Study Guide and Solutions Manual for Organic Chemistry

SIXTH EDITION

Neil E. Schore
University of California, Davis

W. H. Freeman and Company
New York

ISBN-13: 978-1-4292-3136-7
ISBN-10: 1-4292-3136-X

Printed in the United States of America

First printing

W. H. Freeman and Company
41 Madison Avenue
New York, NY 10010
Houndmills, Basingstoke RG21 6XS, England
www.whfreeman.com

Contents

Preface

From One Organic Chemistry Teacher to Another

"I study all the time, I understand what you're saying in the lecture, and I do all the problems. So how come I got a '12' on the exam?" Ouch! We've all heard this from our students, haven't we? (At least I *assume* I'm not the only one.) Why is it that perfectly reasonable students of perfectly reasonable intelligence sometimes wind up being hopelessly buried by this course? More to the point, what, if anything, can *we* do about it? Clearly, in a perfect world, where students have ample time to do everything they're supposed to do and know everything that they need to know from freshman chemistry, things would be better. Typically, however, that is not the case. Students are pressed to budget their time and divide it among their courses. Because they often can't spend sufficient time studying for each course during the term, they sometimes lag so far behind in their studying that, come exam time, they fall into the "Big Trap": they try to memorize everything. And then they get those "12"s and wonder what went wrong.

Well, we're the *teachers,* and we ought to know what's gone wrong and how to help the students do better. My experience has taught me that two critical factors almost always contribute to these predicaments: flawed understanding of basic concepts and lack of ability to apply the concepts to new, unfamiliar situations. The first involves an unsure grasp of mostly descriptive, informational material. Students must learn this fundamental material as surely as they learn the grammar and vocabulary of a foreign language. The basics can usually be mastered by serious study. Continuing emphasis on concepts and mechanisms, including the self-consistent, functional use of color in the textbook, and reemphasis in terms of relationships among topics in the "Introduction" and "Keys" sections of each chapter of this study guide are intended to make this process as manageable as possible for students.

The second factor is the killer for teachers: how to teach struggling students to (1) sort out the concepts and patterns relevant to a given problem and then (2) apply them in a logical way to the development of an answer. We all recognize that what we are trying to teach is not simply a body of information, but a thought *process*. How does one go about teaching a thought process? The most successful way for me has been to lead students through a problem step-by-step, so they can experience the process, even if initially only from the outside looking in. They must be shown the choices that need to be made, why some are wrong and can be dismissed immediately, and how to evaluate the others. My goal in the preparation of the solutions to the end-of-chapter problems in the text was precisely this: to illustrate the thought process involved in getting from the problem to a reasonable answer. I've provided the greatest amount of detail in the earlier chapters, and I've *deliberately* omitted details in answers to some problems toward the end of the book. The learning process almost always requires *direct* experience on the students' part. It isn't enough for students to read an answer *even if it is fully understood!* Students must have opportunities to carry out the mental process for themselves. Therefore in many cases I've begun an answer with a hint, asking students to go back and attempt the problem again if they had difficulty the first time. Getting started is often the hard part, and this ploy at least gives serious students a second chance to make the connections required to proceed to a solution. It's a technique I use in helping students during office hours, and it seems to work.

I've also tried to be as rigorous and as complete as possible in the presentation of mechanisms, even to the extent of showing two-electron arrows in simple proton transfer processes. This might seem excessive to some, but remember, here we are dealing with students who may be in a position to derive clarifying insight from even the most insignificant of points. In the end, we must face the fact that our job is not really to "teach students organic chemistry." Our goal really has to be to teach students how to *learn* what organic chemistry is all about and how it *works*. Teaching students "how to learn" can be a difficult task. I hope the approach taken in this book is helpful in achieving that end.

Note on the Sixth Edition

The sixth edition of *Organic Chemistry* includes many new problems. Additional worked-out exercises ("Working with the Concepts") have been included in each chapter. At the end of each chapter is at least one completely worked-out "Chapter Integration" problem. Finally, the end-of-chapter problem sets have been further expanded throughout the book. The emphasis in all cases is to encourage the student to develop problem-solving skills. Those of you familiar with previous editions of this *Study Guide* will recognize the format used for the solutions to these new problems. They are designed to lead the student to think about each problem in stages—considering possible approaches, seeing where they lead, evaluating how productive they are, and finally going on to the next step. As were many of the original solutions in the *Study Guide,* these new solutions are intended to serve as "thought maps" to demonstrate the way successful strategies may be developed. If students use this volume not as a simple answer manual but indeed as a *guide,* I believe that they will have more success in navigating this challenging course.

Acknowledgments

As always, I am indebted to many individuals who have found and helped me correct mistakes in the earlier editions of the *Study Guide.* Professor K. Peter C. Vollhardt and his students at the University of California, Berkeley, and colleagues and students of mine here at the University of California, Davis, have all provided valuable assistance. Dr. Melekeh Nasiri deserves special credit for always being on the lookout for errors and inconsistencies. A great debt of thanks also goes to the friendly and helpful team at W. H. Freeman and all the reviewers they've provided for this edition.

As usual, my personal thanks go out to my wife, Carrie, and my not-so-small-any-longer children, Mike the computer wizard, and Stef the violin virtuoso, for letting me make my usual mess of things with drafts and proofs and models and journals all over the place. You can all have the house back now, at least for a couple of years.

Neil E. Schore
Davis, California

General Introduction
or
Whose "Brilliant" Idea Was It for Me to Take Organic Chemistry, Anyway?

Good question. What is the problem with organic chemistry that causes so many students to view the class with so much anxiety? I think there are at least two good reasons:

1. Very *bad* experiences in freshman chemistry. Even students *interested* in chemistry find significant stretches of "Chem 1" to be intolerably dull.
2. Comments from students who've just finished taking organic chemistry. For example: "You have to memorize eight hundred million reactions, and then they don't even ask you the ones you've had in class on the tests."

Let's take these reasons one at a time. General chemistry is a little like a tossed salad with many different ingredients: a little bit of theoretical chemistry (electronic structure, bonding), physical chemistry (gas laws, equilibria, kinetics), inorganic chemistry (periodic table, descriptive chemistry of the elements, coordination compounds), organic chemistry (hydrocarbons, other types of compounds, nomenclature), and who knows what else. No wonder so many students finish the first year of chemistry without the slightest trace of an overview of what they've sat through, or the faintest hint of an idea of what's supposed to come next. The problem is that "chemistry" is a very big field that covers a lot of territory. It starts with atoms, but can go in many directions, and each of these can get pretty complicated. For now, all you need to know is that only a portion of what you saw in general chemistry is necessary as background for organic chemistry. This will be the subject of the first chapter of your textbook.

As for the second reason people are afraid of organic chemistry, all that famous "memorizing" you have to do, like most stories heard over and over again, there is truth to it. You *will* have to memorize a lot of organic chemistry. However, you *won't* have to memorize eight hundred million reactions. If you try to do that, you will be lucky to pass the course *even if you succeed*. What you really have to memorize are some basic properties of atoms and molecules, a number of principles that describe why and how reactions take place, and a number of reaction *types* that later can be generalized to include the various reactions of organic compounds that you will see throughout the course. From this framework you will be shown how the various details of organic chemistry are derived from some basic principles or "ground rules." You'll be expected to learn about and *really understand* these ground rules, so that you can apply them in a logical way to completely new kinds of situations and come up with sensible answers. It's little like learning arithmetic. You all learned how to add when you were little. So if someone asked you to add $-1845\frac{2}{3}$ to $793\frac{1}{5}$, you would be able to figure out how to do it, even though it's pretty unlikely that you've added $-1845\frac{2}{3}$ to $793\frac{1}{5}$ ever before in your life. This is because you are familiar with some basic ground rules: what $+$ and $-$ signs mean, how to do fractions, the general methodology for adding (carrying numbers and all that). The difference is that you do arithmetic in elementary school and organic chemistry in college. The principles, the ground rules, and the methods of organic chemistry are going to go by quickly, and you're going to have to learn them well enough to make use of them. . . *quickly.*

That is where this *Study Guide* enters the picture. A textbook has *linear* makeup. It starts at the beginning with page 1, and goes on in a straight line until it gets to the end. Now that might be a decent way to present, say, history, where the book could follow a calendar of events as they occur over a period of time. However, it doesn't work quite as cleanly with chemistry, where the same basic principles operating in Chapter 2 are also cooking in Chapter 12 as well as Chapter 20. In a sense, organic chemistry is three-dimensional: there is a network of interrelationships between the various subtopics, *derived from these basic principles,* but hard to bring out clearly within the framework of a linear textbook. But it's a knowledge of these interrelationships that can make learning organic chemistry a much more reasonable job for a student to undertake. So, what you will find in each chapter of this *Study Guide* will be several features aimed at tying things together, so that you can see at every stage of the course the relationship between the new material, what has gone before, and what will be coming up. Each chapter in this guide will have at least the following four components:

1. A general introduction to the textbook chapter as a whole in the context of previously covered material.
2. An outline of the chapter, with brief comments on the nature and significance of each chapter section.
3. More detailed comments about those features of the chapters that are of greater significance in terms of the course as a whole.
4. Solutions to problems at the end of the chapter, with explanations.

As this book is a solutions manual, a comment on that aspect of it is also appropriate. The problems in the textbook range from "drill" problems, which require you to apply only a single new idea in a repetitive way to several simple cases, to "think" problems, where several ideas, new and old, have to be applied, often to cases that at first glance may look very different from the examples presented in the textbook chapter. This cross-section of problems is intended to illustrate the thought processes involved in analyzing this kind of subject matter, and to resemble the kinds of problems you might encounter in exams.

Try to do the problems!!! Try a couple with the textbook and your notes, to ensure that you are using accurate fundamental principles and information. Then close the book, put away your notes, and try to work the rest on your own. If you can't see how to do a problem at first glance, try to analyze its features: what is involved conceptually and what is its context, *before* looking here for the answer. Then, if you're still stuck, note that in some cases you will often find a short introductory comment in this manual before the actual answer to the problem. This is intended to show you where the problem fits into the chapter material and, perhaps, give you enough of a hint so that you might be able to go back and work it yourself. Then the answer will follow, plus an explanation. If you get a problem wrong, try to do two things: (1) understand the *process* for arriving at the answer, as illustrated in this guide, well enough so that you could answer a similar problem yourself without help, and (2) understand *why* the problem was asked in the first place—what points does it illustrate and what kinds of analogies, interpolations, or extrapolations of the basic subject matter does it involve. If an entire class of problems gives you difficulty consistently, refer to the Interlude on problem solving that follows Chapter 11 in the textbook. Categorize the type of problem. Follow the solution stages described until you can make sense of the problem. This kind of exercise will put you in a much better position to face the kinds of problems you are likely to encounter in exam situations.

Good luck!

1

Structure and Bonding in Organic Molecules

The first chapter of the text covers the basic features associated with the bonding together of atoms to make molecules. Much of the material (at least through Section 1-8) is really a review of topics with which you may have some familiarity from freshman chemistry. In other words, it describes just those topics from freshman chemistry that are the most important to know in order to get off to a good start in organic chemistry: bonds, Lewis structures, resonance, atomic and molecular orbitals, and hybrid orbitals. Read the chapter, try the problems, read the comments below, and, if necessary, look to other supplementary sources for additional problems and examples.

Outline of the Chapter

1-1 Overview

1-2 Coulomb Forces
The simple *physical* basis of bonding between atoms. **Conceptually** important.

1-3 Ionic and Covalent Bonds
Review of principles involved.

1-4 Lewis Structures
Perhaps the most critical section of the chapter. You **must** learn how to draw correct Lewis structures of molecules.

1-5 Resonance Forms
Applies to species for which no single Lewis picture adequately describes the true structure.

1-6 Atomic Orbitals
Review material.

1-7 Molecular Orbitals
Review material.

1-8 Hybrid Orbitals
Simplest way to understand the geometry of molecules in the context of the atomic orbitals involved in bonding.

1-9 Structures and Formulas of Organic Molecules
General information section. Conventions for drawing formulas.

Keys to the Chapter

1-2 and 1-3. Coulomb Forces; Bonds

"Unlike charges attract" and "like charges repel." These consequences of elementary physics dealing with *electrostatics* and *Coulomb's law* are central to a basic understanding of chemistry. Not only do they determine whether, and how strongly, atoms will bond to each other (as described in this chapter), but they also influence an even more complicated process: whether two molecules are likely to react with each other. Time and time again we will return to simple electrostatics, in the context of the properties of the individual elements, to explain the reactions of organic chemistry. Most organic molecules contain *polarized covalent bonds*. In bonds of this type, one or more pairs of electrons are shared between two atoms, but because of an *electronegativity difference* between the atoms, the bonding electrons tend to be closer to the more electronegative atom, thereby creating a partial charge separation. In general, for A less electronegative than B, we have $A^{\delta+} : B^{\delta-}$. See the specific examples in Section 1-3. The computer-generated electrostatic potential maps in Section 1-3 are visual representations of relative charge distribution and, therefore, bond polarity: In any such map, the more red in the vicinity of an atom, the more partial negative charge it contains, and the more blue in its vicinity, the more partial positive charge it bears. (Be careful not to compare colors from different maps, however. The colors have been scaled for each map individually to bring out even small polarity differences as clearly as possible.) As you will see later on, most of the reactions in organic chemistry follow a general pattern. First, two nonbonded atoms with opposite charges or polarities are attracted to each other. Then, electrons move from the "electron-rich" to the "electron-poor" atom to form a new covalent bond between them. Because bonds are made up of electrons, it's necessary to keep track of how many electrons are involved, and where they are located. *Lewis structures* are of **paramount** importance in this bookkeeping process.

1-4. Lewis Structures

Whether you've ever done Lewis structures before or not, follow the rules in Section 1-4 **very closely.** Become familiar with the number of electrons around common atoms and the common arrangements of these electrons in the bonds of molecules. This familiarity, brought about by doing **lots** of examples, is the best way to ensure that you will quickly and confidently be able to picture a Lewis structure for any of the types of species you will encounter later on. As you gain confidence through practice, you will be able to use shorthand notations, such as lines instead of dots for bonding electron pairs.

   Organic chemistry involves reactions between organic compounds and other organic or inorganic species. These reactions can involve both bond-breaking and bond-forming processes, and the key to both is the *movement of electrons*. Lewis structures provide the bookkeeping system to help us keep track of electrons in reactions.

1-5. Resonance Forms

Two important conventions involving arrows are introduced in Section 1-5. The first is the use of *double-headed arrows* between resonance forms. This is a special kind of notation because of the special role resonance forms play in organic chemistry. As shown in this section, many species have structures that cannot be represented by a single Lewis structure. They can only be described as intermediate in nature between two or more contributing forms, each of which **by itself** is an incomplete picture of the molecule's structure. We represent such a molecule by drawing the resonance forms separated by double-headed arrows and enclosed in brackets. The true structure is called the *resonance hybrid*. The only difference between the resonance forms is a different location for the **electrons** from one to the next. The same geometrical arrangement of the atoms is maintained in all the resonance forms. **Caution:** molecules that actually exist as resonance hybrids are still often represented by only one Lewis structure. In cases like this you need to be aware of the fact that this is a shortcut used for convenience purposes only and that the real structure is still the resonance hybrid—the other resonance forms are implied even if they aren't written down.

   The second convention in Section 1-5 is the use of *curved arrows* to show the movement of electron pairs. In this section the only application is in showing how the electron pairs shift in going from one Lewis structure of a resonance hybrid to another. Pictorial descriptions of electron movement using these arrows will be very useful tools to help you learn and understand organic chemistry.

Looking closely at electron shifts that convert one resonance form into another, you will notice several patterns that repeat themselves over and over. The two most common types of electron movement are shown in the resonance forms of carbonate on page 18 of your text.

- **"Bond to atom"**: A pair of electrons moves from being a π bond to becoming a lone pair on one of the atoms involved in the original π bond. The π bond is thus broken.
- **"Atom to bond"**: A lone pair of electrons on an atom moves to become a π bond with a neighboring atom. A π bond is thus formed.

Again referring to carbonate on page 18, notice that violation of the octet rule must be avoided at all costs (not the least of which will be to your grade on an exam): If a pair of electrons moves *toward* an atom already containing an octet, another pair of electrons must at the same time move *away* from that atom.

The guidelines in the textbook for determining the relative degree to which each resonance form contributes to the actual structure are the ones you will use the most when dealing with the most common atoms in organic chemistry. It is somewhat abbreviated. Some additional considerations regarding resonance forms (and some reminders of the basic rules) follow.

ALWAYS TRUE (these are your reminders):

1. Individual contributing resonance forms *do not exist*. Only the resonance *hybrid*, which is a "weighted average" or blend of the contributing forms, is real.
2. All resonance forms of a single chemical species *must* have the same total number of valence electrons and the same total charge.
3. Second-row atoms (i.e., up through neon) can *never* exceed an octet in their valence shell in any resonance form. In other words, the rules for drawing Lewis structures apply to drawing resonance forms.
4. Atom positions and geometries *do not change* from one resonance form to another—there is only one geometry and set of atomic positions for the actual chemical substance being represented.

USUALLY TRUE (these are mostly implied by the guidelines in the textbook—we're just spelling them out):

1. Resonance forms differ only in the positions of π and/or nonbonding electrons. The σ bond electrons normally stay put.
2. We convert one resonance form into another by moving electron pairs from places where there is an excess of electron density to places where there is an electron deficiency.
3. Resonance forms with the most covalent bonds are usually more important contributors than resonance forms with fewer covalent bonds (but don't forget that octets take priority).
4. Resonance forms with fewer charged atoms are usually more important contributors.
5. Atoms in the third row (P, S, Cl) and below (Br, I) are *not* limited to octets of electrons in their outer shells. In fact, Lewis structures with 10 or 12 valence electrons are frequently written for these elements.

1-6, 1-7, and 1-8. Orbitals

Atomic orbitals are a convenient way to represent the distribution of electrons in atoms. Note that the + and − signs associated with parts of these orbitals **do not** refer to electrical charges. They refer to mathematical signs of functions (*wave* functions) associated with the distribution of the electrons. Molecular orbitals are similar but are spread out over more than one atom. They provide an alternative to the Lewis electron-dot method for picturing bonds. The number of molecular orbitals involved in describing a bond is always exactly equal to the number of atomic orbitals contributed by the individual atoms. Overlap of atomic orbitals results in bonding, antibonding, and sometimes also nonbonding molecular orbitals. Bonding orbitals are always lower in energy (more stable) than the original constituent atomic levels, and antibonding orbitals are always higher in energy. Thus bonding electrons will be more stable than electrons in nonbonding atomic orbitals and will give rise to strong bonds. Electrons in antibonding orbitals will reduce bonding.

Hybrid orbitals are derived by mixing atomic wave functions. They are used to explain the geometrical shapes of molecules. Hybridization provides several advantages for bonding. With the larger lobe of the hybrid orbital located in between the bonded atoms, more electron density is located where it can "do some good" by contributing to bonding. The participation of different numbers of s and p orbitals in the hybridization allows a wide range of bond angles, thereby permitting electron pairs to get as far away from each other as possible and minimizing unfavorable electrostatic repulsion.

Keep in mind some points for bookkeeping with hybridization. If an atom starts with one s orbital and three p orbitals, it will always end up with a total of four orbitals, no matter how they have hybridized for bonding purposes. Depending on the ratio of atomic orbitals used in the hybridization, we can describe the resulting hybrid orbitals as consisting of certain percentages of "s character" and "p character." For example, an sp orbital contains 50% s and 50% p character, whereas an sp^2 hybrid is $\frac{1}{3}$ s and $\frac{2}{3}$ p in nature. The total s and p character around an atom after hybridization always equals that which was present in the orbitals before hybridization:

1. sp hybridized atom (linear geometry): contains two sp orbitals (each one is $\frac{1}{2}$ s and $\frac{1}{2}$ p in character) and two ordinary p orbitals

$$\text{two } sp + \text{two } p = \text{two}(\tfrac{1}{2}\, s + \tfrac{1}{2}\, p) + \text{two } p = \text{one } s + \text{three } p$$

2. sp^2 hybridized atom (trigonal planar geometry): contains three sp^2 orbitals (each one is $\frac{1}{3}$ s and $\frac{2}{3}$ p in character) and one ordinary p orbital

$$\text{three } sp^2 + \text{one } p = \text{three}(\tfrac{1}{3}\, s + \tfrac{2}{3}\, p) + \text{one } p = \text{one } s + \text{three } p$$

3. sp^3 hybridized atom (tetrahedral geometry): contains four sp^3 orbitals (each one is $\frac{1}{4}$ s and $\frac{3}{4}$ p in character)

$$\text{four } sp^3 = \text{four}(\tfrac{1}{4}\, s + \tfrac{3}{4}\, p) = \text{one } s + \text{three } p$$

So in all cases exactly four orbitals are present and add up to the equivalent of one s and three p, even though each type of hybridization leads to a form of bonding and molecular shape very different from any of the others. As these examples show, the mathematical nature of hybridization is very flexible, to maximize favorable bonding attractions and minimize unfavorable electron–electron repulsions.

Prepare yourself to use the material in these sections: These are the basics, and everything else will build from them.

Solutions to Problems

25. (and 23 and 29—see below)

(a) $\overset{\delta^+}{:}\overset{..}{\underset{..}{Cl}}:\overset{..}{\underset{..}{F}}:\overset{\delta^-}{}$

(b) $:\overset{\delta^-}{\underset{..}{Br}}:\overset{\delta^+}{C}:::\overset{\delta^-}{N}:$ Triple bond is needed to give octets for C and N atoms.

(c) $\left[:\overset{..}{\underset{..}{Cl}}:\overset{+}{\underset{..}{S}}:\overset{..}{\underset{..}{Cl}}: \longleftrightarrow :\overset{\delta^-}{\underset{..}{Cl}}:\overset{\delta^+}{\underset{..}{S}}:\overset{\delta^-}{\underset{..}{Cl}}: \right]$ Note that the availability of d orbitals allows S to be surrounded by a **fifth** electron pair.

$\quad\quad\quad\quad :\overset{..}{\underset{..}{O}}:\underline{}\quad\quad\quad\quad :\overset{..}{\underset{..}{O}}:\overset{\delta^-}{}$

Major (octets)

(d) $\overset{\delta^+}{H}\ \overset{H}{}\ \overset{H}{}\overset{\delta^-}{}$
$\ \ \ H:\overset{..}{C}:\overset{..}{\underset{..}{N}}:H$
$\quad\quad H$

(e) $\quad\quad H\ \overset{\delta^-}{}\ H$
$\ \ \ H:\overset{..}{C}:\overset{..}{\underset{..}{O}}:\overset{..}{C}:H$
$\quad\quad \overset{\delta^+}{H}\ \ H\overset{}{}_{\delta^+}$

(f) $\overset{\delta^+}{H}\overset{\delta^-}{:}\overset{\delta^-}{N}\overset{}{::}\overset{\delta^+}{N}\overset{}{:}H$ Double bond between nitrogens.

(g) $H:\overset{H}{\underset{}{C}}::\overset{\delta^+}{C}::\overset{\delta^-}{O}:$ A molecule with two double bonds.

(h) $\left[H:\overset{}{N}::\overset{+}{N}::\overset{}{N}: \longleftrightarrow H:\overset{}{N}:\overset{+}{N}:::N: \right]$

(i) $\left[:\overset{}{N}::\overset{+}{N}::\overset{}{O}: \longleftrightarrow :N:::\overset{+}{N}:\overset{}{O}: \right]$
$\overset{\delta^+}{}\overset{\delta^-}{}$

Major (O more
electronegative than N)

26. The symbols δ^+ and δ^- are written above or below the appropriate atoms in the answers to Problem 25 above. In each polar bond, the more electropositive atom is designated δ^+, and the more electronegative atom δ^-, based on the electronegativities in the periodic table.

27. **(a)** $H:^-$ Hydride ion. Contrast H^+ (a proton) and $H\cdot$ (H atom).

(b) $H:\overset{H}{\underset{H}{C}}:^-$ A carb*anion*. C has an octet and a -1 charge.

(c) $H:\overset{H}{\underset{H}{C}}^+$ A carbo*cation*. C has only a sextet and a $+1$ charge.

(d) $H:\overset{H}{\underset{H}{C}}\cdot$ A carbon "radical." C is neutral, bonded to only three other atoms, and surrounded by 7 electrons.

(e) $H:\overset{H}{\underset{H}{C}}:\overset{H}{\underset{H}{N}}^+H$ The methylammonium cation. The product of $CH_3NH_2 + H^+$. Compare $NH_3 + H^+ \longrightarrow NH_4^+$.

(f) $H:\overset{H}{\underset{H}{C}}:\overset{}{O}:^-$ Methoxide ion. The product of ionization of methanol, $CH_3OH \rightleftharpoons CH_3O^- + H^+$. Compare $H_2O \rightleftharpoons HO^- + H^+$.

(g) $H:\overset{H}{\underset{}{C}}:$ A "carbene." A neutral carbon, bonded to two other atoms, with only a sextet of electrons.

(h) $H:C:::C:^-$ Another carb*anion*.

Carbanions [(b) and (h)], carbocations (c), free radicals (d), and carbenes (g) are reactive species of high energy. They can, however, function as reaction "intermediates."

(i) $H:\overset{}{O}:\overset{}{O}:H$ Hydrogen peroxide.

28. How to begin? Look at each atom and compare the bonding pattern with simpler structures that are more familiar to you. Count bonds. Count electrons. This exercise will make it easier for you to recognize similar situations later (as in exams). Then do the formal charge determination as described in the text.

(a) The oxygen has three bonds and a lone pair. What simpler species do you know that is similar? Hydronium ion is one, probably the simplest. We've already seen the formal charge determination $+1$ for the oxygen in hydronium ion, based upon the calculation

6 (group # for O) $-$ [3 (half of shared e^- in bonds) $+$ 2 (unshared e^-)] $= +1$

The oxygen in the species in this problem is analogous.

The carbon has four bonds, just like methane, a "normal" neutral species. Therefore, based on our models, left and right, we can derive the solution in the center:

$$H-\overset{\overset{\displaystyle H}{|}}{\underset{\displaystyle \cdot\cdot}{O}}{}^{+}-H \qquad H-\overset{\overset{\displaystyle H}{|}}{\underset{\displaystyle \cdot\cdot}{O}}{}^{+}-\overset{\overset{\displaystyle H}{|}}{\underset{\displaystyle H}{C}}-H \qquad H-\overset{\overset{\displaystyle H}{|}}{\underset{\displaystyle H}{C}}-H$$

(b) The double bond between C and O changes nothing. The count is the same: carbon has four bonds and will be neutral, as in methane, and oxygen has three bonds and a lone pair, as in hydronium, and will have a +1 formal charge:

$$\overset{\displaystyle H\diagdown}{\underset{\displaystyle H\diagup}{}}C=\overset{+}{\underset{\cdot\cdot}{O}}-H$$

(c) A new system, but not complex. You do not have a simpler species for comparison, so just do the calculation.

$$4\ (\text{group \# for C}) - [3\ (\text{half of shared e}^- \text{ in bonds}) + 2\ (\text{unshared e}^-)] = -1$$

It is a carbon anion, or carbanion. It is isoelectronic—it has the same number of valence electrons (5)—with neutral ammonia and positive hydronium ion.

(d) Proceed as for part **(a)**, by considering simpler, analogous species. Suitable examples are ammonium ion for nitrogen with four bonds, and water for oxygen with two bonds and two lone pairs. Thus, we can write the answer (below, center):

$$H-\overset{\overset{\displaystyle H}{|}}{\underset{\displaystyle H}{N}}{}^{+}-H \qquad H-\overset{\overset{\displaystyle H}{|}}{\underset{\displaystyle H}{N}}{}^{+}-\underset{\cdot\cdot}{\overset{\cdot\cdot}{O}}-H \qquad H-\underset{\cdot\cdot}{\overset{\cdot\cdot}{O}}-H$$

The species is called hydroxylammonium ion.

(e) All three oxygen atoms are "normal": two bonds and two lone pairs, and therefore all three are neutral. What about boron? We have seen two relevant examples: borane, BH_3, neutral boron with three bonds, and borohydride, BH_4^-, boron with four bonds and a negative charge, based upon the calculation

$$3\ (\text{group \# for B}) - [4\ (\text{half of shared e}^- \text{ in bonds})] = -1$$

Therefore, we can arrive at the solution (below, center):

$$H-\underset{\cdot\cdot}{\overset{\cdot\cdot}{O}}-H \qquad \overset{\displaystyle H-\overset{\cdot\cdot}{O}\diagdown}{\underset{\displaystyle H-\underset{\cdot\cdot}{O}\diagup}{}}\overset{-}{B}=\overset{\cdot\cdot}{\underset{\cdot\cdot}{O}} \qquad H-\overset{\overset{\displaystyle H}{|}}{\underset{\displaystyle H}{B}}{}^{-}-H$$

(f) The oxygens are the same as in water, no problem there. The nitrogen is unusual: With two bonds and one lone pair it has no familiar analogs. Let's do the math:

$$5\ (\text{group \# for N}) - [2\ (\text{half of shared e}^- \text{ in bonds}) + 2\ (\text{unshared e}^-)] = +1$$

The answer is $H-\underset{\cdot\cdot}{\overset{\cdot\cdot}{O}}-\overset{+}{\underset{}{\overset{\cdot\cdot}{N}}}-\underset{\cdot\cdot}{\overset{\cdot\cdot}{O}}-H$.

29. (a) (i) and (ii) Don't move any atoms! Resonance forms differ only in the location of the electrons. The forms shown place the negative charge on two of the oxygen atoms. Continue with a Lewis structure that places the charge on the third oxygen:

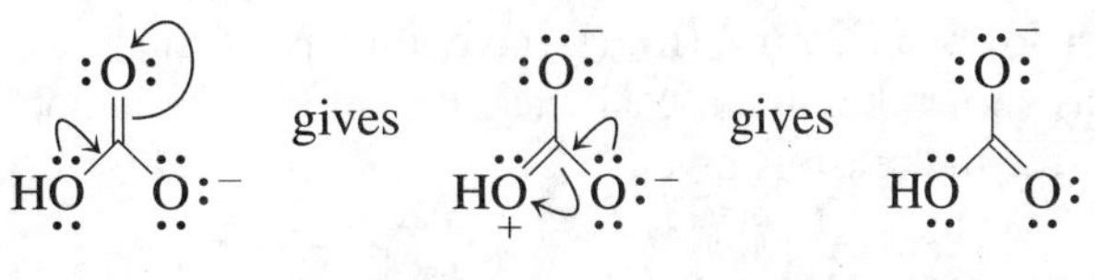

(iii) All three Lewis structures have octets on every large atom, but the middle structure has three charged atoms and two instances of plus–minus charge separations, making it relatively less favorable as a contributor to the hybrid. The first and third forms have only one charged atom and are the major contributors.

(b) Construct one reasonable Lewis structure first: 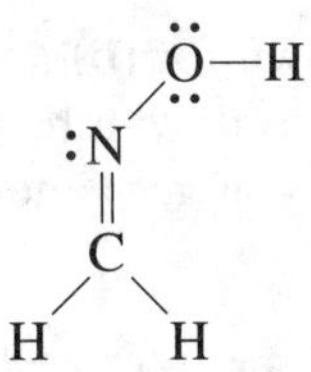 . All the atoms are neutral, so we can move electron pairs in a couple of ways to see what we get. Let's begin by moving a pair of electrons from the double bond. Which way? Doesn't matter—*just move them and see what you get!* If it's something reasonable, fine. If it isn't, it isn't. So, move the pair toward nitrogen:

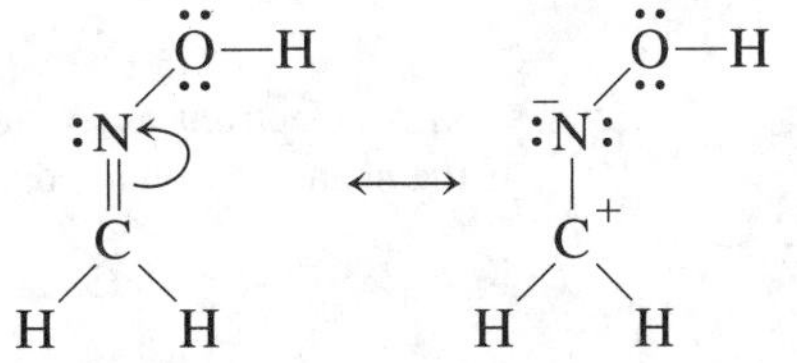

Well, at least the negative charge is on the more electronegative atom (N). But we've separated opposite charges and lost the octet on carbon, so this new resonance form is unlikely to be a major contributor.

What if we shift electrons the other way? Now we get something really hideous, having lost nitrogen's octet and given it a positive charge. However, we can use a lone pair on oxygen to form a double bond with the nitrogen atom, giving the nitrogen back its octet:

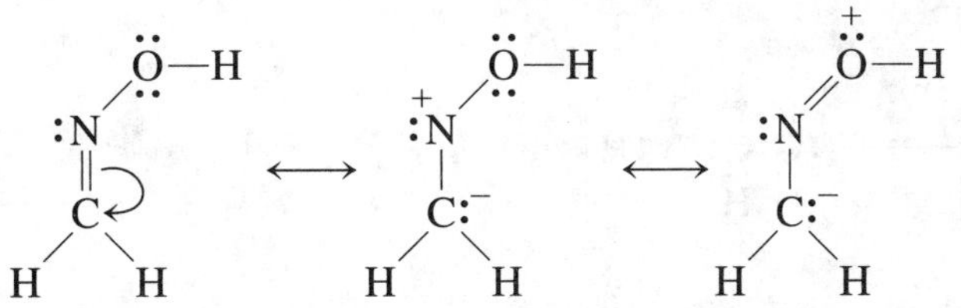

We're not getting anything to write home about here, folks. Just because you can draw resonance forms that don't violate the rules of bonding (like exceeding octets) doesn't mean any of them are going to be any good. The original Lewis structure, with all neutral atoms, represents this compound best. The rest of the resonance forms are at most only marginal contributors.

(c) Now we have a negatively charged atom. Move electrons away from it:

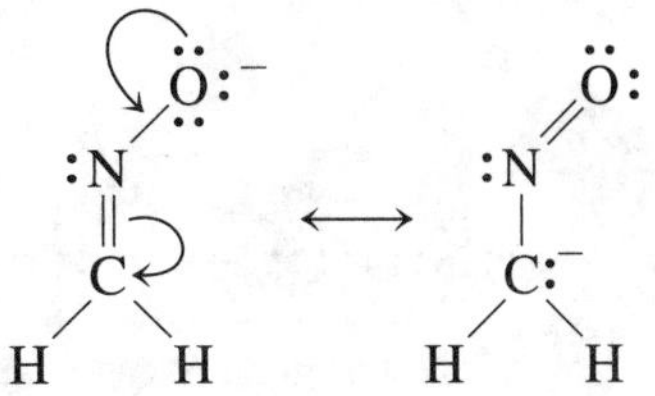

Notice that it's necessary to move an electron pair from the C=N double bond onto the C in order to avoid exceeding an octet on nitrogen. In both forms, all atoms (besides H) have octets. The only difference is the location of the negative charge: it's better on O (more electronegative than C). So the first Lewis structure is better.

30. Resonance structures for Problem 25 (c), (h), (i) have already been shown. Two other species have additional resonance forms, as shown below. In each case the one below is not nearly as good as that shown in the answer to Problem 25, for the reasons given.

(b) 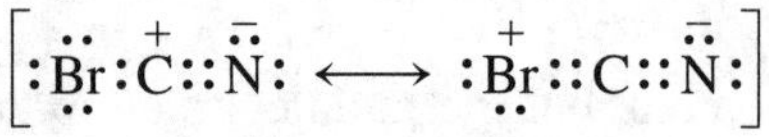 $\left[:\ddot{Br}:\overset{+}{C}::\overset{\bar{}}{N}: \longleftrightarrow :\overset{+}{Br}::C::\overset{\bar{}}{N}: \right]$

   (Carbon sextet)   (Separation of charge)

(g) 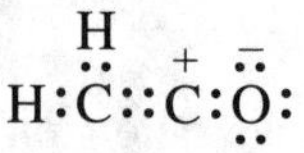 $H:\overset{H}{\underset{}{C}}::\overset{+}{C}:\overset{\bar{}}{\ddot{O}}:$

   (Carbon sextet)

Resonance forms may be drawn for **(b)** and **(e)** of Problem 28—the structures containing double bonds. (It is *always* possible to draw a resonance form for a structure with a multiple bond, although the resonance form you get is not necessarily a major contributor.)

(b) 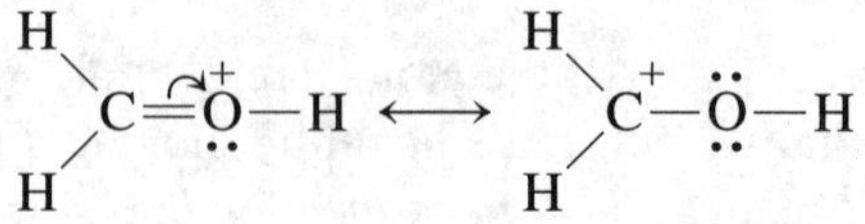 $\overset{H}{\underset{H}{>}}C=\overset{+}{\ddot{O}}-H \longleftrightarrow \overset{H}{\underset{H}{>}}\overset{+}{C}-\overset{..}{\ddot{O}}-H$ Move electrons toward the positively charged atom

     (Minor contributor: carbon lacks an octet)

(e) $\overset{H-\ddot{O}}{\underset{H-\ddot{O}}{>}}B=\overset{\bar{}}{\ddot{O}} \longleftrightarrow \overset{H-\ddot{O}}{\underset{H-\ddot{O}}{>}}B-\ddot{O}:^{-}$ Move electrons away from the negatively charged atom

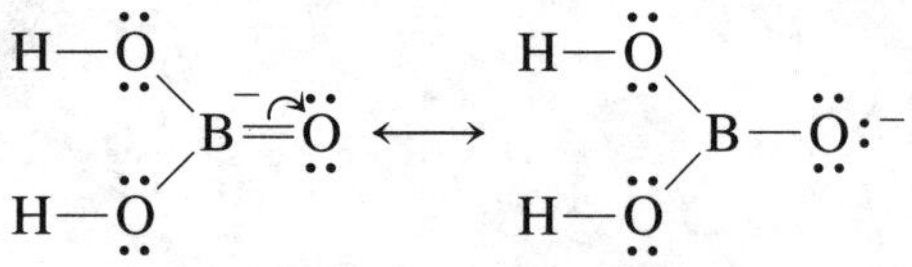

     (Minor contributor: boron lacks an octet)

(f) 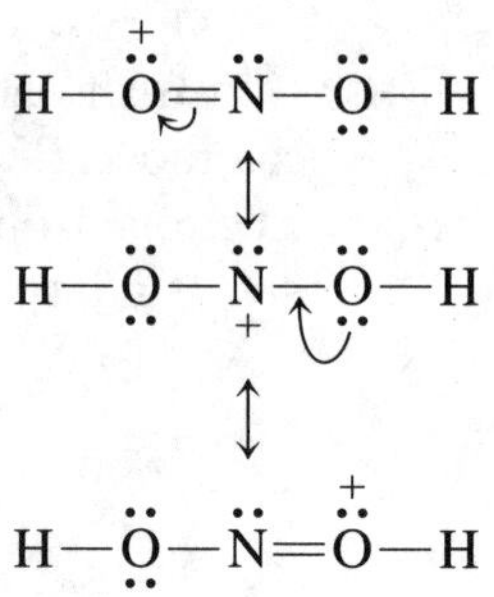 $H-\overset{+}{\ddot{O}}=\overset{..}{N}-\overset{..}{\ddot{O}}-H$

$\updownarrow$

$H-\ddot{O}-\underset{+}{N}-\overset{..}{\ddot{O}}-H$

$\updownarrow$

$H-\overset{..}{\ddot{O}}-\overset{+}{N}=\overset{..}{\ddot{O}}-H$

Move electrons toward the positively charged atom

31. (a) 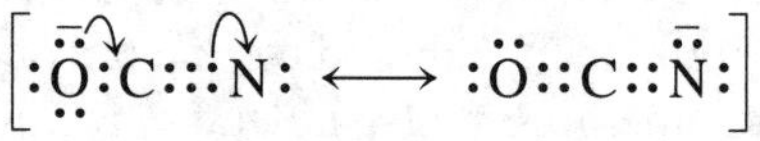 $\left[:\overset{\bar{}}{\ddot{O}}:C:::N: \longleftrightarrow :\ddot{O}::C::\overset{\bar{}}{\ddot{N}}: \right]$

   Major (negative charge prefers the more electronegative atom—oxygen)

In some of these answers, additional, less favorable forms are shown for comparison purposes. Curved arrows show how to convert one form to the next one to its right.

(b) 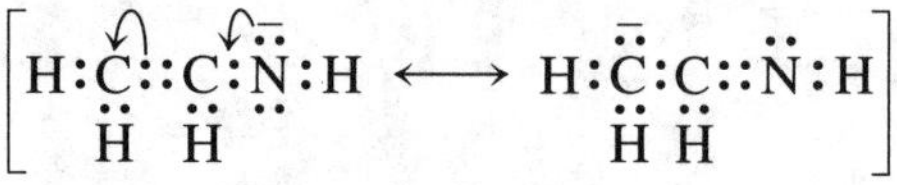 $\left[H:\overset{..}{\underset{H}{C}}::\overset{..}{\underset{H}{C}}:\overset{..}{N}:H \longleftrightarrow H:\overset{\bar{}}{\underset{H}{C}}:\overset{..}{\underset{H}{C}}::\overset{..}{N}:H \right]$

   Major (negative charge prefers the more electronegative atom—nitrogen)

(c) 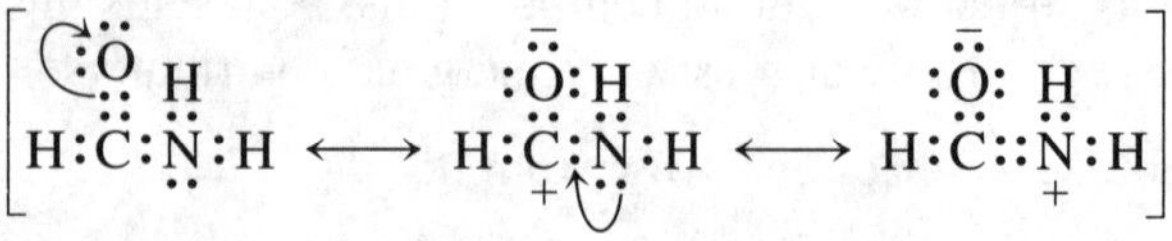 $\left[\overset{\overset{\ddot{O}}{|}}{H:\overset{..}{C}:\overset{..}{N}:H} \longleftrightarrow \overset{:\overset{\bar{}}{\ddot{O}}:H}{H:\overset{..}{C}:\underset{+}{N}:H} \longleftrightarrow \overset{:\overset{\bar{}}{\ddot{O}}:\ H}{H:\overset{..}{C}::\underset{+}{N}:H} \right]$

Major (no separation of charge)

(d) 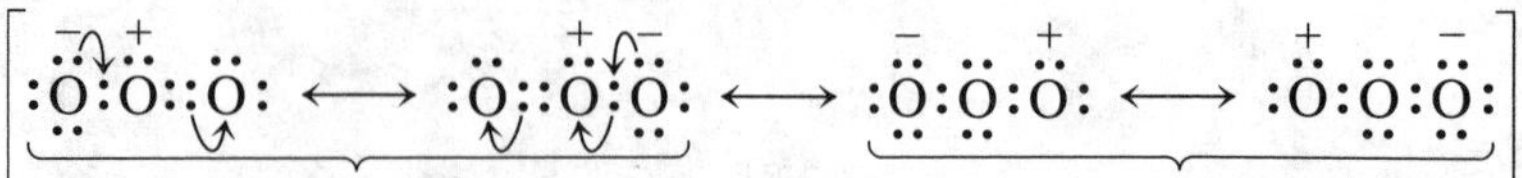 $\left[:\overset{\bar{}}{\ddot{O}}:\overset{+}{\ddot{O}}::\ddot{O}: \longleftrightarrow :\ddot{O}::\overset{+}{\ddot{O}}:\overset{\bar{}}{\ddot{O}}: \longleftrightarrow :\overset{\bar{}}{\ddot{O}}:\overset{+}{\ddot{O}}:\ddot{O}: \longleftrightarrow :\ddot{O}:\overset{+}{\ddot{O}}:\overset{\bar{}}{\ddot{O}}: \right]$

   Equivalent, major   Not as good (oxygen sextet)

(e) 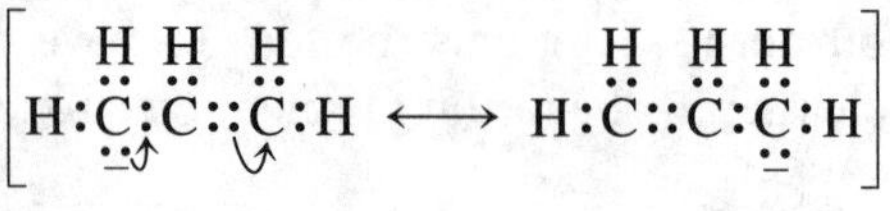 $\left[H:\overset{H}{\underset{}{C}}:\overset{H}{\underset{}{C}}::\overset{H}{\underset{}{C}}:H \longleftrightarrow H:\overset{H}{\underset{}{C}}::\overset{H}{\underset{}{C}}:\overset{H}{\underset{}{C}}:H \right]$ Identical

(f)

$$\left[:\ddot{O}::\ddot{C}l:\ddot{O}:^{-} \longleftrightarrow \ ^{-}:\ddot{O}:\ddot{C}l:\ddot{O}:^{-} \longleftrightarrow \ ^{-}:\ddot{O}:\ddot{C}l::\ddot{O}: \right]$$

Major   Major

(g)

$$\left[\begin{array}{c} H \\ H:\overset{..}{C}:N:H \\ :\ddot{O}: \\ H \end{array} \longleftrightarrow \begin{array}{c} H \\ H:\overset{+}{C}:\overset{..}{N}:H \\ :\ddot{O}: \\ H \end{array} \longleftrightarrow \begin{array}{c} H \\ H:C::N:H \\ :\ddot{O}: \\ H \end{array} \right]$$

(Positive   (Carbon   Major
oxygen)   sextet)

(h)

$$\left[\begin{array}{c} H \\ H:\overset{..}{C}:\overset{+}{C}::N:\ddot{O}:^{-} \\ H \end{array} \longleftrightarrow \begin{array}{c} H \\ H:\overset{..}{C}:C:::\overset{+}{N}:\ddot{O}:^{-} \\ H \end{array} \longleftrightarrow \begin{array}{c} H \\ H:\overset{..}{C}:\overset{..}{C}::N::\overset{+}{O}: \\ H \end{array} \right]$$

(Carbon sextet)   Major (oxygen more
electronegative than carbon)

32. Before starting, notice that the problem tells you how the atoms are attached: both compounds have two N—O bonds. So the N is in the middle of nitromethane. We begin with σ bonds:

$$\begin{array}{ccc} & H & O \\ & | & \diagup \\ H - & C - N \\ & | & \diagdown \\ & H & O \end{array}$$

So far, the valence shells of the carbon and the hydrogens are filled, but the nitrogen and the oxygens are short. But we have 24 electrons to work with (3 from hydrogens + 4 from C + 5 from N + 12 from the oxygens) and we've only shown 12 in these 6 bonds. We could use the remaining 12 to add three lone pairs to each O. Let's do that, and then figure out the formal charges on the atoms:

$$\begin{array}{ccc} & H & \ddot{O}:^{-} \\ & | & \diagup \\ H - & C - \overset{++}{N} \\ & | & \diagdown \\ & H & \ddot{O}:^{-} \end{array}$$

It's a "legal" Lewis structure, we've violated no rules, and we've satisfied the octets for O, but the N is in trouble with only a sextet and a 2+ charge. Can we make this better? Let's move an electron pair away from a negative atom and toward a positive one and see what we get.

$$\begin{array}{ccc} & H & :\ddot{O}:^{-} \\ & | & \diagup \\ H - & C - \overset{++}{N} \\ & | & \diagdown \\ & H & \ddot{O}: \end{array} \longleftrightarrow \begin{array}{ccc} & H & \ddot{O}: \\ & | & \diagup \\ H - & C - \overset{+}{N} \\ & | & \diagdown \\ & H & \ddot{O}: \end{array}$$

Now we're doing better: N has an octet, too. We could, of course, have moved an electron pair from the other oxygen instead. The result is identical to what we just got above, but the N—O single and N=O double bonds are switched, along with the negative charge:

$$\begin{array}{ccc} & H & \ddot{O}:^{-} \\ & | & \diagup \\ H - & C - \overset{++}{N} \\ & | & \diagdown \\ & H & \ddot{O}: \end{array} \longleftrightarrow \begin{array}{ccc} & H & \ddot{O}:^{-} \\ & | & \diagup \\ H - & C - \overset{+}{N} \\ & | & \diagdown \\ & H & \ddot{O}: \end{array}$$

Could we move *two* electron pairs in toward the N, one from each O? Nope: That would violate the octet rule on N and give us an illegal Lewis structure:

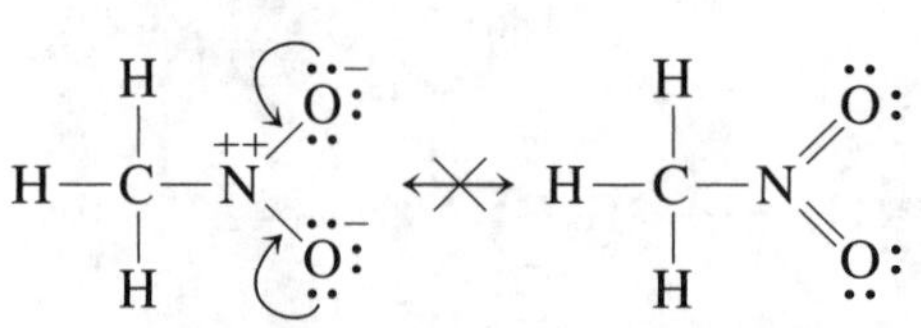

ILLEGAL!
Octet rule violated—
10 electrons on N

So the best two structures are the two we derived above, with octets on all nonhydrogen atoms and a pair of charges. The arrows below show how electron pairs shift to go from one to the other:

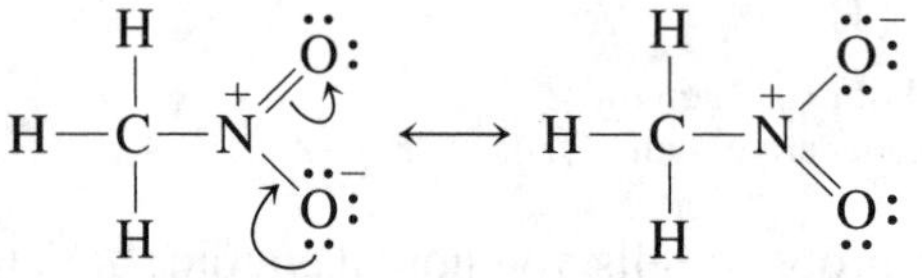

Since these two forms are identical, they contribute equally to the resonance hybrid. The NO bonds are polar, with a full positive charge on N and a negative charge split, half on each O.

You may ask, what would have happened back at the beginning of this exercise if we started by putting one of our extra pairs of electrons on the N, instead of putting all of them on the oxygens? Good question. Our starting Lewis structure (below, left) would then have an octet on the N and one oxygen, but a sextet on the other O. The remedy, shifting the lone pair from the N toward the electron-deficient O, gets us to one of the same final structures that we obtained above:

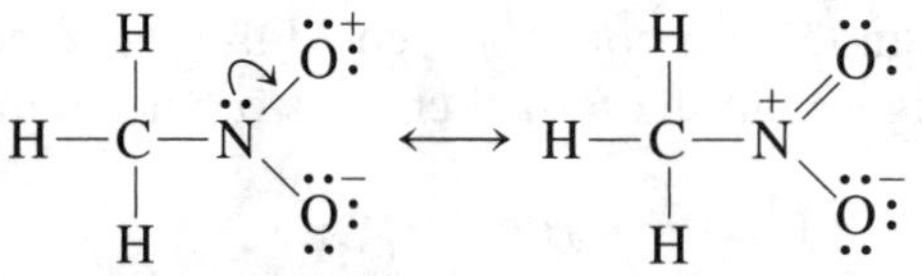

As a rule, as long as all your σ electrons are in place, and you do not violate the octet rule with the rest, any starting structure will eventually get you to the best answer(s).

Now we turn to methyl nitrite. Following the same procedure, we begin with only single bonds, and then arbitrarily add in the remaining electrons as lone pairs, taking care only to avoid violating the octet rule. The structure below left is one result, and it contains a seriously electron-deficient N, just as we found starting out with nitromethane. And, we "fix" it the same way, moving an electron pair "in" from the negatively charged O at the end:

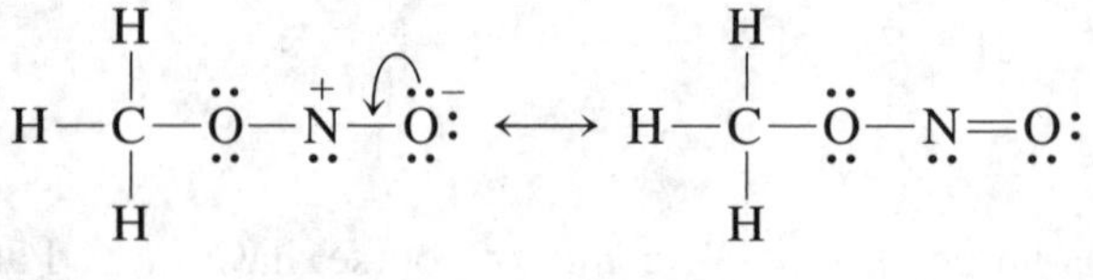

This is pretty good: All nonhydrogen atoms have octets, and none are charged. Can we find any other reasonable resonance forms? There is a common pattern described in the text for systems containing an atom with at least one lone pair attached to one of two atoms connected by a multiple bond. You move the lone pair "in" and move a π bond "out":

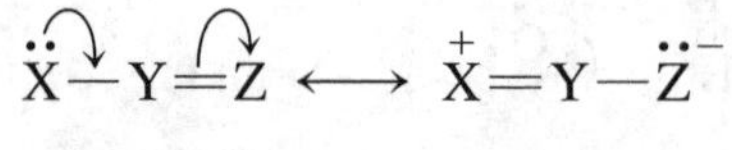

Applying this pattern to methyl nitrite, we get

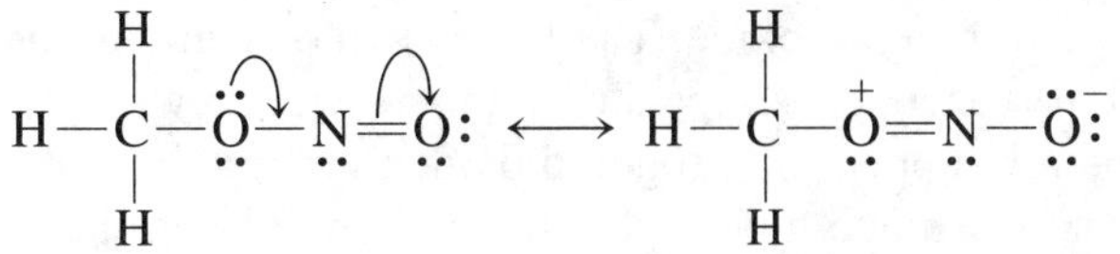

The result is the second-best resonance form—okay for octets, but charges are separated, which makes it less of a contributor than the Lewis structure on the left. The hybrid will more closely resemble the left-hand structure, with two nonequivalent NO bonds. The contribution of the right-hand structure, while small, will tend to make the NO bond at the end the most polar one in the molecule, with the O at the negative end.

33. (a) Chlorine atom is $:\overset{..}{\underset{..}{Cl}}\cdot$ (seven valence electrons, neutral)

Chloride ion is $:\overset{..}{\underset{..}{Cl}}:^{-}$ (eight electrons, negatively charged)

(b) Borane is planar (6e around B), while phosphine is pyramidal (8e around P, like the N in ammonia):

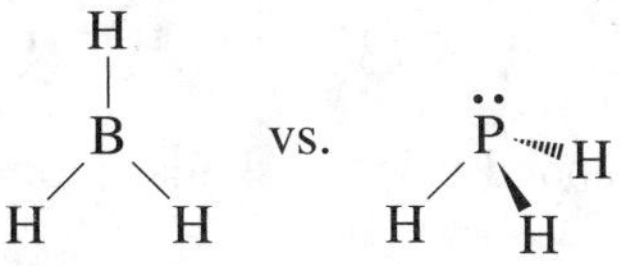

(c) CF_4 is tetrahedral, while BrF_4^-, with *six* electron pairs around Br, is planar with electron pairs above and below.. Notice that all that we need to derive this answer is VSEPR. It is *not necessary* to first try to figure out the hybridization.

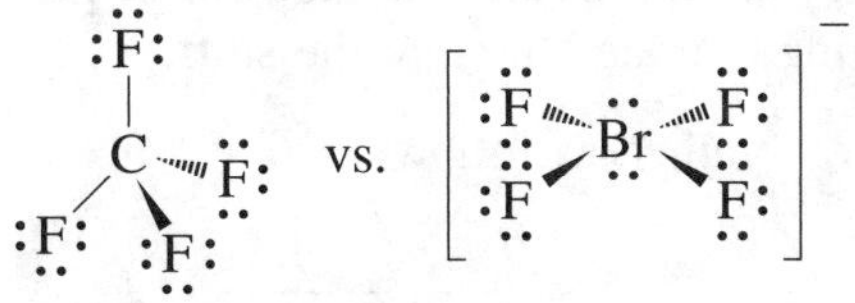

(d) The procedure we follow is the same: Construct the Lewis structures, and then use VSEPR to predict geometries. Do *not* concern yourself with hybridization at first.

   Nitrogen dioxide contains 17 valence electrons (6 from each O and 5 from the N), and nitrite ion contains 18 (the additional electron gives it the −1 charge). N is in the middle, so we have O—N—O to start (4e in σ bonds). For both species we can add 12 of the remaining electrons to oxygens as lone pairs. The last electron (for NO_2) and the last two (for NO_2^-) can go on N, giving $:\overset{-..}{\underset{..}{O}}-\overset{++}{N}-\overset{..}{\underset{..}{O}}:^{-}$ for NO_2 and $:\overset{-..}{\underset{..}{O}}-\overset{+}{N}-\overset{..}{\underset{..}{O}}:^{-}$ for NO_2^-. Each of these Lewis structures is seriously short of an octet on nitrogen and can be improved by resonance delocalization of an electron pair from oxygen toward nitrogen:

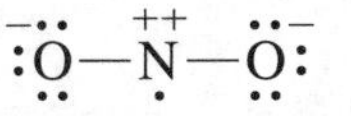

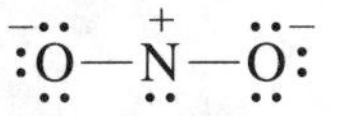

For NO_2,

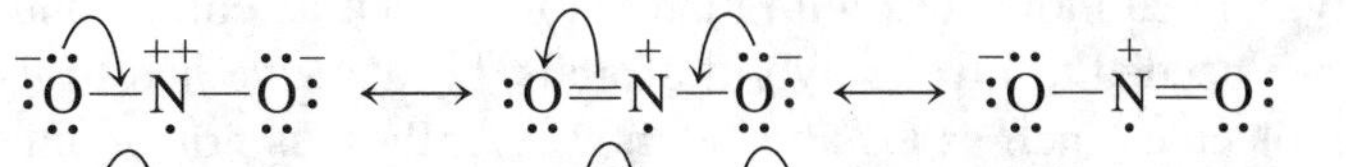

For NO_2^-,

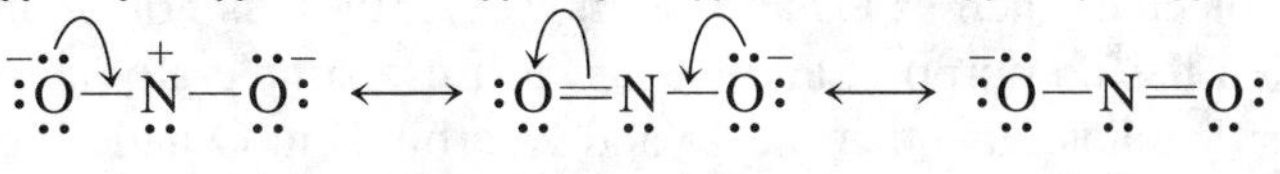

So nitrogen ends up with 7 valence electrons in NO_2 and 8 in NO_2^-.

   How about geometry? Let's begin with NO_2^- because all of its electrons are paired, and VSEPR (valence shell electron *pair* repulsion) can be applied directly. The middle N is surrounded by two σ bonding pairs and a lone pair (π electrons are not considered for VSEPR), and three pairs leads to a bent geometry (which may be rationalized by sp^2 hybridization, if you like). In fact, the O—N—O bond angle in nitrite is 115°. It is a bit smaller than the nominal 120° angle for a trigonal planar structure because the lone pair, being attached to only one atom, exerts more repulsion than do the bonded pairs, closing the bond angle down a little bit.

Now let us examine nitrogen dioxide. The N now bears a single nonbonding electron rather than a lone pair. One electron exerts less repulsion than do two, so we can predict that the O—N—O bond angle should be more open in nitrogen dioxide than it is in nitrite. You do not have enough information to predict how much more open the angle will be. In actual fact it is 134°. The fact that it is larger than 120° means that the two bonding pairs exert more repulsion than does the single nonbonded electron.

You'll no doubt be thrilled to find out that nitrogen dioxide is a significant component of urban atmospheric smog. Smoggy air derives much of its distinctive character from this toxic, smelly, brownish gas.

(e) Now we compare two new dioxides, SO_2 and ClO_2, with the one we've already done, NO_2. Lewis structures and resonance forms first:

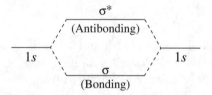

The structures on the extreme right both have expanded valence shells (more than octets), which is OK for third-row atoms.

Based on VSEPR both SO_2 and ClO_2 will be bent structures, because of the lone pair on S and the lone pair + extra single unshared electron on Cl. The actual angle in SO_2 is 129°, and that in ClO_2 is 116°, the difference being due to the extra repulsion of the third nonbonding electron on Cl.

ClO_2, despite the fact that it is smelly, toxic, and tends to blow up, is actually a major industrial chemical used to bleach wood pulp in the manufacture of paper. Prudently, it is prepared just before it is used, eliminating the need to store the stuff.

34. (a) The molecular orbitals are obtained as follows:

$$
\begin{array}{c}
\sigma^* \\
\text{(Antibonding)} \\
1s \qquad\qquad 1s \\
\sigma \\
\text{(Bonding)}
\end{array}
$$

Therefore, the resulting electronic configurations are H_2, $(\sigma)^2$, with two bonding electrons vs. H_2^+, $(\sigma)^1$, with one bonding electron. So H_2 possesses the stronger bond.

(b) Same as Exercise 1-14.

(c) and **(d)** We prepare an orbital diagram similarly. How do we begin? First, out of all the various types of orbitals presented in the chapter, let's see which ones we do and do not have to consider. In multiply-bonded molecules with many atoms, such as ethene and ethyne (Figure 1-21, textbook page 35), we needed to invoke hybrid orbitals, because we had to use them to explain geometry. In *diatomic* molecules such as O_2 and N_2, however, there is no "geometry" to explain, so orbital hybridization has no purpose, and we can just use simple atomic orbitals. That's good—it makes life simpler. We note also that the $1s$ and $2s$ orbitals in O and N are completely filled. In cases such as these, it is customary to ignore the s orbitals, because their overlap will produce no net bonding (just like between two atoms of He)—another welcome simplification. We're down to considering for bonding just the three $2p$ orbitals on each atom, because they are the only ones that are *partly* filled. Referring again to Figure 1-21, we can visualize end-to-end overlap (σ bonding) between the p_x orbitals (one on each atom), which happen to point toward each other, and side-by-side overlap (π bonding) between the remaining p orbitals (two on each atom, p_y and p_z).

Our complete molecular orbital diagram will therefore include three sets of orbital interactions, each of which is shown separately at the left on page 13, and then the three combined at the right. The end-to-end (σ) overlap of the p_x orbitals is shown first, giving σ and σ^* (bonding and antibonding) molecular orbitals, followed by the two π overlaps of the pairs of p_y and p_z orbitals, giving two sets of π and π^* molecular orbitals, respectively. Because σ overlap is generally better than π overlap, the diagrams are shown here with a larger energy gap between the σ and σ^* orbitals than between the π and π^* orbitals—recall from Figure 1-12 that the difference in energy between atomic and molecular orbitals is related to the strength of the bonding—the change in energy going from the atoms to the molecule. (More sophisticated forms of theoretical analysis reveal that the actual ordering of orbital energies is not quite the same as that shown here, but you don't need to worry about that.)

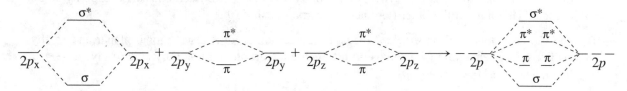

For (c), O_2, $(\sigma)^2(\pi)^2(\pi)^2(\pi^*)^1(\pi^*)^1$, 4 *net* bonding electrons vs. O_2^+, $(\sigma)^2(\pi)^2(\pi)^2(\pi^*)^1$, 5 *net* bonding electrons. So O_2^+ has the stronger bond.

For (d), N_2, $(\sigma)^2(\pi)^2(\pi)^2$, 6 *net* bonding electrons vs. N_2^+, $(\sigma)^2(\pi)^2(\pi)^1$, 5 *net* bonding electrons. So N_2 is better.

35. Use valence shell electron pair repulsion (VSEPR) to predict geometry about any carbon or nitrogen atom. Count the number of other atoms attached to it and add to that the number of lone pair(s) it may contain. Two = linear and *sp* hybridized; 3 = trigonal planar and sp^2 hybridized; 4 = tetrahedral and sp^3. Not complicated.

(a) The four atoms connected to this carbon will be attached by four single bonds and will be arranged in roughly a tetrahedral geometry, which is explained by sp^3 hybridization. It won't be exactly tetrahedral because the four atoms aren't identical (two hydrogens, a carbon, and a Br).

(b) Don't worry about multiple bonds. The carbon in question is attached to three other atoms; therefore it is approximately trigonal planar with sp^2 hybridization.

(c) This carbon is attached to three other atoms. So, as in (b), trigonal planar, sp^2 hybridization.

(d) The nitrogen atom is attached to three other atoms and has one lone pair, so it is sp^3 hybridized. We don't call it tetrahedral, however. When we choose a word to describe the geometry around an atom we usually consider only the atoms to which it's attached, not the lone pairs. So, the nitrogen in CH_3NH_2 is best described as having a *pyramidal* geometry, like the nitrogen in ammonia, NH_3.

(e) This carbon atom sits between two other atoms. Again, VSEPR disregards multiple bonds. The geometry here will be linear, and the carbon *sp* hybridized.

(f) Nitrogen, bonded to three other atoms, is trigonal planar and sp^2 hybridized.

36. (a) This carbon atom uses sp^3 hybrid orbitals for its four sigma (σ) bonds. So does the other carbon atom in the molecule, so the bond between them is formed by overlap between two sp^3 hybrid orbitals. The C—H bonds use overlap between the sp^3 hybrid orbital on carbon and a hydrogen atomic *s* orbital. The C—Br bond forms from overlap of the carbon sp^3 hybrid and an atomic *p* orbital of bromine.

(b) Two of the three sp^2 orbitals on the indicated carbon atom go to ordinary tetrahedral carbons. Those σ bonds therefore involve sp^2–sp^3 overlap. The oxygen atom is bonded to one other atom (our carbon) and has two lone pairs. 1 + 2 = 3. It may be considered to be sp^2 hybridized, in which case the C—O σ bond will be sp^2—sp^2. However, as discussed in Exercise 1-17, oxygen is energetically reluctant to undergo orbital hybridization. It is therefore more accurate to consider this C—O σ bond to involve $C(sp^2)$—$O(p)$ overlap. One of the remaining *p* orbitals on oxygen and the carbon *p* orbital align in a parallel (side-by-side) manner, giving the second pi (π) bond between them.

(c) Each of the three sp^2 orbitals used in sigma (σ) bonds to the indicated carbon atom goes to a different atom. One goes to the other carbon, which is also trigonal planar, so sp^2—sp^2. Each of these carbons has a p orbital left over. They overlap side-by-side to give a p—p π bond. The oxygen is bonded to two atoms and has two lone pairs; we might consider it to be sp^3 hybridized. But as we've just seen, this is not likely for oxygen. The C—O σ bond is C(sp^2)—O(p). The C—H bond is C(sp^2)—H(s).

(d) The C—N bond is sp^3—sp^3. The N—H bonds are sp^3—s.

(e) Both carbons atoms involved in the triple bond are sp hybridized. Therefore the σ bond between them will be sp—sp. There are two pi (π) bonds between them. Each consists of side-by-side p—p overlap. The bond to the other (tetrahedral) carbon is sp—sp^3.

(f) The C—N σ bond will be sp^2—sp^2, because the carbon atom is also trigonal planar and sp^2 hybridized. The remaining (p) orbitals on nitrogen and carbon align side-by-side to give a p—p π bond. The N—H bonds are sp^2—s.

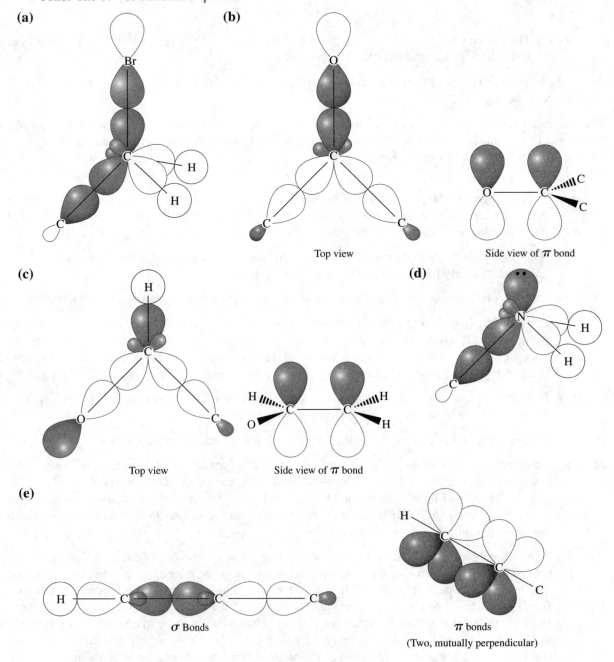

(f)

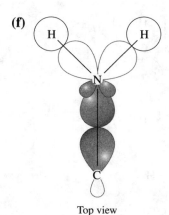

Top view

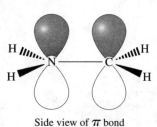

Side view of π bond

38. (a), (b), and **(c)** Each carbon is bonded to four other atoms and, therefore, will possess approximate tetrahedral geometry. Each carbon in these molecules is sp^3 hybridized.

(d) Each carbon is attached to three other atoms (two hydrogens and the other carbon). The bonds to hydrogen are of σ type. One of the carbon–carbon bonds is a σ and the other a π. The result is that each carbon has approximately trigonal geometry (like boron in BH_3) and is sp^2 hybridized. In other words, each carbon uses sp^2 orbitals in three σ bonds, and the leftover p orbital in a π bond.

(e) Each carbon is attached to two other atoms (one hydrogen and the other carbon). The C—H bonds are σ, as is one of the C—C bonds. The other two C—C bonds (of the "triple" bond) are π bonds. Geometry is linear (like beryllium in BeH_2) and each carbon is sp hybridized. Each uses two sp orbitals for σ bonds and two p orbitals for π bonds.

(f)

$$CH_3 - \overset{\overset{\textstyle O}{\|}}{C} - H$$

$$\underset{sp^3}{\uparrow} \qquad \underset{sp^2}{\uparrow}$$

(g) Hybridization must allow both carbons to be doubly bonded (resonance form on the right). Both, therefore, are sp^2.

39. (a) $H-\overset{\overset{\textstyle H}{|}}{\underset{\underset{\textstyle H}{|}}{C}}-C\equiv N\!:$

(b) $H-\overset{\overset{\textstyle H}{|}}{C}-\overset{\overset{\textstyle H \;\; C \;\; H}{|}}{\underset{\underset{\textstyle N}{|}}{C}}-\overset{\overset{\textstyle H}{|}}{C}-\overset{\overset{\textstyle O}{\|}}{C}-\ddot{\underset{..}{O}}-H$

(c) $H-\overset{\overset{\textstyle H}{|}}{\underset{\underset{\textstyle H}{|}}{C}}-\overset{\overset{\textstyle :\!\ddot{O}\!:}{|}}{\underset{\underset{\textstyle H}{|}}{C}}-\overset{\overset{\textstyle H}{|}}{\underset{\underset{\textstyle H}{|}}{C}}-\overset{\overset{\textstyle H}{|}}{\underset{\underset{\textstyle H}{|}}{C}}-H$

(d) $H-\overset{\overset{\textstyle :\!\ddot{Br}\!:\;\;H}{|}}{\underset{\underset{\textstyle H\;\;:\!\ddot{Br}\!:}{|}}{C}}-\overset{}{\underset{}{C}}-\ddot{\underset{..}{Br}}\!:$

(e) $H-\overset{\overset{\textstyle H}{|}}{\underset{\underset{\textstyle H}{|}}{C}}-\overset{\overset{\textstyle :\!\ddot{O}\!:}{\|}}{C}-\overset{\overset{\textstyle H}{|}}{\underset{\underset{\textstyle H}{|}}{C}}-\overset{\overset{\textstyle :\!\ddot{O}\!:}{\|}}{C}-\ddot{\underset{..}{O}}-\overset{\overset{\textstyle H}{|}}{\underset{\underset{\textstyle H}{|}}{C}}-H$

(f) $H-\ddot{\underset{..}{O}}-\overset{\overset{\textstyle H}{|}}{\underset{\underset{\textstyle H}{|}}{C}}-\overset{\overset{\textstyle H}{|}}{\underset{\underset{\textstyle H}{|}}{C}}-\ddot{\underset{..}{O}}-\overset{\overset{\textstyle H}{|}}{\underset{\underset{\textstyle H}{|}}{C}}-\overset{\overset{\textstyle H}{|}}{\underset{\underset{\textstyle H}{|}}{C}}-\ddot{\underset{..}{O}}-H$

Line formulas do **not** as a rule show true bond angles.

40. (a)

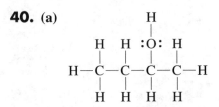

(b)

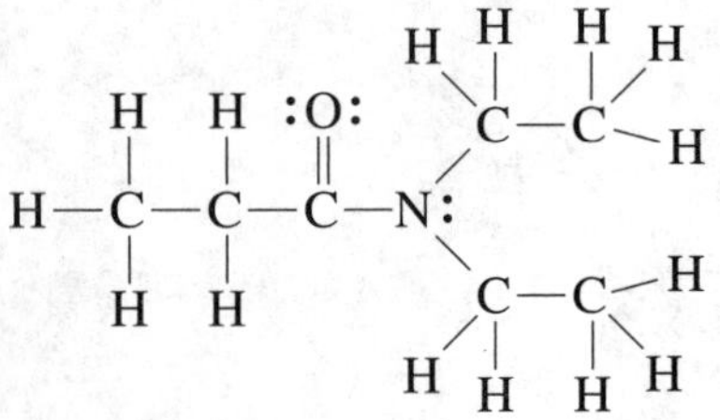

(c)

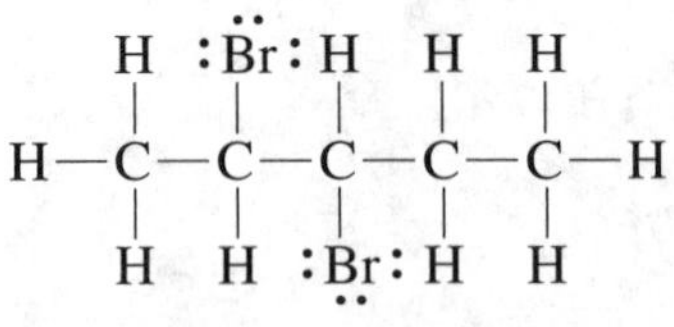

(d)

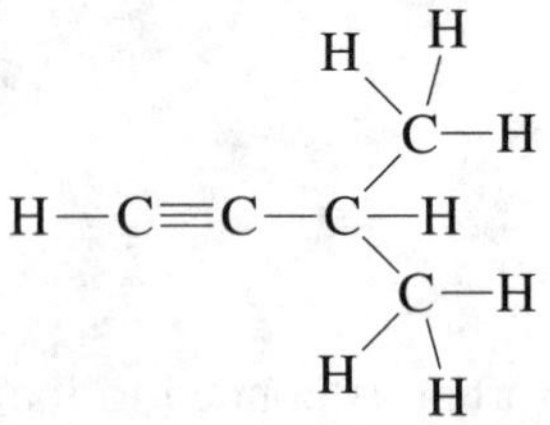

(e)

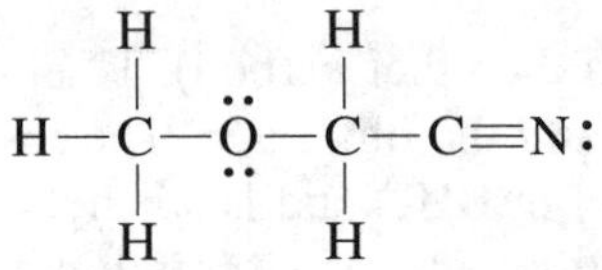

(f)

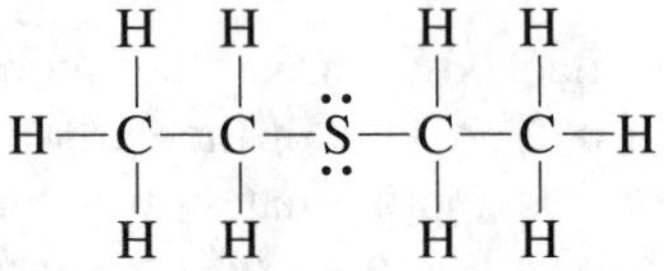

41. (a) $H_2NCH_2CH_2NH_2$ (b) $CH_3CH_2OCH_2CN$ (c) $CHBr_3$

In condensed formulas hydrogens typically **immediately follow** the atoms to which they are connected. This convention is occasionally inverted for the *first* atom at the *left* in a condensed formula, such as (a) which begins with H_2N rather than NH_2, or for a methyl substituent, which you may see written as H_3C rather than CH_3 in that position. However, be careful *never* to separate hydrogens from the atoms to which they are attached by any other atom symbol.

42. (a) $(CH_3)_2NH$ (b) $CH_3\overset{\overset{\text{O}}{\|}}{C}NHCH_2CH_3$ (c) $CH_3CHOHCH_2CH_2SH$

(d) CF_3CH_2OH (e) $CH_3CH{=}C(CH_3)_2$ (f) $CH_2{=}CHC\underset{\underset{\text{O}}{\|}}{C}H_3$

Alternative correct answers exist for several structures in Problems 41 and 42.

43. From Problem 39:

(a) —CN

(b)

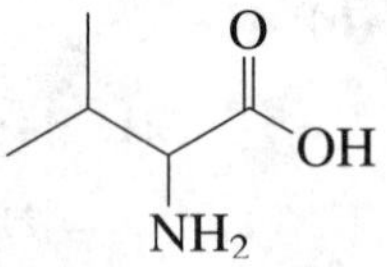

(c)

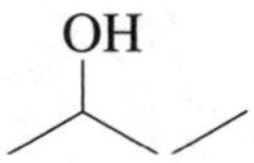

(d)

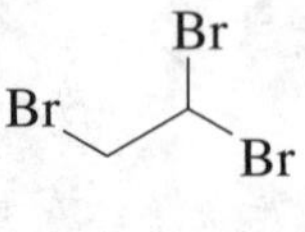

(e)

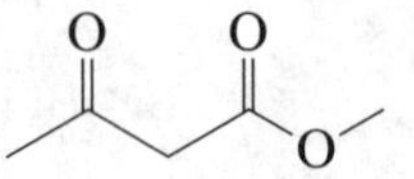

(f)

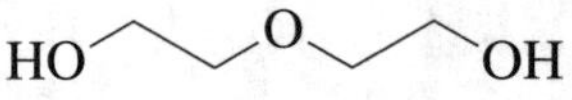

From Problem 42:

(a)

(b)

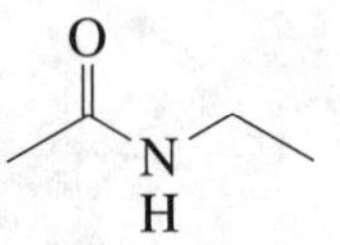

(c)

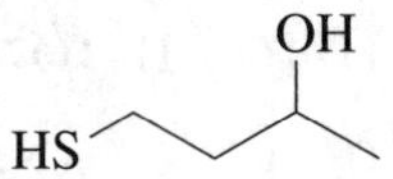

(d)

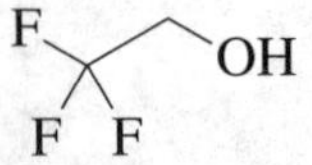

(e)

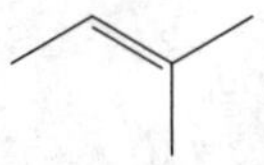

(f)

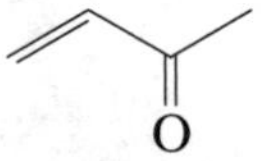

44. **(a)**

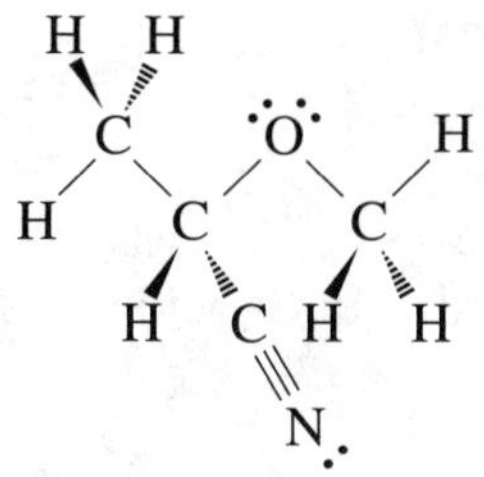

(b)

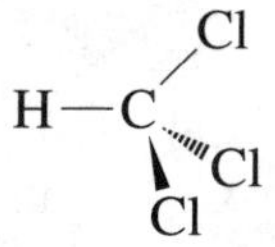

(c)

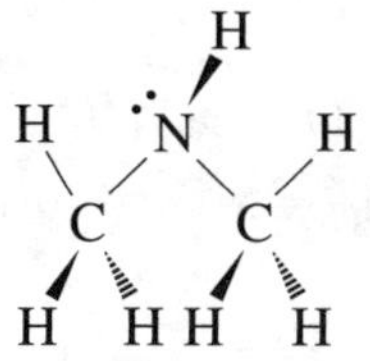

(d)

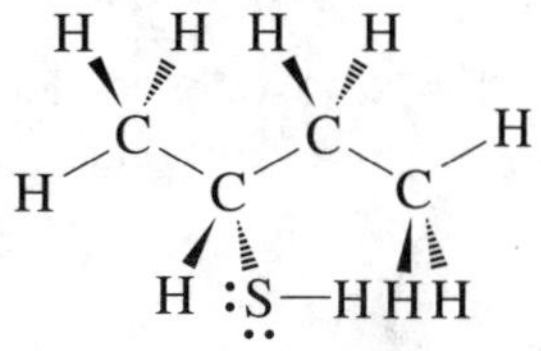

45. **(a)** C_5H_{12}. Begin with the isomer that contains all the carbon atoms connected in a straight chain. Then shorten the chain by one carbon at a time, connecting the removed atom as a substituent to interior positions of the remaining chain until every possibility has been drawn. There are three isomers:

(1) CH_3—CH_2—CH_2—CH_2—CH_3 or $CH_3CH_2CH_2CH_2CH_3$ or $CH_3(CH_2)_3CH_3$. These are all commonly used forms of condensed formulas for the same molecule.

Bond-line:

(2) CH_3—$\overset{\overset{\displaystyle CH_3}{|}}{CH}$—$CH_2$—$CH_3$ CH_3—CH_2—$\overset{\overset{\displaystyle CH_3}{|}}{CH}$—$CH_3$ is the same molecule turned around.

Also, $CH_3\overset{\overset{\displaystyle CH_3}{|}}{CH}CH_2CH_3$ and $(CH_3)_2CHCH_2CH_3$.

Bond-line:

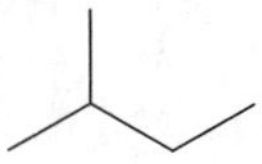

(3) CH_3—$\overset{\overset{\displaystyle CH_3}{|}}{\underset{\underset{\displaystyle CH_3}{|}}{C}}$—$CH_3$ Same as $(CH_3)_4C$.

Bond-line:

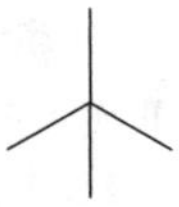

(b) C_3H_8O. Again, there are three:

(1) CH_3—CH_2—CH_2—OH Same as $CH_3CH_2CH_2OH$.

Bond-line:

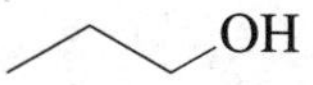

(2) CH_3—$\overset{\overset{\displaystyle OH}{|}}{CH}$—$CH_3$ Same as CH_3—$\overset{\overset{\displaystyle CH_3}{|}}{CH}$—$OH$ Also CH_3CHCH_3, $(CH_3)_2CHOH$.

Bond-line:

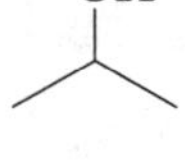

(3) $CH_3\!-\!CH_2\!-\!O\!-\!CH_3$ Same as $CH_3\!-\!O\!-\!CH_2\!-\!CH_3$. Also $CH_3CH_2OCH_3$.

   Bond-line: $\diagup\!\!\diagdown_O\!\diagup$

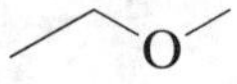

46. Remember: The most important consideration is the presence of electron octets around as many atoms as possible (except for H, of course). All C, N, and O atoms have octets in the structures below.

   (a) $HC\!\equiv\!CCH_3$ and $H_2C\!=\!C\!=\!CH_2$

   (b) $CH_3C\!\equiv\!N\!:$ and $\overset{+}{CH_3N}\!\equiv\!\overset{-}{C}\!:$

   Charges in CH_3NC: for N, (5 valence electrons in atom) $-\frac{1}{2}$(8 e^- in bonds) $= +1$; for C, (4 valence e^- in atom) $-\frac{1}{2}$(6 e^- in bonds) $-$ (2 e^- in lone pair) $= -1$.

   (c) 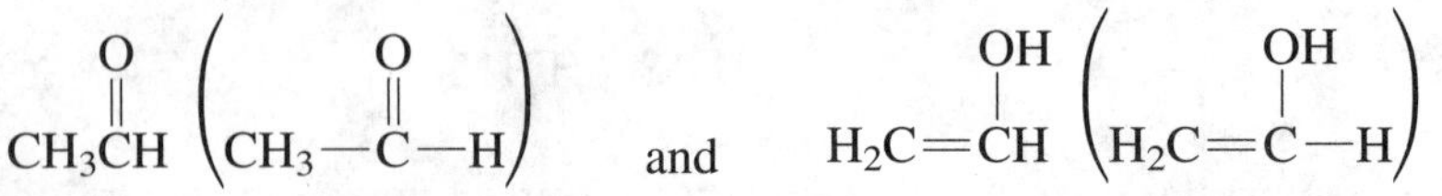 $CH_3\overset{\overset{O}{\parallel}}{C}H$ $\left(CH_3\!-\!\overset{\overset{O}{\parallel}}{C}\!-\!H\right)$ and $H_2C\!=\!\overset{\overset{OH}{|}}{C}H$ $\left(H_2C\!=\!\overset{\overset{OH}{|}}{C}\!-\!H\right)$

   None of the pairs of molecules above consist of resonance forms: In each case the two structures differ in the relative positions of the **atoms**. Resonance forms differ only in the disposition of the **electrons**.

47. (a) 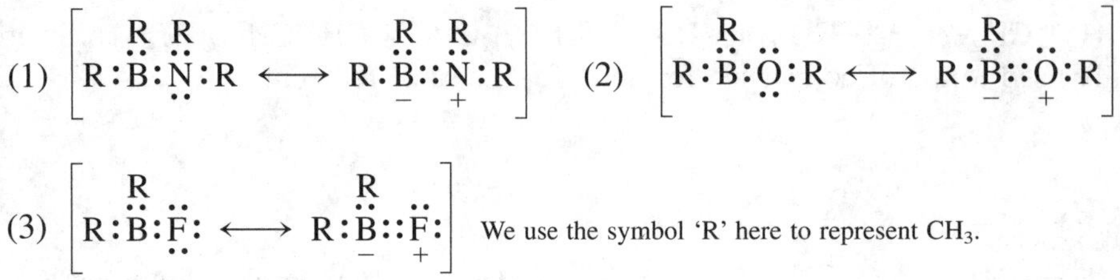 (1) $\left[\begin{array}{c} \overset{R}{\,}\,\overset{R}{\,} \\ R\!:\!\overset{..}{B}\!:\!\overset{..}{N}\!:\!R \\ \overset{..}{\,} \end{array} \longleftrightarrow \begin{array}{c} \overset{R}{\,}\,\overset{R}{\,} \\ R\!:\!\overset{..}{B}\!:\!:\!\overset{..}{N}\!:\!R \\ {}_{-}\quad{}_{+} \end{array}\right]$ (2) $\left[\begin{array}{c} \overset{R}{\,} \\ R\!:\!\overset{..}{B}\!:\!\overset{..}{O}\!:\!R \\ \overset{..}{\,} \end{array} \longleftrightarrow \begin{array}{c} \overset{R}{\,} \\ R\!:\!\overset{..}{B}\!:\!:\!\overset{..}{O}\!:\!R \\ {}_{-}\quad{}_{+} \end{array}\right]$

   (3) $\left[\begin{array}{c} \overset{R}{\,} \\ R\!:\!\overset{..}{B}\!:\!\overset{..}{F}\!: \\ \overset{..}{\,} \end{array} \longleftrightarrow \begin{array}{c} \overset{R}{\,} \\ R\!:\!\overset{..}{B}\!:\!:\!\overset{..}{F}\!: \\ {}_{-}\quad{}_{+} \end{array}\right]$ We use the symbol 'R' here to represent CH_3.

   (b) The octet rule takes precedence over the charge-separation rules, so the double bond-containing structures are preferred in all three cases.

   (c) In each double-bond-containing structure a positive charge resides on an electronegative atom (F, O, or N). As a result of the electronegativity order $F > O > N$, F is least able to accommodate a positive charge. Thus, the charge-separated resonance form of R_2BF will be favored to the smallest extent. This resonance form is more favorable in R_2BOR and is even better in R_2BNR_2, because the ability of the electronegative atom to bear the positive charge increases in the order $F < O < N$.

   (d) Both are sp^2 hybridized to accommodate the resonance form with the double bond.

48. Each marked carbon is attached to its three unmarked neighbors in a *trigonal planar* manner. This arrangement is best explained by sp^2 hybridization at C*, with the three sp^2 hybrids engaged in bonding between each C* and its CH_2 neighbors. The σ bond connecting the two C* atoms is perpendicular to the two planes incorporating the sp^2 hybrids and results from overlap of the pure p orbitals remaining on each C*:

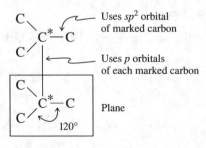

The C*—C* bond, resulting from overlap of unhybridized p orbitals, is longer and much weaker than the usual sp^3—sp^3 linkage.

49. **(a)** (1) The negatively charged carbon is bonded to three other atoms and has a lone pair, similar to N in NH_3: sp^3.

(2) Compare 38(d): Carbon will be sp^2 (double bond requires a p orbital).

(3) Compare 38(e): Carbon will be sp (triple bond requires two p orbitals).

(b) How is orbital energy related to ability to accommodate negative charge? Species containing electrons in lower energy orbitals are more stable than those with electrons in higher energy levels. Given the orbital energy order $sp < sp^2 < sp^3$, it follows that the relative ability to bear negative charge will be $HC{\equiv}C^-$ (charge in an sp orbital) $> CH_2{=}CH^-$ (sp^2) $> CH_3CH_2{}^-$ (sp^3).

(c) From (b), $HC{\equiv}C^-$ is more stable than $CH_2{=}CH^-$, which is more stable than $CH_3CH_2{}^-$. These are formed in the equilibria

$$HC{\equiv}CH \rightleftharpoons H^+ + HC{\equiv}C^- \qquad \text{(Most favorable—gives most stable anion)},$$

$$CH_2{=}CH_2 \rightleftharpoons H^+ + CH_2{=}CH^- \qquad \text{(Less favorable), and}$$

$$CH_3CH_3 \rightleftharpoons H^+ + CH_3CH_2{}^- \qquad \text{(Least favorable—gives least stable anion).}$$

Thus, we have the following order of acidity: $HC{\equiv}CH > CH_2{=}CH_2 > CH_3CH_3$.

50. $e > c > d > a > b$ After the cation, positive character on carbon is related to number of (polarized) bonds to electronegative atoms.

51. The letters in the structure below refer to the letters in the problem. In some cases, only representative bonds or atoms are labeled.

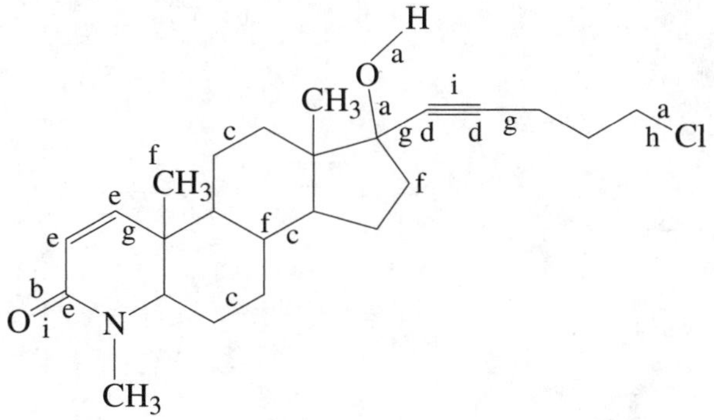

52. **(a)**

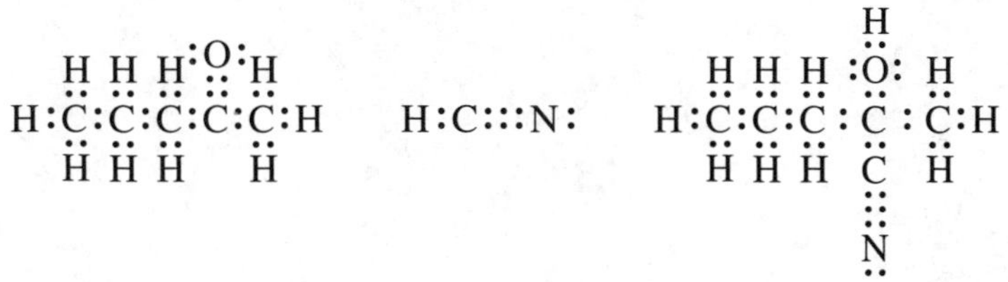

(b) and **(c)**

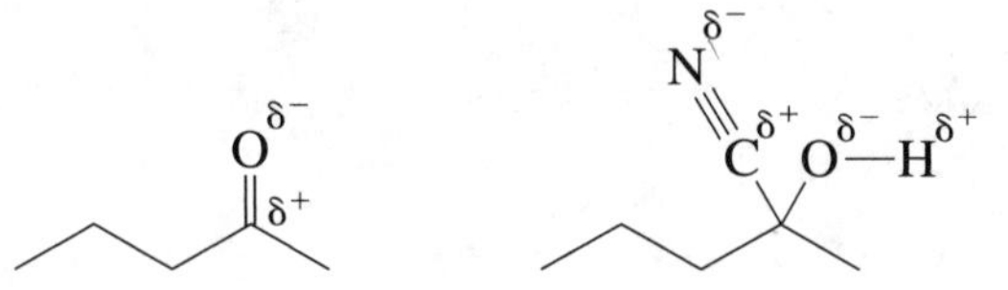

(d)

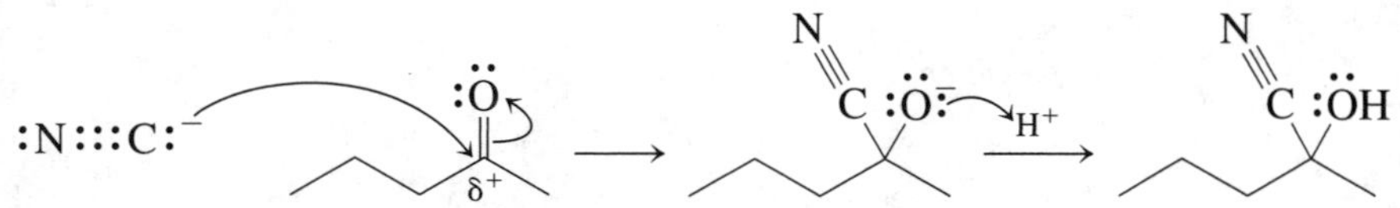

The δ^+ carbon being attacked by the cyanide ion already has an octet. So one electron pair of the two in the double bond is forced to move up to the oxygen atom to avoid violating the octet rule.

53. **(c)** Divide the percentage value by the atomic weight to get the relative number of atoms: carbon, 84/12 = 7   hydrogen, 16/1 = 16

54. **(c)** Aluminum: 3 (group number) − 0 (unshared electrons) − 1/2 (8) (shared electrons) = −1

55. **(a)** Three atoms attached to the carbon: trigonal planar geometry = sp^2 hybridization

56. **(c)**

57. **(a)** Same total number of electrons; all atoms in same locations (same geometric framework)

2

Structure and Reactivity: Acids and Bases, Polar and Nonpolar Molecules

Organic chemistry is largely a study of chemical reactions involving organic molecules. The textbook chapter therefore begins with a review of the principles of kinetics and thermodynamics, which apply to *all* reactions. Then we proceed to discuss acids and bases, which serves two purposes: It provides us with good examples of thermodynamics as applied to reactions, and it illustrates a process that is actually closely related to most of the reactions of polar organic molecules. After an introduction to functional groups and the classes of organic compounds, the chapter turns to a consideration of the simplest of those classes, the nonpolar alkanes. These sections cover (1) how to name organic molecules (*nomenclature*), (2) the relation of the physical properties of molecules to their molecular structure, (3) flexibility and shape of molecules (*conformation*), and (4) applications of kinetics and thermodynamics to changes in molecular shape. In Chapter 3 we cover chemical reactions of alkanes.

Outline of the Chapter

2-1 Kinetics and Thermodynamics
The energetic factors that govern the transformations of molecules.

2-2 Acids and Bases; Electrophiles and Nucleophiles
Reviewing chemistry that you've seen before, with an eye to chemistry you will see a *lot* of in the chapters to come. Also, a first look at the notation used to describe how organic reactions take place.

2-3 Functional Groups
The "business ends" of molecules: where reactions are likely to occur.

2-4 Straight-Chain and Branched Alkanes
Alkanes of various structures. Isomers.

2-5 Nomenclature
The first of a group of rules used to unambiguously name any organic compound.

2-6 Physical Properties
A topic that is usually not emphasized very much but that does reveal several useful, generalizable points about molecules.

2-7 and 2-8 Conformations
A discussion of the spatial arrangements that are possible for atoms in alkanes and the energy changes associated with their interconversions.

Keys to the Chapter

2-1. Kinetics and Thermodynamics

This section introduces ideas associated with energy changes in organic chemistry. Even though some of the terminology may be somewhat familiar to you from general chemistry, a few comments may be useful for orientation purposes. In this course you are going to encounter a lot of discussion concerning the *energy content* of molecules or other species. This term will refer in general to what is called *potential energy* in physics: energy that is stored in some way and can **potentially** be released in some process later on.

Discussions involving energy will often refer to the *stability* or *instability* of various substances or systems. Energy and stability are related in the following way: A species with high-energy content will tend to want to get rid of some of its energy somehow. So, relatively speaking, high-energy species are generally unstable.

Heat and energy are also related, so high-energy species will have a tendency to undergo processes that give off lots of heat. However, that a substance is capable of doing such a thing doesn't necessarily mean that it will do it quickly. The point here is that the *rate* of a process is the subject of *kinetics,* whereas the *energetic favorability* is one of *thermodynamics,* and the two are very different. Energetically favorable processes can take place at fast rates, slow rates, or in some cases, hardly seem to take place at all. A wooden match in the presence of air is a good chemical example of the latter. The reactions with oxygen of the compounds in the wood as well as on the head of the match are all extremely energetically favorable (*thermodynamics*), but nothing perceptible happens at room temperature. Why not? The *rate* of the reaction is too low: The number of molecules actually reacting with the oxygen at room temperature is so small that nothing seems to be happening at all (*kinetics*). However, when we strike the match—heat the match head with friction—it starts to burn and continues until the whole thing has burned up. The reaction of most organic molecules with oxygen requires energy *input* to get started even though it ultimately results in net energy *output* after the reaction has finished. The reason is as follows: In most reactions, old bonds are broken and new ones are formed, but not exactly simultaneously. Some partial breaking of old bonds has to take place before anything else, and that requires an input of energy. Once this process has started, it can lead to the formation of new bonds, and the release of energy—enough to make more old bonds break plus extra in the form of the flame and heat of burning. This initial energy input is the *activation energy* of the reaction, and it is a key factor governing *kinetics: rates* of reactions.

This section provides a brief mathematical description of each of the main concepts involved in *thermodynamics* and *kinetics* as applied to organic chemistry. The equations are generally fairly straightforward in their application. The problems will give you several chances to use them.

2-2. Acids and Bases; Electrophiles and Nucleophiles

The beginning of this text section covers the material you are most likely to have encountered in general chemistry: the mechanics of acid-base chemistry. The guiding principle is that such processes are reversible and are governed by thermodynamics. Notions of strong and weak acids and bases are based upon the positions of equilibria of the general sort

$$\text{stronger acid + stronger base} \rightleftharpoons \text{weaker acid + weaker base} \quad \Delta G° < 0,$$

where the thermodynamic driving force favors conversion of the stronger acid and stronger base into the weaker ones. While this concept may be familiar to you, you may not be quite as used to the *relative* nature of the terms *strong* and *weak.* In other words, a compound that acts as the strong acid in one such equation may be the weak partner in another, or may even play the role of a base. After all, the range of known acid strengths covers sixty orders of magnitude, and in organic chemistry we will encounter examples from every part of the scale. Water is the most familiar of substances to show such varied behavior.

Water acts as a weak acid (the way we normally think of it):

$$HCl \;+\; NaOH \;\rightleftharpoons\; NaCl \;+\; \mathbf{H_2O}$$

stronger	stronger	weaker	**weaker**
acid	base	base	**acid**

But water may act as the strong acid (and conversely, hydroxide as the weak base!):

$$\mathbf{H_2O} \;+\; NaNH_2 \;\rightleftharpoons\; NaOH \;+\; NH_3$$

stronger	stronger	weaker	weaker
acid	base	base	acid

And finally, water acts as a base when it encounters a strong enough acid:

$$HCl \;+\; \mathbf{H_2O} \;\rightleftharpoons\; Cl^- \;+\; H_3O^+$$

stronger	**stronger**	weaker	weaker
acid	**base**	base	acid

In describing acid-base reactions, we define a very simple relationship: the one between an acid and its conjugate base (or, conversely, a base and its conjugate acid). Through this relationship it is possible to estimate the strength of acids and bases that we've never seen before by making structural comparisons with species with which we are more familiar. We use the notion that, *relatively* speaking, strong acids have weak conjugate bases, and vice versa. Through this relationship, we may use an analysis of either component of a conjugate acid-base pair to find the strengths of both, relative to other acids and bases. The most common application is to determine the strength of an acid by evaluating effects that stabilize (and make *weaker*) its conjugate base: increased *size* and increased *electronegativity* of the negatively charged atom, and any effects that disperse negative charge away from the negatively charged atom, such as resonance.* By comparing the degree to which these properties are present in each of a pair of conjugate bases, you can usually tell which of the corresponding conjugate acids is stronger or weaker.

This text section also reviews the definitions of Lewis acids and bases and compares them with their analogs in organic chemistry: *electrophiles* and *nucleophiles*. The latter are the two terms that we use to describe electron-poor and electron-rich atoms in molecules, respectively. Such atoms possess partial or full electrical charges and, as a result, are places where chemical reactivity is usually high. Many of the functional groups are characterized by the presence of electrophilic or nucleophilic carbon atoms, for example. The analogy between a simple inorganic acid-base reaction and an organic nucleophilic substitution is illustrative of these principles. It also utilizes the "curved arrow" notation that we first presented in Chapter 1 when we discussed the shifting of electron pairs to interconvert resonance forms. Here, however, we use the arrows to show the electron movement that takes place when bonds break or form in the course of a chemical reaction. As your first example of a polar organic reaction, pay close attention to the details here: We will be returning to these principles repeatedly. The curved arrow convention is an especially powerful tool to help you understand how and why chemical reactions of organic compounds take place. *The more you understand, the less you will have to memorize.*

2-3. Functional Groups

One look at the 16 classes of organic compounds in Table 2-3 (and these are only some of the most common ones!) will immediately tell you how complicated organic chemistry can become. At the same time, however, closer inspection reveals features of these categories that can greatly simplify learning in this course. Each

---

* You may have learned in general chemistry that the *bond strength* between H and A in H−A relates inversely to its *acid strength.* This correlation is not as general as you may have been led to believe: It holds only when the acids being compared are all from the same column of the periodic table, such as the hydrogen halides. It fails, for instance, in the series CH_4, NH_3, H_2O, HF, where the acid strength increases as the bond strength goes up! The reason? Acidity relates to *heterolytic* bond cleavage to give *ions,* whereas bond strength relates to *homolytic* bond cleavage to give *uncharged* species. The two processes are very different. Differences in atomic *electronegativity* affect heterolytic bond cleavage (and therefore acidity) much more.

compound class is characterized by a specific atomic grouping called a functional group. Notice that only nine different elements are represented: C, H, S, N, O, and the four halogens. In fact, 11 of these classes contain only carbon, hydrogen, and oxygen. Knowledge of the characteristics of these atoms and the bonds between them, as we will see, will tell us the properties of the functional groups in which they appear. The functional groups will, in turn, provide the key to understanding the chemistry of **all** the members of the category. Thus, **all** members of the "alcohol" class of compounds, for instance, have certain common physical and chemical properties, resulting from the presence of the OH group in **all** of them. This kind of generalizable, qualitative similarity among compounds in any given class allows organic chemistry to be learned in a structured, organized, and, above all, logical way.

Functional groups consist either of polarized bonds, whose atoms can attract other polarized or charged species, thereby leading to reactions, or of multiple (double or triple) bonds that also show reactivity for reasons we'll explore later. Functional groups are the parts of molecules that most often take part in chemical reactions of those molecules. They are the "centers of reactivity" of molecules—where the action is.

The most fundamental feature of alkanes relates to this concept of functional groups: Alkanes don't have any. We'll see the consequences of this in the next chapter.

2-4 and 2-5. Structures and Names for Alkanes

There are a **lot** of organic compounds. Table 2-4 lists the numbers of isomers of just alkanes, and only goes up to 20 carbons, and already over half a million structures are possible! Imagine how many more structures can be manufactured when functional groups are present, or when the molecules get larger. Obviously not all these possible structures exist in nature or have been prepared in laboratories. Nonetheless, over 80 million different compounds are known at present, and nomenclature is the language that allows anyone interested in any of these materials to communicate about them in a clear and sensible way.

The text presents brief descriptions of the problems associated with naming compounds before the systematic procedures of the IUPAC were developed. It then goes on to introduce just the rules necessary for naming simple alkanes: molecules containing only carbon and hydrogen atoms and having only single bonds holding the atoms together. Only four rules are needed at this stage:

1. Identify the longest carbon chain (the *parent* chain) in the molecule and name it.
2. Name all groups attached to this chain as *substituents*.
3. Number the carbon atoms of the parent chain from the end that gives the one containing the first substituents the lowest possible number.
4. Assemble the name, using the proper format.

Although examples are given in the text and there are lots of problems for you to practice on, here are four additional worked-out examples to further clarify some fine points of the procedure.

Example 1.

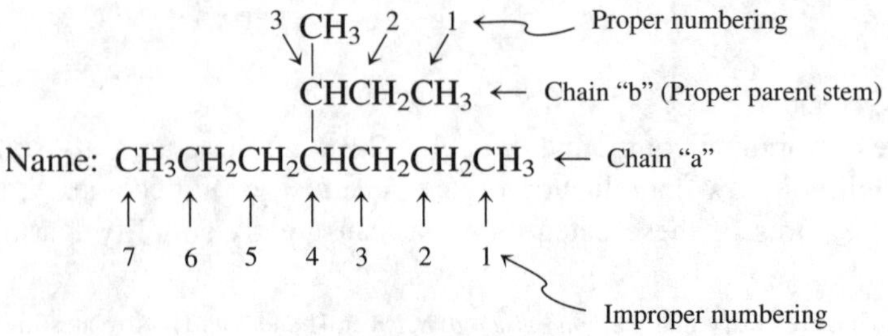

$$\begin{array}{c} \overset{3}{\text{CH}_3} \overset{2}{\diagup} \overset{1}{\diagup} \; \longleftarrow \; \text{Proper numbering} \\ \text{CHCH}_2\text{CH}_3 \; \longleftarrow \; \text{Chain "b" (Proper parent stem)} \\ \text{Name: CH}_3\text{CH}_2\text{CH}_2\text{CHCH}_2\text{CH}_2\text{CH}_3 \; \longleftarrow \; \text{Chain "a"} \\ \overset{\uparrow}{7} \; \overset{\uparrow}{6} \; \overset{\uparrow}{5} \; \overset{\uparrow}{4} \; \overset{\uparrow}{3} \; \overset{\uparrow}{2} \; \overset{\uparrow}{1} \\ \longleftarrow \; \text{Improper numbering} \end{array}$$

Analysis: The longest chain contains seven carbons, so this is a heptane. However, there are two ways to identify a seven-carbon chain (see numbering). Which one is the parent? The rules specify that, in case of a tie for longest chain, the one with the **most** substituents is chosen as the parent. The seven-carbon chain labeled "a" has one substituent (a *sec*-butyl group on carbon 4). The seven-carbon chain labeled "b" has two substituents (a methyl on carbon 3 and a propyl on carbon 4), so it wins. The molecule is called 3-methyl-4-propylheptane.

Example 2.

$$\text{Name: CH}_3\text{CHCH}_2\text{CH}_2\text{CH}_2\text{CHCH}_2\text{CCH}_2\text{C} \longrightarrow \text{CHCHCH}_2\text{CH}_3$$

Analysis: The main chain here is unambiguous and 14 carbons long—the parent is tetradecane. Which is the correct numbering direction, however? Most of the groups are close to the right-hand end and will have low numbers if we number right-to-left. But that is **not** the criterion for determining which way to number the chain. The rule says to number in the direction that gives the carbon containing the **first** substituent the lowest possible number. If we number from right-to-left, the first substituted carbon is C3; if left-to-right, it is C2. So, left-to-right is correct, and the molecule's name is 6,10-diethyl-2,8,8,10,11,12-hexamethyltetradecane. Even though the name that comes from numbering the other way has mostly low numbers, (5,9-diethyl-3,4,5,7,7,13-hexamethyltetradecane), it is wrong—its lowest number is a "3," and the correct name's lowest number is a "2."

Example 3.

$$\text{Name: CH}_3 - \text{CH} - \text{CH} - \text{C} - \text{CH} - \text{CH}_3$$

Analysis: A hexane. Numbering left-to-right gives 2,3,4,4,5-pentamethylhexane; right-to-left gives 2,3,3,4,5-pentamethylhexane. The choice is made by comparing substituent numbers from lowest to highest. The name with the lower number **at the first point of difference** is the winner. So 2,3,3,4,5 is preferred over 2,3,4,4,5.

Example 4.

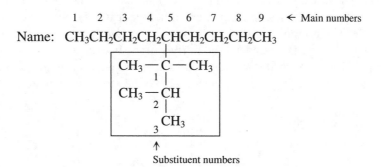

Analysis: A nonane with a complicated substituent on carbon 5. Rule 3 illustrates what to do. Number the substituent carbons from the point of attachment to the main chain, outward along the **substituent's** longest chain. The substituent has three carbons, so it has a name based on propyl. Then add appropriate numbers and names for groups attached to the substituent chain. So, 1,1,2-trimethylpropyl is the complete name **of the substituent.** Now, attach the substituent's name to the name of the main chain to get the name of the whole molecule: 5-(1,1,2-trimethylpropyl)nonane. Note punctuation. It's not hard, but it does take some careful analysis.

The notes above refer to the systematic nomenclature method as it is currently used. Please note, however, that there are many nonsystematic names in common use that are holdovers from the olden days and are still used for convenience or by force of habit. A number of compounds whose systematic names are very complicated have been given names that are well understood by people in the business but may seem random to the uninitiated. Several of these are mentioned in the text. One more example provides perspective in this area. Illustrated below is a compound that we eat every day.

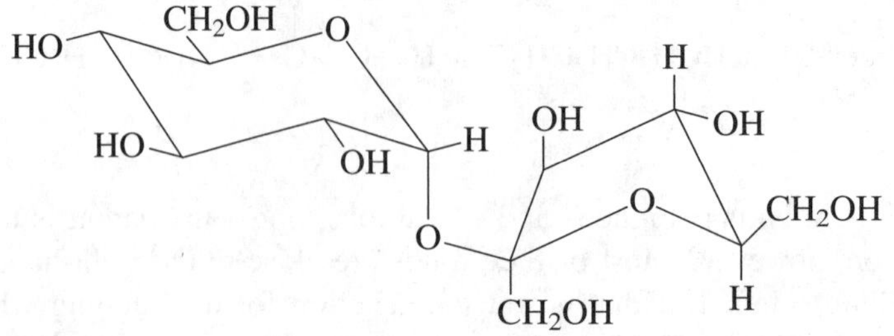

By the end of this course, you could sit down with the handbook of IUPAC rules and come up with the name 1-[3,4-dihydroxy-2,5-bis(hydroxymethyl)oxacyclopent-2-oxy]-3,4,5-trihydroxy-6-(hydroxymethyl)oxacyclohexane (and even this is only partially complete, lacking certain indicators that distinguish it from other known isomers!). Fortunately, the name used by general consent for this molecule, which is none other than ordinary table sugar, is a lot shorter: *sucrose*. See, even chemists use common sense sometimes.

Fear not, odds are you will never, ever, have to give an IUPAC name to a molecule like this. I never did, at least until I had to write this study guide.

2-6. Physical Properties

Every time we encounter a new class of compounds, we will briefly discuss common "physical properties" of members of that compound class. These will include general comments on the nature of the compound under ordinary conditions (e.g., diethylamine, colorless liquid, smells like something died, or, 2-hydroperoxy-2-isopropoxypropane, colorless crystalline solid, blows up like an A-bomb if you look at it cross-eyed). The purpose of these comments is to give you a feeling for what these materials are really like (as well as alerting you to the fact that some organic molecules may not be your friends). For the record, alkanes are colorless gases or liquids, with rather light odors, or white, waxy solids (candle wax is mainly alkanes).

More specific discussion will focus on relationships between molecular structure and physical properties for the class of compounds as a whole. In this chapter a brief summary of the kinds of forces that attract molecules to each other is presented. Alkanes, lacking charged atoms or highly polarized bonds, do not exhibit either ionic or dipolar forces. As **nonpolar** molecules, alkane molecules are attracted to each other by only the rather weak London forces. These can be understood fairly simply. In even a totally unpolarized bond, the electrons are always moving. Even though the average location of the electron pair is exactly half-way between the atoms, at any particular moment in time, the electrons may be closer to one atom or the other:

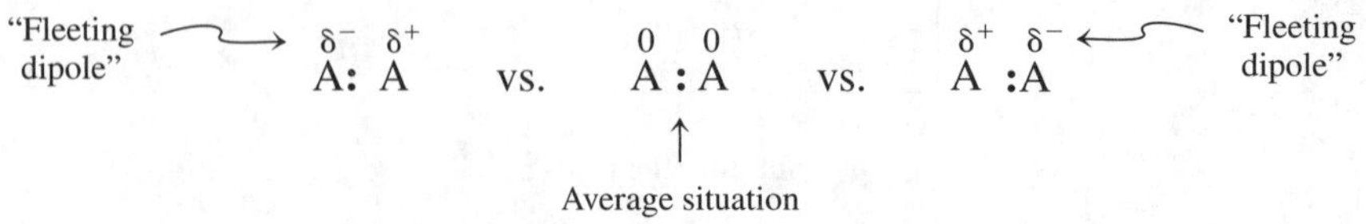

During these moments, the bond is polarized. Because this polarization is not permanent, the partial charges associated with it are only transient, or fleeting in nature, thus the name *fleeting dipoles*. When two nonpolar molecules are close to each other and a bond in one of them exhibits a fleeting dipole, the electrons in a nearby bond of the other molecule will be pushed away from the fleeting dipole's "−" end and attracted toward its "+" end. The positions and movements of all the electrons are said to be "correlated":

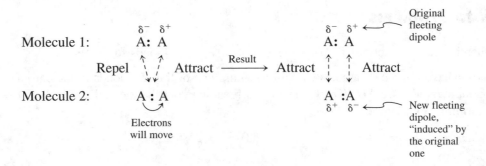

The result will be a new dipole in the second molecule's bond, "induced" by the original fleeting dipole in the first molecule. As the diagram shows, the polarizations that result lead to an attractive force between the molecules—the so-called London forces. Even though the dipoles involved have only transient existence and all the bonds are nonpolar, it turns out that the odds always favor the presence of some fleeting dipoles in a molecule, and the net result is this weak, but real, London attraction.

Because of the weakness of this attraction, alkanes exhibit relatively low melting points and boiling points relative to those of more polar or charged molecules. The nonpolar nature of alkanes results in other physical consequences, such as rather limited ability to serve as solvents for polar compounds (remember "like dissolves like" from freshman chemistry?). Lack of polarized bonds also very much limits the chemistry that alkanes can display. This subject will be taken up in the next chapter.

2-7 and 2-8. Conformations

Although we generally draw pictures of molecules in a single geometrical representation, the fact is that no molecule has a single rigid geometry. The electrons in bonds can be viewed as an elastic glue holding the atoms together. The bonds are therefore somewhat flexible and are subject to some degree of bending or stretching. So, even in the simplest molecules like H_2, the atoms are capable of some degree of movement with respect to one another. In more complicated molecules, additional forms of internal motion become possible. The *conformations* of ethane and larger alkanes are a result of rotation about carbon–carbon single bonds, a relatively easy process. This section describes the energetics associated with this rotation and the names associated with the various shapes of the molecules as this rotation occurs. Newman projections provide an "end-on" view of these conformations:

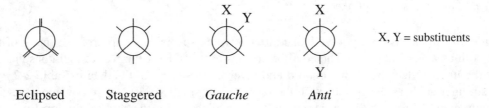

At this point you should take a look at a set of molecular models so that you can become familiar with these conformations in three dimensions.

Conformational energetics can be summarized for alkanes as follows:

1. Eclipsed is 2.9 kcal mol^{-1} higher in energy (less stable) than staggered for ethane.
2. Each CH_3–H eclipsing is 0.3 kcal mol^{-1} worse than an H–H eclipsing (relative to corresponding changes in staggered conformation energies).
3. Each CH_3–CH_3 eclipsing is 2.0 kcal mol^{-1} worse than an H–H eclipsing.
4. Each CH_3–CH_3 *gauche* is 0.9 kcal mol^{-1} worse than CH_3–CH_3 *anti*.

With these individual estimates, the graph of energy vs. rotational angle can be readily sketched for simple alkanes. Note: These "energy" values are actually *enthalpies* (heat content, or $\Delta H°$ values).

Solutions to Problems

26. **(a)** Remember, $\Delta H°$ (reaction) = $\Delta H°$ (bonds broken) − $\Delta H°$ (bonds formed).

(i) To calculate the $\Delta H°$ associated with breaking one of the two bonds in the carbon–carbon double bonds, use $\Delta H°$ (C=C) as a bond-breaking contribution and $\Delta H°$ (C—C) as a bond-forming contribution:

$$\Delta H° = \underset{\substack{\uparrow \\ \text{break} \\ \text{C=C}}}{146} + \underset{\substack{\uparrow \\ \text{break} \\ \text{Br—Br}}}{46} - \underset{\substack{\uparrow \\ \text{form} \\ \text{C—C}}}{83} - \underset{\substack{\uparrow \\ \text{form} \\ \text{2C—Br}}}{2(68)} = -27 \text{ kcal mol}^{-1}$$

(ii) $$\Delta H° = \underset{\substack{\uparrow \\ \text{break} \\ \text{C—H}}}{99} + \underset{\substack{\uparrow \\ \text{break} \\ \text{Br—Br}}}{46} - \underset{\substack{\uparrow \\ \text{form} \\ \text{C—Br}}}{68} - \underset{\substack{\uparrow \\ \text{form} \\ \text{H—Br}}}{87} = -10 \text{ kcal mol}^{-1}$$

(b) In reaction (i), two molecules combine to make one. This concentrates the energy content of the system into fewer particles, resulting in a large negative value for $\Delta S°$ (−35 entropy units) for reaction (i). If you like, the system becomes more "ordered," which is saying much the same thing. In reaction (ii), two molecules react to make two different molecules. The dispersal of the energy content of the system undergoes no major change, resulting in a $\Delta S°$ of approximately zero.

(c) For (i) at 25°C,

$$\Delta G° = \Delta H° - T\Delta S° = -27 - 298(-35 \times 10^{-3}) = -17 \text{ kcal mol}^{-1}$$

For (i) at 600°C,

$$\Delta G° = \Delta H° - T\Delta S° = -27 - 873(-35 \times 10^{-3}) = +4 \text{ kcal mol}^{-1}$$

For (ii) at either 25°C or 600°C, $\Delta G° \approx \Delta H° = -10 \text{ kcal mol}^{-1}$ because $\Delta S° \approx 0$.

Both reactions have negative $\Delta G°$ at 25°C, so both are thermodynamically favorable. At 600°C, the $\Delta S°$ for reaction (i) has made its $\Delta G°$ value positive: The reaction is therefore energetically unfavorable. Reaction (ii) is still just as good as it was at 25°C.

27. Don't identify the acids until you've looked to see which species give up protons; many of the species here can act both as acids and bases! The equilibrium lies to the side of the weaker acid/base pair, as indicated by the unequal lengths of the forward and reverse arrows. From the data in Table 2-2, you can identify the stronger acids as the species with the larger K_a or smaller (less positive or more negative) pK_a. The equilibrium constant for each reaction is found by dividing K_a for the acid on the left by K_a for the acid on the right. How did I know that? Here's how. For the following general reaction:

$$HA_1 + A_2^- \rightleftharpoons HA_2 + A_1^-$$

we have $K_{a_1} = [H^+][A_1^-]/[HA_1]$ and $K_{a_2} = [H^+][A_2^-]/[HA_2]$, right? So, $K_{a_1}/K_{a_2} = [H^+][A_1^-][HA_2]/[HA_1][H^+][A_2^-] = [HA_2][A_1^-]/[HA_1][A_2^-] = K_{eq})$

(a) $\underset{\text{weaker base}}{H_2O} + \underset{\text{weaker acid}}{HCN} \rightleftharpoons \underset{\text{stronger acid}}{H_3O^+} + \underset{\text{stronger base}}{CN^-} \quad K_{eq} = 1.3 \times 10^{-11}$

(b) $\underset{\text{weaker base}}{CH_3O^-} + \underset{\text{weaker acid}}{NH_3} \rightleftharpoons \underset{\text{stronger acid}}{CH_3OH} + \underset{\text{stronger base}}{NH_2^-} \quad K_{eq} = 3.1 \times 10^{-20}$

(c) $\underset{\text{stronger acid}}{HF} + \underset{\text{stronger base}}{CH_3COO^-} \rightleftharpoons \underset{\text{weaker base}}{F^-} + \underset{\text{weaker acid}}{CH_3COOH} \quad K_{eq} = 32$

(d) CH_3^- + NH_3 $\rightleftharpoons$ CH_4 + NH_2^- $K_{eq} = 10^{15}$

    stronger base   stronger acid   weaker acid   weaker base

(e) H_3O^+ + Cl^- $\rightleftharpoons$ H_2O + HCl $K_{eq} \approx 5.0 \times 10^{-7}$

    weaker acid   weaker base   stronger base   stronger acid

(f) CH_3COOH + CH_3S^- $\rightleftharpoons$ CH_3COO^- + CH_3SH $K_{eq} = 2.0 \times 10^5$

    stronger acid   stronger base   weaker base   weaker acid

28. **(a)** $H-\ddot{O}-H$ $H-CN:$ $\rightleftharpoons$ $H-\overset{H}{\underset{}{\overset{|}{O}}}\!\!{}^+\!\!-H$ + $^-:CN:$

(b) $CH_3-\ddot{O}:^-$ $H-\overset{H}{\underset{|}{N}}-H$ $\rightleftharpoons$ $CH_3-\ddot{O}-H$ + $^-:\ddot{N}H_2$

(c) $CH_3\overset{:O:}{\overset{||}{C}}-\ddot{O}:^-$ $H-\ddot{F}:$ $\rightleftharpoons$ $CH_3\overset{:O:}{\overset{||}{C}}-\ddot{O}H$ + $^-:\ddot{F}:$

(d) $H-\overset{H}{\underset{|}{\overset{|}{C}}}\!\!^--H$ $H-\overset{H}{\underset{|}{N}}-H$ $\rightleftharpoons$ CH_4 + $^-:\ddot{N}H_2$

(e) $H-\overset{H}{\underset{|}{\overset{|}{O}}}\!\!{}^+\!\!-H$ $^-:\ddot{Cl}:$ $\rightleftharpoons$ $H-\ddot{O}-H$ + $H-\ddot{Cl}:$

(f) $CH_3\overset{:O:}{\overset{||}{C}}-\ddot{O}-H$ $^-:\ddot{S}CH_3$ $\rightleftharpoons$ $CH_3\overset{:O:}{\overset{||}{C}}-\ddot{O}:^-$ + $H\ddot{S}CH_3$

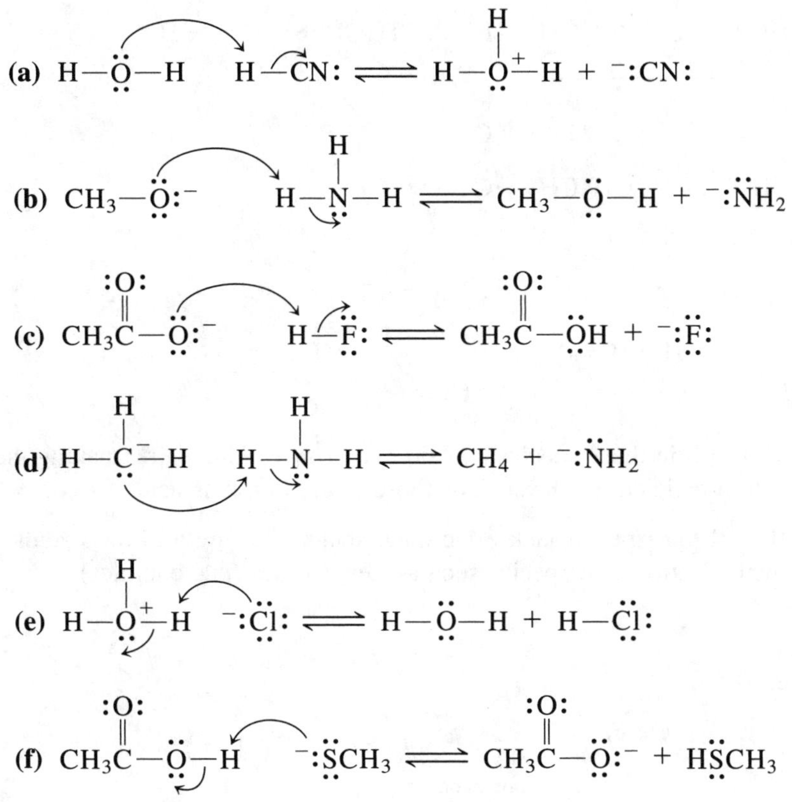

29. **(a)** CN^- is a Lewis base   **(b)** CH_3OH is a Lewis base

(c) $(CH_3)_2CH^+$ is a Lewis acid   **(d)** $MgBr_2$ is a Lewis acid

(e) CH_3BH_2 is a Lewis acid   **(f)** CH_3S^- is a Lewis base

Let's be smart about the second part. We have three Lewis acids and three Lewis bases. So pair them up and answer the question with just three equations. To reduce clutter, the three lone pairs around each halogen atom have been left out:

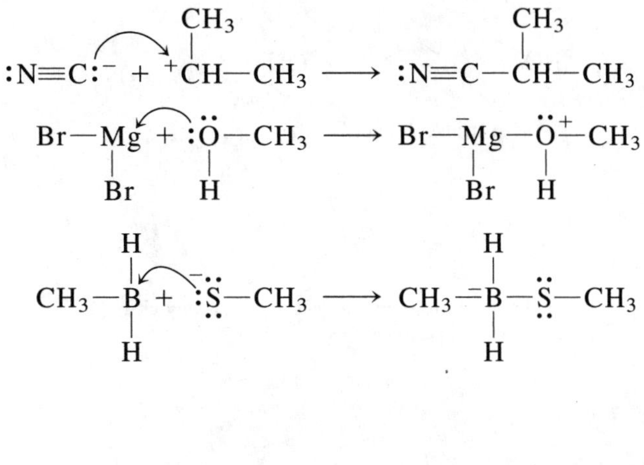

$:N\equiv C:^-$ + $^+\overset{CH_3}{\underset{|}{CH}}-CH_3$ $\longrightarrow$ $:N\equiv C-\overset{CH_3}{\underset{|}{CH}}-CH_3$

$Br-Mg$ + $:\ddot{O}-CH_3$ $\longrightarrow$ $Br-\overset{-}{Mg}-\overset{+}{\ddot{O}}\!\!-CH_3$

$\underset{Br}{|}$ $\underset{H}{|}$ $\underset{Br}{|}$ $\underset{H}{|}$

$CH_3-\overset{H}{\underset{H}{\overset{|}{B}}}$ + $:\ddot{S}-CH_3$ $\longrightarrow$ $CH_3\overset{H}{\underset{H}{=}}B-\ddot{S}-CH_3$

30. Refer to any table of electronegativities to determine bond polarities. Butane, 2-methylpropene, 2-butyne, and methylbenzene lack polarized bonds. The other structures have the polarized bonds shown.

$$CH_3\overset{\delta^+}{C}H_2\overset{\delta^-}{-}Br \qquad (CH_3)_2\overset{\delta^+}{C}H\overset{\delta^-}{-}\overset{}{O}\overset{\delta^+}{-}H \qquad CH_3\overset{\delta^+}{C}H_2\overset{\delta^-}{-}\overset{}{O}\overset{\delta^+}{-}CH_3 \qquad CH_3CH_2\overset{\delta^-}{-}\overset{}{S}\overset{\delta^+}{-}H$$

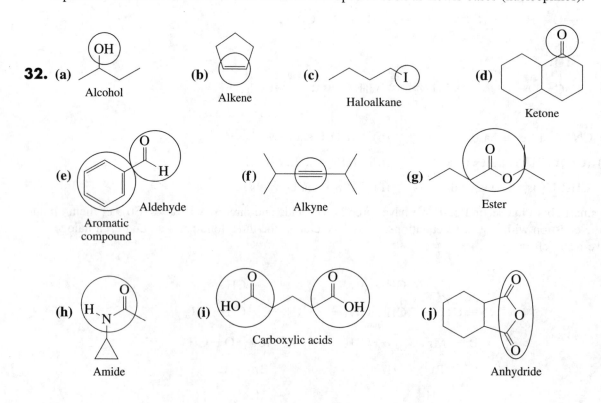

31. Nucleophiles: **(a)** and **(d)**. Both I and S in these species contain one or more lone pairs, making them Lewis bases, capable of attacking electron-deficient atoms such as those found in Lewis acids.

Electrophiles: **(b)**, **(c)**, **(e)**, and **(f)**. All four species lack filled outer shells; they are all Lewis acids capable of chemical interaction with electron-rich species such as Lewis bases (nucleophiles).

32. (a) Alcohol **(b)** Alkene **(c)** Haloalkane **(d)** Ketone

(e) Aromatic compound / Aldehyde **(f)** Alkyne **(g)** Ester

(h) Amide **(i)** Carboxylic acids **(j)** Anhydride

33. (a) $CH_3\overset{\delta^+}{-}CH_2\overset{\delta^-}{-}Br$ The δ^+ carbon (arrow) will attract the negatively charged oxygen atom of hydroxide ion.

(b) $CH_3-CH_2-\overset{\delta+}{C}\overset{\overset{\delta-}{O}}{\overset{||}{}}-H$

The δ^+ carbon will attract the lone pair on the δ^- nitrogen of ammonia. At the same time, the δ^- oxygen will attract a δ^+ hydrogen of ammonia.

(c) $CH_3-\overset{\delta+}{CH_2}-\overset{\delta-}{O}-\overset{\delta+}{CH_3}$

A δ^- oxygen lone pair will bond to H^+.

(d) $CH_3-CH_2-\overset{\delta+}{C}\overset{\overset{\delta-}{O}}{\overset{||}{}}-CH_2-CH_2-CH_3$

The ketone's δ^+ carbon will attract the negatively charged carbon of the carbanion.

(e) $CH_3-\overset{\delta+}{C}\equiv\overset{\delta-}{N}$

The lone pair on nitrogen will be attracted to the positively charged carbon.

(f) No reaction. Butane has no polarized atoms; it is therefore not reactive toward charged or polarized species.

34. **(a)** $H-\ddot{\overset{..}{O}}\!:^-$ H_2C-Br $\longrightarrow$ $HO-CH_2$ + Br^-

$\qquad\qquad\qquad$ $\underset{CH_3}{|}$ $\qquad\qquad\qquad$ $\underset{CH_3}{|}$

Compare to Example 3 in Section 2-2.

(b) $CH_3-CH_2-\overset{\overset{:\overset{..}{O}:}{||}}{C}-H$ $:NH_3$ $\longrightarrow$ $CH_3-CH_2-\underset{\overset{+}{N}H_3}{\overset{\overset{:\overset{..}{O}:^-}{|}}{C}}-H$

Like Example 4(a), except the electron-pair donor is neutral, not negative.

(c) $CH_3-\ddot{\overset{..}{O}}-CH_2CH_3$ H^+ $\longrightarrow$ $CH_3-\underset{..}{\overset{\overset{H}{|}}{\overset{+}{O}}}-CH_2CH_3$

Similar to Example 2, except the donor is neutral, as in part (b).

(d) $CH_3-CH_2-CH_2-\overset{\overset{:\overset{..}{O}:}{||}}{C}-CH_2-CH_3$ $:CH_3$ $\longrightarrow$ $CH_3-CH_2-CH_2-\underset{\overset{|}{CH_3}}{\overset{\overset{:\overset{..}{O}:^-}{|}}{C}}-CH_2-CH_3$

(e) $CH_3-C\equiv N\!:$ $^+CH_3$ $\longrightarrow$ $CH_3-C\equiv\overset{+}{N}-CH_3$ Like part (c) above.

35. Recall that condensed formulas tell you only what atoms are connected to what other atoms, *not* the real three-dimensional shape of a molecule. The longest chain is the chain with the most atoms, not necessarily the one drawn on a single horizontal line in these formulas.

(a) $\overset{5}{C}H_3\overset{4}{C}H_2\overset{3}{C}HCH_3$ $\qquad$ ∴ 2,3-Dimethylpentane

$\qquad\qquad$ $\underset{2}{|}$

$\qquad\qquad$ CH

$\qquad$ $\underset{1}{C}H_3$ CH_3

(b) Parent chain is already horizontal; number left-to-right (nonane): 2-methyl-5-(1-methylethyl)-5-(1-methylpropyl)nonane.

(c) 3,3-Diethylpentane, any way you look at it.

(d) Parent:

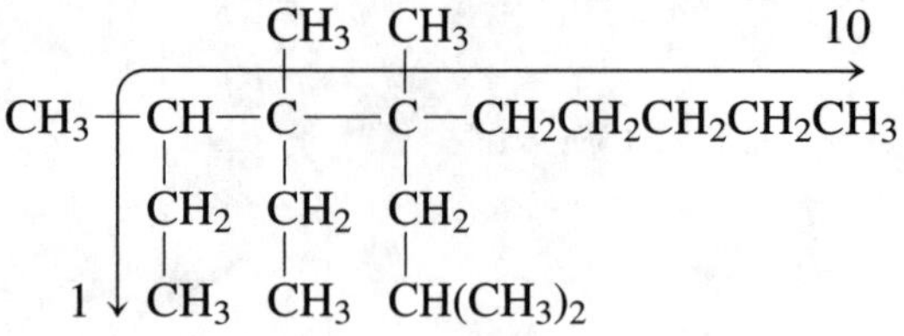

$$CH_3 - CH - C - C - CH_2CH_2CH_2CH_2CH_3$$

with CH_3 CH_3 at top (carbon 10 direction), CH_2 CH_2 CH_2 below, and CH_3 CH_3 $CH(CH_3)_2$ at bottom (carbon 1).

10 carbons, ∴ 4-ethyl-3,4,5-trimethyl-5-(2-methylpropyl)decane.

(e) Redraw:

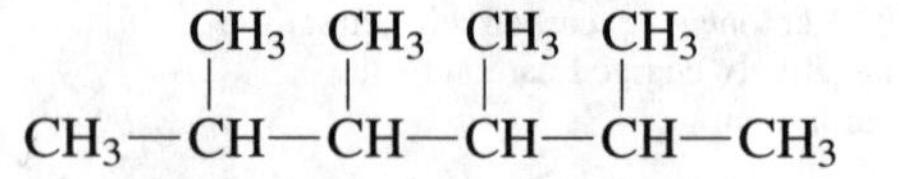

$$CH_3 - CH - CH - CH - CH - CH_3$$

with CH_3 CH_3 CH_3 CH_3 substituents. ∴ 2,3,4,5-Tetramethylhexane

(f) Hexane. Don't be fooled by the way it's drawn.

(g) 2-Methylpropane. For this one as well as the next three, redraw to show all the atoms, if you need to.

(h) 2,2-Dimethylbutane

(i) 2-Methylpentane

(j) 2,5-Dimethyl-4-(1-methylethyl)heptane

36. (a)

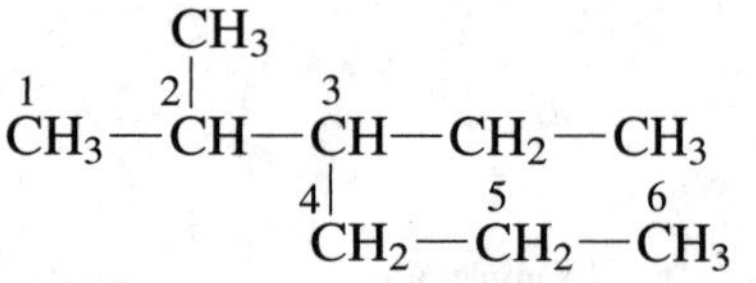

$$\overset{1}{CH_3} - \overset{2}{CH} - \overset{3}{CH} - CH_2 - CH_3$$

with CH_3 at carbon 2 and $\overset{4}{CH_2} - \overset{5}{CH_2} - \overset{6}{CH_3}$ below carbon 3.

Pentane is an incorrect parent name. Correct name is 3-ethyl-2-methylhexane.

(b) 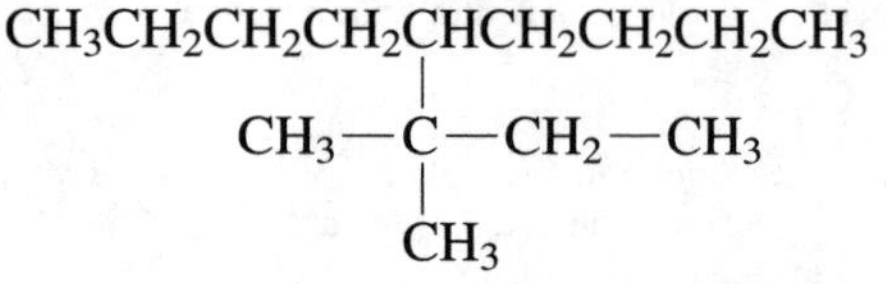 $CH_3CH_2CH_2CH_2CHCH_2CH_2CH_2CH_3$

with $CH_3 - C - CH_2 - CH_3$ and CH_3 below.

Name is correct.

(c)

$$\overset{1}{CH_3} - \overset{2}{CH} - \overset{3}{CH} - \overset{4}{C} - CH_2CH_2CH_3$$

with CH_3 CH_3 CH_3 above carbons 2, 3, 4 and $\overset{5}{CH_2CH_2CH_2CH_3}$ (carbon 8) below carbon 4.

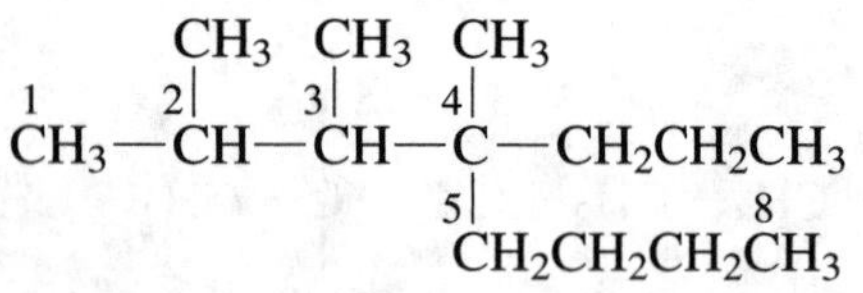

Not a heptane. Should be 2,3,4-trimethyl-4-propyloctane.

(d)

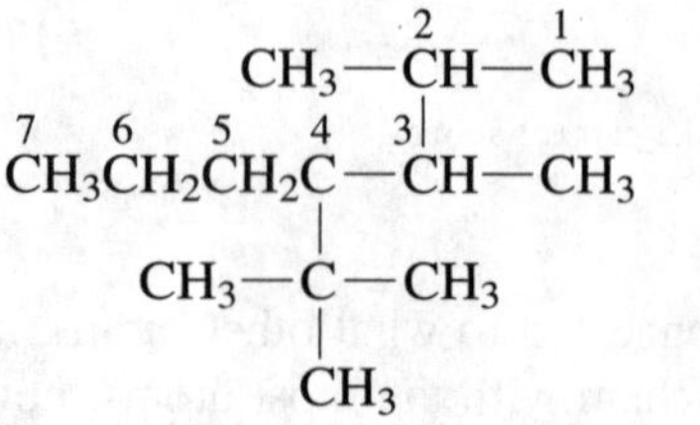

$$\overset{7}{CH_3}\overset{6}{CH_2}\overset{5}{CH_2}\overset{4}{C} - \overset{3}{CH} - CH_3$$

with $\overset{2}{CH} - \overset{1}{CH_3}$ (and CH_3) above carbon 3, and $CH_3 - C - CH_3$ with CH_3 below carbon 4.

Parent and numbering are both wrong. Rename as 4-(1,1-dimethylethyl)-2,3-dimethylheptane.

(e)

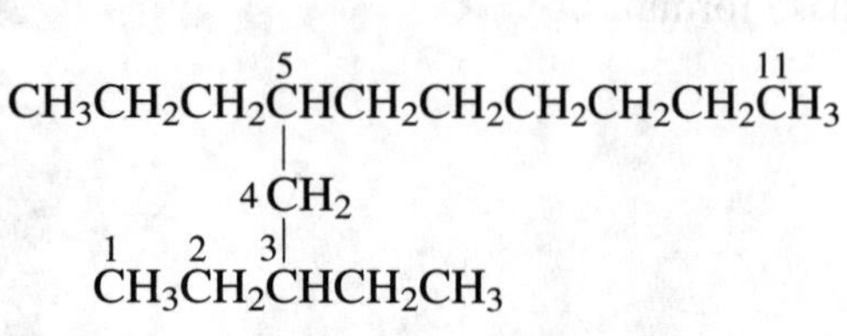

$$CH_3CH_2CH_2\overset{5}{C}HCH_2CH_2CH_2CH_2CH_2\overset{11}{CH_3}$$

with $\overset{4}{CH_2}$ below carbon 5, and $\overset{1}{CH_3}\overset{2}{CH_2}\overset{3}{C}HCH_2CH_3$ below.

Wrong parent chain. This is 3-ethyl-5-propylundecane.

(f) 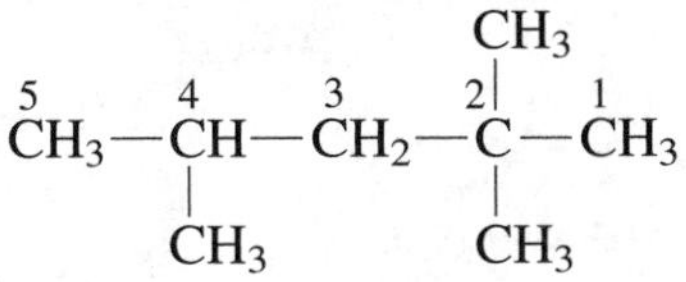 The numbering is backward. It should be 2,2,4-trimethylpentane.

(g) 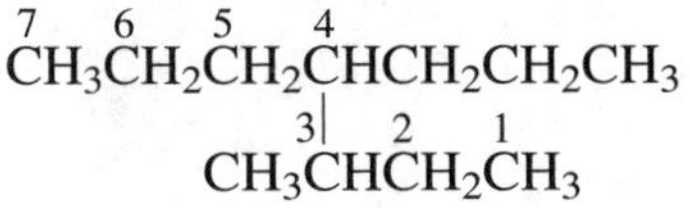 The parent chain is wrong, by the "maximum number of substituents" rule. Call it 3-methyl-4-propylheptane.

(h) 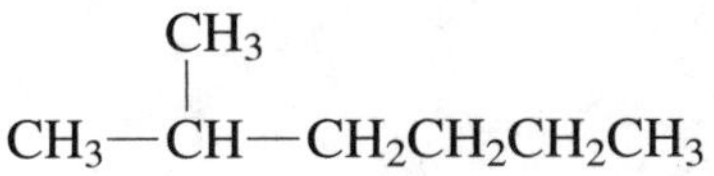 Isoheptane is a common name: Note the $(CH_3)_2CH$ group at the end of an otherwise straight chain. IUPAC name: 2-methylhexane.

(i) 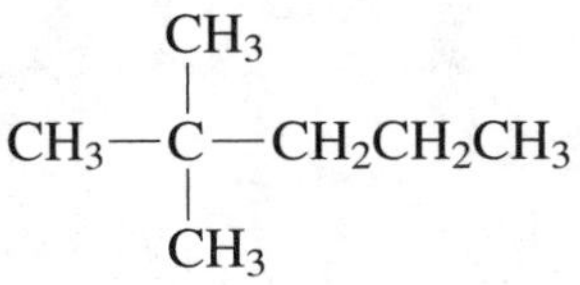 Another common name: *Neo* denotes the $(CH_3)_3C$ group at the end. The IUPAC name is 2,2-dimethylpentane.

37. **(a)** 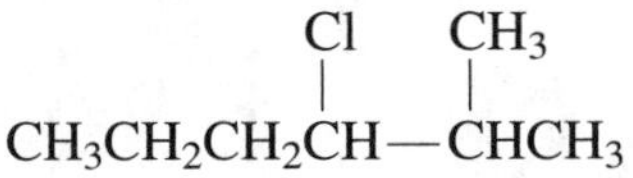 Name is incorrect. Numbering should go the other way (right to left), giving 3-chloro-2-methylhexane.

(b) 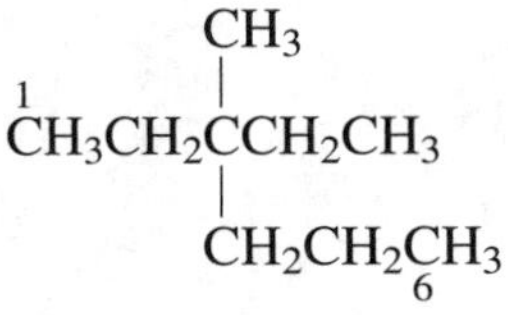 Name is incorrect. The wrong parent chain has been chosen. Number a six-carbon chain as shown, and then name it 3-ethyl-3-methylhexane.

(c) 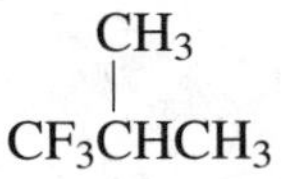 The name is correct.

(d) 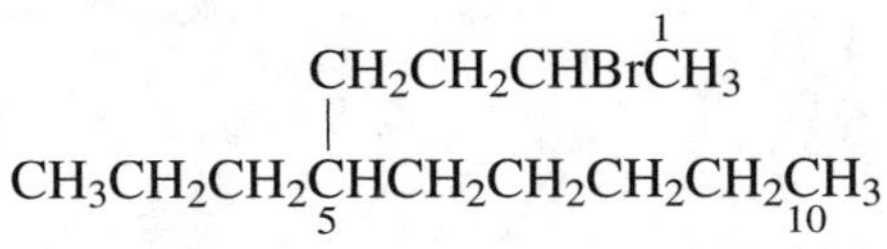 Another incorrect parent chain. Number as shown to give 2-bromo-5-propyldecane.

38. Do not answer questions like this by haphazardly writing down possible structures. You will almost certainly write down some molecules more than once. Do the problem systematically: Write down answers with successively shorter parent chains as shown here. There are nine C_7H_{16} isomers.

(1) $CH_3CH_2CH_2CH_2CH_2CH_2CH_3$ Heptane (7-carbon parent)

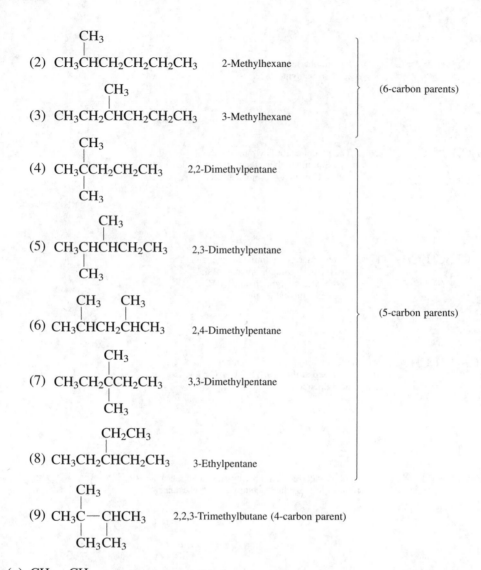

$$\underset{|}{CH_3}$$
(2) $CH_3CHCH_2CH_2CH_2CH_3$ 2-Methylhexane

$$\underset{|}{CH_3}$$
(3) $CH_3CH_2CHCH_2CH_2CH_3$ 3-Methylhexane

(6-carbon parents)

$$\underset{|}{CH_3}$$
(4) $CH_3CCH_2CH_2CH_3$ 2,2-Dimethylpentane
$$\underset{|}{CH_3}$$

$$\underset{|}{CH_3}$$
(5) $CH_3CHCHCH_2CH_3$ 2,3-Dimethylpentane
$$\underset{|}{CH_3}$$

$$\underset{|}{CH_3}\quad\underset{|}{CH_3}$$
(6) $CH_3CHCH_2CHCH_3$ 2,4-Dimethylpentane

(5-carbon parents)

$$\underset{|}{CH_3}$$
(7) $CH_3CH_2CCH_2CH_3$ 3,3-Dimethylpentane
$$\underset{|}{CH_3}$$

$$\underset{|}{CH_2CH_3}$$
(8) $CH_3CH_2CHCH_2CH_3$ 3-Ethylpentane

$$\underset{|}{CH_3}$$
(9) $CH_3C{-}CHCH_3$ 2,2,3-Trimethylbutane (4-carbon parent)
$$\underset{|}{CH_3}\underset{|}{CH_3}$$

39. **(a)** $CH_3{-}CH_3$ Both carbons and all hydrogens are primary.

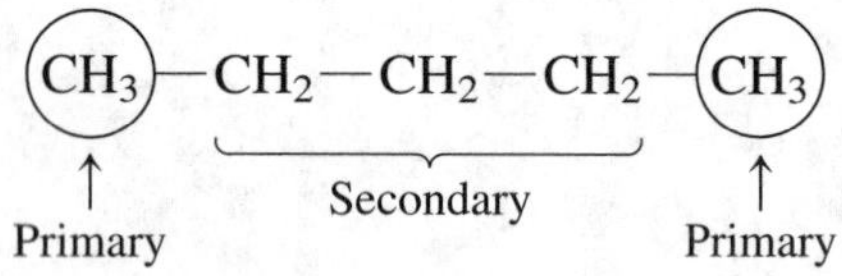

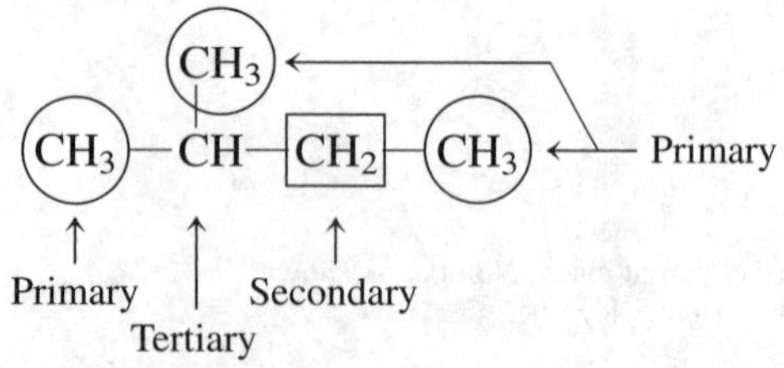

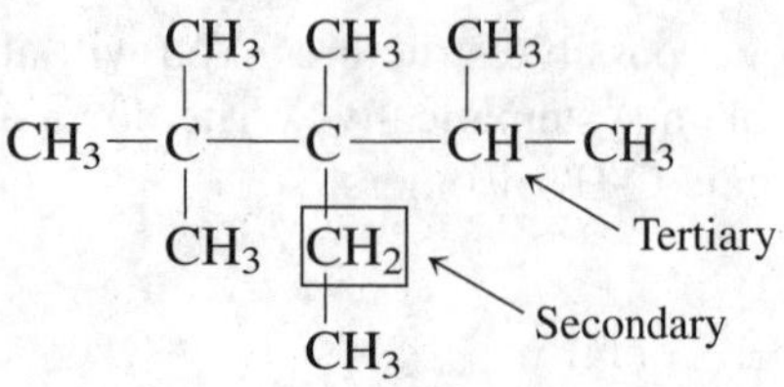

40. The designation is assigned according to the type of carbon at position number 1 (the "point of connection" position, indicated by the "open" bond).

(a)

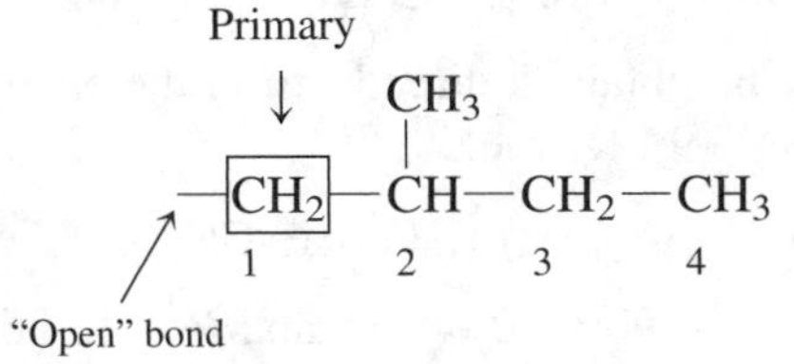

Primary

↓ CH₃
     |
—[CH₂]—CH—CH₂—CH₃   Primary; 2-methylbutyl
     1   2   3   4

↗
"Open" bond

(b) Primary; 3-methylbutyl

(c) Secondary; 1,2-dimethylpropyl

(d) Primary; 2-ethylbutyl

(e) Secondary; 1,2-dimethylbutyl

(f) Tertiary; 1-ethyl-1-methylpropyl

41. These are all isomers (C_8H_{18}) lacking polar functional groups, so the only consideration is that increasing boiling point correlates with increasing molecular surface area: the greater the surface area, the greater the total attractive force between molecules. Therefore straight chain compounds with extended geometric shapes will have the highest boiling points, and more branched isomers with more compact shapes will boil at lower temperatures. The most reasonable order is therefore **(d) < (c) < (a) < (b).**

42. (a) 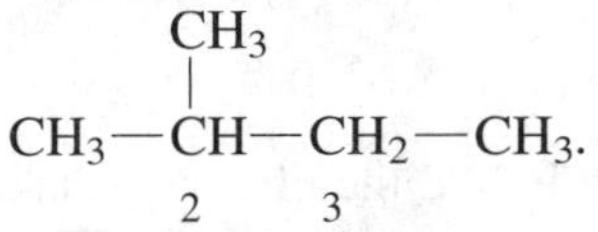
CH₃
 |
CH₃—CH—CH₂—CH₃. Best conformation is
     2   3

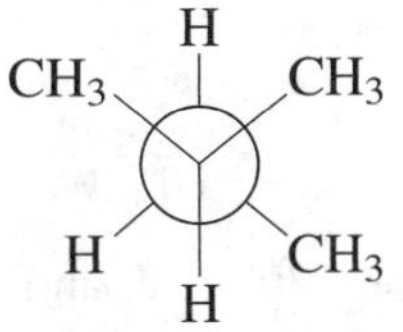

For more details, see Problem 44.

(b) 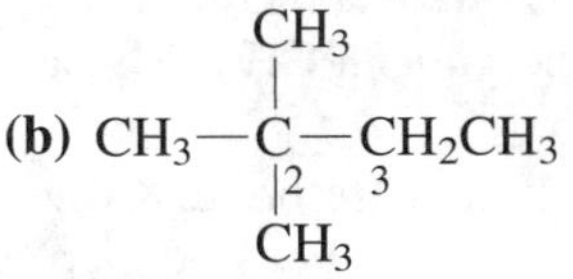
CH₃
 |
CH₃—C—CH₂CH₃
 |2   3
CH₃

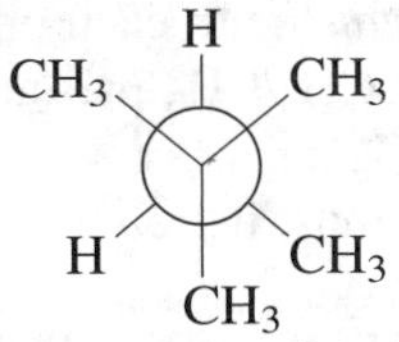

All three staggered conformations are equivalent.

(c) 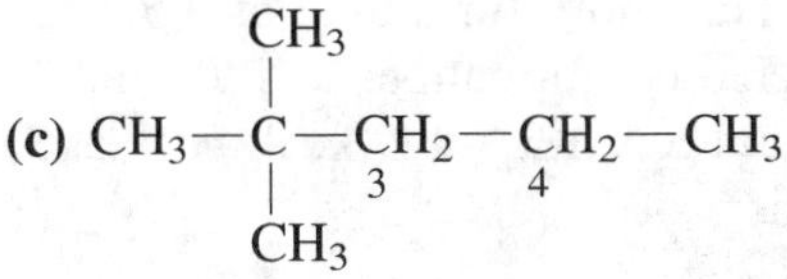
CH₃
 |
CH₃—C—CH₂—CH₂—CH₃
 | 3   4
CH₃

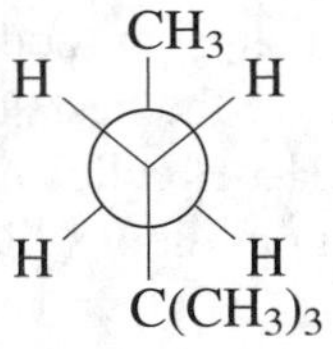

(d) 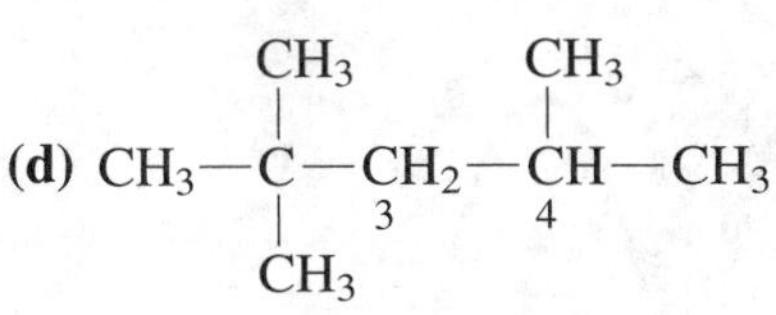
CH₃   CH₃
 | |
CH₃—C—CH₂—CH—CH₃
 | 3   4
CH₃

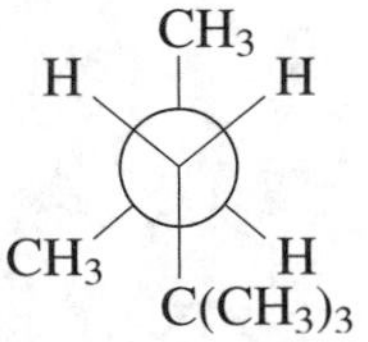

43. **(a)** That would be just ¹/₃ of the energy difference between staggered and eclipsed ethane, or about 1.0 kcal mol⁻¹.

(b) Methyl–hydrogen eclipsing appears in both propane and butane. In propane, the eclipsed conformations are 3.2 kcal mol⁻¹ above the staggered, 0.3 kcal mol⁻¹ higher than the 2.9 kcal mol⁻¹ torsional energy of ethane. This excess corresponds to the difference between CH₃–H eclipsing and H–H eclipsing, giving us an estimate of 1.3 kcal mol⁻¹ for the methyl–hydrogen interaction. We may check this estimate from the 60° and 300° eclipsed conformations of butane in Figure 2-13. Each has one H–H and two CH₃–H eclipsing interactions, and should have an energy of 1.0 + 2(1.3) = 3.6 kcal mol⁻¹, exactly what the plot reveals. It appears from this example that these values are additive, and may be used to predict conformational energies in general.

(c) The CH_3–CH_3 eclipsed conformation in butane (180° in Figure 2-13) is 4.9 kcal mol^{-1} above the most stable conformation. If we assume that 2.0 kcal mol^{-1} of this total derives from the two pairs of eclipsing hydrogens, we arrive at a value of 2.9 kcal mol^{-1} for the methyl–methyl interaction.

(d) A methyl–methyl *gauche* interaction energy value may be obtained directly from the energy difference between the *anti* and *gauche* forms of butane, 0.9 kcal mol^{-1}.

44. The problem deals with conformations about the C_2—C_3 bond of $(CH_3)_2CH$—CH_2CH_3

(a) Use $\Delta G° = -RT \ln K = -2.303RT \log K$. $T = 298$ K, $K = 90\%/10\% = 9$, and $R = 1.986$ cal deg^{-1} mol^{-1}. So, $\Delta G° = -2.303(1.986)(298) \log 9 = -(1.360) \log 9 = -(1.360)(0.954) = -1297$ cal mol^{-1} = -1.30 kcal mol^{-1}.

Do (b) and (c) together: You can't draw the diagram until you know what all the conformations look like! It doesn't matter where you start (what you define as the 0° conformation). Here are four Newman projections showing 180° rotation of C3:

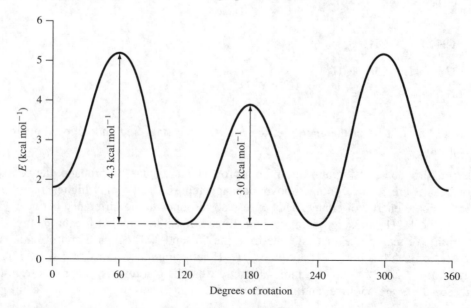

The 240° conformation is like the 120° one, and the 300° conformation is like the 60° one (make a model). Next, calculate relative energies for the diagram. Note that these will be *enthalpies* ($\Delta H°$), not free energies ($\Delta G°$). Determine the energies of each conformation relative to a 0 kcal mol^{-1} reference level that corresponds to a staggered conformation with no groups *gauche* to each other. Of the staggered conformations, the 120°/240° ones are best, with one *anti* CH_3/CH_3 pair and one *gauche* CH_3/CH_3 pair. Set these at +0.9 kcal mol^{-1} (for the one *gauche* CH_3/CH_3 pair).

The 0° conformation has two *gauche* CH_3/CH_3 interactions; its relative energy is therefore $2 \times 0.9 = +1.8$ kcal mol^{-1}.

The 60°/300° conformations are eclipsed with one H–H eclipsing (add 1.0 kcal mol^{-1}), one CH_3–H eclipsing (add 1.3 kcal mol^{-1}), and one CH_3–CH_3 eclipsing (add 2.9 kcal mol), for a total of +5.2 kcal mol^{-1} above our zero reference, 4.3 kcal mol^{-1} above the best staggered conformations at 120° and 240°.

The 180° conformation is eclipsed, with three CH_3–H eclipsings (add $3 \times 1.3 = 3.9$ kcal mol^{-1}), 3.0 kcal mol^{-1} above the 120°/240° conformations. So the graph looks like this:

45. $(CH_3)_2CHCH_2CH_2$—O—C—CH_3
Ester

H—C—CH—CH_2—OH
Aldehyde   Alcohol

Aldehyde

Aromatic compound

(HS)—CH_2—CH—C—OH
Thiol   Amine   Carboxylic acid

Ketone   Alkene

Ether

Alkene

Amine

CH_3—CH=CH—C≡C—C≡C—CH=CH—CH_2—OH
Alkyne   Alkene   Alcohol

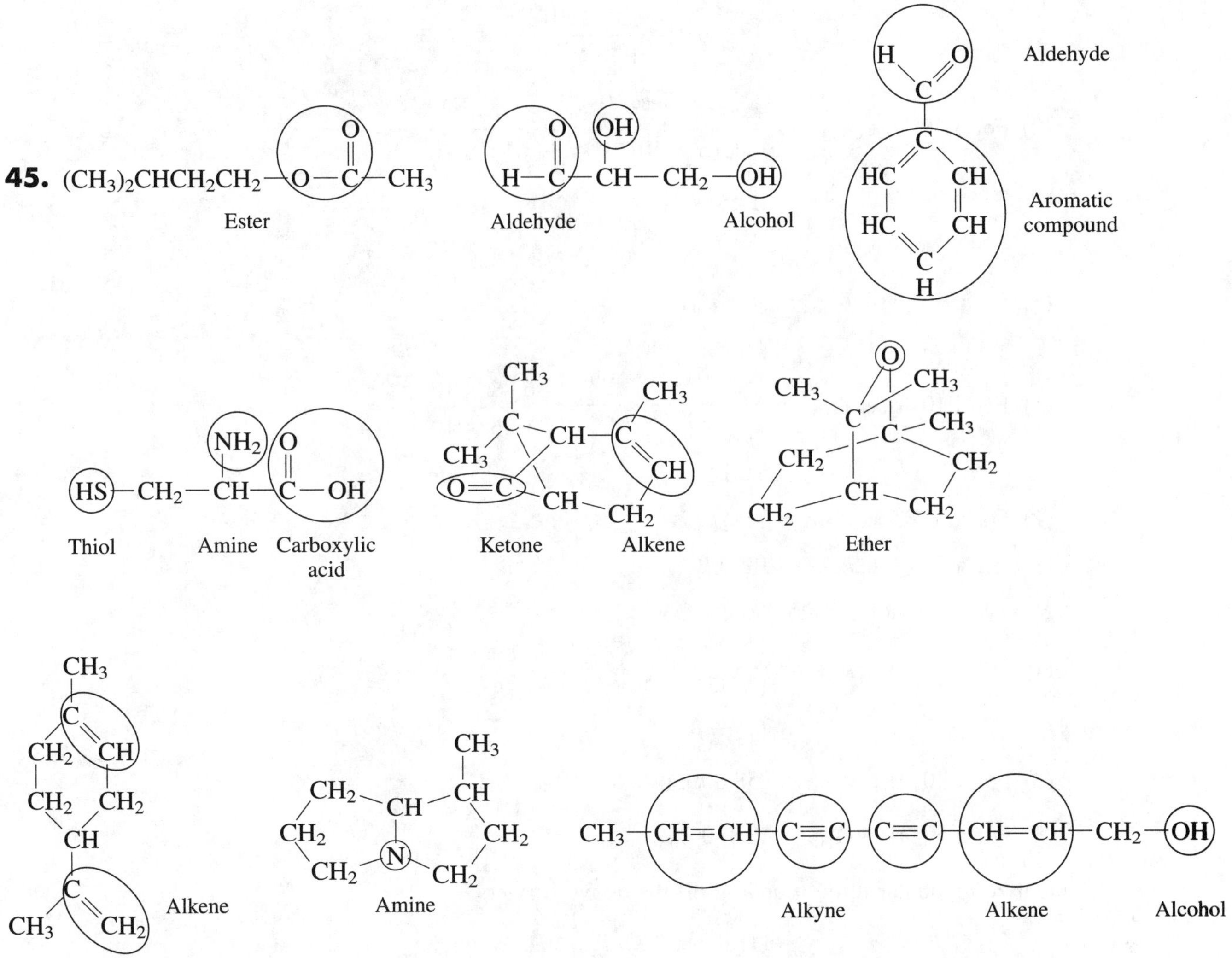

46. In vitamin D_4: 1,4,5-trimethylhexyl (secondary). In cholesterol: 1,5-dimethylhexyl (secondary). In vitamin E: 4,8,12-trimethyltridecyl (primary). In valine: 1-methylethyl (secondary). In leucine: 2-methylpropyl (primary). In isoleucine: 1-methylpropyl (secondary).

47. Before starting, you must understand *quantitatively* just what this question is asking. By "effect on k" of a change in temperature, we mean "how much bigger is k at a higher temperature than at a lower temperature," or the ratio "$k_{\text{higher temp}}/k_{\text{lower temp}}$." You can't answer a question before you know just what the question is asking.

(a) $E_a = 15$ kcal mol^{-1} $k = Ae^{-Ea/RT}$

Set up: It has to be assumed that A is constant at the different temperatures, so it will divide out, giving the general solution

$$\frac{k_{T_2}}{k_{T_1}} = \frac{e^{-E_a/RT_2}}{e^{-E_a/RT_1}} \quad \text{or} \quad k_{T_2} = \left(\frac{e^{-E_a/RT_2}}{e^{-E_a/RT_1}}\right)k_{T_1}$$

Then, remember that $R = 1.986$ *cal* deg^{-1} mol^{-1}, so E_a must be changed from kcal mol^{-1} to cal mol^{-1}. Get out your calculators!

(1) For a 10° rise, $k_{310°} = \dfrac{e^{-15000/(1.986)(310)}}{e^{-15000/(1.986)(300)}} \; k_{300°}$

$$= \frac{e^{-24.36}}{e^{-25.18}} = \frac{2.62 \times 10^{-11}}{1.16 \times 10^{-11}} = 2.26 \, k_{300°}$$

(2) For a 30° rise, $k_{330°} = \frac{e^{-15000/(1.986)(330)}}{1.16 \times 10^{-11}} = \frac{e^{-22.89}}{1.16 \times 10^{-11}}$

$$= \frac{1.15 \times 10^{-10}}{1.16 \times 10^{-11}} = 9.91 \, k_{300°}$$

(3) For a 50° rise, $k_{350°} = 36.6 \, k_{300°}$

(b) $E_a = 30$ kcal mol^{-1} = 30,000 cal mol^{-1}

(1) For a 10° rise, $k_{310°} = \frac{e^{-30000/(1.986)(310)}}{e^{-30000/(1.986)(300)}} = \frac{6.88 \times 10^{-22}}{1.36 \times 10^{-22}}$

$$= 5.06 \, k_{300°}$$

(2) For a 30° rise, $k_{330°} = 96.9 \, k_{300°}$

(3) For a 50° rise, $k_{350°} = 1320 \, k_{300°}$

(c) $E_a = 45$ kcal mol^{-1} = 45,000 cal mol^{-1}

(1) For a 10° rise, $k_{310°} = \frac{e^{-45000/(1.986)(310)}}{e^{-45000/(1.986)(300)}} = \frac{1.80 \times 10^{-32}}{1.58 \times 10^{-33}}$

$$= 11.4 \, k_{300°}$$

(2) For a 30° rise, $k_{330°} = 958.6 \, k_{300°}$

(3) For a 50° rise, $k_{350°} = 48,480 \, k_{300°}$

Let's summarize in tabular form, rounding off the above answers:

$E_a =$	15 kcal mol^{-1}	30 kcal mol^{-1}	45 kcal mol^{-1}
$k_{310°}/k_{300°}$	2	5	10
$k_{330°}/k_{300°}$	10	100	1000
$k_{350°}/k_{300°}$	40	1300	50,000

This problem illustrates the effect of temperature change on rate constants of reactions with three different activation energies. Notice the following:

(1) Reactions with high activation energies are the most sensitive to temperature changes.
(2) Even reactions with lower activation energies show rather significant responses to fairly modest temperature increases. This is relevant because many reactions in organic (and biological) chemistry have E_a values in the 15–30 kcal mol^{-1} range.

48. The general equation for a straight line that plots x vs. y is $y = $ (intercept) + (slope)(x). Compare this with the equation in the problem:

$$\log k = \log A - \frac{E_a}{2.3RT}$$

If we rewrite it slightly, separating out the term $1/T$, we get

$$\log k = \log A - \left(\frac{E_a}{2.3R}\right)\left(\frac{1}{T}\right)$$

Comparing this with the equation for a straight line, we see that a plot of log k vs. $1/T$ will have a slope of $-(E_a/2.3R)$ and an intercept equal to log A. Therefore, multiplying the slope of the line by $-2.3R$ gives E_a. An example of such a plot is shown below.

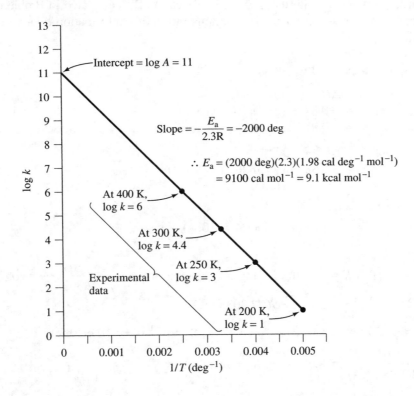

49. **(a)** We have a slight problem here: The positively polarized carbon atom in bromoethane has a closed shell and, therefore, is not a Lewis acid site. See Problem 51 for more on reactions of haloalkanes with Lewis bases.

(b)

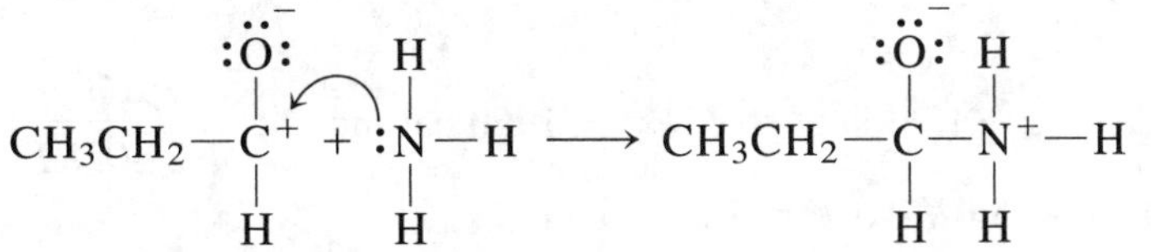

$$CH_3CH_2-\overset{H}{\underset{H}{C^+}} \ + \ :\overset{H}{\underset{H}{N}}-H \longrightarrow CH_3CH_2-\overset{\ddot{\underset{\cdot\cdot}{O}}:^-}{\underset{H}{C}}-\overset{H}{\underset{H}{N^+}}-H$$

(c)

$$H^+ \ + \ :\underset{CH_2CH_3}{\overset{\cdot\cdot}{O}}-CH_3 \longrightarrow H-\underset{CH_2CH_3}{\overset{\cdot\cdot}{O}^+}-CH_3$$

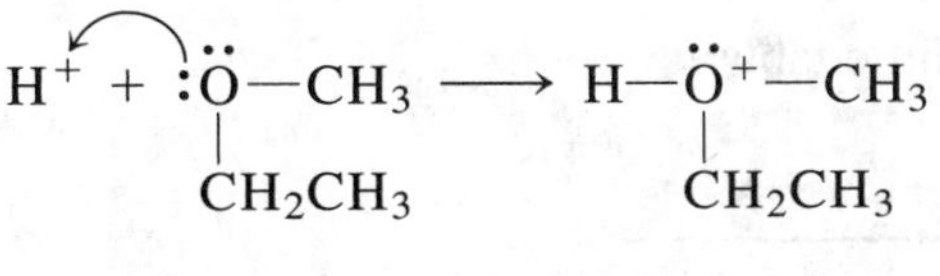

(d)

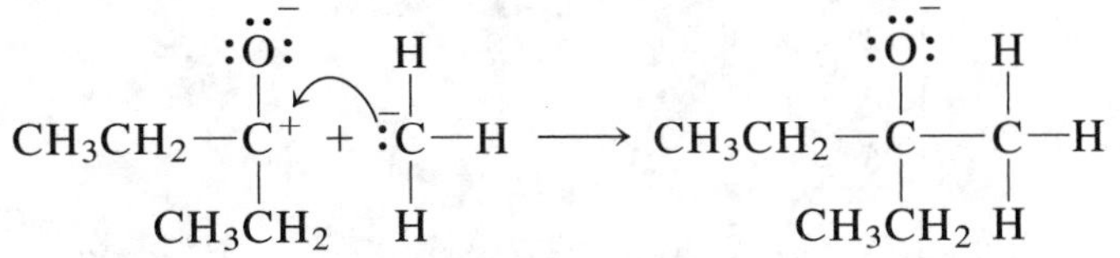

$$CH_3CH_2-\overset{:\ddot{O}:^-}{\underset{CH_3CH_2}{C^+}} \ + \ :\overset{H}{\underset{H}{C}}-H \longrightarrow CH_3CH_2-\overset{:\ddot{O}:^-}{\underset{CH_3CH_2}{C}}-\overset{H}{\underset{H}{C}}-H$$

(e)

$$CH_3-C\equiv N: \ + \ \overset{H}{\underset{H}{\overset{+}{C}}}-H \longrightarrow CH_3-C\equiv N^+-\overset{H}{\underset{H}{C}}-H$$

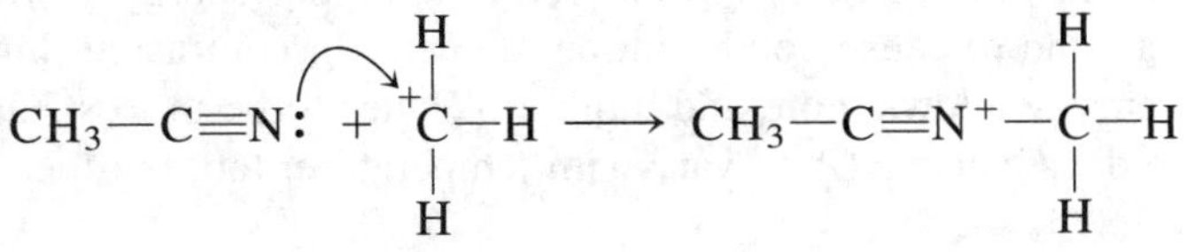

50. (a) From 44(a), we have $\Delta G° = -1.30$ kcal mol^{-1}. $T = 298$ K and $\Delta S° = +1.4$ cal deg^{-1} mol$^{-1} = +1.4 \times 10^{-3}$ kcal deg^{-1} mol^{-1}. So $\Delta G° = \Delta H° - T\Delta S°$ needs to be rearranged to solve for $\Delta H°$:
$\Delta H° = \Delta G° + T\Delta S° = -1.30 + 298(+1.4 \times 10^{-3}) = -1.30 + 0.42$. $\Delta H° = -0.88$ kcal mol^{-1}.

This agrees very nicely with the $\Delta H° = -0.9$ kcal mol^{-1} calculated in Problem 44(b),(c) on page 36 from the number of *gauche* interactions in the 0° conformation relative to the 120° conformation.

(b) Don't forget to change °C to K by adding 273°!

(1) $\Delta G° (-250°C) = \Delta H° - T\Delta S°$
$= -0.88 - (23K)(1.4 \times 10^{-3})$
$= -0.91$ kcal mol^{-1}

(2) $\Delta G° (-100°C) = \Delta H° - T\Delta S°$
$= -0.88 - (1.73K)(1.4 \times 10^{-3})$
$= -1.12$ kcal mol^{-1}

(3) $\Delta G° (500°C) = \Delta H° - T\Delta S°$
$= -0.88 - (773K)(1.4 \times 10^{-3})$
$= -1.96$ kcal mol^{-1}

(c) Use $\Delta G° = -RT \ln K = -2.303 RT \log K$. This rearranges to $-\dfrac{\Delta G°}{2.303RT} = \log K$, or

$K = 10^{(-\Delta G°/2.303RT)} = \text{antilog}(-\Delta G/2.303RT)$

(1) At $T = -250°C = 23$ K, $\Delta G° = -0.91$ kcal mol$^{-1} = -910$ cal mol^{-1}; $-\dfrac{\Delta G°}{2.303RT} =$
$\dfrac{-910}{2.303(1.986)(23)} = 8.65 = \log K$, so $K = 4.5 \times 10^8$.

(2) At $T = -100°C = 173$ K, $\Delta G° = -1.12$ kcal mol$^{-1} = -1120$ cal mol^{-1}; $-\dfrac{\Delta G°}{2.303RT} =$
$-\dfrac{-1120}{2.303(1.986)(173)} = 1.42 = \log K$, so $K = 26$.

(3) At $T = 500°C = 773$ K, $\Delta G° = -1.96$ kcal mol$^{-1} = -1960$ cal mol^{-1}; $-\dfrac{\Delta G°}{2.303RT} =$
$-\dfrac{-1960}{2.303(1.986)(773)} = 0.55 = \log K$, so $K = 3.5$.

We can summarize the results of Problems 50 and 44 in a little table:

T(K)	$\Delta G°$	K
23	−0.91	4.5×10^8
173	−1.12	26
298	−1.30	9
773	−1.96	3.5

These data illustrate two points. The most obvious is the huge effect of temperature on K. At 23 K (which is **very** cold), only two 2-methylbutane molecules out of every billion are in the higher energy (0°) conformation! There is very little thermal energy around to cause bond rotation to occur. In contrast, at higher temperatures K drops as the increased thermal energy allows more and more molecules to be in less stable conformations. Note also that the $\Delta S°$ value does cause $\Delta G°$ to vary with temperature, too, but the effect is small, since $\Delta S°$ is small.

51. It is usually a good idea to make use of the given information—in fact, write it all out so that it's sitting there in front of you—before trying to answer the question.

(a) 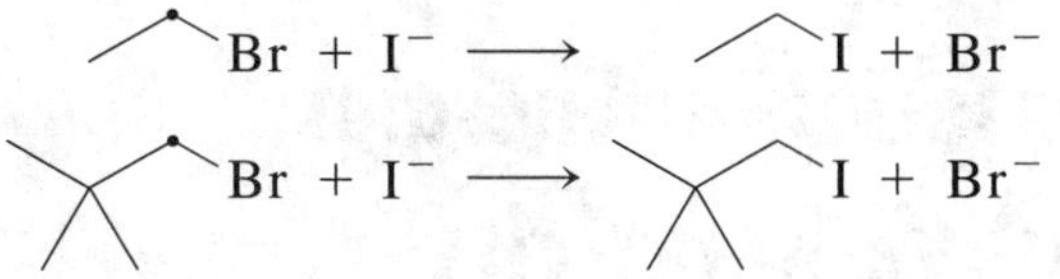   ⌃•Br + I⁻ ⟶ ⌃I + Br⁻

⋎•Br + I⁻ ⟶ ⋎I + Br⁻   This reaction is 10,000 times slower than the one above.

(b) The reaction sites are indicated by dots in the structures above. They are both *primary* carbon atoms because they are each attached directly to exactly one other carbon atom.

(c) Electrostatics suggests that the negative iodide ion will be attracted to the positively polarized carbon atom in the C—Br bond. However, because this carbon atom already has a closed shell, the iodide cannot actually bring in an electron pair for bonding unless some other atom, such as the bromine, leaves and takes an electron pair away. The second-order kinetics does not support a sequence in which the bromide ion leaves before the iodide ion comes in. So most likely both events occur together:

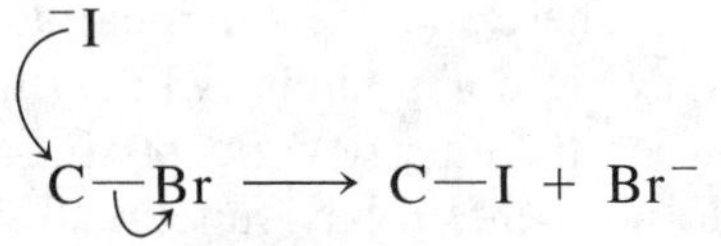

$$\overset{\displaystyle ^-I}{\curvearrowright} \quad C{-}Br \longrightarrow C{-}I + Br^-$$

The big rate reduction in the second example above, relative to the first, suggests that the increased size of the alkyl group has gotten in the way of the iodide ion in its attempt to bond to the carbon atom (an example of steric hindrance; see text Section 2-8, page 84). This makes the most sense if for some reason the iodide ion needs to pass close to this alkyl group to form a bond, perhaps in a trajectory that looks like the following sketch:

(d)

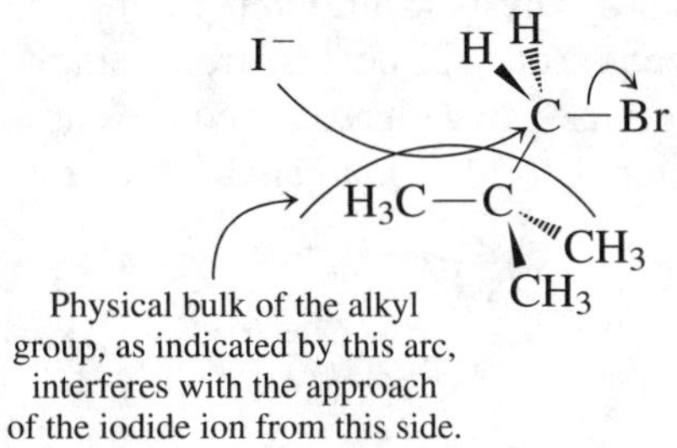

I⁻ H H
   C—Br
H₃C—C
   CH₃ CH₃

Physical bulk of the alkyl group, as indicated by this arc, interferes with the approach of the iodide ion from this side.

52. **(d)** Two secondary hydrogens (on the CH_2 group) and one tertiary (the CH) in $CH_3\overset{\displaystyle CH_3}{\underset{\displaystyle |}{C}}HCH_2CH_3$.

53. **(b)** Products are lower in energy than starting materials, so energy came out.

54. **(b)** It's an alkane. All the angles are 109.5°.

55. **(c)** See Figure 2-12 on text page 84.

56. **(e)** The C=O group is part of an ester.

3

Reactions of Alkanes: Bond-Dissociation Energies, Radical Halogenation, and Relative Reactivity

Discussing reactions of alkanes at the start of an organic chemistry course allows us to learn to work with several concepts that will be useful later. These include the idea of a general *reaction mechanism* that describes how any member of an entire class of compounds is likely to behave under certain reaction conditions. We also see the relationship between the energy concepts introduced in Chapter 2 and reactions that require more than one step. Alkanes do not contain any functional groups: They are made up of nonpolar C—C and C—H bonds, and nothing else. Therefore, alkanes are essentially unreactive toward ionic or polar materials; indeed, alkanes are just about the least reactive of **all** compound classes. Therefore, their chemistry is limited to processes that can lead to cleavage of nonpolar bonds. Thus, the only reasonable way for an alkane bond to cleave is *homolytically,* leaving one electron with each of the formerly bonded atoms: $\text{C—C} \rightarrow \text{C}\cdot + \cdot\text{C}$ or $\text{C—H} \rightarrow \text{C}\cdot + \cdot\text{H}$. This kind of bond cleavage is difficult and only occurs at high temperatures or in the presence of certain especially reactive species like halogen atoms.

This chapter covers three major ways alkane bonds are cleaved: pyrolysis (high temperature), halogenation (by halogen atoms), and combustion (high temperature and oxygen). You will note a strong emphasis on discussions involving bond energies. This should not surprise you, because bond cleavage requires an input of energy. The mechanism presents the reaction in terms of a step-by-step, bond-by-bond analysis that is helpful for spotting trends and analogies.

Outline of the Chapter

3-1 Strength of Alkane Bonds: Radicals
Exactly what **does** it take to cleave bonds in alkanes?

3-2 Alkyl Radicals and Hyperconjugation
The nature of the species obtained upon alkane bond cleavage.

3-3 Conversion of Petroleum: Pyrolysis
A practical example.

3-4 Chlorination of Methane: The Radical Chain Mechanism
Radical substitution of chlorine for hydrogen in methane: mechanism and energetics.

3-5 Other Radical Halogenations of Methane
Similarities and differences.

3-6 Chlorination of Higher Alkanes
What happens when substitution for different hydrogens in an alkane can give different products?

Keys to the Chapter

3-1. Strength of Alkane Bonds: Radicals

A minor but annoying point of confusion is often encountered when one discusses bond strengths. A bond's strength, or more properly, bond-dissociation energy *(DH°)*, is defined as the energy **released** when a bond **forms** or, equivalently, the energy **input** required to **break** a bond:

$$A\cdot \; + \; B\cdot \; \longrightarrow \; A\text{---}B \qquad \Delta H° = -DH° \qquad \text{Energy is released.}$$
$$A\text{---}B \; \longrightarrow \; A\cdot \; + \; B\cdot \qquad \Delta H° = DH° \qquad \text{Energy is put in.}$$

Inspection of these two equations shows that the **bonded** molecule A—B is **more stable—lower** in energy content—than the separated atoms A and B by an amount equal to *DH°*. When the linkages in a molecule are strong (high *DH°*), the molecule is usually relatively low in energy content (e.g., stable). As long as you remember that *DH°* is the energy that **has to be put in** to **break** a bond, you won't fall into the common trap of associating large *DH°* values with high-energy species. Large *DH°* values imply **low** energy, strongly bonded, **stable** species. The tables and figures in this section should further help you develop a comfortable understanding of the meaning of *DH°* values, in preparation for their use later on.

3-2. Alkyl Radicals and Hyperconjugation

Homolytic cleavage of any bond in an alkane generates *radicals:* species with a single unpaired electron where an attached group used to be. The section illustrates four such examples: methyl, $\cdot CH_3$; ethyl, $\cdot CH_2CH_3$; isopropyl, $\cdot CH(CH_3)_2$; and *tert*-butyl, $\cdot C(CH_3)_3$.

Several points are made in the section. First, radical carbons are sp^2 hybridized (planar), not sp^3 hybridized (tetrahedral, as in alkanes). Why should this be? A partial reason goes back to basic electrostatics. The shape of a species will be that which minimizes repulsion between electrons around a central atom (remember valence shell electron pair repulsion, or VSEPR?). In ammonia, $:NH_3$, the four electron pairs around N are best accommodated by a pyramidal shape based on sp^3 hybridization: Repulsion between the lone pair and the electrons in the N—H bonds is important in causing this geometry to be preferred. Reducing the number of non-bonded electrons from two to one as in methyl radical, $\cdot CH_3$, changes the situation. Now, electron repulsion between the pairs in the C—H bonds dominates, a situation leading to sp^2 hybridization and trigonal planar geometry, which allows the C—H bonding electrons to spread out far away from one another.

The second main point in the chapter is the stabilization of a radical center by the presence of alkyl groups attached to the radical carbon. So, *tert*-butyl radical is more stable than isopropyl, which is better than ethyl; methyl radical is the least stable. *Hyperconjugation* is one concept often used to explain this stabilization. Physically, and electrostatically, the radical carbon can be viewed as somewhat *electron deficient* (7 valence electrons instead of an octet). Hyperconjugation provides a means for bonds in neighboring alkyl groups to "lend" a little electron density to the radical center, thereby making it feel a little less electron-poor. In doing so, the alkyl groups effectively take some of the electron deficiency onto themselves, spreading it out, or "delocalizing" it. Delocalization of an electron deficiency or an electron excess over more than just the atom on which it is nominally located is often an energetically favorable, stabilizing process. It effectively allows the "problem" to be diluted over a larger area, rather than being the concentrated burden of a single atom.

The ability of alkyl groups to stabilize electron-deficient centers like radicals is often taken to imply that they are better *donors* of electrons than hydrogen atoms. Alkyl groups are therefore referred to as *electron donating*.

3-3. Conversion of Petroleum: Pyrolysis

The practical, "real-world" aspects of bond cleavage and radical formation are explored. Pyrolysis is a process that often gives mixtures of many products, and methods have been developed (mostly within the petroleum industry) to control this reaction somewhat. We will frequently explore the issue of reaction control: the modification of conditions under which a chemical transformation is carried out in order to give a desired molecule as a major or exclusive product.

3-4. Chlorination of Methane: The Radical Chain Mechanism

In this section the reaction of methane with chlorine molecules is discussed. The process

$$CH_4 + Cl_2 \longrightarrow HCl + CH_3Cl$$

is important because it converts a nonfunctionalized molecule (an alkane) into a molecule containing a functional group (a haloalkane). Once the functional group is present, many more kinds of chemical reactions become possible. This section also presents the mechanism of this reaction in full detail. Pay close attention not only to the steps of the reaction (initiation, propagation, and termination) but also to the finer details relating ΔH°, E_a, and transition state structure for each step. Although some of the terminology introduced here is appropriate only for radical mechanisms and not for the majority of reactions to come later, the type of **information** that the mechanism contains is critical to an understanding of how and why organic reactions occur. Take some time in this section to study each reaction step. What are its energetic circumstances, under what conditions does it occur, what role does it play in the overall process? Try to establish a feeling for the species involved as "stable" or "unstable," "reactive" or "unreactive," relatively speaking. Reaction mechanisms are intended to allow one to make sense out of organic chemistry. Give this one the time to do that for you.

Be sure that you understand the procedure for calculating the ΔH° value for a chemical reaction from the DH° values of the bonds taking part in the transformation. The general formula is

$$\Delta H^\circ_{reaction} = \Sigma DH^\circ \text{ (bonds broken)} - \Sigma DH^\circ \text{ (bonds formed)}$$

$$\uparrow \qquad\qquad\qquad \uparrow$$

$$\text{Energy input} \qquad\qquad \text{Energy output}$$

To illustrate with a reaction different from those in the text, let us calculate ΔH° for the process $C_2H_6 + H_2 \rightarrow 2\,CH_4$. Using the data from Tables 3-1 and 3-2 in the text,

$$CH_3\!-\!CH_3 + H\!-\!H \longrightarrow 2\,CH_3\!-\!H$$
$$DH^\circ = \quad 90 \qquad\quad 104 \qquad\qquad 105 \qquad \text{kcal mol}^{-1}$$
$$H^\circ_{reaction} = [90 + 104] - [2(105)] = -16\,\text{kcal mol}^{-1}$$

$$\uparrow$$

Note 2 methane
C—H bonds

Comment: This is a "hydrocracking" process that, although exothermic, requires very high temperatures to occur (cleavage of C—C bond is necessary).

Note on energetics: The overall enthalpy of a radical chain reaction is the sum of the $\Delta H°$ values for **only** the propagation steps. If we "sum up" the species in these steps, we see that the free atoms and radicals "cancel out," leaving only the molecular species of the overall reaction:

Propagation step 1 $CH_4 + Cl\cdot \rightarrow HCl + \cdot CH_3$ $\Delta H° = +2 \text{ kcal mol}^{-1}$

Propagation step 2 $\cdot CH_3 + Cl_2 \rightarrow CH_3Cl + Cl\cdot$ $\Delta H° = -27 \text{ kcal mol}^{-1}$

Sum: $CH_4 + \cancel{Cl\cdot} + \cancel{\cdot CH_3} + Cl_2 \rightarrow HCl + \cancel{\cdot CH_3} + CH_3Cl + \cancel{Cl\cdot}$ $\Delta H° = -25 \text{ kcal mol}^{-1}$

Removing the $Cl\cdot$ and $\cdot CH_3$ that appear on both sides of the equation leaves just the molecules of the overall process. What about the initiation and termination steps and their $\Delta H°$'s? They are separate. Their $\Delta H°$ values are not a part of the enthalpy change as it is defined for the stoichiometric reaction. When we measure the heat of a radical reaction experimentally, the value we obtain will not be precisely equal to $\Delta H°$ for the propagation steps alone; initiation and termination steps are occurring, too, and their $\Delta H°$'s will introduce an error. This deviation will usually be small, however, because initiation and termination steps occur only infrequently relative to the propagations, and because the $\Delta H°$'s for the endothermic initiation are for the most part canceled out by those of the exothermic termination processes.

3-5. Other Radical Halogenations of Methane

One of the best features of organic chemistry is the fact that one mechanism can hold for many individual reactions. Thus, the same types of steps that occur in the chlorination of methane are followed in its reactions with the other halogens. The similarities are qualitative, however. Differences in energetics are significant, and as a result the reaction-coordinate diagrams differ in appearance in a way that will become important as we continue through the chapter.

3-6. Chlorination of Higher Alkanes

The same mechanism also applies qualitatively to chlorination of other alkanes. The only difference is in the nature of the C—H bonds available in the alkane to be broken. They are generally less strong than those in methane, following a $DH°$ order of $CH_4 > 1° > 2° > 3°$. (Note: $1°$ = primary, $2°$ = secondary, and $3°$ = tertiary. These are commonly used symbols.) The weakest $(3°)$ are the most readily broken; thus, alkanes with different types of C—H bonds display a built-in *selectivity* of $3° > 2° > 1°$ in their reactions with chlorine. This section describes this selectivity quantitatively, illustrating how both reactivity differences and statistical factors combine to produce the observed ratios of products in several representative systems.

3-7. Selectivity with Other Halogens

An extension of the previous sections. The most significant point is that reactivity and selectivity in radical halogenations are inversely related. Simply put, the more reactive halogens are less picky and show less preference for $3°$ vs. $2°$ vs. $1°$ C—H bonds relative to less reactive halogens. The reason lies in the different activation energies associated with the C—H bond-breaking step. The values for fluorine are all very small, and very similar to one another. Fluorine thus reacts very rapidly with any C—H bond in a molecule. The reactions for bromine have large activation energies, with significant differences associated with the different types of C—H bonds present. The result is that bromine is much, much slower than fluorine to react with any alkane, and bromine is much more discriminating (selective) in its reactions, too, greatly preferring $3°$ over $2°$ or $1°$ C—H bonds. The contrasts between reaction-coordinate diagrams for the two halogens provide a pictorial representation of these differences.

3-8. Synthetic Aspects

Synthesis is one primary function associated with organic chemistry. In synthesis, we strive to produce a desired product in good yield with high selectivity, to minimize the effort required to separate this material from side products. In particular, a reaction that gives rise to a hard-to-separate mixture of many components is **synthetically useless.** In this chapter we have seen a large number of possible permutations of a single reaction: alkane halogenation. Not all of the examples shown are equally useful synthetically. The best ones start with an alkane in which all hydrogens are chemically indistinguishable (methane, ethane, neopentane), because they can produce only one monohalogenation product. In the case of most alkanes, synthetic utility will be determined by the number of different types of hydrogens present and whether the desired product derives from substitution of a more reactive or a less reactive hydrogen in the molecule.

In isobutane, for example, there are one 3° and nine 1° hydrogens. If we desire to halogenate at the 3° center, the natural selectivity of bromine makes it the obvious halogen to choose. If we desire to halogenate a 1° carbon, a less selective, more reactive halogen would allow us to take best advantage of the statistical factor of nine possible 1° hydrogens available to be replaced in each molecule. Thus,

$$CH_3-\overset{\overset{\displaystyle CH_3}{|}}{CH}-CH_3 + Br_2 \longrightarrow CH_3-\overset{\overset{\displaystyle CH_3}{|}}{\underset{\underset{\displaystyle Br}{|}}{C}}-CH_3 \quad \text{Major product}$$

$$CH_3-\overset{\overset{\displaystyle CH_3}{|}}{CH}-CH_3 + F_2 \longrightarrow F-CH_2-\overset{\overset{\displaystyle CH_3}{|}}{CH}-CH_3 \quad \text{Major product}$$

3-10. Combustion and Relative Stability

In order to obtain thermodynamic information experimentally, several methods may be used. The measurement of equilibrium constants gives energy differences between species. Directly measuring the heat of a reaction accomplishes the same thing. When the reaction is combustion of a hydrocarbon, the result is a measure of the energy content of the compound relative to that of the product molecules, CO_2 and H_2O. Such data allow comparisons to be made between related compounds, which in turn reveal factors influencing the relative stabilities of different structures.

Solutions to Problems

15. This problem is really a reminder of material from the previous chapter. For shorthand purposes, we use the symbols 1° = primary, 2° = secondary, and 3° = tertiary.

16. (a) $CH_3CH_2\overset{\displaystyle \cdot}{C}HCH_3$ $CH_3CH_2CH_2CH_2\cdot$

1-Methylpropyl (sec-butyl; see Table 2-4)   **Butyl radical**
Secondary (2°), more stable   Primary (1°), less stable

Remember: Identify radicals as 1°, 2°, or 3° by the *radical carbon*. None of the other carbons matter. Two hyperconjugation pictures may be drawn for 1-methylpropyl radical, one with two

C—H bonds overlapping with the radical p orbital, and one with a C—C bond participating instead of a C—H:

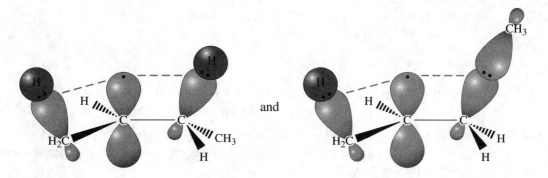

and

(b) In naming these, remember that the radical carbon is **always** C1 (just like the "point of attachment" carbon in alkyl groups). The parent name is based on the longest carbon chain beginning at C1, and all appendages are named as substituents:

$$CH_3{-}CH_2$$
$$|$$
$$CH_3{-}CH_2{-}CH{-}CH_2{\cdot}$$
$$\quad 4 \qquad 3 \qquad 2 \qquad 1$$

2-Ethylbutyl radical
Primary, less stable

$$CH_3{-}CH_2$$
$$|$$
$$CH_3{-}CH_2{-}\overset{\cdot}{C}{-}CH_3$$
$$\qquad 3 \qquad 2 \qquad 1$$

1-Ethyl-1-methylpropyl radical
Tertiary, more stable

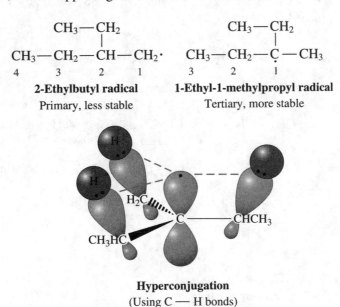

Hyperconjugation
(Using C —— H bonds)

(c) Left to right: 1,2-dimethylpropyl radical, secondary, intermediate in stability; 1,1-dimethylpropyl radical, tertiary, most stable; 3-methylbutyl radical, primary, least stable.

Hyperconjugation in the 1,1-dimethylpropyl radical is the same as in 1-ethyl-1-methylpropyl [(b) above]; in your picture, an H should replace one of the end CH_3 groups.

17. Work problems like this "mechanistically": Proceed via general reaction steps as you have previously seen illustrated in the text until you reach stable molecules. Pyrolysis of propane starts as follows:

(a) $CH_3CH_2{-}CH_3 \longrightarrow CH_3CH_2{\cdot} + {\cdot}CH_3$ C—C bond cleavage

Then there are three possible recombinations:

(b) $2\,CH_3{\cdot} \longrightarrow CH_3CH_3$
Ethane

(c) $2\,CH_3CH_2{\cdot} \longrightarrow CH_3CH_2CH_2CH_3$
Butane

(d) $CH_3{\cdot} + CH_3CH_2{\cdot} \longrightarrow CH_3CH_2CH_3$ (Reverse of first step)
Propane

Two possible hydrogen abstractions can occur:

(e) $CH_3 \cdot + CH_2\overset{H}{—}CH_2 \cdot \longrightarrow CH_4 + CH_2{=}CH_2$

                                            Methane   **Ethene**

(f) $CH_3CH_2 \cdot + CH_2\overset{H}{—}CH_2 \cdot \longrightarrow CH_3CH_3 + CH_2{=}CH_2$

                                            Ethane   **Ethene**

Abstraction only occurs from the carbon **next to** a radical carbon. Methyl radical, $\cdot CH_3$, doesn't have another carbon next to its radical center, so it cannot give up a hydrogen in an abstraction. It can still **accept** a hydrogen, however [reaction (e), above].

So there are four new products formed from cracking of propane: methane, ethane, butane, and ethene (ethylene).

18. **(a)** The weakest bond in butane is the C2—C3 bond, $DH° = 88$ kcal mol^{-1} (Table 3-2). Pyrolysis should therefore proceed as follows:

(1) $CH_3CH_2—CH_2CH_3 \longrightarrow 2\ CH_3CH_2 \cdot$ C—C bond cleavage

(2) $2\ CH_3CH_2 \cdot \longrightarrow CH_3CH_2CH_2CH_3$ Reverse of (1)

(3) $CH_3CH_2 \cdot\ \ H—CH_2—CH_2 \cdot \longrightarrow CH_3CH_3 + CH_2{=}CH_2$ Hydrogen abstraction

                                            Ethane   **Ethene**

(b) The weakest bonds are the three equivalent C—C bonds, $DH° = 88$ kcal mol^{-1}. Therefore,

(1) $(CH_3)_2CH—CH_3 \longrightarrow (CH_3)_2CH \cdot + \cdot CH_3$ Cleavage

(2) $2\ CH_3 \cdot \longrightarrow CH_3CH_3$

            Ethane

(3) $2\ (CH_3)_2CH \cdot \longrightarrow (CH_3)_2CHCH(CH_3)_2$

            2,3-Dimethylbutane

(4) $CH_3 \cdot\ \ \cdot CH(CH_3)_2 \longrightarrow (CH_3)_3CH$ Reverse of (1); recombinations

(5) $CH_3 \cdot\ \ H—CH_2—CHCH_3 \longrightarrow CH_4 + CH_2{=}CHCH_3$ Hydrogen abstractions

                                    Methane   **Propene**

(6) $(CH_3)_2CH \cdot\ \ H—CH_2—CHCH_3 \longrightarrow CH_3CH_2CH_3 + CH_2{=}CHCH_3$

                                    Propane   **Propene**

19. The $DH°$ data are readily found (Tables 3-1 and 3-4). Values in kcal mol^{-1}.

    (a) $104 + 38 - 2(136) = -130$ **(b)** $104 + 58 - 2(103) = -44$

    (c) $104 + 46 - 2(87) = -24$ **(d)** $104 + 36 - 2(71) = -2$

    (e) $96.5 + 38 - (110 + 136) = -111.5$ **(f)** $96.5 + 58 - (85 + 103) = -33.5$

    (g) $96.5 + 46 - (71 + 87) = -15.5$ **(h)** $96.5 + 36 - (55 + 71) = +6.5$

20. **(a)** Two: 1-halobutane and 2-halobutane

(b) Three (see solution to Problem 21)

(c) Four (see solution to Problem 21)

(d) Four:

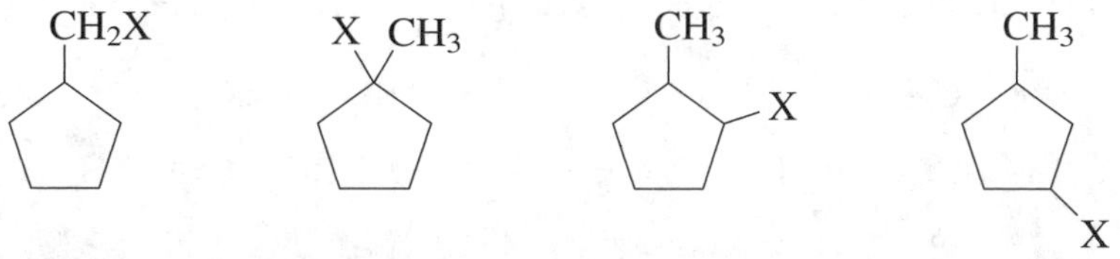

| (Halomethyl)-cyclopentane | 1-Halo-1-methyl-cyclopentane | 1-Halo-2-methyl-cyclopentane | 1-Halo-3-methyl-cyclopentane |

21. **(a)** (i) $CH_3CH_2CH_2CH_2CH_2Cl$ (1-chloropentane),
$CH_3CH_2CH_2CHClCH_3$ (2-chloropentane), and
$CH_3CH_2CHClCH_2CH_3$ (3-chloropentane).

(ii) $CH_3CH_2CH(CH_3)CH_2CH_2Cl$ (1-chloro-3-methylpentane),
$CH_3CH_2CH(CH_3)CHClCH_3$ (2-chloro-3-methylpentane),
$CH_3CH_2CCl(CH_3)CH_2CH_3$ (3-chloro-3-methylpentane), and
$CH_3CH_2CH(CH_2Cl)CH_2CH_3$ [3-(chloromethyl)pentane].

(b) To answer this question, you first have to count up and identify by type (1°, 2°, or 3°) **all** the hydrogens whose removal could lead to **each one** of the products. Then multiply the number of hydrogens you counted by the **relative reactivity** for that **type** of hydrogen in a chlorination reaction at 25°C, the conditions for the reaction stated in the problem. This procedure gives you the relative amount of product corresponding to removal of those hydrogens. After you have done this for all the products, you convert these relative amounts into percentage yields by normalizing to 100% (as shown below).

(i) 1-Chloropentane is formed by chlorination of any of the **six primary hydrogens** (each with a relative reactivity = 1) on carbons 1 and 5; ∴ its relative yield is 6 × 1 = **6.** 2-Chloropentane is formed by chlorination of any of the **four secondary hydrogens** (each with a relative reactivity = 4) on carbons 2 and 4; ∴ its relative yield is 4 × 4 = **16.** 3-Chloropentane is formed by chlorination of any of the **two secondary hydrogens** (each with a relative reactivity = 4) on carbon 3; ∴ its relative yield is 2 × 4 = **8.** The absolute % yield for each product is calculated as follows:

$$\frac{\text{Relative yield of the product}}{\text{Sum of relative yields for all products}} \times 100\% = \% \text{ yield of that product}$$

So, Yield of 1-chloropentane = 100% × 6/(6 + 16 + 8)
= 100% × 6/30 = 20%
Yield of 2-chloropentane = 100% × 16/30 = 53%
Yield of 3-chloropentane = 100% × 8/30 = 27%

(ii) 1-Chloro-3-methylpentane is formed by chlorination of any of the **six primary hydrogens** (relative reactivity = 1) on carbons 1 and 5; ∴ its relative yield is 6 × 1 = **6.** 2-Chloro-3-methylpentane is formed by chlorination of any of the **four secondary hydrogens** (relative reactivity = 4) on carbons 2 and 4; ∴ its relative yield is 4 × 4 = **16.** 3-Chloro-3-methylpentane is formed by chlorination of the **single tertiary hydrogen** (relative reactivity = 5) on carbon 3; ∴ its relative yield is 1 × 5 = **5.** 3-(Chloromethyl)pentane is formed by chlorination of any of the **three primary hydrogens** (relative reactivity = 1) on the methyl group attached to carbon 3; ∴ its relative yield is 1 × 3 = **3.**

So,   Yield of 1-chloro-3-methylpentane = 100% × 6/(6 + 16 + 5 + 3)

= 100% × 6/30 = 20%

Yield of 2-chloro-3-methylpentane = 100% × 16/30 = 53%

Yield of 3-chloro-3-methylpentane = 100% × 5/30 = 17%

Yield of 3-(chloromethyl)pentane   = 100% × 3/30 = 10%

(c) Propagation step 1 [values below are $DH°$ (kcal mol^{-1}) for bonds made or broken]:

$$\underset{96.5}{\overset{\overset{\displaystyle CH_3}{|}}{CH_3CH_2CHCH_2CH_3}} + Cl\cdot \longrightarrow HCl + \underset{103}{\overset{\overset{\displaystyle CH_3}{|}}{CH_3CH_2\overset{\displaystyle .}{C}CH_2CH_3}} \quad \Delta H° = 96.5 - 103 = -6.5$$

Propagation step 2:

$$\underset{58}{\overset{\overset{\displaystyle CH_3}{|}}{CH_3CH_2\overset{\displaystyle .}{C}CH_2CH_3}} + Cl_2 \longrightarrow Cl\cdot + \underset{85}{\overset{\overset{\displaystyle CH_3}{|}}{CH_3CH_2CClCH_2CH_3}} \quad \Delta H° = -27$$

For the reaction overall, $\Delta H° = -33.5$ kcal mol^{-1}.

22. The mechanism is qualitatively the same as that of chlorination of methane (textbook Section 3-4):
Initiation:

$$Br_2 \longrightarrow 2\ Br\cdot$$

Propagation:

(1) $Br\cdot + CH_4 \longrightarrow CH_3\cdot + HBr$

(2) $CH_3\cdot + Br_2 \longrightarrow CH_3Br + Br\cdot$

Termination:

$$Br\cdot + Br\cdot \longrightarrow Br_2$$

$$CH_3\cdot + Br\cdot \longrightarrow CH_3Br$$

$$CH_3\cdot + CH_3\cdot \longrightarrow CH_3CH_3$$

23. The data needed to prepare these diagrams may be found in Table 3-5. Examples to use as models include Figures 3-7 and 3-8.

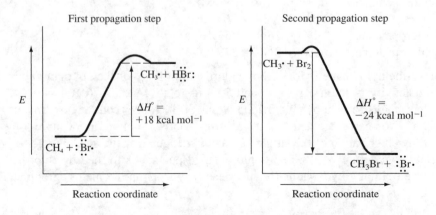

24. Initiation:

$$Br_2 \longrightarrow 2\ Br\cdot \qquad \Delta H° = +46\ \text{kcal mol}^{-1}$$

Propagation:

(1) $Br\cdot + C_6H_6 \longrightarrow HBr + C_6H_5\cdot \qquad \Delta H° = +25\ \text{kcal mol}^{-1}$

(2) $C_6H_5 + Br_2 \longrightarrow C_6H_5Br + Br\cdot \qquad \Delta H° = -35\ \text{kcal mol}^{-1}$

$$\text{Overall} \qquad \Delta H° = -10\ \text{kcal mol}^{-1}$$

The overall $\Delta H°$ is not very different from those of typical alkane C—H bonds: methane, $\Delta H° = -6$ kcal mol^{-1}; 1° C—H, $\Delta H° = -10$; 2° C—H, $\Delta H° = -13.5$; 3° C—H, $\Delta H° = -15.5$. However, the rate-determining first propagation step in the reaction of benzene is **much more endothermic** than any of the alkane reactions, due to the exceptional strength of the C—H bonds in benzene. The result is that bromination of benzene by this mechanism is exceedingly difficult (very slow) and does not compete kinetically with bromination reactions of typical alkanes.

25. Qualitatively the same exercise as in Problem 23, but with different bond strength values.

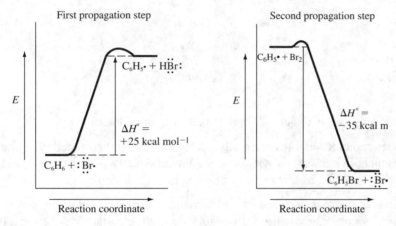

26. The diagram on the left shows that the first propagation step has a late transition state, close to the products along the reaction coordinate. In contrast, the second propagation step (right) has an early transition state, close to the starting materials.

27. Unless otherwise stated, assume that no more than one halogen atom attaches to each alkane molecule.

(a) No reaction. Iodination of alkanes is endothermic.

(b) $CH_3CHFCH_3 + CH_3CH_2CH_2F$ F_2 is not very selective.

(c) CH$_3$ Br   Bromination selects the tertiary position whenever possible. See note in answer to Problem 15(d).

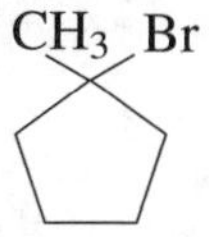

(d) $CH_3{-}\underset{\underset{Cl}{|}}{\overset{\overset{CH_3}{|}}{C}}{-}CH_2{-}\underset{\underset{CH_3}{|}}{\overset{\overset{CH_3}{|}}{C}}{-}CH_3 + Cl{-}CH_2{-}\underset{}{\overset{\overset{CH_3}{|}}{CH}}{-}CH_2{-}\underset{\underset{CH_3}{|}}{\overset{\overset{CH_3}{|}}{C}}{-}CH_3$

$+\ CH_3{-}\underset{\underset{Cl}{|}}{\overset{\overset{CH_3}{|}}{CH}}{-}\underset{\underset{CH_3}{|}}{\overset{\overset{CH_3}{|}}{CH}}{-}\underset{\underset{CH_3}{|}}{\overset{\overset{CH_3}{|}}{C}}{-}CH_3 + CH_3{-}\underset{}{\overset{\overset{CH_3}{|}}{CH}}{-}CH_2{-}\underset{\underset{CH_3}{|}}{\overset{\overset{CH_3}{|}}{C}}{-}CH_2{-}Cl$

A complex mixture is obtained. Cl_2 is more selective than F_2, but still prefers 3° to 1° positions by a factor of only about 5 to 1.

(e) 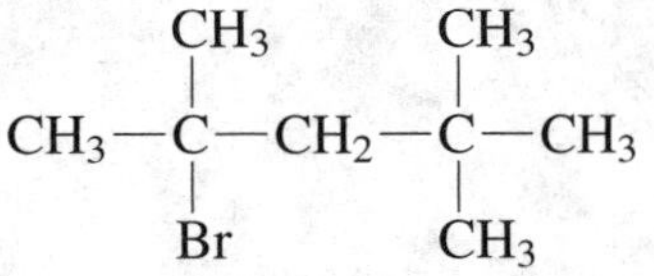
$$CH_3-\overset{\overset{\displaystyle CH_3}{|}}{\underset{\underset{\displaystyle Br}{|}}{C}}-CH_2-\overset{\overset{\displaystyle CH_3}{|}}{\underset{\underset{\displaystyle CH_3}{|}}{C}}-CH_3 \qquad \text{Br}_2 \text{ is very selective for 3° C—H bonds.}$$

28. Relative yield = number of hydrogens of a given type in the starting alkane × relative reactivity

$$\frac{\text{relative yield of one product}}{\text{sum of relative yields of all products}} \times 100\% = \% \text{ yield of that product}$$

Product	Hydrogen type	Number of hydrogens	Relative reactivity	Relative yield	% yield
(b) CH_3CHFCH_3	2°	2	1.2	2.4	29
$CH_3CH_2CH_2F$	1°	6	1	6	71
(d) $(CH_3)_2CClCH_2C(CH_3)_3$	3°	1	5	5	18
$ClCH_2CH(CH_3)CH_2C(CH_3)_3$	1°	6	1	6	21
$(CH_3)_2CHCHClC(CH_3)_3$	2°	2	4	8	29
$(CH_3)_2CHCH_2C(CH_3)_2CH_2Cl$	1°	9	1	9	32

(c) and (e) 3° substitution by Br_2 is about 73% selective in (c) and 91% in (e).

29. Only the bromination reactions (c) and (e) are really acceptable as synthetic methods. The other reactions, giving several products in comparable yields, are not synthetically useful. The fluorination (b) might look good on paper, but use of F_2 as a reagent in practice is very difficult.

30. **(a)** Pentane has two kinds of hydrogens—primary and secondary—but they are distributed in three distinct groups, because three different monobromopentanes are possible, depending on the site of radical bromination. Replacement of any one of the six primary hydrogens at C1 and C5 (group 'a' in the structure below) gives 1-bromopentane. We know that primary hydrogens are 1/80th as reactive as secondary, so this should be the product formed in the lowest yield. The six secondary hydrogens on C2, C3, and C4 are all about equally reactive. Replacement of either of the two on C3 (group 'b') gives 3-bromopentane. Replacement of any one of the four on C2 and C4 (group 'c') gives 2-bromopentane. This will be the major product, because it can form from replacement of the greatest number of the most reactive hydrogens in the starting molecule.

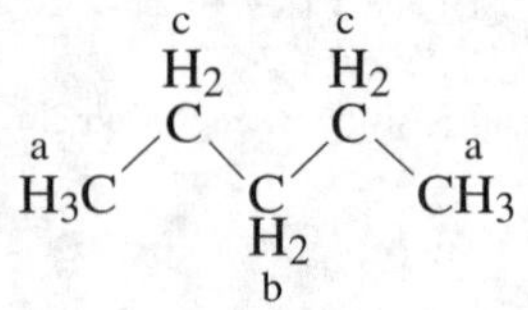

(b) and **(c)** First draw the molecule in a conventional formula and identify the bond of rotation:

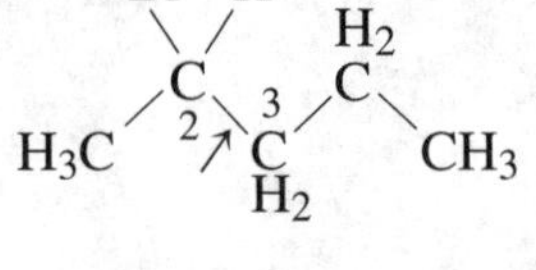

Then choose a sighting direction for preparing the Newman projections. For example, C2 in front and C3 behind. Begin with any arbitrary staggered conformation and proceed with 120° rotations of one carbon (say, the back carbon, C3) with respect to the other, giving all possible staggered conformations arising from rotation about the C2—C3 bond for this product molecule. The potential energy graph may be drawn immediately below the structures.

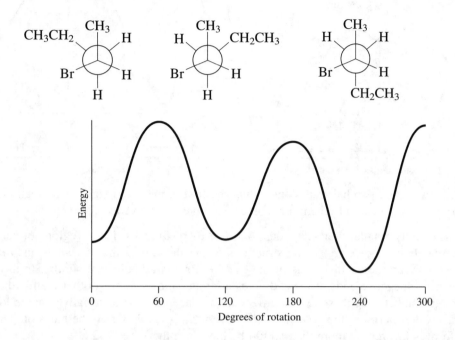

31. First write out the propagation steps and determine the $\Delta H°$ associated with each. As indicated in Problem 30, the major product derives from halogenation at C2. Sketch the graphs, using Figure 3-7 on page 107 of the text as your basic model. The transition states are located at the maxima for each of the steps on the graphs.

Bromination steps:

Propagation step 1

$$CH_3CH_2CH_2CH_2CH_3 + Br\cdot \longrightarrow CH_3\dot{C}HCH_2CH_2CH_3 + HBr \qquad \Delta H°$$

$DH° = 98.5$ kcal mol^{-1} for sec C—H bond $\qquad DH°$ (HBr) $= 87$ kcal mol^{-1} $\quad +11.5$

Propagation step 2

$$CH_3\dot{C}HCH_2CH_2CH_3 + Br_2 \longrightarrow CH_3CHBrCH_2CH_2CH_3 + Br\cdot \qquad \Delta H°$$

$DH°$ (Br$_2$) $= 46$ kcal mol^{-1} $\qquad DH° = 71$ kcal mol^{-1} for sec C—Br bond $\qquad -25$

Overall $\Delta H° = -13.5$ kcal mol^{-1}

Iodination steps:

Propagation step 1

$$CH_3CH_2CH_2CH_2CH_3 + I\cdot \longrightarrow CH_3\dot{C}HCH_2CH_2CH_3 + HI \qquad \Delta H°$$

$DH° = 98.5$ kcal mol^{-1} for sec C—H bond $\qquad DH°$ (HI) $= 71$ kcal mol^{-1} $\quad +27.5$

Propagation step 2

$$CH_3\dot{C}HCH_2CH_2CH_3 + I_2 \longrightarrow CH_3CHICH_2CH_2CH_3 + I\cdot \qquad \Delta H°$$

$DH°$ (I$_2$) $= 36$ kcal mol^{-1} $\qquad DH° = 56$ kcal mol^{-1} for sec C—I bond $\qquad -20$

Overall $\Delta H° = +7.5$ kcal mol^{-1}

Bromination graph: Iodination graph:

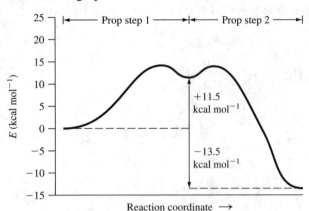   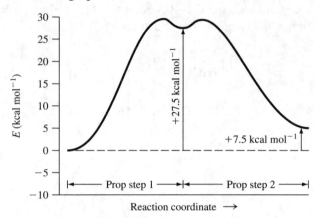

The main difference between the two halogenations is that the first propagation step for iodination is so much more endothermic that the overall reaction becomes endothermic as well.

32. Although Br is sterically smaller than CH_3, the polar C—Br bond gives rise to regions of partial positive and negative charges in the molecule. As the structures below show, in both the *gauche* and *anti* conformations, the distances between the positive ends of the two C—Br bond dipoles are similar. In the *gauche* conformation, however, the negative ends of these dipoles are much closer to each other than is the case in the *anti* form. The repulsion between these two relatively high concentrations of negative charge located near the Br atoms raises the energy of the *gauche* form above what it would be on the basis of sterics alone. Therefore, the *anti* form is favored by more than would be the case otherwise.

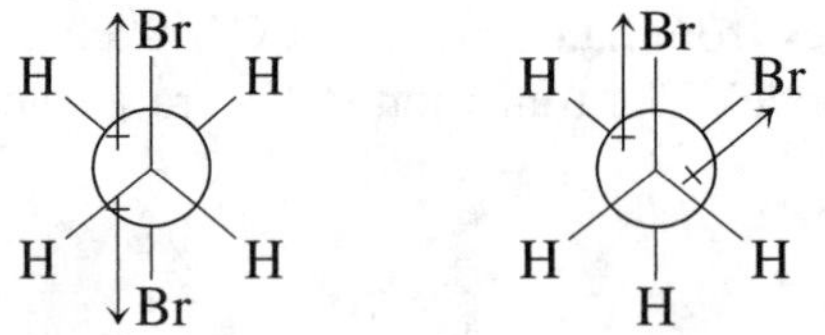

33. (a) $CH_4 + 2\ O_2 \longrightarrow CO_2 + 2\ H_2O$

(b) $C_3H_8 + 5\ O_2 \longrightarrow 3\ CO_2 + 4\ H_2O$

(c) $C_6H_{12} + 9\ O_2 \longrightarrow 6\ CO_2 + 6\ H_2O$

(d) $C_2H_6O + 3\ O_2 \longrightarrow 2\ CO_2 + 3\ H_2O$

(e) $C_{12}H_{22}O_{11} + 12\ O_2 \longrightarrow 12\ CO_2 + 11\ H_2O$

34. (a) $C_3H_6O + 4\ O_2 \longrightarrow 3\ CO_2 + 3\ H_2O$

(b) The energy difference is the difference between the heats of combustion, 6.2 kcal mol^{-1}. Acetone, whose combustion **releases less heat,** must have had the **lower energy content** to begin with.

(c) Acetone, with a lower energy content, is the more stable compound.

35. You need to propose a bond-breaking process for sulfuryl chloride, SO_2Cl_2, which will accomplish the same goal as breaking the Cl—Cl bond in Cl_2. There is really only one option: a sulfur–chlorine bond.

Initiation:

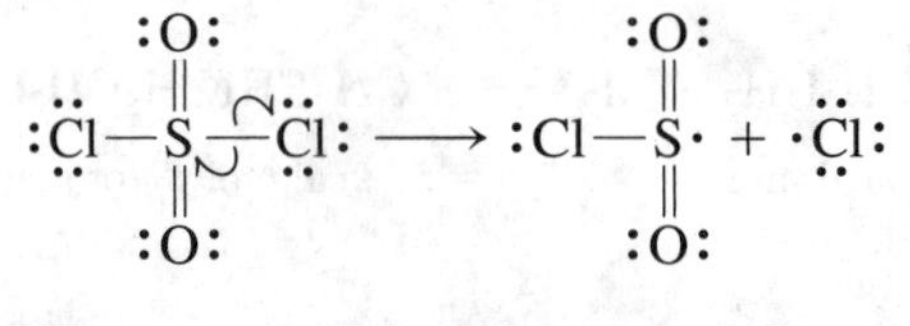

Propagation:

(1)

(2)

We will be stuck unless we can find a way for $ClSO_2\cdot$ to give rise to another propagation step to continue the chain process. Two options are possible, and with the information you have been given, either one is a reasonable proposal.

(a) Unimolecular decomposition of $ClSO_2\cdot$; that is,

, followed by a new

propagation step 1, or

(b) An alternative propagation step that uses $SO_2Cl\cdot$ in place of the $Cl\cdot$ in step 1:

Either is a qualitatively sensible mechanistic possibility.

36. Recall (text Section 3-4) that $E_a = 4$ kcal $mol^{-1} = 4000$ cal mol^{-1} for the reaction between $Cl\cdot$ and CH_4. Therefore,

$$k_{(Cl\cdot\,+\,CH_4)} = A\,e^{-4000/(1.986)(298)}, \text{ and } k_{(Br\cdot\,+\,CH_4)} = A\,e^{-19000/(1.986)(298)}.$$

So, $k_{(Cl\cdot\,+\,CH_4)}/k_{(Br\cdot\,+\,CH_4)} = e^{15000/(1.986)(298)} = e^{25.3} = 9.7 \times 10^{10}$. That's a pretty big rate ratio.

37. As we saw in the case of the reaction of methane with a mixture of Br_2 and Cl_2, the only kinetically viable first propagation steps in this case are reactions of $Cl\cdot$ atoms with propane. Reactions of $Br\cdot$ atoms are far too slow to compete. *It is this first step that determines the ratio of primary to secondary alkyl radicals that form.* Therefore, the selectivity observed is that of $Cl\cdot$ atoms. The two radicals both proceed to react rapidly with either molecular halogen, Cl_2 or Br_2. Because the ratio of radicals present was determined in the prior step, the ratios of chloropropane isomers and of bromopropane isomers obtained are essentially the same, and reflect the selectivity of chlorination.

38. Calculate using the same method that was introduced in Section 3-6. There are three groups of hydrogens with different reactivities: two on C1, two on C2, and three on C3. From the relative yields, it appears that those on C3 are lowest in reactivity. It is sensible therefore to calculate how much more reactive the hydrogens on C2 and C1 are compared to those on C3. First, let's do C2 vs. C3:

$$\frac{\text{Relative reactivity of a hydrogen on C2}}{\text{Relative reactivity of a hydrogen on C3}} = \frac{\left(\begin{array}{c}8.5\%\ \text{C2}\\ \text{chlorination}\end{array}\right) \Big/ \left(\begin{array}{c}2\ \text{C2}\\ \text{hydrogens}\end{array}\right)}{\left(\begin{array}{c}1.5\%\ \text{C3}\\ \text{chlorination}\end{array}\right) \Big/ \left(\begin{array}{c}3\ \text{C3}\\ \text{hydrogens}\end{array}\right)} = 8.5$$

Now C1 vs. C3:

$$\frac{\text{Relative reactivity of a hydrogen on C1}}{\text{Relative reactivity of a hydrogen on C3}} = \left(\begin{array}{c}90\%\ C1\\ \text{chlorination}\end{array}\right) \Bigg/ \left(\begin{array}{c}2\ C1\\ \text{hydrogens}\end{array}\right) \Bigg/ \left(\begin{array}{c}1.5\%\ C3\\ \text{chlorination}\end{array}\right) \Bigg/ \left(\begin{array}{c}3\ C3\\ \text{hydrogens}\end{array}\right) = 90$$

Recalling the results from propane (Section 3-6), the hydrogens at C2 (secondary) were about 4 times more reactive than those at either C1 or C3 (primary). The reason was that secondary alkyl radicals are more stable than primary alkyl radicals. In the molecule in this problem, 1-bromopropane, the most reactive hydrogens are those on the C1, where the Br is attached. Evidently, the bromine atom strongly stabilizes a radical at C1. It is reasonable to speculate that this stabilization may have something to do with the lone pairs on the Br, because we learned that radicals are electron deficient: They are stabilized in alkanes by hyperconjugation, which allows electrons from neighboring bonds to delocalize towards the half-empty radical p orbital. In 1-bromopropane, we can imagine that a p orbital containing a lone pair on the Br could align and overlap with the half-empty p orbital on carbon, an overlap that can also be represented by resonance:

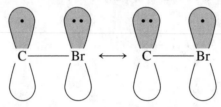

39. All necessary $DH°$ values are in Table 3-1, except for X—X, which are given in Table 3-4.

(a) The answers (kcal mol^{-1}), using $\Delta H° = DH°$(bond broken) $- DH°$(bond formed), are

Reaction	$\Delta H°$ for X = F	Cl	Br	I
(1) X· + CH$_4$ ⟶ CH$_3$X + H·	−5	+20	+35	+48
(2) H· + X$_2$ ⟶ HX + X·	−98	−45	−41	−35
(1) + (2) CH$_4$ + X$_2$ ⟶ CH$_3$X + HX	$\Delta H° = -103$	−25	−6	+13

(b) In every case the $\Delta H°$ for the hypothetical first propagation step above is much **less** favorable than the $\Delta H°$ for the generally accepted, correct step (Table 3-5). Therefore, the E_a values for the first steps above will, in all probability, be much larger than the E_a values for the correct steps. Relative to the correct propagation steps, then, the reaction X· + CH$_4$ ⟶ CH$_3$X + H· will probably be very, very slow, and unlikely to compete kinetically.

40. Inhibition usually comes about by reaction of the inhibitor with one of the reactive "chain-carrying" species in a propagation step. In the case of radical halogenation, the alkyl radical is susceptible to reaction with inhibitors. The products of the inhibition process are not reactive enough to continue on to another propagation step, so the propagation "chain" is broken in much the same way that termination steps break the propagation chain process. We have the following:

Cl$_2$ ⟶ 2 Cl·   Initiation

Cl· + CH$_4$ ⟶ HCl + CH$_3$·   Propagation step 1: $\Delta H° = +2$ kcal mol^{-1}

Then, however, in the presence of I$_2$

CH$_3$· + I$_2$ ⟶ CH$_3$I + I·   Inhibition step: $\Delta H° = -21$ kcal mol^{-1}

The chain started by propagation step 1 is now broken because I· cannot react with CH$_4$ ($\Delta H° = +34$ kcal mol^{-1}, from Table 3-5). A chain carrying CH$_3$· radical has been permanently lost from the reaction system.

41. (a) Divide ΔH°_{comb} by the molecular weight: (kcal mol^{-1}) $\div$ (g mol^{-1}) = kcal g^{-1}.

Methane: $\Delta H^\circ_{comb} = \dfrac{-212.8 \text{ kcal mol}^{-1}}{16 \text{ g mol}^{-1} \text{ (MW of } CH_4)} = -13.3 \text{ kcal g}^{-1}$

Ethane: $\Delta H^\circ_{comb} = -12.4 \text{ kcal g}^{-1}$

Propane: $\Delta H^\circ_{comb} = -12.0 \text{ kcal g}^{-1}$

Pentane: $\Delta H^\circ_{comb} = -11.7 \text{ kcal g}^{-1}$

(b) Ethanol (gas): $\Delta H^\circ_{comb} = -7.3 \text{ kcal g}^{-1}$

(c) Qualitatively, the observation is quite consistent with the much lower heat production by weight from the combustion of ethanol vs. alkanes; oxygen-containing molecules are indeed poorer sources of energy as fuels.

42. (a) $2 \text{ CH}_3\text{OH} + 3 \text{ O}_2 \longrightarrow 2 \text{ CO}_2 + 4 \text{ H}_2\text{O}$

$2 \text{ (CH}_3)_3\text{COCH}_3 + 15 \text{ O}_2 \longrightarrow 10 \text{ CO}_2 + 12 \text{ H}_2\text{O}$

Except for liquid H_2O, all substances are in the gas phase.

(b) The molecular weight of methanol (32) is close to that of ethane (30), but the ΔH°_{comb} value for ethane (-372.8 kcal mol^{-1}) is much greater in magnitude than that of methanol. Similarly, $(CH_3)_3COCH_3$ has a molecular weight of 88, close to that of hexane (86). The ΔH°_{comb} for the latter is -995.0 kcal mol^{-1}, again considerably more than that for 2-methoxy-2-methylpropane. Although addition of oxygenated compounds to the fuel mixtures of internal combustion engines reduces the heat yield per mass of fuel upon combustion, it accomplishes two other goals. First, the oxygen supplied in these additives contributes to permit more complete oxidation of the components of the fuel mix, reducing emission of partly oxidized byproducts such as CO. Second, the fuel mix is less susceptible to so-called pre-ignition, in which the fuel ignites before the piston has had a chance to reach the endpoint of its compression stroke in the cylinder. Pre-ignition results in "knocking" noises in an engine and wastes energy.

43. (a) Calculate ΔH° for both the primary and the secondary hydrogen abstraction reactions first. Data: DH° for primary C—H = 101 kcal mol^{-1}; for secondary C—H, 98.5 kcal mol^{-1}; for HBr, 87 kcal mol^{-1}. Therefore,

$$\Delta H^\circ_{(\text{primary C—H abstraction})} = 101 - 87 = +14 \text{ kcal mol}^{-1}$$

$$\Delta H^\circ_{(\text{secondary C—H abstraction})} = 98.5 - 87 = +11.5 \text{ kcal mol}^{-1}$$

So, we have the following energy diagram as a result:

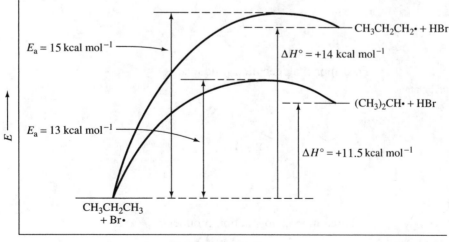

(b) These are both "late" transition states, most resembling the products in energy. (Contrast the very "early" transition states in chlorination.)

(c) These transition states closely resemble the product radicals in structures and therefore have considerable radical character. By comparison, those in Figure 3-9 (for chlorination) show much less radical character, being much "earlier" and much less productlike.

(d) Yes. For bromination the radical-like transition states for primary vs. secondary reaction differ in energy by an amount (2 kcal mol^{-1}) that closely reflects the difference in energy of the radicals themselves (2.5 kcal mol^{-1}). For chlorination, the much less radical-like transition states do not reflect the energies of the product radicals nearly as much, so the difference between them is much less (1 kcal mol^{-1}). Selectivity here is totally determined by the energy difference between transition states of competing pathways; therefore, bromination is much more selective than chlorination.

44. $\Delta H°$ is calculated by subtracting the $DH°$ values for the bonds formed from those of the bonds broken. So,

Propagation step 1: $\Delta H° = 26 - 64 = -38$ kcal mol^{-1}

Propagation step 2: $\Delta H° = 64 - 120 = -56$ kcal mol^{-1}

The overall reaction is just $O_3 + O \rightarrow 2\,O_2$ with $\Delta H° = -94$ kcal mol^{-1}. The reaction is extremely favorable energetically, and, as the equations show, it is **catalyzed** by Cl atoms. A single atom of chlorine is capable of destroying **thousands** of ozone molecules in propagation cycles such as this one.

45. (a) and **(b)**

hexane   Three monohalogenation products:

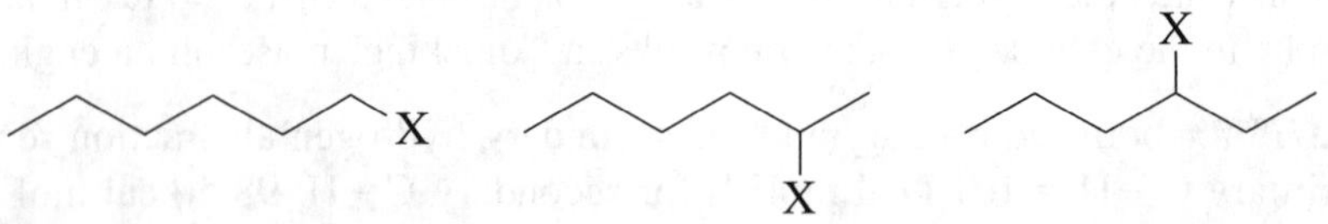

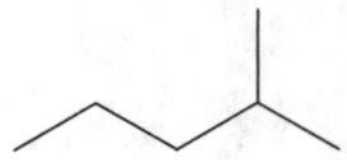

2-methylpentane   Five monohalogenation products:

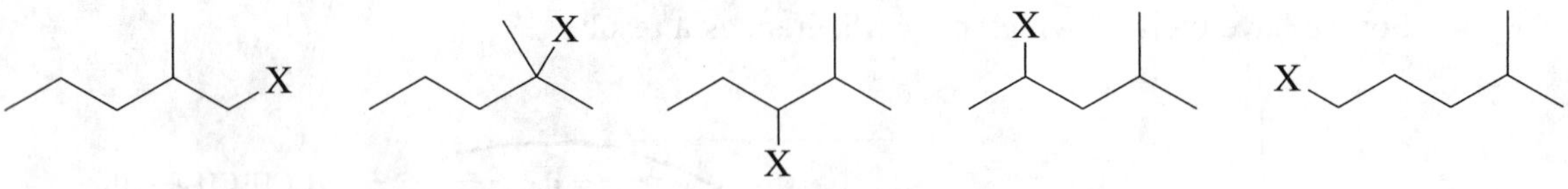

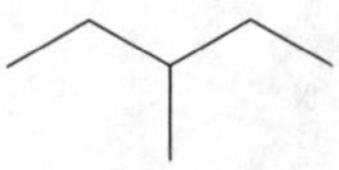

3-methylpentane   Four monohalogenation products:

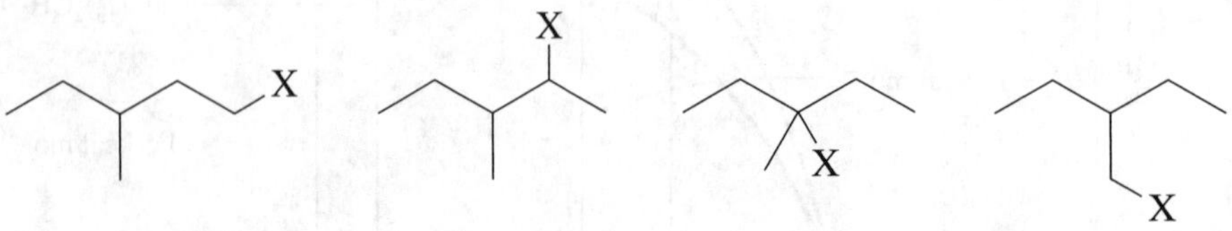

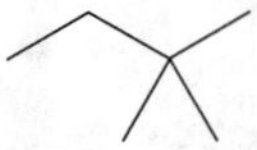

2,2-dimethylbutane   Three monohalogenation products:

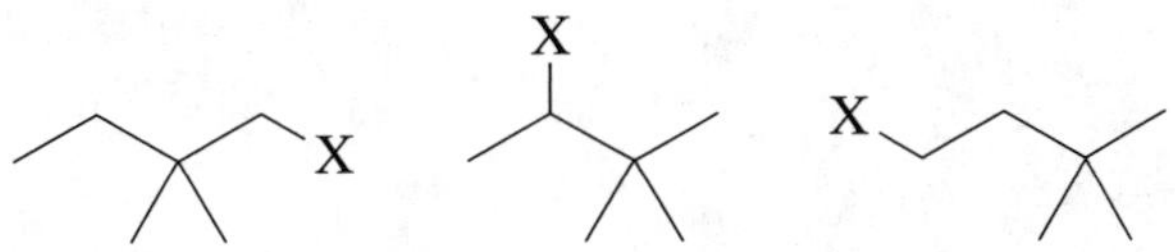

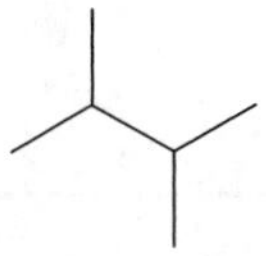

Two monohalogenation products:

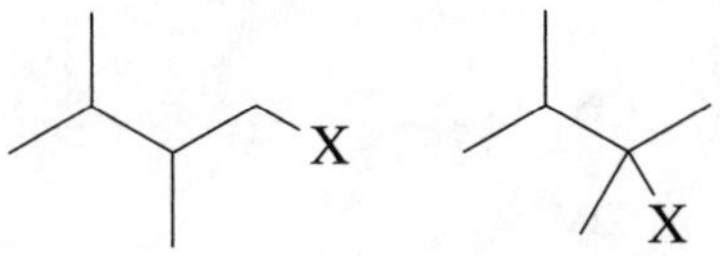

2,3-dimethylbutane

(c) As illustrated above, 2,3-dimethylbutane possesses only two distinct sites for halogenation: two indistinguishable tertiary hydrogens and twelve indistinguishable primary H's. In the case of X = Br, virtually the entire product will arise from bromination at a tertiary position.

46. (e)

47. (c) (1-Chloro-2-methyl-4-propyloctane) Yes, they actually ask questions like this.

48. (a) For chlorination, six secondary hydrogens outcompete the lone tertiary in (b).

49. (d)

4

Cycloalkanes

It is convenient to cover ordinary alkanes and cyclic ones in separate chapters of organic chemistry textbooks. In most cases, however, the presence or absence of a ring in a molecule makes little difference to its physical properties or its chemical behavior. What you have learned in Chapters 2 and 3 can be applied virtually without change to the molecules presented in Chapter 4. Cyclic alkanes are nonpolar, lacking in any functional groups, and therefore are relatively unreactive, like acyclic alkanes. For most of them, the only important reactions are radical reactions. The major topics of concern are those dealing with the shapes (conformations) of the types of ring systems, and the effects of these shapes on the bonding and stability of each size ring. Some new points of nomenclature are presented. On the whole, however, the chapter contains only one new topic that is not a direct extrapolation of what has gone before: the concept of bond angle strain in compounds containing small rings.

Outline of the Chapter

4-1 Nomenclature and Physical Properties
Basic material.

4-2 Ring Strain and Structure
The bonding consequences of closing a chain of atoms into a ring of three, four, or five carbons.

4-3 Cyclohexane
The most common and most important ring size (six carbons). Its shapes, and their consequences.

4-4 Substituted Cyclohexanes
More of the same.

4-5 Larger Cycloalkanes
Very brief overview.

4-6 Polycyclic Alkanes
Ditto.

4-7 Carbocyclic Natural Products
Common ring-containing molecules of biological importance.

Keys to the Chapter

4-1. Nomenclature and Physical Properties
The naming of ring compounds requires two new procedures in addition to those associated with acyclic systems. First, because rings have no "ends," numbering starts at that carbon around the ring giving the lowest

numbers for substituent groups, using the same criteria for "lowest numbers" presented earlier. Second, rings have a "top" and a "bottom" face, relatively speaking. Therefore, substituents on different ring carbons may either be on the same face or on opposite faces, necessitating the cis or trans denotation in the name. All other principles of nomenclature follow unchanged.

4-2.   Ring Strain and Structure

Electron pairs repel each other and try to be as far apart as possible. Rings with only three or four atoms force the electron pairs of the C—C bonds to be closer together than is normal. The repulsion that results is the major cause of the *high-energy* nature of small ring compounds and is the physical cause of the *ring strain* referred to in the text.

To examine the structural aspects of these molecules, you will find your set of models to be indispensable. Cyclopropane is the only flat cycloalkane ring. **All** larger cycloalkanes are nonplanar. Ring distortion away from a planar structure reduces eclipsing interactions between neighboring carbon–hydrogen bonds.

4-3 and 4-4.   Cyclohexanes

Before you do **anything** else, make a model of cyclohexane. Be sure to use the correct atoms and bonds from your kit. The completed model should not be too floppy and should be easily capable of holding the shape shown in Figure 4-5(B). This is the *chair* conformation, with three C—H bonds pointing straight up and three C—H bonds pointing straight down (the *axial* C—H bonds). Starting from this point, you should be able to construct the other important cyclohexane conformations by moving an "end" carbon through the plane of the "middle" four carbons of the ring; that is,

Move up

→ Boat and boatlike conformations (rather floppy, too)

Learn to recognize axial and equatorial positions and their cis/trans interrelationships around the ring. Again, use your model in conjunction with the chapter text and illustrations. Note the congestion associated with large groups in axial positions, a result of *1,3-diaxial* interactions, the main effect that causes differences in energy between the two possible chair conformations of a substituted cyclohexane. This is an example of a *transannular* (literally, "across the ring") interaction, arising in this case from the ring structure forcing groups to adopt *gauche* conformational relationships. Be sure to use your models when trying to do the problems at the end of the chapter.

4-5 and 4-6.   Larger Rings; Polycyclic Molecules

The material in these sections is intended only to give a very brief introduction to areas of organic chemistry that are important in current research but are generally beyond the scope of a course at this level. Only a small number of selected molecules are mentioned with relevant points of structure and nomenclature presented where appropriate.

Solutions to Problems

21. Start with the largest ring and systematically go through successively smaller rings:

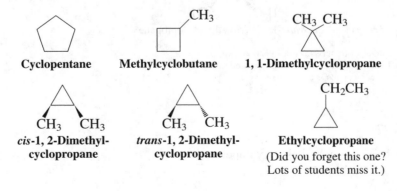

Cyclopentane   Methylcyclobutane   1, 1-Dimethylcyclopropane

cis-1, 2-Dimethyl-
cyclopropane

trans-1, 2-Dimethyl-
cyclopropane

Ethylcyclopropane
(Did you forget this one?
Lots of students miss it.)

22. As before, begin with the largest parent structure (in this case, cyclohexane itself), and work downward in ring size one carbon atom at a time, adding substituents in all possible arrangements to make up the required molecular formula.

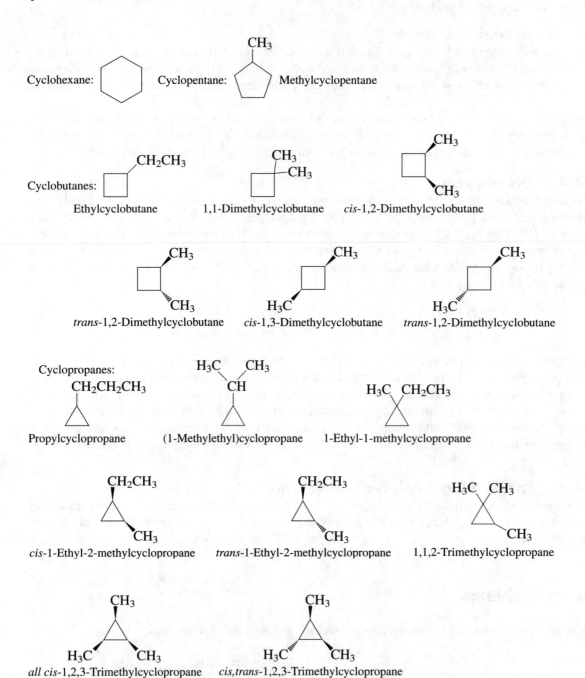

Cyclohexane: Cyclopentane: Methylcyclopentane

Cyclobutanes:

Ethylcyclobutane   1,1-Dimethylcyclobutane   *cis*-1,2-Dimethylcyclobutane

trans-1,2-Dimethylcyclobutane   *cis*-1,3-Dimethylcyclobutane   *trans*-1,2-Dimethylcyclobutane

Cyclopropanes:

Propylcyclopropane   (1-Methylethyl)cyclopropane   1-Ethyl-1-methylcyclopropane

cis-1-Ethyl-2-methylcyclopropane   *trans*-1-Ethyl-2-methylcyclopropane   1,1,2-Trimethylcyclopropane

all cis-1,2,3-Trimethylcyclopropane   *cis,trans*-1,2,3-Trimethylcyclopropane

These two applications of the cis/trans designations are no longer in general use. Chapter 5 will present the modern method for naming compounds such as these.

23. **(a)** Iodocyclopropane

(b) *trans*-1-Methyl-3-(1-methylethyl)cyclopentane

(c) *cis*-1,2-Dichlorocyclobutane

(d) *cis*-1-Cyclohexyl-5-methylcyclodecane

(e) To tell whether this is cis or trans, draw in the hydrogens on the substituted carbons:

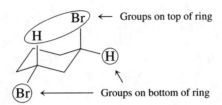

One Br on top, one on bottom, ∴ *trans*-1,3-dibromocyclohexane.

(f) Similarly,

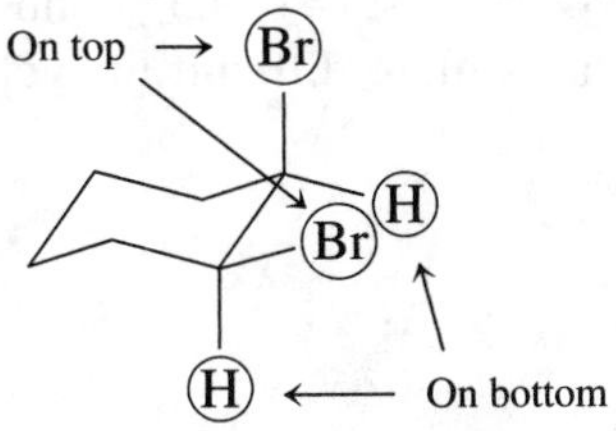

On top → (Br)

(Br)(H)

(H) ← On bottom

∴ *cis*-1,2-dibromocyclohexane

24. **(a)**  The name given is a usable common name but does not follow IUPAC. The correct IUPAC name is (2-methylproply)cyclopentane.

(b) The name is correct.

(c) 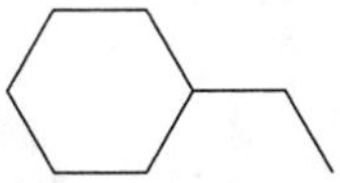 The name given is unambiguous, but IUPAC stipulates that in compounds with a ring attached to a straight chain, the parent should be whichever one has more carbon atoms. Call it ethylcyclohexane.

(d) 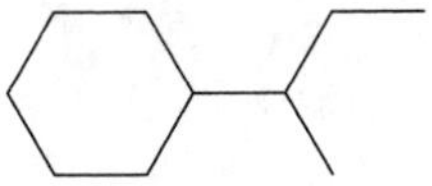 The substituent is named incorrectly. (1-Methylpropyl)cyclohexane is correct.

(e) 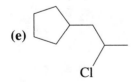 Correct.

Cl

(f) 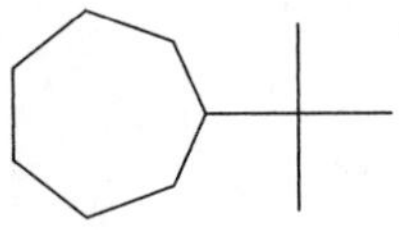 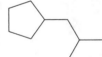 (1,1-Dimethylethyl)cyclohexane.

25.

(a) 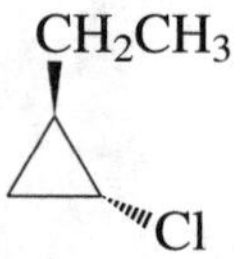

(b) 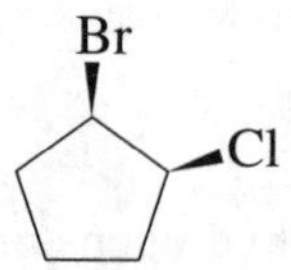

(c) 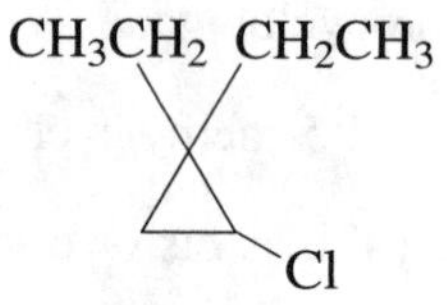

(d) 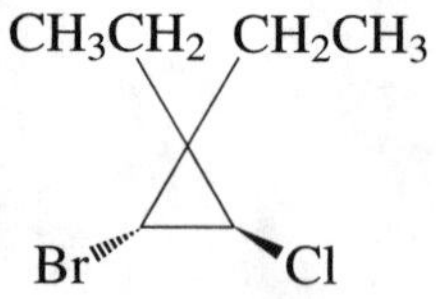

(e) 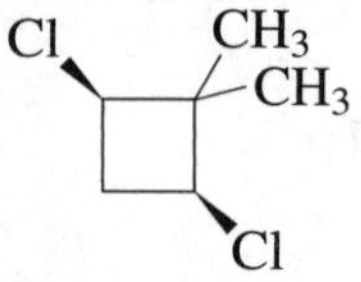

(f) 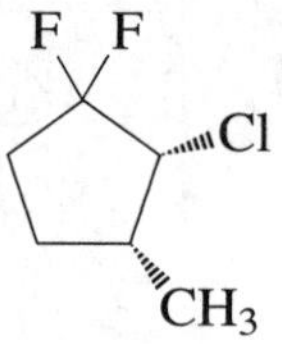

26. **(a)** The very low relative radical chlorination reactivity of cyclopropane implies abnormally **strong** C—H bonds and an abnormally **unstable** cyclopropyl radical.

(b) Radicals prefer sp^2 hybridization, with 120° bond angles. So in the cyclopropyl radical, the bond angle strain at the radical carbon is greater (120° − 60° = 60° bond angle compression) than at a carbon in cyclopropane itself (109.5° − 60° = 49.5° bond angle compression). Forming the radical therefore **increases** ring strain and is more difficult in cyclopropane than in a molecule lacking bond angle distortion to begin with.

27. Initiation:

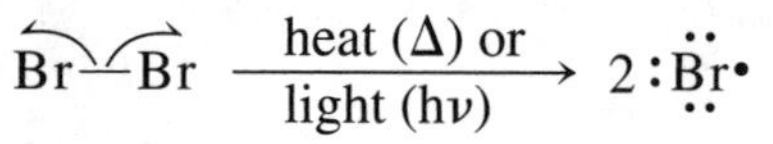

Propagation:

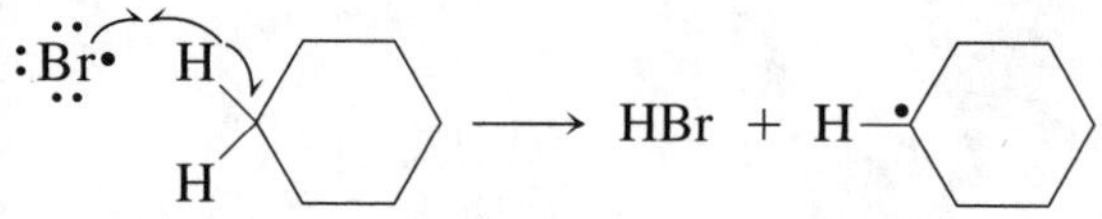

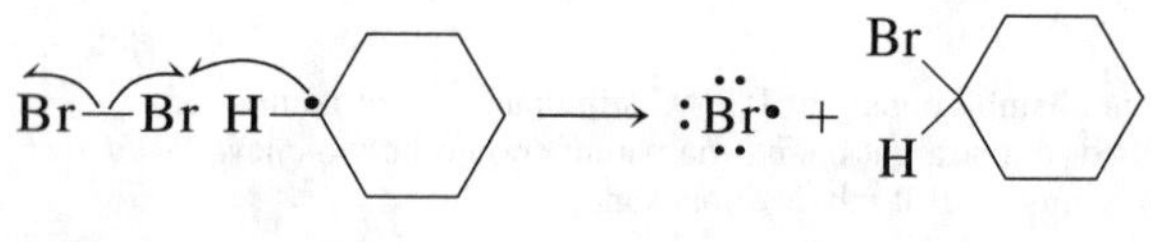

Termination:

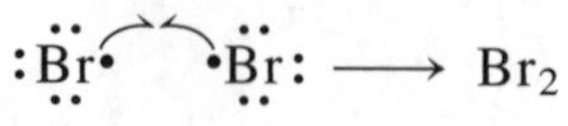

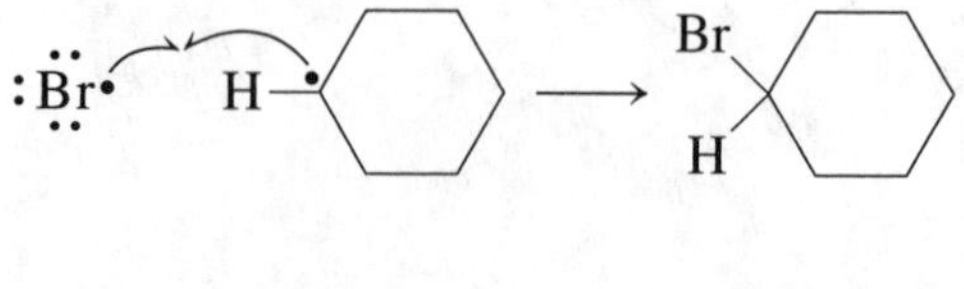

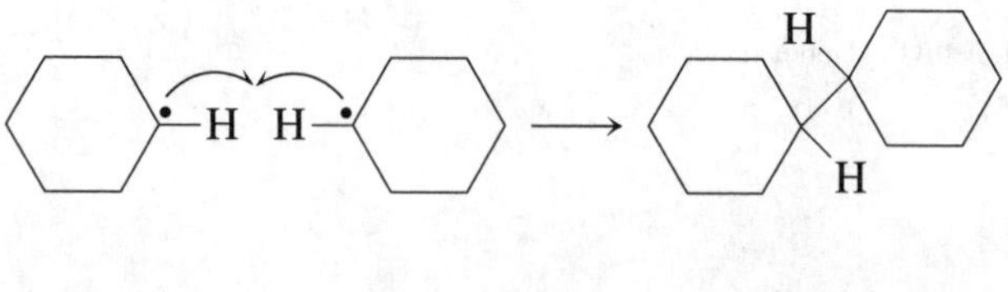

28. In all cases the reference value to begin with is the $DH°$ for the C—C bond between CH_2 (2°) groups, i.e., $DH°$ for CH_3CH_2—CH_2CH_3, 88 kcal mol^{-1} (Table 3-2).

(a) Cleavage of a C—C bond in cyclopropane requires a smaller net energy input because ring strain is relieved in the process. Breaking a "normal" C—C bond would require 88 kcal mol^{-1} input, but because 28 kcal mol^{-1} is recovered as a result of strain relief in opening the three-membered ring, the $DH°$ actually required is $88 - 28 = 60$ kcal mol^{-1}. Note that this is consistent with the E_a of 65 kcal mol^{-1} for ring opening (Section 4-2).

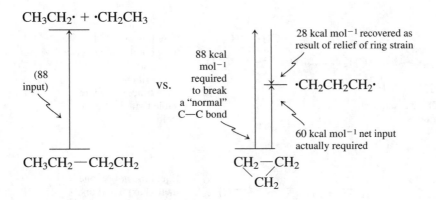

(b) For cyclobutane, our estimated $DH° = 88 - 26 = 62$ kcal mol^{-1}.

(c) $DH° = 88 - 7 = 81$ kcal mol^{-1}

(d) $DH° = 88 - 0 = 88$ kcal mol^{-1}

Thus, the unusual ring-opening reactions of cyclopropane and cyclobutane (relative to other alkanes and cycloalkanes) are thermodynamically reasonable.

29. Here is a drawing of cyclobutane, with axial (a) and equatorial (e) positions labeled.

All the carbons are equivalent, and flipping the puckered form exchanges axial and equatorial positions, exactly as does flipping chair conformations in cyclohexane.

(c) Both CH$_3$'s equatorial; more stable

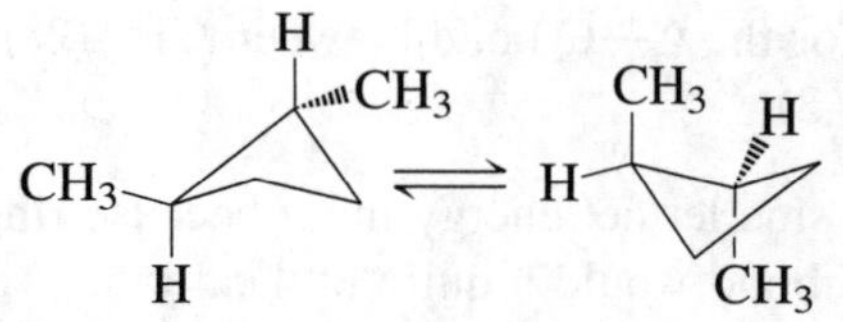

This (the *trans*-1,2 compound) is more stable because both CH$_3$'s can be equatorial at the same time. In the *cis*-1,2 [(b) on previous page] there is always one axial group in either conformation.

(d) Both CH$_3$'s equatorial; more stable

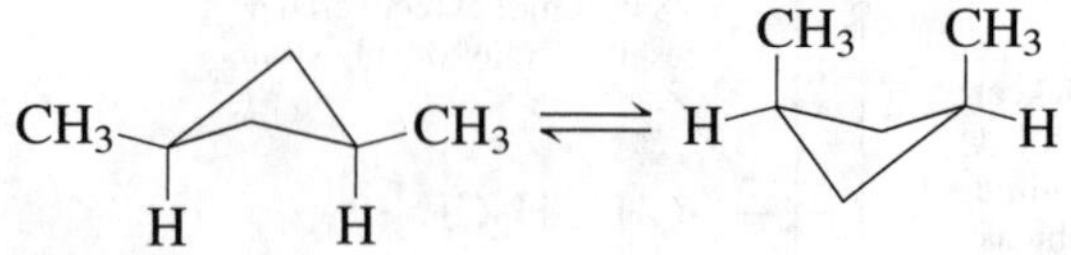

Now it is the *cis*-1,3 compound that can have both CH$_3$'s equatorial at the same time. It is more stable than the trans (below).

(e) Equal in energy: one methyl axial and one equatorial in each conformation.

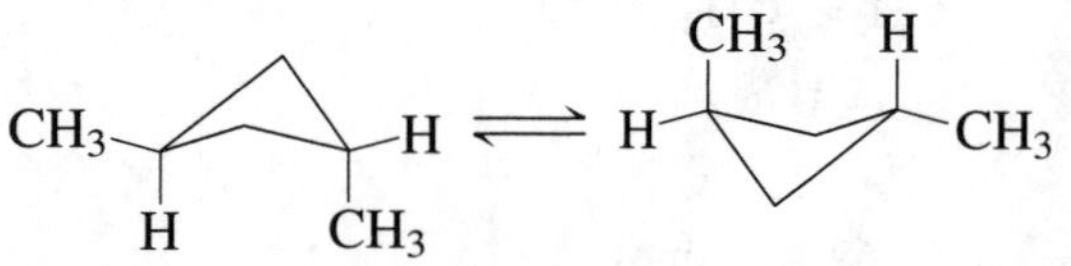

30. Refer to answers to Problems 23(e) and 23(f) for guidelines.

(a) Trans. Not most stable form. Ring flip gives diequatorial conformation:

(b) Trans! (Surprise!) Note positions of hydrogens.

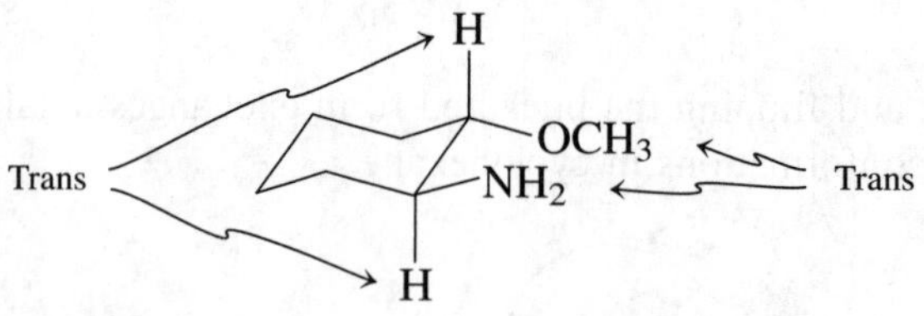

The two hydrogens are trans, so clearly the NH$_2$ and OCH$_3$ groups must be trans, too. The NH$_2$ is cis to the top H, and the OCH$_3$ is cis to the bottom H.

Both groups are equatorial, so this is the most stable conformation.

(c) Cis.

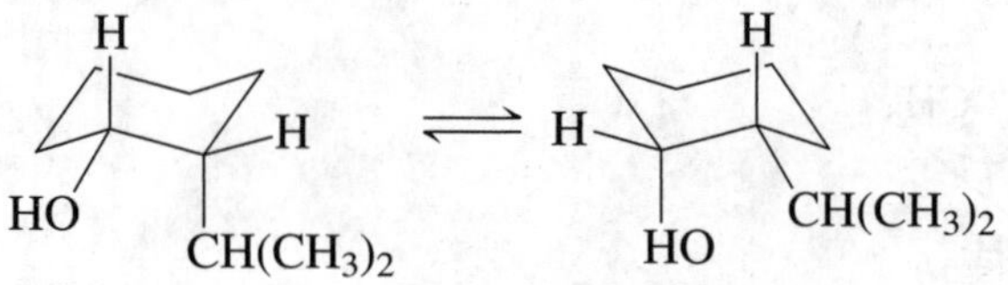

From Table 4-3, we see that $CH(CH_3)_2$ prefers an equatorial position more (2.2 kcal mol^{-1}) than does OH (0.94 kcal mol^{-1}). In the structure drawn, $CH(CH_3)_2$ is axial and OH is equatorial. This is **not** the most stable conformation because the ring can flip to the form on the right, in which $CH(CH_3)_2$ is equatorial and OH axial.

(d) Trans.   Most stable conformation (CH_3 equatorial).

(e) Cis.   Most stable form (CH_3CH_2 equatorial).

(f) Trans.   Most stable form (both groups equatorial).

(g) Cis.   Most stable form (both groups equatorial).

(h) Cis.   Not most stable form. Ring flip makes it diequatorial.

(i) Cis.   Not most stable form. Ring flip makes

$HO-\overset{\overset{\textstyle O}{\|}}{C}$ group equatorial, which is preferable (Table 4-3).

(j) Trans. Most stable form [compare (b), on previous page].

31. The sign for $\Delta G°$ will be negative if the conformation shown in the problem is the less stable one and will be positive if the conformation shown is the more stable one.

(a) $-(1.70 + 0.52) = -2.22$ kcal mol^{-1} **(b)** $1.4 + 0.75 = 2.2$ kcal mol^{-1}

(c) $-(2.20 - 0.94) = -1.26$ kcal mol^{-1} **(d)** $1.70 - 1.29 = 0.41$ kcal mol^{-1}

(e) $1.75 - 0.46 = 1.29$ kcal mol^{-1} **(f)** $1.4 + 0.55 = 2.0$ kcal mol^{-1}

(g) $1.70 + 0.75 = 2.45$ kcal mol^{-1} **(h)** $-(0.94 + 0.25) = -1.19$ kcal mol^{-1}

(i) $-(1.41 - 0.55) = -0.86$ kcal mol^{-1} **(j)** $2.20 + 0.52 = 2.72$ kcal mol^{-1}

32.

Most stable conformation	Least stable conformation

(a) 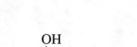 OH

(b) OH — CH$_3$

(c) CH$_3$ — CH(CH$_3$)$_2$

(d) CH$_3$O — CH$_2$CH$_3$

(e) Cl — C(CH$_3$)$_3$

Least stable conformation:

(a) OH

(b) HO — CH$_3$

(c) CH$_3$ — CH(CH$_3$)$_2$

(d) CH$_3$O — CH$_2$CH$_3$

(e) Cl — C(CH$_3$)$_3$

33.

From Table 4-3

(a) 0.94 kcal mol^{-1} (less stable conformation is higher in energy)

(b) 1.7 − 0.94 = 0.8 kcal mol^{-1}

(c) 2.2 + 1.7 = 3.9 kcal mol^{-1}

(d) 1.75 − 0.75 = 1.00 kcal mol^{-1}

(e) 5 + 0.52 = 5.5 kcal mol^{-1}

Ratios, using $\Delta G° = -RT \ln K_{eq}$

$K_{eq} = 4.8$; $\dfrac{4.8}{4.8 + 1} = 0.83$;

∴ 83/17 ratio (in favor of more stable conformation)

$K_{eq} = 3.8$; $\dfrac{3.8}{3.8 + 1} = 0.79$;

∴ 79/21 ratio

$K_{eq} \approx 10^3$; 99.9/0.1 ratio

$K_{eq} = 5.3$; 84/16 ratio

$K_{eq} \approx 10^4$; $\gg$ 99.9/0.1 ratio

In each case the more stable conformation is the one in which the group with the largest $\Delta G°$ value from Table 4-3 is equatorial.

34. The basic idea is that the two extremes of the diagram, the two chair conformations, will no longer be equal in energy: One has the methyl group equatorial, but the other has it axial. You do not have sufficient information to estimate the energies of the twist-boat and boat conformations in the middle of the diagram, except to assume that they will probably be equal to or (more likely) higher in energy than the corresponding conformations of cyclohexane itself, relative to the more stable chair conformations.

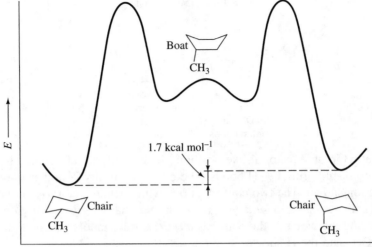

35. Both rings can flip, so there are four possible combinations, two of which are identical:

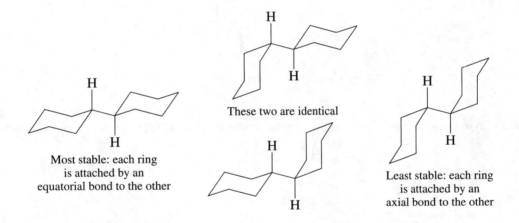

These two are identical

Most stable: each ring
is attached by an
equatorial bond to the other

Least stable: each ring
is attached by an
axial bond to the other

36. Notice how some positions around a boat conformation of cyclohexane are axial-like ("pseudoaxial") and some are equatorial-like ("pseudoequatorial"):

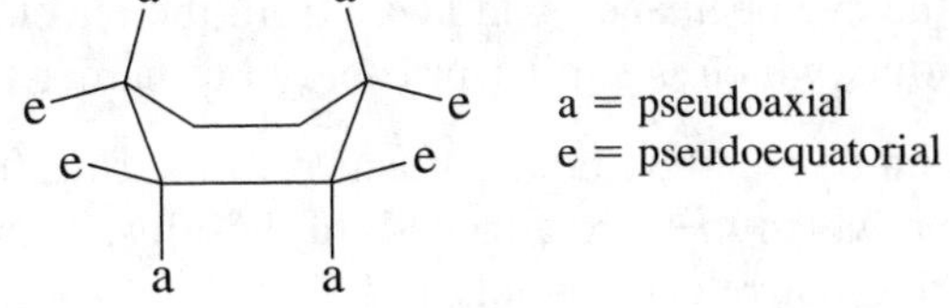

a = pseudoaxial
e = pseudoequatorial

If you draw conformations placing the methyl group in each different type of position and examine each conformation for strain, you will see the following:

Methyl is pseudoequatorial

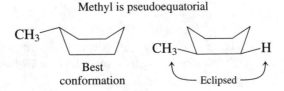

Best
conformation

Eclipsed

Methyl is pseudoaxial

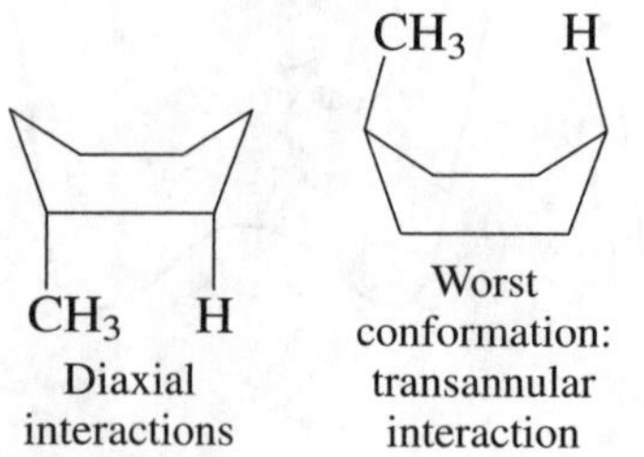

CH_3 H

CH_3 H
Diaxial
interactions

Worst
conformation:
transannular
interaction

Of the two with the methyl in an equatorial-like position, the one on the left has the bond to the CH_3 staggered with respect to neighboring C—H bonds. The other possibility is higher in energy as a result of the eclipsing interaction shown. The two conformations with pseudoaxial CH_3 are both quite high in energy. One actually has three diaxial interactions involving the methyl group (only one is shown; make a model to see the rest!), and the worst of them all has a serious transannular interaction due to the close approach between the CH_3 and the H shown.

37. Only a boat-related conformation permits both bulky groups to avoid axial positions. This molecule will adopt a shape in which both groups are "pseudoequatorial." It will be based on the twist-boat of cyclohexane in order to minimize eclipsing interactions of the true boat conformation (Section 4-3). (Make a model!)

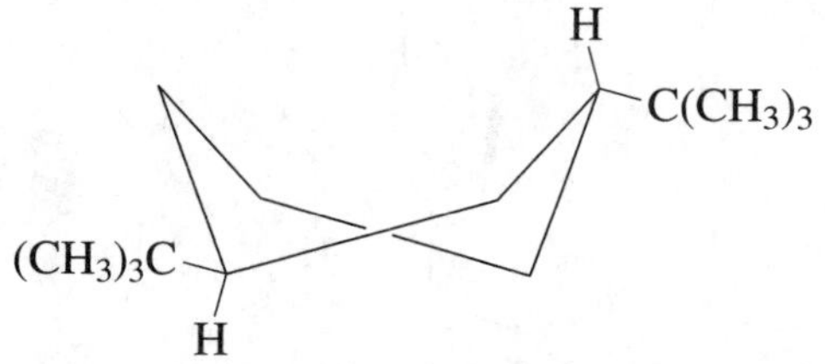

H

$C(CH_3)_3$

$(CH_3)_3C$

H

38. Models may be helpful here. You should be able to construct structures similar to those pictured below.

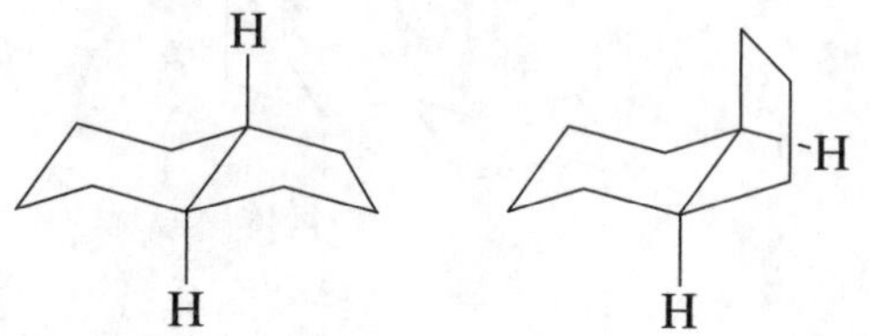

H

H

trans-Hexahydroindane

H

H

cis-Hexahydroindane

Notice how the ring-fusion hydrogens are trans to each other in *trans*-hexahydroindane and cis to each other in *cis*-hexahydroindane. In the drawings, the cyclohexane rings are chairs and the cyclopentane rings are envelopes. A slight twist around the cyclopentane bond away from the envelope "flap" would give rise to the cyclopentane half-chair conformation, which is similar in energy but harder to draw.

39. In *trans*-decalin each of the carbon–carbon bonds attached to the ring fusion occupy an equatorial position with respect to the ring on which they are attached. In the illustration below we have labeled the rings as A and B, and numbered the four relevant bonds 1–4:

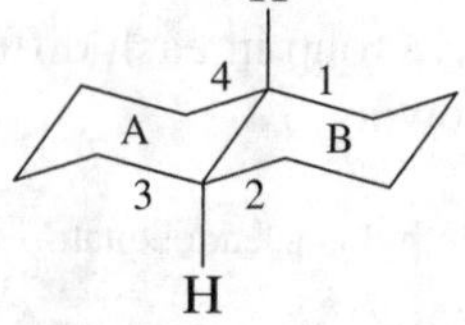

H

4   1
A   B
3   2

H

Bonds 1 and 2 (which are part of ring B) are equatorial substituents with respect to ring A. Bonds 3 and 4 (which are part of ring A) are equatorial substituents with respect to ring B. Both hydrogen atoms on the ring fusion carbons are axial with respect to *both* rings.

In *cis*-decalin the situation is different. Look at the picture:

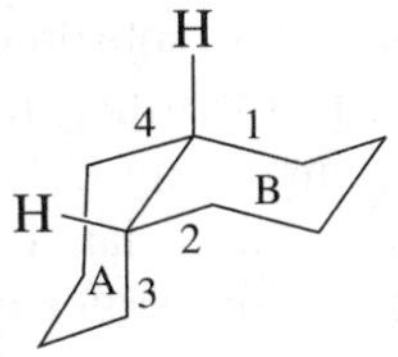

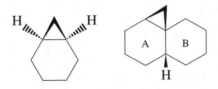

Bond 1 (in ring B) is now *axial* with respect to ring A (rotate the page clockwise by 60° to see this more clearly). Bond 2 is still equatorial. Also, bond 3 (in ring A) is now axial with respect to ring B (bond 4 is still equatorial). So two of these four bonds are axial with respect to the ring on which they are substituents and, as a result, give rise to 1,3-diaxial interactions that raise the enthalpy of the compound. Notice that the ring-fusion hydrogens are now equatorial with respect to one ring and axial with respect to the other.

If we were to assign an energy of about 1.75 kcal mol^{-1} to each axial bond to a carbon in *cis*-decalin, we would conclude that the cis isomer is 3.5 kcal mol^{-1} higher in energy (less stable) than the trans. This turns out to be an overestimate, in part because carbon atoms in rings cannot rotate freely and therefore do not generate as much steric interference as do simple alkyl groups, which can rotate a full 360°. An alternative way to estimate the energy difference is to search for butane structural fragments that possess *gauche* conformations. The cis isomer has three more than does the trans; assuming each *gauche* butane raises the energy content by about 0.9 kcal mol^{-1}, one arrives at an energy difference of 2.7 kcal mol^{-1}.

40. You really do need to go to models to address this question. Start by examining a simpler compound, one in which a cyclopropane ring is fused to a single cyclohexane ring (below, left).

Is the cyclohexane in a chair conformation at all? Actually, it is not. The geometry of the three-membered ring forces the cyclohexane ring bonds associated with it and four of the cyclohexane ring carbons to all be coplanar. Only the two cyclohexane ring carbons farthest from the ring fusion are movable.

In tricyclo[5.4.0^{1,3}.0^{1,7}]undecane (above, right), the ring labeled 'A' is similarly constrained. Only two of its carbons (the one at the bottom and the one at the lower left) are capable of significant motion. This isn't much more than a wiggling motion, one up and one down, relative to the plane of the remaining four carbons in that six-membered ring. As for ring 'B', it is capable of holding one reasonable approximation of a chair conformation, with the appearance shown below.

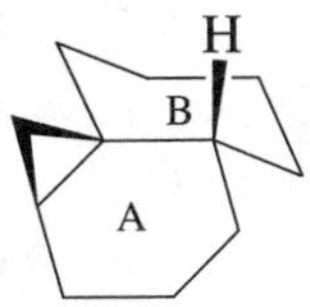

However, if you attempt a chair-chair flip in ring 'B', you will encounter much more resistance from the model set. The flip can be forced, but the conformation that results is a distorted twist-boatlike shape. This is because the two carbons at the ring fusion cannot undergo the corresponding rotation that would finish the chair-chair interconversion. The rigidity of ring 'A' prevents that.

41. (a) The β form is more stable because all of its substituent groups are equatorial. One of the —OH groups (the one on the ring carbon at the right of the picture) in the α form is axial.

(b) $K_{eq} = 64/36 = 1.78$. Using the equation $\Delta G° = -1.36 \log_{10} K_{eq}$ (Section 2-1) for an equilibrium constant at room temperature (25°C), we get $\Delta G° = -0.34$ kcal mol^{-1}. In Table 4-3 the axial/equatorial free energy difference for an —OH substituent on cyclohexane is -0.94 kcal mol^{-1}. The difference derives largely from the fact that the ring in glucose is not a cyclohexane but an *oxa*cyclohexane—a cyclic ether. Replacing a ring CH_2 group with an oxygen has several effects, including removal of the steric interactions associated with the hydrogens on the carbon, and introduction of two polar C—O bonds whose dipoles can give rise to either attractive or repulsive interactions with the C—O bonds of nearby substituted ring carbons.

42. Count carbons in the molecule. If there are 10, the molecule is a monoterpene; if 15, a sesquiterpene; if 20, a diterpene.

   (a) 10 carbons, monoterpene   **(b), (c),** and **(d)** 15 carbons, sesquiterpene

   (e) 11 carbons, but only 10 in the contiguous molecular "skeleton"; monoterpene

   (f) 15, sesquiterpene   **(g)** 10, monoterpene   **(h)** 20, diterpene

43.

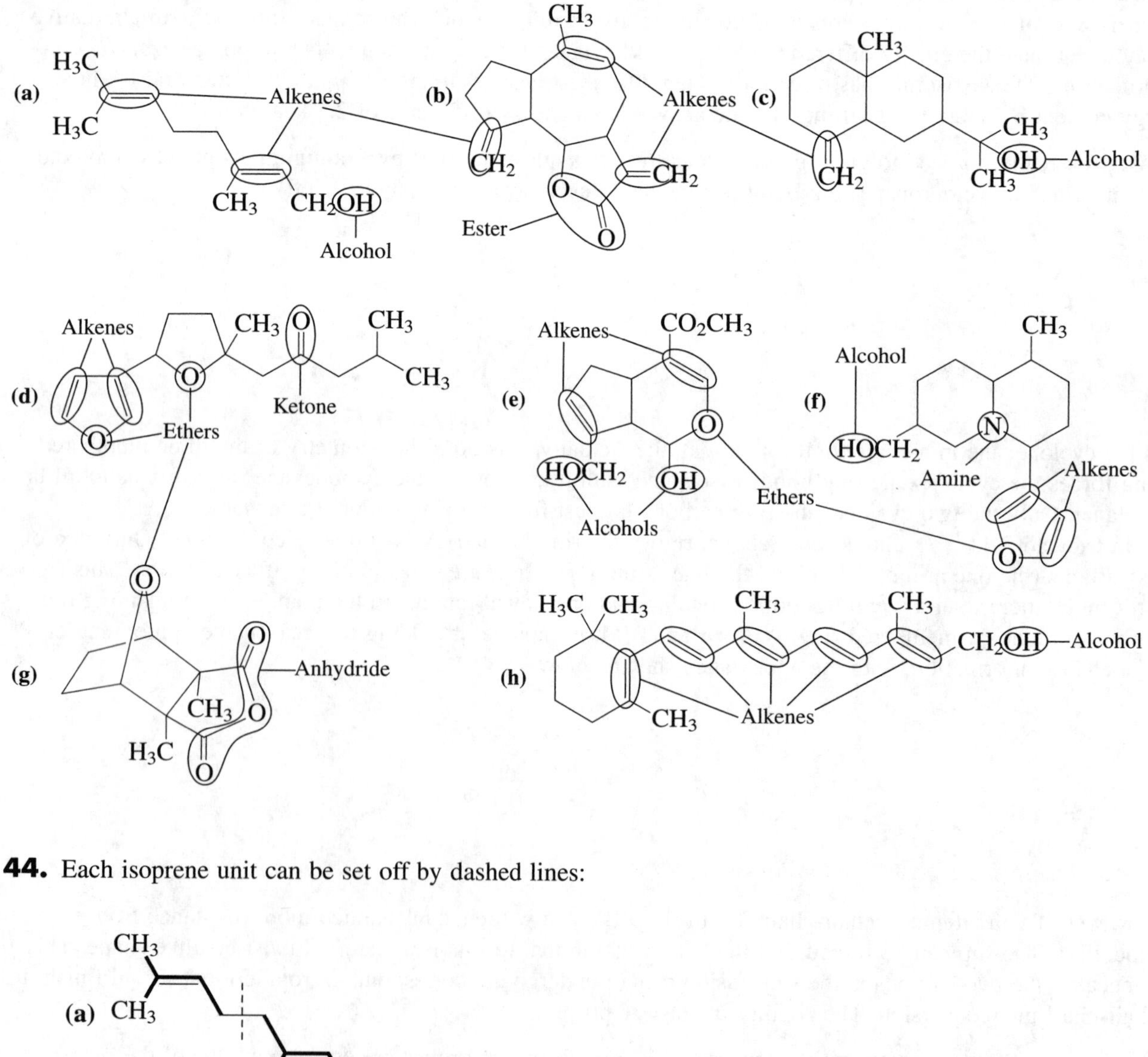

44. Each isoprene unit can be set off by dashed lines:

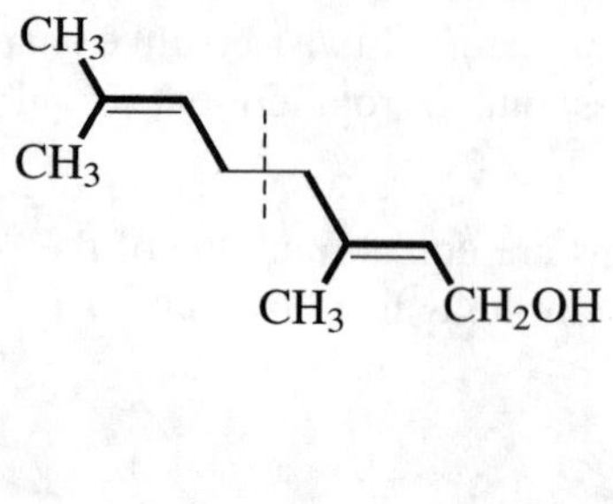

(b)

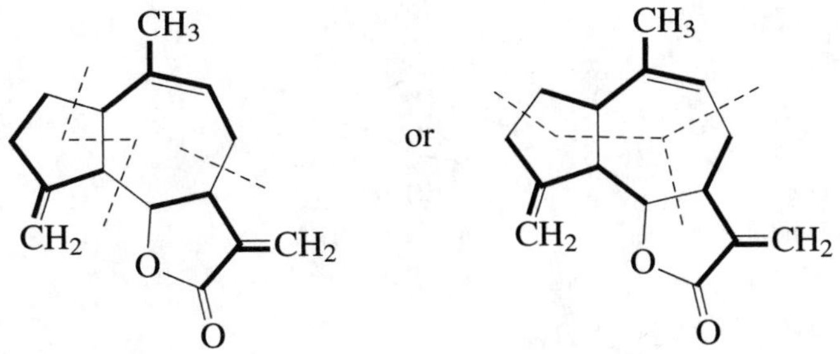

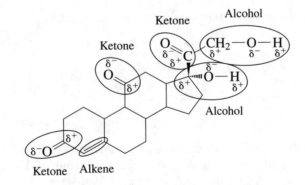

or

(c)

or

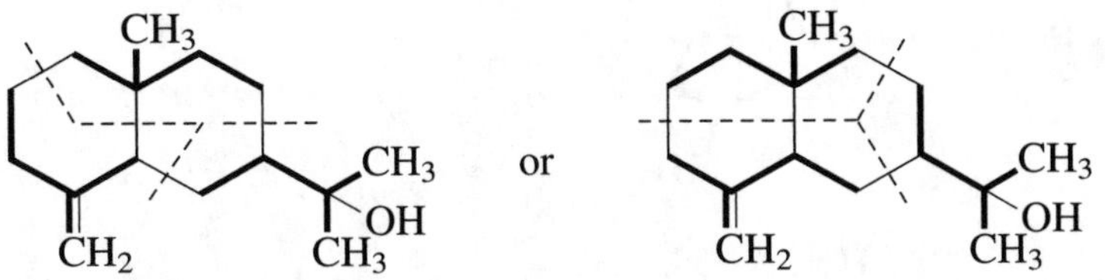

(d)

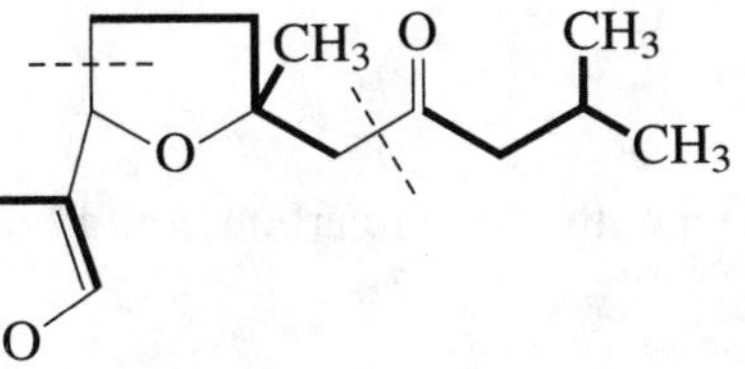

(e)

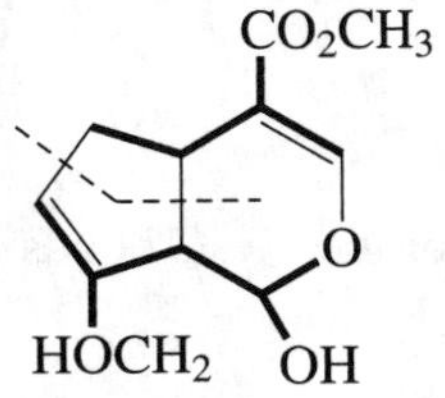

(f)

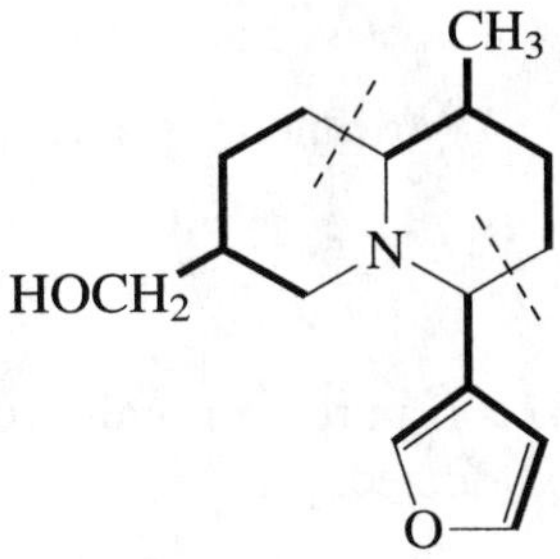

(g)

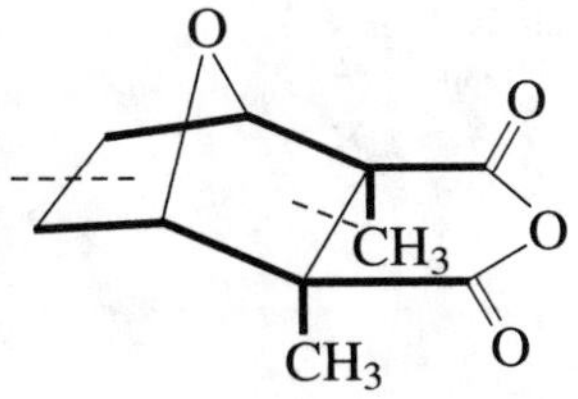

(h)

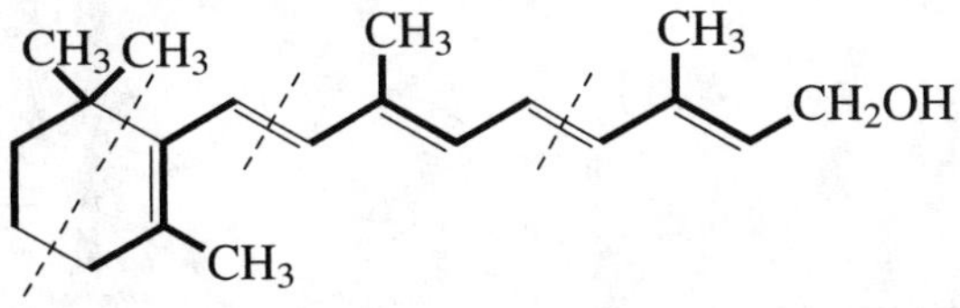

45. Cortisone provides a good example for this exercise:

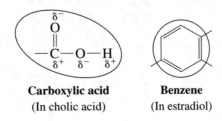

Other functional groups shown in this section include the following:

Carboxylic acid
(In cholic acid)

Benzene
(In estradiol)

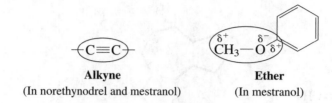

Alkyne
(In norethynodrel and mestranol)

Ether
(In mestranol)

46. α-Pinene is a monoterpene (10 carbons):

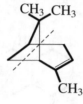

Africanone is a sesquiterpene:

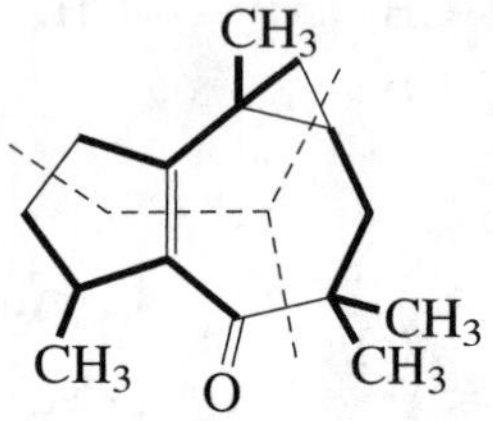

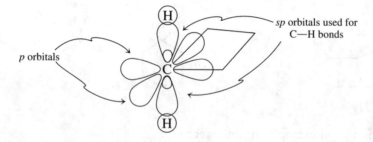

47. If pure *p* orbitals were used for the C—C bonds of cyclobutane, then each carbon could be *sp* hybridized:

sp orbitals used for
C—H bonds

p orbitals

The H—C—H bond angle would be 180° and cyclobutane would look like this:

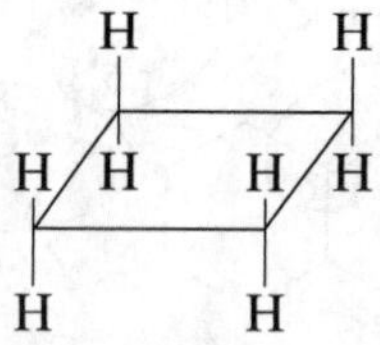

In reality, cyclobutane uses "bent" bonds just like those of cyclopropane, and all four bonds to each carbon involve hybridized orbitals. In addition, cyclobutane is not flat at all, and the H—C—H bond angle is not much different from the normal tetrahedral value of 109° (see Figure 4-3).

48. Notice that making "all-chair" cyclodecane is essentially the same as removing the ring-fusion bond from *trans*-decalin and replacing it with two hydrogens—one on each former ring-fusion carbon:

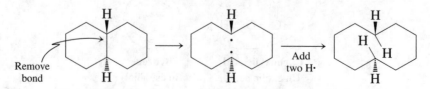

Remove
bond

Add
two H·

The resulting molecule has these two new hydrogens pointing into the center of the ring, in prohibitively close contact with the carbon and hydrogens on the opposite side of the ring. The steric strain of this *transannular* interaction makes this a very high energy conformation.

49. **(a)** From left to right: chair cyclohexane, boat (or twist-boat) cyclohexane, chair cyclohexane, envelope cyclopentane.

(b) All are trans.

(c) α means below and β means above. Therefore, 3α-OH, 4α-CH$_3$, 8α-CH$_3$, 10β-CH$_3$, 11α-OH, 14β-CH$_3$, 16β—OCCH$_3$.

(d) The boatlike cyclohexane ring is the most unusual feature; most steroids have only chair cyclohexanes. The boat shape is a result of the unusual cis relationship of the groups at positions 9 and 10, and also at positions 5 and 8. Note also the unusual number and location of methyl groups: at positions 4, 8, 10, and 14, instead of the more common pair of methyl groups at 10 and 13.

50.

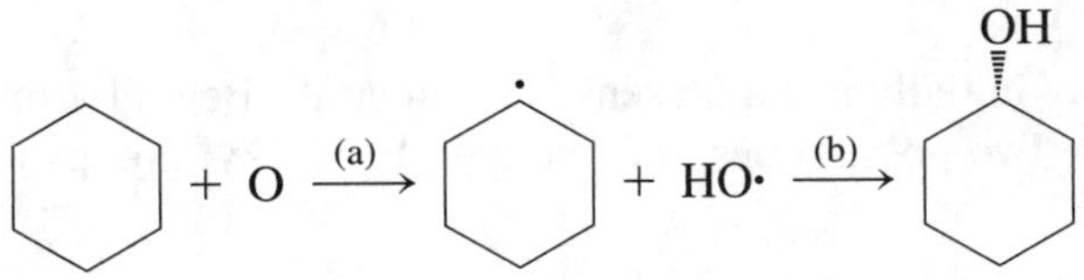

For (a), $\Delta H° = +98.5 - 102.5 = -4.0$ kcal mol^{-1}; for (b), $\Delta H° = -96$ kcal mol^{-1}. Overall, $\Delta H° = -100$ kcal mol^{-1}.

51. **(a)** Propagation:

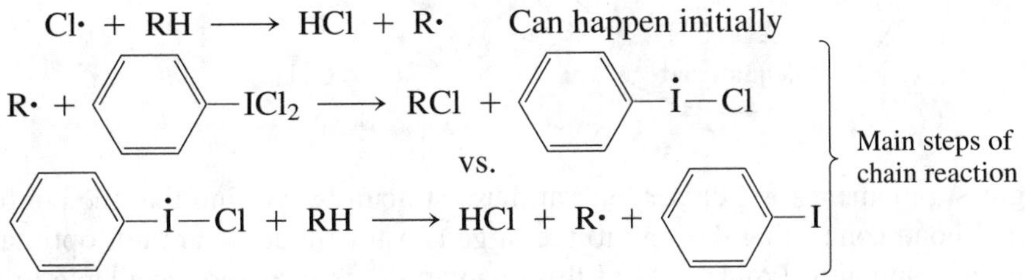

Termination (one possibility):

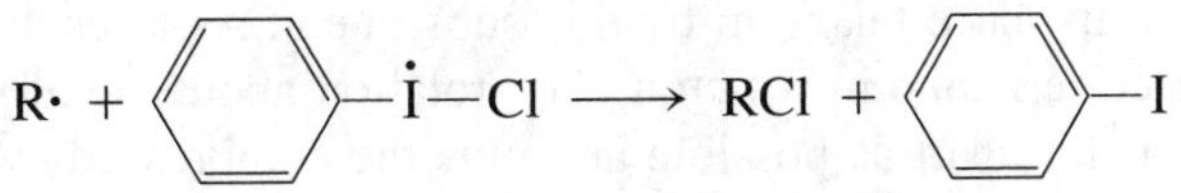

(b) There are four tertiary hydrogens, but the one that is β (up) is too hindered to be chlorinated, because of its 1,3-diaxial interactions with the two β methyl groups. So the sites of chlorination are the three tertiary α (down) hydrogens:

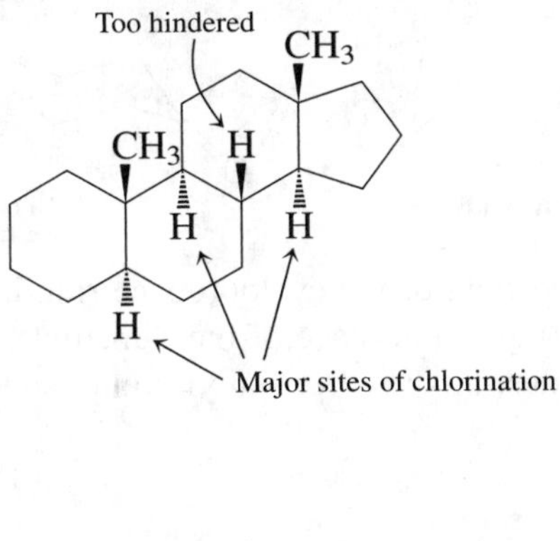

Major sites of chlorination

52. Addition of Cl_2 in each case generates a substituted ⬡-ICl_2 unit. Then, in the presence of light, these become ⬡-İ–Cl groups, which can chlorinate nearby C—H bonds according to the propagation steps shown in the answer to Problem 51(a). The selectivity comes from the fact that in the reaction for (a) the ⬡-İ–Cl group can reach the H at position 9 most easily, whereas in (b) it most easily reaches the H at position 14. Your models should look something like these:

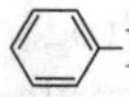

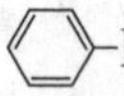

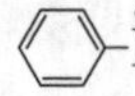

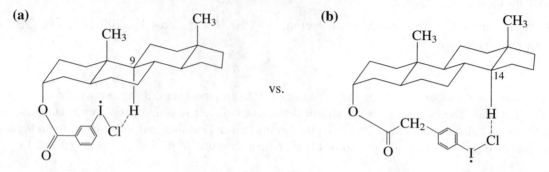

53. **(a)** Without making a model, this problem may initially make no sense to you at all. Begin by finding the better chair conformation for **A.** Of the two possibilities, the one with both alkyl groups equatorial is preferable:

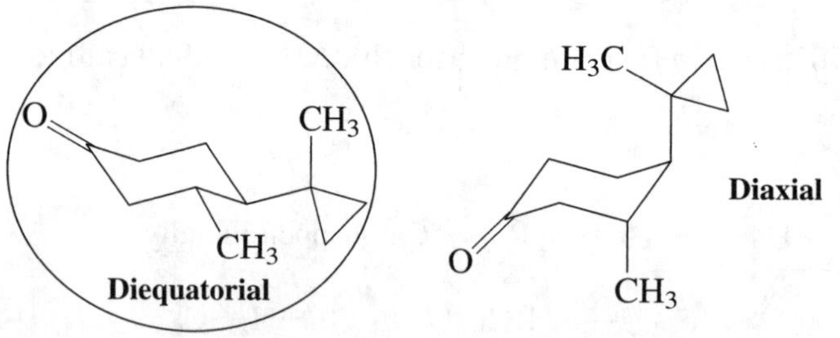

If we digress a moment for a closer look at these structures, we find that the conformations about the external bond connecting the ring to the large substituent at C4 are not optimal. As discussed briefly in the solution to Problem 33 of this chapter, the free rotation available to a simple alkyl group (such as CH_3) causes it to be more sterically demanding than is a comparable molecular fragment contained in a ring. Furthermore, the ring angles in cyclopropanes are only 60°, further shrinking the volume in space taken up by this substituent. As a result, a better conformation for the diequatorial structure is one in which a 120° rotation about the aforementioned bond takes the two methyl groups as far apart as possible and puts the compressed cyclopropane ring on the side of the bond closest to the methyl group at C3. Similar rotation in the diaxial moves the CH_3 group away from the axial hydrogens on the same face of the cyclohexane ring:

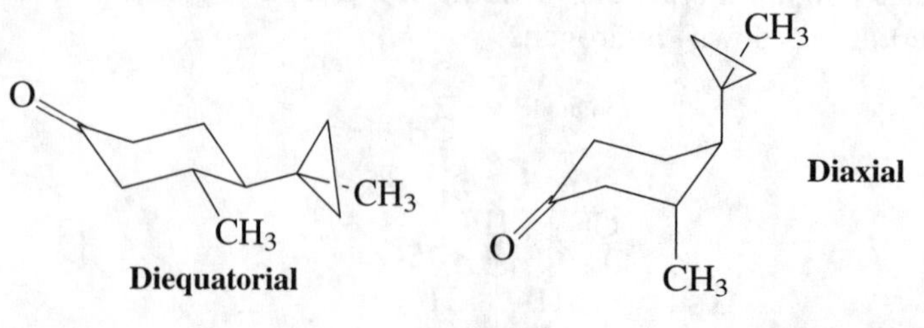

(b) Compound **B** results from opening of the cyclopropane ring in **A.** This process generates a 1,1-dimethylethyl (*tert*-butyl) group in place of the 1-methylcyclopropyl substituent. For purposes of evaluating steric interactions, this new *tert*-butyl group is effectively *much* larger, for two reasons. First of all, the rigid —CH_2—CH_2— fragment of the cyclopropane ring is replaced by two freely spinning CH_3 groups. Secondly, the original 60° angle between the bonds to these carbon atoms widens to the full tetrahedral value of 109.5° upon ring-opening. The increase in steric size of the substituent is considerable and not at all communicated adequately by line drawings. Now that we know what to look for, let's examine conformations of **B** comparable to those of **A** drawn above. The diaxial form is just about impossible, there being no available conformation lacking prohibitive steric compression between one of the CH_3's of the *tert*-butyl group and one or both axial H's on the same ring face. However, as indicated by the intersecting arcs, the diequatorial conformer suffers from a similar interaction between methyl groups, which is unavoidably present in all staggered conformations of the substituent. Basically, hydrogen atoms on these two CH_3's are being forced to try to occupy the same volume of space:

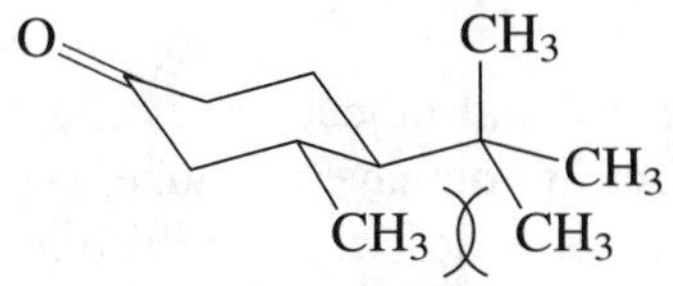

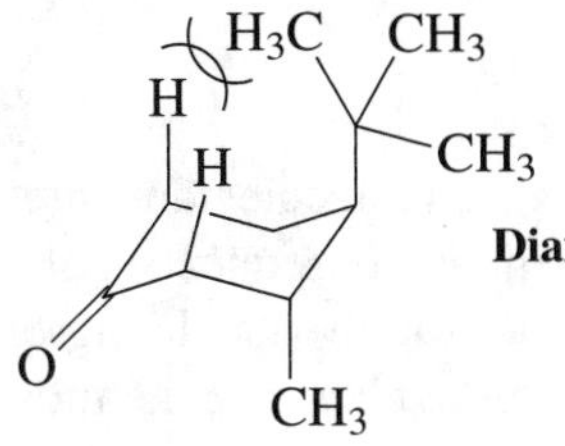

The molecule escapes this dilemma by adopting a shape based upon the boat shown below, but twisted to relieve eclipsing interactions. This unorthodox conformation places the methyl group at C3 in a pseudoaxial position, out of the immediate proximity of the *tert*-butyl group. (The situation is not terribly different from that described in Problem 30.)

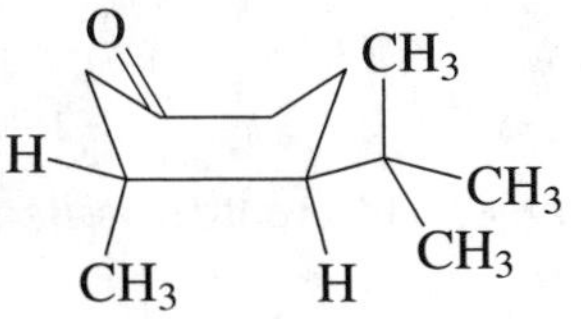

54. (a)

55. (d)

56. (d)

57. (a) Smallest ΔH°_{comb} = most stable (diequatorial)

5

Stereoisomers

By now you are well aware that molecules are three-dimensional objects. This chapter explores some of the more subtle, but extremely critical, consequences of this fact. If you have not done so yet, don't wait any longer to obtain a set of models to aid you in visualizing the structures described in this chapter. For many students, the isomeric relationships discussed here are the most difficult ones encountered in organic chemistry, and they are important later on in descriptions of several types of compounds and reactions. The implications for biological chemistry are especially significant.

Outline of the Chapter

5-1 Chiral Molecules
The key definitions: what makes an object different from its mirror image.

5-2 Optical Activity
Physical properties of chiral molecules.

5-3 The *R–S* Sequence Rules
How to name chiral molecules.

5-4 Fischer Projections
How to represent chiral molecules in two dimensions.

5-5 and 5-6 Molecules Incorporating Several Stereocenters: Diastereomers
Further elaboration, including the introduction of several new terms describing relationships between more complex molecules. Moves toward biological systems.

5-7 Stereochemistry in Chemical Reactions
Ground rules.

5-8 Resolution of Enantiomers
Practical techniques.

Keys to the Chapter

5-1. Chiral Molecules

In this chapter you will be introduced to many new terms. Follow their definitions along with the structures given as examples. The first new term is the chapter title: *stereoisomer*. In brief, stereoisomers are molecules that have the same atoms linked in the same order (i.e., identical connectivity), but that do not have identical three-dimensional shapes. The first example in Section 5-1, 2-bromobutane, is one of a vast number of molecules that are *chiral*. Chiral molecules can exist as either of two stereoisomeric shapes, which are related to each other as an object is to its mirror image. Before going any further, let's make one point clear: Every molecule has a mirror image, obviously. What makes a chiral molecule special is that it is **not identical to its mirror image.** Methane is identical to its mirror image; it is not chiral. The two possible shapes of a chiral molecule differ in the same way that a right-handed object differs from a left-handed object, as gloves, shoes, and hands do. Chirality, therefore, is "handedness" on a molecular level. The two mirror-image shapes of a chiral molecule are called *enantiomers.*

What makes a molecule chiral? The most common of several types of structural features that can make a molecule chiral is the presence of a carbon atom attached to four different atoms or groups (an *asymmetric carbon atom,* an example of what is called a *stereocenter*). At this point it is worthwhile to dust off your set of models and start manufacturing chiral molecules. Prove to yourself that the model of the **mirror image** of one cannot be superimposed on the original. This is the first step toward developing the ability to visualize this relationship clearly.

5-2. Optical Activity

The physical difference between enantiomers is so subtle that they end up for the most part displaying identical physical and chemical properties. They can be distinguished from each other only after interacting with something else that is, itself, already "handed." By analogy, a right and a left glove will have the same weight, color, and texture. However, interaction with, for instance, a right hand will immediately distinguish them. The fact that we can tell them apart just by looking at them reflects the fact that our brain's interpretation of the signals from our binocular visual system is "handed" in its ability to perceive depth as well as left–right relationships. For chiral molecules, the counterpart to this is their interaction with plane-polarized light: The plane of polarization of plane-polarized light is rotated when it passes through a solution of one enantiomer of a chiral molecule. This phenomenon, labeled *optical rotation,* is the most common way of detecting chiral molecules and is described in considerable detail in this section. Molecules possessing the ability to rotate the plane of polarized light are said to be *optically active,* or to display *optical activity;* and another term for enantiomers is *optical isomers.* One further term of importance is the one given to a mixture of equal amounts of the two enantiomers of a chiral molecule: *racemic mixture.* The two enantiomers of a chiral molecule each rotate light by equal amounts, but in opposite directions, so the racemic mixture displays no optical activity because the two components exactly cancel out each other's rotations.

Again, because of the large number of new terms and ideas associated with this material, it merits careful study, with a good set of models close at hand.

5-3. The *R–S* Sequence Rules

Like all the rules of nomenclature, the *R–S* system for molecules containing stereocenters has one purpose: the concise, unambiguous description of a single chemical structure. In most cases the system is not particularly difficult to apply, because the assignment of priorities to the groups on an asymmetric carbon is usually straightforward and use of models to view the stereocenter properly takes care of the rest. Models should be made of the examples in the chapter so that you can confirm their *R* or *S* designations.

Priority rule 2 is an occasional source of trouble until the concept of "first point of difference" is well understood. If this gives you trouble, take it stepwise:

1. Write the substituent groups to be compared side-by-side, with the bond of attachment to the asymmetric atom on the left.
2. Starting at the left, look at the first atom in each substituent chain and identify the atoms attached to it in descending order of priority. If the highest priority atoms in each are the same, work your way down in priority, looking for the first nonidentical atoms (first point of difference). If no differences are found at this stage, move out from the first to the second atom in the chain and repeat the process. If there is branching here, the direction of this move should be chosen to examine the highest priority atom for

a point of difference. If no difference is found, then examine the second highest priority group, and so on.

It is actually easier to **do** than it is to describe, so let's analyze three examples.

Example: Determine *R–S* designation for

$$CCl_3$$
$$H\text{''''}C$$
$$\overset{|}{CH_3}\ \ CH_2Br$$

Procedure: H is obviously lowest priority; the other three need to be compared. Write them out side-by-side:

$$\overset{\textcircled{Cl}}{\underset{Cl}{-C-Cl}}\qquad \overset{\textcircled{Br}}{\underset{H}{-C-H}}\qquad \overset{\textcircled{H}}{\underset{H}{-C-H}}$$
$$\text{–CCl}_3\qquad\qquad \text{–CH}_2\text{Br}\qquad\quad \text{–CH}_3$$

Identify the highest priority atom according to atomic number on each carbon. In —CCl$_3$, it is Cl; in —CH$_2$Br, it is Br; and in —CH$_3$, it is H. All three are different; therefore priorities can be assigned immediately. Br > Cl > H, therefore —CH$_2$Br > —CCl$_3$ > —CH$_3$. The fact that there are **three** Cl's on —CCl$_3$ and only **one** Br on —CH$_2$Br is **irrelevant.** The **first** point of difference is the Br on —CH$_2$Br vs. the **first** Cl on —CCl$_3$, and Br is "bigger" than Cl. Once the first point of difference is identified, **nothing else matters.**

With priorities now assigned, we can redraw the molecule with the highest priority group designated "a," the second "b," the third "c," and the lowest "d":

$$\overset{CCl_3}{\underset{CH_3}{H\text{''''}C}}\ CH_2Br \qquad \text{becomes}\qquad \overset{b}{\underset{c}{d\text{''''}C}}_a \quad\equiv\quad \overset{b}{\underset{c}{C}}_a \quad \text{(d in back)}$$

Counterclockwise = *S*

Example: Determine *R–S* designation for

$$\overset{H}{\underset{(CH_3)_2C}{H_2C\text{''''}C}}\ CH_2CH_3$$

Procedure: The stereocenter is a member of a ring, but the procedure is not really different. The H is lowest priority, and the other three groups have to be compared. The problem is interpreting what it means to have the stereocenter in a ring. One group is ethyl; that's obvious. The other two "groups" are really just the sequence of *ring atoms* attached to the stereocenter. For one group, start at the ring —CH$_2$ and move around the ring; for the other, start at the ring —C(CH$_3$)$_2$, and move around the ring in the other direction. So, we compare the "groups"

$$\text{—CH}_2\text{—CH}_3, \qquad \text{—CH}_2\text{—C(CH}_3)_2\text{—etc.,} \qquad \text{and} \qquad \text{—C(CH}_3)_2\text{—CH}_2\text{—etc.}$$

where the left-hand bond goes to the stereocenter and "etc." means "continuing around the ring." Priorities are assigned in accordance with rule 2 again, because in all three groups the first atom in the "chain" attached to the asymmetric carbon is identical (carbon). We move out to the atoms attached to each of these carbons for comparison (circled):

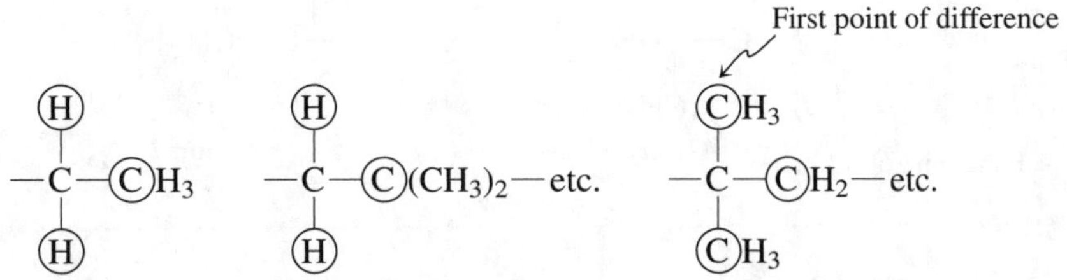

Ethyl **Ring CH_2** **Ring $C(CH_3)_2$**

In each case the largest of these atoms is carbon. No difference. However, in the case of the group at the right, the second largest is also carbon, whereas the second largest for the other two groups is hydrogen. The highest priority of these three groups is therefore the ring-contained —$C(CH_3)_2$—CH_2—etc. Second and third priorities are determined by moving one atom **further** out, as shown (circled):

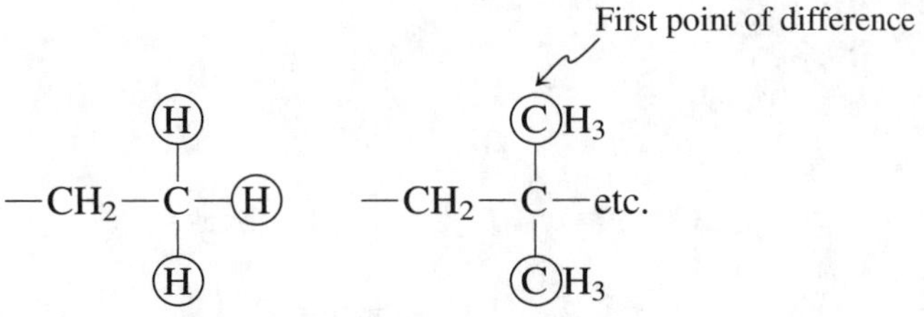

Ethyl **Ring CH_2**

Comparison here is straightforward. The CH_2 in the ethyl is connected only to a simple $\mathbf{CH_3}$ group, whereas the CH_2 of the ring is attached to a —$C(CH_3)_2$—etc. group. So the latter one is higher in priority (**C** larger than **H**). Therefore we have:

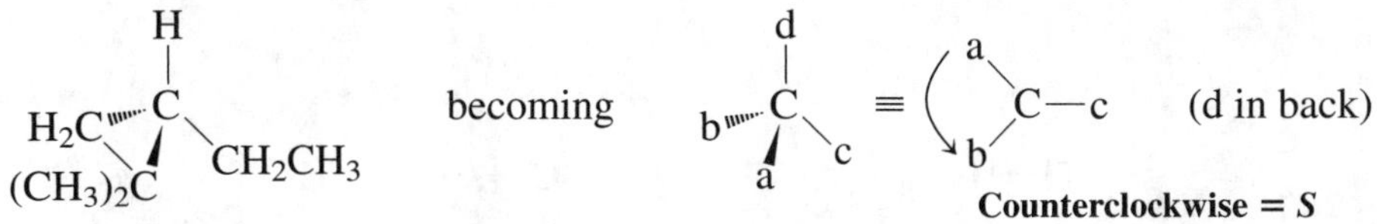

Counterclockwise = *S*

Example: Determine *R–S* designation for the marked carbon in

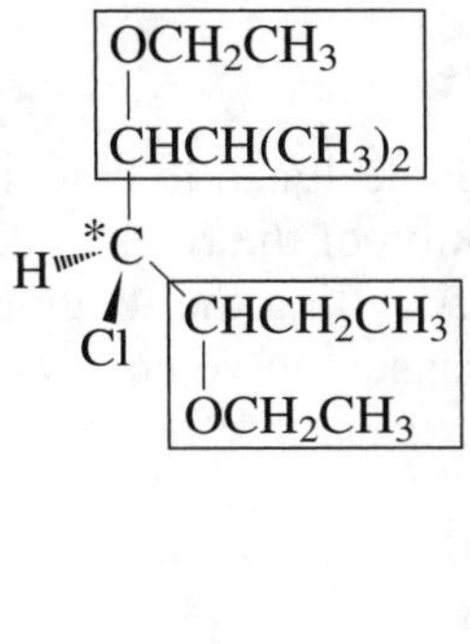

Procedure: Highest priority is Cl, and lowest is H. A priority choice between the two outlined groups needs to be made, however. We write the groups out side-by-side:

$$
\begin{array}{ccc}
\overset{\displaystyle O-CH_2-CH_3}{\underset{\quad a \qquad b}{\big|}} & & \overset{\displaystyle O-CH_2-CH_3}{\underset{\quad a \qquad b}{\big|}} \\
-C-CH-CH_3 & \text{vs.} & -C-CH_2-CH_3 \\
\underset{H}{\big|}\ \underset{CH_3}{\big|} & & \underset{H}{\big|} \\
\text{First point of difference} \quad \textbf{Group 1} & & \textbf{Group 2}
\end{array}
$$

Start at the carbon atom labeled "a" in each group. In both groups, the highest priority atom attached to carbon "a" is oxygen, the second highest is carbon, and the last is hydrogen. No difference is found, so move out to the **highest priority atom** on carbon "a" for the next comparison: Follow the arrow and move to the **oxygen** (**not** carbon atom "b"—**oxygen is higher priority,** so it is evaluated **first**). Both groups 1 and 2 have identical CH_2's attached directly to O, so they are still tied. Turning to the second largest group attached to "a," we look to carbon "b," and see that here the tie can be broken: In group 1, this atom is attached to two carbons and one hydrogen, whereas in group 2, it is attached to just one carbon and two hydrogens. Group 1 therefore is higher in priority than group 2 at atom "b," the first point of difference.

With priorities done, the groups can all be labeled and *R* or *S* assigned:

$$
\underset{a}{\overset{b}{d^{\text{\tiny\|\|\|}}\!\!\text{—}C\text{—}c}} \quad \equiv \quad \underset{a \quad c}{\overset{b}{\big(\,C}} \qquad \text{(d in back)}
$$

Clockwise = R

If, on the other hand, the oxygen in group 1 were attached to an H instead of a C, that would become the first point of difference, and group 2 would be higher in priority. The differences at the "b" carbons would become irrelevant.

$$
\begin{array}{ccc}
\overset{\text{Lower}}{\underset{O-H}{\swarrow}} & & \overset{\text{Higher}}{\underset{O-CH_2-CH_3}{\swarrow}} \\
-C-CH-CH_3 & \text{vs.} & -C-CH_2-CH_3 \\
\underset{H}{\big|}\ \underset{CH_3}{\big|} & & \underset{H}{\big|} \\
\textbf{Group 1} & & \textbf{Group 2}
\end{array}
$$

The above three examples have each been designed to contain a "trick"—an unusual feature that is not often encountered but illustrates the application of the rules in detail. Most chapter and exam problems you see will not be "tricky." However, by understanding the toughies, proper application of the procedure becomes quicker and simpler in **all** cases, because you've now seen what to do when things get complicated.

Before leaving this section, note also that there is a bit of a trick to rule 3 as well. For priority determination purposes, double and triple bonds are changed so that **both atoms involved** are doubled or tripled. For example, a carbon doubly bonded to another carbon is changed to a carbon singly bonded to **two** carbons.

One is the carbon actually present, and the other is invented. So

The procedure is similar for a carbon–oxygen double bond: The carbon bonds are changed and converted to single bonds to two oxygens (one real, one invented), and likewise the oxygen bonds are changed to single bonds to two carbons (one real, one invented). So

5-4. Fischer Projections

The purpose of Fischer projections is to simplify the on-paper drawing of asymmetric carbons by using a simple convention to represent the three-dimensional structural details. The rules are straightforward, but again, following the text with a set of models handy will help you master this material more readily.

5-5 and 5-6. Molecules Incorporating Several Stereocenters: Diastereomers

When a molecule has more than one stereocenter, as 2-bromo-3-chlorobutane does, it will have more than two stereoisomers. In particular, n stereocenters give rise to as many as 2^n stereoisomers. If we consider the case where $n = 3$, how are all the $2^n = 8$ stereoisomers related? Because an object has only one mirror image, if we pick any one of these (stereoisomer A), it may have no more than one enantiomer (stereoisomer B). What about the other six isomers? They are also stereoisomers of A, but they can't be its mirror images. The relationship of any of these other six molecules with A is described by a new term: *diastereomer*. Diastereomers are stereoisomers that are **not mirror images of each other.** Because diastereomers are **not** mirror images of each other, they have **different physical properties and can therefore be separated by standard laboratory techniques.** This is very important. This feature distinguishes diastereomers from enantiomers, which **cannot** readily be separated from one another.

Diastereomer is a very important term, as important as *enantiomer,* and one that you should make a point of understanding well before you leave this section. Notice that both "enantiomer" and "diastereomer" really describe **relationships** between structures. As described above, A is the enantiomer of B; A is also a diastereomer of each of the other six isomers we talked about. It may seem odd at first that a single molecule can be called both an enantiomer and a diastereomer at the same time. However, if you remember that these terms really describe relationships between **pairs** of structures, it makes more sense. The following illustration is not a perfect analogy, but it might help you get the idea.

Imagine a dancer who at different times has various combinations of hands and feet raised. The relationship between right hand–right foot raised and left hand–left foot raised is a mirror-image, enantiomeric one. Neither is the mirror image of, say, right hand–left foot raised, so the latter is diastereomerically related to the first two. The illustration below shows how the four possible combinations of "one hand up, one hand down, one foot up, one foot down" are related in a way similar to the four stereoisomers of a molecule with two chiral carbons.

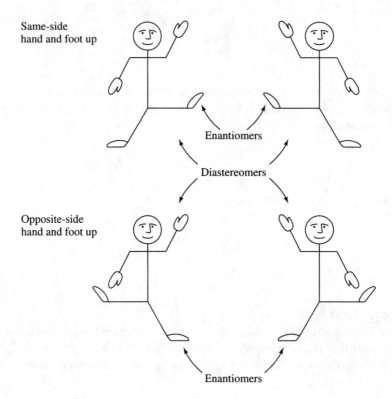

As the figures in this section show, molecules work similarly. Actually the real trouble in determining, for instance, whether two structures are enantiomers or diastereomers of each other stems from the difficulty in drawing and orienting the pictures of the molecules so that the stereocenters can be compared in the first place. Let's look at an example.

Anyone can look at the two compounds,

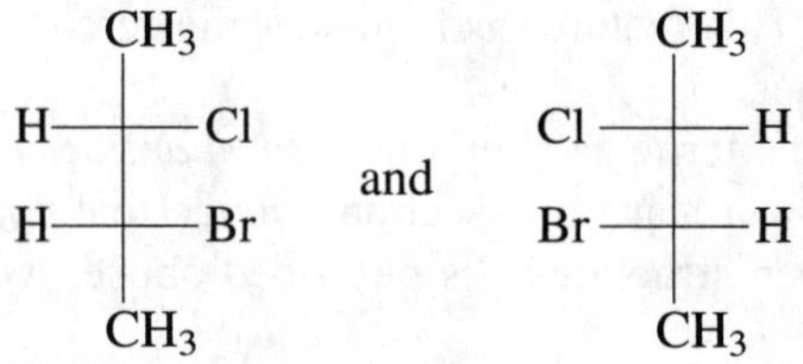

immediately visualize the mirror plane between them, and recognize them as enantiomers. Similarly,

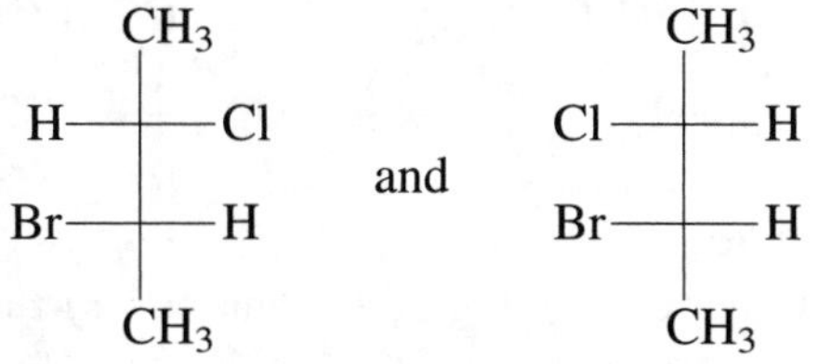

$$
\begin{array}{ccc}
& CH_3 & \\
H\text{---}&|&\text{---}Cl \\
Br\text{---}&|&\text{---}H \\
& CH_3 &
\end{array}
\qquad and \qquad
\begin{array}{ccc}
& CH_3 & \\
Cl\text{---}&|&\text{---}H \\
Br\text{---}&|&\text{---}H \\
& CH_3 &
\end{array}
$$

clearly lack a mirror-image relationship and must be diastereomers.

The hard part is determining, for example, the relationship between

$$
\begin{array}{c}
CH_3 \quad Cl \\
Br^{\text{\tiny w}}C\text{---}C^{\blacktriangleleft}H \\
H \quad CH_3
\end{array}
\qquad and \qquad
\begin{array}{ccc}
& Cl & \\
H\text{---}&|&\text{---}CH_3 \\
CH_3\text{---}&|&\text{---}H \\
& Br &
\end{array}
$$

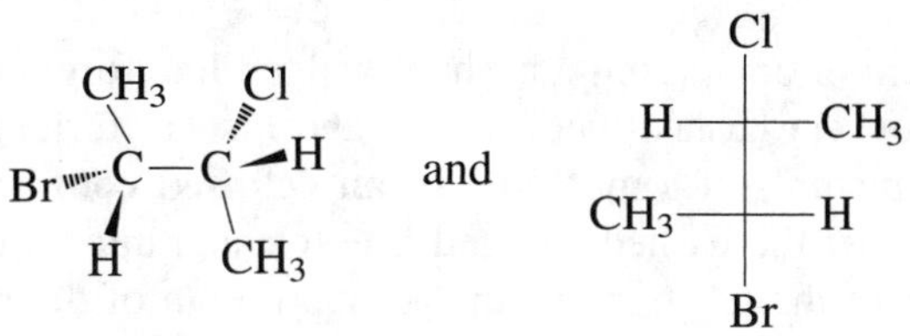

without getting all messed up (or taking forever to do it, and **then** getting all messed up). In short, you need to be able to move quickly and accurately among Newman, hashed-wedged line, and Fischer structures. This capability takes practice at visualizing three-dimensional structures from flat drawings and requires application of a couple of specific techniques. The text covers comparison and interconversion of Fischer projections very thoroughly. The one important precaution here is **never** operate on **more than one** stereocenter at a time when interconverting Fischer projections.

Example: What is the relationship between the following structures: identical, enantiomeric, or diastereomeric?

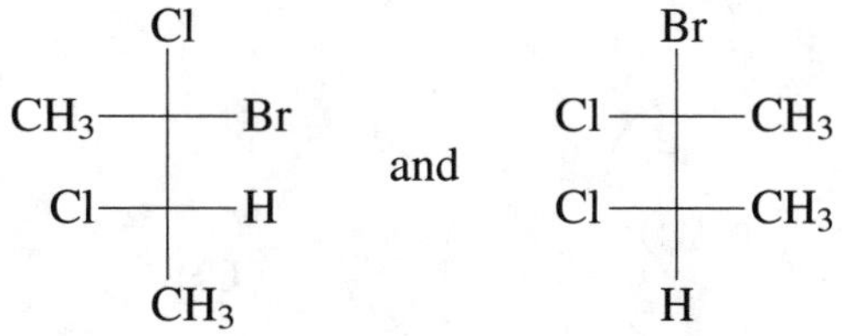

$$
\begin{array}{ccc}
& Cl & \\
CH_3\text{---}&|&\text{---}Br \\
Cl\text{---}&|&\text{---}H \\
& CH_3 &
\end{array}
\qquad and \qquad
\begin{array}{ccc}
& Br & \\
Cl\text{---}&|&\text{---}CH_3 \\
Cl\text{---}&|&\text{---}CH_3 \\
& H &
\end{array}
$$

Structure 1 \qquad **Structure 2**

Procedure:

1. Operate on the top stereocenter.

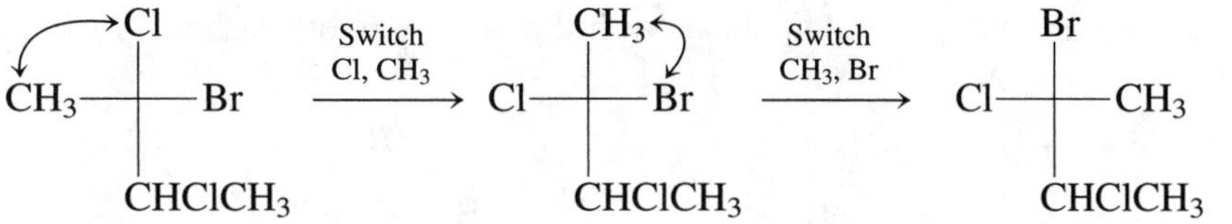

$$
\underset{\textbf{Structure 1}}{\begin{array}{ccc}
& Cl & \\
CH_3\text{---}&|&\text{---}Br \\
& CHClCH_3 &
\end{array}}
\xrightarrow[\text{Cl, CH}_3]{\text{Switch}}
\begin{array}{ccc}
& CH_3 & \\
Cl\text{---}&|&\text{---}Br \\
& CHClCH_3 &
\end{array}
\xrightarrow[\text{CH}_3\text{, Br}]{\text{Switch}}
\underset{\textbf{Same as Structure 2}}{\begin{array}{ccc}
& Br & \\
Cl\text{---}&|&\text{---}CH_3 \\
& CHClCH_3 &
\end{array}}
$$

Two switches made the top stereocenter of 1 identical to that of 2; therefore these carbons are **identical** in configuration.

2. Then do the bottom stereocenter.

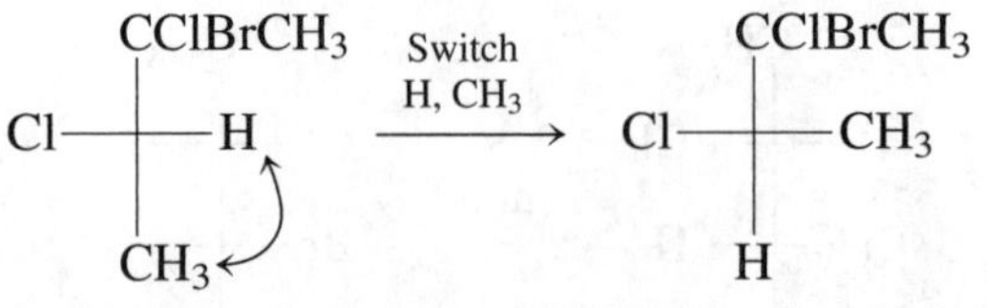

$$
\begin{array}{ccc}
\text{CCIBrCH}_3 & \xrightarrow[\text{H, CH}_3]{\text{Switch}} & \text{CCIBrCH}_3 \\
\text{Cl} \rule[0.5ex]{1.5em}{0.4pt}\!\!\!\!\rule{0.4pt}{2.5ex}\!\!\!\!\rule[0.5ex]{1.5em}{0.4pt} \text{H} & & \text{Cl} \rule[0.5ex]{1.5em}{0.4pt}\!\!\!\!\rule{0.4pt}{2.5ex}\!\!\!\!\rule[0.5ex]{1.5em}{0.4pt} \text{CH}_3 \\
\text{CH}_3 & & \text{H} \\
\textbf{Structure 1} & & \textbf{Same as Structure 2}
\end{array}
$$

One switch made the bottom stereocenter of 1 identical to that of 2; therefore these carbons are **opposite** in configuration.

Answer: The two structures are **diastereomers** (not completely identical and not mirror images).

Let's now examine our problem of comparing a hashed-wedged line formula with a Fischer projection. The actual problem is how to **interconvert** hashed-wedged line and Fischer formulas. The key to this is recognizing that Fischer projections are pictures of a molecule **in an eclipsed conformation.** Very important. So the first step in our comparison is to get the hashed-wedged line formula into an eclipsed conformation. Any 60° rotation will do, such as rotation of the left-hand methyl group up out of the page, toward you:

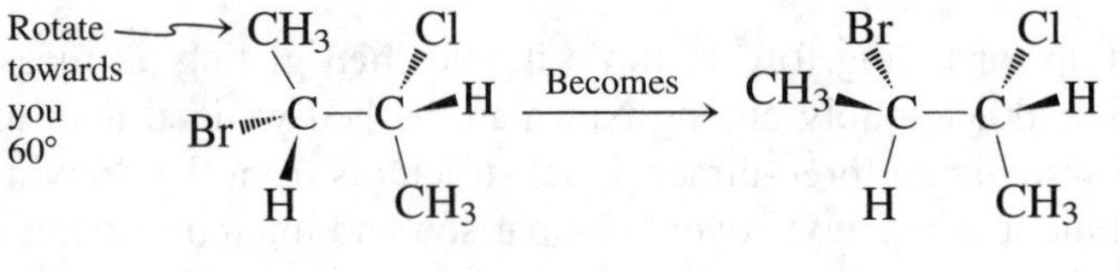

Look at these two conformations of the same, identical molecule **carefully**—with the help of a model, if necessary—to convince yourself that the pictures on the page are what I've said they are.

We can now convert the eclipsed formula, above, into a Fischer projection. Imagine looking at it from a direction such that the carbon–carbon bond is vertical, and the groups horizontal to it point toward us; for example:

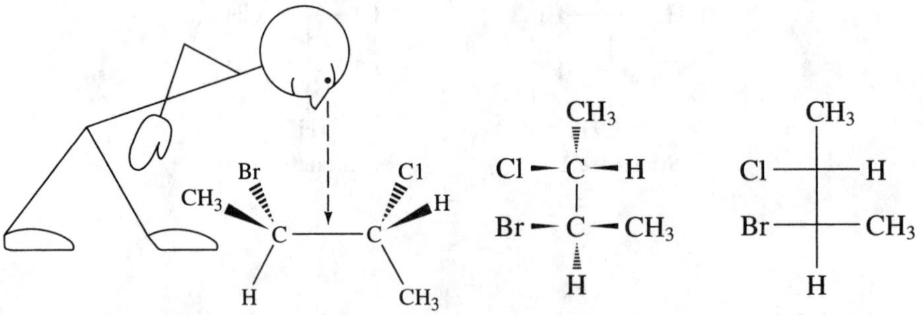

If you look at it like this   **This is what you see**   **Which is this**

Now you can compare the Fischer projection above with the one we wrote earlier:

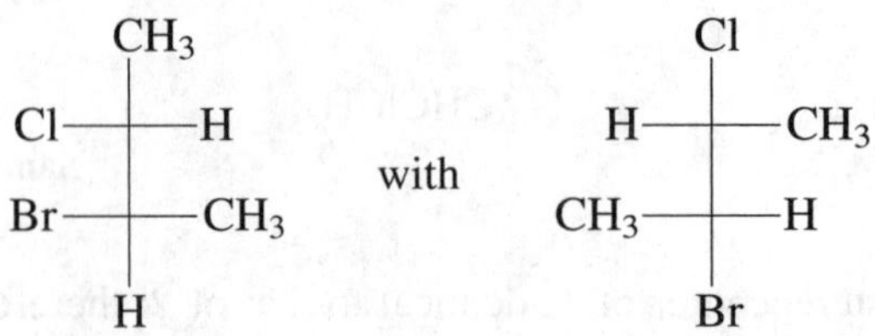

Structure 3   with   **Structure 4**

1. Top stereocenter

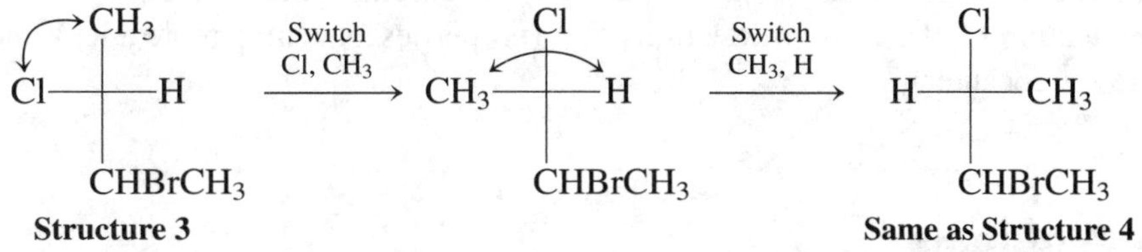

Two switches; therefore identical.

2. Bottom stereocenter

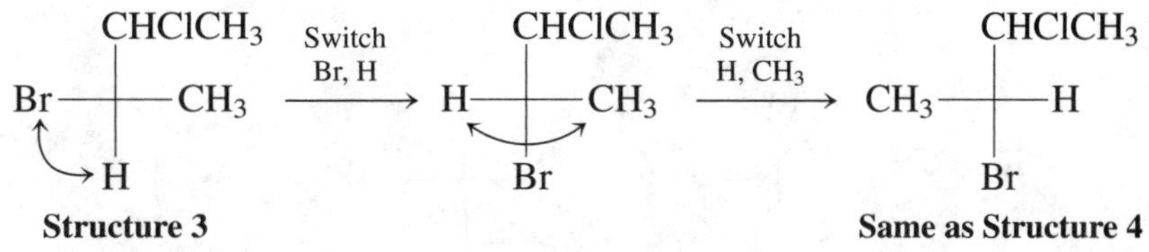

Two switches; therefore also identical.

So the two structures shown earlier were identical to each other.

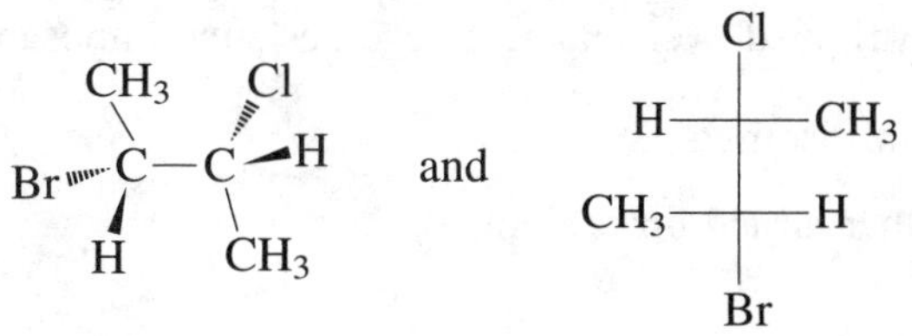

are exactly the same molecule.

It is useful to recognize at this point that comparing pictorial formulas is just one of several ways to determine the relationship between possible stereoisomeric structures. If you have the time, making models is always useful, especially for visualization purposes. An even better way is to apply the rules of nomenclature to each structure and to determine the *R* or *S* configuration of each stereocenter. Once you have become comfortable with this technique as it applies to the different types of structural pictures, you may find that this is the quickest method of all: Once *R–S* assignments have been made to structures under consideration, their stereochemical relationship is obvious. For two stereocenters, *R,R* and *S,S* are enantiomers of each other, *R,S* and *S,R* are enantiomers of each other, and any other combination is a pair of diastereomers.

Notice that I never said this was easy. It takes practice to do and to extend to other common ways of drawing molecules that imply three-dimensional structure. With practice, however, comes experience and confidence, just what you will need come exam time.

5-7. **Stereochemistry in Chemical Reactions**

The apparent complexity of this section is due to the number of situations that are explained in detail. The material actually develops very naturally, with no special pitfalls.

5-8. Resolution of Enantiomers

The most common laboratory method to separate enantiomers is presented. What is required is a pure enantiomer of a molecule that reacts with the molecules of your racemic mixture. The procedure outlined is a straightforward application of the ideas in Sections 5-2 (properties of chiral molecules) and 5-5 (molecules with more than one stereocenter).

Solutions to Problems

30. Chiral: (b)*, (c; fan blades are always twisted!), (d), (e), (h). Achiral: (a), (f), (g), (i), (j), (k), (l). All the achiral objects contain a plane of symmetry:

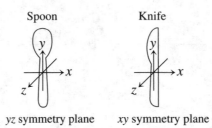

Spoon   Knife

yz symmetry plane   xy symmetry plane

*If you ignore the mounting hinges, then doors are achiral. The plane of a door is a symmetry plane.

31. **(a)** Enantiomers   **(b)** Enantiomers   **(c)** Diastereomers

     (d) Identical (if one pair is flipped over intact, it becomes superimposable with the other pair)

32. **(a)** Constitutional (structural) isomers

     (b) Identical (one is the other turned upside-down)

     (c) Constitutional isomers

     (d) Conformers

     (e) Constitutional (structural) isomers

     (f) Stereoisomers (enantiomers)

     (g) Conformers

     (h) Stereoisomers (enantiomers)

If two conformational isomers are cooled to a temperature low enough to prevent their interconversion by bond rotation or other motion, then they can be described as stereoisomers: structures with the same connectivity but with different atomic arrangements in three dimensions. For example, the conformers in (d) are nonsuperimposable mirror images (enantiomers) if bond rotation is not allowed to occur (make a model!). Those in (g) are not enantiomers, but they are still stereoisomers. The temperatures required to "freeze out" conformational interconversion are very low: on the order of $-200°C$ for substituted ethanes and $-100°C$ for cyclohexanes. Conformers of these types are occasionally referred to as *interconvertible stereoisomers*.

33. The stereocenter has been labeled with an asterisk (*) in each chiral molecule.

(a) 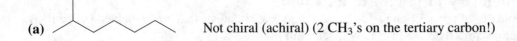   Not chiral (achiral) (2 CH_3's on the tertiary carbon!)

(b) 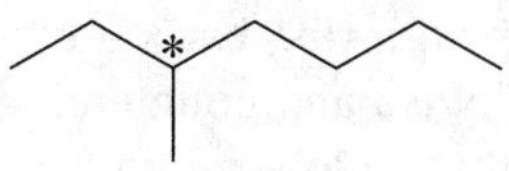 Chiral

(c) 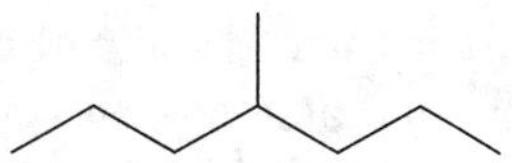 Not chiral

(d) 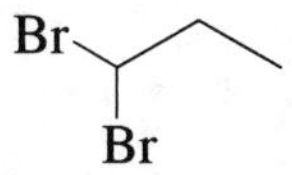 Br Not chiral

(e) 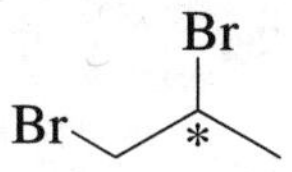 Br Chiral

(f) 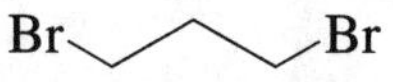 BrBr Not chiral

(g), (h), (i) Not chiral (all are planar molecules)

(j) 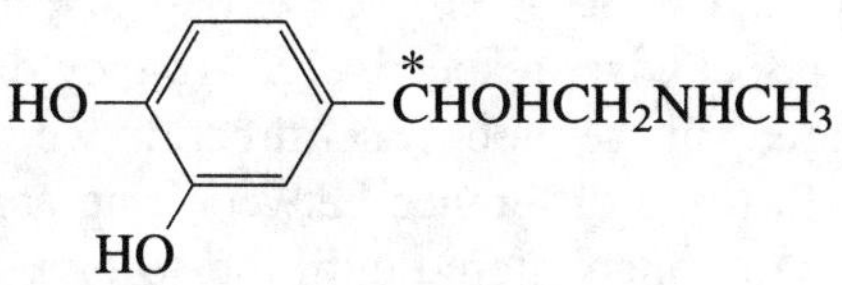 HO——$\overset{*}{C}$HOHCH$_2$NHCH$_3$ Chiral

(k) Not chiral   **(l)** Not chiral

(m) 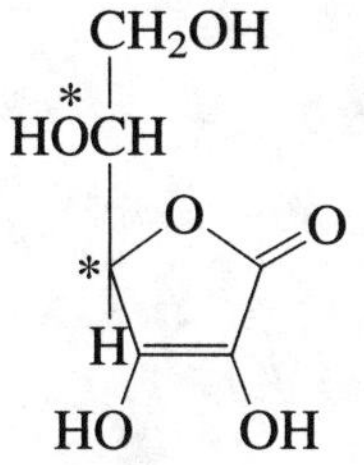 Chiral (has 2 stereocenters)

(n) and **(o)** Not chiral. Both molecules (as drawn) have vertical planes of symmetry bisecting the rings. Make models!

34. Look for a stereocenter: a carbon atom with four different groups. Compounds (a), (c), and (e) are chiral. In (a) and (e) the carbons with the —OH groups are the stereocenters; in (c) the stereocenter is C2, the methyl-substituted carbon atom.

35. As stated in the solution above, molecules **(a)**, **(c)**, and **(e)** are chiral. Both enantiomers of each are shown.

(a)

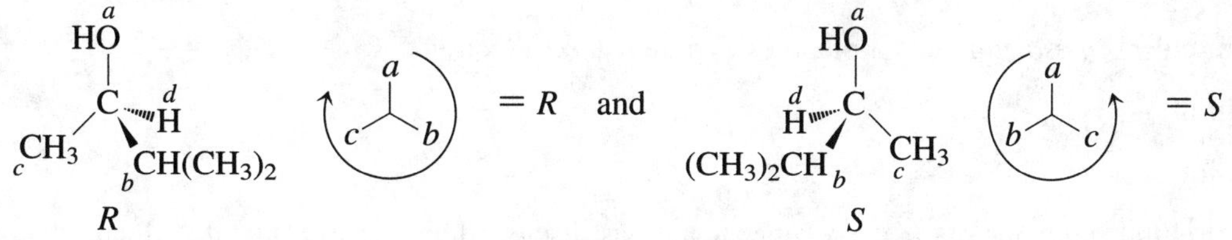

(c)

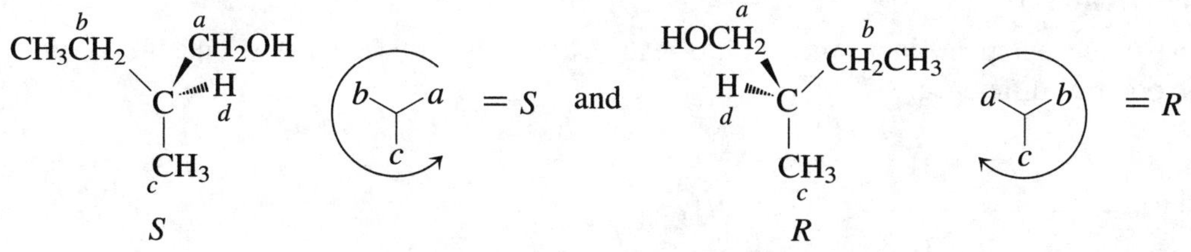

(e)

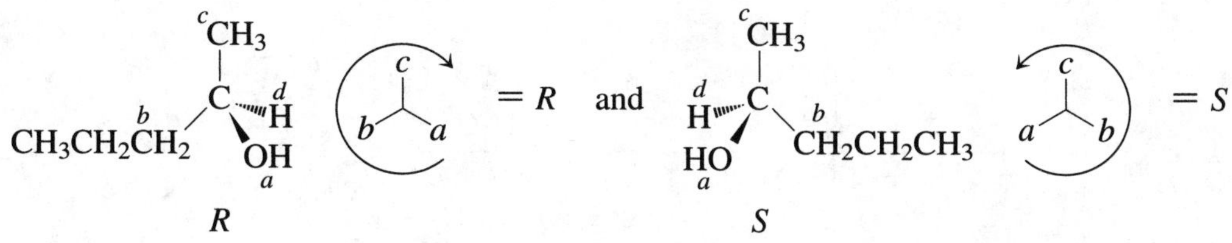

36. Do you know how to recognize chiral carbons in rings? The four groups on a ring carbon include the two external substituents, and the atoms of the cycle itself proceeding in clockwise and counterclockwise directions from the carbon under scrutiny. To take one example, *cis*-1,2-dimethylcyclohexane in (a), consider the topmost ring carbon (marked with a dot in the structure below):

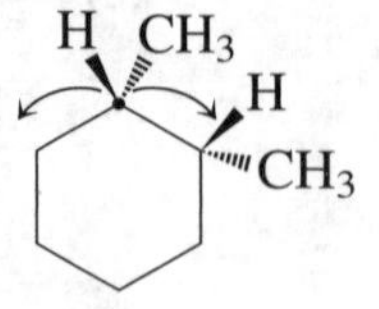

What are its four appendages? It bears an H and a CH_3; that's two. The other two are depicted by the curved arrows: the ring atoms proceeding clockwise and counterclockwise, respectively. Are they different? Yes; proceeding counterclockwise, we first come to a CH_2, whereas in the clockwise direction we encounter a CH—CH_3 first. The same situation is found in (b). In (c), a difference between ring atoms in the clockwise and counterclockwise directions around the ring is not encountered until the second carbon in each direction. All three structures contain chiral carbons (two each). How about (d) below?

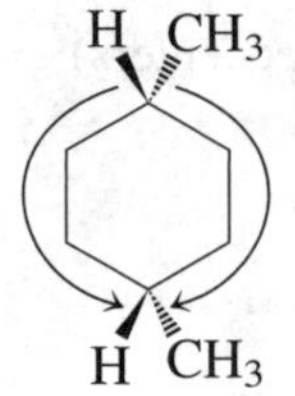

In this case the ring atoms encountered in the two directions around the ring are no longer different: Following the two arrows all the way around reveals that each group we encounter in one direction is identical to the one at the same point in the other. This molecule lacks stereocenters! (Don't be fooled by the way the bottom carbon is drawn: The H is not really on the left, nor the CH_3 on the right. The H is really directly above the CH_3; we only draw them side-by-side so that they can both be seen in the picture.) Now we can answer the questions.

(a) Not chiral (*meso,* contains a plane of symmetry; see Section 5-6)

(b) Chiral

(c) Chiral

(d) Not chiral (contains plane of symmetry, and, as discussed above, the substituted carbons **are not stereocenters**)

37. Labeling stereocenters in rings takes a bit of extra care. Evaluate each stereocenter *separately.* Look for your first point of difference.

(a)

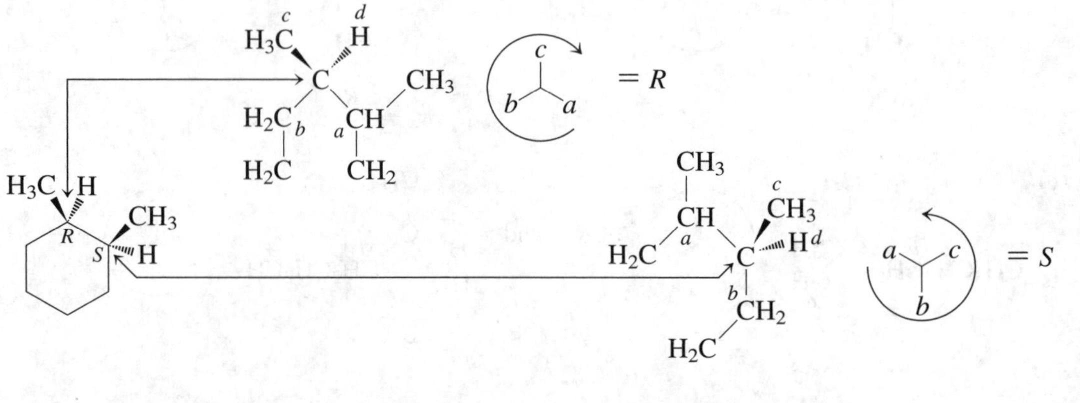

In a *meso* compound the stereocenters must be identically substituted and opposite in configuration.

(b)

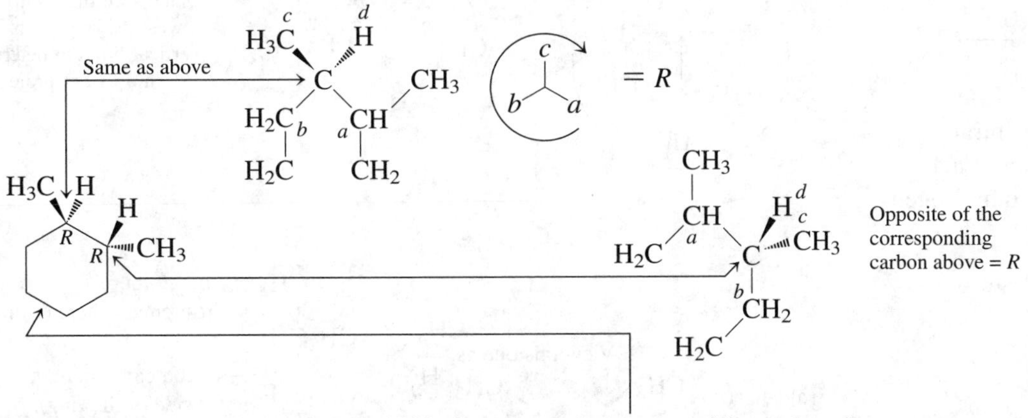

Hint: If you rotate the molecule 180° about an axis here, you'll see that the two stereocenters are identical and, therefore, if the first one is *R* so is the second.

(c)

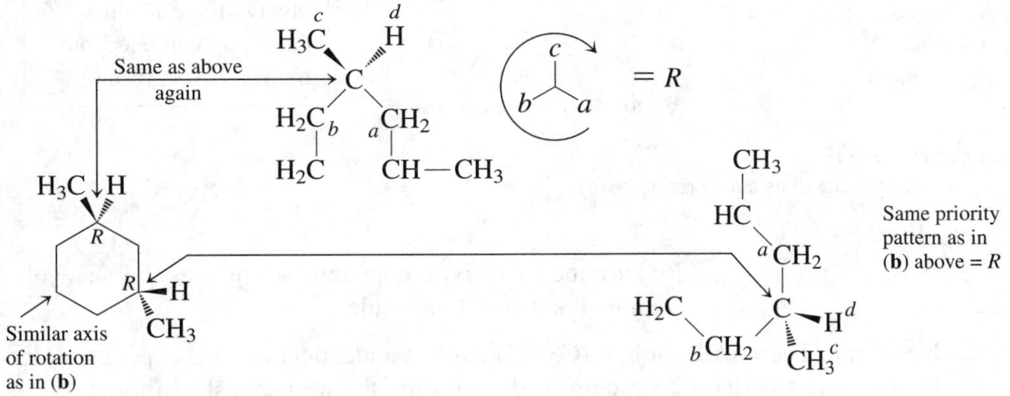

(d) No stereocenters, so no configurations to determine.

38. As in the text, group priorities are given as *a* (highest), *b*, *c*, and *d* (lowest).

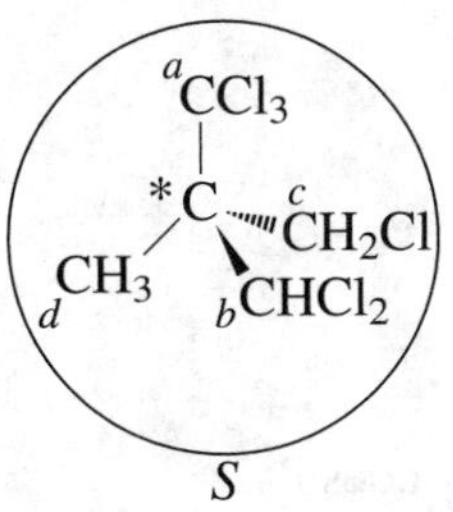

Viewed from the side opposite the '*d*' group you see

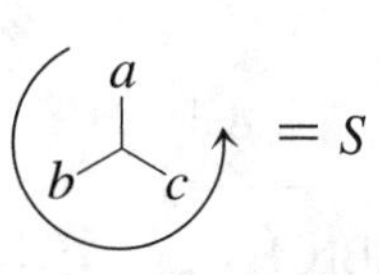

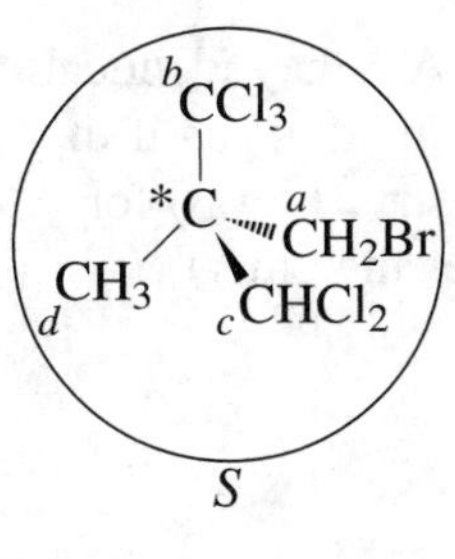

Here the CH_2Br has top priority because Br > Cl (doesn't matter that there are three Cls in the CCl_3 group). In this case, it doesn't change the configuration because the route from *a* to *b* to *c* is still counterclockwise. But let this be your warning to take care with priorities: *first point of difference!* That is, you stop with Br > the first Cl on the CCl_3 group. The other two Cl atoms don't count in comparing CH_2Br with CCl_3. They do count, however, when you make the subsequent comparison of CCl_3 with $CHCl_2$. Then, the first point of difference is the third Cl in CCl_3 > the H in $CHCl_2$.

H
Br
H
H
Cl
H

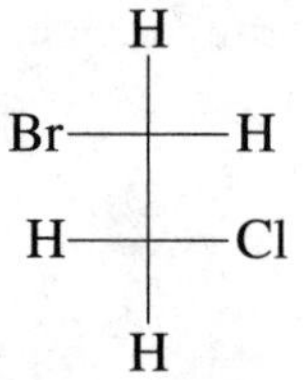

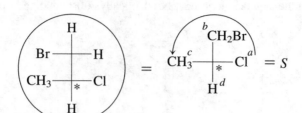

No chiral carbons! Each one has two hydrogens.

H
Br
H
CH3
Cl
H
=
b
CH2Br
c
CH3
a
Cl
H
d
= S

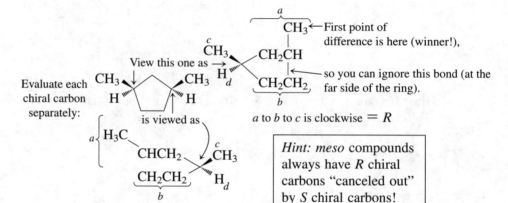

Make sure the group with lowest priority (*d*) is on a vertical bond to determine *R/S* in a Fisher projection!

CH3
CH3
H
H

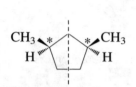

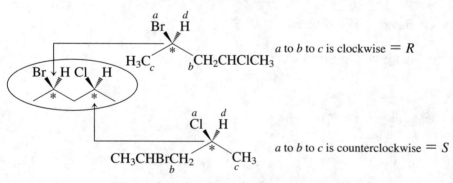

Two chiral carbons, but the perpendicular plane of symmetry (dashed line) makes this *meso*.

Evaluate each chiral carbon separately:

CH3
H
CH3
H
View this one as →
c
CH3
H
d
a
CH3
First point of difference is here (winner!),
CH2CH
CH2CH2
b
so you can ignore this bond (at the far side of the ring).
a to b to c is clockwise = R
is viewed as

a
H3C
CHCH2
CH2CH2
b
c
CH3
H
d
a to b to c is counterclockwise = S

Hint: meso compounds always have *R* chiral carbons "canceled out" by *S* chiral carbons!

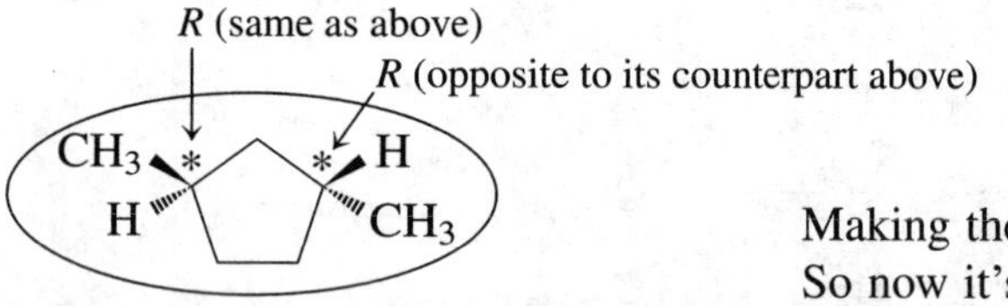

R (same as above)
R (opposite to its counterpart above)
CH3
H
H
CH3

Making the two CH_3 groups trans eliminates the plane of symmetry. So now it's a chiral molecule.

Br
H

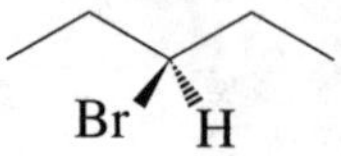

Not chiral. The middle carbon (C3) contains two identical ethyl groups. Don't let yourself be fooled just because a structure is drawn using the wedge-hashed format.

a
Br
d
H
H3C
c
b
CH2CHClCH3
a to b to c is clockwise = R
Br
H
Cl
H
a
Cl
d
H
CH3CHBrCH2
b
CH3
c
a to b to c is counterclockwise = S

39. (a) Two views of the same molecule. Just because we write out a tetrahedral carbon doesn't automatically mean the molecule is chiral. This one has two chlorines on the same carbon.

(b) Now we have a chiral carbon, with the same four substituents in each structure. Are they identical, or are they enantiomers? Hint: Until you are able to rotate these images in your mind to see if they are the same or different, a good way to tell is just to figure out the stereochemistry (*R* or *S*) for both. They are identical (rotation of either one by 180° about a vertical axis gives the other).

(c) Different connectivity now: constitutional isomers.

(d) Enantiomers. **(e)** Identical. **(f)** Identical.

(g) With ring compounds there are some simplifying "tricks" that you can employ. For example, a ring may be treated as if it were flat—planar—for the purpose of determining whether or not it contains a plane of symmetry. These two structures are identical. The compound pictured has

two stereocenters but contains a plane of symmetry, making it meso: ---┼---Plane of symmetry

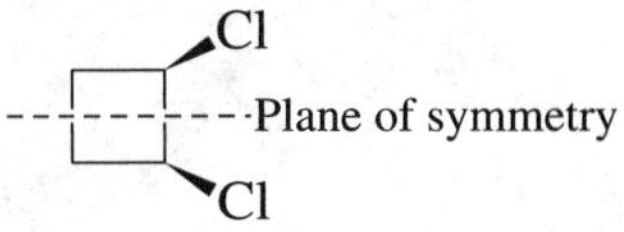

(h) Harder. Not meso. The mirror plane that works for the cis compound in part (g) doesn't in the trans, because it does not reflect one Cl atom into the other. So the trans compound is chiral. Are these structures both of the same enantiomer or of the mirror images? Rotate each by 90°

and put them side by side: 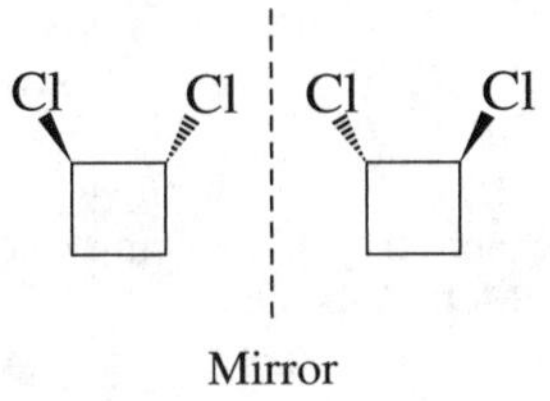 Cl Cl ┊ Cl Cl Enantiomers.

Mirror

(i) Different connectivity: constitutional isomers.

(j) Same thing. Constitutional isomers.

(k) Identical.

(l) Diastereomers (cis/trans isomer pairs are stereoisomers but not mirror images—the definition of diastereomer). Did you notice that the molecules in (k) and (l) do not have stereocenters? The substituted carbons do *not* have four different groups on them.

(m) Look to see if groups have been switched. Switching a pair of groups around a stereocenter changes the stereochemistry at that stereocenter. If both stereocenters are switched, you have two enantiomers. If only one is switched, you have diastereomers. These are diastereomers: Only the Cl-containing center is switched.

(n) Less difficult than it looks: 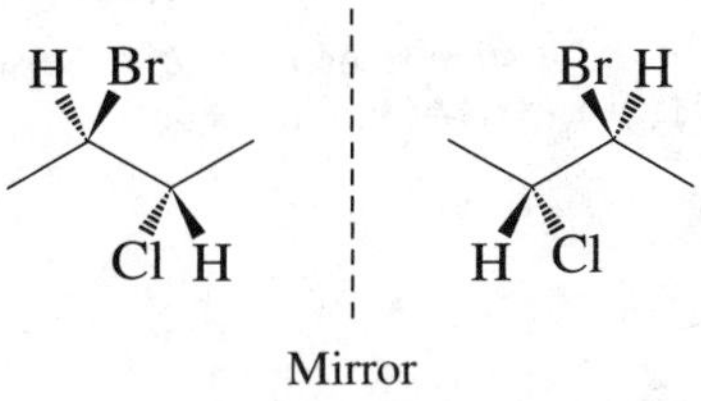 H Br ┊ Br H Enantiomers.
  Cl H ┊ H Cl

Mirror

(o) This one requires some work. One approach: Determine *R/S* for each stereocenter and compare the structures. Easier: Notice that rotation of either one by 180° in the plane of the paper turns it into the other.

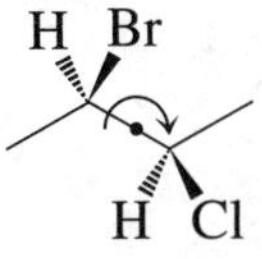

  H Br   H Cl
         becomes
  H Cl   H Br

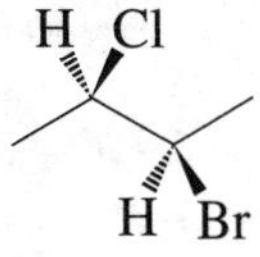

(p) Just as in (n), the structures are mirror images of each other. But now each stereocenter has the same groups, so we could be dealing either with two enantiomers or with two images of a single meso compound. Options: Either determine *R/S* for each stereocenter, or (better) rotate into an eclipsed conformation to see if an internal plane of symmetry is present.

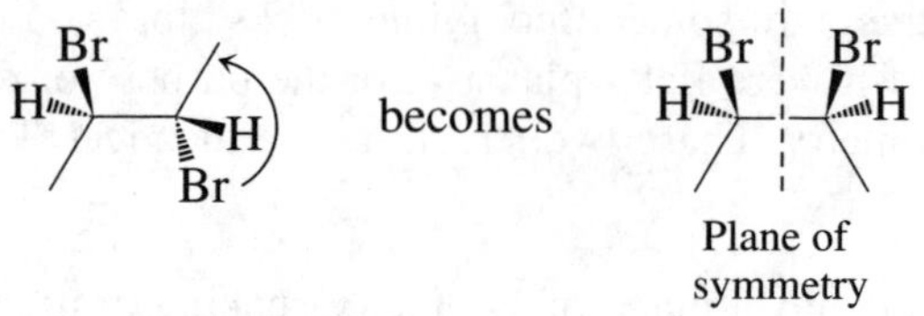

Br   H$_{\text{\tiny III}}$ —— H   Br   becomes   Br | Br   H$_{\text{\tiny III}}$ —|— H

Plane of
symmetry

These are two images of the same meso compound.

40. (a) 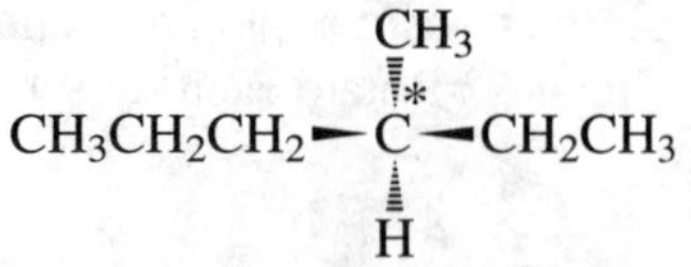
$$CH_3$$
$$CH_3CH_2CH_2 \!-\! \overset{*}{C} \!-\! CH_2CH_3$$
$$H$$

1 stereocenter (*), 2 stereoisomers
(S)-3-Methylhexane

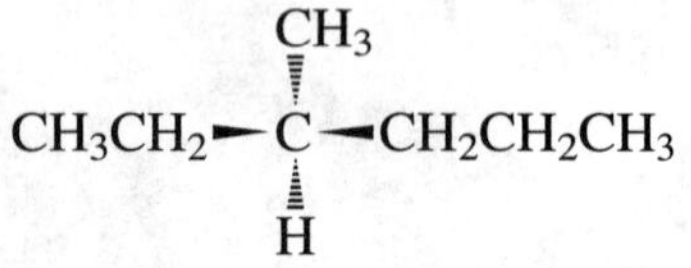

$$CH_3$$
$$CH_3CH_2 \!-\! C \!-\! CH_2CH_2CH_3$$
$$H$$

The enantiomer of the compound above.
(R)-3-Methylhexane

In all the rest of these isomers the R enantiomer would be the mirror image of the structure shown.

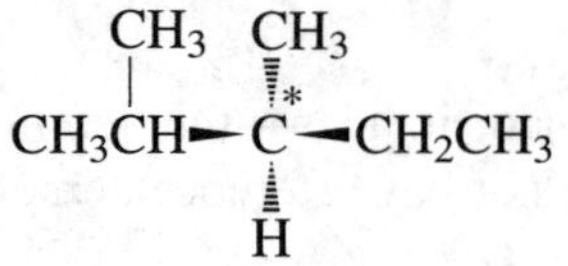

$$CH_3 \quad CH_3$$
$$CH_3CH \!-\! \overset{*}{C} \!-\! CH_2CH_3$$
$$H$$

1 stereocenter (*), 2 stereoisomers
(S)-2,3-Dimethylpentane

Note that there is no need to say "3S" in the name—only carbon 3 is chiral, so "S" is sufficient. Carbon 2 is not a stereocenter because it contains two identical methyl groups. These are the only two chiral isomers of C_7H_{16}.

(b) 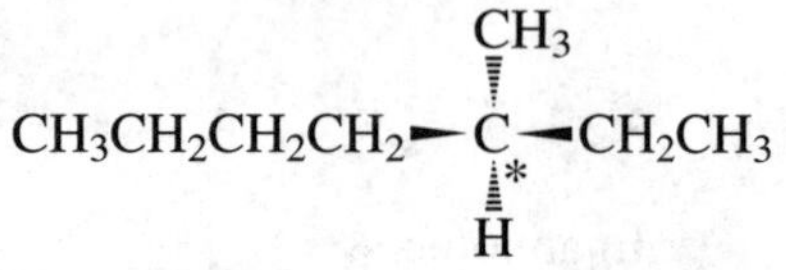
$$CH_3$$
$$CH_3CH_2CH_2CH_2 \!-\! \overset{}{C} \!-\! CH_2CH_3$$
$$\overset{*}{H}$$

1 stereocenter (*), 2 stereoisomers
(S)-3-Methylheptane

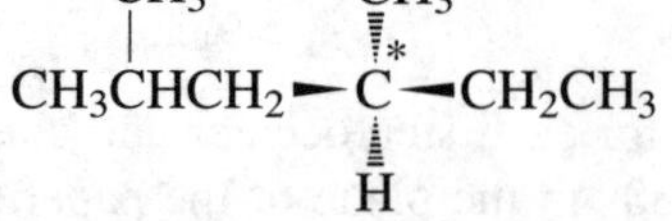

$$CH_3 \qquad CH_3$$
$$CH_3CHCH_2 \!-\! \overset{*}{C} \!-\! CH_2CH_3$$
$$H$$

1 stereocenter (*), 2 stereoisomers
(S)-2,4-Dimethylhexane

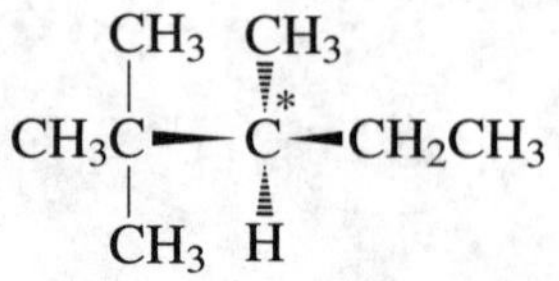

$$CH_3 \quad CH_3$$
$$CH_3C \!-\! \overset{*}{C} \!-\! CH_2CH_3$$
$$CH_3 \quad H$$

1 stereocenter (*), 2 stereoisomers
(S)-2,2,3-Trimethylpentane

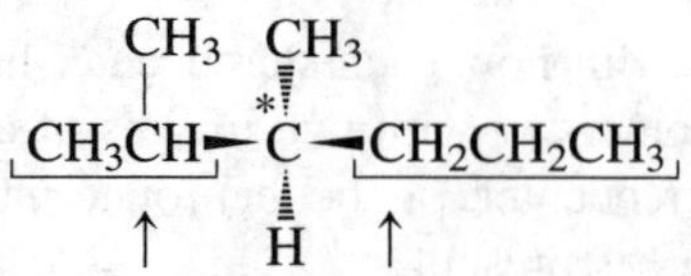

$$CH_3 \quad CH_3$$
$$CH_3CH \!-\! \overset{*}{C} \!-\! CH_2CH_2CH_3$$
$$\uparrow \quad H \quad \uparrow$$

1 stereocenter (*), 2 stereoisomers
(S)-2,3-Dimethylhexane

Note how the isopropyl group has priority over *n*-propyl: At the first point of difference (arrows), the isopropyl carbon is connected to **C**,**C**,H whereas the *n*-propyl carbon is connected to **C**,**H**,H. The boldface atoms (second C for isopropyl vs. first H for *n*-propyl) set the priority.

The structures above are the only ones with exactly one stereocenter. Following is the single isomer with more than one stereocenter.

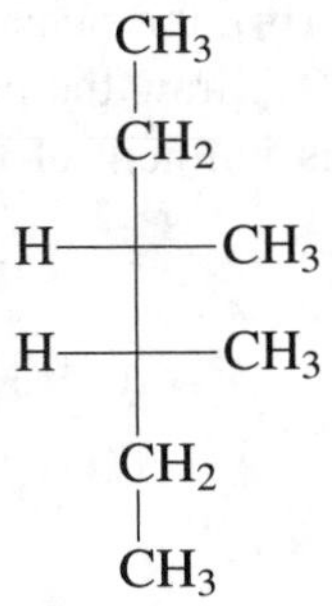

CH₃
|
CH₂
|
H——|——CH₃   2 stereocenters, 3 stereoisomers
| *meso*- or (3*R*,4*S*)-3,4-Dimethyl-
H——|——CH₃   hexane is shown
|
CH₂
|
CH₃

(c) 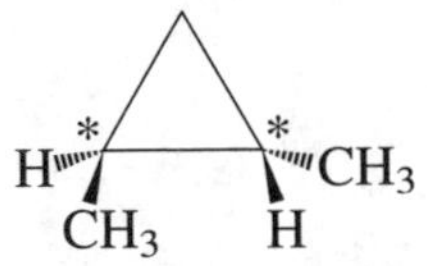 H⁗/‾‾\⁗CH₃   2 stereocenters, 3 stereoisomers
     CH₃ H   (1*S*,2*S*)-1,2-Dimethylcyclopropane

"*Trans*" is implied by the (*S*,*S*) designation. The other two stereoisomers are the (*R*,*R*) compound, its enantiomer (obviously also *trans*), and the cis isomer, a diastereomer, which is a meso (*R*,*S*) compound (Section 5-6). This is the only possible chiral molecule with the formula C₅H₁₀ and one ring.

41. **(a)** Priorities are isopropyl > ethyl > methyl > H. Stereocenter is *R*.

(b) Careful—the highest priority group (OH) is *in back*. Stereocenter is *S*.

(c) Stereocenter is *R* (Cl is highest priority—the Br is beyond the first point of difference and is irrelevant).

(d) *S*.

(e) Hydroxyethyl > hydroxypropyl (oxygen is closer to the stereocenter in hydroxyethyl). Stereocenter is *R*.

(f) Treat as if the molecule were Cl⁑_/\. In other words, consider only the bonds of the ring needed to get you to the first point of difference. Answer is *R*.

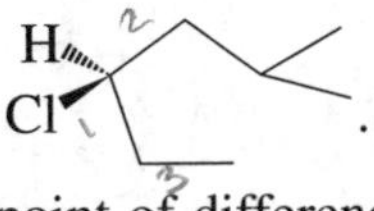

(g) *S*.   **(h)** *R*.

(i) Treat as 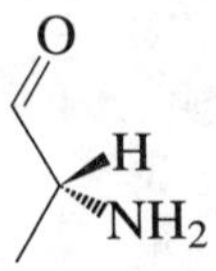   Answer is *R*.

(j) Treat as 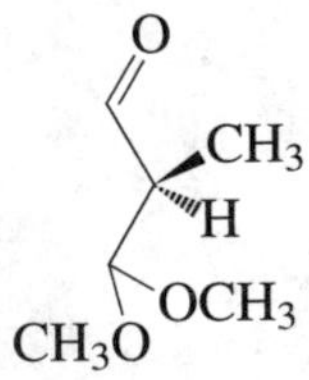   , but it's still the trickiest one of the bunch. Relative priority of C=O versus

$C(OCH_3)_2$ is the issue. The relevant comparison is $-\overset{|}{\underset{|}{C}}-\overset{|}{\underset{|}{O}}$ vs. $-\overset{|}{\underset{|}{C}}-\overset{|}{\underset{|}{O}}$, and the
$\quad\quad\quad\quad\quad\quad\quad\quad\quad\quad\quad\quad\quad\quad\quad\quad\quad O \quad C \quad\quad CH_3O \quad CH_3$

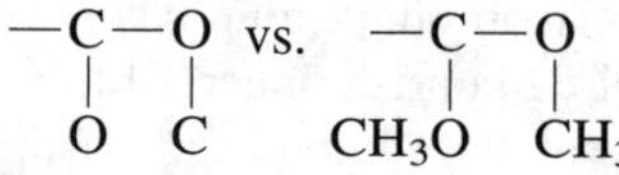

second one is higher by virtue of having carbon atoms attached to *both* oxygens in the functional group. Answer is *R*.

42. Letters correspond to compounds in Problem 33. Consider only (b), (e), (j), and (m), the ones that are chiral. Since you are given a choice of which configuration to draw at each stereocenter, draw the one that has the lowest priority group going "away" from you—on a hashed bond—to make assignment of *R* and *S* easier.

(b)

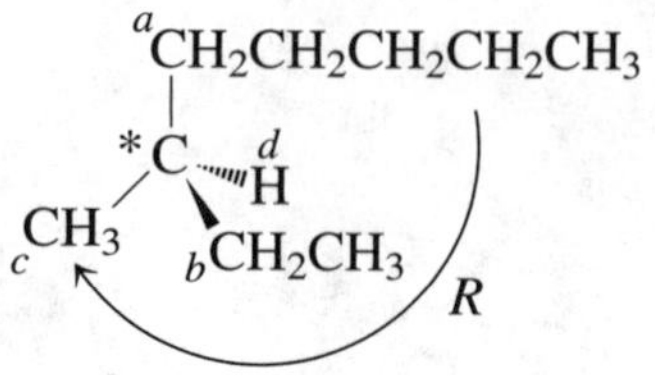

$^a CH_2CH_2CH_2CH_2CH_3$

(e)

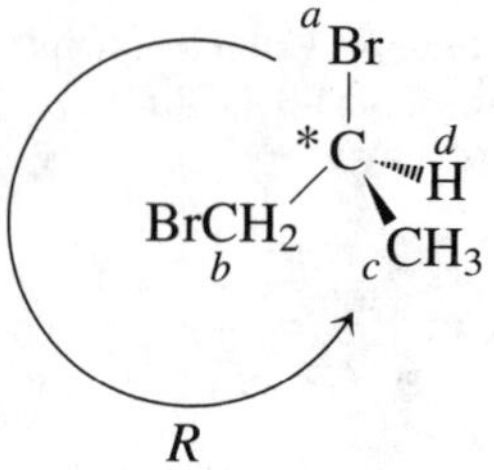

(j)

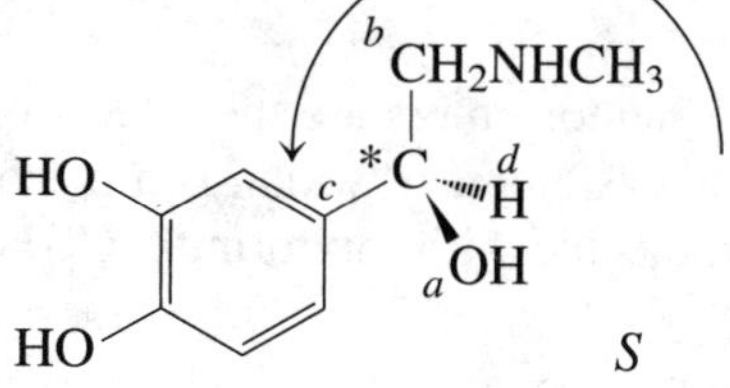

Notice that $-CH_2N$ beats the ring, which is ranked as $-CC_3$, because N > C at the first point of difference.

(m) CH_2OH

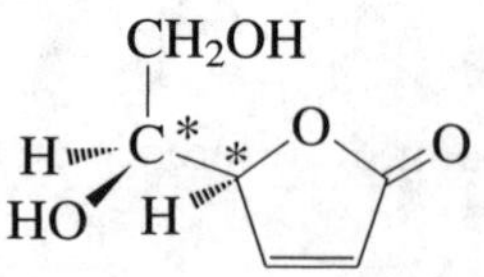

Evaluate each chiral carbon separately:

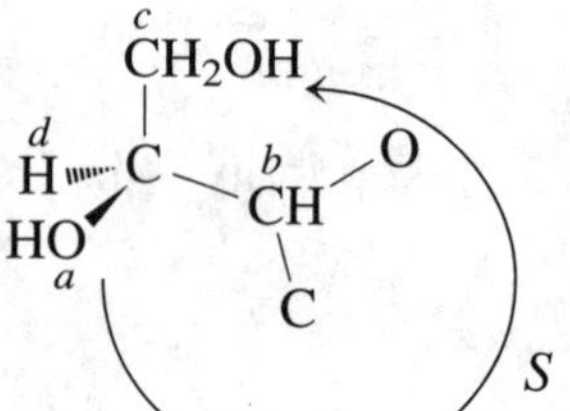

$CHCO > CH_2O$

and

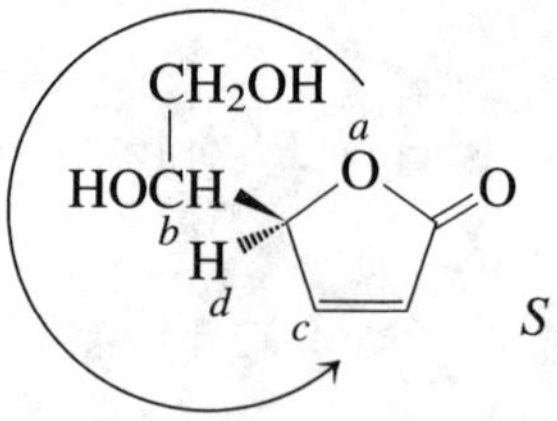

43.

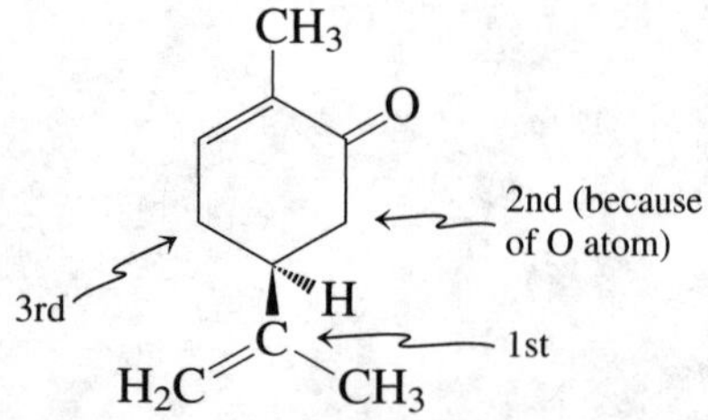

2nd (because of O atom)

3rd

1st

The (*S*) enantiomer (note priorities)

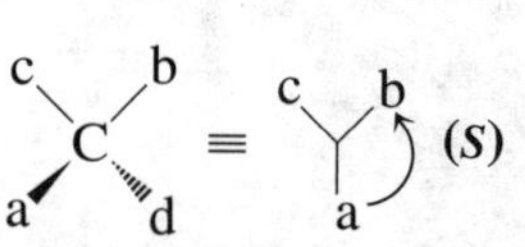

$\equiv$ (*S*)

Conversely, (−)-carvone has the *R* configuration.

44. (a)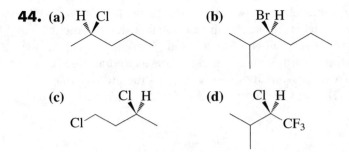

45. The cyclic structures are especially challenging. Again, start by drawing any arbitrary stereoisomer, determine configurations about its stereocenters, and then just switch the two external groups to correct any that are the opposite of what you need.

(a) 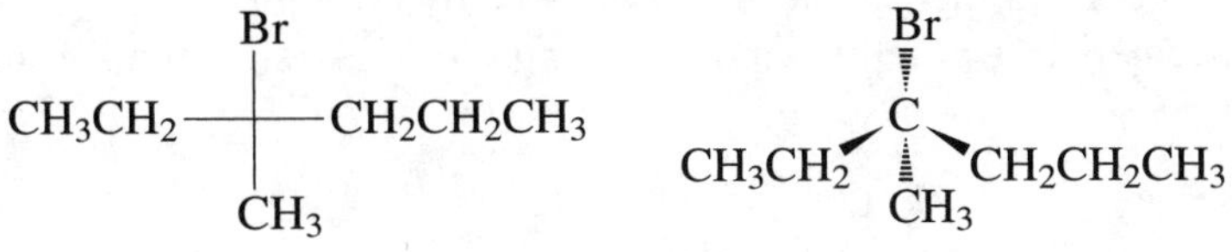 CH_3CH_2—|—$CH_2CH_2CH_3$
      Br
      |
      CH_3

   Br
   ⋮
   C 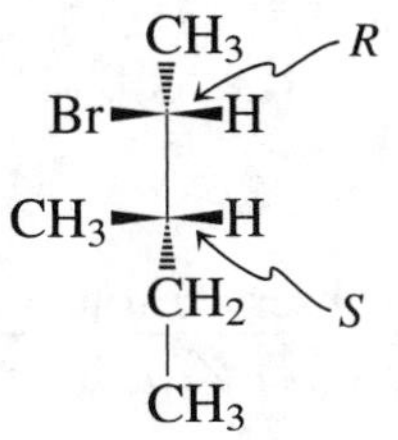
CH_3CH_2⋮ $CH_2CH_2CH_3$
   CH_3

(b) CH_3
    |
    CH_2 ⟋ S
H—|—CH_3
    CH_2 ⟋ R (Meso)
H—|—CH_3
    CH_2
    |
    CH_3

(c) CH_3 ⟍ R
    Br—|—H
    CH_3—|—H
    CH_2 ⟋ S
    CH_3

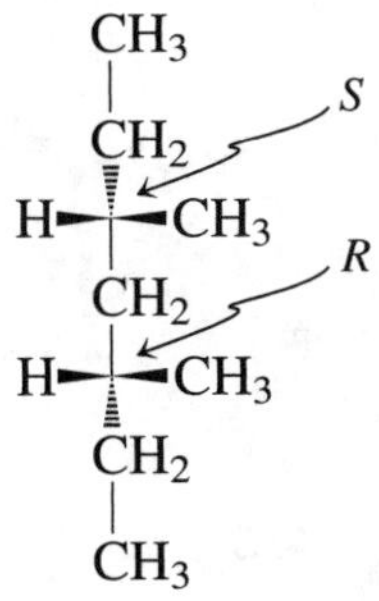

(d) 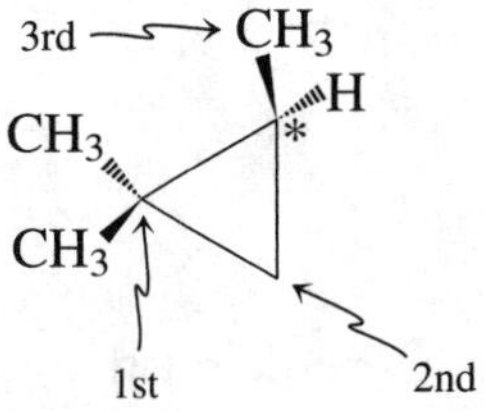 3rd ⟶ CH_3
    CH_3 ⟍ ⫰H
    CH_3 ⫶ *
    CH_3 ↑
    1st   2nd
    (Note priorities)

(e) 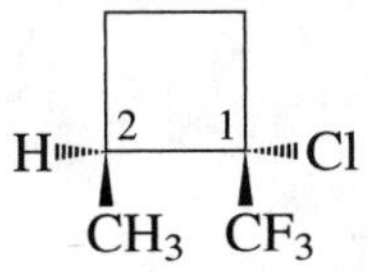
H⫶ 2 1 ⫶Cl
    CH_3 CF_3

Note priorities on C1:
Cl > CF_3 > ring CHCH$_3$ > ring CH$_2$

(f) 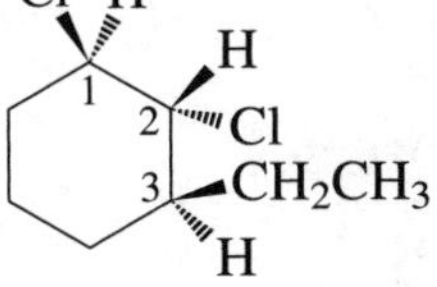 Cl H
        H
    1
    2 ⫶Cl
    3 —CH_2CH_3
        H

46. Begin by expanding the condensed structure so that you can see the connectivity better, give it a name, and identify the location of any stereocenters:

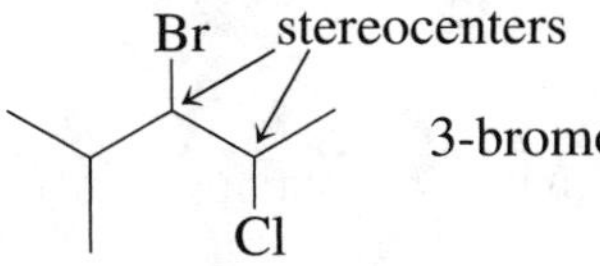

Br stereocenters

3-bromo-2-chloro-4-methylpentane

Cl

Next determine how many stereoisomers can exist. A molecule with two stereocenters can have up to four stereoisomers. This is the case here. If a plane of symmetry were present in any possible stereoisomeric structure, the total number of stereoisomers would drop, because two of them would become identical as the *meso* form. That's not possible with this molecule, so there will be four structures for us to draw and name in full. We make the remainder of the problem as easy as possible by first drawing a structure for which determination of R and S is simplified by putting the H atoms on the stereocenters *below* their carbons (on hashed bonds):

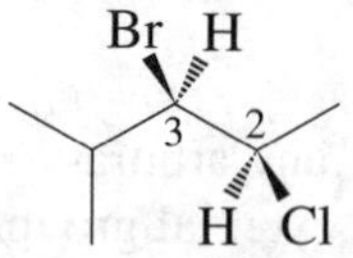

C2 is R
C3 is also R
The name is (2R,3R)-3-bromo-2-chloro-4-methylpentane

We derive the remaining three stereoisomers and their names from this one by switching the stereochemistries of each of the stereocenters one at a time and switching the R/S designations to match:

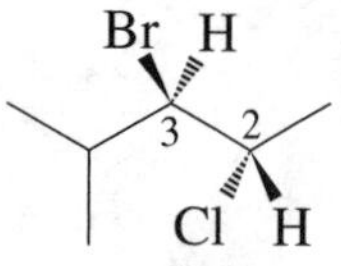

(2S, 3R)   (2R, 3S)   (2S, 3S)

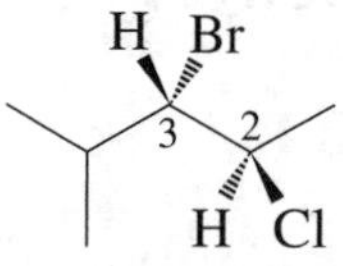

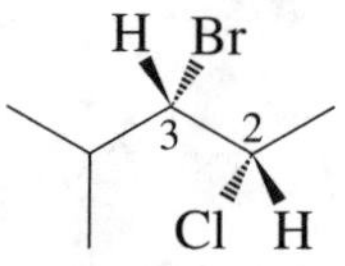

47. Use $[\alpha]_D$ (specific rotation) $= \dfrac{\alpha \text{ (observed rotation)}}{\text{conc. (g mL}^{-1}) \times \text{path length (dm)}}$

(a) $C = 0.4$ g/10 mL $= 0.04$ g mL^{-1}; $\alpha = -0.56°$, and $l = 10$ cm $= 1$ dm; so $[\alpha]_D = -14.0°$

(b) $[\alpha]_D = +66.4°$, $C = 0.3$ g mL^{-1}, $l = 1$ dm; $\therefore \alpha = +19.9°$

(c) Rearranging gives $C = \alpha/[\alpha]_D = 57.3°/23.1° = 2.48$ g mL^{-1}

48. $C = 1$ g/20 mL $= 0.05$ g mL^{-1}, so $\alpha/(C \times l) = -2.5°/0.05 = -50°$, which is identical to the actual $[\alpha]_D$. Therefore the epinephrine is optically pure and, presumably, safe to use.

49. (a)

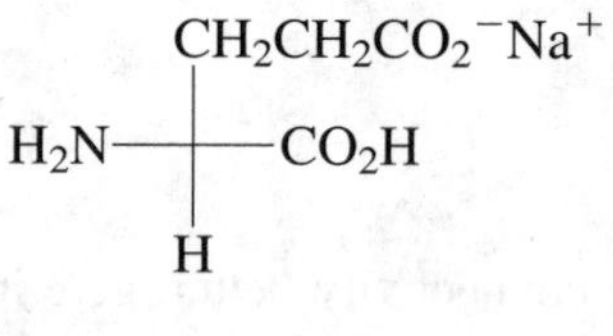

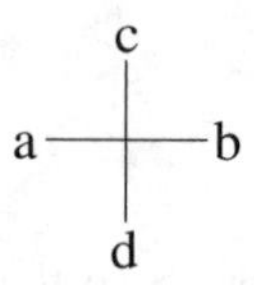

(b) $8°/24° = 0.33$ or 33% optically pure, corresponding to a mixture of 33% pure S + 67% racemate, or 67% S and 33% R.

(c) $16°/24° = 0.67$ or 67% optical purity, which equals 67% pure S + 33% racemate. That's the same as 83% S and 17% R.

50. **(a)** Ring carbons 1, 2, and 5 are stereocenters.

(b) Having three stereocenters, menthol can possess up to $2^3 = 8$ stereoisomers. The menthol structure does not have the characteristics that might lead to achiral meso stereoisomers, so there will be eight stereoisomers in all.

(c) The eight stereoisomers are shown below, in their four pairs of enantiomers, with all stereocenters labeled.

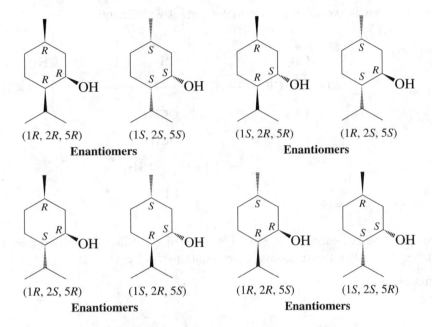

(1R, 2R, 5R)	(1S, 2S, 5S)	(1S, 2R, 5R)	(1R, 2S, 5S)
Enantiomers		**Enantiomers**	
(1R, 2S, 5R)	(1S, 2R, 5S)	(1R, 2R, 5S)	(1S, 2S, 5R)
Enantiomers		**Enantiomers**	

51. **(a)–(c)** See solution to Problem 50, above.

(d) Only menthol possesses a chair conformation in which all three substituents are equatorial. It will be the most stable isomer. Neomenthol's two alkyl groups, like those in menthol, are trans and in a 1,4 relationship in the ring. Neomenthol therefore has a conformation in which both alkyl groups are equatorial and only the relatively "small" hydroxy group is axial. Isomenthol, on the other hand, possesses cis alkyl groups, one of which therefore must be axial in any chair conformation. So the stability order is

$$\text{menthol} > \text{neomenthol} > \text{isomenthol}$$

52. The problem asks for you to solve the equation below for the proportions of menthol and neomenthol:

$$-33° = (-51°)(\text{mole fraction menthol}) + (+21°)(\text{mole fraction neomenthol})$$

That's one equation and two unknowns. However, if we take the information in the problem to mean that menthol and neomenthol together make up virtually 100% of Mentha oil, then their mole fractions will add up to approximately 1:

$$\text{mole fraction menthol} + \text{mole fraction neomenthol} = 1$$

Now we have two equations to use to solve for two unknowns. Using the general symbol X for mole fraction, substitute $X_{neomenthol} = 1 - X_{menthol}$ so as to get

$$-33° = (-51°)(X_{menthol}) + (+21°)(1 - X_{menthol})$$

Solve to get $X_{menthol} = 75\%$ and therefore $X_{neomenthol} = 25\%$.

53. (a) Identical. The carbon in the middle is not a stereocenter; it has two ethyl groups on it.

(b) Enantiomers

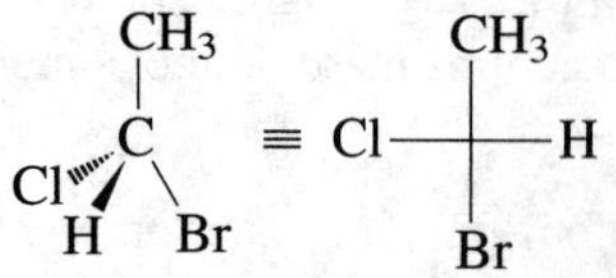

(c) Carry out pairwise switches on one to compare it with the other:

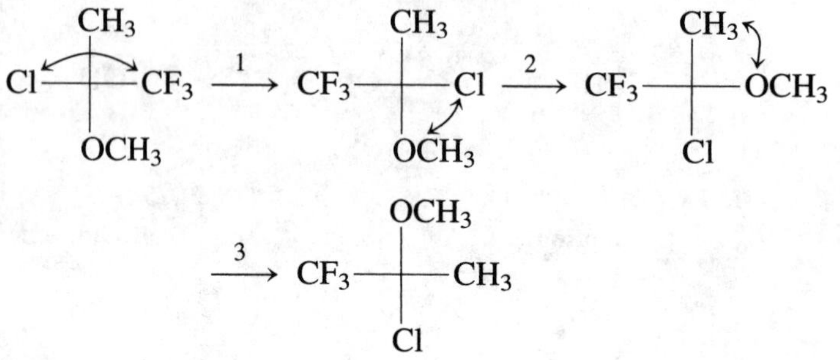

Three pairwise switches turn the first Fischer projection into the second one. The odd number of switches means that the structures were enantiomers of each other.

(d) Similarly,

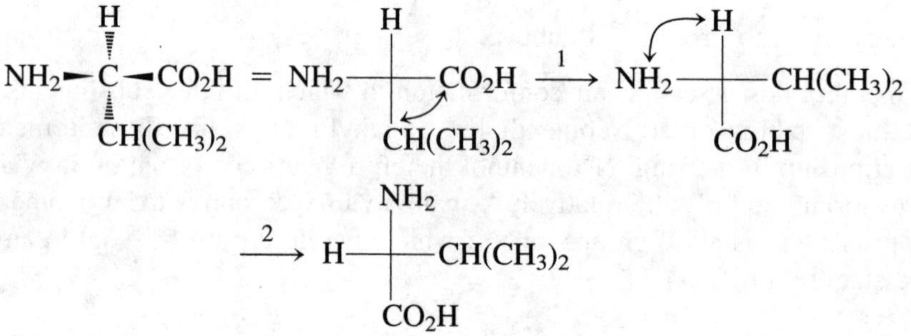

Two pairwise switches are needed, so the structures are identical.

54. (a) No asymmetric carbons.

(b)

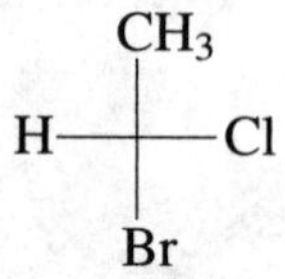

is *R*. Other is *S*.

(c)

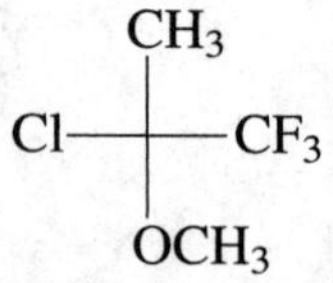

is *S*. Other is *R*.   **(d)** Both are *R*.

55. (a)

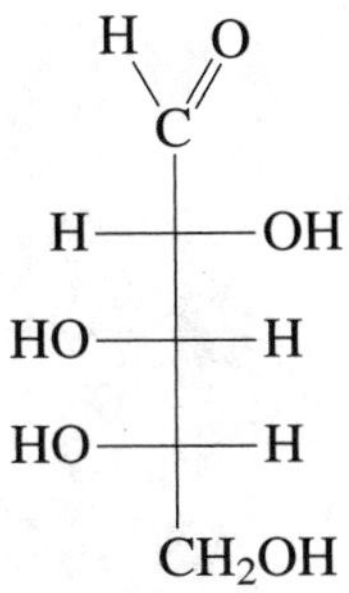

$$
\begin{array}{c}
\text{H} \diagdown \;\; \text{O} \\
\text{C} \\
\text{H} \!\!-\!\!|\!\!-\!\! \text{OH} \\
\text{HO} \!\!-\!\!|\!\!-\!\! \text{H} \\
\text{HO} \!\!-\!\!|\!\!-\!\! \text{H} \\
\text{CH}_2\text{OH}
\end{array}
$$

(b) No. An object can have only one mirror image.

(c) There are several. Here's one:

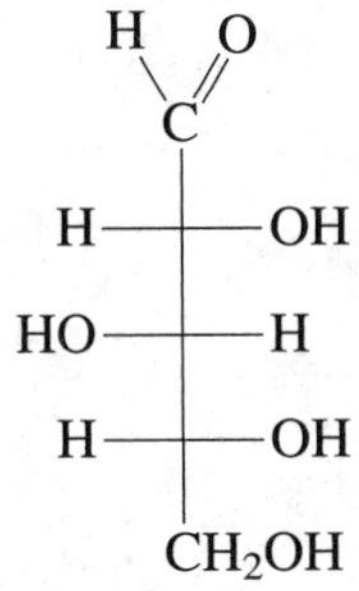

$$
\begin{array}{c}
\text{H} \diagdown \;\; \text{O} \\
\text{C} \\
\text{H} \!\!-\!\!|\!\!-\!\! \text{OH} \\
\text{HO} \!\!-\!\!|\!\!-\!\! \text{H} \\
\text{H} \!\!-\!\!|\!\!-\!\! \text{OH} \\
\text{CH}_2\text{OH}
\end{array}
$$

(d) Yes.

(e) $+105°$

(f) There is no way to predict the optical rotation of a diastereomer of a compound whose optical rotation you know. Diastereomeric compounds usually have very different physical properties.

(g) No. Because the groups on the two end carbons are different, there is no way to have the plane of symmetry that would be required for a meso compound.

56. S-1,3-Dichloropentane is the compound's name.

(a) A single achiral product is formed: $ClCH_2CH_2CCl_2CH_2CH_3$.

(b) Two diastereomers are formed in unequal amounts:

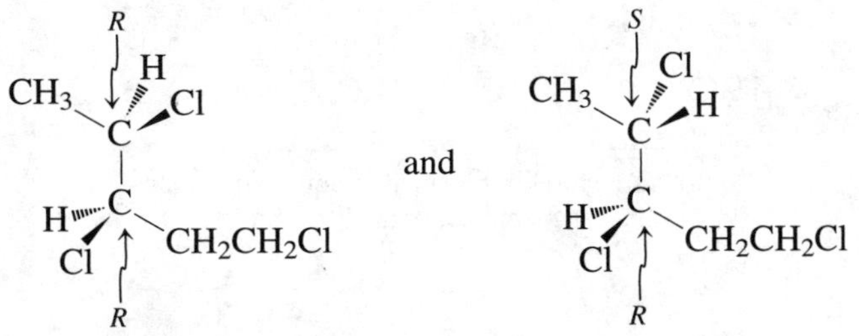

and

Notice that the designation for C3 has changed to R because the priority of the $CHClCH_3$ group is higher than that of the CH_2CH_2Cl group.

(c) A single achiral product is formed: $ClCH_2CH_2CHClCH_2CH_2Cl$.

Notice that in (a) and (c) the chirality at C3 has been eliminated in two slightly different ways. In (a), a bond to C3 has been broken and a second Cl attached. In (c), no bond to C3 has been affected, but a remote change has made two groups on C3 identical.

57. (a) A single achiral product is formed: 1-chloro-1-methylcyclopentane.

(b) Four stereoisomers are formed, as shown below:

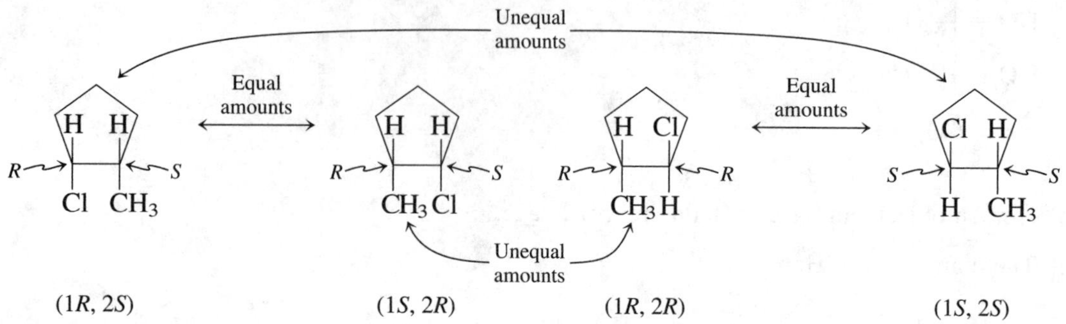

(1R, 2S) (1S, 2R) (1R, 2R) (1S, 2S)

1-Chloro-2-methylcyclopentane

(c) Four stereoisomers are again formed:

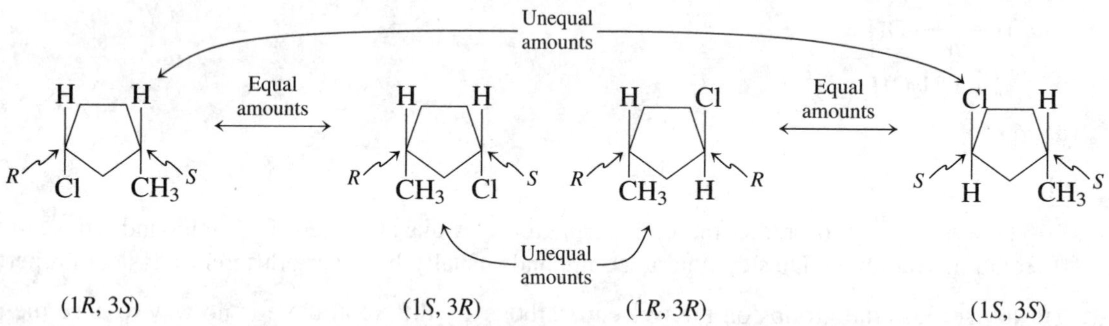

(1R, 3S) (1S, 3R) (1R, 3R) (1S, 3S)

1-Chloro-3-methylcyclopentane

58. Attack at C1:

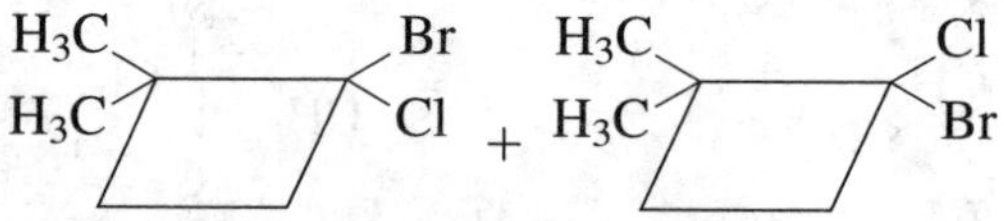

Both are chiral but formed in equal amounts;
this is a racemate, optically inactive.

Attack at the methyl groups:

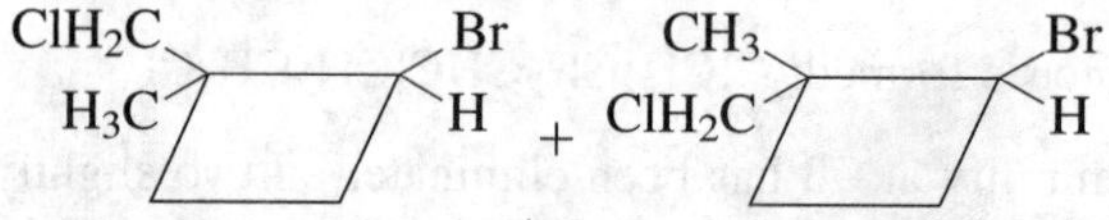

Both are diastereomers, formed in unequal amounts
and in optically active form.

Attack at C3:

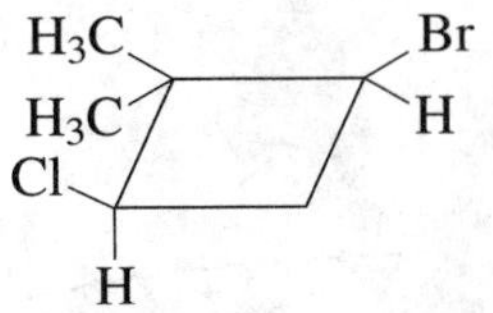

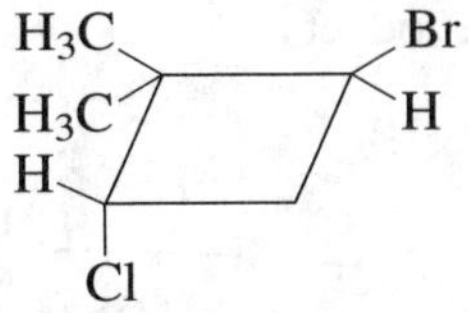

H₃C   Br
H₃C   H
Cl
     H   +

H₃C   Br
H₃C   H
H
     Cl

Chiral cis diastereomer

Formed in amounts different
from the trans dihalide;
optically active

Chiral trans diastereomer

Formed in amounts different
from the cis isomer;
optically active

Attack at C4:

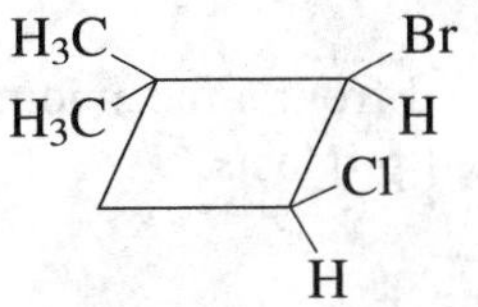

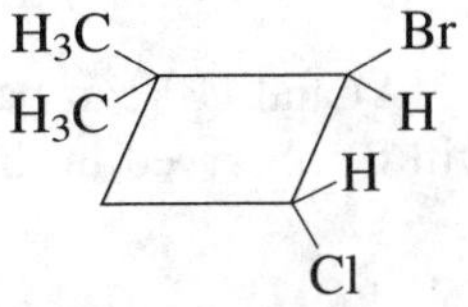

H₃C   Br
H₃C   H
            Cl
     H   +

H₃C   Br
H₃C   H
            H
     Cl

Chiral cis diastereomer

Formed in amounts different
from the trans dihalide;
optically active

Chiral trans diastereomer

Formed in amounts different
from the cis isomer;
optically active

59. In steps:

(1) Racemic amine is neutralized with an equimolar amount of an optically pure acid, such as naturally occurring (*S*)-lactic acid, CH₃CHOHCOOH. This reaction would produce two diastereomeric salts: (*R*)-amine/(*S*)-acid salt and (*S*)-amine/(*S*)-acid salt.

(2) The salt mixture is separated into its two diastereomeric components by recrystallization, typically from an alcohol–water solvent mixture.

(3) The two diastereomeric salts are separately treated with strong base such as aqueous NaOH, thereby removing the lactic acid and freeing the individual enantiomers of the amine, which can be isolated in pure form.

60. Reversing the procedure of Problem 59:

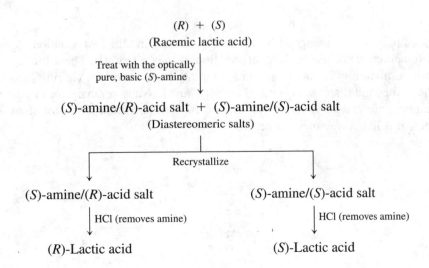

61. Bromination is highly selective, so consider reacting **only** at the tertiary carbons!

(a) Four stereoisomeric products:

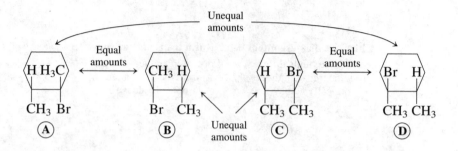

A racemic mixture of Ⓐ and Ⓑ can be separated from a racemic mixture of Ⓒ and Ⓓ, but no net optical activity will be observed in the isolated products.

(b)

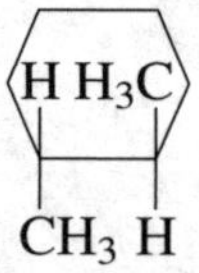

R, R-1, 2-Dimethylcyclohexane

Only two of the four products in (a) are possible: Ⓐ and Ⓒ. They are diastereomers of each other; they are formed in unequal amounts; they can be physically separated; and both are formed optically pure.

Don't forget, radical halogenation gives both possible stereochemical configurations at the **reacting** carbon, but it does not change the configuration of carbons that do not react.

62.

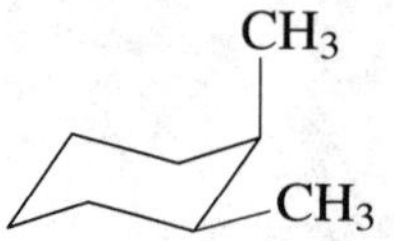

CH₃

CH₃

This molecule lacks a plane of symmetry. It is chiral when locked in this conformation. A ring flip leads to the **mirror image** (enantiomer) of the above conformation, however (try it!). Therefore, cis-1,2-dimethylcyclohexane is actually a mixture of enantiomeric chairs rapidly interconverting via ring flips. Because of this interconversion, the compound displays no optical activity, just like any ordinary meso compound. Although neither chair has a plane of symmetry, notice that one of the boat transition states for chair–chair interconversion does, just like a true meso structure:

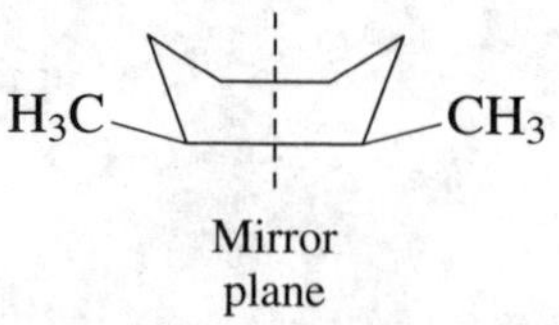

H₃C ⎯⎯⎯ CH₃

Mirror
plane

63. **(a)** There are three (C9, C13, C14):

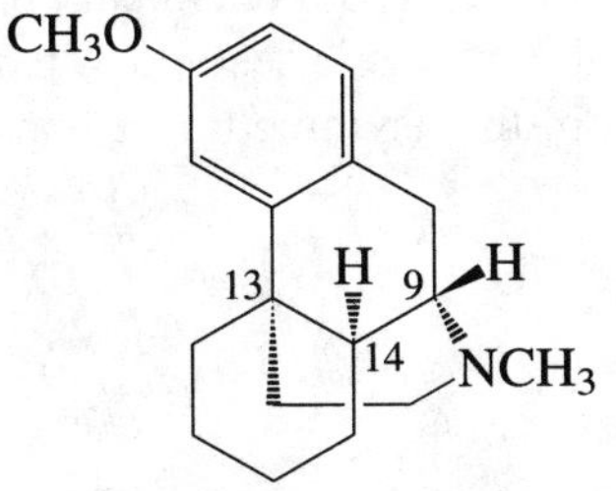

(b) CH₃O

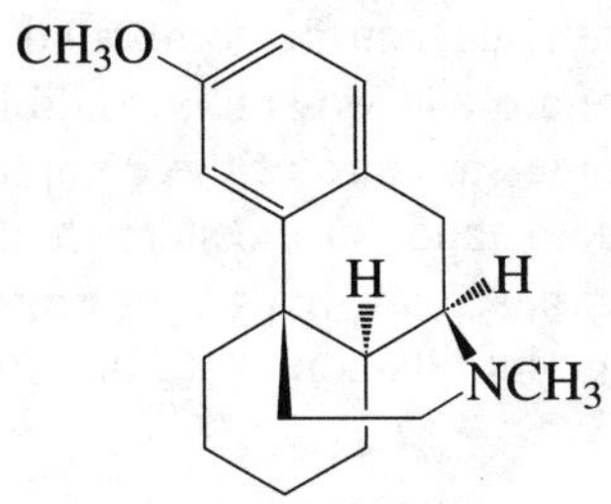

(c) For each of these stereocenters, the groups are drawn out as far as needed to reach the first point of difference.

C9:

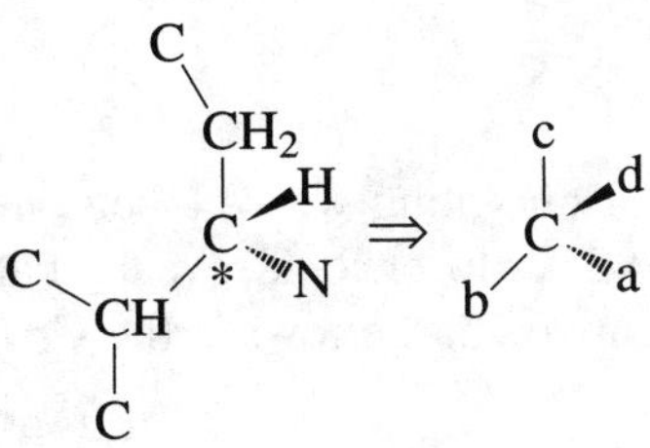

Flip so that the lowest priority group d (the H) is in the back.

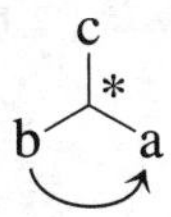

$$= S$$

C13:

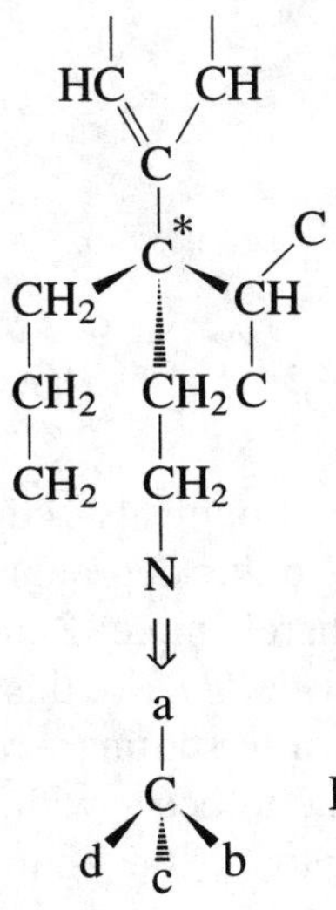

Treat the benzene carbon as any carbon with a double bond:

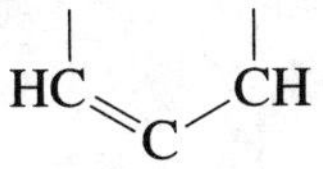

HC═══CH   becomes

HC─C─CH   and is

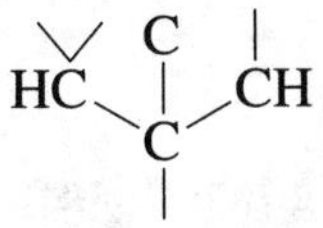

therefore highest in priority over all the other groups.

Rotate d in back: = S again.

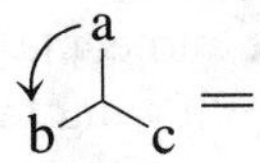

C14:

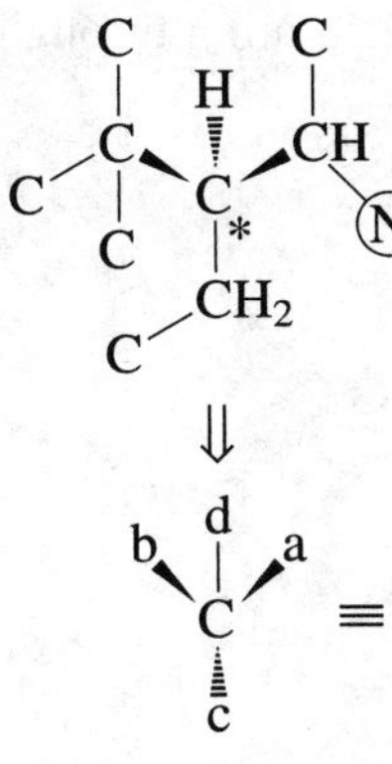

Careful! The C on the right (attached to N, C, H) takes precedence over the C on the left (attached to 3 C's). At the first point of difference, N (circled) > C.

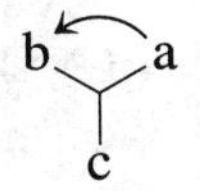

$$\equiv \qquad S \text{ again.}$$

Dextromethorphan therefore has the (9*S*, 13*S*, 14*S*) configuration.

64. The question tells us that the compound loses its optical activity when treated with base. Hypothetically, there are two ways this could happen: The stereocenter could be destroyed by some process that totally changes the structure of the compound, or the compound could undergo a chemical process that causes *R/S* randomization at the stereocenter, leading to a racemic mixture. Let's see which is more likely. Reaction with base (say, hydroxide) removes the acidic proton from the asymmetric carbon and converts the molecule into its conjugate base (Chapter 2):

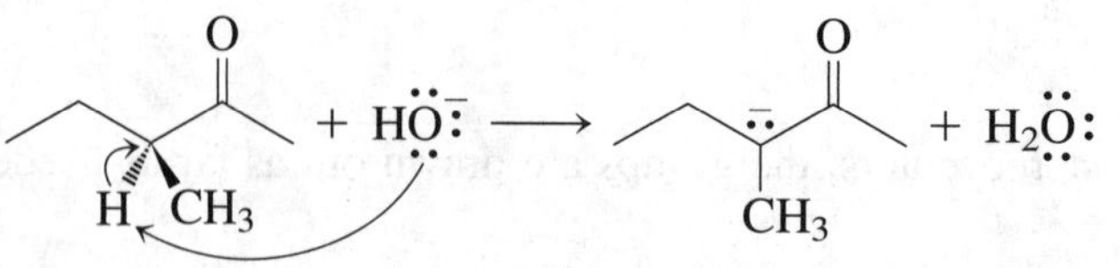

So? Big hulking deal. What has that to do with anything? Well, for one thing we've taken one of the four (different) groups off the asymmetric carbon. What is its geometry? A closer look reveals that we can draw another resonance form, by delocalizing the lone pair and the negative charge toward the C=O double bond:

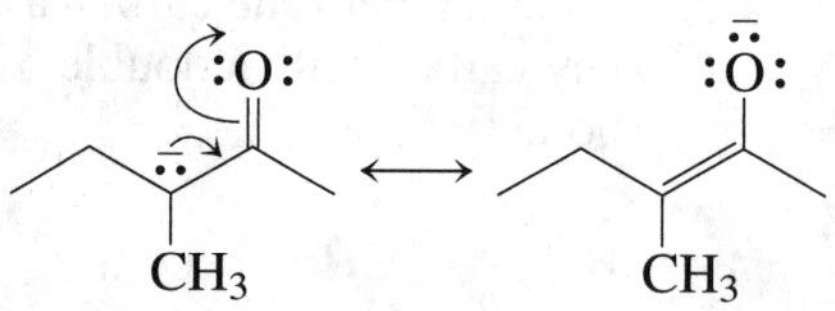

So, our former stereocenter is now planar, meaning that this species is *achiral*—not optically active. Next we recall that the problem said that only a *catalytic* amount of base was present, so this deprotonation happens to only a small percentage of the molecules at any one time. But wait, there's more: Remember (Chapter 2 again) that acid-base reactions are equilibrium processes—they are *reversible*. So, this anionic conjugate base can react to pick up a proton from the molecule of water and return to starting material. A molecule of hydroxide is regenerated in the process, allowing the event to continue to occur with more molecules of starting ketone. Exactly where does this proton attach? Either to the top or the bottom, randomly, just like the attachment of a halogen to a radical in radical halogenation. Therefore, the final outcome is a racemic mixture of the two enantiomers of the starting ketone, equilibrating through this achiral conjugate base:

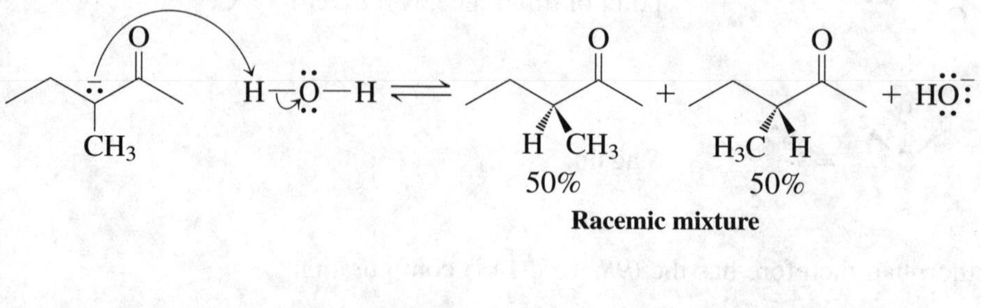

65. (a)

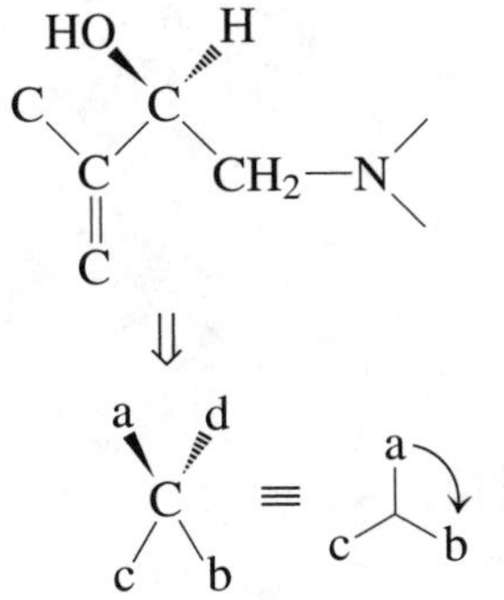

Careful! The —CH$_2$N= group is higher priority than the benzene ring, even though the latter translates into —C$\overset{C}{\underset{C}{-}}$C.

The first point of difference is N > C!

R is the answer.

(b) Equal energy: The transition states are mirror images of each other (they are **enantiomeric**).

(c) The enzyme must be lowering the energy of the transition state that leads to the (−) isomer, relative to the one that leads to the (+) isomer. To do this, the enzyme must be chiral, and optically pure as well, so that the two transition states in the presence of the enzyme become **diastereomeric** and, therefore, different in energy from each other.

66. (a) The four structures on which your models should be based are as follows. In each molecule the chiral carbons are the two ring fusion carbons. For each, the highest priority substituent is the other ring fusion; the ring bond closer to the N is second.

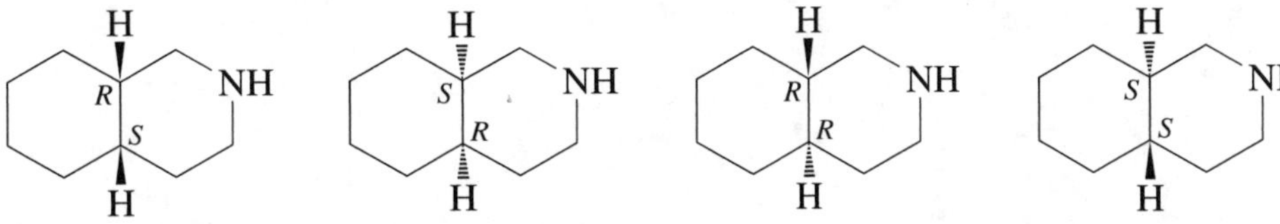

Cis
(enantiomers)

Trans
(enantiomers)

Cis and trans are diastereomers of one another.

(b) Ring bond priorities are not changed by the additional groups because they are beyond the first points of difference with respect to the ring fusion chiral carbon atoms.

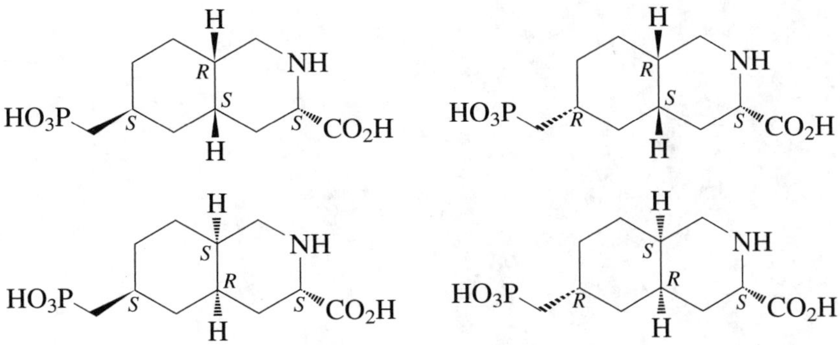

(c)

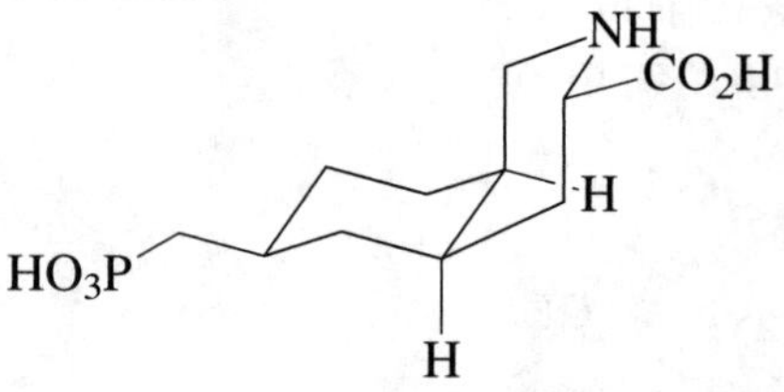

(3S,4aR,6S,8aS)
This is the isomer shown at the
lower left in part (b)

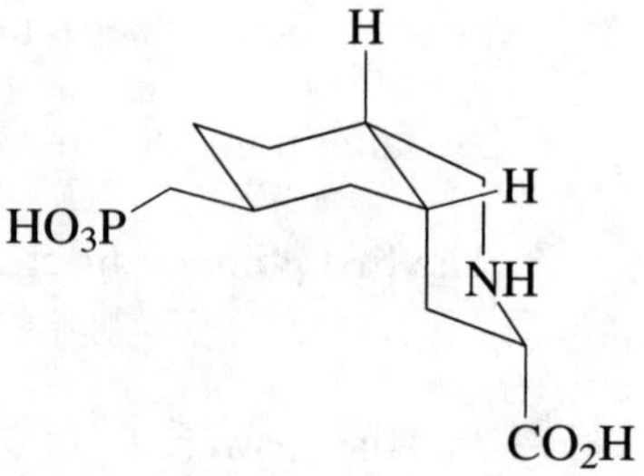

(3S,4aS,6S,8aR)
This is the isomer shown at the
upper left in part (b)

67. **(d)** is meso

68. **(a)** Exchanging any one pair of groups inverts a stereocenter.

69. **(c)**

70. **(b)** lacks a plane of symmetry

6

Properties and Reactions of Haloalkanes: Bimolecular Nucleophilic Substitution

Referring to Chapter 1, recall that polarized covalent bonds are at the heart of most of organic reaction chemistry. Here, for the first time, the properties and chemical behavior of molecules containing a polarized covalent bond are presented in full detail. They are the *haloalkanes,* containing a polarized carbon(δ^+)–halogen(δ^-) bond. Because the reactions here serve as models for many subsequent presentations of the chemistry of organic functional groups, this chapter should be examined particularly closely. Many concepts discussed in detail here are fully applicable later on, even though later presentations may not be as comprehensive. With this and the next chapter we really start covering **typical** organic chemistry. Now you get your first opportunity to see the "big picture" with respect to one portion of this subject. In addition, much of what you've seen up to now will be used to help explain the behavior of haloalkanes. Comprehending this material in a **general overall** sense is in the long run more useful than remembering every last detail.

Outline of the Chapter

6-1 Physical Properties of Haloalkanes

6-2 Nucleophilic Substitution

The nature of this functional group, and an introduction to its most characteristic reaction.

6-3 Reaction Mechanisms

Describing how reactions occur. A very important section.

6-4 Mechanism of Nucleophilic Substitution: Kinetics

The first section in the text to cover in detail the mechanism of a specific polar reaction. Also a very important section.

6-5 Stereochemistry

Tying together this reaction with material in Chapter 5.

6-6 Consequences of Inversion in S_N2 Reactions

Further implications of material in Chapters 4 and 5.

6-7 The Leaving Group

6-8 The Nucleophile

6-9 The Structure of the Substrate

The roles of the major variables on the favorability of the single reaction type under discussion.

Keys to the Chapter

6-1. Physical Properties

The carbon–halogen bond is the focal point of this chapter. Its polarization is the major feature governing the physical and chemical behavior of these molecules. Differences among the four halogens will affect the **degree** to which a haloalkane exhibits any given physical or chemical characteristic. This section reveals many qualitative similarities among the haloalkanes. Understand these first; you will then be in a better position to appreciate the differences that are presented later.

6-2. Nucleophilic Substitution

At this stage we begin a detailed discussion of our first major class of polar reactions. Note the following features:

1. All nucleophilic substitutions have a similar general appearance, with a similar cast of characters, so to speak. Using the first example in the text table, we have

$$\underset{\textbf{Nucleophile}}{HO^-} + \underset{\textbf{Substrate}}{CH_3Cl} \longrightarrow \underset{\textbf{Product}}{CH_3OH} + \underset{\textbf{Leaving group}}{Cl^-}$$

Whenever a new reaction class is presented, analyze the various examples presented from the point of view of the common roles played by the chemical species involved. You should do this now for all the examples in the table in this text section. That is, in each case identify the nucleophile, substrate, product, and leaving group. Start getting used to the variety of species that belong to each category.

2. All nucleophilic substitutions are *electrostatically sensible*. In every case an *electron-rich* atom in the nucleophile ultimately becomes attached via a new bond to an *electrophilic* (positively polarized) carbon atom in the substrate. Opposite charges attract!

Again, whenever a new reaction class is presented, analyze the examples given on the basis of electrostatics. Focus on the logical consequences of oppositely charged or polarized atoms attracting each other. Do this now for the examples in the reaction table in the text section.

When you analyze the components involved in a new reaction class and understand in a fundamental way why the reaction is reasonable, you have taken the first step toward learning organic chemistry by understanding instead of by memorizing.

6-3. Reaction Mechanisms

The mechanism of a reaction describes in detail when and how bonding changes occur. One useful feature of such information is its *predictive* value: Mechanistic understanding can tell us whether an unknown new example of a reaction is likely to work or not. This knowledge is important in *synthesis:* the preparation of new molecules from old ones for, for example, medicinal purposes or theoretical study.

6-4, 6-5, and 6-6. Mechanism of Nucleophilic Substitution

Section 6-4 describes one of the more common methods of deriving mechanistic information: kinetic experiments. Section 6-5 describes another common method: experiments involving the observation of stereochemical changes. By combining information from these and other types of experiments, the mechanistic picture of the bimolecular nucleophilic displacement, or S_N2 reaction, was developed over the first 30-odd years of the twentieth century. The key features, second-order kinetics and inversion of configuration at the substrate carbon, are common to all the reactions in this chapter. The mechanism allows predictions to be made concerning the effects of changing any of a number of variables in the S_N2 reaction. The last four sections of the chapter explore the most important of these. It should be pointed out that many of the observations described in these sections **were predicted ahead of time** on the basis of the logical implications of the S_N2 mechanism. Part of your job as a student in organic chemistry will be to develop the ability to predict the result of a reaction of molecules **you may never have seen before** on the basis of your knowledge of reaction mechanisms associated with the functional groups the molecules contain.

One final comment on mechanisms in general: There exists a common sort of shorthand way of writing organic reaction mechanisms. First, each step is written separately. Second, bonding changes in each step are indicated by **arrows** that represent the movement of **pairs of electrons.** For the S_N2 mechanism, a one-step process, we have:

$$\ddot{H\ddot{O}}:^- + CH_3-Cl \longrightarrow HO-CH_3 + Cl^-$$

The two arrows show (1) movement of a pair of electrons from oxygen to carbon to form the new C—O bond and (2) movement of a pair of electrons from the C—Cl bond to chlorine to form the chloride ion. Mechanism arrows depict the movement of pairs of electrons, **not the movement of atoms.** This is a logical result of the fact that chemical reactions are due to changes in **bonds,** and bonds are made out of electrons. Practice using "electron pushing" in mechanisms as frequently as you can so that you can get used to the technique. Note that proper use of electron-pair arrows **automatically** results in the correct Lewis structures for the products of a reaction, including charges, if any. An example of its application to an even simpler reaction, the reaction of an acid and a base, is shown below.

$$\ddot{H\ddot{O}}:^- + H-Cl \longrightarrow HO-H \; (H_2O) + Cl^-$$

6-7, 6-8, and 6-9. The S_N2 Reaction in Depth

These sections explore the effects of changing three variables on the S_N2 reaction rate: leaving group, nucleophile, and substrate structure. All the material in these sections derives logically from a knowledge of the mechanism and an awareness of the role each of the three variables can play. In each of these sections the effect of the variable on the **rate** of reaction is considered. This should tell you that the point of each discussion will be: How does changing this variable affect the activation energy—the energy of the transition state relative to that of the starting materials? We may not say it in so many words every time, but **that's what we mean.** Even if the discussion is **totally qualitative,** and no actual rate data are given, the focus of such a discussion will still be the effect of the variable in question on relative transition state energy.

Consideration of the *leaving group* is fairly simple. Because the leaving group is "beginning to leave" in the S_N2 transition state, the energy of the transition state will reflect to some extent the stability of the leaving group. Stable leaving groups are therefore better (interpret: "faster") leaving groups. The parallel between leaving-group ability and nonbasic character drawn in the chapter is the easiest one to work with and is fairly reliable. "Good" leaving groups are usually the conjugate bases of strong acids (Table 6-4). "Bad" leaving groups are strong bases, i.e., they are the conjugate bases of weak acids. Later you will learn how to manufacture good leaving groups out of bad ones. Notice how such fundamental concepts as acid/base strength can play pivotal roles in organic chemistry.

Consideration of the *nucleophile* follows. The section discusses **two** characteristics that can give rise to high nucleophilicity: high *basicity,* as in hydroxide ion, and large size of the nucleophilic atom, as in iodide ion. The reasons behind this are explored in detail and are relatively straightforward. In attempting to understand these, consider, again, how the characteristics in question influence the relative stability of the **transition state** of the reaction. Get into the habit of doing this whenever you are faced with totally new concepts relating to reactions and their kinetic favorability.

A third factor can affect the competition between base strength and atom size in determining relative nucleophilicity: the solvent in which the reaction is performed. Some solvents do not affect nucleophiles. These are the **polar, aprotic solvents,** of which there are several including acetone, *N,N*-dimethylformamide (DMF), and dimethylsulfoxide (DMSO). When nucleophilic substitutions are carried out in polar, aprotic solvents, the competition between more basic nucleophiles and larger nucleophilic atoms is close but usually won by the stronger base. Therefore, fluoride, the strongest base among the halides, reacts fastest in S_N2 displacements carried out in polar aprotic solvents such as DMF.

The situation changes when an S_N2 reaction is run in a **protic solvent,** such as water or an alcohol. Protic solvents interact the strongest with (especially) negatively charged **small** nucleophilic atoms. The nature of this interaction is hydrogen bonding, and it effectively **gets in the way** of the nucleophilic atom, inhibiting it from reacting. Fluoride, the smallest halide ion, is most affected by hydrogen bonding in a protic solvent, and is most severely inhibited from reacting. This hydrogen bonding has the effect of canceling out the base-strength advantage of fluoride as a nucleophile. Fluoride goes from being the best halide nucleophile in polar, aprotic solvent, to the worst in polar, protic solvent:

Nucleophilicity order in polar, aprotic solvent (acetone, DMF, DMSO)—basicity dominates:

$$F^- > Cl^- > Br^- > I^-$$

Nucleophilicity order in polar, protic solvent (water, alcohols)—size dominates:

$$I^- > Br^- > Cl^- > F^-$$

Consideration of *substrate structure* is much simpler, as the major consideration for an S_N2 reaction is, simply, How crowded is the "back side" of the atom being attacked? The less crowded, then the less *steric hindrance* to approach of the nucleophile on route to the transition state, so the better will be the situation for an S_N2 reaction. The extreme importance of steric hindrance is emphasized by the fact that *tert*-butyl and neopentyl halides are virtually incapable of reacting by the S_N2 mechanism.

Solutions to Problems

31. (a) Chloroethane   (b) 1,2-Dibromoethane

(c) 3-(Fluoromethyl)pentane   (d) 1-Iodo-2,2-dimethylpropane

(e) (Trichloromethyl)cyclohexane   (f) Tribromomethane

32. (a)
$$\overset{\overset{\displaystyle CH_2CH_3}{|}}{CH_3CHICHCH_2CH_3}$$
        (b) $CHCl_2CH_2CHBrCH_3$

(c)

(d) $\overset{CCl_3}{\triangle}$

(e)
$$CH_2Cl\overset{\overset{\displaystyle Cl}{|}}{\underset{\underset{\displaystyle CH_3}{|}}{C}}CH_2Cl$$

33. Answers for Problem 35 are also given. Stereocenters are marked with an asterisk and numbers of stereoisomers are given in parentheses.

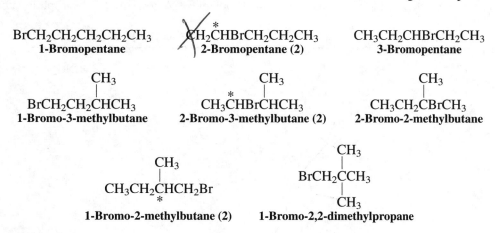

BrClCHCH$_2$CH$_3$
1-Bromo-1-chloropropane (2)

BrCH$_2$CHClCH$_3$
1-Bromo-2-chloropropane (2)

ClCH$_2$CHBrCH$_3$
2-Bromo-1-chloropropane (2)

CH$_3$CBrClCH$_3$
2-Bromo-2-chloropropane

BrCH$_2$CH$_2$CH$_2$Cl
1-Bromo-3-chloropropane

34. Stereocenters are marked with an asterisk and numbers of stereoisomers are given in parentheses.

BrCH$_2$CH$_2$CH$_2$CH$_2$CH$_3$
1-Bromopentane

CH$_2$CHBrCH$_2$CH$_2$CH$_3$
2-Bromopentane (2)

CH$_3$CH$_2$CHBrCH$_2$CH$_3$
3-Bromopentane

CH$_3$
|
BrCH$_2$CH$_2$CHCH$_3$
1-Bromo-3-methylbutane

CH$_3$
|
CH$_3$CHBrCHCH$_3$
2-Bromo-3-methylbutane (2)

CH$_3$
|
CH$_3$CH$_2$CBrCH$_3$
2-Bromo-2-methylbutane

CH$_3$
|
CH$_3$CH$_2$CHCH$_2$Br
1-Bromo-2-methylbutane (2)

CH$_3$
|
BrCH$_2$CCH$_3$
|
CH$_3$
1-Bromo-2,2-dimethylpropane

35. See 33 and 34.

36. In the answers below, the nucleophilic atom in the nucleophile and the electrophilic atom in the substrate are both <u>underlined</u>.

Reaction	Nucleophile	Substrate	Leaving group
1.	H<u>O</u>:⁻	<u>C</u>H$_3$Cl	Cl⁻
2.	CH$_3$<u>O</u>:⁻	CH$_3$<u>C</u>H$_2$I	I⁻
3.	:<u>I</u>:⁻	CH$_3$<u>C</u>HBrCH$_2$CH$_3$	Br⁻
4.	:N≡<u>C</u>:⁻	(CH$_3$)$_2$CH<u>C</u>H$_2$I	I⁻
5.	CH$_3$<u>S</u>:⁻	<u>C</u>HBr	Br⁻
6.	:<u>N</u>H$_3$	CH$_3$<u>C</u>H$_2$I	I⁻
7.	:<u>P</u>(CH$_3$)$_3$	<u>C</u>H$_3$Br	Br⁻

37. (a) 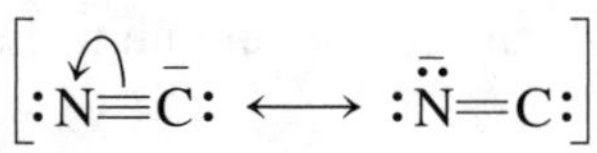 $\left[:N{\equiv}\bar{C}: \longleftrightarrow :\ddot{N}{=}C: \right]$ in reaction 4.

(b) The N may act as a nucleophilic atom in cyanide (CN^-). The reaction would then proceed as follows:

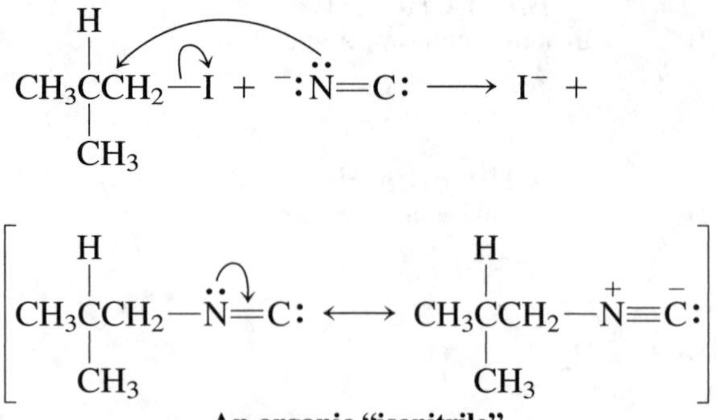

$$CH_3\underset{\underset{CH_3}{|}}{\overset{\overset{H}{|}}{C}}CH_2{-}I + {^-}{:}\ddot{N}{=}C: \longrightarrow I^- +$$

$$\left[CH_3\underset{\underset{CH_3}{|}}{\overset{\overset{H}{|}}{C}}CH_2{-}\ddot{N}{=}C: \longleftrightarrow CH_3\underset{\underset{CH_3}{|}}{\overset{\overset{H}{|}}{C}}CH_2{-}\overset{+}{N}{\equiv}\bar{C}: \right]$$

An organic "isonitrile"

38. Answers are presented in the same manner as for Problem 36. Arrows mark electrophilic atoms.

Reaction	Nucleophile	Substrate	Leaving group	Product
(a)	$^-NH_2$	$\underline{C}H_3I$	I^-	CH_3NH_2
(b)	HS^-	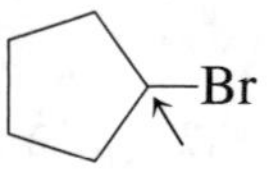	Br^-	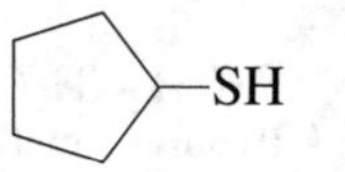
(c)	I^-	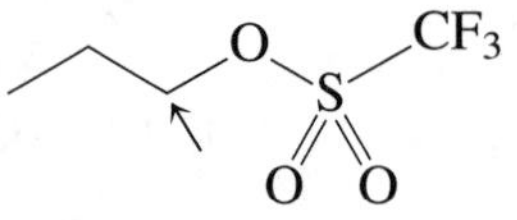	$CF_3SO_3^-$	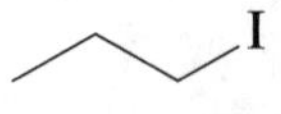
(d)	$\underline{N}_3{}^-$	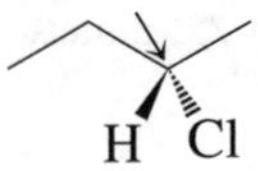	Cl^-	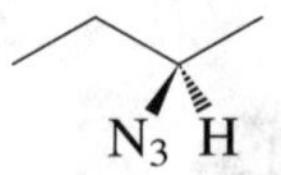
(e)	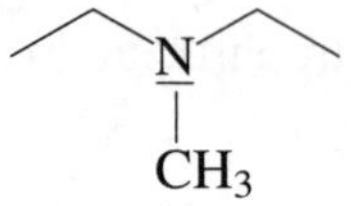	$\underline{C}H_3Cl$	Cl^-	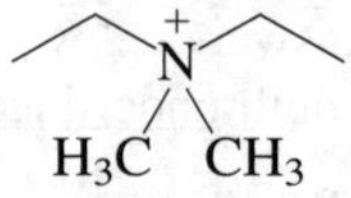
(f)	$^-\underline{Se}CN$	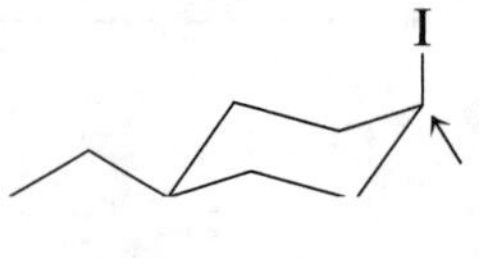	I^-	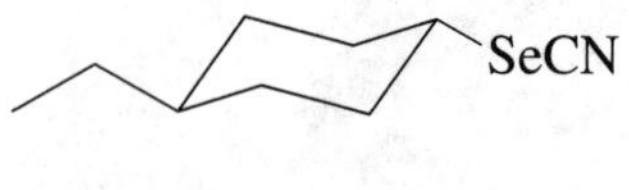

39. Only the curved arrows in the mechanism step itself are shown. The products were given in the solution for Problem 38. The procedure is as follows: (1) identify the electron pair of the nucleophile, (2) draw a curved arrow from it to the carbon bearing the leaving group, and (3) draw a curved arrow from the bond between the carbon and the leaving group and pointing at the leaving group.

(a) 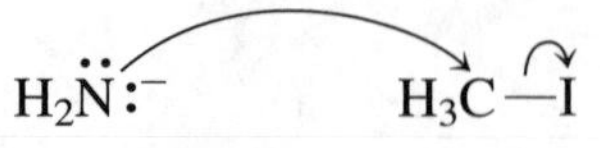 $H_2\ddot{N}{:}^-$ $\qquad$ $H_3C{-}I$

(b) 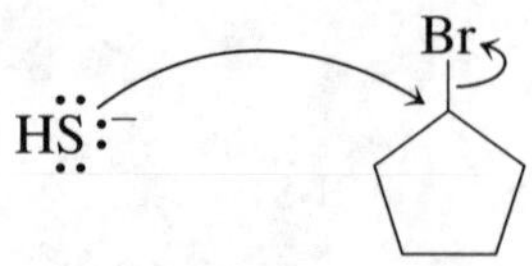 $H\ddot{S}{:}^-$

(c) 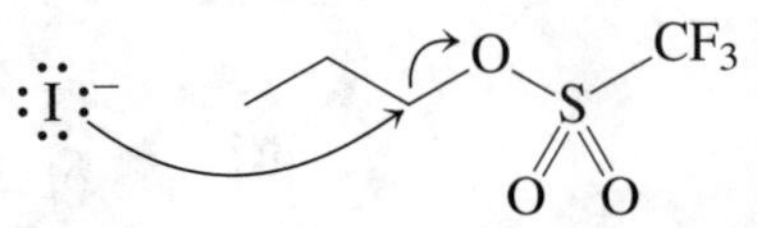 $:\ddot{I}{:}^-$

(d) 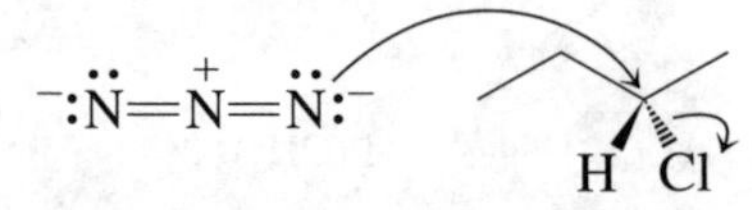 $^-{:}\ddot{N}{=}\overset{+}{N}{=}\ddot{N}{:}^-$

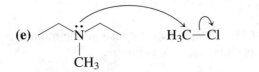

(e) H₃C—Cl   **(f)**   ⁻:SeCN

40. Bimolecular displacement is first order in each component.

(a) Rate = $k[CH_3Cl][KSCN]$: 2×10^{-8} mol $L^{-1}s^{-1} = k(0.1$ M$)(0.1$ M$)$, so $k = 2 \times 10^{-6}$ L mol^{-1}s^{-1}.

(b) The three rates are (i) 4×10^{-8}; (ii) 1.2×10^{-7}; and (iii) 3.2×10^{-7} mol $L^{-1}s^{-1}$, respectively.

41. (a) $CH_3CH_2CH_2I$ (b) $(CH_3)_2CHCH_2CN$ (c) $CH_3OCH(CH_3)_2$

(d) $CH_3CH_2SCH_2CH_3$ (e) ⬠—$CH_2Se(CH_2CH_3)_2{}^+$ Cl^-

(f) $(CH_3)_2CHN(CH_3)_3{}^+$ $^-OSO_2CH_3$

42. (a) Starting material is *R*.

Br—⊢—CH₃   Product is *S*.
   |
H (top), CH₂CH₃ (bottom)

(b) Starting material is (2*S*,3*S*)-2-bromo-3-chlorobutane.

CH₃ ··· / I / CH₃   Product is (2*R*,3*R*)-2,3-diiodobutane.
      I   H

(c) Starting material is (1*S*,3*R*)-3-chlorocyclohexanol (the position of the OH group is understood to be C1).

Product is (1*S*,3*S*)-3-hydroxycyclohexyl acetate.

(d) Starting material is (1*S*,3*S*)-3-chlorocyclohexanol.

Product is (1*R*,3*S*)-3-hydroxycyclohexyl acetate.

Notice that in the product for (c), exchanging the OH and $\overset{\overset{\displaystyle O}{\|}}{O}CCH_3$ groups does not change the molecule. However, in the product for (d), such an exchange turns the molecule into its enantiomer.

43. **(41a)** 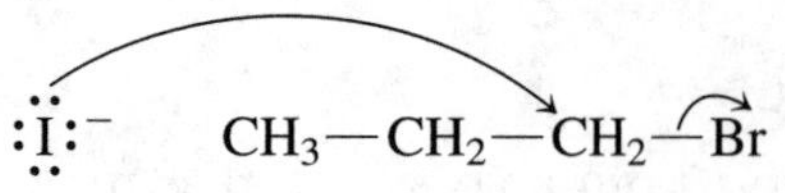 $:\ddot{I}:^-$ $CH_3-CH_2-CH_2-Br$

    (41b) 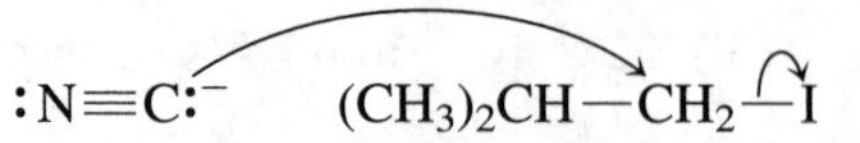 $:N\equiv C:^-$ $(CH_3)_2CH-CH_2-I$

    (41c) 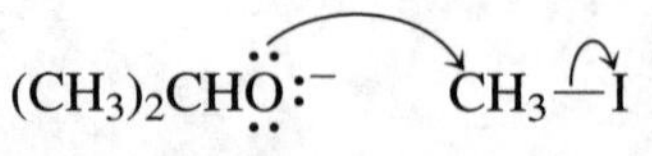 $(CH_3)_2CH\ddot{O}:^-$ CH_3-I

    (41d)  $CH_3CH_2\ddot{S}:^-$ CH_3-CH_2-Br

    (41e) 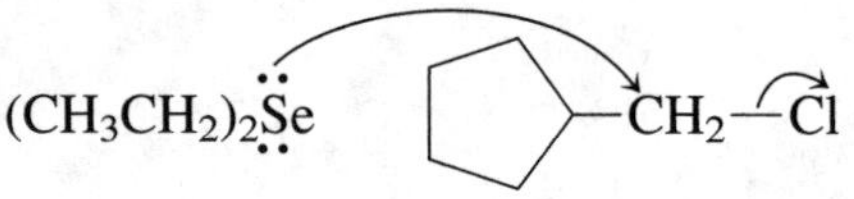 $(CH_3CH_2)_2\ddot{S}e$ $-CH_2-Cl$

    (41f)  $(CH_3)_3N:$ $(CH_3)_2CH-OSO_2CH_3$

    (42a) 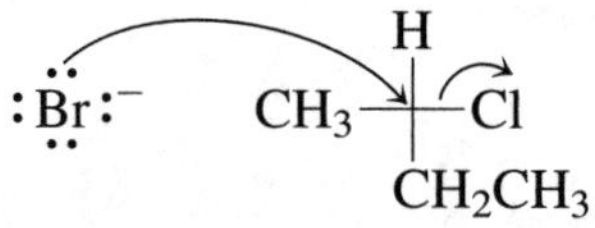 $:\ddot{B}r:^-$ $CH_3-\overset{\overset{\displaystyle H}{|}}{\underset{\underset{\displaystyle CH_2CH_3}{|}}{C}}-Cl$

    (42b) 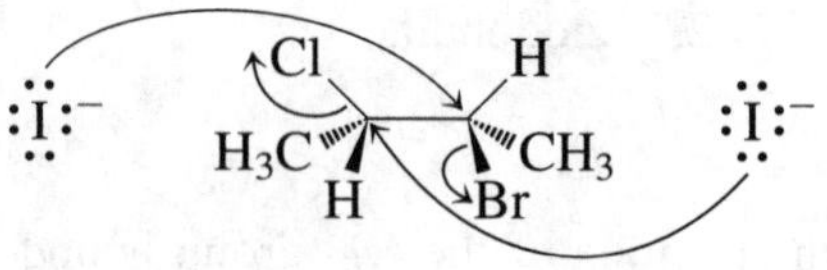 $:\ddot{I}:^-$ $\underset{\underset{\displaystyle H}{}}{H_3C}\overset{\overset{\displaystyle Cl}{}}{C}\overset{\overset{\displaystyle H}{}}{\underset{\underset{\displaystyle Br}{}}{C}}CH_3$ $:\ddot{I}:^-$

    (42c) 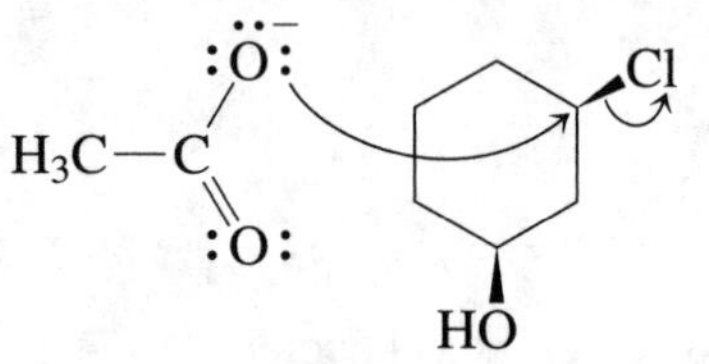 $H_3C-\overset{\overset{\displaystyle :\ddot{O}:^-}{|}}{\underset{\underset{\displaystyle :O:}{\|}}{C}}$ [cyclohexane ring with Cl and HO substituents]

    (42d) 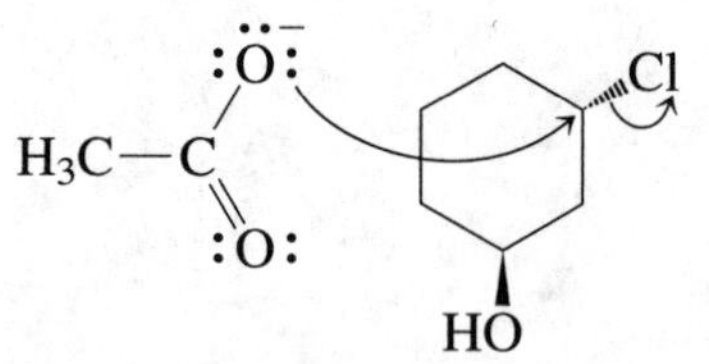 $H_3C-\overset{\overset{\displaystyle :\ddot{O}:^-}{|}}{\underset{\underset{\displaystyle :O:}{\|}}{C}}$ [cyclohexane ring with Cl and HO substituents]

44. **(a)** No reaction, although $CH_3CH_2CH_2OH$ would eventually form (after a few centuries). H_2O is a very poor nucleophile.

    (b) No reaction. H_2SO_4 is a strong acid, a very weak base, and a very, **very** poor nucleophile! Its conjugate base, HSO_4^-, is a very weak nucleophile as well.

    (c) $CH_3CH_2CH_2OH$

    (d) $CH_3CH_2CH_2I$

    (e) $CH_3CH_2CH_2CN$

(f) No reaction. As in (b), HCl is also a very weak base and therefore a dreadful excuse for a nucleophile. Being a strong acid, however, HCl can dissociate in appropriate solvents to release Cl^- ions, which are moderately nucleophilic, so it would not be wrong to answer, "Slow formation of $CH_3CH_2CH_2Cl$."

(g) $CH_3CH_2CH_2S(CH_3)_2{}^+ Br^-$

(h) $CH_3CH_2CH_2NH_3{}^+ Br^-$

(i) No reaction. However, in the presence of heat or light, radical chlorination will occur, giving a mixture of products.

(j) No (or very slow) reaction. F^- is a rather poor nucleophile. (Note: F^- is a much better nucleophile in aprotic solvents.)

45. (a) $CH_3CH_2CH_2CH_2OH$ **(b)** CH_3CH_2Cl **(c)** 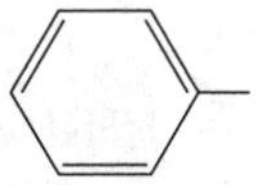 ⬡—$CH_2OCH_2CH_3$

(d) $(CH_3)_2CHCH_2I$ (rather slowly)   **(e)** $CH_3CH_2CH_2SCN$

(f) No reaction. F^- is a very poor leaving group.

(g) No reaction. OH^- is an even worse leaving group.

(h) CH_3SCH_3

(i) No reaction. $^-OCH_2CH_3$ is not a reasonable leaving group.

(j) $CH_3CH_2O\overset{\overset{\displaystyle O}{\|}}{C}CH_3$

The halide ions that are released in reactions (a) through (e), (h), and (j) (Cl^-, Br^-, and I^-) are all good leaving groups.

46. (a) $(R)\text{-}CH_3\overset{\overset{\displaystyle OSO_2CH_3}{|}}{C}HCH_2CH_3 + Na^+N_3{}^- \xrightarrow{CH_3OH} (S)\text{-}CH_3\overset{\overset{\displaystyle N_3}{|}}{C}HCH_2CH_3$

(b) In contrast to (a), where inversion of configuration at the stereocenter was required, you are asked here to substitute Br with CN **with retention.** Because each S_N2 reaction proceeds with inversion, it is necessary to devise a two-inversion scheme made up of two S_N2 reactions to get the proper stereochemical result. The first S_N2 reaction must be done with a nucleophile that is also a good leaving group (I^- fits this description):

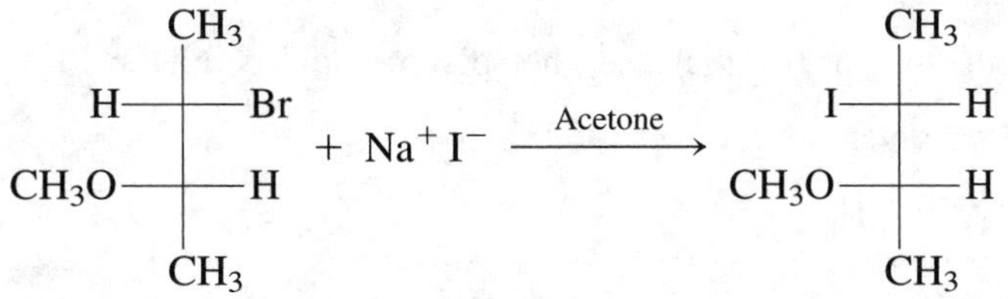

First S_N2 inversion

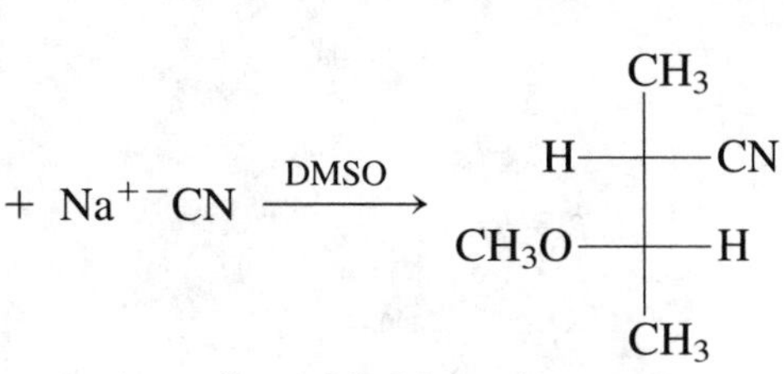

Second S_N2 inversion

(c) Notice that an inversion is required: In the substrate the Br is trans to the ring fusion hydrogens, whereas in the product the SCH_3 group is cis to them. Use one S_N2 reaction:

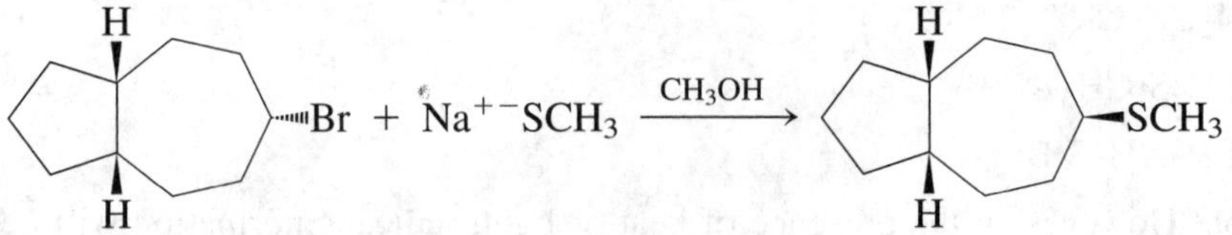

(d) Does this look strange? This shows a *nucleophile,* to which you are asked to attach an alkyl group: an *alkylation* reaction. This is just the same as an S_N2 reaction, but now you need to supply an appropriate *haloalkane substrate* to react with the nucleophile, instead of the other way around:

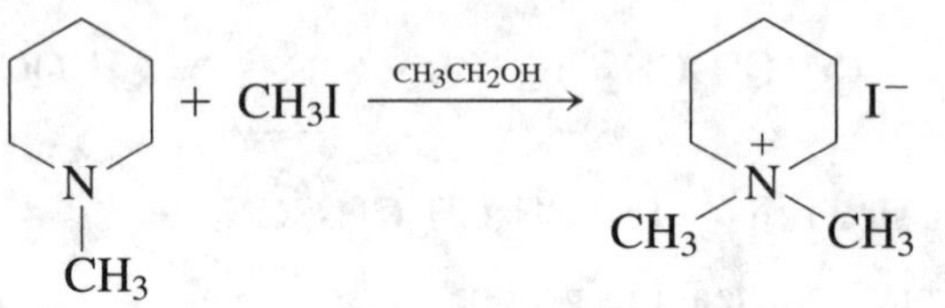

47. (a) (1) $HO^- > CH_3CO_2^- > H_2O$ Basicity increases with charge and decreases with charge stabilization.

(2) $HO^- > CH_3CO_2^- > H_2O$ For a single atom, nucleophilicity parallels basicity.

(3) $H_2O > CH_3CO_2^- > HO^-$ Leaving-group ability is inversely related to basicity.

(b) (1) $F^- > Cl^- > Br^- > I^-$ Larger size stabilizes negative charge, making base weaker.

(2) Depends on the solvent. In polar, aprotic solvent:

$F^- > Cl^- > Br^- > I^-$ Stronger base is better nucleophile.

In polar, protic solvent:

$I^- > Br^- > Cl^- > F^-$ Larger, more polarizable atom is better nucleophile.

(3) $I^- > Br^- > Cl^- > F^-$ Reverse order of (1).

(c) (1) $^-NH_2 > ^-PH_2 > NH_3$ Larger size makes $^-PH_2$ a weaker base than $^-NH_2$; lack of charge makes NH_3 the weakest.

(2) Depends on the solvent. In polar, aprotic solvent:

$NH_2^- > PH_2^- > NH_3$ Stronger base is better nucleophile; lack of charge puts NH_3 last.

In polar, protic solvent (in this case, the solvent could be NH_3):

$PH_2^- > NH_2^- > NH_3$ Size puts PH_2^- first.

(3) $NH_3 > ^-PH_2 > ^-NH_2$ Reverse of (1).

(d) (1) $^-OCN > ^-SCN$ Size (smaller atom is more basic).

(2) Depends on the solvent, again. In polar, aprotic solvent:

$^-OCN > ^-SCN$ Stronger base wins.

In polar, protic solvent:

$^-SCN > ^-OCN$ Size wins.

(3) $^-SCN > ^-OCN$ Reverse of (1).

(e) (1) $HO^- > CH_3S^- > F^-$ HO^- stronger than CH_3S^- because of size, and stronger than F^- because of electronegativity difference. Comparison between CH_3S^- and F^- hard to make as small size favors F^-, whereas lower electronegativity favors CH_3S^-.

    (2) In polar, aprotic solvent:

        $HO^- > CH_3S^- > F^-$ Strongest base.
        In polar, protic solvent:

        $CH_3S^- > HO^- > F^-$ Largest size puts CH_3S^- first; then base strength.

    (3) $F^- > CH_3S^- > HO^-$ They're all bad. Order is reverse of (1).

(f) (1) $NH_3 > H_2O > H_2S$ Electronegativity, then size.

    (2) $H_2S > NH_3 > H_2O$ Size, then base strength.

    (3) $H_2S > H_2O > NH_3$ Reverse of (1).

48. **(a)** No reaction. Starting material is an alkane. Alkanes don't react with nucleophiles.

(b) $CH_3CH_2OCH_3$

(c)

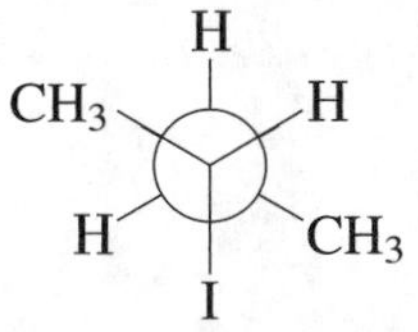

A good conformation of the product after inversion of the reacting carbon. Immediately following S_N2 displacement, the molecule possesses an eclipsed shape. Draw it!

(d)

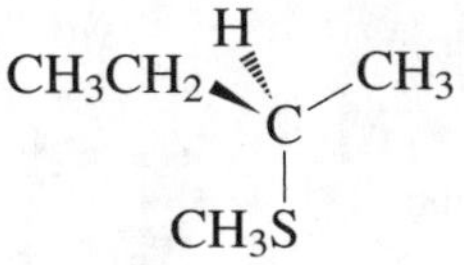

(e) No reaction. The leaving group would be ^-OH, a strong base. Strong bases are very poor leaving groups.

(f) No (or extremely slow) reaction. Now the leaving group is good ($CH_3SO_3^-$), but HCN is a weak acid and therefore a very poor source of CN^- nucleophiles.

(g) Finally, everything is right for a reaction to occur. Both a good leaving group ($CH_3SO_3^-$) and a good nucleophile (CN^-) are present, so the product $(CH_3)_2CHCN$ forms readily.

(h) $(CH_3)_2CHCH_2CH_2SCN$. Leaving group is

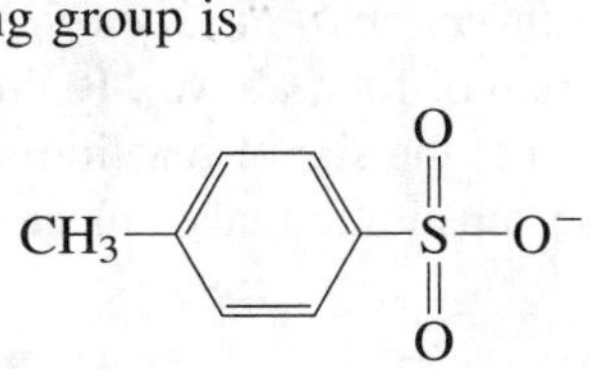

(i) No reaction. NH_2^- is a bad leaving group.

(j) CH_3NH_2

49. If necessary refer to the solution to Problem 39 for guidance.

(b) 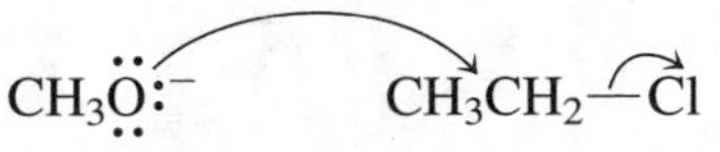   **(c)**

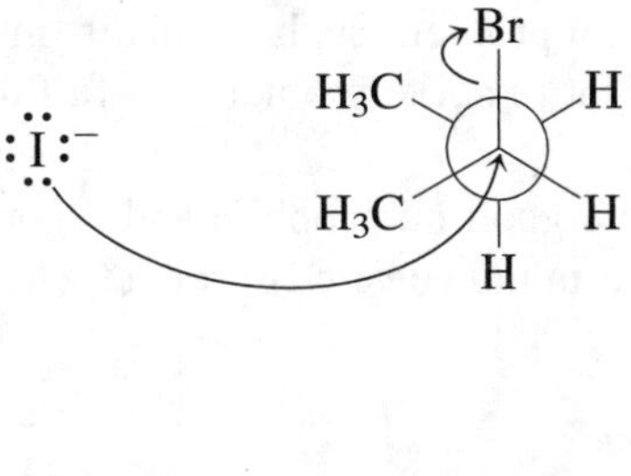

(d) 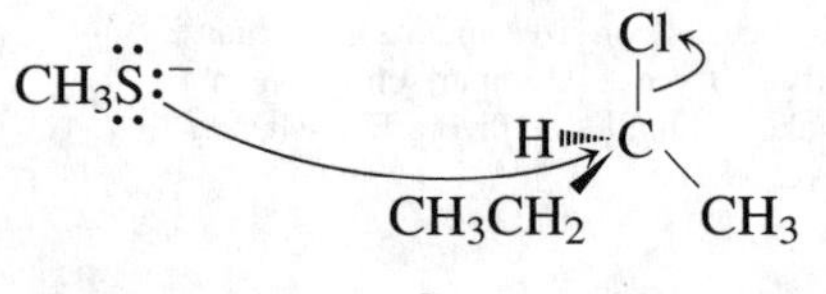 $CH_3\ddot{S}:^-$

(g)

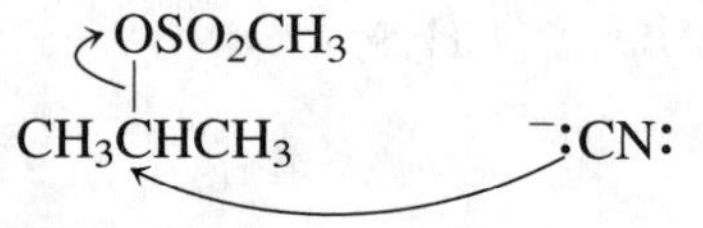

(h) 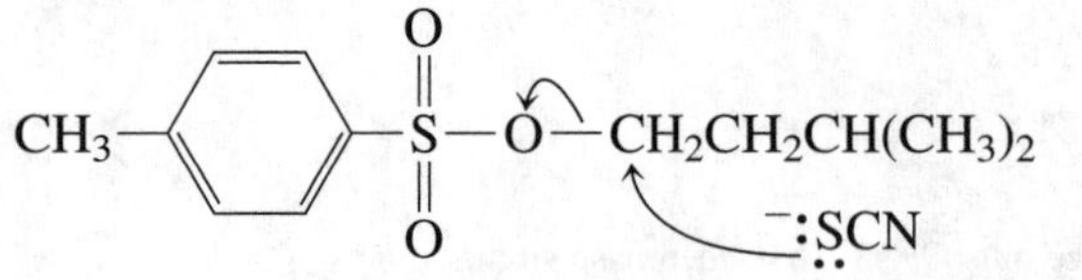 CH_3-

(j)  $H_2\ddot{N}:^-$ H_3C-I

50. (a) BMIM is polar (very! It is an ionic salt!) and aprotic (it lacks hydrogens attached to electronegative atoms), and, therefore, cannot serve as a source of δ^+ hydrogens for hydrogen bonding.

(b) Being a polar, aprotic solvent, BMIM should enhance the rate of any nucleophilic substitution reaction that proceeds via the S_N2 mechanism.

51. The four diastereomers of 2-bromo-3-hydroxy-4-methylpentanoic acid are shown below, with the configuration of each stereocenter identified.

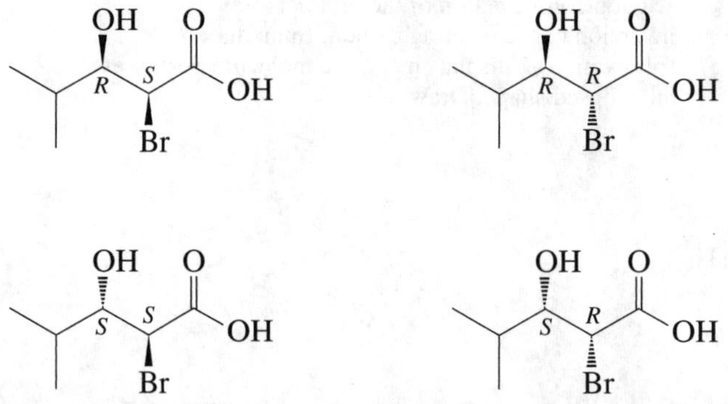

The target molecule is the 2S,3S isomer of 3-hydroxyleucine. To prepare this particular diastereomer from one of the bromo compounds above we need to consider both the reaction that we need to use and its stereochemical consequences. Replacement of bromine by nitrogen is needed; we have been introduced to ammonia, NH_3, as a suitable nitrogen-containing nucleophile. So the necessary reaction is identified. But which stereoisomeric starting material is the best? The mechanism gives us the answer: S_N2, which proceeds with stereochemical inversion. This insight tells us that we must begin with the stereoisomer of the starting compound that will give us the product we need *after stereochemical inversion at the site of reaction*. Therefore, we should choose the 2R,3S isomer (at the lower right from the group of four, above). It already contains the correct configuration at the hydroxy carbon (C3); that stereocenter is not the site of reaction so it remains unchanged. Meanwhile, C2 will undergo inversion from R to S as the substitution takes place, giving the required outcome:

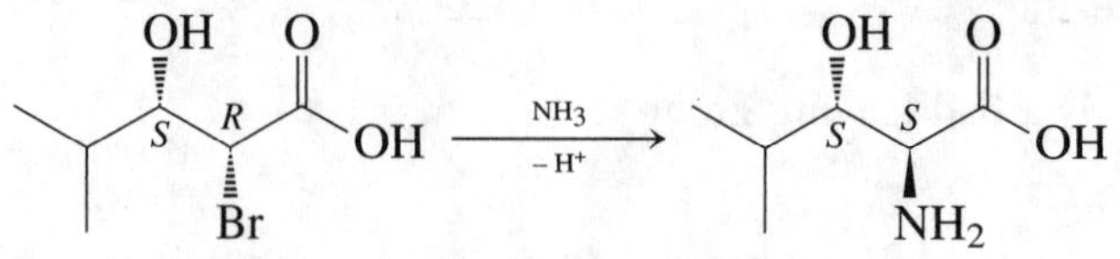

The lesson to take from a problem such as this is one that recurs time and time again: *Always look to the mechanism* for the insight to solve problems in full detail.

52. Iodide ion is both a good nucleophile and a good leaving group. Given the presence of excess iodide ion and lots of time, the initial (desired) product, (R)-2-iodoheptane, can react with iodide via backside displacement *again.* Iodide displaces iodide! The product is the enantiomer (*S*)-2-iodoheptane. This process can occur as long as the reaction mixture, with all its ingredients, is left intact. By the time the reaction is finally stopped by separation of the organic products from iodide ion, all the iodoalkanes have experienced many such displacements. Statistically, about half undergo an even number of iodide substitutions to finish with the *R* stereochemistry (same as they started with), and the other half react an odd number of times, ending up *S*. So a racemic mixture results.

53. (a) $CH_3CH_2CH_3 \xrightarrow{Cl_2,\ 25°C,\ h\nu} CH_3CH_2CH_2Cl + CH_3CHClCH_3$

43%   57%

This is the best way you know, even though a mixture is formed.

(b) Making use of the selectivity of bromination for secondary C—H bonds leads to the best route:

$$CH_3CH_2CH_2 \xrightarrow{Br_2,\ h\nu} CH_3CHBrCH_3 \xrightarrow{KCl,\ DMSO} CH_3CHClCH_3$$

(c) $CH_3CH_2CH_2Cl$ [from (a)] $\xrightarrow{NaBr,\ acetone} CH_3CH_2CH_2Br$

(d) See (b).

(e) $CH_3CH_2CH_2Cl$ [from (a)] $\xrightarrow{NaI,\ acetone} CH_3CH_2CH_2I$

(f) $CH_3CHBrCH_3$ [from (b)] $\xrightarrow{NaI,\ acetone} CH_3CHICH_3$

54. Don't forget: Each S_N2 reaction **inverts** the stereochemistry at the site of reaction.

(a) $\xrightarrow[\text{Acetone}]{CH_3S^-\ Na^+}$ Trans product from one S_N2 inversion

(b) The starting material is already trans. Direct S_N2 reaction with CH_3S^- will give a cis product, which you don't want. Instead, plan out a synthesis involving **two successive** S_N2 inversions: Go from trans to cis first, and then back to trans. The first S_N2 reaction should use a nucleophile that can later function as a leaving group, such as Br^-; CH_3S^- can then be the nucleophile in the second S_N2 step:

$\xrightarrow[\text{DMSO}]{KBr}$ $\xrightarrow[\text{Acetone}]{CH_3S^-\ Na^+}$ Trans product

55. In each pair that follows indicate the one that would be better suited for S_N2 reaction.

(a) PH_3 P and N are in the same group of the periodic table. N, being above P, is more basic, but P, the larger atom, is more polarizable. Polarizability normally wins out in comparing neutral nucleophiles, regardless of solvent, because neutral nucleophiles interact only weakly with solvent molecules. (With negatively charged nucleophiles the solvent plays more of a role. In H-bonding solvents, the larger, more polarizable nucleophile is the more nucleophilic. In polar aprotic solvents, the more basic wins.)

(b) 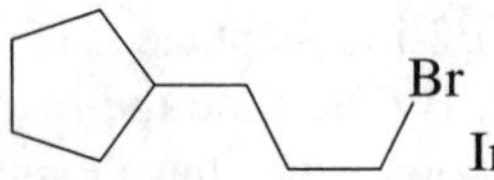 [structure: cyclopentane with CH₂CH₂Br chain] In this substrate the carbon bearing the leaving group—the site of attack—is farther from the nearest substituted (branched) carbon and therefore in a less hindered environment.

(c) 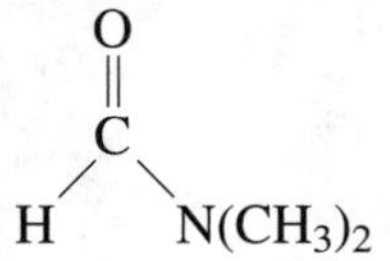 [structure: H—C(=O)—N(CH₃)₂] It is polar and aprotic. [structure: H—C(=O)—NH₂] contains NH bonds and is protic.

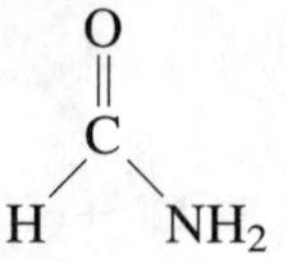

(d) CH_3SH is the weaker base (S below O in the periodic table) and the better leaving group.

56. (a) $CH_3Br > CH_3CH_2Br > (CH_3)_2CHBr$

(b) $(CH_3)_2CHCH_2CH_2Cl > (CH_3)_2CHCH_2Cl > (CH_3)_2CHCl$

(c) $CH_3CH_2I > CH_3CH_2Cl >$ [cyclohexyl]—Cl

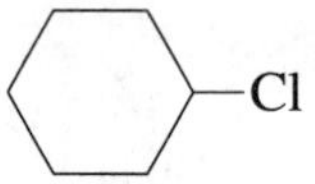

(d) $(CH_3)_2CHCH_2Br > (CH_3CH_2)_2CHCH_2Br > CH_3CH_2CH_2CHBrCH_3$

57. (a) I⁻ is a better leaving group, so the reaction will be faster.

(b) ⁻SCH₃ is a better nucleophile, so the reaction will be faster.

(c) The back side of reacting carbon is strongly sterically hindered to attack, so the reaction will be slowed down greatly.

(d) An aprotic solvent will not hydrogen bond to the nucleophile, so the reaction will be greatly speeded up.

58. The nucleophilicity of the three unsolvated anions is reflected by the rate constants in DMF: Cl⁻ > Br⁻ = ⁻SeCN. This order reflects the higher basicity for Cl⁻. Hydrogen bonding to CH_3OH reduces the reactivity of all three differently. Cl⁻, the smallest ion, is solvated most and becomes the poorest nucleophile in methanol. Solvation is a little less for Br⁻ and much less for ⁻SeCN, as a result of the delocalized charge in the latter.

59. (a) 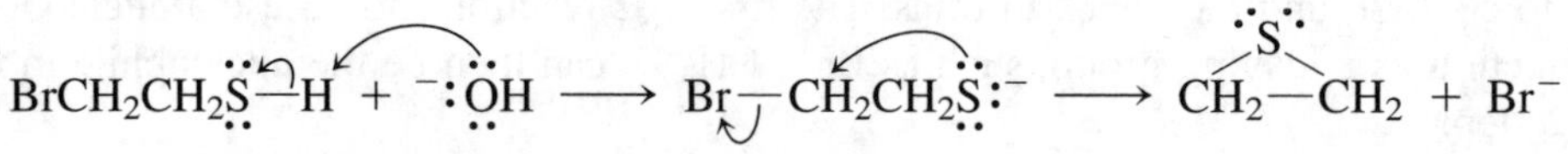 $BrCH_2CH_2\overset{..}{\underset{..}{S}}$—H + ⁻:ÖH ⟶ Br—$CH_2CH_2\overset{..}{\underset{..}{S}}$:⁻ ⟶ CH_2—CH_2 + Br⁻ [with S bridging]

(b) 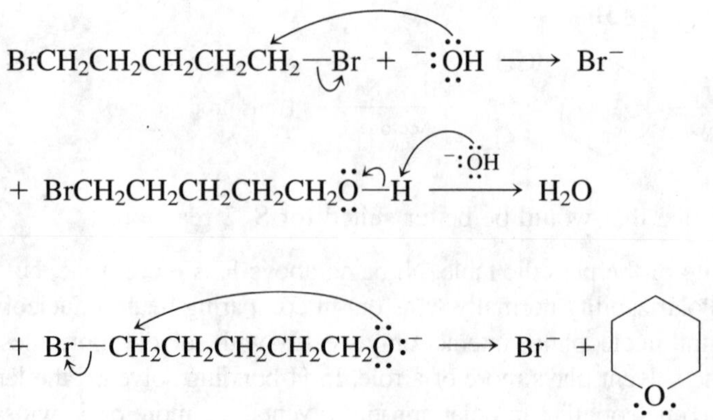 $BrCH_2CH_2CH_2CH_2CH_2$—Br + ⁻:ÖH ⟶ Br⁻

+ $BrCH_2CH_2CH_2CH_2CH_2\overset{..}{O}$—H $\xrightarrow{\text{⁻:ÖH}}$ H_2O

+ Br—$CH_2CH_2CH_2CH_2CH_2\overset{..}{\underset{..}{O}}$:⁻ ⟶ Br⁻ + [tetrahydropyran ring with O]

(c) $BrCH_2CH_2CH_2CH_2CH_2 \overset{\curvearrowright}{-} Br + \overset{..}{N}H_3 \longrightarrow Br^-$

$+ BrCH_2CH_2CH_2CH_2CH_2 \overset{\overset{\overset{H}{\overset{..}{C|}}}{N}H_2}{}{}^+ \;\; \overset{:NH_3}{\rightleftharpoons} \;\; NH_4{}^+$

$+ \overset{\curvearrowright}{Br} \overset{\curvearrowleft}{-} CH_2CH_2CH_2CH_2CH_2\overset{..}{N}H_2 \longrightarrow$

60. In an acyclic (noncyclic) haloalkane the hybridization around the halogen-bearing carbon is sp^3, and the bond angles are about 109°. In the transition state for an S_N2 reaction this carbon has rehybridized to sp^2, and the bond angles involving the three nonreacting groups around this carbon have **expanded** to 120°:

Substrate:
RCR ∠s ≈ 109.5° →

S_N2 transition state:
← RCR ∠s **open** to ≈ 120°

Now take a look at a hypothetical S_N2 reaction on a halocyclopropane (below). The substrate, with ring bond angles of 60°, is already very strained because its carbons would of course prefer to have bond angles of 109°, corresponding to sp^3 hybridization. The amount of strain in the substrate is related to this 109 − 60 = 49° bond angle compression. However, bond angle compression in the S_N2 transition state would be **even worse:** It wants its bond angles to open up even more, to 120°! With a bond angle compression of 120 − 60 = 60°, the S_N2 transition state for a halocyclopropane is much more strained than the halocyclopropane itself, leading to a much higher activation barrier (E_a). A similar analysis fits for halocyclobutanes.

Substrate strain:
∝109 − 60 = 49°
angle compression

→

Transition state strain:
∝120 − 60 = 60°
angle compression
∴ higher activation energy

61. Consider steric hindrance to S_N2 reaction in both of the following situations: (a) Nu⁻ attacks with the leaving group equatorial, and (b) Nu⁻ attacks with the leaving group axial.

(a) Leaving group equatorial. The transition state looks like this:

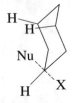

Back side approach of Nu⁻ is impeded by the axial hydrogens on the same side of the ring. **Each** causes steric hindrance similar to that shown in the attack of a nucleophile on a halopropane in an *anti* conformation (Figure 6-9C).

(b) Leaving group axial. Now we have the following:

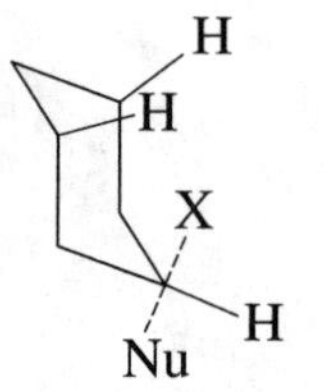

Steric hindrance to approach of Nu⁻ is reduced now: This is like Figure 6-9D. But there is a price to pay. First, X is axial, which costs energy. Second **departure** of X⁻ is now impeded by the axial hydrogens on **its** side of the ring. So this is still a slow reaction.

62. For the answers to the first part of the problem see the answers to Problems 34 and 35. The eight constitutional isomers are listed below by name, classification, and presence or absence of chirality.

Name	Type	Chiral?	Relative S_N2 reactivity
1-Bromopentane	primary	no	A
2-Bromopentane	secondary	yes	D
3-Bromopentane	secondary	no	E
1-Bromo-3-methylbutane	primary	no	B
2-Bromo-3-methylbutane	secondary	yes	F
2-Bromo-2-methylbutane	tertiary	no	H
1-Bromo-2-methylbutane	branched primary	yes	C
1-Bromo-2,2-dimethylpropane	neopentyl	no	G

We can fill in the relative reactivity column by using the information in Section 6-9, together with the hints in the problem. The two unbranched primary haloalkanes should be most reactive (A and B), with the straight-chain isomer slightly superior. The branched primary should be next (C), and this conclusion is supported by the information that C is indeed chiral. Following these should be the two straight-chain secondary halides, of which one is chiral. The branched secondary (F) comes next, followed distantly by the neopentyl (G). Finally, the tertiary (H) is unreactive. The final pieces of information confirm the assignments: The stereocenters in substrates D and F are the carbons bearing the leaving groups; therefore, they undergo inversion of configuration upon S_N2 reaction. In contrast, the stereocenter is unaffected in the substitution reaction of C, *because its stereocenter is not the site of the reaction.*

63. (b)

64. (a)

65. (d)

66. (b)

7

Further Reactions of Haloalkanes: Unimolecular Substitution and Pathways of Elimination

In Chapter 7 we continue and complete a discussion of major reaction types and mechanisms for haloalkanes. Three new processes are discussed. Pay close attention to how the mechanisms of each make electrostatic sense: Just like the S_N2 mechanism in the last chapter, each of these new processes provides a means for an electron pair to move toward the electrophilic carbon, forming a new bond. As we have emphasized before, conceptual understanding is an important step to comprehension of the material.

Of more practical importance and a focal point of the chapter is the fact that similar compounds can undergo several different types of reactions. A goal of both the text and this guide is to show you how to choose which process is most likely to occur under any given set of conditions based on a mechanistic understanding of the reactions. You are asked to apply your knowledge logically and to pay attention to detail—to reason like a scientist.

Outline of the Chapter

7-1 Solvolysis of Tertiary and Secondary Haloalkanes
A surprising reaction of compounds that do not undergo the S_N2 process.

7-2, 7-3, and 7-4 Mechanism of Solvolysis: Unimolecular Nucleophilic Substitution
The explanation, with details and consequences.

7-5 Substrate Structure: The Stability of Carbocations
A new kind of reactive intermediate in organic chemistry. Also, a summary of nucleophilic displacement reactions.

7-6 Unimolecular Elimination: E1

7-7 Bimolecular Elimination: E2
Two mechanisms for a new reaction of haloalkanes.

7-8 Substitution or Elimination?
Factors that allow prediction of favored reaction pathways.

Keys to the Chapter

7-1 through 7-4. Solvolysis: Unimolecular Nucleophilic Substitution
Tertiary haloalkanes, although they do not react via the S_N2 mechanism (Chapter 6), still undergo very rapid nucleophilic displacement under certain reaction conditions. This reaction is due to the appearance of a

completely new mechanism for displacement for which **tertiary** halides are the best suited substrate molecules: unimolecular nucleophilic substitution, or the S_N1 mechanism. Sections 7-2 and 7-3 present the kinetic and stereochemical details of experiments that led to the formulation of this mechanism. Note that this is typically a two- or three-step reaction, in contrast to the single-step S_N2 process. It requires a rate-determining *ionization* of the carbon–halogen (or carbon–leaving group) bond, leading to a new reactive species called a *carbocation* (pronounced car-bo-cat'-ion). You may occasionally encounter the term *carbonium ion,* an older name for these species. Another name currently in use is *carbenium ion.* Note carefully how the various features of this reaction as described in this section (e.g., rate, stereochemistry, sensitivity to solvent polarity and leaving group) are closely and logically derived from the nature of the rate-determining first step.

7-5. Substrate Structure: The Stability of Carbocations

The chief requirement for an S_N1 mechanism is ease of ionization of the bond between the carbon atom and the leaving group. Because the initial result of this is formation of a positively charged (cationic) carbon species (carbocation), it is logical that the ease of the S_N1 mechanism will reflect the ease of generation of the corresponding carbocation. Carbon is not a very electropositive atom, and a carbon with a full positive charge is, in general, not very stable, and, therefore, difficult to generate. As explained here, however, alkyl groups (as opposed to hydrogen atoms) stabilize cationic carbon centers. Thus, cations in which the positive charge is on a **tertiary** carbon atom will be the most stable and the easiest to generate, because **three** alkyl groups are present to help alleviate the electron deficiency of the cationic carbon. Therefore, the reactivity of tertiary substrates in S_N1 reactions boils down to the ease of their ionization to form the **relatively** stable tertiary carbocation intermediate. Table 7-2 summarizes the distinct modes of substitution for methyl and primary halides (S_N2 only) vs. tertiary halides (S_N1 only). For secondary substrates the behavior is more complicated. Depending on the specific situation, either S_N1 or S_N2 reaction, or both, may occur.

Prediction of the pathway a secondary substrate will follow is another direct application of the logical consequences of the two competing mechanisms. Read the last portion of Section 7-5 particularly carefully, because it presents a classic example of the use of mechanistic information to explain (and predict) the effects of reaction variables on possible reaction pathways.

7-6 and 7-7. Elimination Reactions

The positively charged carbon atom in a carbocation is an extremely electron-deficient (electrophilic) carbon. As such, its behavior is dominated by a need to obtain an electron pair from any available source. The S_N1 reaction illustrates the most obvious fate of a carbocation: combination with an external Lewis base, forming a new bond to carbon. However, the electron deficiency of cationic carbon is so great that even under typical S_N1 solvolysis conditions, surrounded by nucleophilic solvent molecules, some of the cations won't wait to combine with external electron-pair sources. Instead, they will seek available electron pairs within their own molecular structures. The most available of these are electrons in carbon–hydrogen bonds one carbon removed from the cationic center (at the so-called β carbon):

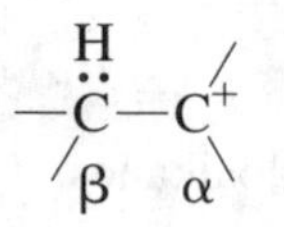

Attraction of this electron pair toward the positively charged carbon leads to two products: an alkene and a proton, the result of E1 elimination. As mentioned in the text, the proton doesn't just "fall off." Actually, it is removed by any Lewis base available in the reaction system, such as solvent or other nucleophile molecules.

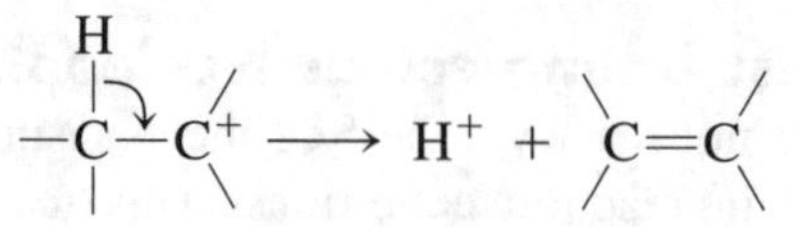

Note that only the β carbon–hydrogen bond (the one next to the positive carbon) is susceptible to cleavage in this manner. Other C—H bonds would not give such stable products if they were to be broken:

α C—H bond: $H-C^+ \longrightarrow H^+ + :C$

A carbene

(Very unstable)

γ C—H bond: $-C-C-C^+ \longrightarrow H^+ + $ a three-membered ring

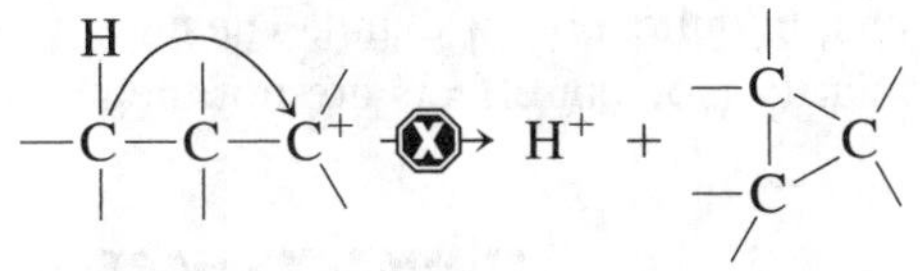

A cyclopropane

(Strained)

Both of these are very uncommon processes. Their occurrence is limited to situations where normal β elimination cannot take place.

Because the E1 process involves the same rate-determining step as the S_N1 reaction, its kinetics are the same: first order. E1 elimination almost always accompanies S_N1 substitution. The difference is simple: In S_N1, the nucleophile attaches to the cationic carbon; in E1, it attaches to and removes a proton. For the practical purposes of synthesis, the presence of the E1 "side reaction" can limit the usefulness of S_N1 substitution.

If, however, elimination is desired, it may be achieved by addition of strong base to tertiary halides. At high concentrations, in fact, a second elimination mechanism with second-order kinetic behavior occurs (E2 reaction). Section 7-7 describes its details pictorially. As in the E1 process, the electrons in a β C—H bond move toward the electrophilic carbon; in the E2 reaction, however, this electron movement occurs **simultaneously** with the loss of the leaving group. Examine Figure 7-8 closely. Note the electron motion toward the chlorine-bearing carbon: It is actually quite similar to the electron motion of an S_N2 process! The tertiary halide cannot undergo S_N2 displacement, but the E2 mechanism is a way for it to move electrons in a similar manner, getting them from a β C—H bond instead of from an external nucleophile.

As the rest of this section shows, **any** haloalkane with a β C—H bond can undergo E2 elimination. In the case of 1° and 2° halides, E2 and S_N2 reactions compete. However, as the following sections in the text and below will show, it is very easy to predict the favored products in these cases.

7-8. Substitution vs. Elimination: General Guidelines for Prediction

The key consideration for synthetic purposes is elimination versus substitution, and, as the text shows, the preference in most cases can be determined by answering the following three questions:

1. Is the nucleophile a strong base?
2. Is the nucleophile sterically very bulky?
3. Is the substrate sterically hindered (i.e., 3°, 2°, or 1° with branching)?

If the answer to at least two out of these three questions is "yes," elimination will be favored. Otherwise substitution will predominate. Check the reactions in the chapter to see how these guidelines work. Here is one more for practice.

$$CH_3CH_2CH_2CH_2I + Na^+ \ ^-NH_2 \xrightarrow{\text{Liquid } NH_3} ?$$

Substitution or elimination? Analysis:

Factor	Favors
SUBSTRATE is primary: unhindered	Substitution
NUCLEOPHILE is $^-NH_2$: strongly basic	Elimination
NUCLEOPHILE is $^-NH_2$: sterically unhindered	Substitution

Result: Substitution, to form $CH_3CH_2CH_2CH_2NH_2$.

The text summarizes the preferences for E1, E2, S_N1, and S_N2 reactions for 1°, 2°, and 3° haloalkanes, as a function of reaction conditions, in quite a bit of detail. The chart that follows repeats the same material, again somewhat oversimplified for clarity (solvent effects are not included, for instance).

SUMMARY CHART
Major reactions of haloalkanes with nucleophiles

Type of halide	Poor nucleophile like H_2O	Weakly basic good nucleophile like I^-	Strongly basic unhindered nucleophile like CH_3O^-	Strongly basic hindered nucleophile like $(CH_3)_3CO^-$
Methyl	No reaction	S_N2	S_N2	S_N2
1°	No reaction	S_N2	S_N2	E2
2°	Slow S_N1	S_N2	E2	E2
3°	S_N1 and E1	S_N1 and E1	E2	E2

Type of nucleophile (spanning the four right columns)

Solutions to Problems

25. (a) $(CH_3)_3COCH_2CH_3$ **(b)** $(CH_3)_2\overset{\displaystyle OCH_2CF_3}{\underset{\displaystyle |}{C}}CH_2CH_3$

(c)

$CH_3CH_2\ \ OCH_3$

(cyclopentane ring with substituents)

(d)

(cyclohexyl)$-\overset{\displaystyle CH_3}{\underset{\displaystyle CH_3}{\overset{\displaystyle |}{\underset{\displaystyle |}{C}}}}-\overset{\displaystyle O}{\overset{\displaystyle \|}{O}CH}$

(e) $(CH_3)_3COD$ **(f)** $(CH_3)_3C-O-$(cyclohexyl with H)

26. (a) For the answer to this part we show each step on a separate line.

$$CH_3-\overset{\displaystyle CH_3}{\underset{\displaystyle CH_3}{\overset{\displaystyle |}{\underset{\displaystyle |}{C}}}}-Br \longrightarrow CH_3-\overset{\displaystyle CH_3}{\underset{\displaystyle CH_3}{\overset{\displaystyle |}{\underset{\displaystyle |}{C^+}}}} + Br^-$$

$$CH_3-\overset{\overset{\displaystyle CH_3}{|}}{\underset{\underset{\displaystyle CH_3}{|}}{C}}{}^+ \quad CH_3CH_2\ddot{O}H \longrightarrow CH_3-\overset{\overset{\displaystyle CH_3}{|}}{\underset{\underset{\displaystyle CH_3}{|}}{C}}-\overset{+}{\underset{\displaystyle CH_2CH_3}{O}}\overset{\displaystyle H}{}$$

$$CH_3-\overset{\overset{\displaystyle CH_3}{|}}{\underset{\underset{\displaystyle CH_3}{|}}{C}}-\overset{+}{\underset{\displaystyle CH_2CH_3}{\ddot{O}}}\overset{\displaystyle H}{} \longrightarrow CH_3-\overset{\overset{\displaystyle CH_3}{|}}{\underset{\underset{\displaystyle CH_3}{|}}{C}}-\ddot{O}-CH_2CH_3 + H^+$$

In the last step of problems like this students frequently make the mistake of removing an alkyl group from the oxygen, and giving an alcohol as the final reaction product. That process does not happen in solvolyses in alcohol solvents: Loss of a proton is far more favorable than loss of an unstable alkyl cation (which would be a primary ethyl cation in the example above).

(b) For the remaining parts, we still show each step separately, but we connect the steps in a continuous sequence. The full structure of each intermediate is still shown.

$$CH_3-\overset{\overset{\displaystyle \curvearrowleft Br}{|}}{\underset{\underset{\displaystyle CH_3}{|}}{C}}-CH_2CH_3 \xrightarrow{-Br^-} CH_3-\overset{\overset{\displaystyle}{|}}{\underset{\underset{\displaystyle CH_3}{|}}{\overset{+}{C}}}-CH_2CH_3 \xrightarrow{CF_3CH_2-\ddot{O}-H}$$

$$CH_3-\overset{\overset{\displaystyle \overset{H}{\underset{\displaystyle :\ddot{O}+}{\diagdown}}\,CH_2CF_3}{}}{\underset{\underset{\displaystyle CH_3}{|}}{C}}-CH_2CH_3 \xrightarrow{-H^+} CH_3-\overset{\overset{\displaystyle OCH_2CF_3}{|}}{\underset{\underset{\displaystyle CH_3}{|}}{C}}-CH_2CH_3$$

(c)

(cyclopentane ring with Cl and CH₂CH₃) $\xrightarrow{-Cl^-}$ (cyclopentane ring with CH₂CH₃ and +) $\xrightarrow{CH_3-\ddot{O}-H}$ (cyclopentane ring with $CH_3-\overset{+}{\underset{\displaystyle\ddot{}}{O}}$, H, CH₂CH₃) $\xrightarrow{-H^+}$ (cyclopentane ring with $CH_3-\ddot{O}:$ CH₂CH₃)

(d)

(cyclohexane with $\overset{\curvearrowleft Br}{\underset{\underset{\displaystyle CH_3}{|}}{C}}-CH_3$) $\xrightarrow{-Br^-}$ (cyclohexane with $\overset{+}{\underset{\underset{\displaystyle CH_3}{|}}{C}}-CH_3$) $\xrightarrow{\overset{:\ddot{O}:}{\underset{HO:}{\diagup}}C-H}$

(cyclohexane with $\overset{\overset{\displaystyle H\diagdown C\overset{\ddot{O}}{\diagup}H}{\underset{\displaystyle :\overset{+}{O}}{\|}}}{\underset{\underset{\displaystyle CH_3}{|}}{C}}-CH_3$) $\xrightarrow{-H^+}$ (cyclohexane with $\overset{\overset{\displaystyle H\diagdown C=\ddot{O}}{}}{\underset{\underset{\displaystyle CH_3}{|}}{\underset{\displaystyle :O:}{C}}}-CH_3$)

OK, you're asking, why did that oxygen attach to the carbocation instead of the other one? Makes it look a lot more complicated, too. The reason is that the species you get is resonance stabilized (see below). When we cover carboxylic acids a dozen chapters from now you'll see more of this. Sorry—reality is reality.

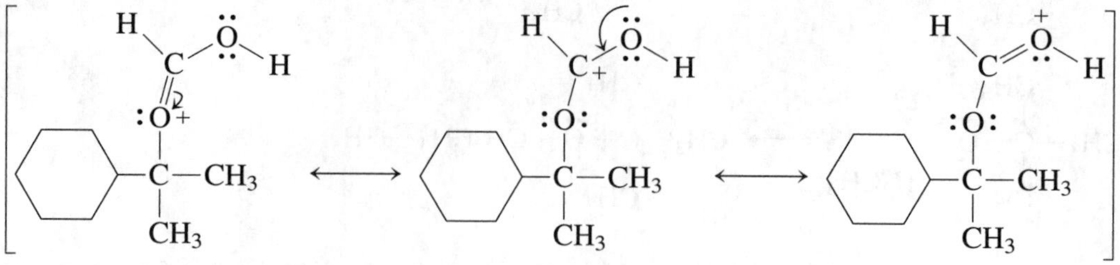

(e)

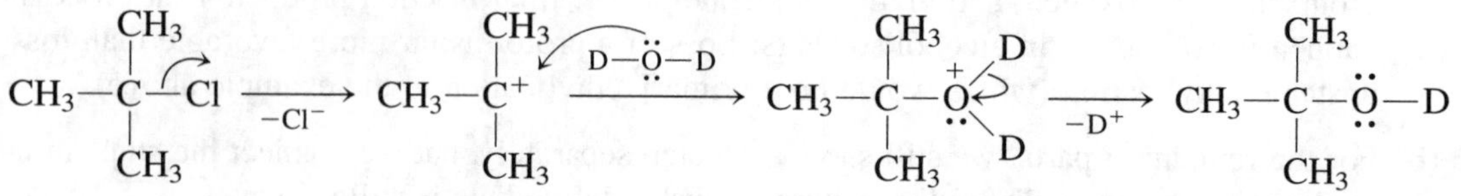

(f)

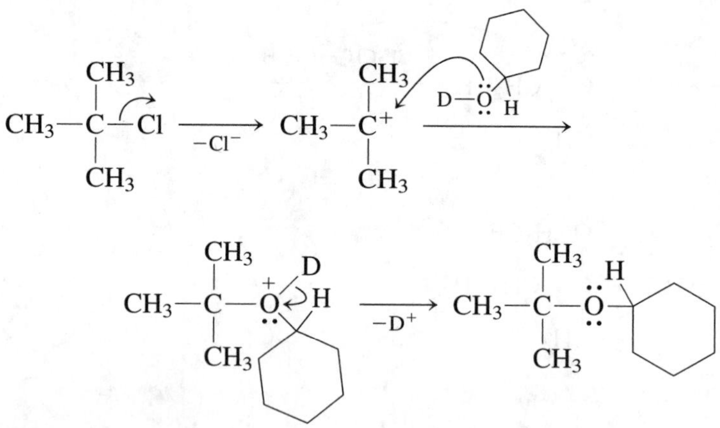

27.

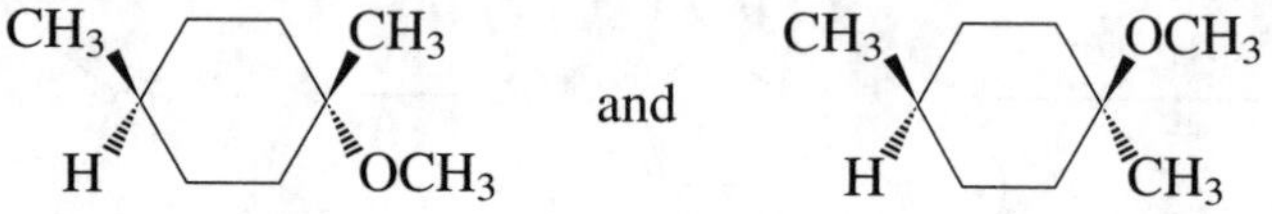

and

(a) Two steps:

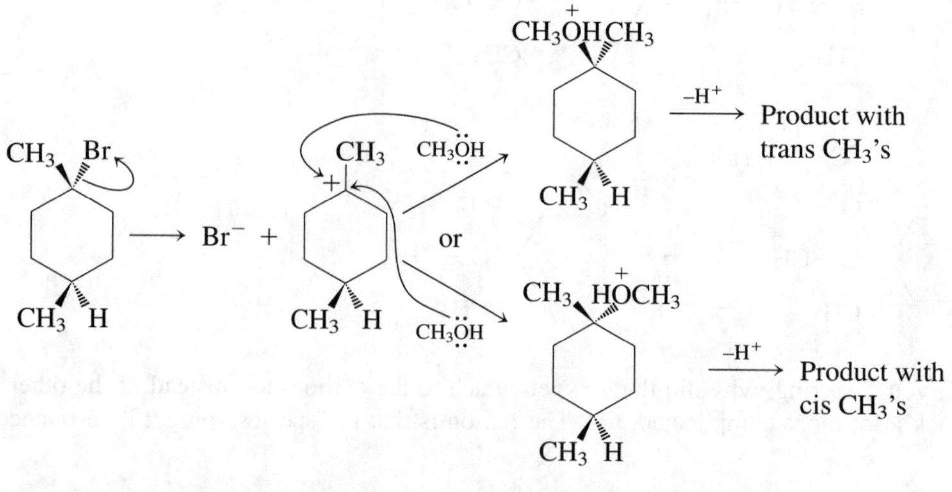

The nucleophile can attach to either face of the planar cationic carbon, reactions yielding the two products shown.

(b) 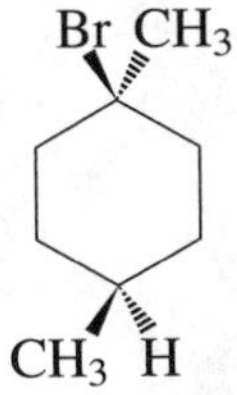 By reattachment of Br^- to the opposite side of cationic carbon. (Reversal of the dissociation step.)

28. Two products: 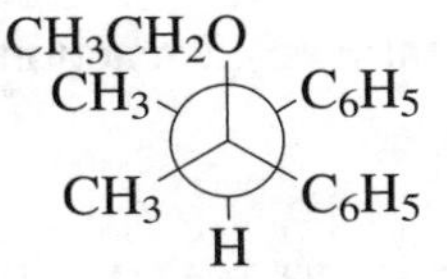 and

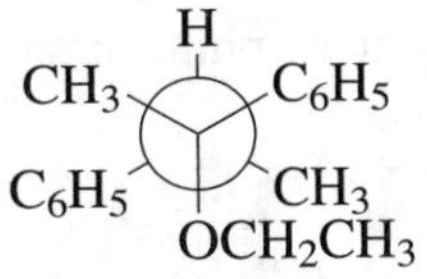

From attachment of nucleophile to either face of the planar cation.

29. (a) H_2O will speed up all of the reactions except for 25(d), because it is more polar than any of the other solvolysis solvents. It will also compete for the carbocations, forming alcohols as products.

(b) Ionic salts strongly increase polarity and accelerate S_N1 reactions (see Problem 51, however). The main products will be iodoalkanes.

(c) Same as (b); azide ion is a strong nucleophile and products will be azidoalkanes (alkyl azides, $R-N_3$).

(d) This solvent should reduce polarity and slow down all the solvolyses.

30.

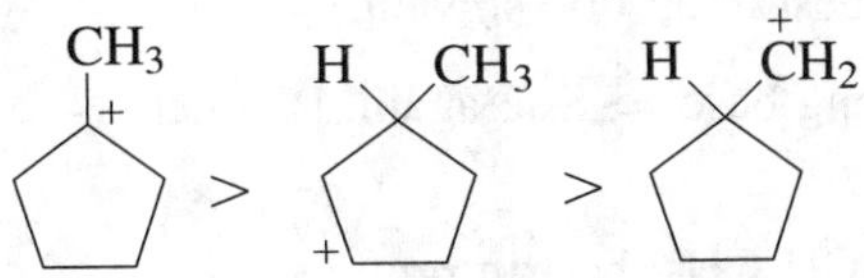

(Tertiary) (Secondary) (Primary)

31. (a) $(CH_3)_2CClCH_2CH_3 > (CH_3)_2CHCHClCH_3 > (CH_3)_2CHCH_2CH_2Cl$ ($3° > 2° > 1°$)

(b) $RCl > ROCCH_3 > ROH$ (Order of leaving group ability)

(c)

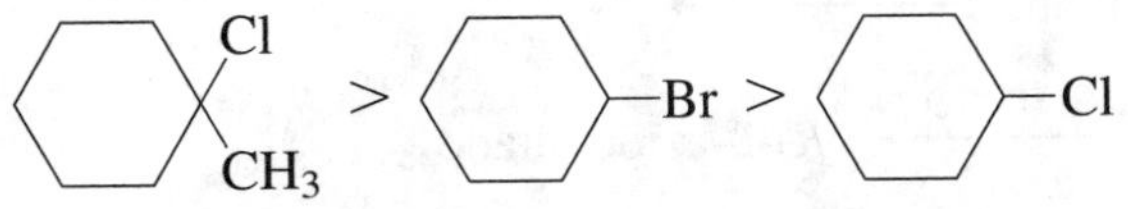

32. (a) A secondary system with an excellent leaving group and a poor nucleophile $\Rightarrow S_N1$ reaction.

$$(CH_3)_2CH-OSO_2CF_3 \longrightarrow {}^-OSO_2CF_3 + (CH_3)_2\overset{+}{CH} \xrightarrow{\quad CH_3CH_2-\overset{..}{O}H \quad}$$

$$(CH_3)_2CH\overset{+}{O}CH_2CH_3 \longrightarrow H^+ + (CH_3)_2CHOCH_2CH_3$$

(b) A tertiary halide in a polar solvent $\Rightarrow$ S$_N$1 reaction

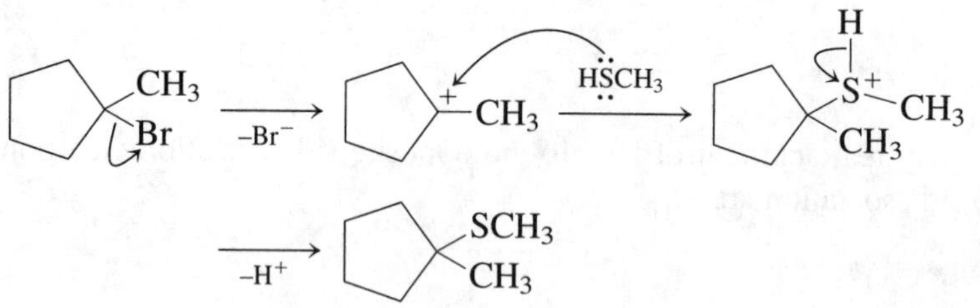

(c) A primary halide with a good nucleophile in an aprotic solvent $\Rightarrow$ S$_N$2 reaction.

$$CH_3CH_2CH_2CH_2{-}Br \ + \ (C_6H_5)_3\ddot{P} \longrightarrow CH_3CH_2CH_2CH_2\overset{+}{P}(C_6H_5)_3 \ Br^-$$

(d) Similar to (c), except a secondary halide $\Rightarrow$ still an S$_N$2 reaction.

$$CH_3CH_2\overset{Cl}{\underset{|}{C}H}CH_2CH_3 \ + \ I^- \longrightarrow CH_3CH_2CHICH_2CH_3$$

33. First decide what the most likely mechanism is for each reaction. Then write the product. Finally, recall that S$_N$2 reactions are faster in polar aprotic solvents, whereas S$_N$1 reactions are faster in polar protic solvents because of the greater stabilization of the transition state for cation–anion dissociation.

  (a) Primary substrate $\Rightarrow$ S$_N$2 to give $CH_3CH_2CH_2CH_2CN$; best in aprotic solvent.

  (b) Branched, but still primary, and nucleophile is not a strong base $\Rightarrow$ S$_N$2 again. Product is $(CH_3)_2CHCH_2N_3$; aprotic solvent is again best.

  (c) Tertiary substrate $\Rightarrow$ S$_N$1 substitution to form $(CH_3)_3CSCH_2CH_3$; best in protic solvent.

  (d) Secondary substrate with an excellent leaving group and a weak nucleophile $\Rightarrow$ S$_N$1 is the most likely mechanism from this combination, forming $(CH_3)_2CHOCH(CH_3)_2$; fastest in protic solvent.

34. **Two successive** S$_N$2 inversion steps are necessary to give the desired net result of stereochemical retention:

$$(R)\text{-2-chlorobutane} \xrightarrow{\text{KBr, DMSO}}$$

$$(S)\text{-2-bromobutane} \xrightarrow{\text{NaN}_3, \text{ DMSO}} (R)\text{-2-azidobutane}$$

35. (1) Racemic $CH_3CH_2CH(O\overset{\displaystyle O}{\overset{\|}{C}}H)CH_3$ will be formed via an S$_N$1 process (a solvolysis). The solvent (a carboxylic acid) is very polar and protic, but a weak nucleophile.

  (2) $(R)\text{-}CH_3CH_2CH(O\overset{\displaystyle O}{\overset{\|}{C}}H)CH_3$ will be formed via an S$_N$2 process (good nucleophile, aprotic solvent). Note the very different conditions.

36. To form the desired product requires stereochemical inversion. The substrate is secondary, so some care should be taken to ensure that the S_N2 mechanism is favored. Use a good nucleophile such as acetate (as opposed to acetic acid, which would be a poor nucleophile) in a polar aprotic solvent like dimethylformamide (DMF):

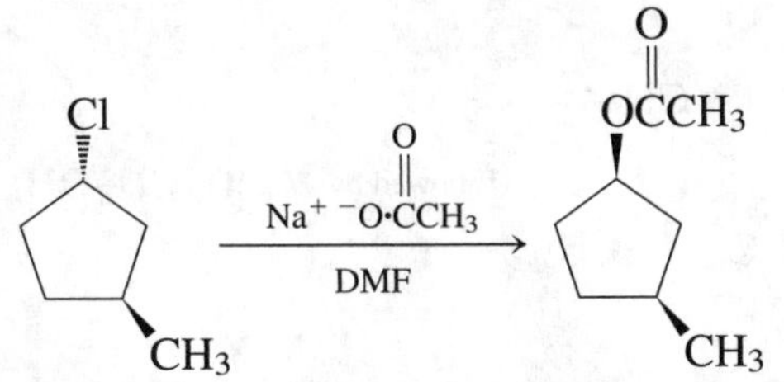

37. The first reaction is an uncomplicated S_N2 displacement. The second is S_N2 as well, but there is a complication. Let's take a look at how its product can react further, and then perhaps we can see why it happens in the second case but not in the first.

The pathway to the byproduct, $(CH_3CH_2CH_2CH_2)_2S$, must involve reaction between the product of the first displacement, $CH_3CH_2CH_2CH_2SH$, and another molecule of the initial starting material $CH_3CH_2CH_2CH_2Br$: This is the only reasonable way to get a second butyl group on the sulfur. Since butyl is primary, we are limited to the S_N2 mechanism. The simplest way to get there is to use the product molecule itself as a nucleophile:

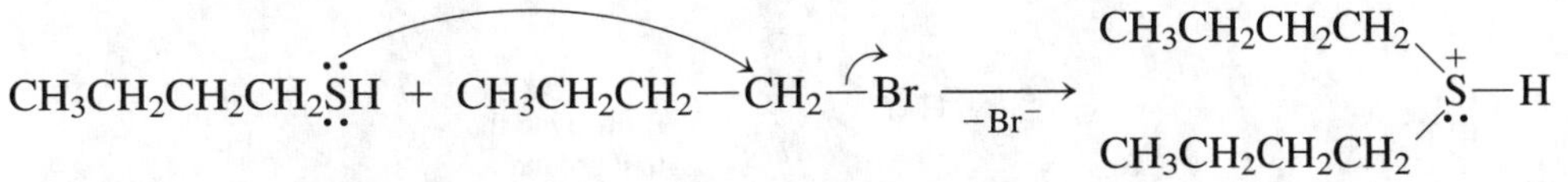

Then loss of H^+ from sulfur completes the sequence. Why doesn't the same thing happen in the first reaction, the formation of the alcohol? What do you know about the differences in nucleophile strength between S and O? Sulfur is far better, especially in protic solvent (Section 6-8). Alcohols are too weak as nucleophiles to carry out S_N2 displacements, while thiols can achieve this transformation readily.

Did you consider an alternative mechanism, one in which the SH group of the thiol is deprotonated *before* nucleophilic attack? That is,

$$CH_3CH_2CH_2CH_2\ddot{S}H \rightleftharpoons CH_3CH_2CH_2CH_2\ddot{S}:^- + H^+,$$

followed by

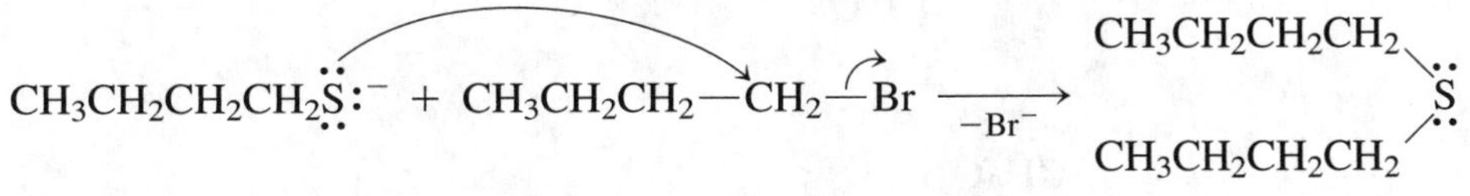

This mechanism is *qualitatively* plausible, but because the pK_a of the thiol SH bond is around 10, in the absence of base the equilibrium of the initial deprotonation is too unfavorable for this sequence to be competitive: The concentration of the conjugate base is too low.

38. (**1a**) $(CH_3)_2C{=}CH_2$

(**1b**) $CH_2{=}C(CH_3)CH_2CH_3$ $CH_3CH{=}C(CH_3)_2$

(**1c**)

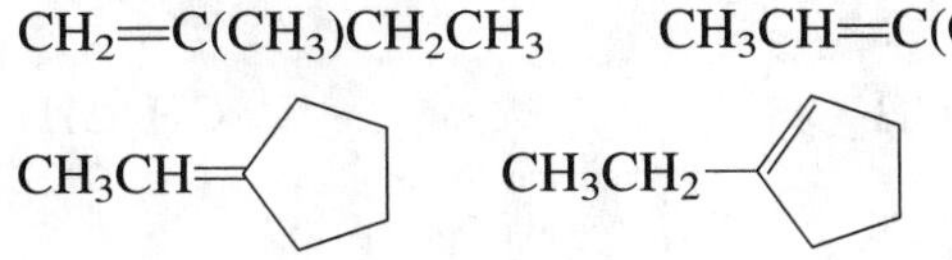

(**1d**)

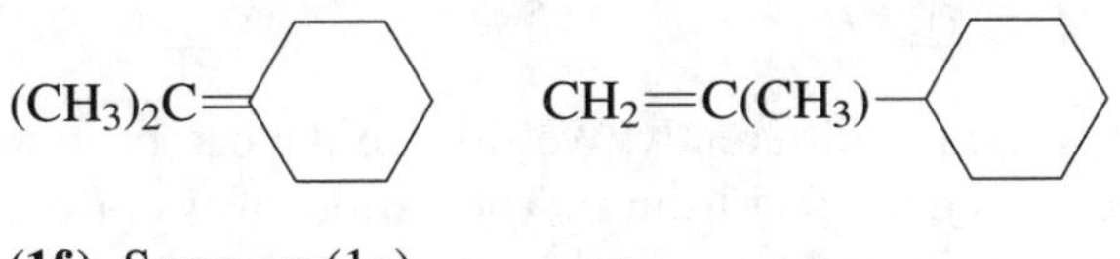

(**1e**), (**1f**) Same as (1a)

39. You are given the product structures. Therefore your task is to identify the mechanistic pathway in each case: How do you get "there" from "here"? Use the information in Section 7-6. Use as a model the mechanism on page 266 for E1 reaction of 2-bromo-2-methylpropane (*tert*-butyl bromide) in methanol.

(a)

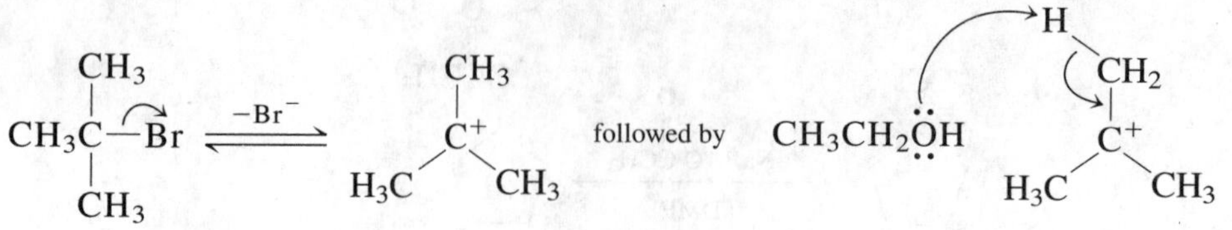

For the remainder of the mechanisms we show the steps, still separately and one after another, but in a continuous scheme. Either way of depicting the separate steps is acceptable. However, observe the following **Caution:** Always indicate the individual steps as being separate when you write any multistep mechanism. *NEVER* combine separate steps in writing multistep mechanisms.

(b) In this case, and in the ones that follow, two products are possible. The reactions diverge in the second step. For clarity, we also show the products of the two pathways for these parts.

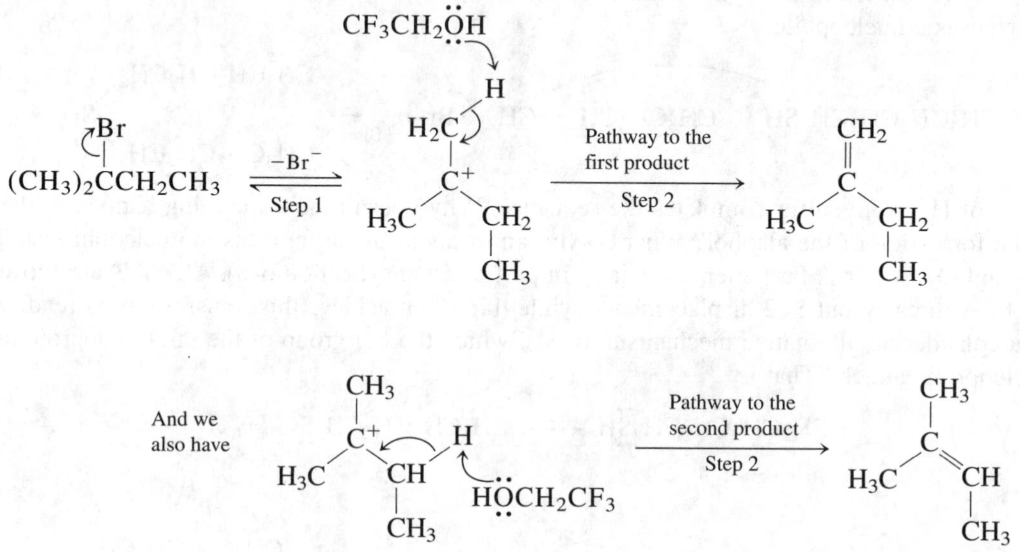

(c) CH₃CH₂

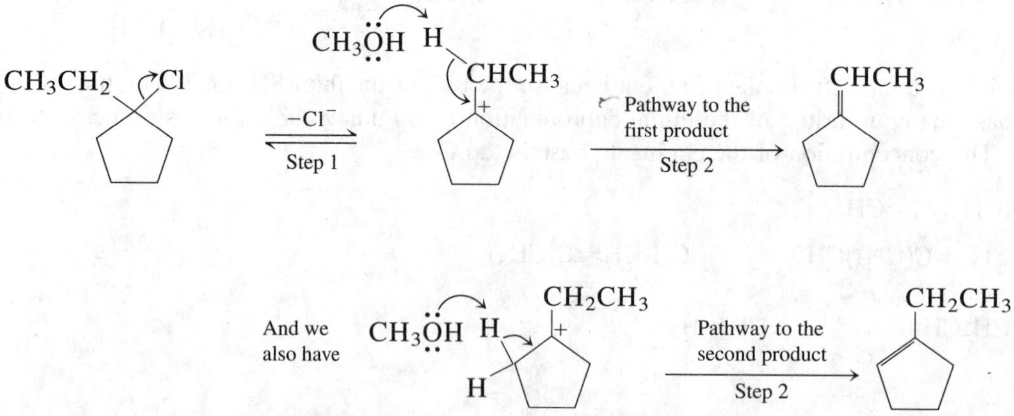

(d) For part **(d)** we'll show the two pathways slightly differently. We'll write the carbocation intermediate just once and show the two E1 products diverging from it. The various ways of depicting divergent pathways are all commonly used, both in the textbook as well as elsewhere.

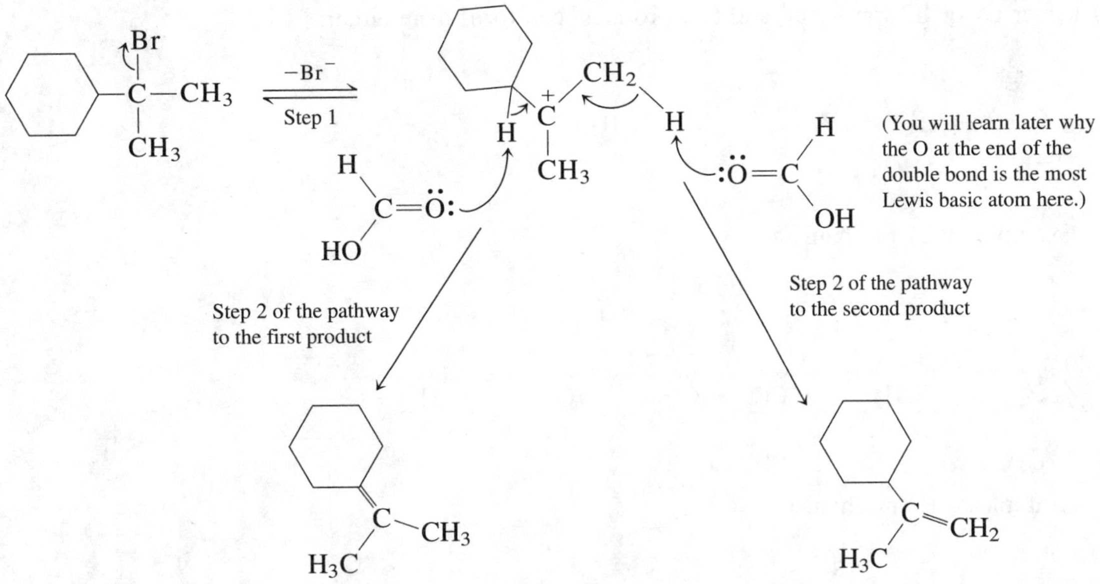

(e) and (f) are essentially the same as (a) in mechanism. The leaving group is changed from Br⁻ to Cl⁻. The nucleophile/base is different but functions in exactly the same way, with the oxygen atom removing a proton from the carbon neighboring the positive center in the intermediate carbocation.

40. **(a)** The base is a very strong one (the pK_a of NH_3 is way up there, around 35; it is a very **weak** acid), and that favors E2 rather than E1. The exclusive product is

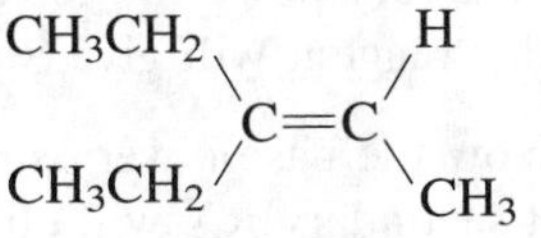

This product forms no matter which of the six possible protons next to the reactive carbon is lost:

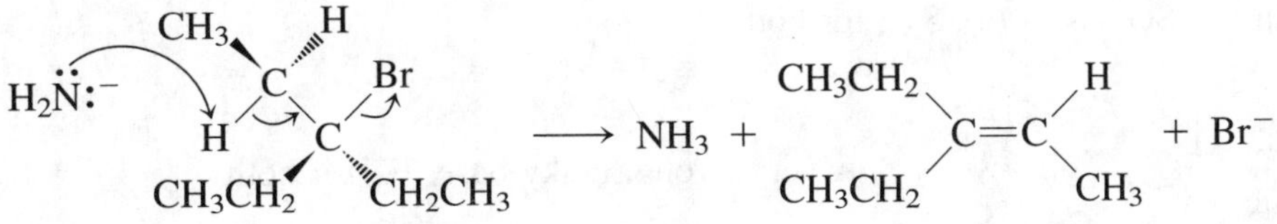

(b) E2

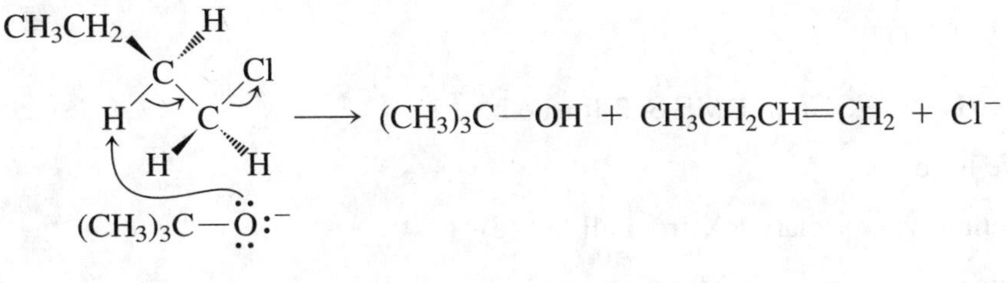

(c) E2

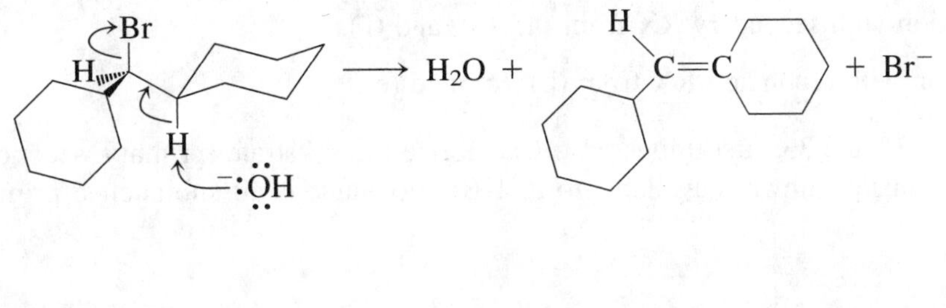

(d) Either E1 or E2 can occur, and two products can form from either:

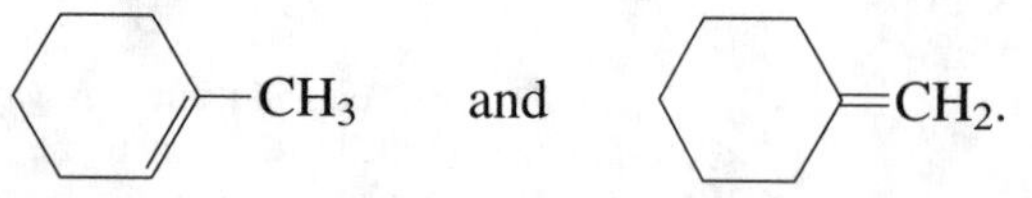

—CH$_3$ and =CH$_2$.

Example of E1 mechanism:

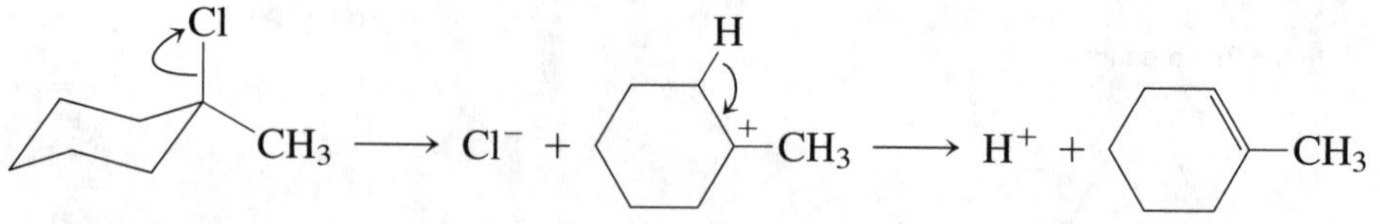

Example of E2 mechanism:

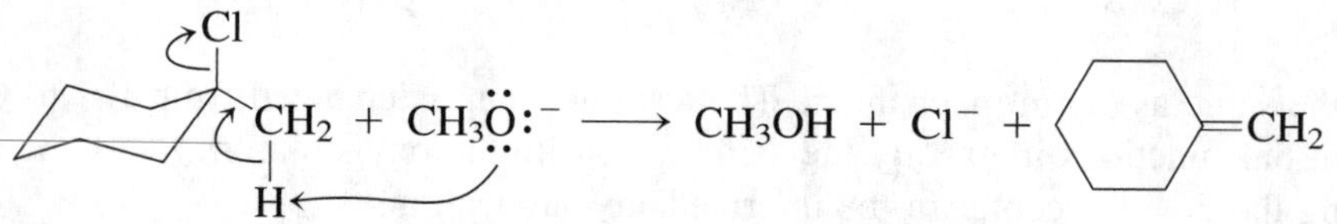

41. Let's first examine the reagents individually. Then we can answer the question as it has been phrased.

(a) NaSCH$_3$ in CH$_3$OH Here we have a very powerful nucleophile which is not a particularly strong base. Despite the protic solvent, this reagent will give S$_N$2 with both primary and secondary RX.

(b) (CH$_3$)$_2$CHOLi in (CH$_3$)$_2$CHOH Now the base is very strong, and somewhat bulky as well. Primary RX will still give mostly S$_N$2, but secondary RX will eliminate via the E2 mechanism.

(c) NaNH$_2$ in liquid NH$_3$ Similar to (b); the base is even stronger but not bulky. S$_N$2 for primary and E2 for secondary.

(d) KCN in DMSO As in (a), S$_N$2 for both.

(e) 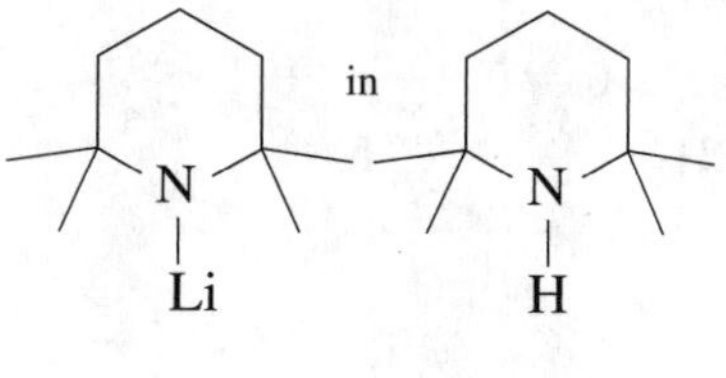 A very strong, bulky base: E2 for both.

(f) CH$_3$CH$_2$CH$_2$—C(=O)—ONa in DMF S$_N$2 for both.

Therefore we have

(i) S$_N$2 reaction with primary RX from all but (e).

(ii) E2 reaction with primary RX from only (e).

(iii) S$_N$2 reaction with secondary RX from (a), (d), and (f).

(iv) E2 reaction with secondary RX from (b), (c), and (e).

42. As in Problems 32 and 33, first things first: Categorize the substrate (primary, secondary, etc.) in order to identify the mechanistic pathways available to it. 1-Bromobutane is an unbranched primary haloalkane;

the mechanistic possibilities are S_N2 and E2. S_N2 will occur with good nucleophiles, but poor nucleophiles don't react. Strong, hindered bases give E2.

(a) and (c) 1-Chlorobutane   (b) 1-Iodobutane

(d) $CH_3CH_2CH_2CH_2NH_3^+$ Br^- (e) $CH_3CH_2CH_2CH_2OCH_2CH_3$

All of the above form via S_N2 mechanisms.

(f) No reaction. Alcohols are poor nucleophiles.

(g) $CH_3CH_2CH{=}CH_2$ (E2 reaction)   (h) $CH_3CH_2CH_2CH_2P(CH_3)_3^+$ Br^- (S_N2)

(i) No reaction. Carboxylic acids are poor nucleophiles. (No, we never told you that, but you do know that nucleophilicity goes along with basicity, and carboxylic *acids* are not terribly likely to be strong *bases*. Make sense?)

43. Now the substrate, 2-bromobutane, is secondary. All mechanisms are possible.

(a) 2-Chlorobutane   (b) 2-Iodobutane

DMF is a polar aprotic solvent. (a) and (b) both are S_N2 reactions.

(c) 2-Chlorobutane, but now by the S_N1 pathway. Nitromethane is a nonnucleophilic solvent with good ionizing power.

(d) $CH_3CH_2CH(NH_3)CH_3^+$ Br^-

S_N2: NH_3 is a good but not *strongly* basic nucleophile.

(e) and (g) $CH_3CH_2CH{=}CH_2$ and $CH_3CH{=}CHCH_2$ (E2 products—the problem states that the reagents are in excess, thus favoring E2 over E1)

(f) $CH_3CH_2CH(OCH_2CH_3)CH_3$ Secondary substrate undergoes S_N1 reaction with the poorly nucleophilic alcohol (solvolysis).

(h) $CH_3CH_2CH[P(CH_3)_3]CH_3^+$ Br^- (S_N2)

(i) $CH_3CH_2CH(O_2CCH_3)CH_3$ Another S_N1 solvolysis, as in (f).

44. The substrate, 2-bromo-2-methylpropane, is now tertiary. The S_N2 pathway is now eliminated as a possibility.

(a) and (b) No reaction. The usual polar aprotic solvents have rather poor ionizing power. DMF is somewhat better than acetone (a really poor S_N1 solvent), but substitution will still be very slow.

(c) 2-Chloro-2-methylpropane (S_N1)

(d) Ammonia is a borderline case in terms of base strength, giving more substitution with secondary substrates (where S_N2 is an option), but more elimination (E2) with tertiary haloalkanes: $(CH_3)_2C{=}CH_2$

(e) and (g) $(CH_3)_2C{=}CH_2$ (E2)   (f) $(CH_3)_3C{-}O{-}CH_2CH_3$

(h) $(CH_3)_3CP(CH_3)_3^+$ Br^- (S_N1), together with $(CH_3)_2C{=}CH_2$ (E1)

(i) $(CH_3)_3CO_2CCH_3$ (S_N1)

45. (a) (1) $(CH_3)_3CSH$ + HCl

(2) $(CH_3)_3CO_2CCH_3$ + $(CH_3)_2C{=}CH_2$ + KCl + CH_3COOH ($CH_3COO^-K^+$ is basic enough to give some elimination from tertiary halides.)

(3) $(CH_3)_2C{=}CH_2$ + KCl + CH_3OH

(b) Rates of (1) and (2) will be the same, or nearly so. (1) is S_N1 and (2) is S_N1 and E1, and they have the same rate-determining step; if the reaction mixtures are similar in polarity, the rates will be very close. Rate of (3) will depend on the concentration of the base because of the occurrence of bimolecular elimination.

46. (a) $=CH_2$ E2   (b) No reaction. Poor leaving group.

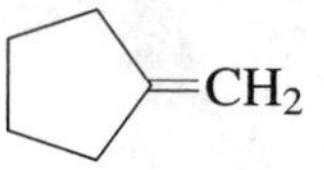

(c) Racemic $CH_3CH_2CHOHCH_3$ S_N1 (d) E2

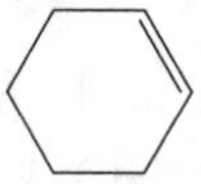

(e) $(CH_3)_2CHCH_2CH_2OCH_2CH_3$ S_N2

(f) Racemic $CH_3CH_2\overset{\overset{\displaystyle I}{|}}{C}(CH_3)CH_2CH_2CH_3$ S_N1

(g) No reaction, except for reversible proton transfer.

(h) $-CH_2CH_2CH_2CN$ and

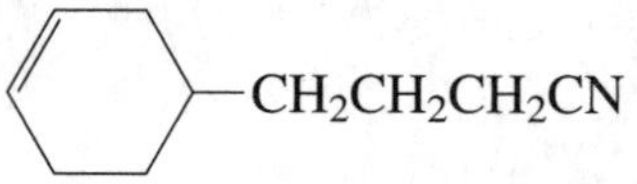

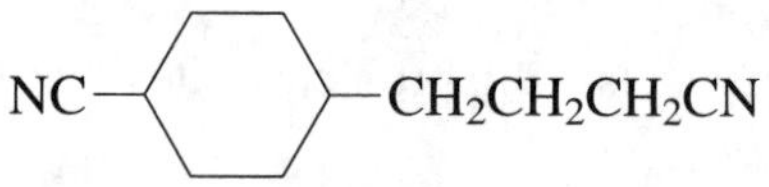

$NC-$$-CH_2CH_2CH_2CN$ E2 and S_N2

(i) $(S)-CH_3CH_2CHSHCH_3$ S_N2 (j) $\overset{OCH_3}{\underset{CH_2CH_3}{}}$ S_N1

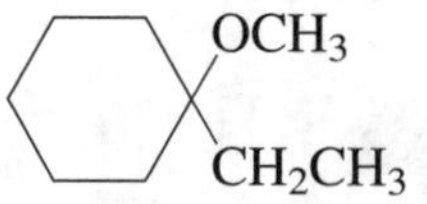

(k) $(CH_3)CCH=CH_2$ E2   (l) No reaction. Poor nucleophile.

47.

Haloalkane	Reagent			
	H_2O	$NaSeCH_3$	$NaOCH_3$	$KOC(CH_3)_3$
CH_3Cl	no reaction	CH_3SeCH_3	CH_3OCH_3	$CH_3OC(CH_3)_3$ S_N2
$CH_3CH_2CH_2Cl$	no reaction	$CH_3CH_2CH_2SeCH_3$ $\rbrace$ S_N2	$CH_3CH_2CH_2OCH_3$ $\rbrace$ S_N2	$CH_3CH=CH_2$
$(CH_3)_2CHCl$	$(CH_3)_2CHOH$ $\rbrace$ S_N1	$(CH_3)_2CHSeCH_3$	$CH_3CH=CH_2$ $\rbrace$ E2	$CH_3CH=CH_2$ $\rbrace$ E2
$(CH_3)_3CCl$	$(CH_3)_3COH$	$(CH_3)_3CSeCH_3$ S_N1	$(CH_3)_2C=CH_2$	$(CH_3)_2C=CH_2$
	and	and		
	$(CH_3)_2C=CH_2\rbrace$ E1	$(CH_3)_2C=CH_2\rbrace$ E1		

48. See answer to Problem 47. Secondary halides give higher E2/E1 ratios than do tertiary halides (secondary carbocations form less readily than tertiary).

49. (a) Poorly. $CH_3CH=CHCH_3$ and $CH_3CH_2CH=CH_3$ are important products.

(b) Not at all. No reaction; poor nucleophile.

(c) Not at all. No appreciable reaction besides S_N1 with the solvent.

(d) Well. An "intramolecular" (internal) S_N1 reaction.

(e) Well, eventually, but very, very slowly.

(f) Well.

(g) Well.

(h) Not at all. No reaction because of the very poor nucleophile.

(i) Not at all. No reaction.

(j) Not at all. No reaction because of the very poor leaving group.

(k) Poorly. $(CH_3)_2CHCH_2CH_2OCH_2CH_3$ forms also.

(l) Poorly. Good nucleophile gives mainly $CH_3CH_2CH_2CH_2OCH_2CH_3$.

50. (a) $CH_3CH_2CH_2CH_3 \xrightarrow{Br_2, \Delta} CH_3CH_2CHBrCH_3 \xrightarrow{KI, DMSO} CH_3CH_2CHICH_3$

(b) $CH_3CH_2CH_2CH_3 \xrightarrow{Cl_2, h\nu}$ some $CH_3CH_2CH_2CH_2Cl \xrightarrow{NaI, acetone} CH_3CH_2CH_2CH_2I$

(c) $CH_4 \xrightarrow{Cl_2, h\nu} CH_3Cl \xrightarrow{KOH, H_2O} CH_3OH$; then $(CH_3)_3CH \xrightarrow{Br_2, \Delta}$

$(CH_3)_3CBr \xrightarrow{CH_3OH} (CH_3)_3COCH_3$

(d)

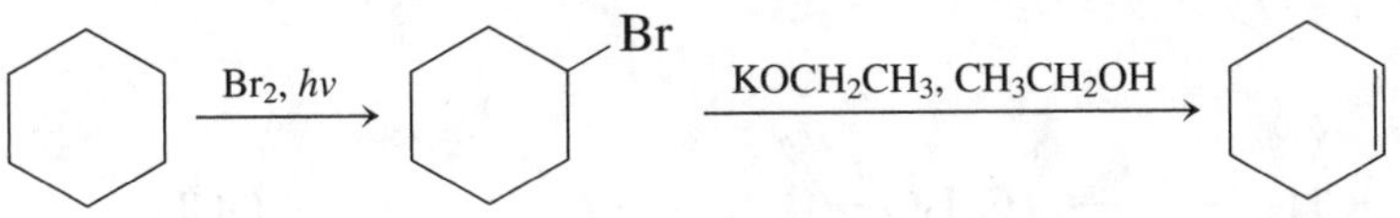

(e) From (d),

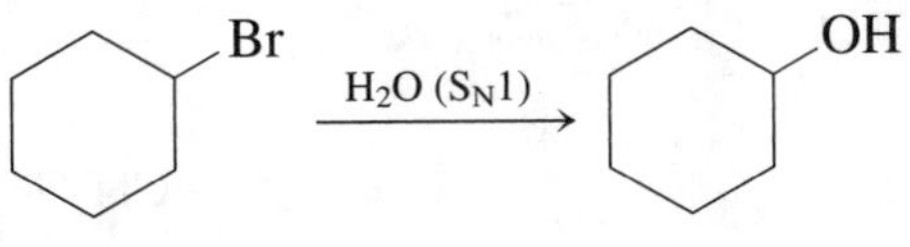

A better method will be presented in Chapter 8.

(f) Na_2S in alcohol:

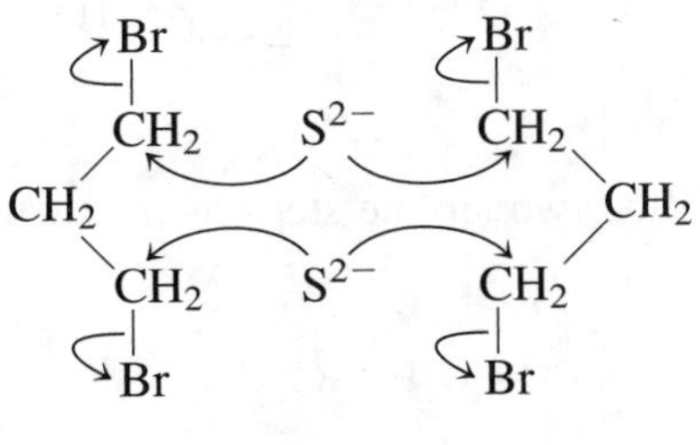

Four S_N2 displacements!

51. (a) Rate $= k[RBr]$, so $2 \times 10^{-4} = k(0.1)$, and therefore $k = 2 \times 10^{-3}$ s^{-1}.

Product is

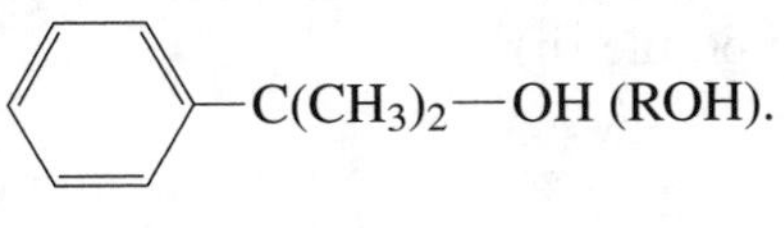

$-C(CH_3)_2-OH$ (ROH).

(b) New "k_{LiCl}" $= 4 \times 10^{-3}$ s^{-1}. Addition of LiCl increases the polarity of the solution by adding ions, and this speeds up the rate-determining ionization step in the solvolysis process.

(c) In this case the added salt contains Br^-, **which is also the leaving group in the solvolysis reaction.** This leads to a **decrease** in rate because the first step in solvolysis is reversible, and **recombination** of R^+ and Br^- is occurring in competition with reaction of R^+ with H_2O:

$$RBr \underset{\text{Recombination}}{\overset{\text{Ionization}}{\rightleftharpoons}} Br^- + R^+ \xrightarrow{H_2O} ROH$$

52. (a) The solvent is an alcohol and, therefore, already not very nucleophilic. In addition, however, the three fluorine atoms in the structure, CF_3CH_2OH, are very electronegative and electron-attracting. The combined effect of their electron-attracting power reduces the ability of the oxygen atom to serve as an electron-donating nucleophilic atom. This is an example of an inductive effect, which we will discuss more fully in the next chapter. For now all we need to recognize is that this alcohol is a much poorer nucleophile than "normal" alcohols (ones lacking halogen substituents).

(b) In the usual S_N1 reaction mechanism the first step is rate-limiting; that is, $rate_1 < rate_2$. In the case of the reaction in this problem, the reduced nucleophilicity of the solvent diminishes $rate_2$ so that it becomes slow and rate-limiting. Thus the carbocation intermediate has the opportunity to build up in concentration, because it is formed faster than it is consumed by attachment to the nucleophile.

(c) As the carbocation stability increases and the solvent nucleophilicity decreases, $rate_1$ should increase further relative to $rate_2$, and the carbocation intermediate should exhibit a more extended lifetime. This principle was used by George Olah of the University of Southern California in his pioneering studies associated with direct observation of carbocations and the direct measurement of their properties in solvent environments of extremely low nucleophilicity.

(d)

53. (a) A tertiary halide $\Rightarrow S_N1$ reaction, which will be two simple steps as in reaction profile (iii).

$$E = (CH_3)_3\overset{\delta^+}{C}\cdots\overset{\delta^-}{C} \qquad F = (CH_3)_3C^+$$
$$G = (CH_3)_3\overset{\delta^+}{C}\cdots\overset{\delta^+}{P}(C_6H_5)_3 \qquad H = (CH_3)_3C\overset{+}{P}(C_6H_5)_3$$

(b) A secondary halide being displaced by another halide $\Rightarrow S_N2$ reaction. The product and starting material are comparable in stability: reaction profile (ii).

$$C = \overset{\delta^-}{Br}\cdots\underset{\underset{CH_3}{|}}{\overset{\overset{H}{|}}{C}}\cdots\overset{\delta^-}{I} \qquad D = (CH_3)_2CHBr$$
with CH_3 CH_3 on the central carbon

(c) A tertiary halide with a poor nucleophile $\Rightarrow$ S_N1 (solvolysis), but with one more step than (a) because of proton loss from an intermediate alkyloxonium ion: reaction profile (iv).

$$I = (CH_3)_3\overset{\delta+}{C}\cdots\overset{\delta-}{Br} \qquad J = (CH_3)_3C^+$$

$$K = (CH_3)_3\overset{\delta+}{C}\cdots\overset{\delta+}{O}CH_2CH_3 \qquad L = (CH_3)_3\overset{+}{C}OCH_2CH_3$$
$$\qquad\qquad\quad | \qquad\qquad\qquad\qquad\qquad |$$
$$\qquad\qquad\quad H \qquad\qquad\qquad\qquad\qquad H$$

$$M = (CH_3)_3C\overset{\delta+}{O}CH_2CH_3 \qquad N = (CH_3)_3COCH_2CH_3$$
$$\qquad\qquad\quad \overset{..}{H}$$
$$\qquad\qquad\quad \underset{\delta+}{}$$

(d) A primary halide and a good nucleophile $\Rightarrow$ S_N2 reaction, but the product is much more stable than the starting material (C—O bonds are stronger than C—Br bonds): reaction profile (i).

$$A = CH_3CH_2\overset{\delta-}{O}\cdots\overset{H\ \ H}{\underset{|}{\underset{CH_3}{C}}}\cdots\overset{\delta-}{Br} \qquad B = CH_3CH_2OCH_2CH_3$$

54. Neutral polar conditions are ideal for an **intra**molecular S_N1 reaction:

$$(CH_3)_2CCH_2CH_2CH_2OH \longrightarrow (CH_3)_2\overset{+}{C}CH_2CH_2CH_2\overset{..}{O}H$$

Basic conditions promote elimination leading to alkenes. Two isomeric alkenes are actually possible: $CH_2=C(CH_3)CH_2CH_2CH_2OH$ and $(CH_3)_2C=CHCH_2CH_2OH$.

55. (a) E1 rate $= (1.4 \times 10^{-4}\ s^{-1})(2 \times 10^{-2}\ M) = 2.8 \times 10^{-6}\ mol\ L^{-1}\ s^{-1}$
E2 rate $= (1.9 \times 10^{-4}\ L\ mol^{-1}\ s^{-1})(2 \times 10^{-2}\ M)(5 \times 10^{-1}\ M)$
$\qquad\quad = 1.9 \times 10^{-6}\ mol\ L^{-1}\ s^{-1}$
E1 rate is faster, so E1 reaction predominates.

(b) E1 rate $= 2.8 \times 10^{-6}\ mol\ L^{-1}\ s^{-1}$ (no change)
E2 rate $= (1.9 \times 10^{-4}\ L\ mol^{-1}\ s^{-1})(2 \times 10^{-2}\ M)(2\ M)$
$\qquad\quad = 7.6 \times 10^{-6}\ mol\ L^{-1}\ s^{-1}$
E2 rate is now faster, so E2 reaction predominates.

(c) Solve for $[NaOCH_3]$ when E1 rate $=$ E2 rate: $2.8 \times 10^{-6}\ mol\ L^{-1}\ s^{-1} = (1.9 \times 10^{-4}\ L\ mol^{-1}\ s^{-1})\ (2 \times 10^{-2}\ M)\ [NaOCH_3]$. $[NaOCH_3] = 0.74\ M$.

56. The first task is to examine both starting materials and products in order to classify the reaction that has taken place. A methyl group has been removed from the substrate and converted to iodomethane. In addition, an oxygen–carbon covalent bond has become an ionic bond between oxygen and lithium. Does this fit any pattern we've seen? Yes. If we go back to Chapter 6 and examine the components of substitution reactions, we can see two familiar elements: a nucleophile (iodide) and an organic substrate to the left of the arrow. Having defined these, it is then possible to recognize iodomethane as the product of substitution, and the lithium salt structure as containing the leaving group.

Having defined the process as a substitution, what remains is to consider experiments to identify whether the operative mechanism is S_N1 or S_N2. A kinetic study to reveal whether or not the iodide ion appears in the rate equation is one possibility. Changing the solvent from pyridine (polar, aprotic) to a protic solvent (e.g., alcohol) and observing the effect on the rate is another. Of course, you are not operating entirely in the dark: Given that the displacement is occurring at a methyl carbon, it is reasonable to predict that the answer will be S_N2.

57. As in Problem 56, first classify the reaction. Not only does a carbon–oxygen bond break, but a hydrogen is lost from the *tert*-butyl group, apparently shifting to the oxygen to give an alcohol (cyclohexanol) as one product, and leaving an alkene as a second organic product. Loss of both hydrogen and oxygen from two adjacent carbon atoms to give a carbon–carbon double bond resembles elimination, but it differs from the eliminations from haloalkanes that we have encountered previously, in which one of the atoms lost was a halogen. Halogens, in the form of halide ions, are good leaving groups. Oxygen is another story, however, because negatively charged oxygen (as in hydroxide, for example) is usually strongly basic and not easily displaced. Where do we turn for a rationalization, then?

Consider the reaction conditions. If this is indeed an elimination, it is an unusual one: The eliminations in this chapter occur in the presence of base, not acid. Taking that anomaly one step further, what role could the acid serve? Where on the substrate could a proton conceivably go? There is only one answer: the oxygen atom. The product is an oxonium ion, an alkyl-substituted analog of the hydronium ion. Looking at the relationship between the structure of this ion and the products, notice that departure of cyclohexanol to leave a *tert*-butyl carbocation is a reasonable next step. Cyclohexanol is a neutral, weakly basic molecule and should be a good leaving group. The *tert*-butyl cation is stable, as carbocations go, and it can lose a proton to give the observed alkene product. Overall, this is a variation on the E1 process, using the acid to turn a poor leaving group into a better one.

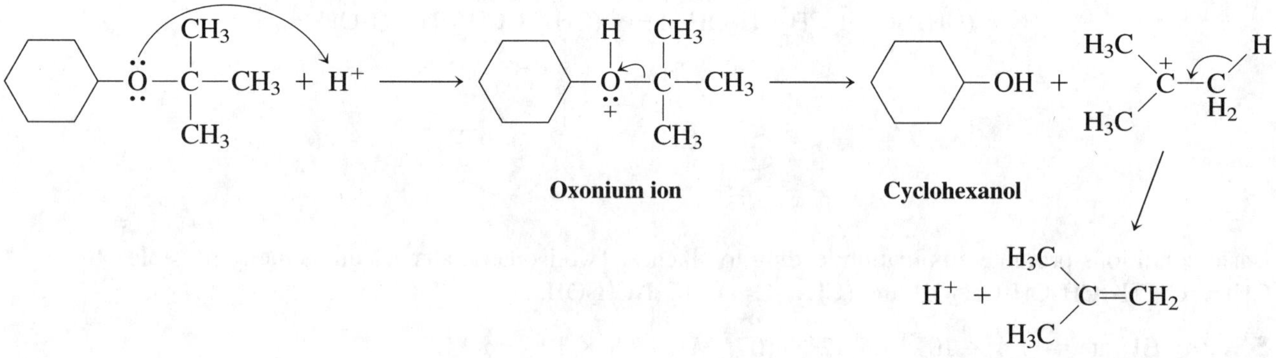

Oxonium ion   **Cyclohexanol**

58. **(a)**, **(d)**, **(e)**, **(g)**, **(h)**, **(i)** Mainly S_N2. These nucleophiles are all weak bases, so elimination is not favored.

(c), **(f)** Mixture of S_N2 and E2.

(b) Mainly E2. Strong bases favor elimination.

59. In both **A** and **B** the necessary *anti* alignment between H and Cl (both of which are in axial positions) is present, so E2 elimination will give the desired alkene. In **C** the Cl is equatorial and *anti* elimination cannot occur (make a model). Instead, very slow eliminations will proceed via *syn* geometries to give a mixture containing the desired alkene together with the isomer shown below.

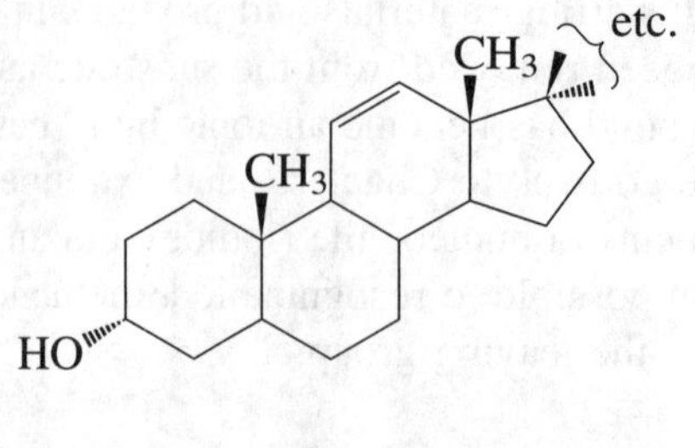

60. Look at the conformations first [the deuterium atoms for (b) are also written in].

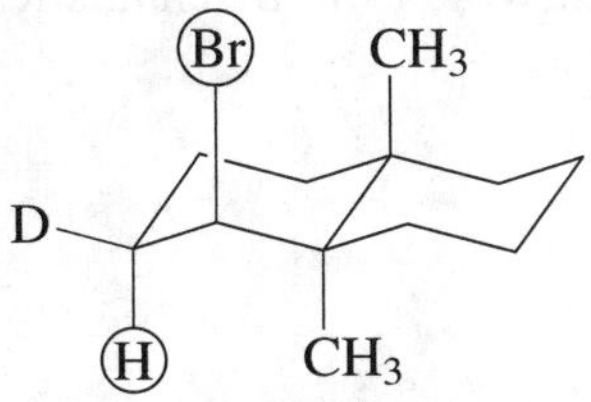

(i-d)   vs.   (ii-d)

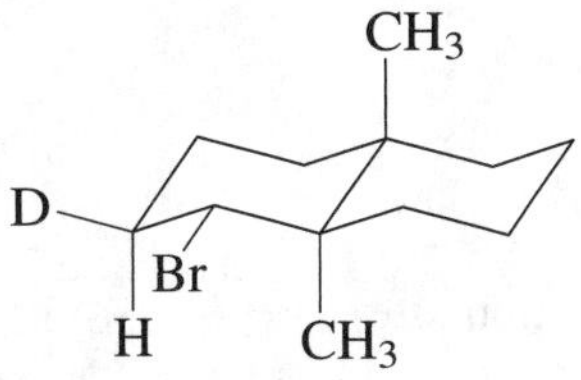

(a) and (b) Compound **i** reacts much faster because it already possesses the necessary *anti* alignment of Br and H (on the neighboring carbon, circled). In the deuterated example shown, the E2 reaction results in loss of HBr only; all the D is retained because the D atom is not capable of adopting an *anti* alignment relative to bromine. The conformation shown is the reactive one.

Compound **ii** possesses no hydrogen *anti* to Br in the conformation shown. However, if the left-hand ring adopts a boatlike conformation resembling that shown below, E2 elimination via an *anti* transition state can readily occur.

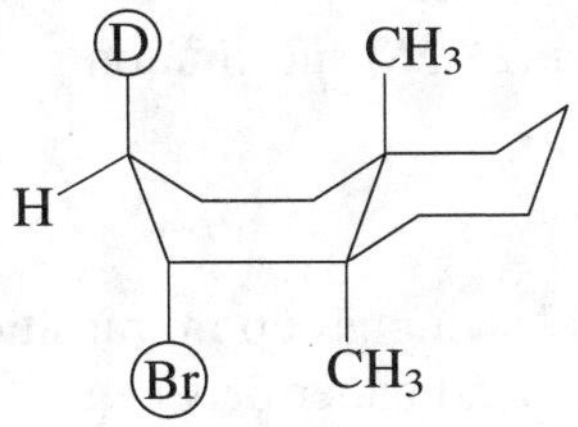

As indicated, all the D should be lost, according to this mechanism, and that is the result observed.

61. As we have been doing, characterize each substrate–nucleophile (base) combination before going any further. We have:

A primary; **B** and **C** secondary; and **D** tertiary bromoalkanes, reacting with (a) a good nucleophile that is a moderately weak base; (b) a strong, bulky base; (c) a good nucleophile that is a moderately strong, nonbulky base; (d) like (a), a good but not particularly strongly basic nucleophile; and (e) a weak and essentially nonbasic nucleophile.

The problem is best solved by taking each bromoalkane in turn and evaluating the most reasonable mechanism that will take place upon its reaction under each of the conditions (a) through (e). We'll do the easy ones first.

A 1-Bromobutane (primary) will give S_N2 products, $CH_3CH_2CH_2CH_2Nu$, under all conditions except (b) (E2 with strong bulky base to give $CH_3CH_2CH=CH_2$) and (e) (no reaction with poor nucleophile).

D 2-Bromo-2-methylpropane (tertiary) with methanol (e) will give the S_N1 product, $(CH_3)_3COCH_3$, accompanied by minor amounts of $(CH_3)_2C=CH_2$, the E1 side product. The S_N1 products $(CH_3)_3CN_3$ and $(CH_3)_3CO_2CCH_3$ will form under conditions (a) and (d), respectively, together with increased amounts of $(CH_3)_2C=CH_2$, which now arises through E2 reaction with the moderately basic azide and acetate nucleophiles. Under strongly basic conditions (c) and (b) only E2 takes place.

B and **C** 2-Bromobutane (secondary) will give S_N2 products under conditions (a) and (d) (good nucleophiles that are not very strong bases), a mixture of S_N2 and E2 with hydroxide (c) (at the borderline of strong base for substitution versus elimination with secondary substrates), and E2 with LDA (b). Methanol (e) gives mostly S_N1, some E1. The presence of two diastereomeric deuterium-substituted 2-bromobutanes raises stereochemical questions, which the mechanistic descriptions that follow explore:

Substitution reactions

S_N2: Inversion at reaction site

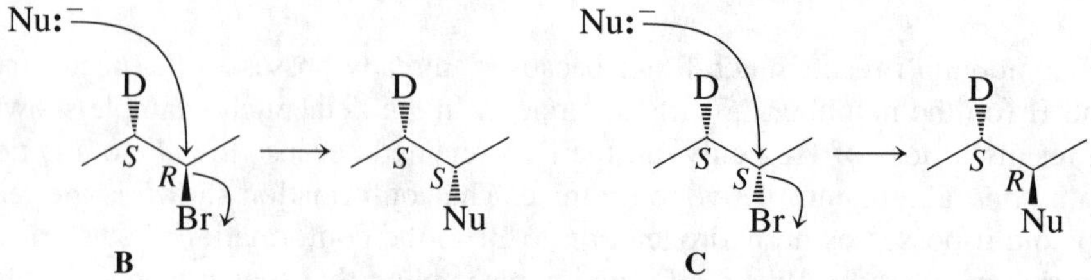

S_N1: Mixture of stereoisomers at reaction site obtained

Either **B** or **C** → mixture of (2S,3S) and (2R,3S) substitution products.

Elimination reactions

E2: *Anti* conformation of hydrogen and leaving group in transition state. Each substrate may undergo reaction to give alkenes with the double bond either between C1 and C2 or between C2 and C3; the latter may form from either of two conformations.

B

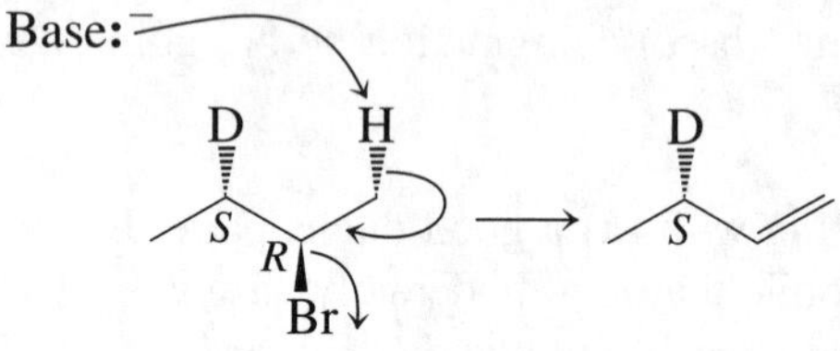

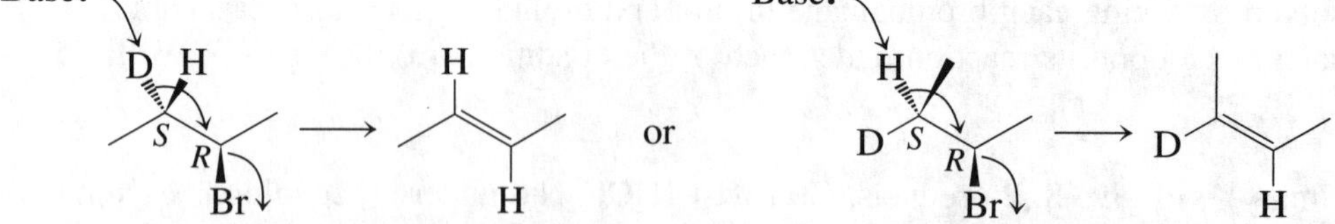

The factors that result in one E2 pathway being preferred over another will be presented in Chapter 11.

C

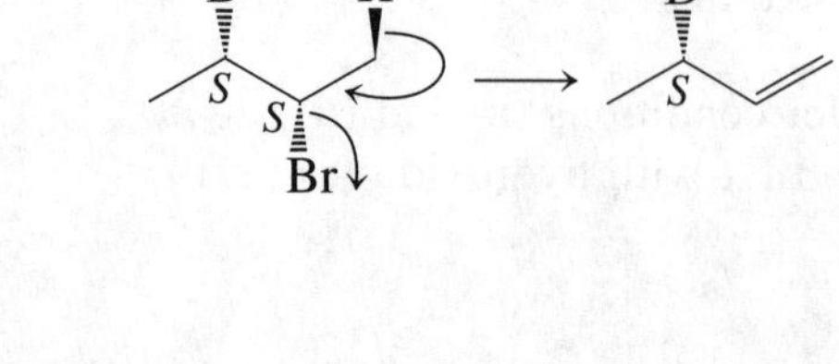

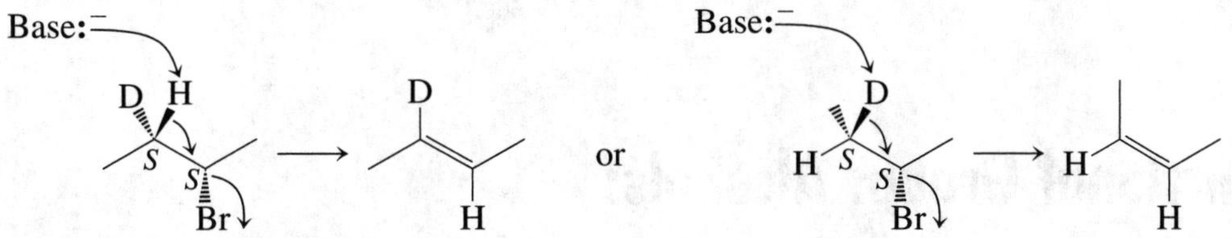

E1: Mixture of all alkene isomers obtained from either substrate

62. (d)

63. (b)

64. (b)

65. (b)

66. (a)

8

Hydroxy Functional Group: Alcohols: Properties, Preparation, and Strategy of Synthesis

With this chapter we begin a detailed examination of alcohols, molecules containing the hydroxy functional group. Apart from carbonyl compounds, alcohols are the most important molecules in organic chemistry. They can be prepared from many different types of compounds, and, in turn, they can be converted into many different types of compounds. They therefore play a central role in organic chemistry. Moreover, their preparations, properties, and reactions serve as excellent illustrations of the logic underlying the behavior of organic compounds.

The approach to Chapters 8 and 9 is similar to that used in Chapters 6 and 7. However—and this is very important—alcohols have much more potential for chemical conversion than haloalkanes do. You must be prepared even more than before to focus on the functional groups and their polar bonds as sites of possible reactivity. A comparison of, first, the bonds and, second, the potential bonding changes available in alcohols and haloalkanes is useful. Haloalkane chemistry involves mainly three bonds:

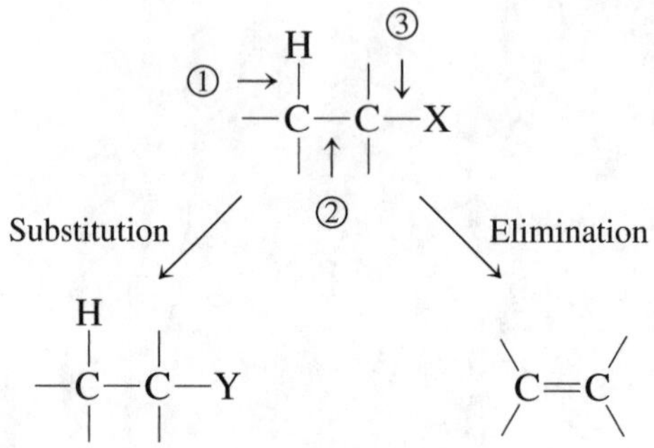

$$
\begin{array}{c}
\text{①} \to \quad \underset{\displaystyle \text{②}}{\overset{\displaystyle \text{H} \quad \text{③}}{-\text{C}-\text{C}-\text{X}}}
\end{array}
$$

Substitution ⟋   Elimination ⟍

$$
-\text{C}-\text{C}-\text{Y} \qquad\qquad \text{C}=\text{C}
$$

Substitution breaks bond 3. Elimination breaks bonds 1 and 3, and "doubles" bond 2. By comparison, **five** bonds may participate in chemical reactions of alcohols:

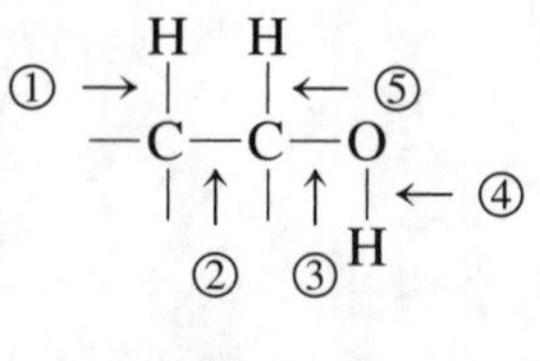

$$
\begin{array}{c}
\text{①} \to \quad \overset{\displaystyle \text{H} \quad \text{H}}{-\text{C}-\text{C}-\text{O}} \ \leftarrow \text{⑤} \\
\underset{\text{②} \quad \text{③}}{} \ \ \overset{}{\text{H}} \ \leftarrow \text{④}
\end{array}
$$

We focus first on properties of alcohols. Then we examine their preparation, using this discussion to examine the general problem of synthetic strategy: how to logically plan a practical and efficient sequence of chemical steps that allows the conversion of a starting material into an organic product.

Outline of the Chapter

8-1   Nomenclature
A description of both systematic and common naming systems for these very common molecules.

8-2   Physical Properties
A new factor is introduced: hydrogen bonding.

8-3   Acidity and Basicity of Alcohols
Similarities and differences with water, the simplest inorganic relative.

8-4   Industrial Preparation of Alcohols
A brief survey of methods for the preparation of commercially useful alcohols.

8-5   Synthesis of Alcohols by Nucleophilic Substitution

8-6   Oxidation and Reduction
Alcohols and carbonyl compounds are interconverted in oxidation–reduction processes, opening up many synthetic possibilities.

8-7   Organometallic Reagents
A detour, introducing compounds containing **negatively polarized, nucleophilic** carbon atoms.

8-8   Organometallic Reagents in the Synthesis of Alcohols
The most important general alcohol syntheses: addition of nucleophilic carbon compounds to carbonyl groups.

8-9   An Introduction to Synthetic Strategy
How to look at a "target" molecule and logically "plan" its synthesis using sequences of several reactions.

Keys to the Chapter

8-1 and 8-2.   Nomenclature and Physical Properties

Two points need to be made concerning nomenclature. First, alcohols have been around for a long time, and the common names given in this section are still in widespread use and need to be learned and understood. Second, there is an **order of precedence** when there are two or more different functional groups in a molecule. This precedence order determines the numbering. The alcohol group is of fairly high ranking on this list, whereas halogens are near the bottom. So,

$$
\underset{CH_3CHCH_2CHCH_3}{\overset{\underset{|}{Cl}\qquad\overset{|}{OH}}{}}
$$

This compound is 4-chloro-2-pentanol (not 2-chloro-4-pentanol).

This compound is 4-bromo-2-chlorocyclopentanol. Note that the OH is understood to be attached to carbon 1 in the ring.

The physical properties of alcohols are strongly influenced by the **hydrogen bonding ability** of the OH group. As in water, the hydroxy hydrogen in an alcohol and a lone pair on a highly electronegative atom (typically F, O, or N) can participate in an unusually strong form of dipole–dipole (electrostatic) interaction. Although much weaker than an ordinary covalent bond, this effect may be worth several kcal mol^{-1} and is much stronger than any other dipole–dipole attraction, which is why it merits the special name *hydrogen bonding*.

Weak dipole–dipole
attractions in
dimethyl ether

Strong "hydrogen bonding" type
dipole–dipole attractions in methanol

Remember, the requirements for hydrogen bonding are δ^+ hydrogens, such as those attached to very electronegative atoms (e.g., N, O, F), and electronegative atoms with lone pairs (again, mainly N, O, and F) to attract them.

8-3. Acidity and Basicity of Alcohols

If you understand the acidic and basic nature of water, then you will need to learn only a little bit that is new here with alcohols. The equilibrium processes are qualitatively similar:

$$H_3O^+ \rightleftharpoons H_2O \rightleftharpoons HO^-$$

As a
base
(adds H^+)

As an
acid
(loses H^+)

$$ROH_2^+ \rightleftharpoons ROH \rightleftharpoons RO^-$$

The differences arise from the presence of the R group (instead of an H), which can affect the relative stabilities of the three species involved. Simple alkyl groups generally destabilize both ROH_2^+ and RO^- in solution, relative to ROH, more than H_3O^+ and OH^- are destabilized, relative to water. So most ordinary alcohols are both weaker acids and weaker bases than water. Electron-withdrawing substituents in R (such as halogens) will **stabilize** RO^-, however, thereby making ROH a stronger acid (see entries in Table 8-2).

   The acidity and basicity of alcohols will play major roles in many of their reactions. When an alcohol acts as a base and is protonated by a strong acid, becoming ROH_2^+, it then contains a good leaving group and is capable of both substitution and elimination reactions (Chapter 9). When an alcohol acts as an acid and loses a proton, it becomes RO^-, a good nucleophile and strong base capable of entering into E2 and S_N2 reactions (Chapters 6 and 9). So this chemistry really serves as a general entry to the more extensive survey of reactions of alcohols coming up later.

8-5. Synthesis of Alcohols by Nucleophilic Substitution

After a section on industrial methods, a review covering S_N2 and S_N1 routes to alcohols is presented. Primary alcohols may be prepared by S_N2 displacement reactions of HO^- with appropriate substrates (e.g., primary haloalkanes). This approach sometimes works for secondary systems, but elimination often interferes. To a limited extent, both secondary and tertiary alcohols may be formed in S_N1 reactions with water as the nucleophile. However, the chemistry described in the remainder of the chapter provides much more versatile and reliable means of synthesizing alcohols.

8-6. Oxidation and Reduction

Carbonyl compounds such as ketones and aldehydes are useful *precursors* (starting materials) for the synthesis of alcohols. Reaction with the hydride reagents $NaBH_4$ and $LiAlH_4$ converts aldehydes to primary alcohols. The same process converts ketones to secondary alcohols. These hydride reductions are the first of many examples that you will see of nucleophilic additions to the electrophilic carbons of carbonyl groups. This is one of the most important classes of reactions in organic chemistry.

The second part of this text section introduces the oxidation of alcohols to aldehydes and ketones, the reverse of reduction. Alcohol oxidation is very useful in that it produces a carbonyl group, the most important functional group of all. Note the two types of reagents based on Cr(VI): PCC (pyridinium chlorochromate, pyH^+ CrO_3Cl^-), which is specifically intended for oxidation of primary alcohols to aldehydes, and aqueous dichromate, which oxidizes secondary alcohols to ketones, but **overoxidizes** 1° alcohols to carboxylic acids.

With these aspects of alcohol preparation and chemistry as background, we now turn to a discussion of molecules that will greatly expand our ability to make bonds: compounds with nucleophilic carbon atoms.

8-7. Organometallic Reagents

So far the only kind of polarized carbons we've looked at in any detail is the δ^+ (electrophilic) carbon that results from its attachment to a very electronegative atom:

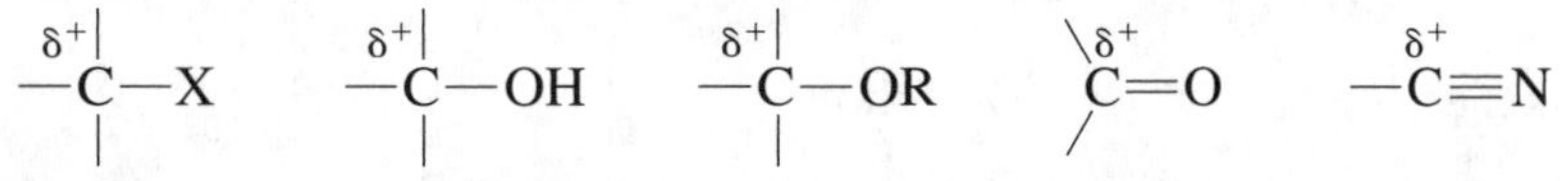

$$\overset{\delta^+}{-}\overset{|}{C}-X \qquad \overset{\delta^+}{-}\overset{|}{C}-OH \qquad \overset{\delta^+}{-}\overset{|}{C}-OR \qquad \overset{\delta^+}{\underset{/}{\overset{\backslash}{C}}}=O \qquad \overset{\delta^+}{-}C\equiv N$$

Haloalkane **Alcohol** **Ether** **Carbonyl** **Nitrile**

If we were to seek a **logical** way to link two carbon atoms in a synthetic process, we would try to take advantage of electrostatics and find molecules with δ^- carbons, which could combine with the δ^+ carbons above. That's very nice, but where can we find δ^- carbon atoms? Logically, if δ^+ carbons result from attachment to electronegative atoms, then δ^- carbons should result from attachment to **electropositive** atoms. **Metals** are the most electropositive elements, so the way to get a δ^- (nucleophilic) carbon would be to attach it to a metal. Compounds with carbon–metal bonds are called organometallic compounds and are sources of nucleophilic carbons. This section describes the preparations of some organometallic compounds. These reactions are easy to do, and the reagents you get are very useful in synthesis. The R groups in RLi and RMgX also act as very strong bases. They are protonated by even weak acids like water or ammonia, giving the hydrocarbon RH as the product:

$$\text{``}R^-\text{''} \quad + \quad \text{``}H^+\text{''} \quad \longrightarrow \quad RH$$

Strong base   From even weakly   Weak acid
(as in R—M)   acidic molecules
                     (e.g., H_2O, ROH, NH_3)

8-8. Organometallic Reagents in the Synthesis of Alcohols

We now come to the primary value of these reagents: their ability to react as nucleophiles toward the electrophilic carbon in carbonyl compounds ($\overset{\delta^+}{>}C=O$). In a reaction mechanistically analogous to the hydride additions of Section 8-6, organometallic reagents add nucleophilic carbon to aldehydes and ketones, resulting in alcohols, and making a new carbon–carbon bond in the process. At the end of the next section, you will find a summary chart of the major types of reactions that convert carbonyl compounds to alcohols.

8-9. An Introduction to Synthetic Strategy

In order to learn how to devise sensible ways to make large organic molecules from small ones (a typical task of synthesis), you need to approach the problem systematically. First, note that the reactions you are learning can be classified into two categories:

1. Reactions that exchange one functional group for another but do not make or break any carbon–carbon bonds. These are called *functional group interconversions,* and a simple example is

$$CH_3I + NaOH \rightarrow CH_3OH + NaI$$

2. Reactions that make or break carbon–carbon bonds. In the text Section 8-8 you saw several very typical examples of this kind of reaction.

There are transformations that overlap the two categories (for instance, S_N2 reactions with cyanide), but let's just keep things simple for now.

Next, you should draw a chart of the general functional group interconversions that we've seen so far. This is how it would look at the moment:

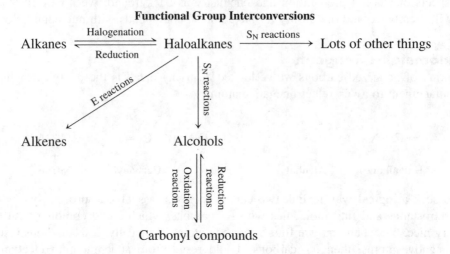

Functional Group Interconversions

It is absolutely necessary to know these interconversion patterns, because they provide the framework for designing synthetic strategy. Suppose we wish to synthesize an alcohol starting with an alkane. From this chart, we see immediately that we have no direct method for converting alkanes to alcohols. We **must first** make a haloalkane and then use it in another reaction to make an alcohol. We set up the proposed synthesis in just that way and insert the specific reagents necessary to carry out the two synthetic steps:

$$CH_4 \; \text{—}\!\otimes\!\text{→} \; CH_3OH \qquad \text{No direct method}$$

$$CH_4 \xrightarrow{Cl_2,\ hv} CH_3Cl \xrightarrow{HO^-} CH_3OH \qquad \text{A sensible synthesis}$$

Note that it is not enough to use the general labels "radical halogenation" or "S_N2 reaction" in the synthesis equations; the actual reagents have to be given.

All right. What about carbon–carbon bond-forming reactions? Syntheses requiring the formation of new bonds are best approached via the *retrosynthetic analysis* described in the text. One works **backward** on paper, imagining **disconnecting** bonds in the desired product molecule, looking for methods that can put together the necessary bonds in an efficient way from reasonable starting molecules. Functional group interconversions are applied as necessary. Notice that by combining the oxidation of alcohols with the one main carbon–carbon bond-forming reaction you've had so far—addition of Grignard reagents to carbonyl compounds to make alcohols—you can now assemble pretty big molecules via synthetic schemes of several steps. The key is the following sequence, made possible by the capability to oxidize alcohols to carbonyl compounds:

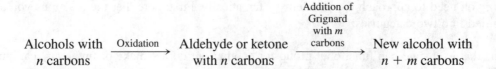

Take some time to look over the examples in the text. Note the preferability of some routes over others, based on efficient incorporation of small molecules into large ones. Finally, apply the techniques to the problems that follow. The practice will not only help you develop and improve your ability to analyze synthesis problems, it will also help you become more and more familiar with all the reactions and reagents. You will have to know them in the end.

SUMMARY CHART
Synthesis of alcohols from carbonyl compounds

Carbonyl compounds	Alcohol product from reaction with	
	$NaBH_4$ or $LiAlH_4$	$R''Li$ or $R''MgX$
Formaldehyde (HCHO)	methanol (CH_3OH)	1° alcohol ($\underline{R''}CH_2OH$)
Aldehyde (RCHO)	1° alcohol (RCH_2OH)	2° alcohol $\left(\begin{array}{c}R\\ \backslash CHOH\\ \underline{R''}\end{array}\right)$
Ketone $\left(\begin{array}{c}R\\ \backslash CO\\ R'\end{array}\right)$	2° alcohol $\left(\begin{array}{c}R\\ \backslash CHOH\\ R'\end{array}\right)$	3° alcohol $\left(\begin{array}{c}R\\ R'-COH\\ \underline{R''}\end{array}\right)$

Solutions to Problems

23. (a) 2-Butanol, 2°   (b) 5-Bromo-3-hexanol, 2°

(c) 2-Propyl-1-pentanol, 1°   (d) (*S*)-1-Chloro-2-propanol, 2°

(e) 1-Ethylcyclobutanol, 3°   (f) (1*R*,2*R*)-2-Bromocyclodecanol, 2°

(g) 2,2-Bis(hydroxymethyl)-1,3-propanediol, 1° ["Bis" is used as the prefix instead of "di" when the name that follows is complicated enough to be in parentheses.]

(h) *meso*-1,2,3,4-Butanetetraol, 1° on C1 and C4, 2° on C2 and C3

(i) (1*R*,2*R*)-2-(2-Hydroxyethyl)cyclopentanol, 2° on ring, 1° on side chain

(j) (*R*)-2-Chloro-2-methyl-1-butanol, 1°

24. (a) $(CH_3)_3SiCH_2CH_2OH$ (b) [structure]

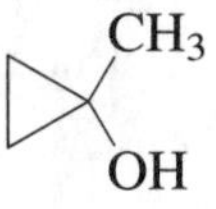

(c) $CH_3CHOHCHCH_2CH_2CH_3$ with $CH(CH_3)_2$ (d) structure

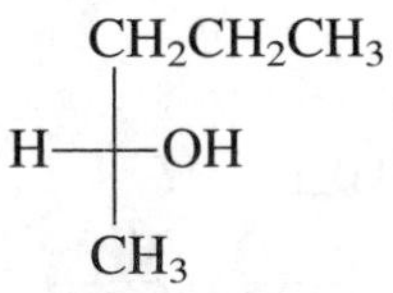

(e) [structure]

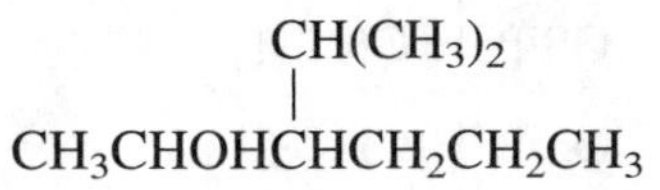

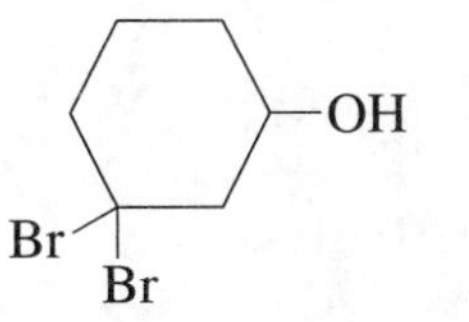

25. (a) Cyclohexanol > chlorocyclohexane > cyclohexane (polarity)

(b) 2-Heptanol > 2-methyl-2-hexanol > 2,3-dimethyl-2-pentanol (branching)

27. **(a)** Ethanol hydrogen bonds to water. Chloroethane is attracted to water by dipole forces. Ethane is nonpolar and attracted least to water.

(b) Solubility decreases as the relative size of the nonpolar portion of a molecule increases.

28. Intramolecular hydrogen bonding stabilizes the *gauche* conformation of 1,2-ethanediol, shown at right, but not the *anti* form. In 2-chloroethanol, similar hydrogen bonding can occur (although it is weaker because of poorer overlap between the large Cl $3p$ lone pair orbital with the small hydrogen):

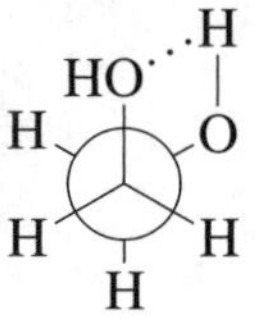

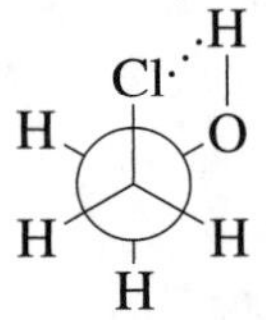

So the conformational ratio of 2-chloroethanol should be more like that of 1,2-ethanediol than 1,2-dichloroethane, in which hydrogen bonding is absent.

29. **(a)** The diequatorial conformation of the diol is shown below, left. It is stabilized in two ways, relative to the diaxial form. First, the —OH groups are larger than —H, and therefore sterics favors the diequatorial. Second, the two hydroxy groups are close enough in the trans-1,2-diequatorial conformation to form an internal hydrogen bond (below, right). The energy of this conformation is further lowered by the strength of this hydrogen bond, about 2 kcal mol^{-1}.

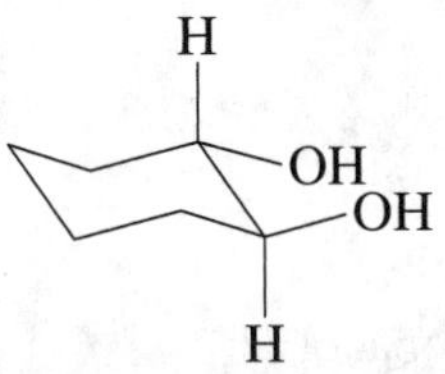

trans-**1,2-Cyclohexanediol**
(both —OH groups equatorial)

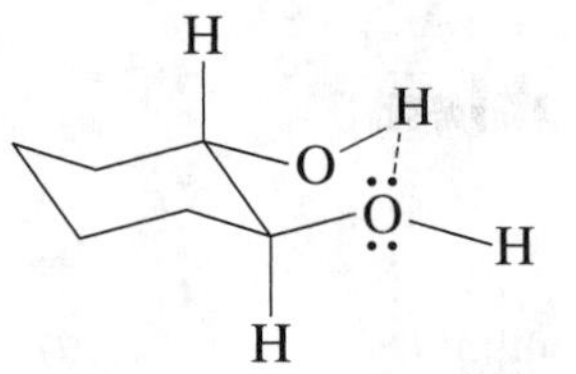

(b) As the models of the diol in part (a) show, adjacent diequatorial substituents are in rather close proximity. A Newman projection view further illustrates this point, revealing that the groups are in a *gauche* relationship. Replacement of the two hydroxy hydrogens with very bulky silyl groups makes the diequatorial form less stable than the diaxial, because the 1,2-silyl-silyl steric interference is greater than the alternative, two pairs of 1,3-silyl-hydrogen diaxial interactions. In addition, with the hydrogens gone from the oxygen atoms no hydrogen bonding is possible to help lower the energy of the diequatorial form. Compare the structures:

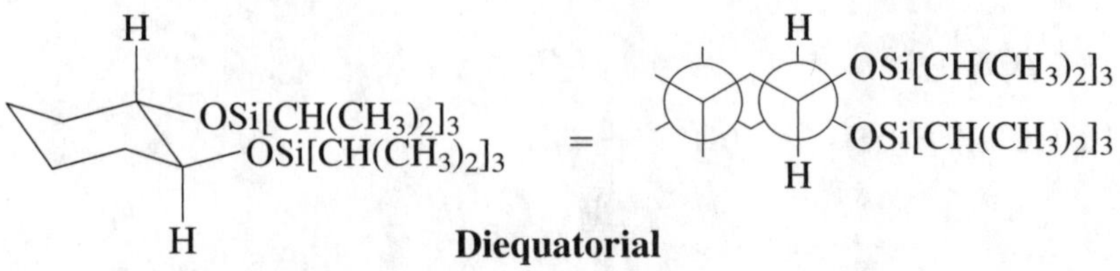

Diequatorial

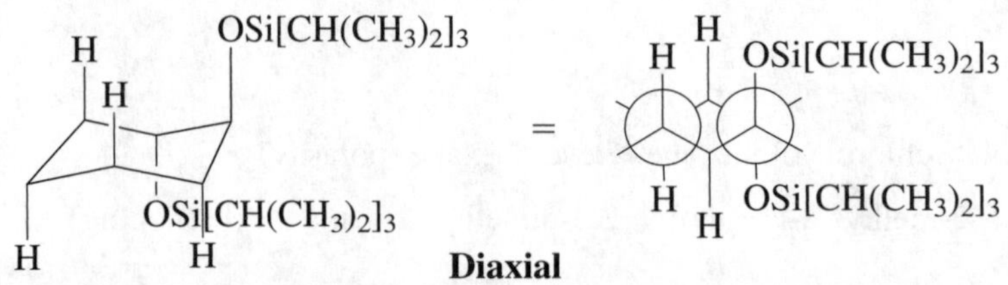

Diaxial

30. Three factors are involved: the electronegativity of the electron-withdrawing atoms, how many there are, and their distance from the hydroxy group.

(a) $CH_3CHClCH_2OH > CH_3CHBrCH_2OH > BrCH_2CH_2CH_2OH$

(b) $CCl_3CH_2OH > CH_3CCl_2CH_2OH > (CH_3)_2CClCH_2OH$

(c) $(CF_3)_2CHOH > (CCl_3)_2CHOH > (CH_3)_2CHOH$

31. (a) $(CH_3)_2C\overset{+}{H}OH_2$ $\xleftarrow{\text{As a base, add } H^+}$ $(CH_3)_2CHOH$ $\xrightarrow{\text{As an acid, add } HO^-}$ $(CH_3)_2CHO^-$

2-Propanol is both a weaker acid and a weaker base than methanol (Tables 8-2 and 8-3).

(b) $CH_3CHFCH_2\overset{+}{O}H_2$ $\xleftarrow{H^+}$ CH_3CHFCH_2OH $\xrightarrow{HO^-, (-H^+)}$ $CH_3CHFCH_2O^-$

(c) $CCl_3CH_2\overset{+}{O}H_2$ $\xleftarrow{H^+}$ CCl_3CH_2OH $\xrightarrow{HO^-, (-H^+)}$ $CCl_3CH_2O^-$

The last two alcohols are both stronger acids and weaker bases than methanol. In each, the alkoxide is stabilized and the alkyloxonium ion destabilized by the electronegative halogens.

32. (a) Halfway between the two pK_a values: pH 6.7. (Compare H_2O: At pH 7, equal amounts of H_3O^+ and HO^- are present.)

(b) pH -2.2 (c) pH $+15.5$

33. No. Recall that stabilization of carbocations by hyperconjugation involves π overlap between a bonding hybrid orbital and an empty p orbital on carbon (Section 7-5). There aren't any empty p orbitals on oxygen in alkyloxonium ions; therefore such overlap is not possible.

34. (a) Worthless. H_2O is a very poor nucleophile in S_N2 reactions.

(b) Good. Excellent S_N2 reaction—CH_3 attached to sulfonate leaving group.

(c) Not so good. Bases give much elimination with 2° haloalkanes.

(d) Good, but slow, via an S_N1 mechanism.

(e) Worthless. ^-CN is a bad leaving group.

(f) Worthless. $^-OCH_3$ is a bad leaving group.

(g) Good. S_N1 first step.

(h) Not so good. Branching reduces S_N2 reactivity, and E2 occurs.

35. (a) Replace the water by hydroxide, a good nucleophile. Run the reaction with NaOH in water or, better yet, run it in a polar aprotic solvent like DMSO.

(c) Two viable options: You can simply heat the substrate in the presence of water and rely on the S_N1 process to give a decent yield. Alternatively, and better, use the acetate substitution—hydrolysis method (Section 8-5). React with sodium acetate in DMF (a high-yield S_N2 displacement because acetate is a weak base but a good nucleophile), and then treat the acetate product with sodium hydroxide in water to effect ester hydrolysis to the alcohol:

(e) and (f) You have not yet encountered any methods to carry out these conversions. Chapter 9 will contain a solution for (f).

(h) The acetate substitution-hydrolysis method given for (c) works better here, too.

36. (a) $CH_3CH_2CHOHCH_3$ (b) $CH_3CHOHCH_2CH_2CHOHCH_3$

(c)

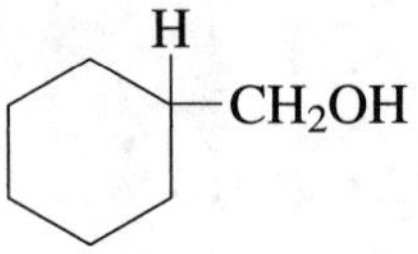

(d)

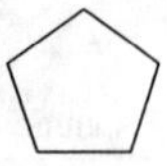

(e) $(CH_3)_2CH$
From addition of hydride to less sterically hindered (bottom) face of ring

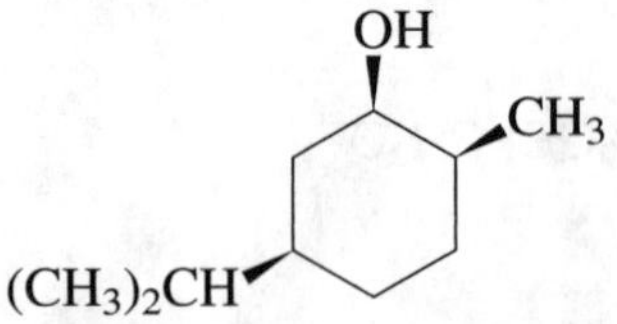

(f) Make a model. Here the main steric interference with hydride addition is by the axial hydrogens on the "top" side of the ring in the chair conformation:

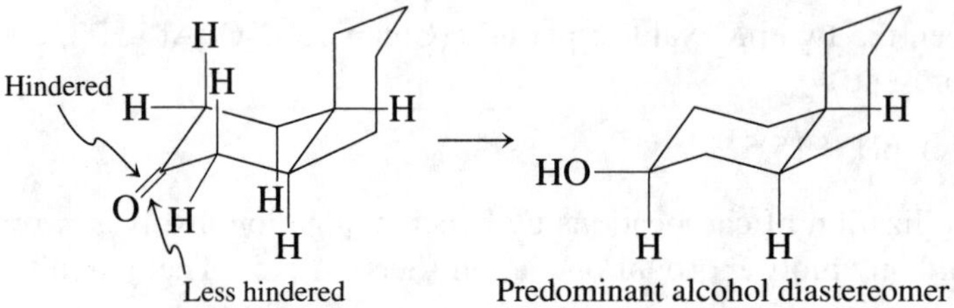

Hindered

Less hindered   Predominant alcohol diastereomer

37. To the right. H_2 is a weaker acid than H_2O, and HO^- is a weaker base than H^-.

38. (a) CH_3CHDOH

(b) CH_3CH_2OD, from reaction of $CH_3CH_2O^-$ with D^+

(c) CH_3CH_2D, from S_N2 reaction. (LiAlD$_4$ serves as a source of "deuteride" nucleophile, D^-, just as LiAlH$_4$ is a source of hydride nucleophile, H^-.)

39. For simplicity we'll just show delivery of a single hydride from one molecule of LiAlH$_4$ (or deuteride from LiAlD$_4$). In reality these reagents can deliver up to all four ions (H:$^-$ or D:$^-$) to as many substrate molecules.

(a)

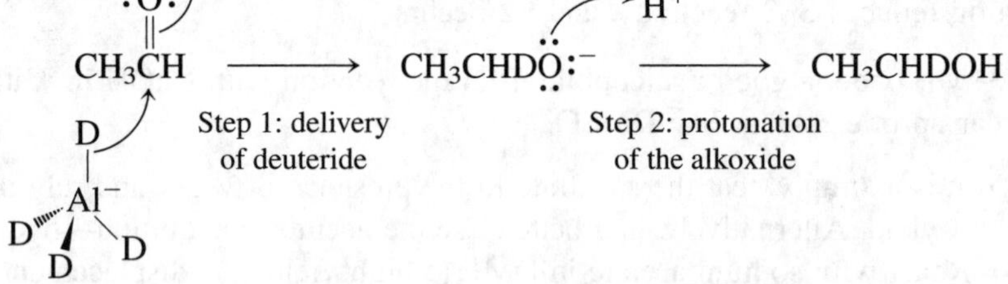

$CH_3CH \longrightarrow CH_3CHDO:^- \longrightarrow CH_3CHDOH$

Step 1: delivery of deuteride   Step 2: protonation of the alkoxide

(b)

$CH_3CH \longrightarrow CH_3CH_2O:^- \longrightarrow CH_3CH_2OD$

Step 1: delivery of hydride   Step 2: alkoxide accepts D^+ from deuterated acid.

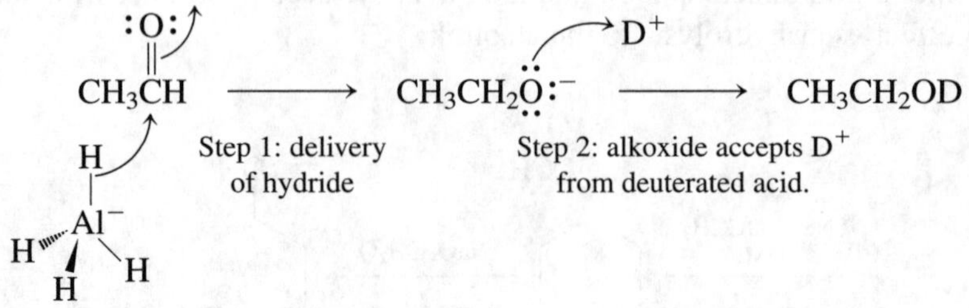

(c) 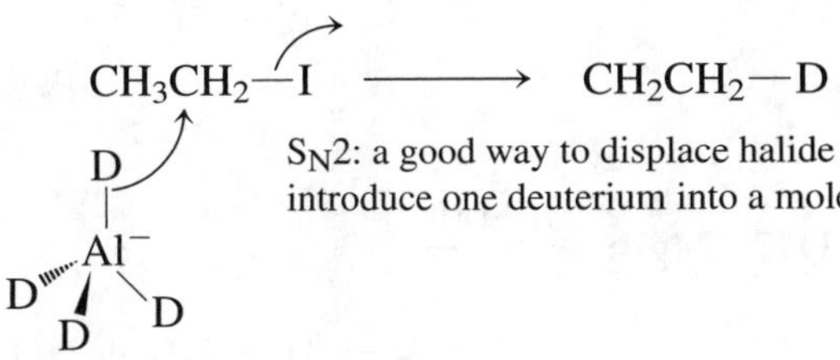 $CH_3CH_2-I \longrightarrow CH_2CH_2-D$

S_N2: a good way to displace halide and thereby introduce one deuterium into a molecule.

40. (a) 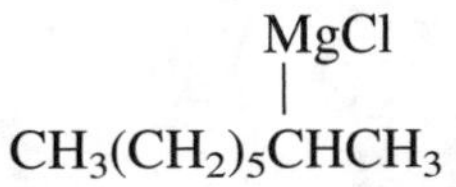 $CH_3(CH_2)_5\overset{\overset{\displaystyle MgCl}{|}}{C}HCH_3$ **(b)** $CH_3(CH_2)_5CHDCH_3$

(c)

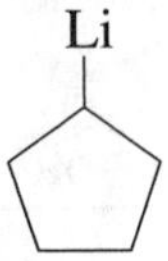

(d)

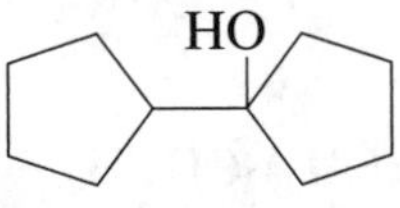

(e) $CH_3CH_2CH_2MgCl$ **(f)**

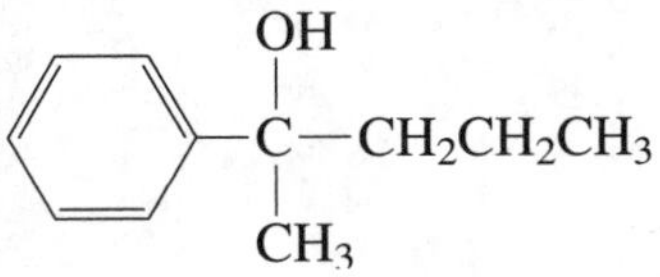

(g) 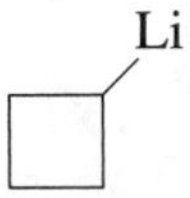   **(h)** 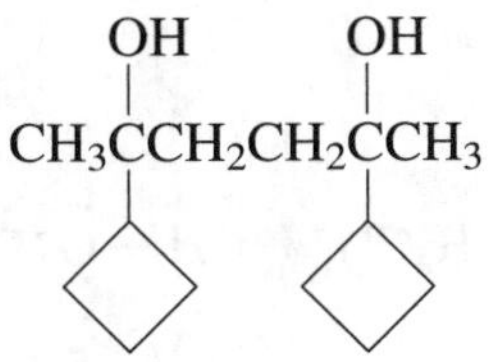 $CH_3\overset{\overset{\displaystyle OH}{|}}{C}CH_2CH_2\overset{\overset{\displaystyle OH}{|}}{C}CH_3$

41. The desired reaction is $CH_3MgI + C_6H_5CHO \rightarrow CH_3CH(OH)C_6H_5$, a synthesis of a secondary alcohol by addition of a Grignard reagent to an aldehyde. The unexpected and undesired side reaction is the addition of methylmagnesium iodide to acetone: $CH_3MgI + CH_3COCH_3 \rightarrow (CH_3)_3COH$, forming 2-methyl-2-propanol (*tert*-butyl alcohol).

42. Only the compounds pictured in parts (a) and (c) will form Grignard reagents in a problem-free manner. In (b) the —OH group contains an acidic hydrogen that will destroy the carbon–metal bond of any Grignard reagent that forms; similarly, the terminal hydrogen on the alkyne function in (e) is acidic enough to do the same (the pK_a is about 25, although all you need to know—from the information in Problem 49 of Chapter 1—is that such hydrogens are much more acidic than hydrogens on alkane carbons). Finally, in (d) the carbonyl function contains a strongly electrophilic carbon and will interfere. The ether function in (c) is not a problem, because it does not contain any acidic hydrogen atoms.

43. Products after hydrolysis are given.

(a) 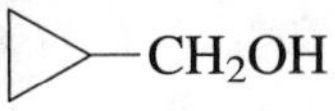 $-CH_2OH$ **(b)** $(CH_3)_2CHCH_2CHOHCH_3$

(c) $C_6H_5CH_2CHOHC_6H_5$ **(d)**

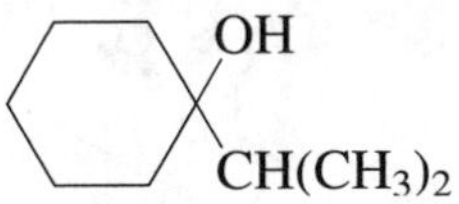

(e)  $-CHOHCH(CH_2CH_3)_2$

44. **(a)**

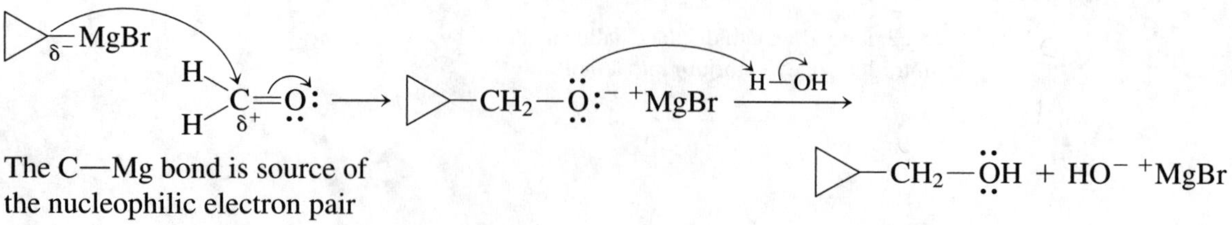

The C—Mg bond is source of
the nucleophilic electron pair

(b)

$(CH_3)_2CHCH_2$—MgCl

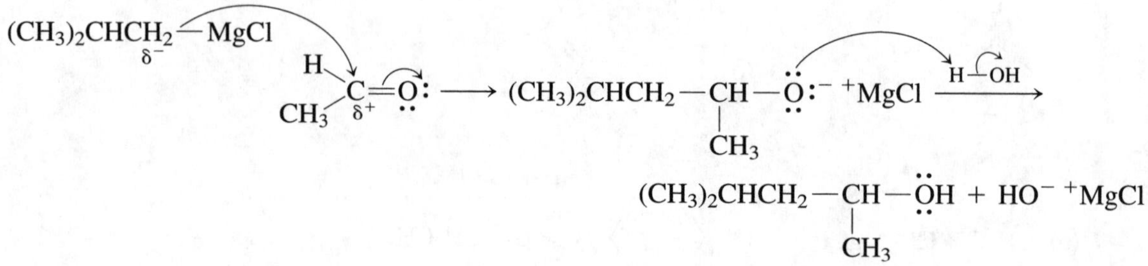

$(CH_3)_2CHCH_2$—CH—$\overset{..}{\underset{..}{O}}$:$^-$ $^+$MgCl

$(CH_3)_2CHCH_2$—CH—$\overset{..}{\underset{..}{O}}$H + HO$^-$ $^+$MgCl

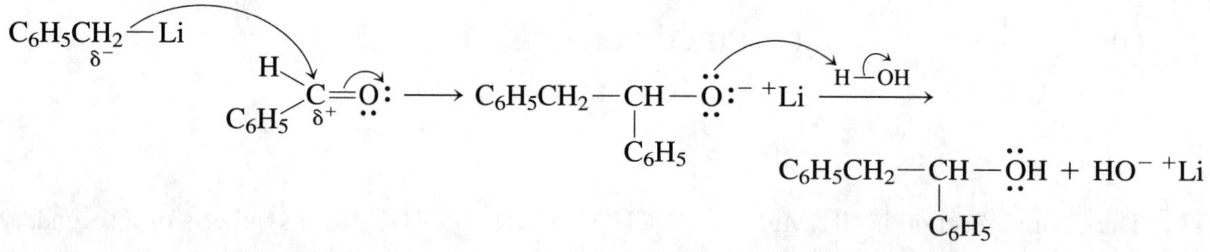

(c)

$C_6H_5CH_2$—Li

$C_6H_5CH_2$—CH—$\overset{..}{\underset{..}{O}}$:$^-$ $^+$Li

$C_6H_5CH_2$—CH—$\overset{..}{\underset{..}{O}}$H + HO$^-$ $^+$Li

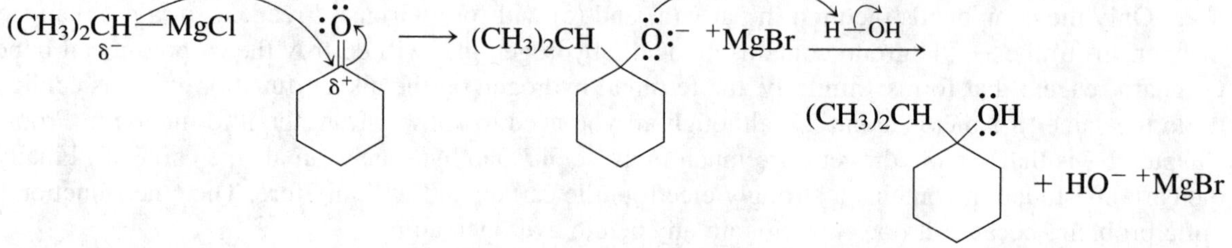

(d)

$(CH_3)_2CH$—MgCl

$(CH_3)_2CH$ $\overset{..}{\underset{..}{O}}$:$^-$ $^+$MgBr

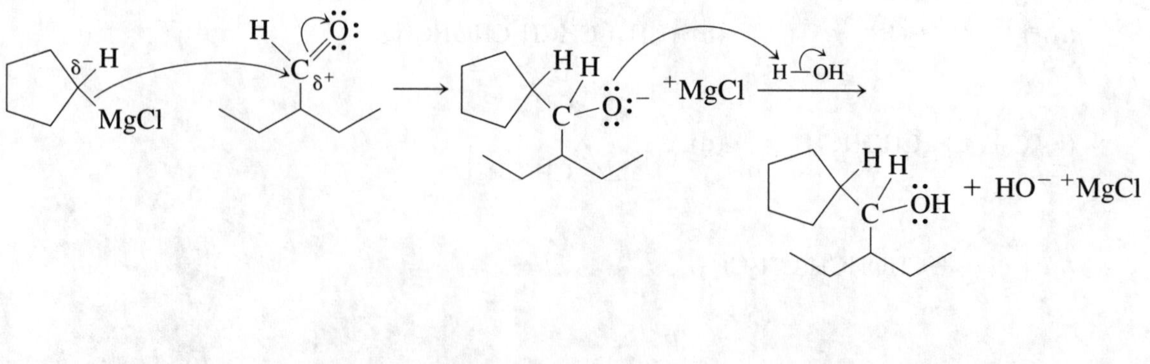

$(CH_3)_2CH$ $\overset{..}{\underset{..}{O}}$H + HO$^-$ $^+$MgBr

(e)

45. (a) 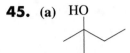 **(b)** 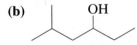 **(c)**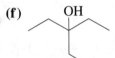

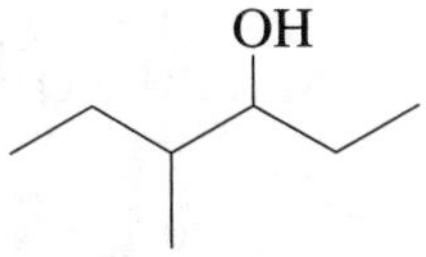

In (b) and (c) the hydroxy groups are located on newly formed stereocenters. The product in (b) forms as a racemic mixture of enantiomers. The starting material in (c) already possessed one stereocenter; addition of a second means that a total of four stereoisomers (two diastereomeric pairs of enantiomers) are formed.

(d) 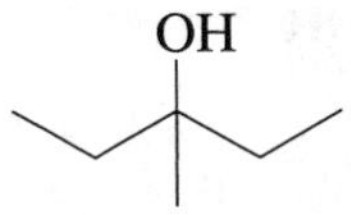 OH **(e)** 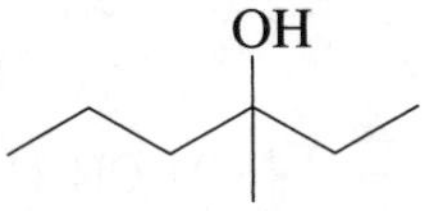 OH **(f)** OH **(g)** 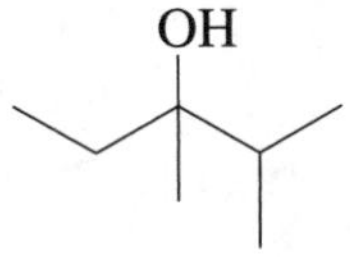 OH

The products in (e) and (g) will both be racemic mixtures of enantiomers.

(h) Two diastereomeric products:

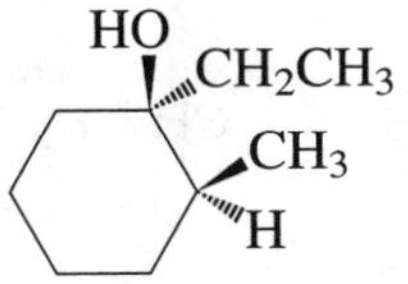

HO
CH$_2$CH$_3$
CH$_3$
H

and

CH$_3$CH$_2$
OH
CH$_3$
H

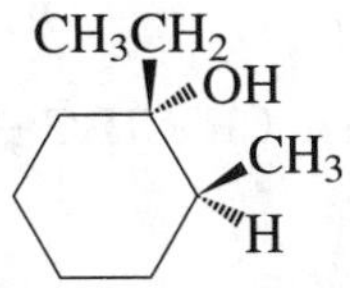

(i) Two enantiomers of

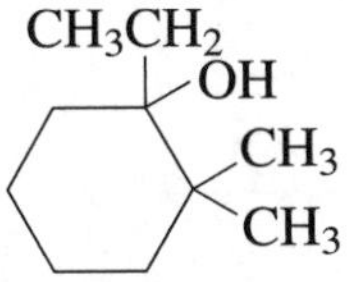

CH$_3$CH$_2$
OH
CH$_3$
CH$_3$

(j)

CH$_3$ H HO
CH$_2$CH$_3$
CH$_3$
H

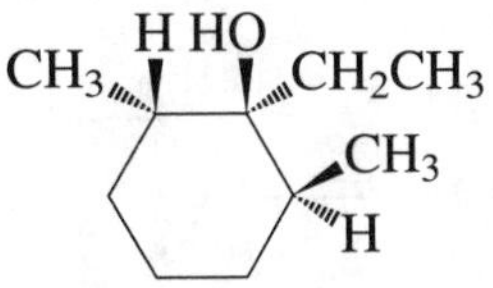

46. (a) CH$_3$CH$_2$CO$_2$H. The carboxylic acid is the major product (from overoxidation) when primary alcohols are treated with Na$_2$Cr$_2$O$_7$ in aqueous H$^+$.

(b) (CH$_3$)$_2$CHCHO

(c)

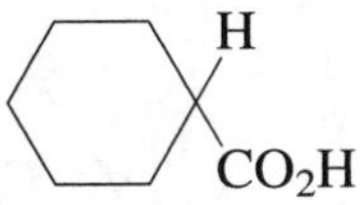

H
CO$_2$H

(d)

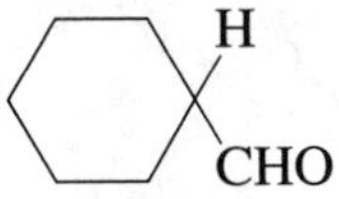

H
CHO

(e)

=O

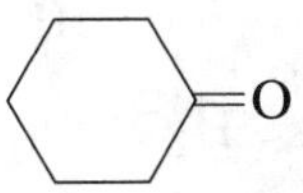

47. Good drill for a mechanism that has a pretty high probability of appearing on an exam. The mechanisms are quite similar regardless of the specific source of Cr(VI)—either Na$_2$Cr$_2$O$_7$ in aqueous acid or PCC in CH$_2$Cl$_2$. In either case the alcohol reacts with the Cr(VI) reagent to give a chromic ester that undergoes elimination. Water is the base using aqueous acidic Na$_2$Cr$_2$O$_7$, as in **(a)** and **(c)**, and pyridine is the base when PCC is the oxidant, as in **(b)**, **(d)**, and **(e)**. Therefore we have:

(a)

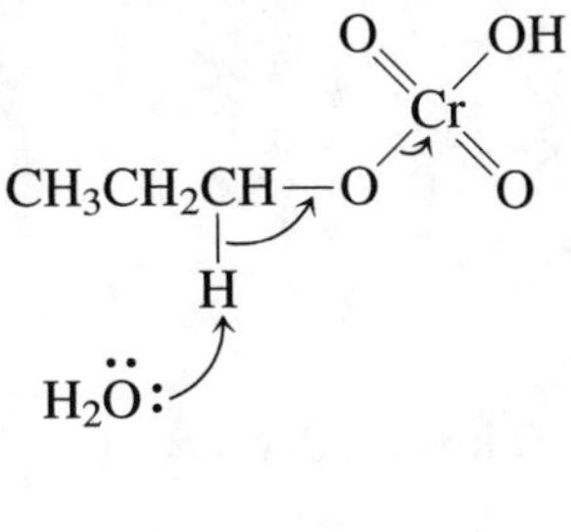

O   OH
Cr
CH$_3$CH$_2$CH—O   O
H

H$_2$Ö:

(b)

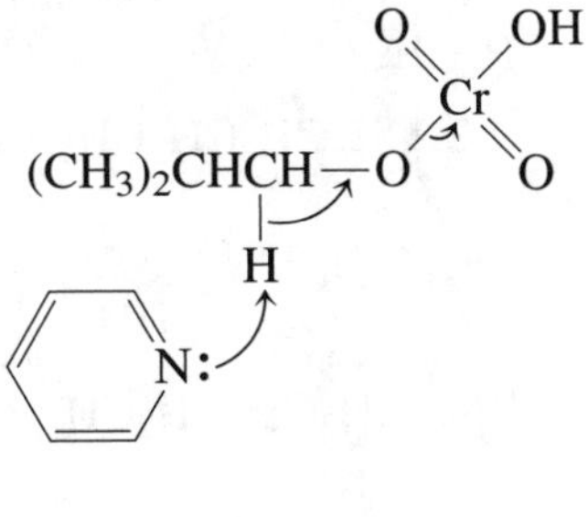

O   OH
Cr
(CH$_3$)$_2$CHCH—O   O
H

N:

(c)

O   OH
H   Cr
CH—O   O
H

H$_2$Ö:

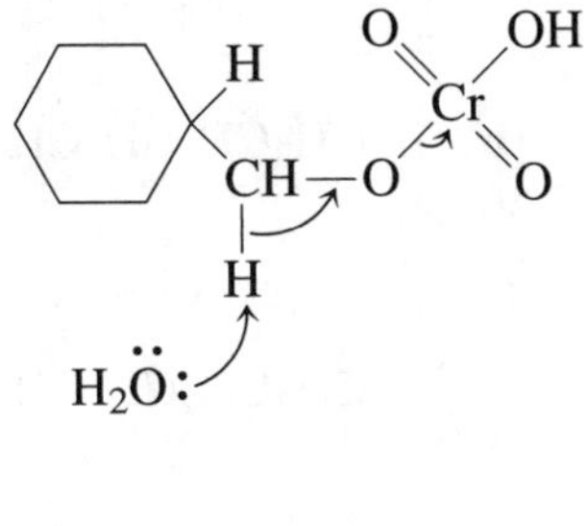

(d)

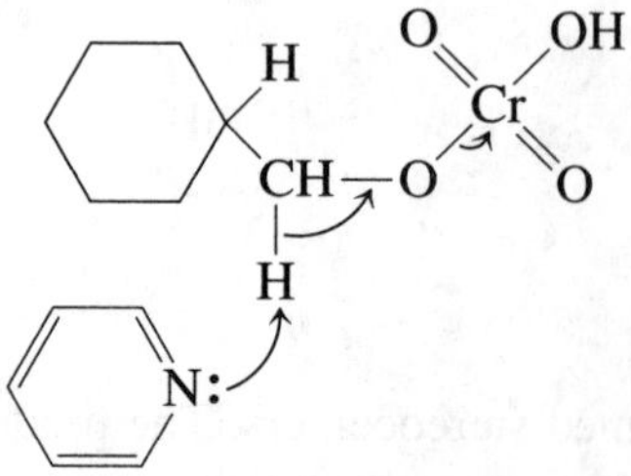

(e)

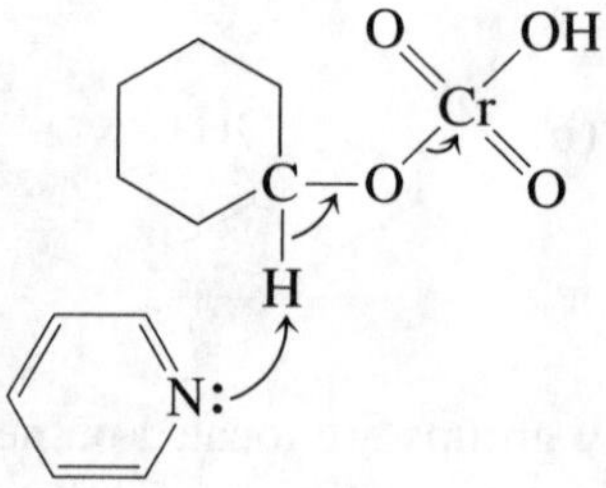

In each case the initial product is the aldehyde or ketone. In (a) and (c) the aldehyde overoxidizes via addition of water to give a 1,1-diol, followed by a second oxidation:

(a)

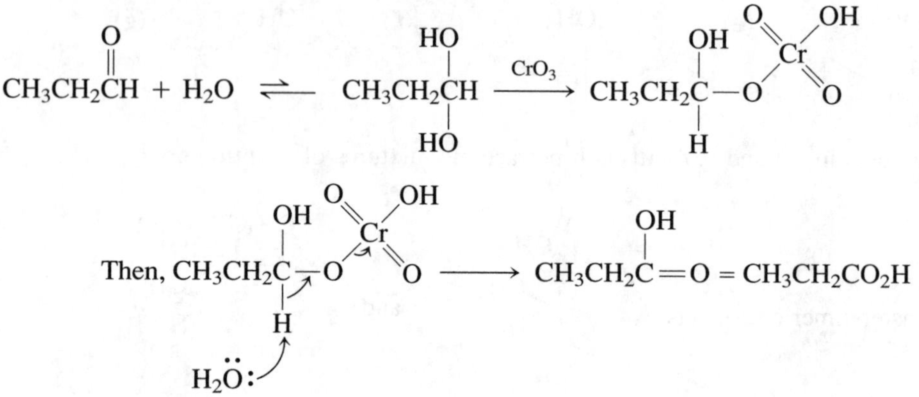

Then,

(c)

followed by

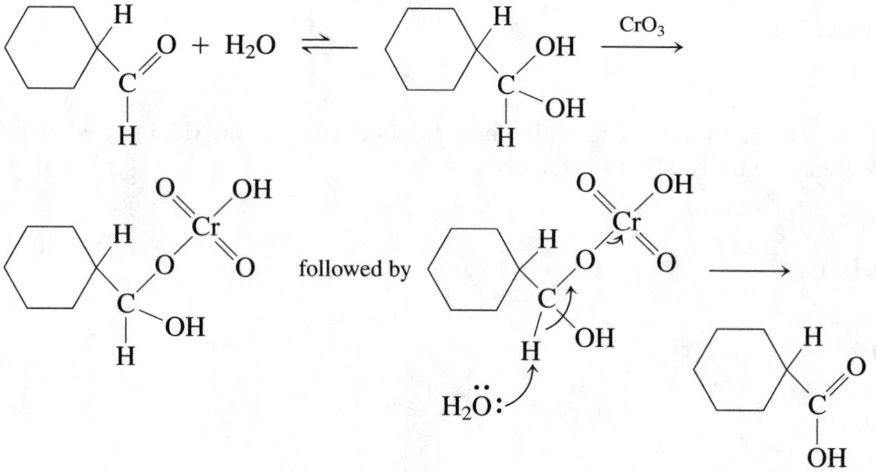

48. No need for you to write anything but the **final** product, but for your information the compounds formed after each step in (a) and (b) are given here.

(a) 1. $(CH_3)_2C=O$; 2. $(CH_3)_2C-OMgBr$; 3. $(CH_3)_2C-OH$ (Final product)
                       | |
                     CH_2CH_3 CH_2CH_3

(b) 1. $CH_3CH_2CH_2CH_2OH$; 2. $CH_3CH_2CH_2CHO$ (Aldehyde);

3. 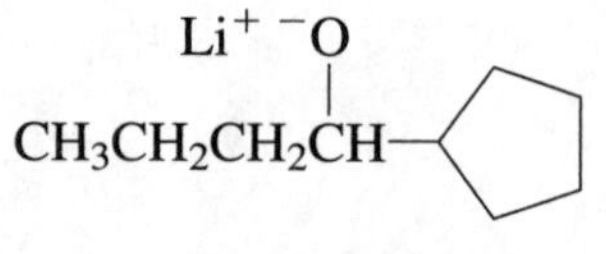 $CH_3CH_2CH_2CH-$⬠ (with Li⁺ ⁻O)   4. 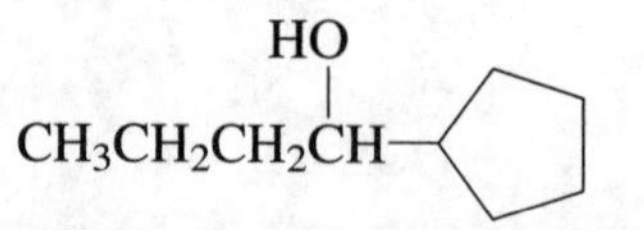 $CH_3CH_2CH_2CH-$⬠ (with HO) (Final product)

(c) 1. $\underset{\text{O}}{\overset{\text{O}}{\text{CH}_3\text{CH}_2\text{CH}_2\text{C}}}$⟨ cyclopentane ⟩   3. $\text{CH}_3\text{CH}_2\text{CH}_2\overset{\text{OH}}{\underset{\text{D}}{\text{C}}}$⟨ cyclopentane ⟩   (Final product)

49. The Wurtz reaction constitutes the direct coupling of the nucleophilic carbon of an alkylsodium compound and the electrophilic carbon in a haloalkane. It is an **unselective** reaction. There is no way to control it to prevent coupling of two "like" alkyl groups while attempting to couple two different ones. In other words, the reaction between chloroethane and 1-chloropropane gives a statistical mixture of butane (from two ethyls), pentane (from an ethyl and a propyl), and hexane (from two propyls).

50. A = $\text{BrMgCH}_2\text{CH}_2\text{CH}_2\text{CH}_2\text{MgBr}$ (A "bis-Grignard" reagent)

B = $\text{CH}_3\text{CHOHCH}_2\text{CH}_2\text{CH}_2\text{CH}_2\text{CHOHCH}_3$ (From addition of each end to aldehyde)

51. (a) $\text{CH}_4 \xrightarrow{\text{Cl}_2,\ hv} \text{CH}_3\text{Cl} \xrightarrow{\text{HO}^-,\ \text{H}_2\text{O}} \text{CH}_3\text{OH}$

(b) Same as (a), starting with ethane.

(c) Same as (a), starting with propane. Not very good.

(d) $\text{CH}_3\text{CH}_2\text{CH}_3 \xrightarrow{\text{Br}_2,\ \Delta} \text{CH}_3\text{CHBrCH}_3 \xrightarrow{\text{H}_2\text{O}} \text{CH}_3\text{CHOHCH}_3$

(e) Same as (a), starting with butane. Not very good.

(f) Same as (d), starting with butane. Much better.

(g) $(\text{CH}_3)_3\text{CH} \xrightarrow{\text{Br}_2,\ \Delta} (\text{CH}_3)_3\text{CBr} \xrightarrow{\text{H}_2\text{O}} (\text{CH}_3)_3\text{COH}$

Beginning with alkanes, the only possible first step is halogenation. Functionalization of 1° carbons is difficult, too: Recall that radical halogenation favors 3° and 2° positions.

52. Note: The halide (X) in all Grignard reagents may be Cl, Br, or I.

(a) From aldehydes, use

$$\text{RCHO} \xrightarrow{\text{NaBH}_4 \text{ or LiAlH}_4} \text{RCH}_2\text{OH}$$

for (a) R = H, (b) R = CH_3, (c) R = CH_3CH_2, and (e) R = $\text{CH}_3\text{CH}_2\text{CH}_2$. Alternatively, for all but the first one, use

$$\text{HCHO} \xrightarrow{\text{RMgX or RLi}} \text{RCH}_2\text{OH}$$

with (b) R = CH_3, (c) R = CH_3CH_2, and (e) R = $\text{CH}_3\text{CH}_2\text{CH}_2$. Finally,

(d) $\text{CH}_3\text{CHO} \xrightarrow{\text{CH}_3\text{MgX or CH}_3\text{Li}} \text{CH}_3\text{CHOHCH}_3$ and

(f) either $\text{CH}_3\text{CHO} \xrightarrow{\text{CH}_3\text{CH}_2\text{MgX or CH}_3\text{CH}_2\text{Li}} \text{CH}_3\text{CHOHCH}_2\text{CH}_3$ or

$\text{CH}_3\text{CH}_2\text{CHO} \xrightarrow{\text{CH}_3\text{MgX or CH}_3\text{Li}} \text{CH}_3\text{CH}_2\text{CHOHCH}_3$

(ii) From ketones, use

$$\overset{\displaystyle O}{\underset{\displaystyle RCR'}{\|}} \xrightarrow{\text{NaBH}_4 \text{ or LiAlH}_4} \text{RCHOHR}'$$

for (d) both R and R' = CH_3 and (f) R = CH_3, R' = CH_3CH_2. Also

(g) $CH_3\overset{\displaystyle O}{\overset{\displaystyle \|}{C}}CH_3 \xrightarrow{\text{CH}_3\text{MgX or CH}_3\text{Li}} (CH_3)_3COH$

A suitable solvent for all these reactions is $(CH_3CH_2)_2O$, and protonation of all alkoxide products with aqueous acid is required to generate each final alcohol.

53. (a) ⬠—OH $\xrightarrow{\text{Na}_2\text{Cr}_2\text{O}_7, \text{H}_2\text{SO}_4, \text{H}_2\text{O}}$

(b) $CH_3CH_2CH_2CH_2CH_2OH \xrightarrow{\text{Na}_2\text{Cr}_2\text{O}_7, \text{H}_2\text{SO}_4, \text{H}_2\text{O}}$

(c) ⬡—$CH_2OH \xrightarrow{\text{*PCC, CH}_2\text{Cl}_2}$

(d) $(CH_3)_2CHCHOHCH_3 \xrightarrow{\text{Na}_2\text{Cr}_2\text{O}_7, \text{H}_2\text{SO}_4, \text{H}_2\text{O}}$

(e) $CH_3CH_2OH \xrightarrow{\text{*PCC, CH}_2\text{Cl}_2}$

*Use PCC to avoid overoxidation.

54. Target molecule is $CH_3\overset{\displaystyle OH}{\underset{\displaystyle CH_3}{\overset{\displaystyle |}{\underset{\displaystyle |}{C}}}}CH_2CH_2CH_2CH_3$.

(a) $CH_3\overset{\displaystyle O}{\overset{\displaystyle \|}{C}}CH_3 + CH_3CH_2CH_2CH_2Li$, followed by H^+, H_2O.

(b) $CH_3\overset{\displaystyle O}{\overset{\displaystyle \|}{C}}CH_2CH_2CH_2CH_3 + CH_3Li$, followed by H^+, H_2O.

(c) This is a bit more tricky. Analyzing retrosynthetically, we look at the bonds in the product that need to be formed if we start with the 5-carbon aldehyde:

$$CH_3\overset{\displaystyle OH}{\underset{\displaystyle CH_3}{\overset{b\,|}{\underset{a\,|}{C}}}}CH_2CH_2CH_2CH_3 \Rightarrow CH_3\overset{\displaystyle O}{\overset{\|}{C}}CH_2CH_2CH_2CH_3 \Rightarrow CH_3\overset{\displaystyle OH}{\overset{|}{C}}HCH_2CH_2CH_2CH_3$$

Starting with an **aldehyde,** we will need to make two bonds ("a" and "b") by adding **two** methyl groups. Working backward, bond "a" is addition of a methyl nucleophile to a ketone. Where does the ketone come from? It must be via **oxidation** of a secondary alcohol. The secondary alcohol then comes from formation of bond "b" to the aldehyde. The answer: $CH_3CH_2CH_2CH_2CHO + CH_3Li$ gives $CH_3CH_2CH_2CH_2CHOHCH_3$. Oxidation of this secondary alcohol with Cr(VI) in aqueous H^+ gives the ketone $CH_3CH_2CH_2CH_2COCH_3$. Addition of a second CH_3Li to this ketone ends in the desired tertiary alcohol. This sequence may be written in the following way:

$$CH_3CH_2CH_2CH_2CHO \xrightarrow[\substack{\text{2. } H^+, H_2O \\ \text{3. } Na_2Cr_2O_7, H_2SO_4, H_2O}]{\text{1. } CH_3Li} CH_3CH_2CH_2CH_2COCH_3 \xrightarrow[\text{2. } H^+, H_2O]{\text{1. } CH_3Li} \text{product}$$

Grignards may be used in place of alkyllithiums in any of these. The solvent in each case is $(CH_3CH_2)_2O$.

55. (a) $CH_3CH_2COCH_2CH_2CH_2CH_2CH_3 + LiAlH_4$ or $NaBH_4$, CH_3CH_2OH

(b) $CH_3CH_2CHO + CH_3CH_2CH_2CH_2CH_2MgBr$ (Lithium reagent okay, too)

(c) $CH_3CH_2CH_2CH_2CH_2CHO + CH_3CH_2MgBr$ (Lithium reagent okay, too)

The solvent in each case is $(CH_3CH_2)_2O$, unless indicated otherwise.

56. Transformation **A** is replacement of a tertiary hydrogen by bromine. There are two tertiary hydrogens in the molecule, but they are at equivalent positions because of symmetry. All the other hydrogens in the compound are secondary. Therefore, this step can be achieved by a simple radical bromination, which we know is highly selective for tertiary over secondary hydrogens (Chapter 3).

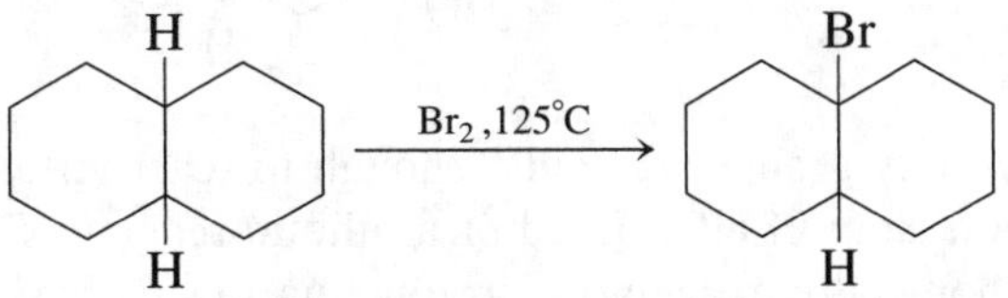

Transformation **B** requires carbon–carbon bond formation, and generation of alcohol functionality, for which we have a new strategy in this chapter: addition of an organometallic reagent (Grignard or lithium) to a carbonyl compound. So the question becomes, which organometallic reagent and which carbonyl compound? The answers are to be found in the structures already given in the problem. The required new carbon–carbon bond goes from the bromine-bearing carbon to a hydroxy-bearing carbon of a new 2-carbon substituent. Based upon the models in this chapter, the solution is to convert the brominated product of transformation **A** into an organometallic reagent, and to react this species with a 2-carbon carbonyl compound:

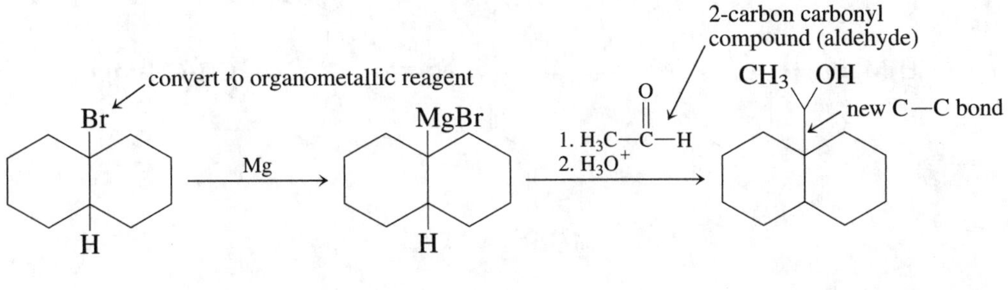

57. $CH_3(CH_2)_{14}\overset{\overset{\displaystyle O}{\|}}{C}O^- + I(CH_2)_{15}CH_3$

58. (a) CH_3CH_2OH

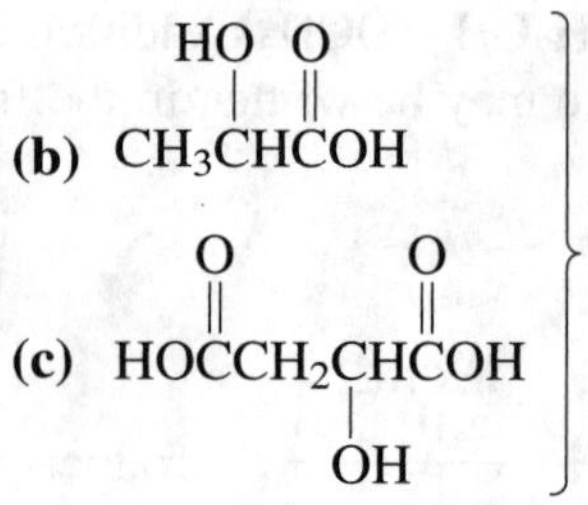

(b) CH$_3$CHCOH with HO and O groups

(c) HOCCH$_2$CHCOH with O, O, and OH groups

Only the ketone carbonyl is reduced in these.

59.

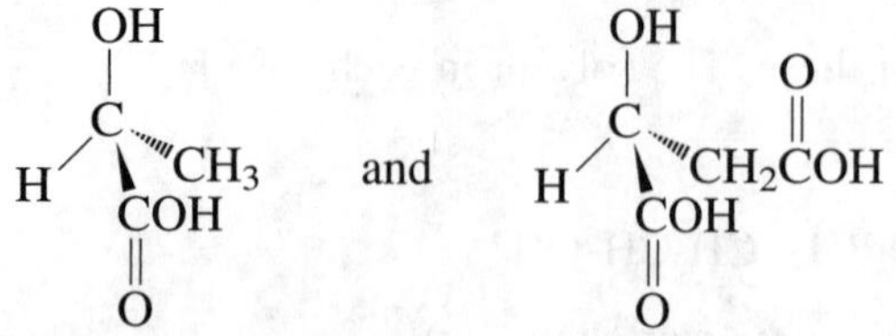

and

60. The bottom (α) face of the steroid is less hindered.

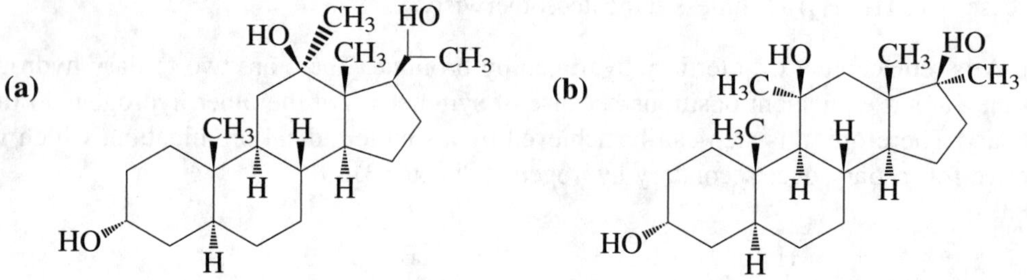

(a)

(b)

61. The hydrogen atoms in hydroxy groups are acidic enough to react virtually instantaneously with the highly basic metal-bearing carbon atom of a Grignard or a lithium reagent. This process, which is an acid-base reaction, destroys the organometallic reagent by protonating this carbon and forming an inert alkane. The alcohol group is converted to its conjugate base, an alkoxide, with a positively charged magnesium or lithium counterion. Such acid-base reactions are much faster than addition reactions of organometallic reagents to carbonyl group C=O double bonds. Therefore, enough excess organometallic species has to be introduced to compensate for the amount that is destroyed by this acid-base process, with enough left over to then add to the carbonyl groups in the substrate molecule.

62. Design each strategy beginning with disconnection of one of the bonds to the alcohol carbon atom (the "strategic bonds").

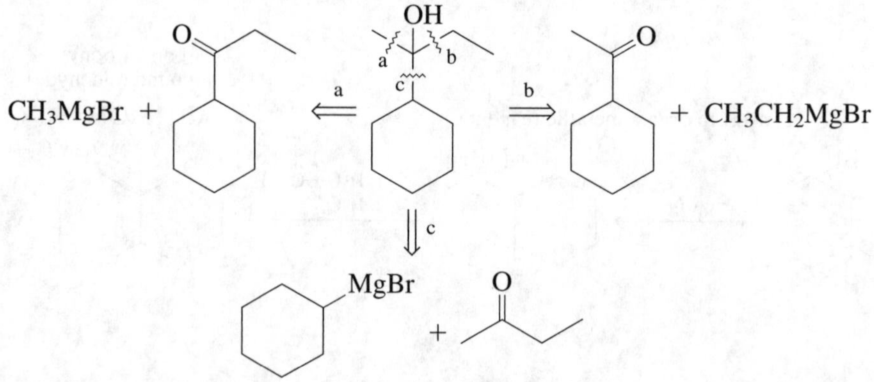

$$CH_3MgBr +$$ ... $$+ CH_3CH_2MgBr$$

Next, consider how readily available (and if available, how expensive) are the starting materials arising from each disconnection. For Grignard reagents, consider the corresponding halides; for ketones, consider the corresponding alcohols:

Disconnection "a"	Disconnection "b"	Disconnection "c"
CH_3Br \$640/kg	CH_3CH_2Br \$36/kg	$CH_3CHOHCH_2CH_3$ \$31/kg

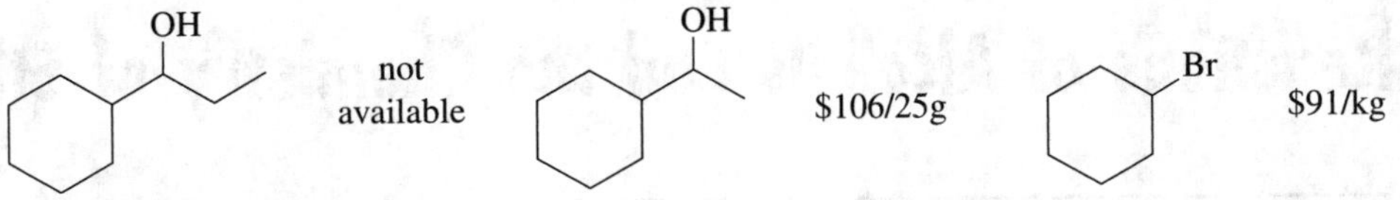
OH
not available
OH
$106/25g
Br
$91/kg

Although 1-cyclohexyl-1-propanol could certainly be prepared by oxidation of cyclohexylmethanol to the aldehyde and addition of CH_3CH_2MgBr, the extra effort and expense makes route "a" a poor choice. Route "b" is not bad, but "c" is certainly the most efficient overall. Thus:

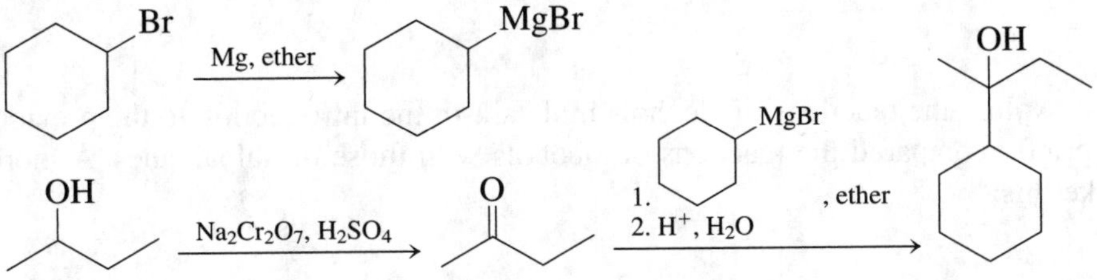
Br
Mg, ether
MgBr
OH
MgBr
OH
Na₂Cr₂O₇, H₂SO₄
O
1.
, ether
2. H⁺, H₂O

63. (b)

64. (b)

65. (c)

66. (b)

9

Further Reactions of Alcohols and the Chemistry of Ethers

In this chapter we explore the reactions of alcohols in detail. In the introduction to the previous chapter in this study guide, we briefly compared the reactions of alcohols with those of haloalkanes. A more detailed analysis might look like this:

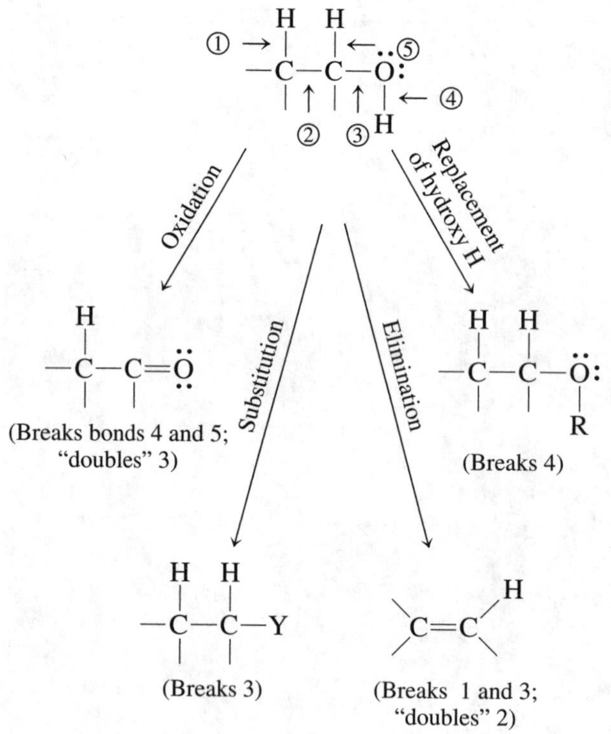

Oxidation
Substitution
Elimination
Replacement of hydroxy H
(Breaks bonds 4 and 5; "doubles" 3)
(Breaks 4)
(Breaks 3)
(Breaks 1 and 3; "doubles" 2)

Indeed, any of **five** bonds may be involved in alcohol chemistry, and if we treat substitution and elimination separately, a total of four types of reactions are possible, as shown. A related class of compounds—ethers—is also presented. Because these lack an oxygen–hydrogen bond, the two reactions of alcohols that involve the O—H bond are not available to ethers. In fact, only substitution reactions turn out to be important in the chemistry of ethers, and those occur only under certain sets of conditions, depending on the nature of the ether. By and large, ethers, in contrast with alcohols, have been found to be very **unreactive** molecules, which results in their usefulness as solvents for a wide variety of reactions in organic chemistry.

Outline of the Chapter

Keys to the Chapter

9-1. Preparation of Alkoxides

We have already seen how the acidity of alcohols resembles the acidity of water. Here two general approaches are presented for the removal of a proton from an alcohol to form an alkoxide ion: reaction with strong bases (such as $[(CH_3)_2CH]_2N^-$ or hydride) and reaction with active metals (especially alkali metals). Alkoxides are readily available species whose reactions will be explored at several places in this chapter.

9-2. Alkyloxonium Ions: Substitution and Elimination Reactions of Alcohols

The other side of the acid-base story for alcohols is their basicity: Just like water, they can be protonated by strong acid, making alkyloxonium ions. These turn out to be very important in the chemistry of alcohols, because they allow reactions that involve cleavage of the carbon–oxygen bond. This bond is hard to break under neutral or basic conditions, and a comparison with haloalkanes shows why: Haloalkanes already possess a good leaving group (halide ion), whereas alcohols do not. For instance, compare the following reactions.

$$\text{Nuc:}^- + \text{R}—\ddot{\text{X}}\text{:} \longrightarrow \text{R}—\text{Nuc} + \text{:}\ddot{\text{X}}\text{:}^- \quad \text{Good leaving group}$$

$$\text{Nuc:}^- + \text{R}—\ddot{\text{O}}\text{H} \longrightarrow \text{R}—\text{Nuc} + \text{H}\ddot{\text{O}}\text{:}^- \quad \text{Bad leaving group}$$

Alcohols require improvement of their leaving group before they can become substrates in substitution reactions. The most common way to do this is protonation of the oxygen atom with strong acid. This reaction converts a bad leaving group (HO^-) into a good one (H_2O, about as good as Br^-). Then, 1° alcohols can undergo S_N2 reactions, and 2° and 3° alcohols can undergo S_N1 reactions. Common nucleophiles in these processes are halide ions, which form haloalkanes, and other molecules of alcohol, which form ethers. As always, eliminations can compete with these substitutions, especially at high temperatures, and alkenes are the products of the very important *acid-catalyzed dehydration* of alcohols.

9-3. Carbocation Rearrangements

So far you have seen two reactions of carbocations: combination with a nucleophile (the second step of the S_N1 process) and loss of a proton (the second step of the E1 process). There are more, as you might expect. After all, carbocations are very reactive species, and they will do just about anything to find sources of electrons. They can even attack unsuspecting atoms or groups in their own molecule, moving the atom or group together with its bonding electrons, from its original location to the positively charged carbon. Such shifts of atoms or groups from one place in a molecule to another are called *rearrangements*. The most common kind of rearrangement is that shown in this text section: shift of a hydride ($H:^-$) or an alkyl group from one atom to another, with the electrons of the breaking bond, to generate a more stable carbocation from a less stable one. The most typical example is a rearrangement that changes a 2° carbocation into a 3° one, a thermodynamically favorable process. Other common types of shifts turn 2° ions into new 2° ions and 3° ions into new 3° ions. All these are liable to occur whenever "rearrangeable" carbocations are formed, namely, in the first steps of S_N1 or E1 reactions of appropriately constructed molecules. In addition, protonated 1° alcohols like 2,2-dimethyl-1-propanol can sometimes change directly to 2° or 3° carbocations via simultaneous ionization and rearrangement, even though they don't undergo simple ionization to 1° ions. A short list of examples of the main types follows.

1. *2° → 2° via hydride shift*

$$CH_3\overset{+}{C}HCHCH_3 \rightleftharpoons CH_3\overset{+}{C}HCHCH_3 \qquad \text{Readily reversible}$$
$$\quad\;\; | \qquad\qquad\qquad\quad |$$
$$\quad\;\; H \qquad\qquad\qquad\quad H$$

2a. *2° → 3° via hydride shift*

$$(CH_3)_2\overset{+}{C}CHCH_3 \rightleftharpoons (CH_3)_2\overset{+}{C}CHCH_3 \qquad \text{Reversible but normally}$$
$$\qquad\quad | \qquad\qquad\qquad\qquad | \qquad\qquad\qquad \text{favored in direction shown}$$
$$\qquad\quad H \qquad\qquad\qquad\qquad H$$

2b. *2° → 3° via alkyl shift*

$$(CH_3)_2\overset{+}{C}CHCH_3 \longrightarrow (CH_3)_2\overset{+}{C}CHCH_3 \qquad \text{Normally not reversible; but product}$$
$$\qquad\quad | \qquad\qquad\qquad\qquad\quad | \qquad\qquad\qquad \text{ion can undergo } 3° \rightleftharpoons 3°$$
$$\qquad\quad CH_3 \qquad\qquad\qquad\quad CH_3 \qquad\qquad\quad \text{interconversion; see next example}$$

3a. *3° → 3° via hydride shift*

$$(CH_3)_2\overset{+}{C}C(CH_3)_2 \rightleftharpoons (CH_3)_2\overset{+}{C}C(CH_3)_2 \qquad \text{Reversible}$$
$$\qquad\quad | \qquad\qquad\qquad\qquad\quad |$$
$$\qquad\quad H \qquad\qquad\qquad\qquad\quad H$$

3b. *3° → 3° via alkyl shift*

$$(CH_3)_2\overset{+}{C}C(CH_3)_2 \rightleftharpoons (CH_3)_2\overset{+}{C}C(CH_3)_2 \qquad \text{Reversible}$$
$$\qquad\quad | \qquad\qquad\qquad\qquad\quad |$$
$$\qquad\quad CH_3 \qquad\qquad\qquad\qquad CH_3$$

4. *"1°" → 2° via hydride shift*

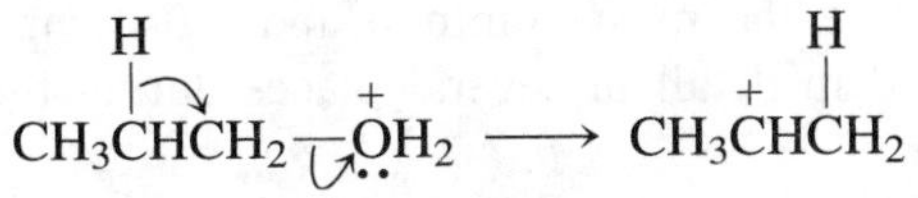
H
CH3CHCH2—OH2
H
CH3CHCH2

5a. *"1°" → 3° via hydride shift*

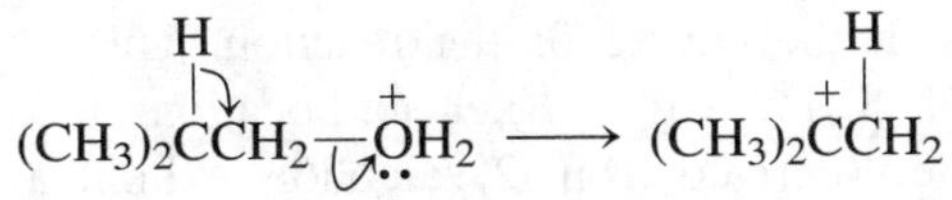
H
(CH3)2CCH2—OH2
H
(CH3)2CCH2

5b. *"1°" → 3° via alkyl shift*

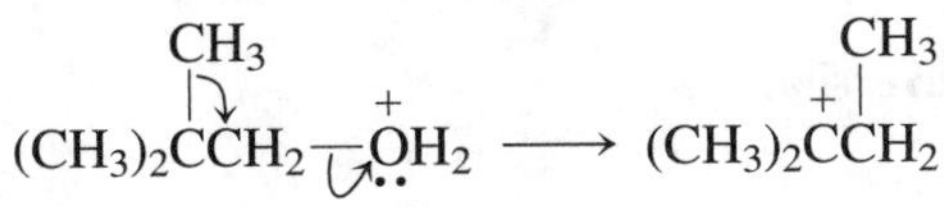
CH3
(CH3)2CCH2—OH2
CH3
(CH3)2CCH2

Notice that in every example of carbocation rearrangement the migrating atom (or group) and the (+) charge switch places.

In some of the problems you will be asked to find products of rearrangement of carbocations in ring compounds. One of the hardest types of shifts to visualize at first is alkyl migration when the alkyl group is part of a ring. This *ring-bond migration* has the effect of **changing the number of atoms in the ring.** Here is an example showing how a secondary cycloheptyl cation can rearrange to become tertiary in two ways.

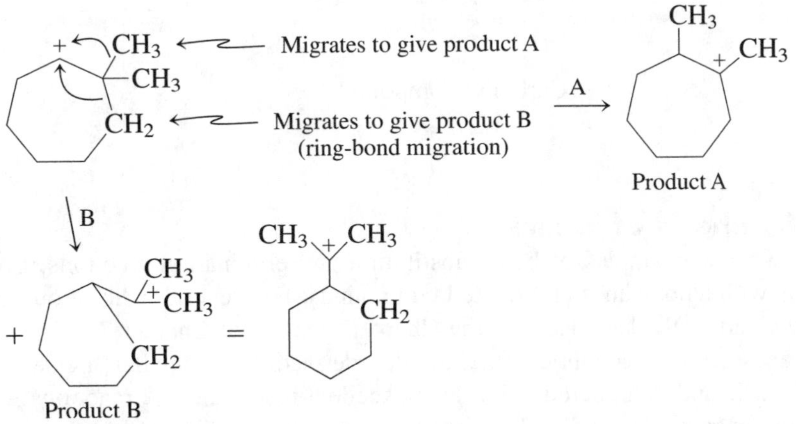
CH3
CH3
CH2
Migrates to give product A
Migrates to give product B
(ring-bond migration)
A
CH3
CH3
Product A
B
CH3
CH3
CH2
Product B
CH3
CH3
CH2

Methyl migration to give A is no different from alkyl shifts you've seen already. To understand the result of migration of the **ring CH_2 group,** follow the bonding change: The bond between the CH_2 and the carbon with the methyls breaks, and the CH_2 forms a new bond to the original carbocation carbon. The result is B. Now if you count the number of atoms in this new, funny-looking ring, it turns out to be six. So then redraw it properly like a normal six-membered ring (shown above). Part of the driving force for this ring-bond shift is the formation of a less strained ring. Now see if you can solve this practice problem: Write the carbocations that can form when the compound shown in the margin is treated with strong acid.

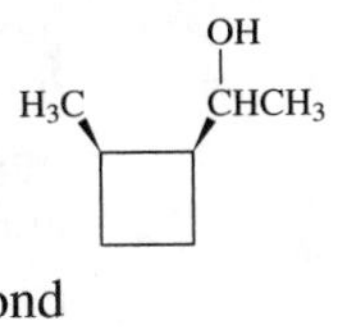
OH
H3C
CHCH3

Finally note that all of these rearranged carbocations can either combine with a nucleophile to give a substitution product or lose a proton to give an alkene (elimination) **just like any other carbocation.**

9-4. Esters from Alcohols

The reversible reaction of alcohols and carboxylic acids to make organic esters is presented here only to alert you to the major connection alcohols have with esters. Esters are the most common and most important *carboxylic acid derivatives,* and their chemistry will be explored in detail in several places during the last third of the course.

Inorganic esters serve useful purposes as synthetic intermediates for certain functional group interconversions. Here, alternative ways to transform alcohols into haloalkanes using these compounds are shown to be often superior to the more "classical" method involving acid-catalyzed substitution. Whereas the latter is frequently susceptible to rearrangement, the phosphorus and sulfur reagents presented here can often allow substitutions to occur without having rearrangements interfere with the course of the reaction. This is most noticeable with 2° alcohols. Upon protonation of a 2° alcohol, S_N1 reactivity (i.e., carbocation chemistry) predominates. In contrast, the leaving groups of inorganic esters derived from 2° alcohols exhibit a more moderate and well-behaved S_N2 reactivity. This can be **very** useful.

With these reactions and the other reactions so far described in this chapter, the chart of functional group interconversions first presented in Chapter 8 has grown to look like this:

Functional Group Interconversions

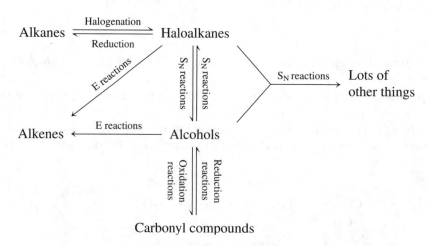

9-6 and 9-7. Synthesis of Ethers

In Chapters 6 and 7 we saw examples of both substitution and elimination reactions involving alcohols and alkoxides in reactions with haloalkanes and related compounds. For general calibration purposes, please refer now to the Summary Chart in the last "Keys to the Chapter" section of Chapter 7 of this study guide ("Major reactions of haloalkanes with nucleophiles"). Alkoxides derived from smaller alcohols are comparable to hydroxide—strongly basic and unhindered—and give excellent results in S_N2 reactions with both methyl and 1° halides (fourth column, rows 1 and 2). These are the prototypical Williamson ether syntheses, and several are illustrated in the text. Increased bulk in either the alkoxide [e.g., $(CH_3)_2CHO^-$ or $(CH_3)_3CO^-$] or the haloalkane (branched 1°, 2°, etc.) tends to increase the amount of E2 reaction at the expense of S_N2 chemistry (refer to the "three questions" for favoring elimination or substitution, also in Chapter 7 of the text and also this study guide). Obviously, 3° halides are worthless in the Williamson ether synthesis because they give only elimination products upon reaction with the strongly basic alkoxide reagent. Normal considerations of kinetics and stereochemistry apply, of course, to these substitution and elimination processes.

In contrast with alkoxides, alcohols are poor nucleophiles, like water (see second column in "Major reactions of haloalkanes with nucleophiles" chart). However, alcohols can act as nucleophiles to make ethers in either of two ways: strongly acidic conditions when no other nucleophiles are present, and solvolytic (S_N1) conditions with 2° and 3° halides. Typical examples of each are presented.

This section concludes the coverage of alcohol chemistry for now. A summary chart concerning the various conditions for substitution and elimination reactions follows.

SUMMARY CHART
Substitution and elimination reactions of alcohols

Type of alcohol	Substitution via inorganic ester, e.g., RSO_2Cl, then $I^{\ominus}$	Strong acid with good nucleophile, e.g., conc. HI	Strong acid with poor nucleophile, e.g., H_2SO_4, in alcohol as solvent	
			Lower temperatures	Higher temperatures
Methyl	S_N2	S_N2	S_N2	S_N2
1°	S_N2	S_N2	S_N2	E2
2°	S_N2	S_N1	S_N1	E1
3°	S_N1	S_N1	S_N1	E1
Rearrangements?	uncommon	common	common	common

9-8. Reactions of Ethers

As mentioned in the introduction to this study guide chapter, the chemistry of ethers is very limited, showing a tendency toward nucleophilic displacement reactivity only under fairly special conditions. As is the case with alcohols, for any kind of nucleophilic displacement to occur to an ether (S_N1 or S_N2), the leaving group (alkoxide in this case) has to be improved. This improvement is again done most simply by protonation with a strong acid. Then reaction can occur with a good nucleophile.

$$\ddot{Nuc}^- + R\overset{\overset{\displaystyle H}{|}}{\underset{\underset{+}{\cdot\cdot}}{O}}R \longrightarrow R-Nuc + H-\ddot{O}-R \qquad \text{Good leaving group}$$

Notice that the nucleophile in such a reaction cannot ever be a strong base! A strong base cannot be present together with the strong acid needed to protonate the ether: They would just neutralize each other. Addition of a strongly basic nucleophile to an already protonated ether is also no good. All that would happen would be loss of the proton from the protonated ether to the base; no nucleophilic displacement would occur.

$$CH_3-\ddot{O}-CH_3 \rightleftharpoons CH_3-\overset{\overset{\displaystyle H}{|}}{\underset{\underset{+}{\cdot\cdot}}{O}}-CH_3 \xrightarrow[\text{(Basic!)}]{\substack{\text{Then add} \\ CH_3CH_2O^-Na^+}}$$

(Acidic)

$$CH_3-\ddot{O}-CH_3 + CH_3CH_2OH$$

(**Not** $CH_3CH_2-O-CH_3$)

For these reasons, nucleophilic ether cleavages are limited to good nucleophiles that are weakly basic like Br^- and I^-, which can exist in the presence of strong acid. (If you look back now at the reactions of alcohols, you'll see the same considerations applying there, too.) Methyl and 1° alkyl ethers react via the S_N2 mechanism, whereas 3° ethers follow an S_N1 pathway. Least reactive are 2° ethers (worse than 1° for S_N2, and worse than 3° for S_N1; the latter mechanism is more typical, however).

9-9. Reactions of Oxacyclopropanes

Strained cyclic ethers (e.g., oxacyclopropanes) react with acid like ordinary ethers do, only faster. Order of reactivity is again $3° > 2° \leqslant 1°$. At a $1°$ carbon, the reaction clearly takes place via an S_N2 mechanism to displace the protonated oxygen.

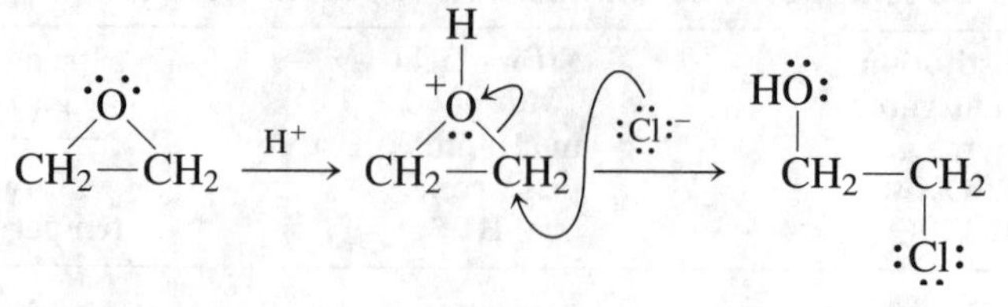

At $2°$ and $3°$ carbons, the reaction may be described as an "S_N2-like S_N1 reaction." To clarify this, let's look at three ways to draw the Lewis structure of protonated trimethyloxacyclopropane.

Trimethyloxacyclopropane

2° Carbocation   **Alkyloxonium ion**   **3° Carbocation**

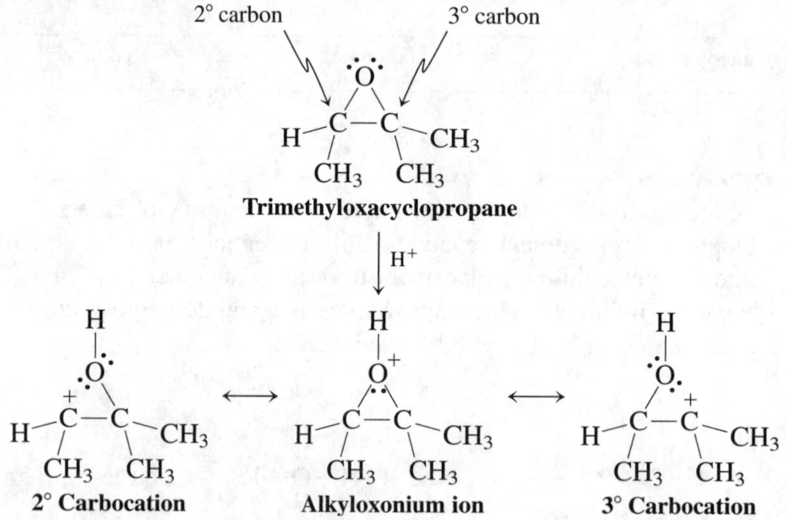

These are actually three resonance forms of the protonated molecule. It may look odd to draw resonance forms where a whole single bond is missing, but such pictures ("no-bond structures") are useful in some cases, provided you recognize that the individual forms are not real and that only the intermediate resonance hybrid really counts. In the case above, the resonance hybrid will probably look more like the alkyloxonium and $3°$ carbocation structures than the $2°$ carbocation structure (because $2°$ carbocations are worse than $3°$ carbocations).

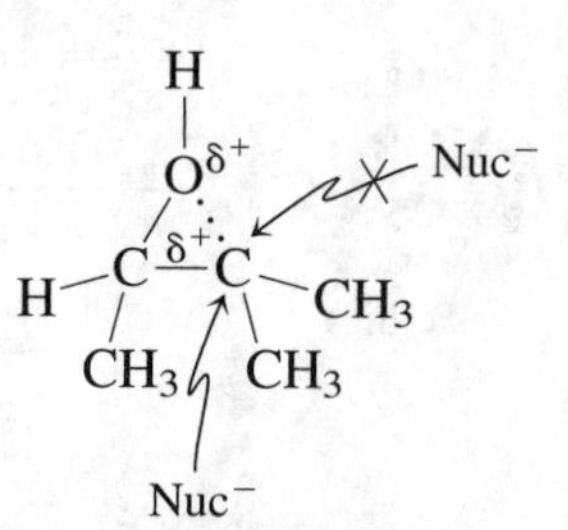

Likely resonance hybrid

Reaction of this protonated molecule with a nucleophile will therefore occur at the $3°$ carbon, which is the most carbocation-like, as you would expect for an S_N1 reaction. However, because of the position of the oxygen leaving group, which is at least partially bonded to the $3°$ carbon, the nucleophile can attach to the $3°$ carbon only from the side **opposite** the oxygen, resulting in **inversion** at that carbon atom, as you would expect for an S_N2 reaction (see illustration above). For these reasons, the S_N1 and S_N2 labels really don't apply in a clear-cut way: S_N1 considerations determine which C—O bond **breaks,** but the direction of approach of the **nucleophile** (back-side attack) is characteristic of an S_N2 process.

  Strained cyclic ethers also react with basic nucleophiles. This is an S_N2 process that displaces an alkoxide, which is a very bad leaving group. The nucleophile has to be a good one because the leaving group (a negatively charged alkoxide ion) is a terrible one. The reaction follows the S_N2 reactivity order of $1° \gg 2° \gg 3°$. Normally alkoxides cannot be displaced in S_N2 reactions. In oxacyclopropanes, however, ring strain raises the energy content such that suitably reactive nucleophiles can displace negatively charged oxygen leaving groups (see graph, below). The only reason this reaction occurs at all is that the displacement reaction breaks open a small, strained ring. Please note that this is a reaction unique to small-ring ethers. **Unstrained ethers are unreactive toward basic nucleophiles.**

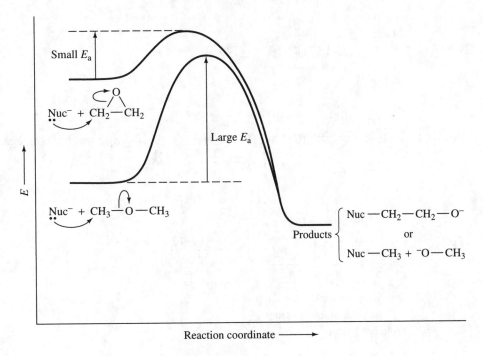

9-10. Sulfur Analogs of Alcohols and Ethers

This short section expands on the obvious parallels between oxygen and sulfur that arise as a result of their relationship in the periodic table. As you saw earlier, the larger sized atoms are more nucleophilic, but less basic. Thus comparisons of the chemical properties of pairs of species like HS^- vs. HO^-, H_2S vs. H_2O, and CH_3SH vs. CH_3OH are straightforward. Larger atoms are also more readily oxidized, and sulfur chemistry includes a variety of oxidized species. Common examples are SO_2 and H_2SO_4. New systems introduced here include sulfonic acids (RSO_3H), sulfoxides ($RSOR'$), and sulfones (RSO_2R').

Solutions to Problems

28. Equilibrium always lies on the side with the **weaker acid-base pair.**

   (a) Left;   **(b)** left;   **(c)** right;   **(d)** right.

29. **(a)** $CH_3CH_2CH_2I$ **(b)** $(CH_3)_2CHCH_2CH_2Br$

   (c)    **(d)** $(CH_3CH_2)_3CCl$

✓**30.** The products are shown above. Except for part (a), only the mechanisms and intermediates will be presented here.

(a)

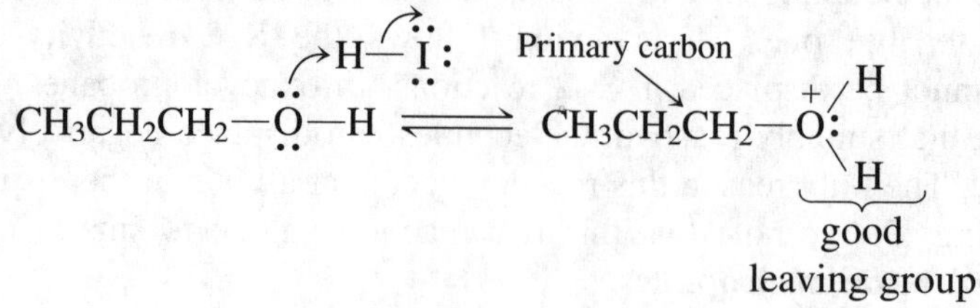

This combination, together with the strong nucleophile I^-, suggests an S_N2 mechanism.

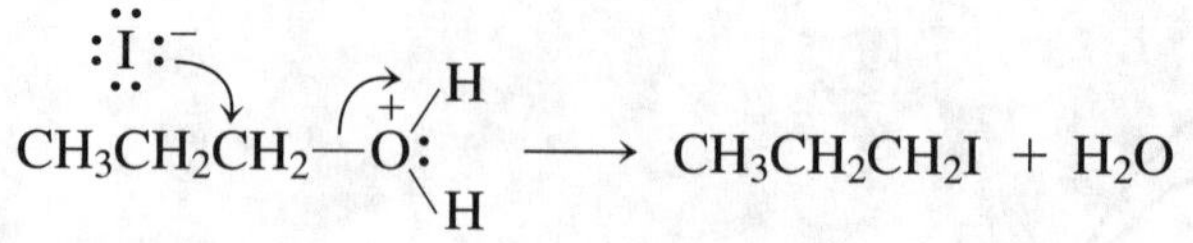

(b)

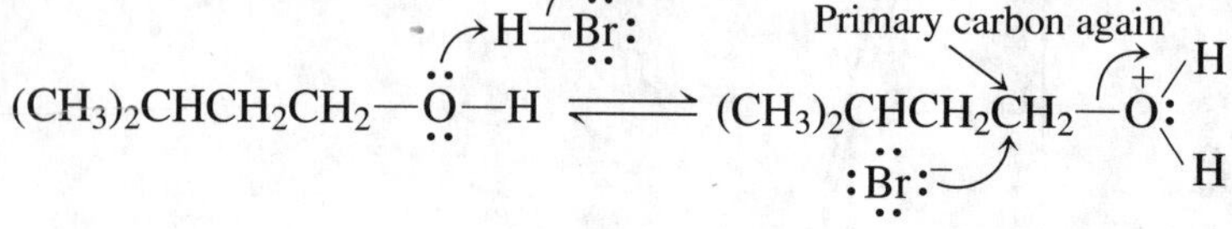

S_N2 mechanism, again

(c)

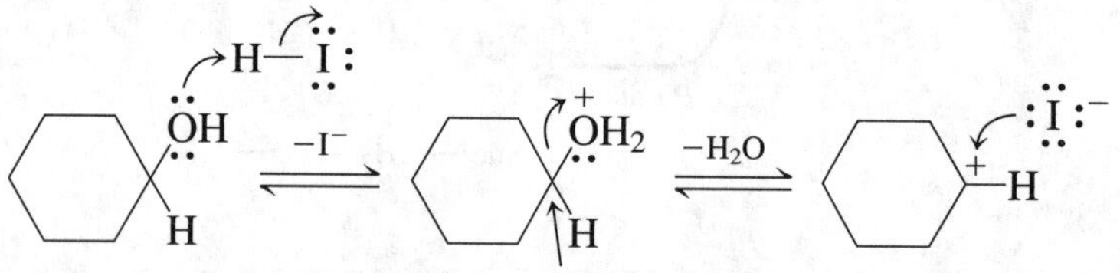

Secondary carbon this time; S_N1 mechanism follows from protonated secondary alcohols

(d)

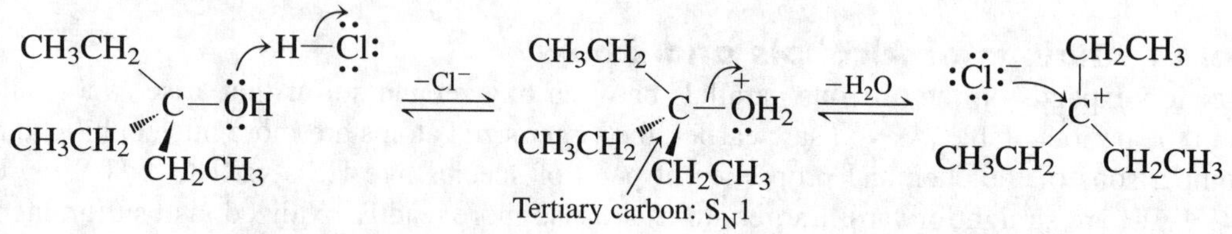

Tertiary carbon: S_N1

31. In each case the species are written in an order reflecting a sequence of rearrangement steps. Rearrangements do not occur to an equal extent under all circumstances. Structures to the right are generally most stable.

(a) $CH_3CH_2CH_2\overset{+}{O}H_2$, $CH_3\overset{+}{C}HCH_3$ (Similar to rearrangement of 2,2-dimethyl-1-propanol in Section 9-3.)

(b) $CH_3CH\overset{+}{O}H_2CH_3$, $CH_3\overset{+}{C}HCH_3$

(c) $CH_3CH_2CH_2CH_2\overset{+}{O}H_2$, $CH_3CH_2\overset{+}{C}HCH_3$

(d) $(CH_3)_2CHCH_2\overset{+}{O}H_2$, $(CH_3)_3C^+$

(e) $(CH_3)_3CCH_2CH_2\overset{+}{O}H_2$, $(CH_3)_3C\overset{+}{C}HCH_3$, $(CH_3)_2\overset{+}{C}CH(CH_3)_2$

Some mechanism arrows are included on the following page to help you find your way.

(f)

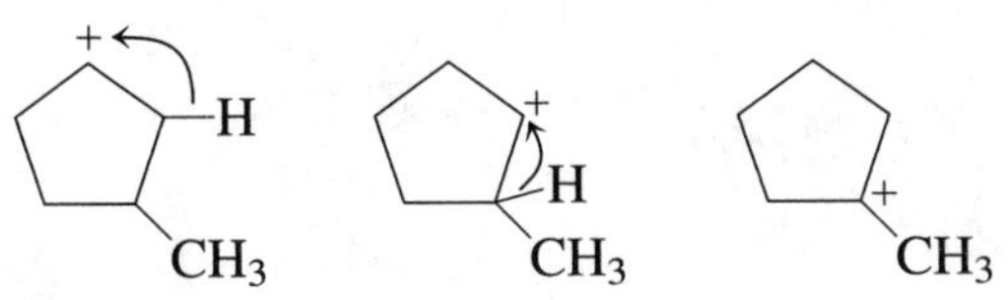

(g)

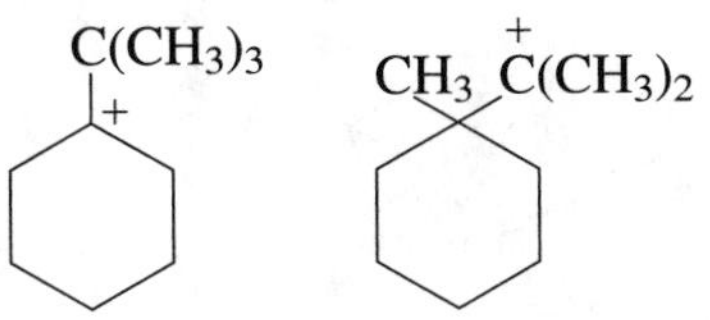

(h) 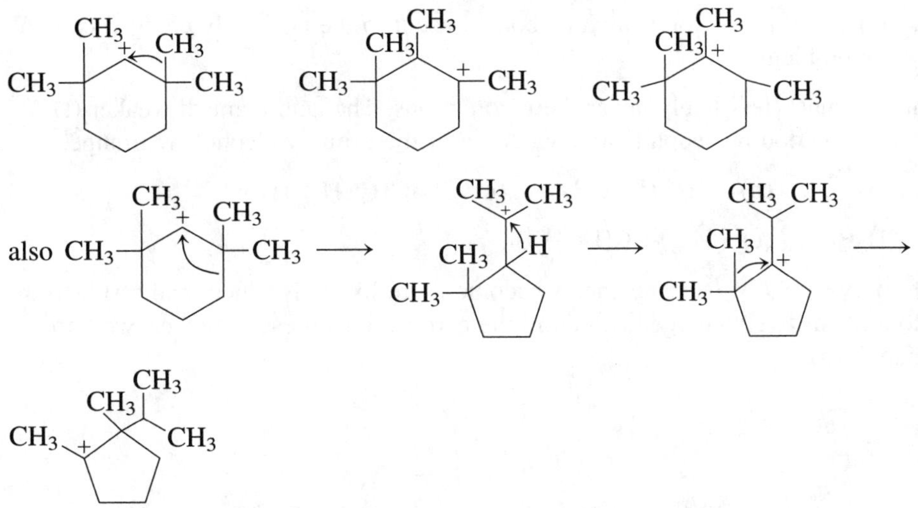 CH₃

also CH₃ ⟶ ⟶ ⟶

✓**32.** These conditions favor rearrangements. They allow carbocations to exist for a long time, because the reaction mixtures are strongly acidic and lack decent nucleophiles.

(a), (b) $CH_3CH{=}CH_2$

(c) $CH_3CH_2CH{=}CH_2$, $CH_3CH{=}CHCH_3$ (major product)

(d) $(CH_3)_2C{=}CH_2$

(e) $(CH_3)_3CCH{=}CH_2$, $(CH_3)_2C{=}C(CH_3)_2$ (major product),

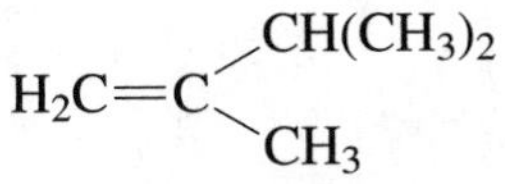

Line formulas will be used for most of the cyclic structures below. Note that methyl groups are understood to be present at the ends of lines even when "CH₃" is not written in.

(f) 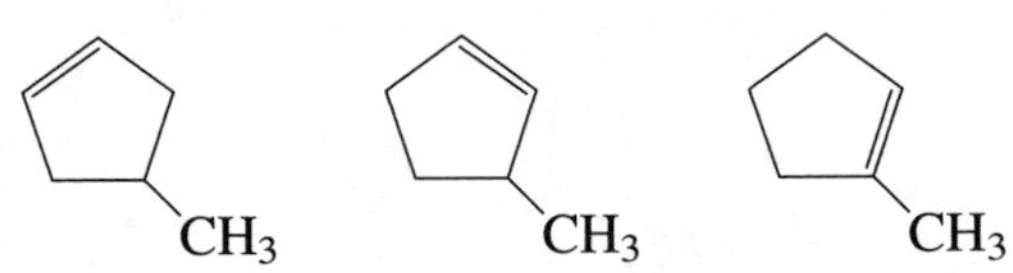   (major product)

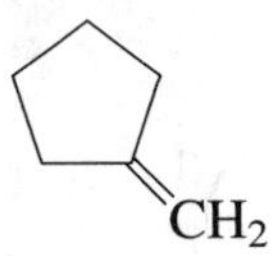

(g)

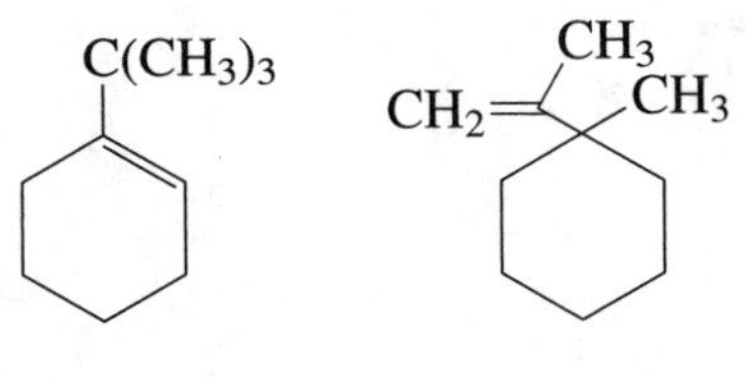

(h)

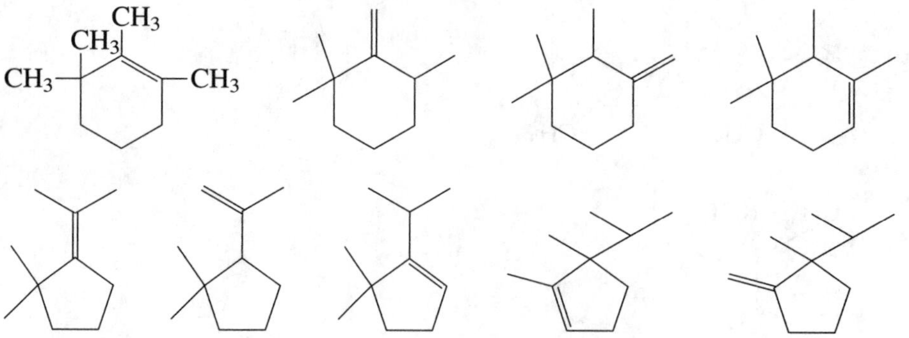

Each product arises from loss of a proton from a carbon adjacent to the positively charged carbon of a structure in the previous problem.

33. Rearrangements are much less likely under these conditions: The acid is much weaker (H_3O^+ rather than H_2SO_4), and there is a good nucleophile around. None of the primary alcohols rearrange.

(a) $CH_3CH_2CH_2Br$ **(b)** $CH_3CHBrCH_3$ **(c)** $CH_3CH_2CH_2CH_2Br$

(d) $(CH_3)_2CHCH_2Br$ **(e)** $(CH_3)_3CCH_2CH_2Br$

With secondary or tertiary alcohols, rearrangements become more likely. Products will result from attachment of Br^- to any positively charged carbon in the carbocations present. See answers to Problem 31(f) through 31(h).

34. (a)

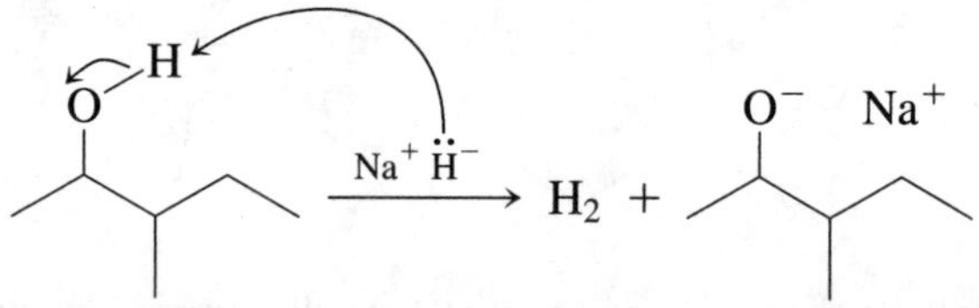

(b)

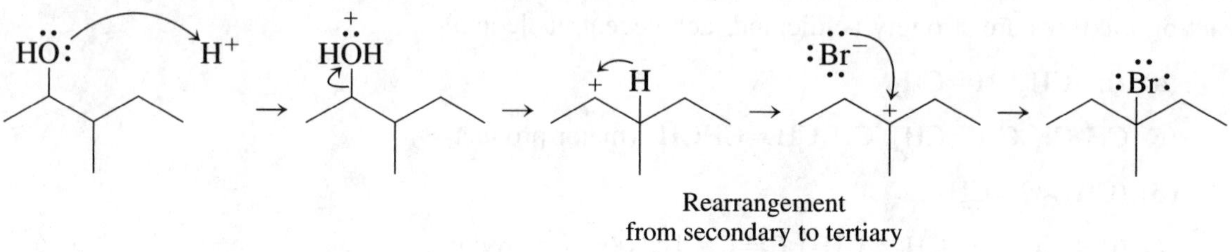

Rearrangement
from secondary to tertiary

(c)

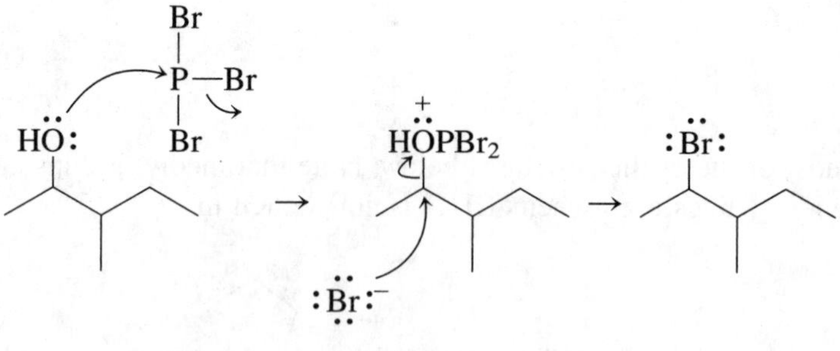

(d)

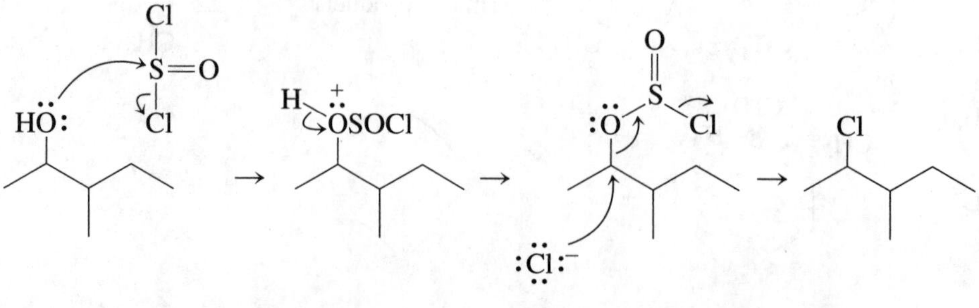

(e)

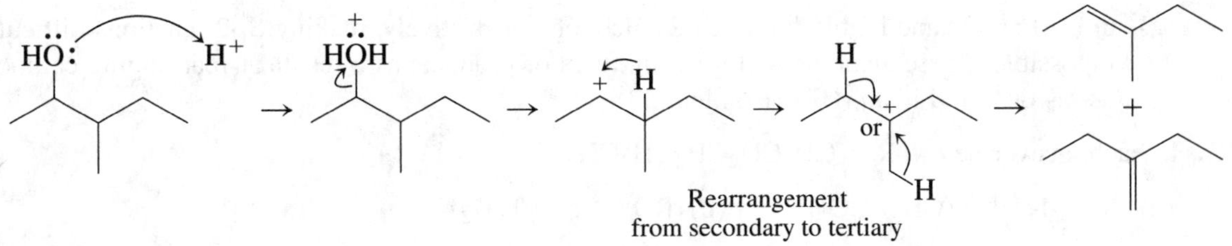

Rearrangement
from secondary to tertiary

(f) 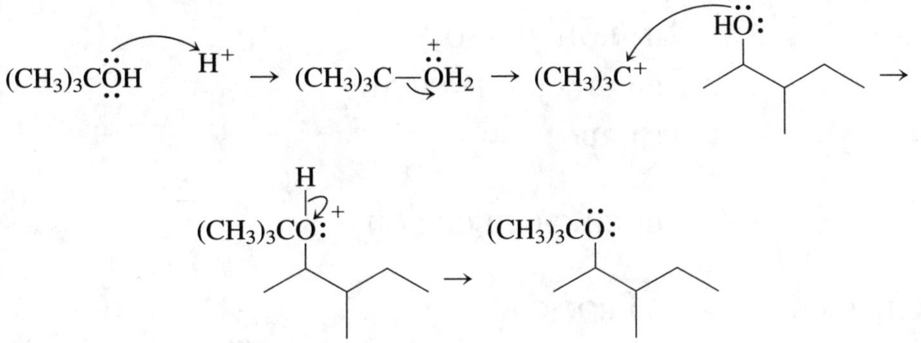 $(CH_3)_3C\ddot{O}H$ → $(CH_3)_3C—\overset{+}{\ddot{O}}H_2$ → $(CH_3)_3\overset{+}{C}$ →

$(CH_3)_3C\overset{+}{\ddot{O}}$ → $(CH_3)_3C\ddot{O}$:

35. The water-free conditions allow **quantitative** conversion of the alcohol to its alkyloxonium ion form, in the presence of high Br^- concentration:

$$RCH_2OH \xrightleftharpoons{\text{Conc. H}_2\text{SO}_4} \underset{\text{Ca. 100\%}}{RCH_2\overset{+}{O}H_2} \xrightarrow{Br^- (S_N2)} RCH_2Br$$

In concentrated aqueous HBr, the main acid present is H_3O^+, which is a **weaker** acid than the alkyloxonium ion. The first equilibrium lies well to the left, so the overall reaction rate is much lower (there is much less protonated alcohol to react with Br^-):

$$\underset{\text{Ca. 99\%}}{RCH_2OH} \xrightleftharpoons{H_3O^+} \underset{\text{Ca. 1\%}}{RCH_2\overset{+}{O}H_2} \xrightarrow{Br^- (S_N2)} RCH_2Br$$

36. (a)

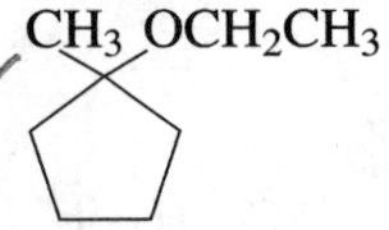

A likely choice, after 2° (secondary)
→ 3° (tertiary) carbocation rearrangement.

(b) Both $(CH_3)_3CCH_2I$ and $(CH_3)_2CICH_2CH_3$.

(c)

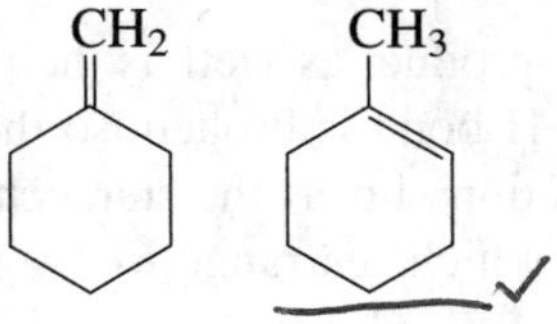

(d)

$(CH_3)_2\overset{OH}{C}—CH(CH_3)_2$

37. (a) through **(e)** are the same, via the S_N2 displacement of a phosphite leaving group by bromide ion.

(f)

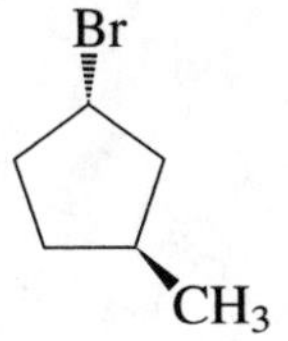

(g) and (h) are 3° and highly hindered 2° alcohols, respectively, making S_N2 reactions difficult or impossible. These alcohols will give mixtures of products as a result of rearranging carbocations, just as occurred in aq. HBr (Problem 33).

38. In each answer below, $R = CH_3CH_2CH_2CH_2CH_2-$.

    (a) $RO^- K^+ [+ (CH_3)_3COH]$ (b) $RO^- Na^+ (+ H_2)$

    (c) $RO^- Li^+ (+ CH_4)$ (d) RI

    (e) No reaction   (f) $R\overset{+}{O}H_2 (+ FSO_3^-)$

    (g) ROR   (h) $CH_3CH_2CH{=}CHCH_3$

    (i) CH_3SO_3R (j) RBr

    (k) RCI   (l) $CH_3CH_2CH_2CH_2\overset{\overset{O}{\|}}{C}OH$

    (m) $CH_3CH_2CH_2CH_2\overset{\overset{O}{\|}}{C}H$ (n) $ROC(CH_3)_3$

39. (a), (b), (c) 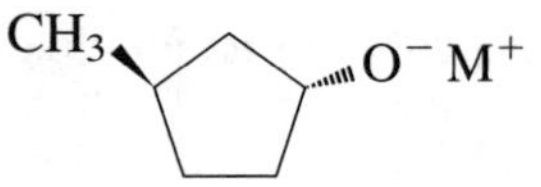 $O^- M^+$ ($M^+ = K^+$, Na^+, or Li^+)   (d)

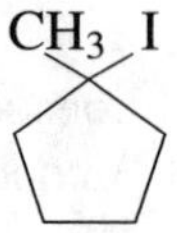

    (e) 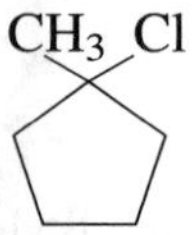   (f)

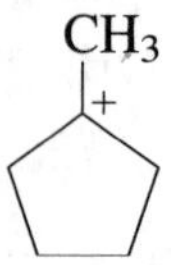

    (g), (h) 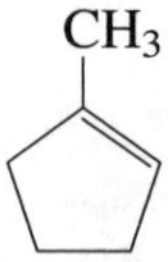   (i) 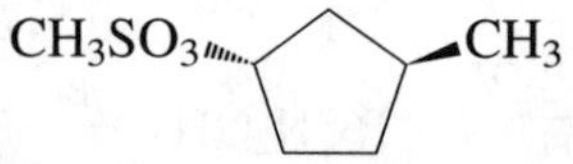 CH_3SO_3

    (j) 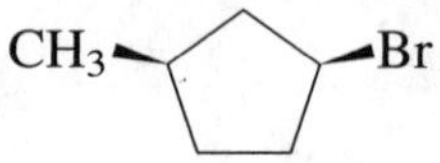 CH_3 Br   (k) 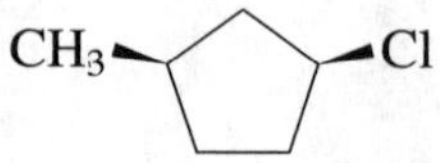 CH_3 Cl

    (l), (m) 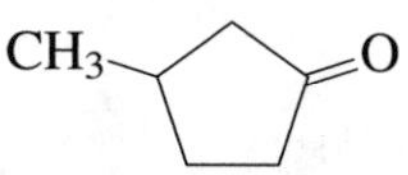 CH_3 O   (n)  CH_3 $OC(CH_3)_3$ [From ROH + $^+C(CH_3)_3$]

Notice how the type of reaction determines the stereochemistry of the product as well as the presence or absence of rearrangement. In (a), (b), (c), (i), and (n), only the O—H bond is broken, so the trans stereochemistry around the ring is maintained. In (j) and (k), S_N2 reactions invert the stereochemistry, giving cis products. In (d) through (h), 2° carbocations are generated, which rearrange to 3°, giving the observed results.

40. (a) $SOCl_2$ (b) PBr_3 (Branching raises the risk of rearrangement.)

    (c) HCl   (d) $P + I_2$ (To avoid rearrangement)

41. (a) 2-Ethoxypropane   (b) 2-Methoxyethanol

    (c) Cyclopentoxycyclopentane   (d) 1-Chloro-2-(2-chloroethoxy)ethane

    (e) 1-Methoxy-1-methylcyclopentane   (f) *cis*-1,4-Dimethoxycyclohexane

    (g) Chloro(methoxy)methane

42. Ethers lack hydrogens attached to oxygen, and therefore, unlike alcohol molecules, ether molecules cannot hydrogen bond to one another. Ethers can hydrogen bond to the hydrogens of water, however, and therefore are more soluble in water than the analogous alkanes but still less soluble than alcohols of comparable molecular weight.

43. The usual options are Williamson ether synthesis (if at least one of the components is a good S_N2 substrate) or acid-catalyzed reaction in the case of symmetrical ethers or ethers containing one fragment suitable for solvolysis (S_N1) chemistry (usually tertiary).

(a) $2\ CH_3CH_2CH_2OH \xrightarrow{H_2SO_4}$

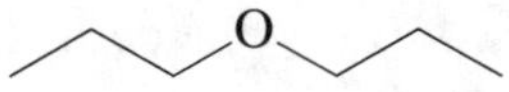

(b) $CH_3CH_2CH_2CH_2CH_2OH \xrightarrow{NaH} CH_3CH_2CH_2CH_2CH_2O^-\ Na^+ \xrightarrow{CH_3I}$

(c) $CH_3CH_2CH_2CH_2OH \xrightarrow{NaH} CH_3CH_2CH_2CH_2O^-\ Na^+ \xrightarrow{CH_3CH_2Br}$

(d) $(CH_3)_2CHCH_2OH \xrightarrow{NaH} (CH_3)_2CHCH_2O^-\ Na^+ \xrightarrow{CH_3CH_2Br}$

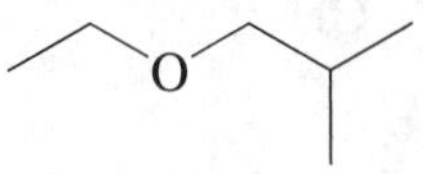

(e) $(CH_3)_2CHOH \xrightarrow{NaH} (CH_3)_2CHO^-\ Na^+ \xrightarrow{CH_3CH_2CH_2Br}$

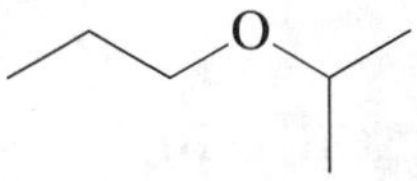

(f) $(CH_3)_2CHCH_2CH_2OH \xrightarrow{NaH} (CH_3)_2CHCH_2CH_2O^-\ Na^+ \xrightarrow{CH_3I}$

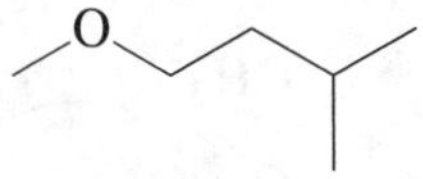

A Williamson ether synthesis would also be usable to synthesize propoxypropane (a), but the mineral acid method is simpler. All the rest are Williamson methods. In the cases of the ethers in (b), (c), and (f), the synthesis would work as well if carried out in the complementary manner, reversing the starting materials' roles, because both alkyl groups are suitable for use as S_N2 substrates. Such is not the case in (d) and (e): Branched primary and secondary haloalkanes are poorer substrates in S_N2 reactions and undergo elimination to a significant degree.

44. (a) $CH_3CH_2CH_2OCH(CH_2CH_3)_2$ (S_N2 okay on 1° haloalkane.)

(b) $CH_3CH_2CH_2OH + CH_3CN{=}CHCH_2CH_3$ (Basic alkoxide gives mainly E2 with 2° haloalkane.)

(c) 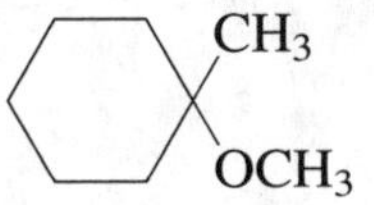   **(d)** $(CH_3)_2CHOCH_2CH_2CH(CH_3)_2$

(e) Cyclohexanol + cyclohexene [Same situation as (b).]

(f)

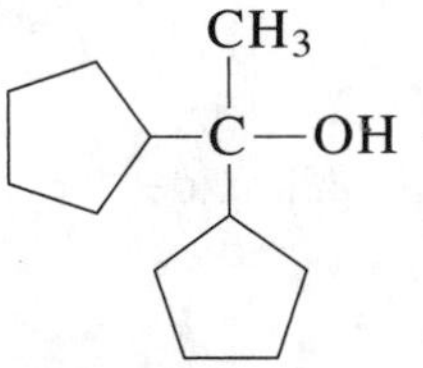

45. The products are shown above. As in the solutions for Problem 30, only the mechanisms and intermediates (if any) will be presented here.

(a) CH$_3$CH$_2$CHCH$_2$CH$_3$

:Ö:⁻

CH$_3$CH$_2$CH$_2$—C̈l:

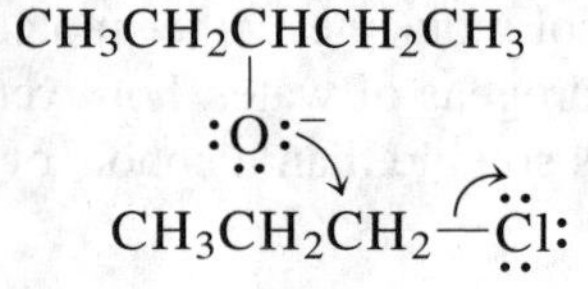

The substrate is a primary haloalkane: S$_N$2 with any but tertiary alkoxides (this one is secondary). This reaction is a good Williamson ether synthesis.

(b) CH$_3$CH$_2$CH$_2$—Ö:⁻

H

CH$_3$CH═CHCH$_2$CH$_3$

:C̈l:

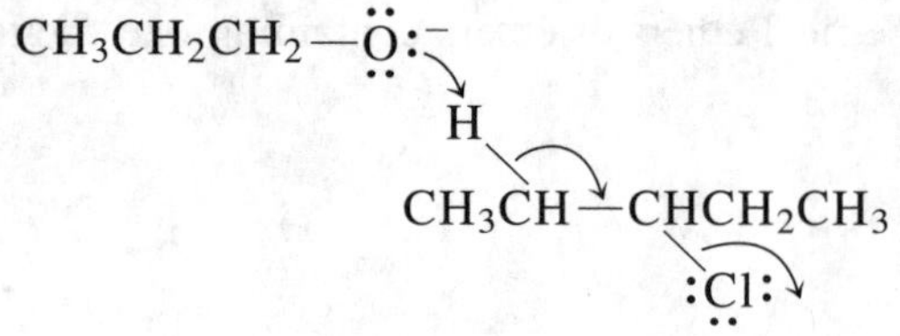

Reverse the roles and the substrate is now a *secondary* haloalkane: S$_N$2 fails with alkoxides because they are strong bases. E2 prevails, and the desired ether is no longer the major product.

(c)

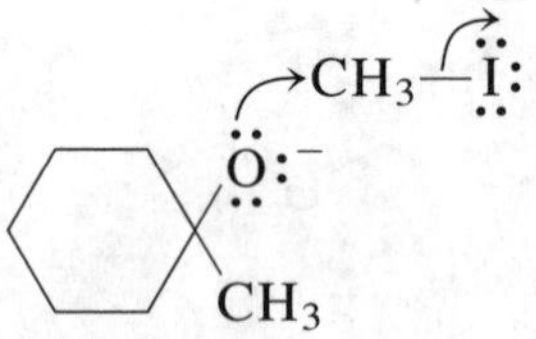

CH$_3$—Ï:

Ö:⁻

CH$_3$

Yes, I see that the alkoxide is tertiary and would give E2 with almost any haloalkane. Except methyl! Iodomethane has only one carbon. It cannot eliminate (where would you put the double bond!?) and instead simply follows the S$_N$2 pathway, its only option.

(d)   CH$_3$CHCH$_3$

:Ö:⁻

(CH$_3$)$_2$CHCH$_2$CH$_2$—B̈r:

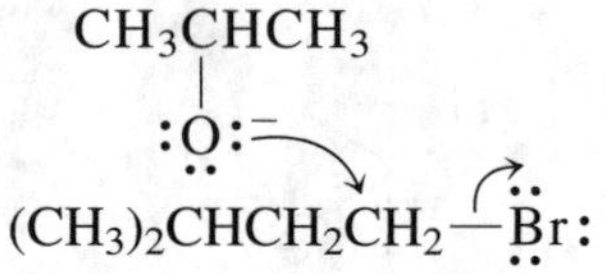

Essentially the same as **(a)**.

(e)

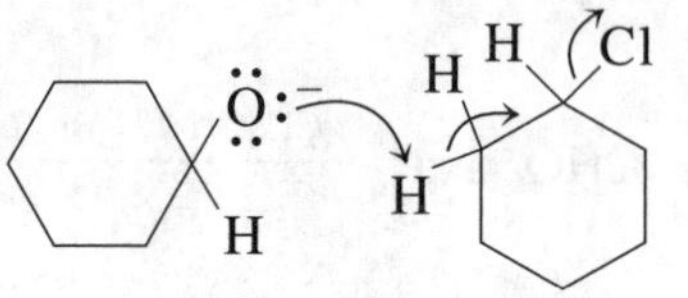

Ö:⁻

H   H   Cl

H

H

Secondary haloalkane + alkoxide; you know the drill. E2, same as **(b)**.

(f)

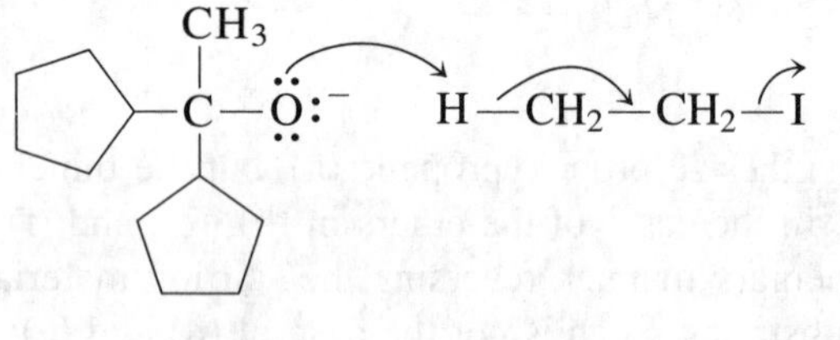

CH$_3$

—C—Ö:⁻   H—CH$_2$—CH$_2$—I

Yes, the substrate is primary. But now the alkoxide is tertiary. E2 results (except when structurally impossible, as was the case in **(c)** above).

46. **(b)** Reaction is given in Problem 44(a).

(e) Use S$_N$1 conditions: chlorocyclohexane in neutral cyclohexanol (solvolysis, without a basic nucleophile).

(f) Use another solvolysis, this time with the 3° haloalkane.

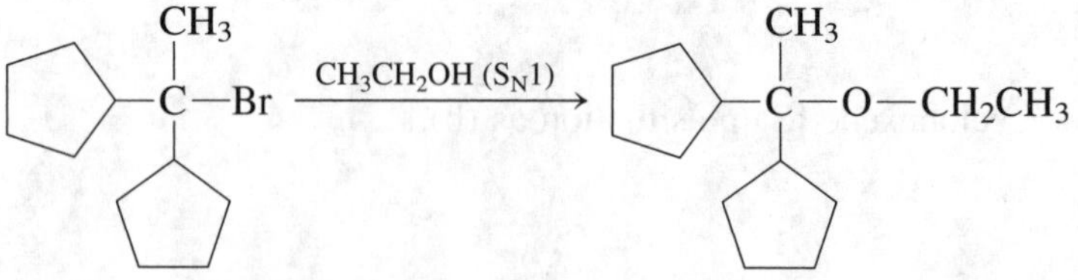

CH$_3$

—C—Br $\xrightarrow{\text{CH}_3\text{CH}_2\text{OH (S}_N\text{1)}}$ —C—O—CH$_2$CH$_3$

CH$_3$

47. **(a)** An oxacyclopropane forms, via an internal displacement of bromide by the alkoxide:

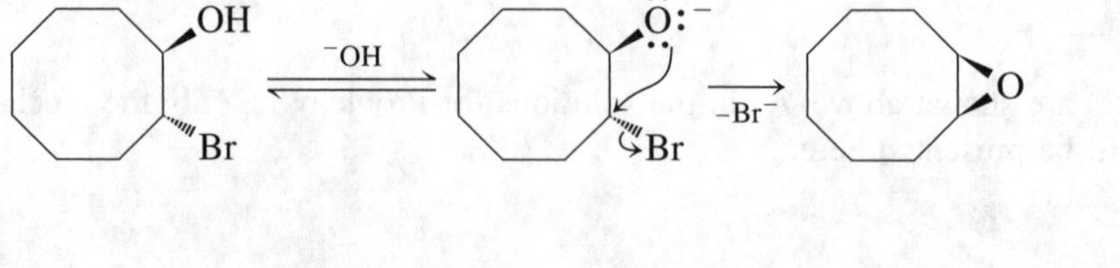

OH $\xrightarrow{\text{}^-\text{OH}}$ Ö:⁻ $\xrightarrow{-\text{Br}^-}$ O

Br   Br

(b) First let us consider how entropy applies to this situation. As discussed in Section 9-6, closing a three-membered ring requires only that the nucleophilic alkoxide oxygen be positioned to carry out a backside displacement on the carbon atom bearing the leaving group. In the open-chain example shown in Figure 9-6, there is free rotation about the bond connecting the oxygen- and bromine-bearing carbons. Restricting this rotation to attain the necessary conformation for backside displacement reduces energy dispersal and is an unfavorable entropy effect. By comparison, the example in Exercise 9-14 places the same carbon–carbon bond in a five-membered ring, in which there is very little range of bond rotation available. Even though the cyclopentane ring has some flexibility (Section 4-2), the trans alkoxide oxygen is never far from the desired position for backside attack. Very little rotational restriction is needed to attain the correct conformation for reaction, and therefore little entropic price needs to be paid.

In the example in this problem, the eight-membered ring has more rotational flexibility than does a five-membered ring, but less than an acyclic compound. Therefore, the entropic cost associated with closing the oxacyclopropane ring will be intermediate: greater than that of the five-membered ring of Exercise 9-14, but less than that of the open-chain compound in Figure 9-6.

48. Take care to limit S_N2 syntheses to 1° haloalkanes. S_N1 reactions are most useful for 3° systems.

(a) $CH_3CH_2CHOHCH_3$ $\xrightarrow[\text{2. } CH_3CH_2Br]{\text{1. NaH}}$ (S_N2)

(b)
+ $CH_3CH_2CH_2CH_2OH$ (solvent) $\longrightarrow$ (solvolysis)

(c) $HOCH_2CH_2CH_2C(CH_3)_2Br \longrightarrow$ (intramolecular S_N1) or

$HOCH_2CH_2CH_2C(CH_3)_2OH \xrightarrow{H^+}$ (same)

(d)
$\xrightarrow{H_2SO_4,\ 40°}$ (S_N1)

49. (a) CH_3CH_2I + $CH_3CH_2CH_2I$ (b) CH_3Br + $(CH_3)_2CHBr$

Reprotonation

(c) 2 CH_3I + ICH_2CH_2I (d)

(e)
(make a model!)

(f)

50. Reactions of strained rings such as oxacyclopropanes with nucleophiles cause opening of the ring. Under acidic, S_N1-like conditions, the nucleophile attaches to the more substituted ring carbon (the one more capable of supporting a positive charge). Under basic, S_N2-like conditions, substitution takes place at the less-substituted, less-hindered ring carbon atom.

(a)
(b)
(c)

(d)
(e)
(f)

51. The problem requests that we figure out how to carry out the following change:

turn [cyclohexanone structure] =O and $BrCH_2CH_2CH_2OH$ into [cyclohexane structure with] $CH_2CH_2CH_2OH$ / OH

We begin by identifying in detail exactly what we are being asked to do: (1) make a new carbon–carbon bond, (2) change a ketone into a tertiary alcohol, and (3) get rid of a carbon–halogen bond in the process. At this stage we don't have a lot of carbon–carbon bond-forming reactions to choose from, and we can settle on some form of addition of an organometallic reagent, derived from the brominated compound, to the ketone carbonyl carbon atom (Section 8-8). For the organometallic species, a Grignard reagent of the structure $BrMgCH_2CH_2CH_2OH$ comes to mind. But we recall (Sections 8-7 and 8-9, and see also Problem 42 in Chapter 8) that alcohol functions and Grignards are incompatible with one another, because the acidity of the O—H group is enough to destroy a C—Mg or C—Li bond by protonation of the carbon atom (which is strongly basic in this type of compound) in a very favorable acid-base reaction, $C—M + H^+ \rightarrow C—H + M^+$.

To do this synthesis, therefore, we cannot have the O—H bond present during the formation or use of the organometallic species. The chemistry we have seen in Chapter 9 (Section 9-8 in particular) gives us ways to solve this problem. Ethers are unreactive toward organometallics, and *tertiary* ethers can be readily formed from alcohols and, later, hydrolyzed by acid to give back alcohols. They make good *protecting groups* for alcohols in situations such as this. Our solution, therefore, is to begin by converting the —OH group of the 3-bromopropanol into a tertiary ether:

$$BrCH_2CH_2CH_2OH \xrightarrow[S_N1]{(CH_3)_3COH,\ H^+} BrCH_2CH_2CH_2OC(CH_3)_3$$

The —OH function is gone now, and we can proceed with the Grignard process:

$$BrCH_2CH_2CH_2OC(CH_3)_3 \xrightarrow{Mg,\ (CH_3CH_2)_2O} BrMgCH_2CH_2CH_2OC(CH_3)_3 \longrightarrow$$

[cyclohexanone structure] =O

[cyclohexane structure with] $CH_2CH_2CH_2OC(CH_3)_3$ / $O^-\ ^+MgBr$ $\xrightarrow{H^+,\ H_2O}$ [cyclohexane structure with] $CH_2CH_2CH_2OH$ / OH

Notice that the aqueous acid work-up, necessary in the Grignard synthesis to protonate the initially formed alkoxide and convert it into the target tertiary alcohol, is also sufficient to hydrolyze the tertiary ether protecting group, releasing the primary alcohol at the end of the three-carbon chain.

52. The leaving group in a hypothetical nucleophilic displacement reaction to cleave any acyclic ether would be an alkoxide. Alkoxides are strong bases. Strong bases are poor leaving groups.

$$Nuc{:}^- + R—O—R \xrightarrow{\quad \otimes \quad} Nuc—R + {}^-O—R$$
Poor
leaving group

The exception is the S_N2 displacement reaction to cleave an oxacyclopropane. The release of bond-angle strain upon opening the three-membered ring overcomes the unfavorable energetics associated with producing a poor leaving group.

53. (a) Ethyloxacyclopropane (racemic); **(b)** (S)-2,3,3-trimethyloxacyclobutane; **(c)** (2R,5S)-2-(chloromethyl)-5-methoxyoxacyclopentane [common name: (2R,5S)-2-(chloromethyl)-5-methoxytetrahydrofuran]; **(d)** 4-oxacyclohexanol; **(e)** oxacycloheptane; **(f)** 2,2-dimethyl-1,3-dioxacycloheptane.

54. **(a)** HOCH$_2$CH$_2$NH$_2$ (S$_N$2 ring opening)

(b) HO—CH(CH$_3$)(H)—SCH$_2$CH$_3$ (Same. Reaction occurs at least substituted carbon of ring.)

(c) BrCH$_2$CH$_2$CH$_2$Br

(d) HOCH$_2$CH$_2$C(CH$_3$)$_2$OCH$_3$ (S$_N$1 ring opening)

(e) CH$_3$OCH$_2$CH$_2$C(CH$_3$)$_2$OH   (S$_N$2 at least substituted ring carbon)

(f) DCH$_2$C(CH$_3$)$_2$OH   (From attack of D$^-$ on less hindered carbon of ring)

(g) (CH$_3$)$_2$CHCH$_2$C(CH$_3$)$_2$OH   (From attack of Grignard at less hindered carbon)

(h) ⬠—CH$_2$CH$_2$OH   (An example of "2-carbon homologation"; formation of an alcohol two carbons longer than a starting organometallic reagent, by addition to oxacyclopropane)

55. Surprisingly, all but methanol can be made by either hydride or organometallic additions to properly designed oxacyclopropanes. Remember: Anionic nucleophiles always add to the less hindered carbon of the ring. We can carry out retrosynthetic analyses as follows:

For any primary alcohol ending in —CH$_2$CH$_2$OH, we can use

$$\text{Nu—CH}_2\text{—CH}_2\text{—OH} \Rightarrow \text{Nu}^- + \triangle\text{O}$$

where Nu = H from LiAlH$_4$ or R from either RLi or RMgX.

For any secondary alcohol ending in —CH$_2$CHOHCH$_3$, we can use

$$\text{Nu—CH}_2\text{—CH(CH}_3\text{)—OH} \Rightarrow \text{Nu}^- + \triangle\text{O}\text{—CH}_3$$

where Nu = H from LiAlH$_4$ or R from either RLi or RMgX.

Do you see the pattern? For a tertiary alcohol ending in —CH$_2$COH(CH$_3$)$_2$, we can use

$$\text{Nu—CH}_2\text{—C(CH}_3\text{)}_2\text{—OH} \Rightarrow \text{Nu}^- + \triangle\text{O}\text{—CH}_3\text{—CH}_3$$

where Nu = H from LiAlH$_4$ or R from either RLi or RMgX, again.

Working forward now, here are reasonable answers. All reactions use (CH$_3$CH$_2$)$_2$O as solvent and are followed by aqueous acid work-up.

$$\triangle\text{O} \xrightarrow{\textbf{(b)} \text{ LiAlH}_4} \text{CH}_3\text{CH}_2\text{OH}$$
$$\xrightarrow{\textbf{(c)} \text{ CH}_3\text{MgX or CH}_3\text{Li}} \text{CH}_3\text{CH}_2\text{CH}_2\text{OH}$$
$$\xrightarrow{\textbf{(e)} \text{ CH}_3\text{CH}_2\text{MgX or CH}_3\text{CH}_2\text{Li}} \text{CH}_3\text{CH}_2\text{CH}_2\text{CH}_2\text{OH}$$

$$\underset{\text{CH}_3}{\triangle\text{O}} \xrightarrow{\textbf{(d)} \text{ LiAlH}_4} \text{CH}_3\text{CHOHCH}_3$$
$$\xrightarrow{\textbf{(f)} \text{ CH}_3\text{MgX or CH}_3\text{Li}} \text{CH}_3\text{CHOHCH}_2\text{CH}_3$$

Also, for **(f)** 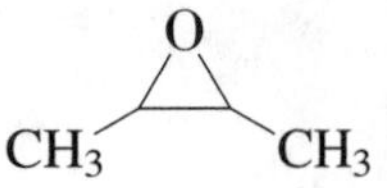 CH_3 ◁O▷ CH_3 $\xrightarrow{\text{LiAlH}_4}$ $CH_3CHOHCH_2CH_3$

Finally **(g)** 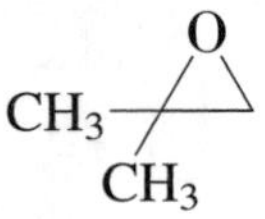 CH_3 ◁O▷ CH_3 $\xrightarrow{\text{LiAlH}_4}$ $(CH_3)_3COH$

56. (a) 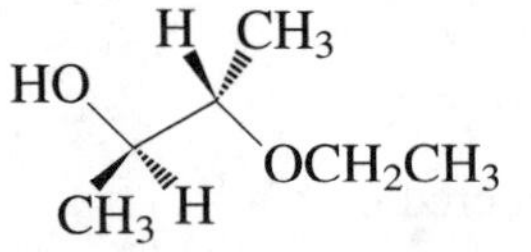 SN2 by ethanol on protonated oxacyclopropane. Attack at either ring carbon gives same product.

(b) 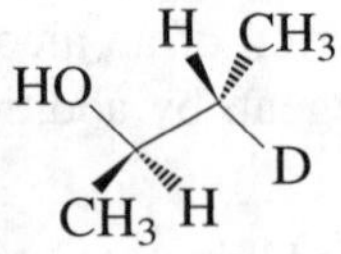 SN2 at either ring carbon.

57. (a) Cyclopropylmethanethiol

(b) 2-(Methylthio)butane, or methyl (1-methyl)propyl sulfide

(c) 1-Propanesulfonic acid

(d) Trifluoromethylsulfonyl chloride

58. In each case, (1) is the stronger acid and (2) is the stronger base.

(a) (1) CH_3SH, (2) CH_3OH

(b) (1) HS^-, (2) HO^-

(c) (1) H_3S^+, (2) H_2S

59. (a)  (Via $^-SCH_2CH_2CH_2CH_2Cl$ intermediate)

(b) 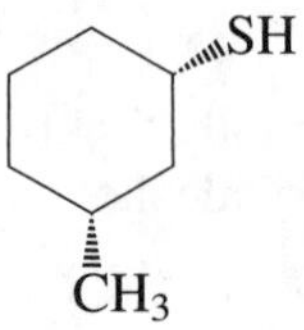 (SN2)

(c) 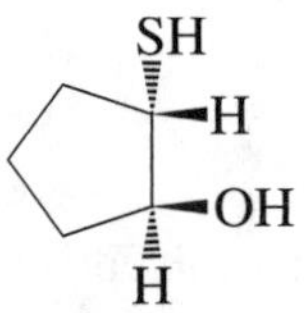 (SN2 again)

(d) $(CH_3CH_2)_3CSCH_3$ (SN1)

(e) $(CH_3)_2CHSSCH(CH_3)_2$

(f) 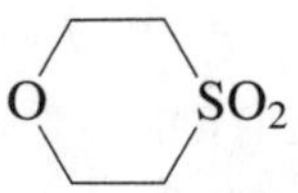 ⬡ SO_2

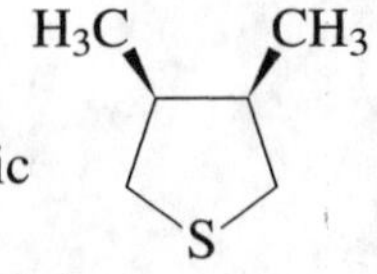

60. A "road-map" problem. Hint: There is "hidden" information. For example, if you just work out the molecular formula of the final product, whose structure is shown, you obtain a useful clue. It is $C_6H_{12}SO_2$, differing from unknown C by only two oxygens. So we can identify structure C as the cyclic sulfide prior to oxidation (see structure in the margin).

Where do we go next? The reaction that produces C involves treatment of B with Na_2S. Look at Problem 58(a). A butane with leaving groups at each end gives thiacyclopentane on reaction with Na_2S. Do the same thing here, but pay attention to the methyl substituents and their stereochemistry. C must be formed by reaction of Na_2S with a *meso*-2,3-dimethylbutane with leaving groups at each end:

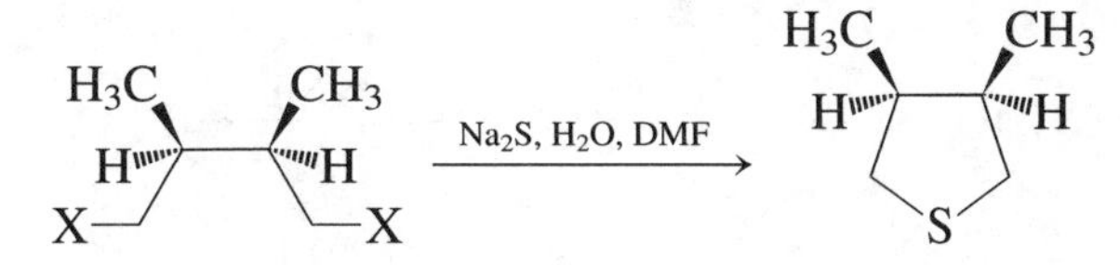

B

(Partial structure; X = unidentified leaving group)

How do we identify the leaving groups X? Look at the precursors to B: an acyclic compound A with the formula $C_6H_{14}O_2$, and two equivalents of a sulfonyl chloride, CH_3SO_2Cl. The simplest solution is to assume X is the sulfonate CH_3SO_3—, and A is the dialcohol corresponding to B. Thus, we arrive at these structures:

A   B

P.S. The final product of the sequence is a (cyclic) sulfone.

61. Everything goes well until the last step. Then—disaster!

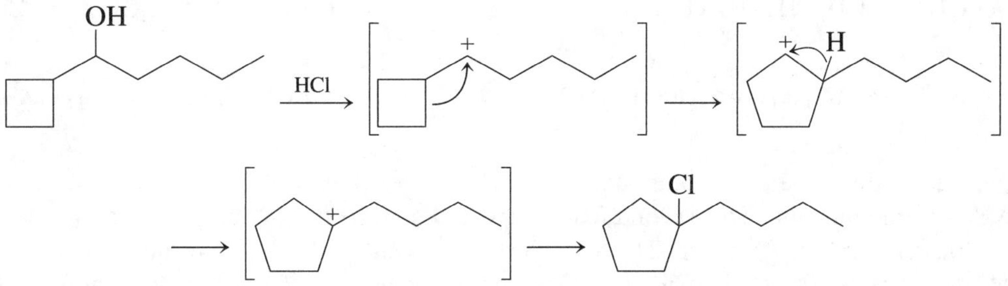

Strained rings are particularly good candidates for carbocation rearrangement if the strain can be relieved in the process.

62. Use thionyl chloride, $SOCl_2$, for the alcohol to chloroalkane conversion. Using thionyl chloride with secondary alcohols shifts the mechanism from S_N1 to S_N2 and, therefore, removes carbocation formation from the process. No carbocation, less risk of rearrangement.

63. The first reaction converts the alcohol and the tosyl chloride into a tosylate. In this process the *alcohol* is the nucleophile, displacing chloride from sulfur. There is no chemical change at the C—O bond of the alcohol, and, therefore, no change in the (R) chirality at C1 of the alcohol:

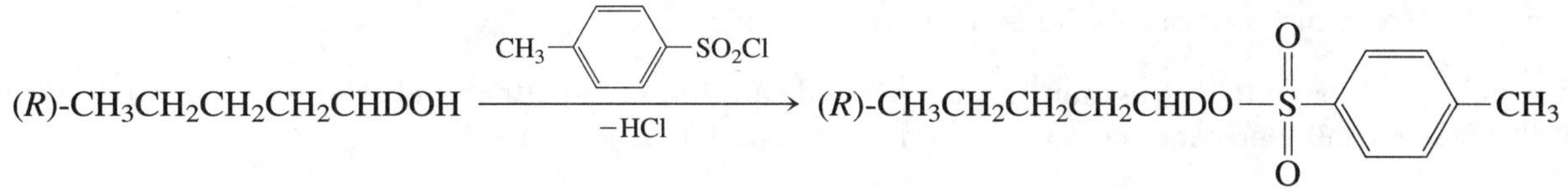

(R)-CH₃CH₂CH₂CH₂CHDOH

(*R*)-1-Deuterio-1-pentanol

(R)-CH₃CH₂CH₂CH₂CHDO—

(*R*)-1-Deuterio-1-pentyl tosylate

The second step is displacement of the tosylate group by the nucleophile ammonia. The reaction occurs at a primary carbon, so it may be safely assumed to proceed by the S_N2 mechanism. Therefore in this step the C—O bond of the original alcohol molecule breaks, with inversion, to give a product with the (S) configuration at C1:

$$(R)\text{-}CH_3CH_2CH_2CH_2CHDO-\overset{\displaystyle O}{\underset{\displaystyle O}{\overset{\|}{\underset{\|}{S}}}}\!-\!\!\left\langle\right\rangle\!\!-CH_3 \xrightarrow{\text{excess } NH_3} (S)\text{-}CH_3CH_2CH_2CH_2CHDNH_2$$

(R)-1-Deuterio-1-pentyl tosylate　　　　　　　**(S)-1-Deuterio-1-pentanamine**

The product should be 100% optically pure (*S*), because S_N2 reactions proceed with 100% inversion. But it is not: It is 70% (*S*) and 30% (*R*). How? One possibility is that some of the molecules reacted by the S_N1 mechanism. (In fact, in a related series of experiments carried out in the 1950s, this was the conclusion reached by the chemists involved.) But this turns out to be *wrong*: Displacements at simple primary carbons do *not* proceed by the S_N1 mechanism, because simple primary carbocations are too unstable to form. How else can we explain the appearance of products with the same (*R*) configuration as was present in the starting material?

Back in Section 6-6 you were shown that *two successive* S_N2 displacements ("double inversion") at the same carbon atom would result in net *retention* of the original configuration. Could something of this sort be happening in the example given for this problem? Using the hint, let us consider the implications of the fact that chloride ion is produced in the course of tosylate formation. Suppose *some* of this chloride ion, a nucleophile, were to react with *some* of the tosylate formed in the first step. This process would be an S_N2 displacement, at C1, with inversion, producing an (*S*) chloroalkane:

$$(R)\text{-}CH_3CH_2CH_2CH_2CHDO-\overset{\displaystyle O}{\underset{\displaystyle O}{\overset{\|}{\underset{\|}{S}}}}\!-\!\!\left\langle\right\rangle\!\!-CH_3 \xrightarrow{Cl^-} (S)\text{-}CH_3CH_2CH_2CH_2CHDCl$$

(R)-1-Deuterio-1-pentyl tosylate　　　　　　　**(S)-1-Chloro-1-deuteriopentane**

Then we would have a mixture containing *both* (*R*) tosylate and (*S*) chloroalkane going into the second reaction, the S_N2 with ammonia. This second reaction would convert the (*R*) tosylate present into (*S*) amine, and any (*S*) chloroalkane into (*R*) amine. This explains how an unequal mixture of amine enantiomers came out of the reaction: About 30% of the first-formed tosylate reacted with chloride before the reaction with ammonia was carried out, giving a second inversion and therefore overall net retention of configuration at C1 for the corresponding 30% of the final product.

64. Product:

$$CH_3\!-\!\!\overset{\displaystyle H}{\underset{\displaystyle CH_2}{\overset{}{C}}}\qquad\overset{\displaystyle D}{\underset{\displaystyle CH_2}{\overset{\displaystyle O}{C}}}\!\!-\!H$$

Note the **inversion** at the carbon that formerly contained the bromine. Make a model if necessary.

The reaction is first order, because the nucleophile and the haloalkane functional groups are in the same molecule. The mechanism is identical to that of the familiar S_N2 reaction, but the "2" here is not applicable because both reacting components are in the same molecule.

65. We'll go through the retrosynthetic analyses in detail, followed by the syntheses. As usual, the solvent for all organometallic preparations and reactions is $(CH_3CH_2)_2O$.

(a) Retrosynthetic analysis

$$CH_3CH_2CH\!-\!CH_2CH_2SO_3H\qquad CH_3CH_2CH\!-\!CH_2CH_2SH$$

$$\bigcirc \qquad\qquad \Rightarrow \qquad\qquad \bigcirc \qquad\qquad \Rightarrow$$

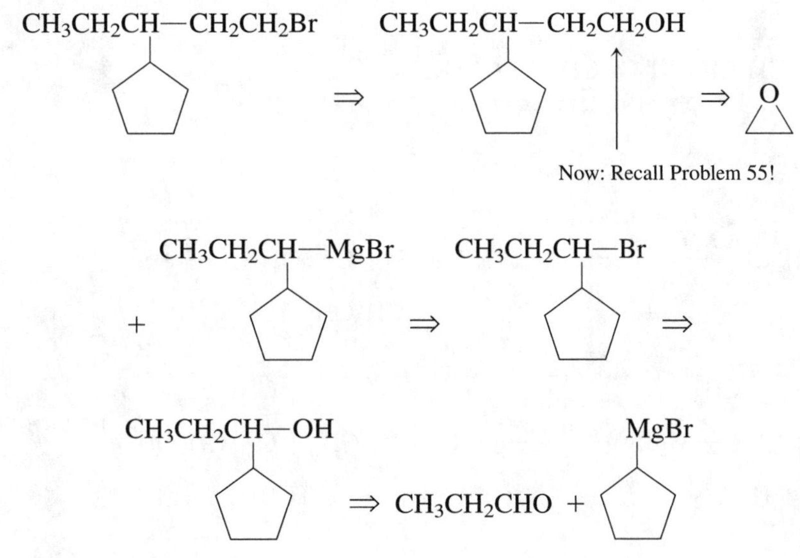

$$CH_3CH_2CH-CH_2CH_2Br \qquad CH_3CH_2CH-CH_2CH_2OH$$

$$\Rightarrow \qquad \qquad \Rightarrow \quad \triangle\!\!O$$

Now: Recall Problem 55!

$$CH_3CH_2CH-MgBr \qquad CH_3CH_2CH-Br$$

$$+ \qquad \qquad \Rightarrow \qquad \qquad \Rightarrow$$

$$CH_3CH_2CH-OH$$

$$\Rightarrow CH_3CH_2CHO \; + \quad \overset{MgBr}{\bigcirc}$$

Synthesis
Starting materials $\quad CH_3CH_2\overset{O}{\overset{\|}{C}}H, \quad \overset{\bigcirc}{\qquad}-Br, \quad CH_2\overset{O}{-}CH_2.$

Br
1. Mg
2. $CH_3CH_2\overset{O}{\overset{\|}{C}}H$
3. H^+, H_2O $\qquad \longrightarrow \qquad CH_3CH_2\overset{OH}{\underset{|}{C}}H-\bigcirc$
1. PBr_3
2. Mg
3. $CH_2\overset{O}{-}CH_2$ $\longrightarrow$

$$CH_3CH_2CHCH_2CH_2OH$$
1. PBr_3
2. NaSH
3. $KMnO_4$ $\longrightarrow$ product

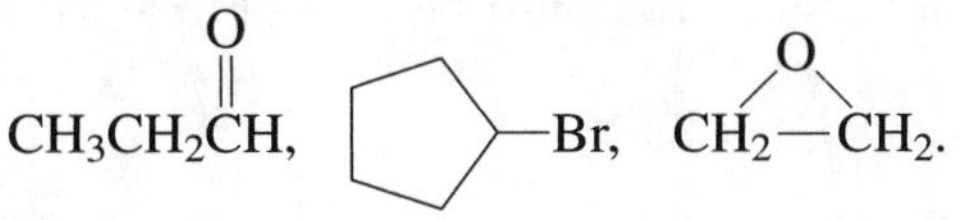

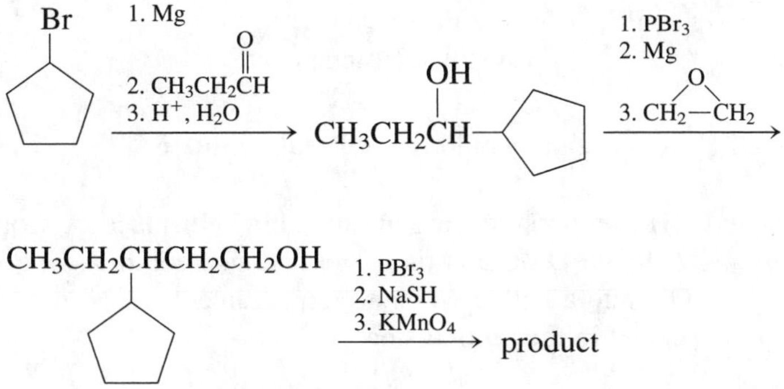

(b) Retrosynthetic analysis

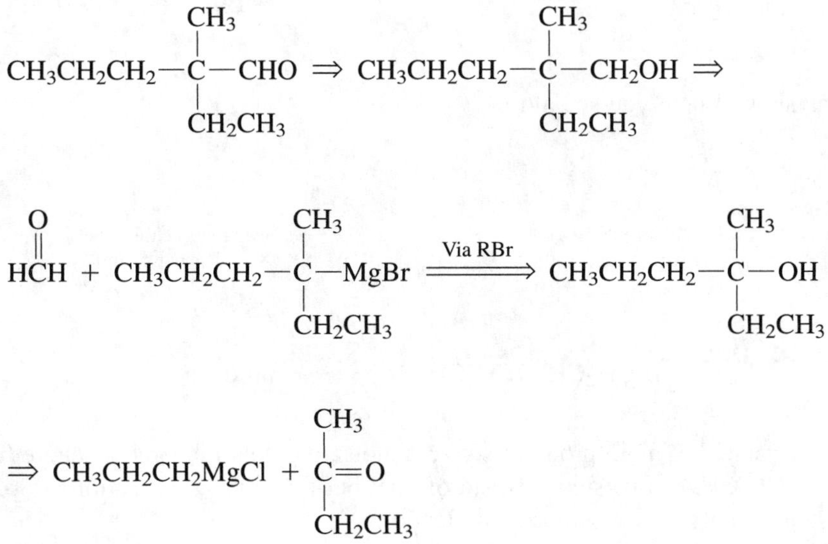

$$CH_3CH_2CH_2-\underset{\underset{CH_2CH_3}{|}}{\overset{\overset{CH_3}{|}}{C}}-CHO \Rightarrow CH_3CH_2CH_2-\underset{\underset{CH_2CH_3}{|}}{\overset{\overset{CH_3}{|}}{C}}-CH_2OH \Rightarrow$$

$$H\overset{O}{\overset{\|}{C}}H \; + \; CH_3CH_2CH_2-\underset{\underset{CH_2CH_3}{|}}{\overset{\overset{CH_3}{|}}{C}}-MgBr \xrightarrow{\text{Via RBr}} CH_3CH_2CH_2-\underset{\underset{CH_2CH_3}{|}}{\overset{\overset{CH_3}{|}}{C}}-OH$$

$$\Rightarrow CH_3CH_2CH_2MgCl \; + \; \underset{\underset{CH_2CH_3}{|}}{\overset{\overset{CH_3}{|}}{C}}=O$$

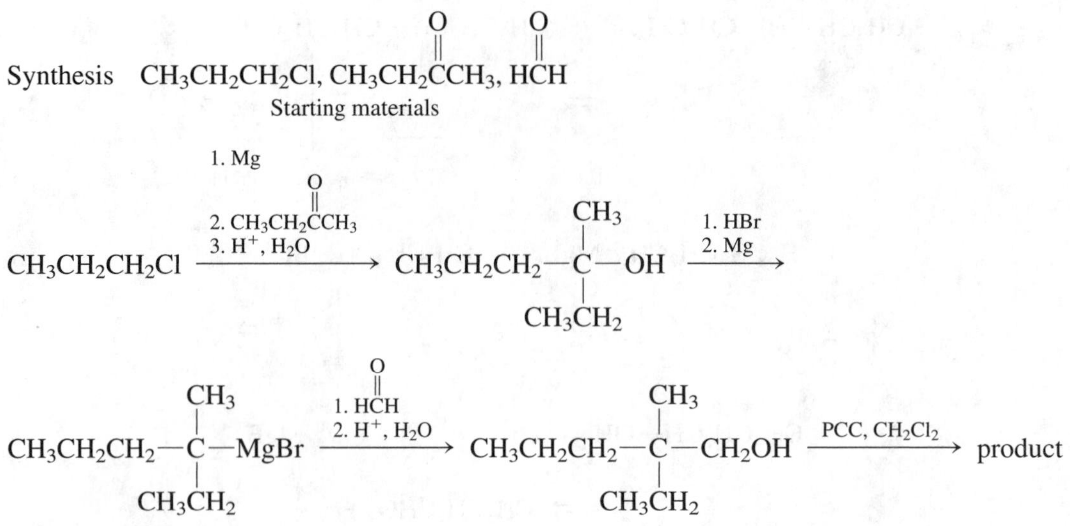

Synthesis $CH_3CH_2CH_2Cl$, $CH_3CH_2\overset{\overset{O}{\|}}{C}CH_3$, $H\overset{\overset{O}{\|}}{C}H$

Starting materials

$$CH_3CH_2CH_2Cl \xrightarrow[\substack{3.\ H^+,\ H_2O}]{\substack{1.\ Mg\\ 2.\ CH_3CH_2\overset{\overset{O}{\|}}{C}CH_3}} CH_3CH_2CH_2-\underset{\underset{CH_3CH_2}{|}}{\overset{\overset{CH_3}{|}}{C}}-OH \xrightarrow[2.\ Mg]{1.\ HBr}$$

$$CH_3CH_2CH_2-\underset{\underset{CH_3CH_2}{|}}{\overset{\overset{CH_3}{|}}{C}}-MgBr \xrightarrow[\substack{2.\ H^+,\ H_2O}]{\substack{1.\ H\overset{\overset{O}{\|}}{C}H}} CH_3CH_2CH_2-\underset{\underset{CH_3CH_2}{|}}{\overset{\overset{CH_3}{|}}{C}}-CH_2OH \xrightarrow{PCC,\ CH_2Cl_2} product$$

66. **(a)** Secondary alcohol to bromide conversion: You must be careful to avoid rearrangement and also control stereochemistry.

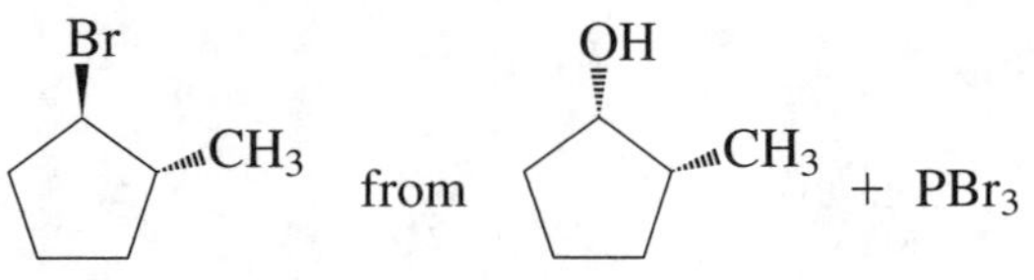

from + PBr$_3$ S$_N$2 inverts at C1; avoids rearrangement.

Rearrangement to 1-bromo-1-methylcyclopentane would be the result with HBr.

(b) Two operations needed: (1) Convert OH to a good leaving group, and (2) displace. A sequence such as 1. CH_3SO_2Cl, $(CH_3CH_2)_3N$; 2. KCN, DMSO is fine. The first step could also be PBr$_3$ or SOCl$_2$ (but not HBr), converting the OH into a halide without rearrangement. KCN, not HCN (a very poor source of cyanide ion), is needed in the second step.

(c) Careful here! Use conc. HCl from **requires** rearrangement.

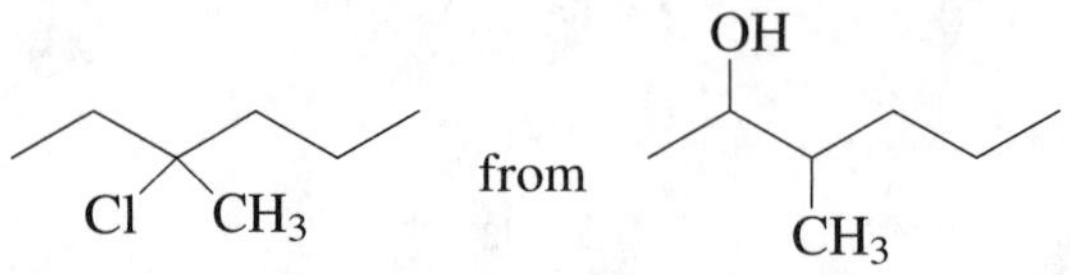

(d) Retrosynthetic analysis will help you see a trick.

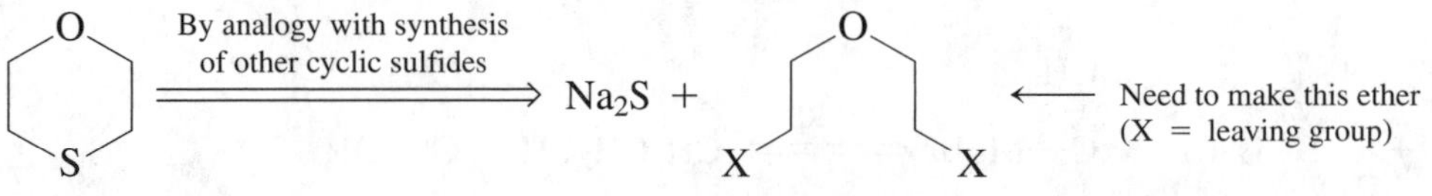

$$2BrCH_2CH_2OH \xrightarrow{H_2SO_4,\ 130°C} BrCH_2CH_2OCH_2CH_2Br \xrightarrow{Na_2S} product$$

67. (1) H$_2$SO$_4$/180°C: shorter route, but more prone to side reactions, e.g., rearrangements and ether formation. (2) PBr$_3$, then K$^+$ $^-$OC(CH$_3$)$_3$: two steps instead of one; but the only side reaction is S$_N$2 to give an ether in the second step, usually only a minor complication.

68. (a) It is a dehydration reaction (an elimination).

(b) The Lewis acid may convert the hydroxy group into a better leaving group.

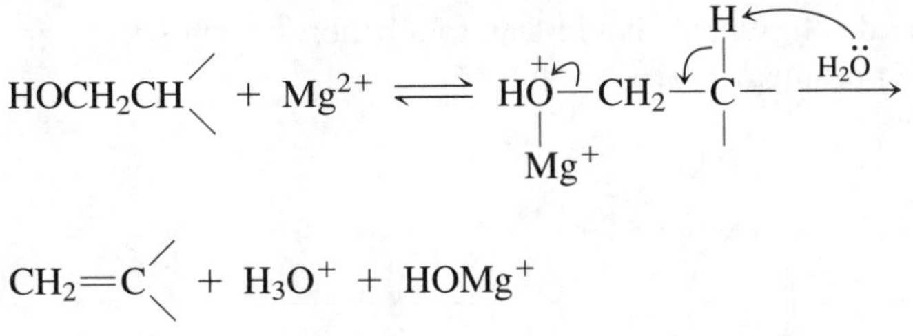

69. (a) Yes!

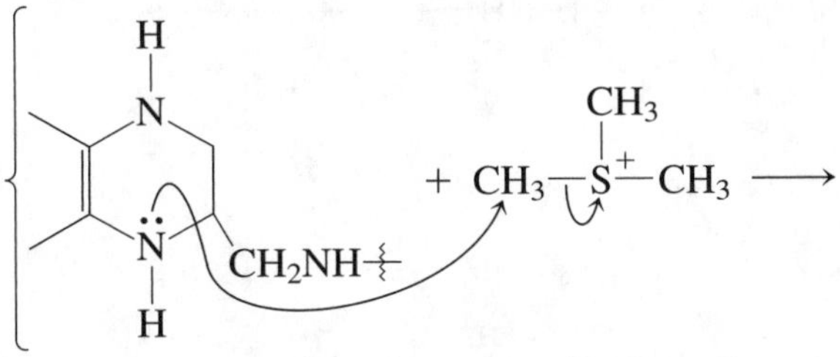

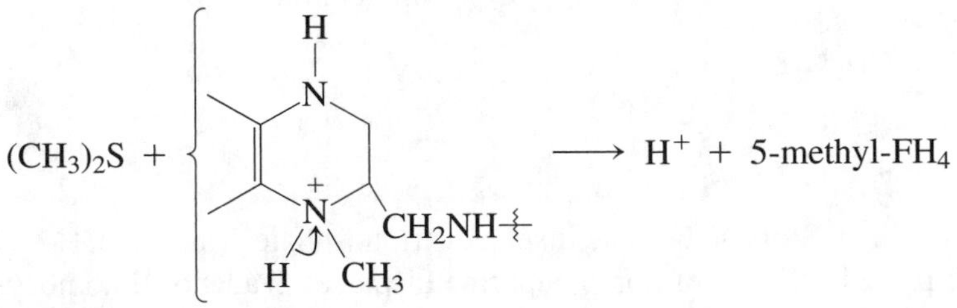

(b) Nucleophile: FH_4. Nucleophilic atom: N5. Electrophilic atom: C in methyl group on $(CH_3)_3S^+$. Leaving group: $(CH_3)_2S$.

(c) Yes! N5 in FH_4 resembles the N in ammonia and is therefore a Lewis base and a reasonable candidate for a nucleophile. The methyls in $(CH_3)_3S^+$ should be polarized δ^+ because of electron attraction by positively charged sulfur. They should be reasonable electrophiles. $(CH_3)_2S$ is neutral and a weak base, and should therefore be a very good leaving group.

70. (a) Yes! A possible mechanism:

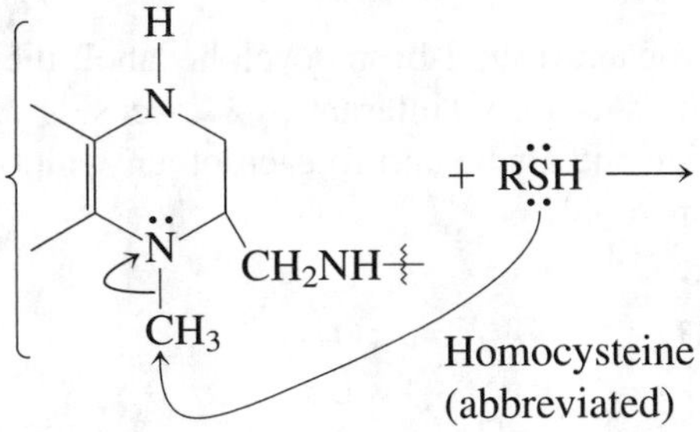

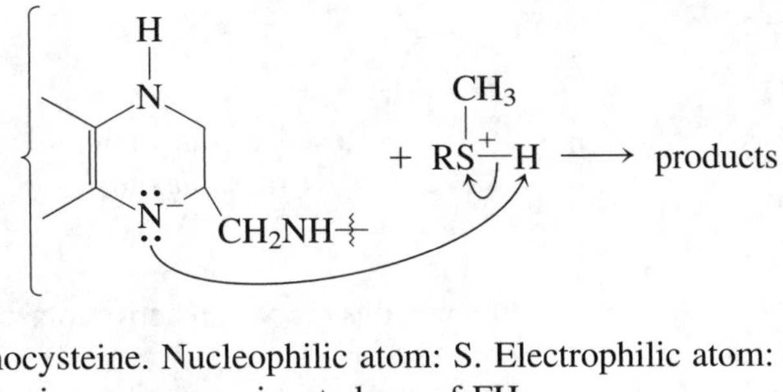

(b) Nucleophile: homocysteine. Nucleophilic atom: S. Electrophilic atom: C of N5 methyl group in 5-methyl-FH_4. Leaving group: conjugate base of FH_4.

(c) Everything is fine except for the leaving group. As the conjugate base of a rather weak acid, it will be a moderately strong base and therefore not a very good leaving group. However, just as protonation of oxygen in alcohols leads to a better leaving group (water), protonation of N5 by acid in 5-methyl-FH$_4$ **before** nucleophilic displacement should lead to a better leaving group (FH$_4$ itself) for the reaction in this problem:

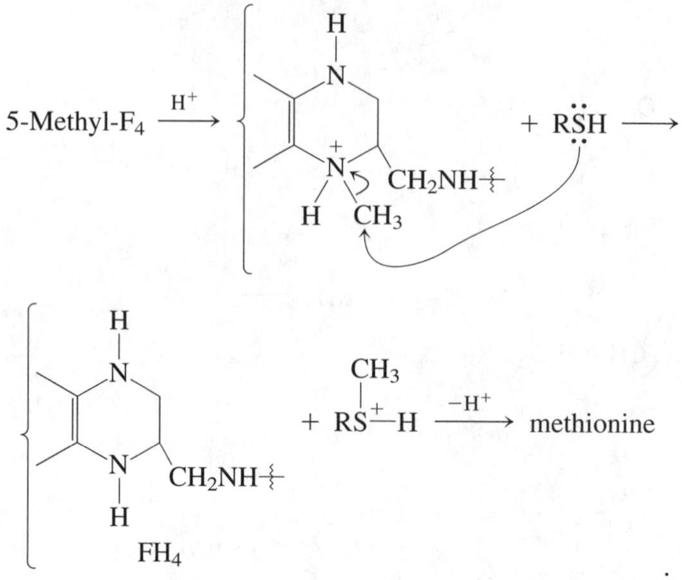

71. (a) Reaction 1 is an S$_N$2 process. The S of methionine displaces triphosphate from the CH$_2$ group in ATP. Reaction 2 is an S$_N$2 process. The N of norepinephrine displaces *S*-adenosyl homocysteine from a CH$_3$ group, making the key CH$_3$—N bond in adrenaline. The ATP makes the second S$_N$2 reaction possible by turning everything that's attached to the CH$_2$ of methionine into one great big leaving group. In other words, *S*-adenosyl methionine is a fancy biological equivalent of CH$_3$I.

(b) No. S$_N$2 reaction on the CH$_3$ doesn't occur because the leaving group (essentially RS$^-$) is not a good one.

(c) Simple! React it with one equivalent of CH$_3$I! (Actually, it's not quite so simple—care is required to prevent extensive reaction of the adrenaline nitrogen, which is still nucleophilic, with additional CH$_3$I.)

72. (a) To make an oxacyclopropane from 2-bromocyclohexanol, the molecule must be able to adapt a geometry that allows the necessary "internal S$_N$2" backside displacement to occur. This geometry requires that the alkoxide and Br be anti to each other, which is possible only from the trans isomer of the starting material.

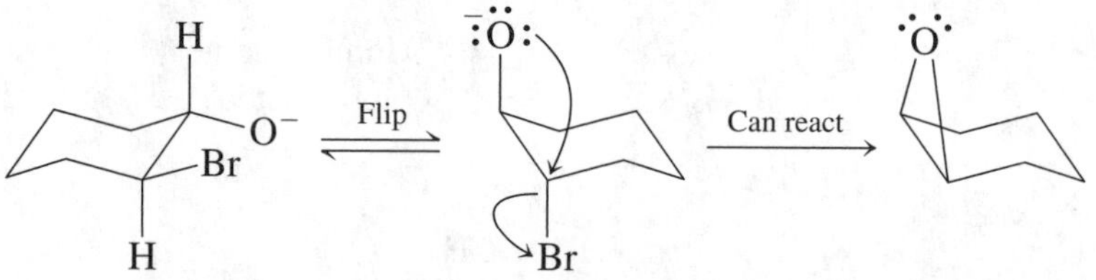

Reactive conformation
(trans, diaxial)

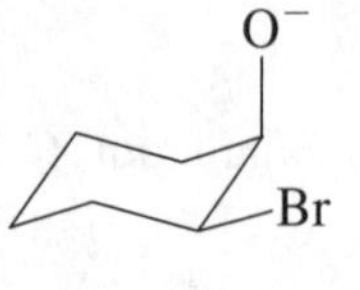

Compare this cis with reactive trans; no good.

(b) 1. NaBH$_4$. Reduces ketone to alcohol. 2. NaOH. Makes alkoxide, leading to internal displacement of Br$^-$ and oxacyclopropane formation.

(c) The first step must form a β-OH, with the new hydroxy group trans to the original Br. Otherwise the second step will not give an oxacyclopropane, for the reason given in (a). Notice that the β-OH and α-Br groups are automatically trans, diaxial as a result of the natural shape of the steroid rings.

73. $CH_2{=}CH{-}CH_2Cl \xrightarrow{\ HS^-Na^+\ } CH_2{=}CH{-}CH_2SH \xrightarrow{\ I_2\ }$

$CH_2{=}CH{-}CH_2{-}S{-}S{-}CH_2{-}CH{=}CH_2 \xrightarrow{\ 1\ \text{equiv.}\ H_2O_2\ }$ allicin

74. Four diastereomers can be drawn based on the general structure given in the problem:

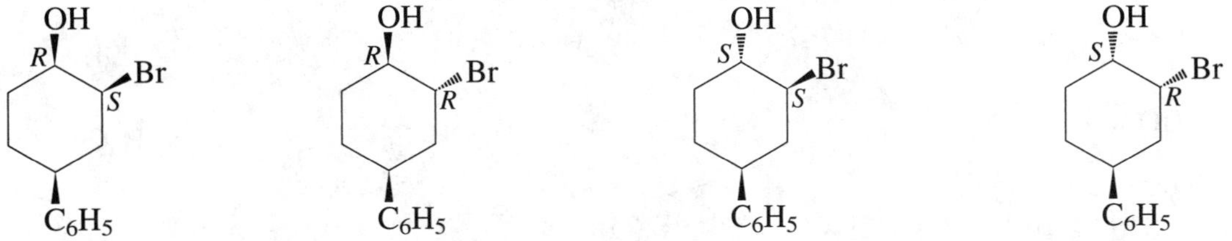

Their most stable conformations are as follows.

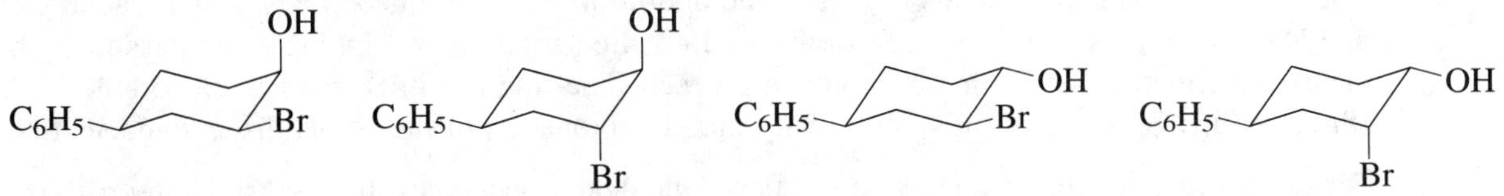

(a) Upon treatment with base, the reaction that occurs will be governed principally by the configuration of the stereocenters in the substrate and the conformation in which the compound is most stable. The main options are E2 reaction and deprotonation of the —OH group followed by internal displacement. For internal (backside, S$_N$2-like) displacement, the —Br and —OH groups must be trans and both axial:

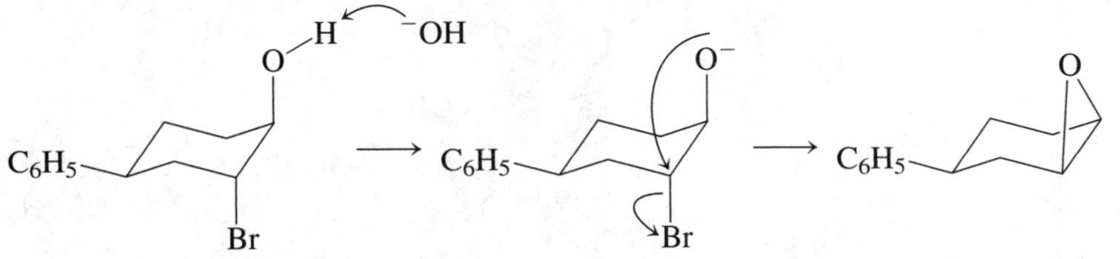

The substrate for this process is therefore identified as compound A.
Next, we look for a substrate that can easily eliminate to furnish the enol pictured in the problem.

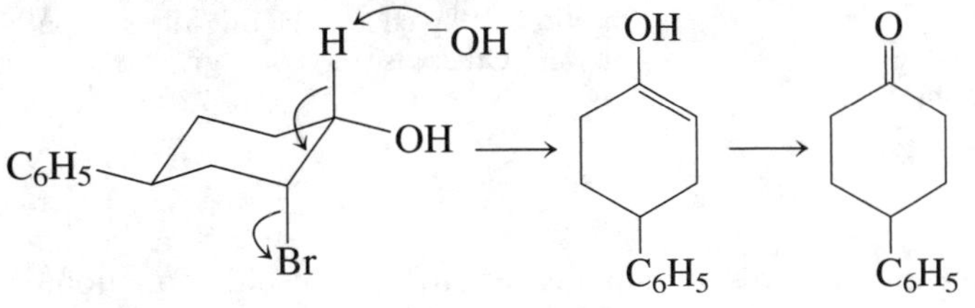

Compound C is identified. You may be wondering exactly why in the world the base removes a proton from carbon, when there is a far more acidic —OH group in the molecule as well. Recall that acid-base reactions typically are fast and reversible. Even if the E2 process shown above takes place only infrequently, under the reaction conditions it is irreversible and will eventually lead to the observed major product.

What remains is to identify compound B, which gives an oxacyclopropane stereoisomeric to that formed from compound A, but at a slower rate, and D, which gives the same product as C, but also more slowly. Because both types of reactions under consideration require an axial leaving group, it makes sense to flip the chairs of the remaining starting compounds (in which the —Br is currently equatorial) and see what we get.

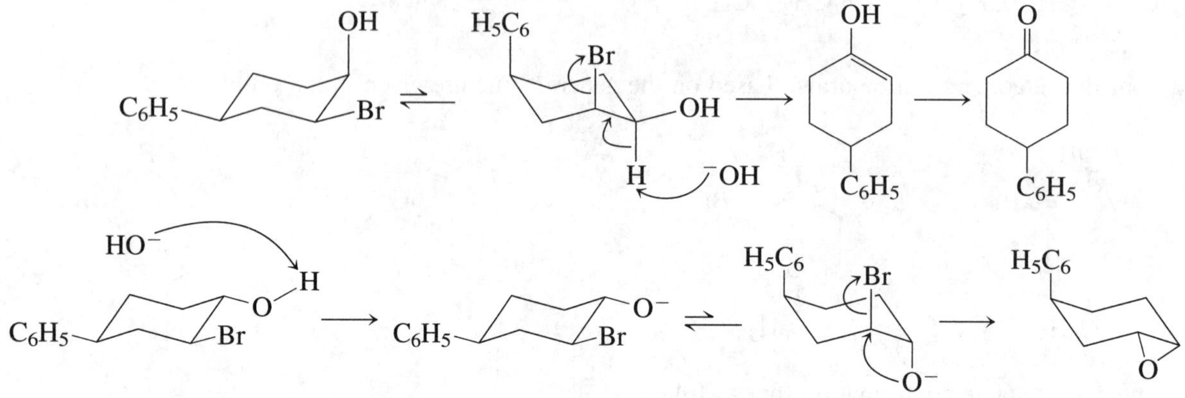

The top compound corresponds to D, and the bottom to B. The reduced rates of the reactions of B and D, relative to A and C, arise from the additional energy cost of flipping into a less favorable chair conformation before the reaction can proceed. The ring flip in B is *very* unfavorable (three groups go axial!), so it's more likely that deprotonation of the —OH group happens first.

(b) Silver ion treatment of secondary and tertiary haloalkanes generates the corresponding carbocations. A quick look at the carbocations derived from compounds A, C, and D confirms that they can be expected to give rise to the same products obtained from base treatment. Notice that the trigonal planar geometry of a carbocation somewhat distorts the cyclohexane chair shape.

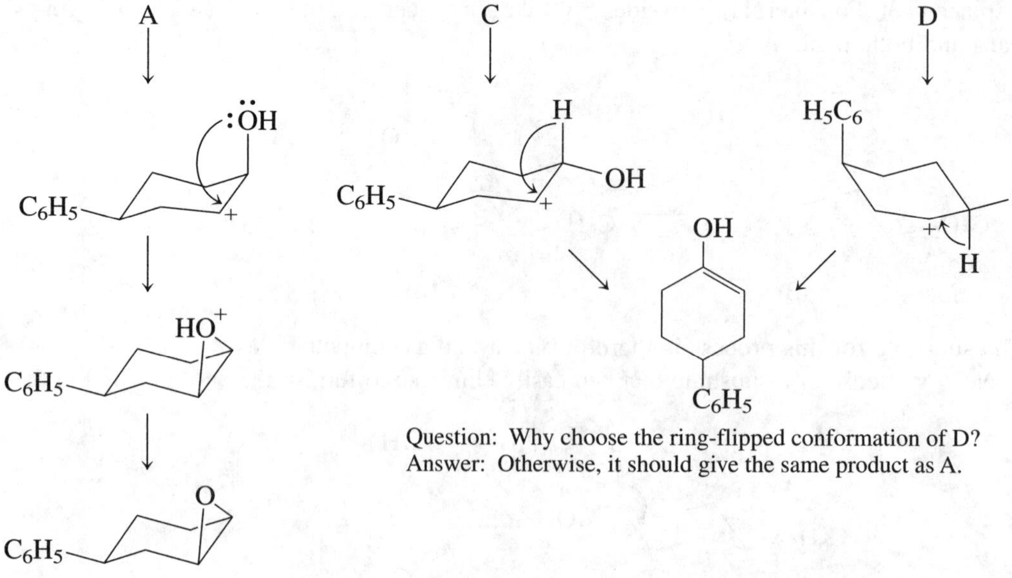

Question: Why choose the ring-flipped conformation of D?
Answer: Otherwise, it should give the same product as A.

(c) Silver ion treatment of B gives a dramatically different result: formation of a ring-contracted aldehyde. Our explanation must include a reason why neither an oxacyclopropane nor a cyclohexenol forms. Any answer that uses one of the intermediate cations shown above is, by definition, wrong! Look again at the result obtained from D in part (b). Had we used the most stable conformation of D, the cation would be identical to that of A and would necessarily give the same product. Since the observed product is *not* the same, *the intermediate must not be the same,* thus the ring flip prior to loss of bromide. Is there any lesson here? It appears that the silver ion-promoted reaction prefers an *axial* Br. Furthermore, inspection of the reactions in part (b) reveals that in each case an adjacent *axial* group reacts with the carbocation. So we return to compound B. Examine its structure. There's something funny going on here. If we flip to the alternative all-axial

chair and have silver remove bromide, we get 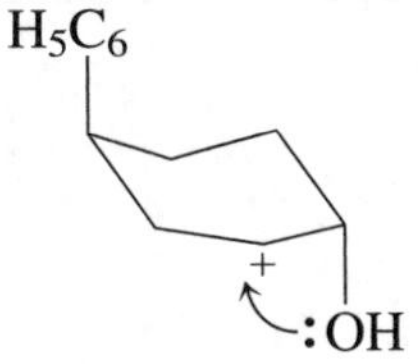 which should ring-close

like a shot to give the same oxacyclopropane product that forms upon treatment of B with base. *Since that doesn't happen,* let's think of something different. We know that this conformation is poor, over 4 kcal mol^{-1} higher in energy than the alternative. According to Table 2-1, less than 0.1% (one in a thousand) of the molecules of B will adopt this conformation at any time. So the likelihood is that the reaction of B with silver ion proceeds through the all-equatorial conformation, contrary to the apparent preference of silver for removing axial bromine. But it would still be wrong to simply suppose that silver removes Br from this conformation, because the result would be the same cation that formed from isomer C, which converted to cyclohexenol. In case you hadn't noticed, we've just ruled out all the carbocations as possibilities. What's left? Perhaps silver can't promote removal of equatorial Br to give a cation, but, instead, rearrangement occurs *simultaneously* with Br loss, bypassing the ring carbocation:

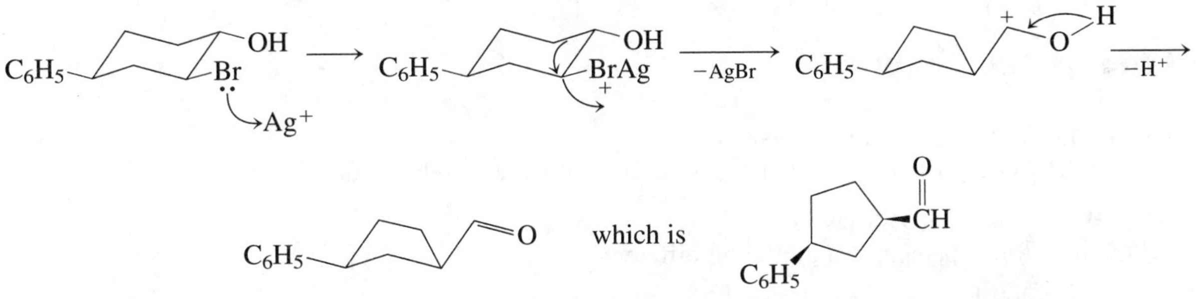

Two points: Notice that the cyclohexane ring bond that migrates is the one located *anti* to the C—Br bond being broken (this *anti* business—backside attack in S$_N$2, the E2 transition state, and now this—keeps showing up, doesn't it?). Furthermore, the cation that results places a positive charge on an HO— substituted carbon atom, permitting stabilization by resonance with oxygen's lone pairs and providing the driving force for the rearrangement.

75. (c)

76. (c)

77. (c)

78. (a)

P.S. Regarding the question in Section 9-3 of this study guide, the alcohol forms three products.

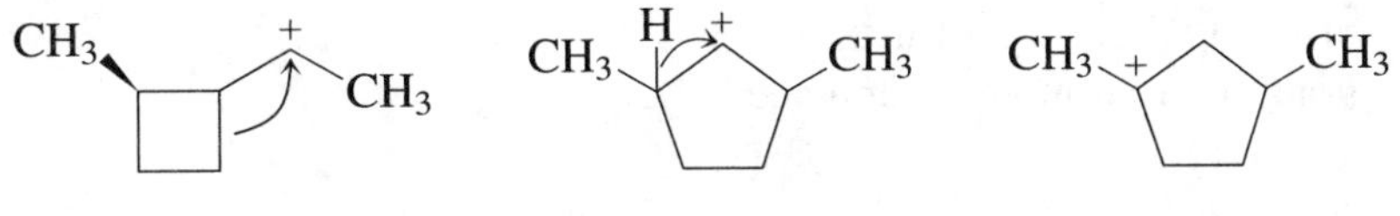

10

Using Nuclear Magnetic Resonance Spectroscopy to Deduce Structure

In this chapter we address the question that faces anyone trying to identify the molecular structure of a substance, namely, "What is it?" To put it another way, this chapter begins to answer the obvious question that you may already have been thinking after the first nine chapters of this book, namely, "How does anyone really know that all those molecules are what we say they are?" In the "olden days," tedious indirect identification methods had to be used, and some are described in the text. Nowadays these questions are answered through the use of spectroscopy, a technique that serves as the "eyes" of an organic chemist with respect to the structures of molecules. The most important and widely used type of spectroscopy, *nuclear magnetic resonance* (NMR), is described in this chapter.

Outline of the Chapter

10-9 Carbon-13 Nuclear Magnetic Resonance

The utility of another magnetic nucleus in NMR.

Keys to the Chapter

10-2. Defining Spectroscopy

Spectroscopy is mainly physics. In spite of that, the material in this section is really very basic and quite understandable: **Spectroscopy detects the absorption of energy by molecules.** The stake that the organic chemist has in this is also simple. Determining the structure of a molecule requires that the chemist be able to **interpret** spectroscopically observed energy absorptions in terms of structural features of an unknown molecule. This chapter describes the physical phenomena associated with nuclear magnetic resonance (NMR). It then describes their logical implications as applied to identifying structural features in a molecule from a nuclear magnetic resonance "spectrum." Don't be frightened. The end result—the ability to interpret spectroscopic data in terms of molecular structure—is actually one of the easiest skills to acquire in this course. It's almost fun!

10-3. Hydrogen Nuclear Magnetic Resonance

In the upcoming text sections the physical basis for the utility of nuclear magnetic resonance in organic chemistry is presented. This text section describes a couple of concepts that are likely to be unfamiliar to you. The first is the idea that a magnet can align with an external magnetic field in more than one way. Most of us are familiar with bar magnet compasses, which orient themselves along the Earth's magnetic field in one direction only. Nuclear magnets are actually similar, except for the fact that the reorientation energies are so small that energy quantization becomes significant. Like the compass, the nuclear magnet does have one energetically preferred alignment (the α spin for a proton). Less favored alignments (*spin states*) are very close in energy to the preferred one on a nuclear scale, however. So, unlike the macroscopic compass, the microscopic nuclei can commonly be observed in less favorable, higher energy orientations in a magnetic field.

   The second new idea is that of *resonance*. This is the term that describes the absorption of exactly the correct amount of quantized energy to cause a species in a lower energy state to move to a higher energy state. For nuclear magnets, this is commonly described as a *spin flip* (α spin state to β spin state for a proton, the nucleus of the hydrogen atom). The amount of energy involved depends on the identity of the nucleus and the size of the external magnet. The text section describes these relationships in detail. It is the observation of resonance energy absorptions by magnetic nuclei at certain values of magnetic field strength and energy input (in the form of radio waves) that constitutes the physical basis for nuclear magnetic resonance spectroscopy.

10-4. The Hydrogen Chemical Shift

The normal NMR spectrum can provide four important pieces of structural information about an unknown molecule. The first two pieces of information are derived from the fact that hydrogens in different chemical environments display separate resonance lines in a high resolution ^{1}H NMR spectrum. Thus, first, by counting the **number of resonance signals** in the spectrum, one knows the **number of sets of hydrogen atoms in different chemical environments** contained in the molecule. Second, the actual **position of each resonance signal** is characteristic of **certain kinds of chemical environments,** e.g., it can imply proximity to a specific type of functional group, or attachment to a certain type of atom. The phenomenon responsible for this is the *shielding* of a magnetic nucleus under observation by the nearby electrons in the molecule. How this comes about physically is described in detail. On a typical NMR spectrum, resonance signals to the left of the chart result from less strongly shielded (*deshielded*) hydrogens, whereas signals to the right are representative of more strongly shielded ones. Deshielded hydrogens require lower external magnetic fields for resonance, whereas shielded ones require higher fields. So we have NMR spectra that have the following qualitative relationships:

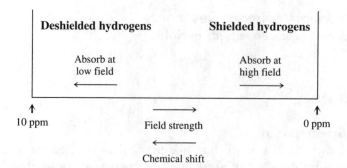

As described in the text, a resonance signal's position is measured as a **field-independent** *chemical shift*, which has units of parts per million (ppm) of the total applied field, often called δ units. These read from right to left, with the usual hydrogen spectrum covering a range from 0 to 10 ppm. Table 10-2 shows typical chemical shifts for common types of hydrogens. There is a lot here, but for most purposes all you really need to know are the types of hydrogens that resonate in several general regions of the NMR spectrum.

General Regions of the NMR Spectrum

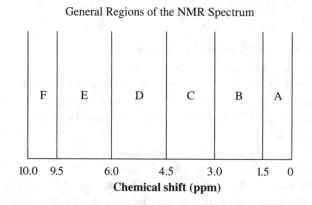

Chemical shift (ppm)

Region	Chemical shift	Hydrogen type
A	0–1.5 ppm	alkane-type hydrogens
B	1.5–3.0 ppm	hydrogens on carbons next to carbon-containing functional groups
C	3.0–4.5 ppm	hydrogens on carbons attached to electronegative atoms
D	4.5–6.0 ppm	alkene-type hydrogens
E	6.0–9.5 ppm	benzene-type hydrogens
F	9.5–10.0 ppm	hydrogens of aldehyde group

With this as a basis, you are ready to start interpreting NMR spectra. For the initial problems of this type, simply **count** the number of signals in the spectrum and note the **position** of each one. Then see if you can come up with a structure that displays the correct number of signals in approximately the observed places, using the material in the text section. If you can, you have probably picked a sensible structure for the unknown molecule.

10-5. Chemical Shift Equivalence

This section presents detailed procedures for determining which hydrogens in a molecule have identical chemical shifts due to *chemical equivalence*. The simplest examples of this are the four hydrogens of methane or the six hydrogens of ethane. There are some minor complications that might require a review of the material toward the end of Chapter 5, but on the whole these procedures are not difficult.

10-6. Integration

Integration provides the third major piece of information available from NMR: the relative number of hydrogens responsible for each separate NMR signal. The integration is measured electronically by the NMR spectrometer and plotted directly on the spectrum. It tells you whether a given NMR signal is due to a single hydrogen or some number of chemical-shift-equivalent hydrogens in the molecule. The integration is critical for the correct interpretation of NMR spectra.

10-7. Spin–Spin Splitting

Because many atomic nuclei are magnets, the NMR signal of a nucleus under observation can, in principle, be affected by the presence of other nearby magnetic nuclei. These neighboring nuclear magnets can align with or against the external magnetic field, so **their fields can add to or subtract from the field generated by the NMR machine.** The result is a slight change in the resonance line position for the nucleus under observation and is called *spin–spin coupling* or *spin–spin splitting*. Again, for spectral interpretation purposes, the theory, presented in the text (although it is not really complicated at all), is secondary to understanding the meaning of the phenomenon in terms of molecular structure. Thus it is often sufficient to rely on two simple rules:

1. Spin–spin splitting is not observed between chemical-shift-equivalent hydrogens.
2. The signal of a hydrogen with N neighboring hydrogens will be split into $N + 1$ lines ("$N + 1$ rule").

There are two important qualifications to rule 2:

a. $N + 1$ lines is a **minimum.** There may be more (Section 10-8).
b. In determining N, you count only neighbors whose chemical shifts are different from that of the hydrogen whose signal is being considered (because of rule 1).

A little careful observation of the text examples will help you get used to the consequences of spin–spin splitting in its most common forms. Table 10-5 and Figures 10-16, 21, and 22 nicely illustrate these situations.

10-8. Spin–Spin Splitting: Complications

The splitting rules outlined in the previous section are idealized for two conditions that are only rarely met in any NMR spectrum. For these rules to hold **exactly:** First, all the signals must be separated from one another by a distance much greater than the coupling constant within each of their patterns (i.e., $\Delta v \gg J$). Second, all the coupling constants (J values) associated with any hydrogen must be identical in size, even if that hydrogen is coupled to more than one group of neighboring hydrogens. If either of these conditions is not fulfilled, the spectrum won't look exactly as you might expect. Fortunately, a lot of the time these conditions are **approached,** especially for hydrogens near electronegative atoms or functional groups. You will therefore have to be aware of the possible effects of "non-first-order" situations; but, for the most part, you won't have to worry a whole lot about them.

10-9. Carbon-13 Nuclear Magnetic Resonance

This is an extension of NMR spectroscopy that is now widely used. There are two reasons. First, modern NMR instruments make these spectra much easier to obtain than originally was the case. Second, the spectra contain useful information that is very easy to interpret, especially under conditions of broad-band hydrogen decoupling, which "wipes out" the spin–spin splitting by neighboring hydrogens. The results are spectra that contain only *singlets,* that is, a single line for each carbon or group of chemically equivalent carbons. Given such a spectrum, you can quickly determine whether or not it corresponds to a proposed structure simply by counting the lines in the spectrum. Of course, more extensive information is also available from the ^{13}C spectrum if desired, in the form of the carbon chemical shifts (Table 10-6), the proton splittings of the "undecoupled" spectrum (e.g., Figure 10-30), or the "DEPT" spectra (Figure 10-33).

Solutions to Problems

25. To do this, you need to tell the difference between **frequencies,** ν, in units of s^{-1}, and **wavenumbers,** $\tilde{\nu}$, in units of cm^{-1}. Section 10-2 shows how they are related: $\nu = c/\lambda$ and $\tilde{\nu} = 1/\lambda$, so $\nu = c\tilde{\nu}$, or $\tilde{\nu} = \nu/c$. For AM radio ($\nu = 10^6 \ s^{-1}$), $\tilde{\nu} = 10^6/(3\times10^{10}) \approx 3 \times 10^{-5} \ cm^{-1}$; and for FM and TV ($\nu = 10^8 \ s^{-1}$), $\tilde{\nu} = 10^8/(3 \times 10^{10}) \approx 3 \times 10^{-3} \ cm^{-1}$. All these are well to the right end of the chart, very low in energy relative to most of the forms of electromagnetic radiation on the chart.

26. The conversion formulas are $\lambda = 1/\tilde{\nu}$ and $\nu = c/\lambda$ (Section 10-2).

(a) $\lambda = 1/(1050 \ cm^{-1}) = 9.5 \times 10^{-4} \ cm = 9.5 \ \mu m$

(b) $510 \ nm = 5.1 \times 10^{-5} \ cm$; $\nu = (3 \times 10^{10} \ cm \ s^{-1})/(5.1 \times 10^{-5} \ cm) = 5.9 \times 10^{14} \ s^{-1}$

(c) $6.15 \ \mu m = 6.15 \times 10^{-4} \ cm$; $\tilde{\nu} = 1/(6.15 \times 10^{-4} \ cm) = 1.63 \times 10^3 \ cm^{-1}$

(d) $\nu = c\tilde{\nu} = (3 \times 10^{10} \ cm \ s^{-1})(2.25 \times 10^3 \ cm^{-1}) = 6.75 \times 10^{13} \ s^{-1}$

27. Use $\Delta E = 28{,}600/\lambda$ (Section 10-2), and use the equations $\lambda = 1/\tilde{\nu}$ and $\lambda = c/\nu$. Be sure to convert the units of λ to nm before calculating ΔE, though!

(a) $\lambda = 1/750 = 1.33 \times 10^{-3} \ cm = 1.33 \times 10^4 \ nm$ (1 cm = 10^{-2} m, and 1 nm = 10^{-9} m, or 1 cm = 10^7 nm), so $\Delta E = (2.86 \times 10^4)/(1.33 \times 10^4) = 2.15 \ kcal \ mol^{-1}$

(b) $\lambda = 1/2900 = 3.45 \times 10^{-4} \ cm = 3.45 \times 10^3 \ nm$, so $\Delta E = (2.86 \times 10^4)/(3.45 \times 10^3) = $ 8.29 $kcal \ mol^{-1}$

(c) $\lambda = 350 \ nm$ (given), so $\Delta E = (2.86 \times 10^4)/350 = 82 \ kcal \ mol^{-1}$

(d) $\lambda = 3 \times 10^{10}/(8.8 \times 10^7) = 3.4 \times 10^2 \ cm = 3.4 \times 10^9 \ nm$, so $\Delta E = (2.86 \times 10^4)/(3.4 \times 10^9) = $ $8.4 \times 10^{-6} \ kcal \ mol^{-1}$

(e) $\lambda = 7 \times 10^{-2} \ nm$, so $\Delta E = (2.86 \times 10^4)/(7 \times 10^{-2}) = 4.1 \times 10^5 \ kcal \ mol^{-1}$

28. Only the value of ν is needed to calculate ΔE. Use $\Delta E = 28{,}600/\lambda$, together with $\lambda = c/\nu$.

(a) $\lambda = (3 \times 10^{10} \ cm \ s^{-1})/(9 \times 10^7 \ s^{-1}) = 333 \ cm = 3.33 \times 10^9 \ nm$, so $\Delta E = (2.86 \times 10^4)/(3.33 \times 10^9) = 8.59 \times 10^{-6} \ kcal \ mol^{-1}$

(b) $\Delta E = 4.76 \times 10^{-5} \ kcal \ mol^{-1}$

29. (a) Increasing radio frequency—to the LEFT

(b) increasing magnetic field strength (moving "upfield")—to the RIGHT

(c) increasing chemical shift—LEFT

(d) increased shielding—RIGHT

30. (a)

(b) Like (a), but with an additional signal at 90 MHz (^{1}H).

(c) This will show all the signals present in both (a) and (b). In addition, signals for ^{79}Br and ^{81}Br will be present (at 22.5 and 24.3 MHz, respectively). At 8.46 T the positions of all lines will be at frequencies 4× greater than at 2.11 T. For example, a ^1H signal will be at 360 MHz.

31. In (c) the high resolution spectrum around 22.6 MHz will show **two** ^{13}C resonance signals, because this molecule contains two nonidentical carbon atoms.

32. **(a)** Divide by 300: 307/300 = 1.02; 617/300 = 2.06; 683/300 = 2.28 ppm.

(b) At 90 MHz: 307 × (90/300) = 92 Hz; 617 × (90/300) = 185 Hz; 683 × (90/300) = 205 Hz.
At 500 MHz: 307 × (500/300) = 512 Hz; 617 × (500/300) = 1028 Hz; 683 × (500/300) = 1138 Hz.

(c)

$$CH_3—\overset{\overset{\displaystyle O}{\|}}{C}—CH_2—C(CH_3)_3$$

$$\delta = \quad 2.06 \qquad 2.28 \quad 1.02$$

33. **(a)** (lowest chemical shift and most upfield), **(d)**, **(f)**, **(c)**, **(b)**, **(e)** (highest chemical shift and most downfield).

34. **(a)** $(CH_3)_2O$ O more electronegative than N, so hydrogens in the ether are less shielded

(b) $CH_3\overset{\overset{\displaystyle O}{\|}}{C}OCH_3$ Hydrogens on carbons next to electronegative atoms are downfield relative to those next to double-bonded functional groups (Table 10-2)

(c) $CH_3CH_2CH_2OH$ Closer to the electronegative atom

(d) $(CH_3)_2S{=}O$ Attachment to oxygen increases the electron-withdrawing nature of the sulfur, deshielding the hydrogens in the sulfoxide more than in the sulfide

35. The number of signals equals the number of nonequivalent sets of hydrogens in the molecule. In the answers below, each nonequivalent hydrogen or set of hydrogens is labeled a, b, c, etc.

(a) Three—the hydrogens cis to the Br (H_b) are different from the hydrogens trans to the Br (H_c)

(b) One—all four H's are equivalent

(c) Three   **(d)** Two   **(e)** Three

36. Chemical shifts have been estimated from the values in Table 10-2, with adjustments for nearby functional groups, and are approximate.

(a) 2 signals: CH_3—CH_2—CH_2—CH_3

↑ ↑
0.9 1.3

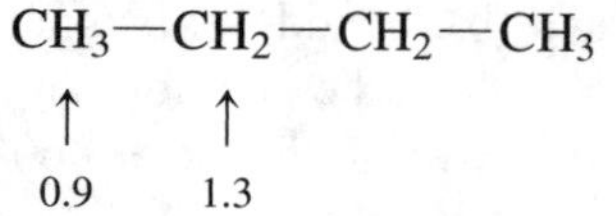

(b) 2 signals: CH_3—$CHBr$—CH_3

↑ ↑
1.5 3.8

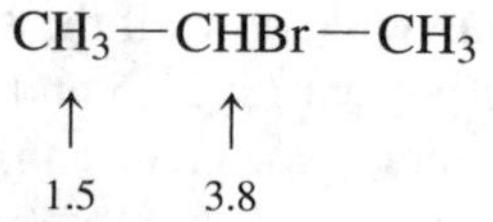

(c) 3 signals: H—O—CH_2—$CCl(CH_3)_2$

↑ ↑ ↑
Variable 4.0 1.4

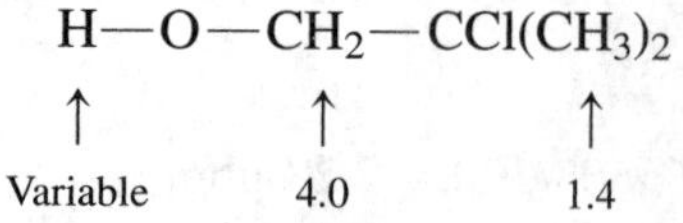

(d) 4 signals: $(CH_3)_2CH$—CH_2—CH_3

↑ ↑ ↑ ↑
0.9 1.5 1.3 0.9

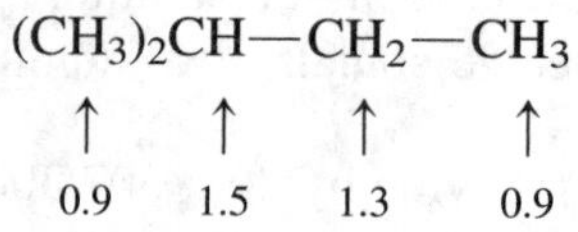

(e) 2 signals: $(CH_3)_3C$—NH_2

↑ ↑
1.3 Variable

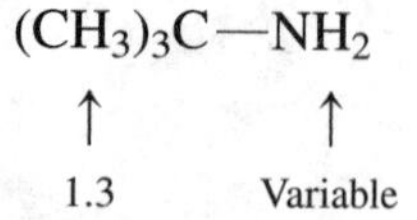

(f) 3 signals: $(CH_3CH_2)_3CH$

↑ ↑ ↑
0.9 1.3 1.5

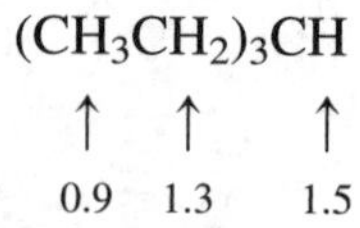

(g) 4 signals: CH_3—O—CH_2—CH_2—CH_3

↑ ↑ ↑ ↑
3.4 3.8 1.7 1.0

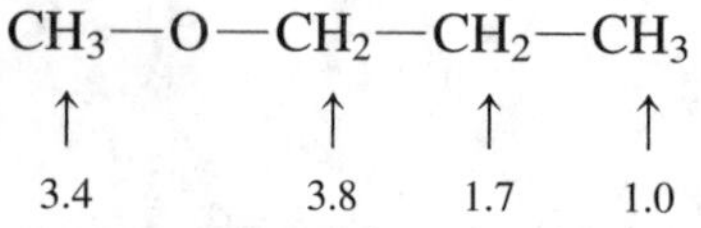

(h) 2 signals:

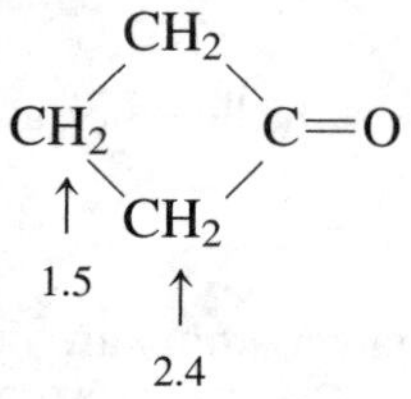

$$CH_2$$

CH_2 C=O

↑ CH_2
1.5 ↑
2.4

(i) 3 signals: CH_3—CH_2—C

O

H ← 9.5

↑ ↑
1.2 2.0

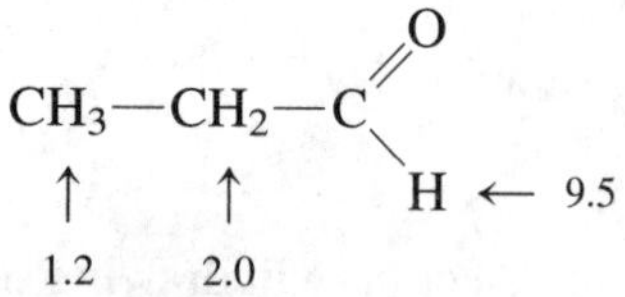

(j) 4 signals:

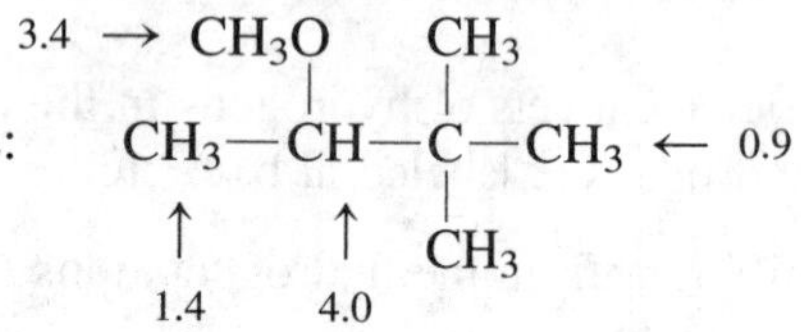

3.4 → CH_3O CH_3

CH_3—CH—C—CH_3 ← 0.9

↑ ↑ CH_3
1.4 4.0

37. As in the previous problem, the chemical shifts are approximations. Chemical boxes are most useful for distinguishing structures. Integrations are given in parentheses. Signals marked with asterisks (*) will be complicated as a result of the presence of a chiral carbon in the molecule (see Chemical Highlight 10-3 for further details).

CH_3 ← 1.1 (3)

(a) $(CH_3)_2CBr$—CH_2—CH_3 $BrCH_2$—CH—CH_2—CH_3

↑   ↑   ↑   ↑   ↑   ↑   ↑
1.5   1.8 1.1   3.5   2.0 1.5   0.9
(6)   (2) (3)   (2)*   (1) (2)*   (3)

$(CH_3)_2CH$—CH_2—CH_2Br

↑   ↑   ↑   ↑
0.9 1.6 1.8 3.5
(6) (1) (2) (2)

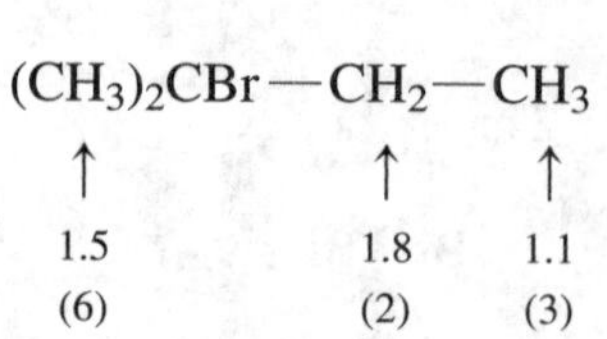

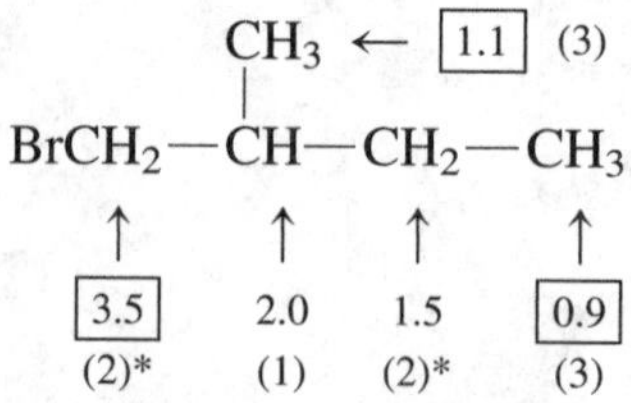

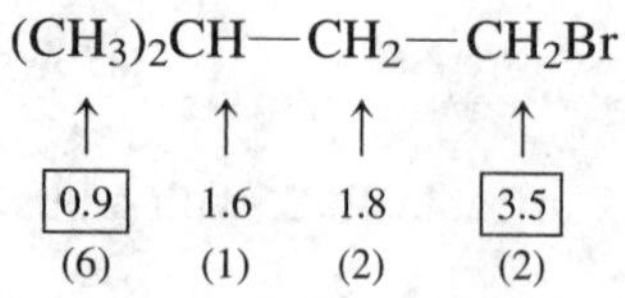

Compounds are readily distinguished by their NMR spectra: The first one has no signals downfield of $\delta = 2$, whereas the others do. The latter will also show different numbers of signals. The second compound has two nonequivalent methyl groups, whereas the third has two identical methyls.

(b) Cl—CH$_2$—CH$_2$—CH$_2$—CH$_2$—OH

$\uparrow$ $\uparrow$ $\uparrow$ $\uparrow$ $\uparrow$

| 3.7 | 1.7 | 1.6 | 3.6 | Var |
| (2) | (2) | (2) | (2) | (1) |

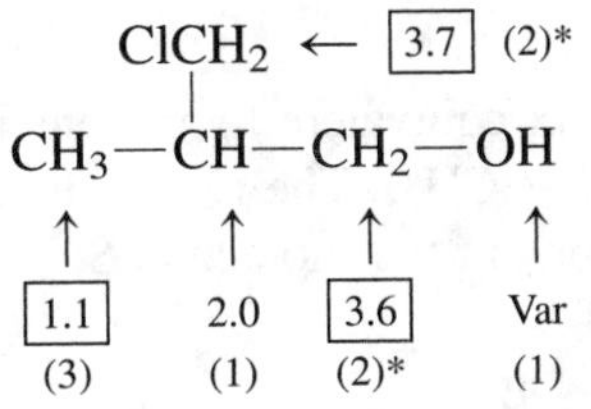

ClCH$_2$ ← 3.7 (2)*
|
CH$_3$—CH—CH$_2$—OH

$\uparrow$ $\uparrow$ $\uparrow$ $\uparrow$

| 1.1 | 2.0 | 3.6 | Var |
| (3) | (1) | (2)* | (1) |

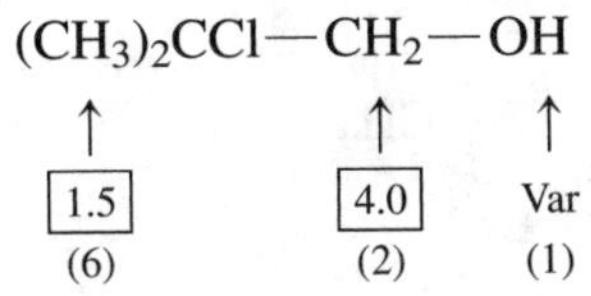

(CH$_3$)$_2$CCl—CH$_2$—OH

$\uparrow$ $\uparrow$ $\uparrow$

| 1.5 | | 4.0 | Var |
| (6) | | (2) | (1) |

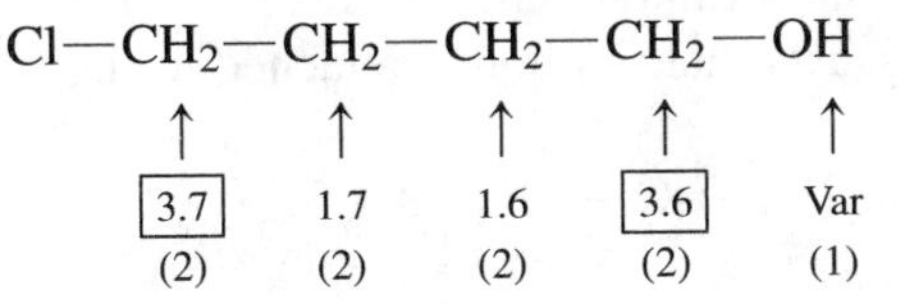

Compounds are again distinguishable. The number and integration of signals downfield of $\delta = 3$ distinguishes the last compound. The other two are distinguished by the presence of an upfield methyl signal in one but not in the other.

CH$_3$ ← 1.5 (3)
|
(c) ClCH$_2$CBrCH—(CH$_3$)$_2$

$\uparrow$ $\uparrow$ $\uparrow$

4.0 2.0 1.1
(2)* (1) (6)*

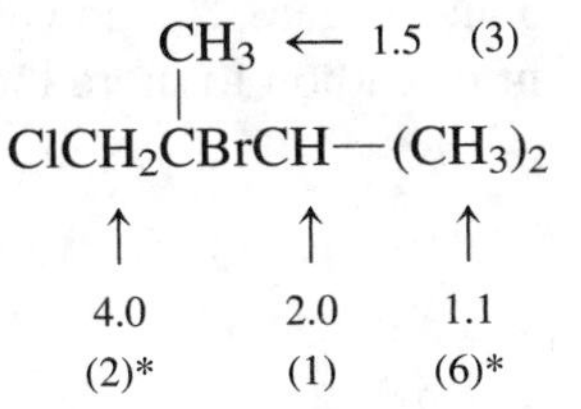

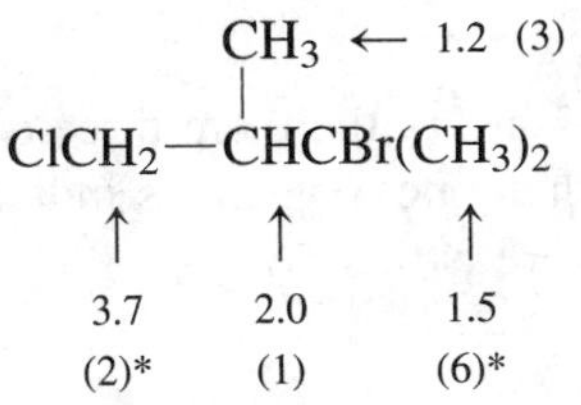

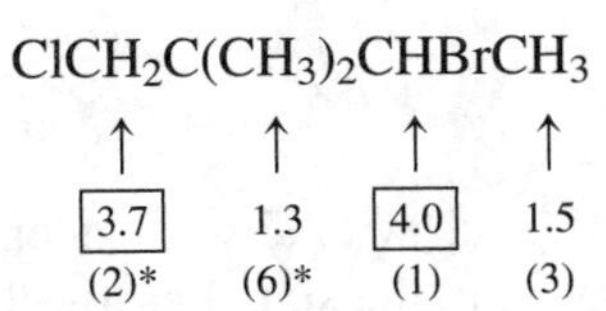

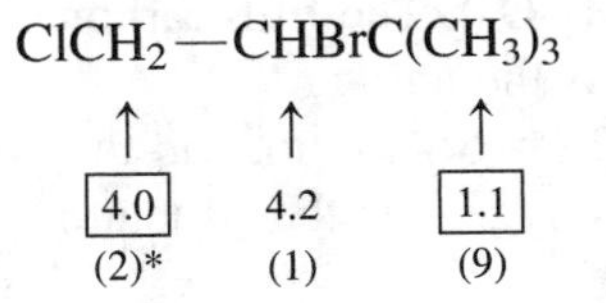

The last two compounds are readily identified. The second of these contains three equivalent methyl groups, giving a single signal of intensity 9. Although the other three compounds all have four signals, only the third one has two signals downfield of $\delta = 3$ (with a total integration of 3). The first two compounds will be difficult to distinguish by NMR—they each have the same number of signals with the same integration ratios. Only minor differences in chemical shifts will be present.

CH$_3$ ← 1.2 (3)
|
ClCH$_2$—CHCBr(CH$_3$)$_2$

$\uparrow$ $\uparrow$ $\uparrow$

3.7 2.0 1.5
(2)* (1) (6)*

ClCH$_2$C(CH$_3$)$_2$CHBrCH$_3$

$\uparrow$ $\uparrow$ $\uparrow$ $\uparrow$

| 3.7 | 1.3 | 4.0 | 1.5 |
| (2)* | (6)* | (1) | (3) |

ClCH$_2$—CHBrC(CH$_3$)$_3$

$\uparrow$ $\uparrow$ $\uparrow$

| 4.0 | 4.2 | 1.1 |
| (2)* | (1) | (9) |

38. **(a)** The spectrum shows two signals, at $\delta = 1.1$ and 3.3. The $\delta = 1.1$ signal is for 9 equivalent hydrogens, and the $\delta = 3.3$ signal is for 2 equivalent hydrogens. A good way to get nine equivalent hydrogens is with a $(CH_3)_3C$ group. The other two hydrogens must be on a separate, single carbon, because the $(CH_3)_3C$ group contains all but one of the five carbons in the formula. The downfield location of these two hydrogens suggests their carbon is attached to the Cl. So,

$$(CH_3)_3C- \; + \; -CH_2- \; + \; -Cl \; \Rightarrow \; (CH_3)_3C-CH_2-Cl$$

$$\uparrow \qquad\qquad \uparrow$$
$$1.1 \qquad\qquad 3.3 \qquad\qquad \text{as a plausible structure.}$$

(b) Somewhat similar: two signals, $\delta = 1.9$ and 3.8. The signal for six equivalent hydrogens is probably due to two methyl groups on the same carbon (CH_3-C-CH_3). The two-hydrogen signal can only be a $-CH_2-$, because there are only four carbons in the molecule. So,

$$CH_3-\overset{\textstyle |}{\underset{\textstyle |}{C}}-CH_3 \; + \; -CH_2- \; + \; (2\times)-Br \; \Rightarrow \;$$

$$\begin{array}{c} Br \\ | \\ CH_2 \;\leftarrow\; 3.8 \\ | \\ CH_3-\overset{\textstyle |}{\underset{\textstyle |}{C}}-CH_3 \\ Br \\ 1.9 \end{array}$$

39. **(a)** The spectrum has two signals in a $3:1$ intensity ratio. The molecule has eight hydrogens, so there must be one group of six equivalent H's and another of two equivalent H's ($6:2 = 3:1$). Two equivalent CH_3's and a CH_2 account for all but the two oxygen atoms in the formula. The larger signal at $\delta = 3.3$ is just right for hydrogens on carbons attached to an oxygen. The downfield location for the small signal ($\delta = 4.4$) is consistent with attachment of carbon to more than one oxygen. Putting it all together we get

$$CH_3-O-CH_2-O-CH_3$$
$$\uparrow$$
$$4.4$$
$$\text{Equivalent, at } 3.3$$

(b) Again two signals, but now in a $9:1$ intensity ratio. Reasoning as in (a), there are three equivalent CH_3's, each attached to an oxygen, and a CH attached to more than one oxygen, as indicated by its downfield chemical shift ($\delta = 4.9$). The only consistent structure is then

$$(CH_3O)_3CH$$
$$\uparrow \qquad \uparrow$$
$$3.3 \qquad 4.9$$

(c) Two signals that are equal in intensity imply two different groups, each with six equivalent H's. The signal at $\delta = 3.1$ could imply two equivalent CH_3-O groups, and the signal at $\delta = 1.2$ suggests two equivalent CH_3 groups not attached to oxygen. These all add up to $C_4H_{12}O_2$, leaving one carbon unaccounted for in the formula of the molecule ($C_5H_{12}O_2$). The fifth carbon can be used to connect the other four groups: $(CH_3O)_2C(CH_3)_2$, which is the answer.

By comparison, 1,2-dimethoxyethane has two signals in a $3:2$ ($= 6:4$) ratio, and they are both in a region consistent with H's on a carbon attached to a **single** oxygen, as the structure $CH_3OCH_2CH_2OCH_3$ requires.

40. **(a)** There are two signals in a 3:1 intensity ratio. There are 12 H's in the formula, so there are 9 H's in one location (not adjacent to a functional group, according to the $\delta = 1.2$, high-field shift) and 3 H's in another (close to a functional group, based on the $\delta = 2.1$ chemical shift). It is simplest to start with CH_3 groups as pieces of a possible structure. There is also a CO group, because the molecule is a ketone. So we have so far

$$CH_3 \quad (3\times)CH_3 \quad CO$$

which adds up to $C_5H_{12}O$. One additional C is needed, but nothing else. If we draw all these pieces with the possible bonds they can form we have

$$-CH_3 \quad \underbrace{-CH_3 \quad -CH_3 \quad -CH_3}_{1.2} \quad \overset{O}{\underset{}{\overset{\|}{-C-}}} \quad -\overset{|}{\underset{|}{C}}-$$
$$\,_{2.1}$$

Attaching the first CH_3 to the CO will explain its chemical shift:

$$CH_3-\overset{O}{\overset{\|}{C}}-$$

None of the other three CH_3's can be directly on the CO because (1) the chemical shifts are wrong and (2) there would be no place left to attach the rest of the pieces. So, attach the unbonded C atom instead:

$$CH_3-\overset{O}{\overset{\|}{C}}-\overset{|}{\underset{|}{C}}-$$

Now the only possible final step is to attach the three CH_3's:

$$CH_3-\overset{O}{\overset{\|}{C}}-C(CH_3)_3 \qquad \text{This is the answer.}$$

(b) Both show two signals in a 3:1 ratio, corresponding to nine H's and three H's again. Let's assume again that the pieces are the same as in (a), plus the extra O atom. Let's also assume that three of the CH_3's are again on the extra C atom; that is, we have as our pieces

$$-CH_3 \quad -C(CH_3)_3 \quad -\overset{O}{\overset{\|}{C}}- \quad -O-$$

For isomer 1, $-CH_3$ is at $\delta = 2.0$ and $-C(CH_3)_3$ is at $\delta = 1.5$. The $-C(CH_3)_3$ is downfield relative to the ketone in (a).

For isomer 2, $-CH_3$ is at $\delta = 3.6$, far downfield relative to the ketone in (a), but the $-C(CH_3)_3$ is at $\delta = 1.2$, almost identical to the ketone in (a).

The CH_3 in isomer 2 must be attached to the extra O, with the rest of the molecule similar to the ketone in (a). This compound is an ester.

$$CH_3-O-\overset{O}{\overset{\|}{C}}-C(CH_3)_3$$

Isomer 1 has the extra O on the other side of the CO, accounting for the slight downfield shift of the $-C(CH_3)_3$. This compound is also an ester.

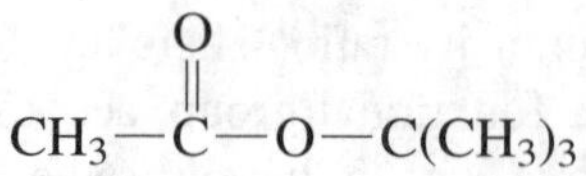

$$CH_3-\overset{\overset{\displaystyle O}{\|}}{C}-O-C(CH_3)_3$$

41. (1) number of signals = number of groups of hydrogen atoms in distinct chemical environments; (2) integrated intensity of each signal ∝ number of hydrogen atoms responsible for each signal; (3) chemical shift of each signal, related to chemical environment of hydrogens responsible for it; and (4) splitting of each signal, related to number and location of neighboring chemical-shift-distinct hydrogens.

42. Taking the last part of the question first, from left to right, the first two compounds are esters, the third is a ketone, and the fourth is an aldehyde.

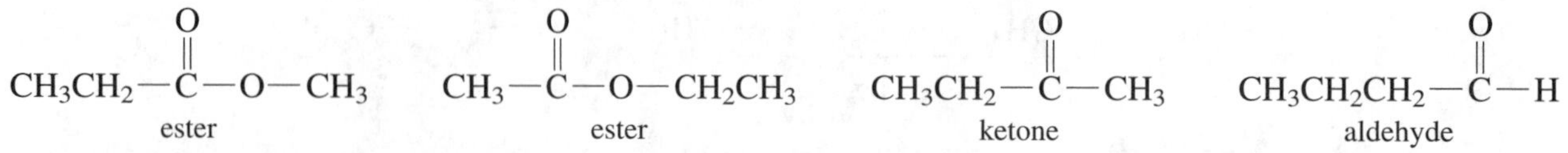

$$CH_3CH_2-\overset{\overset{\displaystyle O}{\|}}{C}-O-CH_3 \qquad CH_3-\overset{\overset{\displaystyle O}{\|}}{C}-O-CH_2CH_3 \qquad CH_3CH_2-\overset{\overset{\displaystyle O}{\|}}{C}-CH_3 \qquad CH_3CH_2CH_2-\overset{\overset{\displaystyle O}{\|}}{C}-H$$

　　　　　ester 　　　　　　　　　ester 　　　　　　　　ketone 　　　　　　　aldehyde

Number of signals, and integration of each: The first three compounds will all have three ^1H NMR signals, two with an integration of 3, and one with an integration of 2. The aldehyde will be different, having an ^1H NMR spectrum with four signals, one with an integration of 3, two with an integration of 2, and one with an integration of 1 (the aldehyde hydrogen attached to the C=O carbon).

　　Chemical shifts will distinguish the first three compounds from each other, because the signals for hydrogens on carbon atoms attached to oxygen appear in the region between about δ 3.0 and δ 4.5 ppm. The ketone lacks such hydrogens; its entire NMR spectrum will appear upfield of δ 2.5 ppm. Splitting follows the $N + 1$ rule, except for the aldehyde. The coupling constant for splitting between the unique ("aldehyde") hydrogen on the C=O carbon and the CH_2 group on the other side is smaller than those for the other splittings. As a result that particular CH_2 shows a more complicated pattern based upon triplet splitting by the neighboring CH_2 group, but with each of the three lines of this triplet doubled by coupling to the aldehyde hydrogen. We describe this pattern as a "doublet of triplets," abbreviated dt. Approximate chemical shifts and splittings are given below (s = singlet, d = doublet, t = triplet, q = quartet, sex = sextet).

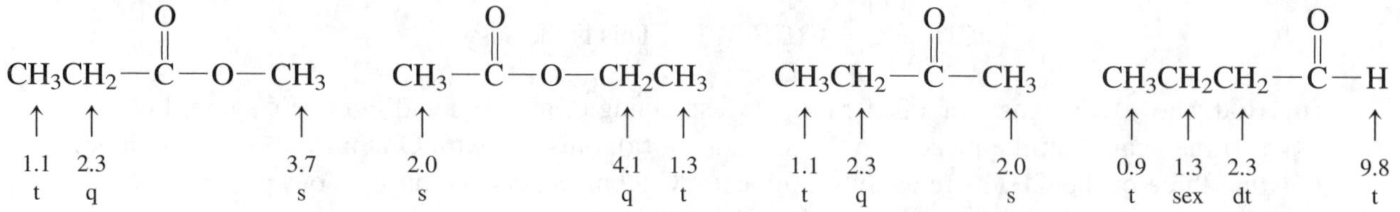

$$CH_3CH_2-\overset{\overset{\displaystyle O}{\|}}{C}-O-CH_3 \qquad CH_3-\overset{\overset{\displaystyle O}{\|}}{C}-O-CH_2CH_3 \qquad CH_3CH_2-\overset{\overset{\displaystyle O}{\|}}{C}-CH_3 \qquad CH_3CH_2CH_2-\overset{\overset{\displaystyle O}{\|}}{C}-H$$

1.1	2.3		3.7	2.0		4.1	1.3	1.1	2.3		2.0	0.9	1.3	2.3		9.8
t	q		s	s		q	t	t	q		s	t	sex	dt		t

43. (a) This compound should show four signals, so it could match either (ii) or (iii). The methyl group in this molecule is far from the nearest Cl, and it has a CH_2 as a neighbor, so it should exhibit a triplet (two neighbors, so $N + 1 = 3$) to relatively high field (low δ value). The answer therefore is (iii), with the triplet at $\delta = 1.0$ ppm.

(b) Symmetry in the structure should simplify the NMR to two signals; (i) is the answer.

(c) Four signals are expected, but unlike (a) the methyl group here is closer to a Cl and has only a CH as a neighbor. Its signal should be a doubler, shifted downfield somewhat; (ii) is the spectrum, with the doublet at $\delta = 1.6$ ppm.

Notice that there was no need to analyze every peak to match the answers in this problem. The remaining signals can be analyzed, of course, as follows:

(a) The CH_2Cl has a CH as a neighbor and is the doublet at $\delta = 3.6$ ppm; the CHCl has four neighbors and is the quintet at $\delta = 3.9$ ppm; and the CH_2 at C3 also has four neighbors and is the quintet at $\delta = 1.9$ ppm.

(b) Each CH_3 has a CH as a neighbor, giving the doublet at $\delta = 1.5$ ppm; and each CHCl has four neighbors, giving the quintet at $\delta = 4.1$ ppm.

(c) The CH_2Cl has a CH_2 as a neighbor and is the triplet at $\delta = 3.6$ ppm; the CH_2 at C2 has three neighbors and is the quartet at $\delta = 2.1$ ppm; and the CHCl has five neighbors and is the sextet at $\delta = 4.2$ ppm.

44. Use the $N + 1$ rule: Number of lines = number of <u>N</u>eighbors $\pm$ 1.

(a) The signal for the CH_3 groups at $\delta = 0.9$ will be split into a triplet by the two hydrogens on the neighboring CH_2 group of each one $(2 + 1 = 3)$. The signal for the CH_2 groups at $\delta = 1.3$ will be split into a quartet by the three hydrogens on the neighboring CH_3 group $(3 + 1 = 4)$. Although each CH_2 group has the other CH_2 group as its neighbor also, splitting is not observed between them because they are equivalent by symmetry and have the same chemical shift. Chemical-shift-equivalent nuclei *do not give rise to observable splitting with one another.*

(b) The signal for the CH_3 groups at $\delta = 1.5$ will be split into a doublet by the single hydrogen on its neighboring CH group $(1 + 1 = 2)$. The signal for the CH group at $\delta = 3.8$ will be split into a septet by the six hydrogens on the two neighboring CH_3 groups $(6 + 1 = 7)$.

(c) All signals will be singlets. The CCl group is in between the CH_3 groups and the CH_2 group. The hydroxy hydrogen does not split its neighbor.

(d) This spectrum will be messy, because three CH_3 groups have their resonances at about the same chemical shift, $\delta = 0.9$, and they are not all equivalent. The signal for the two CH_3 groups with the neighboring CH group will be split by the latter into a doublet. The signal for the remaining CH_3 group at $\delta = 0.9$ will be split into a triplet by its neighboring CH_2 group. The doublet and triplet signals will overlap in the spectrum. The signal for the CH_2 group at $\delta = 1.4$ will be split into a quintet by the combined effect of its neighboring CH_3 group and the CH group on its other side $(4 + 1 = 5)$. Finally, the signal for the CH group at $\delta = 1.5$ will possess nine lines (a nonet) as a result of splitting by a total of eight neighboring hydrogens (two CH_3 groups and one CH_2 group). These latter two patterns, the quintet and the nonet, will also overlap, because of their similar chemical shift positions in the spectrum.

(e) No splitting. All singlets.

(f) The signal for the CH_3 groups at $\delta = 0.9$ will be split into a triplet by the two hydrogens on each one's neighboring CH_2 group. The signal for the CH_2 groups at $\delta = 1.3$ will be split into a quintet by the combined effect of the neighboring CH_3 group and the CH group on the other side $(4 + 1 = 5)$. The signal for the CH group at $\delta = 1.5$ will be a septet (7 lines), arising from splitting by six hydrogens on the three neighboring CH_2 groups.

(g) The signal for the CH_3 group at $\delta = 1.0$ will be split into a triplet by the neighboring CH_2 group. The signal for the CH_2 group at $\delta = 1.7$ will be split into a sextet by the combined effect of its neighboring CH_3 group and the CH_2 group on its other side $(5 + 1 = 6)$. The signal for the CH_2 group at $\delta = 3.8$ will be split into a triplet by its neighboring CH_2 group. The signal for the CH_3 group at $\delta = 3.4$ will be a singlet.

(h) In the simplest possible scenario, the signal for the CH_2 group at $\delta = 1.5$ will be split into a quintet by the combined effect of its two neighboring CH_2 groups on either side $(4 + 1 = 5)$. The signal for the CH_2 groups at $\delta = 2.4$ will be split into a triplet by each one's single neighboring CH_2 group. Unfortunately, simplicity is not likely to be the case here: In cyclic compounds, coupling constants (J values) between nonequivalent hydrogens that are cis to each other are often different in magnitude from the J values between hydrogens that are trans. As discussed in Section 10-8, the usual consequence of this phenomenon is more lines. Taking this situation into account, the signal for the CH_2 group at $\delta = 1.5$ may be split into as many as nine lines—a "triplet of triplets"—by the combined coupling with two neighboring cis hydrogens and two neighboring trans hydrogens ($[2 + 1 = 3] \times [2 + 1 = 3] = 9$). Conversely, the signal for the CH_2 groups at $\delta = 2.4$ may be split into four lines—a "doublet of doublets"—by different size coupling with the cis and trans hydrogens on the neighboring CH_2 group ($[1 + 1 = 2] \times [1 + 1 = 2] = 4$).

(i) The signal for the CH_3 group at $\delta = 1.2$ will be split into a triplet by the neighboring CH_2 group. In the simplest possible situation, the signal for the CH_2 group at $\delta = 2.0$ will be split into a quintet by the combined effect of its neighboring CH_3 group and the CH group on its other side (on the carbonyl carbon). The signal for the CH group at $\delta = 9.5$ will be split into a triplet by its neighboring CH_2 group. As we shall see in Chapter 17, coupling constants to the $H-\overset{\overset{\displaystyle O}{\|}}{C}-$ hydrogen are smaller than usual and lead to more complicated patterns with more lines than the simple $N + 1$ rule predicts. In reality, the signal for the CH_2 group will be split into a quartet of doublets.

(j) The signals at $\delta = 0.9$ and $\delta = 3.4$ will be singlets (no neighbors). The signal for the CH_3 group at $\delta = 1.4$ will be split into a doublet by the neighboring CH group. The signal for the CH group at $\delta = 4.0$ will be split into a quartet by the neighboring CH_3 group.

45. Procedures are similar to those of Problem 44. We show each structure below, and near each group of hydrogens, the multiplicity of its signal (in plain English, the number of lines into which it is split), by using one of the following abbreviations: s, singlet; d, doublet; t, triplet; q, quartet; quin, quintet; sex, sextet; sept, septet; oct, octet; and non, nonet. All multiplicities have been determined by applying the $N + 1$ rule.

(a) $\underset{\overset{|}{\underset{Br}{(q)}}}{\overset{CH_3\ (s)}{\underset{|}{CH_3}}C}CH_2CH_3,\quad BrCH_2CHCH_2CH_3,\quad CH_3CHCH_2CH_2Br$

(b) $ClCH_2CH_2CH_2CH_2OH,\quad CH_3CHCH_2OH,\quad CH_3CCH_2OH$

(c) $ClCH_2C{-}CHCH_3,\quad ClCH_2CH{-}CCH_3,\quad ClCH_2C{-}CHCH_3,\quad ClCH_2CHCCH_3$

46. In some cases assignments of signals with similar chemical shifts cannot be unambiguously made without further information, such as integration data.

(a) Cl_2CHCH_2Cl

$\delta = 5.8\quad 4.0$ ppm

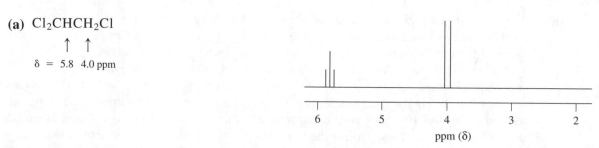

(b) $CH_3CHBrCH_2CH_3$

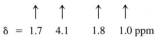

δ = 1.7   4.1   1.8   1.0 ppm

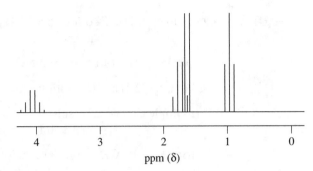

(c) $CH_3CH_2CH_2COOCH_3$

δ = 1.0   1.7   2.3   3.6 ppm

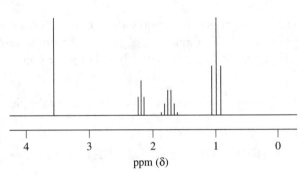

(d) $ClCH_2CHOHCH_3$

δ = 3.4   3.9   3.0   1.2 ppm

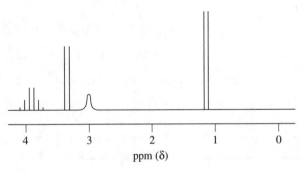

47. In the answers that follow, signals will be described in an abbreviated way, and possible groups to which they correspond will be given. Remember the $N + 1$ rule: N hydrogen neighbors split into $N + 1$ lines!

(C) δ = 0.8 (triplet, 3 H): **CH$_3$,** split by a neighboring CH$_2$ (2 + 1 = 3 = triplet)

δ = 1.2 (singlet, 6 H): **Two identical CH$_3$'s,** with no neighboring H's to split

δ = 1.4 (quartet, 2 H): **CH$_2$,** split by a neighboring CH$_3$ (3 + 1 = 4 = quartet)

δ = 1.9 (singlet, 1 H): Unsplit **CH** or **OH;** OH more likely because molecule is an alcohol

These add up to C$_4$H$_{12}$O. One more C is needed to give C$_5$H$_{12}$O; then we have the pieces

$$CH_3-CH_2- \qquad \underbrace{CH_3- \qquad CH_3-}_{1.2} \qquad -OH \qquad -\overset{|}{\underset{|}{C}}-$$
$$\quad 0.8 \quad 1.4 \qquad\qquad\qquad\qquad\qquad 1.9$$

Putting the first four groups on to the last, unattached C gives the correct answer: 2-methyl-2-butanol.

$$CH_3CH_2-\overset{\displaystyle CH_3}{\underset{\displaystyle CH_3}{\overset{|}{\underset{|}{C}}}}-OH$$

(D) $\delta = 0.9$ (doublet, 6 H): **Two identical CH₃'s,** both split by a neighboring CH ($1 + 1 = 2 =$ doublet)

$\delta = 1.3$ (singlet, 1 H): Most likely the **OH** (the broad signal gives it away)

$\delta = 1.45$ (quartet, 2 H): **CH₂,** split by three neighboring hydrogens ($3 + 1 = 4 =$ quartet)

$\delta = 1.7$ (multiplet, 1 H): **CH,** split by perhaps 7 neighboring hydrogens (eight lines are visible; there could always be more that are not)

$\delta = 3.7$ (triplet, 2 H): **CH₂,** split by a neighboring CH₂ ($2 + 1 = 3 =$ triplet), and connected directly to O, according to the chemical shift

These five groups add up to $C_5H_{12}O$, so we've found all the pieces in clearly separated signals. The **CH** group at 1.7 is reasonably assigned as the group responsible for splitting the 0.9 signal into a doublet. The **CH₂** group at 3.7 is presumably split by the other **CH₂** at 1.45; however, the latter is split by a third neighbor—must be the **CH** group. We can put the pieces together directly to give

$$\begin{array}{c} CH_3 \\ \diagdown \\ \diagup \\ CH_3 \end{array} CH - CH_2 - CH_2 - OH \, .$$
$$\quad\quad 0.9 \quad\; 1.7 \quad 1.45 \quad 3.7 \quad\; 1.3$$

(E) $\delta = 0.8$ (triplet, 3 H): **CH₃,** split by a neighboring CH₂

$\delta = 1.3$ (a mess, 4 H): ???

$\delta = 1.5$ (quintet?, 2 H): **CH₂,** split by four (?) neighboring hydrogens

$\delta = 3.0$ (broad singlet, 1 H): **OH** again

$\delta = 3.5$ (triplet, 2 H): **CH₂,** split by a neighboring CH₂ and on the O

The pieces we can identify add up to C_3H_8O. Some guessing is needed. Let's start by assuming that the CH₂ groups at 1.5 and 3.5 are attached to each other and see if it gets us anywhere. The CH₃ then has to be attached to a CH₂ that is buried in the signal at 1.3. Now our groups total up to $C_4H_{10}O$. There has to be one more CH₂ group hiding at 1.3. So what do we have?

$$CH_3 - CH_2 - \quad\quad -CH_2 - CH_2 - OH \quad\quad -CH_2 -$$
$$\;\; 0.8 \quad\;\; 1.3 \quad\quad\quad\;\; 1.5 \quad\; 3.5 \quad\; 3.0 \quad\quad\quad 1.3$$

There's only one way to put the pieces together to include everything, 1-pentanol: $CH_3 - CH_2 - CH_2 - CH_2 - CH_2 - OH$. By the way, are you noticing how the chemical shifts of the OH groups of these compounds vary all over the place? That's normal.

(F) $\delta = 0.9$ (triplet, although a poor excuse for one, 3 H): **CH₃,** next to a CH₂

$\delta = 1.2$ (doublet, 3 H): **CH₃,** next to a CH

$\delta = 1.4$ (complex signal, 4 H): ???

$\delta = 1.6$ (broad singlet, 1 H): **OH**

$\delta = 3.8$ (four, maybe five lines, 1 H): **CH,** next to O, split by at least 3 and maybe 4 neighboring H's

Let's look at these pieces. The CH₃ at 1.2 could be attached to the CH at 3.8. The CH₃ at 0.9 could be attached to a CH₂ that is part of the 1.4 signal. That gives a total of $C_4H_{10}O$, so we need another CH₂, presumably also at 1.4. The pieces are

$$\begin{array}{c} | \\ -CH_3 - CH - OH \end{array} \quad\quad CH_3 - CH_2 - \quad\quad -CH_2 -$$
$$\;\;\; 1.2 \quad\;\; 3.8 \quad 1.6 \quad\quad\; 0.9 \quad\; 1.4 \quad\quad\quad 1.4$$

Put them together to give 2-pentanol.

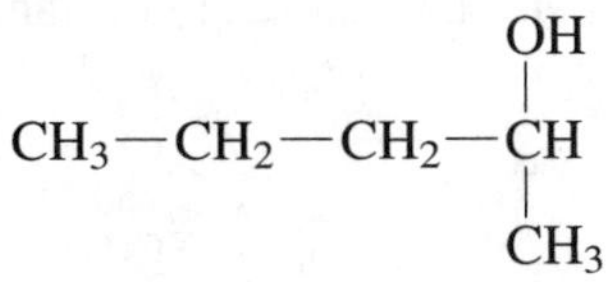

$$CH_3-CH_2-CH_2-\overset{\displaystyle OH}{\underset{\displaystyle CH_3}{CH}}$$

Notice how both spectra E and F have had very distorted triplets around 0.9 for CH_3 groups next to CH_2's. This distorted appearance is very common and is due to the closeness of the chemical shifts of the methyls ($\delta = 0.9$) and the groups splitting them ($\delta = 1.2-1.8$ in E and $\delta = 1.4$ in F).

48. **(a)** The sketch shows typical ethyl group signals (upfield triplet for the CH_3 and downfield quartet for the CH_2) and a very deshielded CH_2 singlet due to attachment to two electronegative atoms (chemical shifts in the spectrum are those actually found).

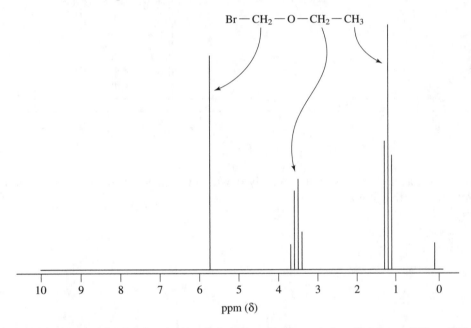

(b) The sketch shows a singlet for the methyl and two close triplets for the $-CH_2-CH_2-$ unit. The actual spectrum would show complications in the splitting of the latter because of the small chemical-shift difference (refer to Section 10-8). We have ignored this effect and shown how the first-order spectrum would look.

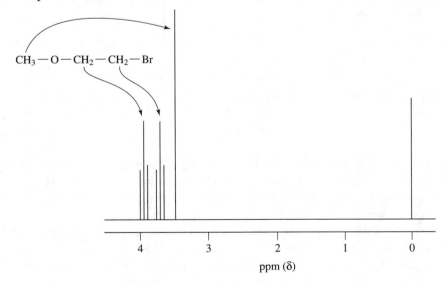

(c) The sketch should look like the spectrum of $CH_3CH_2CH_2Br$ (text Figure 10-27). C2 gives a sextet (five neighbors), assuming similar coupling constants all around.

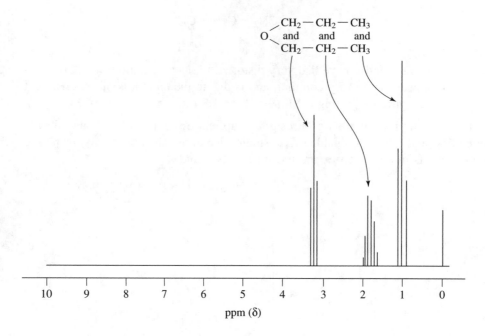

(d) As in (a), we again have a signal relatively downfield because of the group's location between two electronegative atoms.

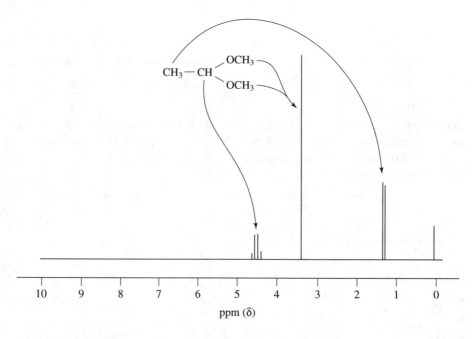

49. Two signals, $\delta = 0.9$ (doublet) and $\delta = 1.4$ (septet?) with an intensity of 12 H for the large signal and 2 H for the small one. Twelve equivalent hydrogens probably mean four equivalent CH_3 groups, which equals C_4H_{12}. That leaves C_2H_2 left to make up the molecule. The only way to do it that makes all the CH_3's identical and split into a doublet is $(CH_3)_2CH—CH(CH_3)_2$, 2,3-dimethylbutane.

   Figure 10-22 shows the NMR of 2-iodopropane, another molecule containing the $(CH_3)_2CH$ group. Again, the methyl signal is a doublet, but the CH signal is better resolved as a clean septet. The larger chemical shift difference between the two sets of signals in 2-iodopropane gives rise to a more nearly "first-order" appearance to its spectrum, compared with the spectrum of 2,3-dimethylbutane.

50. The NMR of the product closely resembles that of the tertiary alcohol 2-methyl-2-butanol (Problem 47, spectrum C), but the signal for the OH group is missing, and the molecular formula has a Br instead of an OH. The product is 2-bromo-2-methylbutane, and the signals in the spectrum are assigned similarly. How does it form? Rearrangements!

51. At 60 MHz, only the CH_2 next to Cl ($\delta = 3.5$) is clearly resolved. The very distorted CH_3 triplet is barely separated ($\delta = 0.9$) from the other three CH_2's, which overlap from $\delta = 1.0-2.0$. At 500 MHz, the separation of signals in hertz is so much greater that the entire spectrum consists of nearly first-order signals: $\delta = 0.92$ (triplet, CH_3), 1.36 (sextet, C4 CH_2), 1.42 (quintet, C3 CH_2), 1.79 (quintet, C2 CH_2), 3.53 (triplet, Cl CH_2). Notice how the multiplets appear so much narrower at 500 MHz. In reality, the coupling constants are unchanged. However, recall that, at 60 MHz, the distance between $\delta = 0$ and $\delta = 4$ is only 240 Hz, whereas, at 500 MHz, the same 4-ppm spectrum width corresponds to 2000 Hz! So the 6- to 8-Hz splittings appear to spread out over a much larger chemical-shift range at 60 MHz than they do at 500 MHz.

52. Neighboring nonequivalent hydrogens split one another. When a hydrogen is split by two or more hydrogens that are themselves not equivalent, the coupling constants of those splittings may be different in magnitude, as illustrated by the spectrum of 1,1,2-trichloropropane (Figures 10-25 and 26).

    (a) H_a—a triplet of triplets, from large splitting by the two cis neighbors (H_c) and smaller splitting by the two trans neighbors (H_b); H_b—a quartet, from geminal splitting to the H_c on the same carbon and trans splittings to H_a and the H_c one carbon over, all of which are approximately the same in magnitude; H_c—a doublet of triplets, the doublet splitting with the (cis) H_a being larger than the triplet splitting to the two H_b (one geminal and one trans).

    (b) No splitting.

    (c) H_a—a quartet, from geminal splitting to H_b and trans splittings to the two H_c; H_b—a doublet of triplets, the doublet geminal splitting with H_a being smaller than the triplet splitting to the two cis H_c; H_c—a doublet of doublets, the doublet splitting with the (cis) H_b being larger than that to the (trans) H_a.

(d) H_a—a doublet of doublets, the doublet splitting with the cis H_b being larger than that to the trans H_b; H_b—also a doublet of doublets, for the same reason: The doublet splitting with the cis H_a is larger than that to the trans H_a.

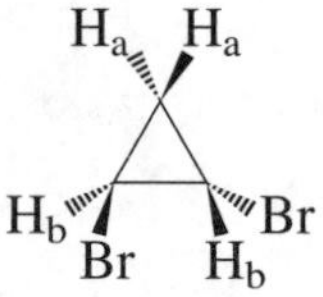

(e) H_a—a triplet, because the geminal splitting with H_b is about the same as splitting with the trans H_c; H_b—a doublet of doublets, the doublet splitting with the cis H_c being larger than the geminal splitting to H_a; H_c—a doublet of doublets, the doublet splitting with the (cis) H_b being larger than that to the (trans) H_a.

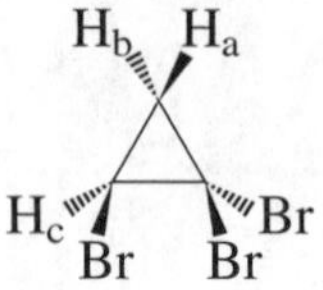

53. Determine how many different signals would be displayed by each isomer.

Pentanes: C—C—C—C—C   3 signals

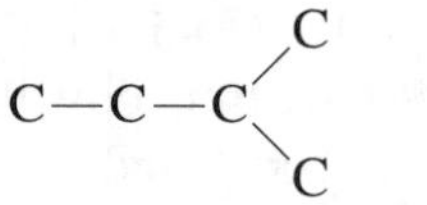

4 signals

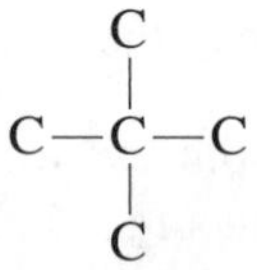

2 signals

All three can be readily identified by ^{13}C NMR.

Hexanes: C—C—C—C—C—C   3 signals

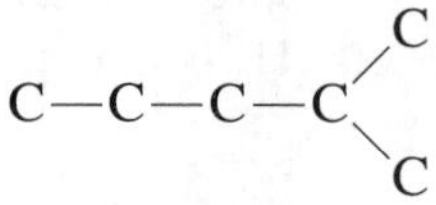

5 signals

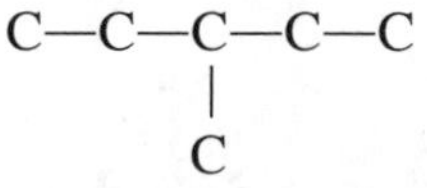

4 signals*

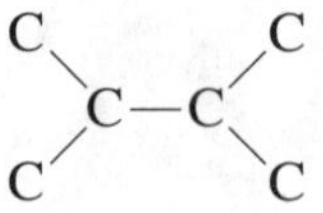

2 signals

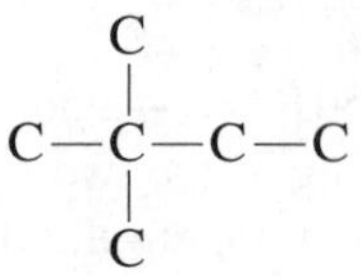

4 signals*

The ^{13}C NMR spectra of 3-methylpentane and 2,2-dimethylbutane (marked with asterisks) are similar, each having four different carbon environments. The two spectra must be distinguished by signal **intensities:** 2,2-dimethylbutane has three equivalent methyl carbons, which give rise to an exceptionally intense signal.

54. Answers include the number of signals and (for the spectrum without hydrogen decoupling) the splitting by directly attached hydrogens ($N + 1$ rule). Chemical shifts (based on Table 10-6) are **very** approximate.

(a) 2 signals: $\delta = 10$ (CH_3, q) and 20 (CH_2, t)

(b) 2 signals: $\delta = 25$ (CH_3, q) and 45 (CHBr, d)

(c) 3 signals: $\delta = 25$ (CH_3, q), 60 (CCl, s), and 65 (CH_2OH, t)

(d) 4 signals: $\delta = 10$ (C4 CH_3, q), 15 (other CH_3's, q), 25 (CH_2, t), and 30 (CH, d)

(e) 2 signals: $\delta = 30$ (CH_3, q) and 50 (CNH_2, s)

(f) 3 signals: $\delta = 10$ (CH_3, q), 25 (CH_2, t), and 30 (CH, d)

(g) 4 signals: $\delta = 10$ (CH_3, q), 30 (CH_2, t), 60 (CH_3O, q), and 65 (CH_2O, t)

(h) 3 signals: $\delta = 15$ (CH_2, t), 45 (CH_2 next to C=O, t), and 200 (C=O, s)

(i) 3 signals: $\delta = 15$ (CH_3, q), 40 (CH_2, t), and 200 (CHO, d)

(j) 5 signals: $\delta = 15$ (3 CH_3, q), 20 (other CH_3, q), 40 (C, s), 60 (CH_3O, q), and 80 (CHO, d)

55. In each group consider the compounds from left to right. Again, chemical shifts are **very rough estimates** based on data in Table 10-6 and closeness of carbons to electronegative atoms.

(a) First: 4 signals: $\delta = 15$ (CH_3), 25 (2 CH_3), 35 (CH_2), 50 (CBr)

Second: 5 signals: $\delta = 10$ (CH_3), 15 (CH_3), 25 (CH_2), 40 (CH_2Br), 45 (CH)

Third: 4 signals: $\delta = 10$ (2 CH_3), 30 (CH_2), 35 (CH_2Br), 40 (CH)

The first and third compounds would be difficult to distinguish on the basis of this spectroscopic information alone.

(b) First: 4 signals: $\delta = 25$ (CH_2), 30 (CH_2), 40 (CH_2Cl), 60 (CH_2OH)

Second: 4 signals: $\delta = 20$ (CH_3), 35 (CH), 45 (CH_2Cl), 60 (CH_2OH)

Third: 3 signals: $\delta = 25$ (2 CH_3), 55 (CCl), 65 (CH_2OH)

The first and second are not readily distinguished.

(c) First: 5 signals: $\delta = 15$ (2 CH_3), 25 (CH_3), 45 (CH), 50 (CBr), 55 (CH_2Cl)

Second: 5 signals: $\delta = 15$ (CH_3), 25 (2 CH_3), 45 (CBr), 50 (CH_2Cl), 55 (CH)

Third: 5 signals: $\delta = 15$ (2 CH_3), 25 (CH_3), 40 (CHBr), 45 (C), 50 (CH_2Cl)

Fourth: 4 signals: $\delta = 10$ (3 CH_3), 45 (CHBr), 50 (C), 55 (CH_2Cl)

The first three are virtually impossible to tell apart.

56. The different versions of the DEPT spectra show carbon atoms with different numbers of attached hydrogens. For example, in 35(a) (bromocyclopropane) there are two chemical-shift distinct carbon signals, one derived from the CH and the other resulting from the two identical CH_2 groups. Thus the CH DEPT spectrum and the CH_2 DEPT spectrum will each show one signal. Another example, 2-bromo-2-methylbutane from 37(a) will show one signal in the CH_2 DEPT spectrum and two signals in the CH_3 DEPT (two of the methyl groups are identical, but differ from the third). These problems are close to being self-explanatory; just be sure to check to make certain that every line in the normal carbon spectrum has a counterpart in *one* of the DEPT spectra.

57. (a) $(CH_3)_2CHCH(CH_3)_2$: The only one that should show only two signals

(b) 1-Chlorobutane: The only one that should show exactly four signals

(c) Cycloheptanone: Same reason as (b); note symmetry in molecule

(d) $CH_2=CHCH_2Cl$: The only example with alkene carbons ($\delta = 100–150$ ppm)

58. (a) ^{13}C: Seven different carbon signals show, six of which are CH_2 groups. ^1H: Five signals, $\delta = 0.6$ (broad singlet, 1 H), 0.8 (triplet, 3 H), 1.2 (broad, 8 H), 1.3 (multiplet, 2 H), 3.3 (triplet, 2 H). At this point, the signals at 1.2 and 1.3 are close to useless, but the others can get us off to a good start. We can propose $—CH_2—\overline{(CH_3)}$ ($\delta = 0.8$) and $—CH_2—\overline{(CH_2)}—OH$ ($\delta = 3.3$; the $—OH$

proton is at $\delta = 0.6$) as molecular fragments. These fragments add up to $C_4H_{10}O$, leaving C_3H_6 unaccounted for. There is no sign of any other signal for a CH_3 group in the ^1H NMR spectrum; indeed, the ^{13}C DEPT spectrum tells us that the remaining carbons are all CH_2 groups. So the answer has to be 1-heptanol. (Alternatives such as 3- or 4-methyl-1-hexanol would show a methyl doublet near $\delta = 0.9$ in the ^1H NMR spectrum in addition to the triplet already present, and one more non-CH_2 peak in the ^{13}C DEPT spectrum.)

(b) ^{13}C: Four signals for the seven carbons, so the molecule must have **symmetry.** ^1H: Four distinguishable signals, $\delta = 0.9$ (triplet, 6 H), 1.3 (quartet, 4 H), 2.8 (singlet, 2 H), 3.6 (singlet, 4 H). A couple of fragments may be discerned:

$$(2\times) \; —\overline{(CH_2)}—\overline{(CH_3)} \qquad (2\times) \; —\overline{(CH_2)}—OH$$
$$ \quad 1.3 \quad\;\; 0.9 \qquad\qquad\quad 3.6 \quad\; 2.8$$

These add up to $C_6H_{16}O_2$; adding in the quaternary carbon that shows up in the ^{13}C (DEPT) spectrum gives the complete molecular formula. What we have, therefore, are the four fragments shown above, plus a carbon atom. Connecting the four fragments to the four bonds of the seventh carbon gives us the solution:

$$\begin{array}{c} CH_2OH \\ | \\ CH_3CH_2—C—CH_2CH_3 \\ | \\ CH_2OH \end{array}$$

59. An assortment of overlapping sharp singlets and doublets for the CH_3 groups is evident between $\delta = 0.6$ and 1.1. Signals for the benzene H's are located between $\delta = 7.2$ and 8.2. Three other signals may be interpreted as follows:

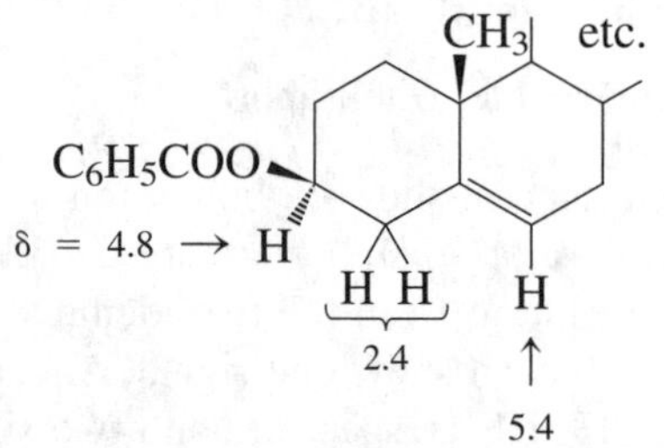

The CH_2 at $\delta = 2.4$ is split into a doublet by the neighboring CH. This CH is the cause of the $\delta = 4.85$ signal, and its complex splitting results from the CH_2 neighbors on both sides. The expansion of this signal shows 9 lines. How can we explain its appearance? Recall Problem 36(h): The CH hydrogen is trans to two of its neighbors and cis to the other two. As in the earlier problem, the cis and trans coupling constants in the ring are different. Applying the type of analysis discussed in Section 10-8, we can attempt to analyze the splitting by means of two successive $N + 1$ rule "stages"—a triplet (from the trans coupling to two of the neighbors) of triplets (from the cis coupling to the other two):

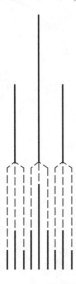

It doesn't work! The predicted line intensities are way off! What could be wrong? Perhaps two coupling constants that we have assumed are the same (such as the cis J values to each side) are also different? Then we are looking at a doublet of doublets for just these two couplings, on top of the triplet for the others. Thus:

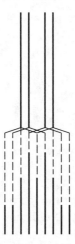

60. The spectrum indicates the presence of two almost equivalent CH_3's, not split by any neighboring H's ($\delta = 1.1$), a third unsplit CH_3, probably attached to a functional group due to its chemical shift ($\delta = 1.6$), and an alkene H showing some (triplet?) splitting ($\delta = 5.3$). The lack of any signal in the $\delta = 3-5$ range means that the alcohol carbon must not have any H's on it. Let's compare the pieces we have with the molecule framework that's supposed to be present.

← Two nearly equal CH_3's, unsplit →	
HO on C lacking hydrogens →	
Triplet split alkene H →	
← Unsplit CH_3 on functional group →	

That's the answer!

61. **(a)** 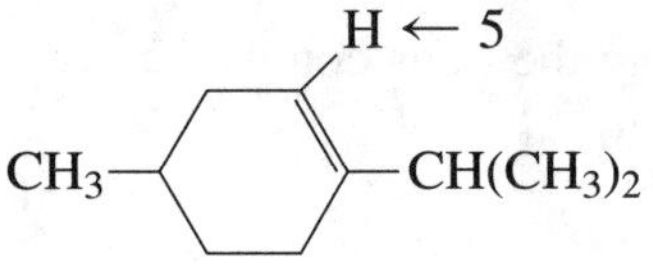 CH$_3$— (cyclohexene ring with H ← 5 at top vinyl position) —CH(CH$_3$)$_2$

(b) 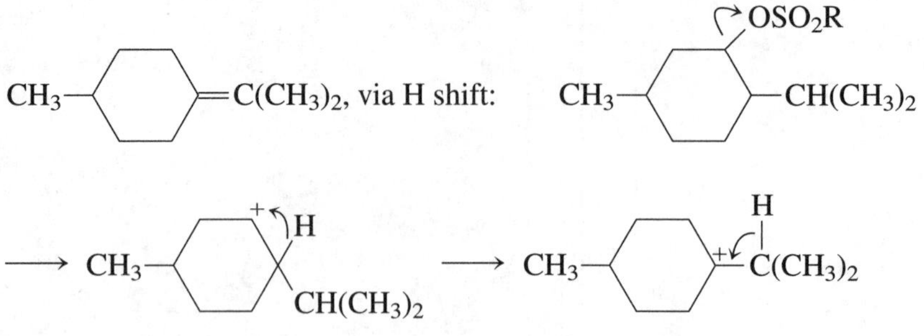 CH$_3$— (cyclohexane ring) =C(CH$_3$)$_2$, via H shift: CH$_3$— (cyclohexane ring with OSO$_2$R) —CH(CH$_3$)$_2$

⟶ CH$_3$— (cyclohexane ring, +, H, CH(CH$_3$)$_2$) ⟶ CH$_3$— (cyclohexane ring, +, H, C(CH$_3$)$_2$)

⟶ product

(c) Follow all possible E1 mechanisms to the first product, with D in the starting molecule:

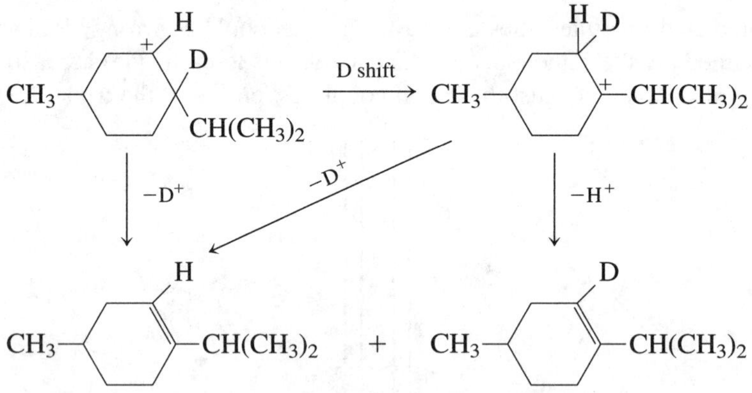

(reaction scheme)

CH$_3$— (ring, +, H, D, CH(CH$_3$)$_2$)   —D shift→   CH$_3$— (ring, H, D, +, CH(CH$_3$)$_2$)

| −D$^+$ −D$^+$ −H$^+$

CH$_3$— (ring with H) —CH(CH$_3$)$_2$ +   CH$_3$— (ring with D) —CH(CH$_3$)$_2$

"Major product" is now a **mixture**
of these two species.

According to these three pathways, some of the major product formed will have an alkene D instead of an alkene H. The molecules with alkene D will **not** show an ^{1}H NMR signal near $\delta = 5$. So, the $\delta = 5$ signal for the major product as a whole will be reduced in intensity compared with the major product from nondeuterated starting material.

   This result was good evidence for the involvement of hydride shifts in what were previously thought to be simple E1 reactions.

62. Write out your given information first. Believe me, it helps.

   A (C$_4$H$_9$BrO) + KOH → E (C$_4$H$_8$O); E has 2 complex ^{1}H NMR signals and 2 ^{13}C peaks

   B (C$_4$H$_9$BrO) + KOH → F (C$_4$H$_8$O); F has 2 ^{1}H NMR singlets and 3 ^{13}C peaks

   C (C$_4$H$_9$BrO) + KOH → G (C$_4$H$_8$O); G has 2 complex ^{1}H NMR signals and 2 ^{13}C peaks

   D (C$_4$H$_9$BrO) + KOH → G (C$_4$H$_8$O); C and D have identical ^{1}H NMR spectra

   Starting materials may be optically active; products are not.

   What can we say *for sure?* Well, if C and D are different compounds but have identical NMR spectra, they're enantiomers (think about it). If they both give G after reaction, either G is achiral, or it is a meso compound. Next question. What kind of chemical reactions are we talking about here? Each proceeds with loss of HBr, but normal elimination processes are ruled out because the ^{1}H NMR spectra of the products all lack peaks for alkene hydrogens between δ 4.6−5.7 ppm. However, internal displacement reactions of the Williamson type to give cyclic ethers accomplish the same change in molecular formula (recall Section 9-6). A little trial and error should convince you that the only way you're going to get compounds from such reactions with the formula C_4H_8O and no double bond is by having a cyclic ether (you'll see a more systematic way of figuring that out in Chapter 11). So draw some cyclic ethers with the formula C_4H_8O. There are six:

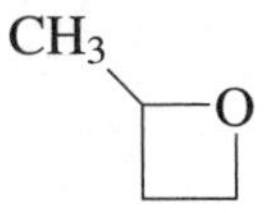

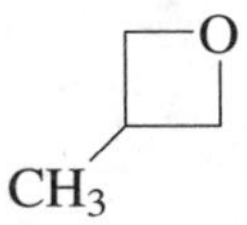

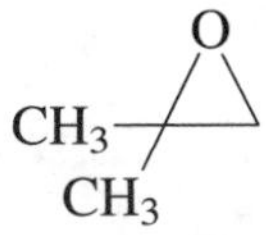

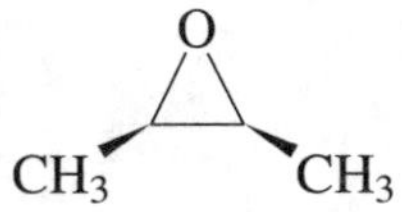

Compounds 2 and 6 are chiral; rule them out. Compound 3 would have 3 ^{1}H NMR signals; it's out, too. Three left, so let's try to match them up with the data we have. Compound 4 will show two singlets in its ^{1}H NMR spectrum, and it has 3 nonequivalent carbons, suggesting that it is F. What could correspond to its precursor, B? An appropriate bromoalcohol:

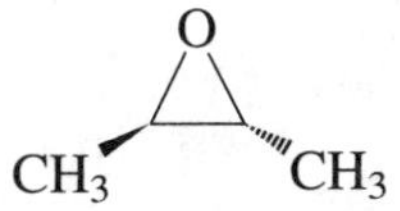

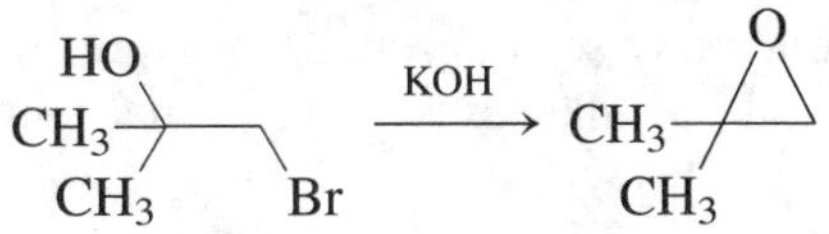

Turning to the remaining product options, compounds 1 and 5, they will both show two complex ^{1}H NMR signals and 2 ^{13}C peaks. Compound 1 has only one possible precursor, as illustrated by the following reaction, identifying it as E:

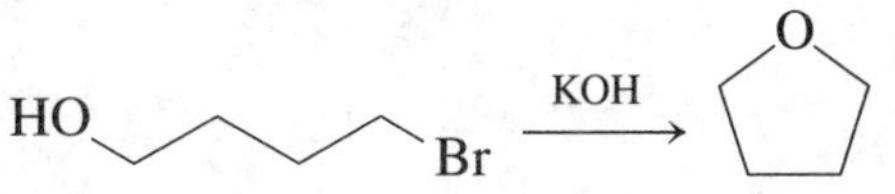

In contrast, compound 5 (which is meso) can be formed from either of two enantiomeric precursors (recall that backside displacement inverts the site of attack):

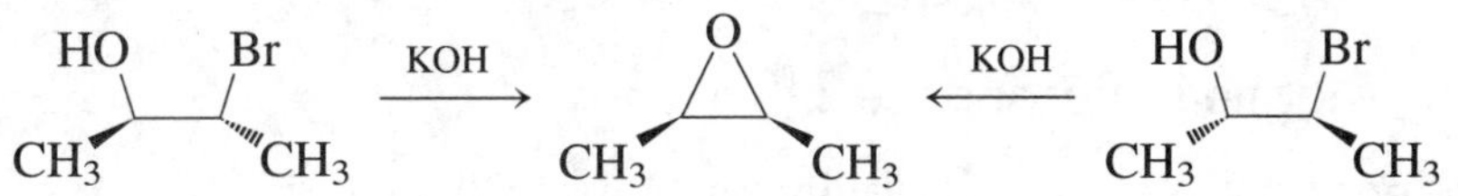

One of the precursor molecules is C, the other D. Chemical shifts for compounds E, F, and G may be estimated directly from Tables 10-2 and 10-6. The DEPT spectra also will be quite distinctive.

63. (c)

64. (b)

65. (d)

66. (e)

11

Alkenes; Infrared Spectroscopy and Mass Spectrometry

In Chapters 11 and 12 we return to the presentation of a new functional group: the carbon–carbon double bond. This functional group differs from those seen so far in that it lacks strongly polarized covalent bonds. Instead, its reactivity arises from special characteristics of electrons in so-called π bonds. The properties of these electrons and their consequences are discussed in the next chapter. Chapter 11 is restricted to a general description of alkenes as a compound class and a presentation of methods of preparation of double bonds. Most of the reactions are ones you have already seen because the major methods of alkene syntheses are the same elimination reactions of alcohols and haloalkanes that were presented in Chapters 7 and 9. Only some finer details have been added.

This chapter also introduces infrared spectroscopy, a useful tool for qualitative identification of functional groups, and mass spectrometry, the best method for determining molecular composition.

Outline of the Chapter

11-1   Nomenclature

11-2   Structure and Bonding in Ethene

11-3   Physical Properties of Alkenes

11-4   Nuclear Magnetic Resonance of Alkenes

11-5   Hydrogenation: Relative Stability of Double Bonds
Comparing alkenes and alkanes.

11-6   Preparation of Alkenes: Elimination Revisited

11-7   Alkenes by Dehydration of Alcohols
Mostly review material in these two sections.

11-8   Infrared Spectroscopy
Another useful spectroscopic technique.

11-9 and 11-10 Mass Spectrometry
A different sort of technique for molecular characterization.

11-11 Degree of Unsaturation
More information for solving structure problems.

Keys to the Chapter

11-1 through 11-4. Nomenclature and Physical Properties

Little needs to be added to the text descriptions for these two topics. The nomenclature rules are straightforward. Again, a small number of common names are still in use and must be learned. However, the systematic nomenclature is logical and easy to master. Note that alkenes, like cyclic alkanes, have two distinct "sides," and therefore substituents may be either cis or trans to each other. For alkenes, however, the cis and trans designations should be restricted to molecules with exactly two substituents, one on each of the two doubly bonded carbons. If more substituents are present, the *E,Z nomenclature* system should always be applied.

The two sides of alkenes are due to the nature of the four-electron double bond. In the simplest picture, we assign two of its electrons to a basic, garden-variety σ bond between the atoms. The other two electrons are then placed in two parallel *p* orbitals, overlapping "sideways" to form the π bond. This π-type overlap prevents the carbons at each end of the double bond from rotating with respect to one another. Ethene, therefore, is a perfectly flat molecule, and, in general, the carbons of the alkene functional group and all the atoms attached to them will lie in a plane, with the π electrons above and below.

One other significant consequence of the enforced planarity of double bonds and the cis-trans relationships of attached groups is seen in the NMR spectra of alkenes. A molecule's alkene hydrogens do not all have to be chemical-shift equivalent. When they aren't, coupling will be observable, sometimes leading to very complicated patterns as a result of *J* values that vary widely as a function of the structural relationships between the hydrogens involved (see Table 11-2). Figure 11-11 illustrates this feature. Even in complex spectra, however, you will still be able to derive the information you need for structure determination as long as you remember to look separately for the four basic pieces of information the spectrum contains: number of signals, chemical shift of each one, integration, and splitting patterns. If the splitting is too complicated to interpret, you can still use the other three pieces of data to come up with an answer.

11-5. Hydrogenation: Relative Stability of Double Bonds

The stability order of different kinds of alkenes is a well-established feature of this compound class: More substituted alkenes are more stable than less substituted ones, and trans are more stable than cis. This topic does not exist in isolation, however. In fact, it has important consequences for both reactions that form alkenes as well as reactions that alkenes undergo. Learn this stability order. You will need to use it later.

11-6. Preparation of Alkenes: Elimination Revisited

This is a review of the material from Sections 7-6 and 7-7. There are two new considerations. First, many haloalkanes can give rise to several alkenes upon elimination, each with the double bond in a different position in the carbon chain. These products arise when there are several different β-hydrogens that can be lost in the elimination process together with the leaving group. The rule to remember is as follows: All E1 and, with one main exception, all E2 processes tend to produce the most highly substituted, **most stable alkene** (Saytzev elimination). The major **exception** is that **very bulky bases** will favor production of the least substituted, **least stable alkene** in E2 processes (Hofmann elimination).

The second new consideration relates to stereochemistry. As was briefly mentioned in Chapter 7, the E2 elimination mechanism strongly prefers an *anti* **conformation** between the leaving group and the β-hydrogen being removed. The result is that E2 eliminations will tend to give alkenes arising from the best available *anti* conformation. E1 eliminations are not as restricted and will simply tend to give the most stable alkene (i.e., trans in preference to cis) as the major product. Stereochemistry is important when considering the use of elimination reactions for alkene synthesis. Certain kinds of haloalkanes possess only one reactive conformation for E2 elimination (see Problem 40) and will therefore give only a single stereoisomer upon reaction. This can be very useful. E1 eliminations, however, are more prone to yield mixtures of stereoisomeric products.

11-7. Alkenes by Dehydration of Alcohols

Again, this is mainly a review of earlier material (Chapters 7 and 9). Note that, unlike the situation with base-promoted E2 eliminations, under the reaction conditions for alcohol dehydration, the usual result is formation of the **most stable alkene** (the thermodynamic product). Alcohol dehydrations are susceptible to rearrangement processes. A classic example is encountered in attempted syntheses of terminal alkenes such as 1-butene. The **only** 100% reliable method is base-promoted E2 elimination of a suitable 1-butyl compound (e.g., 1-bromo-butane, 1-butyl tosylate). Any other method will give mixtures:

$$CH_3CH_2CH_2CH_2Br \xrightarrow{\text{K}^+{}^-\text{OC(CH}_3)_3, \text{ (CH}_3)_3\text{COH}} CH_3CH_2CH=CH_2$$

The only elimination product

$$CH_3CH_2CHBrCH_3 \xrightarrow{\text{K}^+{}^-\text{OC(CH}_3)_3, \text{ (CH}_3)_3\text{COH}} CH_3CH_2CH=CH_2 + CH_3CH=CHCH_3$$

Major   Minor, cis and trans

$$CH_3CH_2CHBrCH_3 \xrightarrow{\text{Na}^+{}^-\text{OCH}_2\text{CH}_3, \text{ CH}_3\text{CH}_2\text{OH}} CH_3CH_2CH=CH_2 + CH_3CH=CHCH_3$$

Minor   Major, cis and trans

$$\text{Either 1- or 2-butanol} \xrightarrow{\text{Conc. H}_2\text{SO}_4, \Delta} CH_3CH_2CH=CH_2 + CH_3CH=CHCH_3$$

Minor   Major, cis and trans

Additional information pertaining to this process is presented in Chapter 12.

11-8. Infrared Spectroscopy

Once the most important spectroscopic technique, infrared spectroscopy is now used to complement NMR data. The IR technique helps confirm the presence or absence of common functional groups in a molecule. It is most diagnostic for the following: HO, $C\equiv N$, $C\equiv C$, $C=O$, and $C=C$. Different types of C—H bonds can be readily identified, helping to confirm information obtained by NMR. Although occasionally the detailed data in Table 11-4 and in the text may be necessary to solve a problem, for the most part you will only need to look for bands in certain **general** regions of the IR spectrum, much the same way you have learned to divide up the NMR spectrum into rather general segments (e.g., alkane C—H and alkene C—H). The following illustration, derived from the data in Table 11-4, shows these regions.

For example, a compound exhibiting a strong band somewhere between 1680 and 1800 cm^{-1} contains a $C=O$ group. However, IR data tells us about both the presence **and the absence** of functional groups in a molecule. Don't neglect the usefulness of the latter! For instance, a molecule lacking absorption between 3200 and 3700 cm^{-1} **cannot** be an alcohol. Combining information from a molecular formula with NMR and IR data often permits complete determination of the structure of an unknown molecule. The text problems give you opportunities to practice.

Regions of the Infrared Spectrum

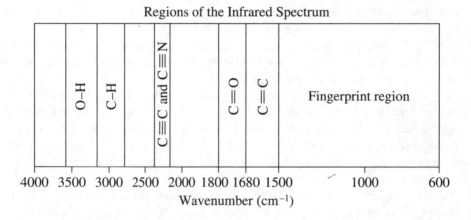

11-9 and 11-10. Mass Spectrometry

The kind of information available from mass spectrometry falls into two categories. First, the m/z value for the molecular ion provides information useful in calculating the molecular formula of the molecule. Second, the lower molecular weight fragments that appear in the mass spectrum contain clues concerning structural features of the molecule in question. Be sure that you understand how to extract these kinds of information from mass spectral data.

11-11. Degree of Unsaturation

When you are faced with the problem of coming up with a reasonable structure to match some spectroscopic or chemical data, it is possible to waste a lot of time writing answers that are incompatible with the molecular formula of the molecule. Determining the number of rings + π bonds ahead of time (the degree of unsaturation) can make solving these problems go much more smoothly: You automatically know whether or not you need to consider these structural elements as possible parts of an unknown molecule.

In practice, you should try to reconcile your IR and NMR data with the degree of unsaturation **before** you start writing down possible structures. For example, if the IR and NMR indicate the absence of π bonds (no IR bands around 1650 cm^{-1}; no NMR signals downfield of about δ = 5) but the formula indicates one degree of unsaturation, then your answer **must** contain one ring. On the other hand, if the IR and NMR do show such signals in a molecule with exactly one degree of unsaturation, then the unsaturation **must** be a π bond and a ring **cannot** be present. Thus by the process of elimination, you get closer to the right answer more quickly. Try it yourself, beginning with Problem 34. In some cases you will be given an additional piece of data: the result of hydrogenation. Use it to distinguish π bonds from rings, because, in general, after hydrogenation the π bonds will be gone, but the rings will still be there. The degree of unsaturation will be given in the answers in this study guide.

Solutions to Problems

29. (a) (b)

(c) (d) (e)

30. (a) *cis-* or *Z*-2-Pentene

 (b) 3-Ethyl-1-pentene

 (c) *trans-* or *E*-6-Chloro-5-hexen-2-ol

 (d) *Z*-1-Bromo-2-chloro-2-fluoro-1-iodoethene (Priorities are I > Br on C1 and Cl > F on C2.)

 (e) *Z*-2-Ethyl-5,5,5-trifluoro-4-methyl-2-penten-1-ol

 (f) 1,1-Dichloro-1-butene

 (g) *Z*-1,2-Dimethoxypropene

 (h) *Z*-2,3-Dimethyl-3-heptene

 (i) 1-Ethyl-6-methylcyclohexene. This name is better than 2-ethyl-3-methylcyclohexene, because the **first** number is smaller.

31. (a) *E*-2-chloro-2-butene (*E* because the highest-priority groups on the two alkene carbon atoms are Cl and CH$_3$, respectively, and they are trans to each other; note that *cis/trans* may not be used in the name because the alkene carbons collectively have more than two substituents—three in this case); (b) *E*- or *trans*-1-chloro-2-butene; (c) 1-chloro-2-methylpropene.

32. In each case both the higher dipole moment and the higher boiling point are found in the compound that has the fluorine atoms cis to each other. (a) *cis*-1,2-difluoroethene; (b) *Z*-1,2-difluoropropene; (c) *Z*-2,3-difluoro-2-butene.

33.

The two alcohols are considerably more acidic ($pK_a \approx 17$) than the alkene (pK_a 40 or so), which in turn is more acidic than the alkane ($pK_a > 50$). 3-Cyclopenten-1-ol is slightly more acidic than cyclopentanol because of a small electron-withdrawing inductive effect exerted by the sp^2-hybridized alkene carbon atoms.

34. (a) $H_{sat} = 8 + 2 - 1 = 9$; degree of unsaturation = $(9 - 7)/2 = 1$ π bond or ring present. The integrated intensities reveal the pieces.

$\delta = 1.8$ (s, 3 H): **CH$_3$,** attached to an unsaturated functional group

$\delta = 4.0$ (s, 2 H): **CH$_2$,** most likely attached to Cl

$\delta = 4.9$ and 5.1 (singlets, each 1 H): Two alkene hydrogens

Thus, you have CH$_3$—, —CH$_2$—Cl, and $\overset{\diagdown}{\diagup}C{=}C\overset{\diagup}{\diagdown}$ attached to two H's.

There are three ways to attach the four groups around the double bond.

In the first two compounds, all the NMR signals should show substantial couplings.

Only the third compound will show a spectrum as simple as A (remember that $={=}C\overset{\diagup \text{H}}{\diagdown \text{H}}$ couplings are typically very small whereas cis and trans H—C=C—H couplings are large); it is the correct answer.

(b) $H_{sat} = 10 + 2 = 12$; degree of unsaturation = $(12 - 8)/2 = 2$ π bonds and/or rings. The NMR shows the following:

$\delta = 2.1$ (s, 3 H): **CH$_3$,** next to an unsaturated functional group

$\delta = 4.5$ (d, 2 H): **CH$_2$,** attached to oxygen, split by one H

$\delta = 5.3$ and 5.9 (m, 2H and 1H): **—CH=CH$_2$,** the internal H downfield of the other two, typical of a terminal ethenyl group; the extensive splitting of the δ 5.9 signal suggests an adjacent CH$_2$ [compare Figure 11-11(b) in the text]

The pieces are CH$_3$— and CH$_2$=CH—CH$_2$—O— so far: C$_4$H$_8$O, leaving a C and an O to add

in, and one more π bond (a ring would be impossible). So let them be $\overset{\diagdown}{\diagup}C{=}O$, giving the final solution:

$$CH_3-\overset{\overset{\displaystyle O}{\|}}{C}-O-CH_2-CH=CH_2$$

(c) $H_{sat} = 8 + 2 = 10$; degree of unsaturation = $(10 - 8)/2 = 1$ π bond or ring.

$\delta = 1.3$ (doublet, 3 H): Most likely **CH$_3$—CH**

$\delta = 1.6$ (broad singlet, 1 H): Perhaps **OH**?

$\delta = 4.3$ (quintet w/fine splitting, 1 H): Probably **CH—OH,** four immediate neighbors to the CH (not including the OH)

$\delta = 5.1$ and 5.3 (one narrow doublet and one wider doublet, 1 H each): Alkene hydrogens, one cis and one trans to a third alkene hydrogen

δ = 5.9 (multiplet, 1 H): The third alkene hydrogen, in a $H_2C=CH-$ fragment

The methyl group (at δ = 1.3) must be attached to the CH (at δ = 4.3), giving as the only possible solution

$$CH_3-CHOH-CH=CH_2$$

(d) Same formula as (c), so again 1 π bond or ring.

δ = 1.4 (broad singlet, 1 H): Perhaps **OH** again?

δ = 2.3 (quartet w/fine splitting, 2 H): A **CH₂** with three immediate neighbors

δ = 3.7 (triplet, 2 H): Almost certainly CH_2-**CH₂**$-OH$ (no splitting to OH)

δ = 5.2 (multiplet, 2H): and 5.7 (multiplet, 1 H): Alkene hydrogens, **H₂C**=**CH**— again

As in (c), with only four carbons in the molecule everything is accounted for, and the two CH₂ groups must be attached to each other; thus,

$$H_2C=CH-CH_2-CH_2OH$$

(e) $H_{sat} = 6 + 2 - 2 = 6$; degree of unsaturation $= (6 - 4)/2 = 1$ π bond or ring.

δ = 1.7 (doublet, 3 H): **CH₃**, next to CH

δ = 5.7 (quartet, 1 H): alkene **CH**, next to CH₃

All that's left is one C and two Cl's, so the pieces, $CH_3-CH=C\diagup$ and two Cl's, combine to give the answer, $CH_3-CH=CCl_2$.

35. The multiplet at δ = 2.3 is a quartet for the C2 CH₂ group, because of the three neighbors, the CH₂ of C1 and the CH of C3. The signal shows additional very fine splitting, indicating that there is a small coupling to the alkene hydrogens on C4 ($J < 1$ Hz). The multiplet for the hydrogen signal at δ = 5.7 can be interpreted in detail. It corresponds to the CH at C3, which is split by the alkane CH₂ on one side and the alkene CH₂ on the other. The result is a pattern that can be described as a doublet (for trans alkene coupling) of doublets (for cis alkene coupling) of triplets (for coupling to the alkane CH₂), 12 lines total, of which 10 are seen. If we construct a plausible splitting diagram we find that the two "missing" lines are actually "hidden" under the tallest lines in the center of the pattern:

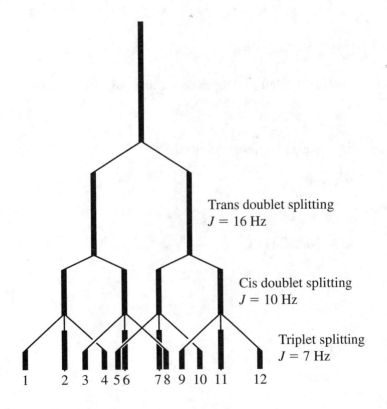

Trans doublet splitting
$J = 16$ Hz

Cis doublet splitting
$J = 10$ Hz

Triplet splitting
$J = 7$ Hz

1   2 3   4 5 6   7 8 9 10 11   12

In other words, lines 5 and 6 overlap to give one tall line in the center of the pattern, and lines 7 and 8 overlap to give the other.

36. (a) Yes. 1-Butene > *trans*-2-butene (which should be zero)

(b) No

(c) Yes. cis > trans (which, again, is zero)

37. The two orders are opposite each other because very stable alkenes are low in energy to start with and give off less heat upon hydrogenation than do less stable alkenes. The stability orders are given below.

(a) $CH_2{=}CH_2 < (CH_3)_2C{=}CH_2 < (CH_3)_2C{=}C(CH_3)_2$ Increasing substitution

(b)

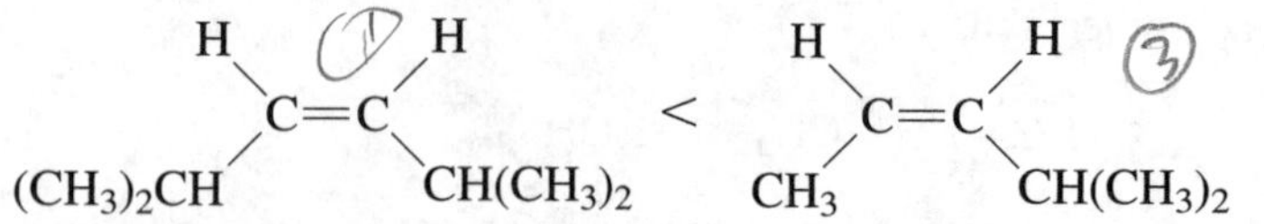

(Two large groups cis)   (A small and a large group cis)

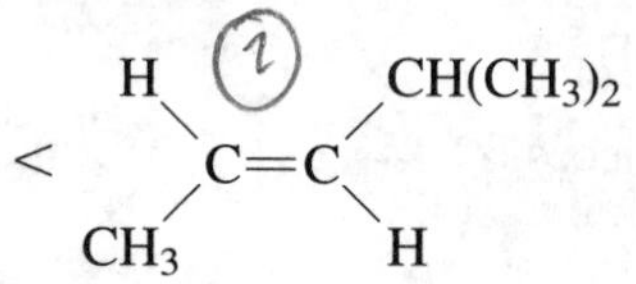

(Trans)

(c)

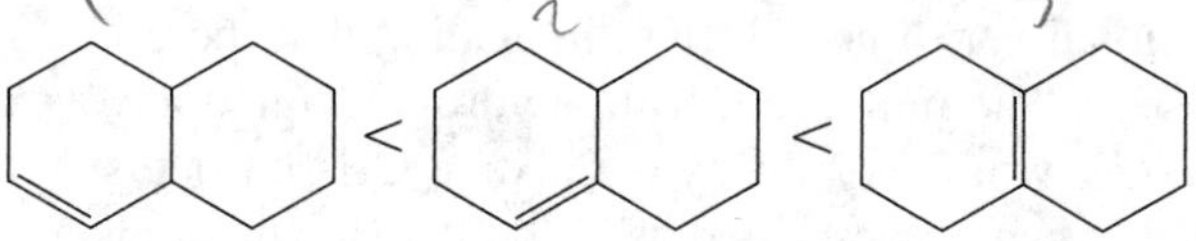

Similar to (a)

(d)

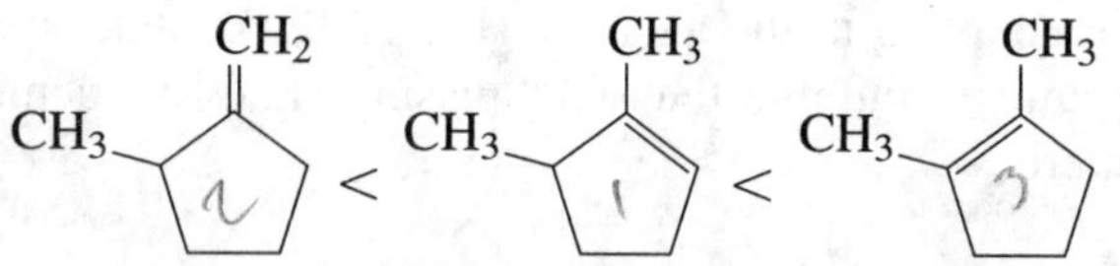

Similar to (a) and (c)

(e)

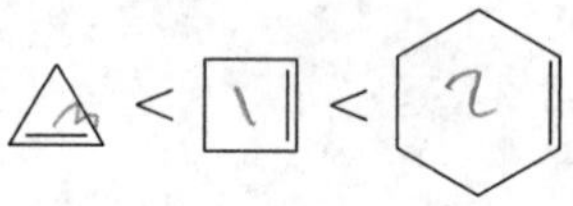

Ring strain

38. Generally the alkenes are written left-to-right in decreasing order of stability.

(a)

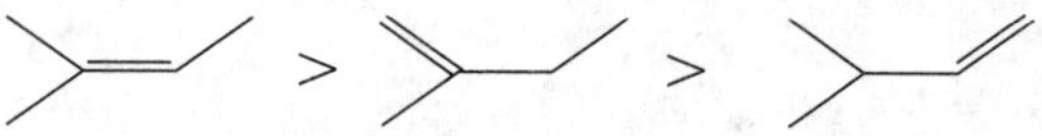

(trisubstituted)   (disubstituted)   (monosubstituted)

(b)

(tetrasubstituted)   (disubstituted)

(c)

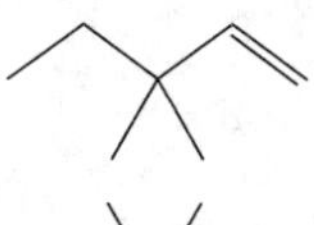

(only possibility)

(d)

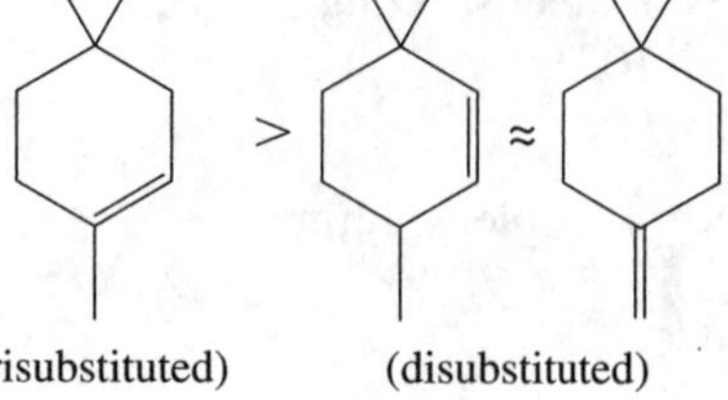

(trisubstituted)   (disubstituted)

39. $CH_3CH_2CH=CH_2$

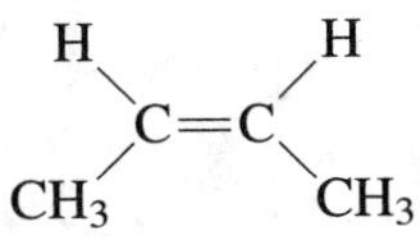

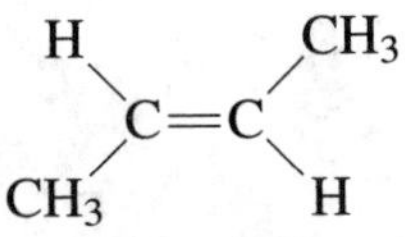

Smallest amount   Major product

40. A haloalkane of general structure $R-CH_2-CHX-R'$ will have **two** conformations with H *anti* to X; one gives the cis and the other gives the trans alkene as product. For example, consider 2-bromobutane:

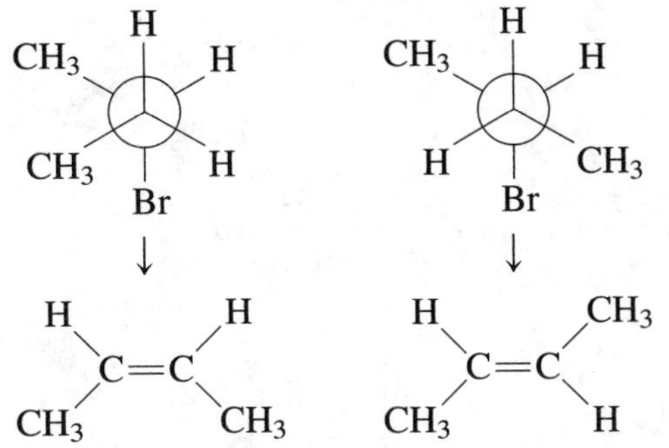

A haloalkane of general structure $RR'CH-CHX-R''$ will have only one conformation with H *anti* to X. Therefore, only a single alkene stereoisomer can form. Its stereochemistry will be determined by the stereochemistry of the two chiral carbons in the haloalkane.

41. The more hindered $(CH_3)_3CO^- K^+$ will favor removal of protons toward the less crowded "ends" of molecules, thus giving less stable products. Ethoxide eliminations will favor more stable products.

Product with $NaOCH_2CH_3$ in CH_3CH_2OH	Structure of starting material	Product with $(CH_3)_3COK$ in $(CH_3)_3COH$
(a) $CH_3OCH_2CH_3$	CH_3Cl	$CH_3OC(CH_3)_3$
(b) $CH_3CH_2CH_2CH_2CH_2OCH_2CH_3$ + 1-pentene	$CH_3CH_2CH_2CH_2CH_2Br$	1-Pentene
(c) *trans-* and *cis*-2-pentenes	$CH_3CH_2CH_2CHBrCH_3$	1-Pentene
(d) 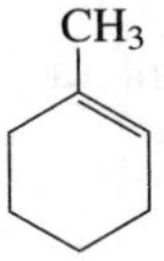	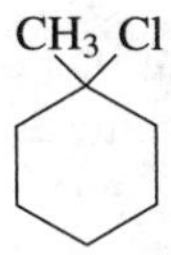	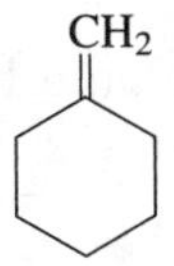
(e) 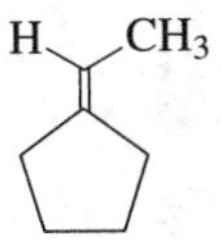	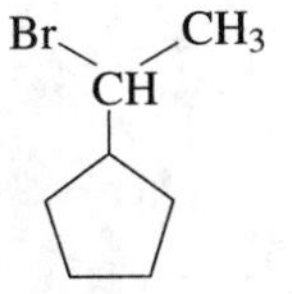	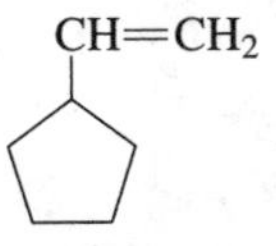
(f) 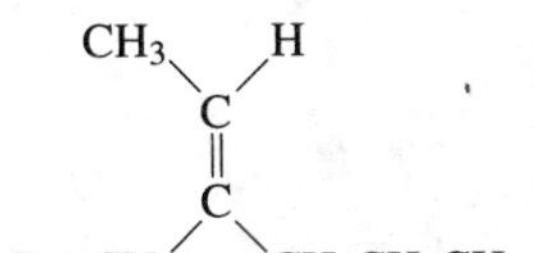	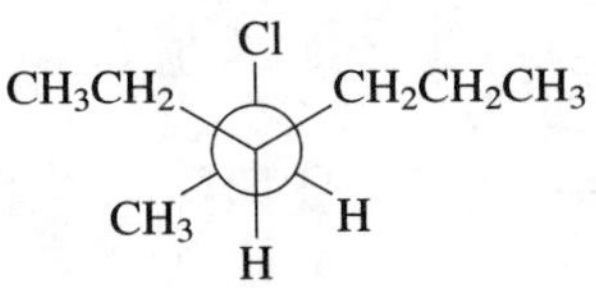 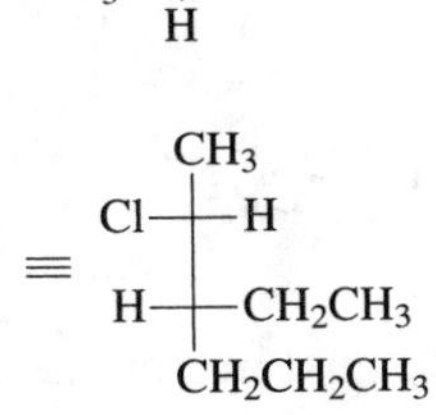	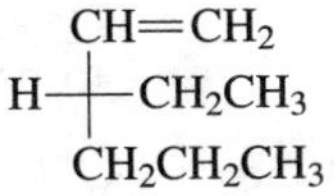

(g)

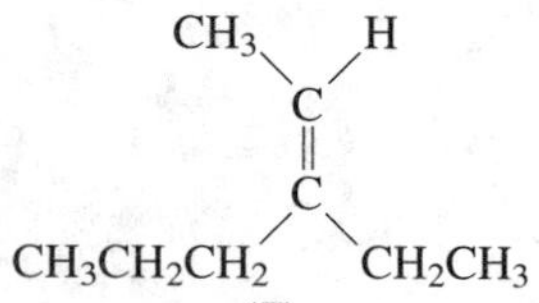

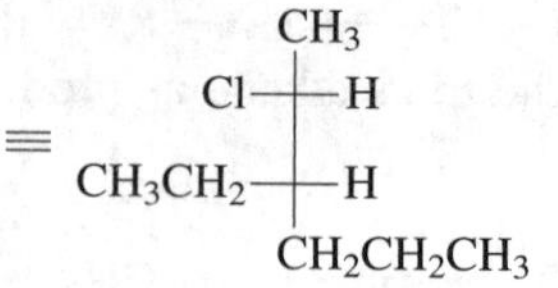

(h)

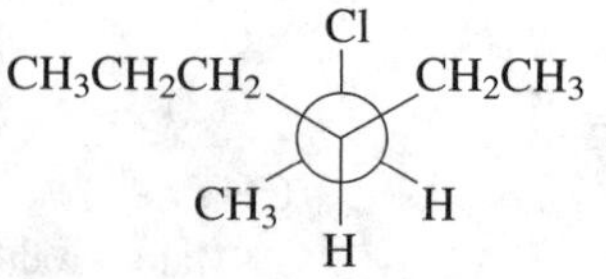

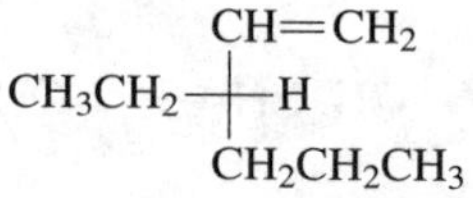

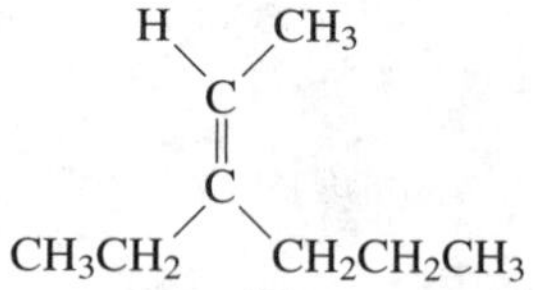

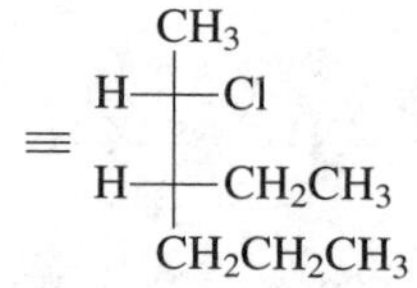

42. The energy of the E2 transition state for 1-bromopropane is similar to that for 2-bromopropane, because the products are the same (propene). So the E2 rate constants for these two compounds will be similar. However, 1-bromopropane, being primary, will have a faster competing S_N2 reaction than will the secondary 2-bromopropane. Overall, it is reasonable to estimate that the actual E2 rate for 1-bromopropane will be intermediate between the rates for bromoethane and 2-bromopropane.

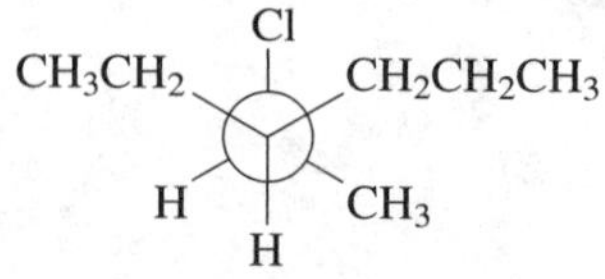

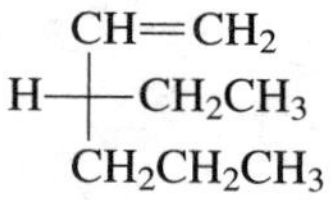

43.

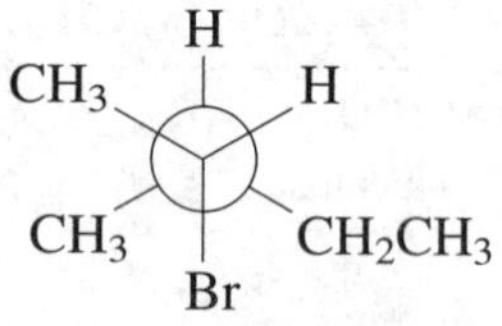

These are probably not the most stable conformations because all contain three *gauche* interactions between alkyl groups or bromine. For the *R,R* isomer, a better conformation would probably be

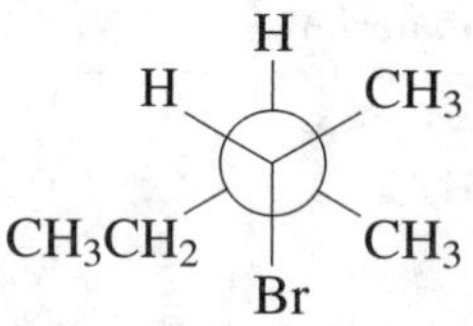

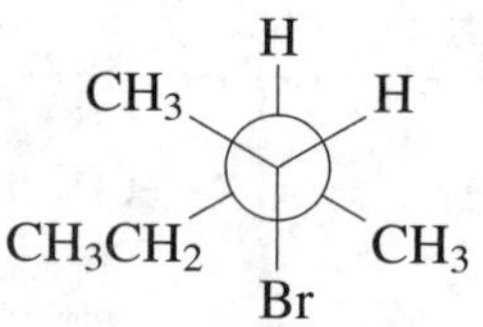

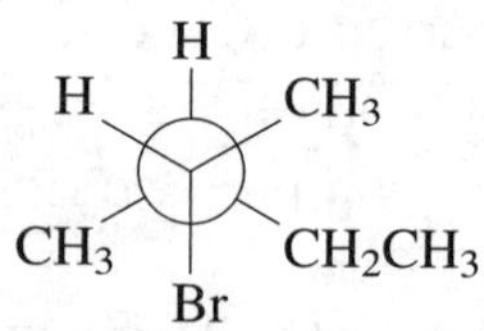

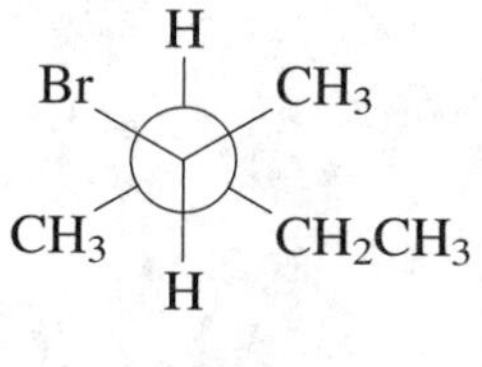

The others would be similar.

44. Referring to problem numbers in Chapter 7:

(**25b**) $CH_3CH=C(CH_3)_2 > CH_2=C(CH_3)CH_2CH_3$

(**25c**) Comparable amounts

(**25d**) 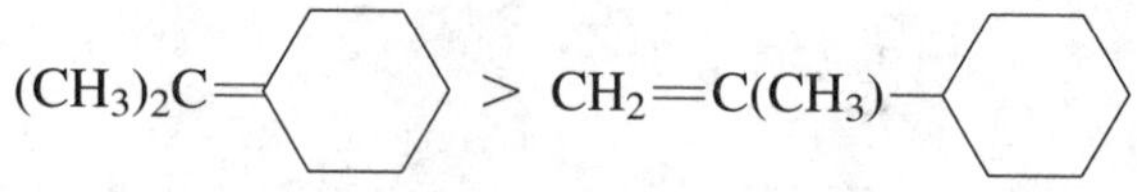 $(CH_3)_2C=$ $> CH_2=C(CH_3)-$

These are the only elimination product mixtures in these problems.

45. Major products are labeled for (a) through (f). Generally, more highly substituted alkene isomers form in highest yields.

(g)

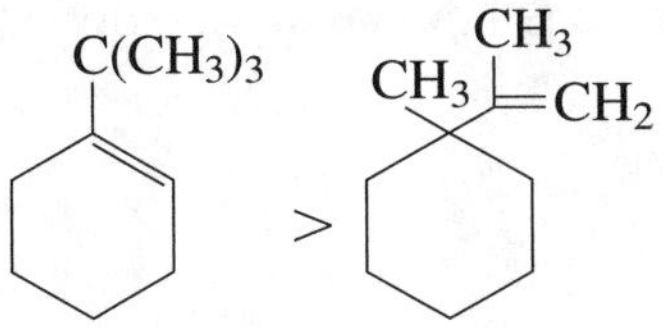

(h)

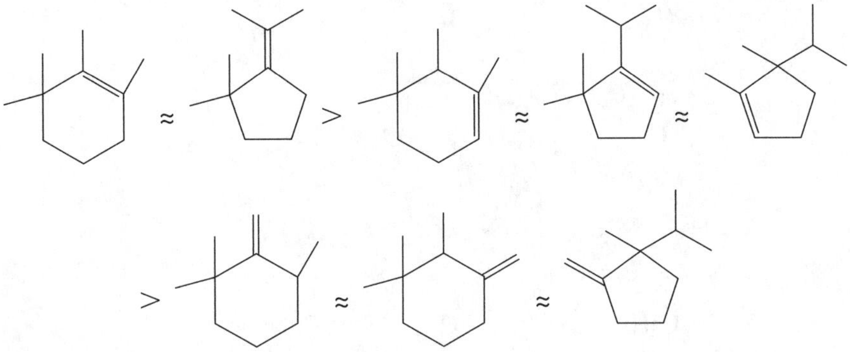

46. (a) 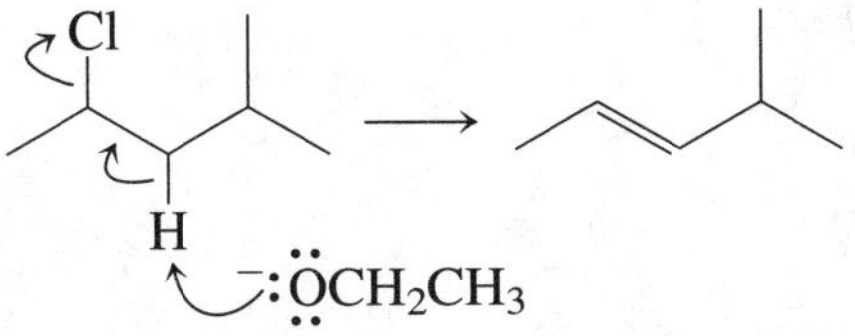   **(b)**

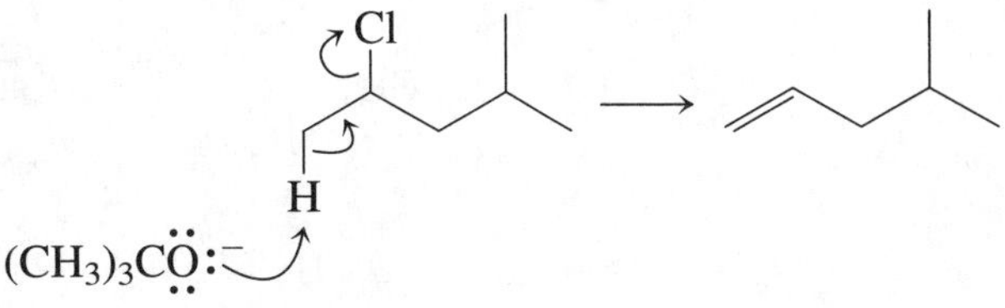

(c)

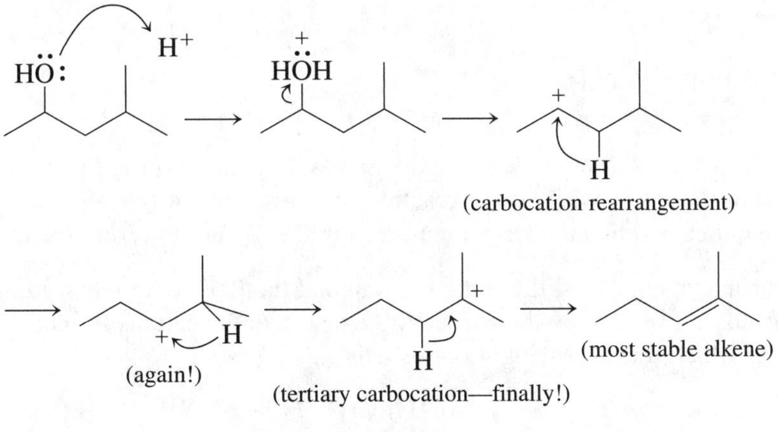

(carbocation rearrangement)

(again!)   (tertiary carbocation—finally!)   (most stable alkene)

47. A would be expected to give the tetrasubstituted alkene

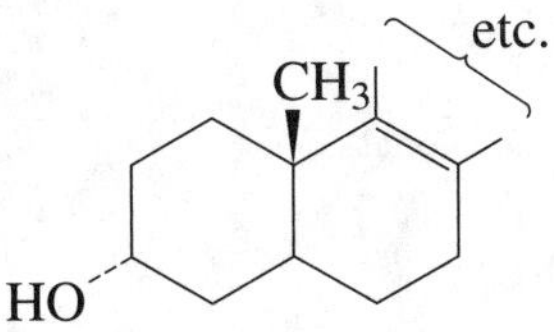

CH₃   etc.

HO

B gives the (trisubstituted) alkene shown in the problem in the text. C is a special situation.

48. 1-Methylcyclohexene contains a trisubstituted double bond and is therefore more stable than methyl-enecyclohexane, whose double bond is only disubstituted. The stability order in the three-membered rings is affected by angle strain because the geometry compresses bond angles to about 60°: This compression is worse for sp^2 carbons (which prefer 120° angles) than for sp^3 carbons (109°). 1-Methylcyclopropene, with two sp^2 carbons, will be much more strained than methylenecyclopropane, which has only one sp^2 carbon in the ring.

49. Rewrite each compound in a Newman projection with the Br and the β-H in an *anti* conformation.

(a)

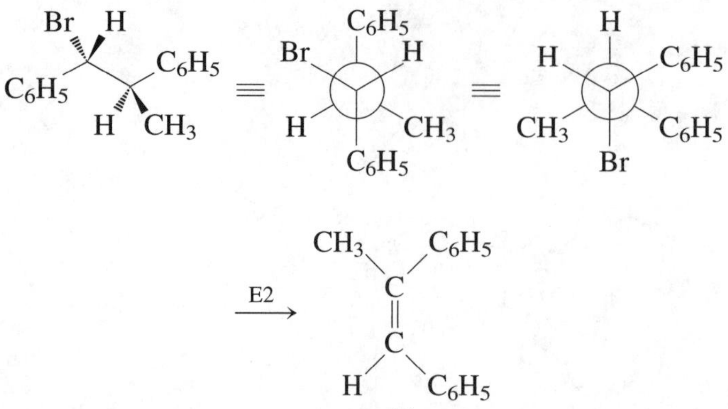

(b)

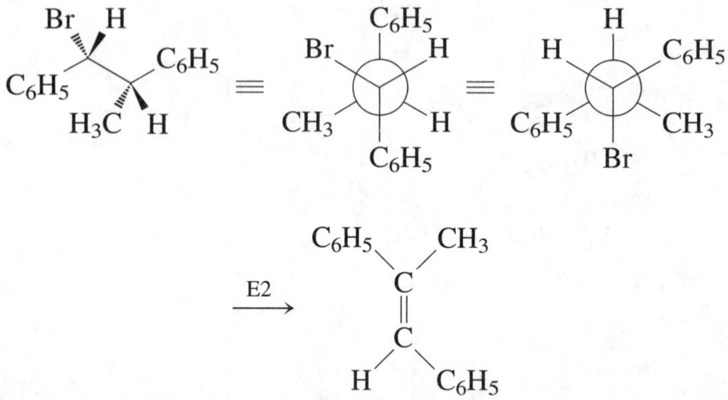

Elimination of (b) should be faster than that of (a) because the necessary conformation for reaction in (b) has the large C_6H_5 groups *anti* to one another, whereas the reactive conformation for (a) requires the C_6H_5 groups to be *gauche*. The latter requires additional energy input, raising the E_a and lowering the rate.

50. When E2 eliminations occur in cyclohexanes, the leaving group and the β-H to be removed must have a 1,2-trans **diaxial** relationship. So, first, draw the two possible chair conformations for each starting compound, and analyze the one in which the Cl leaving group is axial.

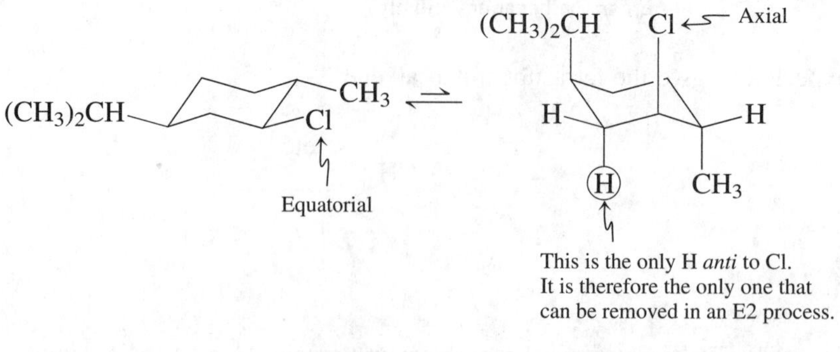

This is the only H *anti* to Cl. It is therefore the only one that can be removed in an E2 process.

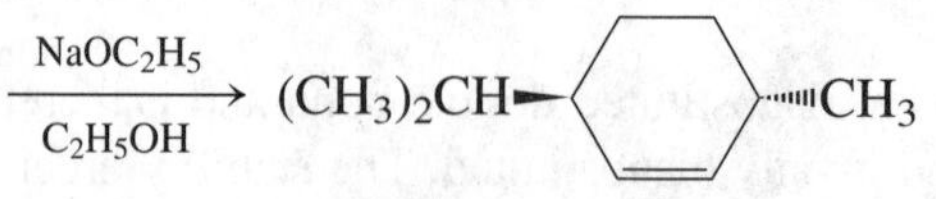

Only possible E2 product

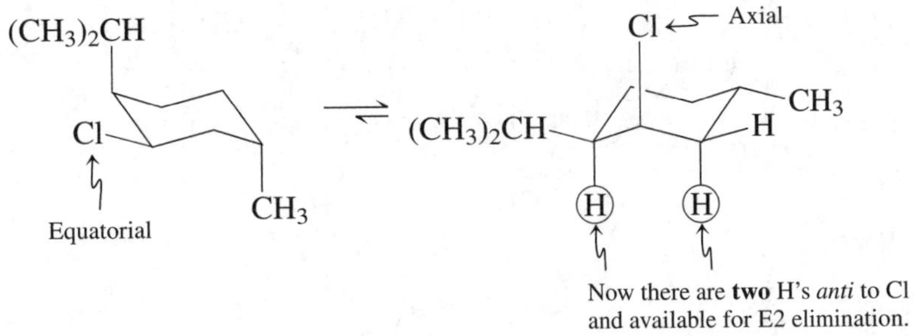

(CH3)2CH
Cl
Equatorial
CH3
Cl
Axial
CH3
(CH3)2CH
H
H
H
Now there are **two** H's *anti* to Cl and available for E2 elimination.

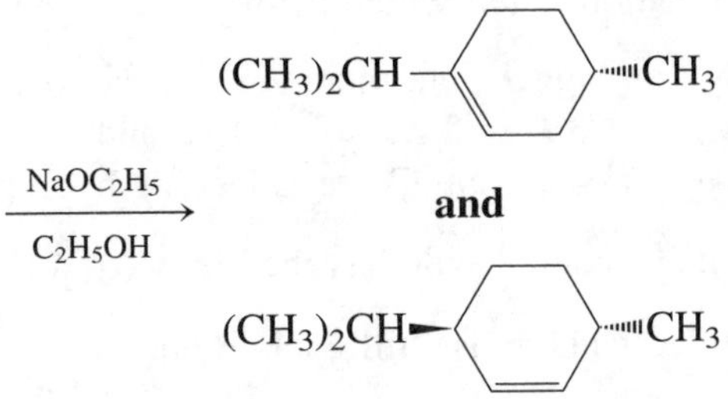

NaOC2H5
C2H5OH
(CH3)2CH
CH3
and
(CH3)2CH
CH3

51. Use the carbon types together with the chemical shifts to choose between alternative possibilities.

(a) $H_{sat} = 8 + 2 = 10$; degree of unsaturation $= (10 - 6)/2 = 2$ π bonds or rings present. $\delta =$ 30.2 is a CH_2 group; $\delta = 136.0$ is an alkene CH. Because those alone only add up to C_2H_3, there must be two of each. Two —CH_2—'s, plus —CH=CH—, which can only combine to make

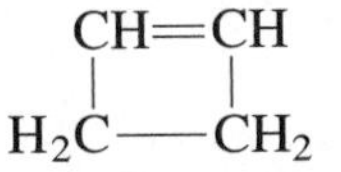

Cyclobutene (1 π bond and 1 ring)

(b) $H_{sat} = 8 + 2 = 10$; degree of unsaturation $= (10 - 6)/2 = 2$ again. $\delta = 18.2$ is a CH_3, **not** attached to the oxygen; $\delta = 134.9$ and 153.7 are alkene CH's; $\delta = 193.4$ is in the C=O region, and therefore, must be a —CH(=O) group because it is a doublet. The answer is therefore

CH_3—CH=CH—C(=O)—H (2 π bonds). You do not have enough information about ^{13}C NMR to determine the stereochemistry.

(c) $H_{sat} = 8 + 2 = 10$; degree of unsaturation $= (10 - 8)/2 = 1$.

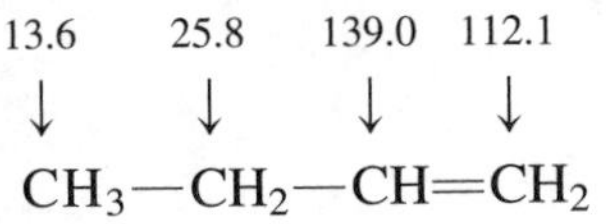

13.6
25.8
139.0
112.1
CH3—CH2—CH=CH2

Answer directly available from the carbon types.

(d) $H_{sat} = 10 + 2 = 12$; degree of unsaturation $= (12 - 10)/2 = 1$. This one has two CH_3 groups ($\delta = 17.6$ and 25.4), a CH_2 downfield enough ($\delta = 58.8$) to be attached to the O, and two alkene carbons, of which only one ($\delta = 125.7$) has an H on it.

The pieces: (2×)CH_3—, —CH_2—O—, —CH=C<. There is one H unlocated. Because it is not attached to one of the carbons, it must be on the oxygen. So, the possible answers are

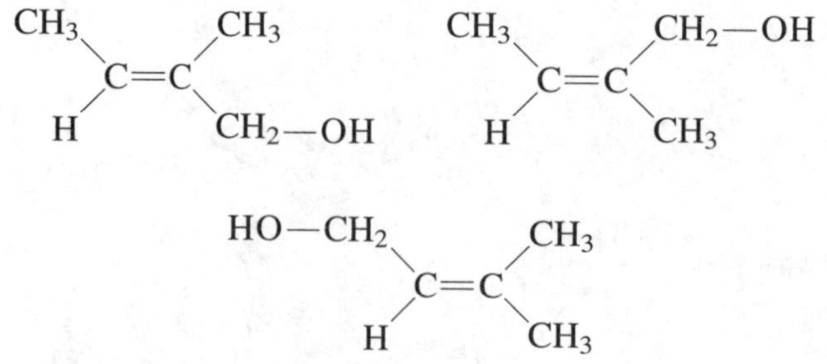

You do not have the information to tell which of the three is the actual compound.

(e) Notice that there are only four signals, but there are five carbons. Be careful. $H_{sat} = 10 + 2 = 12$; degree of unsaturation $= (12 - 8)/2 = 2$ now. $\delta = 15.8$ and 31.1 are CH_2 groups; $\delta = 103.9$ is an alkene CH_2, whereas $\delta = 149.2$ is an alkene C lacking hydrogens.

What do you have so far? The molecule has the piece $CH_2{=}C\diagdown^{\diagup}$, leaving three C's and six H's to make up the formula, which must still contain one more element of unsaturation (a ring?).

Because the highfield signals are triplets, these can only be CH_2 groups: three of them.

Combining $CH_2{=}C\diagdown^{\diagup}$, with three CH_2's can only give

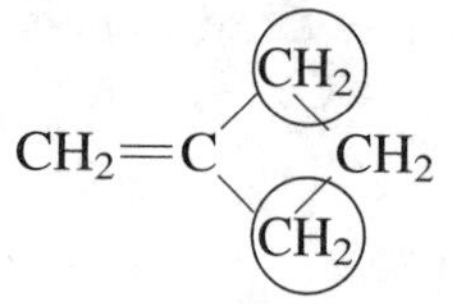

The $\delta = 31.1$ signal accounts for the two **equivalent** CH_2 groups (circled).

(f) $H_{sat} = 14 + 2 = 16$; degree of unsaturation $= (16 - 10)/2 = 3$, or 1 π bond and **2** rings. Again, be careful. Now there are four signals, but **seven** carbons in the molecule. Upfield, there are two different kinds of CH_2's ($\delta = 25.2$ and 48.5) and one kind of CH ($\delta = 41.9$). There is one kind of alkene carbon ($\delta = 135.2$). Because a double bond must connect **two** alkene carbons, this signal must represent two equivalent alkene CH groups: $-CH{=}CH-$. So you have at least two CH_2's, an alkane CH, and $-CH{=}CH-$, for a total of C_5H_7. So two carbons and three H's are still required: One more CH_2 and one more CH would do, and these each must be equivalent to groups already identified in order to keep the NMR spectrum as simple as it is. In other words, here are the pieces you have for the molecule: 2 equivalent $-CH_2-$'s, 2 equivalent $-\overset{|}{\underset{|}{C}}H$'s, a single unique $-CH_2-$, and the $-CH{=}CH-$ group, for a total of C_7H_{10}.

How do you put this all together? Remembering that symmetry can make groups equivalent, you can write these groups in symmetrical arrangements and connect them in a trial-and-error manner.

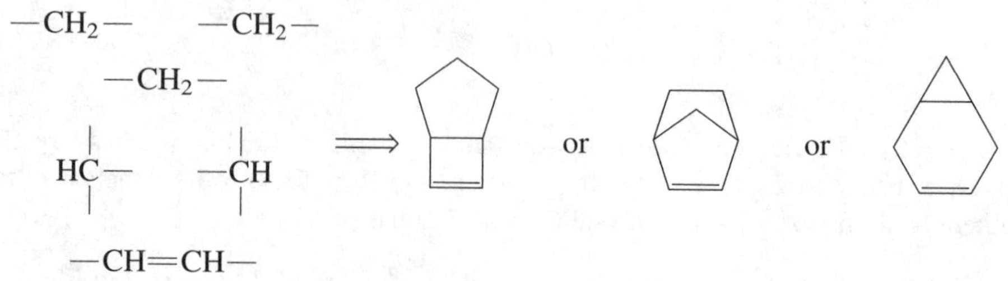

Each is a reasonable possibility (the second one, norbornene, is actually correct).

52. $H_{sat} = 10 + 2 = 12$; degree of unsaturation $= (12 - 10)/2 = 1$.

(a) The only way for five carbons to be equivalent is to make a ring: ⬠ is the answer.

(b) Three CH$_3$'s, and a $-CH=C{\Large\diagup}\diagdown$:

$$CH_3-CH=C\genfrac{}{}{0pt}{}{\diagup CH_3}{\diagdown CH_3}$$

(c) Two CH$_3$'s, one CH$_2$, and $-CH=CH-$: CH$_3-CH_2-CH=CH-CH_3$ is the answer (stereochemistry is ambiguous).

53. Lower, because the vibrational frequency varies **inversely** with the square of the "reduced" mass involving the atoms about the bond. So, bonds involving heavier atoms have lower energies associated with vibrational excitation. Typically, $\tilde{\nu}_{C-Cl} \approx 700$ cm^{-1}, $\tilde{\nu}_{C-Br} \approx 600$ cm^{-1}, and $\tilde{\nu}_{C-I} \approx 500$ cm^{-1}.

54. $10{,}000/\tilde{\nu} = \mu$m. (a) 5.81 μm; (b) 6.06 μm; (c) 3.03 μm; (d) 11.24 μm; (e) 9.09 μm; (f) 4.42 μm.

55. A—(b) (saturated alkane)

B—(d) (alcohol band at 3300 cm^{-1})

C—(a) (alkene band at 1640 cm^{-1} in addition to alcohol band)

D—(c) (alkene band at 1665 cm^{-1} but no alcohol band)

56. (i) Both alkene (1660) and alcohol (3350) products have formed.

(ii) Only alkene (1670) forms. (iii) Only alcohol (3350) forms.

(a) Conclusions: Isomer C is probably a primary bromoalkane, which gives a primary alcohol product (S$_N$2). Isomer B is probably a tertiary bromoalkane, which gives only alkene as product (E2). Isomer A is probably a secondary bromoalkane, which gives a mixture of S$_N$2 and E2 products.

(b) A possibilities: CH$_3$CHBrCH$_2$CH$_2$CH$_3$, CH$_3$CHBrCH(CH$_3$)$_2$, or CH$_3$CH$_2$CHBrCH$_2$CH$_3$
B possibilities: (CH$_3$)$_2$CBrCH$_2$CH$_3$ (only tertiary isomer)
C possibilities: CH$_3$CH$_2$CH$_2$CH$_2$CH$_2$Br, (CH$_3$)$_2$CHCH$_2$CH$_2$Br, or CH$_3$CH$_2$CH(CH$_3$)CH$_2$Br, but probably not (CH$_3$)$_3$CCH$_2$Br (too hindered to give S$_N$2 reaction)

57. Process of elimination: we begin by noticing the *absence* of absorptions in certain regions of the spectrum, and conclude that the corresponding functional groups cannot be in the actual molecule. So, no O—H (no strong broad band around 3300 cm^{-1}), no C=O (nothing around 1700 cm^{-1} or so), no C≡C—H (bands near 2100 and 3300 cm^{-1}), and no C=C (1680 cm^{-1}). All that are left as possibilities are the alkane and the ether. The spectrum shows strong bands between 1000 and 1200 cm^{-1}, strongly suggesting the presence of C—O bonds. The fact that such absorptions do not appear in the sample spectra for alkanes in the chapter (see Figures 11-18 and 19) confirms that the correct answer is the ether, $\diagup\diagdown_{\!\!\!O}\diagup\diagdown$.

58. Begin by noting the masses of the most prominent ions in each mass spectrum. Then try to predict how each of the alkanes might be most likely to fragment, using as a guiding principle a preference for forming more rather than less stable carbocations upon bond cleavage. The three compounds are constitutional isomers, all with the molecular formula C$_6$H$_{14}$ and a molecular weight of 86.

Spectrum A shows a base peak at $m/z = 57$ (C_4H_9) and other significant ions with $m/z = 56, 41, 29$ (C_2H_5) and 27. The molecular ion at $m/z = 86$ is weak.

Spectrum B also shows a base peak at $m/z = 57$ (C_4H_9). The peak at $m/z = 43$ (C_3H_7) is bigger than in spectrum A, while that at $m/z = 29$ (C_2H_5) is smaller. The molecular ion is more intense.

Spectrum C shows a base peak at $m/z = 43$ (C_3H_7). The molecular ion is weak, but $m/z = 71$ (C_5H_{11}) is prominent in this spectrum.

Now consider the three structures and the bonds in the molecular ion most likely to fragment in each:

Hexane: $[CH_3CH_2{-}CH_2{-}CH_2{-}CH_2CH_3]^{+\cdot} \longrightarrow$ $\begin{cases} CH_3CH_2^+ \ (m/z = 29) \\ CH_3CH_2CH_2^+ \ (m/z = 43) \\ CH_3CH_2CH_2CH_2^+ \ (m/z = 57) \end{cases}$

Cleavage is most favorable between CH_2 groups (avoiding formation of methyl cation), but since the best cations that you can get are only primary—and not very stable—fragmentation as a whole will be less likely. Spectrum B seems the best match because of the prominent molecular ion.

2-Methylpentane: $\overset{\displaystyle CH_3}{\underset{}{[CH_3{-}CH{-}CH_2CH_2CH_3]^{+\cdot}}} \longrightarrow$ $\begin{cases} [CH_3CHCH_3]^+ \ (m/z = 43) \\ [CH_3CH_2CH_2CHCH_3]^+ \ (m/z = 71) \end{cases}$

Cleavage will occur mainly about the CH to give secondary cations. The best match is spectrum C.

3-Methylpentane: $\overset{\displaystyle CH_3}{\underset{}{[CH_3CH_2{-}CH{-}CH_2CH_3]^{+\cdot}}} \longrightarrow [CH_3CH_2CHCH_3]^+ \ (m/z = 57)$

Fragmentation occurs at the indicated bond to form mainly *sec*-butyl cation and, to a lesser extent, ethyl cation ($m/z = 29$). Spectrum A fits best.

As is typical in mass spectrometry, the large amount of energy imparted to molecules in the process induces rearrangements and other modes of fragmentation as well. Fortunately, ions arising from such processes do not usually dominate the spectrum.

59. Major peaks: m/z 43 $(CH_3CH_2CH_2)^+$ from M—Br

  m/z 41 $(CH_2CH{=}CH_2)^+$ from M—HBr—H

Minor peaks: m/z 109 $(CH_2CH_2{}^{81}Br)^+$ $\Big\}$ from M—CH$_3$
  m/z 107 $(CH_2CH_2{}^{79}Br)^+$

  m/z 42 $(CH_3CH{=}CH_2)^+$ from M—HBr
  m/z 29 $(CH_3CH_2)^+$ from M—Br—CH$_2$
  m/z 28 $(CH_2{=}CH_2)^+$ from M—Br—CH$_3$
  m/z 27 $(CH_2{=}CH)^+$ from M—Br—CH$_3$—H

60. Compound is saturated (see Section 11-9). Try to use the general guidelines that intense fragment peaks either result from the loss of relatively stable neutral species or are due to relatively stable cations.

So, looking at the high intensity m/z 73 peak for isomer C, it corresponds to $(M - 15)^+$, or loss of CH_3. This is most likely if the remaining fragment is a very stable cation, for example,

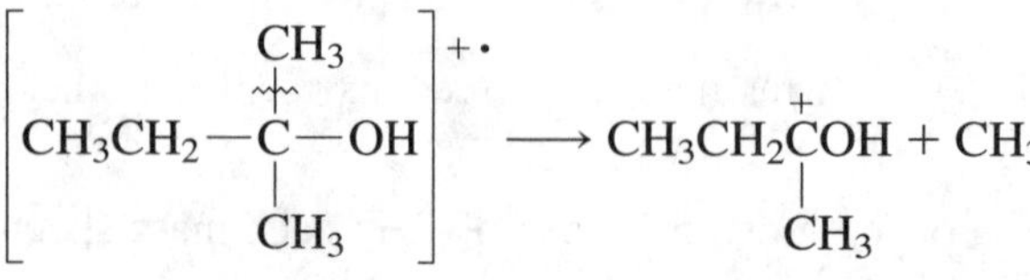

$$\left[\overset{\displaystyle CH_3}{\underset{\displaystyle CH_3}{CH_3CH_2{-}\overset{|}{\underset{|}{C}}{-}OH}} \right]^{+\cdot} \longrightarrow CH_3CH_2\overset{+}{\underset{\displaystyle \underset{|}{CH_3}}{C}}OH + CH_3 \cdot$$

m/z 73
cation, stabilized by
oxygen lone pair.

Looking at the rest of the spectrum, the base peak is at m/z 59, or $(M - 29)^+$, loss of CH_3CH_2.

$$
\left[\begin{array}{c} CH_3 \\ | \\ CH_3CH_2 \overset{\cdot}{\dashv} \overset{+}{C} - OH \\ | \\ CH_3 \end{array} \right]^{+\cdot} \longrightarrow (CH_3)_2\overset{+}{C}OH + CH_3CH_2 \cdot
$$
$$m/z\ 59$$

This is, all together, good evidence for isomer C being 2-methyl-2-butanol, as shown.

Isomer B also has a peak at m/z 73 for loss of CH_3. Its base peak (m/z 45) corresponds to loss of 43, or $CH_3CH_2CH_2$. These signals are what you might expect for 2-pentanol.

$$
\left[\begin{array}{c} OH \\ \overset{b}{} | \overset{a}{} \\ CH_3CH_2CH_2 \overset{\cdot}{\dashv} CH \overset{\cdot}{\dashv} CH_3 \end{array} \right]^{+\cdot}
$$

a →
$$
\begin{array}{c} OH \\ | \\ CH_3CH_2CH_2\overset{+}{CH} + CH_3 \cdot \\ m/z\ 73 \end{array}
$$

b →
$$
\begin{array}{c} OH \\ | \\ CH_3\overset{+}{CH} + CH_3CH_2CH_2 \cdot \\ m/z\ 45 \end{array}
$$

Both fragmentations give cations stabilized by resonance from an oxygen lone pair. This is, in fact, the correct answer.

Isomer A does **not** lose CH_3 or CH_3CH_2 (no peaks at m/z 73 or 59). That pretty much rules out any tertiary or secondary alcohol structure as a possibility. (Any example that you can write should show those fragmentations.) How about possible primary alcohol structures? Look again to intense fragment peaks for clues. The m/z 70, loss of water, doesn't really help much, except to rule out $(CH_3)_3CCH_2OH$, which has no β-hydrogens and therefore cannot dehydrate. That leaves three possibilities for A:

$$CH_3CH_2CH_2CH_2CH_2OH \qquad \begin{array}{c} CH_3 \\ | \\ CH_3CH_2CHCH_2OH \end{array} \qquad \begin{array}{c} CH_3 \\ | \\ CH_3CHCH_2CH_2OH \end{array}$$

The data that you have are in fact quite consistent with either of the first two (the third is difficult to maneuver into a fragment with m/z 42). If you got this far, you are doing **well!** (The actual spectrum of isomer A is that of 1-pentanol, by the way.)

61. (a) C_7H_{14}
  $H_{sat} = 2(7) + 2 = 16$; degree of unsaturation $= (16 - 14)/2 = 1$

  (b) C_3H_5Cl
  $H_{sat} = 2(3) + 2 - 1$ (for the Cl) $= 7$; degree of unsaturation $= (7 - 5)/2 = 1$

  (c) C_7H_{12}
  $H_{sat} = 2(7) + 2 = 16$; degree of unsaturation $= (16 - 12)/2 = 2$

  (d) C_5H_6
  $H_{sat} = 2(5) + 2 = 12$; degree of unsaturation $= (12 - 6)/2 = 3$

  (e) $C_6H_{11}N$
  $H_{sat} = 2(6) + 2 + 1$ (for the N) $= 15$; degree of unsaturation $= (15 - 11)/2 = 2$

  (f) C_5H_8O
  $H_{sat} = 2(5) + 2 = 12$; degree of unsaturation $= (12 - 8)/2 = 2$

  (g) $C_5H_{10}O$
  $H_{sat} = 2(5) + 2 = 12$; degree of unsaturation $= (12 - 10)/2 = 1$

(h) $C_{10}H_{14}$

$H_{sat} = 2(10) + 2 = 22$; degree of unsaturation $= (22 - 14)/2 = 4$

62. (a) $H_{sat} = 2(7) + 2 = 16$; degree of unsaturation $= (16 - 12)/2 = 2$

(b) $H_{sat} = 2(8) + 2 + 1$ (for the N) $= 19$; degree of unsaturation $= (19 - 7)/2 = 6$

(c) $H_{sat} = 2(6) + 2 - 6$ (for the Cl's) $= 8$; degree of unsaturation $= (8 - 0)/2 = 4$

(d) $H_{sat} = 2(10) + 2 = 22$; degree of unsaturation $= (22 - 22)/2 = 0$

(e) $H_{sat} = 2(6) + 2 = 14$; degree of unsaturation $= (14 - 10)/2 = 2$

(f) $H_{sat} = 2(18) + 2 = 38$; degree of unsaturation $= (38 - 28)/2 = 5$

63. Begin by using the molecular mass to determine the molecular formula. Given that the unknown is a hydrocarbon, first determine which combinations of carbon and hydrogen can give an approximate mass of 96. Although 96 is the mass of 8 carbon atoms, that option is not possible (no mass left for hydrogen atoms). Reduce the number of carbon atoms and add the necessary number of hydrogens to arrive at a feasible formula. Thus, 7 carbons + 12 hydrogens also gives a mass of 96, and by using the data in Table 11-5, we quickly find a good match for the exact mass measurement: $7(12.000) + 12(1.0078) = 96.0936$. Thus we are working with a formula of C_7H_{12}: two degrees of unsaturation. The unknown compound contains a total of two rings or π bonds. The IR is useful: The peaks marked indicate at least one of the degrees of unsaturation is from a $C=C$ bond. Also, the sharp band at 888 cm^{-1} is diagnostic for a $R_2C=CH_2$ group. Can you decide whether the unknown has two double bonds, or one double bond and one ring? Hydrogenation gives you C_7H_{14}: One degree of unsaturation still remains, suggesting one ring in the original compound. The ^{1}H NMR supports this conclusion: The integration of the alkene H signal (at $\delta = 4.8$) is 2 H. That limits the structure of one $R_2C=CH_2$ group. What else can you learn from the ^{1}H NMR? The alkene signal is a quintet (five lines). On the basis of the $N + 1$ rule, you can try to make a structure where four equivalent neighbor H's split the alkene H's:

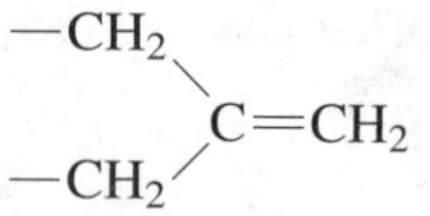

This structure fits the signal pattern, and the J of 3 Hz is right for allylic coupling, too (Table 11-2). The two CH_2 groups fit the ^{1}H NMR signal for 4 H at $\delta = 2.2$ as well. This takes care of four carbons and six hydrogens, leaving C_3H_6 unaccounted for. Try the simplest way to make a ring: add three CH_2 groups to get

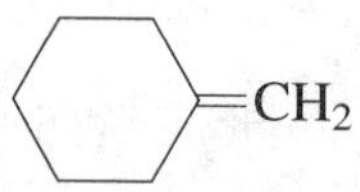

Does this structure fit the rest of the spectrum? The structure contains two more equivalent CH_2 groups, and the unique CH_2 at the opposite side of the ring from the double bond, fitting the ^{1}H NMR, and the ^{13}C NMR reflects the symmetry with five peaks. This is indeed the answer.

64. A saturated 60-carbon alkane has the formula $C_{60}H_{122}$. Therefore, "bucky-ball" possesses $122/2 = 61$ degrees of unsaturation. In its hydrogenation product, $C_{60}H_{36}$, there are $(122 - 36)/2 = 43$ degrees of unsaturation. Therefore, there are **at least** $61 - 43 = 18$ π bonds in C_{60}. (As you will see later, there are actually 30 π bonds, but not all undergo hydrogenation.)

65. First list the information you can derive from the data:

(a) Molecular formula: There is one degree of unsaturation.

(b) ^{1}H NMR: There are three methyl groups; two (at $\delta = 1.63$ and 1.71 ppm) probably are attached to alkenyl carbons, the other split by one neighbor, as CH_3-CH. A $-CH_2-CH_2-O-$ group is likely due to the triplet at $\delta = 3.68$ ppm. There is one alkene hydrogen, whose signal is also a triplet, so the molecule would appear to contain a $RR'C'{=}CH{-}CH_2{-}$ grouping.

(c) ^{13}C NMR: There are signals for an alcohol carbon and two alkene carbons.

(d) IR: Alkene $C{=}C$ and alcohol $O{-}H$ stretches are apparent.

(e) Oxidation eliminates signals for the

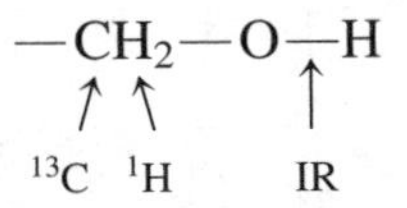

$$-CH_2-O-H$$

$$^{13}C \quad {}^1H \qquad IR$$

group, replacing them with IR and ^{13}C evidence for $C{=}O$, which the ^1H NMR indicates is an aldehyde (signal at $\delta = 9.64$ ppm), as expected for the product of oxidation of a primary alcohol.

(f) Hydrogenation gives the same product as from hydrogenation of geraniol.
Put it all together: The last item gives us the skeleton,

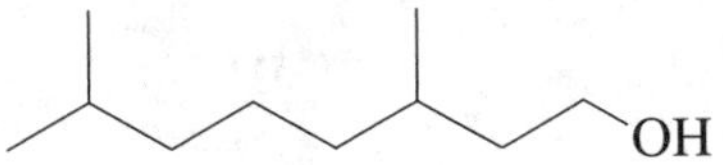

You still need to find the correct place for the double bond. The ^1H NMR spectrum shows only one alkenyl H, so the double bond must be trisubstituted. That means that there are only three possible locations:

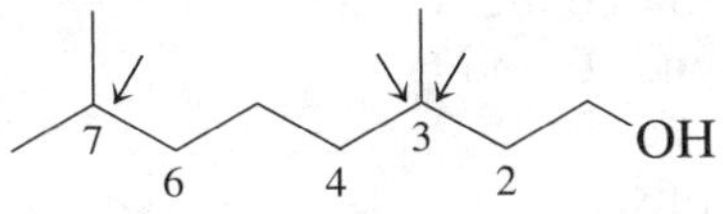

Again, the ^1H NMR helps you, because it contains **two** signals for methyl groups in the region characteristic of attachment to unsaturated functional groups. The double bond must therefore be between C6 and C7 to be consistent with this observation. The answer is

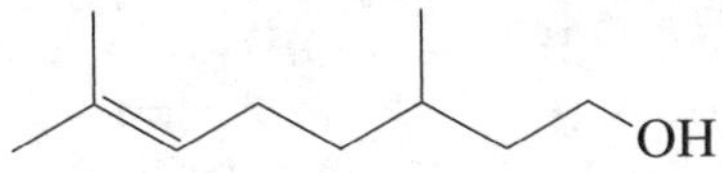

Its name is citronellol, and both it and geraniol are isolated from the oily extract of citronella, a fragrant grass native to southern Asia. The oil has long been used as an insect repellent and a liniment as well as for perfume.

66. Remember that the IR is not used to elucidate an entire structure in detail now that fancy NMR techniques are available. Just match IR bands with functional groups.

(a) Camphor has one functional group, the $C{=}O$ (carbonyl) group; a single IR band in the range $1690{-}1750$ cm^{-1} is expected, and this is found in (d), 1738 cm^{-1}.

(b) Menthol is a simple alcohol; the single band in (a) is the expected $O{-}H$ stretch.

(c) Chrysanthemic ester contains two functional groups, an alkene $C{=}C$ bond and an ester $C{=}O$ group; the alkene should give two bands, one around 1650 cm^{-1}, the other around 3080 cm^{-1}; the ester should give a peak around 1740 cm^{-1}. The match with (b) is not perfect, but it is the correct answer. The cyclopropane affects the precise band locations.

(d) Epiandrosterone contains an alcohol and a ketone; (c) fits perfectly.

67.

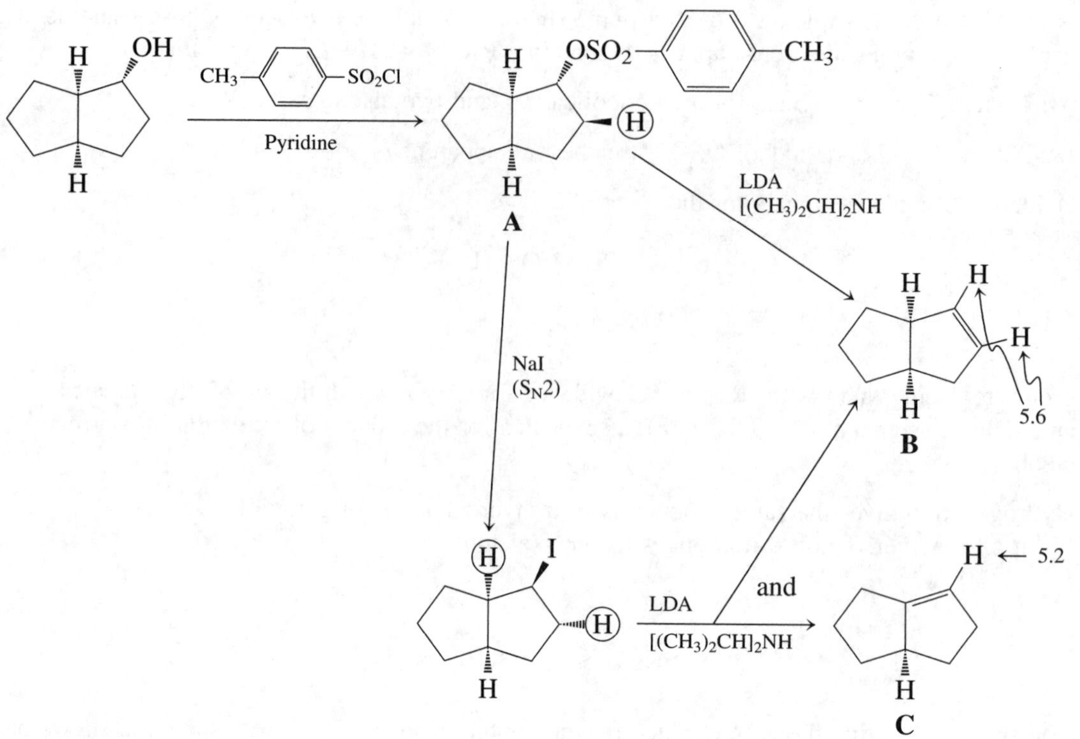

E2 *anti* elimination on A can lead to only one alkene B via removal of the circled hydrogen. Conversion to the iodo-derivative via S_N2 inversion allows either of the two circled hydrogens to be removed in *anti* elimination, leading to a mixture of alkenes B and C.

68. (a) Newman projections help for this.

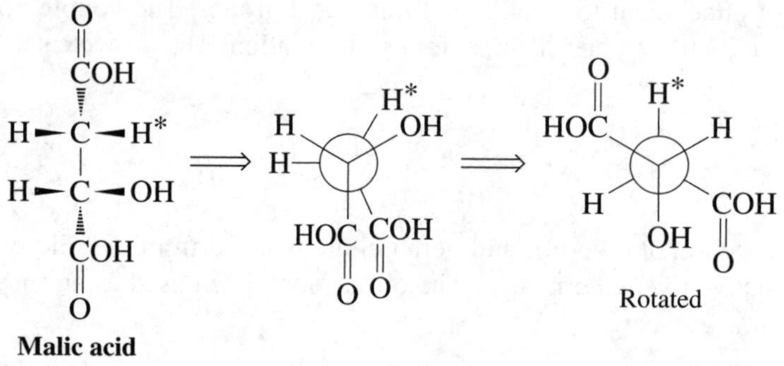

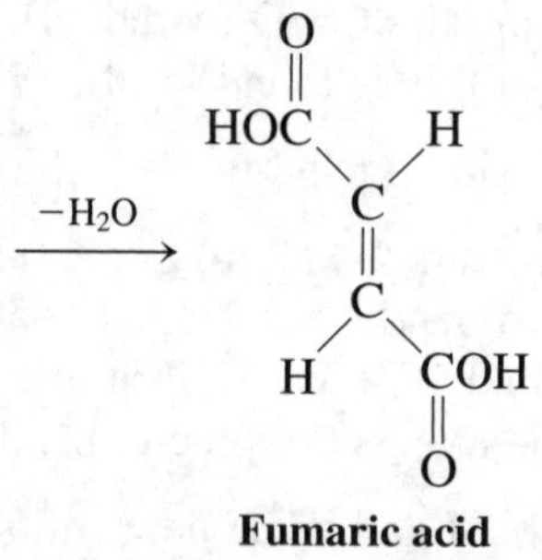

Malic acid

Fumaric acid

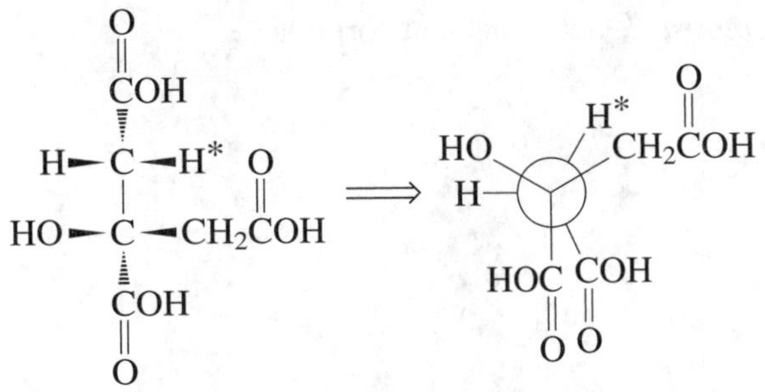

Citric acid

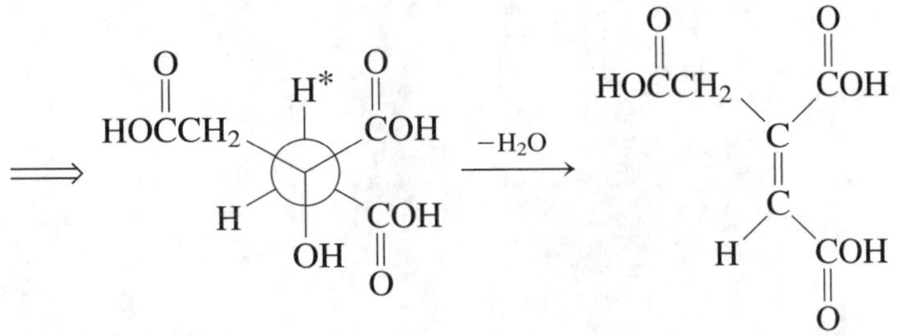

Aconitic acid

(b) Fumaric acid is *E*; aconitic acid is *Z*.

(c) Four, because it contains two stereocenters that do not have identical groups attached (so there is no possibility of an achiral *meso* isomer). Two of them will give Z-aconitic acid via *anti* elimination (circled groups are removed).

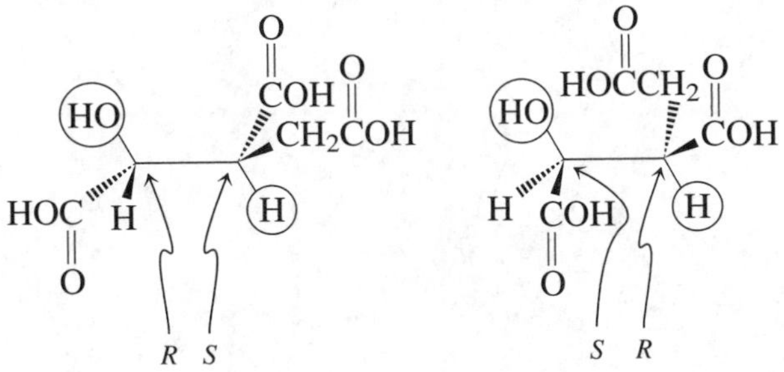

69. The chiral carbon common to all four compounds 1a–1d has the *S* configuration. The chiral carbon to its left has the indicated configuration: **1a,** *R*; **1b,** *S*; **1c,** *R*; **1d,** *S*. Work backwards from the stereochemistry of the alkene function to a Newman projection of the required configuration in the necessary anti (for E2) conformation. Make sure that the specified chiral carbon is depicted **exactly** the way it appears in the formula of the substrate so that you don't inadvertently change its configuration. If necessary, determine its configuration in both the structure given in the problem (it is *S*) and in your Newman projection—they better both be the same!

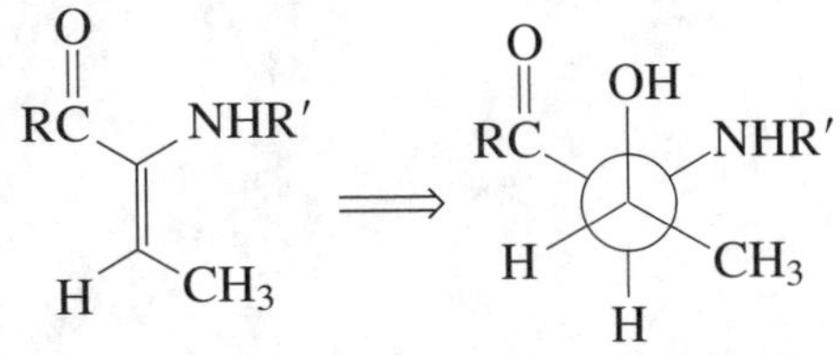

Now work out the configuration of the carbon bearing the hydroxy group: It is *S*.

70. **(d)** (Question asks for *empirical*, not molecular formula)

71. **(b)**

72. **(b)**

73. **(a)**

74. **(d)**

12

Reactions of Alkenes

Alkenes are reactive and synthetically useful molecules. Their reactivity originates from the electrons in the π bond. They are on the average further away from (and therefore electrostatically less tightly held to) the carbon nuclei than are electrons in σ bonds. As a result, they are "available" in a "Lewis base" sense, somewhat like the lone pair electrons on the oxygen of water or the nitrogen of ammonia. As you will see, attachment of an electrophile to these basic π electrons has the effect of breaking the π bond and is the first step in many reactions of alkenes, *additions*. Many of these addition reactions are known, and they allow conversion of alkenes to other organic molecules, including ones we've seen before (haloalkanes and alcohols) as well as new ones. These additions expand the scope of our synthetic possibilities significantly. Therefore, an updated Functional Group Interconversions chart is presented in this chapter of the study guide.

Outline of the Chapter

12-1   Thermodynamic Feasibility of Addition Reactions

12-2   Hydrogenation of Alkenes

12-3 through 12-7 Electrophilic Additions to Alkenes
>   The largest group of reactions characteristic of alkenes.

12-8   Regio- and Stereoselective Hydration of Alkenes by Hydroboration
>   Another functional group interconversion involving addition.

12-9   Diazomethane, Carbenes, and Cyclopropane Synthesis
>   Preparing three-membered rings.

12-10, 12-11, 12-12 Oxidation of Alkenes
>   Additions of oxygen-containing electrophiles.

12-13   Radical Additions to Alkenes
>   A new kind of mechanism for alkene reactions: radical additions.

12-14, 12-15 Dimerization, Oligomerization, and Polymerization of Alkenes
>   Reactions of, mainly, industrial use. Or, "where everything in your kitchen that you don't eat comes from."

12-16   Ethene in Industry

12-17   Alkenes in Nature

Keys to the Chapter

12-1. Thermodynamic Feasibility of Addition Reactions

Carbon–carbon π bonds are weaker than σ bonds. Addition reactions to alkenes are generally exothermic because one of the bonds broken is weak (the π bond), but both new bonds formed are strong.

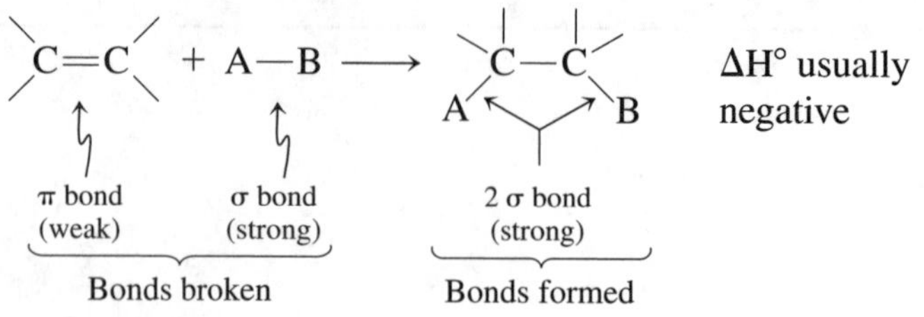

12-2. Hydrogenation of Alkenes

The simplest reaction of alkenes is the addition of hydrogen to form alkanes. *Hydrogenation* is a nonpolar process, unlike many of the addition reactions to be described later, which involve electrophiles attaching to the nucleophilic π electrons of the alkene. Hydrogenation reactions require catalysts, such as platinum oxide (Adams's catalyst), palladium on carbon (Pd-C), or Raney nickel (Ra-Ni).

12-3 through 12-7. Electrophilic Additions to Alkenes

In these sections we turn to the largest and most typical class of addition reactions of alkenes. These all proceed in two steps. The first step is attachment of an electrophilic atom to one of the alkene's double-bonded carbons, a reaction forming a cationic intermediate. Combination of this cation with any available nucleophile gives the final addition product:

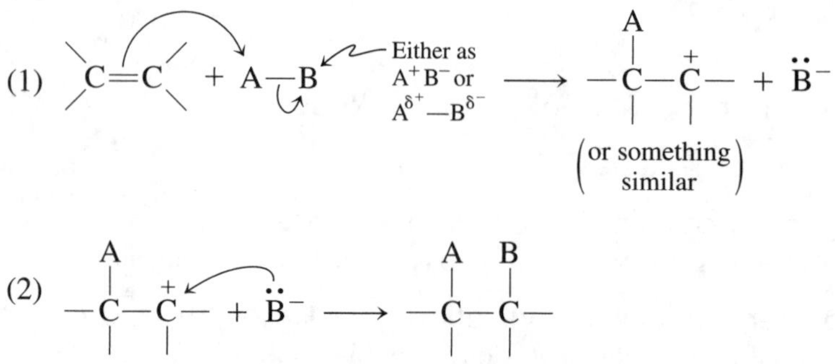

From a synthetic point of view, addition reactions allow conversion of alkenes to molecules containing new functional groups at one or both of the original double-bonded carbons. These sections present many such examples.

Most of the "A–B"-type molecules that participate in electrophilic addition to alkenes are strongly polarized, such as $H^{\delta+}$—$Cl^{\delta-}$. Others are not but can still serve as sources of electrophilic atoms. Halogens (Cl_2, Br_2) fall into this category. Although nonpolar molecules like Br_2 do not have permanent dipoles, recall that electron movement can give rise to "fleeting dipoles," thus allowing the halogen molecule to behave as if it contained an electrophilic atom, namely, "Br^+".

The details of the addition mechanism vary somewhat with the nature of the electrophile. Thus, protons add regiospecifically to give the more stable of the two possible carbocations in the case of unsymmetrical double bonds. The cation then can do any of its usual reactions (e.g., attach to a nucleophile or rearrange).

Larger electrophiles, especially ones with lone pairs (e.g., "$:\overset{..}{Br}:^+$" from Br_2), add to give positively charged three-membered rings (bromonium ion, chloronium ion, etc.). Nucleophiles react with these ions in much the same way that they open the rings of the cyclic alkyloxonium ions of Chapter 9. Addition occurs at the most highly substituted carbon (Markovnikov orientation) in an *anti* manner.

12-8. Regio- and Stereoselective Functionalization of Alkenes by Hydroboration

This section describes the special characteristics and utility of the reaction between an alkene and a *borane,* a molecule containing B—H bonds. Boranes are electrophilic, and their reactions with alkenes are therefore very reasonable (note carefully, however, that *borohydrides* like BH_4^- are very different—they are anions, not electrophiles, and **don't react with alkenes**).

Borane B—H bonds add across alkenes. The reaction (1) is regioselectively anti-Markovnikov, (2) is stereospecifically a *syn* addition, (3) goes by a concerted, one-step mechanism, and (4) is synthetically very useful because the new carbon–boron bond can be oxidized by basic H_2O_2 to give an alcohol.

$$-\overset{|}{\underset{|}{C}}-B\diagup \longrightarrow -\overset{|}{\underset{|}{C}}-OH$$

The chart that follows summarizes the new reactions of Sections 12-3 through 12-8, using 1-methylcyclohexene as a substrate. Note the three methods for adding H_2O to a double bond, each with its own specific characteristics.

1. Hydroboration–oxidation (anti-Markovnikov), e.g.,

2. Oxymercuration–demercuration (Markovnikov), e.g.,

3. Acid-catalyzed hydration (Markovnikov, with carbocations that may rearrange), e.g.,

CHART OF EXAMPLES
Electrophilic Additions to 1-Methylcyclohexene

Example	Typical reagent	Electrophile	Nucleophile	Intermediate	Regiochemistry Stereochemistry	Major Product
Hydroboration	BH_3	$B^{\delta+}$	$H^{\delta-}$	None	anti-Markovnikov *syn* addition	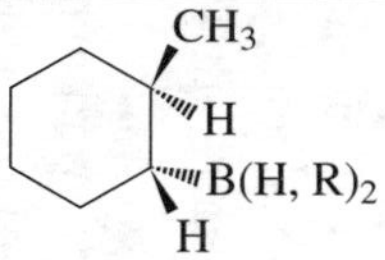
HX addition	HCl	H^+	Cl^-	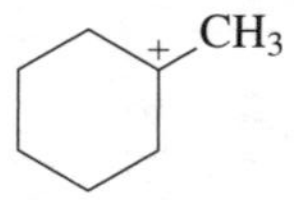	Markovnikov	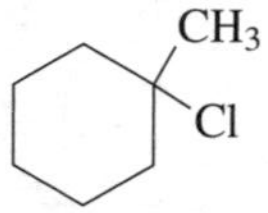
Hydration	H_2SO_4, H_2O	H^+	H_2O	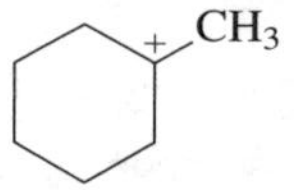	Markovnikov	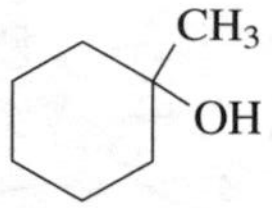
Halogenation	ICl	$I^{\delta+}$	Cl^-	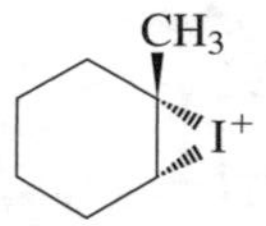	Markovnikov *anti* addition	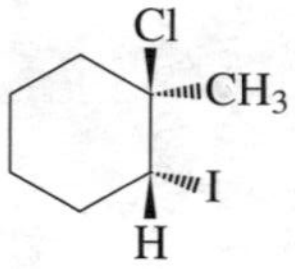
Haloalcohol formation	Cl_2, H_2O	$Cl^{\delta+}$	H_2O	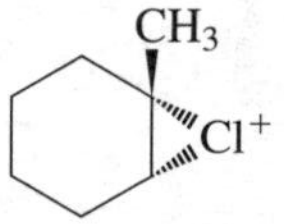	Markovnikov *anti* addition	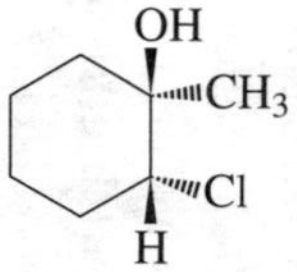
Chlorosulfenylation	CH_3SCl	$CH_3S^{\delta+}$	Cl^-	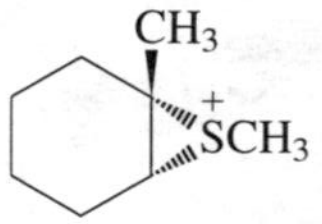	Markovnikov *anti* addition	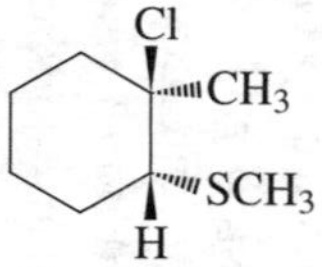
Oxymercuration	$Hg(OCCH_3)_2$, H_2O	$CH_3COHg^{\delta+}$	H_2O		Markovnikov *anti* addition	

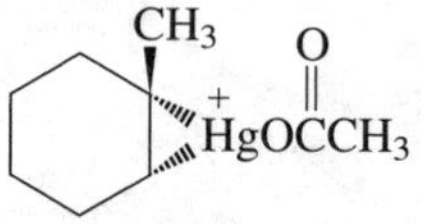

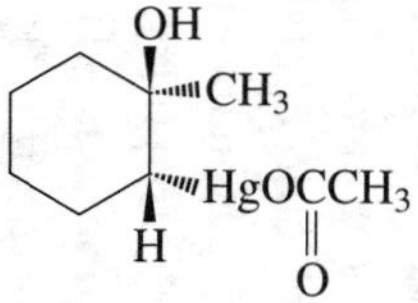

12-9. Diazomethane, Carbenes, and Cyclopropane Synthesis

This section introduces an unusual electrophile, the carbene, which has a specific use in synthesis. Carbenes, which have the general structure $R_2C\colon$, have a neutral but electron-deficient carbon capable of electrophilic additions. Lacking an octet, carbenes are high-energy species that cannot be synthesized and stored but have to be prepared in the presence of whatever substance they are intended to react with. In the presence of alkenes, carbenes react rapidly to form bonds with both alkene carbons, giving a cyclopropane as a result. In the text section, look at the arrows in the mechanism for this process. The electron pair in the π bond of the alkene moves to make one bond between one of the alkene carbons and the electrophilic carbene carbon, and the electron pair on the carbene carbon itself moves to make the second bond with the other alkene carbon. Typically these two bond-forming processes happen together, making this an example of a concerted reaction, one in which multiple bonding changes happen simultaneously.

Some alternative species are encountered that also make cyclopropanes from alkenes; these are called carbenoids because they act like carbenes without actually being carbenes. Like carbenes they contain a special carbon atom capable of reacting as an electrophile with the π bond of the alkene to form one carbon–carbon bond and also supplying an electron pair to form a second carbon–carbon bond.

12-10, 12-11, and 12-12. Oxidation of Alkenes

These sections present several reactions that attach oxygen to both the double-bonded carbons of alkenes. Each reaction is an example of a concerted process, in which several bonding changes occur in one step. Peroxycarboxylic acids such as MCPBA contain an electrophilic oxygen

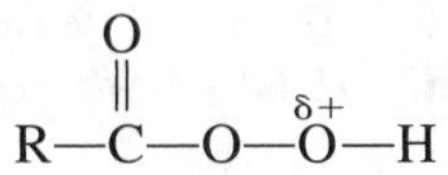

$$
\begin{array}{c}
\text{O} \\
\parallel \\
\text{R}-\text{C}-\text{O}-\overset{\delta+}{\text{O}}-\text{H}
\end{array}
$$

that adds to give oxacyclopropanes. Osmium tetroxide and ozone both participate in a process where three pairs of electrons move in a circle to simultaneously form new C—O bonds at both the double-bonded carbons. The products of these *concerted cycloadditions* are ring compounds:

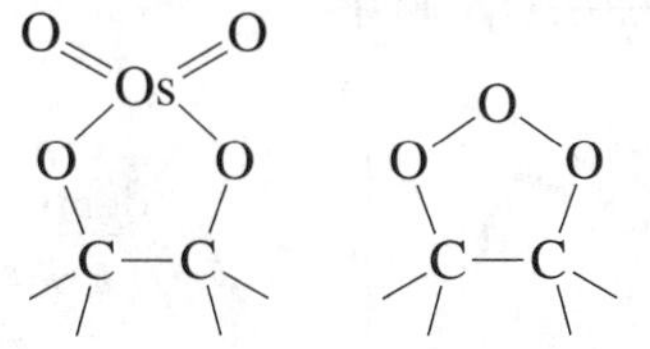

Notice, for synthetic purposes, that ozonolysis is the first process you've seen that can lead to complete cleavage of carbon–carbon bonds. Ozonolysis followed by reduction breaks double bonds, giving two carbonyl compounds:

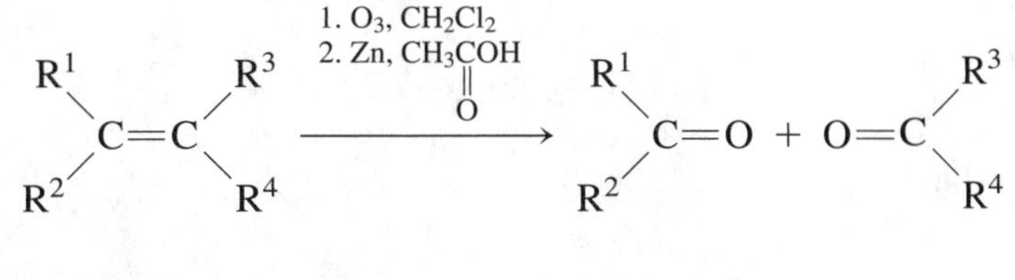

12-13. Radical Additions to Alkenes

Although typical radicals are neutral, they are electron deficient in the sense that they are one electron short of having a full octet. Reaction between the π electrons of an alkene double bond can take place, completing the octet around the radical atom. Unlike additions of positively charged electrophiles, however, the result of this addition is not a carbocation but another radical. Like the processes we studied in Chapter 3, these reactions follow radical chain mechanisms.

The text describes a brief history of the story entitled "What's the Matter with HBr?" In the radical addition of HBr to an alkene, the reactive species is $:\overset{..}{\underset{..}{Br}}\cdot$ instead of the H^+ in ionic additions. So the radical addition appears to "turn around" the direction of reaction to unsymmetrically substituted alkenes. Notice that **both** the radical **and** the ionic additions regiospecifically give the **most stable intermediate.**

Ionic Addition of HBr

$$CH_3CH=CH_2 \xrightarrow{H^+} CH_3\overset{+}{C}HCH_3 \xrightarrow{:\overset{..}{\underset{..}{Br}}:^-} CH_3\overset{\overset{\displaystyle :\overset{..}{\underset{..}{Br}}:}{|}}{C}HCH_3$$

2° carbocation

Only product
(Markovnikov)

Radical addition of HBr
(Requires presence of initiators
such as peroxides or UV radiation.)

$$CH_3CH \overset{..}{\cdots} CH_2 \xrightarrow{Br\cdot} CH_3\overset{\cdot}{C}HCH_2Br \xrightarrow{H\cdot\cdot Br} CH_3CH_2CH_2Br$$

2° radical

Only product
(anti-Markovnikov)

For radical chain reactions to be kinetically feasible, **both** propagation steps must have relatively low activation barriers. Such is the case for addition of HBr, but **not** for HCl or HI. **Only HBr** addition "turns around" in the presence of peroxides. **HCl and HI additions remain ionic, with Markovnikov orientation, whether peroxides are present or not!**

As promised earlier, the updated chart of synthetic transformations that follows includes the reactions of alkenes you have just seen.

Functional Group Interconversions

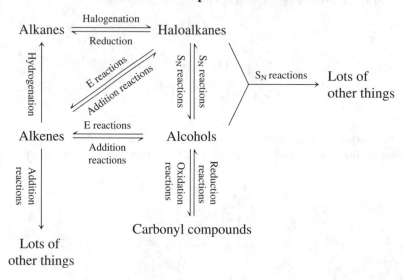

12-14 and 12-15. Dimerization, Oligomerization, and Polymerization of Alkenes

These are addition reactions of carbocations, carbanions, or radicals with alkenes. In each case the addition gives a new carbocation, carbanion, or radical (as the case may be), which can add to another alkene molecule. This process can go on and on, making the molecule bigger and bigger: a *polymer*. Polymers can contain units that are all identical (like Teflon, which is a polymer of $CF_2=CF_2$) or can contain two or more different *monomer* units (the original Saran Wrap was a *copolymer* of $CH_2=CHCl$ and $CH_2=CCl_2$).

Solutions to Problems

33. Careful! Use $DH°$ for CH_3CH_2—X, not CH_3—X, from Table 3-1.

    (a) $C_2H_4 + Cl_2 \longrightarrow Cl$—$CH_2CH_2$—$Cl$
    Heat in: 65 + 58   Heat out: 2 × 84
    ∴ $\Delta H° = 65 + 58 - (2 × 84) = -45$ kcal mol^{-1}

    (b) $C_2H_4 + IF \longrightarrow I$—$CH_2CH_2$—$F$; $\Delta H° = 65 + 67 - (56 + 111) = -35$ kcal mol^{-1}

    (c) $C_2H_4 + IBr \longrightarrow I$—$CH_2CH_2$—$Br$; $\Delta H° = 65 + 43 - (56 + 70) = -18$ kcal mol^{-1}

    (d) $C_2H_4 + HF \longrightarrow H$—$CH_2CH_2$—$F$; $\Delta H° = 65 + 135 - (101 + 111) = -12$ kcal mol^{-1}

    (e) $C_2H_4 + HI \longrightarrow H$—$CH_2CH_2$—$I$; $\Delta H° = 65 + 71 - (101 + 56) = -21$ kcal mol^{-1}

    (f) $C_2H_4 + HOCl \longrightarrow HO$—$CH_2CH_2$—$Cl$; $\Delta H° = 65 + 60 - (94 + 84) = -53$ kcal mol^{-1}

    (g) $C_2H_4 + BrCN \longrightarrow Br$—$CH_2CH_2$—$CN$; $\Delta H° = 65 + 83 - (70 + 124) = -46$ kcal mol^{-1}

    (h) $C_2H_4 + CH_3SH \longrightarrow CH_3S$—$CH_2CH_2$—$H$; $\Delta H° = 65 + 88 - (60 + 101) = -8$ kcal mol^{-1}

34. The structure of the product shows that the process leads to attachment of two hydrogen atoms to the face of the double bond *opposite* to the cyclopropane ring. Thus we can infer that the starting cyclic alkene is able to complex to the catalyst surface from only one face, the face *opposite* to the fused three-membered ring (below, left). The steric bulk of the cyclopropane makes approach of the other face of the alkene very difficult (below, right), interfering with complexation and leading to high stereoselectivity in the hydrogenation process.

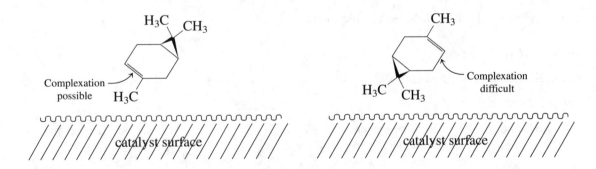

35. In all cases identify the face of the double bond that can complex to the catalyst surface with the least steric interference. Add H_2 to that side of the molecule.

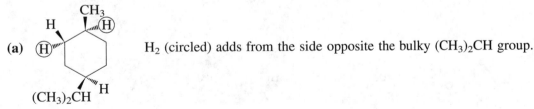

(a) H_2 (circled) adds from the side opposite the bulky $(CH_3)_2CH$ group.

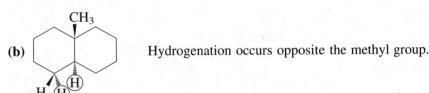

(b)   Hydrogenation occurs opposite the methyl group.

(c) 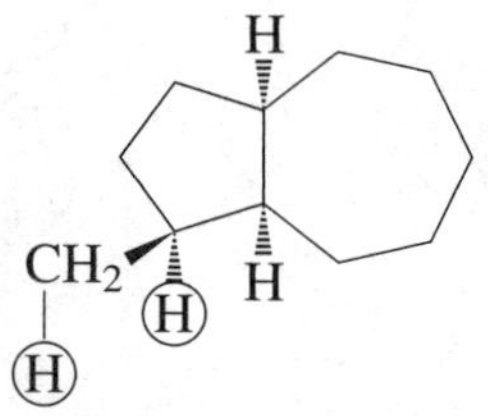 Hydrogenation occurs from the more exposed (bottom) side of the folded molecule.

36. More exothermic. There is essentially no bond angle strain in either cyclohexane or cyclohexene. Heat of hydrogenation of the latter is essentially the same as that for an acyclic cis-disubstituted alkene. Both cyclobutane and cyclobutene are strained, but bond angle compression is greater in the alkene ($120° - 90° = 30°$) compared with the alkane ($109° - 90° = 19°$). The greater strain in cyclobutene increases the energy difference between it and cyclobutane, thus resulting in greater energy release upon hydrogenation.

37.

	(i) Peroxide-free HBr (Markovnikov addition)	**(ii) HBr + peroxides** (anti-Markovnikov addition)
(a)	2-Bromohexane	1-Bromohexane
(b)	2-Bromo-2-methylpentane	1-Bromo-2-methylpentane
(c)	2-Bromo-2-methylpentane	3-Bromo-2-methylpentane
(d)	3-Bromohexane	3-Bromohexane
(e)	Bromocyclohexane	Bromocyclohexane

All chiral products are formed as racemic mixtures.

38. **(a)** 1,2-Dibromohexane

(b) 1,2-Dibromo-2-methylpentane

(c) 2,3-Dibromo-2-methylpentane

(d) (*R,R*) and (*S,S*)-3,4-Dibromohexane. *Anti* addition to a cis compound gives a racemic mixture of chiral products; a trans substrate gives the meso isomer.

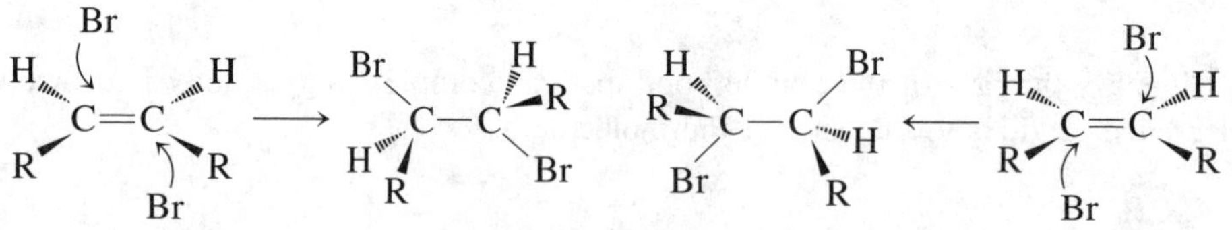

Racemic mixture

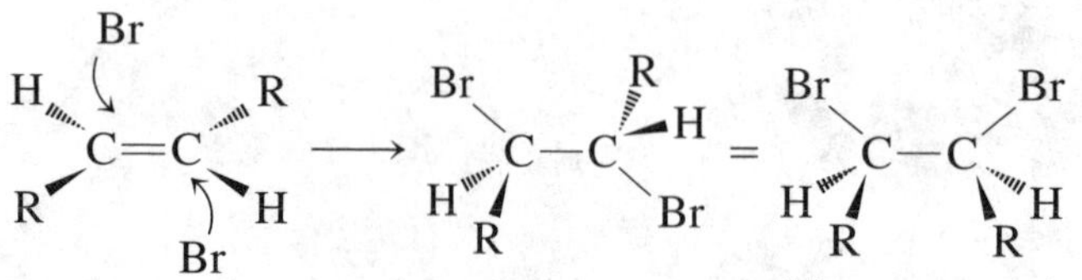

Meso compound

(e) *trans*-1,2-Dibromocyclohexane

All products in this problem (except for the meso compound) are chiral; all are racemic.

39.

H$_2$SO$_4$ + H$_2$O (Markovnikov hydration)	BH$_3$, THF; then NaOH, H$_2$O$_2$ (anti-Markovnikov hydration)
(a) 2-Hexanol	1-Hexanol
(b) 2-Methyl-2-pentanol	2-Methyl-1-pentanol
(c) 2-Methyl-2-pentanol	2-Methyl-3-pentanol
(d) 3-Hexanol	3-Hexanol
(e) Cyclohexanol	Cyclohexanol

Oxymercuration–demercuration gives the same products as aqueous sulfuric acid. The carbocations formed from these substrates and H$^+$ are not particularly prone to rearrangement. All chiral products are formed as racemic mixtures.

40. (a) Hot, concentrated H$_2$SO$_4$

(b) Cold, aqueous H$_2$SO$_4$

(c) NaOCH$_2$CH$_3$ in CH$_3$CH$_2$OH

(d) HCl in CCl$_4$

Additions [reactions (b) and (d)] are normally favored by thermodynamics (Section 12-1). For elimination to occur, conditions have to be established to drive the equilibria the opposite way. In (a) the water lost in the reversible E1 process is protonated by the concentrated H$_2$SO$_4$, removing it from the equilibrium. No good nucleophiles are present; therefore, the carbocation undergoes loss of a proton to form the alkene. In (c) the strongly basic ethoxide ion induces bimolecular elimination and neutralizes the liberated HCl, forming ethanol and NaCl. No species electrophilic enough to add to the alkene are present in the reaction mixture.

41. The transformation proceeds through *anti* addition. The possible products arising from addition to the *trans* and the *cis* isomers of the starting compound are shown below. Addition to the *trans* isomer gives the required 2S,3S isomer, but as a racemic mixture with its 2R,3R enantiomer, because initial attack by bromine on the double bond has an equal chance of occurring from the top and bottom faces of the double bond.

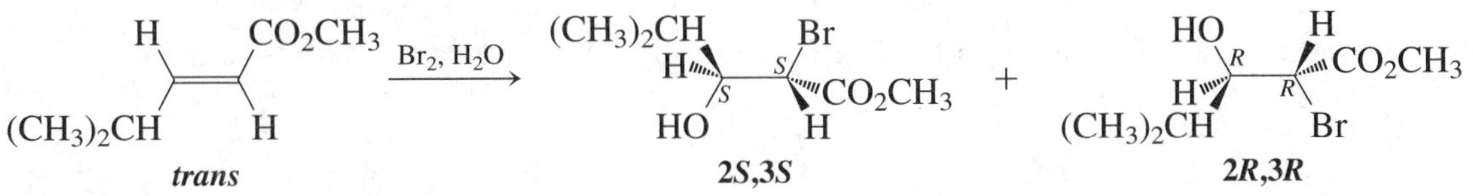

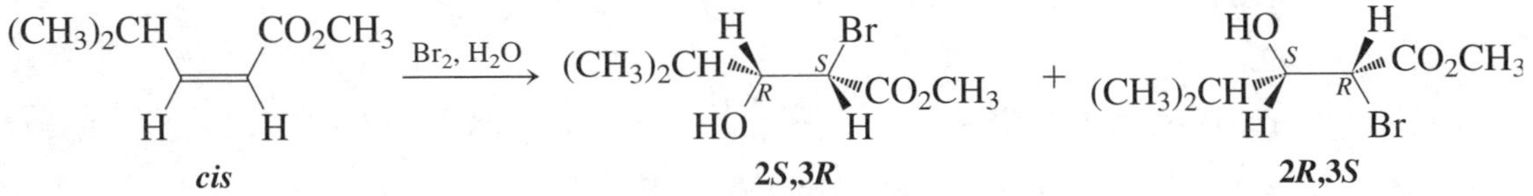

42. All chiral products are formed as racemic mixtures.

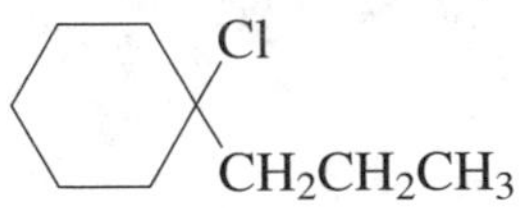

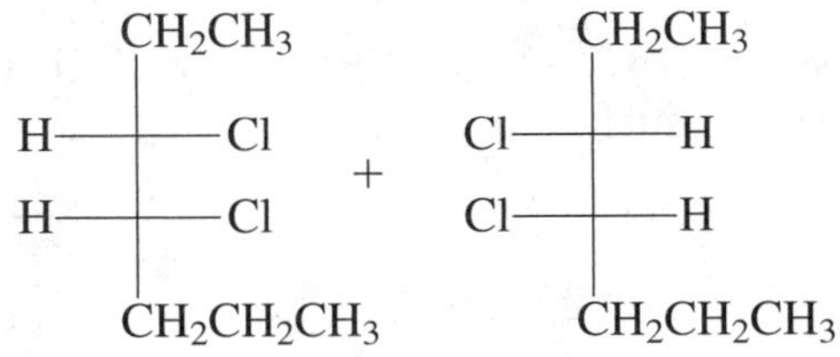

(c) 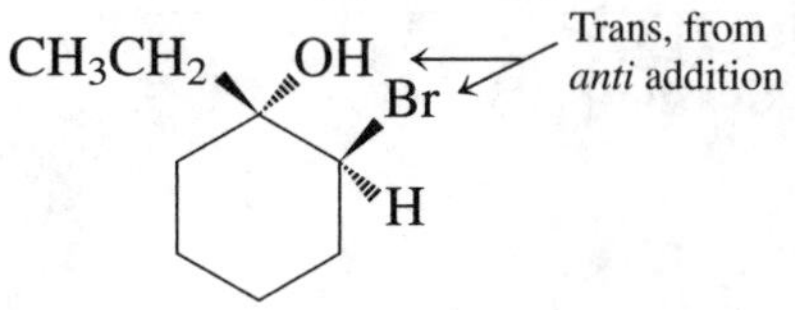 CH₃CH₂ OH ← Trans, from *anti* addition

(d) 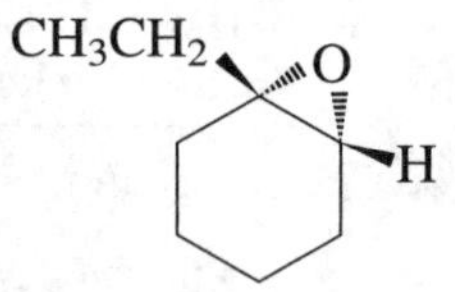 CH₃CH₂

(e) 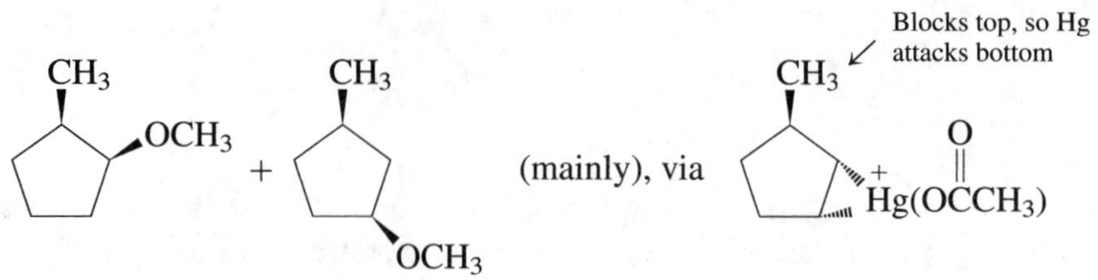 CH₃ OCH₃ + CH₃ OCH₃ (mainly), via

Blocks top, so Hg attacks bottom ↙
CH₃
+ Hg(OCCH₃)

All products are formed as racemic mixtures.

(f) 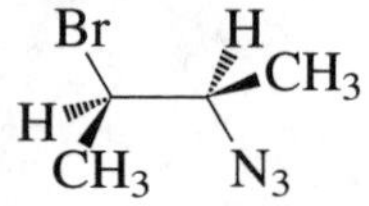 Br H CH₃ H CH₃ N₃ (+ enantiomer)

(g) 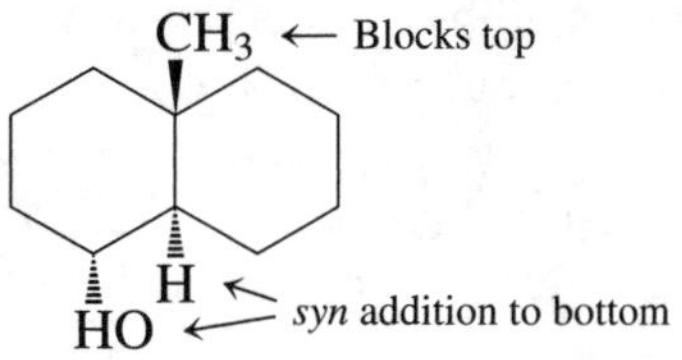 CH₃ ← Blocks top

H
HO ← *syn* addition to bottom

Note the anti-Markovnikov regiochemistry for hydroboration.

43. A brief analysis of possible choices is provided for each problem.

OH H
↓   ↓
(a) Need Markovnikov hydration of $(CH_3)_2CHCH{=}CH_2$, or anti-Markovnikov hydration of

H   OH
↓   ↓
$(CH_3)_2C{=}CHCH_3$. Either can be done.

$$(CH_3)_2CHCH{=}CH_2 \xrightarrow[\text{2. NaBH}_4\text{, NaOH, H}_2\text{O}]{\text{1. Hg(OAc)}_2\text{, H}_2\text{O}} (CH_3)_2CHCHOHCH_3$$

$$(CH_3)_2C{=}CHCH_3 \xrightarrow[\text{2. NaOH, H}_2\text{O}_2\text{, H}_2\text{O}]{\text{1. BH}_3\text{, THF}} (CH_3)_2CHCHOHCH_3$$

(b) Need addition to propene of "Cl⁺" and "$(CH_3)_2CHO^-$." Cl_2 is the source of "Cl⁺", and $(CH_3)_2CHOH$ provides the nucleophile.

$$CH_2{=}CHCH_3 \xrightarrow{\text{Cl}_2\text{, (CH}_3)_2\text{CHOH solvent}} ClCH_2CH(CH_3)OCH(CH_3)_2$$

(c), (d) These involve addition of Br$_2$ to *cis-* or *trans-*4-octene isomers. Addition is *anti,* so *trans-*octene gives meso and *cis-*octene gives a racemic mixture.

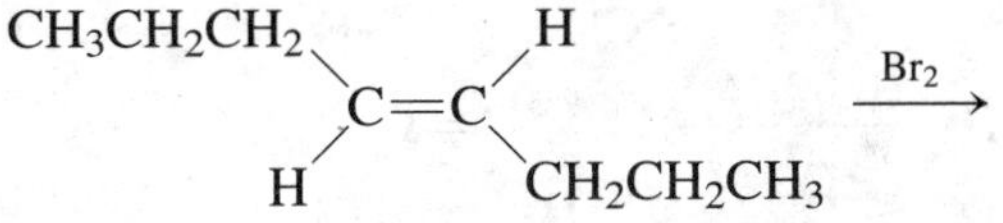

*meso-*CH$_3$CH$_2$CH$_2$CHBrCHBrCH$_2$CH$_2$CH$_3$

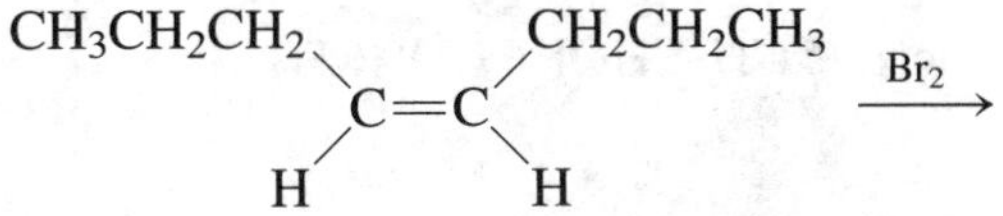

racemic CH$_3$CH$_2$CH$_2$CHBrCHBrCH$_2$CH$_2$CH$_3$

(e) This synthesis is easier—the methyl of 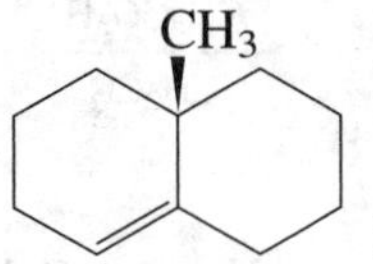 blocks the top face, allowing reaction with a peroxycarboxylic acid to occur from below:

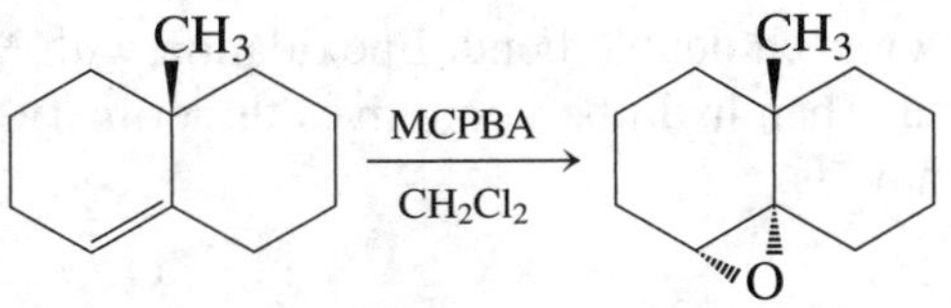

(f) This synthesis is harder. How do you get the oxygen attached on the more crowded side? It has to be done stepwise, with an inversion step following initial attack of an electrophile on the less crowded bottom face. Apply the general oxacyclopropane synthesis sequence

Alkene $\xrightarrow{\text{X}_2,\ \text{H}_2\text{O}}$ haloalcohol (*anti* addition) $\xrightarrow{\text{Base}}$ oxacyclopropane (internal S$_N$2)

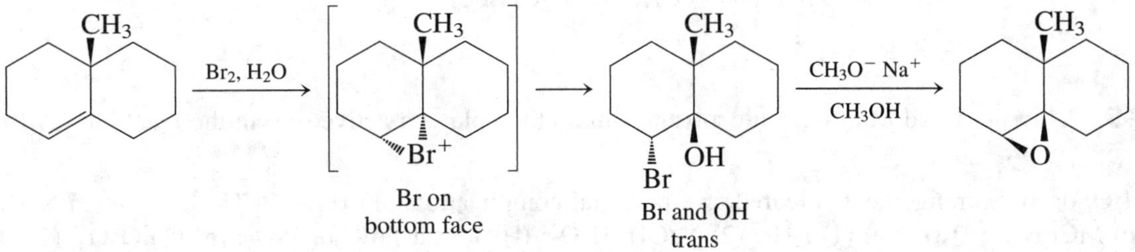

44. (a) Need to proceed via anti-Markovnikov addition to 1-butene. A bulky base is necessary to make 1-butene by elimination.

CH$_3$CH$_2$CHBrCH$_3$ $\xrightarrow{(\text{CH}_3)_3\text{CO}^-\ \text{K}^+,\ (\text{CH}_3)_3\text{COH}}$ CH$_3$CH$_2$CH=CH$_2$

We cannot add HI directly to an alkene with anti-Markovnikov orientation. One indirect possibility is shown.

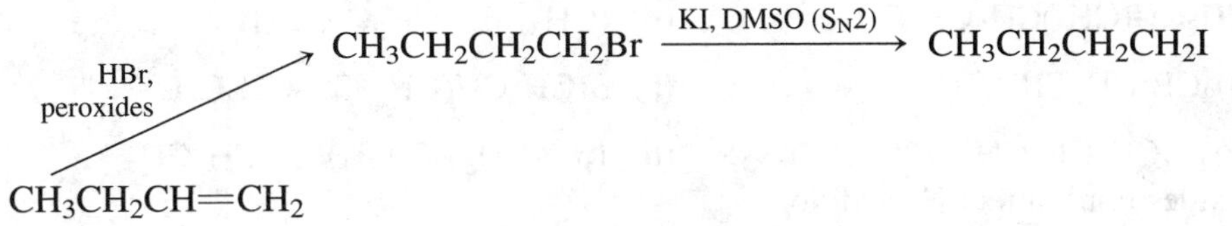

(b), **(c)** Because dehydration gives mainly trans alkenes, these problems involve finding ways to add two OH's either *anti* (⟶ meso) or *syn* (⟶ racemic). *Syn* is easy.

$$CH_3CHOHCH_2CH_3 \xrightarrow{H_2SO_4, \Delta}$$

$$\xrightarrow[\text{(syn)}]{\substack{1.\ OsO_4,\ THF \\ 2.\ H_2S}} \text{racemic } CH_3CHOHCHOHCH_3$$

Anti is also readily achieved.

$$\xrightarrow[\text{(with inversion)}]{H^+,\ H_2O} meso\text{-}CH_3CHOHCHOHCH_3$$

(d) Need different reactions on each double bond. Epoxidation with MCPBA is very selective for the trisubstituted double bond. Then hydroboration gives the primary alcohol. Finally, oxidation to an aldehyde finishes the synthesis.

$$(CH_3)_2C{=}CHCH_2CH_2CH{=}CH_2 \xrightarrow[CH_2Cl_2]{MCPBA} (CH_3)_2\overset{O}{\overset{\diagdown\diagup}{C{-}}}CHCH_2CH_2CH{=}CH_2$$

$$\xrightarrow[]{\substack{1.\ BH_3, \\ THF \\ 2.\ H_2O_2, \\ HO^-}} (CH_3)_2\overset{O}{\overset{\diagdown\diagup}{C{-}}}CHCH_2CH_2CH_2CH_2OH$$

$$\xrightarrow{PCC,\ CH_2Cl_2} (CH_3)_2\overset{O}{\overset{\diagdown\diagup}{C{-}}}CHCH_2CH_2CH_2\overset{O}{\overset{\|}{C}}H$$

45. As in the Road Map we omit solvents unless they play an active role in the reaction. **(a)** Br$_2$; **(b)** H$^+$,

H$_2$O, or, if rearrangement is feared as a potential complication, (1) Hg(O$\overset{O}{\overset{\|}{C}}CH_3$)$_2$, H$_2$O, (2) NaBH$_4$;
(c) MCPBA; **(d)** HI; **(e)** (1) BH$_3$, (2) NaOH, H$_2$O$_2$; **(f)** H$_2$, cat Pd/C or PtO$_2$; **(g)** (1) OsO$_4$, (2) H$_2$S, or cat
OsO$_4$, H$_2$O$_2$; **(h)** HBr, ROOR (R = *tert*-butyl, for example); **(i)** CH$_2$N$_2$, *hv* or Δ or Cu; or CH$_2$I$_2$, Zn-Cu;

(j) (1) Hg(O$\overset{O}{\overset{\|}{C}}CH_3$)$_2$, CH$_3$OH, (2) NaBH$_4$; **(k)** HBr; **(l)** Br$_2$, H$_2$O; **(m)** HCl; **(n)** H$^+$ or ROOR or base;
(o) Br$_2$, CH$_3$OH; **(p)** (1) O$_3$, (2) H$_2$S or Zn; **(q)** CH$_3$CH$_2$SH, ROOR.

46. Starting compound is CH$_2$=C(CH$_3$)CH$_2$CH$_2$CH$_3$.

(a) (CH$_3$)$_2$CHCH$_2$CH$_2$CH$_3$

(b) CH$_2$DCD(CH$_3$)CH$_2$CH$_2$CH$_3$

(c) HOCH$_2$CH(CH$_3$)CH$_2$CH$_2$CH$_3$

(d) (CH$_3$)$_2$CClCH$_2$CH$_2$CH$_3$

(e) (CH$_3$)$_2$CBrCH$_2$CH$_2$CH$_3$

(f) BrCH$_2$CH(CH$_3$)CH$_2$CH$_2$CH$_3$

(g) (CH$_3$)$_2$CICH$_2$CH$_2$CH$_3$
(peroxides don't affect HI addition)

(h), **(p)** (CH$_3$)$_2$C(OH)CH$_2$CH$_2$CH$_3$

(i) ClCH$_2$CCl(CH$_3$)CH$_2$CH$_2$CH$_3$ **(j)** ICH$_2$CCl(CH$_3$)CH$_2$CH$_2$CH$_3$

(k) BrCH$_2$C(OCH$_2$CH$_3$)(CH$_3$)CH$_2$CH$_2$CH$_3$

(l) CH$_3$SCH$_2$CH(CH$_3$)CH$_2$CH$_2$CH$_3$ **(m)**

$$CH_2 \overset{O}{\overset{\diagup \diagdown}{-}} C(CH_3)CH_2CH_2CH_3$$

(n) HOCH$_2$C(OH)(CH$_3$)CH$_2$CH$_2$CH$_3$

(o) H$_2$C=O + CH$_3$$\overset{\overset{\textstyle O}{\|}}{C}CH_2CH_2CH_3$ **(q)** (CH$_3$)$_2$C=CHCH$_2$CH$_3$

47. Starting compound is

$$\underset{CH_3}{\overset{CH_3CH_2}{}} C=C \underset{CH_2CH_3}{\overset{H}{}}$$

All chiral products will be formed as racemic mixtures.

(a) CH$_3$CH$_2$CH(CH$_3$)CH$_2$CH$_2$CH$_3$

(b) *Syn* addition: CH$_3$CH$_2$$\overset{\overset{\textstyle D \quad D}{|\quad |}}{C-C}$H with CH$_3$ and CH$_2$CH$_3$

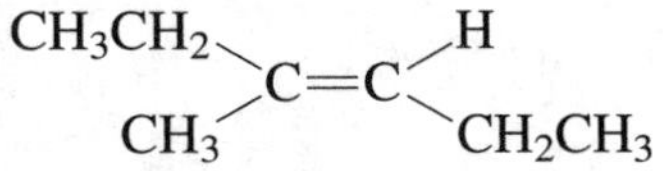

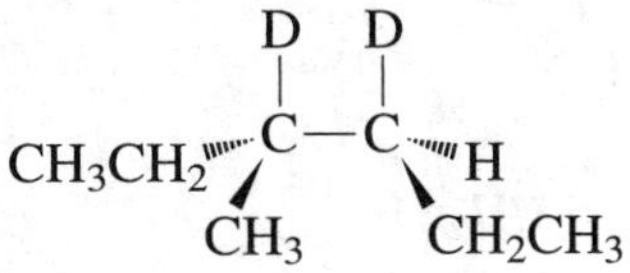

(c) *Syn:* CH$_3$CH$_2$$\overset{\overset{\textstyle H \quad OH}{|\quad |}}{C-C}$H with CH$_3$ and CH$_2$CH$_3$

(d) CH$_3$CH$_2$CCl(CH$_3$)CH$_2$CH$_2$CH$_3$

(e) CH$_3$CH$_2$CBr(CH$_3$)CH$_2$CH$_2$CH$_3$

(f) CH$_3$CH$_2$CH(CH$_3$)CHBrCH$_2$CH$_3$ (mixture of stereoisomers)

(g) CH$_3$CH$_2$CI(CH$_3$)CH$_2$CH$_2$CH$_3$

(h), (p) CH$_3$CH$_2$C(OH)(CH$_3$)CH$_2$CH$_2$CH$_3$

(i) *Anti* addition: CH$_3$CH$_2$$\overset{\overset{\textstyle Cl \quad H}{|\quad }}{C-C}CH_2CH_3$ with CH$_3$ and Cl

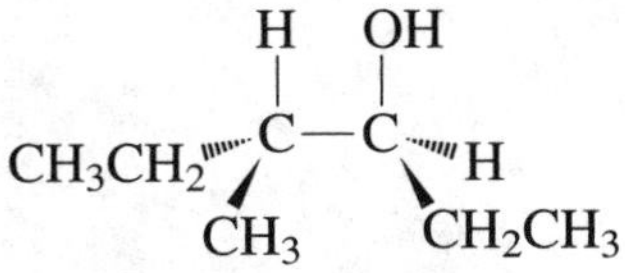

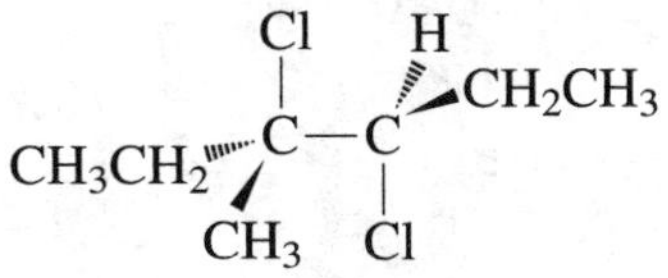

(j) *Anti:* CH$_3$CH$_2$$\overset{\overset{\textstyle Cl \quad H}{|\quad }}{C-C}CH_2CH_3$ with CH$_3$ and I

(k) *Anti:* CH$_3$CH$_2$$\overset{\overset{\textstyle CH_3CH_2O \quad H}{|\quad }}{C-C}CH_2CH_3$ with CH$_3$ and Br

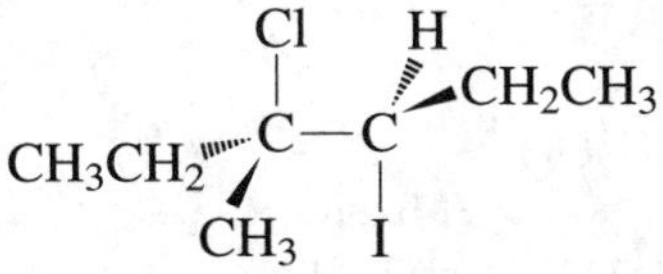

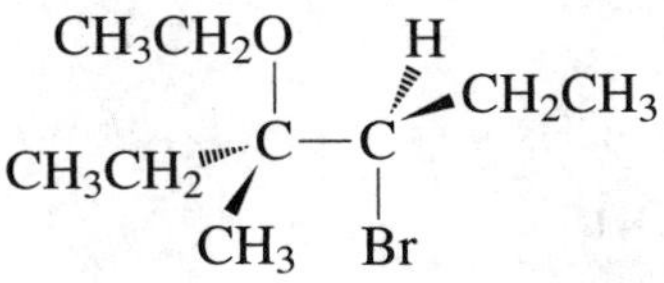

(l) CH$_3$CH$_2$CH(CH$_3$)CH(SCH$_3$)CH$_2$CH$_3$ (mixture of isomers)

(m) 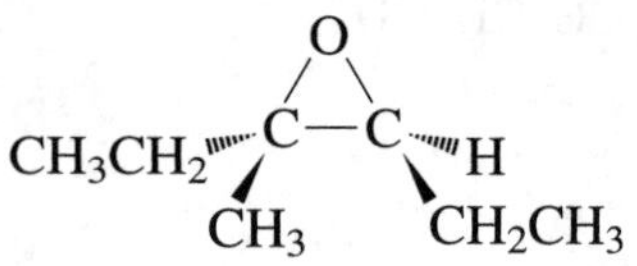 CH₃CH₂ —C—C— H (with epoxide O bridge), CH₃, CH₂CH₃

$$\text{(m)} \quad CH_3CH_2\text{''''}\overset{\displaystyle O}{\underset{\displaystyle CH_3}{C}}\!\!-\!\!\overset{}{\underset{\displaystyle CH_2CH_3}{C}}\text{''''}H$$

(n) *Syn:*

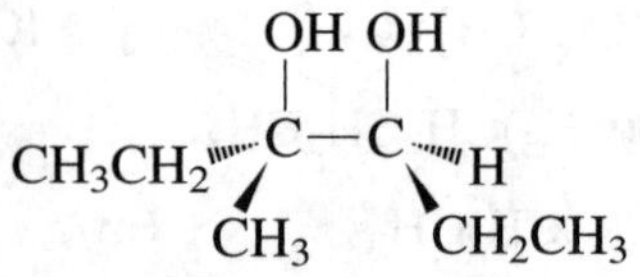

$$CH_3CH_2\text{''''}\overset{\displaystyle OH}{\underset{\displaystyle CH_3}{C}}\!\!-\!\!\overset{\displaystyle OH}{\underset{\displaystyle CH_2CH_3}{C}}\text{''''}H$$

(o)

$$CH_3CH_2\overset{\displaystyle O}{\overset{\|}{C}}CH_3 \;+\; CH_3CH_2\overset{\displaystyle O}{\overset{\|}{C}}H$$

(q)   Mixture of *E* + *Z* isomers of starting compound, and *E* + *Z* isomers of CH₃CH=C(CH₃)CH₂CH₂CH₃ (which is also trisubstituted)

48. Starting compound is ⌬—CH₂CH₃ (cyclopentene with CH₂CH₃ substituent)

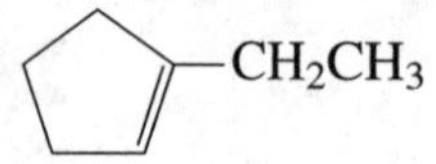

(a) 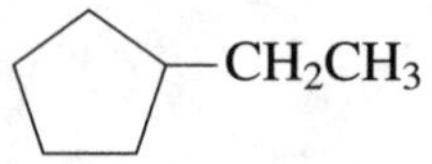 ⌬—CH₂CH₃ (cyclopentane with ethyl)

(b) 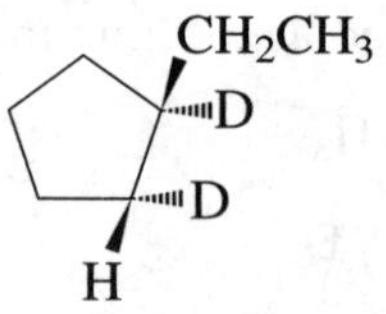 cyclopentane with CH₂CH₃, ''''D, ''''D, H

(c) cyclopentane with CH₂CH₃, ''''H, ''''OH, H

(d) cyclopentane with CH₂CH₃, Cl

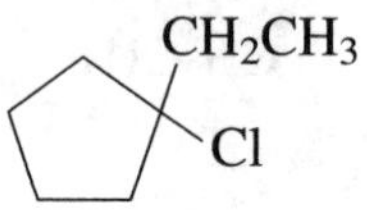

(e) 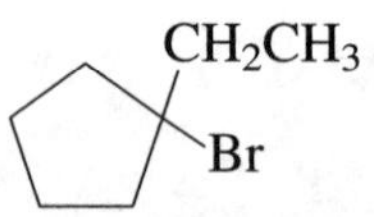 cyclopentane with CH₂CH₃, Br

(f) 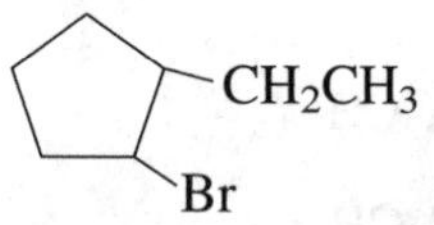 cyclopentane with —CH₂CH₃, Br   (Mixture of isomers)

(g) 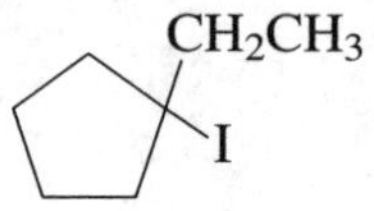 cyclopentane with CH₂CH₃, I

(h), (p) 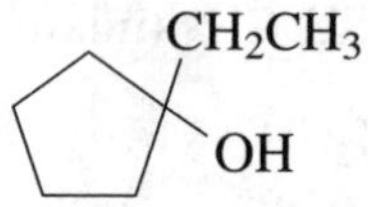 cyclopentane with CH₂CH₃, OH

(i) 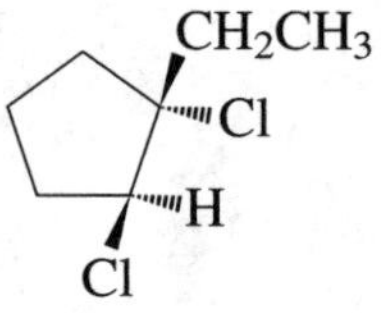 cyclopentane with CH₂CH₃, ''''Cl, ''''H, Cl

(j) 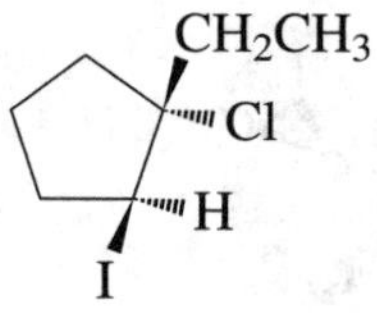 cyclopentane with CH₂CH₃, ''''Cl, ''''H, I

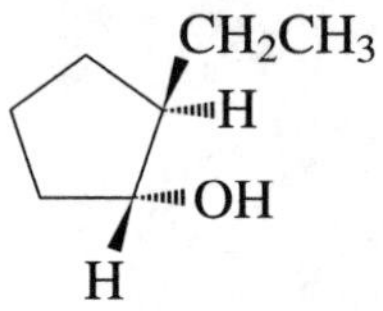

(k) 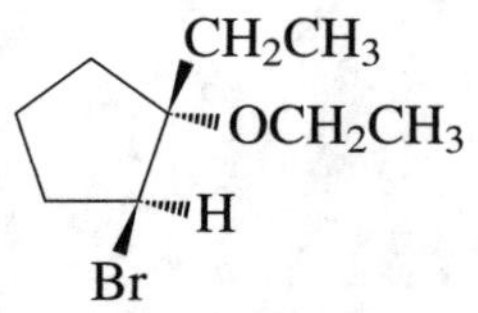 cyclopentane with CH₂CH₃, ''''OCH₂CH₃, ''''H, Br

(l) 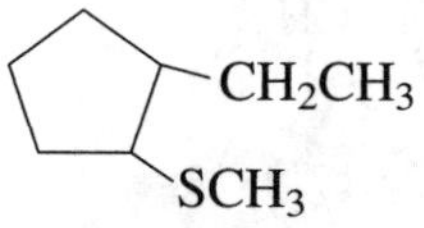 cyclopentane with —CH₂CH₃, SCH₃   (Mixture of isomers)

(m) 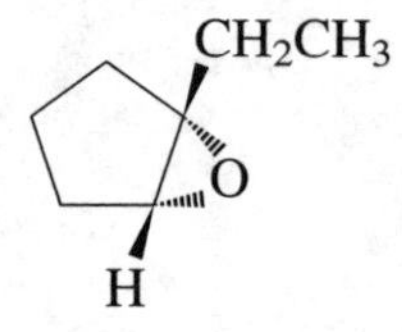 cyclopentane with CH₂CH₃, O (epoxide), H

(n) 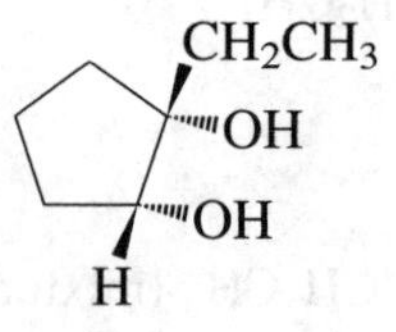 cyclopentane with CH₂CH₃, ''''OH, ''''OH, H

(o) $\overset{O}{\overset{\|}{H}C}CH_2CH_2CH_2\overset{O}{\overset{\|}{C}}CH_2CH_3$

(q) Mixture of starting compound and

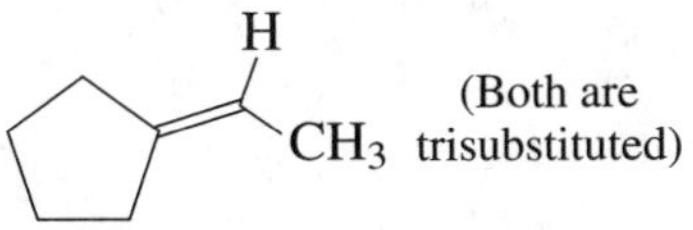

(Both are trisubstituted)

49. The products are shown above; only the mechanisms are given here.

(c)

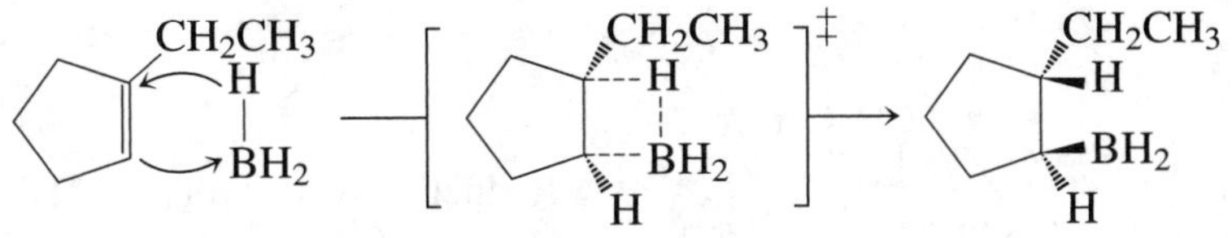

For brevity we show only one BH bond reacting with one molecule of alkene.

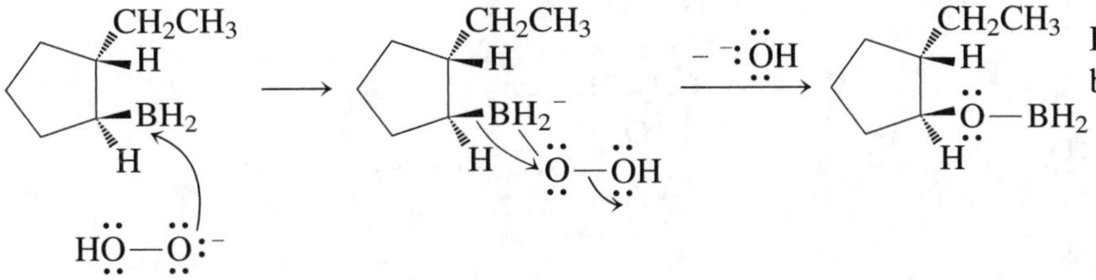

Hydrolysis of the O–B bond gives the product.

(e)

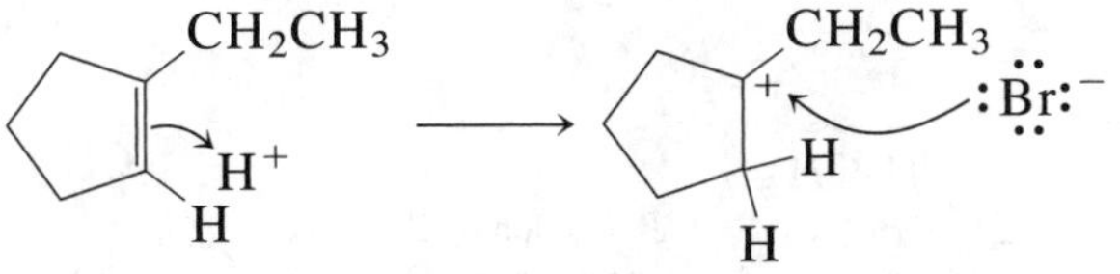

(f) Initiation steps: $RO\!-\!OR \longrightarrow 2\ RO\cdot$ then $RO\cdot \ \ H\!-\!Br\!: \longrightarrow ROH + :Br\cdot$

Propagation steps:

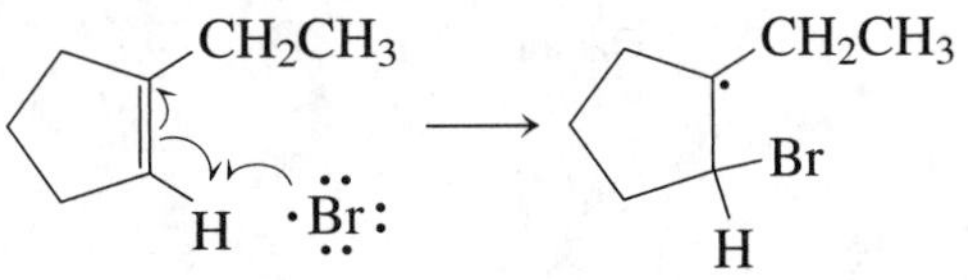

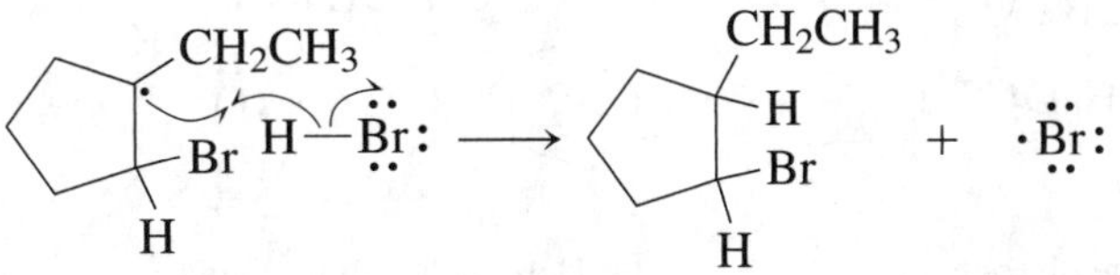

(h)

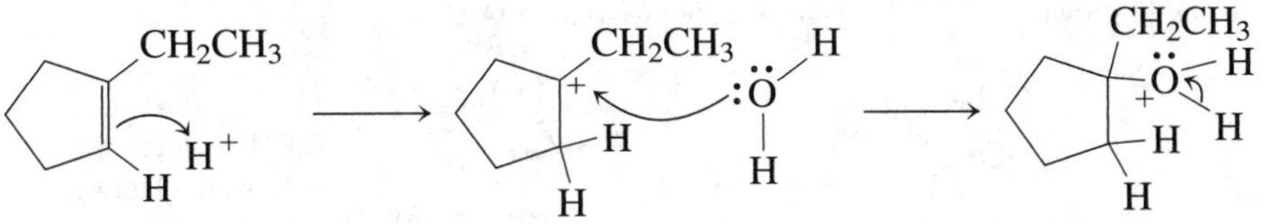

(j)

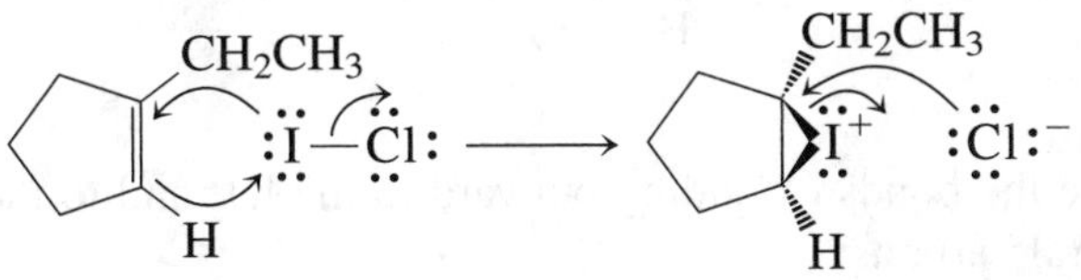

(k)

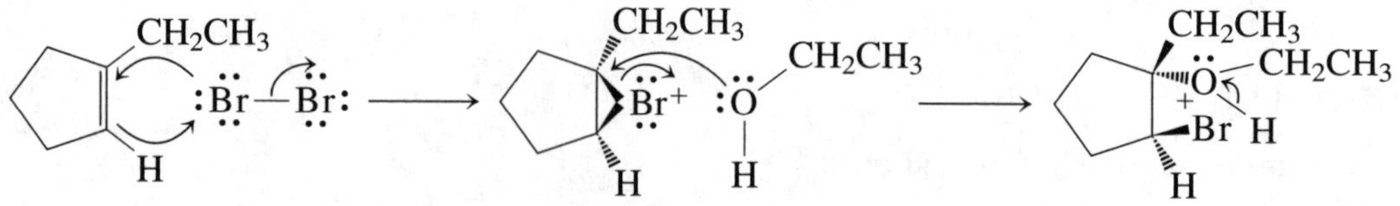

(m)

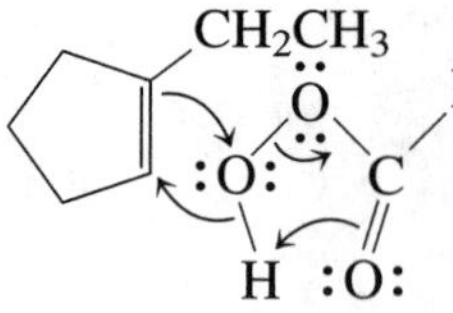

(n)

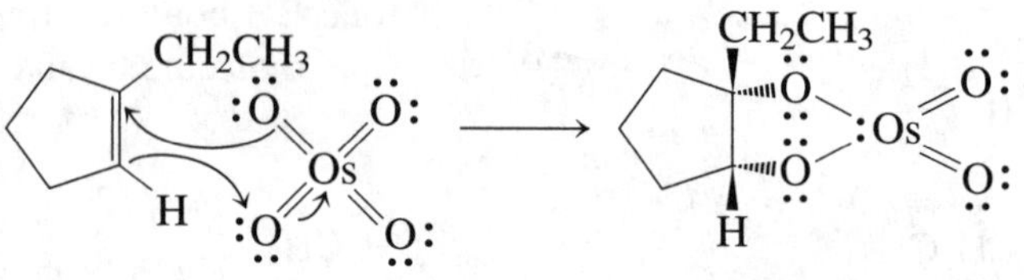

Reductive work-up using H_2S cleaves off Os to give the diol.

(o)

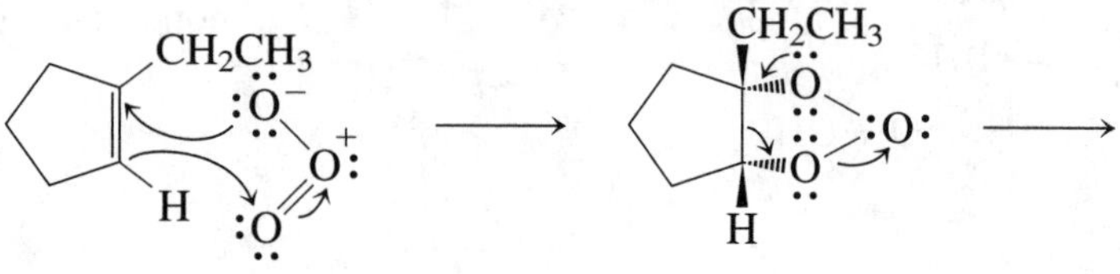

Cycloaddition of O_3
to give molozonide

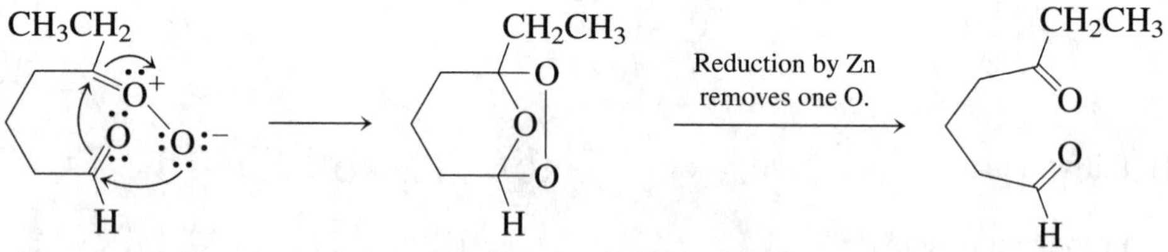

Negative oxygen adds
to carbonyl carbon; then
carbonyl adds to form second
new O—C bond of ozonide.

Ozonide

Reduction by Zn
removes one O.

(p)

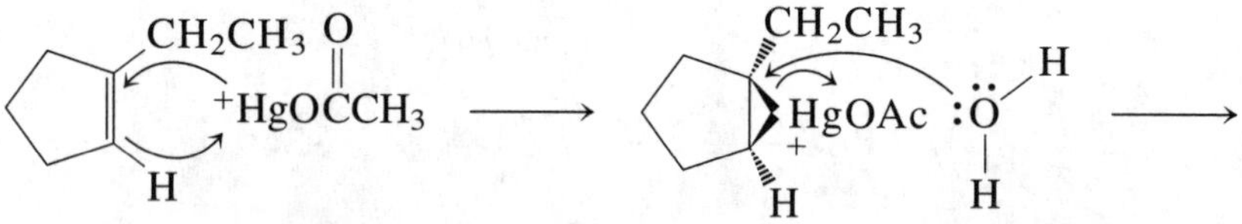

Mercuric acetate dissociates
to give the electrophile shown.

Here we've abbreviated
the acetate group as "OAc."

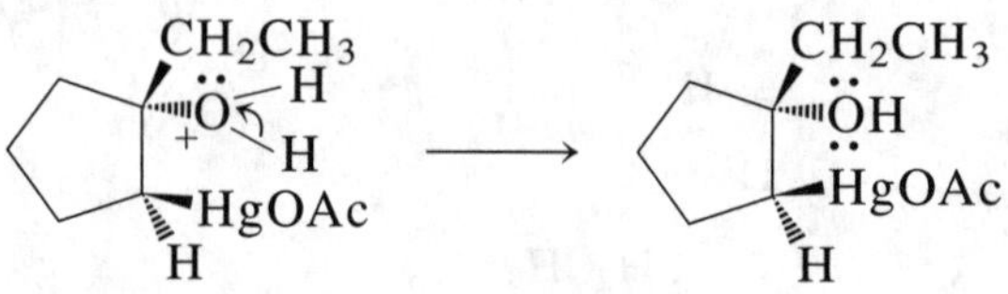

NaBH$_4$ reduction
replaces the HgOAc
by hydrogen.

50. Take the structure in the parentheses, delete the bonds extending outward to the left and to the right, and reform a double bond between the two central carbons:

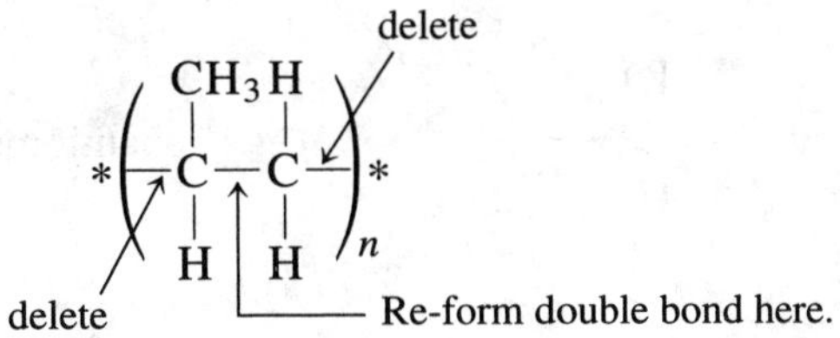

delete

delete

Re-form double bond here.

The answer is propene,

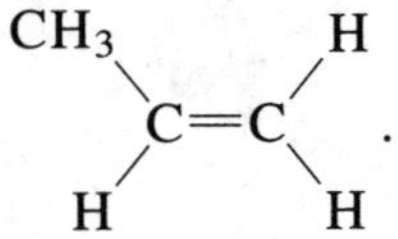

.

51. **(a)** Initial protonation at C1 gives a secondary carbocation that can rearrange to a more stable tertiary cation by a hydride shift. The product is a tertiary alcohol.

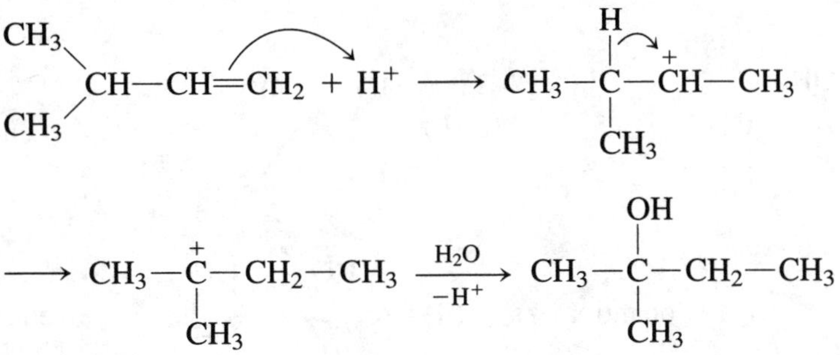

(b) Markovnikov hydration occurs, without rearrangement, giving

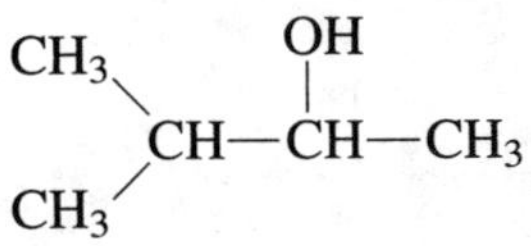

(c) The anti-Markovnikov product is formed.

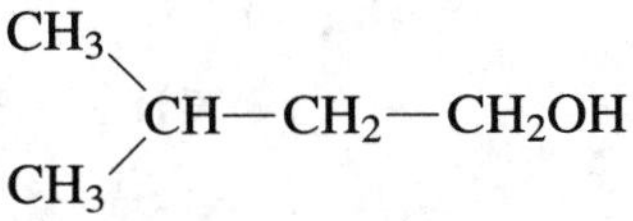

52. **(a)**

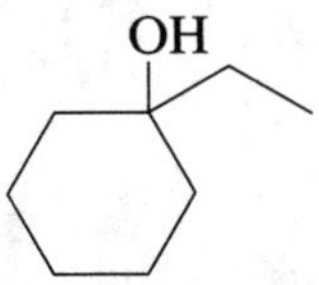

(b)

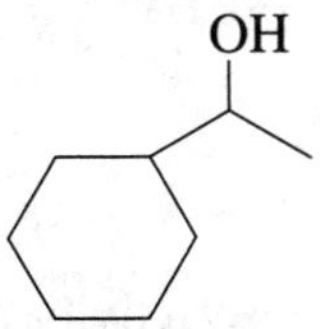

(c)

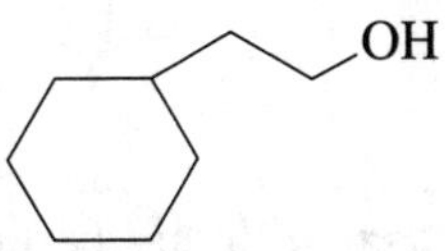

53. **(a)**

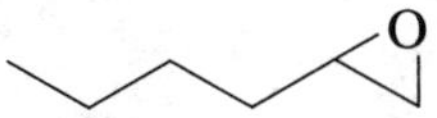

and

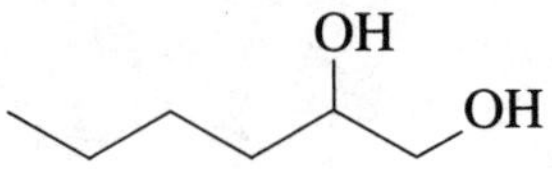

(b)

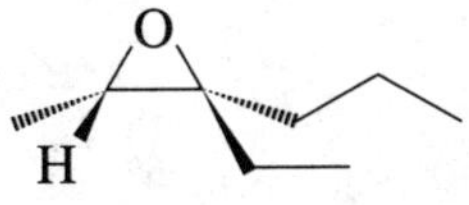

and 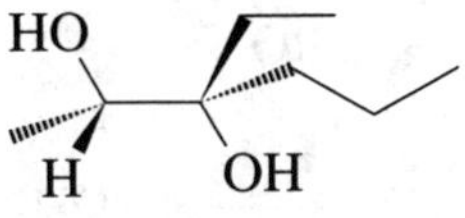 (+ enantiomer)

(c) 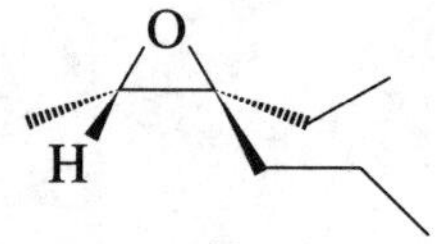 and 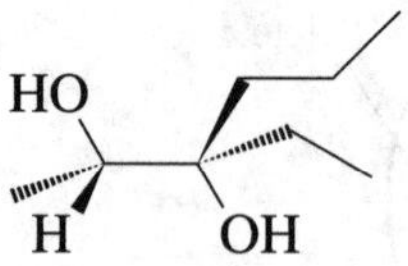 (+ enantiomer)

(d) 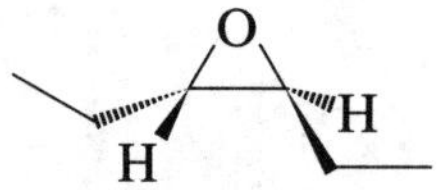 and

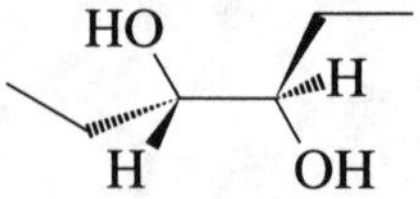

(e) 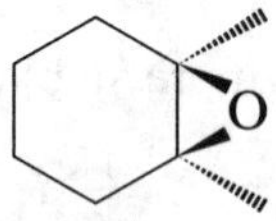 and 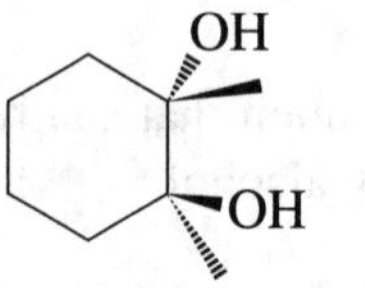 (+ enantiomer)

54. (a) 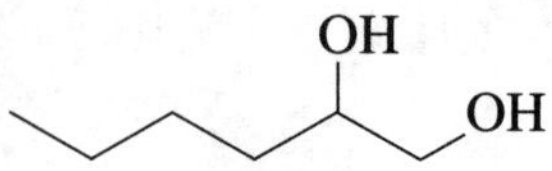   **(b)** 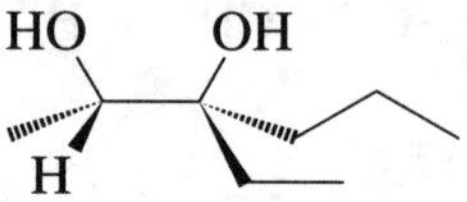 (+ enantiomer)

(c) 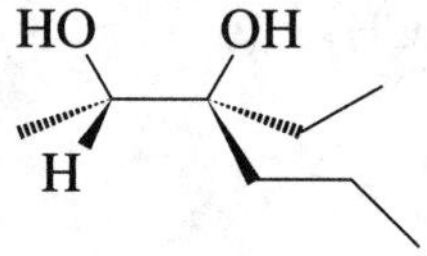 (+ enantiomer)   **(d)** 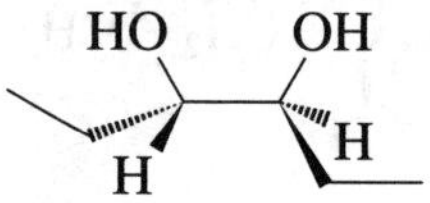 (+ enantiomer)

(e)

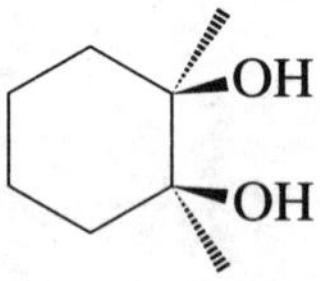

55. (a)    **(b), (c)**

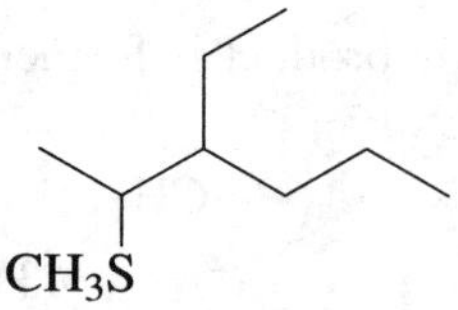

(d) 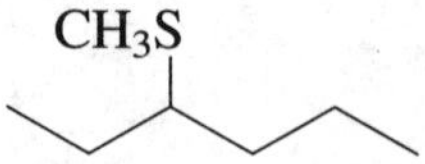   **(e)**

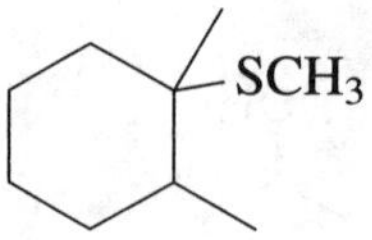

Radical additions do not proceed with any particular stereoselectivity, because the radical intermediates can freely rotate about all C—C bonds.

56. Initiation sequence

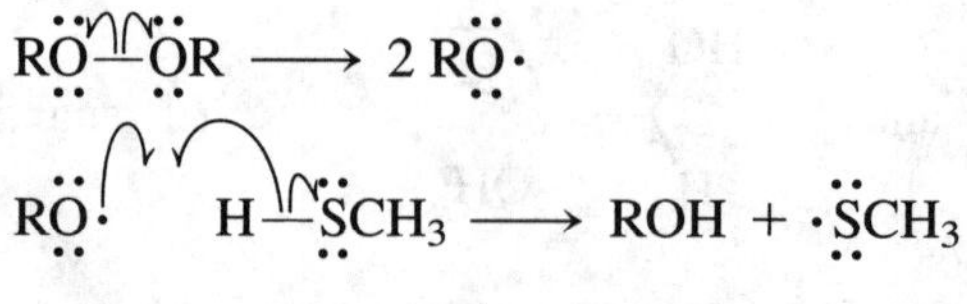

$$RO\!-\!OR \longrightarrow 2\ RO\cdot$$

$$RO\cdot \quad H\!-\!SCH_3 \longrightarrow ROH + \cdot SCH_3$$

Propagation steps

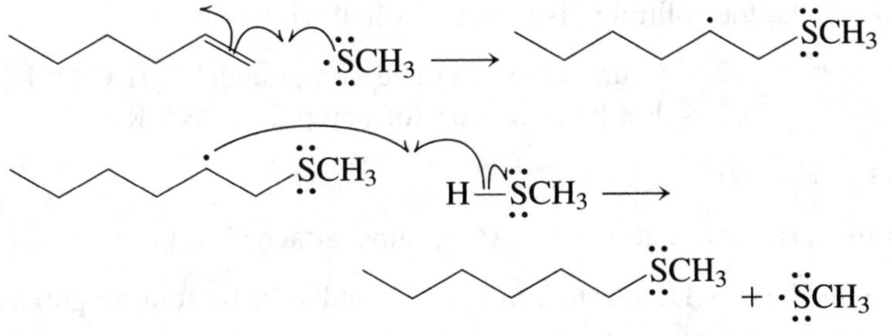

57. All cyclopropane-forming reactions. Watch stereochemistry! In all these reactions the original stereochemical relationships around the double bond are preserved.

(a)

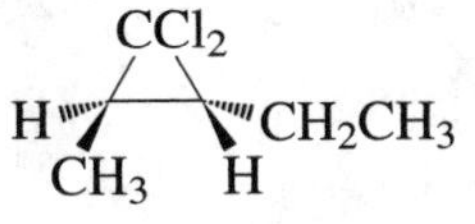

(b)

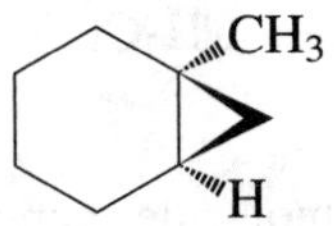

(c)

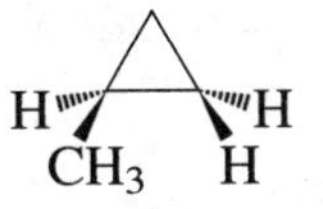

(d)

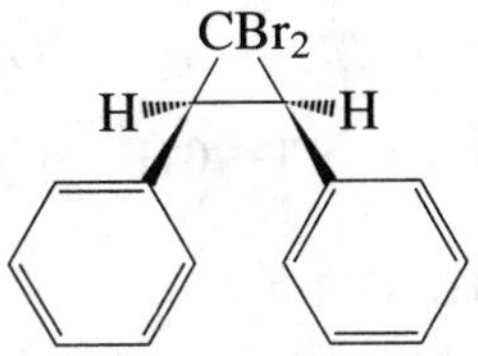

(e)

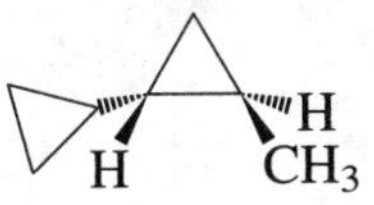

(f) from 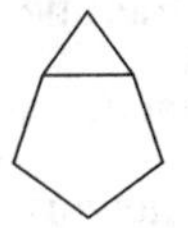 from

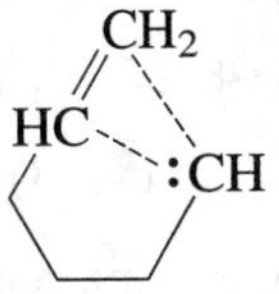

58. **(a), (b)** $H_{sat} = 6 + 2 - 1 = 7$; degree of unsaturation $= (7 - 5)/2 = 1$ π bond or ring. Review Problem 34(b) of Chapter 11 for help, if necessary. The answer is

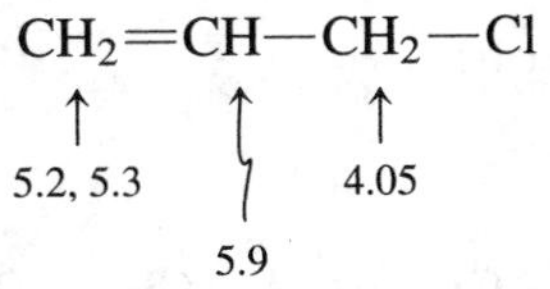

$$CH_2{=}CH{-}CH_2{-}Cl$$

5.2, 5.3   4.05

5.9

IR assignments: 730 (C—Cl); 930 and 980 (terminal alkene); 1630 (C=C); 3090 (alkene C—H).

(c) Yes. This is an expected size of coupling for the following structure (see Table 11-2).

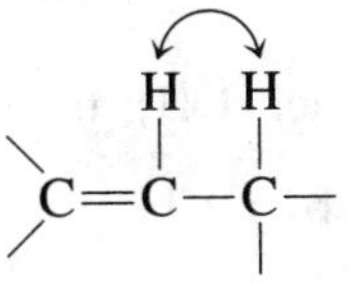

(d) This is a "long-range" (allylic) coupling between the "far" alkene hydrogens and the saturated CH_2 hydrogens:

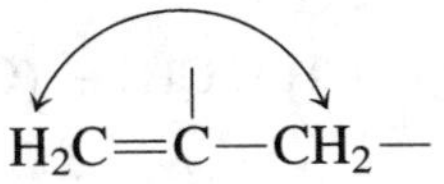

Because of the distance between the hydrogens (that's what we mean by "long-range"), the size of the splitting is small (again, see Table 11-2). And because there are **two** such alkene hydrogens splitting the CH_2 with very similar J values, the splitting gives rise to small triplets.

59. $C_3H_6Cl_2O$: $H_{sat} = 6 + 2 - 2 = 6$; these are saturated compounds. C_3H_5ClO: $H_{sat} = 6 + 2 - 1 = 7$; degree of unsaturation = $(7 - 5)/2 = 1$ π bond or ring for compound "NMR-D."

(a) B $\delta = 2.9$ (s, 1 H): **OH**?

$\delta = 3.65$ (d, 4 H): Two equivalent CH_2 groups, attached to O or Cl and split by a CH

$\delta = 4.0$ (quin, 1 H): **CH,** attached to O or Cl and split by four neighboring H's

Since there are two Cl's but only one oxygen, the two identical CH_2's must be attached to Cl's. So the pieces are

$$(2\times)-CH_2Cl \qquad -\overset{|}{\underset{|}{CH}} \qquad -OH$$

which equals $C_3H_6Cl_2O$. The molecule therefore must be

$$Cl-CH_2-\overset{\overset{\displaystyle OH}{|}}{CH}-CH_2-Cl$$

C $\delta = 2.0$ (broad s, 1 H): **OH**

$\delta = 3.7$ (d, 2 H): CH_2, attached to O or Cl and split by a CH

$\delta = 3.8$ (d, 2 H): CH_2, like the one above, but not equivalent to it

$\delta = 4.1$ (quin, 1 H): **CH,** attached to O or Cl, split by four neighbors

The molecule again has the framework

$$-CH_2-\overset{|}{CH}-CH_2-,$$

but is different from B, so it must be

$$Cl-CH_2-\overset{\overset{\displaystyle Cl}{|}}{CH}-CH_2-OH$$

The extra splitting in the signal at $\delta = 3.7$ derives from the fact that C2 in this compound is a stereocenter.

D All the signals are upfield of $\delta = 5$ ppm. The reaction with base must have led to an oxacyclopropane, not an alkene:

$\delta = 2.7$, 2.9, and 3.2 (multiplets, each 1 H): three **CH**'s?

$\delta = 3.6$ (doublet, 2 H): CH_2, split by a CH

Wait a minute! That adds up to four carbons, but there are only three carbons in the molecule. Two of the three upfield H's must therefore be on the same carbon. So, again, the structure seems to have the framework

$$-CH_2-\overset{|}{CH}-CH_2- \ (C_3H_5),$$

with one Cl and one O to be attached.

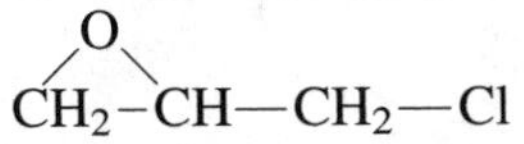

$$\underset{CH_2-CH-CH_2-Cl}{\overset{O}{\triangle}}$$

As expected, the IR spectrum indicates that D has no C=C or O—H bonds; 720 is for the C—Cl; 1260 is new for you, a special oxacyclopropane band for its C—O stretches. C—O stretches are more normally seen around 1000–1100 cm^{-1}.

(b) Electrophilic addition of Cl_2 to the double bond in CH_2=CH—CH_2—Cl gives

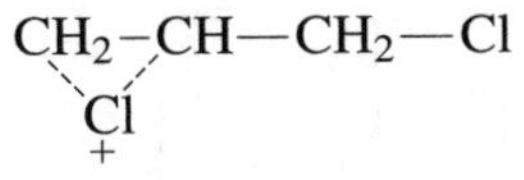

$$\underset{\overset{|}{\underset{+}{Cl}}}{CH_2-CH-CH_2-Cl}$$

Attack of H_2O would ordinarily be favored at the middle (2°) carbon, which should be the site of the more stable cation. However, the other (electron-withdrawing) Cl inductively reduces the preference of 2° cation over 1°, so some chloronium ions react with H_2O at the primary carbon instead.

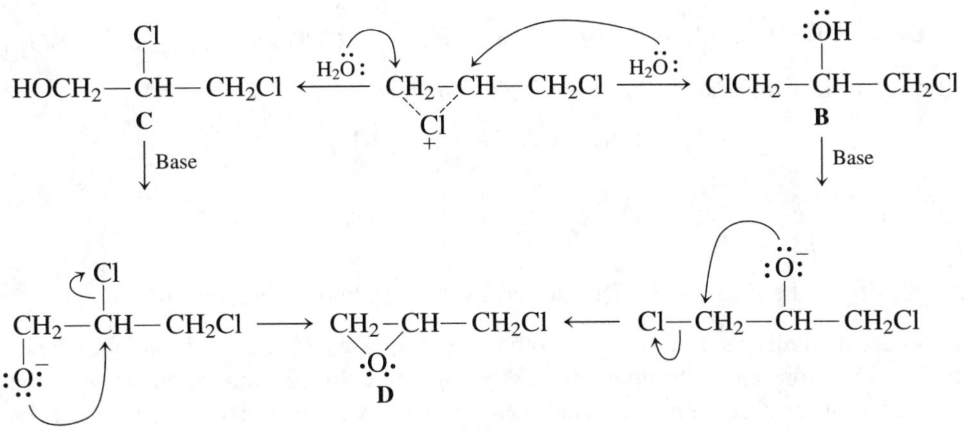

60. **(a)** $H_{sat} = 8 + 2 = 10$; degree of unsaturation $= (10 - 8)/2 = 1$ π bond or ring.

δ = 1.2 (d, 3 H): **CH$_3$**, split by a CH

δ = 1.5 (broad s, 1 H): **OH** (IR at 3360 cm^{-1})

δ = 4.3 (quin, 1 H): **CH,** split by four hydrogens, attached to O

δ = 5.0–6.0: Terminal alkene, —**CH**=**CH$_2$** (IR at 945, 1015, 1665, and 3095 cm^{-1})

The molecule is

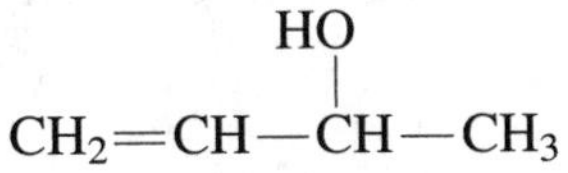

$$\underset{CH_2=CH-CH-CH_3}{\overset{HO}{\overset{|}{}}}$$

This is the only way to put the pieces together.

(b) Upfield signals are assigned above. The alkene assignments are

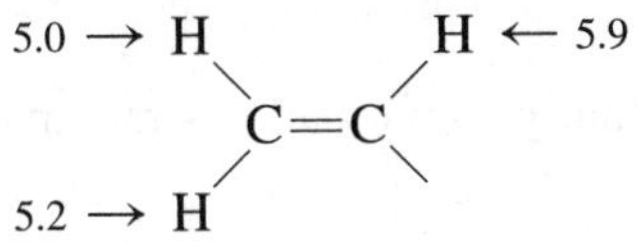

(c) Upfield splittings are assigned in (a). The δ = 5.9 hydrogen shows an 8-line pattern due to coupling to **three nonequivalent hydrogens,** each with a **different** coupling constant.

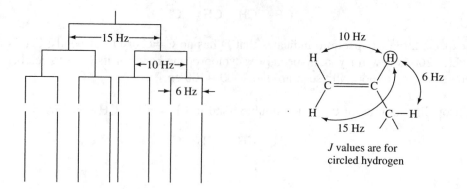

J values are for circled hydrogen

61. Replacement of OH by Cl has occurred:

$$CH_2{=}CH{-}\underset{\underset{\displaystyle Cl}{|}}{CH}{-}CH_3 \qquad \text{(Reaction is ROH + SOCl}_2 \rightarrow \text{RCl + HCl + SO}_2.)$$

F Reaction is just hydrogenation, forming 2-chlorobutane.

$$CH_3{-}CHCl{-}CH_2{-}CH_3$$

    ↑   ↑   ↑   ↑
  1.5(d)   3.9   1.7   1.0(t)
            (sex)   (m)

(The complexity of the signal at δ1.7 is caused by the neighboring stereocenter.)

62. Chlorine occurs naturally as a mixture of isotopes with masses 35 and 37 in an abundance ratio of approximately 3:1. Therefore, in all compounds possessing one chlorine atom about 75% of the molecules contain Cl^{35} and 25% of them contain Cl^{37}. The mass spectrum will thus display two molecular ion peaks two mass units apart, with the lower mass peak (due to the molecules containing Cl^{35}) three times as intense as the higher mass peak (from those containing Cl^{37}).

63. Ozonolysis cleaves double bonds: $C{=}C \rightarrow C{=}O + O{=}C$. So, to determine the alkene from which two carbonyl compounds are derived by ozonolysis, reverse the process in your mind and reattach the carbonyl carbons: $C{=}O + O{=}C \rightarrow C{=}C$.

(a) When only a single carbonyl compound is obtained in ozonolysis, it just means that the alkene precursor was symmetrical, with both "halves" the same. The answer is 2-butene, $CH_3CH{=}CHCH_3$. Both cis and trans isomers give the same ozonolysis products, two molecules of CH_3CHO.

(b) 2-Pentene, $CH_3CH{=}CHCH_2CH_3$. Again, stereochemistry is irrelevant in ozonolysis of alkenes to give carbonyl compounds.

(c) 2-Methylpropene, $(CH_3)_2C{=}CH_2$

(d)

$$\underset{\displaystyle CH_3 \quad\quad H}{\overset{\displaystyle CH_3CH_2 \quad\quad CH_3}{\diagdown C{=}C \diagup}}$$

(Or stereoisomers)

(e)

a cyclopentane ring attached to $C{=}C$ with substituents CH_2CH_3 and H

64. Assume that chiral products are actually to be made as racemic mixtures.

(a) Best bond to make:

$$CH_3CH_2\overset{\overset{\displaystyle O}{\|}}{C}\!\!\overset{\downarrow}{-}\!\!CH(CH_3)_2$$

You need to connect the end carbon of one propene with the middle carbon of the other, so you need to functionalize accordingly.

$$CH_3CH{=}CH_2 \xrightarrow{\text{HCl}} CH_3CHClCH_3 \xrightarrow{\text{Mg, }(CH_3CH_2)_2O} \overset{\overset{\displaystyle MgCl}{|}}{CH_3CHCH_3}$$

$$CH_3CH{=}CH_2 \xrightarrow[\text{2. H}_2\text{O}_2\text{, HO}^-]{\text{1. BH}_3\text{, THF}} CH_3CH_2CH_2OH$$

$$\xrightarrow{\text{PCC, CH}_2\text{Cl}_2} CH_3CH_2\overset{\overset{\displaystyle O}{\|}}{C}H$$

$$\text{product} \xleftarrow{\text{CrO}_3\text{, CH}_2\text{Cl}_2} CH_3CH_2\overset{\overset{\displaystyle OH}{|}}{C}H{-}CH(CH_3)_2$$

(b) Analysis: $CH_3CH_2CH_2{-}\overset{\overset{\displaystyle Cl}{|}}{C}H{-}CH_2CH_2CH_3$

The final product can come from reaction of a suitable reagent such as $SOCl_2$ with 4-heptanol. This, in turn, can come from a Grignard synthesis. Here is an outline of the retrosynthetic plan.

$$CH_3CH_2CH_2\overset{\overset{\displaystyle Cl}{|}}{C}HCH_2CH_2CH_3 \Rightarrow CH_3CH_2CH_2\overset{a}{\underset{}{\Big\{}}\overset{\overset{\displaystyle OH}{|}}{C}H\overset{b}{\underset{}{\Big\}}}CH_2CH_2CH_3$$

$$\Rightarrow H{-}\overset{\overset{\displaystyle O}{\|}}{C}\Big\{CH_2CH_2CH_3 \Rightarrow H_2C\Big\{CH_2CH_2CH_3$$

As mentioned, the synthesis ends easily enough: changing the OH of 4-heptanol to Cl with $SOCl_2$. To form bonds "a" and "b", you must make $CH_3CH_2CH_2MgBr$. That will require an **anti-Markovnikov** addition to propene. Working backward, bond "a" is addition of this Grignard to an aldehyde. Where does the aldehyde come from? It must be from **oxidation** of a primary alcohol with PCC. The alcohol then comes from formation of bond "b" from addition of the same Grignard to formaldehyde. So you have

$$CH_3CH{=}CH_2 \xrightarrow{\text{HBr, peroxide}} CH_3CH_2CH_3Br \xrightarrow{\text{Mg,}(CH_3CH_2)_2O}$$

$CH_3CH_2CH_2MgBr$, first; then,

$$CH_3CH_2CH_2MgBr \xrightarrow[\substack{\text{3. PCC, CH}_2\text{Cl}_2}]{\substack{\text{1. H}_2\text{C}{=}\text{O, }(CH_3CH_2)_2O \\ \text{2. H}^+\text{, H}_2\text{O}}} CH_3CH_2CH_2CHO$$

$$\xrightarrow[\substack{\text{2. H}^+\text{, H}_2\text{O}}]{\substack{\text{1. CH}_3\text{CH}_2\text{CH}_2\text{MgBr,} \\ (CH_3CH_2)_2O}} CH_3CH_2CH_2\overset{\overset{\displaystyle}{}}{-}\underset{\underset{\displaystyle OH}{|}}{C}H{-}CH_2CH_2CH_3 \xrightarrow{\text{SOCl}_2} \text{product}$$

(c) This one requires some extrapolation. You need to figure out how to add a methyl group and an OH to the double bond of an alkene. Thinking back to Section 9-9, oxacyclopropane rings are opened by Grignard reagents to give alcohols. The alkyl group of the organometallic reagent attaches to the carbon next to that of the alcohol functional group.

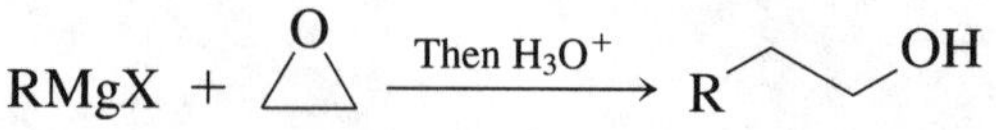

$$RMgX + \triangle O \xrightarrow{\text{Then } H_3O^+} R\diagup\diagup OH$$

If you imagine that the oxacyclopropane was made from an alkene, then the final product contains an alcohol and an R group attached to each end of the original double bond, just what you want. Apply this approach here, using the oxacyclopropane derived from cyclohexene:

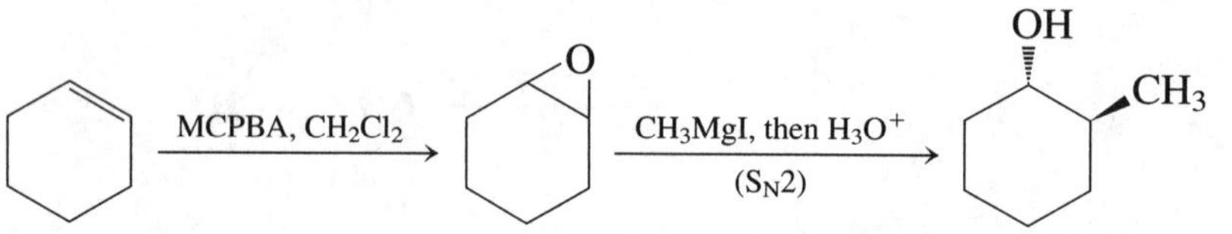

A common (and very wrong) answer seen for problems like this goes as follows: A student will add a halogen and water to an alkene to make a 2-haloalchohol. Then (s)he will react this compound with an organometallic reagent in an attempt to displace the halogen with an alkyl group. This sequence doesn't work because (1) the OH of the haloalcohol instantly destroys the organometallic reagent and (2) even if (1) didn't happen, Grignards and organolithiums **do not form C—C bonds with haloalkanes.** Use the oxacyclopropane route!

65. First 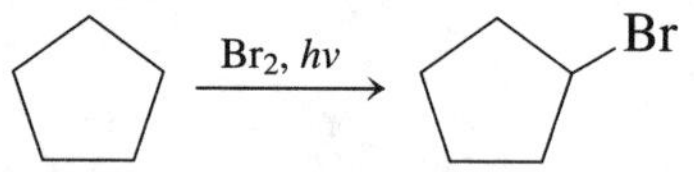 is necessary; you must functionalize the alkane before doing anything else! Note: Later sections of this problem use molecules made in earlier sections.

(a)

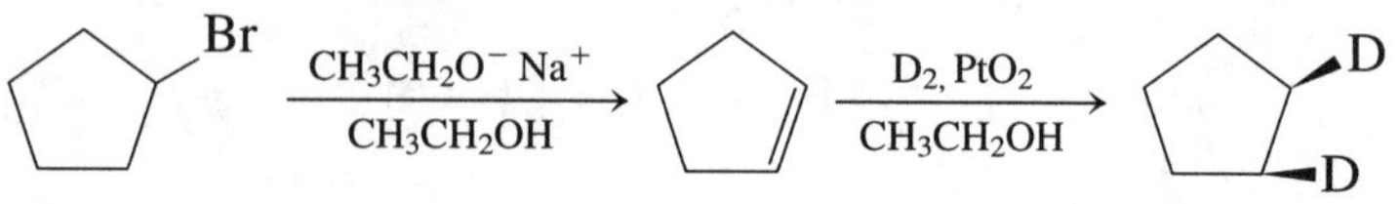

(b) 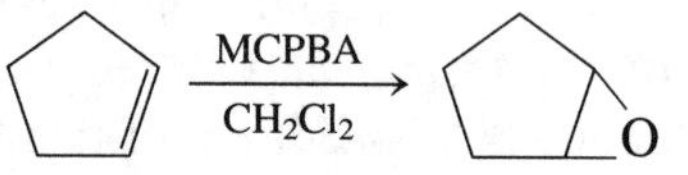 (Stop! If you didn't get this one, try to complete the problem yourself from here before looking at the rest of the answer.) The rest:

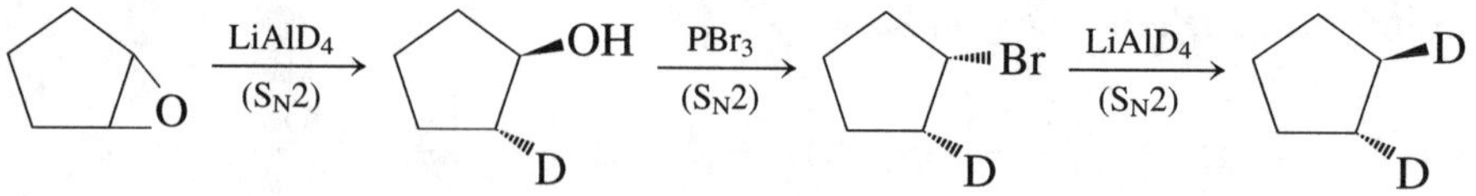

(c)

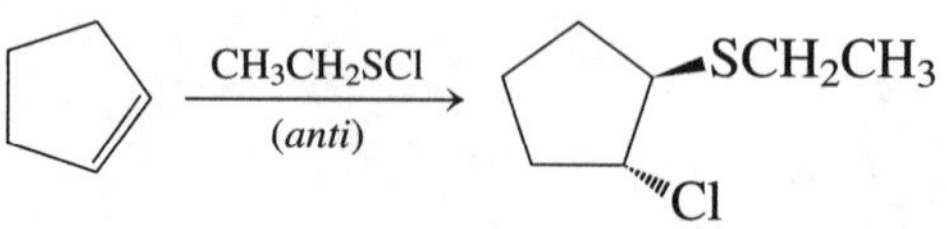

(d)

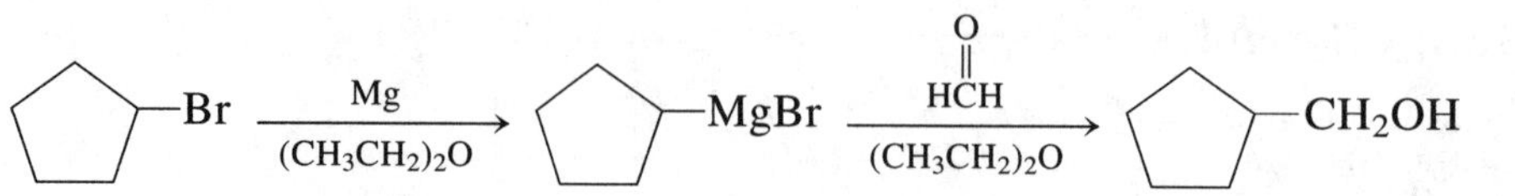

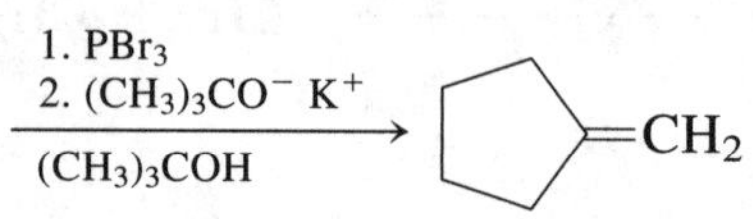

(e) 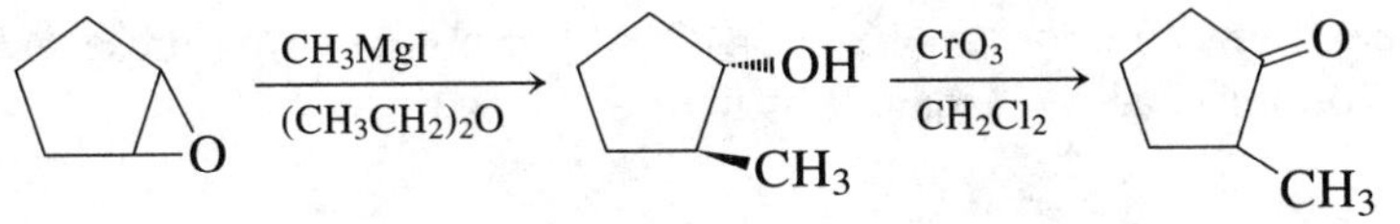 [structure] $\xrightarrow[\text{(CH}_3\text{CH}_2)_2\text{O}]{\text{CH}_3\text{MgI}}$ [structure with ""OH and CH$_3$] $\xrightarrow[\text{CH}_2\text{Cl}_2]{\text{CrO}_3}$ [ketone with CH$_3$]

(f) 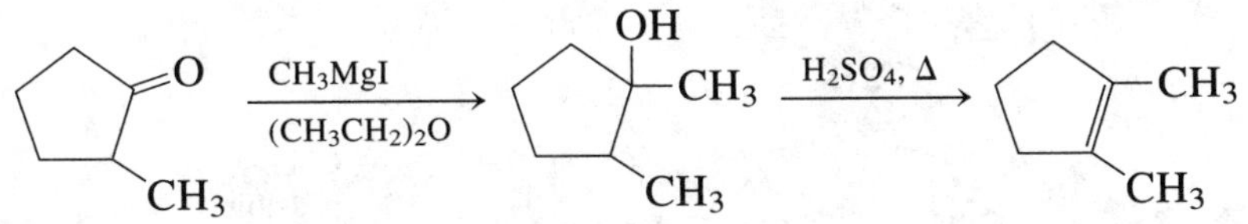 [cyclopentanone with CH$_3$] $\xrightarrow[\text{(CH}_3\text{CH}_2)_2\text{O}]{\text{CH}_3\text{MgI}}$ [structure with OH, CH$_3$, CH$_3$] $\xrightarrow[]{\text{H}_2\text{SO}_4,\ \Delta}$ [cyclopentene with CH$_3$, CH$_3$]

(g) 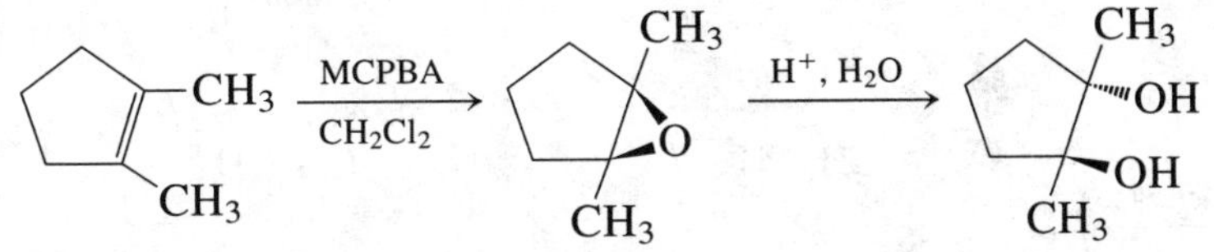 [cyclopentene with CH$_3$, CH$_3$] $\xrightarrow[\text{CH}_2\text{Cl}_2]{\text{MCPBA}}$ [epoxide with CH$_3$, O, CH$_3$] $\xrightarrow[]{\text{H}^+,\ \text{H}_2\text{O}}$ [diol with CH$_3$, ""OH, "OH, CH$_3$]

66. **(a)** $CH_3OCH_2CH_2CH(OCH_3)CH_3$ (Markovnikov ether synthesis)

(b) $HOCH_2\overset{\overset{\displaystyle OH}{|}}{\underset{\underset{\displaystyle CH_3}{|}}{C}}CH_2OH$ (Oxacyclopropane → ring opening)

(c) Rearranges: [cyclobutane with $\overset{+}{C}HCH_3$] $\longrightarrow$ [cyclopentyl cation $\overset{+}{}CHCH_3$] $\longrightarrow$ [cyclopentane with CH$_3$ and I]

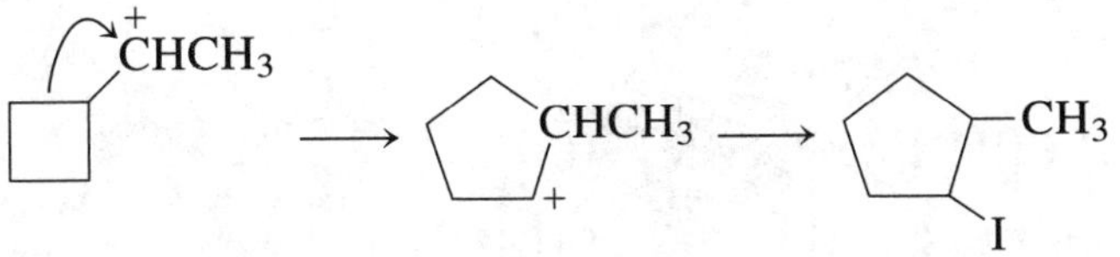

Product is a cis/trans mixture.   (Strain relieved)

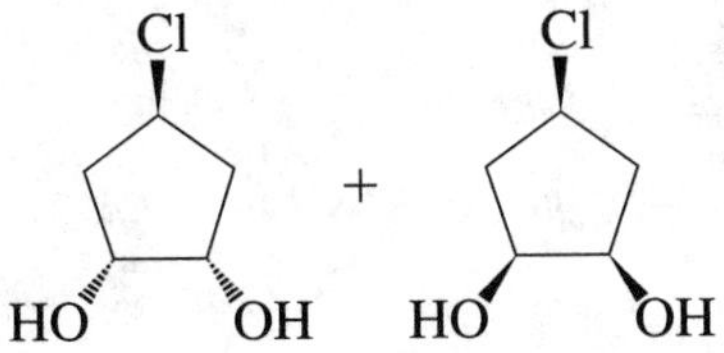

(d) $CH_3CH_2\overset{\overset{\displaystyle O}{\|}}{C}H$ + $H\overset{\overset{\displaystyle O}{\|}}{C}CH_2CH_2\overset{\overset{\displaystyle O}{\|}}{C}CH_2CH_2CH_2CH_2\overset{\overset{\displaystyle O}{\|}}{C}H$

(e) Adds as $Br^+\ {}^-CN$, *anti* stereochemistry: [structure: CH$_3$""C—C"CH$_3$ with CN, H, CH$_3$CH$_2$, Br]

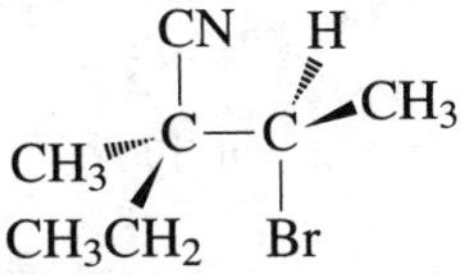

(f) [cyclopentane with Cl, HO, OH] + [cyclopentane with Cl, HO, OH]

(g) $-(CH-CH_2)_{\overline{n}}$ "Polypropylene" with CH$_3$

(h) Lewis structure: $CH_2=CH-\overset{+}{N}\overset{\displaystyle O}{\underset{\displaystyle O^-}{}}$ Positive N implies NO$_2$ group is electron withdrawing.

Therefore, this alkene should be readily polymerized by base, just like super glue (Section 12-15).

Polymer structure will be $-(CH_2-CH)_{\overline{n}}$ with NO$_2$

67. **(a)** $H_{sat} = 14 + 2 = 16$; degree of unsaturation $= (16 - 14)/2 = 1$ π bond or ring present. First do the mechanism, to help you arrive at a reasonable structure:

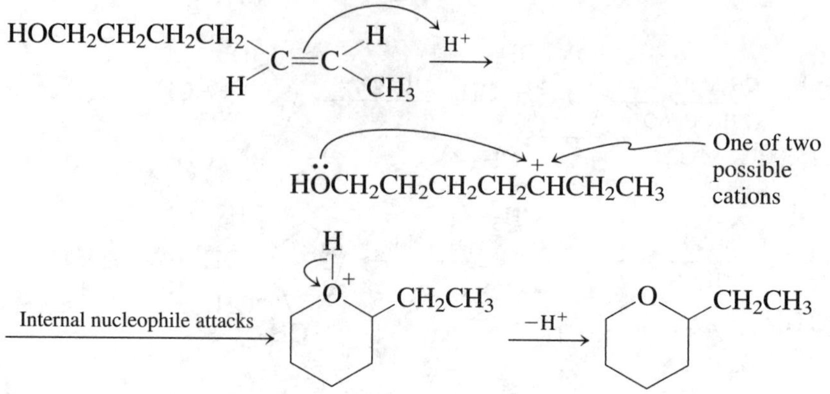

(b) $H_{sat} = 14 + 2 - 1 = 15$; degree of unsaturation $(15 - 13)/2 = 1$ π bond or ring present; IR; no double bonds, no OH groups.

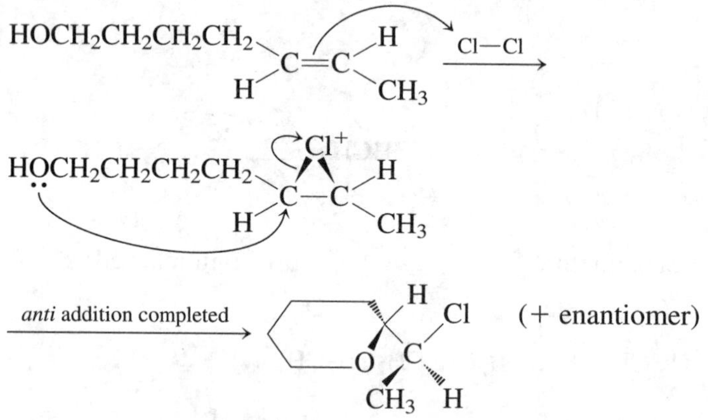

68. Heat or light causes $I_2 \rightarrow 2$ I; then you can have

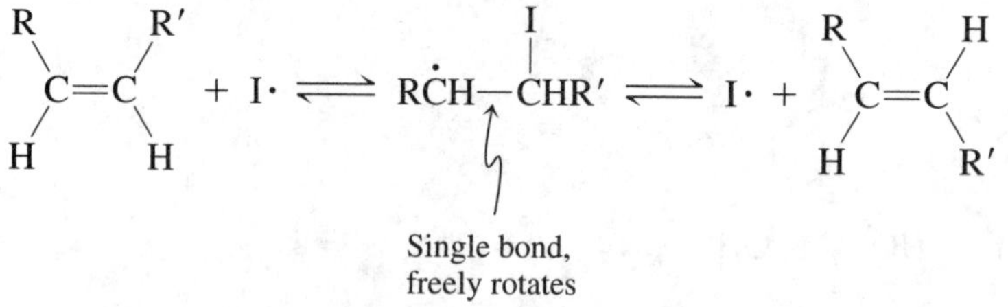

Remember: I_2 addition is endothermic because C—I bonds are weak ($DH° = 52$–53 kcal mol^{-1}). As a consequence, after one I· adds, a second does not attach; the weak C—I linkage simply breaks, regenerating the π bond.

69. $H_{sat} = 20 + 2 = 22$; degree of unsaturation $= (22 - 18)/2 = 2$ π bonds or rings still present. These must both be rings, because the IR spectrum lacks a C=C stretching band. Note only seven signals in the ^{13}C NMR: The product must have greater symmetry than the starting material. Notice, also, **two** signals for C attached to oxygen ($\delta = 69.6$ and 73.5), but the product's formula only contains **one** oxygen. The only conclusion possible is that the product contains an **ether:** C—O—C. How? Think mechanistically.

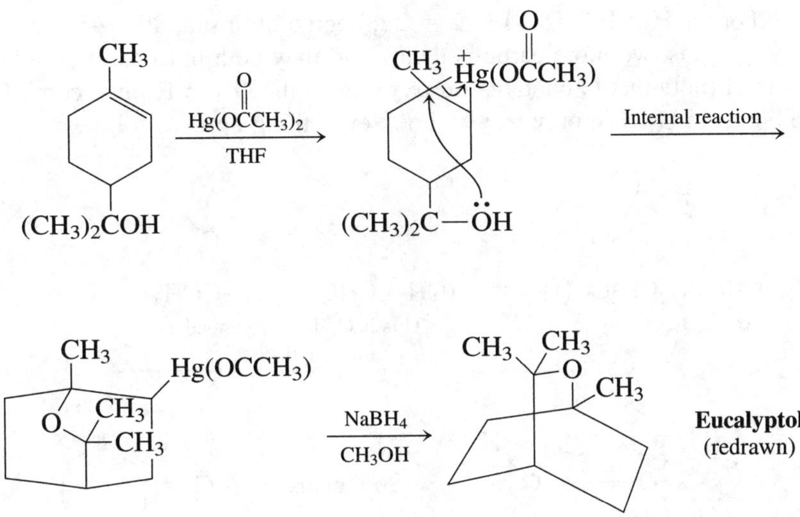
CH3
(CH3)2COH
Hg(OCCH3)2
THF
CH3
Hg(OCCH3)
(CH3)2C—OH
Internal reaction
CH3
Hg(OCCH3)
CH3
CH3
NaBH4
CH3OH
CH3
CH3
O
CH3

Eucalyptol (redrawn)

By the way, eucalyptol is just another name for cineole (Chapter 2, Problem 45).

70. Situation similar to that in problem 44(d).

(a)

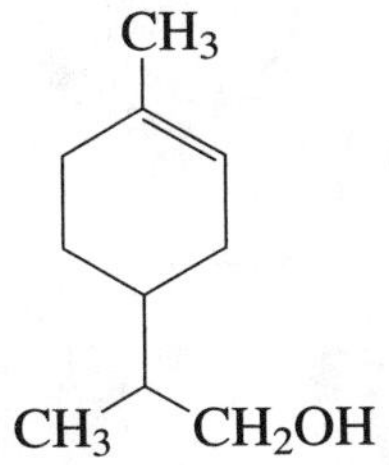
CH3
CH3
CH2OH

BH_3 prefers less substituted double bonds because of their reduced steric crowding.

(b)

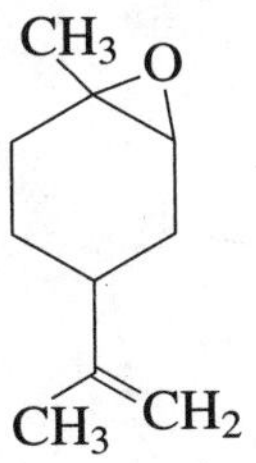
CH3
O
CH3
CH2

Electrophiles such as MCPBA prefer more substituted double bonds because they are more nucleophilic (electron-rich), and their alkyl groups help stabilize both carbocations and carbocation-resembling transition states.

71. A "roadmap" problem. Organize your given information and work from it, stepwise, toward the answers.

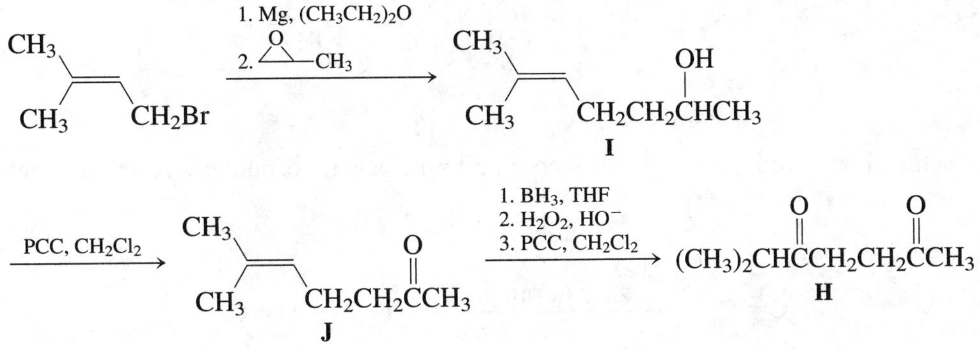
CH3
CH3
CH2Br
1. Mg, (CH3CH2)2O
2. CH3
CH3
CH3
OH
CH2CH2CHCH3
I
PCC, CH2Cl2
CH3
CH3
CH2CH2CCH3
J
1. BH3, THF
2. H2O2, HO−
3. PCC, CH2Cl2
(CH3)2CHCCH2CH2CCH3
H

Now, how about G? For $C_{10}H_{16}$, $H_{sat} = 20 + 2 = 22$; degree of unsaturation $= (22 - 16)/2 = 3\,\pi$ bonds or rings. G only contains two more carbons than H, so they both must form double bonds with the two carbonyl carbons in H (note that G contains **no oxygen,** so the O's in H must come from the ozonolysis of double bonds in G). The only way to hook everything up is as follows:

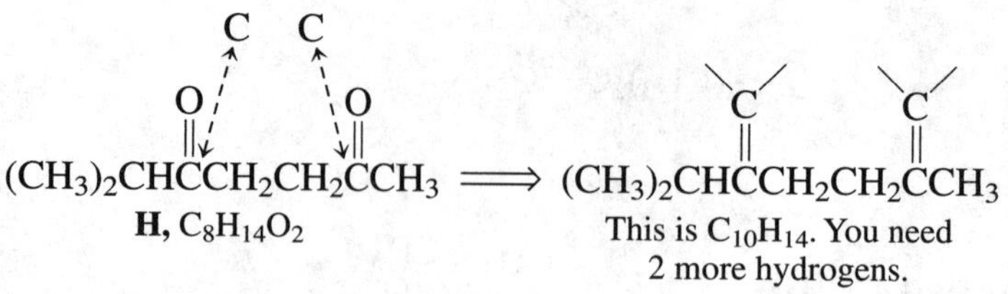

$(CH_3)_2CHCCH_2CH_2CCH_3$ $\Longrightarrow$ $(CH_3)_2CHCCH_2CH_2CCH_3$

H, $C_8H_{14}O_2$

This is $C_{10}H_{14}$. You need 2 more hydrogens.

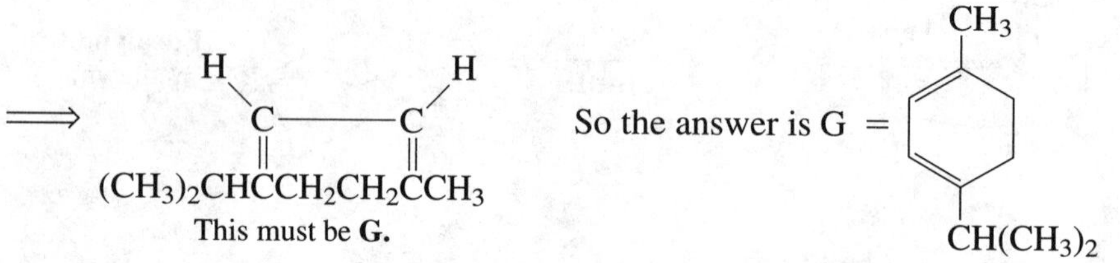

$\Longrightarrow$

$(CH_3)_2CHCCH_2CH_2CCH_3$

This must be **G.**

So the answer is G =

This compound is known as α-terpinene.

72.

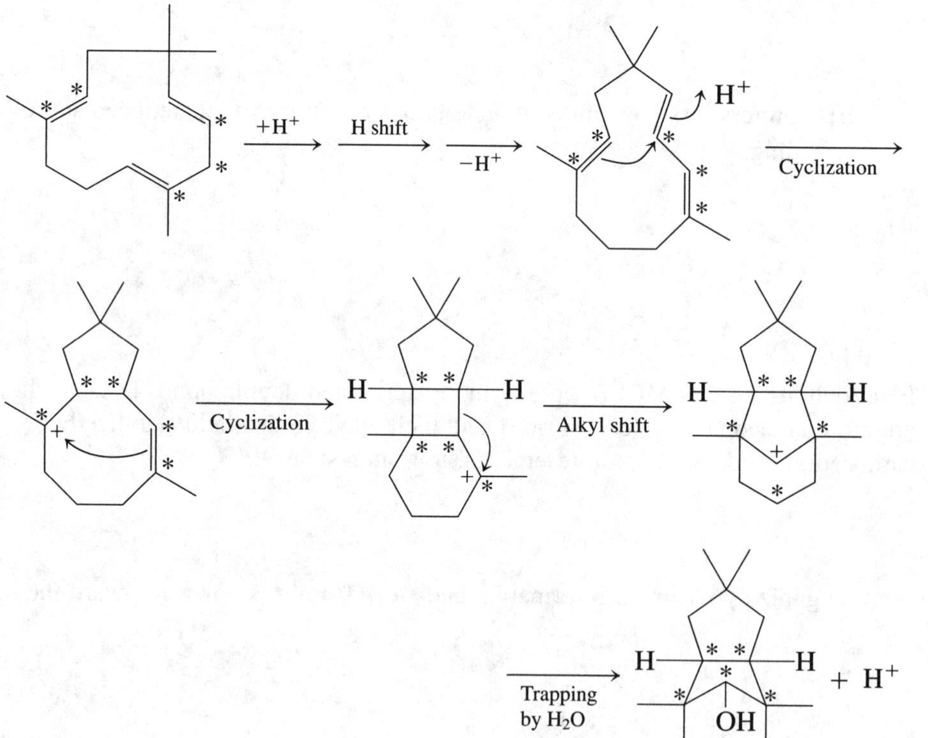

73. Cleave each carbon–carbon double bond according to the general ozonolysis reaction pattern:

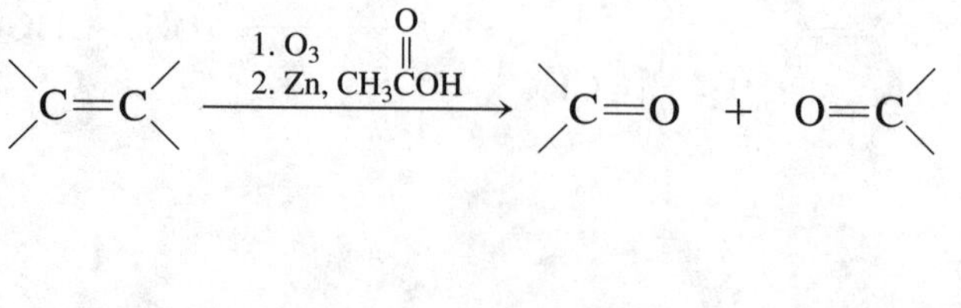

Thus, for humulene we cleave at the three double bonds, giving three "dicarbonyl" compounds that we may draw by copying the structural portions from the starting molecule, as follows:

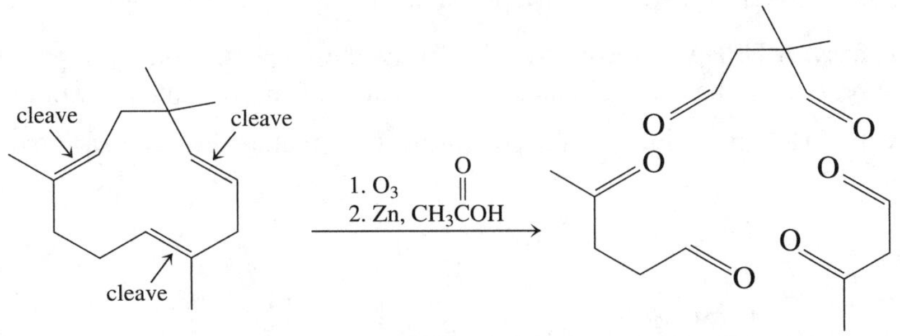

74. $H_{sat} = 30 + 2 = 32$; degree of unsaturation $= (32 - 24)/2 = 4$ π bonds or rings.

Reaction 1 tells you two π bonds are present (hydrogenation only adds two H_2), so there must be two rings. Reaction 2 forms two pieces: formaldehyde (CH_2O) and the triketone shown, which is $C_{14}H_{22}O_3$. These account for all the carbons and hydrogens in caryophyllene. The oxygens came from the ozonolysis. All that's left to decide is how the four carbonyl carbons originally connected up into two alkene double bonds.

Reaction 3 gives the answer to this last question. Hydroboration converts one of the caryophyllene double bonds into an alcohol. Then, ozonolysis cleaves the other to give the diketo-alcohol shown. Working backward from this structure, you can write the following:

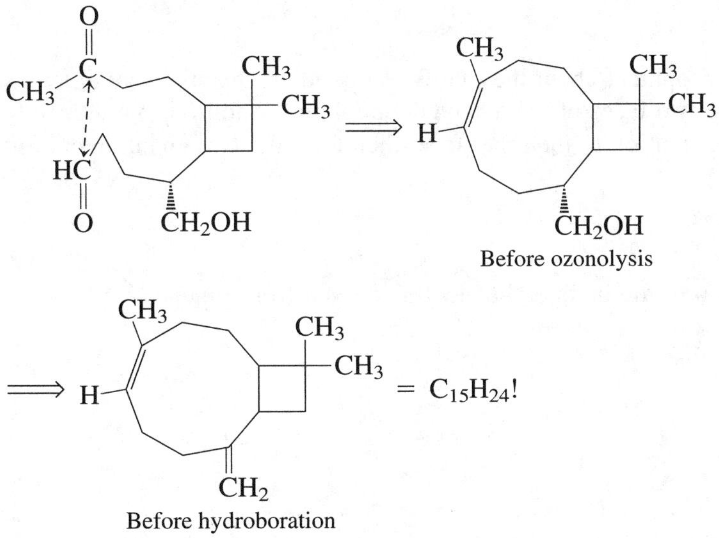

Before ozonolysis

$= C_{15}H_{24}!$

Before hydroboration

The only question left is whether the double bond in the nine-membered ring is cis or trans (more properly, Z or E)—it could be either. In fact, this is the difference between caryophyllene (the E isomer) and the isocaryophyllene (Z):

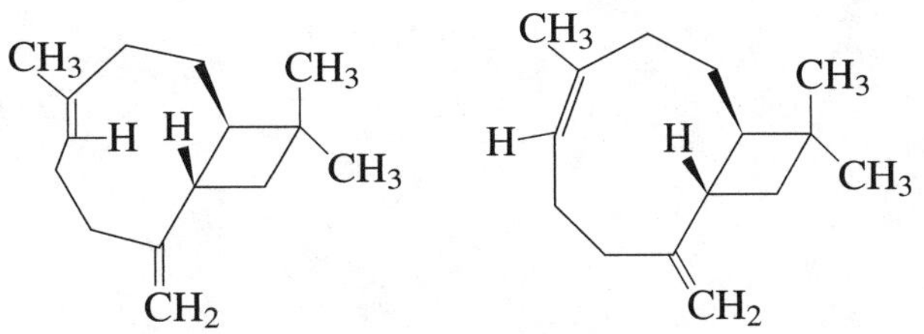

Caryophyllene　　**Isocaryophyllene**

Notice that hydroboration of the bottom double bond does not affect the *E-Z* relationship between these two, but once O_3 cleaves the other alkene, the products are identical.

75. As usual, consider the goal to be a **racemic** product.

(a) This part is easy: MCPBA stereospecifically forms the necessary oxacyclopropane, with the same *Z* geometry present in the original alkene. A haloalcohol–base sequence would work equally well.

(b) Reaction with CH_3MgCl, although nonstereoselective, makes the necessary oxacyclopropane directly.

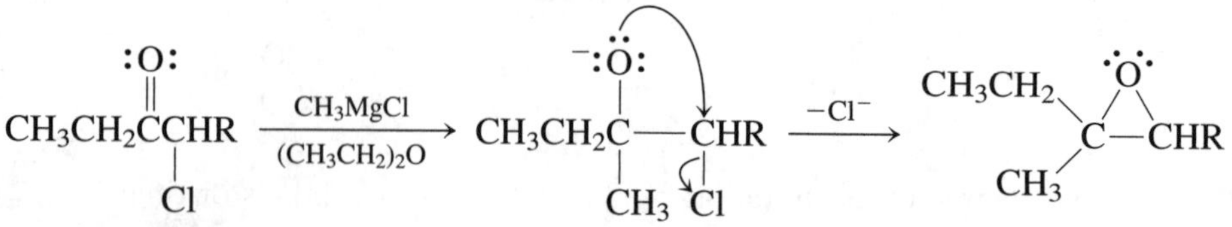

76. (a) Bis(1,2-dimethylpropyl) borane ("Disiamylborane") is the hydroboration product of 2-methyl-2-butene, 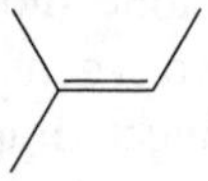 . One molecule of borane normally can add to only two molecules of trisubstituted alkenes because of steric hindrance, leaving one B—H free to add to a third alkene molecule. 9-BBN is the product formed upon hydroboration of either 1,4-cyclooctadiene, , or 1,5-cyclooctadiene, .

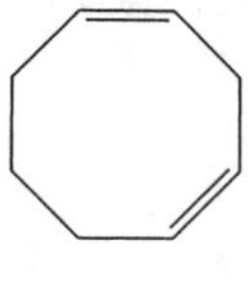

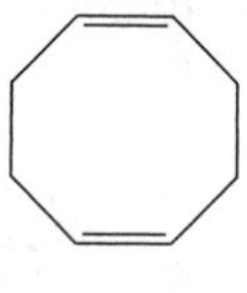

(b) The chiral environment about the free B—H bond creates different amounts of steric hindrance to addition to the two faces of the alkene double bond. Addition to one face forms the enantiomer of the product of addition to the other face; therefore, the two enantiomers form in unequal amounts.

Upon oxidation of the borane, the alcohol derived from pinene, 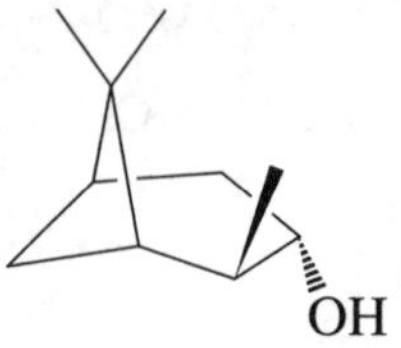 , is also formed.

77. (b)

78. (b)

79. (c)

80. (a)

81. (d)

13

Alkynes: The Carbon–Carbon Triple Bond

Now that the properties of carbon–carbon double bonds have been examined in detail, it's time to have a brief look at their relatives, carbon–carbon triple bonds. Not surprisingly, what you will find will be very similar to what you've just seen. Addition reactions will make up the main portion of triple bond chemistry. An important feature is the hydrogen attached to triply bonded carbon: It is unusually acidic, allowing ready removal by strong bases to form a new and synthetically useful class of carbanions called alkynyl anions.

Outline of the Chapter

13-1 Nomenclature

13-2 Structure and Bonding

13-3 Spectroscopy of the Alkynes
An unusual effect in the NMR due to electron motion.

13-4 Preparation of Alkynes by Double Elimination

13-5 Preparation of Alkynes by Alkylation of Alkynyl Anions
A synthesis of internal alkynes starting from terminal alkynes.

13-6, 13-7, 13-8 Reactions of Alkynes

13-9 Alkenyl Halides
A brief description of these compounds, and an introduction to reactions of organic compounds catalyzed by transition metals.

13-10 Ethyne as an Industrial Starting Material

13-11 Naturally Occurring and Physiologically Active Alkynes

Keys to the Chapter

13-1 and 13-2. Nomenclature, Structure, and Bonding

The alkyne functional group has a geometry simpler than that of the alkenes: It is linear. Thus, no cis-trans isomerism is possible; and in naming the alkynes it is necessary only to indicate the position of the triple bond. When the triple bond is at the end of a chain, it is said to be terminal, and it thus possesses a —C≡C—H unit. This hydrogen is characterized by unexpectedly high-field absorption in the proton NMR spectrum and by a very strong, polarized carbon–hydrogen bond, making it relatively acidic. Section 13-5 examines some consequences of that acidity.

13-3. Spectroscopy of the Alkynes

The high-field position of the hydrogen resonance is surprising at first, but it is the logical result of the cylindrical symmetry of the triple bond, which allows electrons to rotate in a tight circle about its axis. These alkyne signals are often easy to spot because of their characteristic splitting patterns, which arise from long-range coupling to the neighboring nuclei across the triple bond (see, for instance, the alkynyl hydrogen triplet in Fig. 13-5). Other NMR properties of these compounds are pretty normal. Diagnostic bands for $C\equiv C$ and $\equiv C-H$ in the IR spectra of terminal alkynes complement the NMR data, especially when the latter become more complex as a result of overlapping signals. Take note, however, of the weakness or even absence of an IR band for *internal* $C\equiv C$ triple bonds.

13-4. Preparation of Alkynes by Double Elimination

The elimination route to alkynes involves removal of **two** moles of hydrogen halide by strong base from a **di**haloalkane. The most common kind of sequence involves forming the dihaloalkane by addition of halogen to an alkene double bond. Because you already know how to make alkenes, you now have access to alkynes in the following way:

$$-CH_2-CH_2- \xrightarrow[\text{e.g., Br}_2,\, hv]{\text{Halogenation,}} -CH_2-CHBr- \xrightarrow[\text{e.g., KOC(CH}_3)_3]{\text{Elimination,}}$$

Alkane   **Haloalkane**

$$-CH=CH- \xrightarrow[\text{e.g., Br}_2]{\text{Addition,}} -CHBr-CHBr- \xrightarrow[\text{e.g., NaNH}_2]{\substack{\text{Double} \\ \text{elimination,}}} -C\equiv C-$$

Alkene   **1,2-Dihaloalkane**   **Alkyne**

13-5. Preparation of Alkynes from Alkynyl Anions

The other major alkyne preparation is based on the easy accessibility of nucleophilic carbanions from terminal alkynes (Section 13-2). So, a wide variety of internal alkynes may be made from any terminal alkyne, via the scheme

$$R-C\equiv C-H \xrightarrow[\text{base}]{\text{Strong}} R-C\equiv C:^- \xrightarrow[\text{"E}^+\text{"}]{\substack{\text{Any} \\ \text{electrophile}}} R-C\equiv C-E$$

Nucleophile

where "E" is the electrophilic carbon in a primary haloalkane, a strained cyclic ether, or a carbonyl compound.

13-6, 13-7, and 13-8. Reactions of Alkynes

Just as you saw with alkenes, alkynes are subject to a variety of addition reactions. These can occur in two stages: a single addition to make an alkene, and then a second addition to give an alkane derivative. The mechanisms, stereochemistry and regiochemistry, are essentially analogous to those you've already seen. So, you can look back to the reactions of the previous chapter as points of reference. Only an occasional detail or two will be different. When you choose to stop the addition reaction at the alkene stage, there is often the possibility of picking reagents and conditions to allow for specific formation of either the trans or cis alkene product. This flexibility further contributes to the usefulness of alkynes in synthesis.

A special note should be made concerning hydration reactions of alkynes. Remember that hydration of alkenes leads to alcohols. Alkynes can be hydrated, too. Markovnikov addition is achieved with an aqueous acidic Hg(II) catalyst. Anti-Markovnikov addition occurs *via* a modified hydroboration–oxidation sequence. Both initially give *vinylic alcohols* (or *enols*) as products, but these are kinetically and thermodynamically unstable and isomerize to carbonyl compounds in a reaction called *tautomerism.*

The latter is thermodynamically favorable because of the very strong carbon–oxygen double bond that is formed. It is kinetically rapid because the enol O—H bond is acidic and readily deprotonated, allowing the proton eventually to find its way to the nearby carbon. More details concerning the process will be upcoming when carbonyl compounds are discussed. Note that the hydration of alkynes is a new synthesis of aldehydes and ketones.

13-9. Alkenyl Halides

Although the carbon–halogen bonds in alkenyl halides do not participate in the usual substitution reactions, they may be converted into carbon–metal bonds, thus allowing formation of lithium and Grignard reagents that can be used in the synthesis of allylic alcohols by exposure to ketones and aldehydes.

The alkenyl halide carbon–halogen bond is also reactive in a variety of transformations involving transition metals. The Heck reaction utilizes this property to enable the coupling of the halogen-bearing carbon of an alkenyl halide to one of the double-bonded carbons of a second alkene, giving rise to dienes as products. We won't be exploring this kind of chemistry in much depth, partly because it falls outside the scope of mainstream mechanistic organic chemistry. However, take note of the fact that such processes now make up an important part of the so-called synthetic toolbox of the pharmaceutical and medicinal chemist: Reactions such as these have made the syntheses of many types of compounds of therapeutic value much easier than was the case before their discovery and development.

Solutions to Problems

27. (a) (b)

(c)

28. (a) 3-Chloro-3-methyl-1-butyne

(b) 2-Methyl-3-butyn-2-ol

(c) 4-Propyl-5-hexyn-1-ol

(d) *trans*-3-Penten-1-yne

(e) *E*-5-Methyl-4-(1-methylbutyl)-4-hepten-2-yne

(f) *cis*-1-Ethenyl-2-ethynylcyclopentane

29. Bond strengths: ethyne > ethene > ethane. Ethyne C—H bond uses the *sp* orbital from carbon, which overlaps best with the 1*s* orbital on hydrogen. The high (50%) *s* character strongly attracts the bonding electrons to the carbon nucleus. This effect shifts these electrons closer to carbon, thereby enhancing the bond polarity, which follows the same order: $^{\delta-}$C—H$^{\delta+}$ greatest in ethyne. In turn, the greater bond polarity in ethyne contributes to the acidity of the hydrogen (together with the enhanced stability of the conjugate base, the ethynyl anion, also a result of hybridization effects).

It may seem paradoxical to you that the strongest C—H bond is the easiest one to deprotonate. Remember, however, that bond strength relates to homolytic cleavage (to C· and H·), whereas acidity refers to heterolytic cleavage (to C$^-$ and H$^+$).

30. The bond strength should be greatest and the bond length smallest in propyne, again as a result of the *sp* (50% *s* character) orbital at C2.

31. In analogy to alkynes, alkenes, and alkanes, the three compound types differ in the hybridization of the nitrogen atom. Acid strengths vary in the same way:

$$CH_3C\equiv NH^+ > CH_3CH=NH_2^+ > CH_3CH_2NH_2^+$$

32. Stability order is cyclopentene > 1,4-pentadiene > 1-pentyne. Cyclopentene has the most σ bonds, which are generally stronger than π bonds. 1,4-Pentadiene and 1-pentyne both have two π bonds, but the alkyne is of higher energy. Notice (Section 13-2) that heats of hydrogenation for alkynes are 65–70 kcal mol^{-1}, or 32.5–35 kcal mol^{-1} per π bond, whereas the normal range for alkenes is 27–30 kcal mol^{-1} (Section 11-10).

33. (a) 3-Heptyne > 1-heptyne (internal more stable than terminal)

(b) Stability decreases from left to right. The first two, isomers of propynylcyclopentane, follow the order "internal more stable than terminal." The last, cyclooctyne, despite being internal, is less stable than either as a result of bond angle strain. Make a model: The alkyne carbons cannot have 180° bond angles. The compound has actually been made, but it doesn't hang around very long and has a strain energy in excess of 20 kcal mol^{-1}.

34. Degree of unsaturation is calculated for each compound.

(a) $H_{sat} = 12 + 2 = 14$; degree of unsaturation = $(14 - 10)/2 = 2$ π bonds or rings. NMR looks like an ethyl group: **CH$_3$** (t, δ = 1.0) next to **CH$_2$** (q, δ = 2.0). Because the molecule has 10 H's, there must be two equivalent ethyl groups, 2 CH$_3$CH$_2$—, adding up to C$_4$H$_{10}$. Two C atoms are all that are left to account for. To get two degrees of unsaturation, make a triple bond between them (2 π bonds):

$$2\ CH_3CH_2—\ \text{and}\ —C\equiv C—\ \Rightarrow\ CH_3CH_2—C\equiv C—CH_2CH_3$$

(b) $H_{sat} = 14 + 2 = 16$; degree of unsaturation = $(16 - 12)/2 = 2$ π bonds or rings. IR: terminal —C≡CH. NMR:

δ = 0.9 (t, 3 H) ⇒ **CH$_3$**, next to CH$_2$

δ = 1.3 (m, 4 H) ⇒ ?

δ = 1.5 (quintet, 2H) ⇒ **CH$_2$**, with CH$_2$ groups on both sides

δ = 1.7 (t, small *J*, H) ⇒ Aha! How about

$$\overset{\frown}{\text{Split by}}$$
$$H—C\equiv C—CH_2— \quad \text{(Compare Figure 13-5.)}$$

δ = 2.2 (m, 2H) ⇒ Perhaps the CH$_2$ referred to here?

So far, you have CH$_3$—CH$_2$— and —CH$_2$—C≡CH, or C$_5$H$_8$; you need C$_2$H$_4$ more. The simplest way is CH$_3$CH$_2$CH$_2$CH$_2$CH$_2$C≡CH, 1-heptyne.

(c) The molecular formula is C_5H_8O. How? Briefly:

Carbon, 71.4% of 84 = 60; 60/12 (atomic mass of C) = 5

Hydrogen, 9.6% of 84 = 8; 8/1 (atomic mass of H) = 8

Oxygen, 19% (the remainder) of 84 = 16; 16/16 (atomic mass of O) = 1

Double check using exact masses from Table 11-5:

5(12.00000) + 8(1.00783) + 15.9949 = 84.05754

H_{sat} = 10 + 2 = 12; degree of unsaturation = (12 − 8)/2 = 2 π bonds or rings. IR: —C≡C— stretch shows at 2100 cm^{-1}, broad band from 3200–3500 cm^{-1} suggest —O—H. NMR, focus on the signals with the simplest splitting patterns first:

δ = 1.8 (broad s, 1 H) $\Rightarrow$ **OH, broad** singlet gives it away

δ = 3.7 (t, 2 H) $\Rightarrow$ **CH$_2$**, next OH (chemical shift tells you this), and also next to another CH$_2$ (triplet splitting tells you that)

δ = 1.9 (t, 1 H) $\Rightarrow$ **C≡CH** (narrowness of splitting is typical), "long-range" coupled to a CH$_2$ on the other side of the triple bond.

Let's see what we know so far. We have figured out that the molecule contains the two pieces HO—CH$_2$—CH$_2$— and —CH$_2$—C≡CH. Add them up and you get C_5H_8O: that's all there is in the formula, so just put them together: HO—CH$_2$—CH$_2$—CH$_2$—C≡CH. The two middle CH$_2$ groups are responsible for the two signal sets that we didn't bother to try to interpret because they were more complicated. Try to figure out on your own why they look the way they do.

35. ≡C—H of terminal alkyne has $\tilde{v}_{C-H} \approx$ 3300 cm^{-1}.

(a) D—C≡CCH$_2$CH$_2$CH$_2$CH$_2$CH$_2$C≡C—D   **(b)** C≡C—D ($\tilde{v}_{C-D}$)

(c) Before reaction, m_1 is H (mass = 1) and m_2 is C$_9$H$_{11}$ (mass = 119). Rewrite the Hooke's law equation as $\tilde{v}^2 = k^2 f(m_1 + m_2)/m_1 m_2$. So $(3300)^2 = k^2 f(120/119)$, or $k^2 f = 1.1 \times 10^7$. Because k and f are assumed to be constant, use this value for $k^2 f$ to predict $\tilde{v}^2$ for the product. Now m_1 is D (mass = 2), so $\tilde{v}^2 = (1.1 \times 10^7)(122/240) = 5.6 \times 10^6$ and predicted $\tilde{v}_{C-D}$ = 2366 cm^{-1}. The discrepancy of about 10% is typical and due to changes in k and f.

36. (a) CH$_3$CH$_2$CH(CH$_3$)C≡CH   **(b)** CH$_3$OCH$_2$CH$_2$CH$_2$C≡CCH$_3$

(after aqueous work-up)

(c)

(d) The reverse of the *meso* compound, this gives

37. (a) *trans*-3-octene, via two sequential one-electron reductions as described in Section 13-6:

(b) Upon sodium/liquid ammonia reduction of a triple bond, the trans stereochemistry of the double bond that normally results is determined in the first two steps of the mechanism. Addition of one electron gives the alkyne radical anion, in which the two substituents on the original alkyne carbons can exist in either a cis or a trans arrangement. In the reduction of an acyclic alkyne, the trans radical anion is the more stable because it suffers from less steric hindrance than the cis isomer. Beginning with cyclooctyne, however, the reverse is the case: The cis radical anion is more stable because angle, torsional, and other strain effects combine to raise the energy of its trans counterpart:

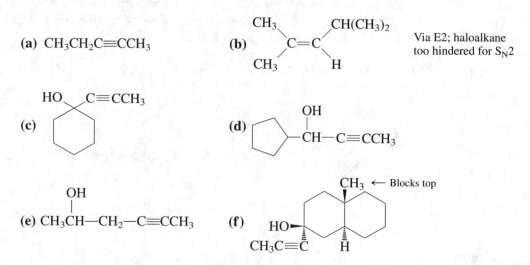

The *cis*-cyclooctenyl radical is formed preferentially, and ultimately goes on via reduction by a second electron to give *cis*-cyclooctene.

38. All products are those obtained after aqueous work-up.

(a) $CH_3CH_2C{\equiv}CCH_3$

(b)

$$CH_3 \diagdown C{=}C \diagup CH(CH_3)_2$$
$$CH_3 \diagup \qquad \diagdown H$$

Via E2; haloalkane too hindered for S_N2

(c)

$$HO \diagdown \; C{\equiv}CCH_3$$

(d)

$$\text{(cyclopentyl)}{-}\underset{\displaystyle OH}{CH}{-}C{\equiv}CCH_3$$

(e) $CH_3\underset{\displaystyle \overset{|}{OH}}{CH}{-}CH_2{-}C{\equiv}CCH_3$

(f)

CH₃ ← Blocks top

HO—

$CH_3C{\equiv}C$ H

39. Since you are not told otherwise, assume a racemic mixture for the (chiral) oxacyclopropane starting material. Write out the answer for one enantiomer; then indicate that the product is racemic. The two carbons of the oxacyclopropane are equivalent (rotate a model of the molecule 180° about a vertical axis to prove this to yourself), so it makes no difference which of the two carbons you attack with the nucleophilic carbon of the lithium reagent. This is ring-opening under basic conditions and follows an S_N2 mechanism with inversion at the site of nucleophilic attack (Section 9-9).

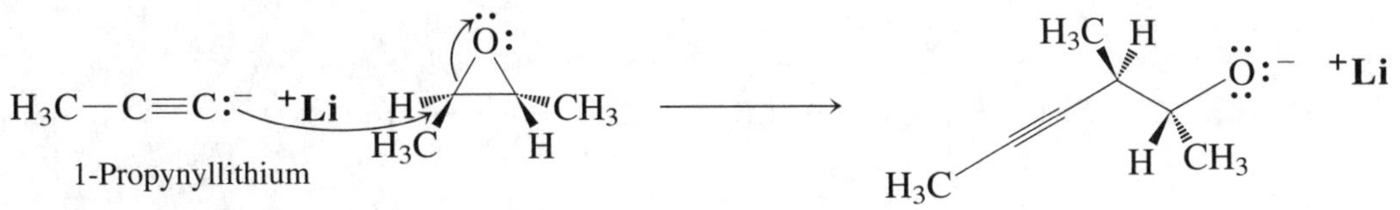

1-Propynyllithium

trans-2,3-Dimethyloxacyclopropane
(only one of the two enantiomers is shown)

(+ Enantiomer)

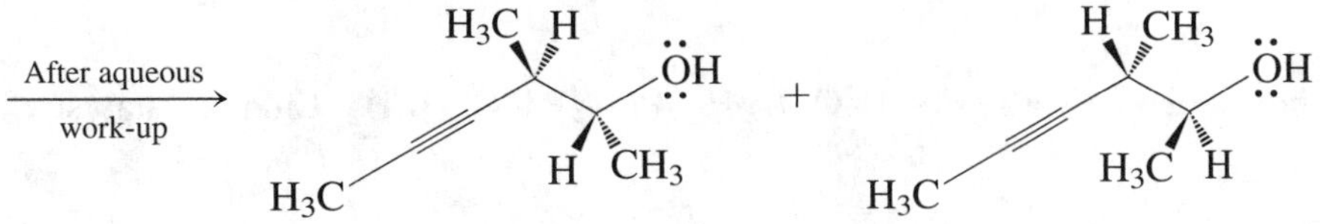

After aqueous
work-up

Racemic mixture

40. Method (d) is the only one that will give a high yield of the correct molecule. Methods (b) and (c) will give some of the target molecule, together with some regioisomeric alkynes. Method (e), and S_N2 with a basic nucleophile on a secondary halide, works to a certain extent but gives much of the E2 product as well. Method (a) is totally bogus.

41. In most cases the answer given is just one of several correct ones.

(a) $HC\equiv CLi \xrightarrow{CH_3CH_2CH_2Br, \text{ DMSO}} HC\equiv CCH_2CH_2CH_3 \xrightarrow[\substack{\text{2. } CH_3CH_2Br, \text{ DMSO}}]{\text{1. } NaNH_2, NH_3}}$ product

(b) $HC\equiv CLi + CH_3CH_2\overset{\overset{\displaystyle O}{\|}}{C}CH_3 \longrightarrow$ product

(c) To get the triple bond one carbon away from the alcohol carbon, use the ring opening of an oxacyclopropane by an alkynyl anion.

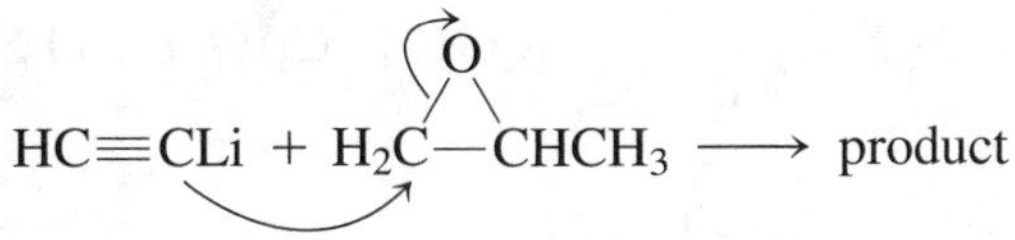

$HC\equiv CLi + H_2C\!-\!CHCH_3 \longrightarrow$ product

(d) Reaction of the basic alkynyl anion with a tertiary halide would give elimination. Instead, use a tertiary Grignard reagent to make the necessary carbon–carbon bond by addition to an aldehyde. An elimination–halogenation–double-dehydrohalogenation sequence generates the triple bond.

$(CH_3)_3CCl \xrightarrow[\substack{\text{2. } CH_3\overset{\overset{\displaystyle O}{\|}}{C}H}]{\text{1. Mg}} (CH_3)_3C\overset{\overset{\displaystyle OH}{|}}{C}HCH_3 \xrightarrow[\substack{\text{2. LDA, THF}}]{\text{1. } PBr_3} (CH_3)_3CCH\!=\!CH_2 \xrightarrow[\substack{\text{2. } NaNH_2, NH_3}]{\text{1. } Br_2, CCl_4}$ product

42. Priority of D is higher than H, but lower than anything else. Structure is

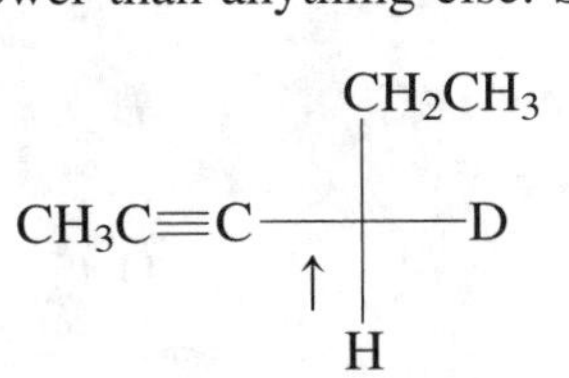

Best bond to make is marked (arrow). Synthesis of the optically active product could be achieved if the enantiomerically pure haloalkane could be obtained.

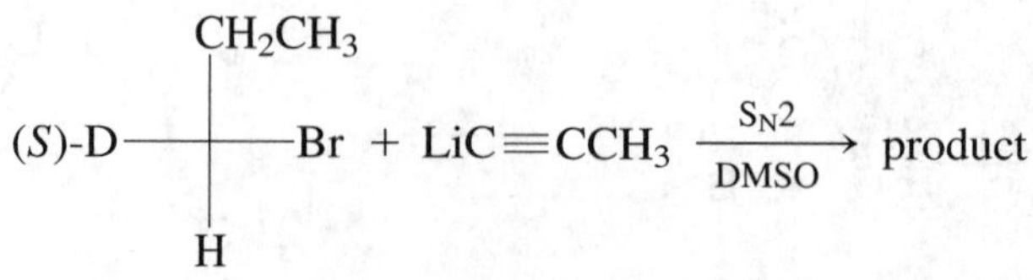

$$(S)\text{-D}\underset{\overset{|}{H}}{\overset{\overset{CH_2CH_3}{|}}{-}}Br + LiC\equiv CCH_3 \xrightarrow[\text{DMSO}]{S_N2} \text{product}$$

43. **(a)** Br_2 (1 equivalent), LiBr, CH_3COOH (Section 13-7); **(b)** Br_2 (2 equivalents), CCl_4; **(c)** $HgSO_4$, H_2O, H_2SO_4; **(d)** HI (excess); **(e)** strong base, e.g., R′Li, R′MgX, or $NaNH_2$; **(f)** (1) strong base as in **(e)**, (2) R′COR″, (3) H^+, H_2O; **(g)** (1) strong base as in **(e)**, (2) R′X (R′ = methyl or primary alkyl);

(h) (1) strong base as in **(e)**, (2) 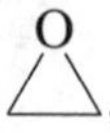 ◁ O, (3) H^+, H_2O; **(i)** H_2, Pt or Pd–C; **(j)** H_2, Lindlar catalyst

(Pd–CaCO₃, Pb(OCCH₃)₂, quinoline); **(k)** (1) 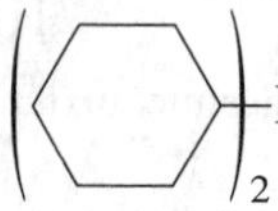 ⬡–BH)₂, (2) NaOH, H_2O_2.

44. **(a)** 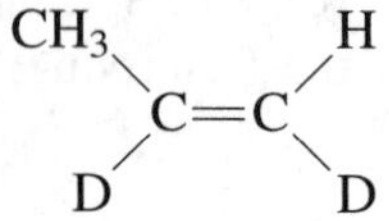 CH₃\C=C/H with D, D   **(b)** 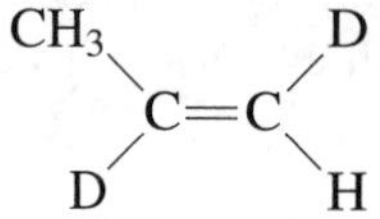 CH₃\C=C/D with D, H   **(c)** $CH_3Cl=CH_2$

(d) $CH_3Cl_2CH_3$ **(e)** 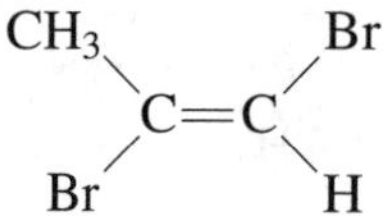 CH₃\C=C/Br with Br, H   **(f)** 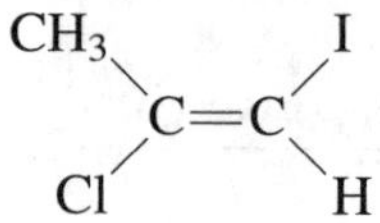 CH₃\C=C/I with Cl, H

(g) $CH_3CCl_2CHI_2$ **(h)** $CH_3\overset{O}{\overset{||}{C}}CH_3$ **(i)** $CH_3CH_2\overset{O}{\overset{||}{C}}H$

45. In the structures below R = cyclohexyl.

(a) 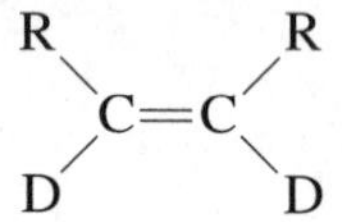 R\C=C/R with D, D   **(b)** R\C=C/D with D, R   **(c)** RCl=CHR (*E* and *Z*)

(d) RCl_2CH_2R **(e)** 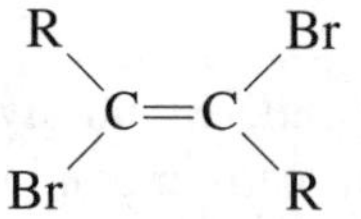 R\C=C/Br with Br, R   **(f)** 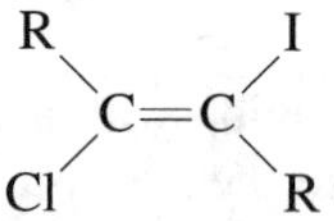 R\C=C/I with Cl, R

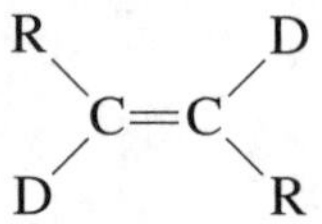

(g) $RCCl_2Cl_2R$ + $R\overset{I}{\overset{|}{C}}Cl\overset{I}{\overset{|}{C}}ClR$ **(h)** and **(i)** $R\overset{O}{\overset{||}{C}}CH_2R$

46. The reaction adds H_2O to the triple bond with anti-Markovnikov regiochemistry:

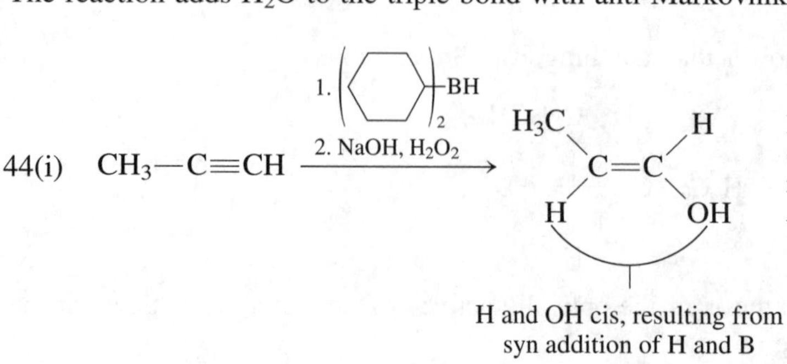

44(i) $CH_3—C\equiv CH$ $\xrightarrow[\text{2. NaOH, H}_2O_2]{\text{1. ⬡–BH)}_2}$ H₃C\C=C/H with H, OH

Hydroboration is syn; oxidation occurs with retention; therefore the *E* enol results.

H and OH cis, resulting from syn addition of H and B

45(i)

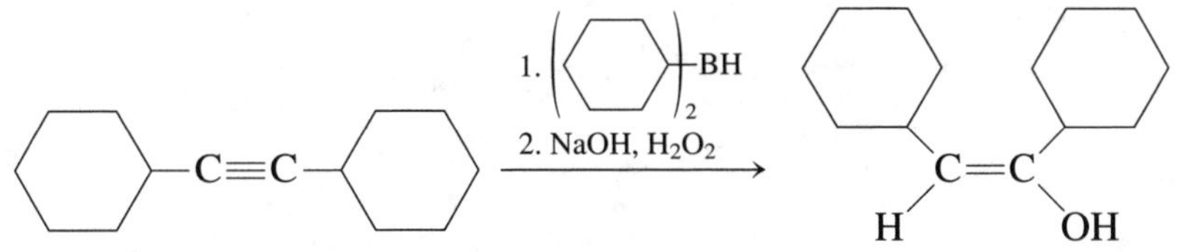

47. In these, "racemic" means a racemic mixture of *R,R* and *S,S* stereoisomers.

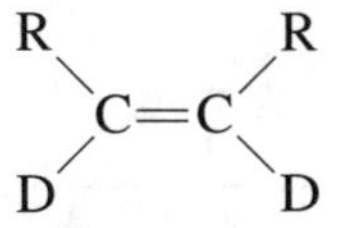

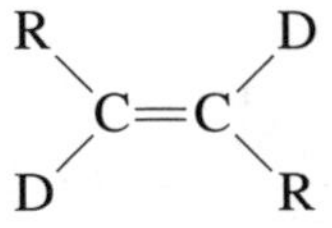

(a) *meso*-RCHDCHDR

(b) *racemic* RCDCDR
    | |
    Br Br

(c) *racemic*

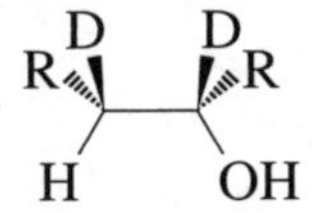

(d)

(e) *meso*-

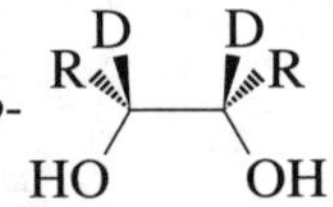

(a) *racemic* RCHDCHDR

(b) *meso*-RCDCDR
    | |
    Br Br

(c) *racemic*

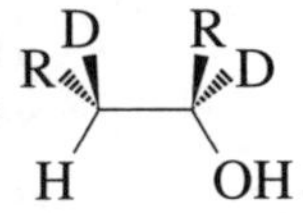

(d)

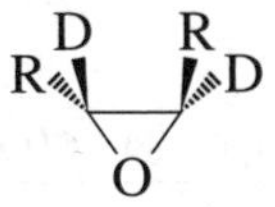

(e) *racemic*

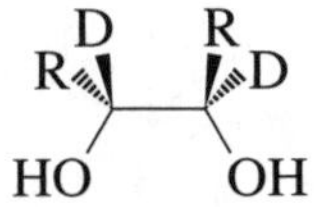

48. The only high-yield precursor is 3-heptyne (g), which gives *cis*-3-heptene with both good yield and selectivity upon hydrogenation over Lindlar's catalyst. Eliminations of 3-chloroheptane (a) with base and 3-heptanol (d) with acid are the poorest choices because both will give regioisomeric and stereoisomeric mixtures of *cis*- and *trans*-2- and 3-heptenes. Similar eliminations of 4-chloroheptane (b) and 4-heptanol (e) are better because at least the regioisomer problem is gone: Only *cis*- and *trans*-3-heptenes can form. Double elimination from 3,4-dichloroheptane (c) gives mostly 3-heptyne (g), but other unsaturated regioisomers are obtained as minor products. Finally, addition of chlorine to *trans*-3-heptene gives 3,4-dichloroheptane.

49. (a) $CH_3CH_2C{\equiv}CH$ $\xrightarrow{\text{1. HCl} \atop \text{2. HBr}}$ product

(b) $CH_3CH_2CH_2CH_2C{\equiv}CH$ $\xrightarrow{\text{2HI}}$ product

(c) 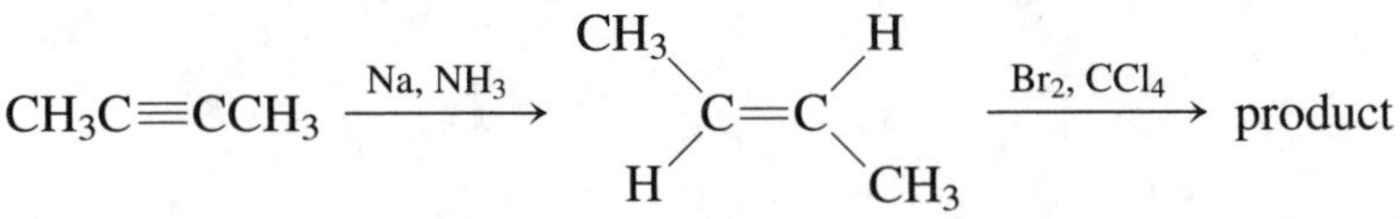 $CH_3C{\equiv}CCH_3$ $\xrightarrow{\text{Na, NH}_3}$

$\xrightarrow{\text{Br}_2, \text{CCl}_4}$ product

(d) 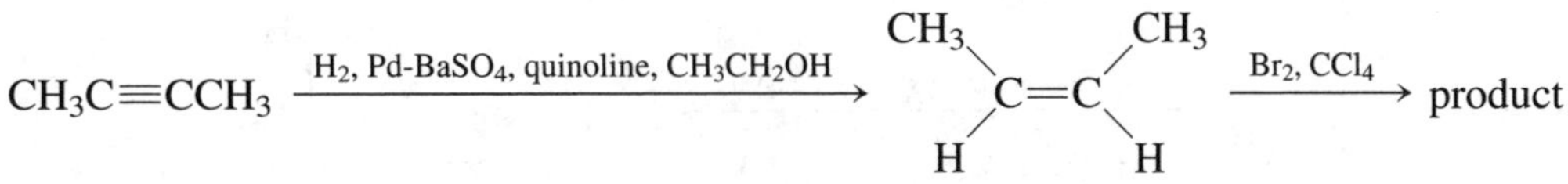 $CH_3C{\equiv}CCH_3$ $\xrightarrow{\text{H}_2, \text{Pd-BaSO}_4, \text{quinoline, CH}_3CH_2OH}$

$\xrightarrow{\text{Br}_2, \text{CCl}_4}$ product

(e) 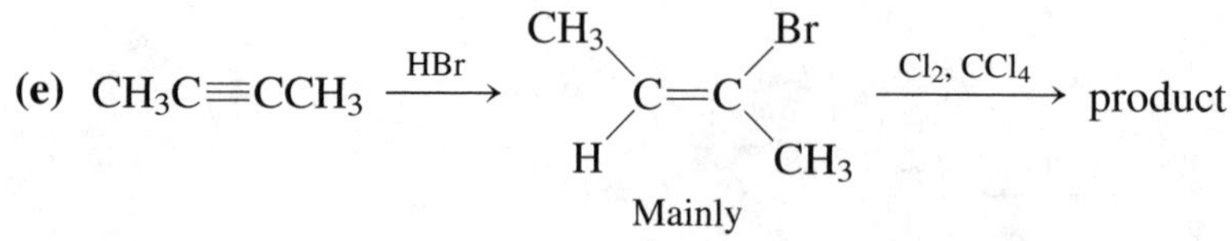 $CH_3C\equiv CCH_3$ $\xrightarrow{HBr}$

$$\underset{\text{Mainly}}{\underset{H}{\overset{CH_3}{C}}=\underset{CH_3}{\overset{Br}{C}}}$$

$\xrightarrow{Cl_2,\ CCl_4}$ product

(f) $CH_3CH_2CH_2C\equiv CCH_2CH_2CH_3$ $\xrightarrow{HgSO_4,\ H_2SO_4,\ H_2O}$ product

(g) $HC\equiv C\overset{\overset{OH}{|}}{C}HCH_3$ $\xrightarrow{H_2,\ Pd\text{-}BaSO_4,\ quinoline,\ CH_3CH_2OH}$ $H_2C=CH\overset{\overset{OH}{|}}{C}HCH_3$ $\xrightarrow[\text{2. } H_2O_2,\ HO^-]{\text{1. } BH_3,\ THF}$ product

(h) 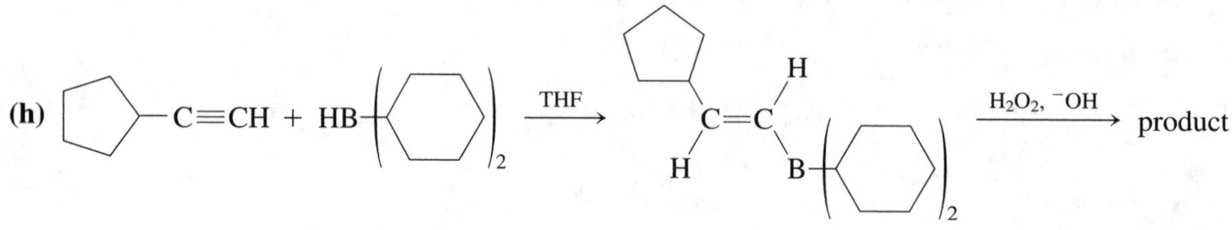 (cyclopentyl)$-C\equiv CH$ + HB(cyclohexyl)$_2$ $\xrightarrow{THF}$ $\underset{H}{\overset{\text{cyclopentyl}}{C}}=\underset{B\text{(cyclohexyl)}_2}{\overset{H}{C}}$ $\xrightarrow{H_2O_2,\ ^-OH}$ product

(i) 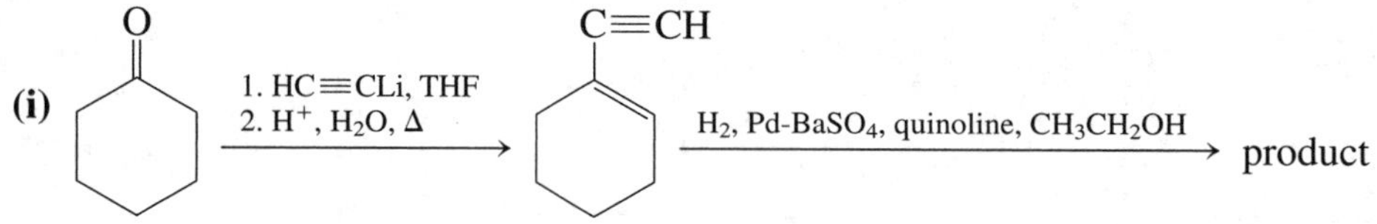 (cyclohexanone) $\xrightarrow[\text{2. } H^+,\ H_2O,\ \Delta]{\text{1. } HC\equiv CLi,\ THF}$ (cyclohexene with $C\equiv CH$) $\xrightarrow{H_2,\ Pd\text{-}BaSO_4,\ quinoline,\ CH_3CH_2OH}$ product

50. **(a)** 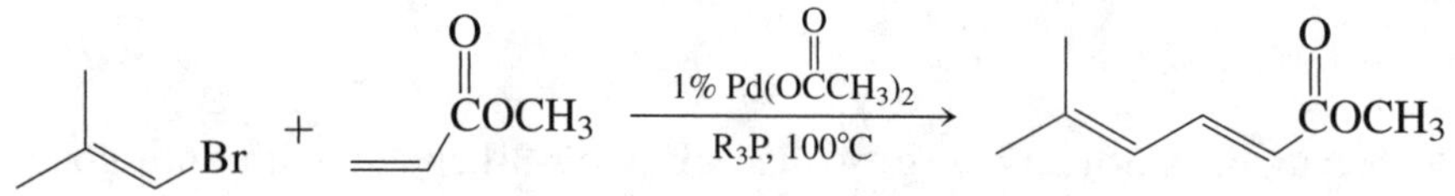 (2-methyl-1-propenyl bromide) + $\underset{\overset{\|}{O}}{COCH_3}$ (vinyl) $\xrightarrow[R_3P,\ 100°C]{1\%\ Pd(OCCH_3)_2}$ (product with $COCH_3$)

(b) 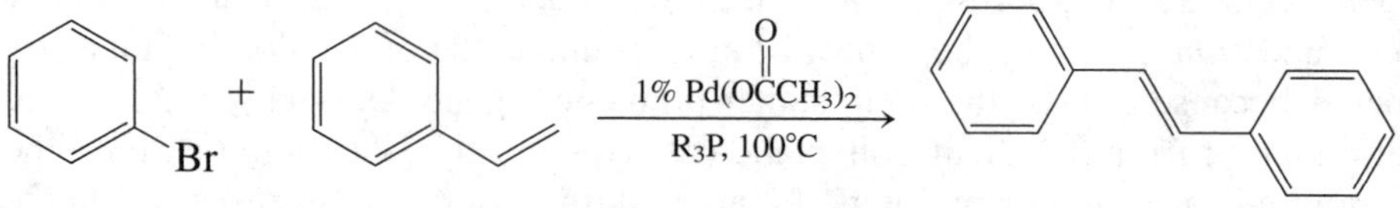 (bromobenzene) + (styrene) $\xrightarrow[R_3P,\ 100°C]{1\%\ Pd(OCCH_3)_2}$ (stilbene)

51. $Ca^{2+}\ ^-:C\equiv C:^-$, a calcium salt of ethyne, is consistent with its reaction with water to form $HC\equiv CH$. One could call this material "calcium acetylide" or, perhaps "ethynediylcalcium," with the "di" referring to the doubly deprotonated ethyne.

52.

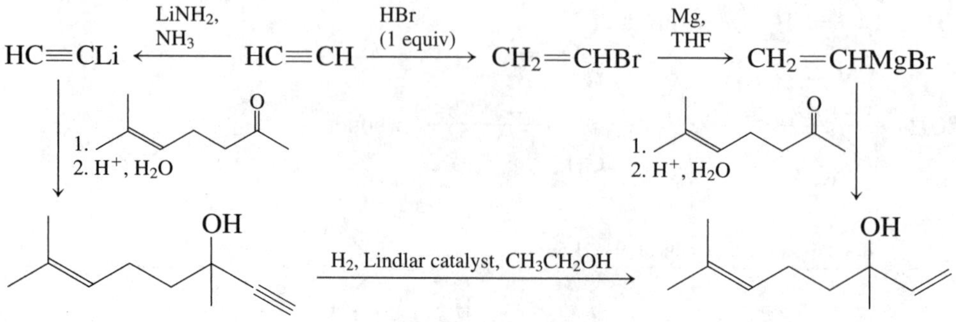

$HC\equiv CLi$ $\xleftarrow[NH_3]{LiNH_2,}$ $HC\equiv CH$ $\xrightarrow[\text{(1 equiv)}]{HBr}$ $CH_2=CHBr$ $\xrightarrow[THF]{Mg,}$ $CH_2=CHMgBr$

$\downarrow$ 1. (ketone structure) 2. $H^+,\ H_2O$

$\downarrow$ 1. (ketone structure) 2. $H^+,\ H_2O$

(alcohol with alkyne) $\xrightarrow{H_2,\ Lindlar\ catalyst,\ CH_3CH_2OH}$ (alcohol with vinyl)

53.

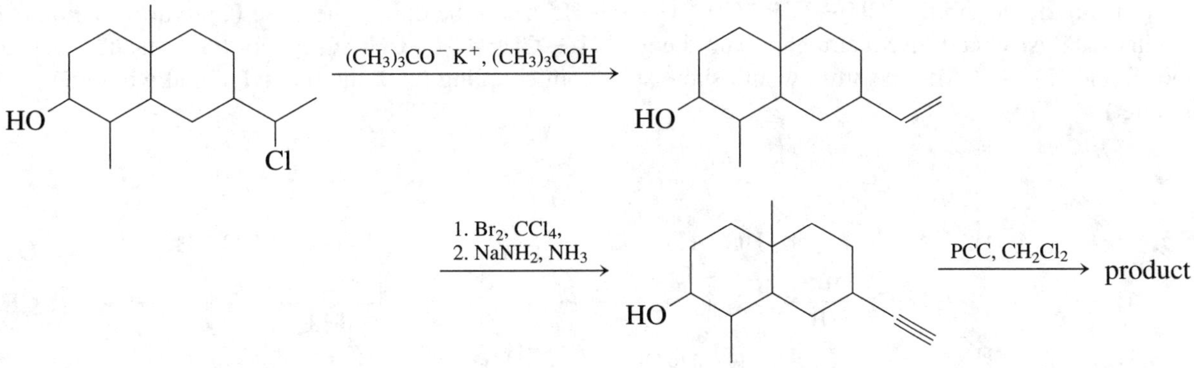

54.

$$RCH_2OH \xrightarrow[\text{2. NaI, HMPA}]{\text{1. } CH_3-\langle\bigcirc\rangle-SO_2Cl\text{, py}^*} RCH_2I \xrightarrow{\text{LiC}\equiv\text{CH, DMSO}} RCH_2C\equiv CH$$

$$\xrightarrow[\text{2. } H_2O_2\text{, } HO^-]{\text{1. } (\langle\bigcirc\rangle)_2BH\text{,}^{**}\text{THF}} RCH_2CH_2\overset{O}{\overset{\|}{C}}H \xrightarrow[\text{2. } PBr_3]{\text{1. } NaBH_4\text{, } CH_3OH} RCH_2CH_2CH_2Br$$

$$\xrightarrow[\text{2. } CH_3\overset{O}{\overset{\|}{C}}CH_3]{\text{1. Mg, THF}} RCH_2CH_2CH_2\overset{OH}{\overset{|}{\underset{CH_3}{\underset{|}{C}}}}CH_3 \xrightarrow{H_2SO_4,\ \Delta} RCH_2CH_2CH{=}C(CH_3)_2 \equiv \text{bergamotene}$$

*R is so hindered that this special sequence (tosylate ⟶ iodide) is necessary to allow S_N2 reaction with alkynyl anion to be carried out.
**The unhindered triple bond is hydroborated much faster than the hindered (trisubstituted) double bond in the R group.

55. Work backward from the ozonolysis result.

$$2\,H\overset{O}{\overset{\|}{C}}H + CH_3\overset{O}{\overset{\|}{C}}-\overset{O}{\overset{\|}{C}}H \xleftarrow[\text{2. Zn, } H^+\text{, } H_2O]{\text{1. } O_3\text{, } CH_2Cl_2} \text{must have come from } CH_3-\overset{CH_2}{\overset{\|}{C}}-\overset{CH_2}{\overset{\|}{C}}-H.$$

Now look at the spectra: The IR shows bands at 1615 cm^{-1} (C=C) and at 2110 cm^{-1} (C≡C). Notice, too, that absorptions at 3100 cm^{-1} and 3300 cm^{-1} are present, and may be assigned to alkenyl and alkynyl C—H bonds, respectively. The NMR shows four types of hydrogens, with signals in an intensity ratio of 3 : 1 : 1 : 1. A total of six hydrogens are present in the unknown molecule, presumably a CH_2 (δ = 1.9), an alkynyl H (δ = 2.8), and two alkenyl hydrogens (δ = 5.2 and 5.3).

Can we combine this information with the ozonolysis data? The hydrocarbon shown above, which arises from hydrogenation of the unknown, has the formula C_5H_8. Our unknown must therefore be C_5H_6. What pieces have we identified? A CH_3—, a —C≡C—H, and two alkenyl H's, adding up to C_3H_6; so two more carbons are required by the molecular formula. The answer is

$$CH_3-\overset{CH_2}{\overset{\|}{C}}-C\equiv C-H$$

An isomeric possibility, CH_3—CH=CH—C≡C—H, cannot be correct because (1) hydrogenation over Lindlar catalyst would give a straight chain diene, CH_3—CH=CH—CH=CH_2, not the branched one shown above, and (2) the NMR spectrum would show spin–spin coupling (e.g., the methyl signal would be a doublet).

56.

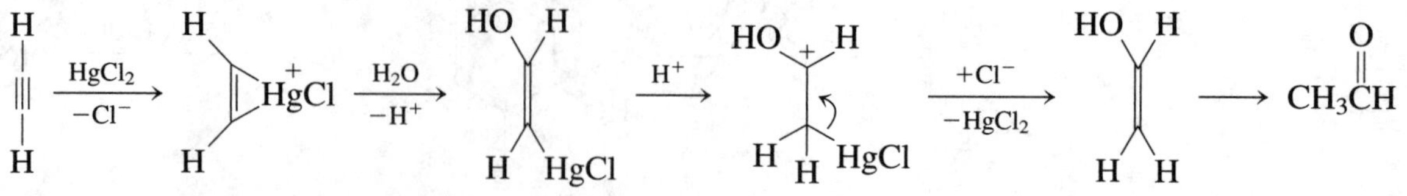

57.

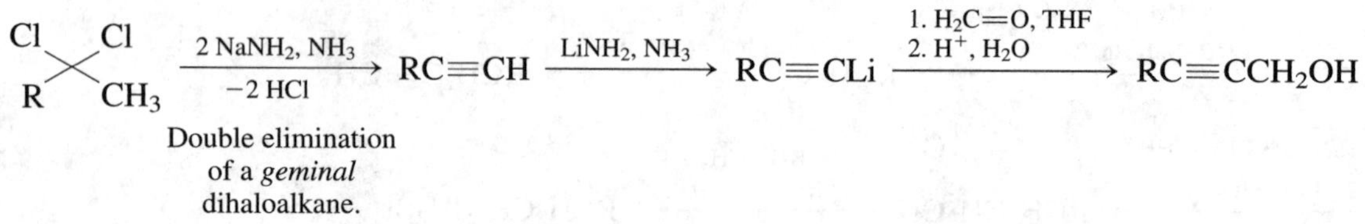

Double elimination
of a *geminal*
dihaloalkane.

58. A The recipe for formation of a sulfonate (an inorganic ester):

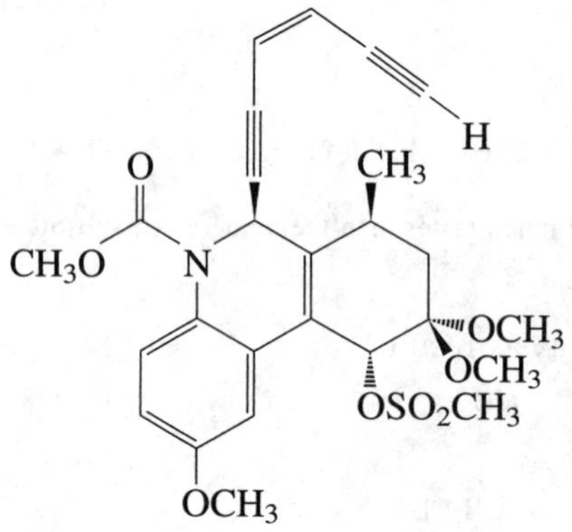

B Oxacyclopropane formation from a C=C double bond: MCPBA or any other peroxycarboxylic acid may be used.

C Similar to A:

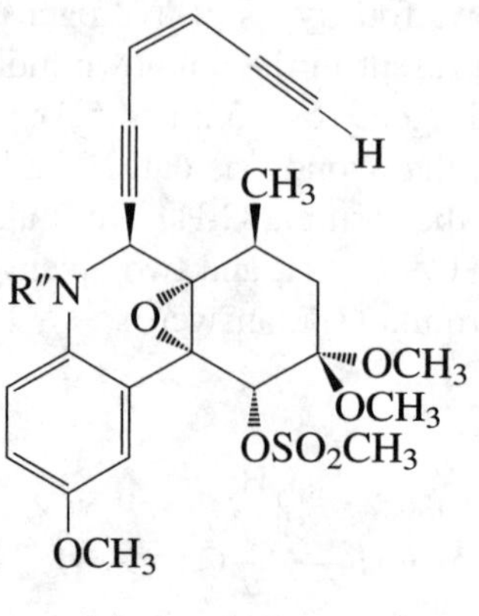

D An alternative oxacyclopropane synthesis:

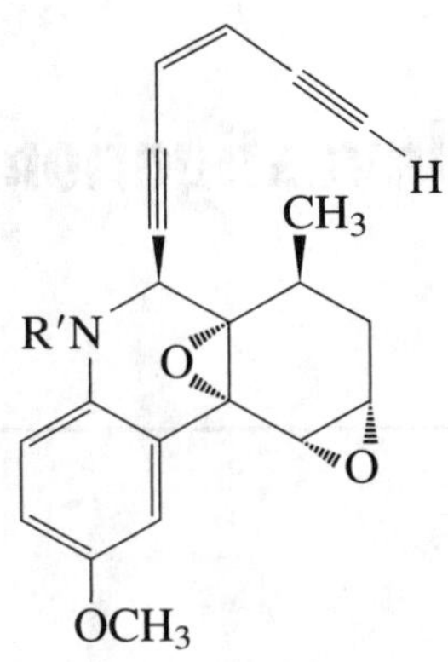

The model study suggests that addition to a carbonyl group is superior to simple displacement reactions for formation of the required medium-sized ring. So it would be reasonable to try something like the following:

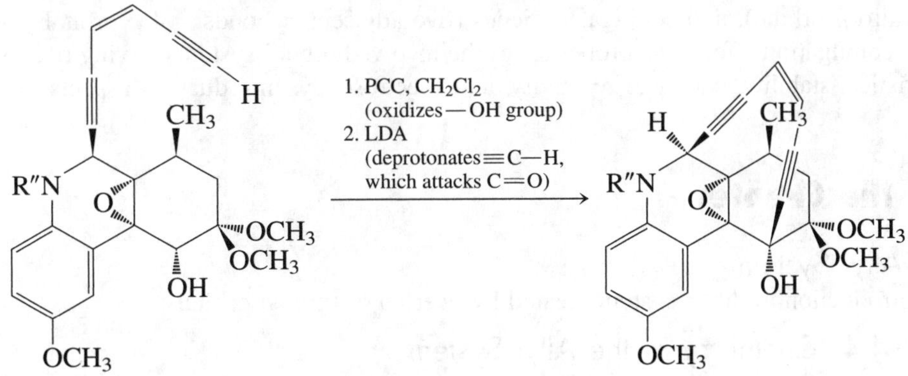

1. PCC, CH_2Cl_2
   (oxidizes — OH group)
2. LDA
   (deprotonates $\equiv$C—H, which attacks C$=$O)

59. **(b)** 5-Chloro-1-pentyne would be equally correct.

60. **(d)**

61. **(a)**

62. **(a)**

63. **(c)**

14

Delocalized Pi Systems: Investigation by Ultraviolet and Visible Spectroscopy

This chapter covers an assortment of topics derived from a single concept: *conjugation*. Conjugation refers to π overlap of three or more *p* orbitals on adjacent atoms in a molecule. The *allyl* systems are the simplest (one π bond plus a third *p* orbital), and conjugated dienes (two adjacent π bonds = 4 *p* orbitals) are next in line. As you will see, conjugation affects the properties of the involved orbital systems, giving rise to modified electronic characteristics, stability, chemical reactivity, and spectroscopy. Introductory aspects of all of these are presented here.

Outline of the Chapter

Keys to the Chapter

14-1.   The Allyl System

Delocalization generally results in stabilization. The experimental results cited in Section 14-1 illustrate the relative ease of generating allylic radicals, cations, and anions, compared with ordinary 1° radicals, cations, or anions. The origins of allylic stabilization are presented in two different but equivalent ways: using resonance and using molecular orbitals. Both viewpoints offer useful insights into the allyl system. You should pay special attention to the electrostatic consequences of conjugation as implied by these resonance and molecular-orbital pictures. Electrons can move freely through conjugated π systems, either toward an electron-deficient atom or away from an electron-rich one. This delocalization obviously is electrostatically desirable and, again, results in overall stabilization.

14-2, 14-3, 14-4.   Chemistry of the Allyl System

The presence of an allyl system gives rise to the possibility of easily formed, stabilized radicals, cations, and anions. It also introduces a new regiochemical factor, because the reactive character of each of these intermediates is now shared by the two carbons at the ends of the allylic system. A reaction sequence involving any allylic radical, cation, or anion can and usually does give two isomeric products, derived from attachment of a group at either of these two "ends."

Notice that **none of these reactions is fundamentally new.** All you are seeing is the modified outcome of a nucleophilic displacement, a radical halogenation, or a Grignard-type reaction when the substrate leads to an allylic intermediate as it follows the **ordinary mechanistic course** of any of these reactions. Learning to understand and handle situations like this requires that you "think mechanistically." That is, you need to apply what you've learned earlier about a reaction mechanism directly to a new type of molecule. You have to follow the mechanism one step at a time, see what you get, and analyze the consequences of any unusual new structural types that turn up. This is a cornerstone of organic chemistry, allowing some degree of extrapolation and predictability in new situations. You haven't been asked to do a whole lot of this up until now, but you will need to develop these skills from now on. Much of what is coming up will involve molecules with multiple functional groups that may affect each other's behavior. Mechanistically oriented thinking is indispensable in deciding just what these molecules are likely to do.

14-5 and 14-6.   Conjugated Dienes

With dienes, you see the first situation where interacting functional groups affect chemical behavior. Conjugated dienes possess p orbitals on four adjacent atoms. They are more stable than the other two alternatives: isolated dienes, where the double bonds are separated by one or more atoms, and "cumulated" dienes (like allene), where the double bonds share a common atom.

Cumulated	**Conjugated**	**Isolated,** $n \geq 1$
("1, 2")	("1, 3")	("1, 4"; "1, 5"; "1, 6"; etc.)

As you saw with allyl systems, the presence of conjugation leads to stabilization. The result is lower energy for conjugated dienes relative to the others. Again, both resonance and molecular-orbital explanations are applicable.

In their qualitative chemistry, conjugated dienes behave very much like alkenes: They readily react with electrophiles in addition reactions. Just as in the case of alkenes, this addition proceeds to give the most stable intermediate. For conjugated dienes, this normally turns out to be a resonance stabilized allylic cation:

$$CH_2=CH-CH=CH_2 + E^+ \nearrow CH_2=CH-\overset{+}{C}H-CH_2-E$$
$$\searrow CH_2=CH-CH-\overset{+}{C}H_2$$
$$|$$
$$E$$

That represents the basic story. The rest of the section deals with details, mainly associated with the fact that the allylic cation can attach a nucleophile at either of two positions. Attachment to give 1,2-addition is usually fastest (kinetic), although the 1,4 product, when it possesses a more highly substituted double bond, is usually more stable.

14-7.   Extended Conjugation and Benzene

Further extrapolation on the same themes.

14-8 and 14-9. Special Reactions of Conjugated π Systems

Up until now we haven't made any special presentations concerning syntheses of rings, because the ring-forming processes you've seen so far were nothing more than intramolecular versions of ordinary reactions, such as

Now, however, a new set of ring-forming reactions are presented separately because they represent a totally new mechanistic class, sometimes collectively called *pericyclic* reactions. Mechanisms for these involve movement of two or more pairs of electrons **in a circle** and the simultaneous breaking and forming of σ and π bonds. They are therefore examples of *concerted* processes. These generally do not involve radicals or ions and don't need polarized bonds to take place, although dipole–dipole attractions between reacting atoms can speed things up. Because reactive species like radicals, ions, or polar bonds are not involved, you might ask why these reactions should happen at all. There are two reasons: kinetic and thermodynamic. Certain special properties of circularly moving groups of electrons give these transformations low activation barriers, and the products are more stable than the starting materials. (That was simple, wasn't it?) To convince yourself of the latter, take a look at all the examples of those thermal reactions given in the New Reaction section of the text. **In every case** the products contain more σ bonds and fewer π bonds than the starting material.

Points to take particular note of have to do with stereochemistry: In particular, stereochemical (e.g., cis-trans) relationships in the starting materials are preserved through the reaction transition states and on into the products. You may need some practice visualizing the reactants to do the problems. For instance, for the Diels-Alder cycloaddition, it may be useful to make models of two reacting molecules and to hold them in an arrangement resembling the cycloaddition transition state (Figure 14-9), to see in three dimensions where all the original groups will wind up relative to the two newly formed σ bonds. You should be able to follow readily the positions of the atoms during the course of the reaction of a molecule like 1,3-cyclopentadiene.

The electrocyclic reactions in Section 14-9 present a more complex situation, where the stereochemistry of the process is a function both of the reaction conditions (heat or light) and of the number of electrons involved. The details are beyond the scope of the course, so only introductory material has been presented.

14-10. Polymerization of Conjugated Dienes

Polymers composed of diene units are significant for two reasons. Just like polymers of simple alkenes, they are industrially important (and have been for a much longer time, by the way). In addition, they are closely related to several major classes of biological molecules formally derived from isoprene (2-methyl-1,3-butadiene) as the monomeric unit. Some of the variety in this biochemistry is illustrated in this section.

14-11. Electronic Spectra: Ultraviolet and Visible Spectroscopy

The principles behind electronic spectroscopy are very simple and, in fact, are really direct extensions of the spectroscopy of atoms, a freshman chemistry topic. Remember how absorption of light by atoms promotes electrons to higher energy levels? Here, you're seeing the same thing, but with molecules; so, the energy levels involved are best described as molecular orbitals.

The experimental techniques for observing these light absorptions are straightforward. UV-vis spectroscopy (as it is often abbreviated) was once very important in determining the presence or absence of conjugation, and so on, in an organic molecule, and therefore in structure determination. Much of its past importance has been reduced by the development of sophisticated NMR equipment and techniques. UV-vis spectroscopy is used to confirm structural assignments made on the basis of NMR and IR spectroscopy and to identify conjugated systems in compounds such as complex biomolecules, whose NMR and IR spectra are more difficult to interpret.

Solutions to Problems

32 and **33.** Major contributing resonance forms are labeled.

(a)

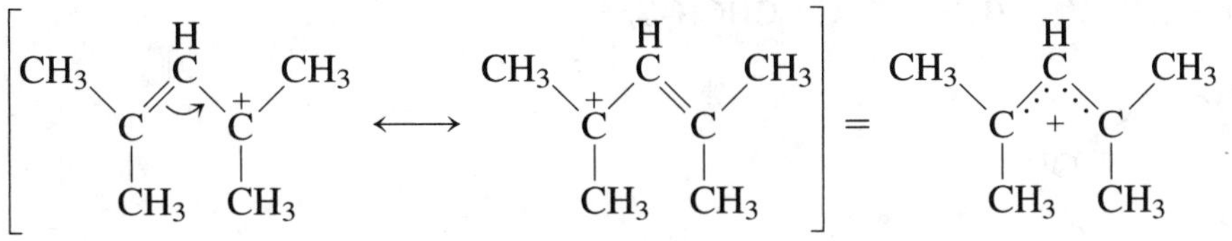

Equal contributors

(b)

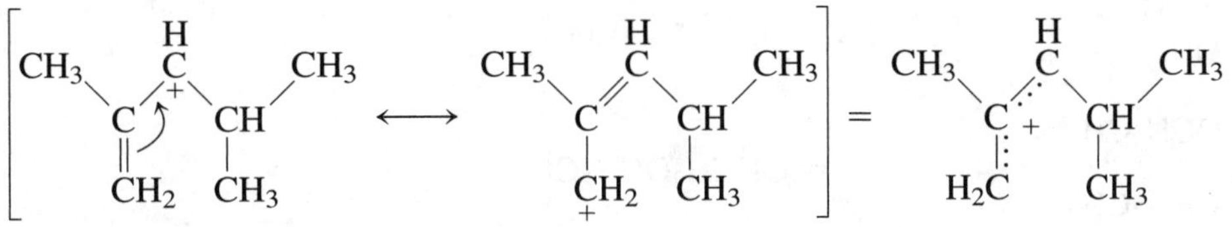

Major contributor
(charge on secondary carbon)

(c)

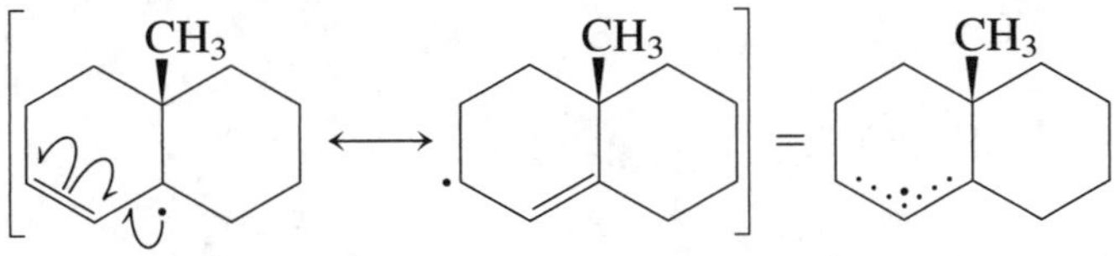

Major contributor
(tertiary radical-like)

(d)

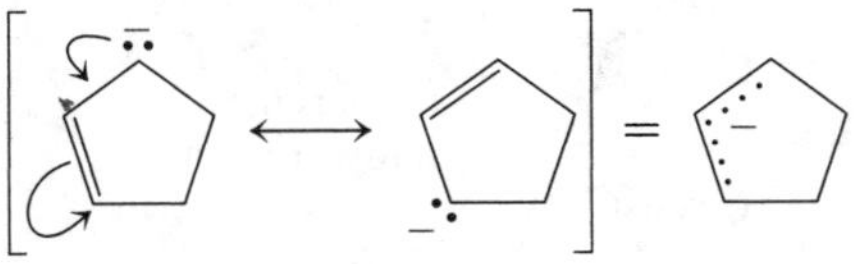

Equal contributors

(e)

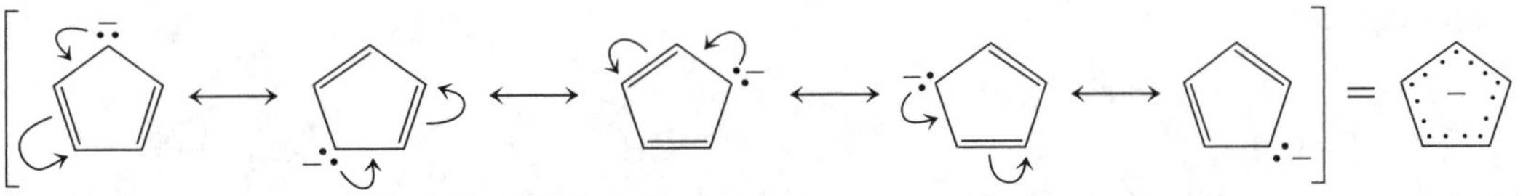

All contributors are equal

34. (a) $\left[CH_3\dot{C}HCH{=}CH_2 \longleftrightarrow CH_3CH{=}CH\dot{C}H_2 \right]$

(b)

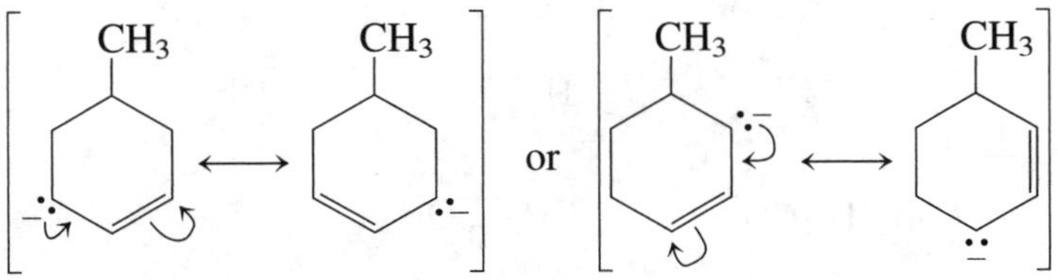

or

(c)

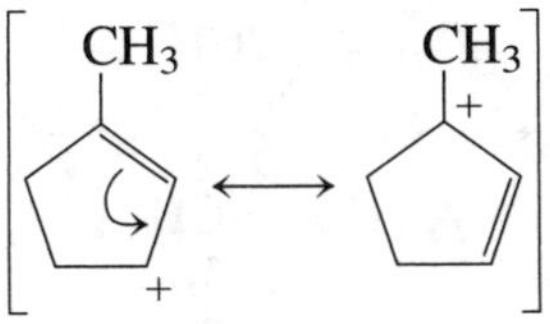

35. Radicals: allylic > tertiary > secondary > primary
Cations: tertiary > allylic ≈ secondary > primary

Hyperconjugation, which is at least partially responsible for the tertiary > secondary > primary stabilization order, is more important for cations than for radicals. The effect is large enough for tertiary cations to exceed resonance-stabilized allylic cations in stability (the reverse of the order for radicals).

36. (a) 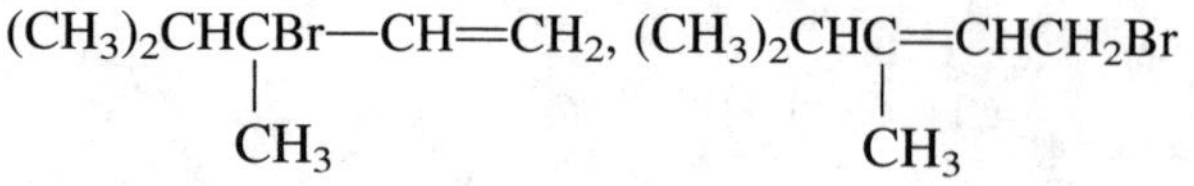 $(CH_3)_2CHCBr-CH=CH_2$, $(CH_3)_2CHC=CHCH_2Br$
with CH_3 substituents

(b)

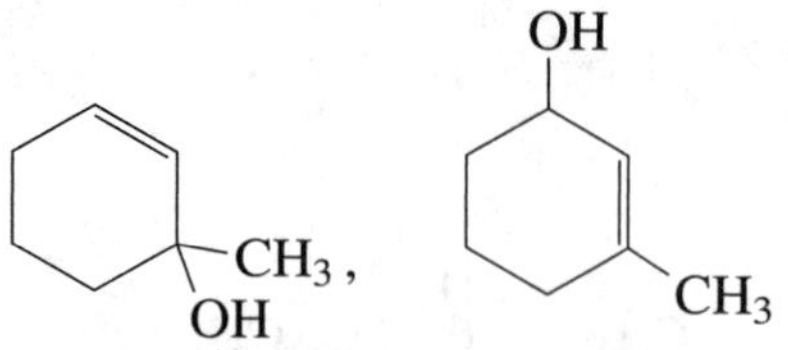

(c)

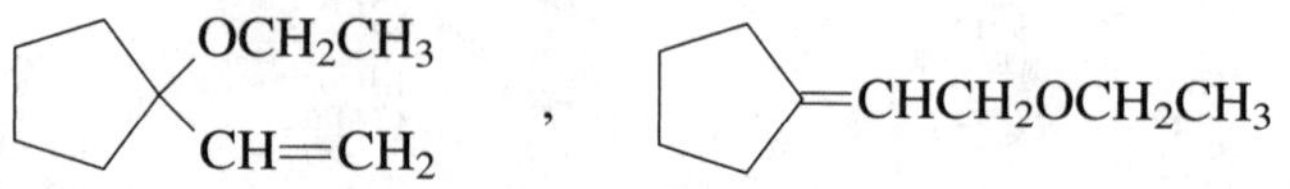

(d)

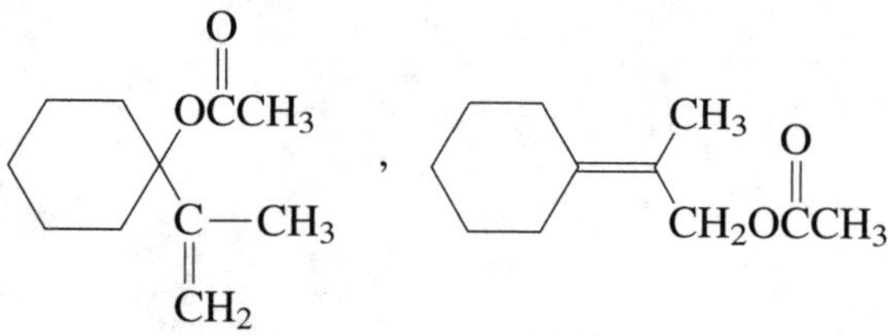

(e) Different! S_N2, not S_N1 conditions:

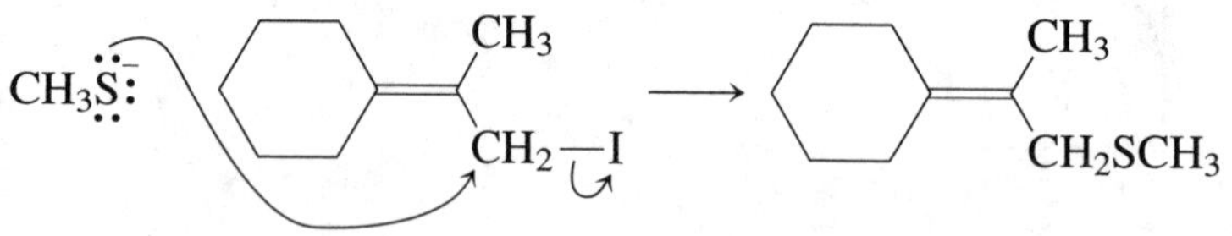

This is the only product.

(f) Intramolecular version:

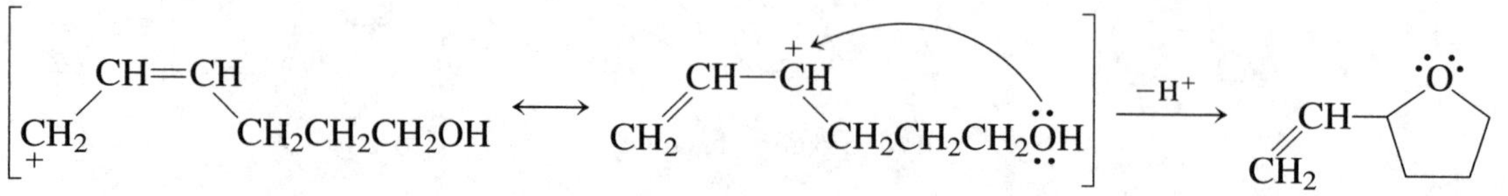

Again, only one product; bond formation at the other end would produce a more strained seven-membered ring.

37. (a)

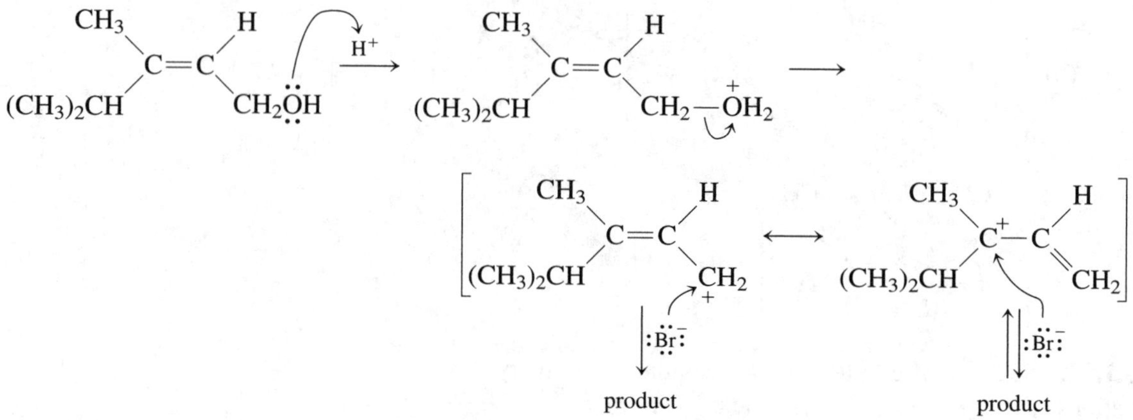

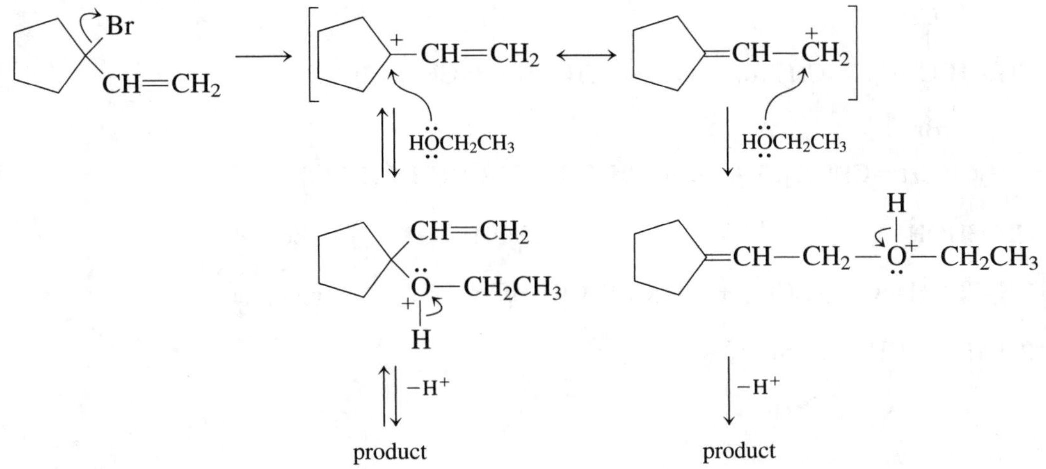

(e), (f) See answers to Problem 36.

38. **(a)** Tertiary > allylic ≈ secondary ≫ primary (order of cation stability)

(b) Allylic > primary > secondary ≫ tertiary

39. S_N1 reactivity: e (allylic and tertiary) > a (allylic and secondary) > d (forms same cation as e, but requires ionization at primary carbon, so will be slower) > c > b > f (these follow cation stability order)

$\quad$ S_N2 reactivity: Steric hindrance predominates, so f > b > d > c > a > e

40. S_N2 reactivities: Data in this chapter (Section 14-3) reveal that allylic halides are about 10^2 more reactive than their non-allylic counterparts in S_N2 displacements. Therefore, all the primary allylic systems (b, c, d, and f) will be more reactive than a saturated primary halide—even the branched system (c) will possess higher reactivity, because branching reduces reactivity only by about a factor of 20 (Table 6-9). The secondary allylic system (a) will be similar in reactivity but perhaps a bit slower than a saturated primary— secondary systems are more than 10^2 slower than primaries (Table 6-8), so the steric hindrance of the greater substitution just about cancels the acceleration due to the allylic system. Both allylic and saturated tertiaries are quite dead to S_N2 displacement.

$\quad$ S_N1 reactivities: Follow cation stabilities together with nature of the position of the leaving group. So (e) is fastest, followed by a saturated tertiary halide. Then comes (a), followed by the primary allylic systems (probably in the order d, c, b, and f), which are comparable to the saturated secondary. Saturated primary halides do not react by the S_N1 mechanism.

41. Write all possible allylic isomers in each case and pay attention to stereochemistry.

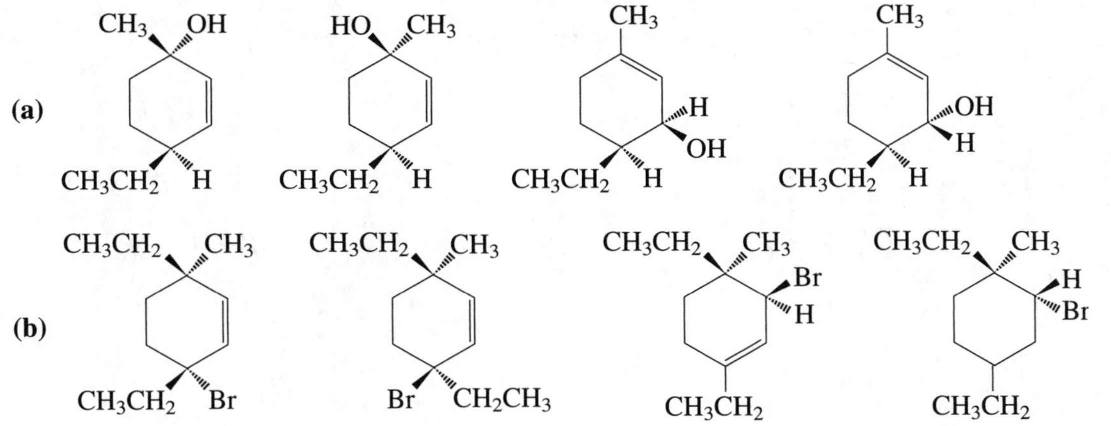

(c) 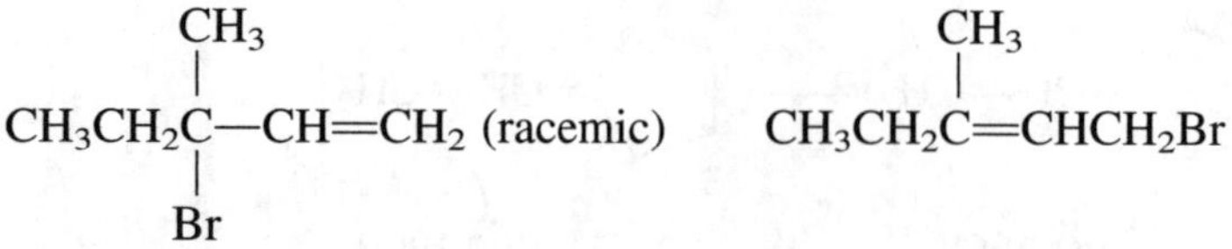
$$CH_3CH_2\overset{\displaystyle CH_3}{\underset{\displaystyle Br}{C}}—CH=CH_2 \text{ (racemic)}$$
$$\overset{\displaystyle CH_3}{CH_3CH_2C}=CHCH_2Br$$

(d) $[CH_3\bar{C}HCH=CHCH_2CH_3 \longleftrightarrow CH_3CH=CH\bar{C}HCH_2CH_3]\ Li^+$

(e) 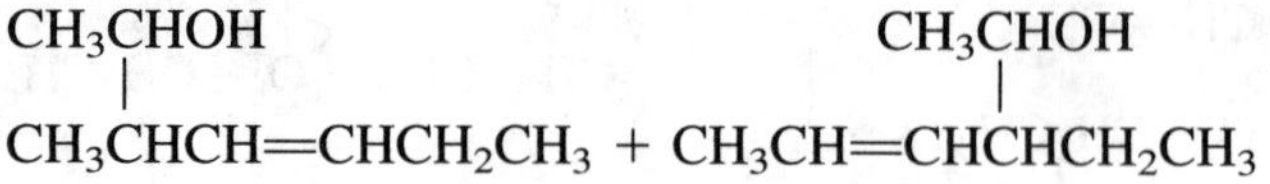 $\overset{\displaystyle CH_3CHOH}{CH_3CHCH}=CHCH_2CH_3 + \overset{\displaystyle CH_3CHOH}{CH_3CH}=CHCHCH_2CH_3$ (All possible stereoisomers for each structure)

(f) 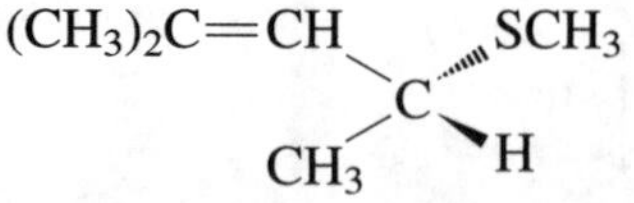
$$(CH_3)_2C=CH\underset{\displaystyle CH_3}{\overset{\displaystyle}{\underset{}{}}}\overset{\text{\scriptsize}}{C}\begin{smallmatrix}SCH_3\\H\end{smallmatrix}$$

42. The substrate is a tertiary, allylic halide, and the leaving group (iodide) is a good one. The solvent is water, an excellent one for formation of ions, suggesting that we should begin the process in the manner of an S_N1 substitution: departure of the leaving group. The result is a carbocation which, being allylic, has two resonance forms:

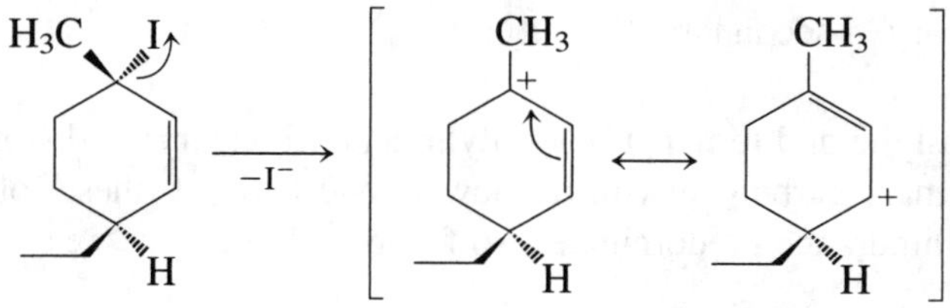

When we proceed to the next step, we will therefore have two possible positively charged positions to which the nucleophile, water, can add.

Notice that the carbon that originally contained the halogen is now trigonal planar. Thus we no longer depict the methyl group with a wedged bond; it has moved into the same plane as the other two bonds to this trigonal carbon atom. This carbocation still contains one stereocenter, however: the carbon with the ethyl substituent. Because of this stereocenter, the two faces of the cyclohexane ring are stereochemically distinct. The addition of water can take place to the same face or the opposite face of the one containing the ethyl group. For clarity, we first draw addition of water to the carbon that originally contained the leaving group (followed by proton loss to give the final products):

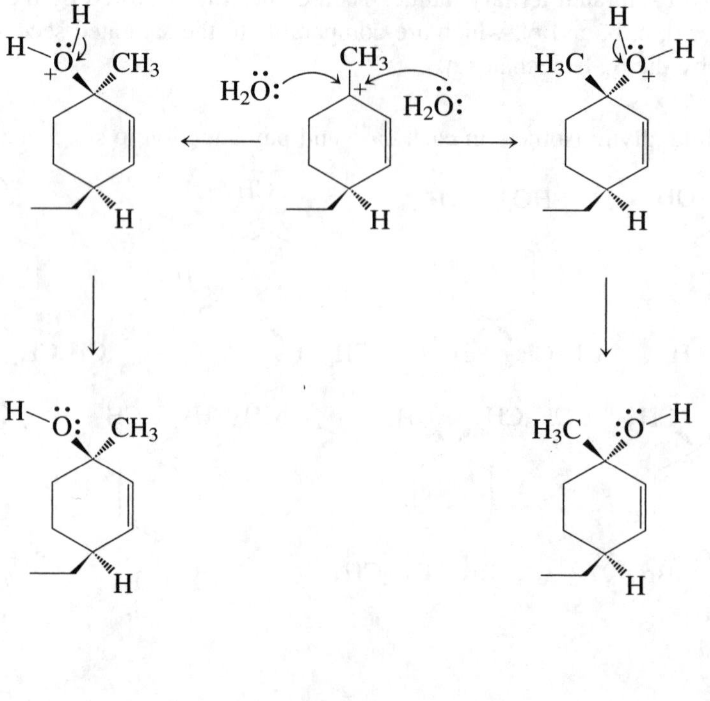

Water can similarly add to the other end of the allylic carbon, the other position with positive charge, and again to either face of the ring, relative to the ethyl group:

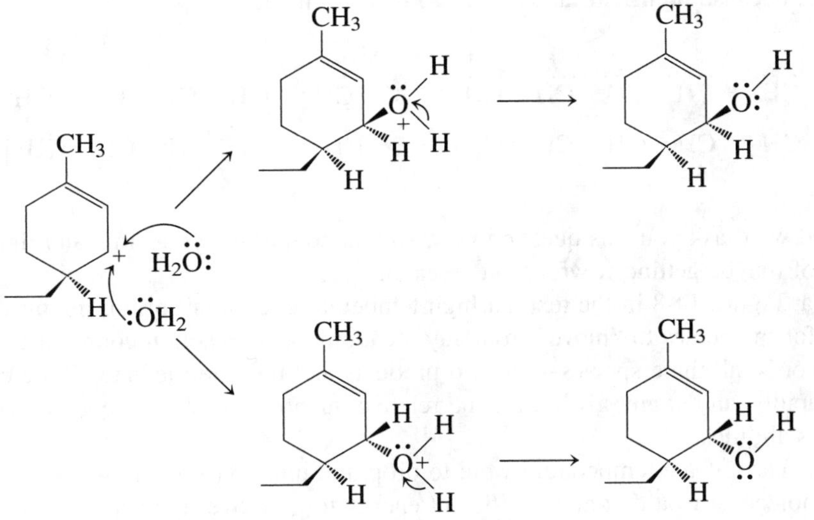

Thus, in all, four isomeric products are possible.

43.

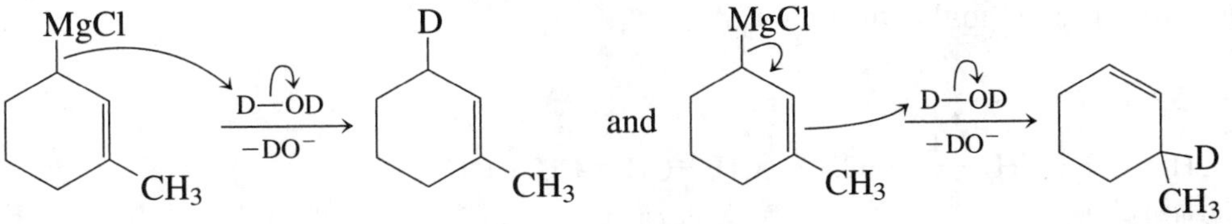

and

44.

or

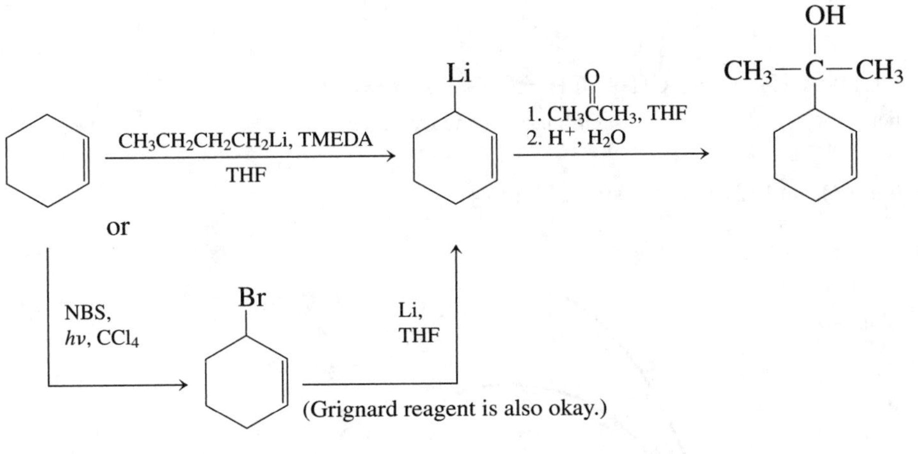

(Grignard reagent is also okay.)

45. (a) *cis*-2-*trans*-5-Heptadiene, or (2*E*, 5*Z*)-2,5-heptadiene

   (b) 2,4-Pentadien-1-ol

   (c) (5*S*, 6*S*)-5,6-Dibromo-1,3-cyclooctadiene

   (d) 4-Ethenylcyclohexene

46. CH$_2$=CH—CH—CH=CH$_2$ 1,4-Pentadiene has the weakest C—H bond (arrow), a bond that is **doubly** allylic ($DH° \sim 77$ kcal mol^{-1}); this isomer will therefore be brominated fastest. Because only a very weak C—H bond needs to be broken, its first propagation step has a much smaller E_a relative to 1,3-pentadiene, where a stronger methyl C—H bond needs to be broken. However, both will give identical product mixtures because **identical** radicals are formed from each:

$$\overset{\cdots\cdots\cdots\overset{\cdot}{}\cdots\cdots\cdots}{CH_2-CH-CH-CH-CH_2} \equiv [\overset{\cdot}{C}H_2-CH=CH-CH=CH_2 \longleftrightarrow$$

$$CH_2=CH-\overset{\cdot}{C}H-CH=CH_2 \longleftrightarrow CH_2=CH-CH=CH-\overset{\cdot}{C}H_2]$$

47. We figured we'd ask you this question now, so you could take your time and figure out the right answer instead of maybe getting it wrong on an exam.

Have a look at Figure 14-8 in the text. At high temperature, an equilibrium mixture exists because there is enough energy for molecules to "move" from any location on the reaction coordinate to any other location on it. In other words, all three species—the two products and the intermediate allylic cation—are interchanging rapidly, and at any given time the relative quantities of each are governed by their relative thermodynamic stabilities.

That being the case, if the temperature were to drop, the interconversion processes would slow down because fewer molecules would contain sufficient energy to pass over the activation barriers. This would mainly affect conversion of the two product molecules into the intermediate carbocation because those processes possess the highest activation barriers. The result is that the thermodynamic ratio of products originally established at high temperature would remain pretty much unchanged (frozen) upon cooling of the reaction mixture. It will **not** revert to the kinetic ratio!

48. $CH_2=CH-CH=CH-CH_3 \xrightarrow{H^+} CH_3-\overset{+}{C}H-CH=CH-CH_3$

     (1) Conjugated diene   (2) Allylic cation,
                                              secondary at each end

$CH_2=CH-CH_2-CH=CH_2 \xrightarrow{H^+} CH_2=CH-CH_2-\overset{+}{C}H-CH_3$

     (3) Isolated diene   (4) Ordinary secondary cation

(1) is more stable than (3), and (2) is more stable than (4).

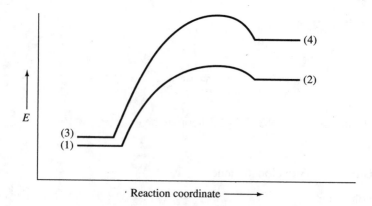

Reaction (1) + $H^+ \rightarrow$ (2) is faster and leads to the more stable cation. Note: When the text says that allylic and secondary cations are similar in energy, it is referring to the ease of formation of the simplest allylic cation, $\overset{+}{C}H_2-CH=CH_2$, which is primary at each end. Additional alkyl groups on allylic cations increase their stability and their ease of formation, as you might expect.

49. Expect 1,2- and 1,4-addition to occur in each case. Note that the 1,2-additions in (b) and (c) might be expected to show *anti* stereochemistry, similar to additions to ordinary alkenes.

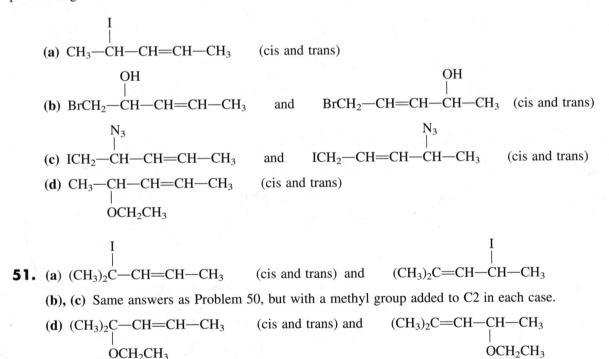

(a) 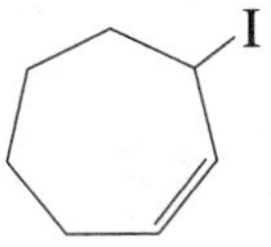 1, 2 and 1, 4 products are the same!

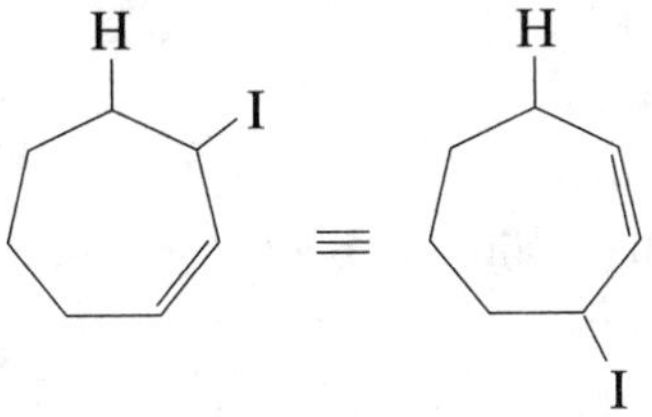

(b) 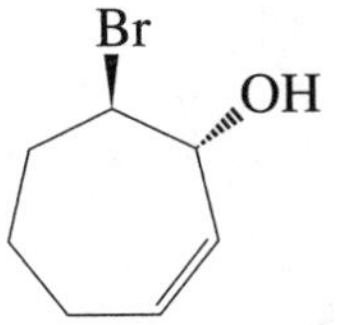 and

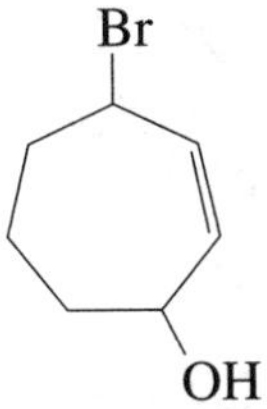

(c) 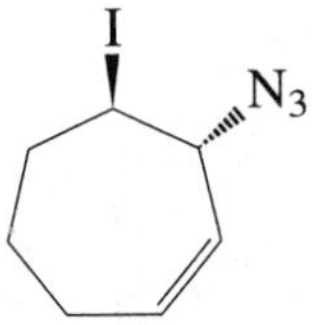 and

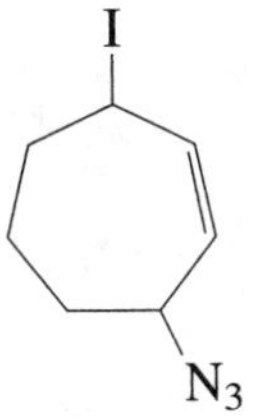

(d) 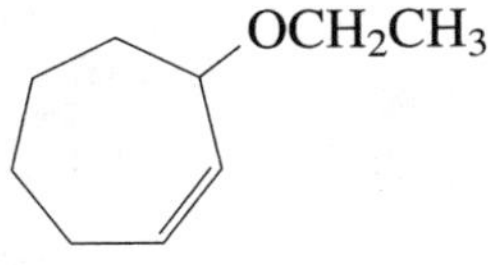 From both 1, 2- and 1, 4-additions

Note in parts (b) and (c) that, while 1,4-addition to conjugated dienes is stereorandom, 1,2-addition is often stereospecific, suggesting a halonium ion-like intermediate. See structures B and C in Exercise 14-13.

50. Addition of the electrophile will always be at C1, generating the best allylic cation. The 1,2-addition product is given first.

(a) CH_3—$\overset{\overset{\displaystyle I}{|}}{CH}$—$CH{=}CH$—$CH_3$ (cis and trans)

(b) $BrCH_2$—$\overset{\overset{\displaystyle OH}{|}}{CH}$—$CH{=}CH$—$CH_3$ and $BrCH_2$—$CH{=}CH$—$\overset{\overset{\displaystyle OH}{|}}{CH}$—$CH_3$ (cis and trans)

(c) ICH_2—$\overset{\overset{\displaystyle N_3}{|}}{CH}$—$CH{=}CH$—$CH_3$ and ICH_2—$CH{=}CH$—$\overset{\overset{\displaystyle N_3}{|}}{CH}$—$CH_3$ (cis and trans)

(d) CH_3—$\underset{\underset{\displaystyle OCH_2CH_3}{|}}{CH}$—$CH{=}CH$—$CH_3$ (cis and trans)

51. **(a)** $(CH_3)_2\overset{\overset{\displaystyle I}{|}}{C}$—$CH{=}CH$—$CH_3$ (cis and trans) and $(CH_3)_2C{=}CH$—$\overset{\overset{\displaystyle I}{|}}{CH}$—$CH_3$

(b), (c) Same answers as Problem 50, but with a methyl group added to C2 in each case.

(d) $(CH_3)_2\underset{\underset{\displaystyle OCH_2CH_3}{|}}{C}$—$CH{=}CH$—$CH_3$ (cis and trans) and $(CH_3)_2C{=}CH$—$\underset{\underset{\displaystyle OCH_2CH_3}{|}}{CH}$—$CH_3$

52. In each case the electrophile adds to give the more stable allylic cation. Thus, addition to C1 is observed:

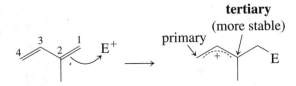

in preference to addition to C4:

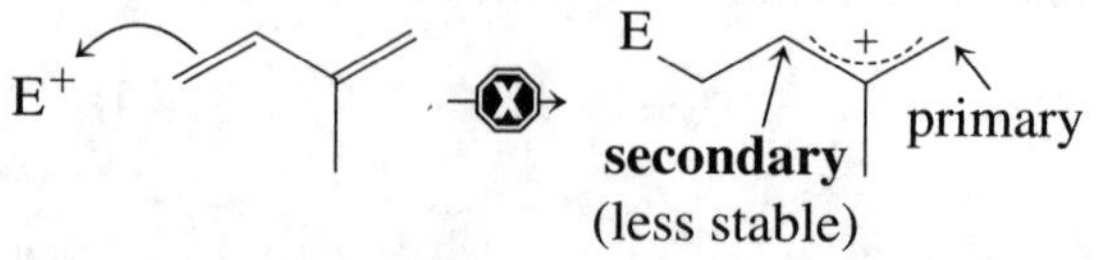

Final products in each case have already been given in the solution to Problem 51.

(a)

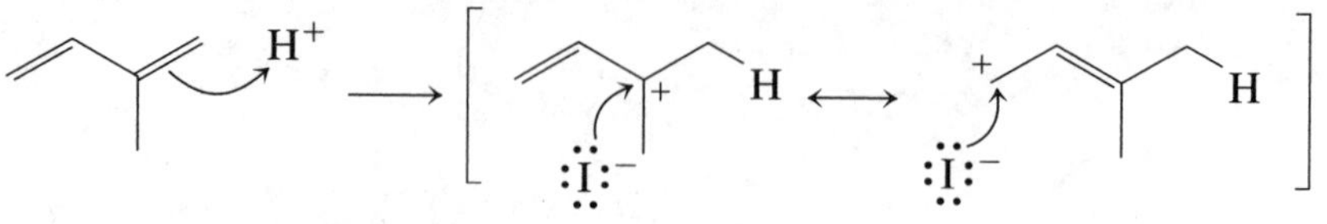

(b)

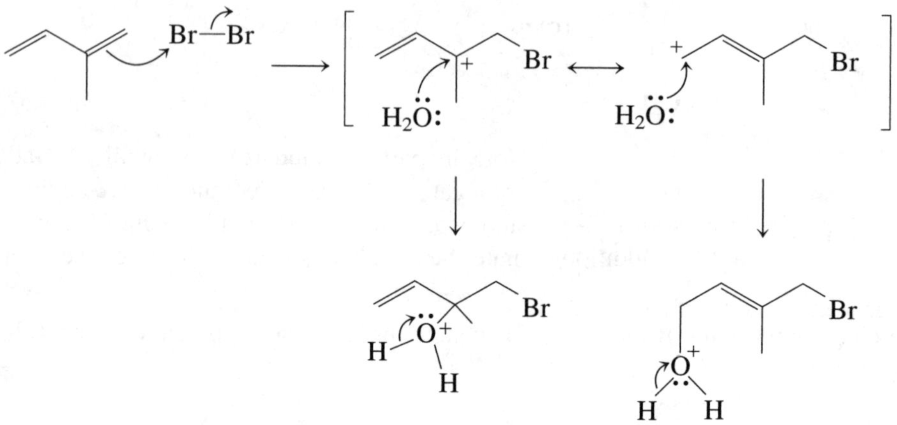

(c)

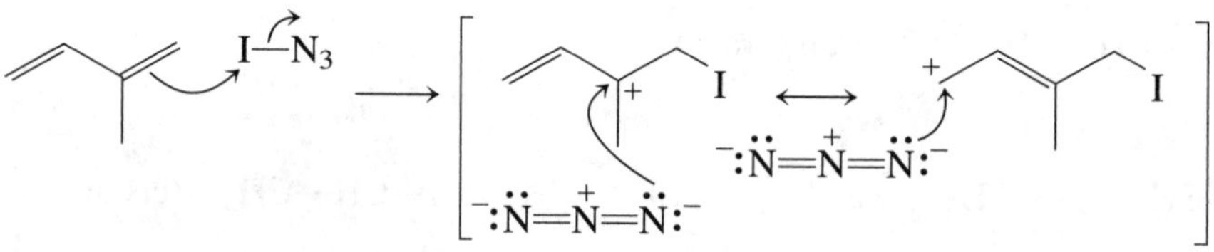

(d)

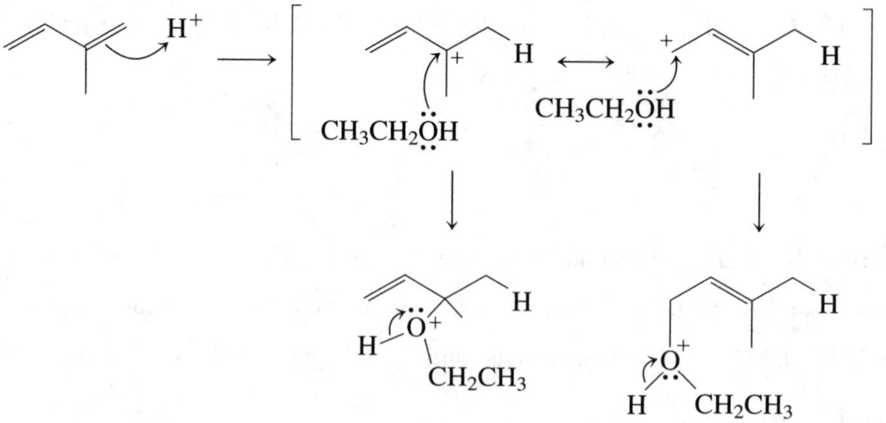

53. (a)

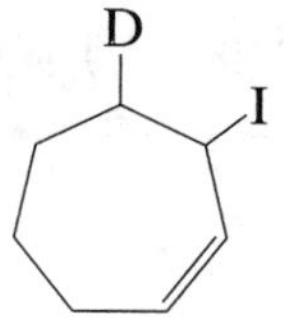

and

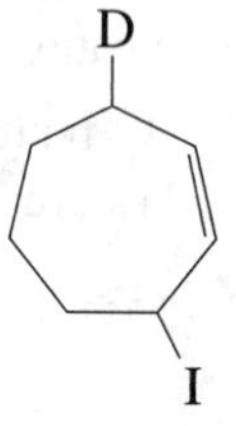

(b) $\overset{\text{D}}{\underset{|}{\text{CH}_2}}-\overset{\text{I}}{\underset{|}{\text{CH}}}-\text{CH}=\text{CH}-\text{CH}_3$ $\text{CH}_3-\overset{\text{I}}{\underset{|}{\text{CH}}}-\text{CH}=\text{CH}-\overset{\text{D}}{\underset{|}{\text{CH}_2}}$

(c) $\overset{\text{D}}{\underset{|}{\text{CH}_2}}-\overset{\text{I}}{\underset{\underset{\text{CH}_3}{|}}{\text{C}}}-\text{CH}=\text{CH}-\text{CH}_3$ and $\overset{\text{D}}{\underset{|}{\text{CH}_2}}-\overset{}{\underset{\underset{\text{CH}_3}{|}}{\text{C}}}=\text{CH}-\overset{\text{I}}{\underset{|}{\text{CH}}}-\text{CH}_3$

With DI it is easy to distinguish between 1,2- and 1,4-addition in the case of the cyclic diene in Problem 44 and the unbranched acyclic diene in Problem 45.

54. **(e)** $[\text{CH}_2=\text{CH}-\overset{+}{\text{CH}}-\text{CH}=\text{CH}_2 \longleftrightarrow \overset{+}{\text{CH}_2}-\text{CH}=\text{CH}-\text{CH}=\text{CH}_2 \longleftrightarrow$

$\text{CH}_2=\text{CH}-\text{CH}=\text{CH}-\overset{+}{\text{CH}_2}] > $ **(d)** $[\text{CH}_3-\overset{+}{\text{CH}}-\text{CH}=\text{CH}-\text{CH}_3 \longleftrightarrow$

$\text{CH}_3-\text{CH}=\text{CH}-\overset{+}{\text{CH}}-\text{CH}_3]$ (secondary allylic at both ends) $>$

(a) $\overset{+}{\text{CH}_2}-\text{CH}=\text{CH}_2 > $ **(c)** $> $ **(b)**

55.

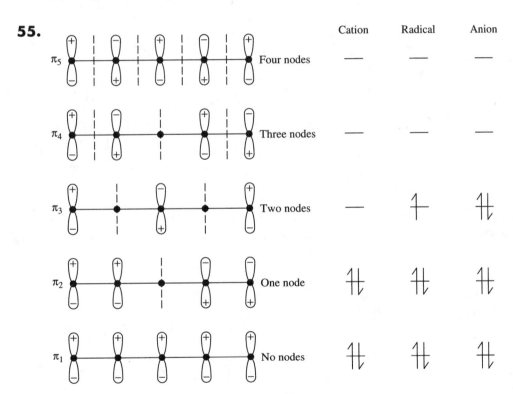

	Cation	Radical	Anion
π_5 Four nodes	—	—	—
π_4 Three nodes	—	—	—
π_3 Two nodes	—	↿	↿⇂
π_2 One node	↿⇂	↿⇂	↿⇂
π_1 No nodes	↿⇂	↿⇂	↿⇂

See the answer to problem 48(e) for the resonance forms of the cation and the answer to Problem 41 for the resonance forms of the radical.

56. $(\text{CH}_3)_2\text{C}=\text{CH}-\text{CH}_2-\overset{..}{\underset{..}{\text{O}}}\text{H} \xrightarrow{\text{H}^+} (\text{CH}_3)_2\text{C}=\text{CH}-\text{CH}_2-\overset{+}{\text{O}}\text{H}_2$

$\xrightarrow{-\text{H}_2\text{O}} \overset{\text{H}}{\text{CH}_2}\overset{}{-}\overset{\text{CH}_3}{\underset{|}{\text{C}}}=\text{CH}-\overset{+}{\text{CH}_2} \xrightarrow{-\text{H}^+}$ product

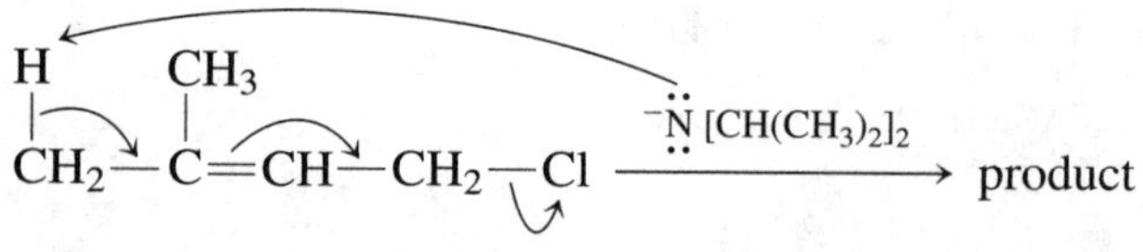

57. Any allylic hydrogen may be lost from the intermediate cation (arrows):

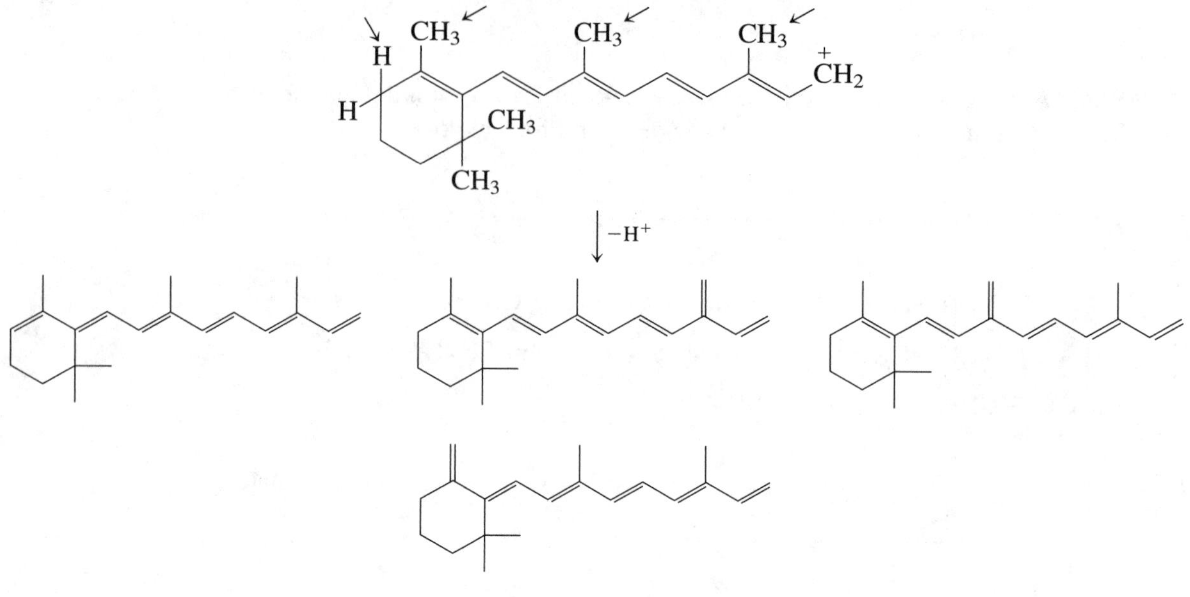

58. (a) 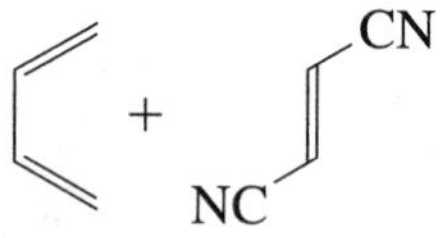 **(b)** 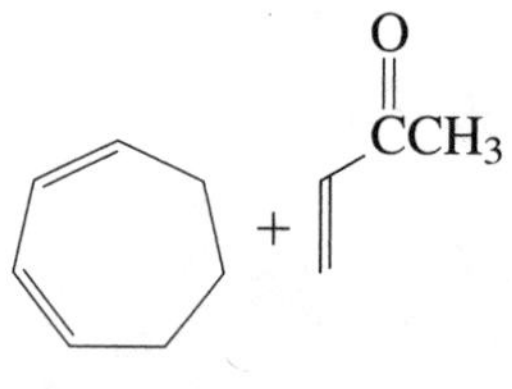 **(c)**

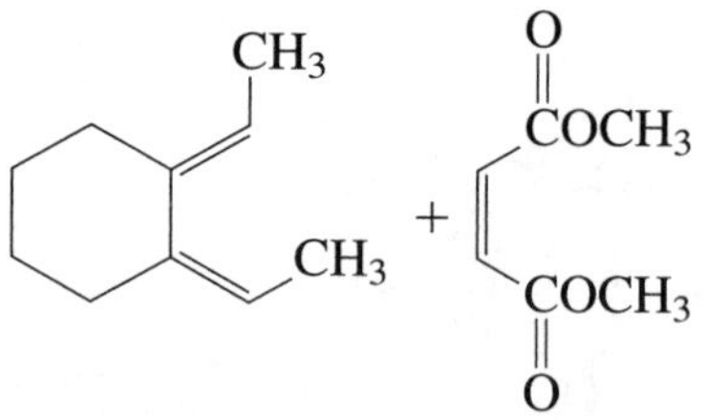

59. (a) Identify the bonds in the products formed in the Diels-Alder reaction. Then work backward to the starting molecules.

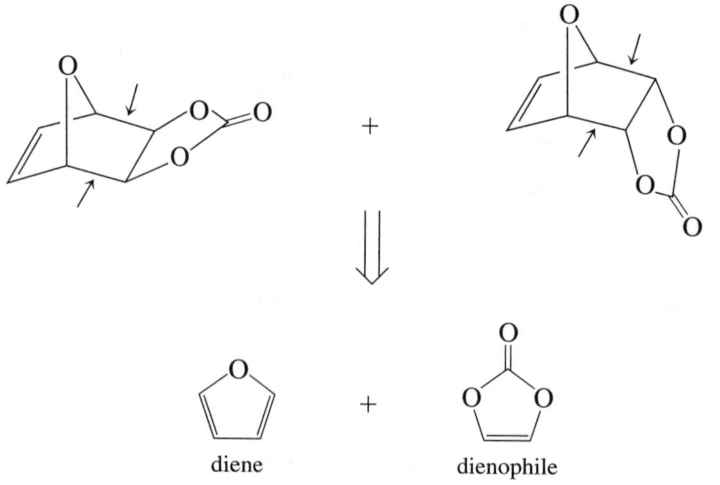

diene   dienophile

(b) See above.

(c) B is the *endo* adduct and is favored, as is expected for the typical Diels-Alder process.

60. (a) $H_2C\!=\!CH\!-\!CH_2\!-\!OCH_3$

(b) $H_2C\!=\!CH\!-\!CHBr\!-\!CH_3 + CH_3\!-\!CH\!=\!CH\!-\!CH_2Br$ (cis and trans)

(c)

(d) $CH_3CH_2CHClCH\!=\!CHCH_3 + CH_3CH_2CH\!=\!CHCHClCH_3$
(both as mixtures of stereoisomers)

(e) $CH_3CHBrCHOHCH\!=\!CHCH_3 + CH_3CHBrCH\!=\!CHCHOHCH_3$
(both as mixtures of stereoisomers)

(f)

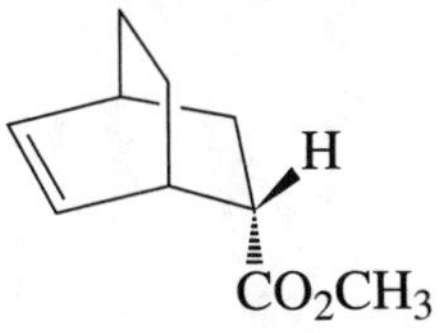

(g)

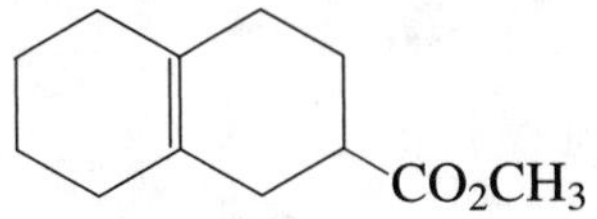

61. Taking note of the hint, recall from Section 14-8 in particular that alkenes containing electron-withdrawing groups are especially good dienophiles in Diels-Alder reactions. So,

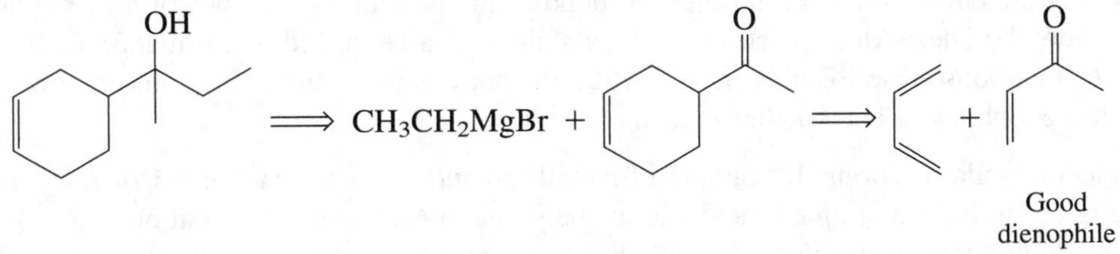

The synthesis, therefore, would go as follows:

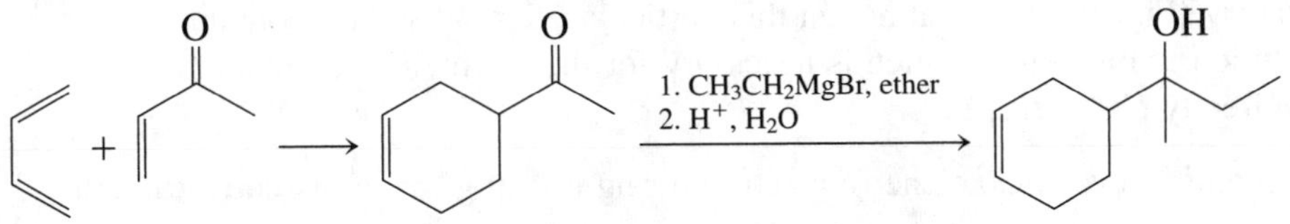

62. (a)

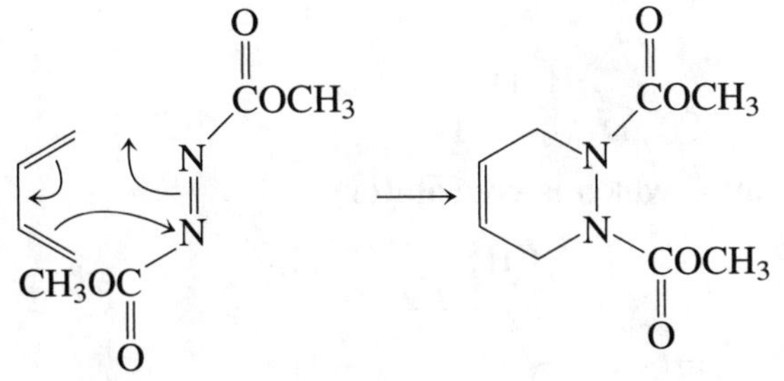

(b)

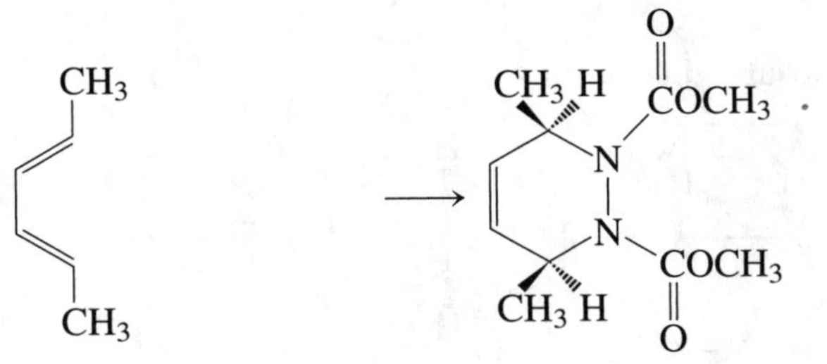

(c)

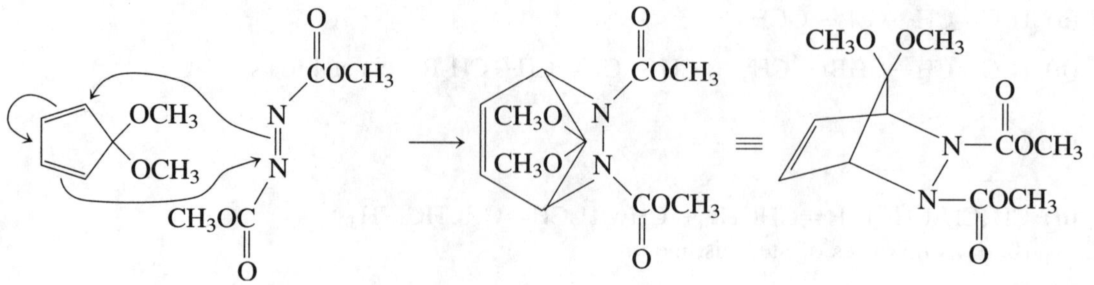

(d)

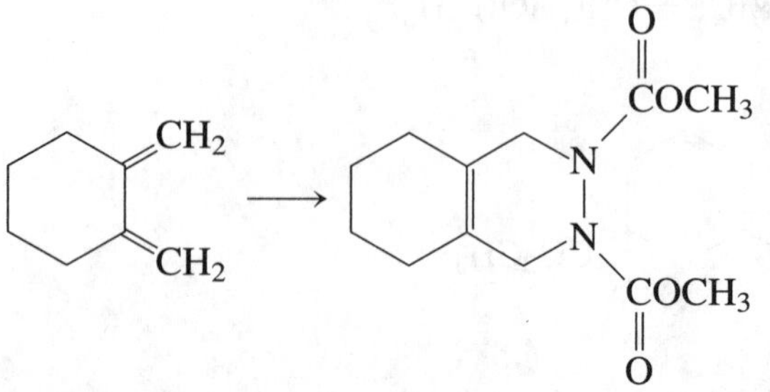

63. Diene A is similar in structure to 1,3-cyclohexadiene, which reacts well in Diels-Alder cycloadditions (Table 14-1). In diene B, however, the ends of the diene are locked in a zigzag conformation (called "s-trans") that puts the end carbons too far apart to bond to the two alkene carbons of a dienophile. Figure 14-9 illustrates how the Diels-Alder reaction involves a diene in a U-shaped conformation (called "s-cis"), which puts its end carbons close together. Dienes (like B) that cannot achieve this conformation will not participate with dienophiles in Diels-Adler reactions.

64. All are electrocyclic reactions. If you are a football fan and know that a **T**ouch**D**own is worth **6** points, then you are eligible to learn a simple mnemonic, namely, that a **6**-electron **T**hermal electrocyclic reaction is **D**isrotatory. Changing either from thermal to photochemical or the number of electrons by ± 2 changes the rotation direction.

(a) A photochemical 1,3-diene ring closure (4 electrons). [Mechanism is disrotatory (Figure 14-11), although this particular diene does not contain substitution at both ends, which is necessary for the disrotatory outcome to be directly observed.]

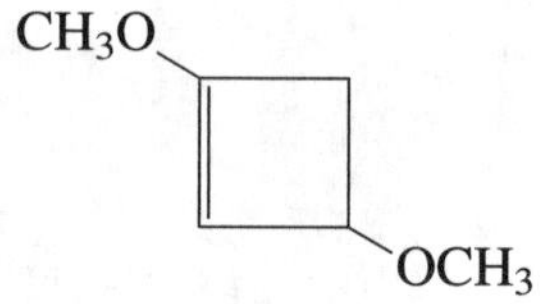

(b) Photochemical cyclohexadiene (6 electrons) ring opening, which is **conrotatory.**

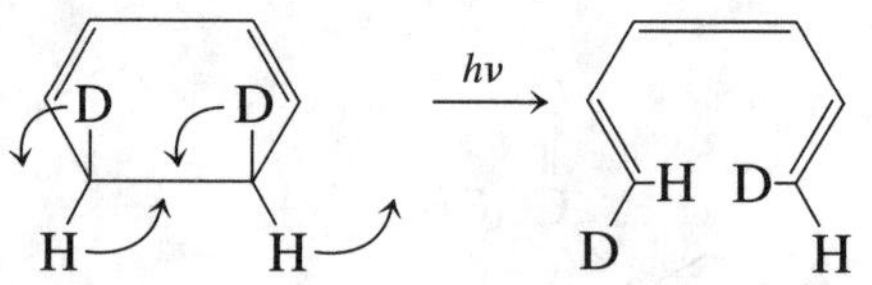

(c) Thermal cyclobutene (4 electrons) ring opening, which is **conrotatory.**

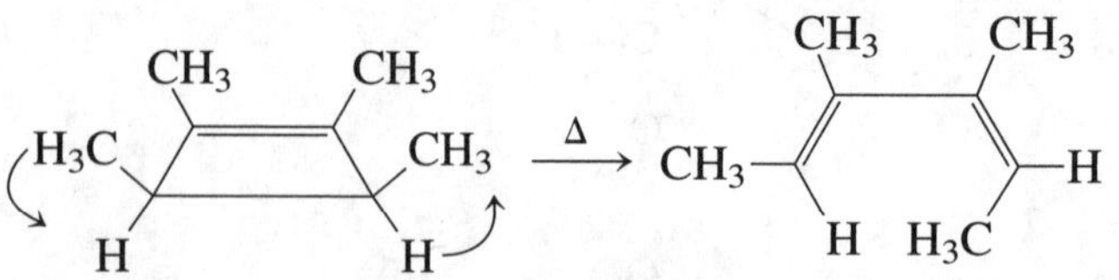

(d) Thermal hexatriene (6 electrons) ring-closure: **disrotatory.**

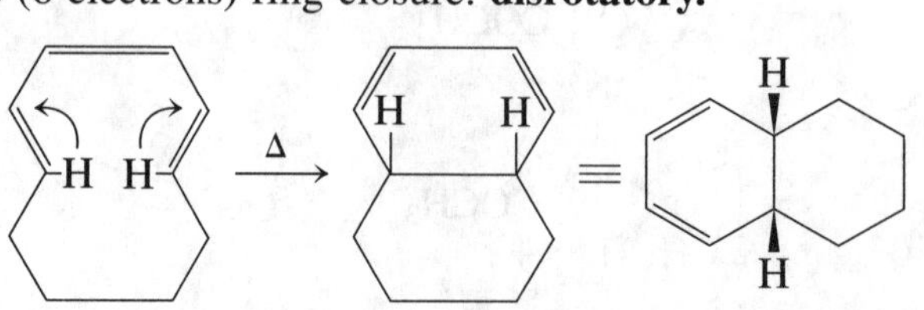

65. **(a)** The first reaction is an electrocyclic ring closure involving the six π electrons of a conjugated triene. In order to determine whether the process shown requires the input of either heat or light, we must examine its stereochemistry to see if it involves a conrotatory or a disrotatory motion. Remember our American football mnemonic: TD = 6, or Thermal/Disrotatory corresponds to six electrons. Therefore, if the six-electron process shown is disrotatory, it will be thermal; if conrotatory, photochemical. Examining the process in detail and using the movement of the two hydrogens at the ends of the conjugated triene as a guide, we have

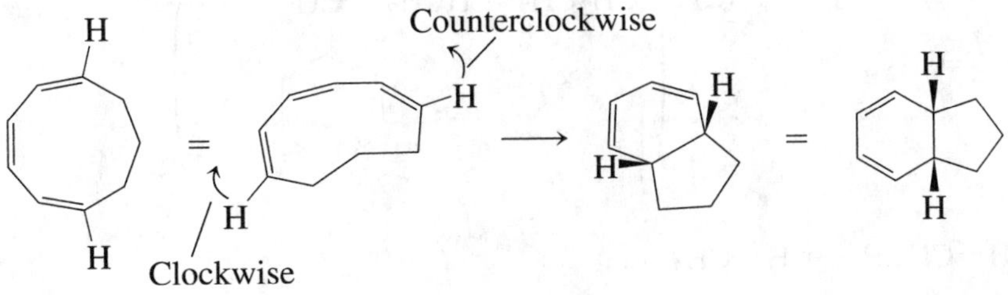

with the arrows showing the motions of the two relevant hydrogen atoms. They move *toward* each other. To accomplish this motion, the C–H bond on the left rotates *clockwise* relative to the plane of the triene, while the C–H bond on the right rotates *counterclockwise*. Opposite directions of rotation correspond to a disrotatory process; thus, according to the mnemonic given above, this process is thermal, requiring the input of heat.

(b) The second transformation gives an isomer of the original all-*cis* triene, in this case *cis, cis, trans*-1,3,5-cyclononatriene. It is a ring-opening process, but still one involving six electrons, four in the two π bonds and two in the cleaving σ bond. The rules and procedure are the same: We must ascertain whether it is con- or disrotatory.

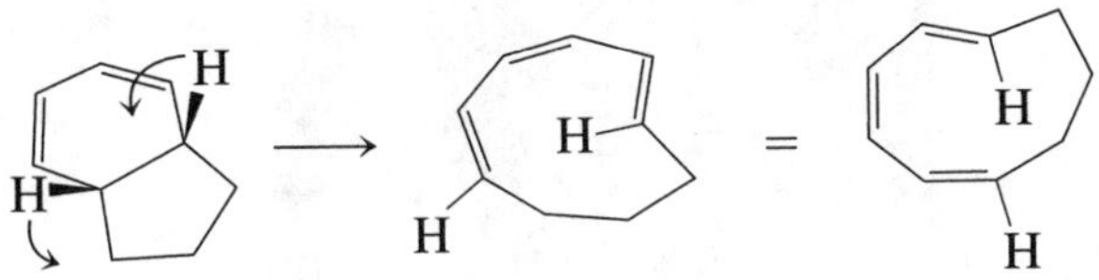

Here we see that *both* C–H bonds undergo counterclockwise rotation in order to place the two relevant hydrogen atoms in their final positions. This is therefore a conrotatory six-electron process, requiring the input of light to permit it to occur.

(c) For the final reaction, we have a second six-electron ring closure. Conrotatory or dis-?

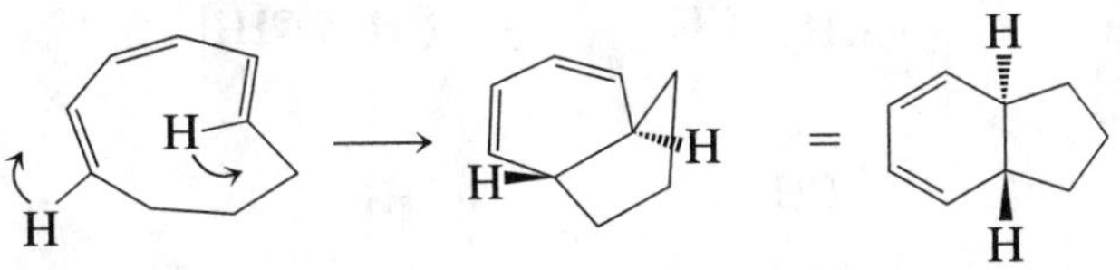

As in reaction (a), the C–H bonds rotate opposite one another: This process is disrotatory in nature and therefore another thermal transformation.

66.

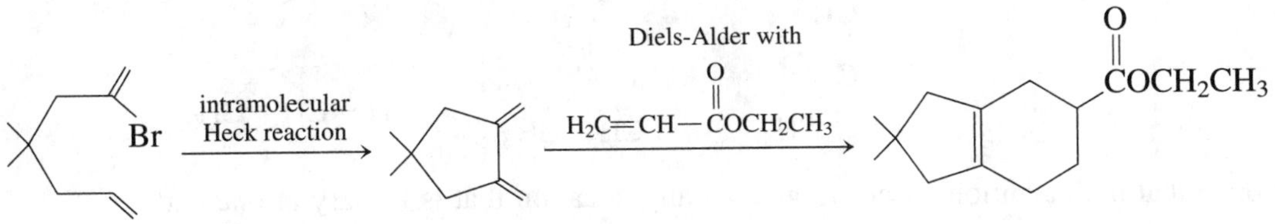

67. (a)

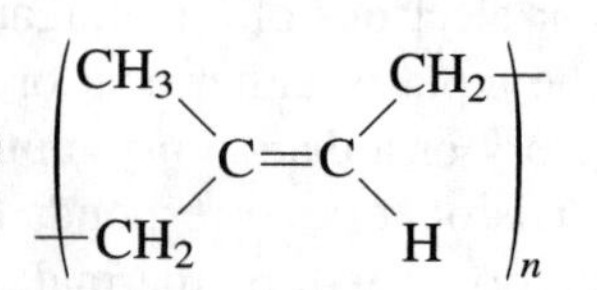

$$\left(\begin{array}{c} CH_3 \quad\quad CH_2 \\ \quad C=C \\ CH_2 \quad\quad H \end{array} \right)_n$$

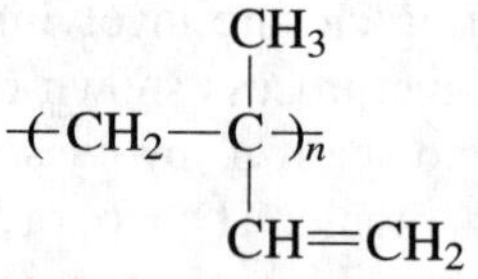

(b)

$$\left(CH_2 - \overset{\overset{\displaystyle CH_3}{|}}{\underset{\underset{\displaystyle CH=CH_2}{|}}{C}} \right)_n$$

(c)

$$\left(CH_2 - \overset{\overset{\displaystyle }{}}{\underset{\underset{\displaystyle CH_3-C=CH_2}{|}}{CH}} \right)_n$$

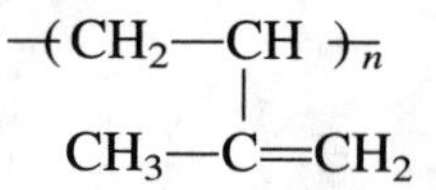

(d)

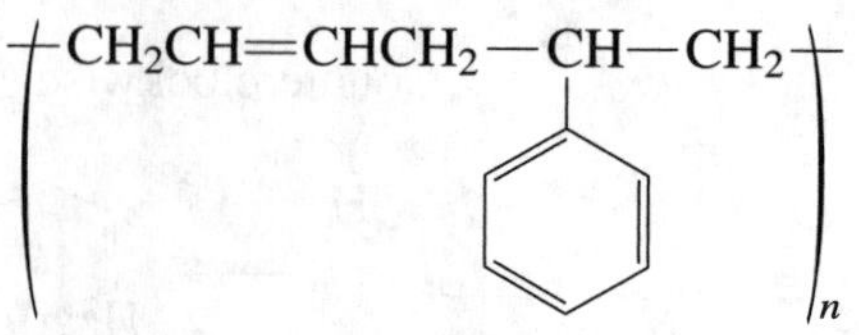

$$\left(CH_2CH=CHCH_2 - \overset{\overset{\displaystyle }{}}{\underset{\underset{\displaystyle \text{(phenyl)}}{|}}{CH}} - CH_2 \right)_n$$

(e)

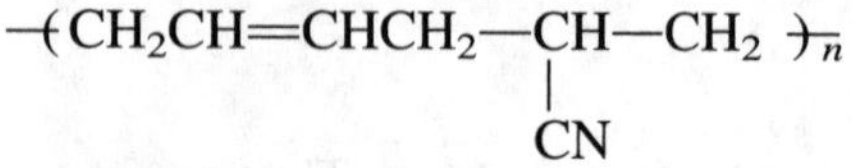

$$\left(CH_2CH=CHCH_2 - \overset{\overset{\displaystyle }{}}{\underset{\underset{\displaystyle CN}{|}}{CH}} - CH_2 \right)_n$$

(f)

$$\left(CH_2\overset{\overset{\displaystyle CH_3}{|}}{C}=CHCH_2 - \overset{\overset{\displaystyle CH_3}{|}}{\underset{\underset{\displaystyle CH_3}{|}}{C}} - CH_2 \right)_n$$

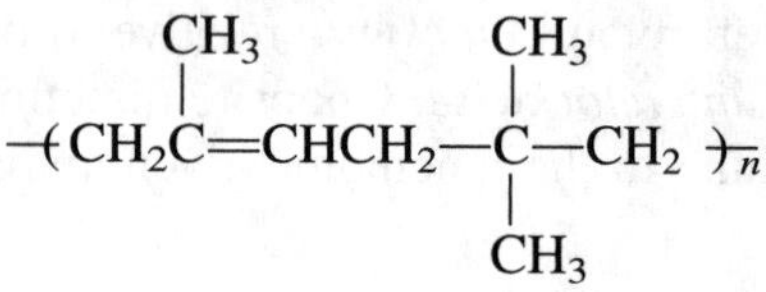

68.

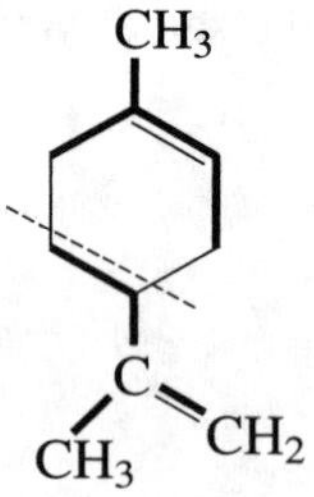

(a)

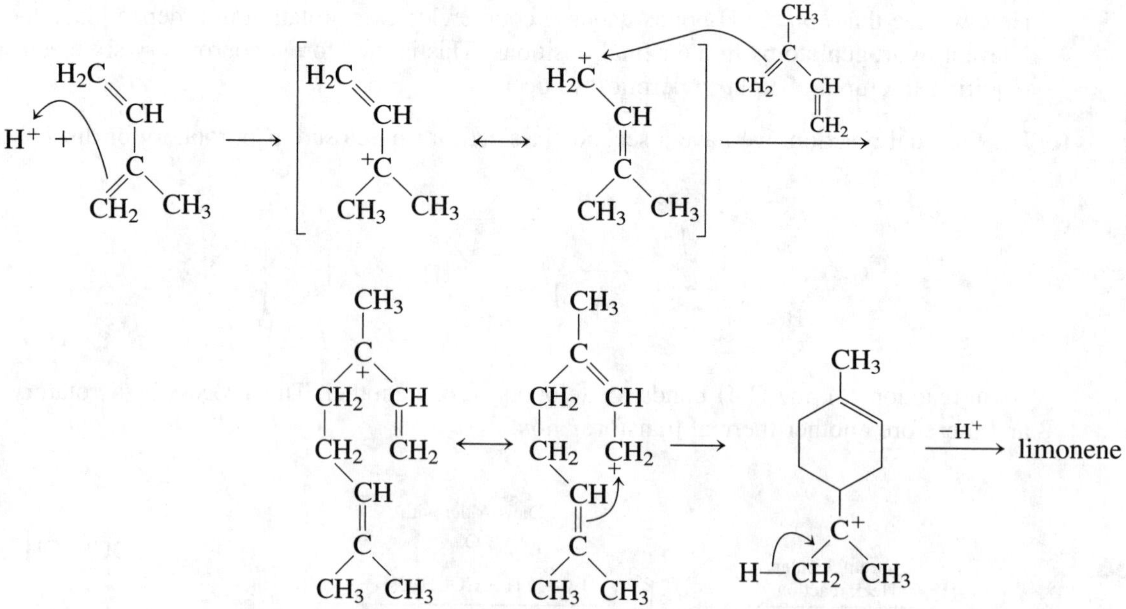

Notice that both additions occur to give an allylic cation that is tertiary at one end.

(b) 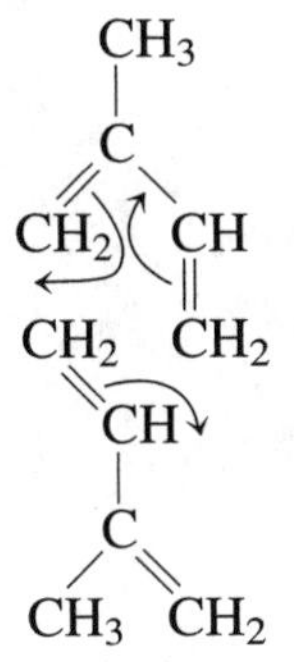 $CH_2 \quad CH_2 \longrightarrow$ limonene   Via a Diels-Alder concerted cycloaddition reaction

69.

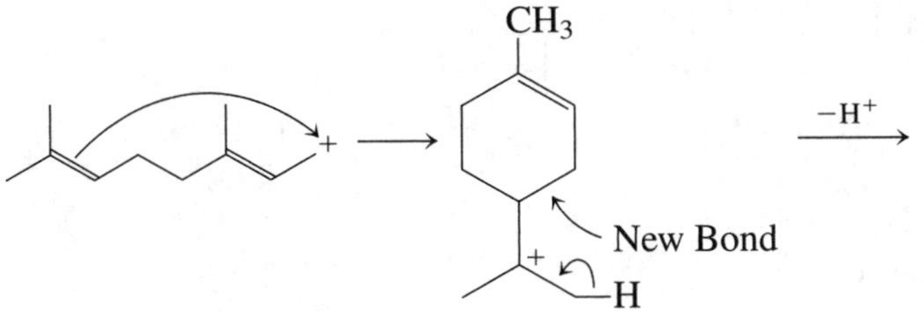

$\xrightarrow{-H^+}$ limonene

New Bond

Alternatively,

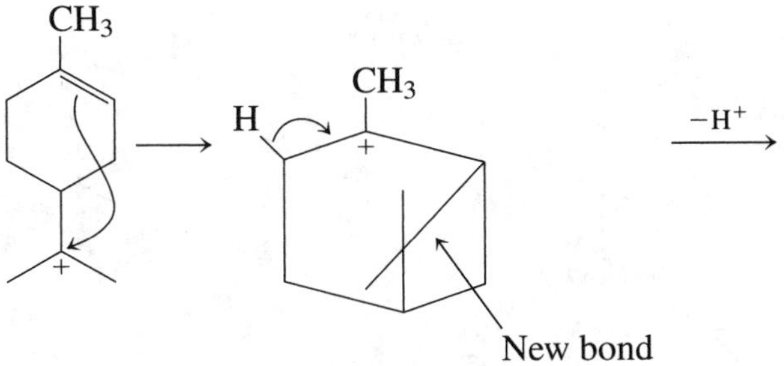

$\xrightarrow{-H^+}$ pinene

New bond

70. In each case the process involves promotion of an electron in the highest occupied molecular orbital (n or π) to the lowest unoccupied molecular orbital (e.g., π^*). Use subscripts when there is more than one orbital of a given type. **(a)** $\pi \rightarrow n$ (alternatively, $\pi_1 \rightarrow \pi_2$); **(b)** $n \rightarrow \pi^*$ (alternatively, $\pi_2 \rightarrow \pi_3$); **(c)** $n \rightarrow \pi^*$; **(d)** $n \rightarrow \pi^*$ (more specifically, $n \rightarrow \pi_3^*$); **(e)** $n \rightarrow \pi^*$ (alternatively, $\pi_3 \rightarrow \pi_4^*$); **(f)** $\pi_3 \rightarrow \pi_4^*$

71. These molecules contain only σ and n electrons, and their lowest unoccupied molecular orbitals are σ^*. The energy gaps between these orbitals are large, resulting in UV absorptions only at wavelengths shorter than 200 nm.

72. $1.95/(2 \times 10^{-4}) = 9750$; $0.008/(2 \times 10^{-4}) = 40$

73.

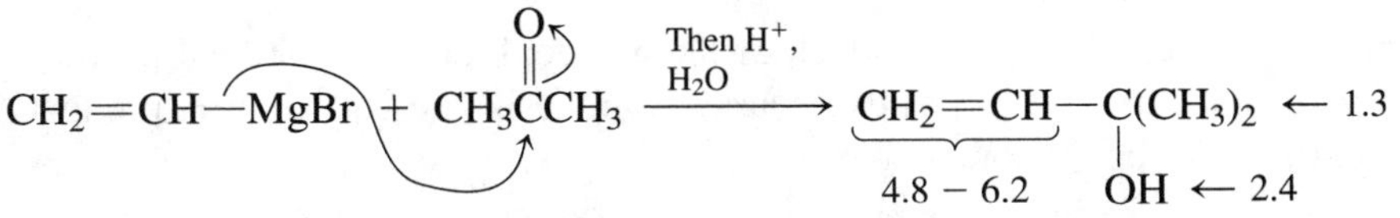

The new compound may be derived from ionization of the allylic alcohol upon extended contact with acid:

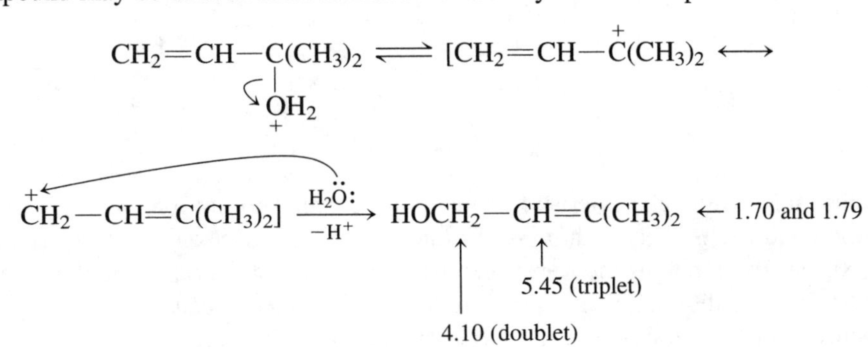

74.

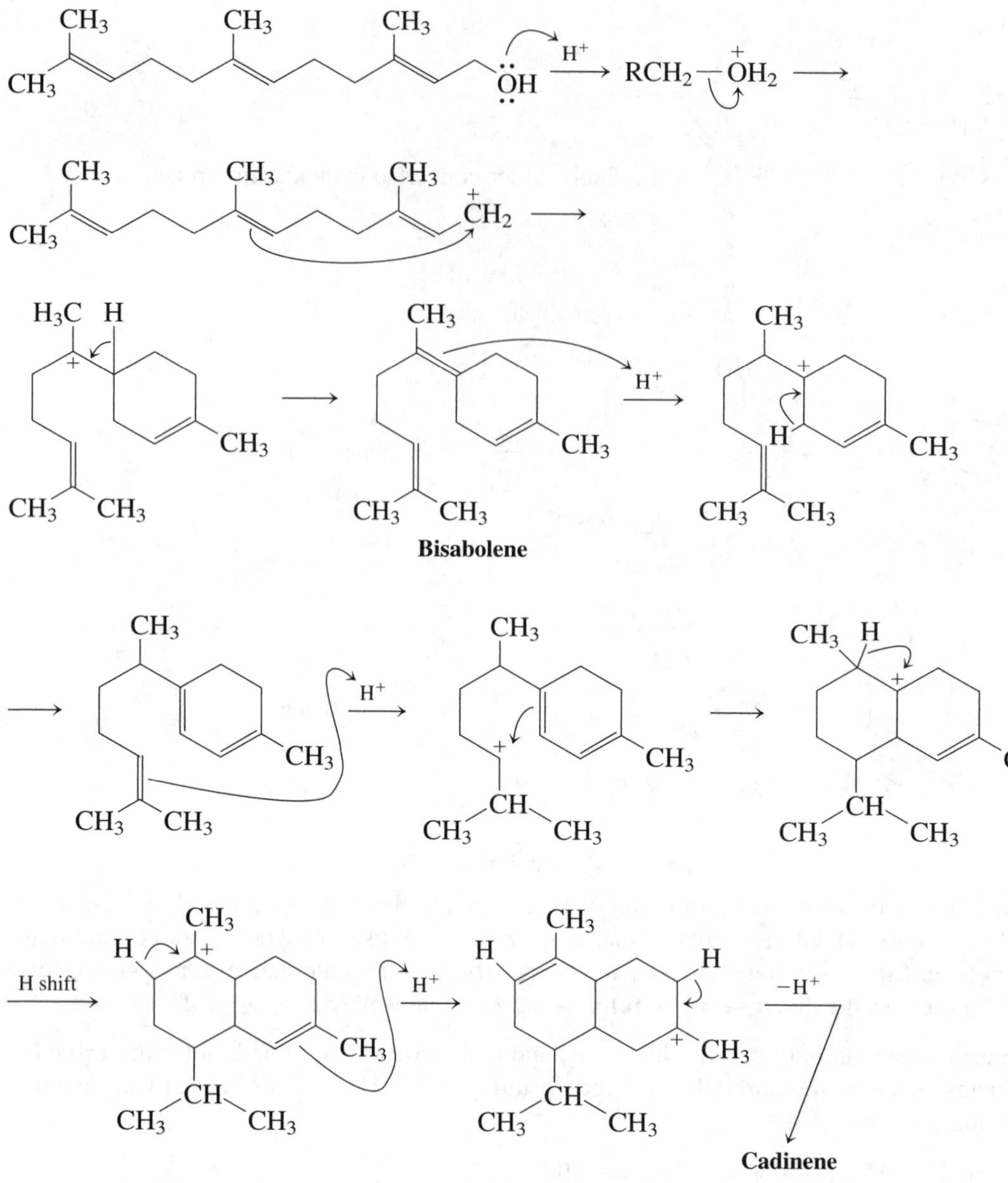

Bisabolene

Cadinene

75. The 1,2-addition product is kinetic, arising from attack of the nucleophile at the internal position of the intermediate allylic cation, where concentration of positive charge is highest. The 1,4-addition product contains an internal double bond, rendering it more stable.

76.

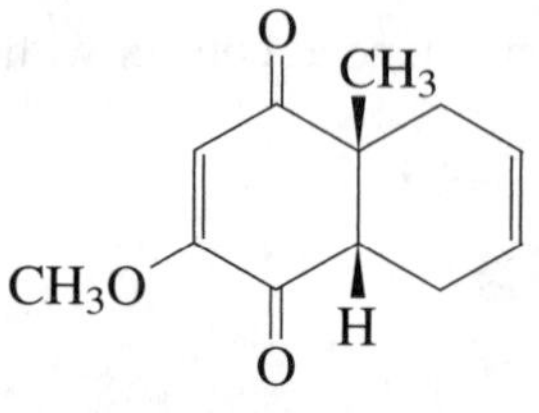

This example illustrates several characteristic features of the cycloaddition process. Only the methyl-substituted double bond reacts. The methoxy-substituted one is too electron rich (see Exercise 14-14) to compete. The cis ring fusion retains the stereochemistry of the double bond in the dienophile. Finally, further cycloadditions of additional diene to the product are inhibited because neither of the double bonds in the latter is sufficiently electron deficient to react at an appreciable rate.

77. Ring closure of a 1,3-butadiene to a cyclobutene is the operative process. In order to bring the ends of the diene together to form the central bond in Dewar benzene, the mode of rotation must be disrotatory; otherwise, an impossibly strained product, with one cyclobutene ring substituted in a trans fashion by the other, would result. Disrotatory ring closure of a 1,3-diene is a photochemically promoted process (a photocyclization; see Figure 14-11. The resistance of this particular substituted Dewar benzene to reversion by ring-opening back to the corresponding benzene is in part a consequence of the reluctance of two tertiary alkyl groups to occupy adjacent (i.e., ortho) positions on a planar benzene ring. Dewar benzenes are severely bent structures (make a model!), and the adjacent *tert*-butyl groups in this one are well out of each other's way as a result.

78. (c)

79. (b)

80. (b)

81. (b)

15

Benzene and Aromaticity: Electrophilic Aromatic Substitution

The most obvious feature of benzene is its superficial similarity to a molecule containing ordinary double bonds. Benzene and its derivatives are special, however, as a result of the new concept of aromatic stabilization covered in the early sections of this chapter. The electronic structure of benzene, which gives this compound unusual stability relative to that of an ordinary triene, profoundly affects benzene's chemistry. Each time you encounter a new property or chemical reaction of a benzene derivative, ask yourself, "How would the properties or reactions of a simple alkene (or diene or triene) compare with this?" See if the differences in behavior of benzene and alkenes make sense to you on thermodynamic grounds. After you've done that, you will be in a better position to learn the material in this and the next chapter more thoroughly, with a more balanced overview of the entire topic.

Benzene and its derivatives make up one of the most important classes of organic compounds (after carbonyl compounds and alcohols). There is, accordingly, a lot of relatively significant material presented in these chapters concerning them.

Other kinds of aromatic compounds exist besides simple benzene derivatives. Two common types will be covered in this chapter: polycyclic fused benzenoid hydrocarbons and other cyclic conjugated polyenes with either more or less than six carbons in the ring. In Chapter 25, a third common class, heterocyclic aromatic compounds, will be presented.

Outline of the Chapter

Keys to the Chapter

15-1. Nomenclature

The systematic (IUPAC) naming system for benzenes coexists with an assortment of names that have been in common use for over a century and show no signs of going away. So, whether anyone likes it or not, terms like aniline (for benzenamine) and styrene (for ethenylbenzene) are, and will probably always be, the ones people use, both in speaking as well as writing about these compounds. In addition to these common names for simple substituted benzenes, a special system exists exclusively for benzenes with **exactly two** substituents on the benzene ring: the "ortho/meta/para" system. Note, carefully, that this method is **never** to be used for benzenes with more than two ring substituents: For those, a name with proper numbering is required. (It is okay, however, to use numbering instead of ortho/meta/para for a benzene with two substituents.) Finally, when numbering around the benzene ring, be careful where you start: If you use one of the special names for a monosubstituted benzene as your parent name (e.g., phenol, aniline, benzaldehyde, toluene), C1 is **always** the carbon containing the substituent implied by the parent (even if starting somewhere else would result in smaller numbers). For example, this compound is 3,4-dibromotoluene (1,2-dibromotoluene would be nonsense, although 1,2-dibromo-4-methylbenzene is the correct **IUPAC** name).

15-2 and 15-3. Structure of Benzene: A First Look at Aromaticity; Molecular Orbitals

Aromaticity as a special property of molecules like benzene is covered in these sections. Structural, thermodynamic, and electronic considerations are presented and serve as an introduction to the more general discussion that is presented in Section 15-6. For now, note simply that the aromaticity of benzene is reflected in (1) its symmetrical structure (as a resonance hybrid), (2) its unexpectedly enhanced thermodynamic stability, and (3) its unusual electronic structure with a completely filled set of strongly stabilized bonding molecular orbitals.

15-4. Spectral Properties

The spectroscopy of benzene is a logical consequence of its structural and electronic properties. Special features in the spectra make the identification of benzenes relatively easy. As has been the case with other compounds, NMR is most useful, followed by IR and UV spectroscopy. A special feature in the infrared spectrum of a benzene derivative is the C—H out-of-plane bending vibration pattern, which provides information concerning the arrangement of substituents around the ring. This information is not always as easily derived from the NMR spectrum. Ultraviolet spectroscopy can be suggestive but not definitive regarding the presence of a benzene derivative. If the presence of the benzene ring is already known from, e.g., NMR, the UV spectrum can be useful in deciding whether it is conjugated with one or more π bond-containing groups.

15-5. Polycyclic Aromatic Hydrocarbons

The fusion of benzene rings leads to a large class of polycyclic hydrocarbons, beginning with naphthalene as the simplest (and only bicyclic) example.

15-6 and 15-7. Other Cyclic Polyenes: Hückel's Rule

Really very simple. For a molecule to be aromatic, it must have $(4n + 2)$ π electrons contained in a complete, unbroken circle of p orbitals.

15-8, 15-9, and 15-10. Electrophilic Aromatic Substitution: Halogenation, Nitration, and Sulfonation of Benzene

These sections introduce the main type of chemistry displayed by benzene rings. The keys to this material are in Section 15-8. You are given a general mechanism (which is repeated later in specific form for each type of reaction that is presented). More important for understanding the concepts, you are given information concerning the energetics of this reaction. Pay close attention to Figure 15-20, Exercise 15-22, and the data in Section 15-8. After you understand this basic material, you will find the specific reactions a little easier to learn. Remember that benzene is stabilized by its special aromatic form of resonance. It is **less reactive** toward electrophiles than are ordinary alkenes; and therefore only **strong electrophiles,** sometimes generated in strange ways, will attack the π electrons in benzene rings. These electrophiles, and the ways they are generated, have to be memorized so that you can use these reactions later on in synthesis problems.

15-11, 15-12, and 15-13. Friedel-Crafts Reaction

Because carbon–carbon bond formation is such an important part of organic synthesis, the attachment of carbon electrophiles to benzene rings is of special significance among the reactions in this chapter. Ordinary carbocations and carbocationlike species are the simplest types of carbon electrophiles, but their use in Friedel-Crafts alkylation has limitations. Rearrangements of the carbon electrophile often occur, and it is hard to prevent multiple alkylation from happening to a single benzene ring. Friedel-Crafts *acylation,* via the acylium ion $[R—\overset{+}{C}=\overset{..}{O}: \longleftrightarrow R—C≡\overset{+}{O}:]$, is not subject to these drawbacks and, whenever possible, is the preferred method for attachment of a carbon unit to a benzene ring.

Solutions to Problems

36. **(a)** 3-Chlorobenzenecarboxylic acid, *m*-chlorobenzoic acid

(b) 1-Methoxy-4-nitrobenzene, *p*-nitroanisole

(c) 2-Hydroxybenzenecarbaldehyde, *o*-hydroxybenzaldehyde

(d) 3-Aminobenzenecarboxylic acid, *m*-aminobenzoic acid

(e) 4-Ethyl-2-methylbenzenanime, 4-ethyl-2-methylaniline

(f) 1-Bromo-2,4-dimethylbenzene

(g) 4-Bromo-3,5-dimethoxybenzenol, 4-bromo-3,5-dimethoxyphenol

(h) 2-Phenylethanol

(i) 3-Ethanoylphenanthrene, 3-acetylphenanthrene

37. **(a)** 1,2,4,5-Tetramethylbenzene   **(b)** 4-Hexyl-1,3-benzenediol

(c) 2-Methoxy-4-(2-propenyl)benzenol

38. **(a)** Name is acceptable. (IUPAC: 2-chlorobenzenecarbaldehyde)

(b) Name is numbered incorrectly; call it 1,3,5-benzenetriol.

(c) 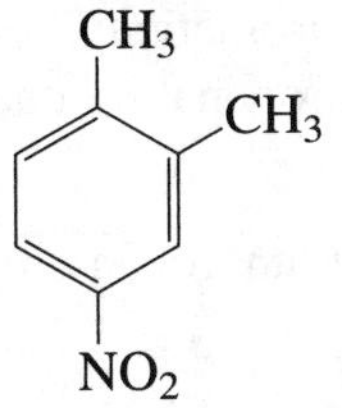 Name is incorrect. Never mix *o, m, p* with numbers; call it 1,2-dimethyl-4-nitrobenzene.

(d) 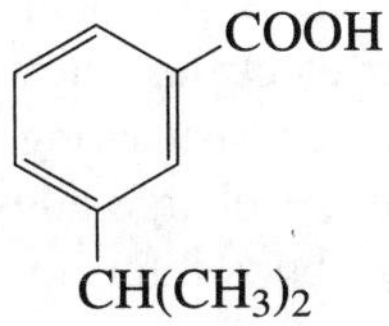 Name is acceptable. (IUPAC: 3-(1-methylethyl)-benzenecarboxylic acid)

(e) 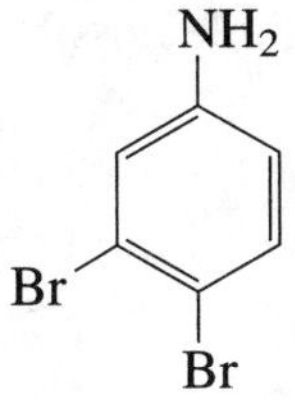 Wrong numbers; 3,4-dibromoaniline or 3,4-dibromobenzenamine.

(f) 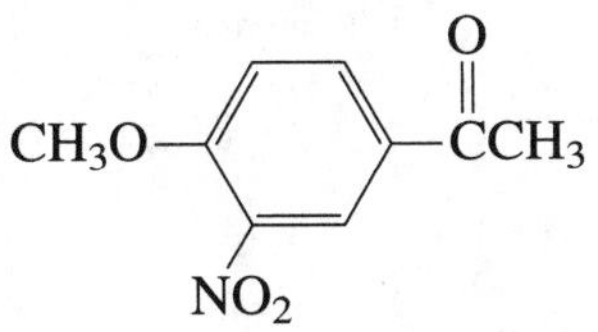 CH₃O— Use **only numbers:** 4-methoxy-3-nitroacetophenone or 1-(4-methoxy-3-nitrophenyl)ethanone.

39. Benzene would be higher in energy by about 30 kcal mol^{-1}, so ΔH_{comb} would be -819 kcal mol^{-1}.

40.

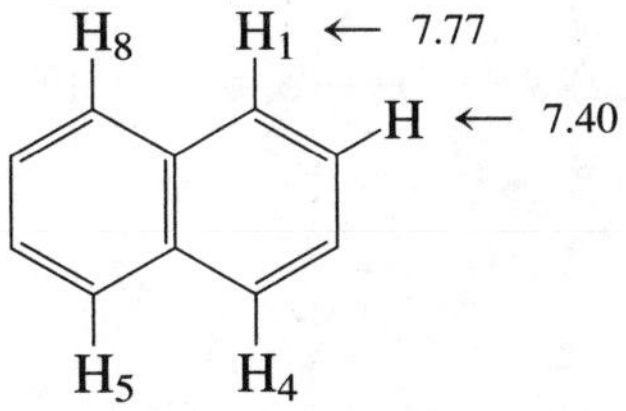

H_8 H_1 ← 7.77

H ← 7.40

H_5 H_4

The hydrogens at carbons 1, 4, 5, and 8 are deshielded because they are closer to the **other** benzene ring in the molecule. They feel the deshielding effects of π-electron ring currents three ways: the ring current around the whole molecule (i), the ring current of their own benzene ring (ii), and the ring current of the **adjacent** benzene ring (iii):

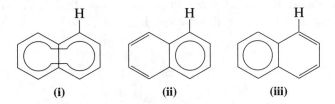

(i)   (ii)   (iii)

The hydrogens at carbons 2, 3, 6, and 7 are too far away to feel a significant amount of the ring current of the other benzene ring.

41. Yes. Cyclooctatetraene is described as lacking any special stabilization such as aromaticity. As a result, hydrogenation of its four double bonds should give off four times the energy that hydrogenating one double bond releases. The data indicate that this is approximately the case.

42. Rule: Aromaticity requires [a] $(4n + 2)$ π electrons contained in [b] a complete, unbroken circle of p orbitals.

(a) No. 3 π electrons. (b) Yes. Benzene is intact; extra double bond is an irrelevant substituent, not being part of the circle. (c) No. The saturated carbon is sp^3, breaking the circle of p orbitals; without a circle of p orbitals, the number of π electrons is irrelevant. (d) Yes. 10 π electrons; the sp^3 carbon here is bridging and does not interrupt the p orbital circle. (e) No. 12 π electrons; wrong number. (f) No. 9 π bonds = 18 π electrons, which would be fine, except that the 2– charge adds 2 more for a total of 20 electrons; wrong number. (g) No. Saturated ring fusion carbons interrupt circle.

43. (a) UV spectrum supports the presence of benzene ring. ^{13}C NMR: three peaks, so the molecule must have symmetry. ^{1}H NMR: two sets of signals in equal intensity. Look at the three possible dibromobenzenes. The answer is clear:

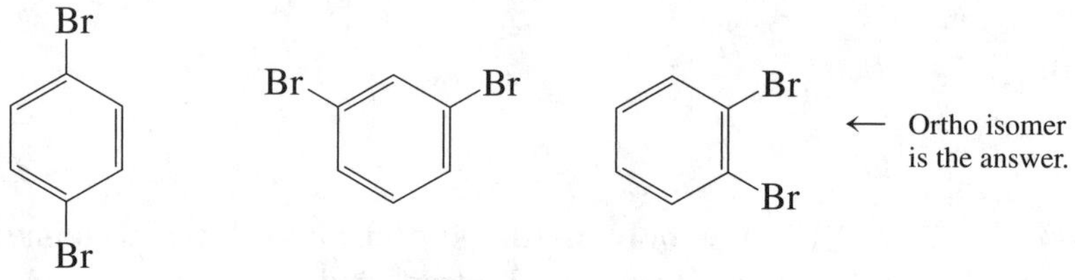

p	**m**	**o**
^{13}C: 2 kinds of carbons	4 kinds of carbons	3 kinds of carbons
^{1}H: All equivalent	3 kinds of hydrogens	2 kinds of hydrogens

The IR (single band at 745 cm^{-1}) agrees with the conclusion.

(b) ^{1}H NMR: four benzene H's (one quite different from the other three), CH$_3$O— (δ = 3.7). IR: meta disubstituted benzene. The answer is

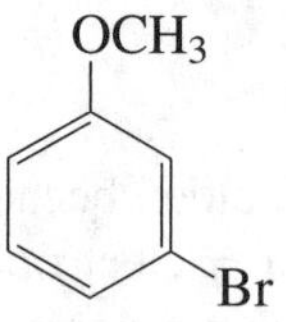

(c) ^{1}H NMR: two benzene H's, three CH$_3$'s (two equivalent; note just **two** methyl carbons in ^{13}C NMR). Also, there are just four benzene carbons in ^{13}C spectrum, so the molecule has some symmetry to it. The answer (use trial and error) is

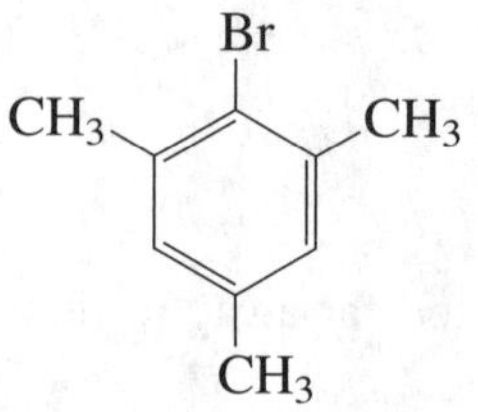

44. The left-hand spectrum corresponds to the benzene derivative. Its general appearance, a strong molecular ion and few peaks of lower mass, reflects a compound that does not undergo ready fragmentation, a characteristic of aromatic compounds. The only significant fragmentation is loss of a single hydrogen atom to

give the resonance-stabilized carbocation $\langle\!\!\!\bigcirc\!\!\!\rangle\!-CH_2^+$ (a system that will be discussed at length in

Chapter 22). In contrast, the mass spectrum at the right shows extensive fragmentation, as one frequently sees with alkynes (Section 13-3). Note, for example, the strong peak at $m/z = 39$ for the resonance-stabilized cation $HC\equiv CCH_2^+$.

45. (a) Yes it is. Compare the answer to Problem 43 (a) and you'll see that each disubstitution pattern gives a unique number of ^{13}C peaks for the ring carbons. Adding the peak for the two equivalent methoxy carbons in each molecule, we expect the para isomer to have three ^{13}C peaks, the meta five, and the ortho four.

(b) The ten possible isomers and the number of ^{13}C peaks (ring carbons + methoxy carbons, which are no longer necessarily equivalent) for each are given below.

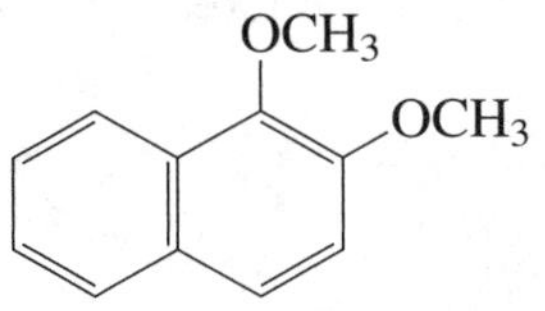

10 ring + 2 CH$_3$O   10 ring + 2 CH$_3$O   5 ring + 1 CH$_3$O   5 ring + 1 CH$_3$O

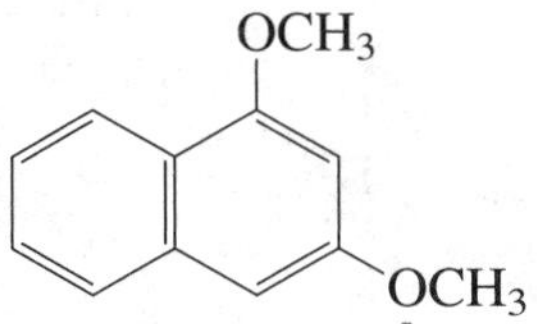

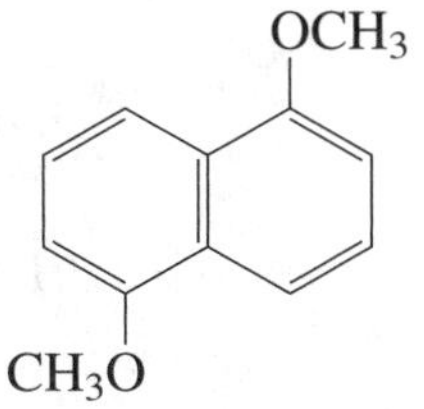

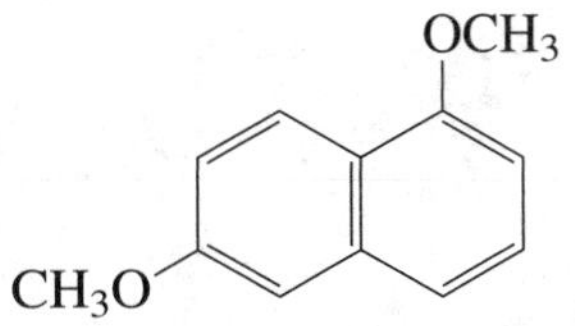

10 ring + 2 CH$_3$O   10 ring + 2 CH$_3$O   6 ring + 1 CH$_3$O

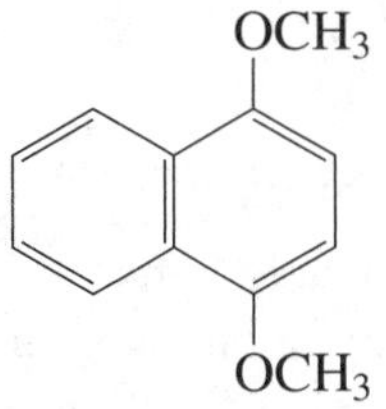

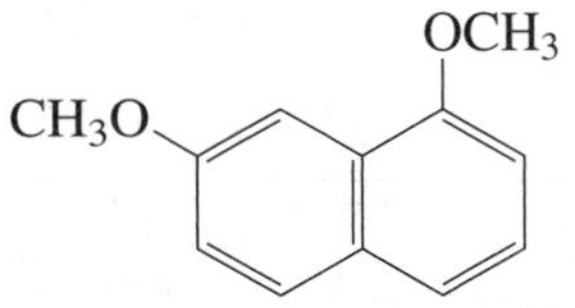

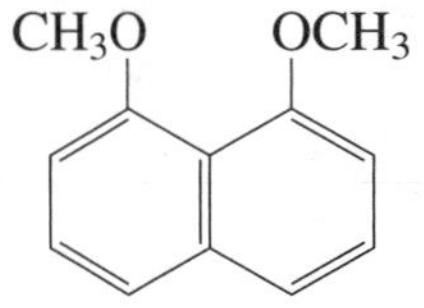

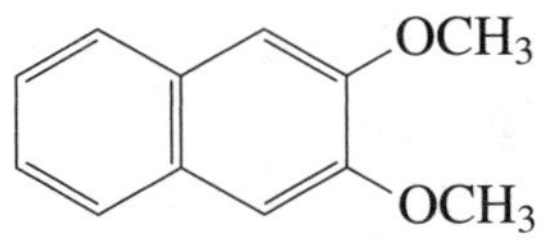

5 ring + 1 CH$_3$O   5 ring + 1 CH$_3$O   6 ring + 1 CH$_3$O

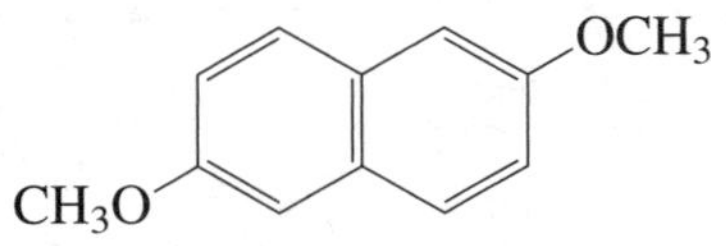

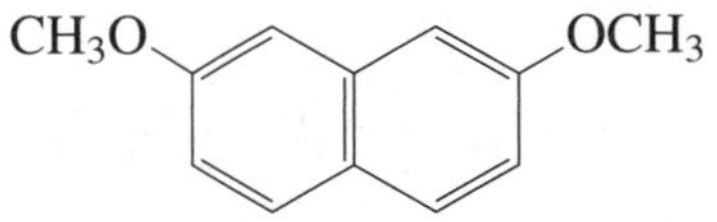

46.

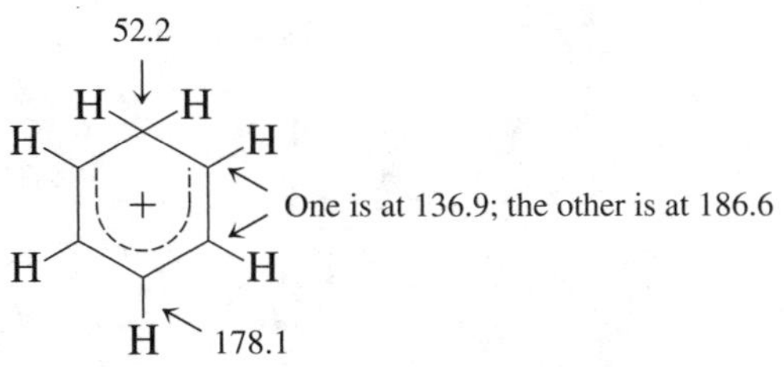

52.2

One is at 136.9; the other is at 186.6

178.1

Assign the signals at 136.9 and 186.6 ppm more precisely by looking at the resonance forms:

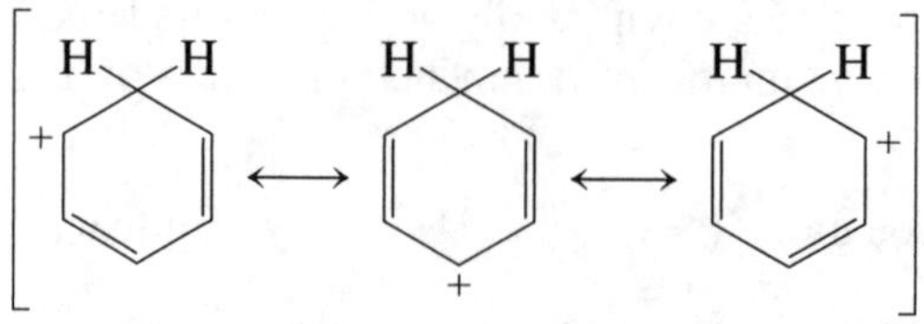

In the resonance forms, positive charges are located at only three of the carbons: They should be the most deshielded. That explains the δ = 178.1 chemical shift of the "bottom" carbon. The δ = 186.6 signal must therefore correspond to the positively charged "end" carbons of the delocalized cation:

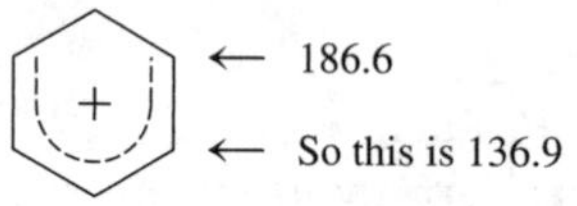

← 186.6

← So this is 136.9

47. (a) $(CH_3)_3CCl$, $AlCl_3$; (b) Cl_2, $FeCl_3$; (c) H_2, Pt; (d) HNO_3, H_2SO_4; (e) CH_3COCl, $AlCl_3$; (f) CH_3CH_2Cl, $AlCl_3$; (g) SO_3, H_2SO_4; (h) Br_2, $FeBr_3$.

48. (a) 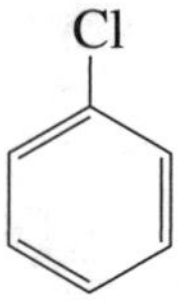 (b) 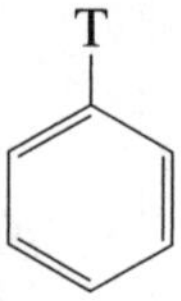 initially; 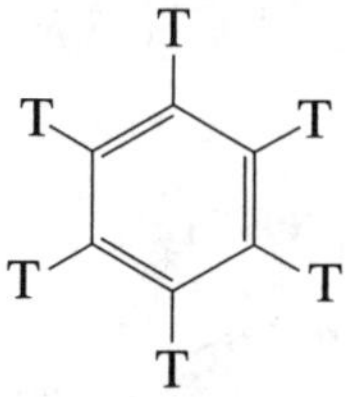 eventually (Compare Exercise 15-23)

(c), (e) 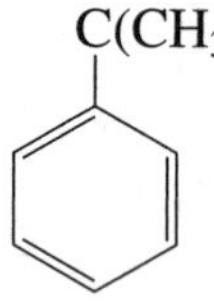 $C(CH_3)_3$ Friedel-Crafts alkylation involving $(CH_3)_3C^+$ cation

(d) 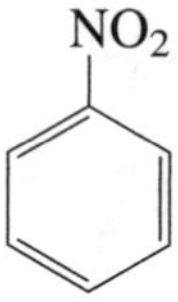 NO_2

(f) Careful!

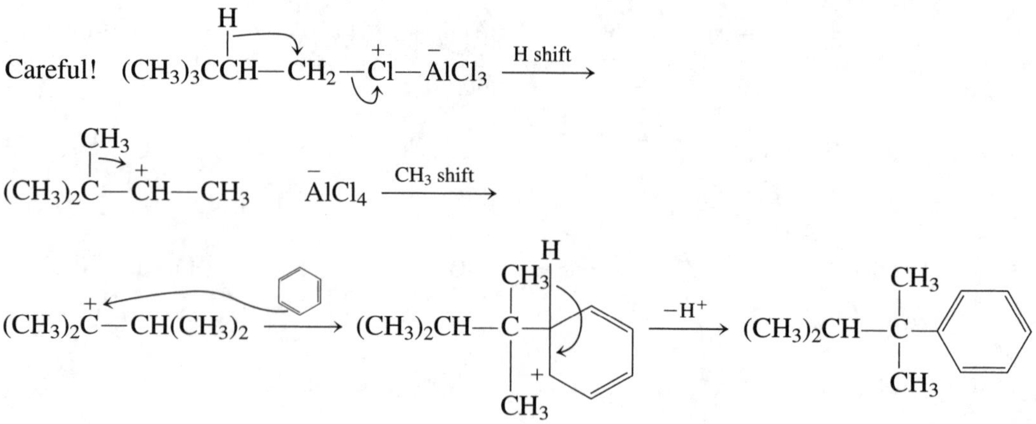

(g) 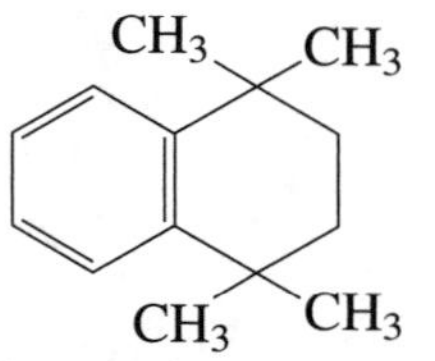   (h) 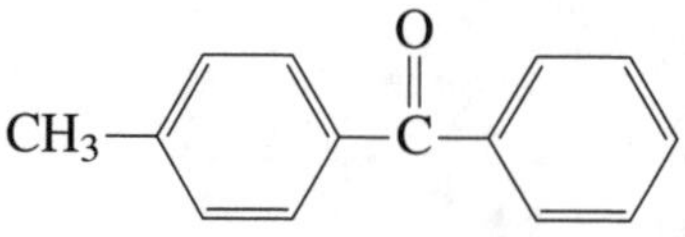 CH_3—

49. (c) $(CH_3)_3C—\ddot{O}H$ H^+ ⟶ $(CH_3)_3C—\overset{+}{\underset{\cdot\cdot}{O}}H_2$ $\xrightarrow{-H_2O}$ $(CH_3)_3\overset{+}{C}$

⟶ $(CH_3)_3C$ $\underset{H}{\big|}$ $\xrightarrow{-H^+}$ $C(CH_3)_3$

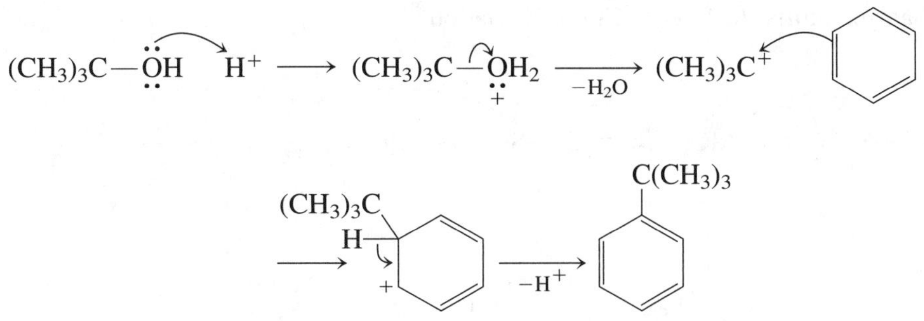

(f) is shown in the answer to Problem 48.

50. The problem is twofold: What should we make C_6D_6 from, and how should we do it? The chapter does not give us any practical way to construct benzene rings from nonbenzenoid starting materials, but it does present methods to replace groups on benzene rings—electrophilic aromatic substitution. We consider what we want to accomplish: the attachment of deuterium, D, to every carbon. A plausible electrophile to consider is the deuterium ion, D^+. In Section 15-10, we see that hydrogen ions, H^+, are sufficiently electrophilic to attack benzene rings (in that case, benzenesulfonic acid, leading to replacement of the —SO_3H group by —H). We could therefore expect that D^+ would behave similarly, attacking benzenesulfonic acid and replacing the —SO_3H group by —D. Indeed, it does, but is that really the best way to solve our problem? That reaction in Section 15-10 was presented to show how to get rid of the —SO_3H group. But do we need that group in the first place for what we're trying to accomplish here—replace H by D? The chemistry in Section 15-10 tells us that D^+ can attack a benzene ring as an electrophile. Therefore, we can assume that it can attack benzene itself:

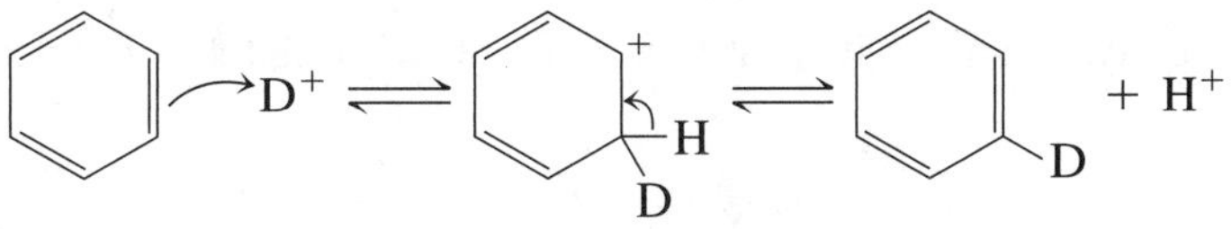

Run in the forward direction, we've replaced an H with a D! But the reaction is an equilibrium: We need to drive it from left to right. The solution is to treat benzene with a large excess of an acidic solution containing D^+ instead of H^+, dilute D_2SO_4 in D_2O, for example. The reaction goes to equilibrium, replacing most of the benzene C—H bonds with C—D bonds, according to the equation above. We repeat the process several times, each time treating the partially deuterated benzene with fresh D_2SO_4 in D_2O, until the amount of residual H in the benzene has been reduced to an acceptably low level (typically well below 1%).

51. Identify a likely electrophilic atom and follow a reasonable mechanistic pattern.

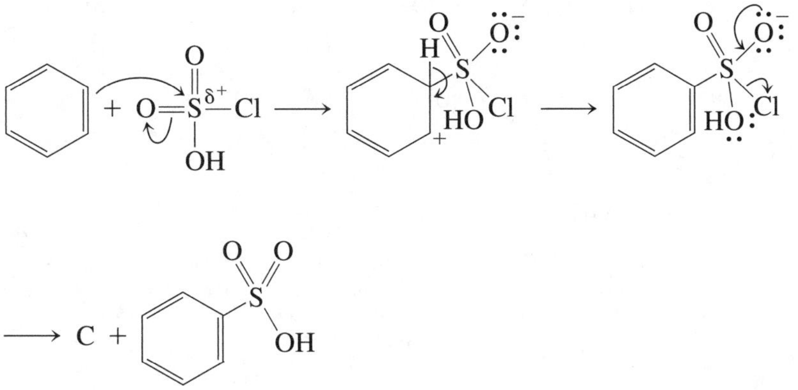

52. Consider a mechanism similar to Friedel-Crafts alkylation. Twice.

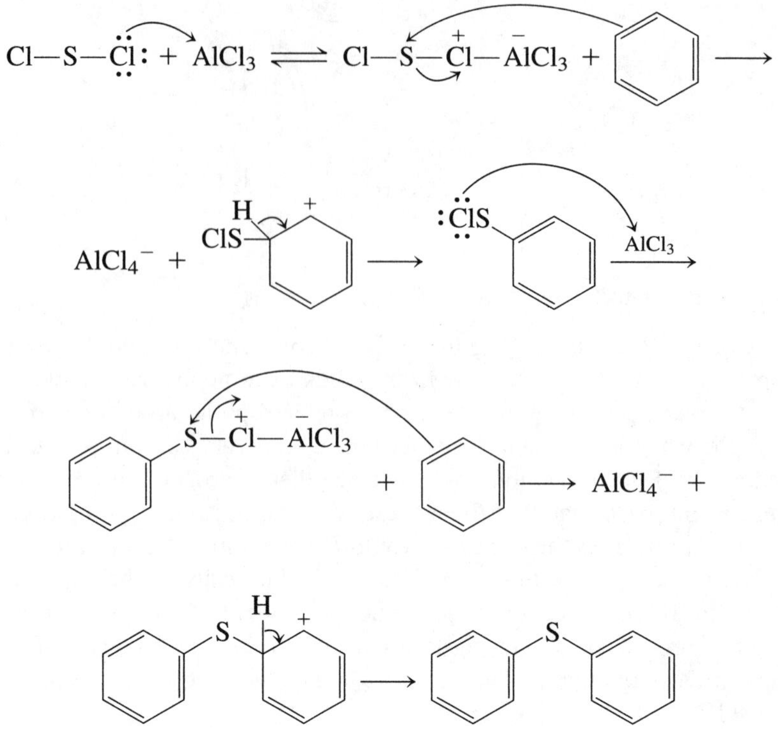

53. **(a)** Think mechanistically. Then see if what you come up with matches the data.

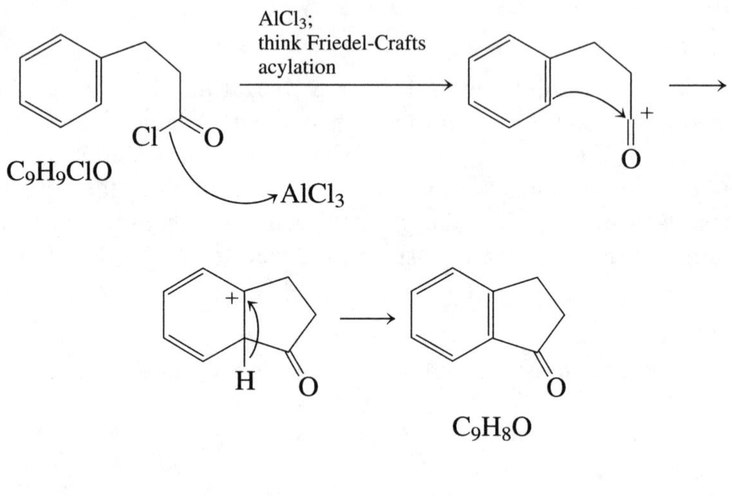

(b)

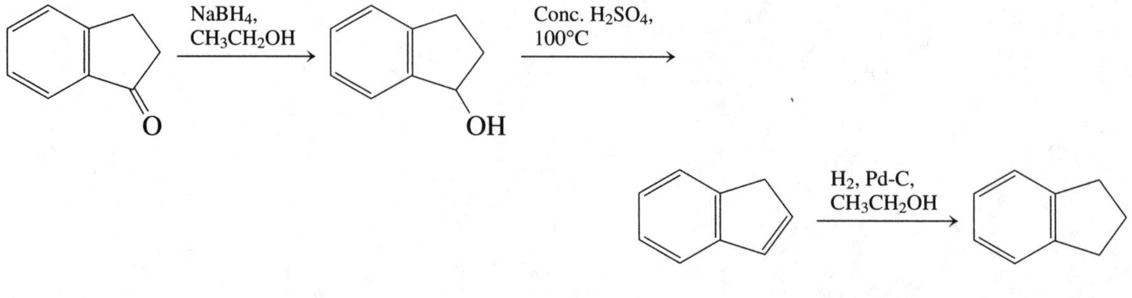

54.

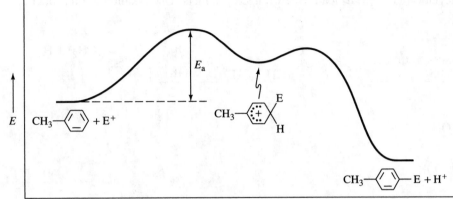

The reaction of methylbenzene should proceed with a lower energy of activation than that of benzene ($E_a^{\text{methylbenzene}} < E_a^{\text{benzene}}$), and the intermediate cation should be more stable.

55. (a) $\underset{\underset{\text{C}_6\text{H}_5\text{CHCH}_3}{|}}{\overset{\text{OH}}{}}$ (b) $\text{C}_6\text{H}_5\text{CH}_2\text{CH}_2\text{OH}$

56. More than one approach can be used for both (a) and (b): one based on Friedel-Crafts acylation (acylation), which uses an alkanoyl chloride, and another using Grignard addition to an aldehyde or a ketone.
If you got one but not the other, try now to give a second answer before peeking at the solutions below.

(c) Friedel-Crafts acylation is liable to lead to rearranged products. Use alkanoylation-reduction.

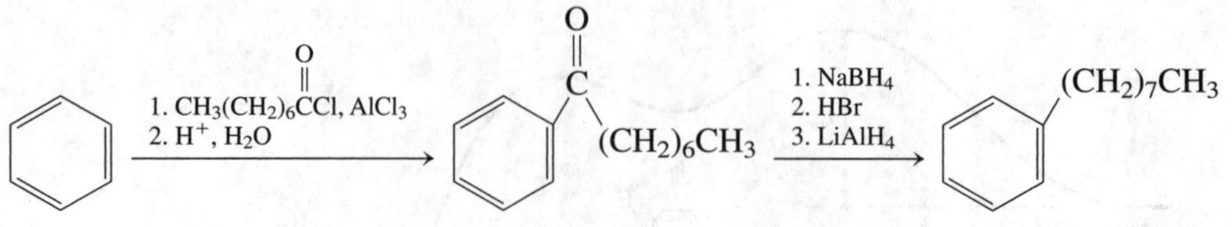

57. (a)

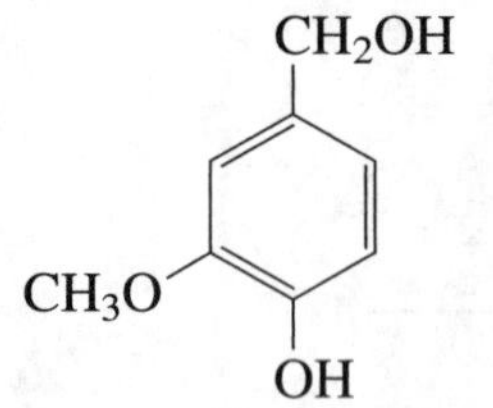

(b)

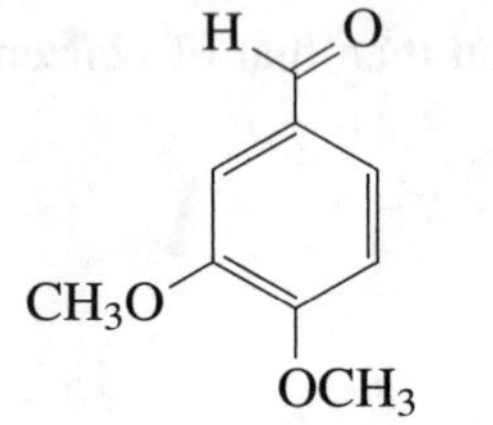

(c)

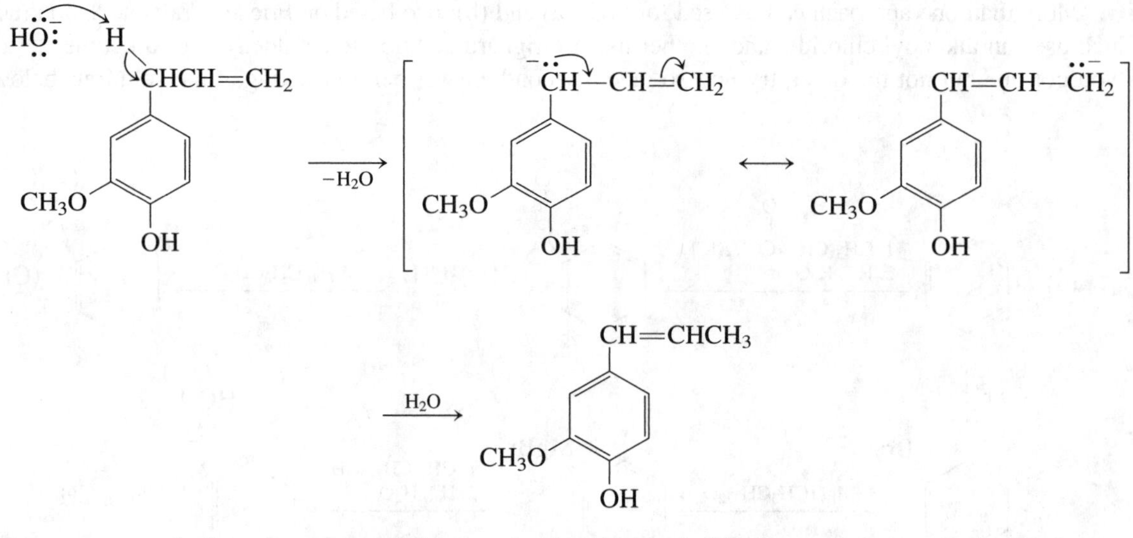

58. Cyclooctatetraene lacks resonance stabilization; its double bonds behave as if they were isolated, not conjugated. This is in fact the case. As a result of the geometry of the molecule (which is **not** planar; see Figure 15-17), the double bonds do not overlap into conjugated systems. So, resonance does **not** occur:

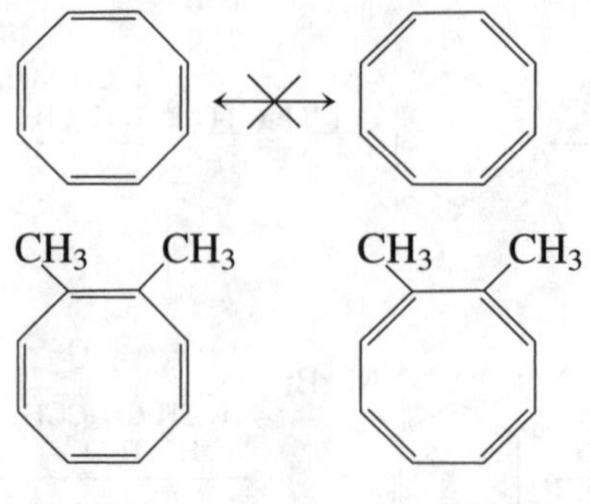

The two structures above actually represent **different** molecules! Their names are 1,2-dimethylcyclooctatetraene and 1,8-dimethylcyclooctatetraene, respectively.

59. (a) Draw the molecular orbitals by referring to the ones in Figure 15-4 and putting the **nodes** in your orbital illustrations the same way.

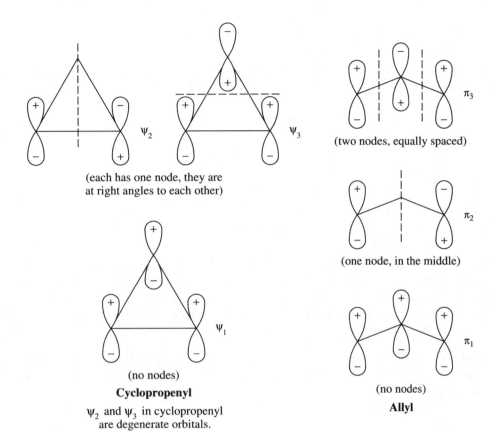

ψ_2
ψ_3
(each has one node, they are at right angles to each other)
ψ_1
(no nodes)
Cyclopropenyl
ψ_2 and ψ_3 in cyclopropenyl are degenerate orbitals.
π_3
(two nodes, equally spaced)
π_2
(one node, in the middle)
π_1
(no nodes)
Allyl

(b) **Two** electrons are best. They would fill the cyclopropenyl ψ_1 orbital and be stabilized relative to electrons in the π_1 allyl orbital. Electrons in ψ_2 or ψ_3 of cyclopropenyl are destabilized relative to π_2 electrons in allyl. Lewis structures for the two-electron cyclopropenyl and allyl systems (both are cations):

and

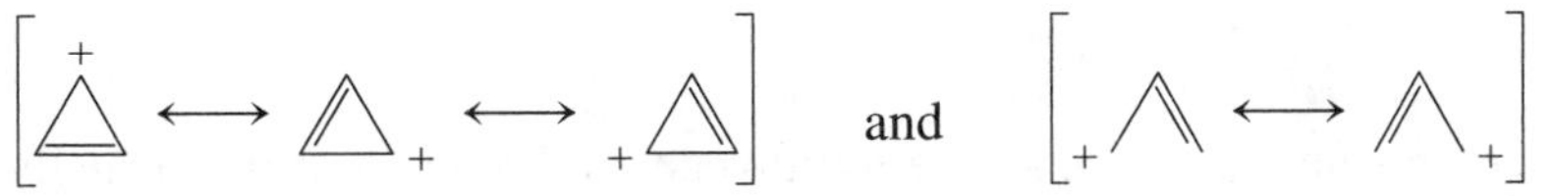

(c) Yes! The electrons in the cyclopropenyl cation can be represented as being delocalized in a **cyclic manner,** and the system is **stabilized** relative to the best acyclic analog, the 2-propenyl (allyl) cation.

60. Although there are several locations in this molecule that are reasonable sites for protonation, one in particular, the Lewis basic carbonyl oxygen, is unusual: Upon protonation, a resonance-stabilized cation is formed, and one of the resonance forms can be recognized as a substituted cyclopropenyl cation, an **aromatic** species (Problem 59). In fact, with the additional phenyl substitution, this species is remarkably stable.

O
C_6H_5
C_6H_5
+ HBr
H
O
C_6H_5
C_6H_5
H
O
C_6H_5
C_6H_5
Br^-

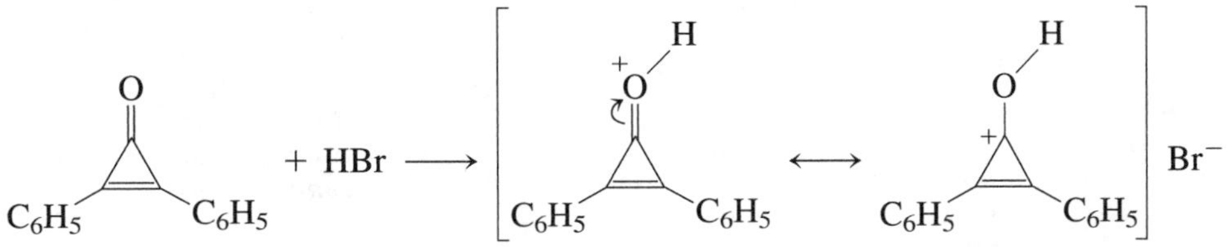

61. The cation is derived by removing two π electrons from the neutral diene:

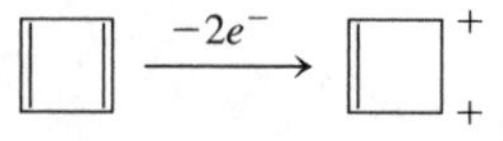

So the π system of the ring contains 2π electrons total. That makes it aromatic according to Hückel's rule. The MO levels look like this:

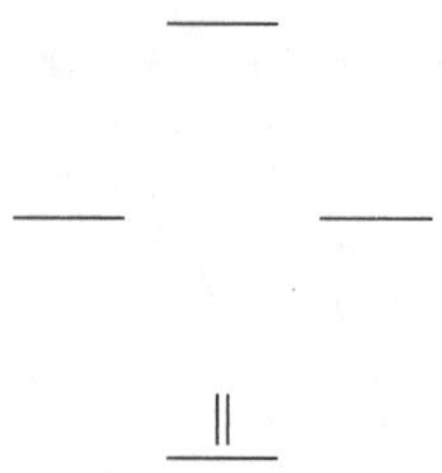

62. Recall that unusually strong acidity correlates with unusual stability of the conjugate base. A common way to increase the acidity of a hydrogen atom in a molecule containing only carbon and hydrogen is for the conjugate base to be stabilized by aromaticity. Use this principle and find the structure in which removal of a proton from one of the carbon atoms gives rise to an aromatic conjugate base. (**Hint:** Look for aromatic five-carbon rings in Section 15-7.)

(a) The cyclopentadienyl anion is aromatic. Only one of the four fulvenes pictured gives an aromatic cyclopentadienyl anion upon loss of a proton, by means of resonance delocalization of the lone pair left behind:

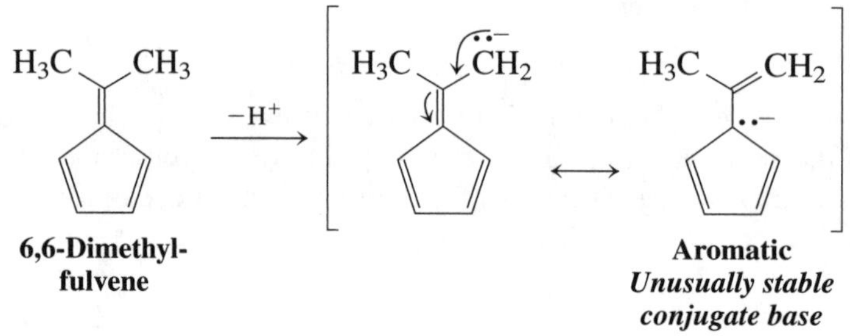

Therefore, the compound shown, 6,6-dimethylfulvene, is the one that is unusually acidic.

(b) The corresponding seven-membered ring compound shown below would give an 8π-electron, *antiaromatic* anion upon deprotonation of one of its methyl groups. This would be a very unstable conjugate base and, therefore, the parent hydrocarbon is a very *weak* acid.

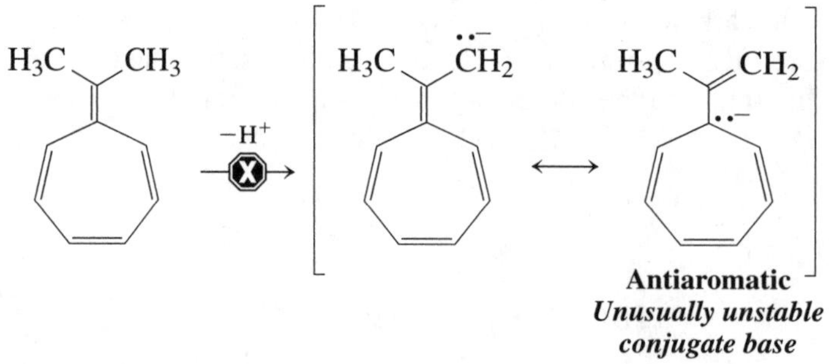

63. Again, as in Problem 62, look for aromaticity to play a role. Addition of a nucleophile to the non-ring carbon gives an aromatic result:

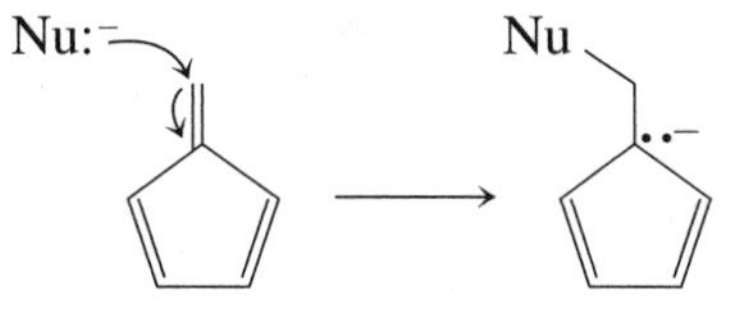

Aromatic

In contrast, addition of a nucleophile to any of the five ring carbons makes that carbon tetrahedral, breaking up the continuous circle of p orbitals necessary for aromaticity. Therefore such additions are normally not observed.

64. Figure 15-9 illustrates the local magnetic field generated by the ring current of an aromatic molecule. As described in Section 15-4, h_{local} reinforces H_0 outside the ring, a reinforcement leading to deshielding of protons on the external periphery. In contrast, h_{local} points in the **opposite** direction at the **center** of the ring and therefore reduces the net field in this region. In aromatic rings large enough to possess hydrogens inside the ring of circulating π electrons, such hydrogens should be strongly shielded, requiring increased values of the external field H_0 to attain resonance. Their NMR signals should be shifted to high field (to the right).

  (a) The structure of [18]annulene (Section 15-6) shows 12 hydrogens outside the ring (blue), which will be deshielded as are those of benzene. The inner six (green) are strongly shielded, a situation resulting in the signal nearly 3 ppm upfield from that of $(CH_3)_4Si$.

  (b) Yes. The eight hydrogens on the π periphery show benzenelike deshielding, whereas the two hydrogens on the bridge over the ring are shielded, similar to those "inside" the [18]annulene ring.

65. Structure A contains four cis and three trans double bonds; structure B contains three cis and four trans double bonds. The NMR indicates the presence of four "inside" and ten "outside" hydrogens, a pattern consistent with structure B only.

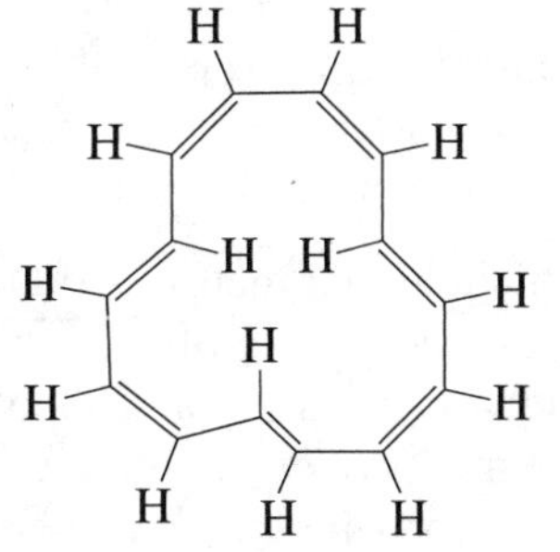

A	B
3 "inside" and 11 "outside" H's	4 "inside" and 10 "outside" H's

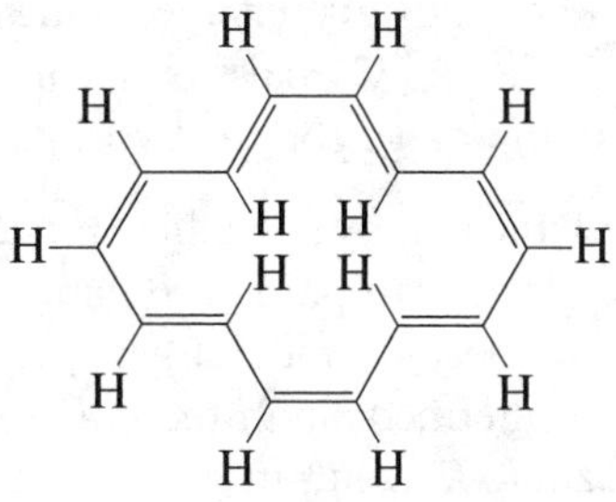

66. Look at $AlCl_3$ as a Lewis acid and benzene as a nucleophile. Recall (Section 9-9) that acid-catalyzed ring openings of oxacyclopropanes give S_N1 regiochemistry (most stable carbocation) but S_N2 stereochemistry (backside nucleophilic attack). So,

$$\text{H}_2\text{C} \overset{:\text{O}:}{\underset{\text{H}}{-\text{C}}} \text{CH}_3 \quad \xrightarrow{\text{AlCl}_3} \quad \left[\text{H}_2\text{C} \overset{^-\text{AlCl}_3 \atop :\text{O}^+}{\underset{\text{H}}{-\text{C}}} \text{CH}_3 \right] \longrightarrow$$

$$\underset{\text{CH}_2-\text{C}}{\overset{^-\text{Cl}_3\text{AlO} \quad \text{CH}_3}{\underset{+}{\quad}}} \overset{\text{H}}{\underset{\text{H}}{}} \quad \xrightarrow[\text{}]{-\text{H}^+} \quad \xrightarrow[-\text{Al}^{3+}\text{ salts}]{\text{After H}^+,\ \text{H}_2\text{O}} \quad \text{HOCH}_2-\overset{\text{CH}_3}{\underset{}{\text{C}}}\text{H}$$

Exclusive
product

67. How about a simple electrophilic substitution? Hg^{2+} in $\text{Hg}(\text{OCCH}_3)_2$ is electrophilic, so try

$$\bigcirc + \text{Hg}(\overset{\text{O}}{\overset{\|}{\text{OCCH}_3}})_2 \quad \xrightarrow{\text{THF}} \quad \bigcirc\text{HgOCCH}_3 \ (\overset{\text{O}}{\|}) \quad + \ \text{CH}_3\text{COOH}$$

68. (a) In analogy to the reaction of alkyl halides, such as RCl, with AlCl_3, a reaction between HCl and AlCl_3 might be expected to occur, too. Recall (Section 15-11) that either of two types of reactions takes place with RCl: A Lewis acid-Lewis base complex forms with primary RCl, but secondary and tertiary RCl ionize to give carbocations. We know that HCl is easily ionized, so it makes sense to suggest that ionization occurs upon reaction with AlCl_3: $\text{HCl} + \text{AlCl}_3 \rightarrow \text{H}^+ + \text{AlCl}_4^-$. The solution of ions conducts electricity.

(b) The largest chemical shifts belong to C1, C3, and C5, which make sense because these happen to be the ones that bear positive charge in the three resonance forms of this cation, and should be most deshielded. C2 and C4 have pretty normal chemical shifts for benzene carbons. C6 has a chemical shift in the range for sp^3-hybridized carbon atoms, as expected from the structure.

(c) Two effects are indicated by the table: First, each additional methyl group added to the ring increases the rate of electrophilic aromatic substitution by a significant amount. This observation is easy to rationalize, because methyl groups are electron releasing and should inductively stabilize the carbocationic intermediate in the reaction and, in turn, lead to a lower-energy transition state leading to the cation. Second, in the three instances for which rate data are given for two *different* isomers with the *same number* of methyl groups, the rates for the two isomers are not the same. The effect is small, about a factor of two, for di- and trimethylbenzenes. In the case of the dimethylbenzenes, the reason is steric: In each molecule the carbocations are similarly stabilized. However, all positions in 1,4-dimethylbenzene are adjacent to a methyl group, while substitution at positions 4 or 5 in 1,2-dimethylbenzene avoids the steric crowding of an adjacent CH_3. For the trisubstituted benzenes, the 1,2,3-isomer is faster because substitution at either of two positions, C4 or C6, gives a carbocation in which two of the three resonance forms benefit by placing the positive charge on a tertiary ring carbon. In contrast, in the 1,2,4-isomer, substitution at C3 is sterically difficult because it lies between methyl groups on either side. Finally, the spectacular rate difference for the two tetrasubstituted benzenes requires explanation. The difference here is that substitution in the 1,2,3,5-isomer proceeds through an intermediate in which all three positions that bear positive charge are tertiary. In the 1,2,3,4-isomer, no intermediate has more than two tertiary sites with positive charge.

(d) The salt may be best formulated as ⟨structure: cyclohexadiene cation with H, CH₂CH₃, H₃C, CH₃ substituents and CH₃ below⟩ BF_4^-. Heating causes elimination of

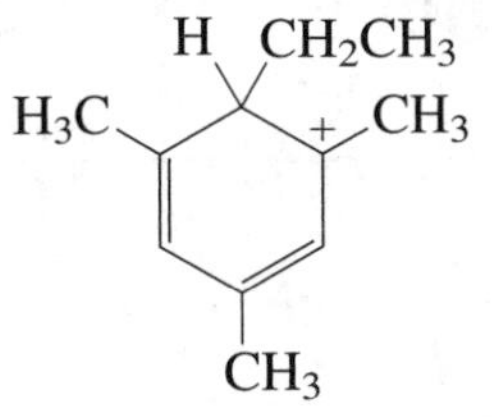

HBF_4 and generation of the neutral, tetrasubstituted benzene. Why is the salt stable in the absence of heat? Make a model. The sp^3 carbon atom holds the bulky ethyl group out of the plane of the methyl groups on either side. Upon loss of the proton, the ethyl group moves into the same plane as these methyl groups and experiences steric hindrance as a result. The benzene aromaticity is enough to thermodynamically overcome the unfavorable steric effect, but the latter clearly shows up in a higher than usual activation barrier for proton loss and aromatization.

69. (a)

70. (c)

71. (d)

72. (b)

73. (d)

16

Electrophilic Attack on Derivatives of Benzene: Substituents Control Regioselectivity

---

The preceding chapter introduced you to the properties and chemistry of benzene itself. In this chapter the chemistry is expanded to include several new reactions not only of benzene but also of derivatives of benzene containing various substituents. The text focuses on the effects of substituents on the chemistry of the benzene ring. The first and rate-determining step of electrophilic substitution is the formation of a delocalized cation. Different types of substituents can affect the stability of this cation and make its formation easier or harder. Substituents on the ring also direct the attack of an electrophile to specific positions. This study guide chapter contains information that may help you avoid a lot of memorization. You will see that the effect of a substituent on benzene chemistry can be predicted from a relatively straightforward examination of its electronic characteristics.

Outline of the Chapter

16-1 Activation or Deactivation by Substituents on the Benzene Ring

16-2 Directing Inductive Effects of Alkyl Groups

16-3 Effects of Substituents in Conjugation with the Benzene Ring

16-4 Electrophilic Attack on Disubstituted Benzenes
Effects of substituents on the reactivity and regioselectivity of further electrophilic substitution on a benzene ring.

16-5 Synthetic Strategies
General principles and useful new reactions.

16-6 Reactivity of Polycyclic Benzenoid Aromatics
What happens when two or more benzene rings are fused together.

16-7 Polycyclic Aromatic Hydrocarbons and Cancer

Keys to the Chapter

16-1 through 16-3. Directing Effects: Activation and Deactivation
Don't forget the **basics: Electrophiles** seek **electrons!** So, reactivity of a benzene ring toward electrophiles will be enhanced by substituents that donate electron density to the ring, and it will be reduced by substituents that withdraw electron density from the ring. "Reactivity" in this context is the **rate** at which substitution occurs, which is determined by the rate-determining step—electrophilic attack to form a carbocation. Electron-donating

groups (*activating groups*) will stabilize the cation, lowering the transition state energy for its formation and speeding up the reaction, and electron-withdrawing groups (*deactivating groups*) will do just the opposite. Furthermore, these effects will differ at different positions around the ring relative to the location of the substituents already present, giving rise to *directing effects*.

Both inductive and resonance effects are involved. The favored reaction proceeds through the most stabilized (or least destabilized) intermediate carbocation. Study **carefully** the resonance forms pictured for the possible cations derived from electrophilic attack on methylbenzene and (trifluoromethyl)benzene (Section 16-2), and on benzenamine (aniline), benzoic acid, and a halobenzene (Section 16-3). Notice the **types** of groups that fall into the different categories in Table 16-1. In particular, notice the following two general trends:

1. **All** substituents with **lone pairs** on the atom attached to the benzene ring are **ortho, para** directors.
2. **All** substituents that have a **positively polarized (δ^+) atom** lacking a lone pair attached to the benzene ring are **meta** directors.

These generalizations always hold and can be used as aids to help you remember whether a group belongs in one category or the other.

16-5. Synthetic Strategies

This section shows how to plan syntheses of multiply substituted benzenes. It also illustrates the interconversion of NH_2 and NO_2, reduction of $C=O$ to CH_2, use of SO_3H as a para-blocking group, and several other valuable synthetic "tricks."

16-6. Reactivity of Polycyclic Benzenoid Aromatics

Fused polycyclics such as naphthalene, anthracene, and phenanthrene exhibit both similarities and differences with respect to benzene. For example, electrophilic aromatic substitution occurs but is generally easier with the polycyclics and proceeds with milder reagents. The activation energies for attack are lower because the intermediate cations are more highly delocalized (more resonance forms!) and, therefore, lower in energy. If the rings are not identically substituted, reaction occurs at the most activated ring. Substituents have the same directing and activating/deactivating effects you saw with benzene.

Fused polycyclics also have an increased tendency to undergo **addition** reaction, which are rare for benzene itself. As a general rule of thumb, additions occur in such a way as to preserve intact benzene rings as much as possible. So, additions to both anthracene and phenanthrene occur at the 9,10 positions, leaving two simple benzene rings intact.

Solutions to Problems

30. Order of **decreasing** reactivity; for brevity, only the substituents are listed.

(a) $-CH_3 > -CH_2Cl > -CHCl_2 > -CCl_3$. Electronegative Cl's make carbon δ^+, so the inductive effect becomes increasingly electron withdrawing and deactivating.

(b) $-O^- Na^+ > -OCH_3 > -O\overset{\displaystyle O}{\overset{\displaystyle \|}{C}}CH_3$. Resonance activators. Compare the availability of oxygen's lone pair electrons, because they activate the ring by resonance donation to stabilize the cationic intermediate. The negatively charged oxygen atom is the best donor, the neutral ether oxygen second best, and the oxygen in the ester worst because it is attached to a δ^+ carbonyl carbon that pulls the electrons away from the ring.

(c) $-CH_2CH_3 > -CH_2CCl_3 > -CH_2CF_3 > -CF_2CH_3$. Inductive effects again, combined with distance to ring.

31. Compare the —CH₂Cl substituent with the simplest similar one you know, the —CH₃ group. We know from Section 16-2 that the methyl group is activating because it is an electron-donating group by induction. The methyl group is also an *ortho,para*-directing group, because its donating effect best stabilizes the carbocations (and therefore the transition states leading to the carbocations) resulting from substitution in the ortho and para positions.

Adding a chlorine atom to the methyl carbon reduces the inductive electron-donating capability of the group because chlorine is very electronegative and inductively electron-withdrawing. The effect is to make the —CH₂Cl group slightly electron-withdrawing and, therefore, a weakly deactivating group. It also is less strongly *ortho,para*-directing, with now a significant amount of *meta*-substitution being observed, at least when nitration is the reaction being carried out. Table 16-2 provides a useful overview of nitrations of a variety of substituted benzenes. Notice that toluene (methylbenzene) nitrates mostly in the ortho position (58%). In (chloromethyl)benzene the amount of ortho nitration is cut almost in half (32%). It is likely that sterics play a role here: The —CH₂Cl substituent is considerably more bulky than methyl, and therefore obstructs ortho substitution to a greater extent.

32. Activated: (c), (d), (e), (g)

33. (a)

(b)

In each series of compounds, the benzene ring with two alkyl group substituents is most activated, the one with two carbonyl group substituents is most deactivated, and the one with one of each is intermediate.

34. The two methyl substituents in 1,3-dimethylbenzene reinforce each other's activating and directing effects: They both direct subsequent electrophilic attack to positions 4 and 6 on the ring. For example, position 4 is ortho to the methyl group at C3 and para to the methyl group at C1, so it is doubly activated for subsequent substitution. In either 1,2- or 1,4-dimethylbenzene (*o*- or *p*-xylene) the two methyl groups direct to different carbons in the ring. For example, in 1,2-dimethylbenzene, the methyl group at C1 directs to positions 4 and 6, while the methyl group at C2 directs to positions 3 and 5.

Another way of looking at the situation is to notice that the intermediate cation resulting from attack at C4 on 1,3-dimethylbenzene has *two* resonance contributors in which the cationic carbon is methyl-substituted:

Therefore the energy of this cation is lower, and the activation barrier associated with its formation is reduced. Its formation is relatively fast. In contrast, none of the cations derived from electrophilic attack on either 1,2- or 1,4-dimethylbenzene are directly stabilized by more than one of the methyl groups. They are all higher in energy, and their formation requires surmounting a larger activation barrier.

35. (a) 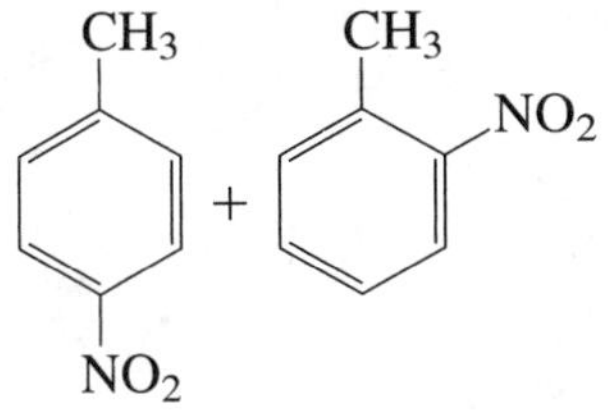 (b)

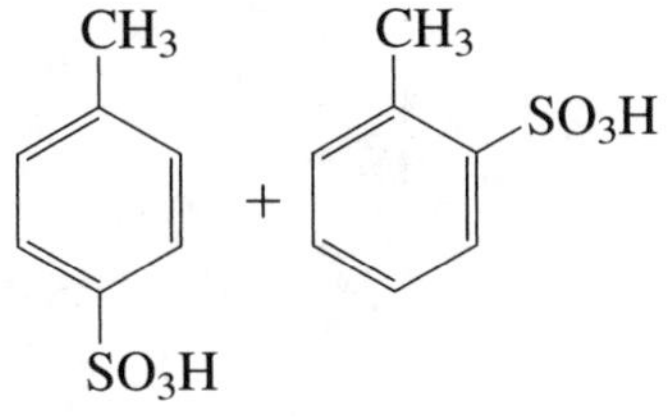

(c) 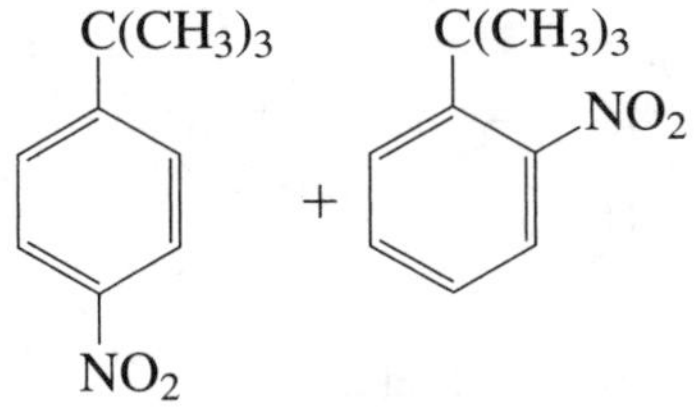 (d)

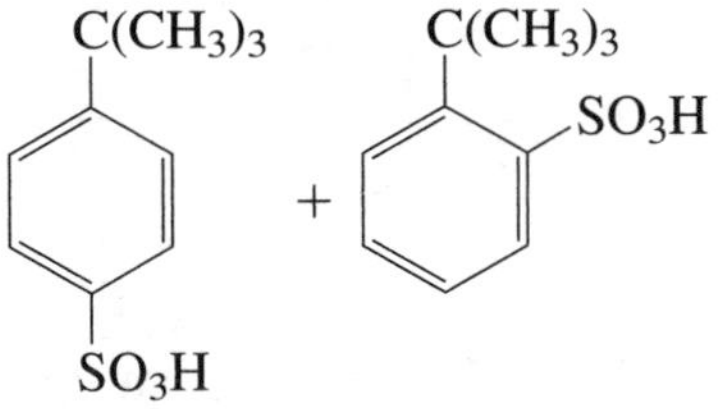

Changing from a small (methyl) group to a very bulky (1,1-dimethylethyl) substituent will strongly sterically inhibit ortho substitution. Reactions (c) and (d) give rise to almost entirely para-disubstituted products.

36. The position at which substitution occurs is determined by the directing effect of the substituent that is already present in the starting compound. Methoxy and chloro are ortho, para-directing groups. Most of the time they direct incoming electrophiles to the para position, especially if the electrophile is somewhat bulky, as is the case in (a) and (d). Nitro and carboxylic acid are meta-directing.

(a) 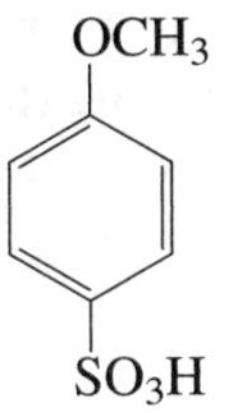 (b) 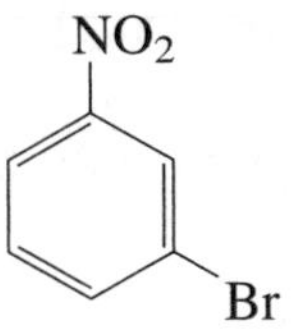 (c) 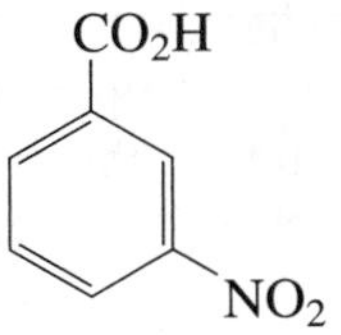 (d)

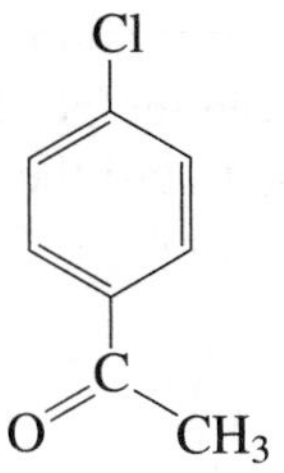

37. Ortho attack:

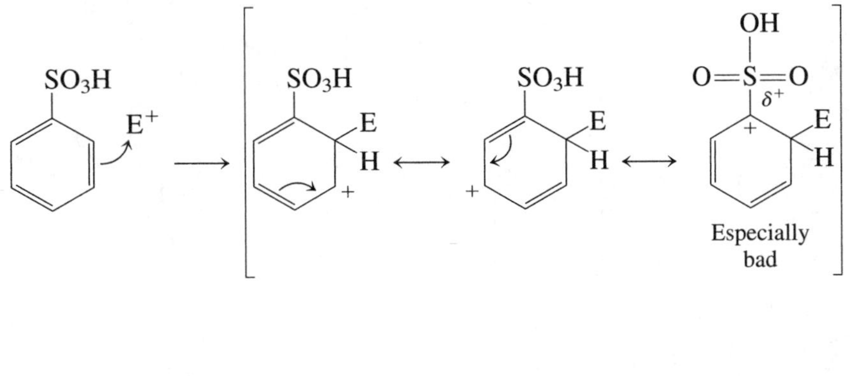

Para attack:

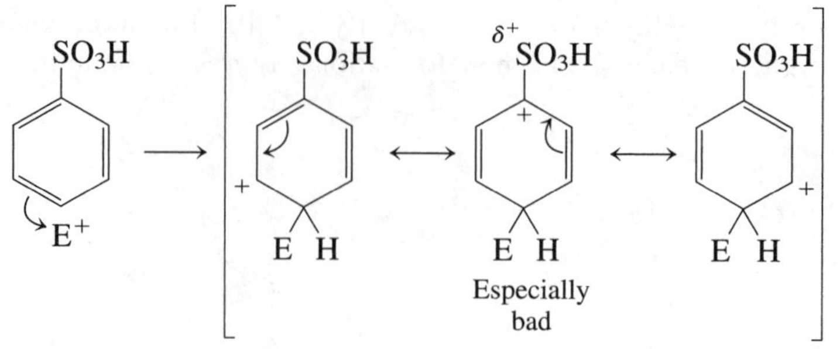

Meta attack:

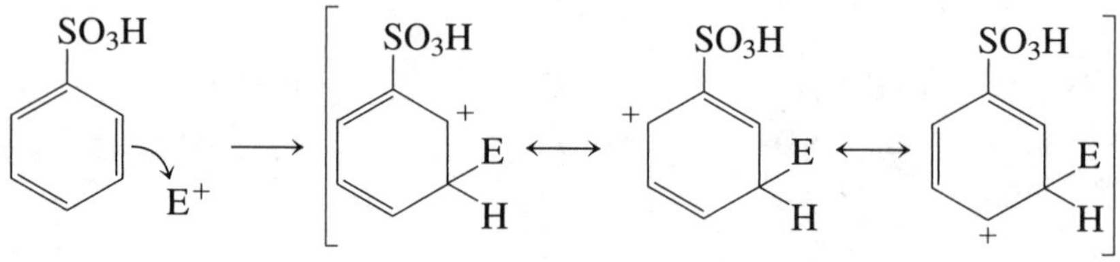

There are no resonance forms with $+$ adjacent to δ^+ sulfur when attack occurs meta.

38. Statement is correct. All meta-directors deactivate the entire ring by inductive electron withdrawal. Deactivation at the ortho and para positions is most intense as a result of resonance (see, for example, the answer to Problem 37). Meta substitution occurs only because the deactivation is felt least strongly at that position.

39. Solve the problem by examining the structures, in particular the resonance forms, of the carbocation that is produced by attachment of a generic electrophile, E^+, to each of the three possible positions on one of the benzene rings: ortho, meta, and para, with respect to the other ring. Use principles developed in this chapter to evaluate the relative stabilities of these carbocations. The more stable carbocations should form faster and will determine the major product(s).

(a) Substitution ortho: attach the electrophile to either ring (they are equivalent), at the position adjacent to the bond connecting the two rings. Initially, we see the same three resonance forms that we can draw for substitution on benzene itself—the top three forms in the brackets below. However, having delocalized the positive charge to the position of attachment of the second ring (top right), we can continue to use the double bonds of this ring to delocalize further, giving us three more resonance forms (bottom three in brackets), for a total of six.

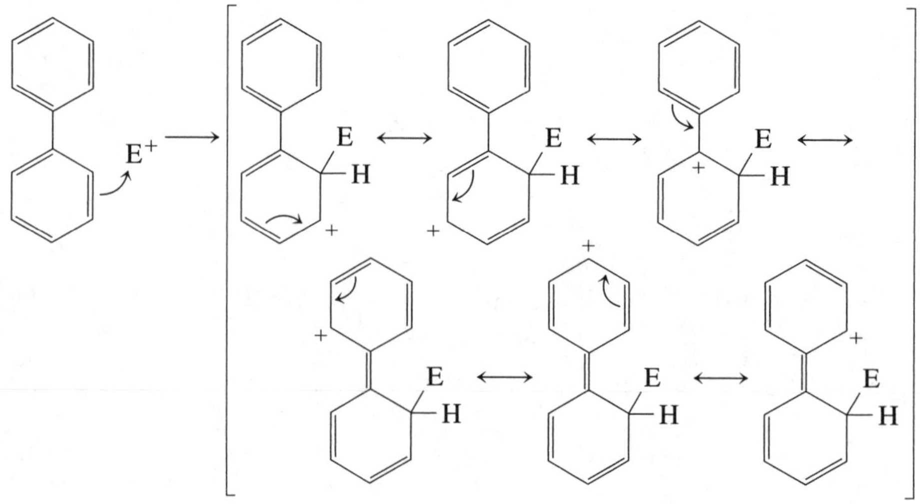

(b) Substitution meta: attach the electrophile one position further removed from the bond between the rings, the meta position. Draw the resonance forms for the carbocation. Notice now that the positive charge is no longer delocalized to the position of attachment to the second ring. This ring is incapable of providing additional resonance stabilization beyond that of the three original forms present in the ring on which the substitution took place.

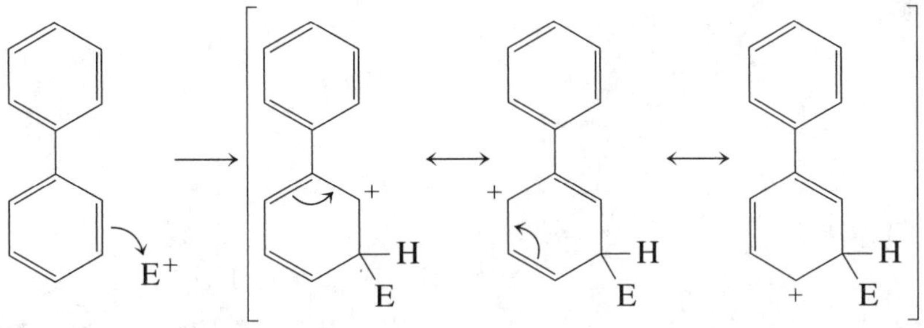

(c) Substitution para: attach the electrophile and proceed as before. As was the case for substitution ortho, one of the three initial forms places the positive charge at the ring connection position, permitting three more resonance forms to participate in stabilizing the intermediate. (The second and third resonance forms in the top group of three have been written in reverse order—and the electron-pushing arrows modified accordingly—so that the form at the right leads naturally to the three forms in the bottom group of three.)

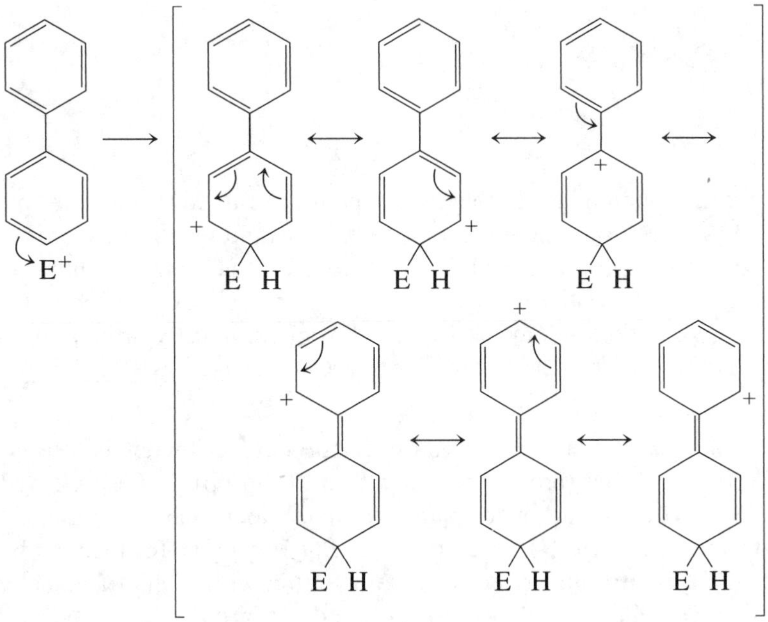

Additional resonance forms provide additional stabilization, and therein lies the answer. Substitution either ortho or para leads to a carbocation delocalized throughout both rings, with a total of six resonance forms, while meta substitution gives a cation limited to delocalization in only one of the rings, with only three contributing resonance forms. The meta cation is less stable, higher in energy, requires surmounting a higher activation energy barrier to form, and therefore forms more slowly. The phenyl ring as a substituent is thus an ortho, para-directing group with respect to electrophilic aromatic substitution. It is also activating, again a consequence of the additional stabilization of the cation formed upon ortho or para substitution, relative to the cation formed from substitution on benzene itself.

40. Where both ortho and para products are formed, the para is generally the major one.

41. **Note:** Write out the solutions for Problems 41 and 42 in full equation form for yourself.
(a) Nitrobenzene + Br_2, $FeBr_3$ (nitration of bromobenzene would give mostly ortho and para products);
(b) iodobenzene + Br_2, $FeBr_3$; **(c)** methoxybenzene + CH_3Cl, $AlCl_3$; **(d)** methoxybenzene + CH_3COCl,
$AlCl_3$; **(e)** nitrobenzene + SO_3, H_2SO_4 or benzenesulfonic acid + HNO_3, H_2SO_4; **(f)** toluene + HNO_3,
H_2SO_4 (Table 16-2: This reaction is somewhat unusual in giving mostly ortho rather than para product);
(g) nitrobenzene + HNO_3, H_2SO_4; **(h)** acetophenone + Cl_2, $FeCl_3$.

42. **Note:** These problems rely on the material in Section 16-5. Review before you begin.
(a) Nitrobenzene + 1. Br_2, $FeBr_3$ (brominates meta to nitro group), 2. Fe, HCl (reduces NO_2 to NH_2);
(b) iodobenzene + SO_3, H_2SO_4 (sulfonates para to iodine; blocks para position), 2. Br_2, $FeBr_3$
(brominates ortho to iodine), 3. H^+, H_2O, Δ (removes SO_3H group); **(c)** methoxybenzene + 1.
$CH_2CH_2CH_2COCl$, $AlCl_3$ (**Caution:** Friedel-Crafts alkylation with 1-chlorobutane would give a rearranged
product; use instead Friedel-Crafts acylation para to methoxy group), 2. H_2, Pd or Zn(Hg), HCl (reduce
C=O to CH_2); **(d)** ethylbenzene + 1. CH_3COCl, $AlCl_3$ (acetylates para to ethyl group), 2. CrO_3, H_2SO_4,
H_2O (oxidizes original ethyl group to acetyl); **(e)** aniline + 1. CH_3COCl, pyridine (protects amino group
as amide), 2. HNO_3 (nitrates para), 3. H^+, H_2O, Δ and 4. ^-OH, H_2O (frees NH_2 group); **(f)** ethylbenzene
+ SO_3, H_2SO_4 (sulfonates para to iodine; blocks para position), 2. Br_2, $FeBr_3$ (brominates ortho to ethyl
group), 3. H^+, H_2O, Δ (removes SO_3H group) (**Caution:** You cannot begin with bromobenzene, because
after sulfonation para you cannot do the necessary subsequent Friedel-Crafts reaction. Friedel-Crafts reactions do not normally proceed when a *meta*-directing group, such as SO_3H, is present); **(g)** nitrobenzene
+ 1. Fe, HCl (reduces NO_2 to NH_2), 2. CH_3COCl, pyridine (protects amino group as amide), 3. HNO_3
(nitrates para), 4. H^+, H_2O, Δ and 5. ^-OH, H_2O (frees NH_2 group), 6. CF_3CO_3H (oxidizes NH_2 to NO_2);
(h) acetophenone + 1. SO_3, H_2SO_4, 2. H_2, Pd.

43. Orientation of reaction is determined by more activating (or less deactivating) substituent (marked in each case below). Again, the para product can be expected to predominate where a choice of ortho or para exists.

(a)

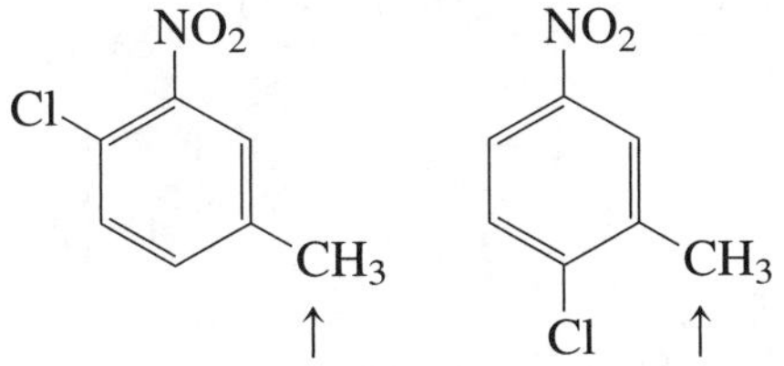

(b)

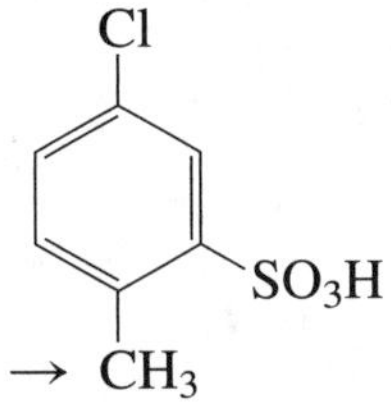

(c)

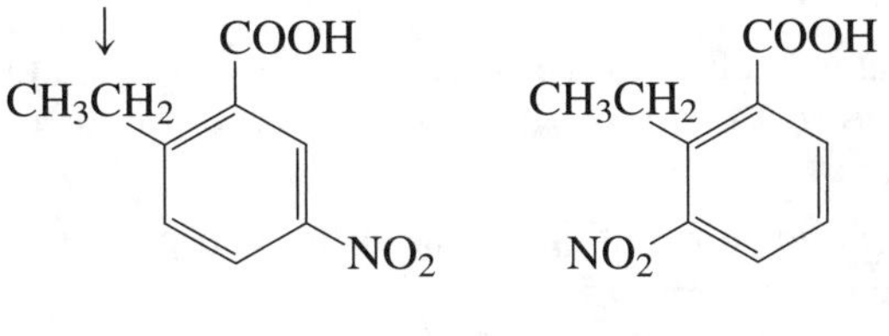

(d)

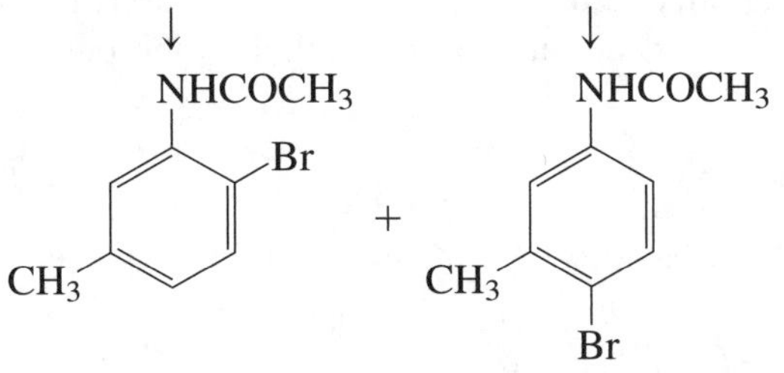

(e)

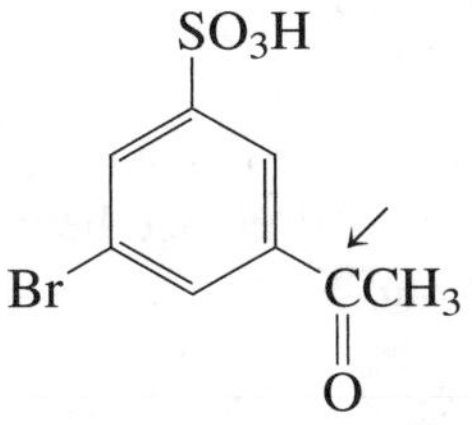

(Both meta directing)

(f)

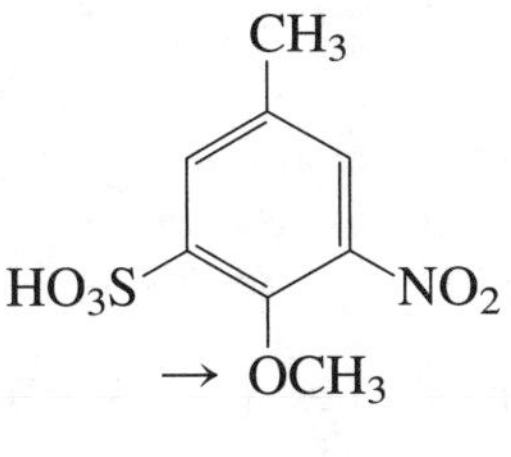

(g)

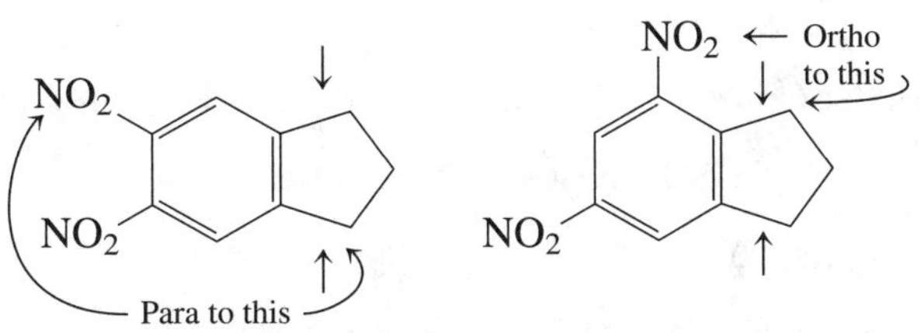

The saturated ring is **not** special.
Treat it just like two alkyl groups.

(h)

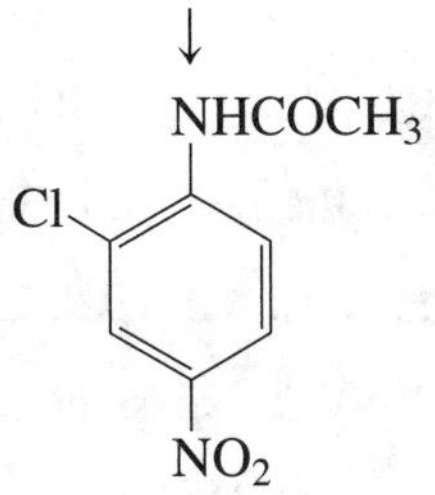

(i) No reaction. Friedel-Crafts reaction does not occur on rings containing meta-directing groups: The ring is too deactivated.

44. (a) Activation effects are additive. 1,3-Dimethylbenzene (*m*-xylene) differs from its 1,2 and 1,4 isomers because it is the **only** one that contains ring positions activated by **both** methyl substituents. Electrophilic substitution at C2, C4, and C6 (which is equivalent to C4) gives intermediate cations whose positive charges are delocalized to two methyl-substituted positions around the ring. For attack at C4, you have

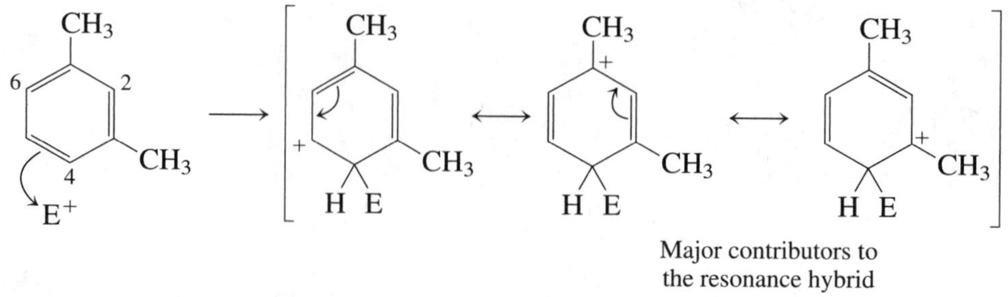

Major contributors to
the resonance hybrid

The cations are more stable, the transition states leading to them are lower in energy, and they form faster. Substitution mainly occurs at C4/6 (C2 is a sterically hindered position because of the methyls on either side).

You should check for yourself that electrophilic substitution on the ortho and para isomers cannot lead to such a doubly stabilized intermediate cation.

(b) Three methyl groups, but the principle to be applied is the same: Look for positions that are doubly or triply activated. In each structure count up the number of methyls ortho or para to each open position.

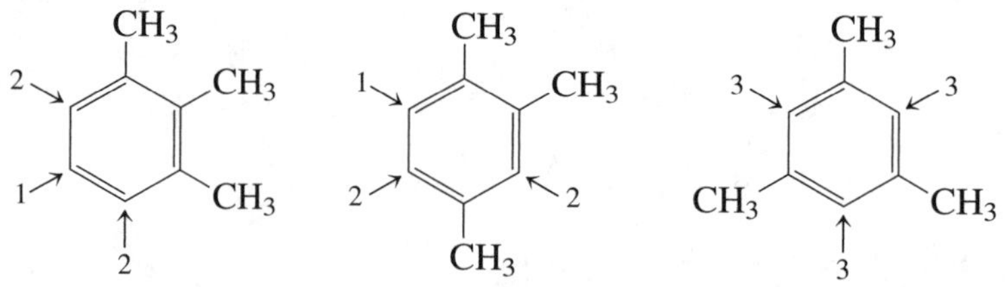

The last should be the most reactive toward electrophiles: Each vacant position is activated by being ortho or para to **all three** methyl groups. By the way, the reactivity difference between the 1,3,5 isomer and the other two is quite large, about a factor of 200 in rate.

45. Friedel-Crafts reactions are understood to be followed by H^+, H_2O work-up.

(a) 1. CH_3CH_2Cl, $AlCl_3$;   2. CH_3COCl, $AlCl_3$

(b) 1. HNO_3, H_2SO_4;   2. Cl_2, $FeCl_3$

(c) 1. CH_3COCl, $AlCl_3$;   2. SO_3, H_2SO_4

Can't do it in the reverse order. The Friedel-Crafts reaction fails with the SO_3H group there.

(d) 1. HNO_3, H_2SO_4;   2. HCl, Zn(Hg);   3. SO_3, H_2SO_4, heat;   4. CF_3CO_3H

(e) 1. Cl_2, $FeCl_3$;   2. excess conc. HNO_3, H_2SO_4, Δ

(f) 1. Br_2, $FeBr_3$;   2. HNO_3, H_2SO_4, separate para from ortho;

3. HCl, Zn(Hg) $\left(\text{makes } Br\!-\!\bigcirc\!-\!NH_2\right)$;   4. Cl_2, $CHCl_3$, 0°C (chlorinates once, ortho to

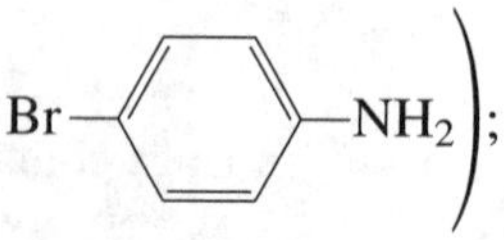

NH_2; see Section 16-3); 5. CF_3CO_3H

(g) 1. Br_2, $FeBr_3$;   2. SO_3, H_2SO_4 (blocks para);   3. Cl_2, $FeCl_3$;   4. H_2O, Δ

(h) 1. CH_3Cl, $AlCl_3$;   2. SO_3, H_2SO_4;   3. excess Br_2, $FeBr_3$;   4. H_2O, Δ

46.

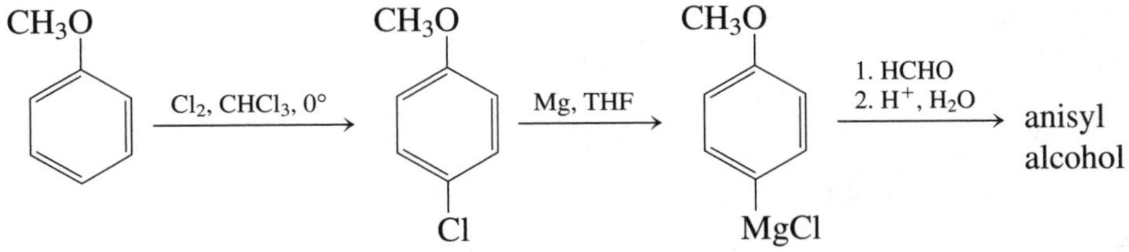

47. The synthesis of URB597 was extremely challenging, more so for the positions of its substituents than for their structure. Both groups are located in meta positions relative to the bond between the rings. Problem 39 revealed that phenyl rings as substituents are ortho, para directing. Therefore electrophilic aromatic substitution reactions performed on biphenyl cannot serve as an entry to the preparation of this important compound.

48. **(a)** NMR shows two sets of benzene hydrogens in a 2 : 3 integration ratio, and the IR bands at 685 and 735 cm^{-1} indicate a monosubstituted benzene. This analysis leads to the only logical answer,

bromobenzene, C_6H_5Br,  —Br, which is made from benzene + Br_2 and $FeBr_3$.

(b) NMR shows three benzene hydrogens, a 1 H triplet and a 2 H doublet; what you would expect for three hydrogens in adjacent positions around a ring, with the end ones equivalent: CH—CH—CH. What else is there? A 2 H lump at $\delta = 4.5$ ppm together with two bands in the IR at 3382 and 3478 cm^{-1} strongly suggest an NH_2 group. How about an atom count? You're up to six carbons for the benzene ring, five H and an N. The only formula that will fit is $C_6H_5Br_2N$. Putting the NH_2 on the benzene ring and then attaching the two Br's one each side of it fits the data: 2,6-dibromobenzenamine is the answer.

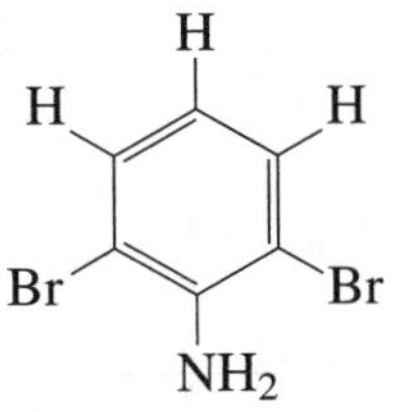

(c) The NH_2 is again present, but the NMR now shows four benzene hydrogens in a complicated pattern. The IR helps: The band at 745 cm^{-1} is in the range for ortho disubstituted rings. 2-Bromobenzenamine is the answer: C_6H_6BrN.

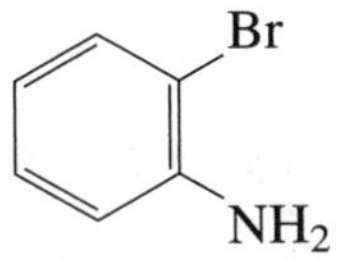

(d) Similar, but the NMR (two doublets, each 2 H) and the IR (820 cm^{-1}) indicate that this is the

para isomer of C: 4-bromobenzenamine: Br——NH_2

Synthesis. (Assume any ortho + para mixtures can be readily separated to give the para product in good yield.)

A $\xrightarrow{\text{1. HNO}_3, \text{H}_2\text{SO}_4}_{\text{2. H}_2-\text{Ni}}$ D

A $\xrightarrow{\text{SO}_3, \text{H}_2\text{SO}_4}$ [structure: Br-substituted benzene with SO₃H] mainly $\xrightarrow[\text{2. H}_2-\text{Ni}]{\text{1. HNO}_3, \text{H}_2\text{SO}_4}$ [structure with Br, NH₂, SO₃H] $\xrightarrow[\substack{\text{Removes} \\ \text{SO}_3\text{H}}]{\text{H}_2\text{O}, \Delta}$ C

A $\xrightarrow{\text{(again)}}$ C $\xrightarrow[\substack{\text{Blocks position} \\ \text{para to amine}}]{\text{SO}_3, \text{H}_2\text{SO}_4}$ [structure with Br, NH₂, HO₃S]

$\xrightarrow{\text{Br}_2, \text{FeBr}_3}$ [structure with Br, NH₂, HO₃S, Br] $\xrightarrow[\text{Removes SO}_3\text{H}]{\text{H}^+, \text{H}_2\text{O}, \Delta}$ B

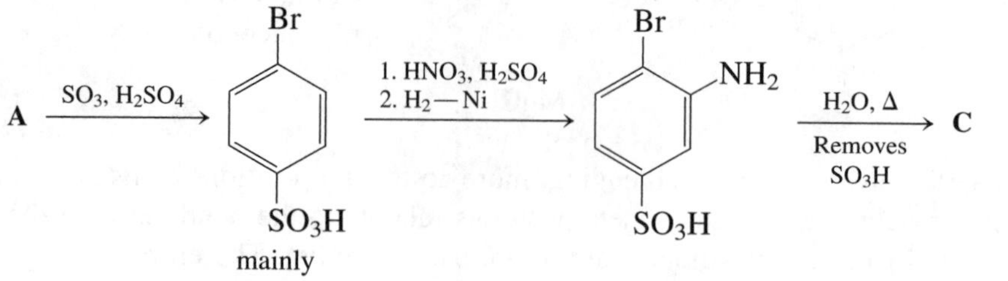

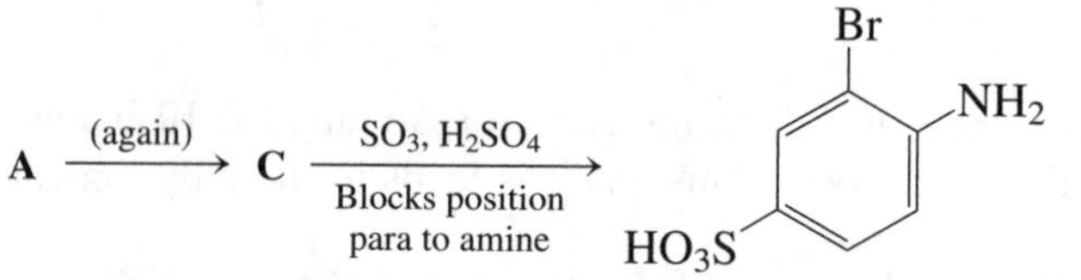

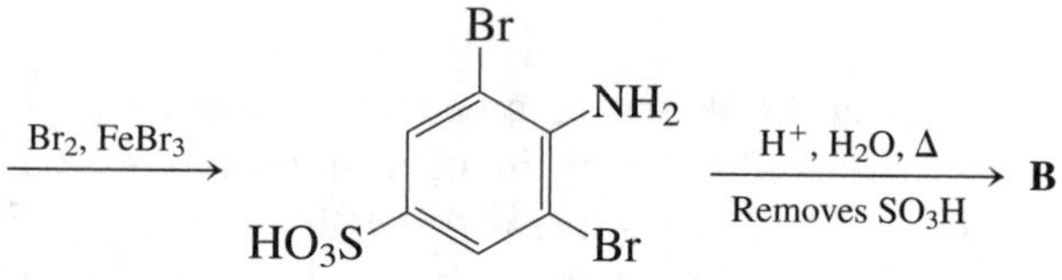

49. 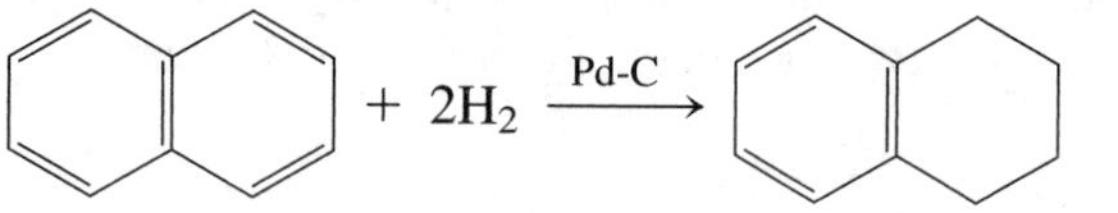 [naphthalene] $+ 2\text{H}_2 \xrightarrow{\text{Pd-C}}$ [tetralin]

Although aromatic, naphthalene does undergo some addition reactions like this one. Addition occurs in such a way that one aromatic benzene ring is left fully intact.

50. Reaction should occur on the most activated (or least deactivated) ring, directed by the groups present.

(a) 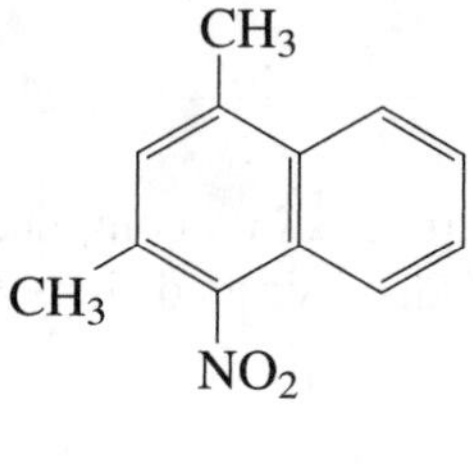 [naphthalene with CH₃, CH₃, NO₂ substituents]

(b) 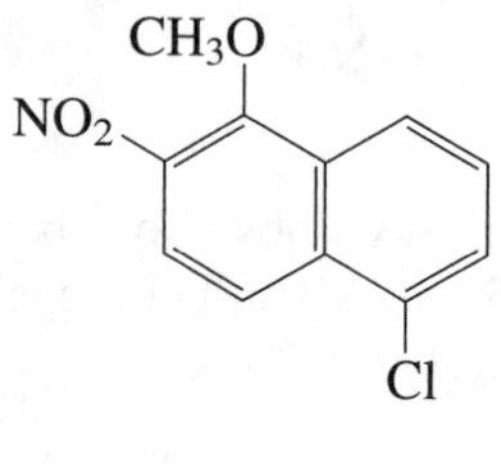 [naphthalene with CH₃O, NO₂, Cl substituents]   Major: [structure]   is sterically hindered

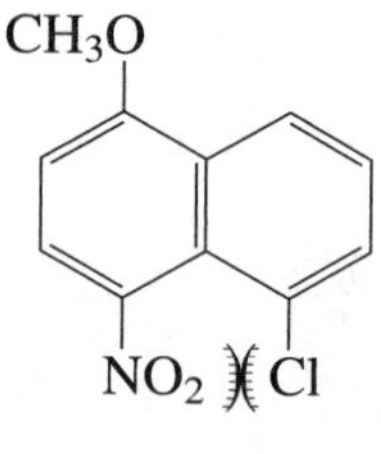

(c) 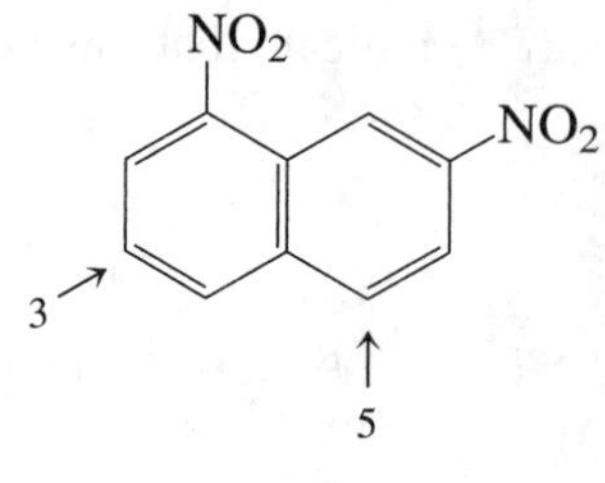 [naphthalene with NO₂, NO₂ substituents, positions 3 and 5 marked]

Two choices: C3 (meta to C1 NO_2) and C5 (meta to C7 NO_2). If all other factors are equal, substitution next to a ring fusion is preferred because one benzene ring is left intact in the intermediate. (See structures for (b) of the solution to Problem 46.) So, nitration occurs at C5.

(d) 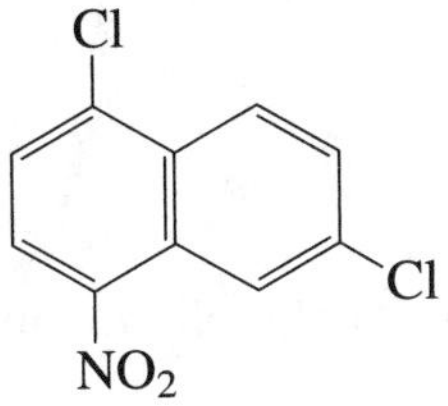   Reasons: para to a C1, next to a ring fusion, and relatively unhindered

51. (a) 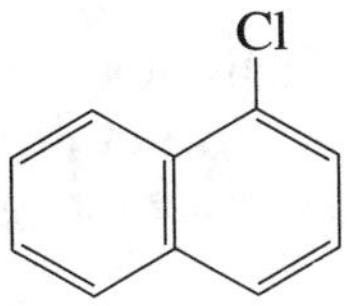   **(b)**

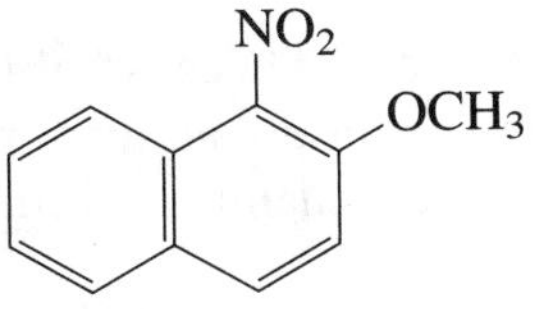

(See below)

(c) 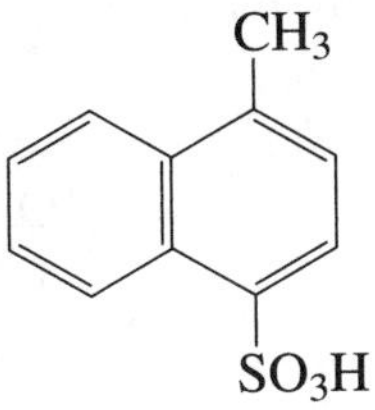   **(d)**

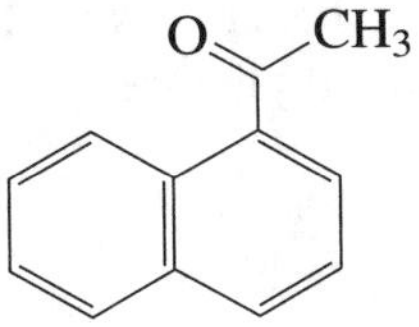

(e) 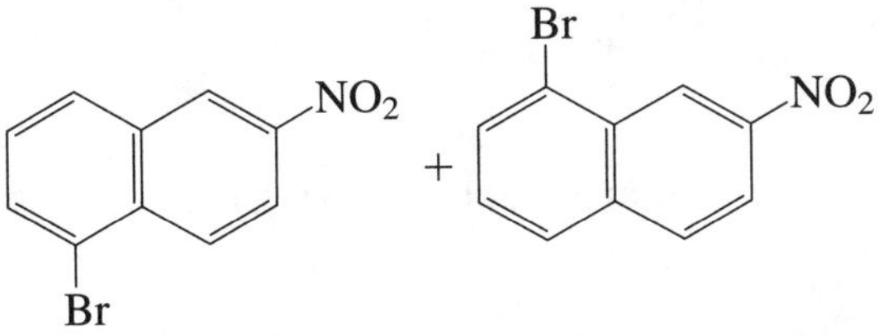 +

In (b), substitution at C1 is favored because the most important resonance form for the intermediate carbocation has an intact 6-π electron benzene ring (see structure, below left). The cation from substitution at C3 does not have this (below, right).

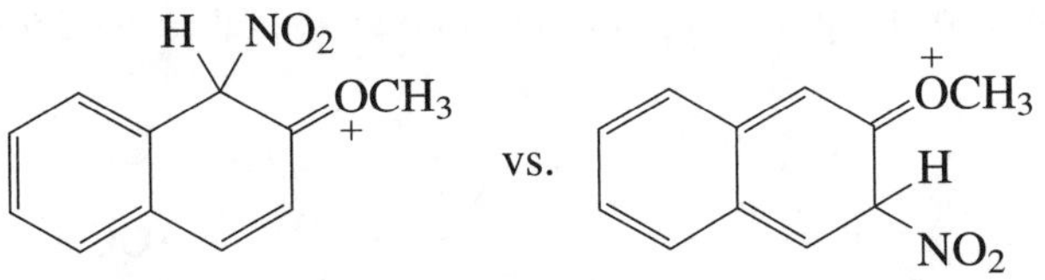

vs.

52. 1-Naphthalenesulfonic acid is the *kinetic* product: It forms faster because the cationic intermediate leading to its formation is more stable, and therefore is produced via a pathway with a lower activation barrier. However, 1-naphthalenesulfonic acid itself is less stable than 2-naphthalenesulfonic acid because in 1-naphthalenesulfonic acid the sulfonic acid group is sterically crowded by the hydrogen atom at C8:

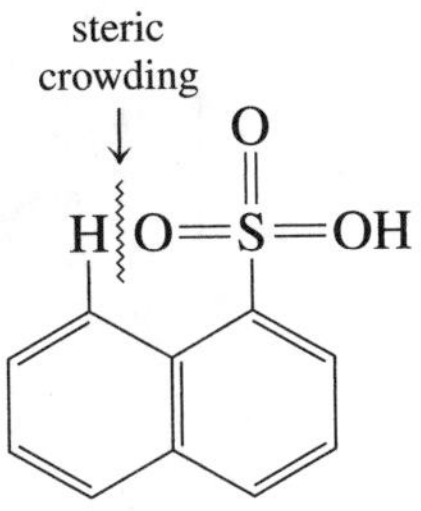

This fact would be completely irrelevant if not for the fact that *sulfonation is reversible*. So, at higher temperature desulfonation from C1 of the kinetic product can take place and (slower) sulfonation at C2 can occur, ultimately leading to build-up of the more stable product, 2-naphthalenesulfonic acid.

53.

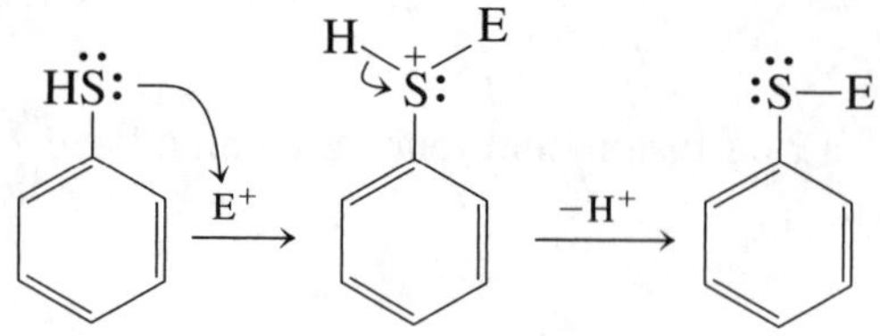

Favored because of high nucleophilicity of sulfur atom

54. The methoxy group is *inductively electron withdrawing* because of the electronegative oxygen atom. The overwhelming resonance effect strongly activates the ortho and para positions, but it is **not directly felt** at the meta positions (see related resonance forms for benzenamine, Section 16-3). The deactivating inductive effect wins at the meta positions.

55. See the answer to Problem 43(g). The saturated six-membered ring is nothing special. You can view it just like two alkyl substituents in adjacent positions on the benzene ring. For example, the starting hydrocarbon in (a) can be treated just as if it were 1,2-dimethylbenzene (*o*-xylene).

(a)

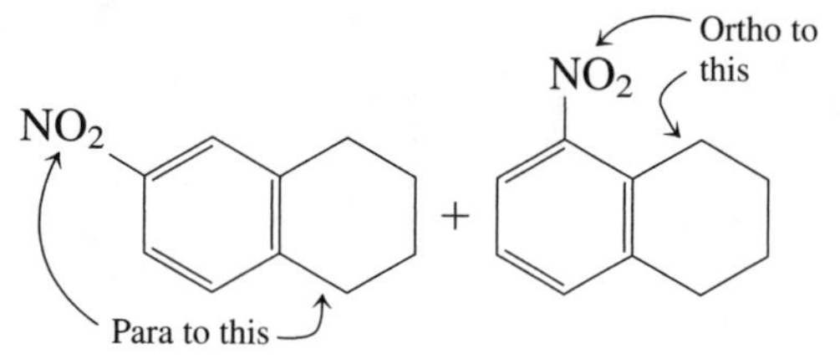

(b)

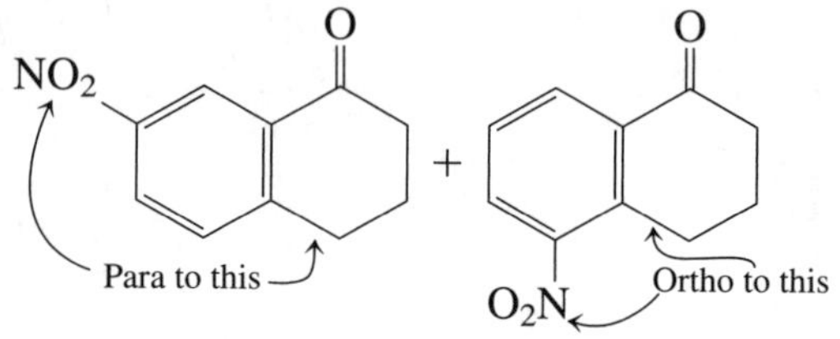

Major
(Actual yield = 43%; 3:1 ratio)

(c)

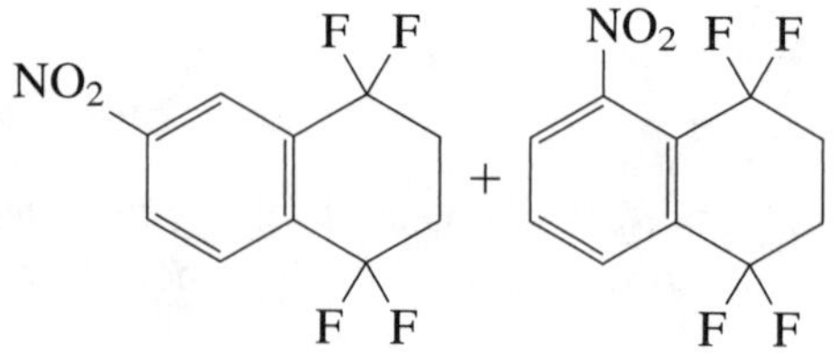

Each position is meta to one of the CF_2 groups.

In (d) and (e), monosubstitution occurs in the more activated or less deactivated ring.

(d)

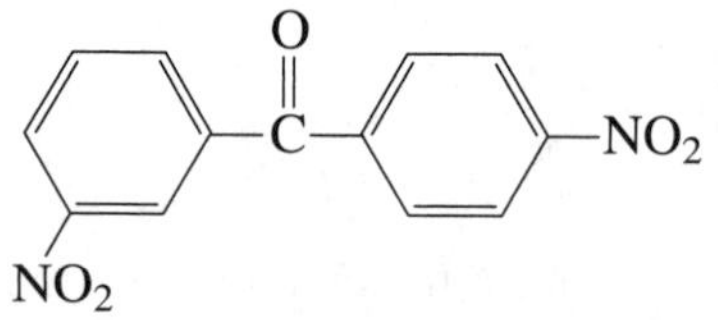

(e)

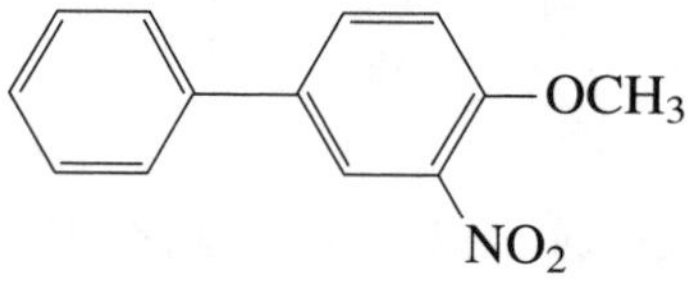

56. Lewis structure: $-\overset{..}{\text{N}}=\overset{..}{\text{O}}:$. The lone pair on N will favor ortho and para substitution through the following resonance forms:

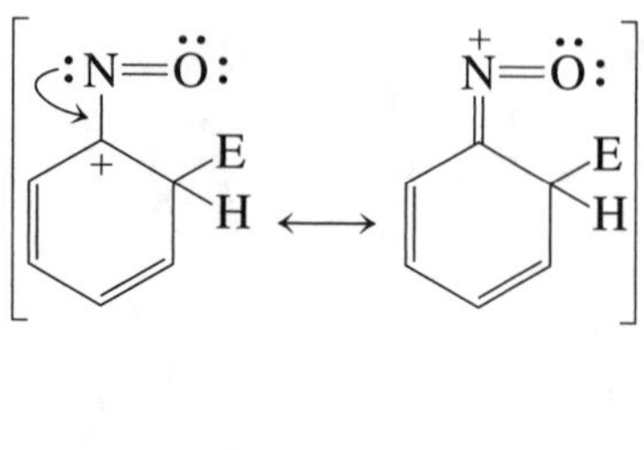

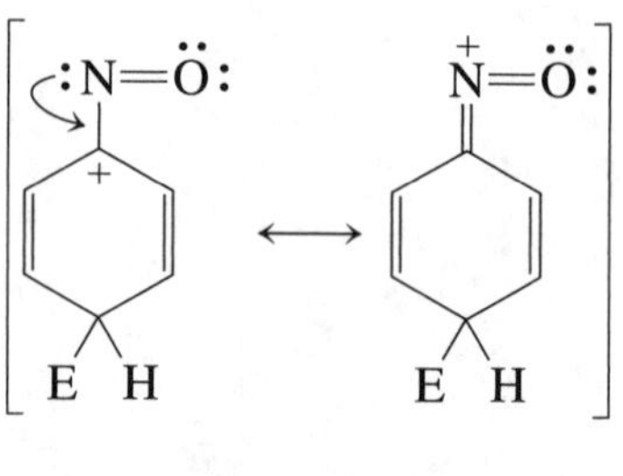

However, the —NO group is inductively electron withdrawing. As is the case with halogen substituents, this deactivating inductive effect is on the average stronger than the resonance effect, so nitrosobenzene is deactivated overall, even though substitution is preferred (by the resonance effect) at the ortho and para positions.

57. The electrophile:

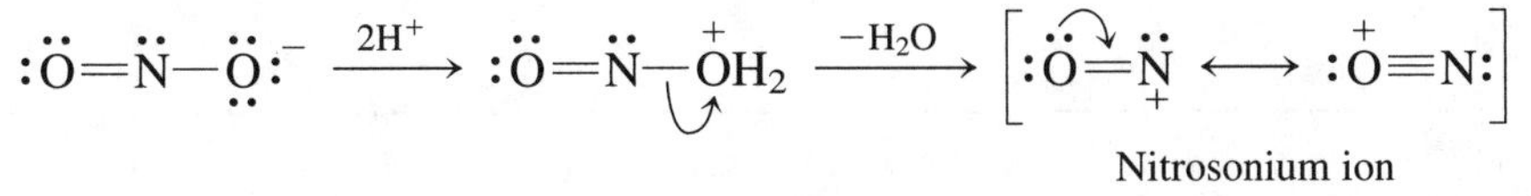

Nitrosonium ion

Then

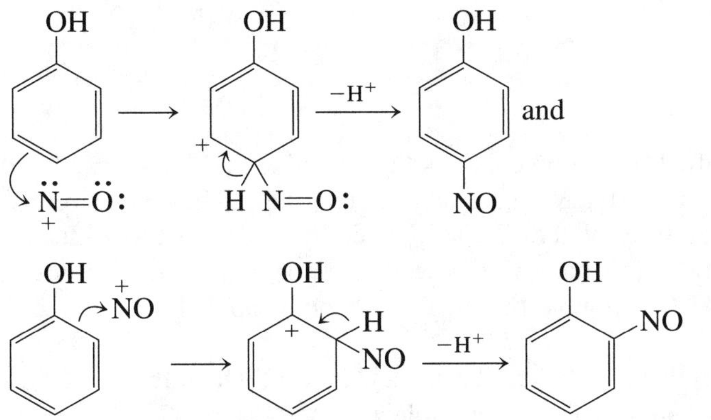

58.

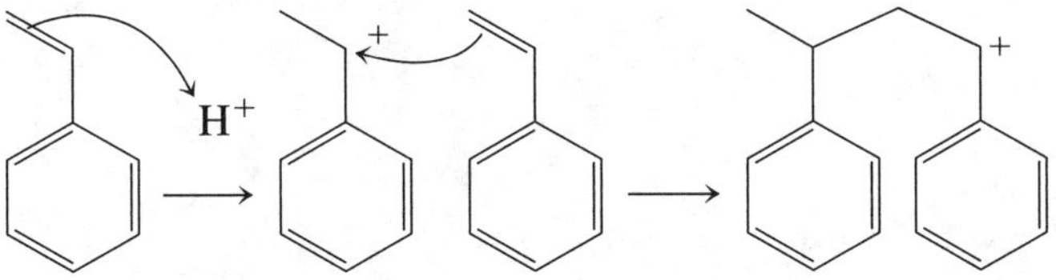

At this point the system has two options: add another styrene and proceed with polymerization, or close a ring by intramolecular Friedel-Crafts reaction.

Polymerization:

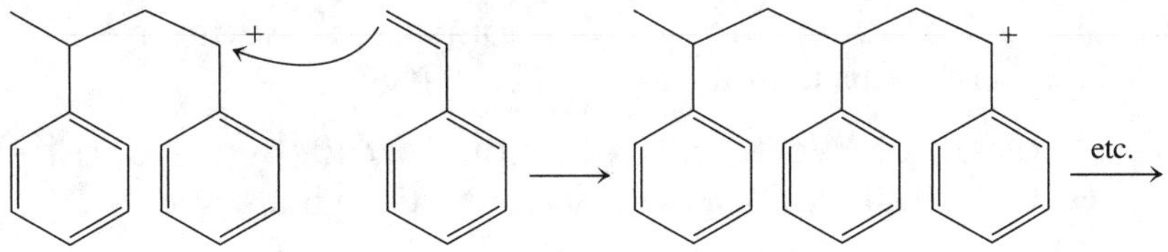

Friedel-Crafts:

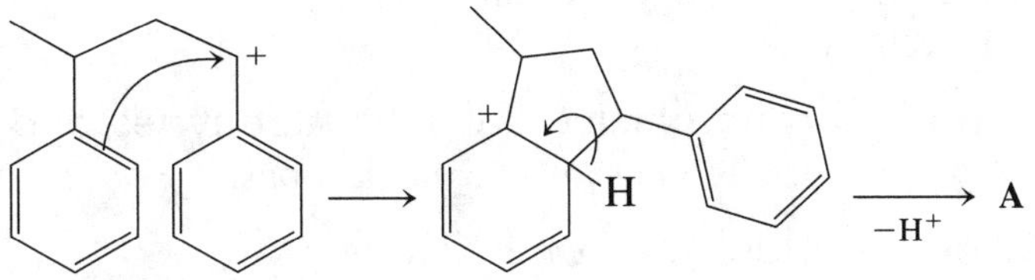

59. (c)

60. (b)

61. (c)

62. (c)

63. (d)

17

Aldehydes and Ketones: The Carbonyl Group

Congratulations! You have finally gotten to the first of several chapters that examine the chemistry of carbonyl compounds: the most important ones in organic chemistry. Why are they so important? They are extremely versatile in carbon–carbon bond formation, and, therefore, synthesis. Carbonyl compounds contain an electrophilic carbonyl carbon (which you already know about from Chapter 8) as well as a potentially nucleophilic carbon next to it (which you will learn about shortly). This "double-barreled" functional capability is unique among the simple compound classes.

The importance of carbonyl compounds extends to biological chemistry as well, where the carbonyl group plays a central role in biochemical synthesis of naturally occurring molecules.

Outline of the Chapter

17-1   Naming the Aldehydes and Ketones

17-2   Structural and Physical Properties

17-3   Spectroscopy

17-4   Preparation of Aldehydes and Ketones: A Review

17-5   Reactivity of the Carbonyl Group: Mechanisms of Addition
          One of two major reaction patterns for carbonyl compounds.

17-6, 17-7, 17-8   Addition of Water and Alcohols to Aldehydes and Ketones
          Hydrates, hemiacetals, acetals, and protecting groups.

17-9   Addition of Amines to Aldehydes and Ketones

17-10   Deoxygenation of the Carbonyl Function
          Reduction of C=O to CH_2.

17-11, 17-12   Addition of Carbon Nucleophiles to Aldehydes and Ketones
          With mechanistic details and synthetic applications.

17-13, 17-14   Oxidations of Aldehydes and Ketones

Keys to the Chapter

17-1. Naming the Aldehydes and Ketones

Most of the material in these sections is of a relatively routine nature, so only a few points of special interest, or with special implications, will be mentioned.

Nomenclature of carbonyl compounds presents a bit of a problem in that several alternative names may be possible for almost any compound.

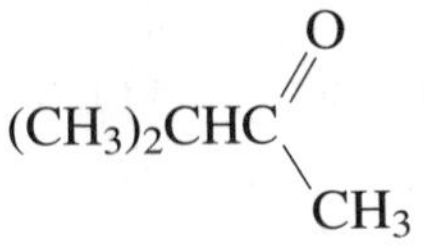

$$(CH_3)_2CHC \overset{\displaystyle O}{\underset{\displaystyle CH_3}{\big\|}}$$

For example, one can give any of several names to the structure above: 3-Methyl-2-butanone (IUPAC) and iso-propyl methyl ketone (common) are just two that are still in current use. Be prepared for some variety, especially in common names! The old (and I mean **old**) naming system for phenyl ketones is especially entertaining. That for C_6H_5COR was derived by combining the (common) name of the carboxylic acid RCO_2H (dropping the final -*ic acid* and adding an *o* if one wasn't already there) with the suffix -*phenone*. So, $C_6H_5COCH_3$ is acet(ic acid) + o + phenone = acetophenone; $C_6H_5CO(CH_2)_4CH_3$ is valer(ic acid) + o + phenone = valerophenone; $C_6H_5COC_6H_5$ is benzo(ic acid) + phenone = benzophenone; and so on. No, it doesn't make much sense to me, either, but that's what people called them.

17-2. Structural and Physical Properties

The polarized carbonyl group is the key to the physical and chemical properties of carbonyl compounds. Although they have polarizations comparable to those found in haloalkanes, carbonyl compounds have a negatively polarized oxygen capable of hydrogen bonding to protic solvents. They are therefore much more water soluble than haloalkanes, for instance. The carbon–oxygen double bond is also considerably stronger than the carbon–carbon double bond in alkenes. One consequence of the relatively strong C=O bond is its tendency to form whenever possible. Whereas additions to C=C bonds are usually quite exothermic, many additions to C=O bonds are not and, indeed, are often reversible, with equilibrium constants near 1. Later you will see a number of reactions that generate carbonyl groups in ways that, at first glance, seem rather surprising.

17-3. Spectroscopy

A few noteworthy points are brought up here. Infrared spectroscopy is very useful for characterization of carbonyl compounds: The band for the C=O stretch is very intense and located in a region of the spectrum (usually 1690–1750 cm^{-1}) that is free from strong absorptions of other functional groups. The precise location of this peak is also useful in establishing the nature of the groups attached to the carbonyl carbon. ^{13}C NMR spectroscopy exhibits signals near $\delta = 200$ ppm for the carbonyl carbon, and the aldehydic (also called 'formyl') hydrogen (—CHO) resonates in the range $\delta = 9.5$–10.0 ppm in the ^1H NMR.

17-4. Preparation of Aldehydes and Ketones: A Review

The number of methods that exist to synthesize carbonyl compounds is impressive to the point of being intimidating. But don't be put off by this situation. Each reaction is presented in an appropriate, logical context. This section mainly reiterates carbonyl syntheses that you've seen before, perhaps just with a few new examples to help reinforce those original presentations.

17-5. Reactivity of the Carbonyl Group: Mechanisms for Addition

This chapter concentrates solely on one type of reaction: addition across the carbon–oxygen double bond of a carbonyl group. Carbonyl additions are polar reactions that exactly follow pathways that would be expected by electrostatics.

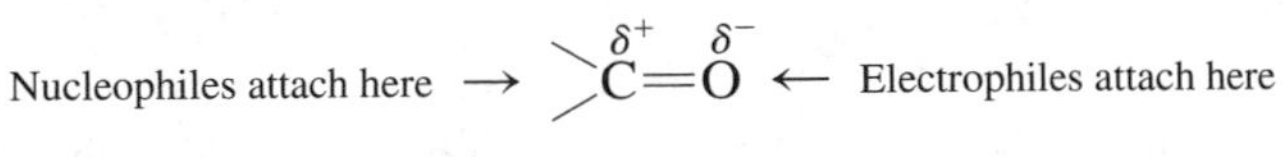

$$\text{Nucleophiles attach here} \rightarrow \overset{\delta^+}{\underset{}{C}}{=}\overset{\delta^-}{\underset{}{O}} \leftarrow \text{Electrophiles attach here}$$

The two main mechanisms described in this section are distinguished by the order of addition. Strong nucleophiles, Nuc$^-$ (which may be added directly or formed by the reaction of a base with Nuc—H), add to the carbonyl carbon first, followed (usually) by protonation of oxygen (which may be a separate step). Conversely, addition of weaker nucleophiles, especially neutral $\ddot{N}$ucH, is helped by prior protonation of the carbonyl oxygen to give the highly electrophilic $>$C$=\overset{+}{O}$H group. Whichever way the reaction proceeds in any given case, its scope is very broad, and many useful types of addition products are known. These are the subjects of the remaining sections of the chapter.

17-6 through 17-9. Addition of Water, Alcohols, and Amines to Aldehydes and Ketones

There are actually two fundamentally different types of reactions in these sections. The first is the reversible addition of a nucleophile to a carbonyl group, which can generally be catalyzed by either base or acid.

1. 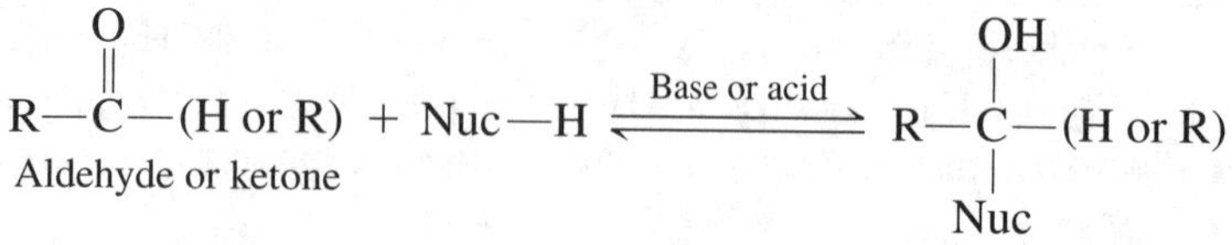
$$\underset{\text{Aldehyde or ketone}}{R-\overset{\displaystyle O}{\overset{\|}{C}}-(H \text{ or } R)} + Nuc-H \xrightleftharpoons{\text{Base or acid}} R-\overset{\displaystyle OH}{\underset{\displaystyle Nuc}{\overset{|}{\underset{|}{C}}}}-(H \text{ or } R)$$

If Nuc—H is	Product is
H_2O	Aldehyde or ketone hydrate
R'OH	Hemiacetal
R'SH	Hemithioacetal
R'NH$_2$ or R$_2$'NH	Hemiaminal

The second is really *carbocation chemistry*. When Nuc—H is R'OH, or R'SH, the product in the reaction shown above may have its OH group replaced by a second Nuc. This reaction occurs by the S_N1 mechanism and leads to relatively stable acetal or thioacetal products. As the text section shows, these latter two kinds of compounds are resistant to attack by many kinds of basic or nucleophilic reagents, such as RLi, Grignards, and hydrides. Conversion of an aldehyde or ketone C$=$O to an acetal or thioacetal is a handy way of protecting it when you need to react some other functional group in a molecule with a strong base or nucleophile.

2a. 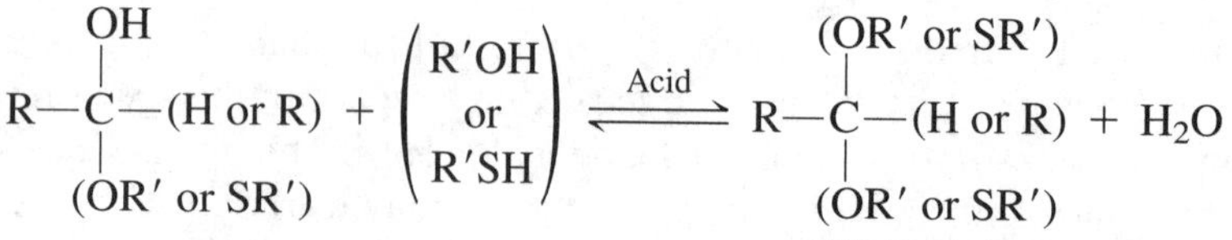
$$R-\overset{\displaystyle OH}{\underset{\displaystyle (OR' \text{ or } SR')}{\overset{|}{\underset{|}{C}}}}-(H \text{ or } R) + \begin{pmatrix} R'OH \\ \text{or} \\ R'SH \end{pmatrix} \xrightleftharpoons{\text{Acid}} R-\overset{\displaystyle (OR' \text{ or } SR')}{\underset{\displaystyle (OR' \text{ or } SR')}{\overset{|}{\underset{|}{C}}}}-(H \text{ or } R) + H_2O$$

If R'OH: Acetal
If R'SH: Thioacetal

When Nuc—H is a primary amine, R'NH$_2$, further reaction of the hemiaminal gives an *imine,* containing a carbon–nitrogen double bond. Alternatively, when the nucleophile is R'$_2$NH, an *enamine* results.

2b. 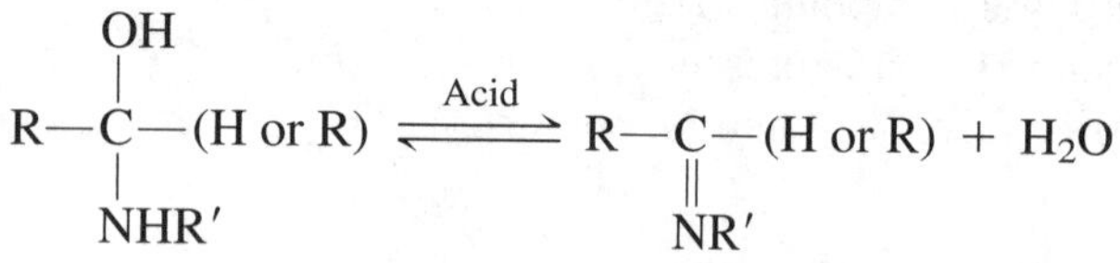
$$R-\overset{\displaystyle OH}{\underset{\displaystyle NHR'}{\overset{|}{\underset{|}{C}}}}-(H \text{ or } R) \xrightleftharpoons{\text{Acid}} R-\overset{\displaystyle }{\underset{\displaystyle NR'}{\overset{\|}{\underset{\|}{C}}}}-(H \text{ or } R) + H_2O$$

Imine

2c. 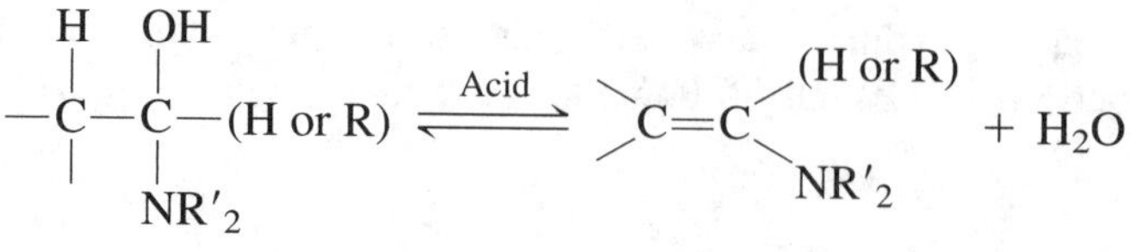
$$-\overset{\displaystyle H}{\underset{\displaystyle }{\overset{|}{\underset{|}{C}}}}-\overset{\displaystyle OH}{\underset{\displaystyle NR'_2}{\overset{|}{\underset{|}{C}}}}-(H \text{ or } R) \xrightleftharpoons{\text{Acid}} >C=C\overset{(H \text{ or } R)}{\underset{NR'_2}{}} + H_2O$$

Enamine

For all practical purposes, acid-catalyzed reaction of an aldehyde or a ketone with any of the nucleophiles listed in reaction 1 will continue straight through to the product of reactions 2a, 2b, and 2c, as the case may be. Conversely, acid-catalyzed hydrolysis of any of the latter products will proceed back all the way to the original aldehyde or ketone and the free nucleophile.

17-10. Deoxygenation of the Carbonyl Function

The reduction of C=O to CH_2 may be achieved in three ways: Raney nickel desulfurization of thioacetals (Section 17-8), Clemmensen reduction (Section 16-5), and, as described here, Wolff-Kishner reduction. Because there are many ways to make carbonyl compounds, and carbonyl compounds are great places to start for making new carbon–carbon bonds, you will find that carbonyl groups are often present when complex molecules are made from simpler ones. This will be especially evident in Chapters 18–20 and 23. Deoxygenation will be useful if you need to get rid of a carbonyl group that has been used to construct bonds in a large molecule but is not wanted in the final product. The example of Friedel-Crafts acylation followed by deoxygenation to give an alkylbenzene is a case in point.

17-11 and 17-12. Addition of Carbon Nucleophiles to Aldehydes and Ketones

Grignard reagents and alkyllithiums are examples of highly reactive "carbanionic" reagents (i.e., they behave like carbanions). In this section less highly energetic carbanionic reagents are introduced. Cyanide ion is the simplest, and the products of its addition to aldehydes and ketones, cyanohydrins, have some specialized synthetic utility. Ylide reagents containing phosphorus are much more useful, particularly in the *regiospecific* synthesis of alkenes, because the double bond is fixed in a single position determined entirely by the starting compounds in the reaction.

$$RCH{=}P(C_6H_5)_3 \; + \; O{=}C\overset{R'}{\underset{R''}{\diagdown}} \; \longrightarrow \; RCH{=}CR'R'' \; + \; (C_6H_5)_3P{=}O$$

Typical phosphorus ylide   **Exclusive regioisomer**

(Mixture of *E* and *Z* stereoisomers)

17-13. Baeyer-Villiger Oxidation of Ketones

Baeyer-Villiger oxidation of ketones to carboxylic esters is significant for two reasons. Mechanistically, it involves a nucleophilic addition to a carbonyl group, which is no big deal. The nucleophile, however, is a peroxidic species, which can lead to rearrangement, forming a new oxygen–carbon bond.

You might recall a somewhat related mechanistic rearrangement, the migration of a group from boron to oxygen during the oxidation of alkylboranes with basic hydrogen peroxide.

The second reason the Baeyer-Villiger reaction is significant is that it **cleaves a carbon–carbon bond.** You have seen only a very small number of reactions capable of this (e.g., ozonolysis; Chapter 12). The power of this method to be a synthetic tool lies both in its high selectivity when the carbonyl compound is not symmetrical (see discussion of migratory aptitudes) and in the subsequent chemistry available from the ester or acid, which will be covered shortly.

Solutions to Problems

25. (a)

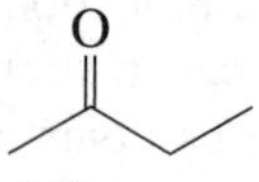

2-Butanone

(b)

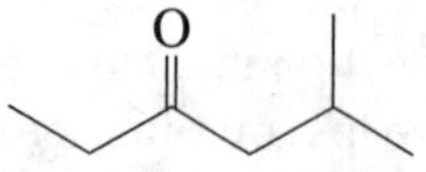

5-Methyl-3-hexanone

(c)

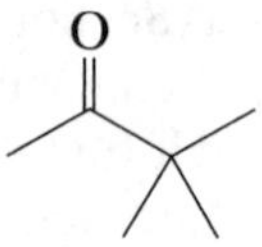

3,3-Dimethyl-2-butanone

(d)

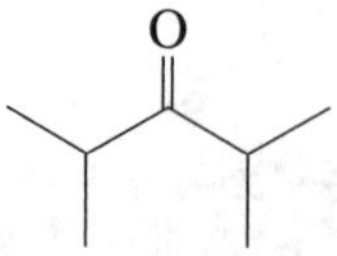

2,4-Dimethyl-3-pentanone

(e)

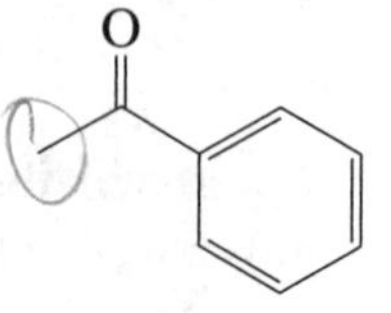

1-Phenylethanone

(f)

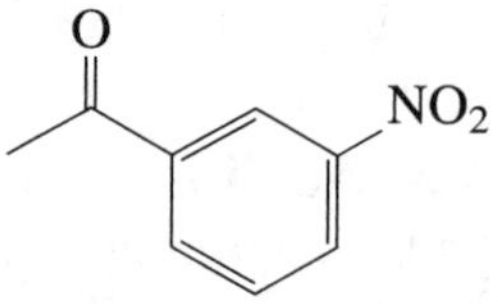

1-(3-Nitrophenyl)ethanone

26. (a) 2,4-Dimethyl-3-pentanone　　**(b)** 4-Methyl-3-phenylpentanal

(c) 3-Buten-2-one　　**(d)** *trans*-4-Chloro-3-butenal

(e) 4-Bromo-2-cyclopentenone　　**(f)** *cis*-2-Ethanoyl-3-phenylcyclohexanone
　　　　　　　　　　　　　　　　　　(*cis*-2-acetyl-3-phenylcyclohexanone)

(g)

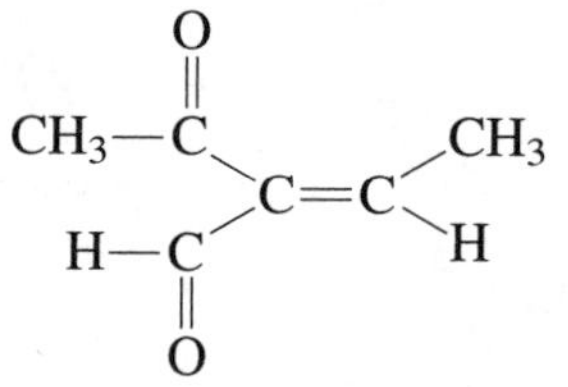

$$CH_3-C \quad CH_3$$

(h)

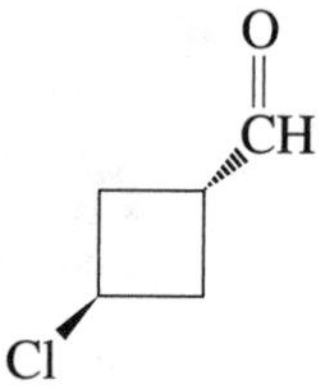

27. Degree of unsaturation for $C_8H_{12}O$, $H_{sat} = 16 + 2 = 18$; degree of unsaturation $= \frac{(18 - 12)}{2} = 3\ \pi$ bonds or rings present.

(a) ^{13}C: Molecule contains $>C=O$ ($\delta = 198.6$) and $>C=C<$ ($\delta = 139.8$ and 140.7).

^1H: Obvious features include $\delta = 2.15$ (s, 3 H), for $CH_3\overset{\overset{\displaystyle O}{\|}}{C}-$, and $\delta = 6.78$ (t, 1 H), for

$>C=C<\overset{CH_2-}{\underset{H}{}}$. Note the absence of any other alkene hydrogens. So we can start with these pieces:

$$CH_3-\overset{\overset{\displaystyle O}{\|}}{C}- \quad \text{and} \quad >C=C<\overset{CH_2-}{\underset{H}{}} \text{, adding up to } C_5H_6O$$

Still needed are three more C's, six more H's, and another degree of unsaturation, which will have to be a ring. A simple way to put together a trial structure would be to finish a six-membered ring with three CH_2 groups, and then attach the acetyl group:

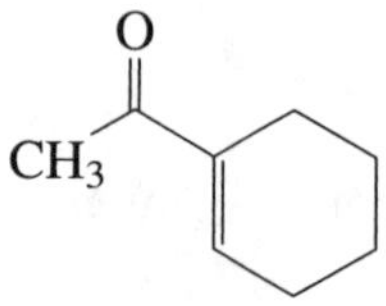

This is not the only possible answer, but it is the actual molecule that gives the indicated spectra.

(b) ^{13}C NMR: One C=O (δ = 193.2) and **two** C=C groups (δ = 129.0, 135.2, 146.7, and 152.5) this time.

^{1}H NMR: The carbonyl group is an **aldehyde** $\left(\delta = 9.56 \text{ for } \overset{\overset{\text{O}}{\|}}{-\text{C}-\text{H}} \right)$. At the other end, we have

$$\underset{\underset{0.94 \quad 1.48 \quad 2.21}{\uparrow \qquad \uparrow \qquad \uparrow}}{\text{CH}_3-\text{CH}_2-\text{CH}_2-}$$

This adds up to C_4H_8O, leaving C_4H_4 to account for. All four of these H's are alkene hydrogens (δ = 5.8–7.1), so these could most simply be two —CH=CH— groups. The result:

$$\text{CH}_3\text{CH}_2\text{CH}_2\text{CH}=\text{CHCH}=\overset{\overset{\text{O}}{\|}}{\text{CHCH}}$$

28. Each is a conjugated carbonyl compound, giving an intense UV absorption with $\lambda_{max} > 200$ nm. The first spectrum matches

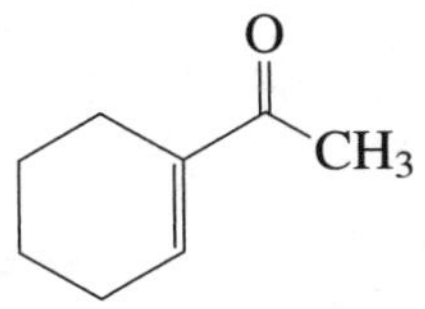

with $\pi \rightarrow \pi^*$ absorption at 232 nm, and carbonyl $n \rightarrow \pi^*$ absorption at 308 nm. The second spectrum matches the diene-aldehyde, with the longer wavelength 272 nm band corresponding to the $\pi \rightarrow \pi^*$ absorption of the more extended conjugated system.

29. (a) MS: $M^{+\bullet}$ of 128 confirms that $C_8H_{16}O$ is the molecular formula as well

$H_{sat} = 16 + 2 = 18$; degree of unsaturation $= \frac{(18-16)}{2} = 1$ π bond or ring present

IR, UV: A ketone C=O appears to be present

NMR:

$$\underset{\underset{0.9(t)}{\uparrow}}{\text{CH}_3-\text{CH}_2-}, \quad \underset{\underset{2.0(s)}{\uparrow}}{\text{CH}_3-\overset{\overset{\text{O}}{\|}}{\text{C}}-}\underset{\underset{2.2(t)}{\uparrow}}{\text{CH}_2-\text{CH}_2-}$$

are likely pieces, adding up to $C_6H_{12}O$; only C_2H_4 are left to add in. Is 2-octanone a reasonable answer?

MS: Base peak (m/z 43) is $\left[\overset{\overset{\text{O}}{\|}}{\text{CH}_3\text{C}} \right]^+$; next largest is m/z 58, consistent with *McLafferty rearrangement* as follows:

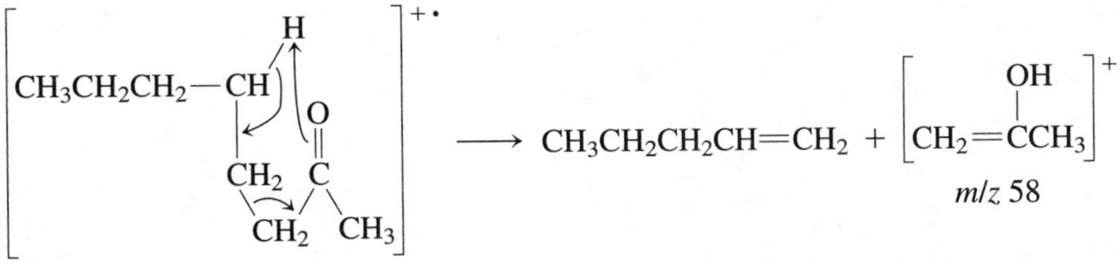

2-Octanone

This answer seems quite reasonable.

30. Be sure you've read the problem correctly. The given structures are your *starting materials,* and you are asked to identify a reagent that will turn each one of them *into* 3-hexanone. **(a)** PCC, CH_2Cl_2 or CrO_3, H_2SO_4, acetone; **(b)** H_2O, H^+; **(c)** $HgCl_2$, $CaCO_3$, CH_3CN, H_2O; **(d)** H_2O, H^+ or ^-OH; **(e)** 1. O_3; 2. $(CH_3)_2S$ or Zn, CH_3COOH; **(f)** H_2O, H^+ or ^-OH; **(g)** Hg^{2+}, H_2O or 1. R_2BH (R = cyclohexyl), 2. H_2O_2, H_2O, ^-OH; **(h)** H_2O, ^-OH.

31. **(a)** CrO_3, H_2SO_4, acetone, or MnO_2, CH_2Cl_2 (better);

(b) PCC, CH_2Cl_2; **(c)** 1. O_3, CH_2Cl_2, 2. Zn, CH_3COOH, H_2O;

(d) $HgSO_4$, H_2O, H_2SO_4; **(e)** same as (d);

(f) 1.

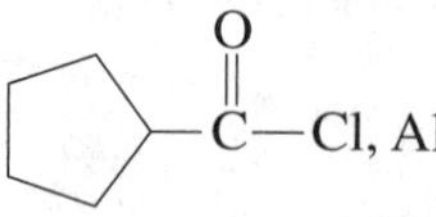

32. **(a)** $CH_3CH_2CH_2CH + HCH$ (with two carbonyl groups) **(b)** 2

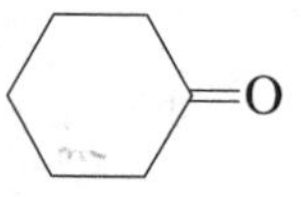

(c) $HCCH_2CH_2CH_2CH_2CH$ (with two carbonyl groups) **(d)**

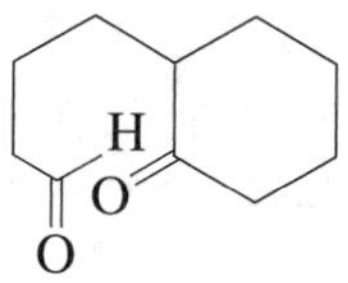

33. **(a)** Ask "How electrophilic is the carbon in question?":

$(CH_3)_2C=\overset{+}{O}H > (CH_3)_2C=O > (CH_3)_2C=NH$ Order of ketone and imine determined by electronegativity.

(b) $CH_3CCCCH_3 > CH_3CCCH_3 > CH_3CCH_3$ (with carbonyl groups)   Adjacent carbonyl groups enhance each other's reactivity by reinforcing the δ^+ character of their respective carbons.

(c) $BrCH_2CHO > CH_3CHO > BrCH_2COCH_3 > CH_3COCH_3$ Aldehydes are more reactive than ketones; halogen substituents increase reactivity.

34. **(a)**

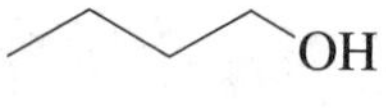

(b)

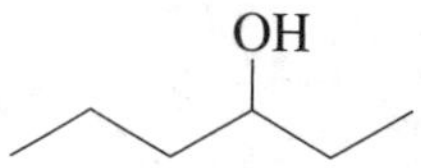

(c)

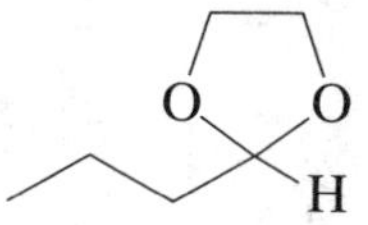

35. **(a)**

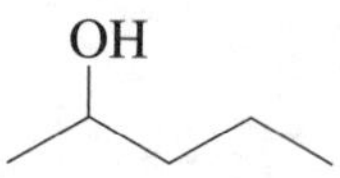

(b)

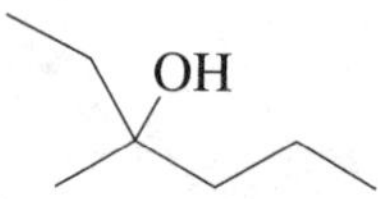

(c)

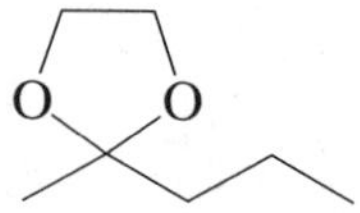

36. **(a)**

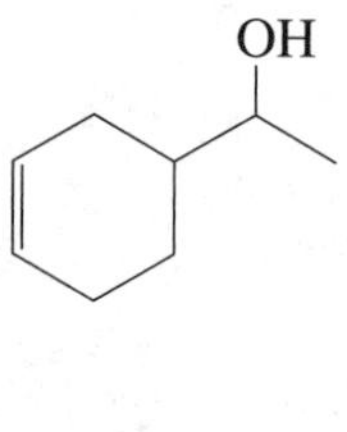

(b)

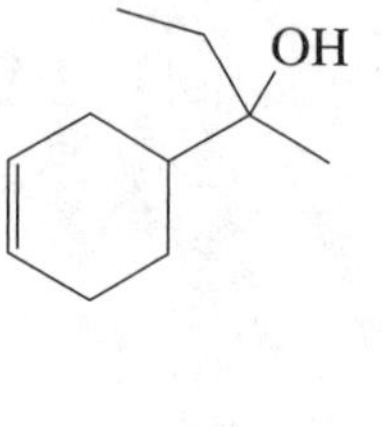

(c)

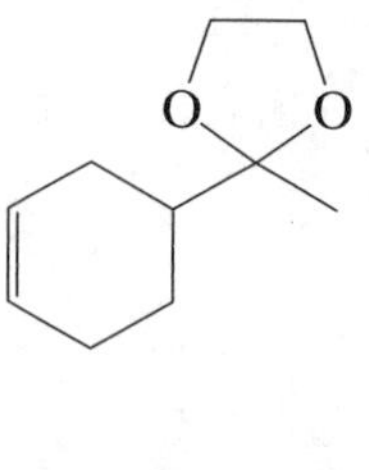

37. **(a)**

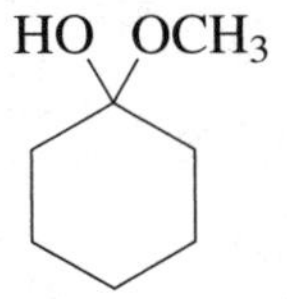

HO OCH$_3$ on cyclohexane

The starting material equilibrates with this product.

(b)

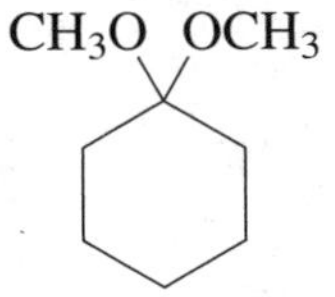

CH$_3$O OCH$_3$ on cyclohexane

Only acid catalyzes acetal formation.

(c)

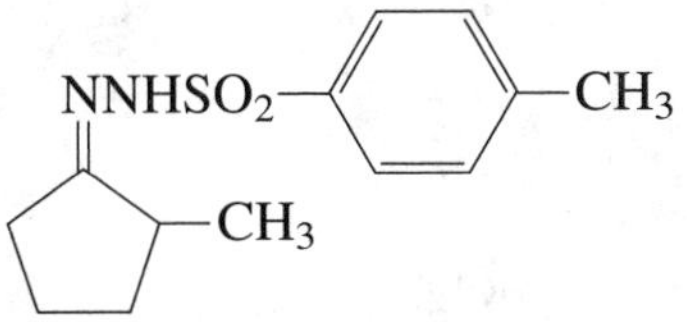

NNHSO$_2$—⟨benzene⟩—CH$_3$ with CH$_3$ on cyclopentane ring

(d)

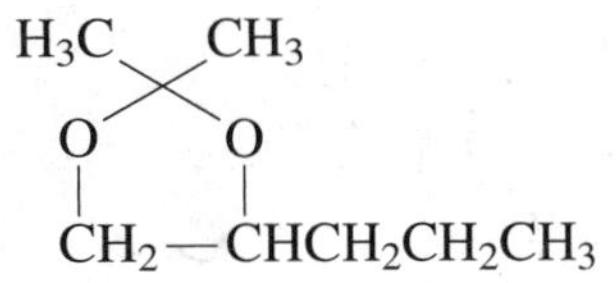

H$_3$C   CH$_3$
     C
    O   O
CH$_2$—CHCH$_2$CH$_2$CH$_3$

(e)

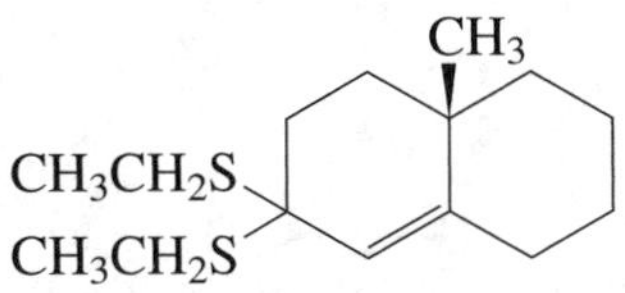

CH$_3$CH$_2$S
            decalin ring with CH$_3$
CH$_3$CH$_2$S

(f)

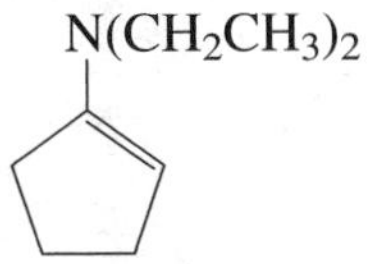

N(CH$_2$CH$_3$)$_2$ on cyclopentene

38. **(a)** In acid:

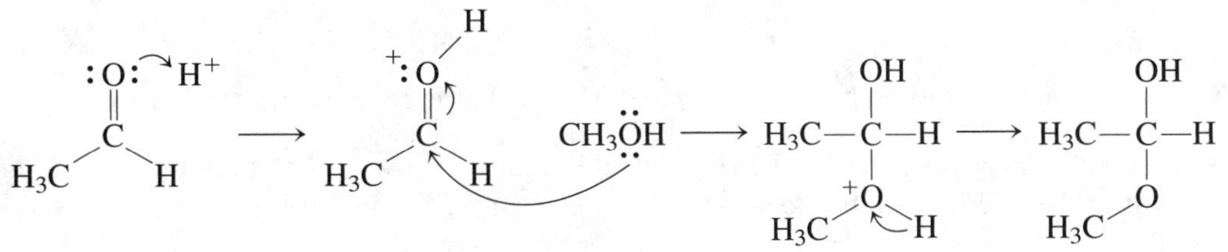

:O: ⤴ H$^+$ / H$_3$C—C—H → $^+$:O—H / H$_3$C—C—H   CH$_3$ÖH → H$_3$C—C(OH)—H with $^+$O(H)—CH$_3$ → H$_3$C—C(OH)—H with O—CH$_3$

In base:

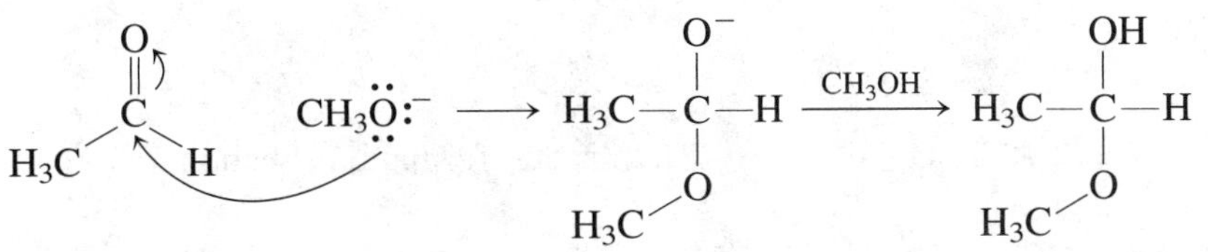

O / H$_3$C—C—H   CH$_3$Ö:$^-$ → H$_3$C—C(O$^-$)—H with O—CH$_3$ $\xrightarrow{CH_3OH}$ H$_3$C—C(OH)—H with O—CH$_3$

(b) In acid:

:O: ⤴ H$^+$ / H—C—CH$_2$CH$_2$CH$_2$CH$_2$OH → $^+$:O—H / H—C—CH$_2$CH$_2$CH$_2$CH$_2$ÖH → HO—C(H) pyran ring with $^+$O(H) → HO—C(H) pyran ring with O

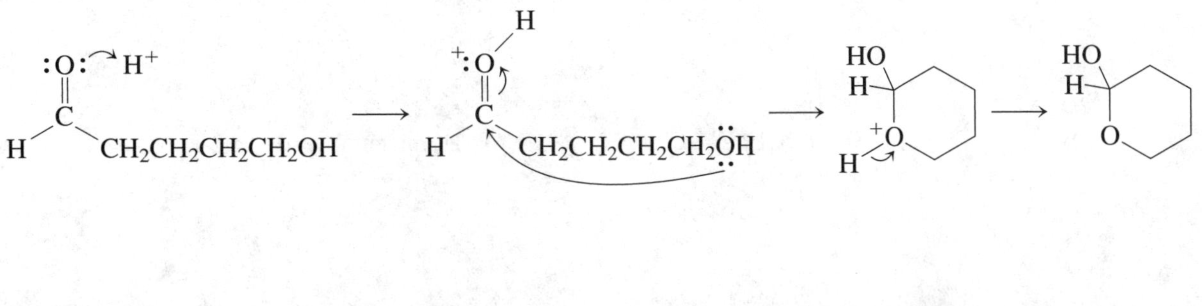

In base:

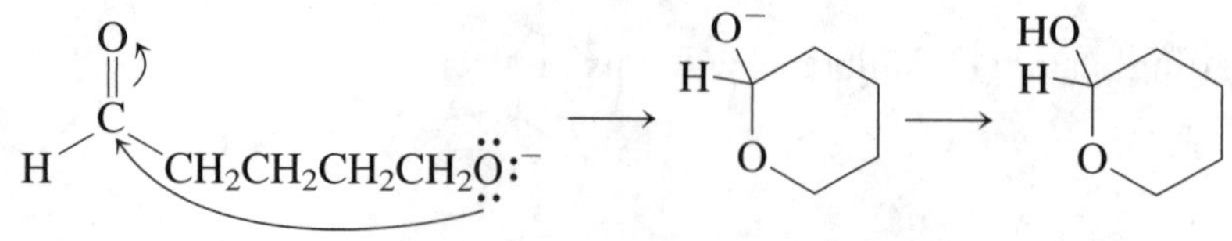

39.

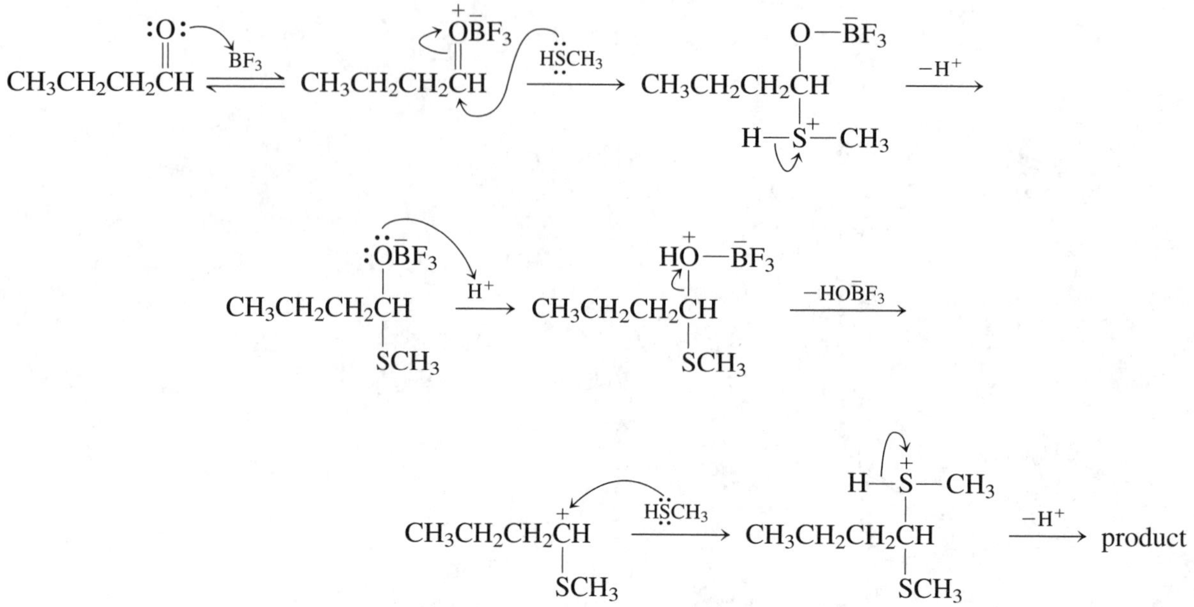

40. (a) Ketone hydrates do **not** contain the H—C—OH unit that is required for further oxidation to occur. Oxidation beyond the ketone stage would require cleavage of a carbon–carbon bond, a difficult process (compare the Baeyer-Villiger oxidation; Section 17-13).

(b) Think about it: If CrO_3 is added to an alcohol, what will be present in the mixture? There will be some aldehyde formed from oxidation, in the presence of an excess of **unreacted alcohol.** What reaction occurs between aldehydes and alcohols?

$$RCH_2OH + RCH \rightleftharpoons RCH_2O{-}\overset{\overset{\displaystyle OH}{|}}{\underset{\underset{\displaystyle H}{|}}{C}}{-}R \qquad \text{Hemiacetal formation (Section 17-7)}$$

(c) (1)

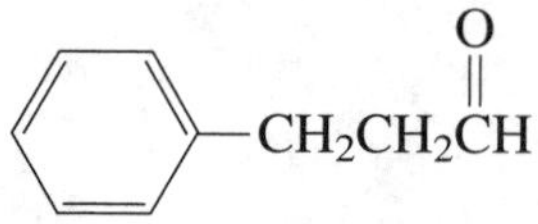

Proper procedure for oxidation of primary alcohols to aldehydes

(2)

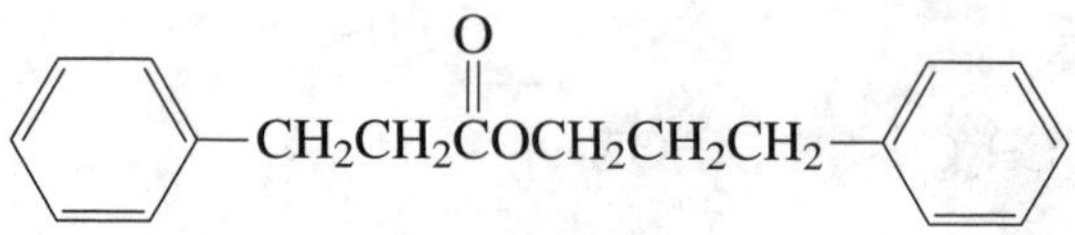

Hemiacetal forms and is oxidized; actual yield = 54%

41. (a)

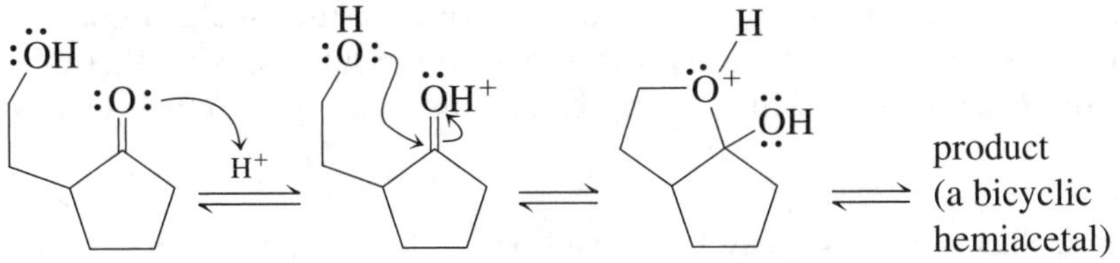

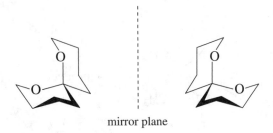

(b) HO ...

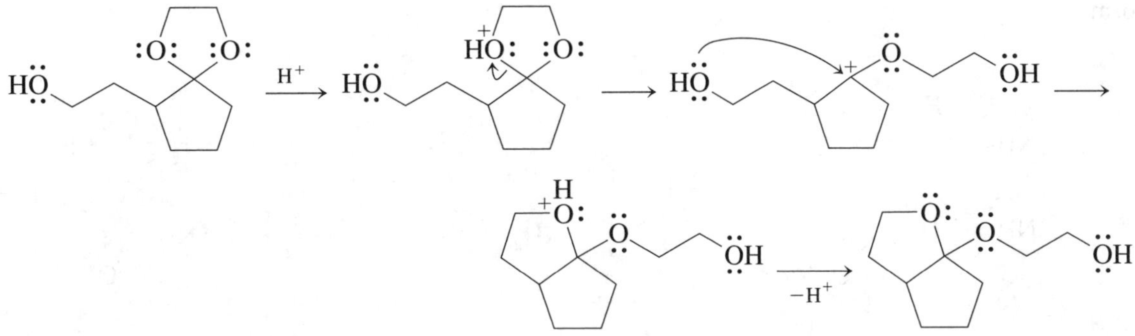

(c) See text sections 17-6 and 17-7 and *Study Guide* p. 332.

42. (a) The structures are presented in a way that should remind you of one particular isomeric relationship. Imagine a vertical mirror placed between them. Do you see each as the reflection of the other? These are enantiomers.

mirror plane

(b) Both molecules contain acetal functions: two oxygen atoms attached by single bonds to a common carbon. (Neither is an OH group; otherwise we'd be looking at hemiacetals.)

(c) The acetal function undergoes hydrolysis under aqueous acidic conditions to a carbonyl compound. The common carbon referred to above becomes double bonded to oxygen. The two oxygens are released as hydroxy functions. First, consider a simpler example:

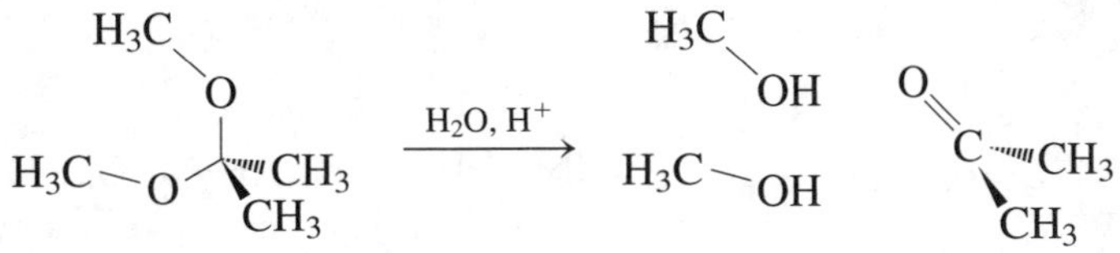

We drew this example in just this orientation to make clearer its resemblance to the molecule in the problem. Now take our molecule and proceed similarly:

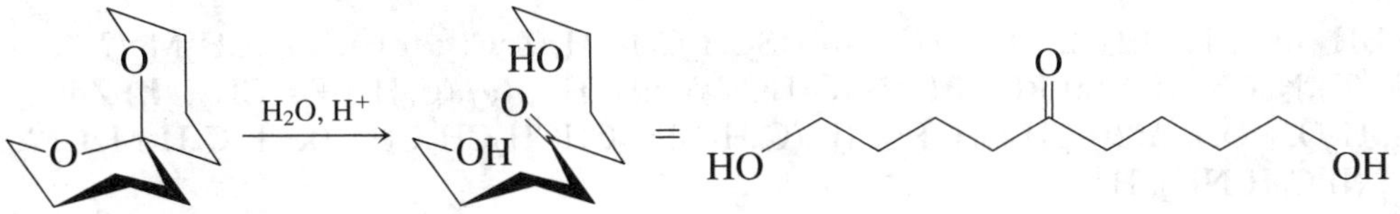

Do you see how these two examples parallel each other? Together they make a good exercise in pattern recognition, a very valuable tool in solving reaction problems. Note that the final product is achiral; both enantiomers of the starting compound hydrolyze to give this same molecule.

43. A double imine formation. Such a process may be carried out without catalysis, but mild acid is usually helpful. The mechanism for the first condensation is shown in detail, the second in abbreviated form.

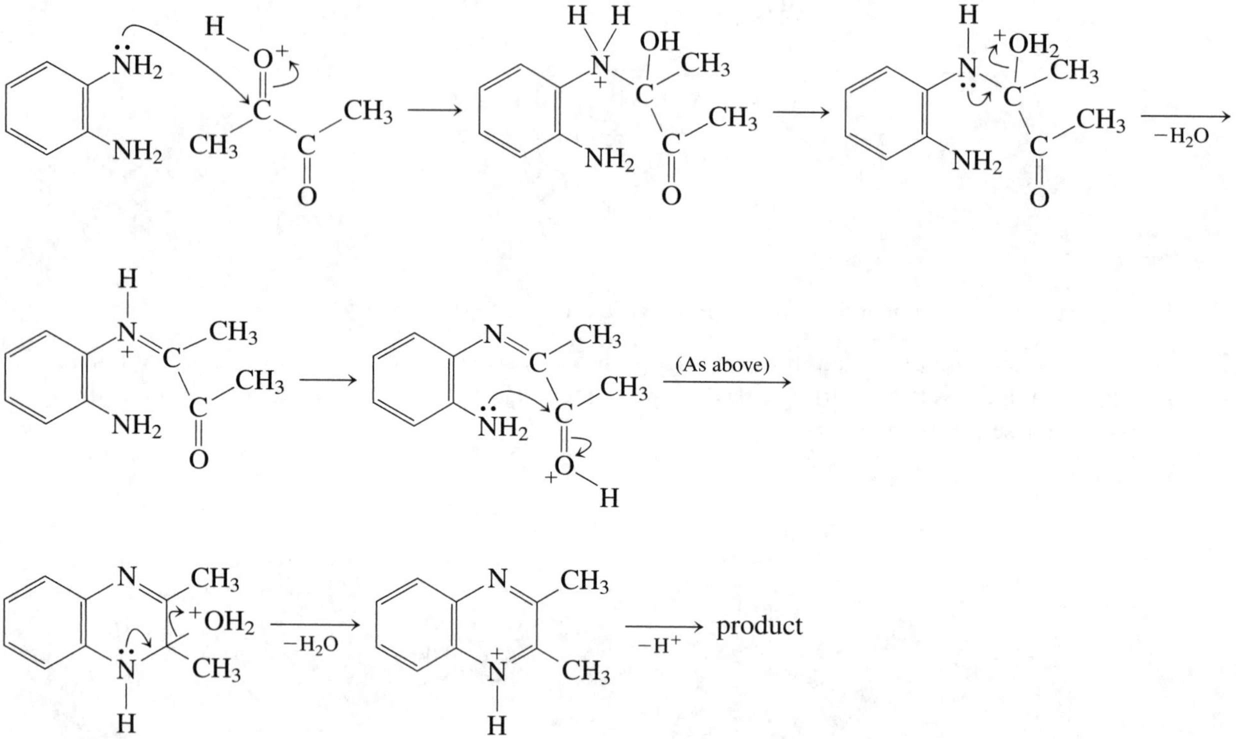

44.

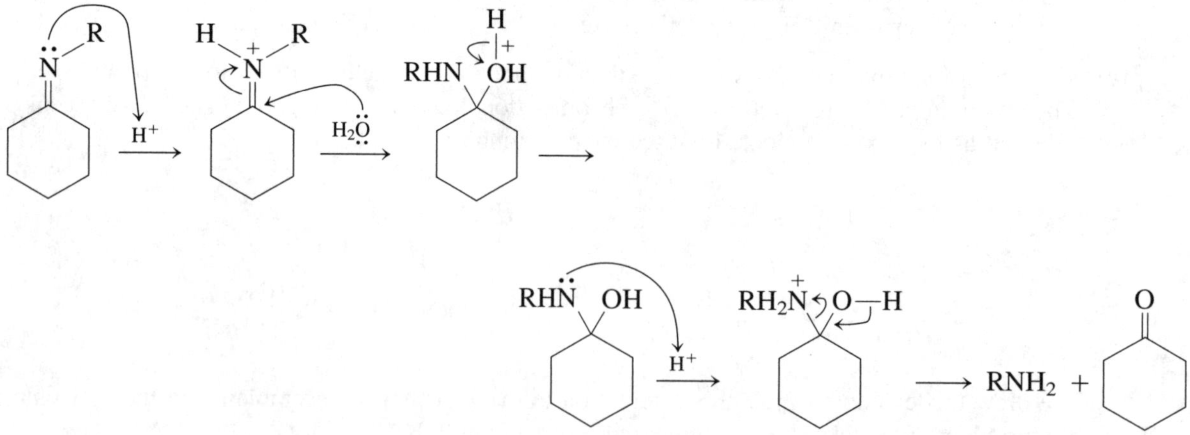

45. (a) NaBH₄ or 1. LiAlH₄, 2. H₂O, H⁺; (b) HSCH₂CH₂SH, ZnCl₂; (c) 1. CH₃CH₂MgCl (or CH₃CH₂Li), 2. H₂O, H⁺; (d) NaCN, H⁺; (e) RCO₃H; (f) (CH₃CH₂)₂NH, H⁺; (g) (C₆H₅)₃P=CH₂; (h) Zn(Hg), HCl, Δ, or H₂NNH₂, H₂O, ⁻OH, Δ; (i) CH₃OH, H⁺; (j) (C₆H₅)₃P=CHCH₂CH₂CH₃; (k) 1. C₆H₅MgCl (or C₆H₅Li), 2. H₂O, H⁺; (l) C₆H₅NH₂, H⁺.

46. **(a)**

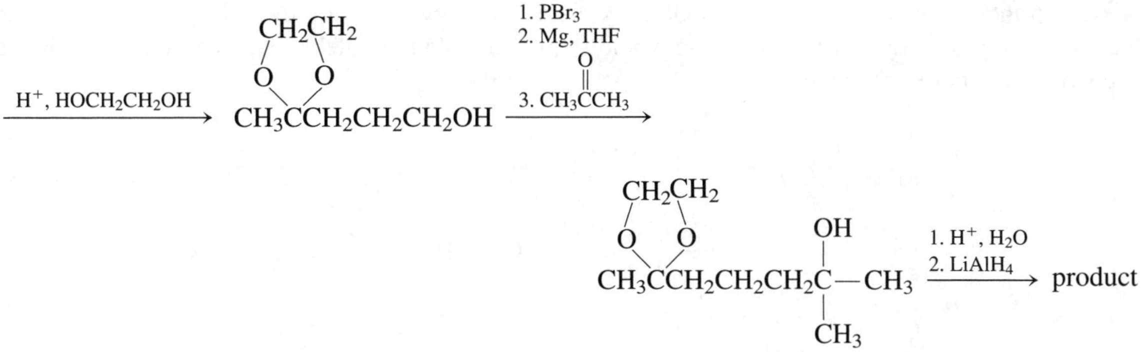

$$\xrightarrow{H^+, HOCH_2CH_2OH}$$ CH₃CCH₂CH₂CH₂OH $$\xrightarrow[\substack{\text{2. Mg, THF}\\ \text{3. CH}_3\text{CCH}_3}]{\text{1. PBr}_3}$$

CH₃CCH₂CH₂CH₂C—CH₃ $$\xrightarrow[\substack{\text{2. LiAlH}_4}]{\text{1. H}^+,\text{ H}_2\text{O}}$$ product

(b) $$\xrightarrow{CrO_3, H_2O, H_2SO_4}$$ 3-pentanone $$\xrightarrow[\substack{CH_3CH_2OH}]{H^+,\ C_6H_5NH_2}$$ product

(c) $$\xrightarrow{2PCC,\ CH_2Cl_2}$$ 1,5-pentanedial $$\xrightarrow{H_2O}$$ product, via

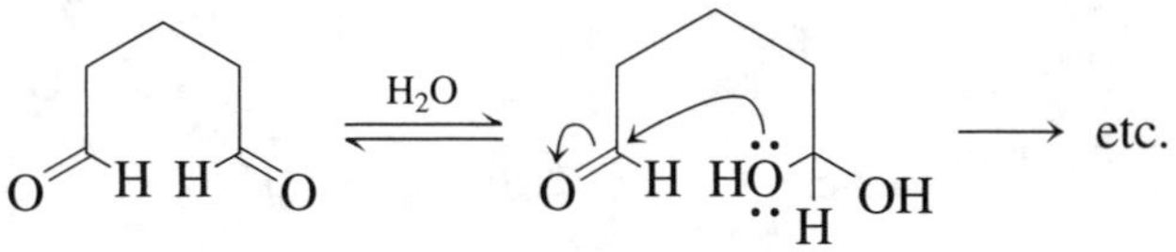

(d) 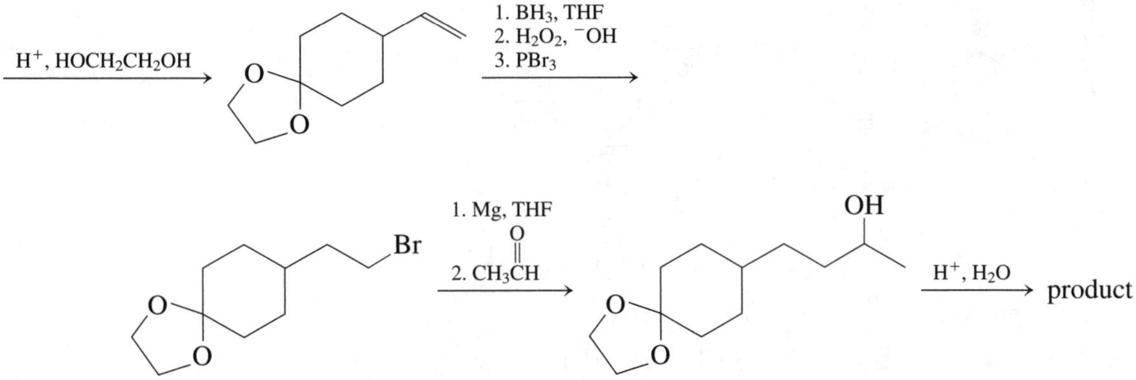 $$\xrightarrow{H^+, HOCH_2CH_2OH}$$

$$\xrightarrow[\substack{\text{2. H}_2\text{O}_2,\ ^-\text{OH}\\ \text{3. PBr}_3}]{\text{1. BH}_3,\text{ THF}}$$

47. Extended conjugation would be expected to shift the absorptions of these hydrazones to longer wavelength, just as it does with the parent ketones. So,

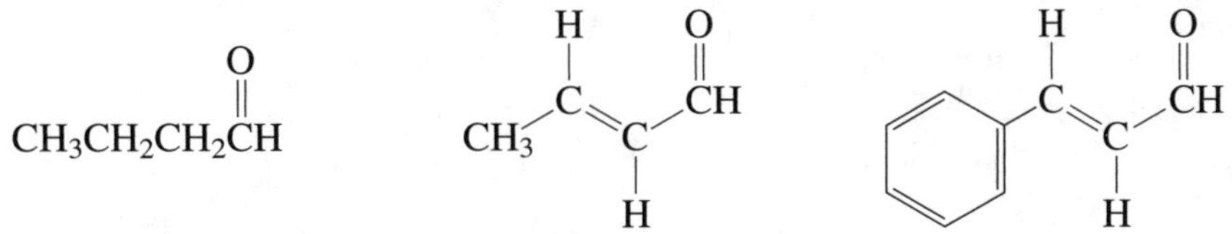

Hydrazone λ_{max} = 358 nm
(yellow)
Bottle 2

Hydrazone λ_{max} = 377 nm
(orange)
Bottle 1

Hydrazone λ_{max} = 394 nm
(red)
Bottle 3

48. **(a)** H₂NNH₂, H₂O, HO⁻, Δ (Wolff-Kishner reduction of both carbonyl groups)

(b) H₂, Pd-C, CH₃CH₂OH (Selective reduction of alkene)

(c) LiAlH₄, (CH₃CH₂)₂O (Selective reduction of aldehyde)

(d) H⁺, cycloheptanone (Formation of an unusual acetal, nothing more)

49. In principle the double bond may be constructed in two ways using the Wittig reaction. The solution shown is the more likely. The alternative would require an alkynyl aldehyde, a compound that is difficult to prepare and is not very stable.

$$CH_3CH_2CH_2C \equiv CCH_2Br \xrightarrow{P(C_6H_5)_3} CH_3CH_2CH_2C \equiv CCH_2\overset{+}{P}(C_6H_5)_3 \quad Br^-$$

$$\xrightarrow{CH_3CH_2CH_2CH_2Li} CH_3CH_2CH_2C \equiv CCH = P(C_6H_5)_3$$

$$\downarrow OHC(CH_2)_8CO_2CH_2CH_3$$

$$CH_3CH_2CH_2C \equiv CCH = CH(CH_2)_8CO_2CH_2CH_3$$

50. (a)

$$\underset{O}{\overset{\|}{CH_3CH}} + (C_6H_5)_3P = CHCH_2CH(CH_3)_2 \xrightarrow[THF]{(1)}$$

$$product \xleftarrow[THF]{(2)} CH_3CH = P(C_6H_5)_3 + \underset{O}{\overset{\|}{HCCH_2CH(CH_3)_2}}$$

(b)

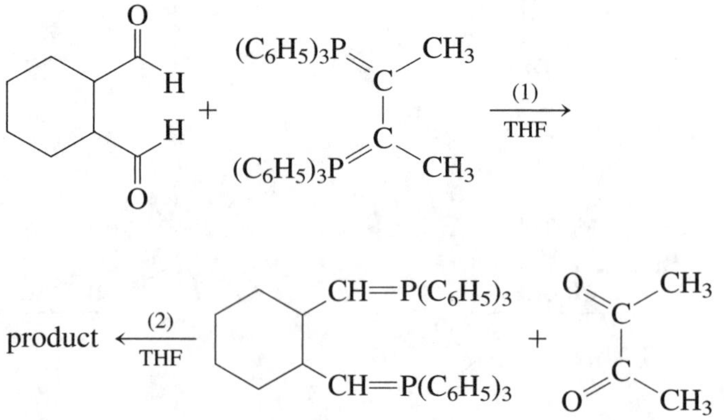

51. There are only three possible ketones: 2-heptanone, 3-heptanone, and 4-heptanone. So this is really a matching problem.

$$\underset{\textbf{2-Heptanone}}{CH_3CH_2CH_2CH_2CH_2\overset{\overset{O}{\|}}{C}CH_3} \xrightarrow{\text{Baeyer-Villiger}}$$

$$\underset{\substack{\textbf{Major product} \\ \text{(From migration of 1° alkyl)}}}{CH_3CH_2CH_2CH_2CH_2O\overset{\overset{O}{\|}}{C}CH_3} + \underset{\substack{\textbf{Minor product} \\ \text{(From migration of methyl)}}}{CH_3CH_2CH_2CH_2CH_2\overset{\overset{O}{\|}}{C}OCH_3}$$

$$\underset{\textbf{3-Heptanone}}{CH_3CH_2CH_2CH_2\overset{\overset{O}{\|}}{C}CH_2CH_3} \xrightarrow{\text{Baeyer-Villiger}}$$

$$CH_3CH_2CH_2CH_2O\overset{\overset{\displaystyle O}{\|}}{C}CH_2CH_3 \ + \ CH_3CH_2CH_2CH_2\overset{\overset{\displaystyle O}{\|}}{C}OCH_2CH_3$$

Two products in approximately equal amounts
(both from migration of primary alkyl groups)

$$CH_3CH_2CH_2\overset{\overset{\displaystyle O}{\|}}{C}CH_2CH_2CH_3 \xrightarrow[\text{Villiger}]{\text{Baeyer-}} CH_3CH_2CH_2O\overset{\overset{\displaystyle O}{\|}}{C}CH_2CH_2CH_3$$

4-Heptanone　　　　　　　　Only product (starting
　　　　　　　　　　　　　　ketone is symmetrical)

Compound A must be 4-heptanone; B is 2-heptanone; C is 3-heptanone.

52. (a) 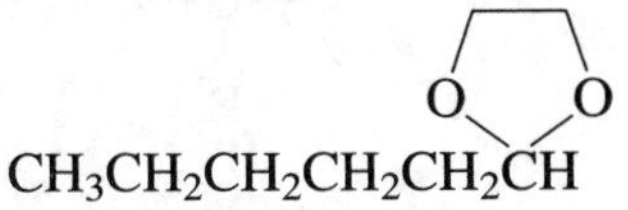 $CH_3CH_2CH_2CH_2CH_2CH$〈O—O ring〉　　(b) 1-Hexanol

(c) $CH_3CH_2CH_2CH_2CH_2CH\text{=}NOH$　　(d) Hexane

(e) $CH_3CH_2CH_2CH_2CH_2CH\text{=}CHCH_2CH(CH_3)_2$ (*E* and *Z*)

(f) 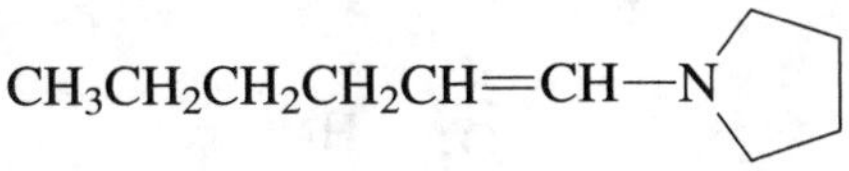 $CH_3CH_2CH_2CH_2CH\text{=}CH\text{—}N$〈pyrrolidine ring〉　　(g) Hexanoic acid + Ag metal

(h) Hexanoic acid　　(i) 2-Hydroxyheptanenitrile $\left(\text{via } R\overset{\overset{\displaystyle O}{\|}}{C}H \ \rightarrow \ R\overset{\overset{\displaystyle OH}{|}}{C}HCN \right)$

53. (a) 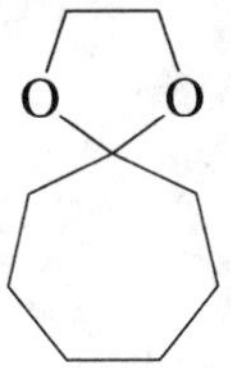 〈spiro dioxolane structure〉　　(b) Cycloheptanol

(c) 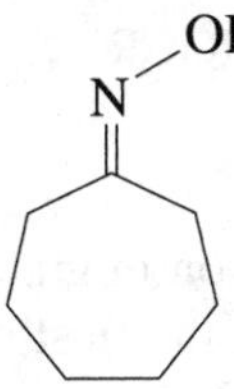 〈cycloheptanone oxime, N—OH〉

(d) Cycloheptane

(e) 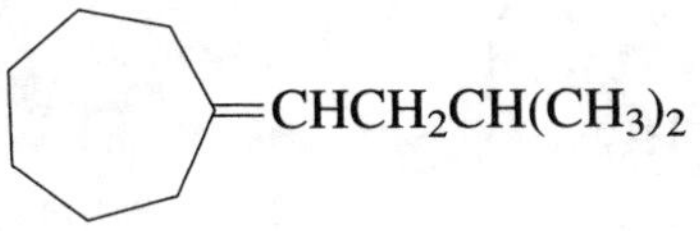 〈cycloheptylidene〉$=CHCH_2CH(CH_3)_2$　　(f) 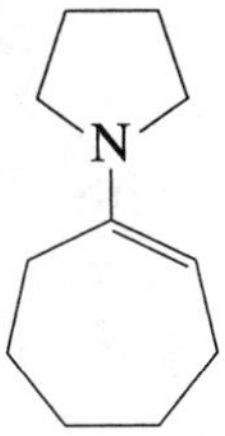 〈enamine with pyrrolidine N〉

(g) and (h) No reaction. These are reactions of aldehydes only.

(i)  〈cycloheptane ring with HO and CN〉

54.

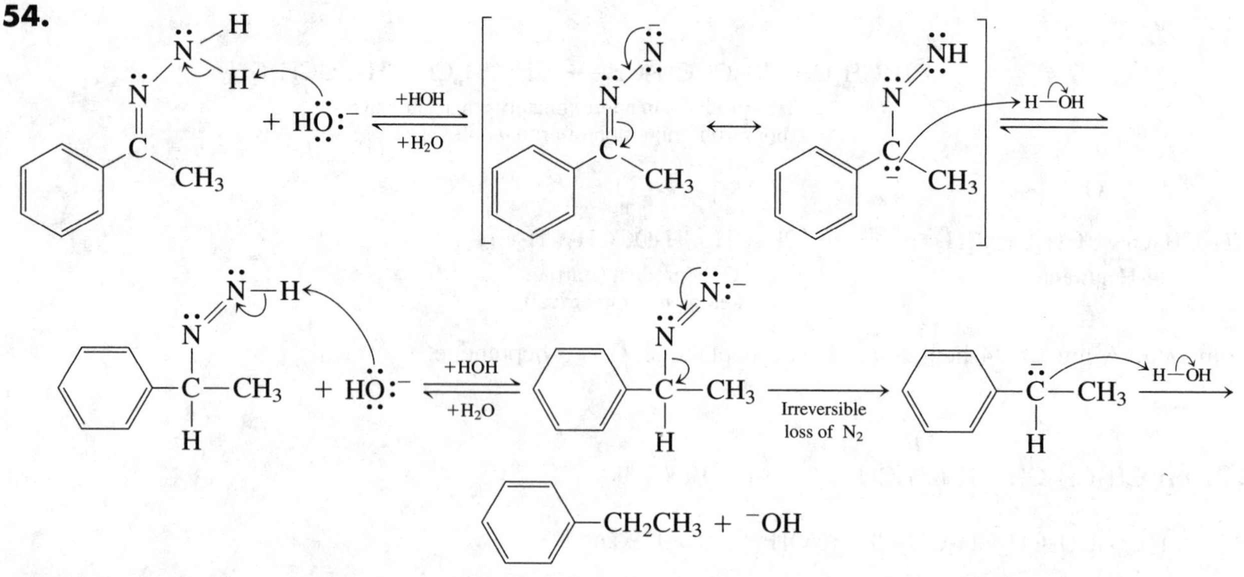

55. The reaction occurs under acidic conditions. Therefore:

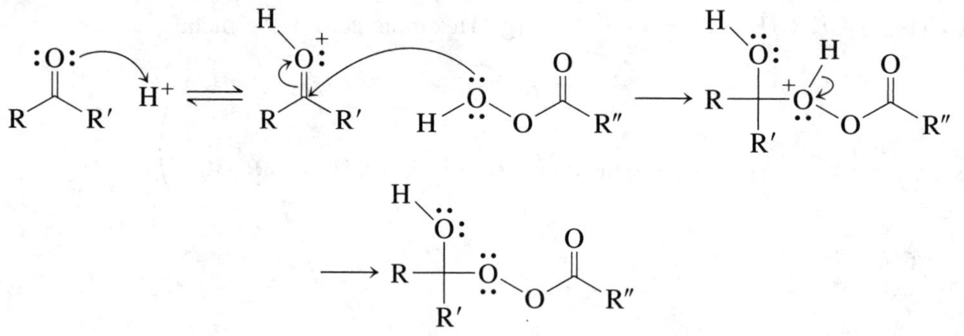

56.

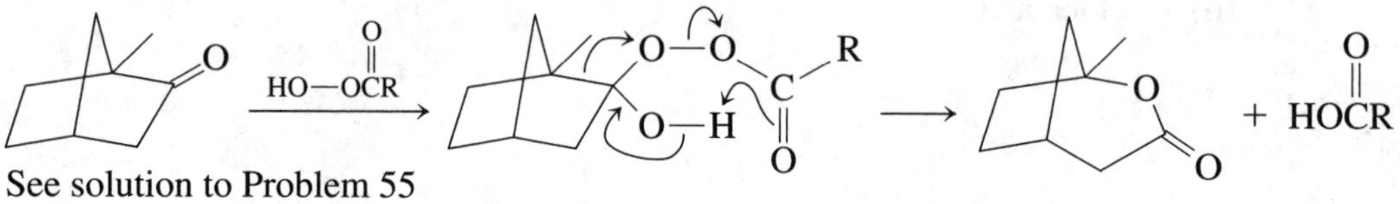

See solution to Problem 55

57. In each case insert O on either side of the carbonyl C. Then refer to the information on migratory aptitutes in Section 17-13 to help you choose the preferred product. In the answers below, the **first** structure is the preferred one

(a)

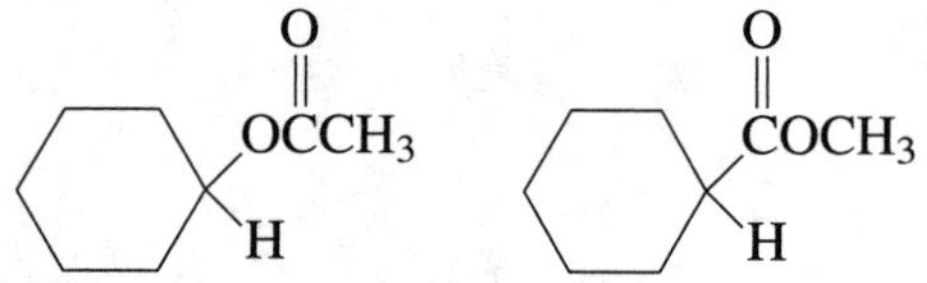

(b)

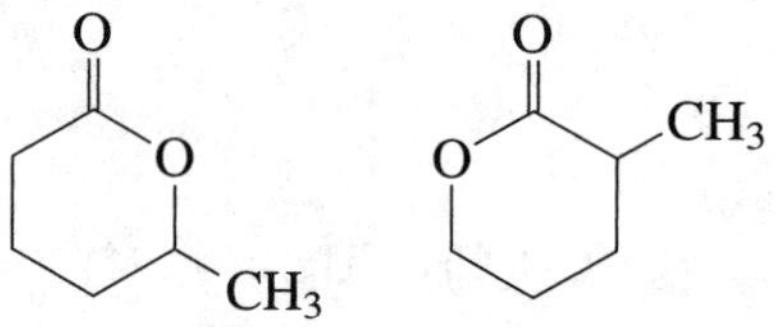

(c) $(CH_3)_2CHOCCH_2CH(CH_3)_2$ $(CH_3)_2CHCOCH_2CH(CH_3)_2$

(d)

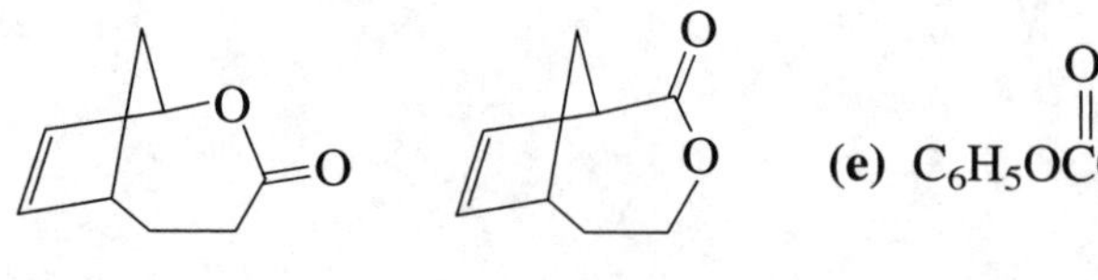

(e) $C_6H_5OCCH_3$ $C_6H_5COCH_3$

58. **(a)** Approach: You need a reaction between CH_3MgI and a ketone carbonyl group, so the aldehyde will have to be protected.

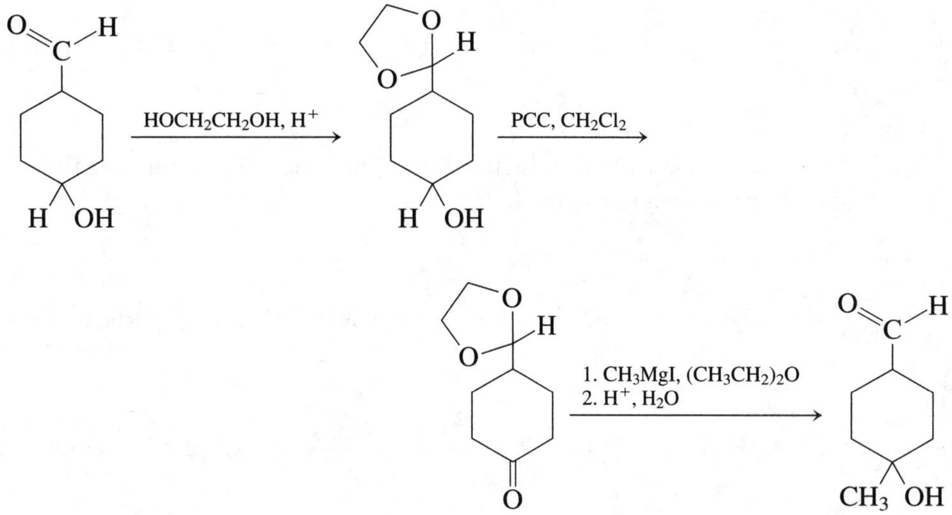

(b) Careful! Better protect the aldehyde again, or it will die when you apply any of several methods to get rid of the alcohol. Here's one:

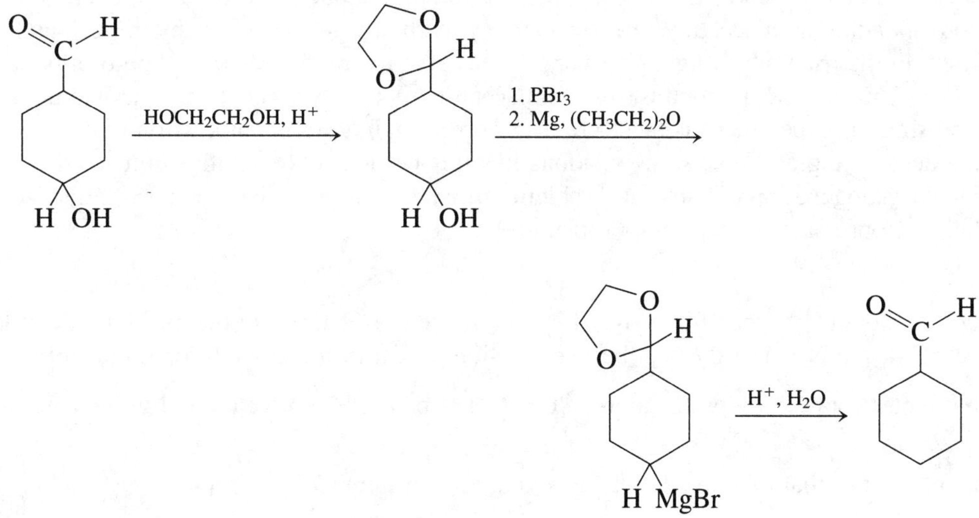

Now go to work on the rest.

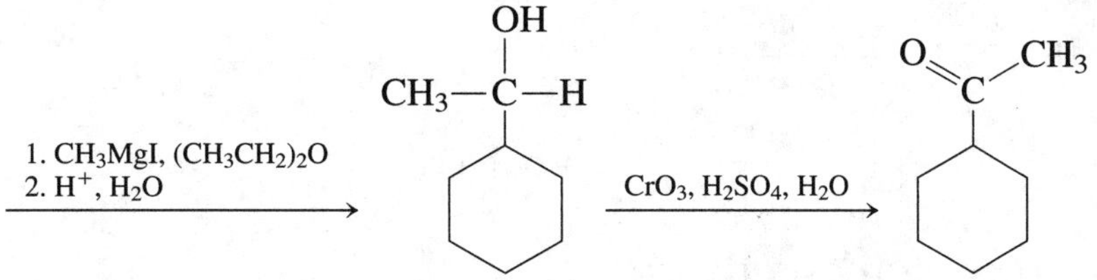

Oxidation of the alcohol to a ketone, followed by Wolff-Kishner reduction works as well.

(c) The "target" molecule is a hemiacetal. Rewrite it in "open" form (that is, as its acyclic isomer).

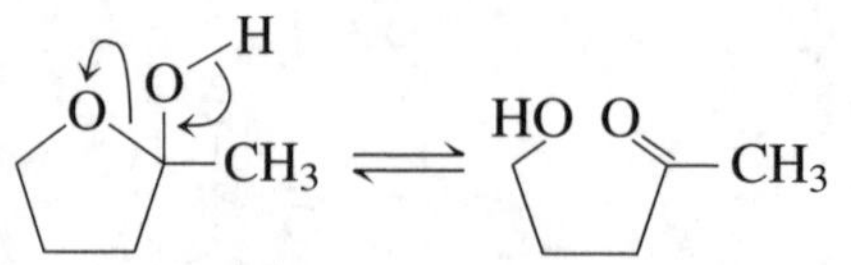

Synthesis of the open form will automatically result in formation of the desired product. So, with care in protecting the alcohol,

$$ClCH_2CH_2CH_2OH \xrightarrow{\text{H}^+, \text{(CH}_3\text{)}_3\text{COH}} ClCH_2CH_2CH_2OC(CH_3)_3 \xrightarrow[\text{2. CH}_3\text{CHO}]{\text{1. Mg, (CH}_3\text{CH}_2\text{)}_2\text{O}}$$

$$\underset{\underset{\text{OH}}{|}}{CH_3CHCH_2CH_2CH_2OC(CH_3)_3} \xrightarrow{\text{PCC, CH}_2\text{Cl}_2} \underset{\underset{\text{O}}{\overset{\text{O}}{||}}}{CH_3CCH_2CH_2CH_2OC(CH_3)_3} \xrightarrow{\text{H}^+, \text{H}_2\text{O}}$$

$$\underset{\underset{\text{O}}{\overset{\text{O}}{||}}}{CH_3CCH_2CH_2CH_2OH} \rightleftharpoons \text{product}$$

59. The answer is actually not that hard: Cyclopropanone has more bond angle strain than does the corresponding hemiacetal. Why? The carbonyl carbon wants to be sp^2 hybridized and have 120° angles. Being stuck in the triangular ring with an angle near 60°, it experiences the strain of a bond angle compression of $120° - 60° = 60°$. In contrast, the hemiacetal carbon only wants to be tetrahedral, with 109° bond angles. The strain it experiences is therefore less, corresponding to a compression of $109° - 60° = 49°$ relative to the desired angle. We've seen situations like this before, in the relative unreactivity of cyclopropane in radical halogenation (Chapter 4, Problem 26), and in the sluggishness of S_N2 displacements on halocyclopropanes (Chapter 6, Problem 60).

60. Below pH 2, the N in NH_2OH is protonated ($^+NH_3OH$), so the nucleophilic atom is effectively gone. At pH 4, the N is mostly free, but the solution is still acidic enough for some carbonyl groups to be made more electrophilic by protonation: $\underset{+}{>C=OH}$. Above pH 7, no carbonyl groups are protonated, so the rate is reduced to that of free NH_2OH attacking unactivated $>C=O$ groups.

61. Work backward from given structural information.

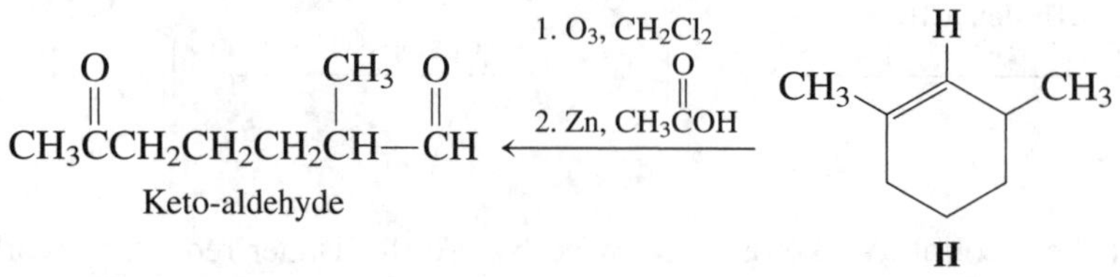

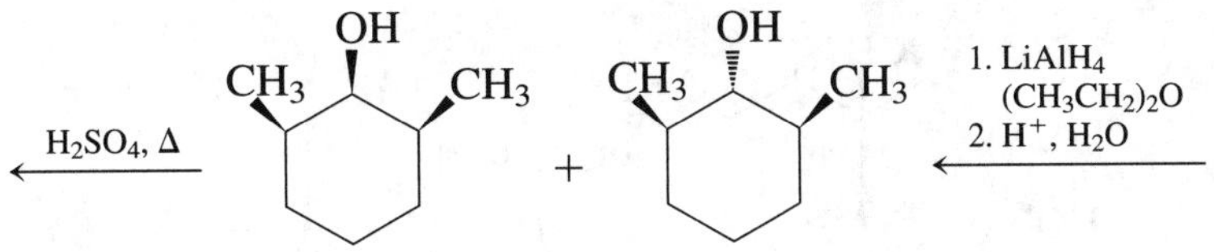

F and G

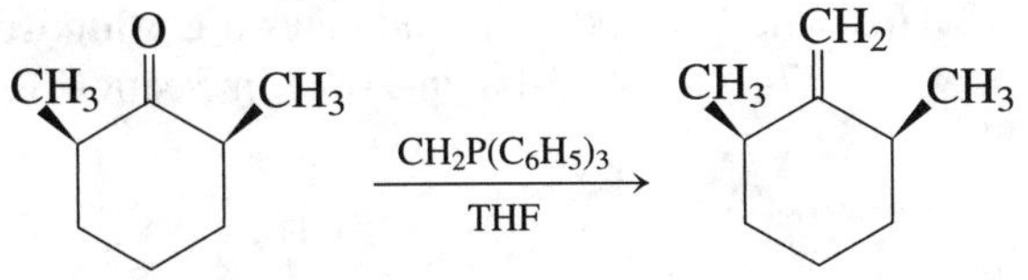

D   **E**

Note: F and G must be **secondary** alcohols because they have been formed by LiAlH$_4$ reduction of something (here, a ketone). The methyls in D must be cis. If they were trans, only one LiAlH$_4$ reduction product would be obtained.

62. The conclusions that may be reached from each piece of information are given to you.

(i) Coprostanol is an alcohol, and J is a ketone.

(ii)

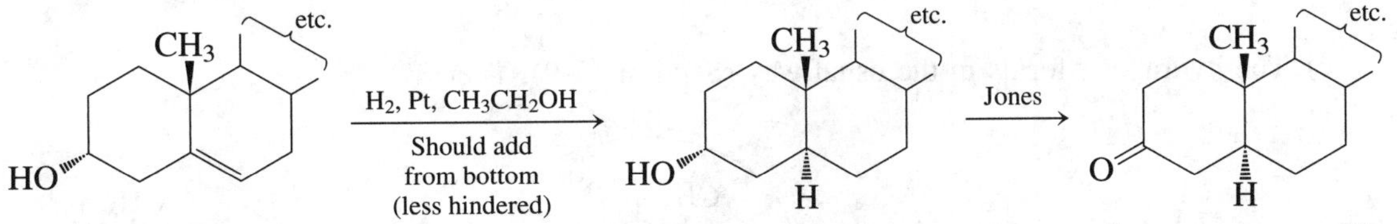

Relevant portion
of cholesterol

H$_2$, Pt, CH$_3$CH$_2$OH
Should add
from bottom
(less hindered)

K, a stereoisomer
of coprostanol

Jones

L, a stereoisomer of J

It appears likely, then, that J is

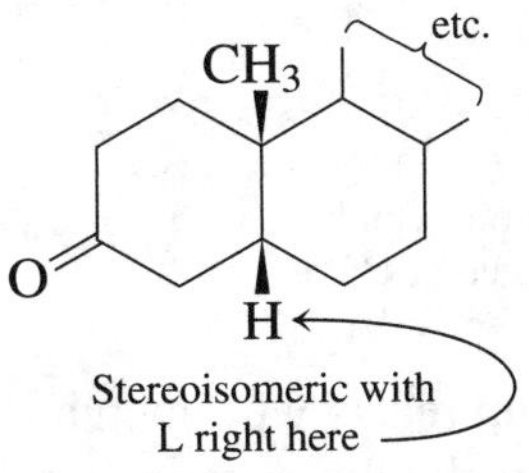

Stereoisomeric with
L right here

(iii) Cholesterol $\xrightarrow{\text{Cr}^{6+}, \text{ acetone}}$

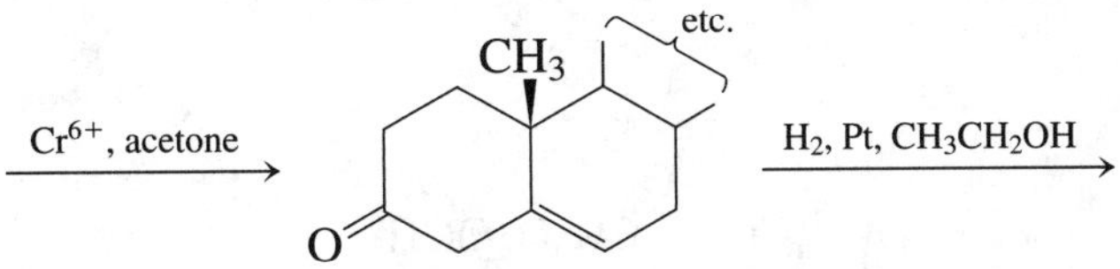

M

$\xrightarrow{\text{H}_2, \text{ Pt, CH}_3\text{CH}_2\text{OH}}$ **L again**

This is what you would expect, based on the information in (ii). So, what's coprostanol? Because it gives J on oxidation with Jones reagent, it must be one of the two alcohols

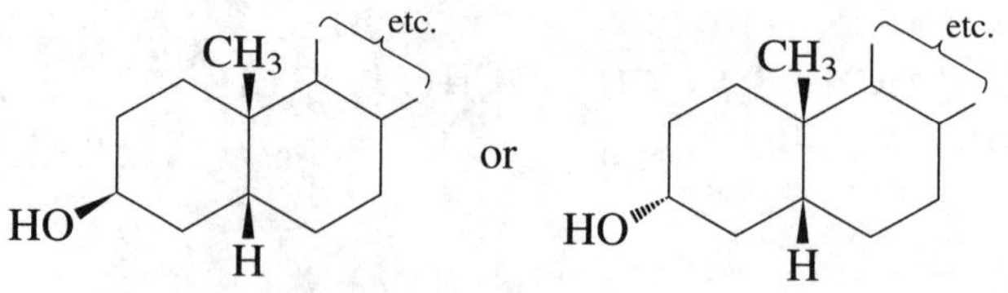

In fact, both are known and are labeled 3β- and 3α-coprostanol, respectively.

63. (a) The UV spectrum indicates a **conjugated** ketone (compare UV data of compounds in Exercise 17-4 and of 3-buten-2-one in Section 17-3). So N would appear to result from

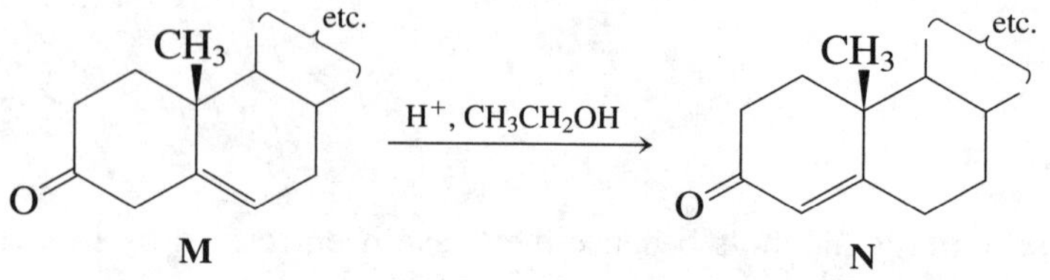

(b) N $\xrightarrow{\text{H}_2, \text{Pd}, \text{CH}_3\text{CH}_2\text{OH}}$

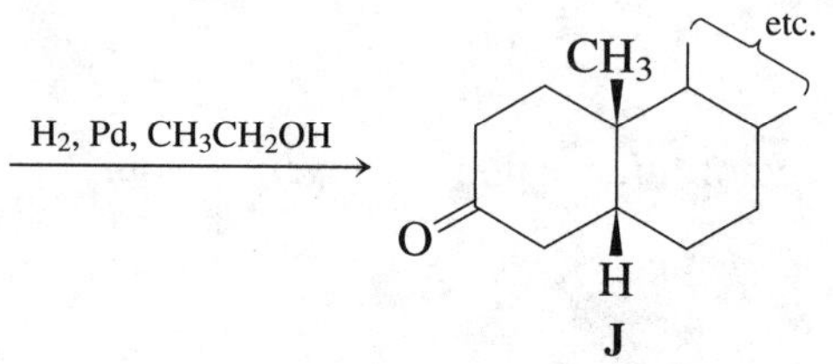

This is indeed odd, as the H$_2$ adds from the **more** hindered top face in this particular case.

(c) The hydrazone forms in the usual way (Section 17-9). Then

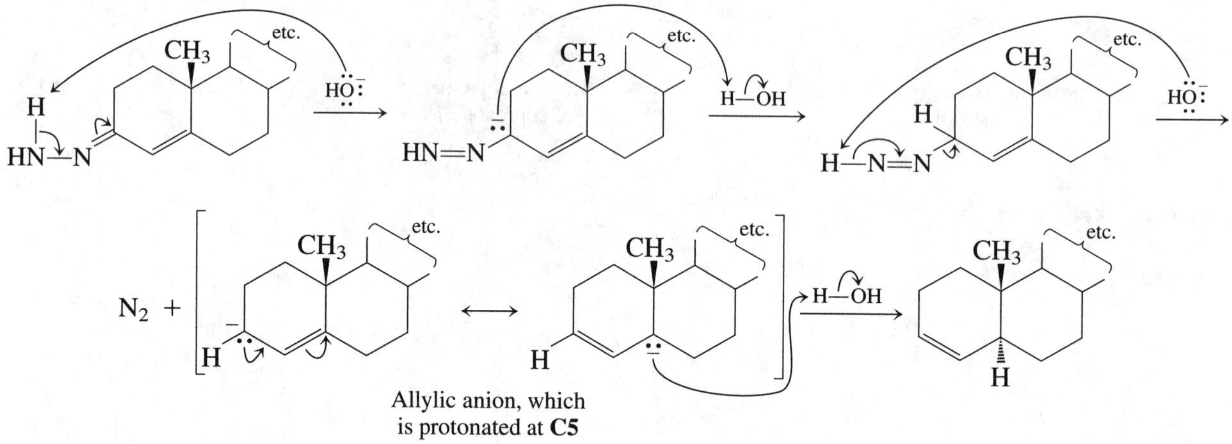

Allylic anion, which
is protonated at **C5**

64. The main issue is which carbonyl group reacts with methanol preferentially. On the basis of the material in Section 17-6, the aldehyde group should be more reactive. Thus, as a working hypothesis, propose a product with an acetal function derived from addition of methanol to the aldehyde carbon, and see if the NMR agrees.

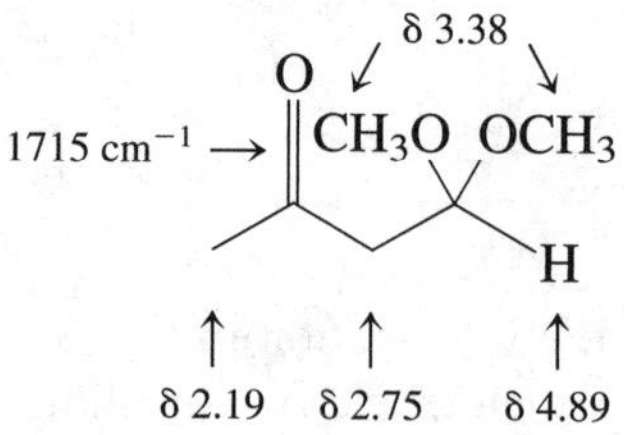

One of the giveaways is the absence of a signal for a formyl (aldehydic) hydrogen in the δ 9.5–10 region of the spectrum. The mechanism of formation is precisely that outlined in Section 17-7 of the textbook.

65. (c)

66. (b)

67. (d)

68. (b)

18

Enols, Enolates, and the Aldol Condensation: α, β-Unsaturated Aldehydes and Ketones

As the text continues to develop the chemistry of aldehydes and ketones, you will now see how the carbon adjacent to a carbonyl group can become nucleophilic. First, reactions of these new nucleophiles with common electrophiles like haloalkanes will be covered: alkylation reactions. More important are reactions of the nucleophilic *α-carbons* of one carbonyl compound with electrophilic carbonyl carbons of another. They are generically termed *carbonyl condensation reactions.* You see them here for aldehydes and ketones: the *aldol condensation.* (In a later chapter you will be introduced to the analogous reaction of carboxylic esters: the Claisen condensation.) The products of aldol condensations are α, β-unsaturated aldehydes and ketones, which contain additional sites of electrophilic and potential nucleophilic character.

Pay close attention to the structural relationships between starting materials and products. These are the reactions of real-world organic synthesis, and they are among the most important ones you will see in this course.

Outline of the Chapter

18-1 The Acidity of α-Hydrogens: Enolate Ions
Making a nucleophile at the carbon α to a carbonyl group.

18-2 Keto-Enol Equilibria

18-3, 18-4 Halogenation and Alkylation at the α-Carbon

18-5, 18-6, 18-7 Aldol Condensation
A very important new reaction for synthesis.

18-8 through 18-11 α, β-Unsaturated Aldehydes and Ketones
The properties of the versatile products of aldol condensation.

Keys to the Chapter

18-1. The Acidity of α-Hydrogens: Enolate Ions

Hydrogens on carbons adjacent to carbonyl or related functional groups are, as a class, the most acidic of alkanelike hydrogens. A short list of representative pK_a values follows. Notice how acidity is enhanced by a second carbonyl group.

Compound	pK_a	Compound	pK_a
RCH_2COR' ‖O	25	$ROCCH_2COR$ ‖O ‖O	13
RCH_2CR' ‖O	20	$RCCH_2COR'$ ‖O ‖O	11
RCH_2CH ‖O	17	$RCCH_2CR$ ‖O ‖O	9
RCH_2CCl ‖O	16	$HCCH_2CH$ ‖O ‖O	5

Any of the boldface protons in the structures above may be removed by an appropriate base, thus giving rise to the central participant in the reactions of this chapter, the enolate ion.

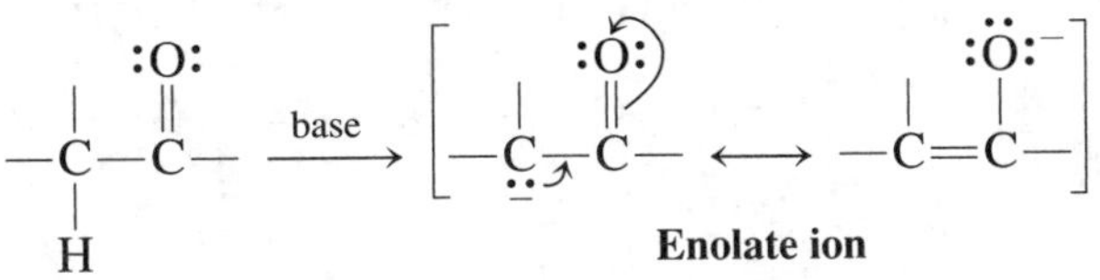

$$\underset{H}{-C}-\overset{:O:}{\overset{\|}{C}}- \xrightarrow{\text{base}} \left[-\underset{\cdot\cdot}{C}\curvearrowright\overset{:\overset{\frown}{O}:}{\overset{\|}{C}}- \longleftrightarrow -C=\overset{:\overset{\cdot\cdot}{O}:^-}{C}- \right]$$

Enolate ion

The enolate is nucleophilic, capable of reaction with the typical electrophiles you've seen before. Their reaction with haloalkanes (alkylation) is the most general way to introduce alkyl substituents adjacent to carbonyl carbons.

18-2. Keto-Enol Equilibria

In this section more mechanistic detail is provided concerning the isomerization of vinyl alcohols (enols) to carbonyl compounds. This isomerization, although it favors the keto structure, allows small concentrations of enols to be in equilibrium with carbonyl compounds, and these enols can lead to productive chemical reactions.

18-3 and 18-4. Halogenation and Alkylation at the α-Carbon

The reactions in this section are mechanistically very similar to ones you know. The nucleophilic carbon of an enolate (or, in acid, an enol) attacks either a halogen or a haloalkane through an S_N2 pathway. Therefore, alkylations succeed best when the haloalkane is methyl or primary. Enamines can be used instead of enolates for alkylation and usually give superior results.

18-5, 18-6, and 18-7. Aldol Condensation

These sections cover the first, and probably the most important carbonyl condensation reaction, the aldol condensation. Its significance is due to its ability to make relatively complex structures, including rings, from much simpler starting molecules. It proceeds via addition of the deprotonated α-carbon of one carbonyl group to the carbonyl carbon of another. As you read this text section, notice the location of the new carbon–carbon bond in each aldol product. Cover up the starting materials and see if you can deduce their structures solely by looking at their aldol products. Although many aldol reactions between two identical molecules are useful, the most important applications fall into two categories: the so-called crossed aldols, and intramolecular condensations of dicarbonyl compounds. Pay particular attention to the examples in this section, because they provide excellent illustrations for the various permutations of enolate and carbonyl "partners" that give successful and useful results in aldol condensations.

18-8 through 18-11. α, β-Unsaturated Aldehydes and Ketones

You see here logical extensions of carbonyl and alkene chemistry. The carbon–carbon double bond in an α, β-unsaturated carbonyl compound generally shows the same addition reactions typical of simple alkenes. However, the highly polar nature of the carbonyl strongly affects the reactivity of the alkene functional group. This is best seen by the resonance forms in these compounds: Note in particular the presence of positive character at the β-carbon as well as at the carbonyl carbon.

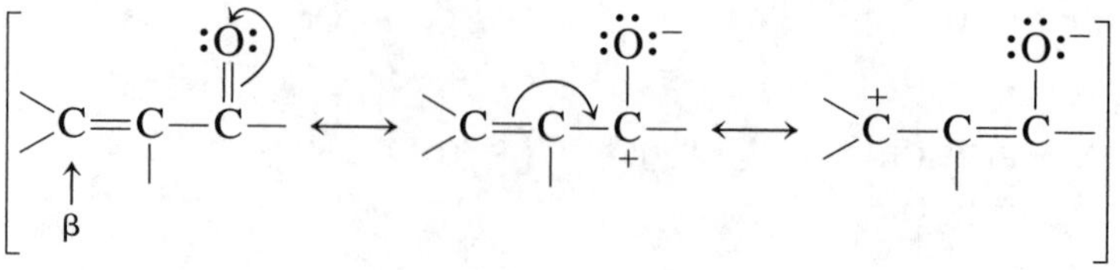

The wide range of *enone* chemistry is understandable in terms of just three fundamental mechanistic processes.

1. Electrophilic addition to carbonyl oxygen, e.g.,

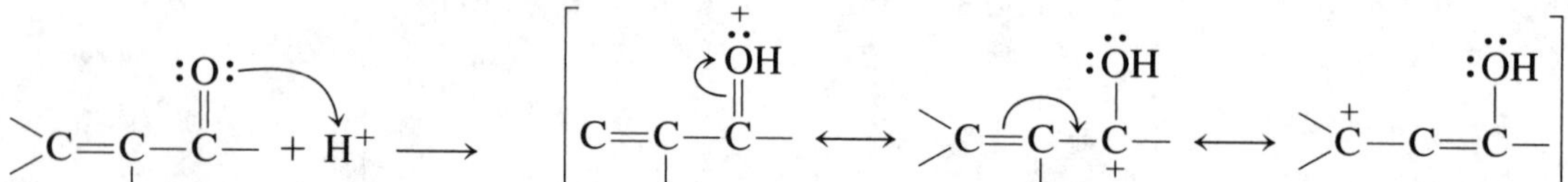

2. Nucleophilic addition to carbonyl carbon, e.g.,

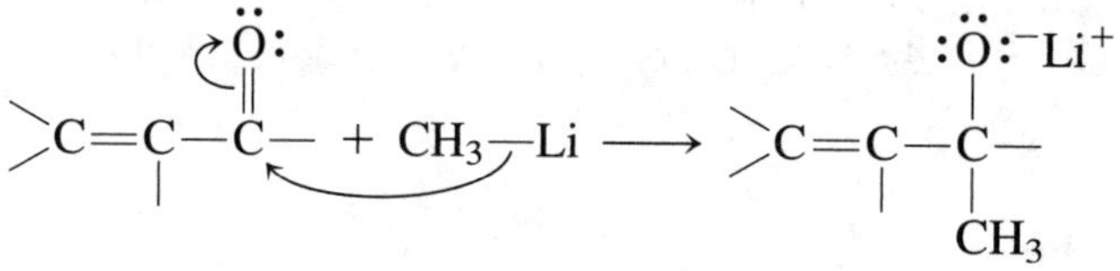

3. Nucleophilic addition to the β-carbon, e.g.,

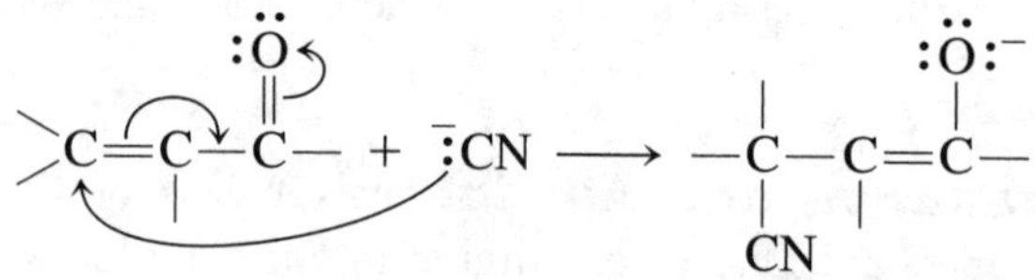

Practice by writing mechanisms for examples in the text for which specific mechanisms are not pictured. Use the appropriate mechanism above to start, and follow it with the behavior expected for the product of that step.

Once you understand the mechanisms, concentrate on the synthetic applications of the process. Focus on the carbon–carbon bond-forming examples, with particular emphasis on the Michael addition, the 1,4-addition of enolates to enones or enals. The combination Michael addition–aldol condensation provides a powerful means of synthesis of six-membered rings, the Robinson annulation. Don't worry about all these people's names; learn the retrosynthetic analysis for compounds containing six-membered rings.

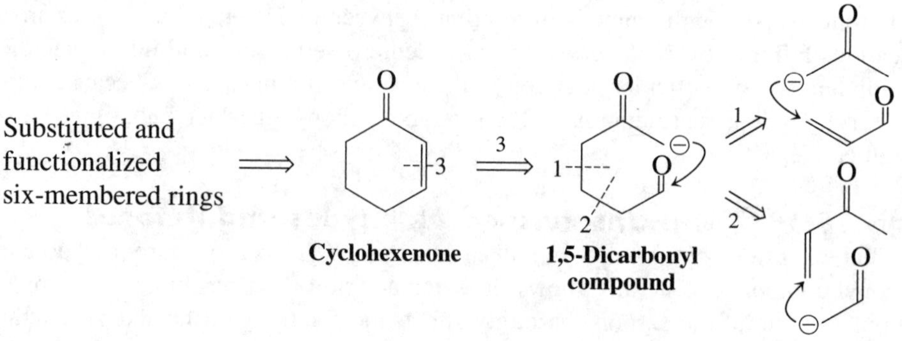

Substituted and functionalized six-membered rings

Cyclohexenone

1,5-Dicarbonyl compound

The first step is Michael addition to give a 1,5-dicarbonyl compound. This, in principle, can be done in either of two ways (formation of either bond 1 or bond 2). Then intramolecular aldol condensation forms bond 3 of a cyclohexenone, which may be converted to a wide range of new compounds using the reactions of these sections.

Solutions to Problems

32. **(a)** $CH_3\underline{C}(H_2)\overset{O}{\overset{\|}{C}}\underline{C}(H_2)CH_3$ **(b)** $\underline{C}(H_3)\overset{O}{\overset{\|}{C}}\underline{C}(H)(CH_3)_2$ **(c)** 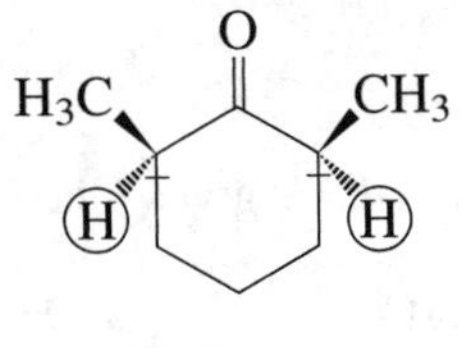

(d) 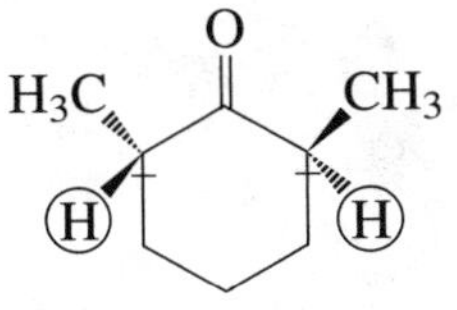 **(e)** 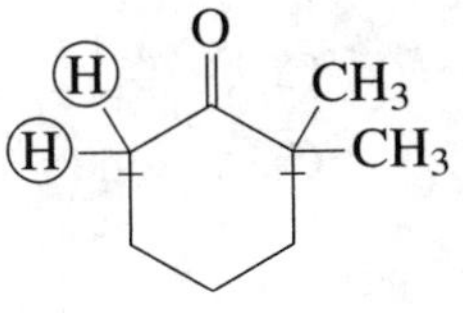 **(f)** 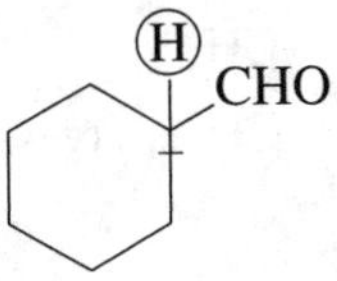

(g) $(CH_3)_3\underline{C}\overset{O}{\overset{\|}{C}}H$ **(h)** $(CH_3)_3C\underline{C}(H_2)\overset{O}{\overset{\|}{C}}H$

33. **(a)** (i) $CH_3CH{=}\overset{OH}{\overset{|}{C}}CH_2CH_3$

(ii) $\left[CH_3\ddot{\bar{C}}H\overset{O}{\overset{\|}{C}}CH_2CH_3 \longleftrightarrow CH_3CH{=}\overset{\ddot{O}^-}{\overset{|}{C}}CH_2CH_3\right]$

For the rest, only one of the enolate resonance forms will be written.

(b) (i) $CH_2{=}\overset{OH}{\overset{|}{C}}CH(CH_3)_2,\ CH_3\overset{OH}{\overset{|}{C}}{=}C(CH_3)_2$

(ii) $\ddot{\bar{C}}H_2\overset{O}{\overset{\|}{C}}CH(CH_3)_2,\ CH_3\overset{O}{\overset{\|}{C}}\ddot{\bar{C}}(CH_3)_2$

(c) (i) 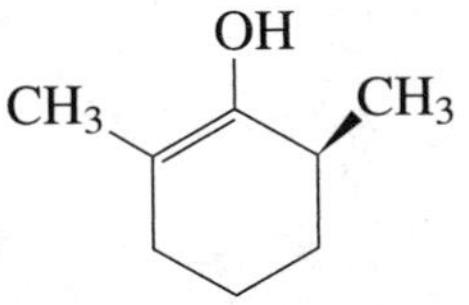 (ii) 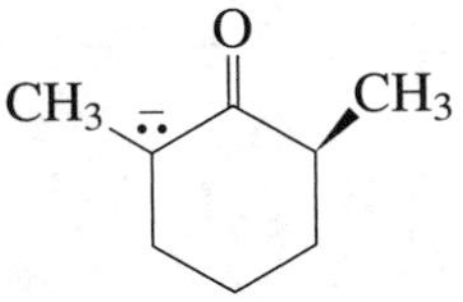

(d) Same as (c). Stereochemistry will be lost as α-carbon becomes sp^2.

(e) (i) 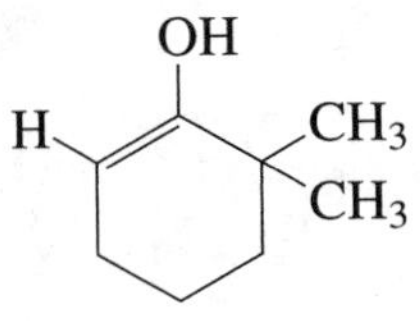 (ii) 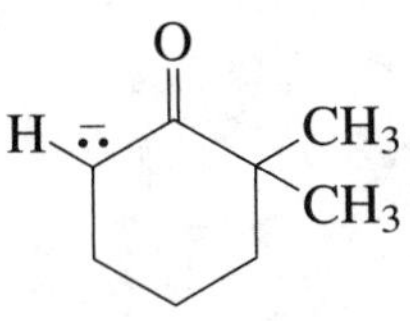

(f) (i) 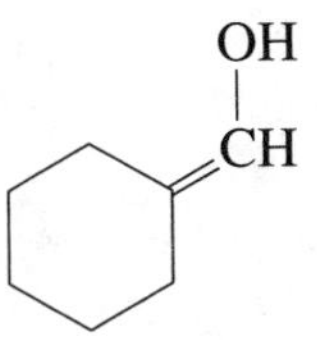 (ii) 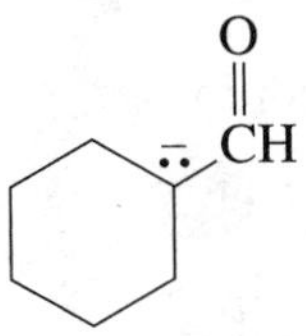

(g) None possible (no α-hydrogens!)

(h) (i) $(CH_3)_3CCH=CH$ with OH on the CH (ii) $(CH_3)_3C\overset{..}{C}HCH$ with O double bond on CH

34. (a) Replace **all** α-hydrogens with D; for instance,

$CH_3CH_2CCH_2CH_3$ (with C=O) gives $CH_3CD_2CCD_2CH_3$,

gives

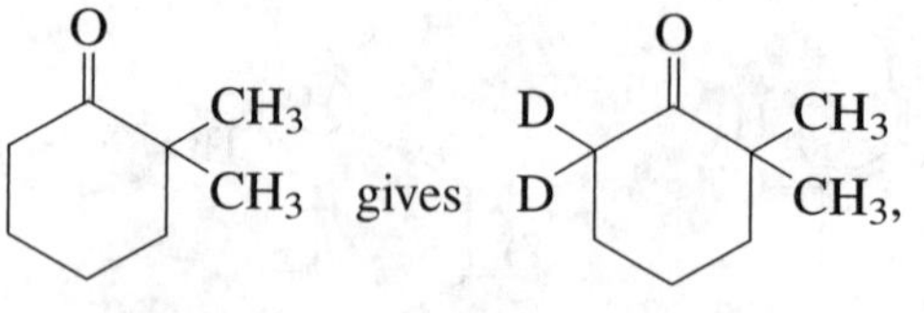

and

$(CH_3)_3CCH_2CH$ (with C=O) gives $(CH_3)CCD_2CH$ (with C=O).

(Note how the aldehyde hydrogen is **not** replaced—it is **not** acidic.)

(b) Conditions for introduction of a **single** α-halogen. In order, the products are

1(a) $CH_3CHBrCCH_2CH_3$ (with C=O)

1(b) A mixture of $CH_3CCBr(CH_3)_2$ and $BrCH_2CCH(CH_3)_2$ (both with C=O)

1(c), (d) 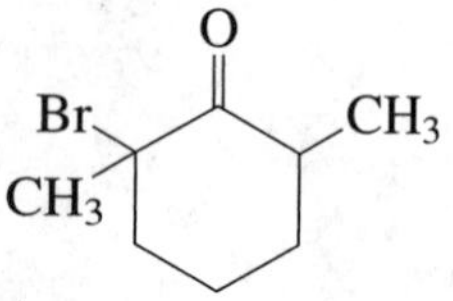   (Mixture of stereoisomers from either cis or trans starting ketone)

1(e)

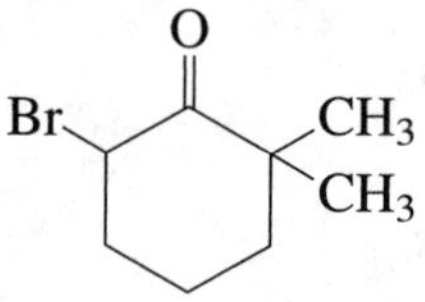

1(f)

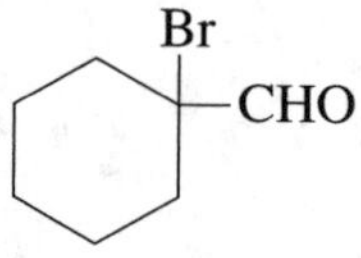

1(g) No reaction

1(h) $(CH_3)_3CCHBrCH$ (with C=O)

(c) All α-hydrogens are replaced by Cl under these conditions.

35. (a) An equivalent of Br_2 in acetic acid (CH_3CO_2H) solvent

(b) Excess Cl_2 in aqueous base

(c) One equivalent of Cl_2 in acetic acid

36.

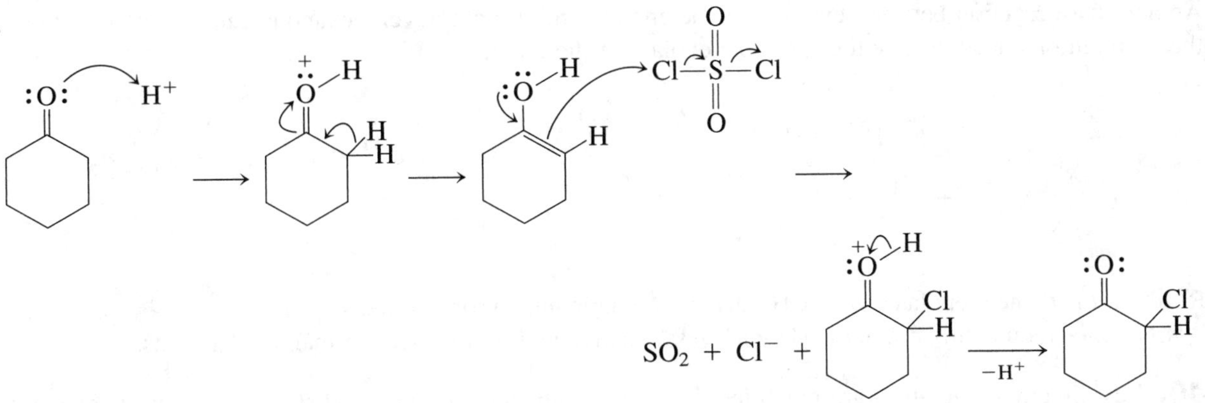

37. (a)

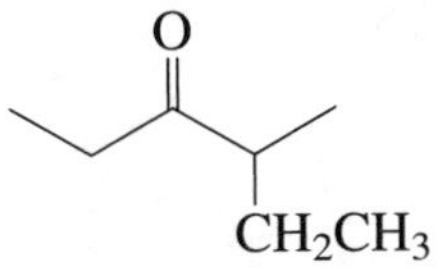

(b) 3-Pentanone + $H_2C=CHCH_3$

(c)

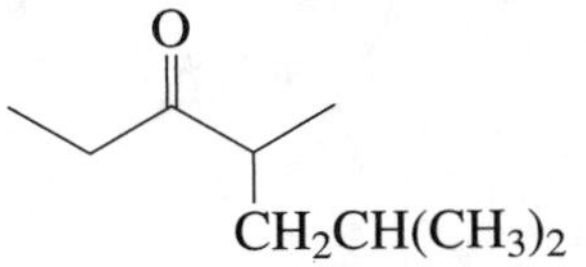

(d) 3-Pentanone + $H_2C=C(CH_3)_2$

(b) and (d) show the results of E2 elimination: Alkylation requires an S_N2 process and you already know that these are poor with secondary and impossible with tertiary haloalkanes.

38. Both are aldehyde → enamine → alkylation sequences. The new carbon–carbon bond is marked with an arrow.

(a) $(CH_3)_2C=CHCH_2\overset{\downarrow}{—}CH_2CHO$ $\left(\text{via } CH_2=CH—N\bigcirc\right)$

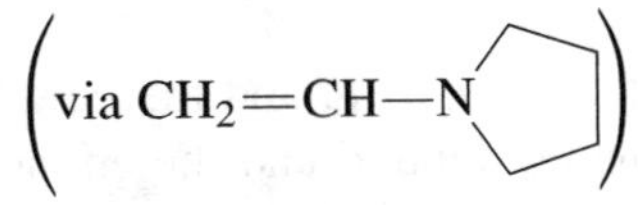

(b)

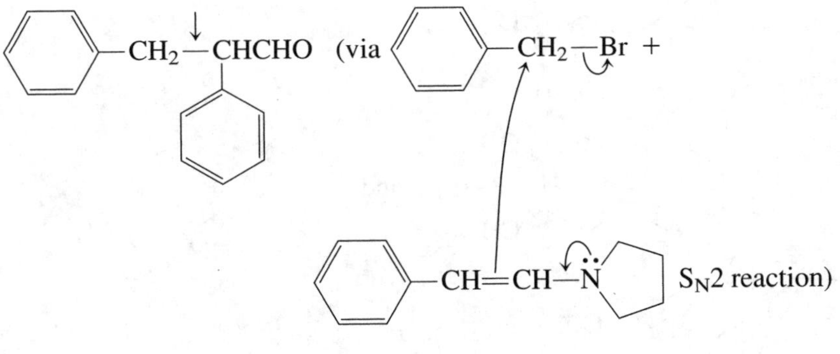

39. Illustrated with cyclohexanone: Before the reaction between cyclohexanone enolate and iodomethane has gone to completion, a mixture is present containing

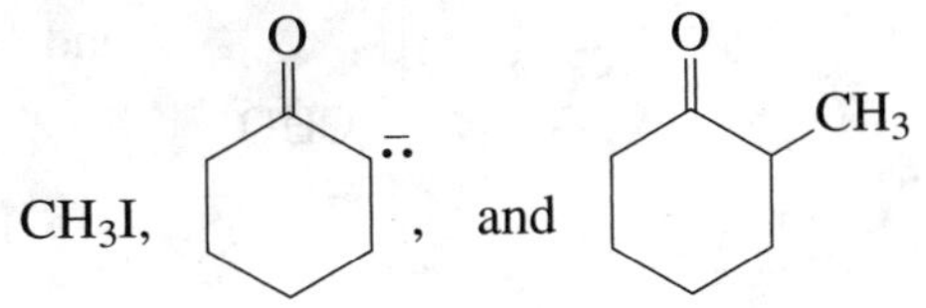

CH_3I,   , and

An acid-base reaction between cyclohexanone enolate and 2-methylcyclohexanone can take place under these conditions, leading to either possible enolate of the latter.

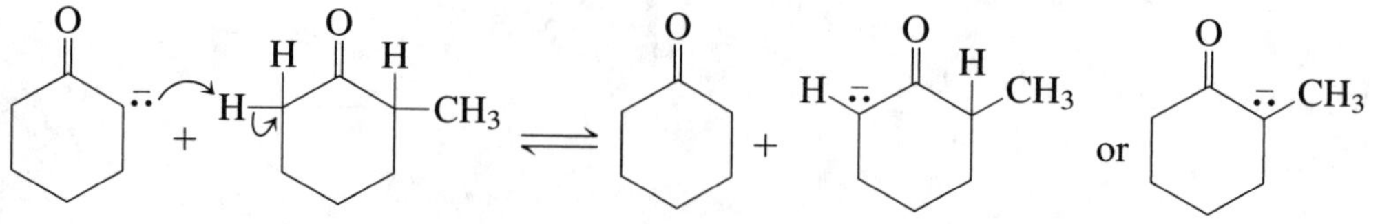

Reaction of the new enolates with CH_3I leads to double alkylation products.

Enamines reduce this problem. They are less reactive and more selective than enolate ions.

40. Yes. Enamines (neutral) are much less basic than enolates (anionic) and show much less tendency to cause E2 elimination reactions.

41. Use the catalyst! Then it's straightforward:

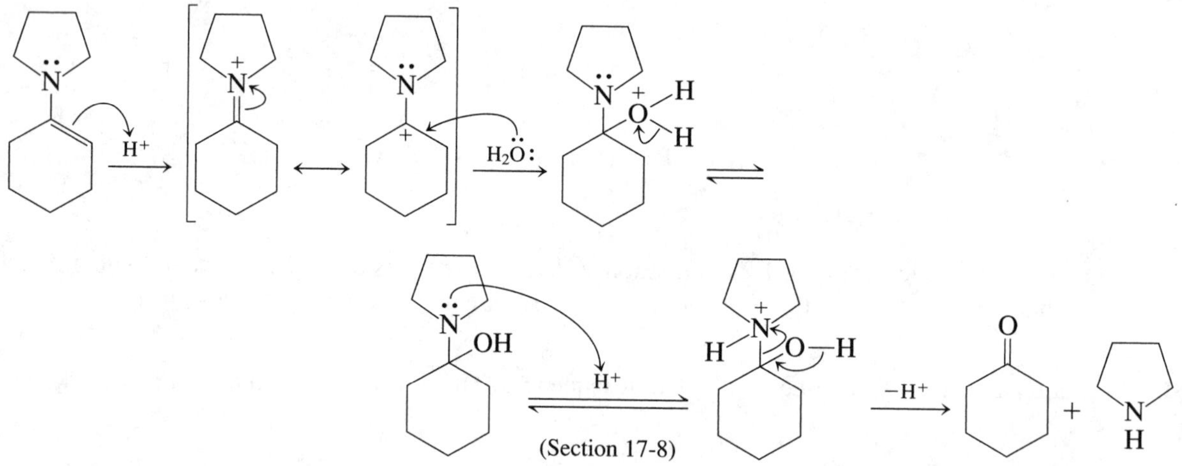

(Section 17-8)

42. The direction of nucleophilic attack is shown, and the new bond is marked in the product with an arrow. Both the initial hydroxycarbonyl compound and the enone that forms upon dehydration are shown.

(a)

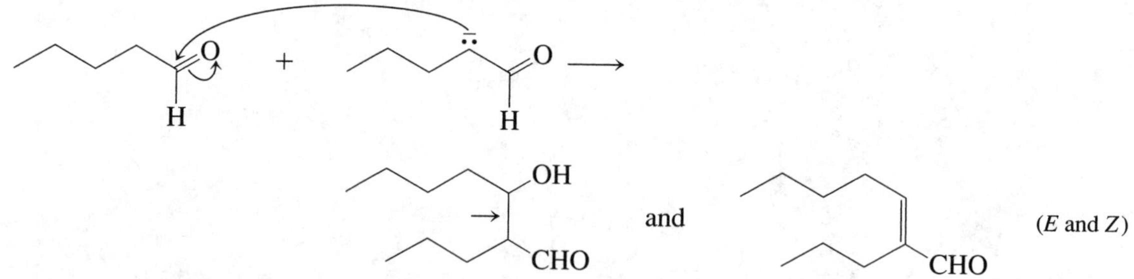

and   (*E* and *Z*)

(b)

and   (*E* and *Z*)

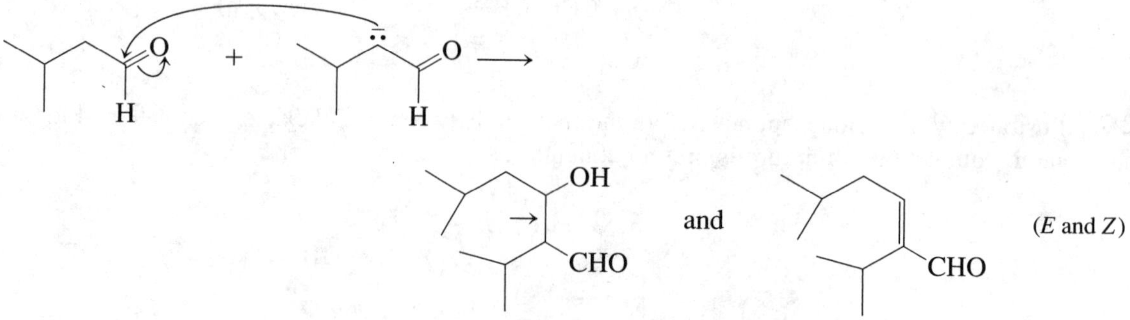

(c) 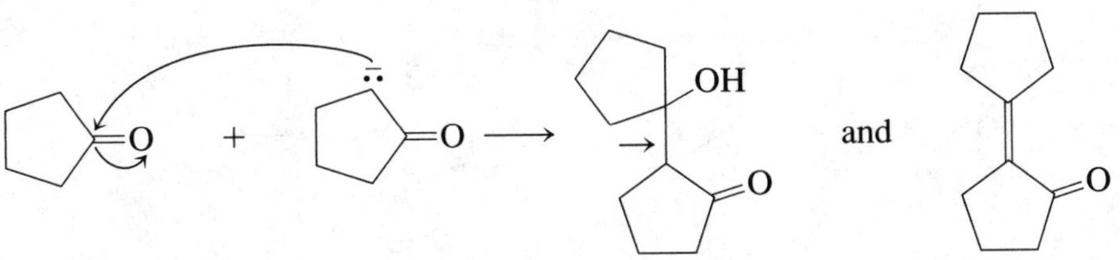 + ⟶ and

43. The direction of nucleophilic attack is shown, and the new bond is marked in the product with an arrow. Both the initial hydroxycarbonyl compound and the enone that forms upon dehydration are shown.

(a) 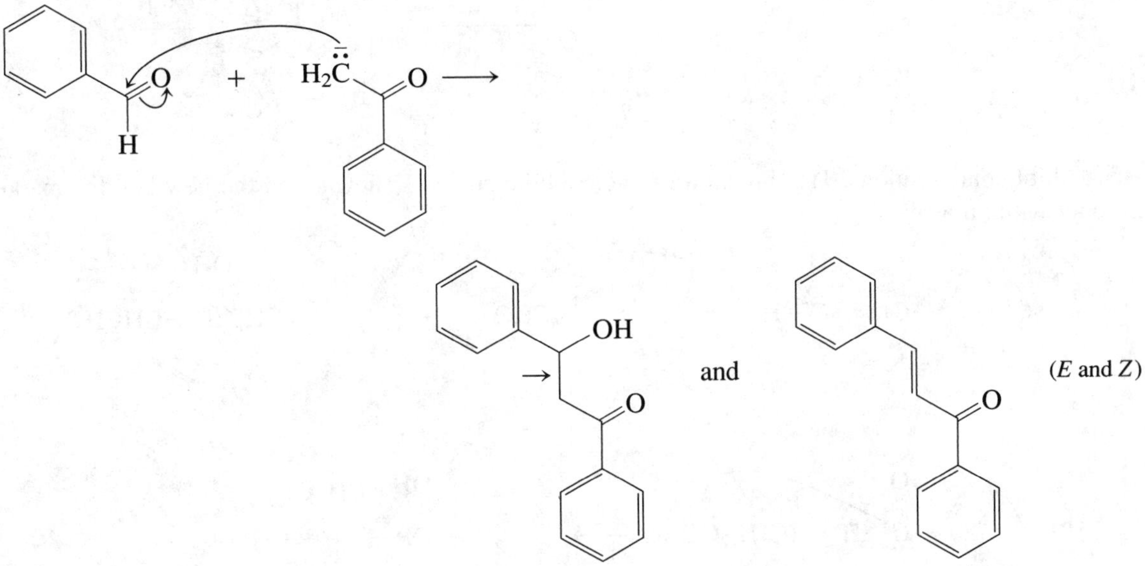 + H₂C̈ ⟶

and (*E* and *Z*)

(b) 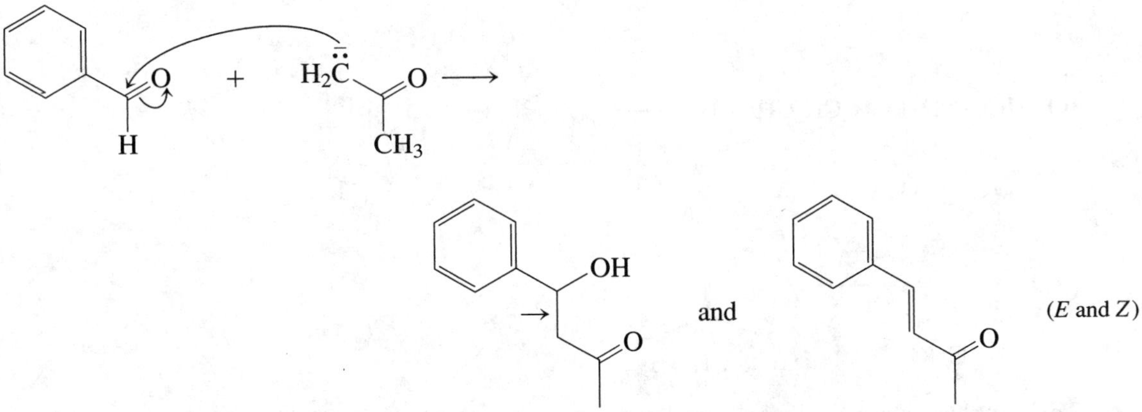 + H₂C̈ CH₃ ⟶

and (*E* and *Z*)

(c) 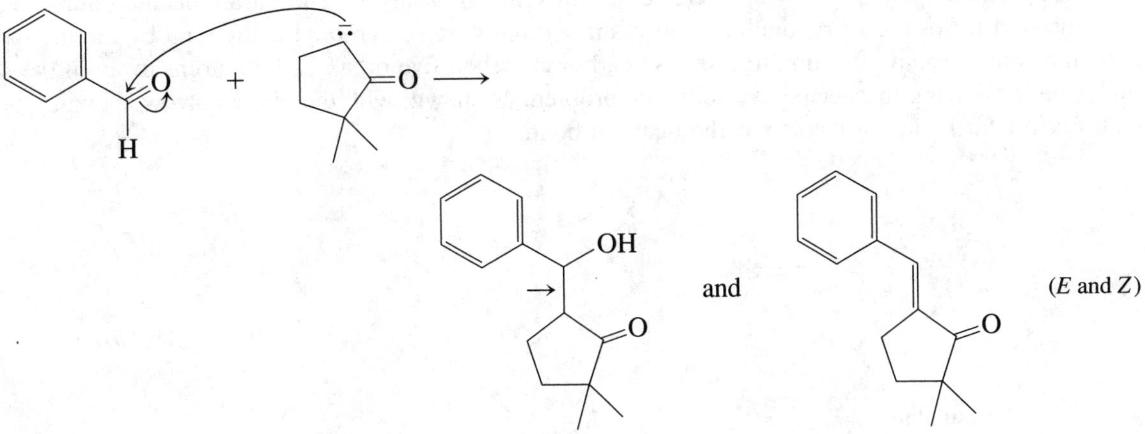 + ⟶

and (*E* and *Z*)

44.

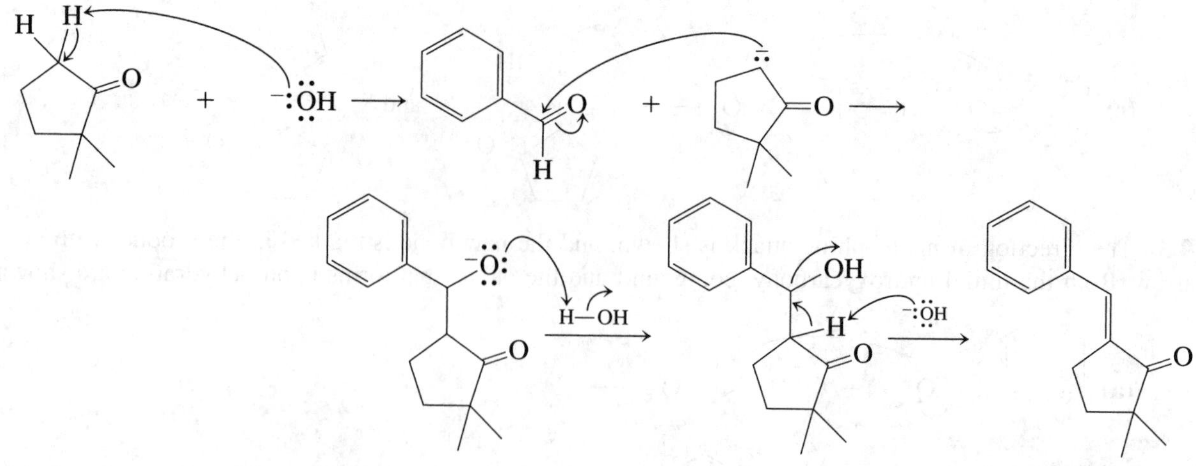

45. Aldol condensations. The direction of nucleophilic attack is shown, and the new bond is marked in the product with an arrow.

(a)

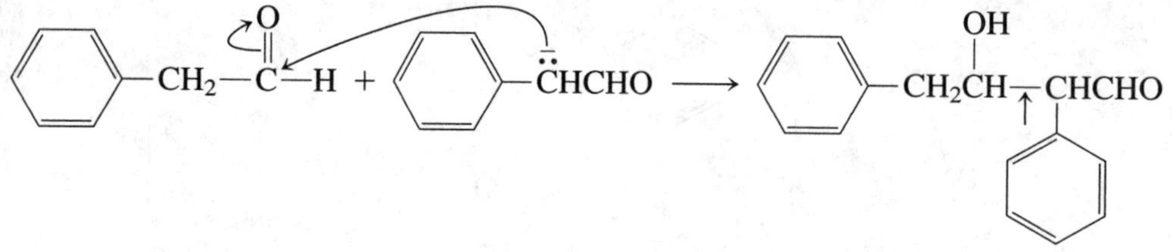

(b)

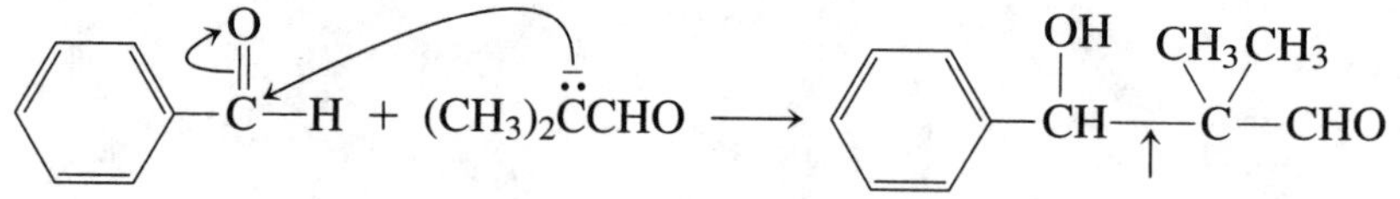

(c) HCC(CH₃)₂CH₂CH₂CHCCH₃ ⟶

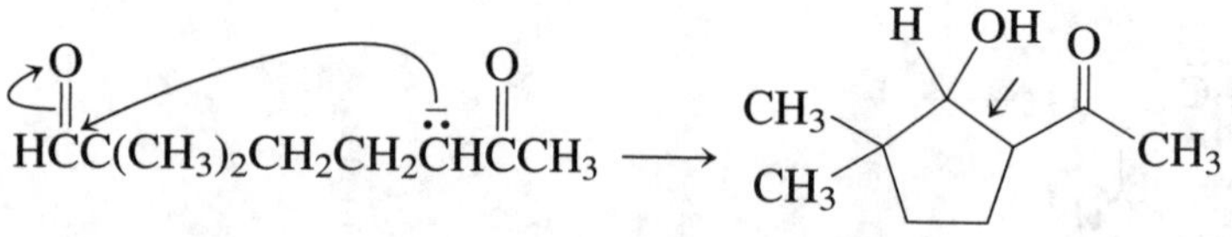

(d)

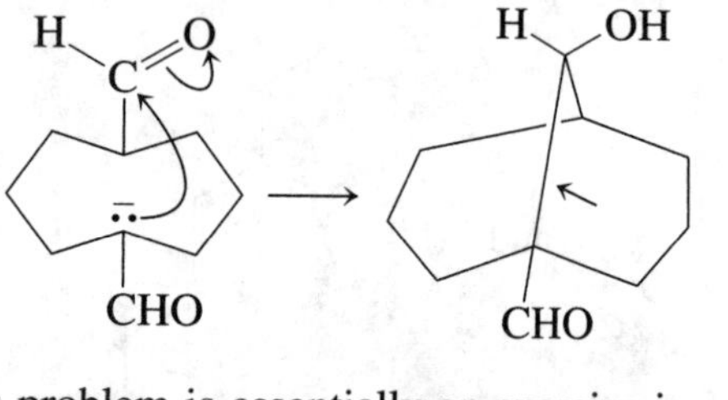

46. This problem is essentially an exercise in retrosynthetic analysis. The intramolecular aldol condensation forms the C=C double bond of an α,β-unsaturated carbonyl compound by linking the α carbon of one carbonyl function to a second carbonyl carbon five or six carbon atoms away in the same molecule. Following that recipe, we solve the problem as shown, with the curved arrow showing the pathway for formation of the new carbon-carbon bond:

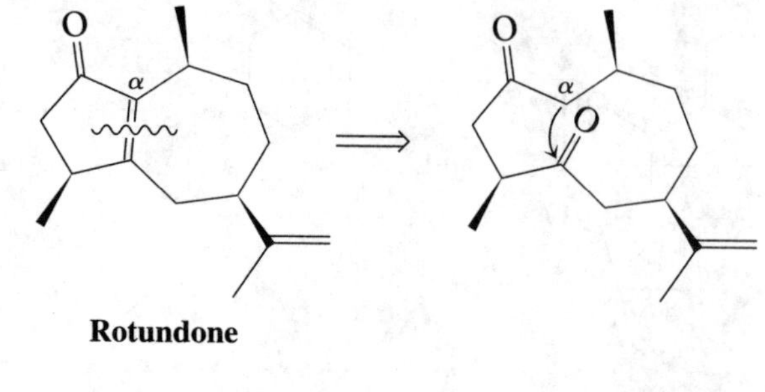

Rotundone

47 and 48. (a) $\overset{\alpha}{C}H_3CHO + CH_3CH_2\overset{\alpha}{C}H_2CHO$

$$CH_3-\underset{\underset{\displaystyle OH}{|}}{CH}-CH_2-CHO$$

$$CH_3CH_2CH_2-\underset{\underset{\displaystyle OH}{|}}{CH}-\underset{\underset{\displaystyle CH_2CH_3}{|}}{CH}-CHO$$

$$CH_3-\underset{\underset{\displaystyle OH}{|}}{CH}-\underset{\underset{\displaystyle CH_2CH_3}{|}}{CH}-CHO$$

$$CH_3CH_2CH_2-\underset{\underset{\displaystyle OH}{|}}{CH}-CH_2-CHO$$

None of these compounds will be a "major" product, but the first one may form in somewhat higher yield than the others because it is the least sterically crowded.

(b) $(CH_3)_3\overset{\alpha}{C}CHO$ + $\overset{\alpha}{C}H_3-C\overset{\displaystyle O}{\parallel}-$⟨phenyl⟩

No α hydrogens

$$(CH_3)_3C-\underset{\underset{\displaystyle OH}{|}}{CH}-CH_2-\overset{\displaystyle O}{\overset{\parallel}{C}}-$$⟨phenyl⟩

Major product

$$CH_3-\underset{\underset{\displaystyle \text{⟨phenyl⟩}}{|}}{\overset{\overset{\displaystyle OH}{|}}{C}}-CH_2-\overset{\displaystyle O}{\overset{\parallel}{C}}-$$⟨phenyl⟩

Very minor
(ketone + ketone)

(c) ⟨phenyl⟩$-\overset{\displaystyle O}{\overset{\parallel}{C}}-H$ + $\overset{\alpha}{C}H_3-\overset{\displaystyle O}{\overset{\parallel}{C}}-\overset{\alpha}{C}H_2CH_3$

No α hydrogens

⟨phenyl⟩$-\underset{\underset{\displaystyle OH}{|}}{CH}-CH_2-\overset{\displaystyle O}{\overset{\parallel}{C}}-CH_2CH_3$

Major product

⟨phenyl⟩$-\underset{\underset{\displaystyle OH}{|}}{CH}-\underset{\underset{\displaystyle CH_3}{|}}{CH}-\overset{\displaystyle O}{\overset{\parallel}{C}}-CH_3$

$$CH_3CH_2-\underset{\underset{\displaystyle CH_3}{|}}{\overset{\overset{\displaystyle OH}{|}}{C}}-CH_2-\overset{\displaystyle O}{\overset{\parallel}{C}}-CH_2CH_3$$

$$CH_3CH_2-\underset{\underset{\displaystyle CH_3}{|}}{\overset{\overset{\displaystyle OH}{|}}{C}}-\underset{\underset{\displaystyle CH_3}{|}}{CH}-\overset{\displaystyle O}{\overset{\parallel}{C}}-CH_3$$

Very little of these
(ketone + ketone)

49. Because enolates are not feasible intermediates, consider an alternative—a neutral enol, which should be a nucleophile somewhat similar to an enamine.

$$\left[\overset{\displaystyle}{C}=C-\overset{..}{\underset{..}{O}}H \longleftrightarrow \overset{-}{\underset{..}{C}}-C=\overset{+}{\underset{..}{O}}H \right], \quad \text{Compare with} \quad \left[C=C-\overset{..}{N}R_2 \longleftrightarrow \overset{-}{\underset{..}{C}}-C=\overset{+}{N}R_2 \right]$$

How can acid play a role? You already know (Section 17-5) that acid can catalyze carbonyl addition reactions by attachment to the carbonyl oxygen, thereby generating a better electrophile. So (using acetaldehyde for illustration),

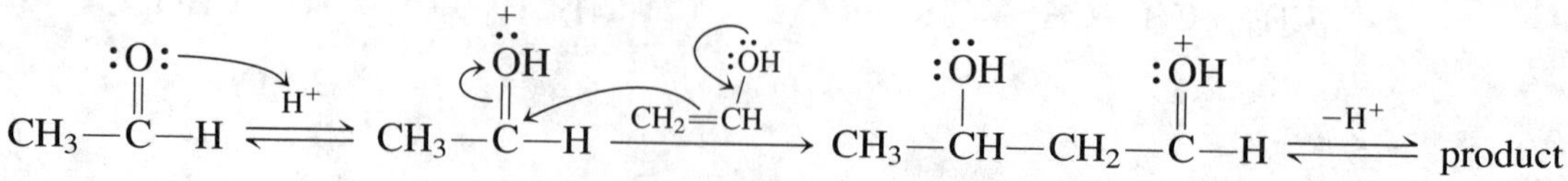

50. (a) Cl_2, CH_3CO_2H, H_2O; (b) Br_2, NaOH, H_2O; (c) 1. R_2NH, H^+, 2. CH_3I; (d) 1. R_2NH, H^+, 2. $C_6H_5CH_2Br$; (e) 1. R_2NH, H^+, 2. CH_2=$CHCH_2Br$; (f) NaOH, H_2O, 5°C (aldol reaction); (g) NaOH, H_2O, heat (aldol condensation).

51. (a) 1. R_2NH, H^+, 2. CH_3CH_2Br; (b) Cl_2, NaOH, H_2O; (c) CH_3CH_2CHO (excess), NaOH, H_2O, 5°C (crossed aldol reaction); (d) 1. R_2NH, H^+, 2. CH_2=$CHCH_2Br$; (e) CH_3CH_2CHO (excess), NaOH, H_2O, heat (crossed aldol condensation); (f) Br_2, CH_3CO_2H, H_2O.

52. (a) 1. $(C_6H_5)_2CuLi$, 2. H^+; (b) 1. $(CH_3)_2CuLi$, 2. H^+; (c) H_2, Pd; (d) 1. $(CH_3)_2CuLi$, 2. CH_3I; (e) $LiAlH_4$; (f) H_2O, ^-OH; (g) CH_3MgI; (h) cyclohexanone (excess), $NaOCH_2CH_3$, CH_3CH_2OH (conjugate addition [Michael addition] of the enolate of cyclohexanone).

53. Retrosynthetically,

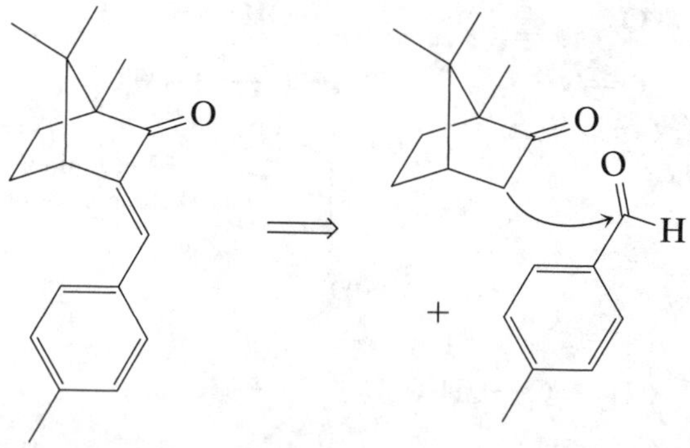

54. (a) The desired reaction is a crossed (or mixed) aldol condensation, but one that suffers from competition with at least two other aldol processes: condensation between two molecules of aldehyde, and condensation of the aldehyde with the methyl group (C1) of the ketone:

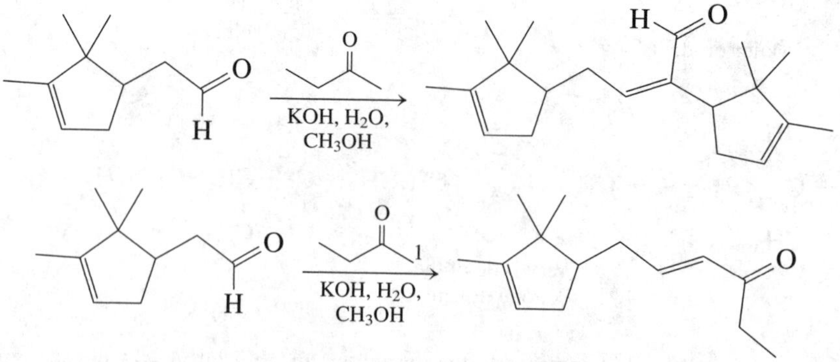

In addition, both of these processes, as well as the condensation that affords the desired compound, give a mixture of E and Z stereoisomers.

(b) This solution forces the formation of just a single nucleophilic enolate carbon, C3 of the ketone. The side reactions described above derive from reaction of nucleophilic enolate ions at C2 of the aldehyde and C1 of the ketone, respectively. By restricting nucleophilic behavior to just the single carbon atom required to allow access to the desired product structure, the two side reactions above are eliminated. In addition, dehydration of the product of this sequence can be controlled to give the necessary stereoisomer, because that isomer places the two larger substituents on the side-chain double bond trans to each other, making it the more stable and the likely product of dehydration under equilibrium (thermodynamic) conditions.

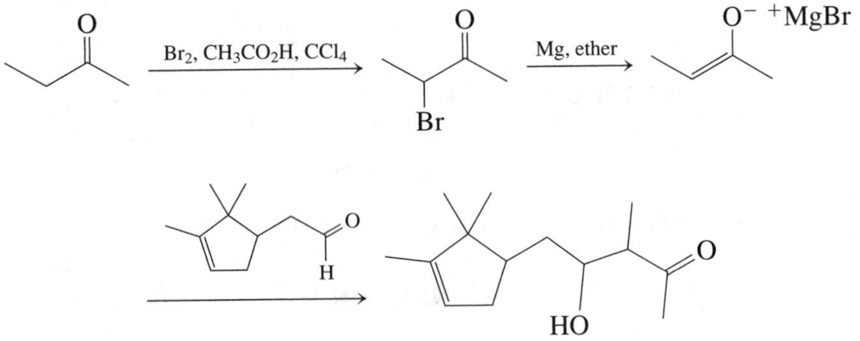

55. 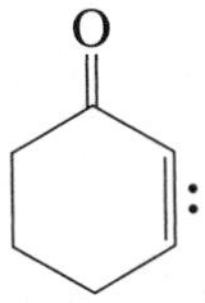 : **(a)** Cyclohexanone **(b)** 2-Cyclohexenol

(c) 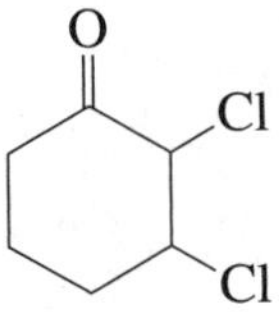

(d) 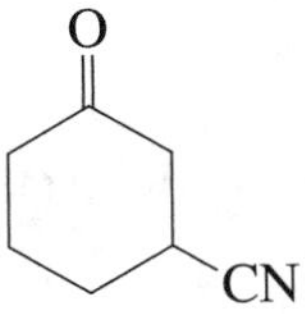

(e) 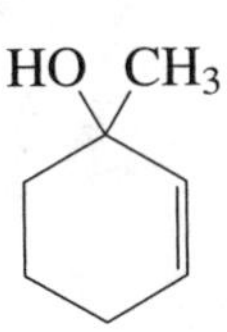

(f) 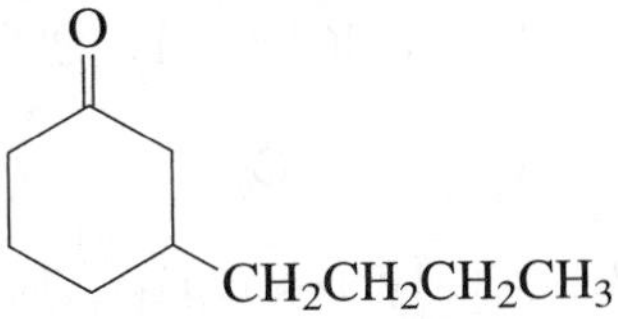

(g) 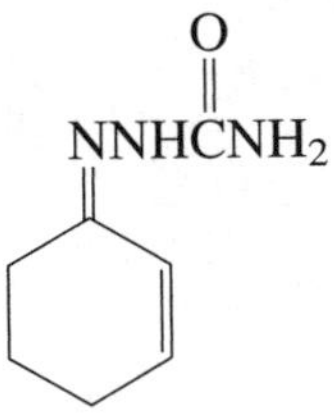

(h) 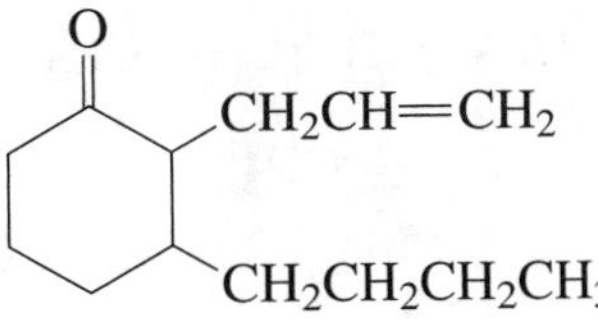

(first step forms an enolate, which is alkylated)

$CH_3CH{=}C(CH_2CH_2CH_3)(CHO)$:

(a) $CH_3CH_2CH(CHO)CH_2CH_2CH_3$

(b) $CH_3CH{=}C(CH_2OH)CH_2CH_2CH_3$

$$\underset{\substack{| \\ \text{CHO}}}{(c)\ CH_3CHClCClCH_2CH_2CH_3}$$

(c) $CH_3CHClC\overset{\overset{\displaystyle CHO}{|}}{Cl}CH_2CH_2CH_3$

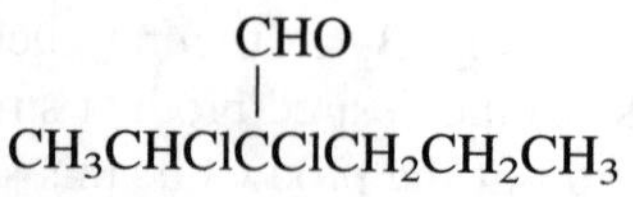

(d) 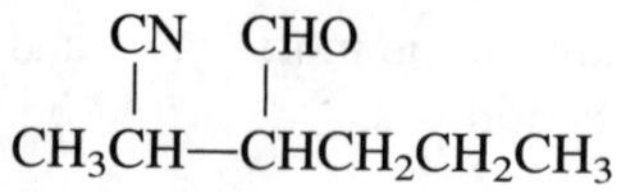 $CH_3CH\overset{\overset{\displaystyle CN}{|}}{{}}\!-\!\overset{\overset{\displaystyle CHO}{|}}{CH}CH_2CH_2CH_3$

(e) 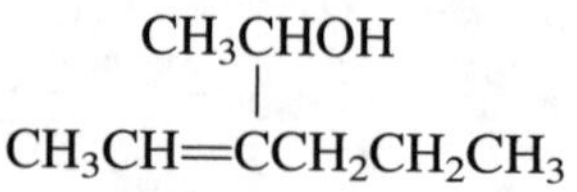 $CH_3CH\!=\!\overset{\overset{\displaystyle CH_3CHOH}{|}}{C}CH_2CH_2CH_3$

(f) 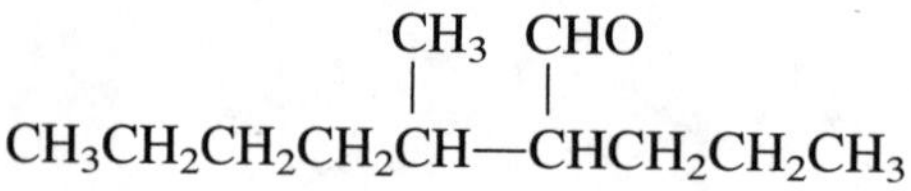 $CH_3CH_2CH_2CH_2CH\overset{\overset{\displaystyle CH_3}{|}}{{}}\!-\!\overset{\overset{\displaystyle CHO}{|}}{CH}CH_2CH_2CH_3$

(g) 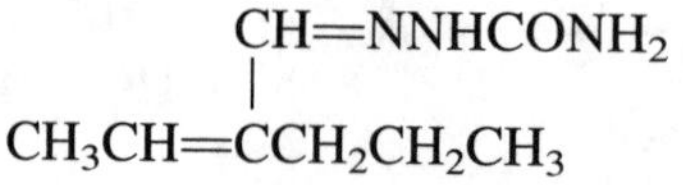 $CH_3CH\!=\!\overset{\overset{\displaystyle CH=NNHCONH_2}{|}}{C}CH_2CH_2CH_3$

(h) 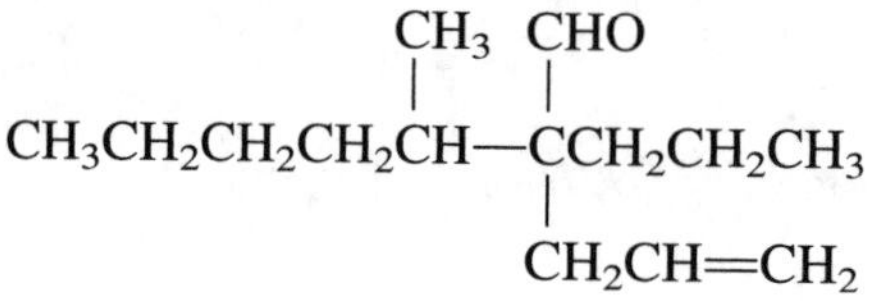 $CH_3CH_2CH_2CH_2CH\overset{\overset{\displaystyle CH_3}{|}}{{}}\!-\!\underset{\underset{\displaystyle CH_2CH=CH_2}{|}}{\overset{\overset{\displaystyle CHO}{|}}{C}}CH_2CH_2CH_3$

$\overset{\overset{\displaystyle O}{\|}}{CH_2\!=\!CHC(CH_2)_4CH_3}:$ **(a)** 3-Octanone **(b)** 1-Octen-3-ol

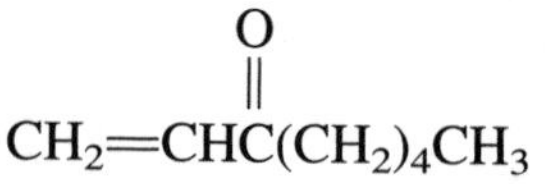

(c)  $ClCH_2CHClC\overset{\overset{\displaystyle O}{\|}}{{}}(CH_2)_4CH_3$

(d)  $NCCH_2CH_2C\overset{\overset{\displaystyle O}{\|}}{{}}(CH_2)_4CH_3$

(e) 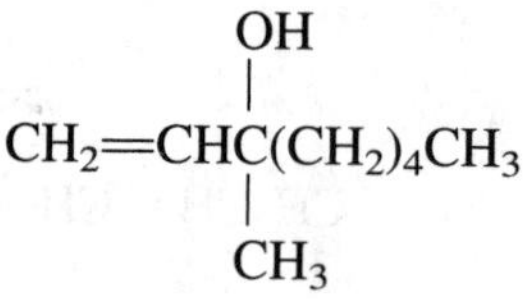 $CH_2\!=\!CHC\overset{\overset{\displaystyle OH}{|}}{\underset{\underset{\displaystyle CH_3}{|}}{{}}}(CH_2)_4CH_3$

(f) 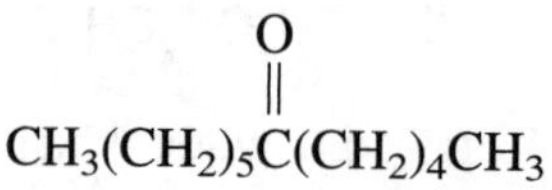 $CH_3(CH_2)_5C\overset{\overset{\displaystyle O}{\|}}{{}}(CH_2)_4CH_3$

(g)  $CH_2\!=\!CHC\overset{\overset{\displaystyle NNHCONH_2}{\|}}{{}}(CH_2)_4CH_3$

(h) 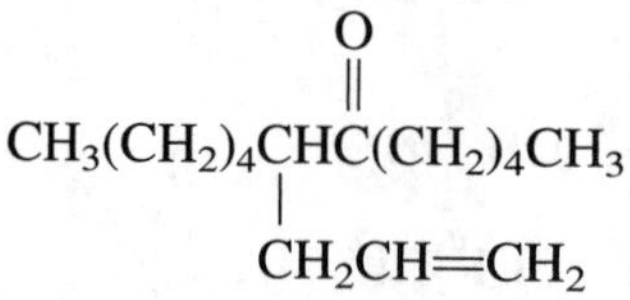 $CH_3(CH_2)_4CHC\overset{\overset{\displaystyle O}{\|}}{\underset{\underset{\displaystyle CH_2CH=CH_2}{|}}{{}}}(CH_2)_4CH_3$

56. (a) $\underset{\displaystyle \overset{O}{\|}}{C_6H_5C}-\underset{\displaystyle \overset{CH_2CH_3}{|}}{CHCH_2CH_3}$ (b)

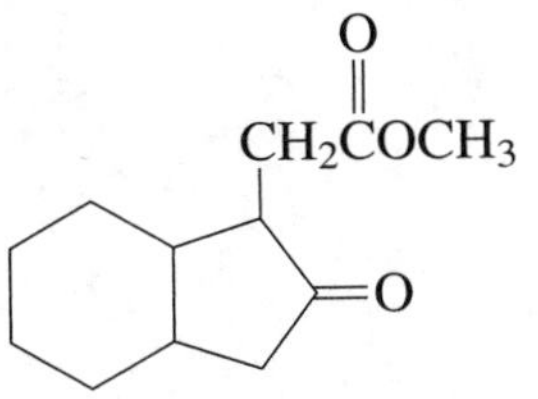

(c)

$\left(\text{via} \quad \text{(structure)} \quad + \ C_6H_5CH_2Cl \right)$

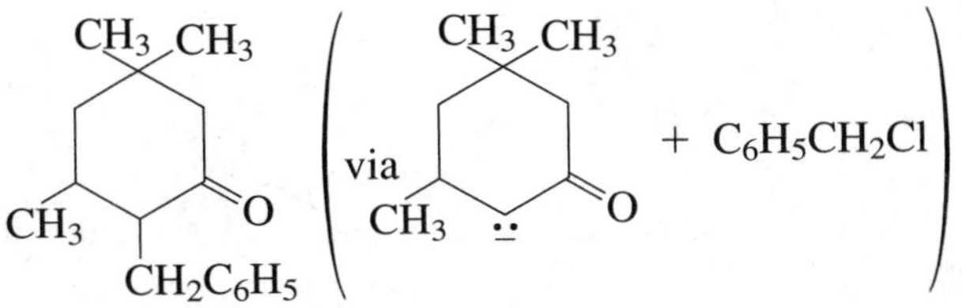

(d)

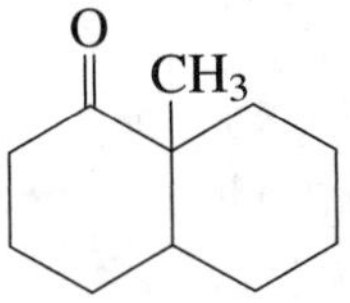

57. Michael additions: 1,4-additions of enolate ions to α, β-unsaturated aldehydes or ketones. New bond(s) indicated by arrow(s).

(a) $\underset{\displaystyle \overset{O}{\|}}{C_6H_5CCH_2}\overset{\downarrow}{-}CH_2CH_2\underset{\displaystyle \overset{O}{\|}}{CC_6H_5}$ (b)

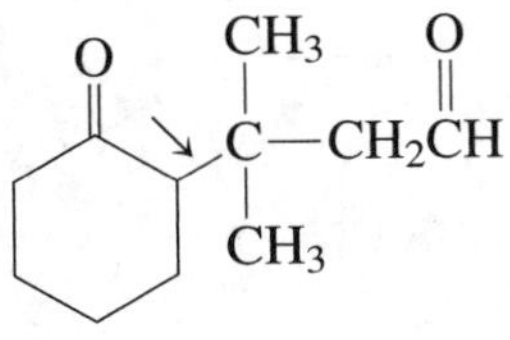

(c)

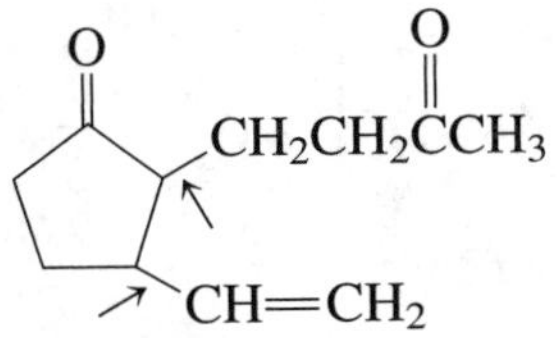

(d)

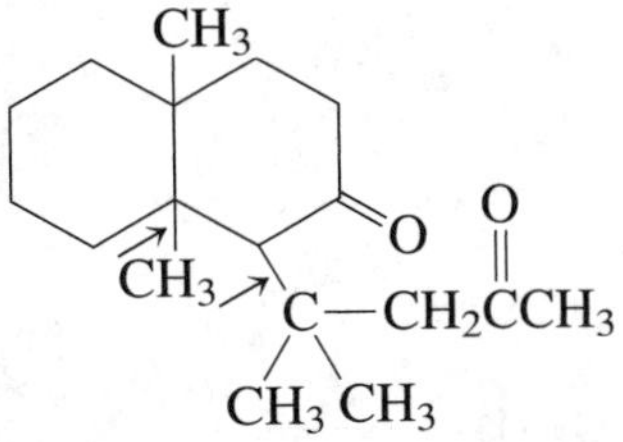

(e) Intramolecular aldol condensations, leading to

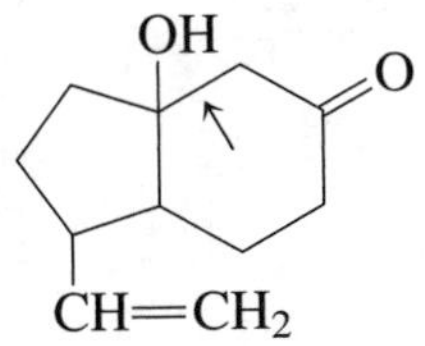

from (c) and 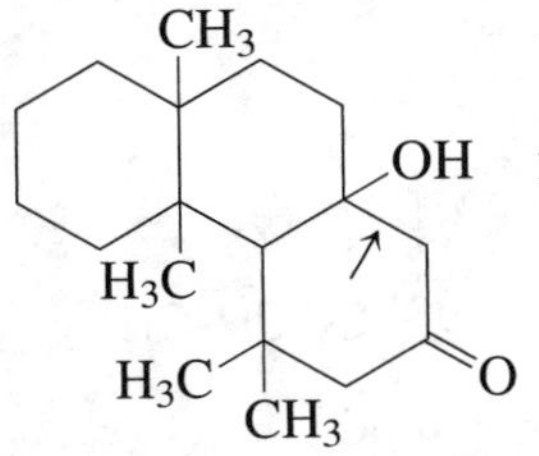   from (d)

58. Robinson annulation sequences (Michael addition followed by aldol).

(a)

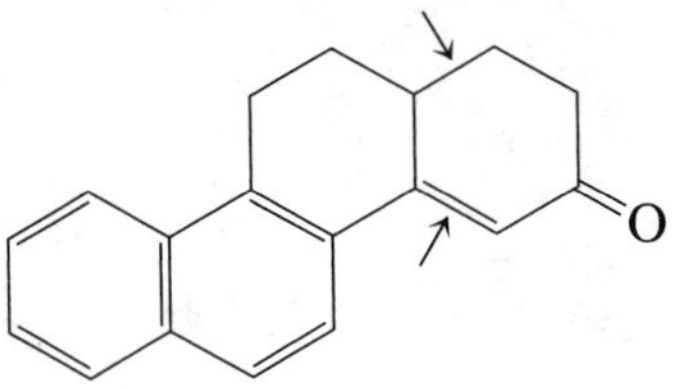

(b)

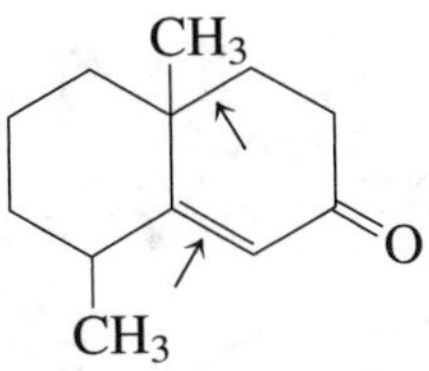

(c)

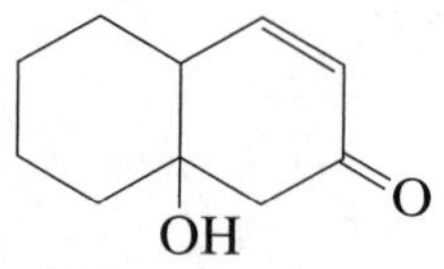

(d)

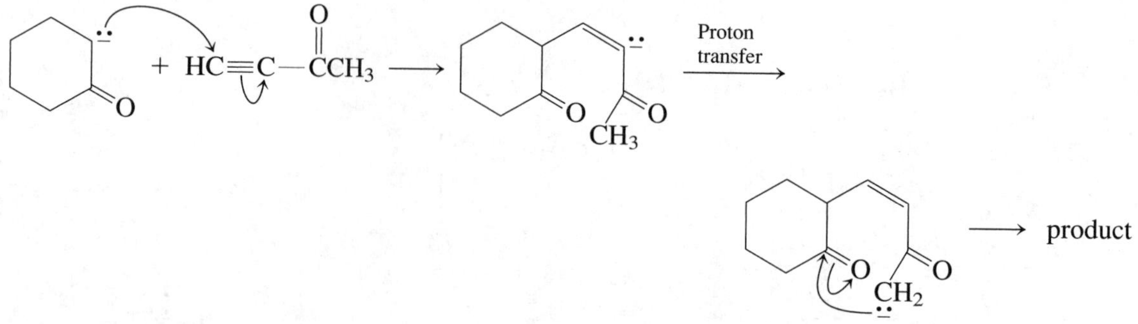

Proton transfer →

→ product

Heating in (a) and (b) tends to drive dehydration to give the α, β-unsaturated ketone product.

59. Retrosynthetically:

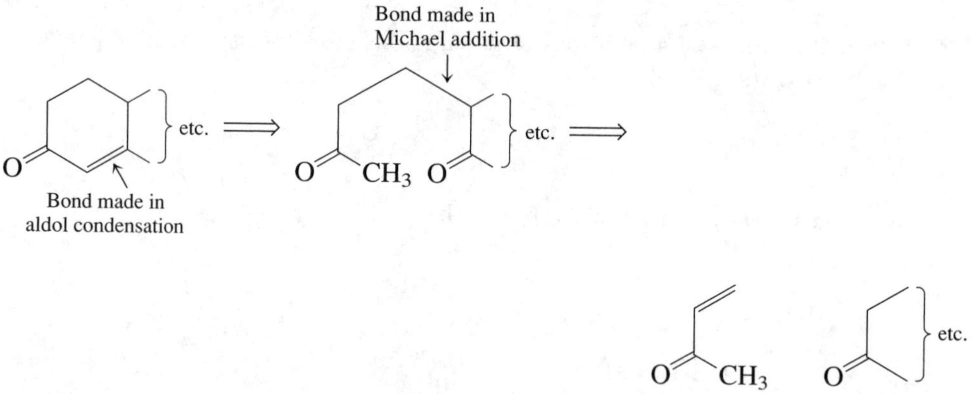

Bond made in Michael addition

Bond made in aldol condensation

etc. ⟹ etc. ⟹

etc.

(a)

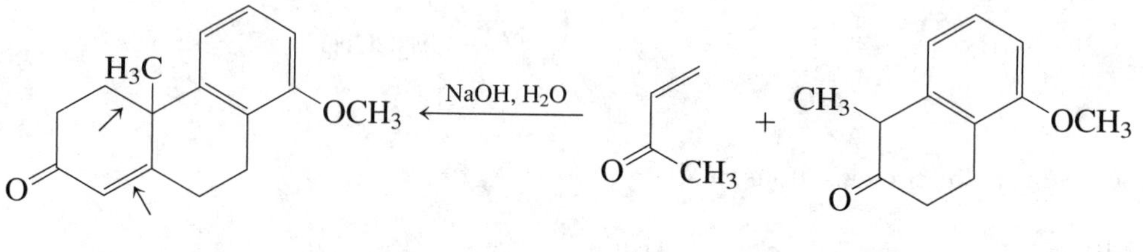

NaOH, H₂O ←

(b)

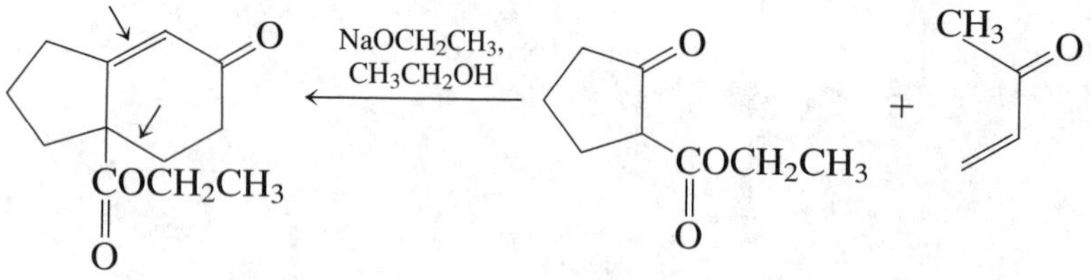

NaOCH₂CH₃, CH₃CH₂OH ←

(c)

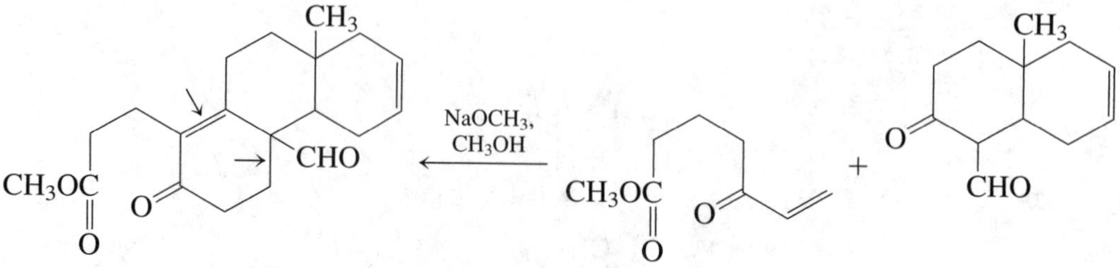

NaOCH₃, CH₃OH ←

It's not a big deal, but use of NaOH, H_2O in (b) and (c) would hydrolyze the ester groups to carboxylic acids ($-CO_2H$). Using NaOR, ROH where R corresponds to the alkyl group of the ester, prevents this (Chapter 20).

(d)

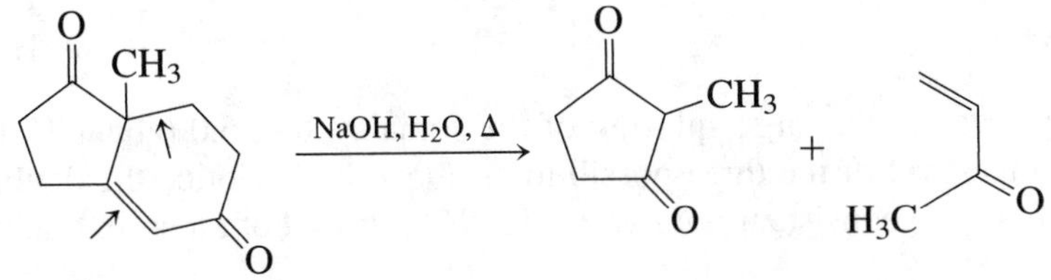

60. No! The carbonyl group changes the mechanism.

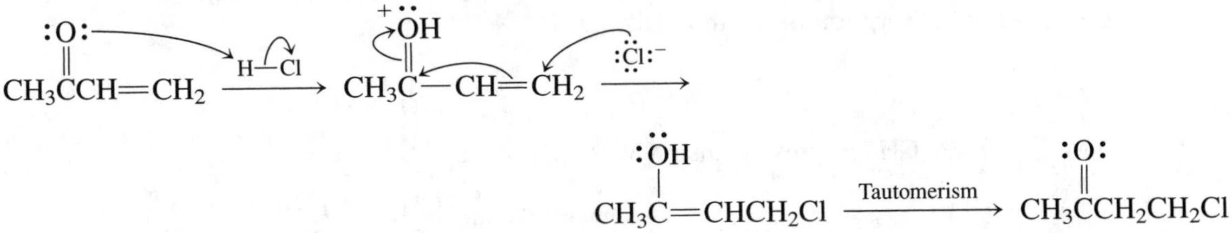

Protonation occurs on oxygen, not carbon, and the result **looks** like it's anti-Markovnikov. It is really 1,4-addition.

61. Calculate your degrees of unsaturation first!

(a) $H_{sat} = 10 + 2 = 12$; degrees of unsaturation $= \frac{12 - 10}{2} = 1$ π bond or ring.
UV: n $\rightarrow$ π^* absorption of carbonyl.

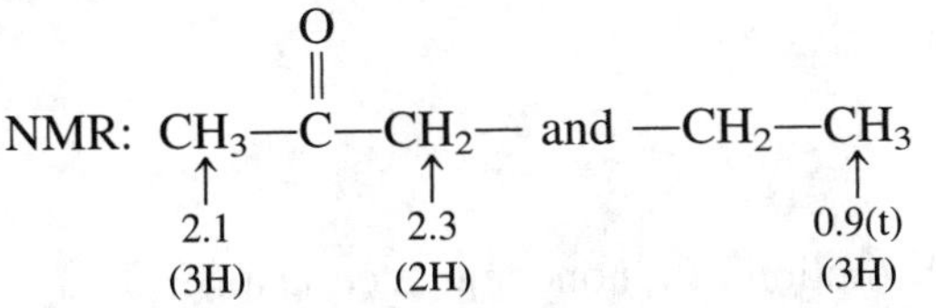

These assignments are clear, giving as the answer 2-pentanone, $CH_3COCH_2CH_2CH_3$ (A). The sextet at $\delta = 1.5$ corresponds to the C4 CH_2 group.

(b) $H_{sat} = 10 + 2 = 12$; degrees of unsaturation $= \frac{12 - 8}{2} = 2$ π bonds or rings.
UV: $\pi \rightarrow \pi^*$ absorption of α, β-unsat'd carbonyl at 220 nm.

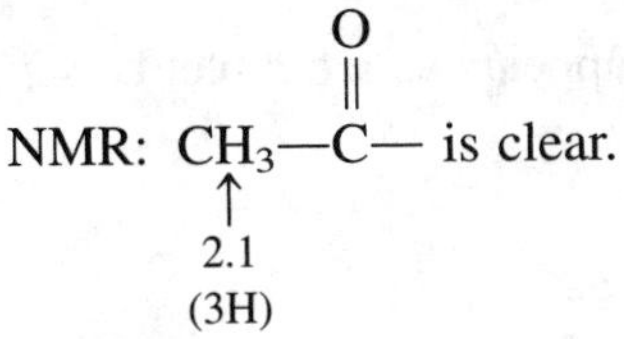

Also two alkene hydrogens ($\delta = 5.8$–6.9) and another three H's at $\delta = 1.9$ (CH_3—?). Because UV indicates conjugation, there are three possibilities:

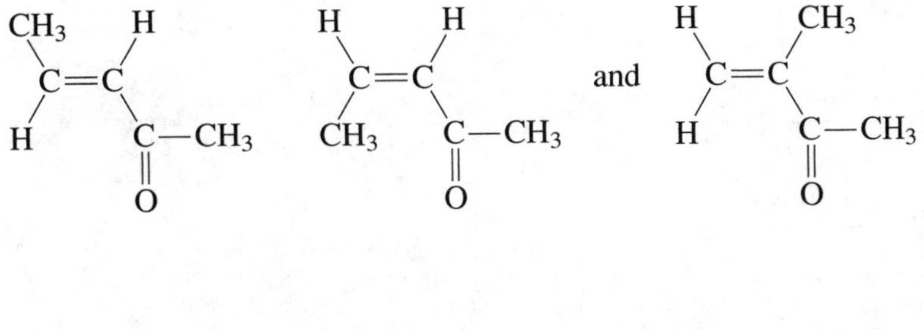

The last is ruled out because the signal at δ = 1.9 has a large splitting, indicating that a $\underset{H}{\overset{CH_3}{\diagdown}}C=$ piece is present. In the alkene region, the large splitting of the signal at δ = 6.0 (about 15 Hz) suggests trans alkene H's, so the **first** of the three possibilities above is B. Notice the doublet of quartets for the signal at δ = 6.7, consistent with an alkene H split by both a second alkene H and a methyl group.

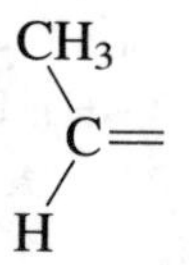

(c) $H_{sat} = 12 + 2 = 14$; degrees of unsaturation = $\frac{14 - 12}{2} = 1$ π bond or ring.
UV: π → π* absorption of a simple alkene.

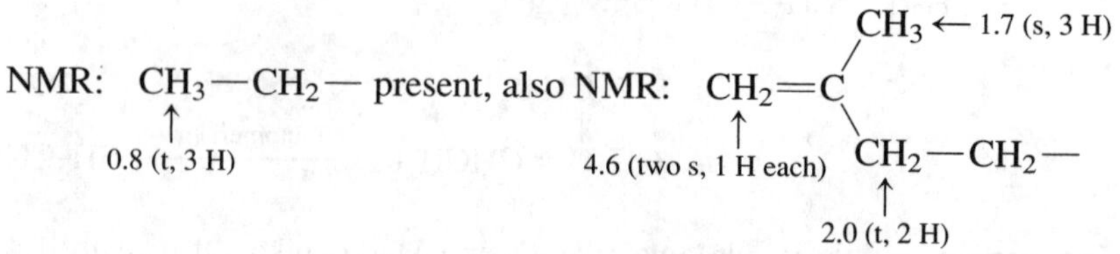

NMR: CH_3—CH_2— present, also NMR:
$\underset{0.8 \text{ (t, 3 H)}}{\uparrow}$

CH₃ ← 1.7 (s, 3 H)
$CH_2=C$
$\underset{4.6 \text{ (two s, 1 H each)}}{\uparrow}$ $\underset{2.0 \text{ (t, 2 H)}}{\overset{CH_2—CH_2—}{\uparrow}}$

can be identified by chemical shifts and splittings. These add up to C_7H_{14}, so one CH_2 group has been duplicated in the two pieces. After correcting for that, the answer for C becomes

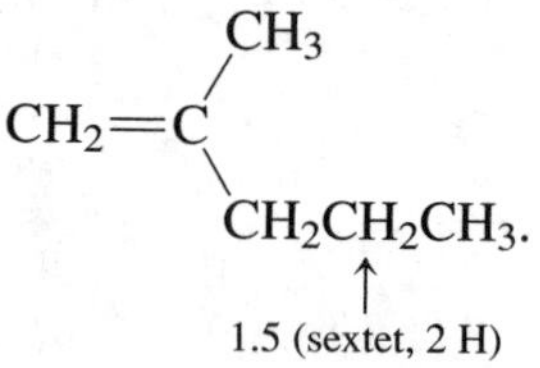

$$CH_2=\underset{\underset{\underset{1.5 \text{ (sextet, 2 H)}}{\uparrow}}{CH_2CH_2CH_3.}}{\overset{CH_3}{\overset{|}{C}}}$$

(d) Also 1 degree of unsaturation. UV: n → π* absorption of a nonconjugated ketone.

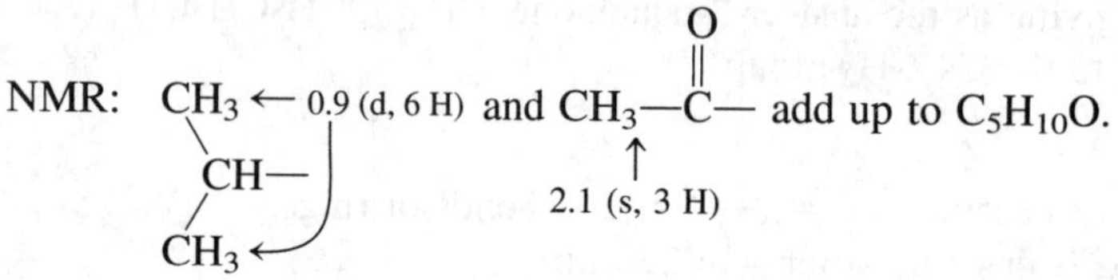

NMR: $\underset{CH—}{\overset{CH_3}{\diagup}}\underset{CH_3}{\diagdown}$ ← 0.9 (d, 6 H) and $CH_3—\overset{\overset{O}{||}}{C}—$ add up to $C_5H_{10}O$.
$\underset{2.1 \text{ (s, 3 H)}}{\uparrow}$

All that's left to arrive at the molecular formula is CH_2, which appears as a doublet at δ = 2.3; insert it between the above two pieces to get the answer:

$$(CH_3)_2CH—CH_2—\overset{\overset{O}{||}}{C}—CH_3 \text{ is D}$$

The signal for the CH group (nine lines!) appears at the base of the CH_3 singlet at δ = 2.1.

(e) A + $CH_2=P(C_6H_5)_3 \xrightarrow{\text{THF}}$ C

(f) B + $(CH_3)_2CuLi \xrightarrow{\text{THF}}$ D

(g) B + H_2, Pd—C $\xrightarrow{CH_3CH_2OH}$ A

62. (a)

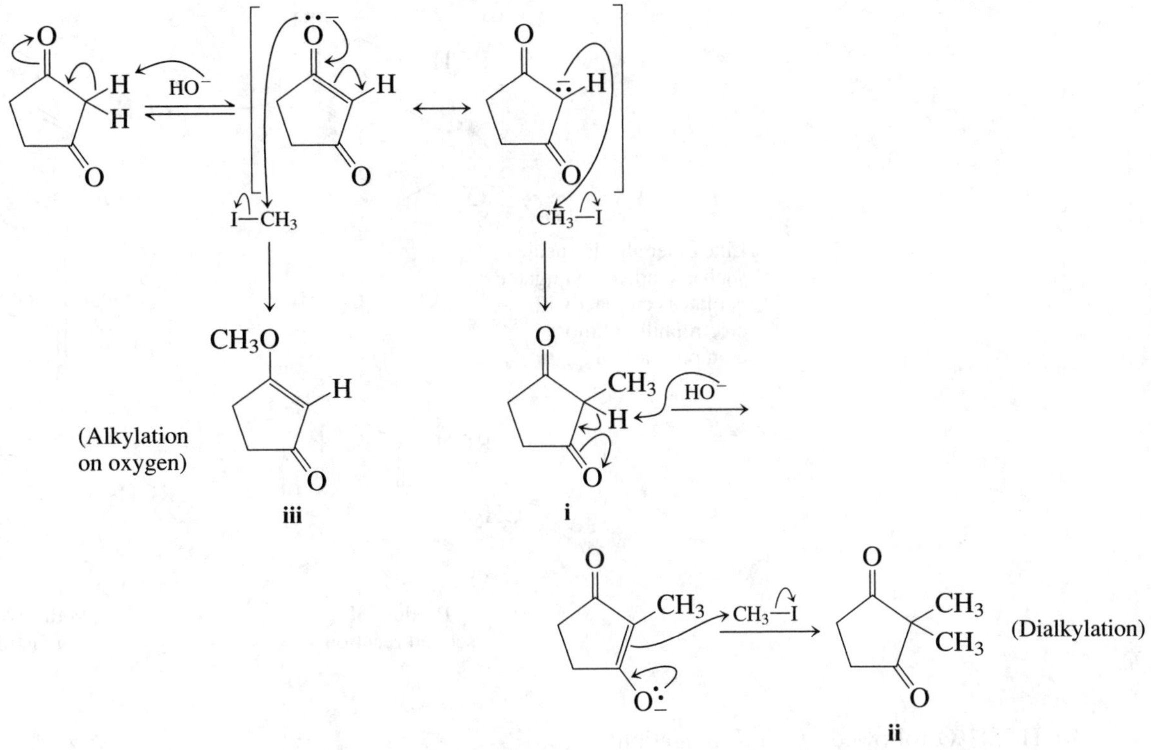

(Alkylation on oxygen)

iii

i

(Dialkylation)

ii

(b)

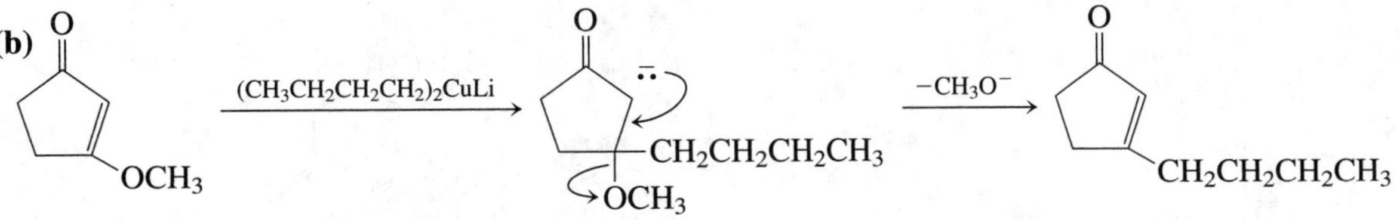

The presence of a leaving group adjacent to the enolate diverts the reaction to elimination.

63. (a) Deprotonate "allylic" to give a conjugate enolatelike anion (see Section 18-11):

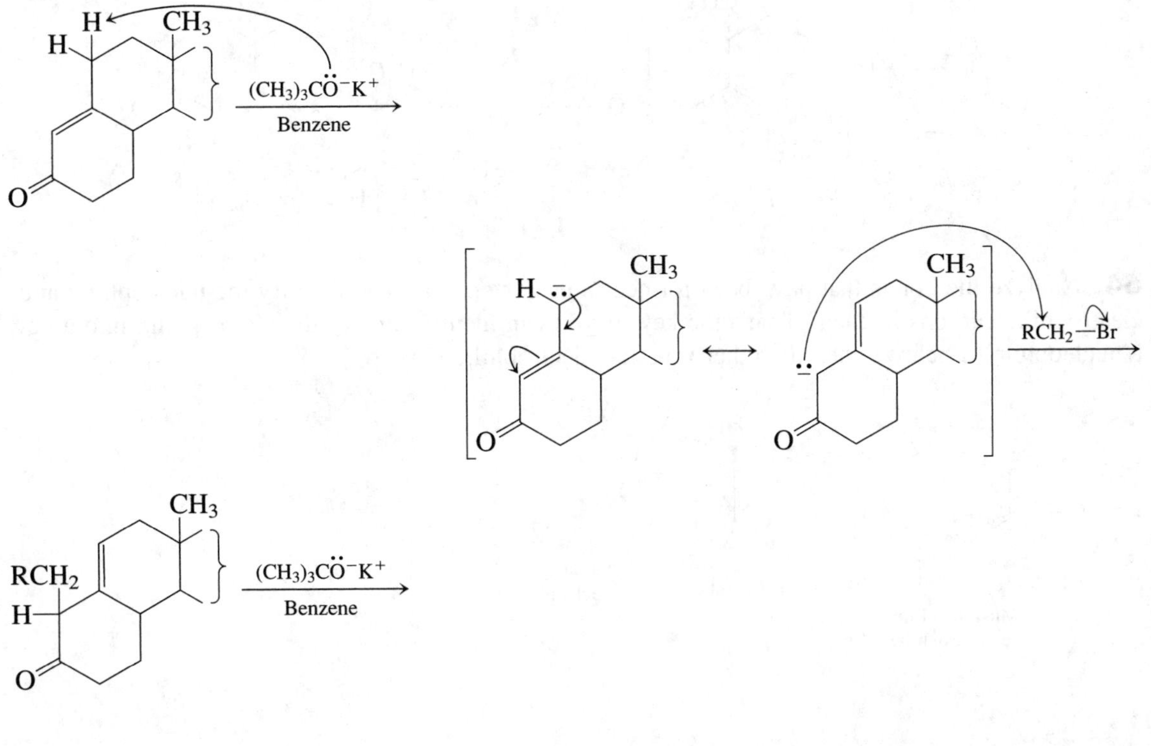

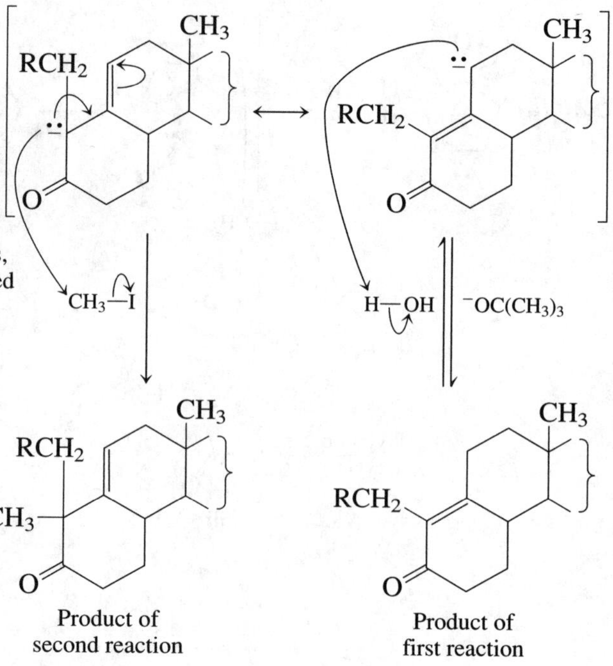

Like other allylic species, such extended, conjugated enolates can react with electrophiles at more than one carbon.

Product of second reaction

Product of first reaction

(b) H^+, H_2O followed by Cr^{6+} oxidation gives

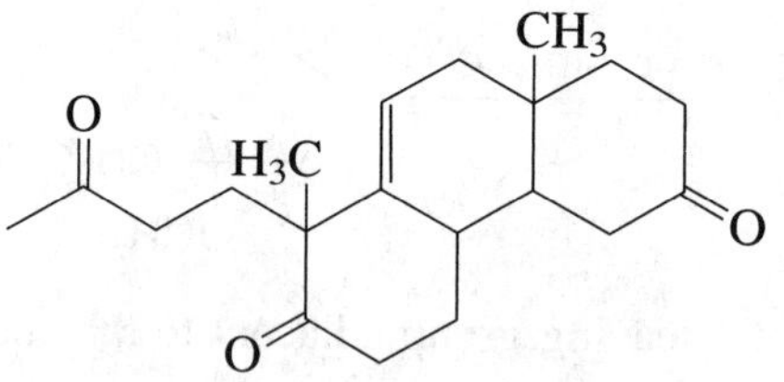

Aldol condensation is then possible:

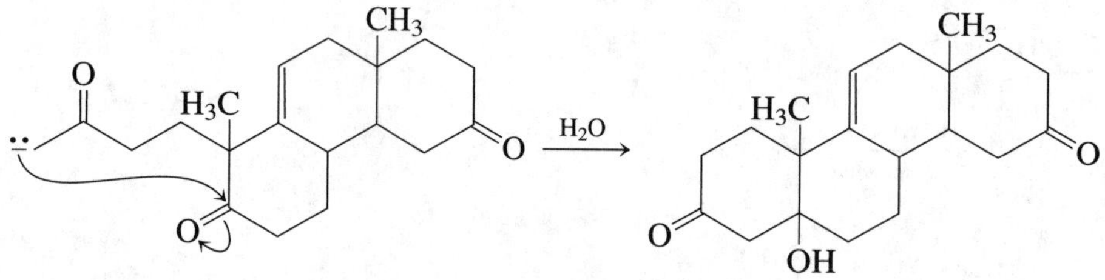

64. Analyze the bonds that have been formed in each step. Can you identify the nucleophilic and electrophilic carbons in each? That's the key: If you can identify one of the atoms going into a new bond as a nucleophile (*a* below), then the other **must** be electrophilic (*b*). So,

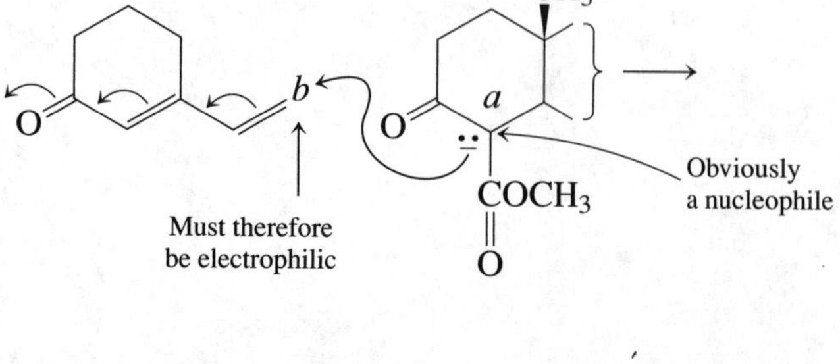

Must therefore be electrophilic

Obviously a nucleophile

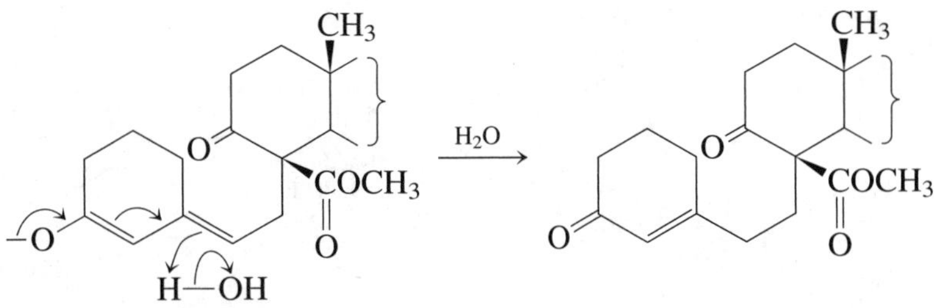

This, therefore, is a version of nucleophilic addition to an unsaturated ketone, but here the ketone is doubly (α, β, γ, δ) unsaturated, and the process is a *1,6-addition*. The product enolate protonates on the δ-carbon, giving an α, β-unsaturated ketone as the product. In the second reaction, remove an allylic proton with base to give an *extended* enolate.

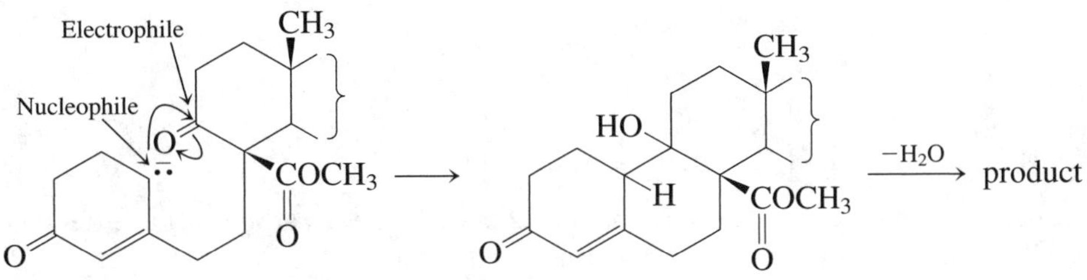

This is just a version of aldol condensation, where the nucleophile is the extended enolate ion at the γ-carbon of an α, β-unsaturated ketone, instead of the simple enolate at the α-carbon of a saturated ketone.

65. Work backward.

(a)

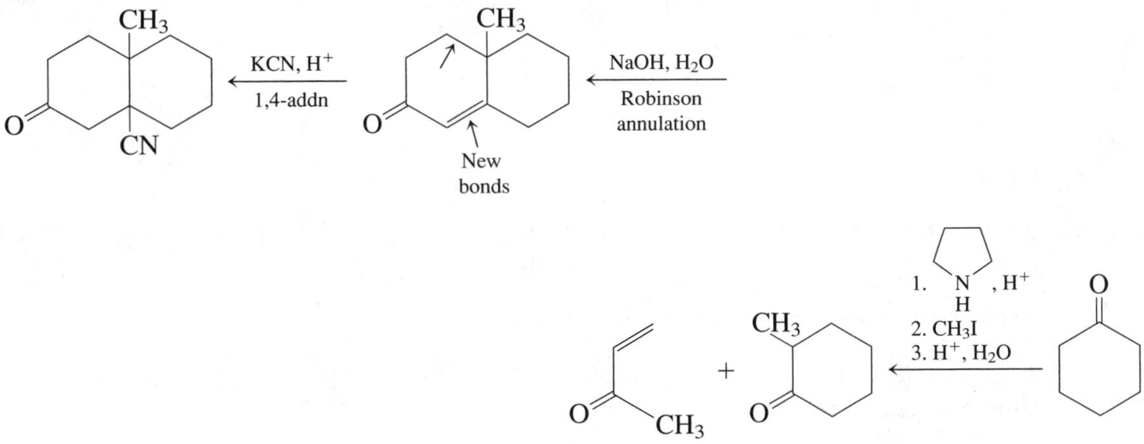

(b) Use the **hint!** Find the carbons of

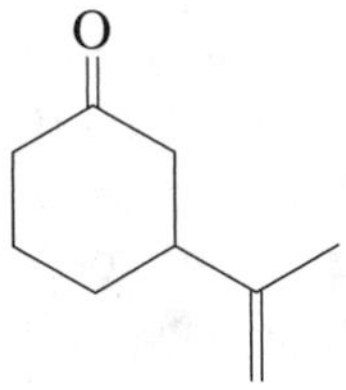

in your final target molecule.

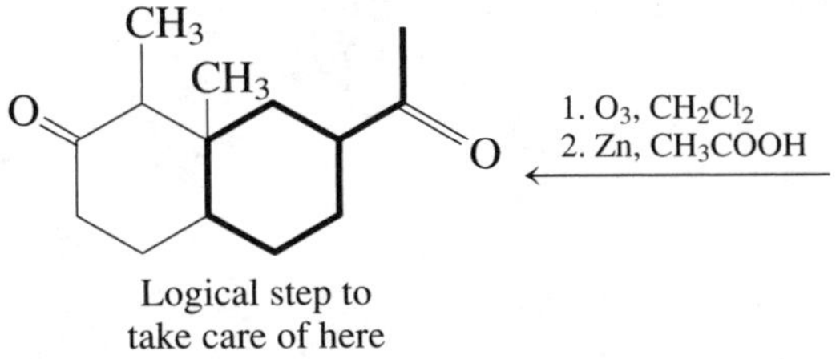

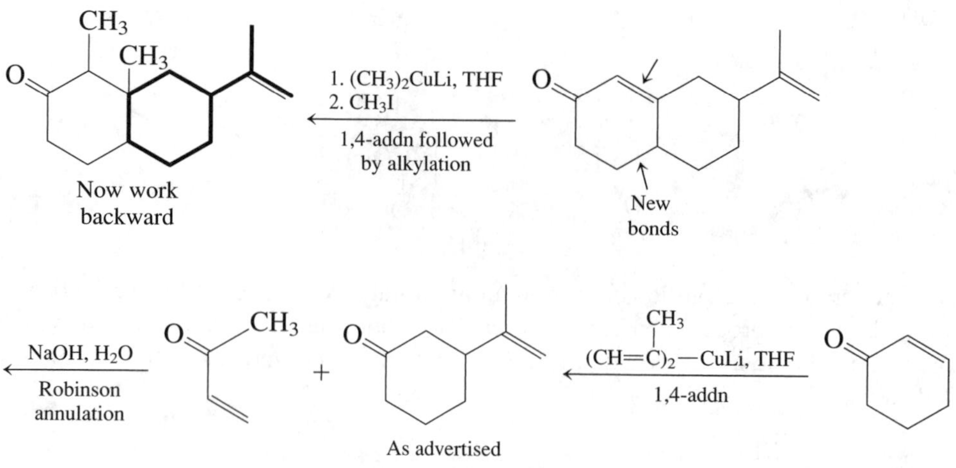

66. (a) $CH_3CH_2CCH=CH_2$, NaOH, H_2O (Robinson annulation)

(b) Exactly the same reagents as (a), but here the nucleophile is the α-carbon of the extended enolate formed by allylic deprotonation of the α, β-unsaturated ketone:

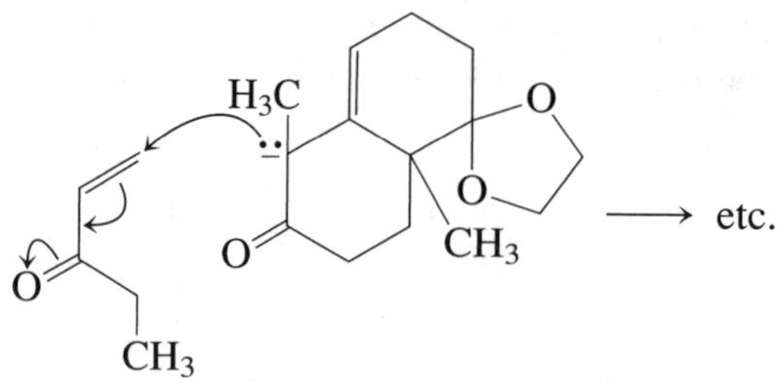

(c) CH_3I

(d) 1. H_2, Pd (reduces C=C bond), 2. $NaBH_4$ (reduces C=O), 3. H^+, H_2O (hydrolyzes acetal),

4. CH_3CCl (makes ester of newly formed alcohol)

(e) 1. Cl_2, CH_3COOH (chlorinates α to ketone), 2. K_2CO_3, H_2O (eliminates HCl to make α, β-unsaturated ketone)

67. Kinetically, the bulky base will prefer to deprotonate the ketone at the less hindered, unsubstituted α-carbon to the left of the carbonyl group. Thermodynamically, however, the more stable enolate ion will be the one on the right, containing a tetrasubstituted double bond (Section 11-7). Under condition B, the excess ketone that is present can reversibly protonate enolate ions, providing a mechanism for enolate equilibration. Under condition A, excess base is always present, and therefore there is never a significant concentration of neutral ketone around to allow enolate equilibration to occur. Your potential energy diagram should show a lower activation barrier leading to the less-stable, higher energy "kinetic" enolate (the product on the left in the problem).

68. (b)

69. (c)

70. (d)

71. (c) is conjugated

19

Carboxylic Acids

The carboxylic acids and their derivatives contain a carbonyl group as one primary source of reactivity and therefore have a lot in common with aldehydes and ketones. However, two major features make carboxylic acids different from aldehydes and ketones. First, the HO proton is made much more acidic by the neighboring C=O function. Second, the HO may behave as a leaving group under the right conditions. You will discover the importance of this ability when you study nucleophilic additions to the carboxy group.

Outline of the Chapter

Keys to the Chapter

19-1 through 19-3. Nomenclature and Physical Properties of Carboxylic Acids

The one characteristic that is unique to carboxylic acids is the strong tendency to form hydrogen-bonded dimers, a process resulting in much higher melting and boiling points relative to comparable compounds (Table 19-2). Notice also the very low field position of the COOH proton in the NMR.

19-4. Acidity and Basicity of Carboxylic Acids

Recall that acid and base strengths are determined by the charge-stabilizing ability in the structures of the acid-base conjugate pair. There are three points to consider in evaluating the stability of a charged species: (1) the electronegativity of the charged atom(s), (2) the size of the charged atom(s), (3) stabilization of charge by either inductive effects or resonance. For carboxylic acids, the text sets up two comparisons.

Carboxylic acids vs. alcohols

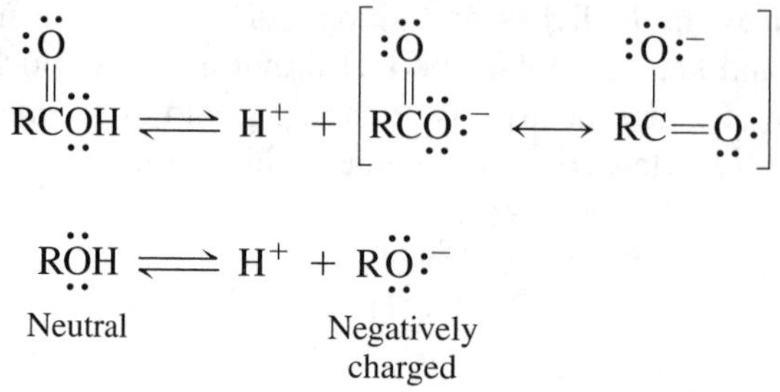

Both negatively charged conjugate bases have the same negative atom (O), but the carboxylate ion is stabilized by inductive effects and resonance. Because the carboxylate is more stable, the carboxylic acid is a stronger acid than the alcohol.

Carboxylic acids vs. ketones or aldehydes

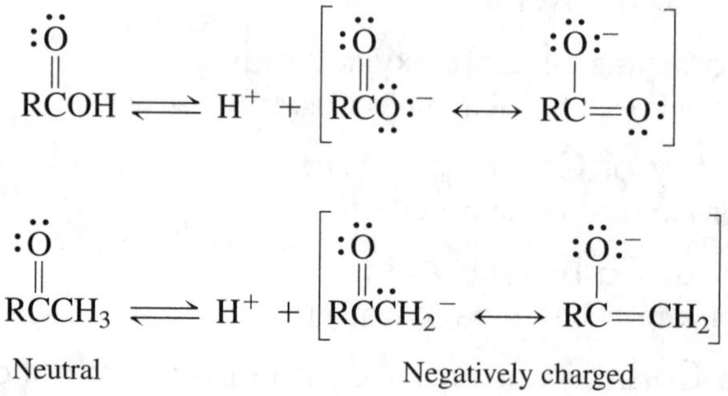

Both negatively charged conjugate bases are stabilized by inductive effects of a δ^+ carbonyl carbon. Both are also stabilized by resonance, but the carboxylate ion distributes its negative charge between **two** electronegative oxygens, whereas the enolate distributes its charge between a carbon and only one electronegative oxygen. The carboxylate ion is more stable, so the carboxylic acid is more acidic.

Similar analyses can be applied to base strength as well. See chapter-end Problems 30 and 49 and their answers for examples and explanations.

19-6. Preparation of Carboxylic Acids

Even though the most obvious syntheses of carboxylic acids are simple functional group interconversions (mainly oxidations), there are quite a few that involve breaking or forming carbon–carbon bonds. You might take the time to do a quick review and then begin organizing these synthetic methods into appropriate categories.

19-7. Reactivity of the Carboxy Group: Addition and Elimination

The mechanistic discussion at the start of this section is central to both the chemistry of carboxylic acids and the chemistry of their derivatives. Read it **very** carefully, and copy down some of the schemes on your own, simply to have the practice of writing them down. Here are two key points to keep in mind.

1. Carboxylic acids and their derivatives contain **potential leaving groups** attached to their carbonyl carbon. After a nucleophile adds, the leaving group may leave, giving a net substitution reaction overall. This is a two-step, *addition–elimination,* mechanism. It is **not** the same as either an S_N2 or an S_N1 reaction, mechanistically: S_N1 and S_N2 reactions **do not apply to sp^2 hybridized carbons.**

2. For carboxylic acids, this new type of substitution process occurs best with weakly basic nucleophiles in the presence of an acid catalyst. Strong bases will deprotonate the acid faster than nucleophilic addition can take place. Therefore, if the nucleophile is a strong base and deprotonation is essentially irreversible, nucleophilic addition will be **very difficult** and will occur only with extraordinarily powerful reagents, such as $LiAlH_4$.

In some of these reactions you will see some unexpected species act as leaving groups, including strong bases such as HO^- and RO^-. Although these were generally far too basic to be leaving groups in ordinary S_N2 reactions (Chapter 6), they can function that way here, because the tetrahedral intermediate is relatively high in energy compared with the carbonyl product, which has a very strong, very stable carbon–oxygen double bond. So, even HO^- and RO^- can leave in the elimination step and still result in an energetically favorable process overall. You saw a similar effect in the nucleophilic ring-opening of oxacyclopropanes with bases (Chapter 9). The high energy of the strained three-membered ring ether made expulsion of an alkoxide possible. To summarize,

Energetically

1. $\ddot{Nu}^-$ + $-\overset{|}{\underset{|}{C}}-OH$ $\longrightarrow$ $Nu-\overset{|}{\underset{|}{C}}-$ + HO^- Unfavorable

2. $\ddot{Nu}^-$ + $>C-C<$ (with O bridging) $\longrightarrow$ $Nu-\overset{|}{\underset{|}{C}}-\overset{|}{\underset{|}{C}}-O^-$ Favorable (Strain is relieved)

3. $Nu-\overset{:\ddot{O}:^-}{\underset{R}{\overset{|}{C}}}-OH$ $\longrightarrow$ $Nu-\overset{:O:}{\overset{||}{C}}-R$ + HO^- Favorable (Stable $C{=}O$ bond is regenerated)

(From Nu^- + $R-\overset{O}{\overset{||}{C}}-OH$)

Note how acid catalysis can be beneficial to all these reactions: Adding a proton to each of these oxygens before it leaves should help by converting it from a strongly basic alkoxide (i.e., bad) leaving group to a neutral alcohol or water (weakly basic, good leaving group).

19-8, 19-9, and 19-10. Acyl Halides, Anhydrides, Esters, and Amides

In these sections the general mechanisms of the previous section will be applied in illustrating the transformations of carboxylic acids into their four most important derivatives. For later synthetic applications, carefully note the reagents involved in each case. To understand these reactions, pick out a few whose mechanisms are not given in detail in the text and try to write them out, step-by-step. The practice will prove to be worthwhile.

19-11. Nucleophilic Attack at a Carboxylate Group

Strongly basic nucleophiles irreversibly deprotonate carboxylic acids, forming carboxylate anions. Addition–elimination reactions on carboxylate anions are hard to do because (1) the addition is hard to do and (2) the elimination is hard to do. More specifically, (1) it is difficult to add one anion to another due to electrostatic repulsion and (2) oxide anions are extremely bad leaving groups. Nevertheless, LiAlH$_4$ is capable of addition to RCOO$^-$, and the product of addition can go on to eliminate, formally, an aluminum oxide anion leaving group. The product of an acid + LiAlH$_4$ is a primary alcohol.

19-12. α-Bromination

A useful method of α-substitution in carboxylic acids is via the α-bromo derivative, synthesized using the Hell-Volhard-Zelinsky reaction.

Solutions to Problems

27. **(a)** 2-Chloro-4-methylpentanoic acid;   **(b)** 2-ethyl-3-butenoic acid

   (c) *E*-2-bromo-3,4-dimethyl-2-pentenoic acid;

   (d) cyclopentylacetic acid;

   (e) *trans*-2-hydroxycyclohexanecarboxylic acid;

   (f) *E*-2-chlorobutenedioic acid;

   (g) 2,4-dihydroxy-6-methylbenzoic acid;

   (h) 1,2-benzenedicarboxylic acid;   **(i)** H$_2$NCH$_2$CH$_2$CH$_2$COOH

   (j)

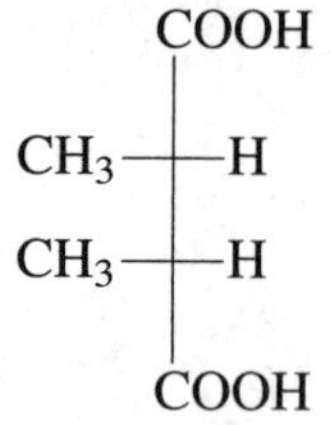

   (k) CH$_3$CCOOH (with O double-bonded above the C)

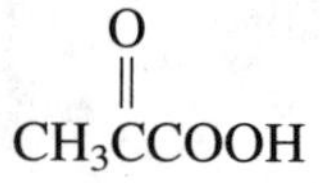

   (l)

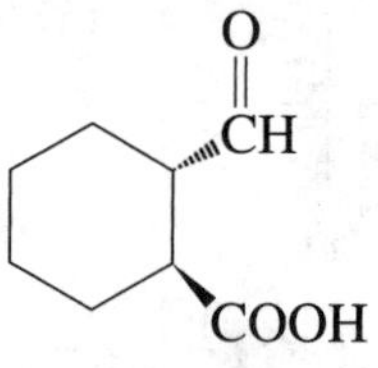

   (m)

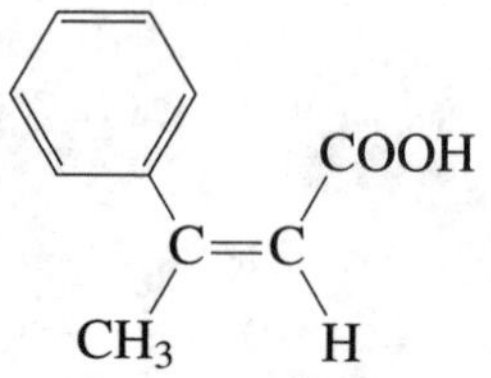

   (n)

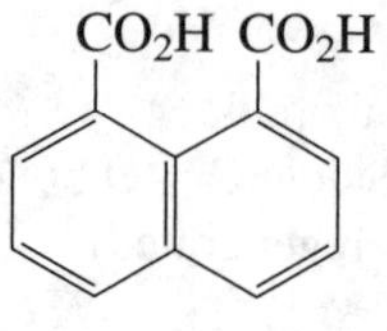

28. **(a)** 3-Chloro-4-hydroxy-2-butanone; **(b)** 2-ethenyl-3-formylbutanoic acid;
(c) 2-hydroxycyclopent-1-enecarboxylic acid or 2-hydroxy-1-cyclopentenecarboxylic acid;
(d) 3-acetyl-5-nitrobenzenecarboxylic acid or 3-acetyl-5-nitrobenzoic acid.

29.

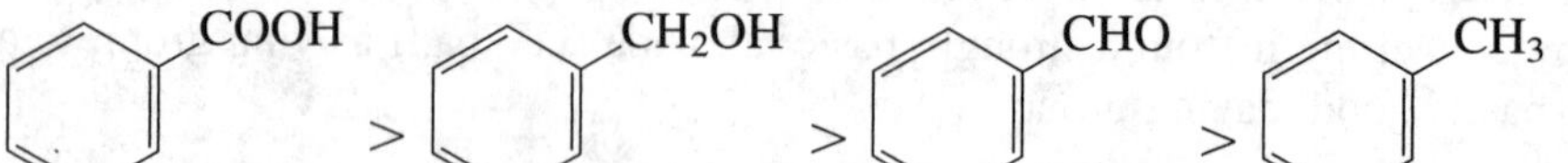

The order is the same for both boiling point and solubility in water. The acid has the most hydrogen-bonding capability and will have the highest boiling point (249°C) because of its hydrogen-bonded dimer formation. The alcohol, which can also hydrogen bond, is next (205°C), the polar aldehyde third (178°C), and the nearly nonpolar hydrocarbon last (115°C). Solubilities follow similar considerations, except that the acid and alcohol are quite similar in water solubility because they can both hydrogen bond with H$_2$O.

30. **(a)** Order is as written;   **(b)** order is reverse of that given;

   (c) CH$_3$CH$_2$CHClCO$_2$H > CH$_3$CHClCH$_2$CO$_2$H > ClCH$_2$CH$_2$CH$_2$CO$_2$H;

   (d) order is as written;   **(e)** 2,4-dinitro > 4-nitro > unsubstituted > 4-methoxybenzoic

31. Begin by scanning the given information to look for data that stand out for being particularly diagnostic. The mass information lends itself only to trial-and-error narrowing down of molecular formulas. Pass on that for starters. The IR, on the other hand, is unambiguously diagnostic for the CO_2H functional group of a carboxylic acid. The ^1H NMR confirms, with a signal near $\delta = 12$ ppm. Three other ^1H NMR signals, and four ^{13}C NMR signals total, suggest a four-carbon acid. The simplest, and a reasonable first guess, is butanoic acid, $CH_3CH_2CH_2CO_2H$. But its mass is only 88. We are missing 28 mass units, suggesting we still have work to do.

The ^1H NMR splitting patterns confirm: *two* equivalent CH_3 groups appear as a 6 H triplet near $\delta = 0.9$ ppm, which can only mean that the molecule contains two equivalent CH_3CH_2 groups. The two CH_2 groups give the 4 H signal at $\delta = 1.6$ ppm. Finally, a 1 H quintet near $\delta = 2.4$ ppm must be a CH attached to both CH_2 groups, and by process of elimination, the COOH as well, giving as our solution the compound

$$CH_3CH_2$$
$$CH{-}CO_2H$$
$$CH_3CH_2$$

A check of the mass: $C_6H_{12}O_2$ gives $72 + 12 + 32 = 116$.

32. (a) $H_{sat} = 14 + 2 = 16$; degrees of unsaturation $= \frac{(16 - 12)}{2} = 2$ π bonds or rings are present.

Compare Figure 19-3. This is a carboxylic acid ($\tilde{\nu} = 1704$ and 3040 cm^{-1}).

(b) For **B**, $H_{sat} = 12 + 2 = 14$; degree of unsaturation $= \frac{(14 - 10)}{2} = 2$ π bonds or rings. From the ^{13}C NMR (three signals), the molecule would seem to have twofold symmetry, with two sets of equivalent pairs of alkyl carbons and a pair of equivalent alkene carbons. The ^1H NMR shows two equal-area upfield signals (4 H each) and a 2 H alkene signal. So the pieces seem to be

which can be simply put together to give [cyclohexene structure], cyclohexene!

Then **C** = [cyclohexanol structure] $\delta = 69.5(^{13}C)$ via oxymercuration–demercuration;

D = [cyclohexanone structure] $\delta = 208.5(^{13}C)$ ($\tilde{\nu}_{C=O} = 1715$ cm^{-1});

E = [methylenecyclohexane structure] ($\tilde{\nu}_{C=C} = 1649$ cm^{-1} and $\tilde{\nu}_{C=CH_2} = 888$ cm^{-1}) via Wittig reaction:

$$F = \text{(cyclohexane ring with H and CH}_2\text{OH, labeled 3.4(d))} \qquad (\tilde{\nu}_{O-H} = 3328 \text{ cm}^{-1}) \text{ via hydroboration–oxidation;}$$

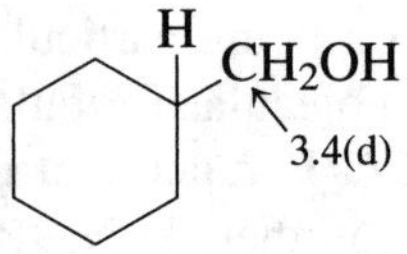

$$A = \text{(cyclohexane ring with CO}_2\text{H)}$$

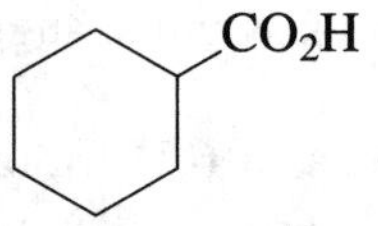

(c) For **G**, $H_{sat} = 16 + 2 = 18$; degrees of unsaturation $= \frac{(18 - 14)}{2} = 2 \pi$ bonds or rings.

IR: $\tilde{\nu} = 1742$ cm^{-1} is C=O, very possibly an ester because of the high value and the large number of oxygens in the formula.

NMR: Only three signals, with integrations of 4, 4, and 6 H. The molecule must be symmetrical,

with pieces such as 2 CH$_3$—O— ($\delta = 3.7$), 2 —CH$_2$—$\overset{\displaystyle O}{\overset{\|}{C}}$—(?) ($\delta = 2.4$), and two more equivalent —CH$_2$—'s ($\delta = 1.7$). The splitting between the upfield signals suggests that the sets of CH$_2$'s are connected, so a reasonable answer is

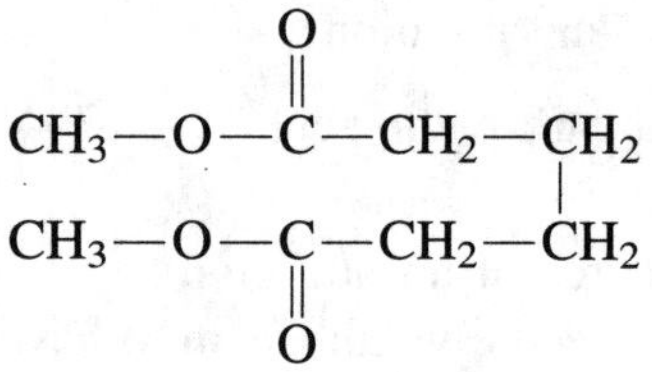

$$\begin{array}{c}\text{CH}_3-\text{O}-\overset{\displaystyle O}{\overset{\|}{\text{C}}}-\text{CH}_2-\text{CH}_2\\ \text{CH}_3-\text{O}-\underset{\displaystyle O}{\underset{\|}{\text{C}}}-\text{CH}_2-\text{CH}_2\end{array}$$

(d)

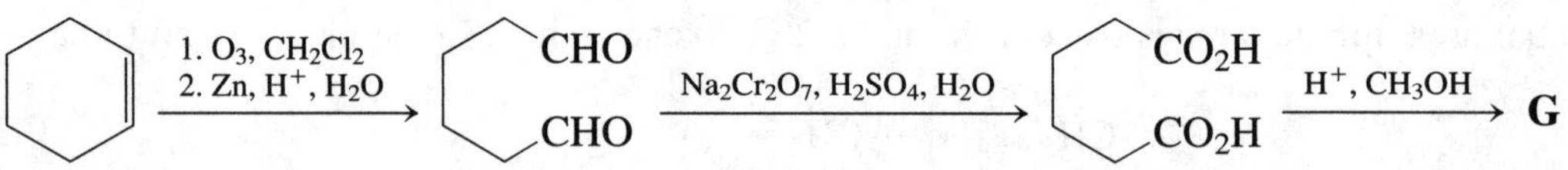

(e) How about

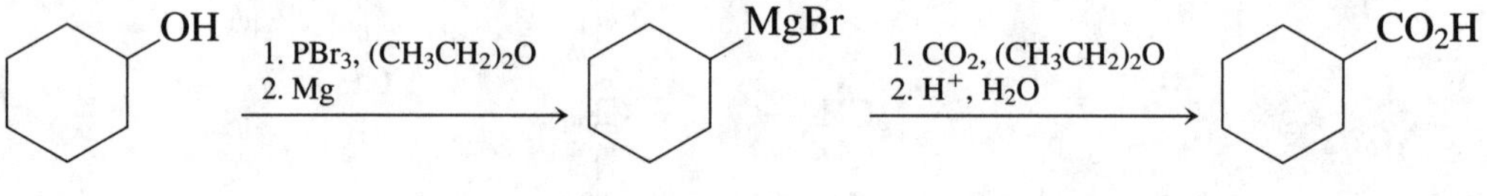

(f)

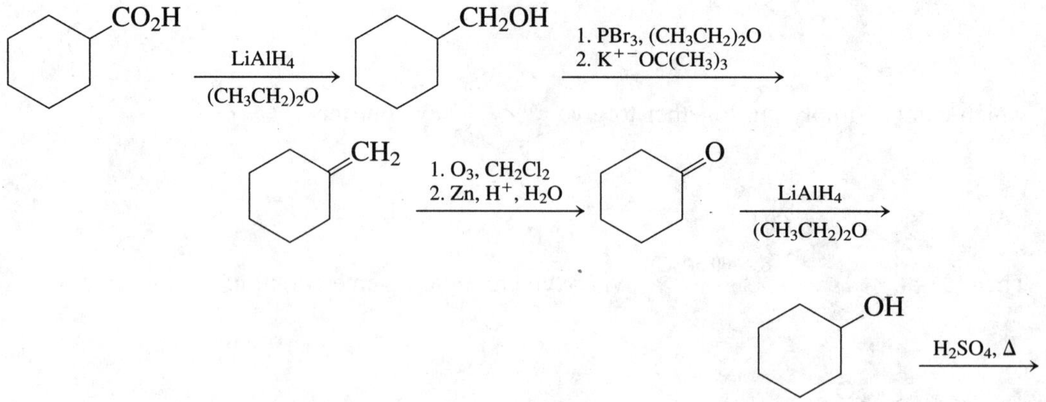

33. (a) (CH$_3$)$_2$CH$_2$CH$_2$COCl (alkanoyl chloride)

(b) (CH$_3$)$_2$CHCH$_2$$\overset{\displaystyle O}{\overset{\|}{\text{C}}}$—O—$\overset{\displaystyle O}{\overset{\|}{\text{C}}}CH_3$ (mixed anhydride)

(c) [cyclopentane ring with] CO$_2$CH$_2$CH$_3$ (ethyl ester)

(d) CH$_3$O—[benzene ring]—COO$^-$ $^+$NH$_4$ (ammonium salt)

(e) CH$_3$O—[benzene ring]—CONH$_2$ (carboxylic amide)

(f) [phthalic anhydride structure] (cyclic anhydride)

34. Recall (Section 19-8) that carboxylic acids may react with alkanoyl halides to produce anhydrides. Upon conversion of one carboxylic acid function of a 1,4- or 1,5-diacid to a halide, the other acid function may react with it intramolecularly to form the cyclic anhydride. For example:

[reaction mechanism scheme showing HO-diacid chloride forming cyclic anhydride]

Are you surprised that I used the oxygen of the carbonyl group to form the ring instead of the O of the carboxy —OH group? Which oxygen is more basic (and therefore more nucleophilic)? Check out Section 19-4 in the textbook.

35. **(a)** any of numerous oxidants such as CrO$_3$, normally in aqueous solvent; **(b)** H$^+$ or $^-$OH, H$_2$O, heat; **(c)** 1. Mg, ether, 2. CO$_2$, 3. H$^+$, H$_2$O or 1. $^-$CN, 2. H$^+$ or $^-$OH, H$_2$O, heat; **(d)** CrO$_3$, H$_2$SO$_4$, H$_2$O; **(e)** H$^+$ or $^-$OH, H$_2$O, heat.

36. **(a)** LiAlH$_4$; **(b)** hexanoyl chloride (see 'c'); **(c)** SOCl$_2$; **(d)** Br$_2$, P; **(e)** CH$_3$CH$_2$OH, H$^+$; **(f)** 1. SOCl$_2$, 2. Excess NH$_3$.

37. **(a)** Na$_2$Cr$_2$O$_7$, H$_2$O, H$_2$SO$_4$;   **(b)** 1. NaCN, H$_2$O, H$_2$SO$_4$, 2. H$^+$, H$_2$O, Δ;   **(c)** 1. Mg, (CH$_3$CH$_2$)$_2$O, 2. CO$_2$, 3. H$^+$, H$_2$O;   **(d)** 1. NaCN, DMSO, 2. KOH, H$_2$O, Δ, 3. H$^+$, H$_2$O, Δ; **(e)** 1. SOCl$_2$ (makes alkanoyl chloride), 2. Add one more mole of starting acid, Δ; **(f)** (CH$_3$)$_2$CHOH, H$^+$;   **(g)** CH$_3$COOH, Δ

38. **(a)** CH$_3$(CH$_2$)$_5$Br

Either
1. Mg, (CH$_3$CH$_2$)$_2$O
2. CO$_2$
3. H$^+$, H$_2$O

or
1. NaCN, DMF
2. KOH, H$_2$O, Δ
3. H$^+$, H$_2$O, Δ

$\longrightarrow$ product

(b) CH$_3$CH=CH$_2$ $\xrightarrow{\text{Cl}_2,\ \text{H}_2\text{O}}$ CH$_3$CHCH$_2$Cl [with OH on the CH]
(O.K. to start here, actually)

1. NaCN, DMSO
2. KOH, H$_2$O, Δ
3. H$^+$, H$_2$O, Δ

$\longrightarrow$ product

(c) $(CH_3)_3CCl$ $\xrightarrow[\text{3. } H^+, H_2O]{\substack{\text{1. Mg, }(CH_3CH_2)_2O \\ \text{2. } CO_2}}$ product

39. (a) Acid catalysis is understood for esterification. So,

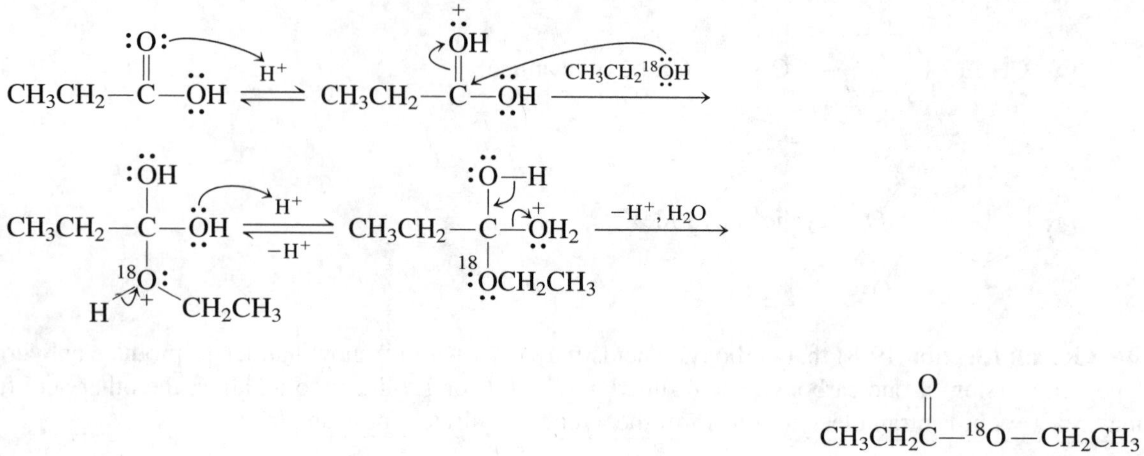

$CH_3CH_2-\overset{:O:}{\overset{\|}{C}}-\ddot{O}H$ $\xrightarrow{H^+}$ $CH_3CH_2-\overset{+\overset{..}{O}H}{\overset{\|}{C}}-\ddot{O}H$ $\xrightarrow{CH_3CH_2{}^{18}\ddot{O}H}$

$CH_3CH_2-\overset{\overset{..}{O}H}{\underset{\underset{H\overset{+}{\overset{18}{O}:}CH_2CH_3}{|}}{C}}-\ddot{O}H$ $\underset{-H^+}{\overset{H^+}{\rightleftharpoons}}$ $CH_3CH_2-\overset{\overset{..}{O}-H}{\underset{\underset{:OCH_2CH_3}{{}^{18}|}}{C}}-\overset{+}{\ddot{O}}H_2$ $\xrightarrow{-H^+, H_2O}$

$$CH_3CH_2\overset{O}{\overset{\|}{C}}-{}^{18}O-CH_2CH_3$$
This is the product.

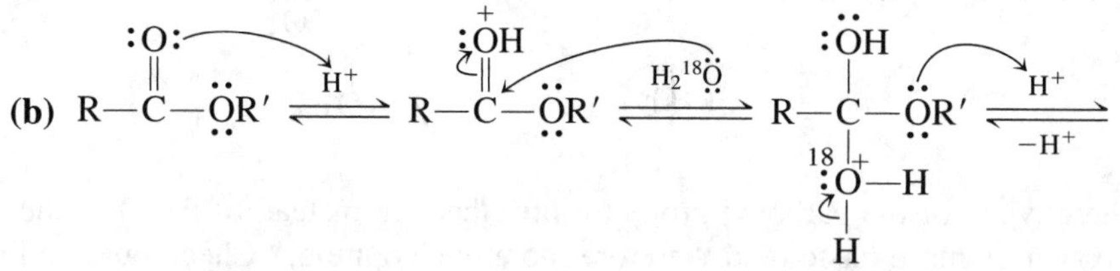

(b) $R-\overset{:O:}{\overset{\|}{C}}-\ddot{O}R'$ $\underset{}{\overset{H^+}{\rightleftharpoons}}$ $R-\overset{\overset{+}{\overset{..}{O}}H}{\overset{\|}{C}}-\ddot{O}R'$ $\underset{}{\overset{H_2{}^{18}\ddot{O}}{\rightleftharpoons}}$ $R-\overset{\overset{..}{O}H}{\underset{\underset{H}{\overset{18}{\overset{+}{\ddot{O}}-H}}}{C}}-\ddot{O}R'$ $\underset{-H^+}{\overset{H^+}{\rightleftharpoons}}$

Intermediate 2

$R-\overset{\overset{..}{O}-H}{\underset{\underset{:\ddot{O}-H}{{}^{18}|}}{C}}-\overset{+}{\overset{..}{O}}R$ $\underset{H}{}$ $\xrightarrow{-H^+, H_2O}$ $R-\overset{:O:}{\overset{\|}{C}}-{}^{18}\ddot{O}H$

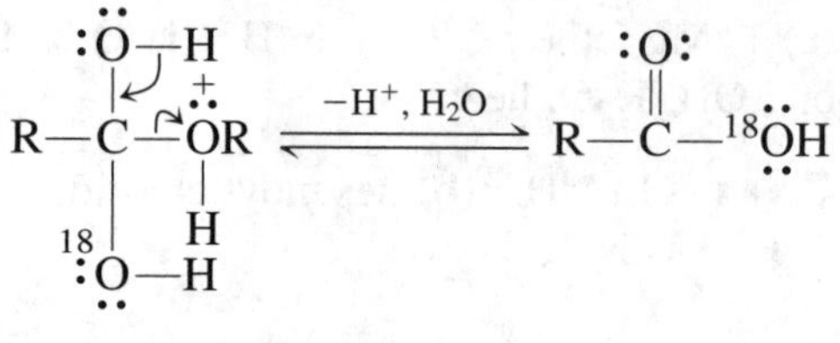

Alternatively, intermediate 2 could protonate on the unlabeled OH group instead of the OR′ group:

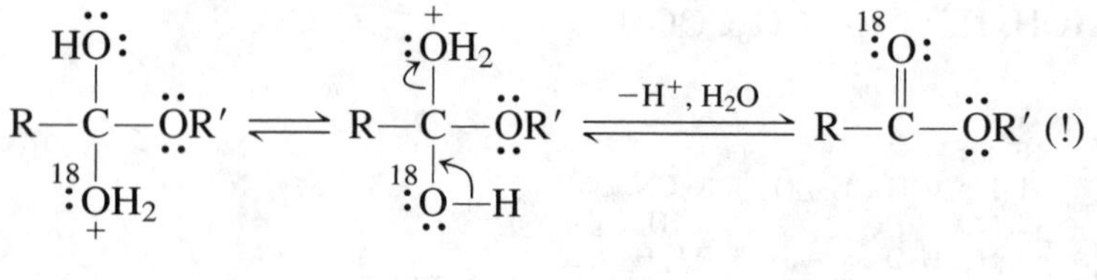

$R-\overset{\overset{..}{H\ddot{O}:}}{\underset{\underset{+}{\overset{18}{:}\ddot{O}H_2}}{C}}-\ddot{O}R'$ $\rightleftharpoons$ $R-\overset{\overset{+}{\overset{..}{O}}H_2}{\underset{\underset{:\overset{18}{O}-H}{}}{C}}-\ddot{O}R'$ $\xrightarrow{-H^+, H_2O}$ $R-\overset{\overset{18}{:}\ddot{O}:}{\overset{\|}{C}}-\ddot{O}R'$ (!)

Ester containing ^{18}O in the carbonyl oxygen. Now if you follow the hydrolysis mechanism written above, the product will be

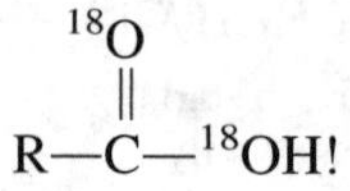

$$R-\overset{{}^{18}O}{\overset{\|}{C}}-{}^{18}OH!$$

40. (a) $CH_3CH_2\overset{\overset{\displaystyle O}{\|}}{C}Cl$; **(b)** $CH_3CH_2\overset{\overset{\displaystyle O}{\|}}{C}Br$; **(c)** $CH_3CH_2\overset{\overset{\displaystyle O}{\|}}{C}O\overset{\overset{\displaystyle O}{\|}}{C}CH_2CH_3$;

(d) $CH_3CH_2\overset{\overset{\displaystyle O}{\|}}{C}OCH(CH_3)_2$; **(e)** 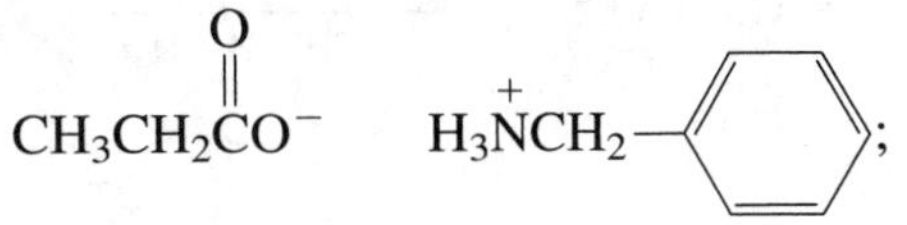 $CH_3CH_2\overset{\overset{\displaystyle O}{\|}}{C}O^-$ $H_3\overset{+}{N}CH_2-$⟨benzene ring⟩;

(f) 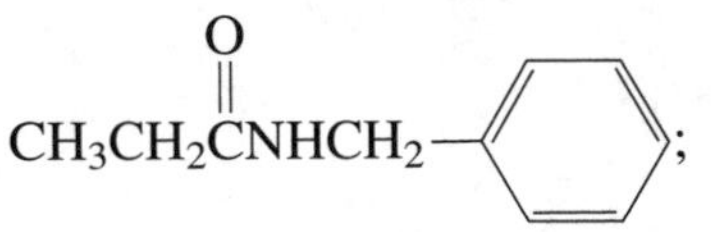 $CH_3CH_2\overset{\overset{\displaystyle O}{\|}}{C}NHCH_2-$⟨benzene ring⟩; **(g)** $CH_3CH_2CH_2OH$;

(h) $CH_3\overset{\overset{\displaystyle Br}{|}}{C}HCO_2H$

41. (a) 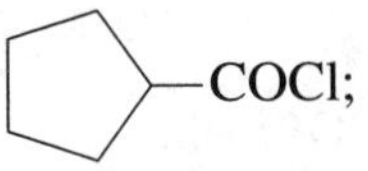 ⟨cyclopentyl⟩$-COCl$; **(b)** 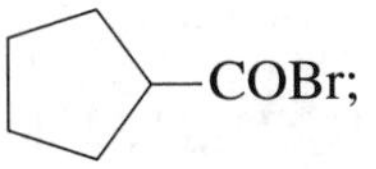 ⟨cyclopentyl⟩$-COBr$;

(c) 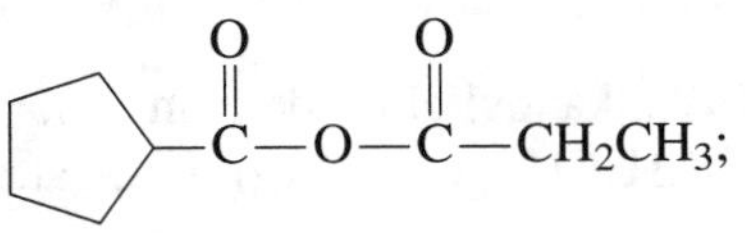 ⟨cyclopentyl⟩$-\overset{\overset{\displaystyle O}{\|}}{C}-O-\overset{\overset{\displaystyle O}{\|}}{C}-CH_2CH_3$; **(d)** 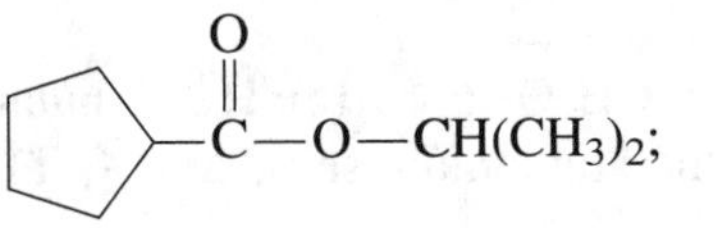 ⟨cyclopentyl⟩$-\overset{\overset{\displaystyle O}{\|}}{C}-O-CH(CH_3)_2$;

(e) 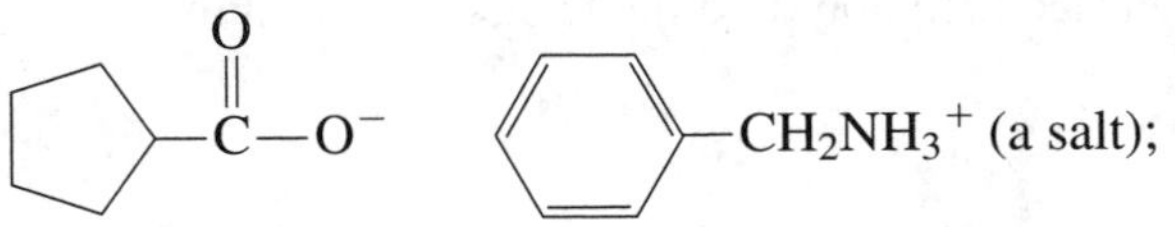 ⟨cyclopentyl⟩$-\overset{\overset{\displaystyle O}{\|}}{C}-O^-$ ⟨benzene ring⟩$-CH_2NH_3^+$ (a salt);

(f) 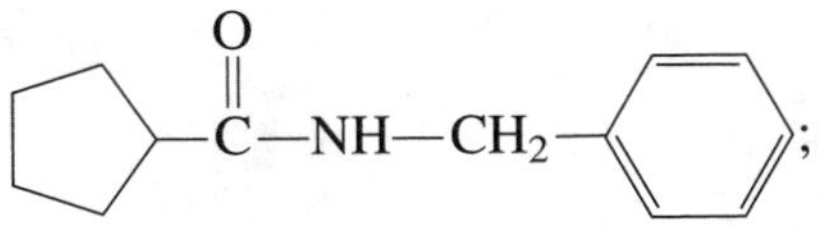 ⟨cyclopentyl⟩$-\overset{\overset{\displaystyle O}{\|}}{C}-NH-CH_2-$⟨benzene ring⟩; **(g)** 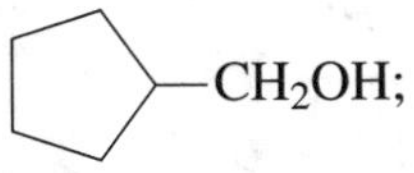 ⟨cyclopentyl⟩$-CH_2OH$;

(h)  ⟨cyclopentyl with $-CO_2H$ and $-Br$⟩

42. The only sensible way to start is by addition of the main nucleophile present (hydroxide) to the carbonyl carbon of the ketone. The resulting tetrahedral intermediate can be taken in the direction of the product by eliminating a novel leaving group, a tribromomethyl carbanion:

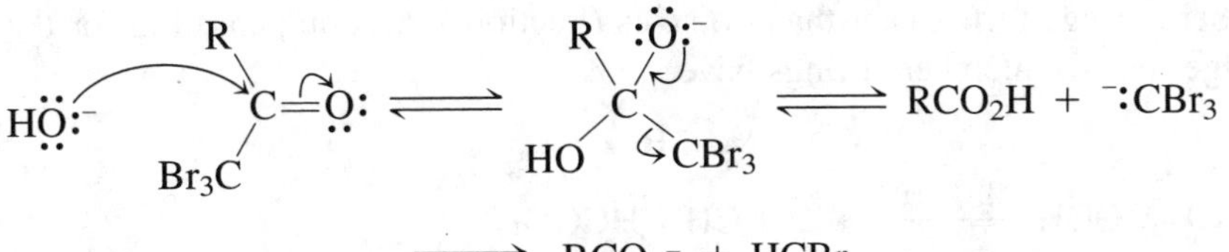

This carbanion is capable of leaving because its negative charge is stabilized by the combined inductive effects of the three halogen atoms. It is still a moderately strong base with a pK_a in the mid-teens. Proton transfer between this anion and the carboxylic acid gives the final products of the reaction, the salt of the acid and the haloform, bromoform ($CHBr_3$). The same process occurs with trichloro- and triiodo-substituted methyl ketones, to give chloroform and iodoform, respectively.

43. Here are the processes that give rise to the labeled peaks.

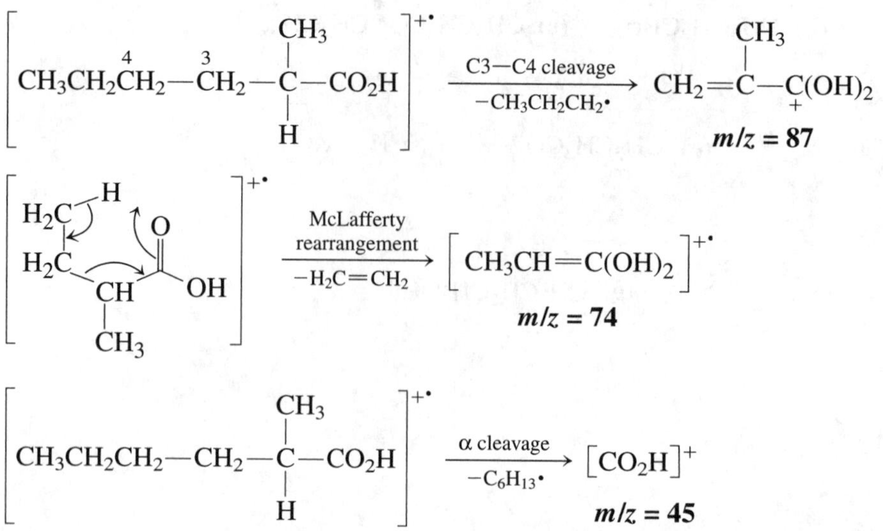

$$\left[CH_3CH_2CH_2 \overset{4}{-} \overset{3}{CH_2} -\underset{\underset{H}{|}}{\overset{\overset{CH_3}{|}}{C}} -CO_2H \right]^{+\cdot} \xrightarrow[-CH_3CH_2CH_2\cdot]{\text{C3—C4 cleavage}} CH_2{=}\underset{\overset{|}{CH_3}}{C} -\underset{+}{C(OH)_2}$$

$m/z = 87$

$$\xrightarrow[-H_2C=CH_2]{\substack{\text{McLafferty} \\ \text{rearrangement}}} \left[CH_3CH{=}C(OH)_2 \right]^{+\cdot}$$

$m/z = 74$

$$\left[CH_3CH_2CH_2 -CH_2 -\underset{\underset{H}{|}}{\overset{\overset{CH_3}{|}}{C}} -CO_2H \right]^{+\cdot} \xrightarrow[-C_6H_{13}\cdot]{\alpha\text{ cleavage}} \left[CO_2H \right]^+$$

$m/z = 45$

44. 1. LiAlH$_4$, (CH$_3$CH$_2$)$_2$O (makes 1-pentanol); 2. KBr, H$_2$SO$_4$, Δ; 3. KCN, DMSO (makes hexanenitrile); 4. KOH, H$_2$O, Δ; 5. H$^+$, H$_2$O, Δ

45. **(a)** SOCl$_2$;   **(b)** H$^+$, CH$_3$OH;   **(c)** H$^+$, 2-butanol;   **(d)** alkanoyl chloride from (a);   **(e)** CH$_3$NH$_2$ (via the ammonium salt), Δ;   **(f)** LiAlH$_4$, (CH$_3$CH$_2$)$_2$O;   **(g)** Br$_2$, cat. P

46. As usual, think mechanistically. The first intermediate should be the cyclic bromonium ion:

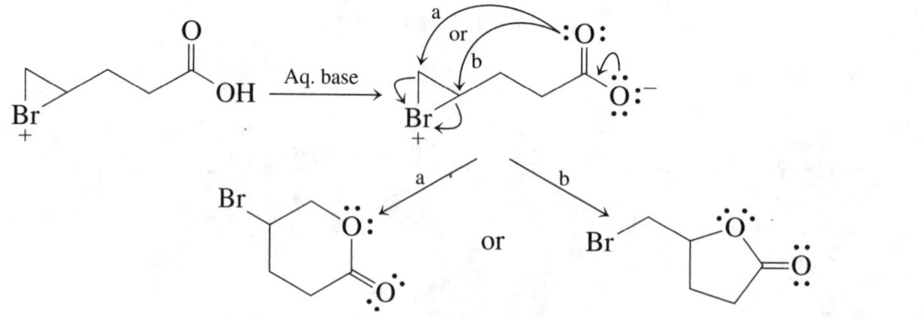

The bromolactones shown are formed by intramolecular attack by a carboxylate oxygen on one of the carbon atoms of a cyclic bromonium ion, the initial product of Br$_2$ addition to the double bond (Section 12-6). Which pathway, a or b, should be favored? The five-membered ring should form preferentially, for two reasons. First, attack by a nucleophile on a brominium ion normally takes place at the more highly substituted carbon atom (which is more positively polarized—see Section 12-6 again). Second, five-membered rings form faster than do sixes (Section 9-6), compensating for the slight thermodynamic edge that six-membered rings have.

47. **(a)** $CH_3CH_2CH_2COOH \xrightarrow{Br_2,\ cat.\ P} CH_3CH_2\underset{\overset{|}{Br}}{\overset{\overset{Br}{|}}{C}}HCOOH$

$$\left(\text{via } CH_3CH_2CH_2\overset{\overset{O}{\|}}{C}Br \rightleftharpoons CH_3CH_2CH{=}\underset{}{\overset{\overset{O-H}{|}}{C}}{-}Br + Br{-}Br \longrightarrow \right.$$

$$\left. CH_3CH_2CHBr\overset{\overset{O}{\|}}{C}Br \xrightarrow{\text{exchange}} \right).$$

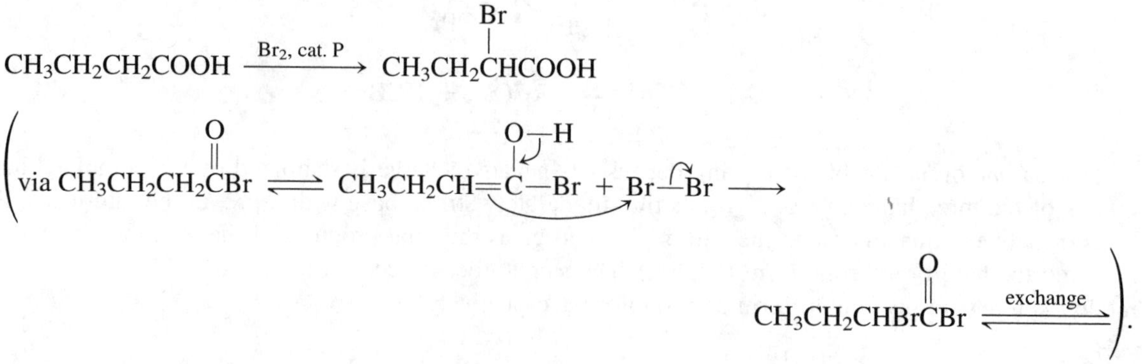

Then

$$CH_3CH_2\overset{\curvearrowleft Br}{\underset{|}{C}}HCOOH + NH_3 \xrightarrow[S_N2]{-Br^-} CH_3CH_2\overset{+NH_3}{\underset{|}{C}}HCOOH \xrightarrow{-H^+} product$$

(b)
$$\langle\!\!\!\bigcirc\!\!\!\rangle-CH_2CO_2H \xrightarrow[\text{2. KCN, DMSO}]{\text{1. Br}_2,\text{ cat. P}} \langle\!\!\!\bigcirc\!\!\!\rangle-\overset{CN}{\underset{|}{C}}HCO_2H \xrightarrow[\text{2. H}^+,\text{ H}_2O]{\text{1. HO}^-,\text{ H}_2O} product$$

(c) $CH_3CH_2CH(CH_3)CH_2CH_2COOH \xrightarrow{Br_2,\text{ cat. P}} \xrightarrow[\text{2. H}^+,\text{ H}_2O]{\text{Then 1. K}_2CO_3,\text{ H}_2O,\ \Delta} product$

(d) $CH_3COOH \xrightarrow{Br_2,\text{ cat. P}} BrCH_2COOH \xrightarrow[\text{2. I}_2]{\underset{\text{1. Excess KSH, CH}_3CH_2OH}{\text{(Section 9–10)}}} product$

(e) $BrCH_2COOH + (CH_3CH_2)_2NH \rightarrow product$

(f) $CH_3CH_2COOH \xrightarrow[\text{2. (C}_6H_5)_3P,\text{ (CH}_3CH_2)_2O]{\text{1. Br}_2,\text{ cat. P}} product$

48. No tricks here! The mechanism is almost exactly as shown in Section 19-12, with only minor differences. Because the method starts with an alkanoyl halide, step 1 is unnecessary. Step 2 (enolization) occurs as written. The third step uses Cl_2, generated in low concentration from NCS, or Br_2, formed in the same way from NBS, or I_2. The fourth step in the mechanism doesn't apply because only alkanoyl halides are present, not carboxylic acids.

49. Acidity: $CH_3\overset{O}{\overset{||}{C}}OH > CH_3\overset{O}{\overset{||}{C}}NH_2 > CH_3\overset{O}{\overset{||}{C}}CH_3$. The most acidic hydrogens in $CH_3\overset{O}{\overset{||}{C}}NH_2$ are on the nitrogen. Acidity order is determined by electronegativity. Two protonation possibilities: on N, giving

$CH_3\overset{O}{\overset{||}{C}}\overset{+}{N}H_3$, and on O, giving $\left[CH_3\overset{+\overset{\curvearrowleft}{O}H}{\overset{||}{C}}\overset{..}{N}H_2 \longleftrightarrow CH_3\overset{OH}{\underset{|}{C}}=\overset{+}{N}H_2\right]$.

Resonance stabilization causes protonation on O to be favored.

50. See Problem 40 in Chapter 17.

$$HOCH_2CH_2CH_2CH_2OH \xrightarrow{CrO_3} H\overset{..}{\underset{..}{O}}CH_2CH_2CH_2\overset{\overset{\overset{\curvearrowleft}{\overset{..}{O}:}}{||}}{C}H \rightleftharpoons \underset{\substack{\text{Hemiacetal} \\ \text{form}}}{\langle\overset{:\overset{..}{O}\quad\overset{..}{O}H}{}\rangle} \xrightarrow{CrO_3} \langle\overset{:\overset{..}{O}\quad\overset{..}{O}:}{}\rangle$$

Initially

51. (a) $CH_3CH_2\ddot{O}H$ + $Cl—\overset{\displaystyle :O:}{\underset{\displaystyle ||}{C}}—$⟨phenyl⟩ $\xrightarrow{\text{Addition}}$

$CH_3CH_2—\overset{+}{\underset{\underset{Cl}{|}}{\overset{|}{\underset{H}{O}}}}—\overset{:\ddot{O}:}{C}—$⟨phenyl⟩ $\xrightarrow{-H^+}$ $CH_3CH_2—\ddot{O}—\overset{\overset{\displaystyle -:\ddot{O}:}{|}}{\underset{Cl}{C}}—$⟨phenyl⟩ $\xrightarrow{H^+}$

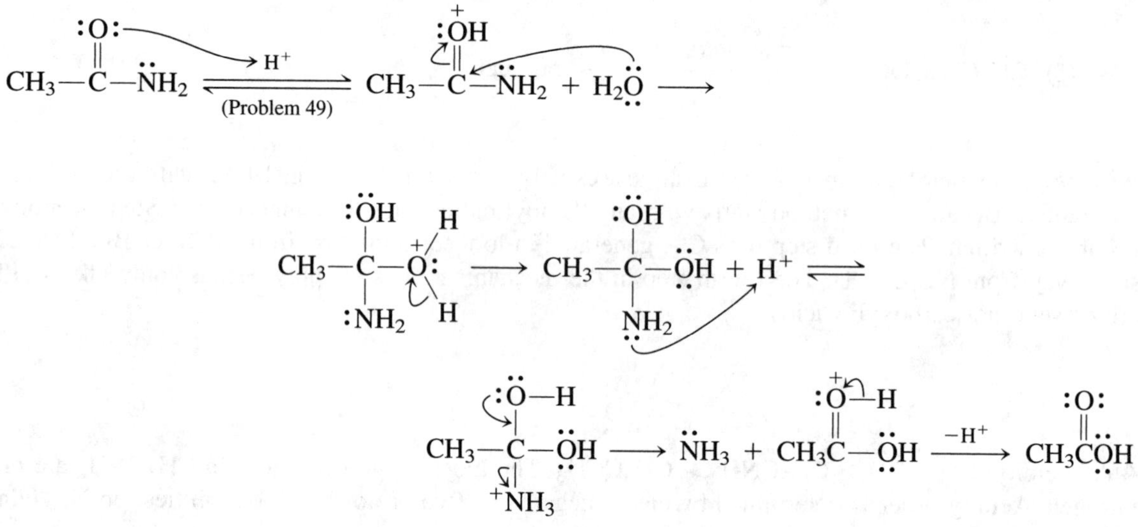

$CH_3CH_2—\ddot{O}—\overset{\overset{\displaystyle :\ddot{O}H}{|}}{\underset{Cl}{C}}—$⟨phenyl⟩ $\xrightarrow[\text{Elimination}]{-Cl^-}$

Tetrahedral
intermediate

$CH_3CH_2—\ddot{O}—\overset{\overset{\displaystyle +\ddot{O}H}{||}}{C}—$⟨phenyl⟩ $\longrightarrow$ $CH_3CH_2—\ddot{O}—\overset{\overset{\displaystyle :O:}{||}}{C}—$⟨phenyl⟩

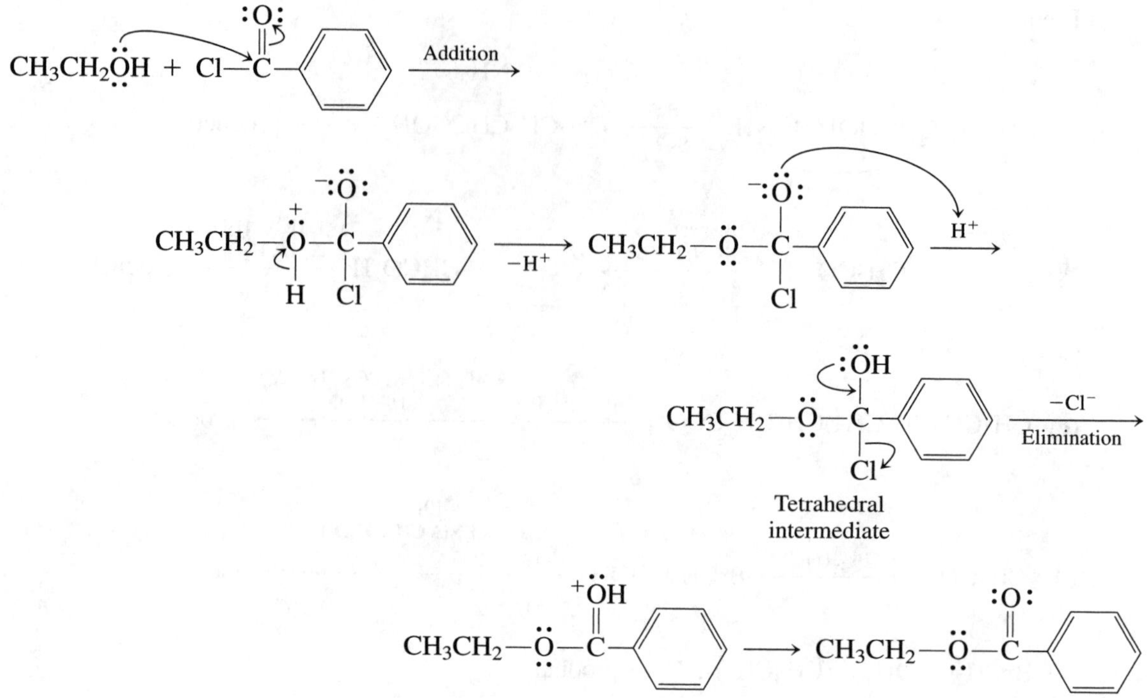

(b) $CH_3—\overset{\overset{\displaystyle :O:}{||}}{C}—\ddot{N}H_2$ $\underset{\text{(Problem 49)}}{\overset{H^+}{\rightleftharpoons}}$ $CH_3—\overset{\overset{\displaystyle +\ddot{O}H}{||}}{C}—\ddot{N}H_2$ + $H_2\ddot{O}$ $\longrightarrow$

$CH_3—\overset{\overset{\displaystyle :\ddot{O}H}{|}}{\underset{:NH_2}{C}}—\overset{+}{\underset{H}{\overset{H}{\ddot{O}}}}:$ $\longrightarrow$ $CH_3—\overset{\overset{\displaystyle :\ddot{O}H}{|}}{\underset{NH_2}{C}}—\ddot{O}H$ + H^+ $\rightleftharpoons$

$CH_3—\overset{\overset{\displaystyle :\ddot{O}—H}{|}}{\underset{+NH_3}{C}}—\ddot{O}H$ $\longrightarrow$ $\ddot{N}H_3$ + $CH_3\overset{\overset{\displaystyle :O:}{||}}{C}—\ddot{O}H$ $\xrightarrow{-H^+}$ $CH_3\overset{\overset{\displaystyle :O:}{||}}{C}\ddot{O}H$

52. H = ⟨bicyclic structure with H, =O, and CH₂CH=C(CH₃)₂ substituent⟩

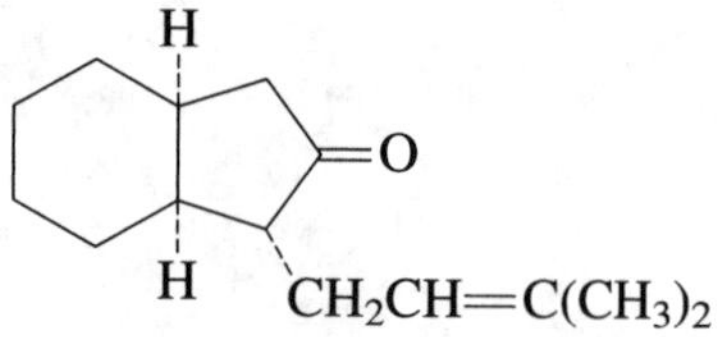

(Alkylation on less
hindered bottom face)

I = ⟨bicyclic structure with dioxolane and CH₂CH=C(CH₃)₂⟩

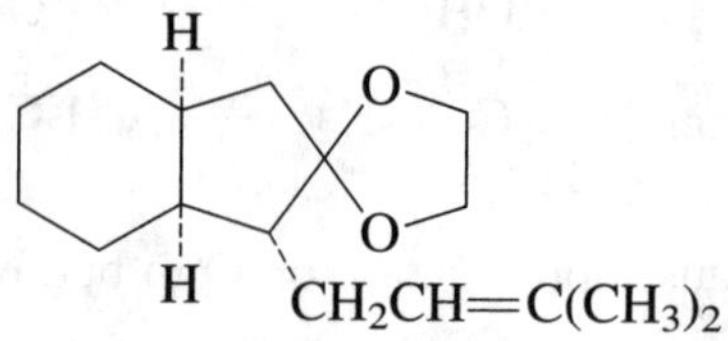

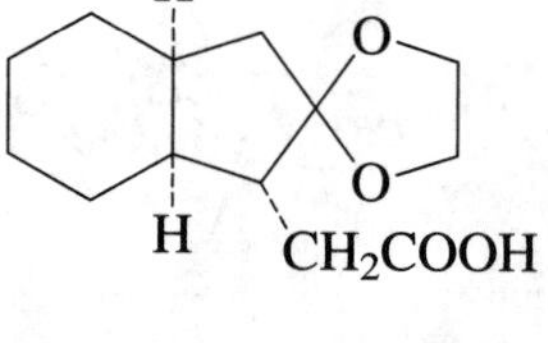

J = ⟨bicyclic structure with dioxolane and CH₂COOH⟩

K = ⟨bicyclic structure with OH and CH₂COOH⟩ (via ketone);

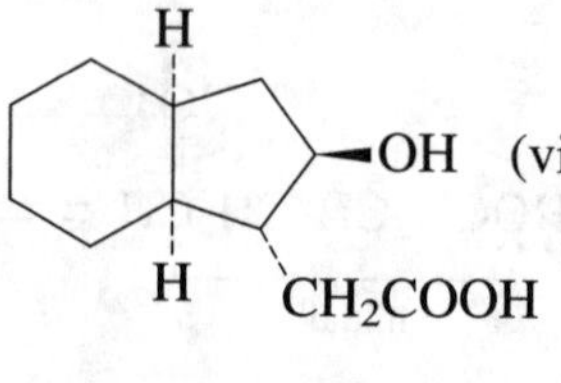

L = A lactone ($\tilde{\nu}_{C=O} = 1770$ cm^{-1})

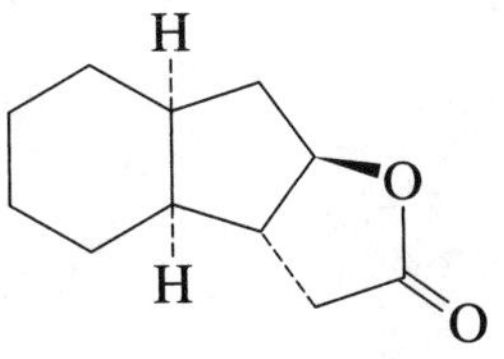

53. **(a)** Haloalkanes are generally not very soluble in water (too much polarity difference). The reaction mixture is heterogeneous, which prevents good mixing of reactants. Water also forms hydrogen bonds to the nucleophile, which doesn't help.

(b) Acetic acid is a better solvent for the haloalkane, so the system is homogeneous, allowing for better mixing of reacting molecules.

(c) Sodium dodecanoate is a soap, and it dissolves in water to make *micelles*. The less polar interior regions of the micelles form good solvents for molecules of low polarity like haloalkanes. The iodobutane dissolves **in the micelles** and is therefore in close proximity to the nucleophilic carboxylate groups, allowing the S$_N$2 reaction to proceed.

54. Work forward and backward from the given structure. The first reaction looks like an aldol condensation.

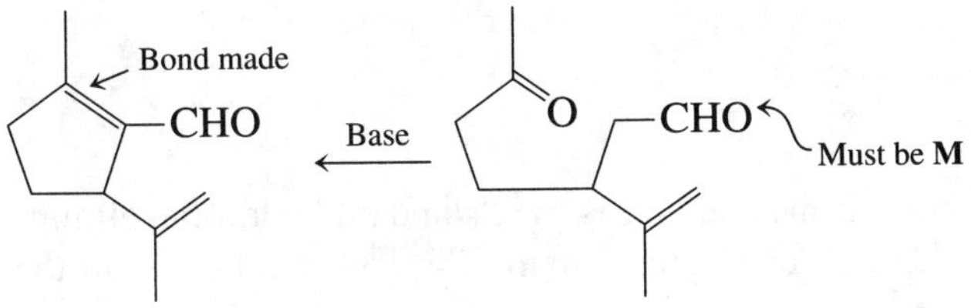

Then **N** 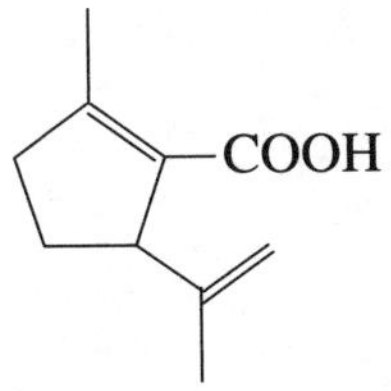 = **O** =

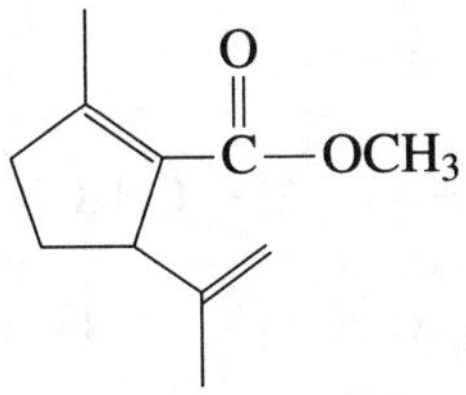

P 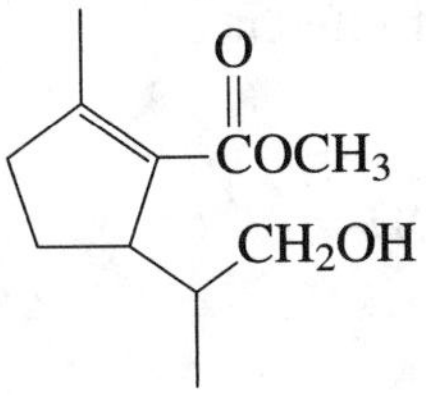 = (Note that only the least hindered alkene is hydroborated.)

Neonepetalactone =

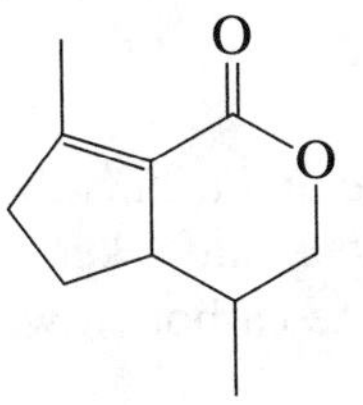

55.

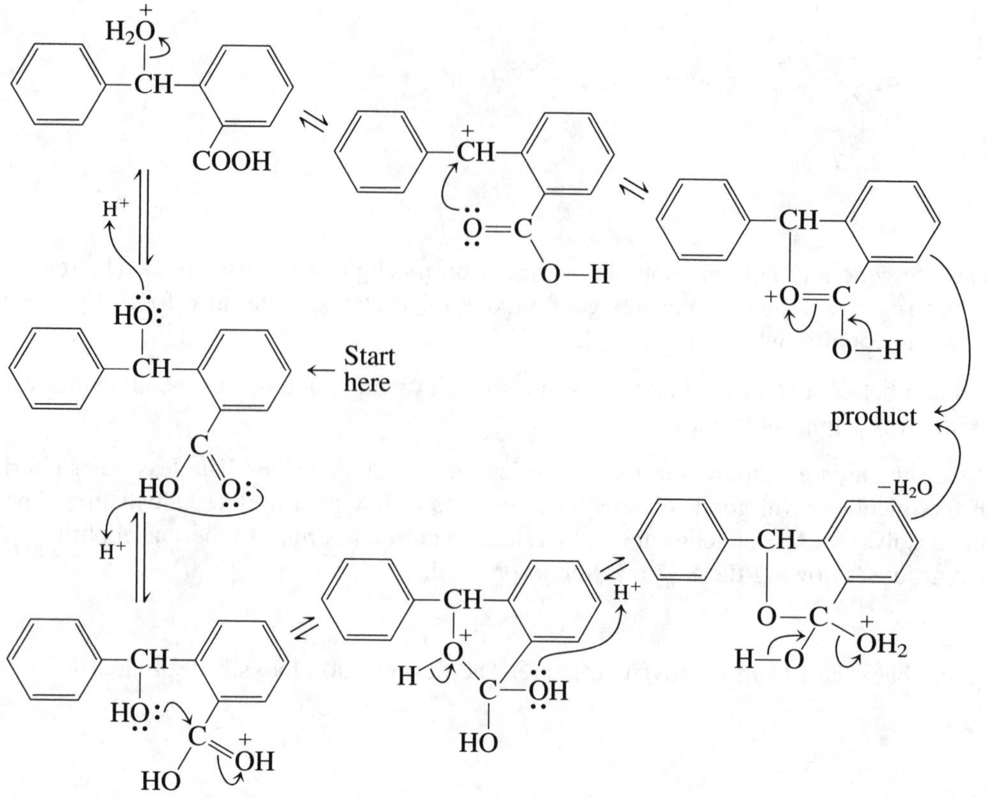

The top mechanism is an S_N1 on the alcohol; the bottom mechanism is the "standard" addition–elimination on the carboxyl group. Labeling the **alcohol** with ^{18}O will distinguish them. The ^{18}O will be lost in the top mechanism, but it will be retained in the bottom one.

56. $CH_3C\equiv CH$ $\xrightarrow{\begin{array}{l}1.\ CH_3CH_2CH_2CH_2Li,\ THF,\ hexane,\ -30°C\\ 2.\ CO_2,\ 0°C\\ 3.\ H^+,\ H_2O\end{array}}$ $CH_3C\equiv CCOOH$
Propyne
$\quad\quad\quad\quad\quad\quad\quad\quad\quad\quad\quad\quad\quad\quad\quad\quad\quad\quad\quad\quad$ 98%
$\quad\quad\quad\quad\quad\quad\quad\quad\quad\quad\quad\quad\quad\quad\quad\quad\quad\quad\quad$ **2-Butynoic acid**

57. Start by identifying the carbon atoms in the product that correspond to those in the starting material. Each has a methyl group (*a*) and a carboxy carbon (*b*), so use these as points of reference.

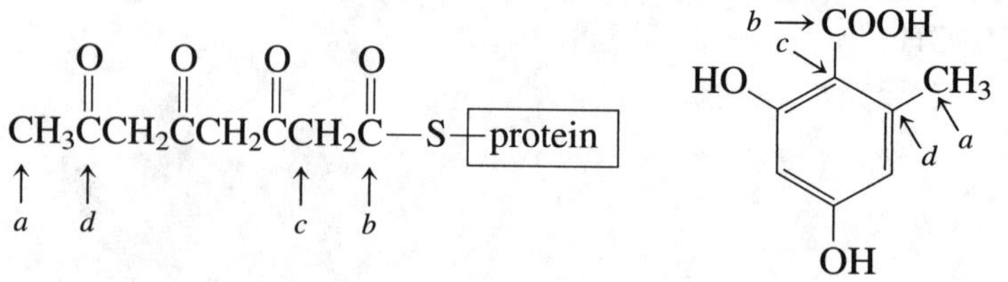

In the structure of the product (above right), note that the carbons *c* and *d* must be the ones that became connected in a ring-forming reaction. They correspond to a methylene and a ketone in the starting material. Because the methylene is an α-carbon with respect to another ketone carbonyl, we know that an aldol condensation can form the bond we need.

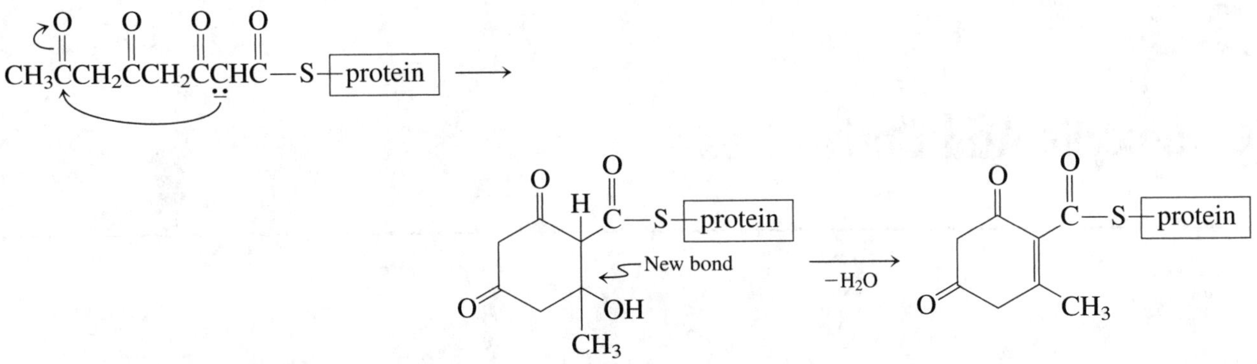

All that remains to generate the benzene ring is to enolize both ketones. This step is favored by aromatic stabilization. Finally, hydrolysis of thioester gives the acid.

58. The first part of the problem is just a slight extension of the question in Problem 46: Stereochemistry has been added by extending the chain of the carboxylic acid from five to six carbons. Again, the kinetic advantage in forming a five- as opposed to a six-membered ring is illustrated. The stereochemical outcome results from the carboxylate oxygen attacking the intermediate bromonium ion in a backside manner (did you make a model?!). The second reaction is a double esterification. First, the hydroxy group of one molecule esterifies the carboxy group of another. Then the intermediate (pictured in the middle) undergoes lactonization. Both processes follow the acid-catalyzed mechanism depicted in text Section 19-9.

59. (a)

60. (d)

61. (c)

20

Carboxylic Acid Derivatives

Derivatives of carboxylic acids share two properties in common with carboxylic acids themselves: (1) They have a carbonyl group that is susceptible to *nucleophilic addition*, and (2) they have a potential *leaving group* attached to the carbonyl carbon. In addition, they may possess *acidic hydrogens* α to the carbonyl group. However, they **lack** an acidic —OH group on the carbonyl carbon itself. Therefore, it is the α-hydrogens in most carboxylic acid derivatives that are the most readily removed by strong bases.

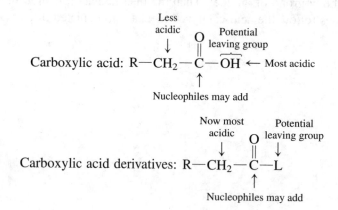

L = halide (acyl halide)
$L = O\overset{\overset{O}{\|}}{C}R'$ (anhydride)
$L = OR'$ (ester)
$L = NR_2'$ (amide)

The mechanisms associated with reactions of these compounds have already been introduced (Section 19-7), so this chapter mainly presents additional examples of reactions that follow familiar patterns. The most important new points involve (1) the relative reactivities of these derivatives and (2) how to convert each kind of derivative into any of the others, for synthetic purposes.

Outline of the Chapter

20-1 Relative Reactivities, Structures, and Spectra of Carboxylic Acid Derivatives

Comparisons of physical properties.

20-2 Acyl Halides

   Details of the chemistry of the four major carboxylic acid derivatives, plus nitriles, which are relatives of amides.

Keys to the Chapter

20-1. Relative Reactivities, Structures, and Spectra of Carboxylic Acid Derivatives

In this section, differences in the physical properties of the four main carboxylic acid derivatives are highlighted. The principal factor determining these differences is the electronegativity of the atom attached to the carbonyl carbon. These differences are reflected in a consistent way throughout all the reactions of these compounds: deprotonation of the α-carbon, addition to the carbonyl carbon, and protonation of the carbonyl oxygen. Make sure you understand the conceptual arguments here before moving ahead to the specific sections on each derivative. Then, refer back to this text section from time to time as you go through the chapter to see how the reaction details presented in the upcoming sections reflect the general principles outlined here.

20-2. Acyl Halides

Acyl halides are very useful in synthesis because (1) they are easily made from carboxylic acids (Section 19-8) and (2) they are easily turned into every major type of carbonyl compound, including aldehydes, ketones, and all the other carboxylic acid derivatives. Of all the carboxylic acid derivatives, acyl halides possess the most reactive carbonyl group toward nucleophilic addition and the best leaving group (halide).

20-3. Carboxylic Anhydrides

Carboxylic acid anhydrides are used in synthesis less often than are acyl halides. They are generally not as easily prepared, and they are less reactive. A major drawback to using an anhydride in synthesis is that only one of its carbonyl groups can add a nucleophile; the other falls off as part of the carboxylate ion leaving group (see "Nucleophilic Addition–Elimination of Anhydrides"). Cyclic anhydrides are more useful because the two ends still stay stuck together by a carbon chain after an addition takes place (see "Nucleophilic Ring Opening of Cyclic Anhydrides").

20-4 and 20-5. Esters

Esters are the most prevalent carboxylic acid derivatives in nature. They also are convenient compounds for synthetic purposes because their intermediate reactivity makes them easy to prepare and store for later use. Acyl halides, in contrast, are so reactive toward water that it takes some care to prevent them from hydrolyzing to some extent upon extended storage. Esters are easily prepared and readily converted into many types of compounds by reactions at both their carbonyl and α-carbons (Chapter 23).

20-6 and 20-7. Amides

Amides are much less reactive than esters toward nucleophilic acyl substitution. They have greater resonance stabilization by the nitrogen lone pair, and they have a poorer leaving group (NH_3 in acid, and NH_2^- in base). Amides do show some special reactions: the possibility of deprotonation of the nitrogen in 1° or 2° amides, reduction reactions that may form either amines or aldehydes, and a new process, the Hofmann rearrangement, which forms an amine with one carbon less than the original amide. Consider the following two amine syntheses.

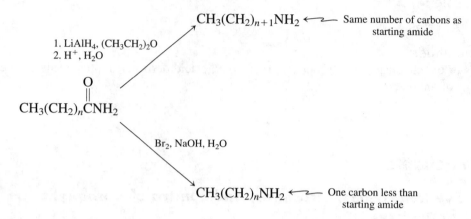

This Hofmann rearrangement has a mechanism that is related conceptually to some you've seen before but has some very new and unusual details. It merits a careful look, especially the final steps.

20-8. Alkanenitriles

The nitrile functional group is, **formally,** a dehydrated primary amide. In fact, a little-used nitrile synthesis involves amide dehydration:

$$R-\overset{\overset{\textstyle O}{\|}}{C}-NH_2 \xrightarrow[(-H_2O)]{P_2O_5} R-C\equiv N$$

In this sense, therefore, it is related to carboxylic acids and their derivatives. The importance of nitriles in synthesis derives from the fact that the nitrile carbon may be introduced into a molecule as the **nucleophilic** cyanide ion in an S_N2 reaction. Then, by the reactions in this text section, the nitrile may be converted to carboxylic acid derivatives, ketones, or aldehydes, in which the originally nucleophilic cyanide carbon has become an **electrophilic** carbonyl carbon. This is a good way to use a nucleophilic substitution to introduce a carbonyl carbon into a molecule.

$$R-CH_2-X \xrightarrow[S_N2]{^-CN} R-CH_2-CN \xrightarrow{\text{Various reactions}} \begin{cases} R-CH_2-CONH_2 \\ R-CH_2-CO_2H \\ R-CH_2-CO_2R' \\ R-CH_2-CHO \\ R-CH_2-COR' \end{cases}$$

Solutions to Problems

30. **(a)** 3-Methylbutanoyl iodide;   **(b)** 1-methylcyclopentanecarbonyl chloride;

(c) 2,2,2-trifluoroacetic anhydride;   **(d)** benzoic propanoic anhydride;

(e) ethyl 2,2-dimethylpropanoate;   **(f)** N-phenylacetamide

(g) $CH_3CH_2CH_2\overset{\overset{\textstyle O}{\|}}{C}OCH_2CH_2CH_3$ **(h)** $CH_3CH_2\overset{\overset{\textstyle O}{\|}}{C}OCH_2CH_2CH_2CH_3$

(i) [structure: benzene ring]—C(=O)OCH$_2$CH$_2$Cl **(j)** [structure: benzene ring]—C(=O)N(CH$_3$)$_2$

(k) CH$_3$CH$_2$CH$_2$CH$_2$CH(CH$_3$)CN **(l)** [cyclopentane structure]—CN

31. **(a)** Methyl 2-chloro-3-butenoate or methyl 2-chlorobut-3-enoate; **(b)** ethyl 1-hydroxycyclobutanecarboxylate; **(c)** *N*-methyl-2-oxobutanamide; **(d)** 5-formyl-1,3-benzenedicarboxamide.

32. The guiding principle is that the strength of an acid is related to the ability of its conjugate base to accommodate the negative charge that remains after loss of a proton (Section 2-2). We are given (Section 20-1) that the order of acidity of α-hydrogens in carboxylic acid derivatives increases in the series carboxamide (weakest acid) < ester < anhydride < alkanoyl halide (strongest acid). **(a)** We know (Section 18-1) that resonance delocalization of the negative charge of an enolate from the α-carbon into the carbonyl group is partly responsible for the acidity of the α-hydrogen. Resonance donation of an electron pair from

L in RC(=O)L to the carbonyl group essentially competes with possible donation from the deprotonated α-carbon. The net effect is destabilization of the anionic enolate. Therefore, as the resonance donation ability of L increases, the acidity of the α-hydrogen decreases. **(b)** Inductively, as the electronegativity of L increases, stabilization of the negative charge of the enolate function should increase accordingly, as is observed.

33. **(a)** Acetyl chloride (Cl is bigger than F, and the bonds to it are longer);

(b) CH$_2$(COCH$_3$)$_2$ (H's α to ketones are more acidic than H's α to esters);

(c) imide (The lone pair on N is shared in resonance by two carbonyl groups, so it does not reduce their electrophilicity as much as the N in an amide. Note that the relationship between an imide and an amide is similar to that between an anhydride and an ester.)

(d) Ethenyl acetate (CH$_3$C(=O)—Ö—CH=CH$_2$ ⟷ CH$_3$C(=O)—Ö$^+$=CH—CH$_2^-$

resonance **reduces** electron donation from oxygen toward the carbonyl carbon. So the resonance form CH$_3$C(O$^-$)=O$^+$—CH=CH$_2$ is reduced in importance, strengthening the C—O double bond and raising the carbonyl stretching frequency, actually to about 1760 cm^{-1}.)

34. **(a)** [decalin structure]—NHC(=O)CH$_3$ + [decalin structure]—N$^+$H$_3$Cl$^-$

(b) CH$_3$(CH$_2$)$_4$C(=O)—[benzene ring] **(c)** (CH$_3$)$_3$CC(=O)H

(d) 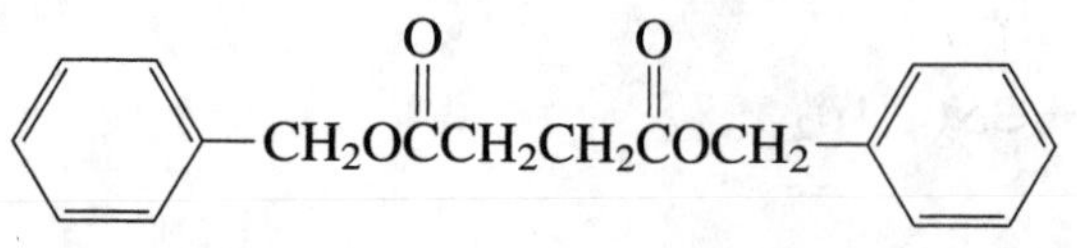 **(e)** 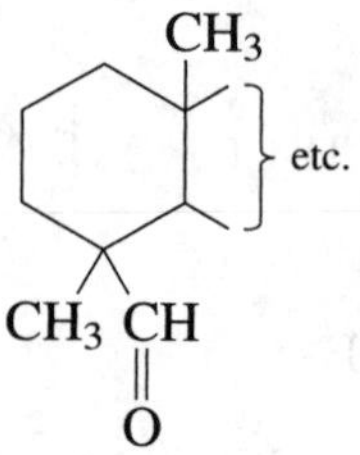

35. 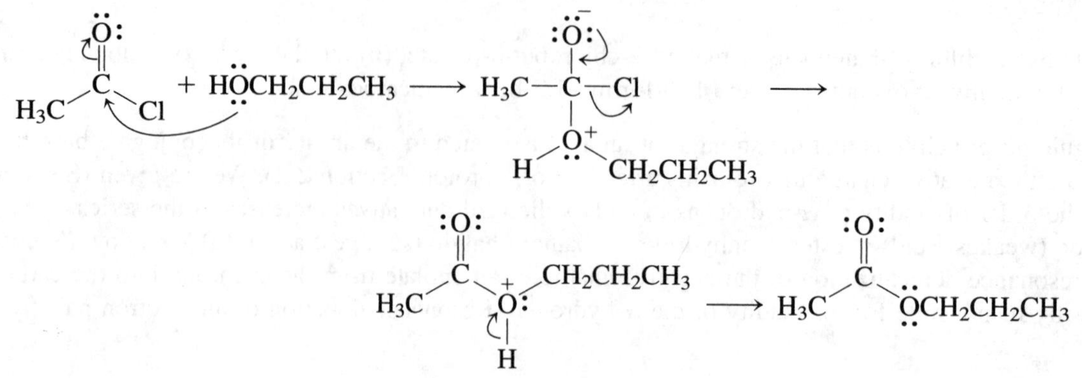

36. Assume aqueous acid work-up for both this and the next problem.

(a) $CH_3COCH(CH_3)_2 + CH_3COH$ **(b)** $CH_3CNH_2 + CH_3COH$

(c) 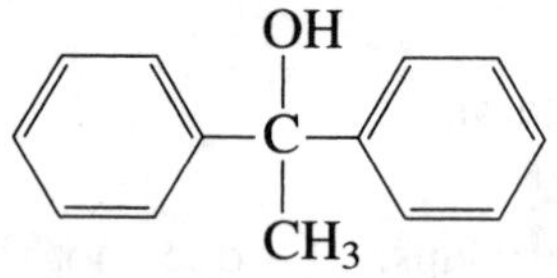 $+ CH_3COH$ **(d)** $2\ CH_3CH_2OH$

37. **(a)** $(CH_3)_2CHOCCH_2CH_2COH$ **(b)** $HOCCH_2CH_2CNH_2$

(c) 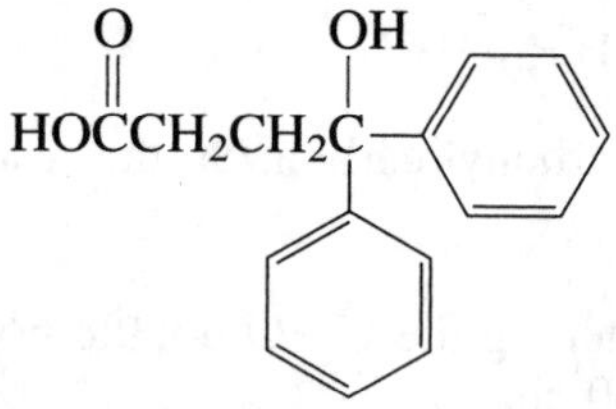 **(d)** $HOCH_2CH_2CH_2CH_2OH$

38. 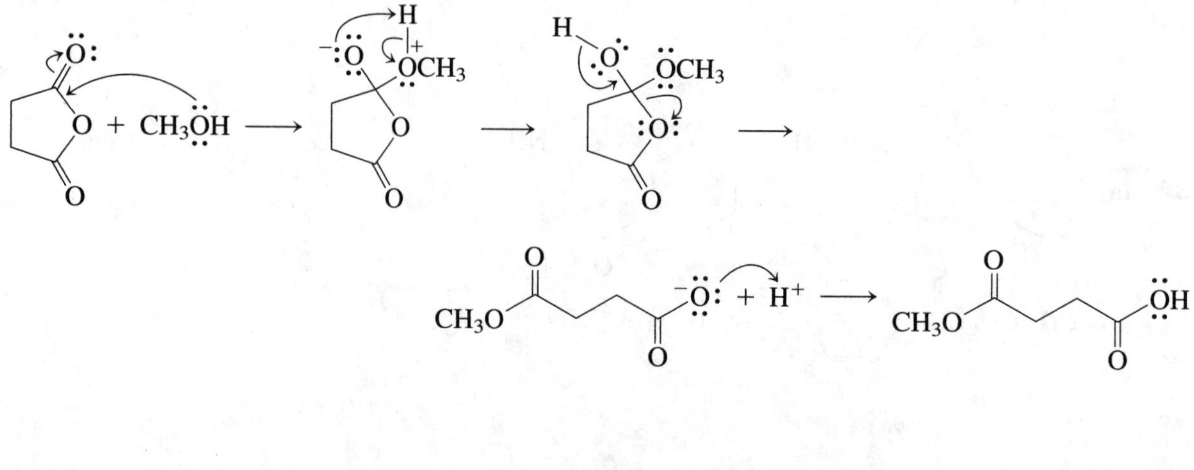

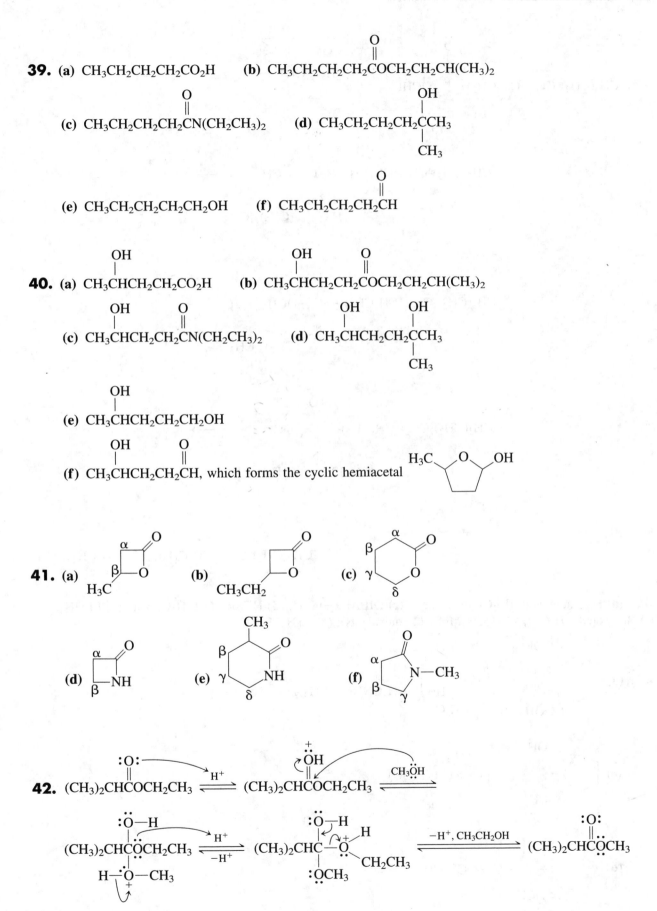

43. 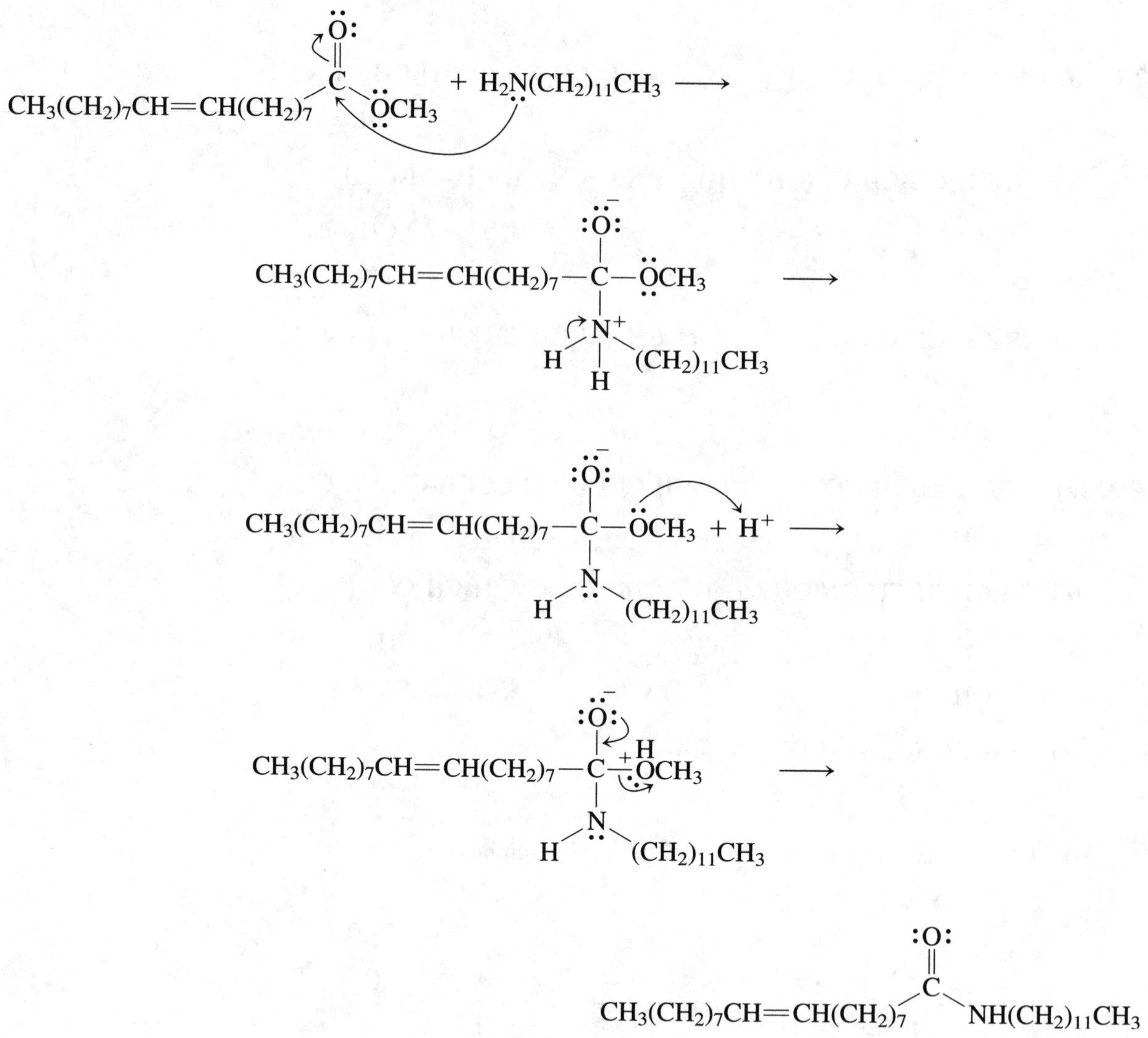 $CH_3(CH_2)_7CH=CH(CH_2)_7$ —C—$\overset{..}{\underset{..}{O}}CH_3$ $\quad + H_2N(CH_2)_{11}CH_3 \longrightarrow$

$CH_3(CH_2)_7CH=CH(CH_2)_7$ —C—$\overset{..}{\underset{..}{O}}CH_3 \quad \longrightarrow$

$\qquad\qquad\qquad\overset{+}{N}$ H $(CH_2)_{11}CH_3$
$\qquad\qquad\qquad\quad$ H

$CH_3(CH_2)_7CH=CH(CH_2)_7$ —C—$\overset{..}{\underset{..}{O}}CH_3 \quad + H^+ \longrightarrow$

$\qquad\qquad\qquad N$
$\qquad\qquad\quad$ H $\quad (CH_2)_{11}CH_3$

$CH_3(CH_2)_7CH=CH(CH_2)_7$ —C—$\overset{+}{\underset{..}{O}}CH_3 \quad \longrightarrow$

$\qquad\qquad\qquad N$
$\qquad\qquad\quad$ H $\quad (CH_2)_{11}CH_3$

$CH_3(CH_2)_7CH=CH(CH_2)_7$ —C—$NH(CH_2)_{11}CH_3$

44. (a) Hexanoic acid; (b) CH_3NH_2, heat; (c) DIBAL, $-60°C$; (d) H^+ or ^-OH, H_2O, heat; (e) $LiAlH_4$; (f) Br_2, NaOH, H_2O; (g) CH_3CH_2MgBr (2 equivalents); (h) C_6H_5MgBr.

45. (a)

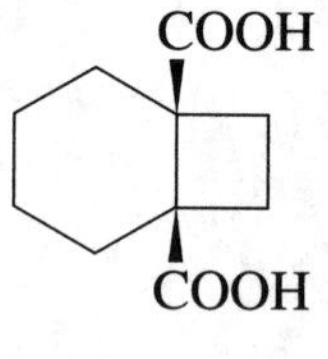

(b)

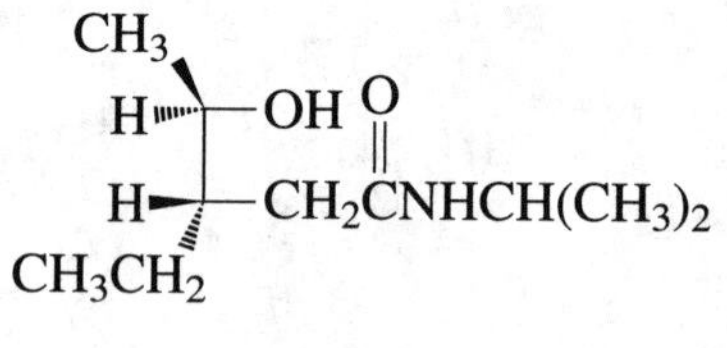

(c) 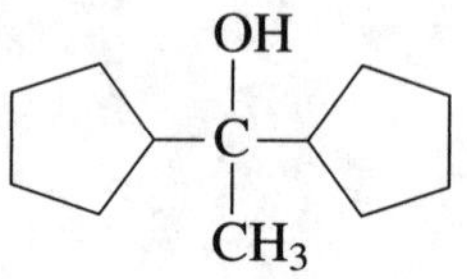 (After H^+, H_2O)

(d)

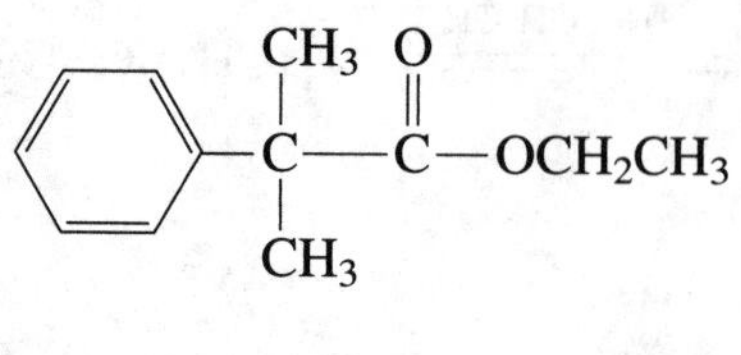

(e)

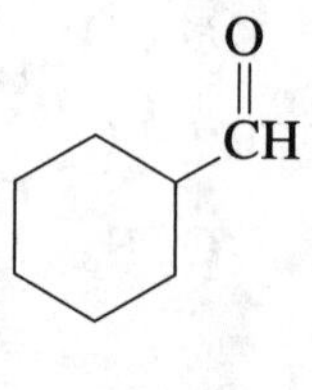

46. Each of the first three answers assumes subsequent acidification to convert the carboxylate salt into the neutral acid.

(a)

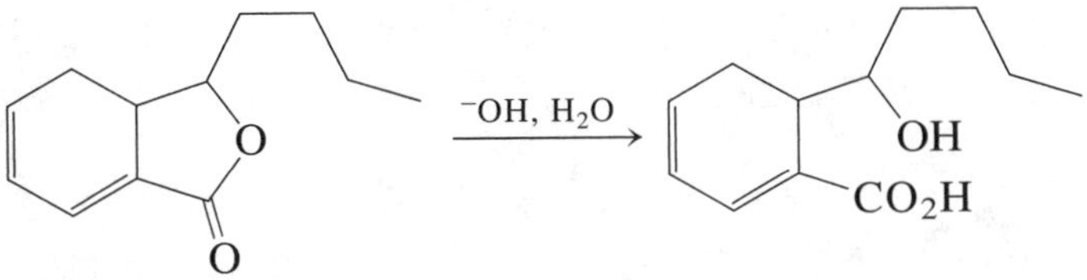
O
O
⁻OH, H₂O
OH
CO₂H

(b)

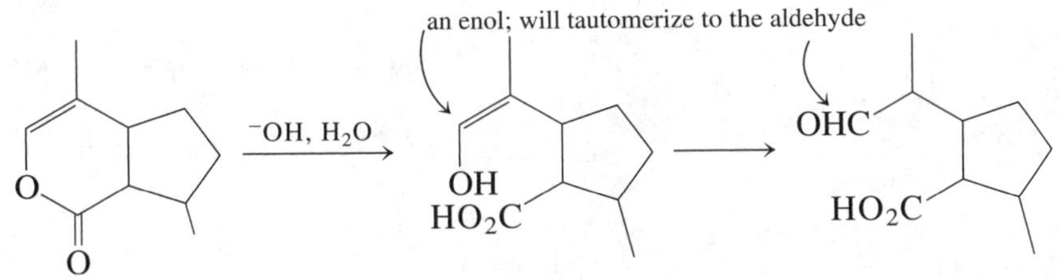
an enol; will tautomerize to the aldehyde
O
O
⁻OH, H₂O
OH
HO₂C
OHC
HO₂C

(c)

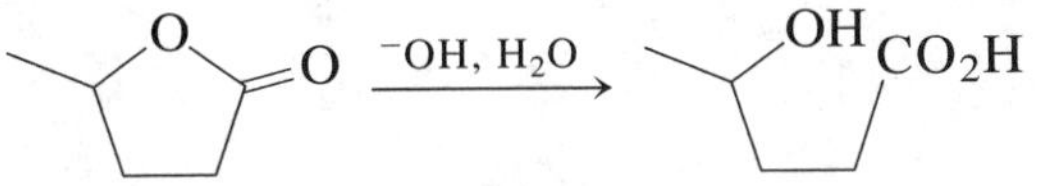
O
O
⁻OH, H₂O
OH
CO₂H

(d)

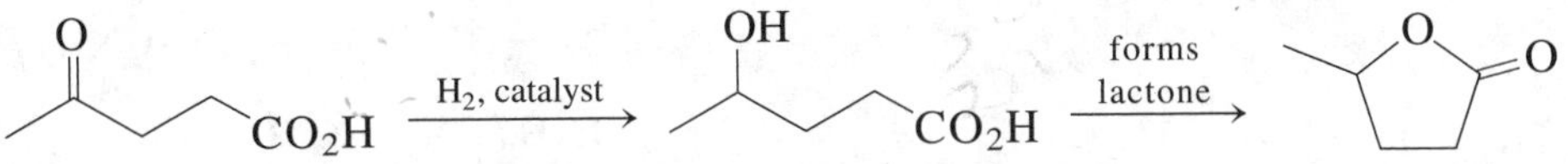
O
CO₂H
H₂, catalyst
OH
CO₂H
forms
lactone
O
O

47. Amides may be made directly from the acids, or indirectly via alkanoyl halides, anhydrides (less efficiently, because only one of the two alkanoyl groups is converted into an amide), or esters. Thus we have the following options:

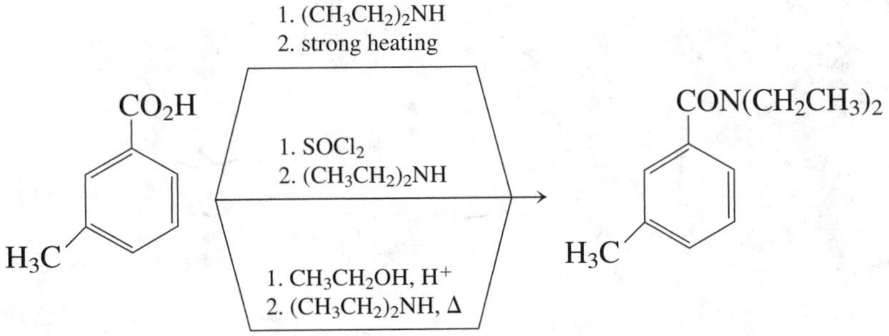
1. (CH₃CH₂)₂NH
2. strong heating
CO₂H
1. SOCl₂
2. (CH₃CH₂)₂NH
H₃C
1. CH₃CH₂OH, H⁺
2. (CH₃CH₂)₂NH, Δ
CON(CH₂CH₃)₂
H₃C

Method 1: Direct reaction between an amine and an acid first gives a salt; strong heating converts the salt to the amide with loss of water. Method 2: Converts to alkanoyl chloride; reaction with amine gives amide with loss of HCl. Often done in the presence of excess amine to consume the HCl. Method 3: Esterification with any inexpensive alcohol (ethanol chosen here), then reaction with the amine under mild heating.

48. Dimethyl sulfate is a highly reactive S_N2 substrate with respect to attack by a nucleophile on a methyl group (similar to CH_3I in reactivity). The process displaces the methyl sulfate anion, a good leaving group. The starting cyclic amide (a lactam) possesses two possible nucleophilic atoms, nitrogen and oxygen.

Reaction of the oxygen with an electrophile gives a resonance-stabilized intermediate; reaction of the nitrogen does not. Therefore, we can proceed as follows:

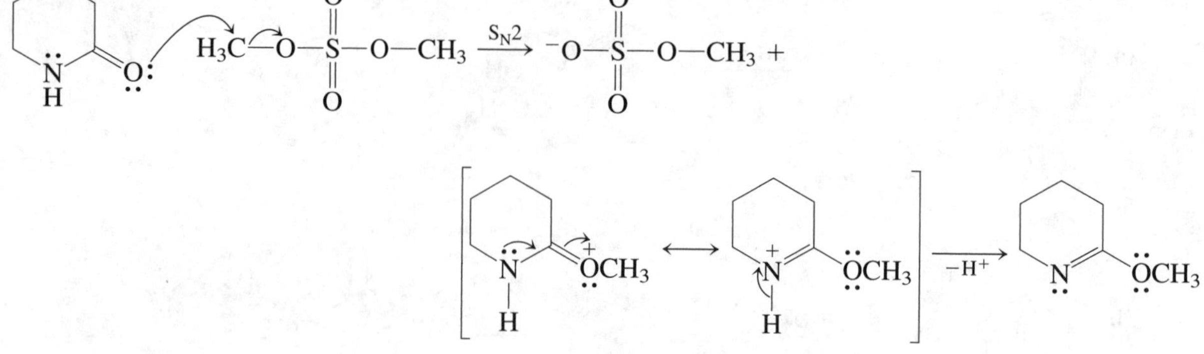

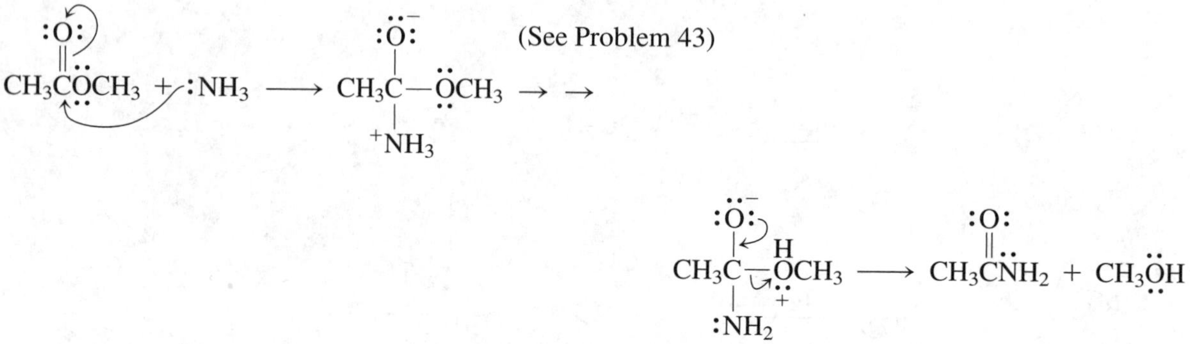

(See Problem 43)

49. $CH_3COCH_3 + :NH_3 \longrightarrow CH_3C-OCH_3 \rightarrow \rightarrow$

$CH_3C-OCH_3 \longrightarrow CH_3CNH_2 + CH_3OH$

50. Pentanamide gives (a) pentanoic acid, (e) pentanamine, and (f) pentanal; N,N-dimethylpentanamide gives the same products for (a) and (f), whereas $LiAlH_4$ reduction in this case gives N,N-dimethylpentanamine, $CH_3CH_2CH_2CH_2CH_2N(CH_3)_2$.

51.

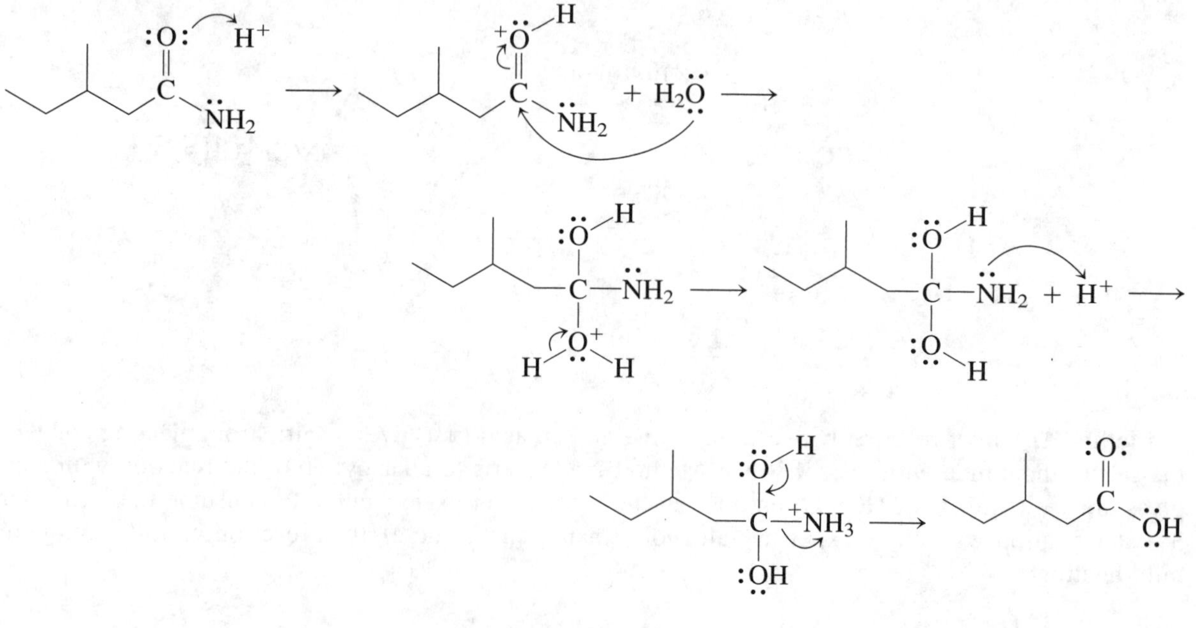

52. **(a)** $(CH_3CH_2CH_2CH_2)_2CuLi$; then H^+, H_2O; **(b), (d)** $LiAlH_4$, $(CH_3CH_2)_2O$; then H^+, H_2O;
(c) $LiAl[OC(CH_3)_3]H$; then H^+, H_2O; **(e), (f)** $CH_3CH_2CH_2MgBr$, $(CH_3CH_2)_2O$; then H^+, H_2O

53. (a)

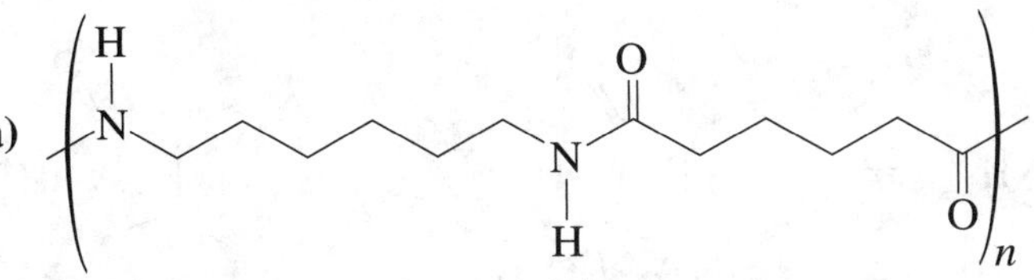

(b) 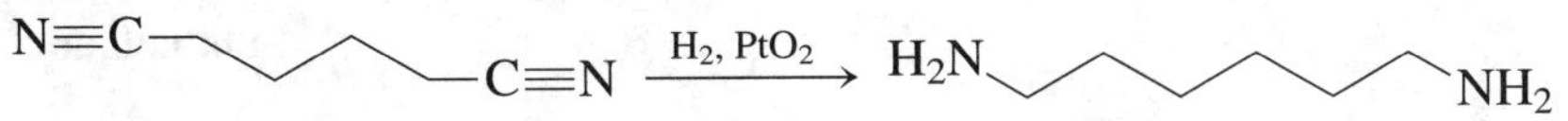 N≡C—⟨⟩—C≡N $\xrightarrow{\text{H}_2,\ \text{PtO}_2}$ H₂N—⟨⟩—NH₂

N≡C—⟨⟩—C≡N $\xrightarrow{\text{H}^+,\ \text{H}_2\text{O}}$ HO—⟨⟩—OH

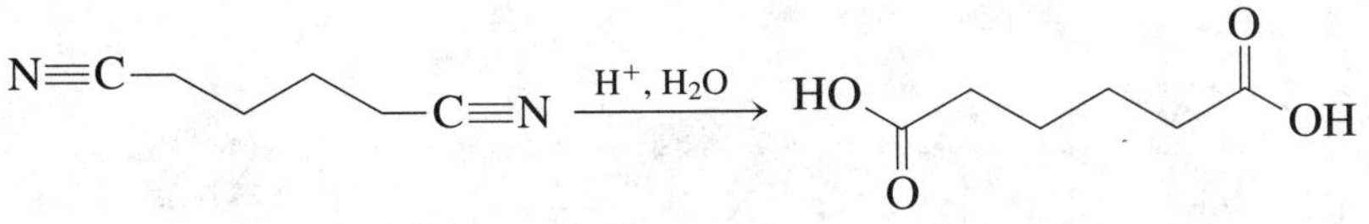

(c)

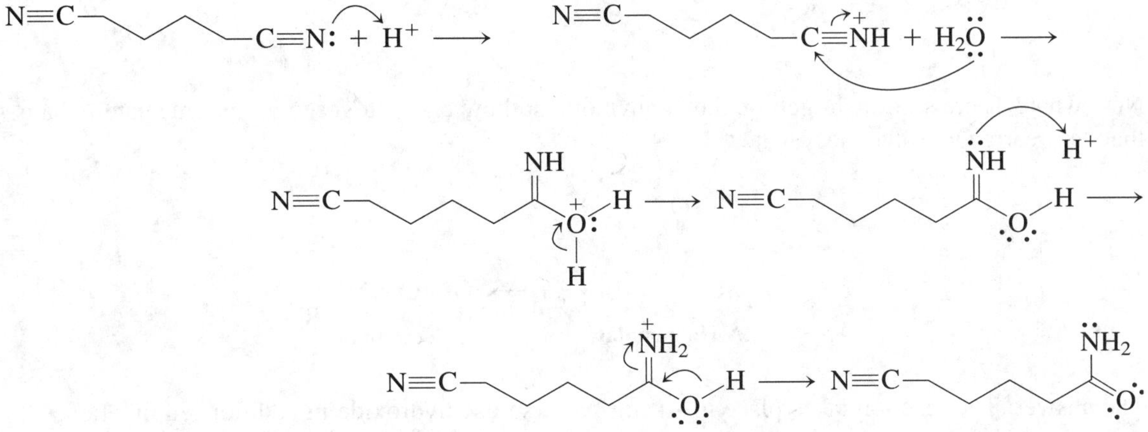

Then continue as in Problem 51 and repeat at the other nitrile group.

54. In A and B, the most acidic hydrogens are α-hydrogens. Deprotonation followed by protonation provides a mechanism for isomerization.

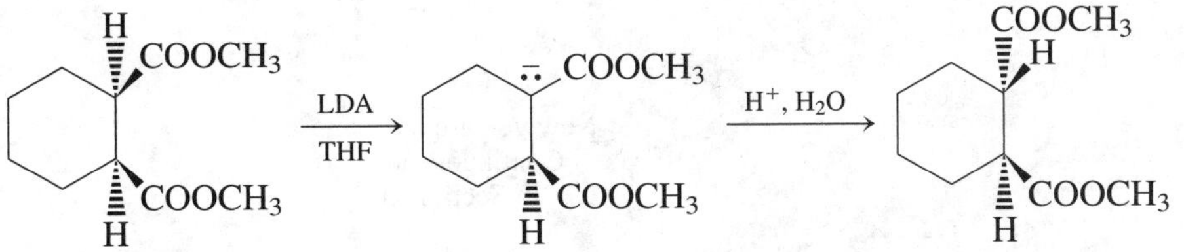

In C the most acidic hydrogens are on nitrogen. The α-hydrogens are not removed, so no isomerization is observed.

55. The mechanisms are related to amide formation and Hofmann rearrangement, respectively. In the formation of phthalimide, only those proton transfers that occur along the pathway to the product are shown.

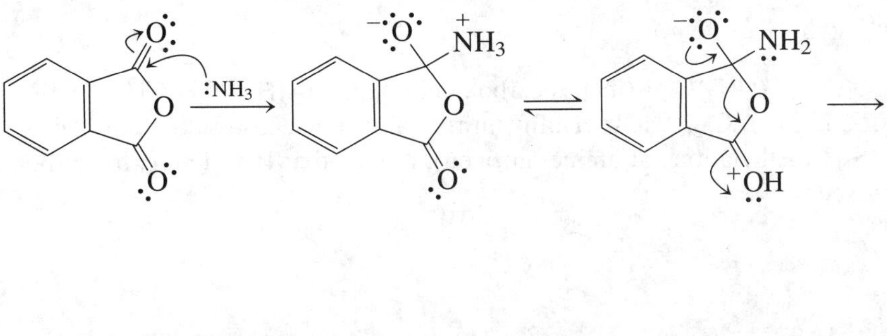

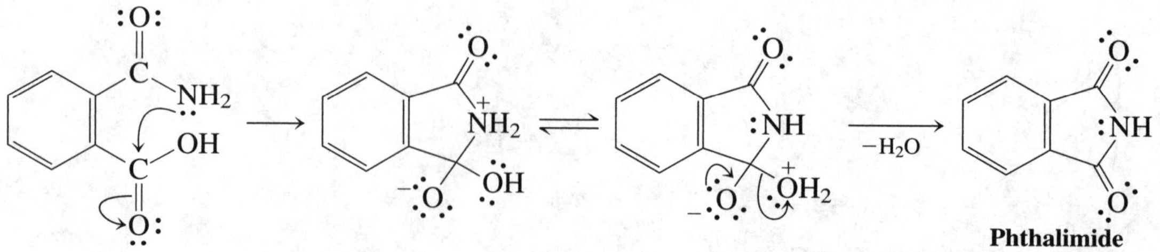

Then

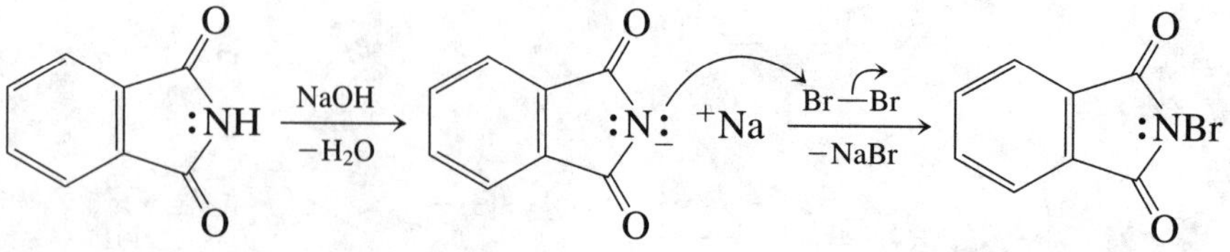

Now what? There's no hydrogen on the N anymore, so how can you get to an intermediate *N*-haloamidate that can rearrange to the isocyanate?

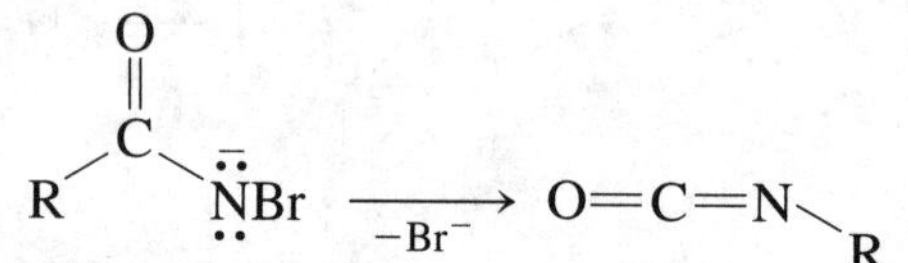

N-Haloamidate   **Isocyanate**

The answer: The reaction takes place in **strong base,** so use hydroxide in addition–elimination.

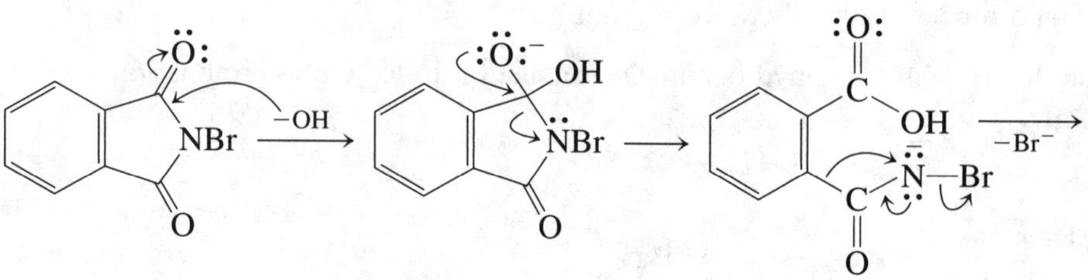

Now you are in business.
Continue exactly as
in Section 20-7.

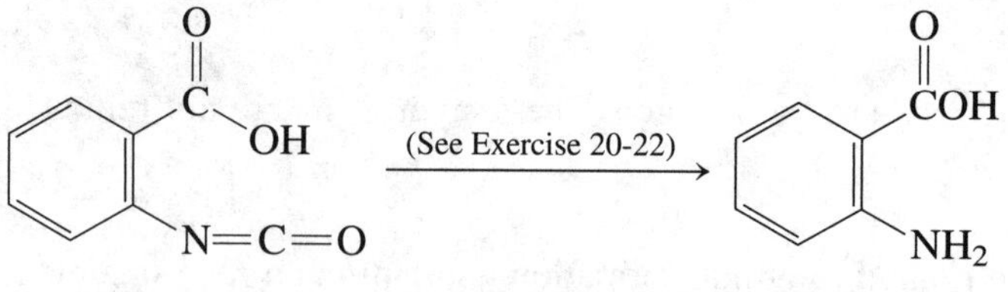

(See Exercise 20-22)

56. Self-explanatory.

57. From A: 1. SOCl$_2$, 2. (CH$_3$)$_2$NH (makes carboxamide), 3. LiAlH$_4$, (CH$_3$CH$_2$)$_2$O, then H$^+$, H$_2$O.
From B: 1. SOCl$_2$, 2. NH$_3$, 3. Cl$_2$, NaOH (Hofmann rearrangement, loses CO$_2$ and makes simple amine),
4. 2 CH$_3$I, NaOH (S$_N$2 methylations of amine nitrogen). Notice that B had an extra carbon that the Hofmann rearrangement removed.

58. The dipolar resonance forms that weaken the carbonyl C—O bond and reduce its stretching frequency are less and less favorable in small rings because of the increased strain associated with a second sp^2 atom.

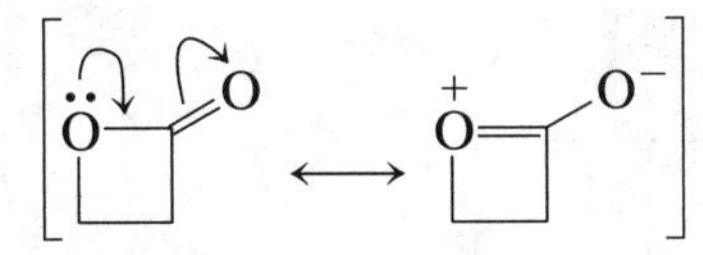

Very strained,
relatively unimportant
resonance form

59. (a) Bad idea. Residual hexanoyl chloride will be converted to hexanoic acid, which smells like a herd of goats on a hot day.

(b) Wash out the glassware with an alcohol like ethanol. Reaction with hexanoyl chloride produces the ester ethyl hexanoate, which smells like fresh fruit. Much better.

60. React with H^+ and CH_3OH! The methyl ester group at the upper right will remain unchanged because the only nucleophile around for attacking carbonyl groups is methanol. However, the acetate function at the lower right will exchange to give methyl acetate, and the steroid alcohol group will be displaced.

$$CH_3\overset{O}{\overset{\|}{C}}-O-\text{steroid} \xrightarrow{H^+, CH_3OH} CH_3\overset{O}{\overset{\|}{C}}-O-CH_3 + HO-\text{steroid}$$

61. Baeyer-Villiger reaction, followed by ester hydrolysis.

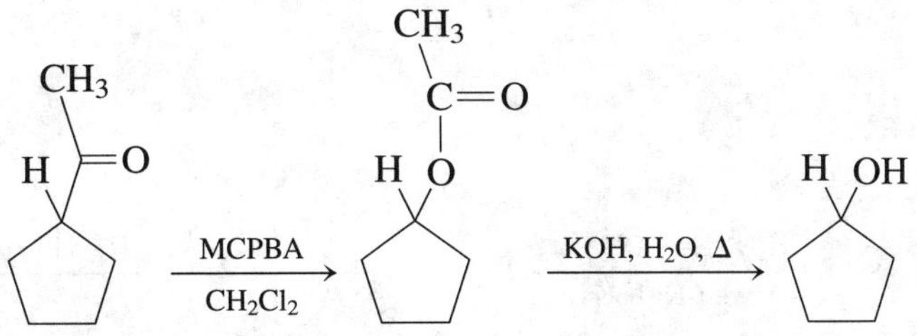

62. Several ways will work. One that uses a reaction from this chapter (the actual route used) is shown below.

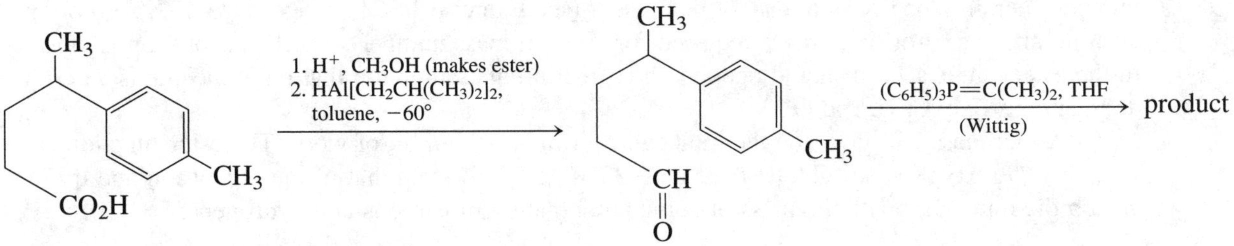

63. 1. $(CH_3)_2NH$, Δ (makes N,N-dimethyl carboxamide), 2. $LiAlH_4$, $(CH_3CH_2)_2O$ (makes amine), 3. H^+, H_2O

64. Pay attention to the stereochemistry.

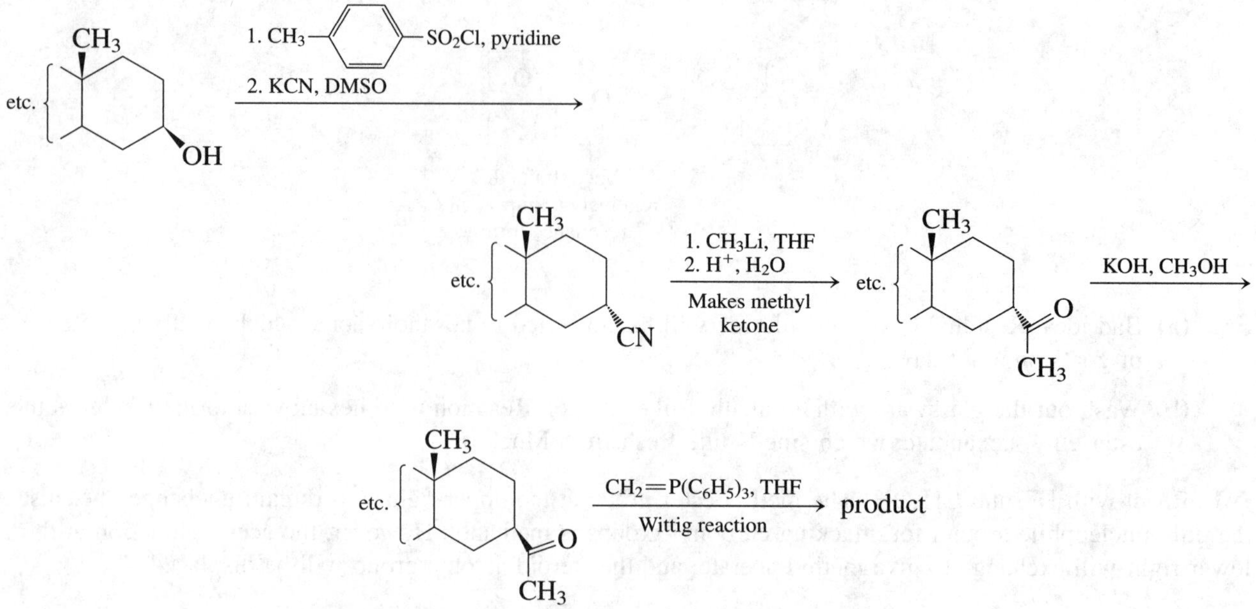

Deprotonation at α-carbon
allows isomerization to more
stable equatorial stereoisomer

65.

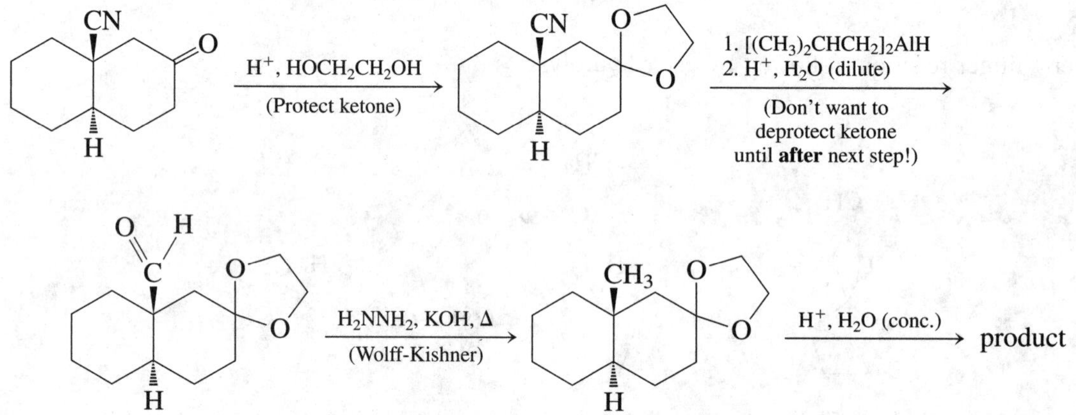

66. **(a)** Collect your information and decide what it tells you. Begin with the molecular ion. It is a single peak, ruling out Cl or Br, which would give two peaks two mass units apart. The value is an even number, therefore one N atom cannot be present (see Exercise 11-21 on text page 477; zero or an even number of nitrogen atoms is required for an even mass number). The IR absorption is closest to the ester range, so you should operate under the initial assumption that the molecule is an ester and contains only C, H, and O.

The exact mass will give you the molecular formula, with a bit of work. The ester function contains two oxygens, so subtract their mass (2 × 15.9949) from that of the parent ion and try to match the remainder with the mass of some combination of carbons and hydrogens:

$$\begin{array}{r} 116.0837 \\ -(2 \times 15.9949) \\ \hline 84.0939 \end{array}$$

The only reasonable combination of C and H that has a mass of 84 is C_6H_{12}. Does it match **exactly?** $(6 \times 12) + (12 \times 1.00783) = 84.0940$—yes, indeed it does. (If that had not been the case, you would have had to explore possibilities containing oxygen, such as C_5H_8O, etc.) The molecular formula is $C_6H_{12}O_2$, giving one degree of unsaturation, the ester carbonyl group.

Now turn to the proton NMR. You have the following absorptions:

$\delta = 1.1$: Triplet, integration of 3 H; a CH_3—CH_2— group

$\delta = 1.2$: Doublet, integration of 6 H; a $(CH_3)_2CH$— group

$\delta = 2.3$: Quartet, integration of 2 H; indicates a CH_3—CH_2— group, possibly attached to a carbonyl carbon due to the slightly deshielded position in the spectrum

$\delta = 5.0$: A septet (see amplified scan), integration of 1 H; seven lines imply two methyl groups as neighbors; chemical shift indicates attachment to oxygen: a $(CH_3)_2CH$—O— group

Do you have your answer? Yes: $CH_3CH_2\overset{\displaystyle O}{\overset{\displaystyle \|}{C}}$—$OCH(CH_3)_2$

Does this fit the rest of the data? The upfield region of the NMR spectrum can now be interpreted as consisting of a large doublet for the methyls of the $CH(CH_3)_2$ group, overlapping one peak of a smaller triplet, which arises from the CH_3 of the ethyl group. How about the mass spectrum? The base peak at $m/z = 57$ arises from a type of α cleavage at the ester C—O bond, giving the very stable acylium ion $CH_3CH_2C{\equiv}O^+$. See how many of the other fragments you can assign.

(b) In this case you see two parent ions, two mass units apart, of equal intensity; a bromine atom is present. About half of the molecules contain ^{79}Br and a molecular weight of 180; the other half contains ^{81}Br with total $m/z = 182$. Again, the IR spectrum tells you that an ester is present.

Using either exact mass, subtract the mass of the appropriate Br isotope and that of the two oxygens of the ester function:

$$\begin{array}{r} 179.9886 \\ -78.9183 \\ \hline 101.0703 \\ -(2 \times 15.9949) \\ \hline 69.0805 \end{array}$$

The only reasonable combination of C and H that has a mass of 69 is C_5H_9. It has the correct exact mass: $(5 \times 12) + (9 \times 1.00783) = 69.0705$. The molecular formula is $C_5H_9O_2Br$, giving one degree of unsaturation, the ester carbonyl group.

You could turn directly to the NMR, but first try to extract all you can from the mass spectral data. The base peak has $m/z = 29$; two possibilities you have seen in this chapter are $CH_3CH_2^+$ and $HC{\equiv}O^+$. A pair of peaks at $m/z = 107$ and 109 imply a Br-containing fragment. What could it be? If you subtract the atomic mass of the appropriate Br isotope from either, you get 28, which could correspond either to C_2H_4 or CO. You can therefore write possible structures for the 107/109 fragment such as CH_3CHBr^+ and $BrC{\equiv}O^+$. The IR told you that you were dealing with an ester, not an alkanoyl halide, so the first of these two possibilities is more plausible. You are getting very close. So look at the proton NMR. You have the following absorptions:

$\delta = 1.3$: Triplet, integration of 3 H; a CH_3—CH_2— group

$\delta = 1.8$: Doublet, integration of 3 H; a CH_3—CH— group

$\delta = 4.2$ and 4.4: Quartets, with integrations of 2 H and 1 H, respectively.

These must be the CH_2 and CH groups inferred from the two signals, above! Because these protons are so deshielded, you can assume that one of these groups is attached to the ester oxygen and the other to bromine. Which is attached to which? If you try connecting the CH_3—CH_2— to bromine, you make CH_3CH_2Br, bromoethane. That's no good because there's no place left to attach all the rest of the atoms. So, instead connect CH_3—CH_2— to the ester oxygen and CH_3—CH— to bromine, giving the answer

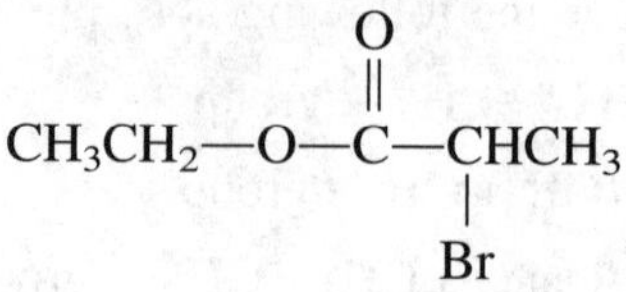

$$CH_3CH_2-O-\overset{\overset{\displaystyle O}{\|}}{C}-\underset{\underset{\displaystyle Br}{|}}{C}HCH_3$$

67. First reaction: The "mixed" anhydride can release either of two different acylium ions.

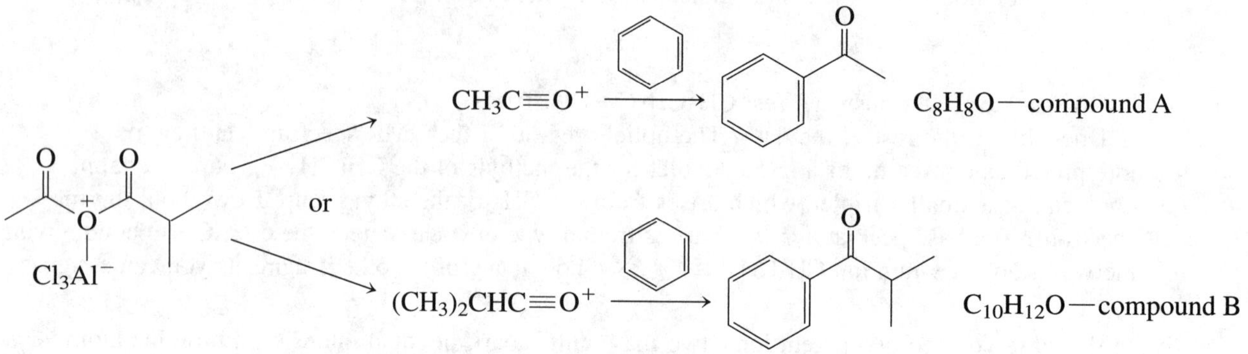

$CH_3C{\equiv}O^+ \longrightarrow$ C$_8$H$_8$O—compound A

$(CH_3)_2CHC{\equiv}O^+ \longrightarrow$ C$_{10}$H$_{12}$O—compound B

Second reaction: An ester can serve as a source of acylium ion, too.

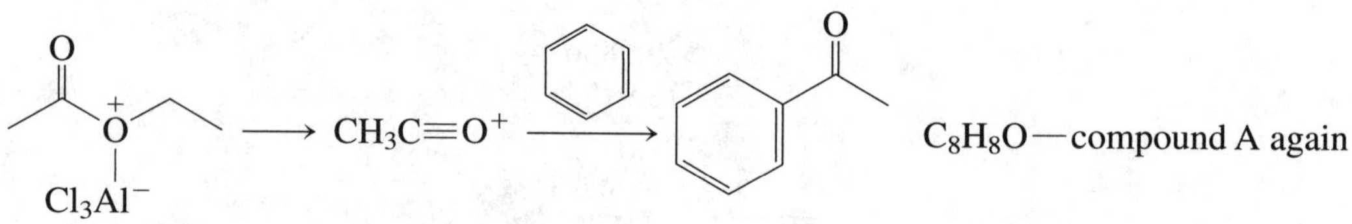

$CH_3C{\equiv}O^+ \longrightarrow$ C$_8$H$_8$O—compound A again

But what else can happen? What is this product C, with the formula C$_8$H$_{10}$? Look at the NMR: Five benzene hydrogens, and CH$_2$ and CH$_3$ groups upfield—looks like ethylbenzene! How? The Lewis acid complexed oxygen is a good leaving group, so we can propose the following:

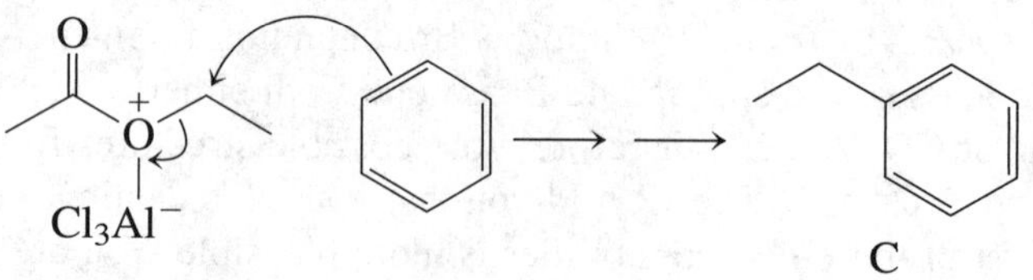

which, of course, can go on to undergo another acylation to give

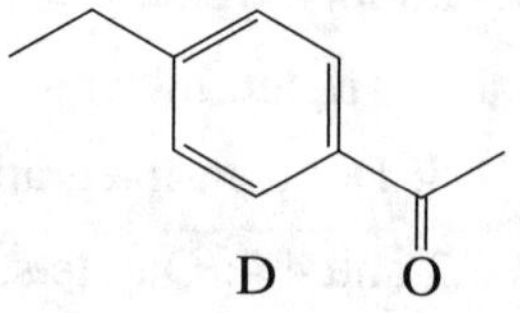

Third reaction sequence:

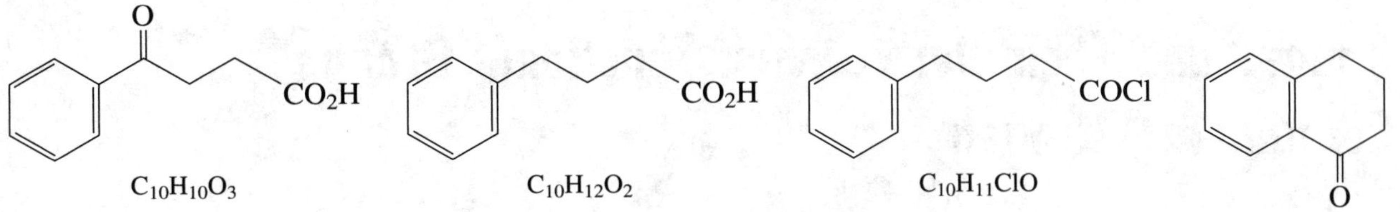

$C_{10}H_{10}O_3$

$C_{10}H_{12}O_2$

$C_{10}H_{11}ClO$

$C_{10}H_{10}O$

68. (e)

69. (a)

70. (d)

71. (b)

21

Amines and Their Derivatives: Functional Groups Containing Nitrogen

Amines represent the last of the simple functional groups that you will encounter in organic chemistry. They are not entirely new to you, of course. They popped up as early as Chapter 6, as the results of nucleophilic substitution reactions between ammonia and haloalkanes. More recently, you have been introduced to amine syntheses starting from carboxamides and nitriles (Sections 20-6 through 20-8). As usual, the chapter begins by presenting the usual body of descriptive information concerning the properties of amines as a class of compounds. It then focuses in more detail on the limitations and variations associated with several amine syntheses, finishing with sections on their reactions. Amines are of substantial importance biologically. But unlike other biologically important compound classes, amines are not involved in a very wide range of distinctly different types of reactions. Thus, you should find this aspect of amine chemistry relatively manageable.

Outline of the Chapter

21-1 Nomenclature

21-2, 21-3, 21-4 Physical, Spectroscopic, and Acid-Base Properties of Amines
> Qualitative and quantitative characteristics of the functional group.

21-5, 21-6, 21-7 Synthesis
> Guidelines for choosing synthetic strategies.

21-8 through 21-10 Reactions
> Mostly extensions of earlier material; a couple of special reactions, too.

Keys to the Chapter

21-1. Nomenclature

To an even greater extent than with most other compounds, common names for amines are still almost universally used. It is therefore necessary to be able to recognize structures from either their alkylamine- or aniline-based name. The systematic alkanamine method looks tricky at first but becomes simple once you recognize that it works in much the same way that the IUPAC alcohol (alkanol) naming system works.

21-2 through 21-4. Properties of Amines

Amines are related to alcohols in the same way that ammonia is related to water. This parallel makes the properties of amines easy to predict, because the main qualitative difference is simply that N is less electronegative than O. So, hydrogen bonding in amines is present, but it is weaker than in alcohols; deshielding of nearby ^{1}H and ^{13}C NMR signals is observed but to a lesser extent than in alcohols; IR spectra are similar; mass spectra are rather predictable. Tertiary amines (R_3N) may be viewed as nitrogen analogs of ethers (R_2O).

By the way, all these nice, tidy analogies do not extend to the **smells** possessed by most amines. Whereas alcohols tend to have, at worst, somewhat heavy, sweetish odors, amines, at best, smell like ammonia, and, at worst, richly deserve the common names that have been bestowed upon some of their representatives. These include names like cadaverine, putrescine, and skatole. Dead fish would be an improvement.

The acid-base properties of amines are an extension of what you know about ammonia: They are weaker acids and stronger bases than are water or alcohols. It will repay you many times over, however, to go over the information involving pK_a's of these molecules. The qualitative ability to handle acid-base concepts is one of the more useful capabilities you can take out of a course in organic chemistry.

21-5, 21-6, and 21-7. Synthesis

By the time you finish reading these sections of the text, you will be aware of the fact that the first amine synthesis you learned, alkylation of NH_3 via S_N2 reaction, is also generally the worst amine synthesis. It is much better to use any one of the special N-containing nucleophiles that give clean monoalkylation products, which can then be turned into amines. For simple systems lacking sensitive functional groups, all the methods will work comparably well. The choice becomes more critical if the molecule is more sensitive. For example, 1. N_3^-, 2. $LiAlH_4$ would be a poor choice for the conversion of $Br(CH_2)_3COCH_3$ into the corresponding amine, because the ketone will be reduced by the hydride reagent along with the azide group. The ketone could be protected before starting, but a better solution would be the Gabriel sequence, which involves hydrolysis instead of reduction in the second step.

An important additional amine synthesis in this section is reductive amination of aldehydes and ketones. In particular, reductive amination of a ketone is a better way to make an amine attached to a 2° alkyl group than is S_N2 reaction with a 2° haloalkane.

You've already seen how to make an amine with one more carbon by using S_N2 reaction with CN^-, followed by reduction. What would you do, however, if you faced the particularly nasty problem of needing to make an amine attached to a 3° alkyl group (R_3CNH_2)? That's tricky, because S_N2 reactions won't work. If you could add a carbon somehow to get to R_3CCONH_2, then you would be all set to use one of the above rearrangements, right? O.K., think about that one for awhile. You will find help in Section 19-6. The solution will be given in the answer to Problem 40(b).

21-8 through 21-10. Reactions

Beyond simple displacement reactions in which an amine behaves as a nucleophile, there is a small group of specialized reactions of amines, each of which has a very specific use. The Hofmann elimination has been more important in structure determination than anything else. It is frequently combined with the Mannich reaction, however, to make an important synthetic entry to cyclic carbonyl compounds with methylene groups next to the carbonyl (see Problem 52). Such structures are present in many naturally occurring (plant-derived) antitumor agents.

The nitrosation reactions and the chemistry of diazoalkanes are also used in synthesizing just certain specific types of compounds, like cyclopropanes. The mechanisms here are more involved, however, and an understanding of their steps at this point is helpful because similar types of chemistry will be seen again later, in Chapter 22. As you plow through this, try to focus on the relationship of each mechanistic step to processes you've seen before. Almost all of this chemistry is based on relatively fundamental sorts of events, like protonation and deprotonation, combined with elimination or addition.

Solutions to Problems

27. **(a)** 3-Hexanamine, 3-aminohexane; **(b)** *N*-methyl-2-propanamine, 2-(methylamino)propane, isopropylmethylamine; **(c)** 2-chlorobenzenamine, *o*-chloroaniline;

(d) *N*-methyl-*N*-propylbenzenamine; *N*-methyl-*N*-propylaniline; (e) *N,N*-dimethylmethanamine (common: trimethylamine); *N,N*-dimethylaminomethane; (f) 4-(*N,N*-dimethylamino)-2-butanone (only satisfactory name); (g) 6-chloro-*N*-cyclopentyl-*N*,5-dimethyl-1-hexanamine (numbers refer to substituents on parent hexane chain); 1-chloro-6-(*N*-cyclopentyl-*N*-methylamino)-2-methylhexane; (h) *N,N*-diethyl-2-propen-1-amine, 3-(*N,N*-diethylamino)-1-propene

28. (a)

N(CH₃)₂

(b)

CH₂CH₂NHCH₂CH₃

(c) HOCH₂CH₂NH₂   **(d)**

NH₂

Cl

29. (a) 1-Cyclohexyl-*N*-methyl-2-propanamine; **(b)** 1-phenyl-2-propanamine; **(c)** 2-(3,4,5-trimethoxyphenyl)ethanamine; **(d)** *R*-4-(1-hydroxy-2-[*N*-methylamino]ethyl)benzene-1,2-diol.

30. (a) 5–7 kcal mol^{-1}, approximately equal to E_a for inversion. **(b)** Methyl anion is isoelectronic with ammonia and, likewise, is tetrahedral (sp^3 hybridized). Methyl radical and cation, with one and two fewer electrons, respectively, are more stable when trigonal planar. They gain bond strength by rehybridizing to use sp^2 orbitals in the σ bonds, with either a singly occupied or vacant *p* orbital "left over." The sp^2 scheme is not as good for the anion or for the ammonia because two electrons in an unhybridized *p* orbital is quite unfavorable in the absence of other stabilizing influences: Such electrons are far from and attracted only poorly by the atom's nucleus.

31. The odd atomic weights suggest that each contains a single nitrogen. The total number of hydrogens is available from the NMR, so the number of carbons can be determined by difference: m/z 129 = 14 (one N) + 19 (19 H's) + weight of carbons. Weight of carbons = 96 → 8 carbons; $C_8H_{19}N$ is the formula for each of these unknowns. Degrees of unsaturation (see Section 11-8): $H_{sat} = 16 + 2 + 1$ (for the N) = 19; compounds are saturated.

A NMR:

$$CH_3—CH_2— \text{ and } —CH_2—CH_2—NH_2$$

0.9(t)   2.7(t)   2.3

Notice that the signal at δ = 2.7 is **not** split by the —NH₂ hydrogens (as is the case for alcohols as well).

The splittings nicely reveal the number of neighboring H's. MS: m/z 30 for [$\overset{+}{C}H_2—\overset{..}{N}H_2 \longleftrightarrow CH_2\overset{+}{=}NH_2$] fragment. All that remains is to insert C_4H_8, and the simplest way to do that is as $CH_3(CH_2)_7NH_2$ (1-octanamine). Other isomers would show additional methyl signals in the NMR near δ = 0.9–1.0.

B NMR:

$$(CH_3)_3C— \qquad 2 CH_3—,$$

1.0(s)   1.2
Likely,   Also two equivalent
                      methyl groups

Perhaps a CH_2 and an NH_2? (Signals at $\delta = 1.3$ and 1.4). MS: m/z 114 is $[M - CH_3]^+$, 72 is $[M - (CH_3)_3C]^+$, and 58 is most likely an iminium ion. Before guessing, notice that there are no NMR signals in the $\delta = 2.7$ region, where you might expect to find $-\overset{\textcircled{H}}{\underset{|}{C}}-N\diagdown$ signals. So, most likely the N is attached to a **tertiary** carbon. Possible pieces:

$$(CH_3)_3C- \qquad 2\ CH_3- \qquad -CH_2- \qquad -\overset{|}{\underset{|}{C}}-NH_2$$

All atoms in the formula are present, so put it together

$$(CH_3)_3C-CH_2-\overset{\displaystyle CH_3}{\underset{\displaystyle CH_3}{\overset{|}{\underset{|}{C}}}}-NH_2$$

The m/z 58 fragment therefore is $[(CH_3)_2C{=}NH_2]^+$.

32. As you do each of these, keep the $C_6H_{15}N$ formula in mind.

    (a) NMR: The $\delta = 23.7$ peak may correspond to one or more than one *equivalent* CH_3- groups, and the peak at $\delta = 45.3$ is one or more equivalent $\diagup CH-$ units (attached to N due to chemical shift).

    IR: A **secondary** amine, $-NH-$. No other signals are present, so attach as many of each as are necessary. The answer is

$$\underset{CH_3}{\overset{CH_3}{\diagdown}}CH-\overset{H}{\underset{}{N}}-CH\underset{CH_3}{\overset{CH_3}{\diagup}}$$

    (b) NMR: Now you have only CH_3- and $-CH_2-$ groups (the latter attached to N); IR: A **tertiary** amine. So, the answer is $(CH_3CH_2)_3N$.

    (c) NMR: CH_3- groups, $-CH_2-$ groups **not** attached to N, and $-CH_2-$ groups that **are** attached to N. IR: Amine is **secondary.** So the answer is $CH_3CH_2CH_2NHCH_2CH_2CH_3$.

    (d) NMR: One CH_3- and five $-CH_2-$'s. IR: **Primary** amine ($-NH_2$). This is $CH_3(CH_2)_5NH_2$.

    (e) NMR: Two different CH_3- types, one ($\delta = 38.7$) attached to N; also a quaternary C attached to N ($\delta = 53.2$). IR: A **tertiary** amine. Remembering the $C_6H_{15}N$ formula, you can construct the molecule:

$$25.6 \longrightarrow (CH_3)_3C-N\underset{CH_3}{\overset{CH_3}{\diagup}} \diagdown\ 38.7$$
$$\underset{53.2}{\uparrow}$$

33. Figure 21-5 is $(CH_3CH_2)_3N$ for comparison purposes. Look in each case for important fragments from $C{+}C-N$ cleavage to make iminium ions.

    (a) m/z 72 is important, which is $[M - 29]^+$ or loss of CH_3CH_2-. The only amine that should easily lose an ethyl group from those in Problem 28 is $CH_3CH_2{+}CH_2-NH-CH_2CH_2CH_3$ [see (c)]. This is the answer.

(b) m/z 86 is rather large, corresponding to the loss of CH_3—. Three amines in Problem 32 should lose CH_3— easily: (a), (b), and (e). *N,N*-Diethylethanamine [(triethylamine, b)] is ruled out because its MS (Figure 21-5) doesn't match. The m/z 58 peak is loss of 43, or C_3H_7. That's easy to visualize from (a): $(CH_3)_2CH{+}NH—CH(CH_3)_2$ (the correct answer), but not from amine (e).

34. The base in question (call it "B^1:") is stronger. If its conjugate acid (B^1H^+) has a high pK_a, then that acid is weak; consequently, its corresponding conjugate base is strong. Here's the equation:

$$B^1\text{:} \quad + \quad B^2H^+ \quad \rightleftharpoons \quad B^1H^+ \quad + \quad B^2\text{:}$$

stronger base   *(stronger acid)*   *weaker acid*   *(weaker base)*

(higher pK_a)

35. (a) To the left. NH_3 and ^-OH are weaker acids and bases, respectively, than are H_2O and NH_2^-.

     (b) To the left. CH_3NH_2 is a weaker base than ^-OH, and H_2O is a weaker acid than $CH_3NH_3^+$.

     (c) To the right. CH_3NH_2 is a stronger base (see textbook Section 21-4) than $(CH_3)_3N$.

36. (a) Weaker bases because lone pair on N is "tied up" by resonance:

$$\left[\begin{array}{c} \overset{O}{\underset{\|}{RC}}\overset{\cdot\cdot}{NH_2} \end{array} \longleftrightarrow \begin{array}{c} \overset{O^-}{\underset{|}{RC}}=NH_2^+ \end{array} \right]$$

Stronger acids because the conjugate base is stabilized by both the inductive effect of the carbonyl group and by resonance:

$$\overset{O}{\underset{\|}{RCNH_2}} \rightleftharpoons H^+ + \left[\overset{O}{\underset{\|}{RC}}\overset{\cdot\cdot}{NH^-} \longleftrightarrow \overset{O^-}{\underset{|}{RC}}=\overset{\cdot\cdot}{NH} \right]$$

     (b) Same as carboxamides, only to a greater extent for both acidity and basicity due to the two carbonyl groups.

     (c) Somewhat weaker bases because of resonance:

$$\left[\overset{\diagdown}{\diagup}C=C\overset{\diagup}{\underset{|}{N}}\overset{\diagdown}{\cdot\cdot} \longleftrightarrow \overset{\diagdown}{\diagup}\overset{\cdot\cdot}{C}-C=\overset{+}{N}\overset{\diagup}{\diagdown} \right]$$

Not acidic, because of lack of H's on nitrogen.

     (d) Weaker bases and stronger acids, for the same reasons given in (a) for carboxamides.

37. Protonate a doubly bonded nitrogen in each case to get a resonance stabilized cation.

DBN-derived cation (cation from DBU is similar)

$$\left[\begin{array}{c} \overset{+}{N}H_2 \\ \ddot{N}H_2-\overset{\|}{C}-\ddot{N}H_2 \end{array} \longleftrightarrow \begin{array}{c} :NH_2 \\ \ddot{N}H_2-\overset{|}{\underset{+}{C}}-\ddot{N}H_2 \end{array} \longleftrightarrow \begin{array}{c} :NH_2 \\ \ddot{N}H_2-\overset{|}{C}=\overset{+}{N}H_2 \end{array} \longleftrightarrow \begin{array}{c} :NH_2 \\ \overset{+}{N}H_2=\overset{|}{C}-\ddot{N}H_2 \end{array} \right]$$

Guanidine

Resonance stabilization of conjugate acids enhances base strengths.

38. **(a)** Excess NH_3 is the easiest way, also 1. N_3^-, 2. $LiAlH_4$, or Gabriel (text, p. 988); **(b)** NH_3, $NaBH_3CN$; **(c)** $CH_3CH_2NH_2$, $NaBH_3CN$; **(d)** 1. ^-CN, 2. $LiAlH_4$; **(e)** $(CH_3)_2NH$, $NaBH_3CN$; **(f)** 1. $CH_2{=}O$, NH_3, HCl, heat, 2. ^-OH, H_2O (Mannich); **(g)** H_2, PtO_2, or $LiAlH_4$; **(h)** Br_2, NaOH, H_2O, heat (Hofmann rearrangement).

39. **(a)** Not at all. This process **adds a carbon** (the ^-CN group), making 1-pentanamine. **(b)** Not at all. S_N2 reactions with tertiary haloalkanes are not possible. **(c)** Well. **(d)** Poorly. Further alkylation can

occur, making $\left(\text{(cyclopentyl-(CH}_2)_3) \right)_2 NCH_3$.

(e) Poorly. The haloalkane, although primary, is highly branched and will not react well in S_N2 reactions. **(f)**, **(g)** Well. **(h)** Poorly. Four-membered rings are strained and difficult to form. The method would work well for a five- or six-membered ring. **(i)** Not at all. Reaction shown is for a benzene, not a cyclohexane compound. **(j)** Well.

40. $(CH_3CH_2)_2O$ solvent and H^+, H_2O work-up are understood for Grignard and $LiAlH_4$ reactions.

(a) 1. NaN_3, DMSO, 2. $LiAlH_4$. **(b)** You need to be devious. Add a carbon and then take it out again!

$$(CH_3)_3CCl \xrightarrow[\text{2. }CO_2]{\text{1. Mg}} (CH_3)_3C\overset{O}{\overset{\|}{C}}OH \xrightarrow[\text{2. }NH_3]{\text{1. }SOCl_2} (CH_3)_3C\overset{O}{\overset{\|}{C}}NH_2 \xrightarrow[\substack{\text{(Hofmann} \\ \text{rearrangement)}}]{Br_2,\ NaOH,\ H_2O} (CH_3)_3CNH_2$$

(d) 1. NaN_3, 2. $LiAlH_4$ (makes primary amine), 3. $H_2C{=}O$ [makes imine; use one equivalent only, to avoid dimethylation (Problem 56)], 4. $NaBH_3CN$, CH_3CH_2OH (completes reductive amination). Note how CH_3 group is introduced as **formaldehyde,** with a subsequent reduction step.

(e) Same as **(b).** **(h)** No simple way to improve the situation. **(i)** Start with (phenyl)—Br

and do reactions shown, making Br—(cyclohexyl)—NH_2. The H_2, Pd (Sections 14-7 and 15-2) will slowly hydrogenate the ring.

41. $CH_3CH_2NH_2$, $(CH_3CH_2)_2NH$, $(CH_3CH_2)_3N$, and $(CH_3CH_2)_4N^+$.

42. Make pseudoephedrine from phenylpropanolamine using reductive amination. As suggested in Problem 40(d), use one equivalent of $H_2C{=}O$ to avoid **di**methylation.

$$RNH_2 \xrightarrow{H_2C{=}O,\ NaBH_3CN} RNHCH_3$$

43. Secondary (general structure RR′NH). **(a)** 1. $CH_3CH_2NH_2$, H^+, 2. $NaBH_3CN$, CH_3CH_2OH; **(b)** 1. NaN_3, DMF, 2. $LiAlH_4$, THF, 3. CH_3CHO, H^+, 4. $NaBH_3CN$, CH_3CH_2OH; **(c)** 1. $SOCl_2$, 2. NH_3, 3. Br_2, NaOH, H_2O, 4. CH_3CHO, H^+, 5. $NaBH_3CN$, CH_3CH_2OH

44. (a) Lots of ways! Bromobutane + azide, then $LiAlH_4$ reduction, *or* bromopropane + cyanide, then $LiAlH_4$, *or* bromobutane + phthalimide salt, then hydrolysis.

(b) First make methanamine from iodomethane and either azide or phthalimide salt (similar to butanamine syntheses in (a), and then carry out reductive amination using the methanamine and butanal:

$$CH_3I \xrightarrow[\text{2. } LiAlH_4]{\text{1. } N_3^-} CH_3NH_2 \xrightarrow[\text{2. } NaBH_3CN, \text{ pH} = 3]{\text{1. } CH_3CH_2CH_2CHO} CH_3NHCH_2CH_2CH_2CH_3$$

(c) Make butanamine **(a)** and then go with double reductive amination using excess formaldehyde and $NaBH_3CN$. Best way to make any *N,N*-dimethyl tertiary amines (see Problem 56).

45. (a) 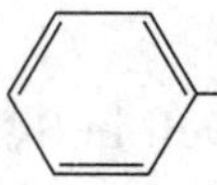 C₆H₅—CH=CHCH₃ (*Z* and *E*) **(b)** 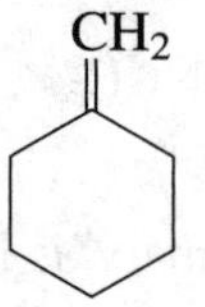 [structure: methylenecyclohexane] and [structure: 1-methylcyclohexene]

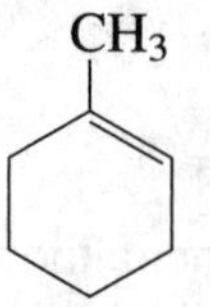

(c) First cycle:

$CH_2=CH(CH_2)_3N(CH_3)_2$ $CH_3CH=CH(CH_2)_2N(CH_3)_2$

$$\underset{\displaystyle CH_3CHCH_2CH=CH_2}{\overset{\displaystyle N(CH_3)_2}{|}}$$

Second cycle:

$CH_2=CHCH_2CH=CH_2$ $CH_3CH=CHCH=CH_2$

(d) 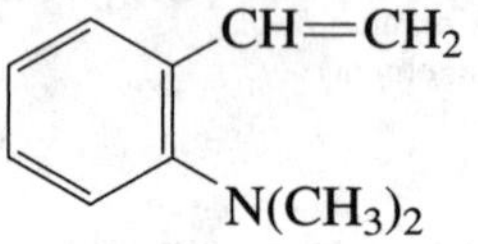 [structure: benzene ring with CH=CH₂ and N(CH₃)₂ substituents]

(e) First cycle:

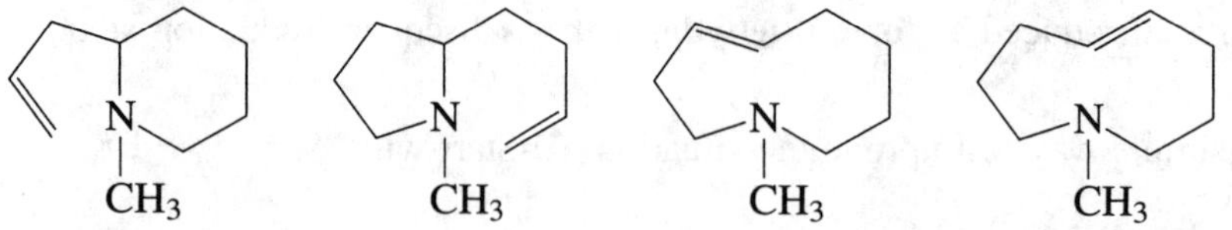

[four bicyclic structures with N–CH₃]

Second cycle:

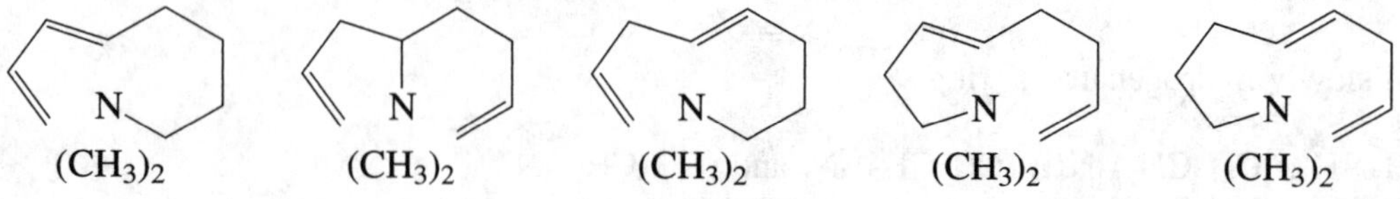

[five structures with N(CH₃)₂]

Third cycle:

[two diene structures]

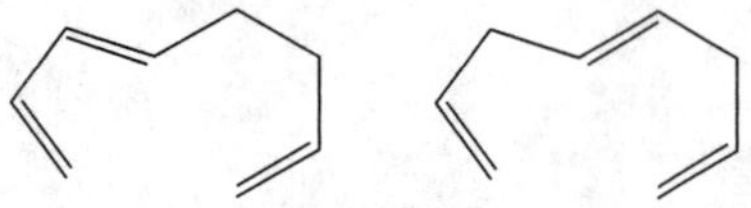

46. (a) 4-Heptanamine; **(b)** 3-heptanamine; **(c)** 1-heptanamine; **(d)** 2-heptanamine.

47. The reaction takes place under *acidic* conditions.

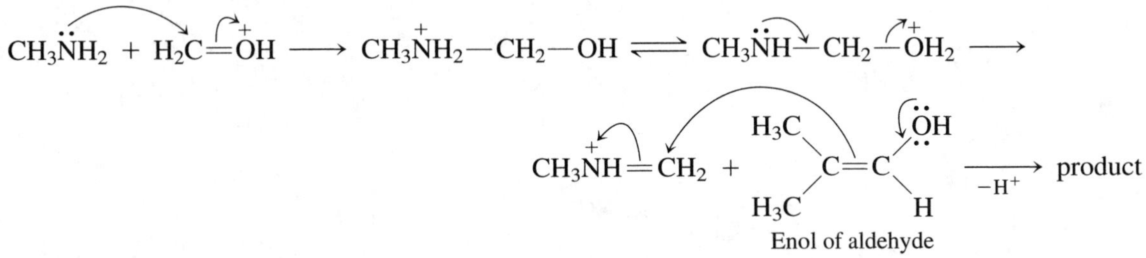

$$CH_3\ddot{N}H_2 + H_2C=\overset{+}{O}H \longrightarrow CH_3\overset{+}{N}H_2—CH_2—OH \rightleftharpoons CH_3\ddot{N}H—CH_2—\overset{+}{O}H_2 \longrightarrow$$

$$CH_3\overset{+}{N}H=CH_2 + \underset{H_3C}{\overset{H_3C}{\diagdown}}C=C\underset{H}{\overset{\ddot{O}H}{\diagup}} \xrightarrow[-H^+]{} product$$

Enol of aldehyde

48. Tropinone is a tertiary amine. Alkylation of nitrogen can occur from either the "left" or "right" (arrows, below), giving stereoisomeric products.

(a)

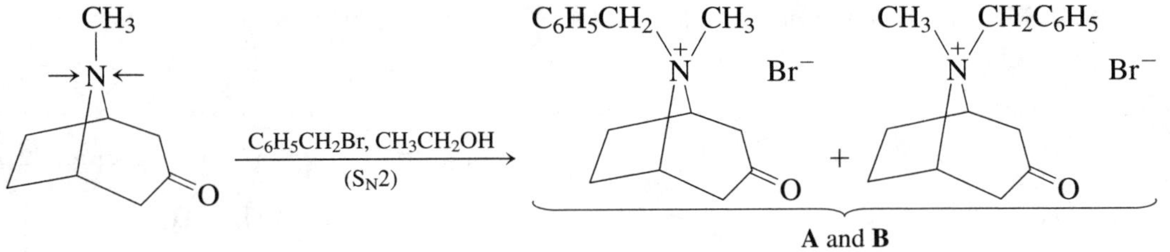

$$\xrightarrow[(S_N2)]{C_6H_5CH_2Br, CH_3CH_2OH}$$

A and B

(b) Diastereomers (they are not mirror images)

(c) Where are acidic hydrogens in A and B? At the carbons α to the ketone carbonyl.

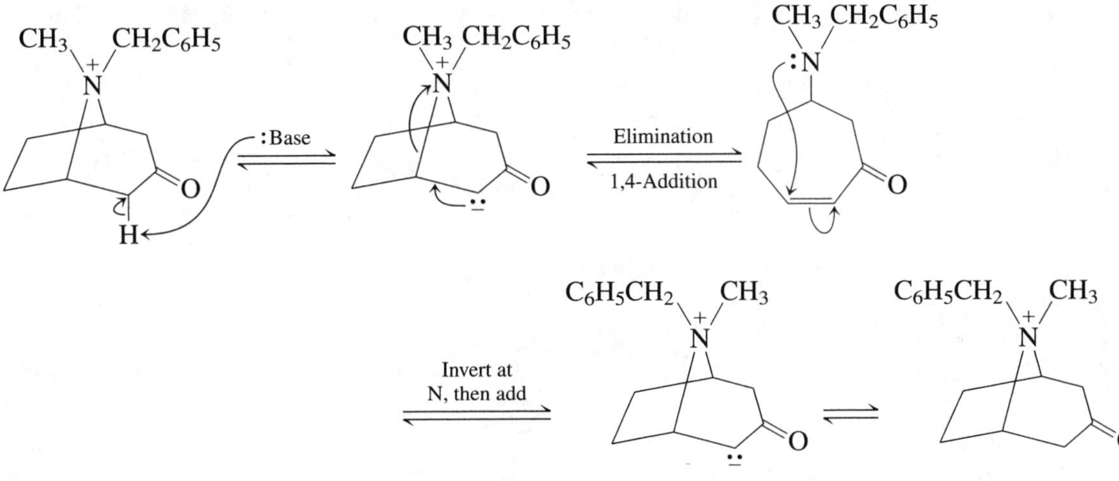

:Base ⇌

$$\xrightarrow[\text{1,4-Addition}]{\text{Elimination}}$$

Invert at
N, then add

Deprotonation and elimination gives an enone, C. The amine is free to add back, re-forming the original ketone, or it can invert at nitrogen first and then add back, which gives the stereoisomeric product (CH$_3$ and C$_6$H$_5$CH$_2$ groups switched places).

49. (a) $HOCH_2CH_2NH_2$ $\xrightarrow[(S_N2's)]{\overset{\text{Excess}}{CH_3I}}$ $HO\ddot{:}\underset{CH_2}{\overset{CH_2}{\diagdown}}\overset{+}{N}(CH_3)_3 \; I^-$ $\xrightarrow{\text{Internal } S_N2}$

$$\underset{H}{\overset{CH_2—CH_2}{\underset{\diagdown}{O}}}\overset{+}{} + (CH_3)_3N\ddot{:} \longrightarrow \overset{CH_2—CH_2}{\underset{O}{}} + (CH_3)_3NH^+I^-$$

Final products

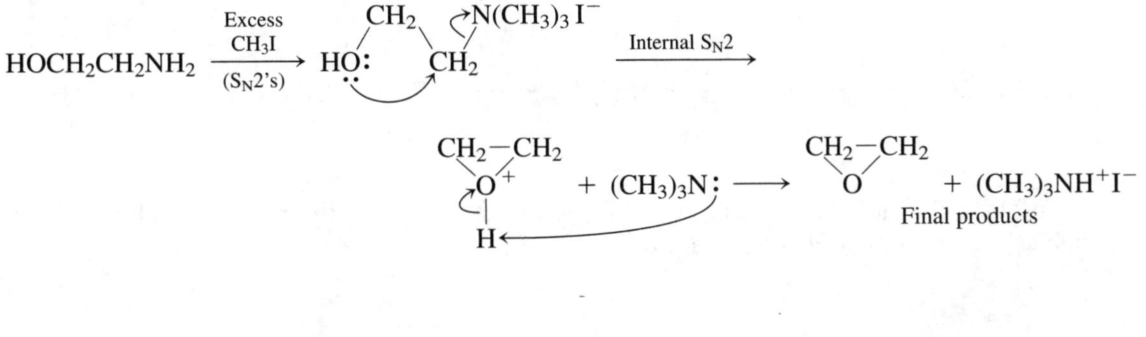

(b) Work backward.

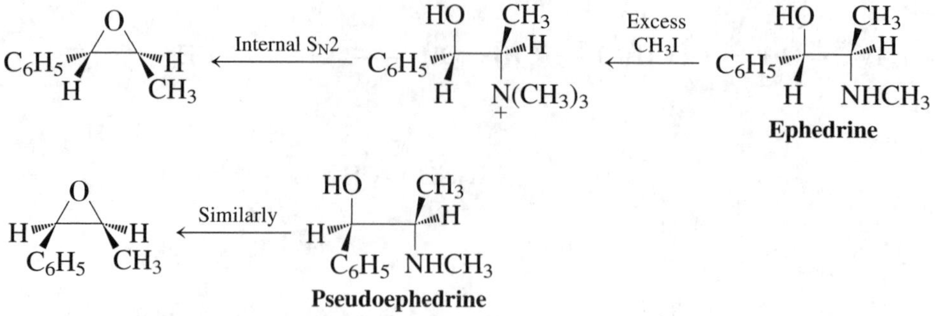

Ephedrine and pseudoephedrine are diastereomers!

50. Analyze the Mannich reaction in terms of the functional unit it constructs.

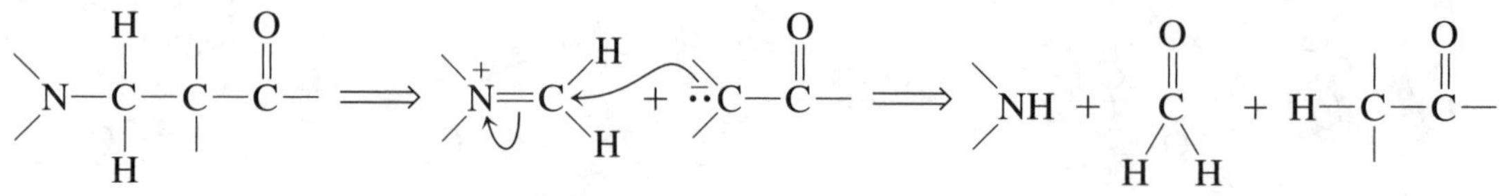

Relevant bonds are emphasized in the answers below.

(a) CH_3CCH_2—CH_2—$N(CH_2CH_3)_2$ $\xleftarrow{\text{1. HCl} \\ \text{2. HO}^-}$ CH_3CCH_3 + $CH_2{=}O$ + $HN(CH_2CH_3)_2$

(b)

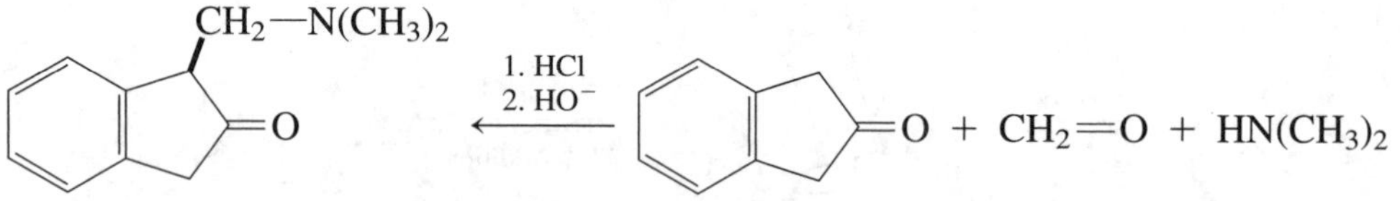

(c) H_2N—CH—CN $\longleftarrow$ NH_3 + CH_3CHO + HCN
$\quad\quad\quad\overset{|}{CH_3}$

Here the nucleophile is different; it is the cyanide ion that adds to the iminium carbon.

(d) $CH_3CH_2CH_2CCH$—CH_2—$N(CH_3)_2$ $\xleftarrow{\text{1. HCl} \\ \text{2. HO}^-}$
$\quad\quad\quad\quad\quad\overset{|}{CH_2CH_3}$

$\quad\quad\quad\quad\quad\quad CH_3CH_2CH_2CCH_2CH_2CH_3$ + $CH_2{=}O$ + $HN(CH_3)_2$

(e) CH_3CCH_2—CH_2—N—CH_2—CH_2CCH_3 $\xleftarrow{\text{1. HCl} \\ \text{2. HO}^-}$ $2\ CH_3CCH_3$ + $2\ CH_2{=}O$ + H_2NCH_3
$\quad\quad\quad\quad\quad\quad\quad\overset{|}{CH_3}$

Two Mannich reactions are involved in this example. Note the **primary** amine and the presence of **two** moles each of formaldehyde and acetone.

51. A double Mannich reaction, similar to Problem 50(e):

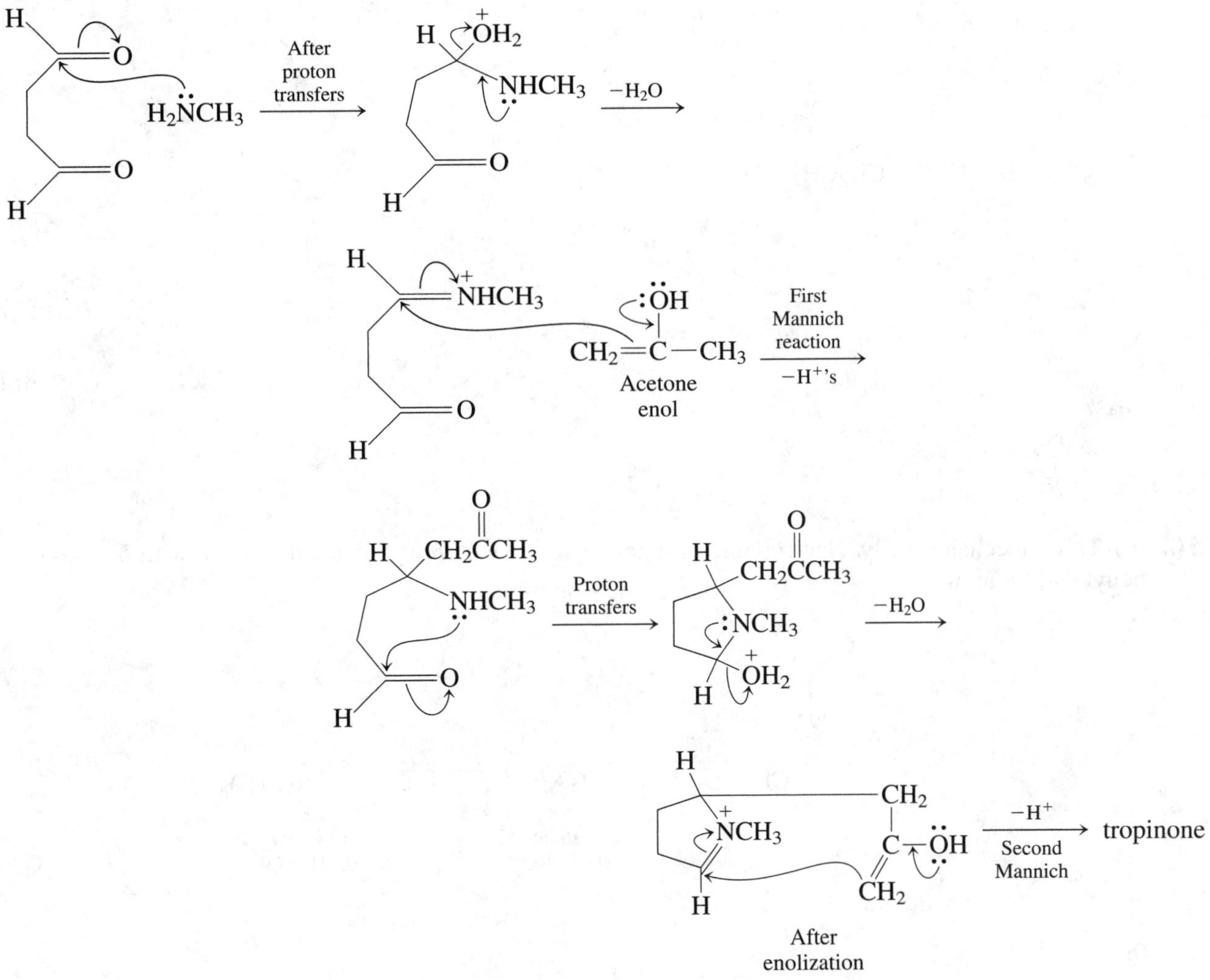

52.

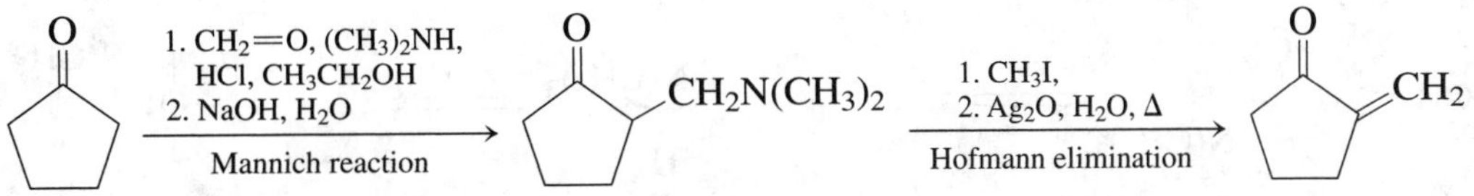

The Mannich reaction–Hofmann elimination sequence is a useful synthesis of α, β-unsaturated ketones.

53. (a) All possible products of

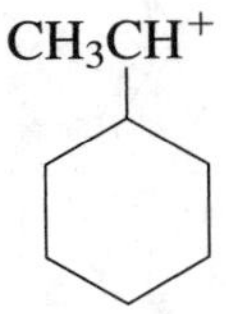

!

So,

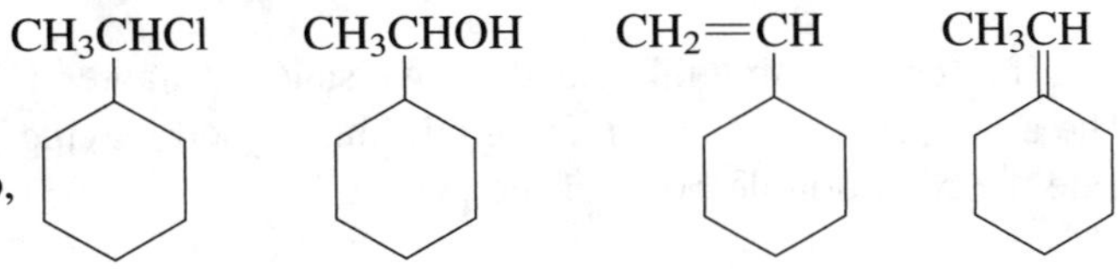

And, after hydride shift to

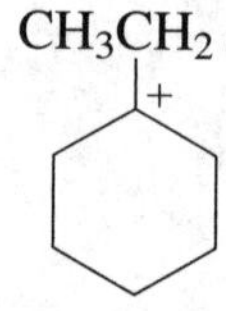

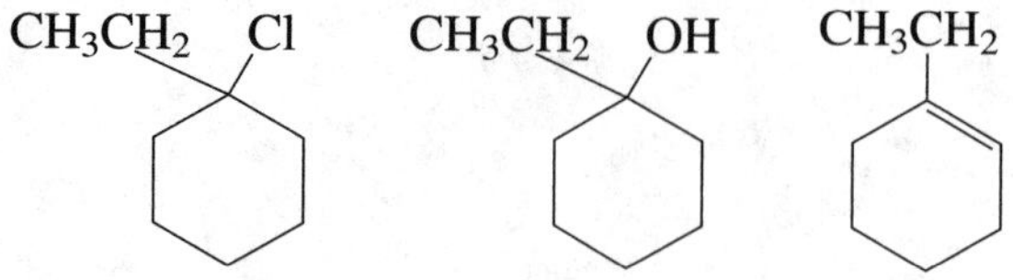

CH$_3$CH$_2$

CH$_3$CH$_2$ Cl   CH$_3$CH$_2$ OH   CH$_3$CH$_2$

(d)

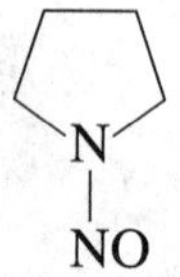

54. (a) Think mechanistically. Nucleophilic nitrogen + reactive carboxylic acid derivative with δ$^+$ carbonyl carbon atom:

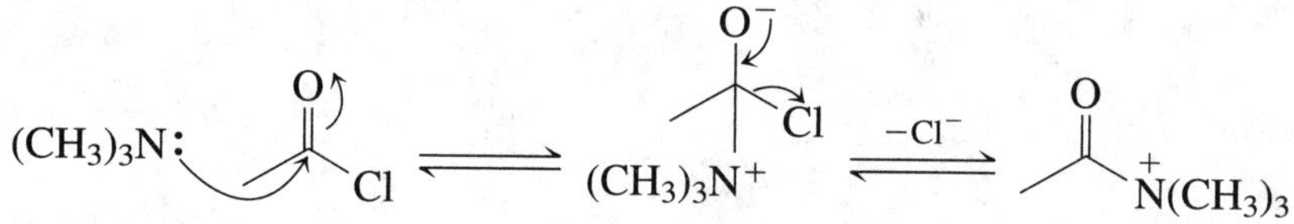

(CH$_3$)$_3$N: ⇌ Intermediate 'A' C$_5$H$_9$ClNO $\xrightarrow{-Cl^-}$ Intermediate 'B' (C$_5$H$_9$NO)$^+$

(b)

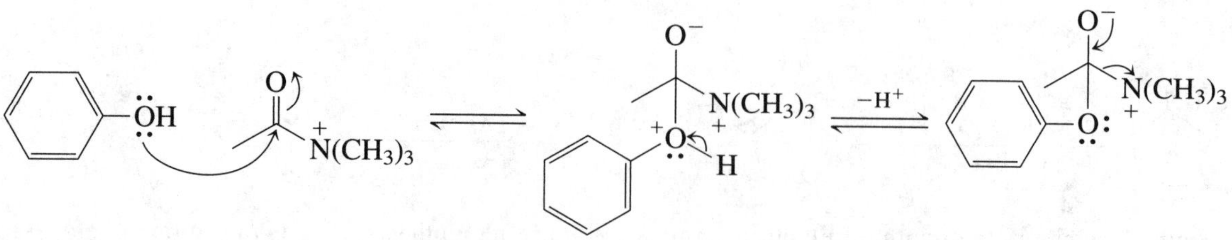

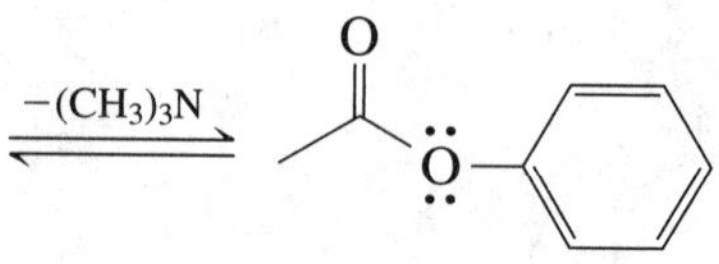

$\xrightarrow{-(CH_3)_3N}$

55. (a) Follow the hint to add the amine. The result is an enolate and, as we saw in Chapter 17, enolates add to aldehydes (aldol reaction). The aldol product has an ammonium ion, a good leaving group, in position to be eliminated to regenerate the α, β double bond of the product:

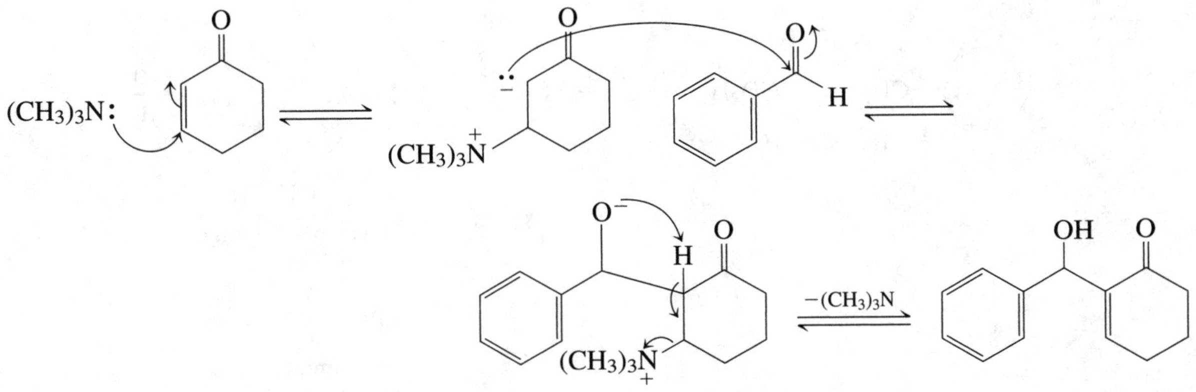

The Baylis-Hillman pattern is connection of the α carbon of the α, β-unsaturated ketone to the carbonyl carbon of the aldehyde, making an alcohol.

(b)

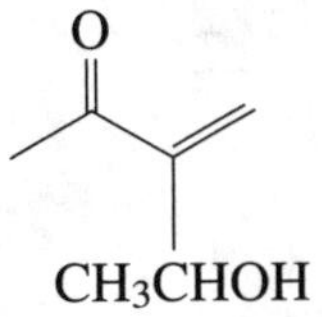

CH₃CHOH

(c)

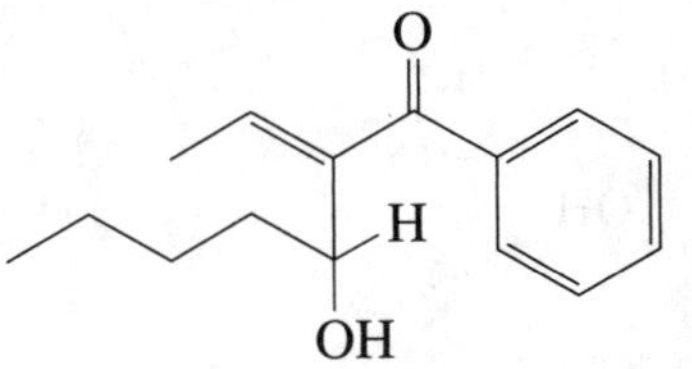

OH

56. Follow the mechanistic pattern **twice**. The first reductive amination forms a secondary amine with one methyl group on nitrogen. A second reductive amination can still take place, in accordance with the pattern shown in Section 21-6, to give the final dimethylated product.

First is (reversible) imine formation:

$$CH_2{=}O + H_2NR \longrightarrow CH_2{=}NR + H_2O$$

Then reduction gives the secondary amine:

$$NaBH_3CN + CH_2{=}NR \longrightarrow CH_3{-}NHR$$

Another $CH_2{=}O$ reacts, forming an iminium ion:

$$CH_2{=}O + HN(CH_3)R \longrightarrow CH_2{=}\overset{+}{N}(CH_3)R + HO^-$$

Reduction of this then gives the final product, a dimethylated amine:

$$NaBH_3CN + CH_2{=}\overset{+}{N}(CH_3)R \longrightarrow (CH_3)_2NR$$

57.

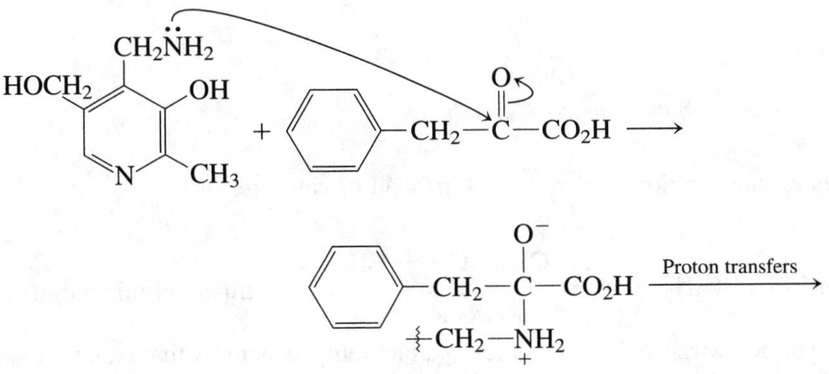

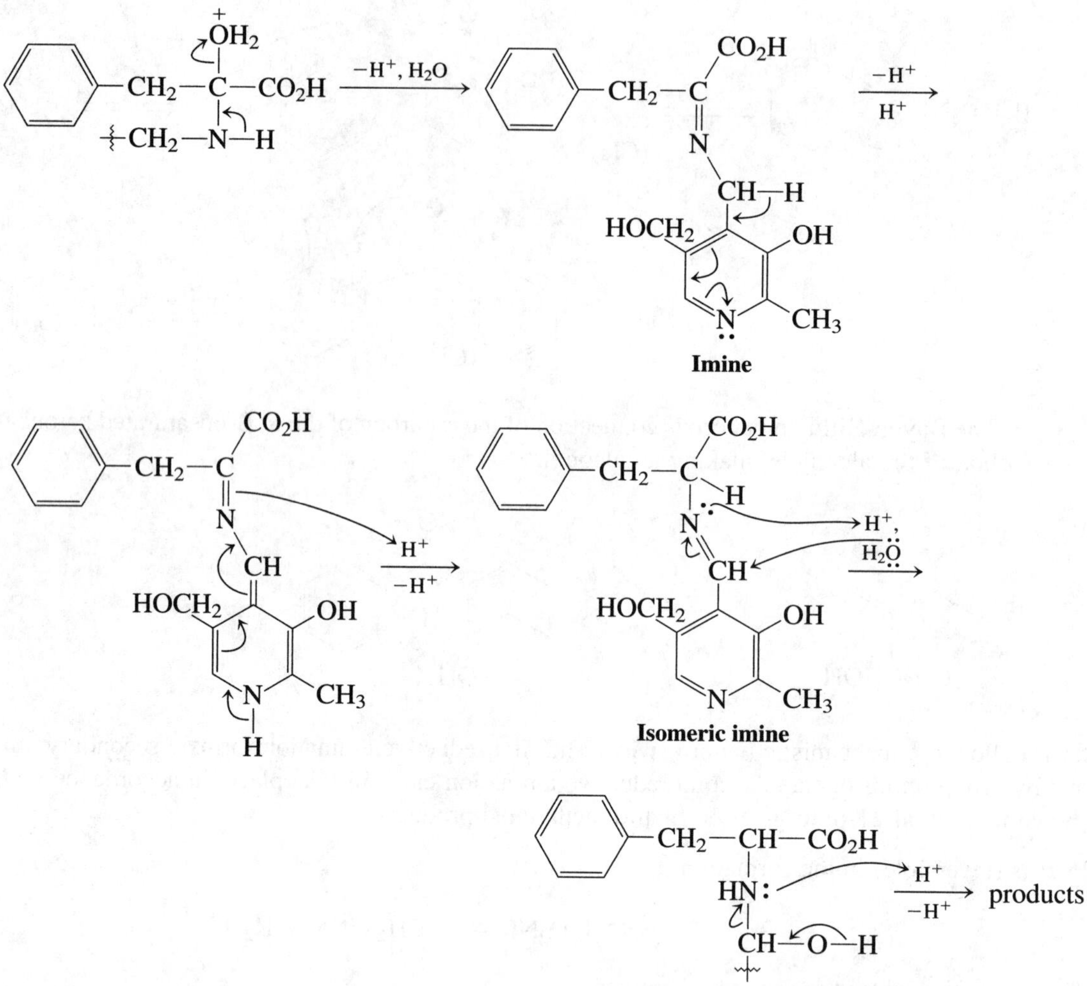

Imine

Isomeric imine

58. IR: Secondary amine.

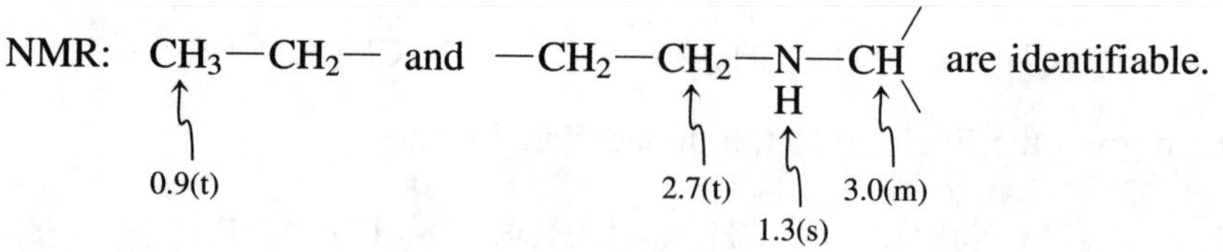

NMR: CH_3-CH_2- and $-CH_2-CH_2-N-CH\diagdown$ are identifiable.

$$0.9(t) \qquad 2.7(t) \quad \underset{H}{|} \quad 3.0(m)$$

$$1.3(s)$$

Total of 17 hydrogens in the molecule.

MS: m/z 127 − 17 (H's) − 14 (N) = 96 or 8 carbons, so $C_8H_{17}N$.

$H_{sat} = 16 + 2 + 1 = 19$; degrees of unsaturation $= \frac{(19 - 17)}{2} = 1$ π bond or ring present.

MS: Base peak is $[M − 43]^+$, or loss of C_3H_7, perhaps $\underbrace{CH_3-CH_2-CH_2-}_{\text{From NMR}}$. Hofmann elimination results:

Reattach N to alkene carbons in various ways to see what's reasonable. Only structures that can eliminate to give **both** 1,4- and 1,5-octadiene are worth further consideration.

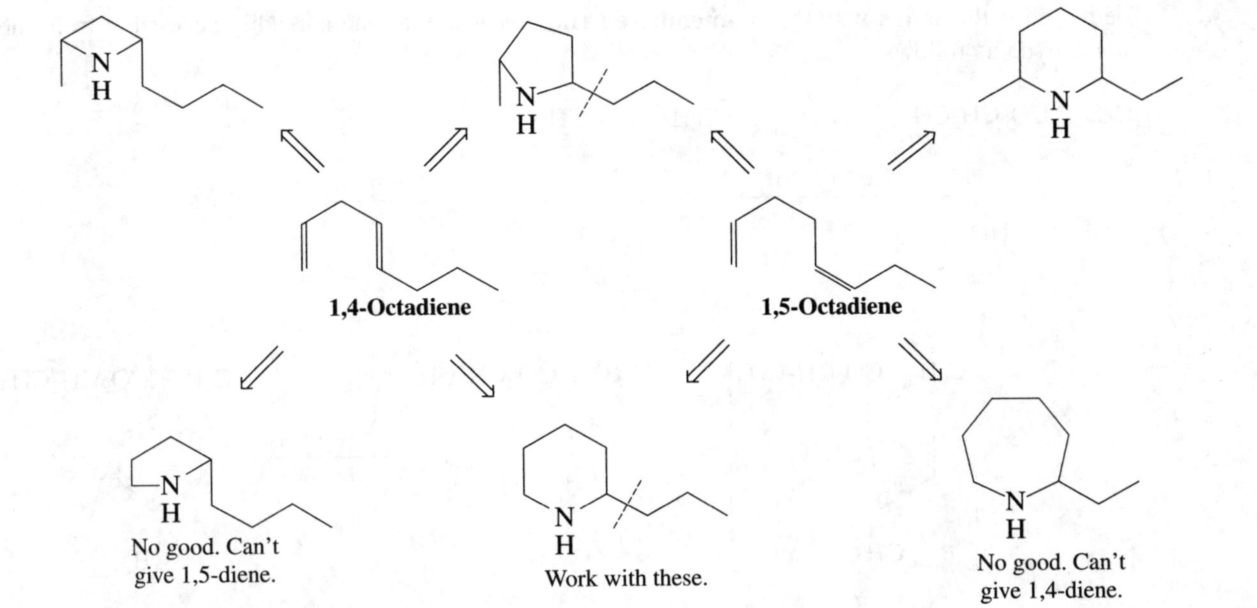

1,4-Octadiene

1,5-Octadiene

No good. Can't give 1,5-diene.

Work with these.

No good. Can't give 1,4-diene.

Both structures in the center should lose C_3H_7 in the MS (see dashed lines). The top one doesn't fit the NMR though; it should have 2 H's on the carbons attached to N. It also should show a strong $[M - 15]^+$ peak in the MS (loss of CH_3), which is not seen, and it should give another Hofmann product: 2,4-octadiene. The only possible correct structure therefore is

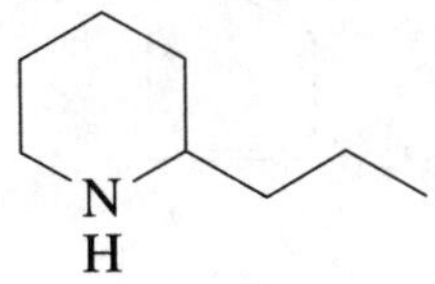

59. Work backward; make sure you include all 15 carbons in your answer.

(a)

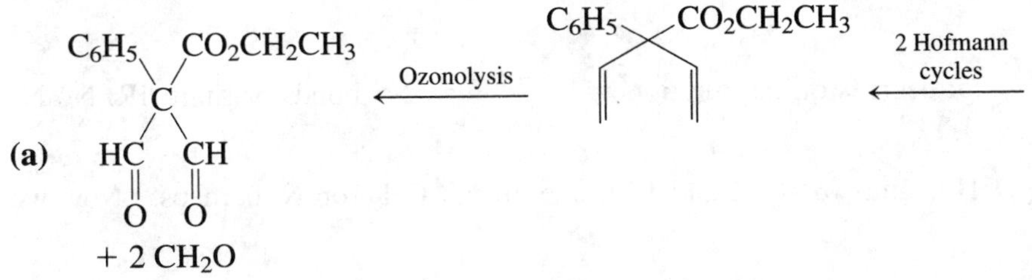

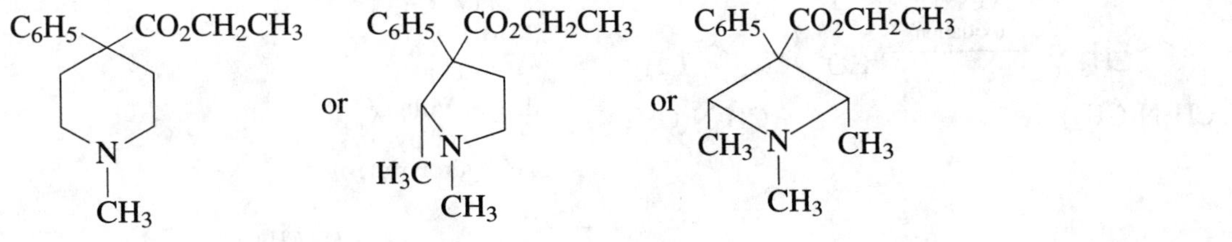

C_6H_5 $CO_2CH_2CH_3$ $\xleftarrow{\text{Ozonolysis}}$ C_6H_5 $CO_2CH_2CH_3$ $\xleftarrow{\text{2 Hofmann cycles}}$

$+ 2 CH_2O$

(The extra CH_3 on nitrogen is needed to give the correct molecular formula.) No way yet to choose which of the three is the correct structure.

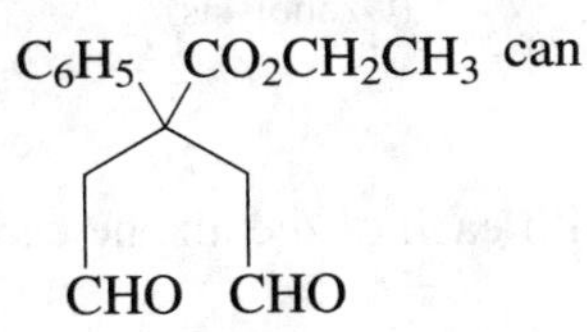

(b) Big hint: C_6H_5 $CO_2CH_2CH_3$ can be converted into pethidine. This strongly suggests that pethidine is the amine with the **six-membered ring** because that one is readily accessible from the dialdehyde as follows:

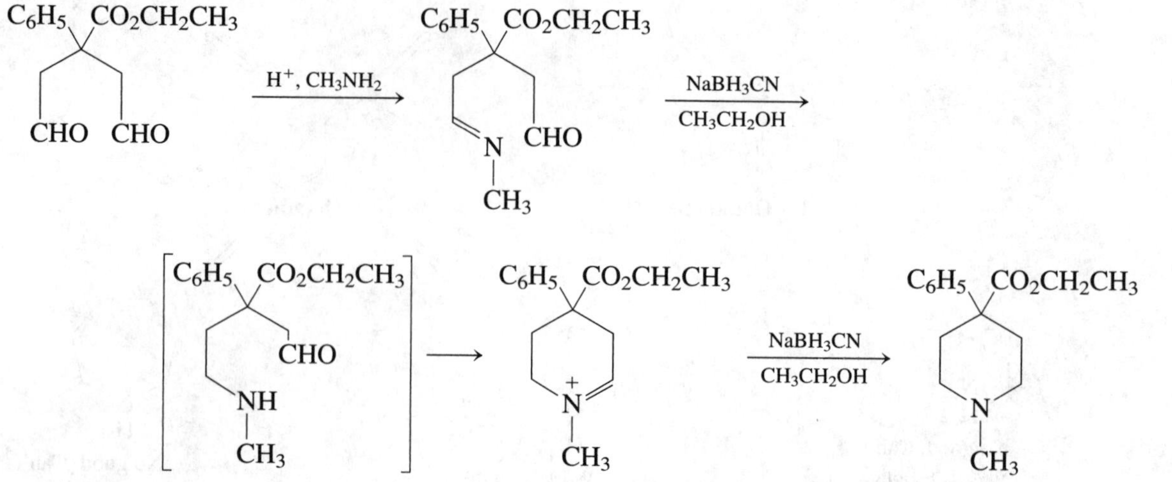

The sequence is actually carried out all at once, by mixing amine, dialdehyde, and $NaBH_3CN$ together.

The dialdehyde synthesis:

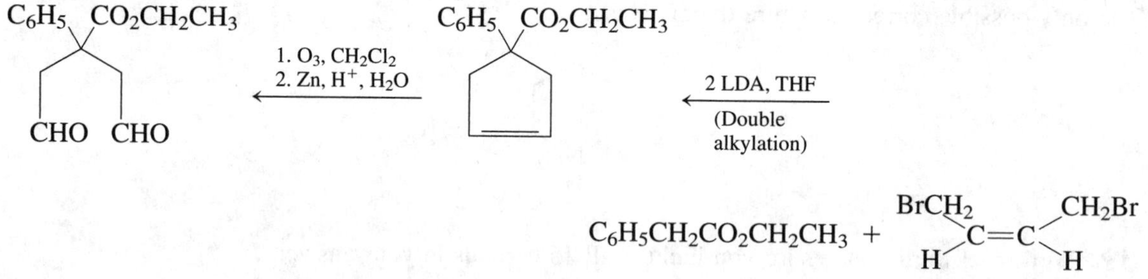

60. $H_{sat} = 22 + 2 + 1(N) = 25$; degrees of unsaturation $= \frac{(25-21)}{2} = 2$ π bonds or rings. IR: No N—H bonds, so amine is tertiary.

NMR: Two dissimilar CH_3—$CH\big<$ units ($\delta = 1.2$ and 1.3); one unsplit CH_3 (on N, perhaps?) Now work backward.

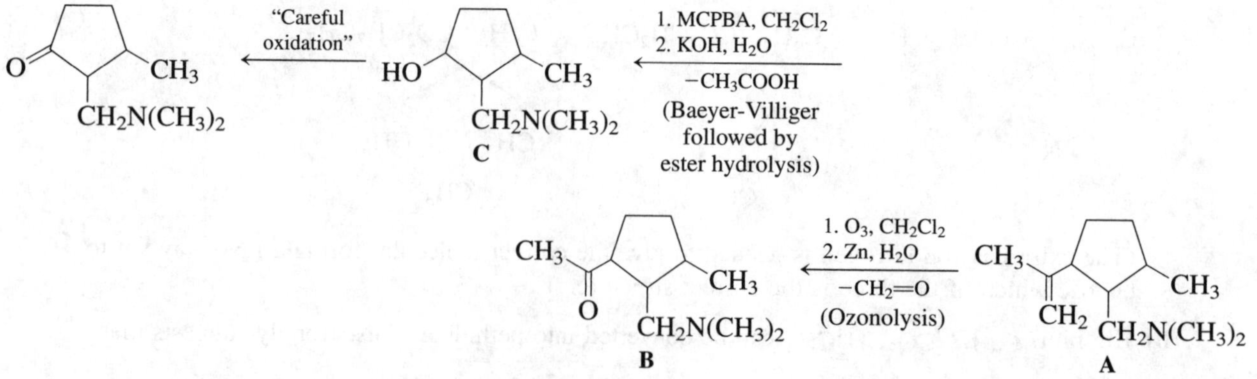

Following everything so far? Now reconnect the nitrogen in A with each of the alkene carbons to establish possible structures before Hofmann elimination.

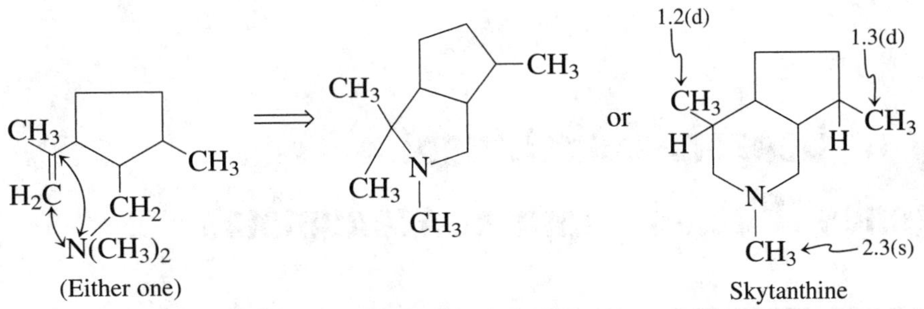

(Either one)

Both are $C_{11}H_{21}N$

Skytanthine

The methyl signals in the NMR match only the second structure, which is the correct one.

61.

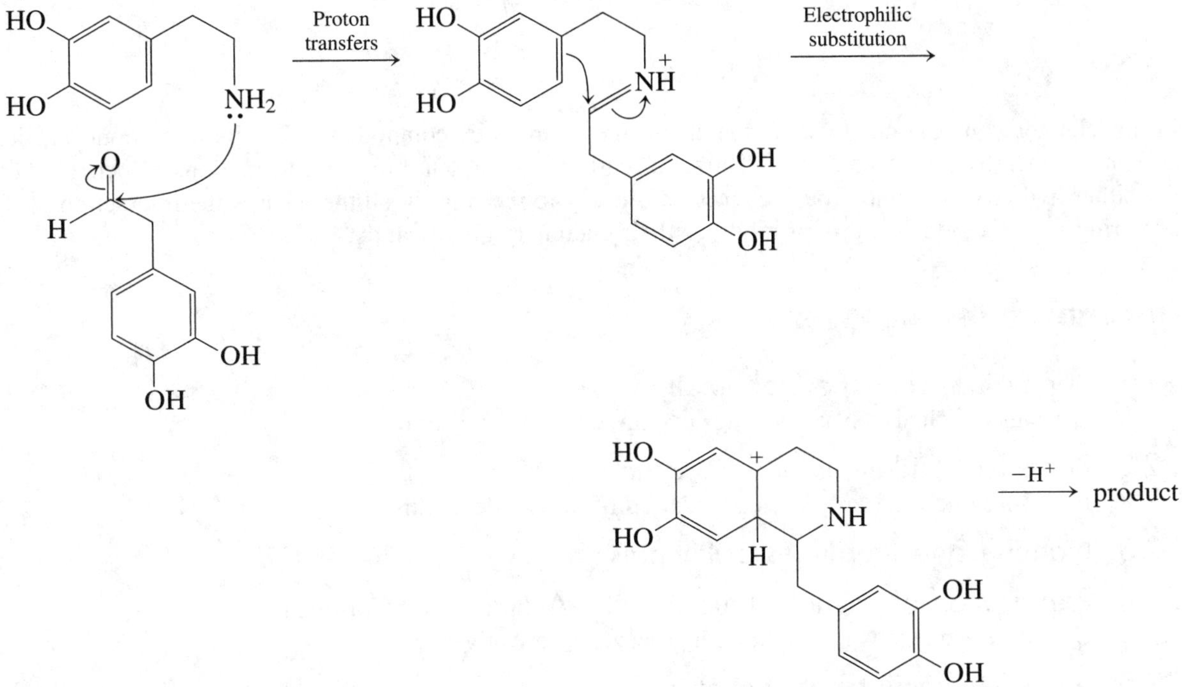

The electrophilic substitution is in fact another variation of the Mannich reaction as well, with the electron-rich benzene ring acting as the nucleophile.

62. (a) The catalyst, being a salt, should have some solubility in the water. Because it contains several hydrocarbon substituents on nitrogen, it should also possess some solubility in decane as well.

(b) The ionic salt NaCN is essentially insoluble in decane. Thus the concentration of nucleophilic cyanide ion in solution is exceedingly tiny, and S_N2 reaction is correspondingly slow.

(c) The ammonium cation may exchange counterions (cyanide for chloride) when it is in the water layer, and upon transfer to the decane, cyanide may accompany it. Thus, a significant concentration of cyanide may develop in the decane to participate in S_N2 reaction with the chlorooctane.

63. (c)

64. (c)

65. (e)

66. (a)

67. (d)

22

Chemistry of Benzene Substituents:
Alkylbenzenes, Phenols, and Benzenamines

In this chapter you continue to study the chemistry of aromatic compounds. You find out about reactions at a carbon attached to a benzene ring (a "benzylic" carbon) and about more transformations of benzenes containing either hydroxy or amino groups connected directly to the ring. In all these cases the interaction of the benzene ring with the attached groups modifies their chemistry significantly.

Outline of the Chapter

Keys to the Chapter

22-1. Benzylic Resonance Stabilization

Chemical reactivity at a saturated carbon atom attached to a benzene ring is greatly enhanced over that of ordinary alkyl carbons. Delocalization stabilizes both reaction transition states and intermediates. After reading this section once, go back and take a good look at the sets of resonance forms for the phenylmethyl (benzyl) radical, cation, and anion. There are four forms for each one, leading to increased stability and ease of formation.

Most often encountered is nucleophilic substitution on compounds containing potential leaving groups on the benzylic carbon. Both the carbocation intermediate for an S_N1 mechanism and the transition state for the S_N2 mechanism are stabilized, so both pathways may be followed. Just as you saw a long time ago with secondary haloalkanes (Section 7-5), and more recently with allylic systems (Sections 14-1 through 14-4), the choice of mechanism is determined by the specific conditions and reagents. Also, in analogy with allylic compounds, both free radical and anionic reactions take place at benzylic carbons as well. Notice, however, that the products of these reactions never involve attachment of groups to the benzene ring itself: Double bond migrations, which are often seen with allylic systems, do not occur with benzylic ones, because the aromaticity of the benzene ring would be lost (see Problem 39).

22-2. Benzylic Oxidations and Reductions

Of the several reactions in this section, the one you are most likely to encounter is the oxidation of alkyl side chains on benzene rings with $KMnO_4$ to form benzoic acids. Notice two features: First, all carbons but the benzylic carbon are chewed off by the reagent and disappear, and second, an *o,p*-directing alkyl group is converted into an *m*-directing COOH group.

22-4. Preparation of Phenols: Nucleophilic Aromatic Substitution

The two **nucleophilic** aromatic (*ipso*) substitution reactions are useful synthetic processes. The addition–elimination version is a common reaction of benzenes containing a leaving group together with good electron-withdrawing (anion-stabilizing) groups like NO_2. The elimination–addition ("benzyne") mechanism occurs when benzenes containing a leaving group, but no other anion-stabilizing groups, are treated with strong bases. Although these nucleophilic substitution reactions are indeed useful, you should keep them in perspective; they are encountered only rarely relative to the electrophilic reactions described earlier. You have already seen the reasons for this: Arenes normally contain electron-rich π systems and are naturally most easily attacked by electrophiles. Only in the presence of strong electron-withdrawing groups or very strong bases will nucleophilic attack be likely to occur.

22-3, 22-5 through 22-7. Reactions of Phenols

The behavior of aromatic alcohols is so different from that of ordinary alcohols that the properties of the aromatic alcohols merit special coverage. Most of the differences are due to the ability of the benzene ring to delocalize a lone pair of electrons from the phenol oxygen. The immediate result is that phenols are **more acidic** and **less basic** than ordinary alcohols. The conjugate bases of phenols, phenoxide ions, are much less basic than alkoxide ions and can therefore be generated by reaction of phenols with HO^-. Being less basic, phenoxide ions are much better leaving groups than are hydroxide or alkoxide ions (see Problem 59 for a practical consequence of this), but they are still good nucleophiles, especially useful in either synthesis via S_N2 reaction.

The resonance between phenol oxygen and the benzene ring also affects benzene reactivity. The extra electron density in the ring makes it susceptible to electrophilic attack by even rather weak electrophiles: Reactions with formaldehyde and CO_2 in the presence of base are examples.

22-8. Oxidation of Phenols: Benzoquinones

The reversible redox relationship between 1,4-benzenediols and *p*-benzoquinones, or just quinones, is a special one because of its relative ease. Conjugate addition and Diels-Alder cycloaddition are common reactions of quinones; conjugate additions have biological importance. The isomeric but less stable 1,2-compounds are encountered much less frequently.

22-9. Oxidation–Reduction Processes in Nature

After a discussion of types of reactions that can damage biological molecules such as lipids in cell membranes, this section illustrates the ways antioxidant molecules such as vitamin E inhibit these processes. These antioxidants share one property in common with 1,4-benzenediols: They are easily oxidized, although they usually aren't oxidized to quinones (their structures don't usually permit that). Read carefully to find out about some molecules that are good for you!

22-10 and 22-11. Arenediazonium Salts and Diazo Coupling

Like aromatic alcohols, aromatic amines are also special. In particular, resonance delocalization of the lone pair on nitrogen into the benzene ring makes aromatic amines much less basic than their alkyl relatives. However, most of the reactions of aromatic amines are similar enough qualitatively to those of alkanamines that there's really no need to rehash all that stuff. Instead, this section presents just one class of reactions of these compounds, chosen because of its special versatility in synthesis. Diazotization of a primary benzenamine produces an arenediazonium salt containing an —N$_2^+$ substituent on the benzene ring. The value of these salts lies in the ease of replacement of this group by any of the following (reagents are given in parentheses):

$$-H \ (H_3PO_2) \qquad -OH \ (H_2O, \Delta) \qquad -Cl, Br \ (CuX, \Delta)$$
$$-I \ (I^-, \Delta) \qquad -CN \ (CuCN, CN^-, \Delta)$$

Arenediazonium salts are also electrophilic enough to react with phenols or benzenamines to form so-called azo dyes via the *diazo coupling* process.

Solutions to Problems

35. (a)

(b)

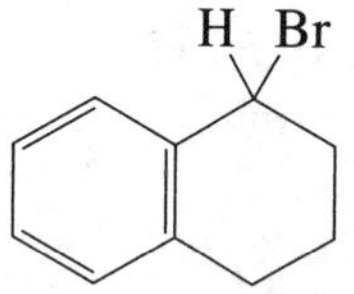

36. Radical chemistry: initiation, propagation, and termination (which we will ignore).

INITIATION:

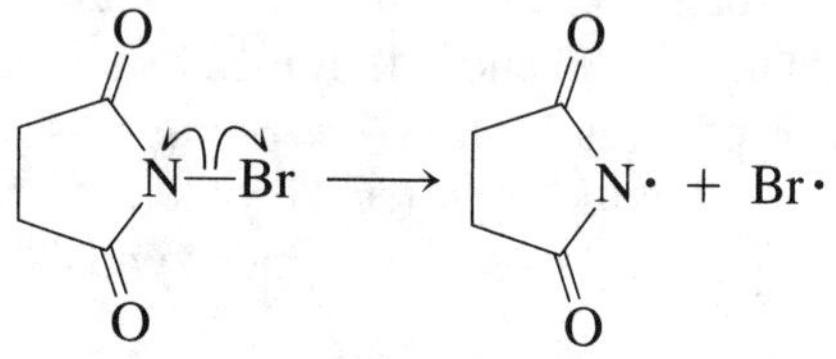

PROPAGATION:

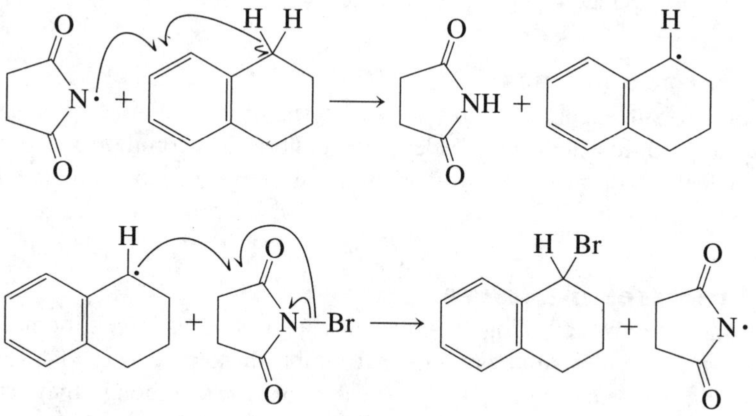

37. Several reactions derive from fundamental material presented in earlier chapters.

(a)

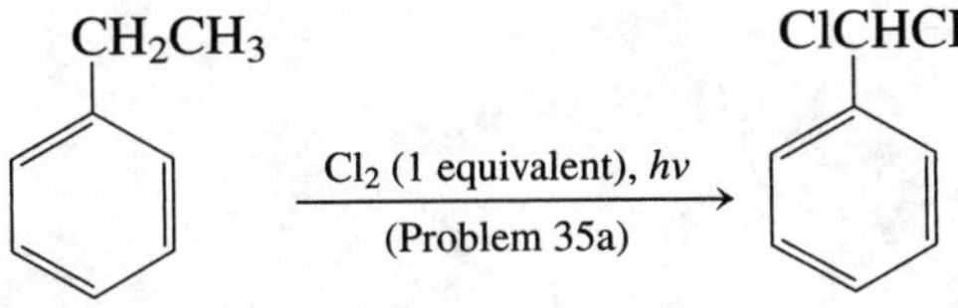

$$\xrightarrow[\text{(Problem 35a)}]{\text{Cl}_2 \text{ (1 equivalent)}, \, hv}$$

CH$_2$CH$_3$ → ClCHCH$_3$

(b)

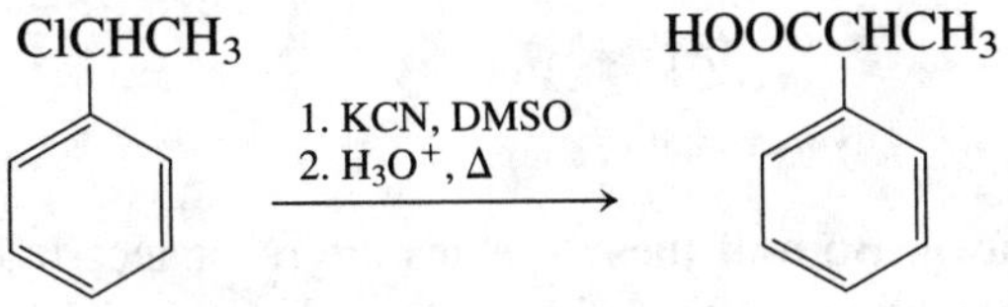

ClCHCH$_3$ $\xrightarrow[\text{2. H}_3\text{O}^+, \, \Delta]{\text{1. KCN, DMSO}}$ HOOCCHCH$_3$

(c)

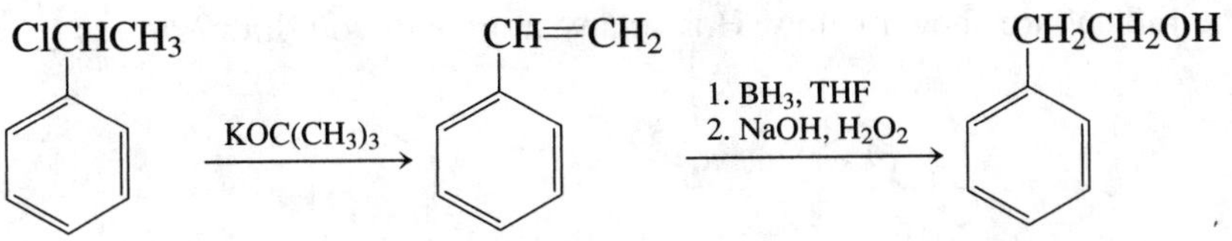

ClCHCH$_3$ $\xrightarrow{\text{KOC(CH}_3)_3}$ CH=CH$_2$ $\xrightarrow[\text{2. NaOH, H}_2\text{O}_2]{\text{1. BH}_3, \text{THF}}$ CH$_2$CH$_2$OH

(d)

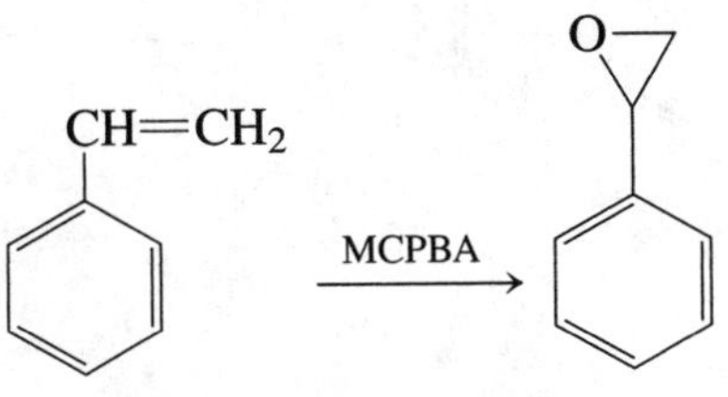

CH=CH$_2$ $\xrightarrow{\text{MCPBA}}$ (epoxide)

38.

ClCH$_2$ $\xrightarrow{-\text{Cl}^-}$ [CH$_2^+$ ↔ CH$_2$ ↔ CH$_2$ ↔ CH$_2$]

ClCH$_2$ / OCH$_3$ $\xrightarrow{-\text{Cl}^-}$ [CH$_2^+$ ↔ CH$_2$ ↔ CH$_2$ ↔ CH$_2$ ↔ CH$_2$] (OCH$_3$ ... :OCH$_3$ ... $^+$OCH$_3$ ... OCH$_3$)

ClCH$_2$ / NO$_2$ $\xrightarrow{-\text{Cl}^-}$ [CH$_2^+$ ↔ CH$_2$ ↔ CH$_2$ ↔ CH$_2$] (NO$_2$)

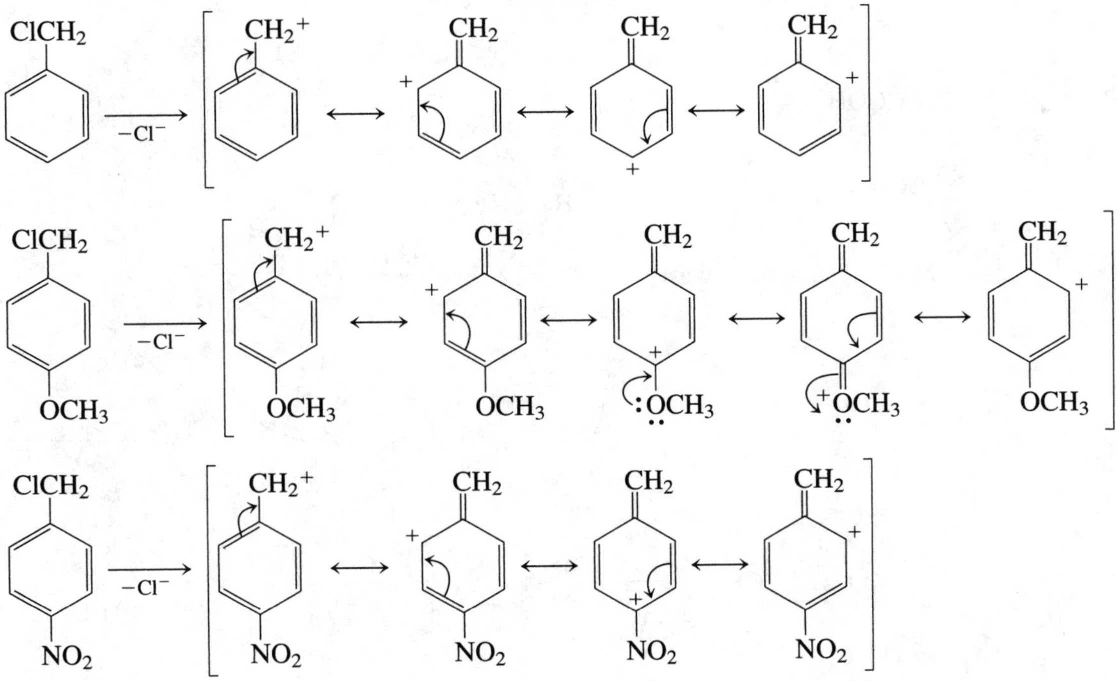

Poor

The cation derived from chloromethylbenzene (benzyl chloride, top) is stabilized by four resonance contributing forms. The cation from 1-(chloromethyl)-4-methoxybenzene (4-methoxybenzyl chloride, middle) is more stable because of the added contributor in which a lone pair on oxygen is delocalized into the ring (second Lewis structure from the right). The cation from 1-(chloromethyl)-4-nitrobenzene

(4-nitrobenzyl chloride, bottom) is least stable because one of its resonance forms, the one with the + charge next to the electron-withdrawing nitro group, is poor.

39.

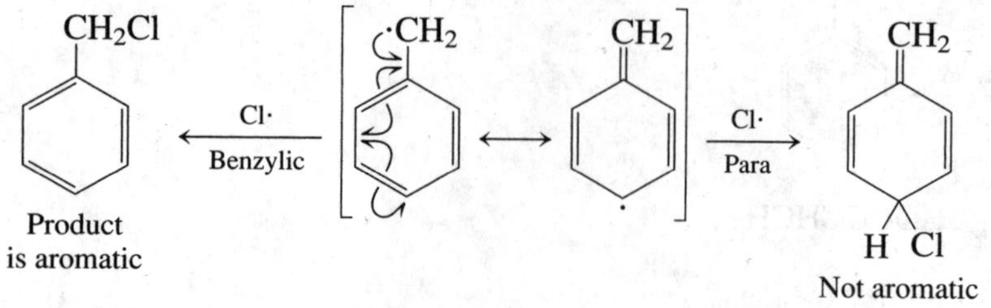
CH₂Cl
Cl·
Benzylic
·CH₂
CH₂
Cl·
Para
CH₂
H Cl
Product
is aromatic
Not aromatic

40. In each case numerous resonance forms can be drawn. Both of these species are resonance hybrids strongly stabilized by delocalization of the charge (in the cation) or the odd electron (in the radical). Three of the resonance forms for the radical are shown below. How many more can you draw?

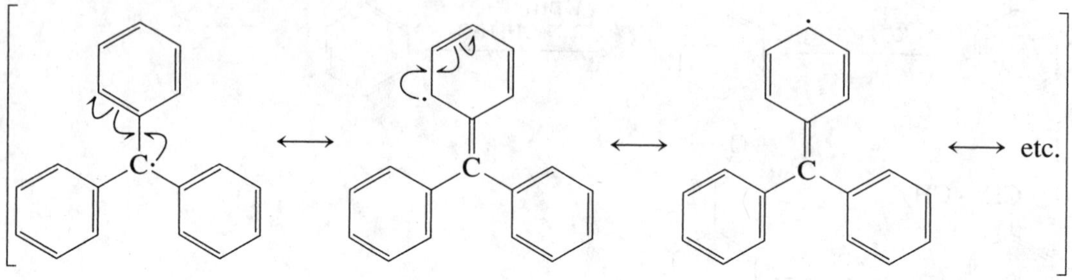
C
C
C
etc.

41. (a) $BrCH_2CH_2CH_2$–C_6H_4–CH_2OH Benzylic position is most reactive in nucleophilic substitution.

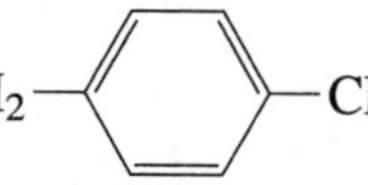

(b)

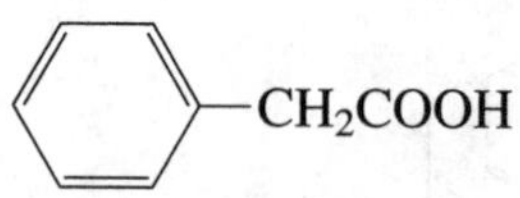
CH₂COOH

(c)

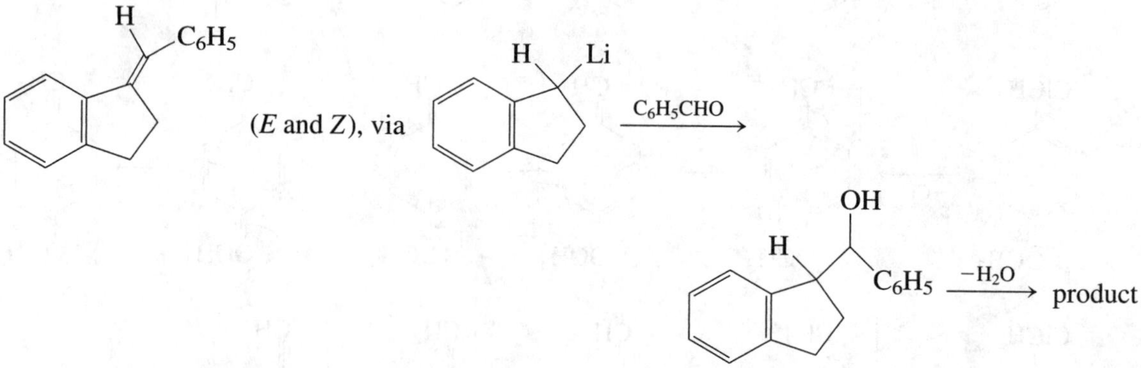
H
C₆H₅
(E and Z), via
H
Li
C₆H₅CHO
OH
H
C₆H₅
−H₂O
product

42.

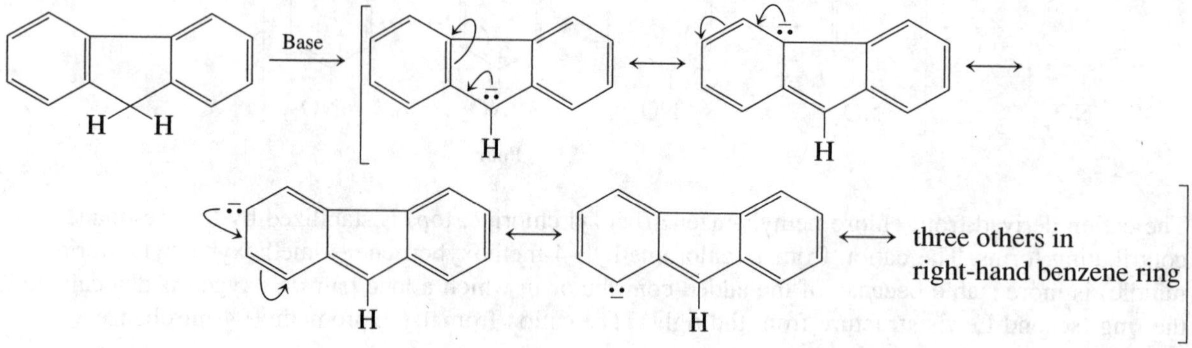
H H
Base
H
H
H
H
three others in
right-hand benzene ring

Seven resonance forms make this carbanion especially stable. It is also aromatic, having 14 π electrons in an unbroken loop of p orbitals.

43. **(a)** One solution, perhaps a bit roundabout:

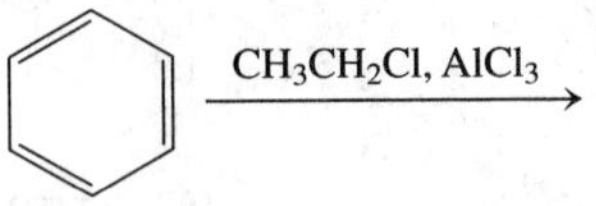

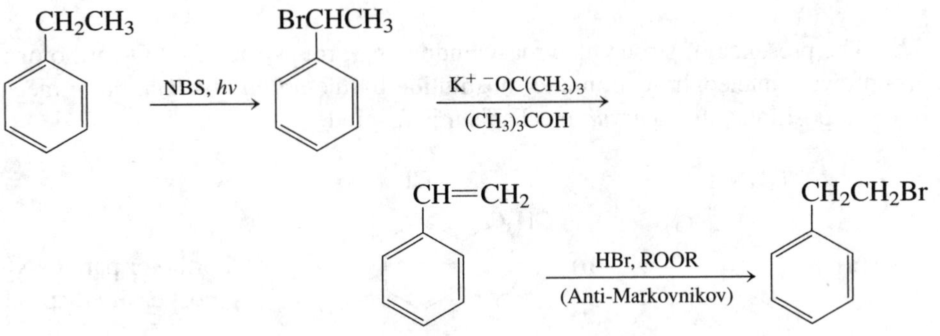

(b)

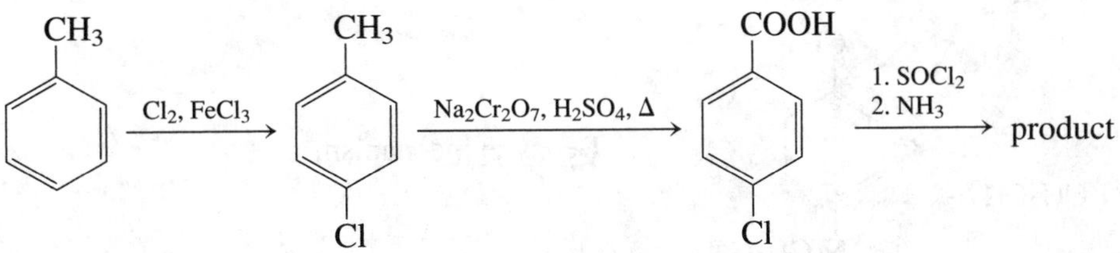

(c)

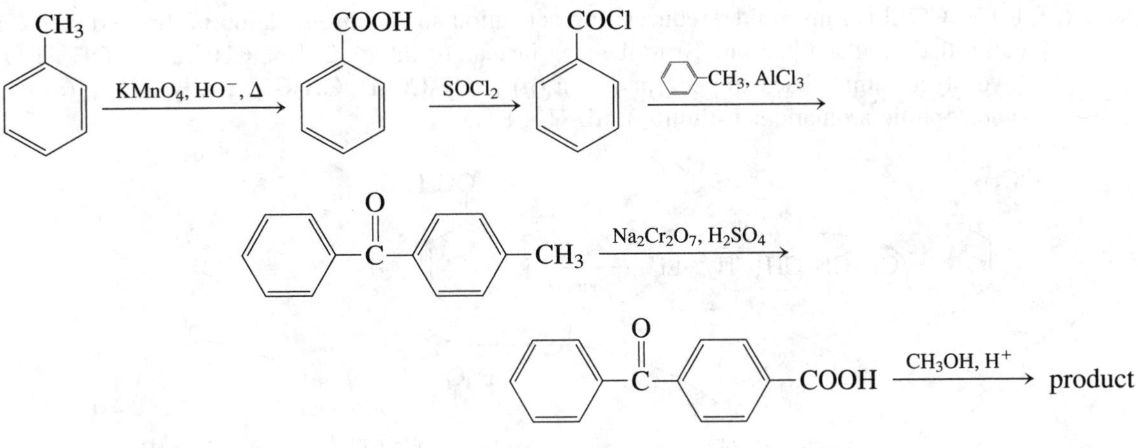

(d)

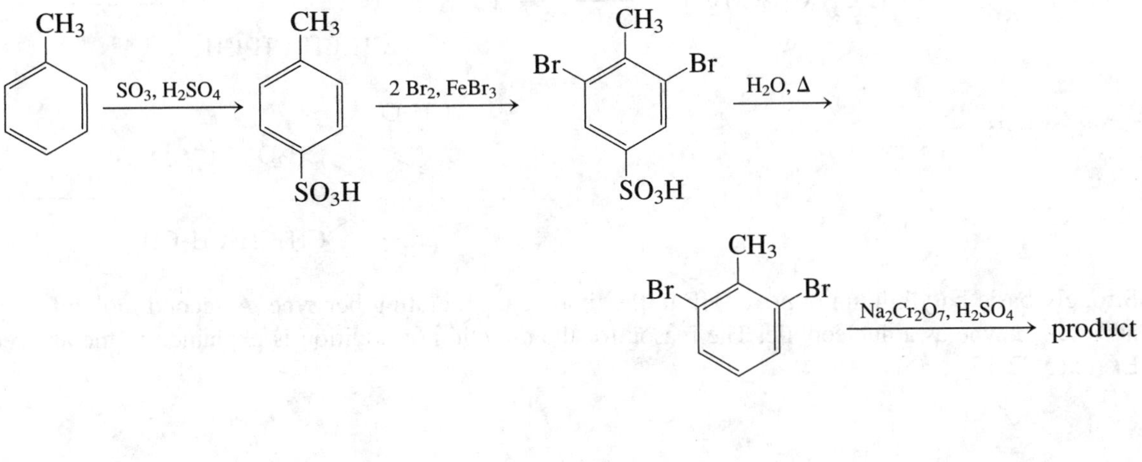

44. Most reactive ones have NO_2 groups ortho or para to the leaving group. So,

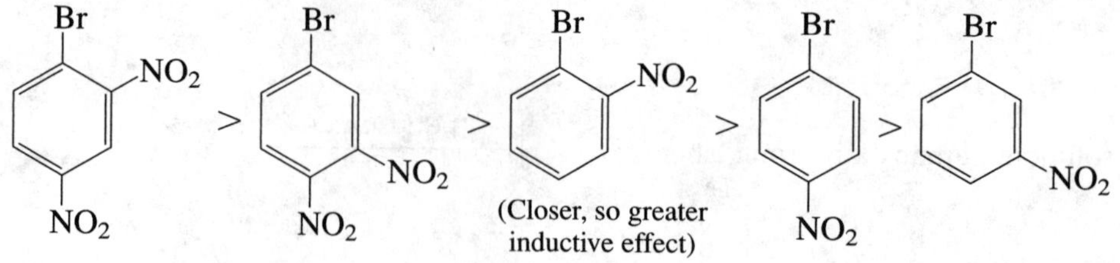

(Closer, so greater inductive effect)

45. The presence of strong electron-withdrawing groups such as NO_2 ortho or para to a potential leaving group favors nucleophilic aromatic substitution by the *addition–elimination* mechanism. Without such groups in these positions, the *benzyne* mechanism is favored.

(a)

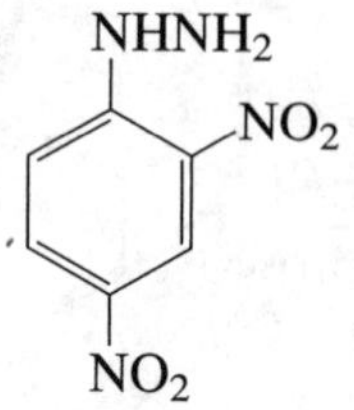

(b)

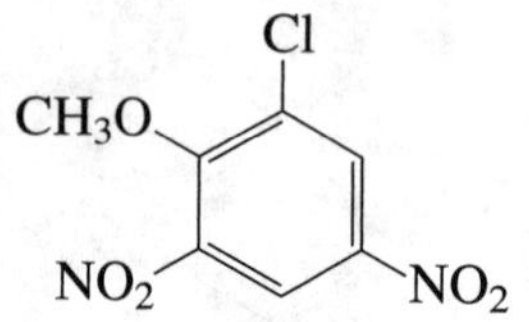

Cl ortho or para to NO_2's is most easily displaced

(c)

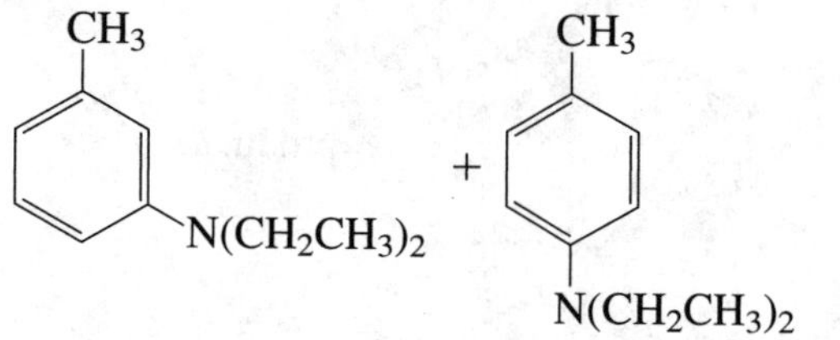

Benzyne mechanism

46. (a) 1. CH_3COCl (forms amide, reduces ring activation so that bromination in the next step can be controlled to attach just one Br to the ring instead of three), 2. Br_2, $CHCl_3$, 3. KOH, H_2O, Δ (hydrolyzes amide back to benzenamine); **(b)** 1. CF_3CO_3H, CH_2Cl_2, 2. Cl_2, $FeCl_3$; **(c)** KCN (nucleophilic aromatic substitution); **(d)** H^+, H_2O, Δ

47.

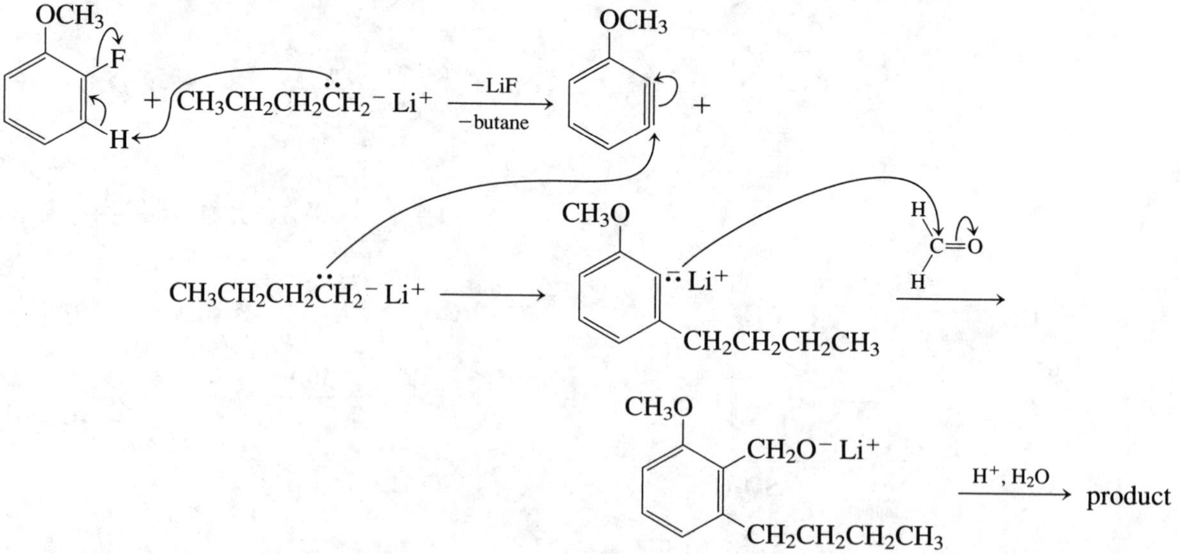

Strongly basic butyllithium removes HF in the first step, generating benzyne. A second mole of butyllithium adds to benzyne as a nucleophile. The reason for the direction of addition is explained in the answer to Exercise 22-12.

48. In nucleophilic aromatic substitution reactions that proceed by the addition-elimination mechanism, the *addition* step is rate-limiting. Fluorine is the most electronegative and, therefore, the most electron-withdrawing of the halogens. Therefore, the presence of F lowers the transition-state energy for nucleophilic addition the most of all the halogens, and stabilizes the resulting anion by an inductive effect to the greatest degree. While it is true that F^- is by far the worst leaving group among halide ions, expulsion of the leaving groups takes place *after* the rate-determining addition and is fast because it benefits from reestablishment of the aromaticity of the benzene ring. In Chapter 25 we will see even worse leaving groups than this expelled to generate aromatic rings.

49.

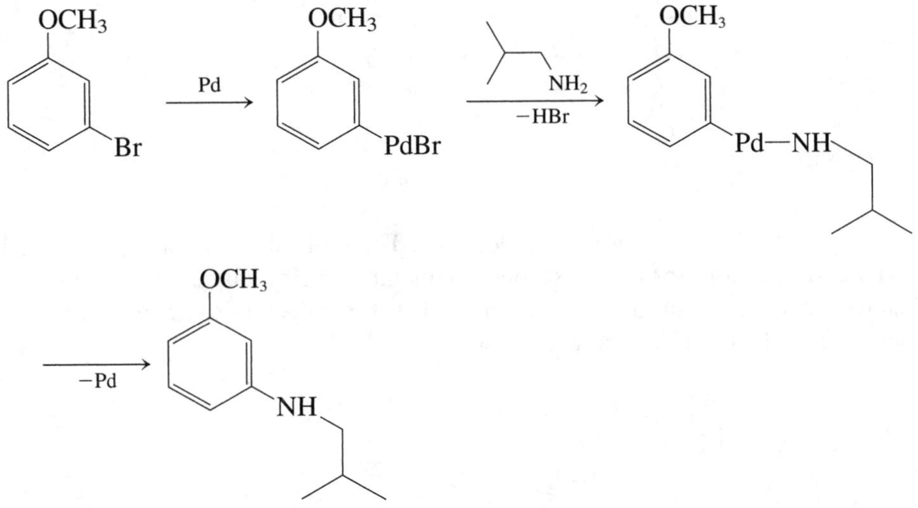

50. (a) F_3C—⟨benzene⟩—N⟨piperidine⟩   (b) ⟨benzene⟩—SCH_2CH_2CH_3

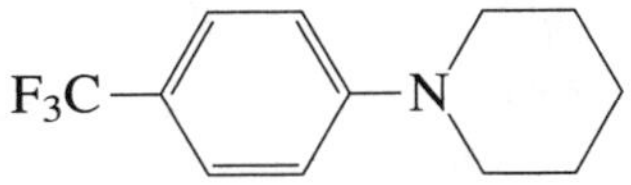

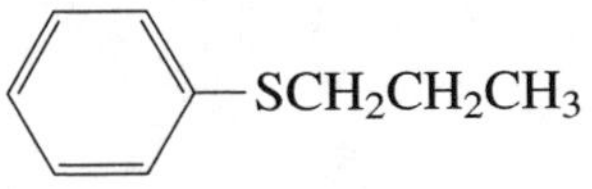

(c) CH_3O—⟨benzene⟩—CN   (d) ⟨structure⟩

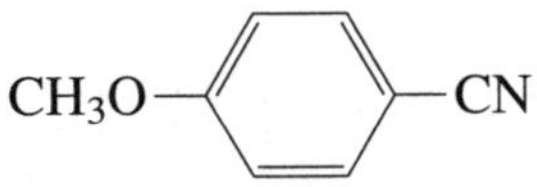

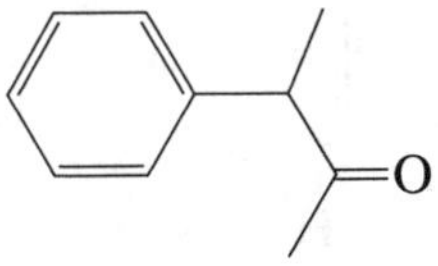

51. (a) $Ph_3P=CH_2$ (Wittig); (b) CH_3COCl, NaOH, H_2O; (c)  Br—⟨benzene⟩—OCOCH_3, $Pd(OCOCH_3)_2$, R_3P, 100°C (Heck); (d) NaOH, H_2O.

52. The first step introduces oxygen into the ring. Nucleophilic aromatic substitution is the most reasonable method. The resulting phenol must be used as a nucleophile itself to attack an appropriate substrate in order to give the final product.

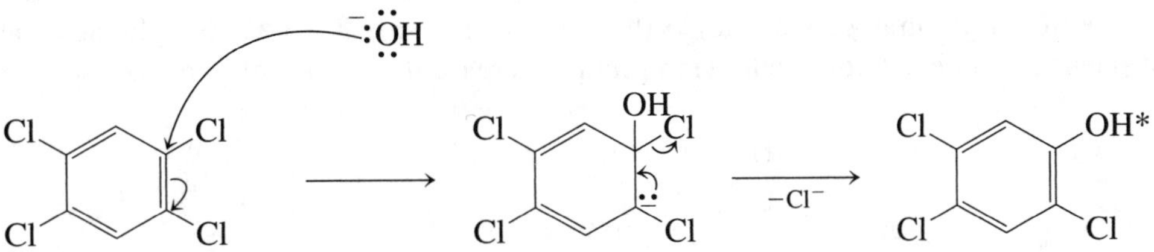

*Acidification is necessary because this phenolic hydroxy group will be deprotonated at the pH used for the initial nucleophilic aromatic substitution.

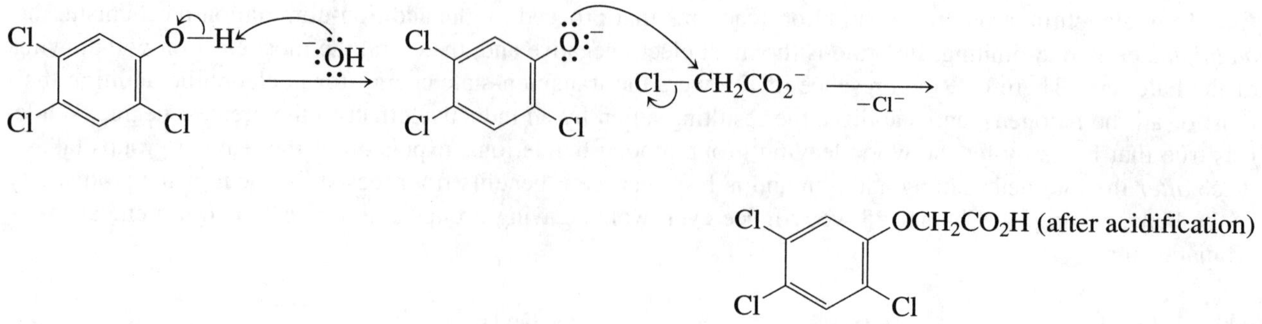

Cl—OCH₂CO₂H (after acidification)

53. (a)

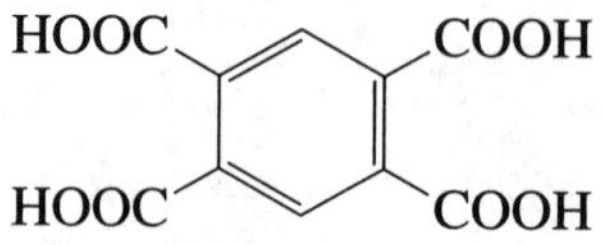

(b) The first step oxidizes the benzylic alcohol to an aldehyde. The molecule is then capable of an intramolecular aldol condensation, forming a second six-membered ring that contains an α, β-unsaturated ketone. This ketone, in turn, is unusual in that it is a cyclohexadienone, the ketone tautomer of a more stable phenol. Using equations, we have

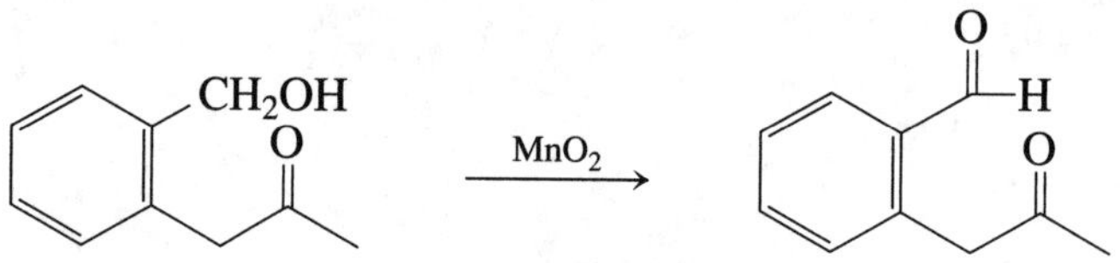

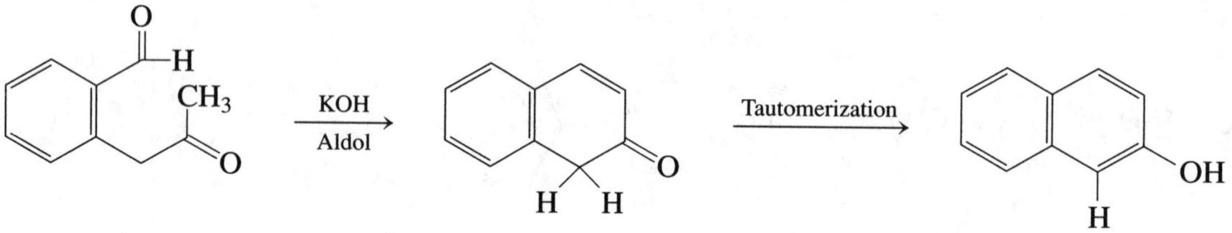

(c)

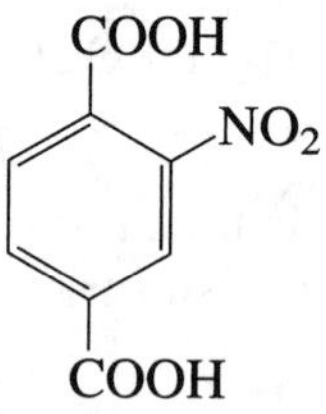

54. (c) (The SO₃H group is very acidic) > (b) > (e) > (f) > (d) > (a). Carboxylic acids are more acidic than most phenols, and electron-withdrawing groups increase the acidity of phenols.

55. (a)

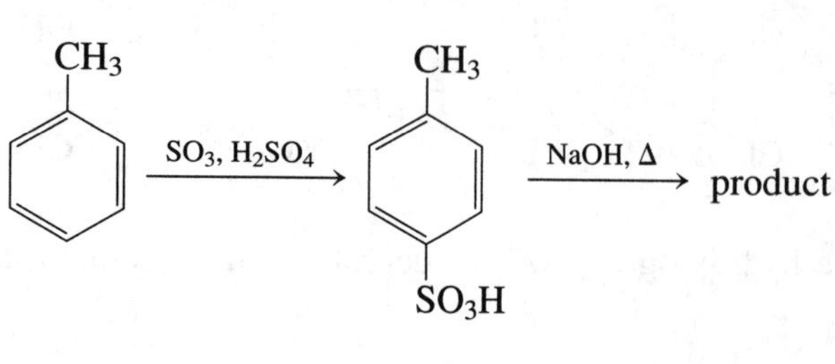

product

(b)

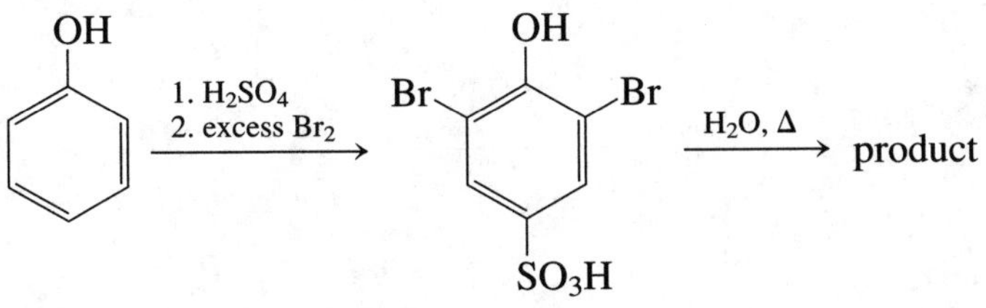
OH
1. H2SO4
2. excess Br2
OH
Br
Br
SO3H
H2O, Δ
product

(c)

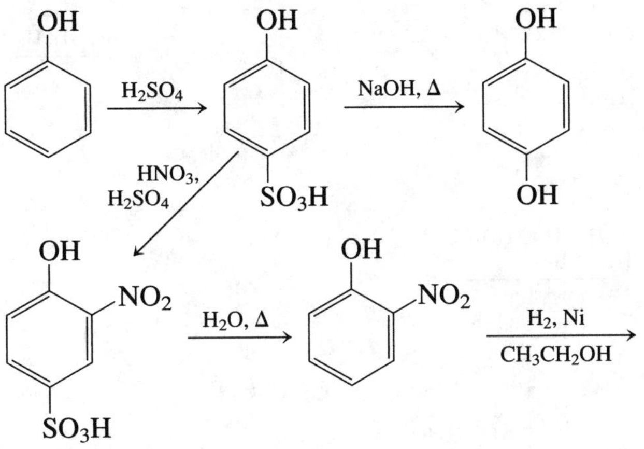
OH
H2SO4
OH
SO3H
NaOH, Δ
OH
OH
HNO3,
H2SO4
OH
NO2
SO3H
H2O, Δ
OH
NO2
H2, Ni
CH3CH2OH

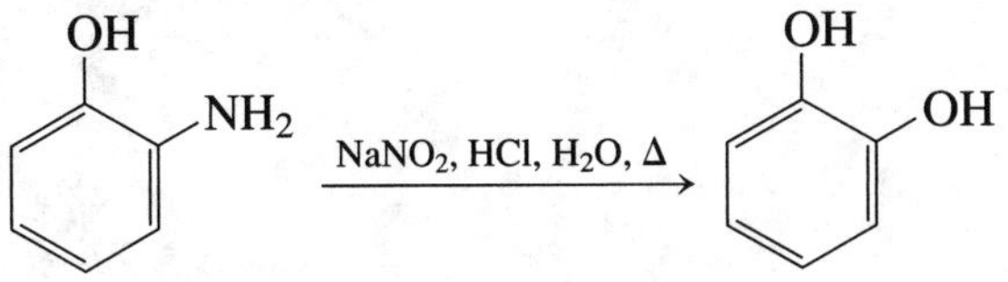
OH
NH2
NaNO2, HCl, H2O, Δ
OH
OH

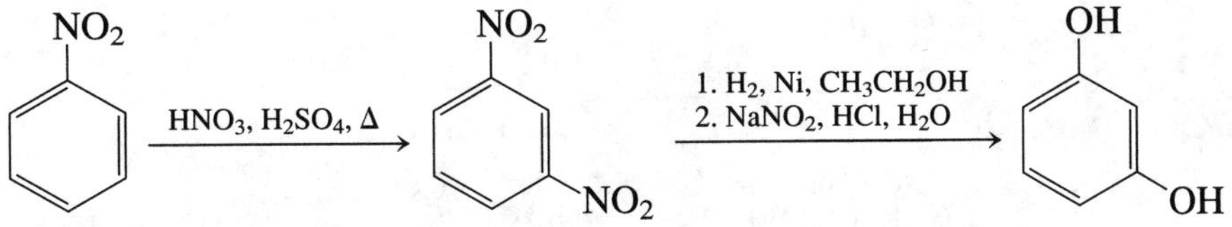
NO2
HNO3, H2SO4, Δ
NO2
NO2
1. H2, Ni, CH3CH2OH
2. NaNO2, HCl, H2O
OH
OH

(d)

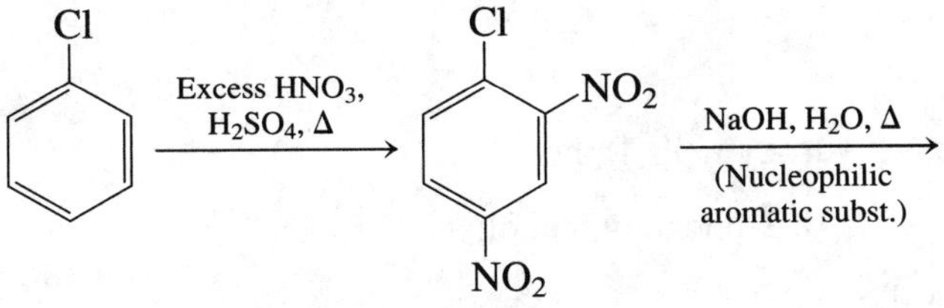
Cl
Excess HNO3,
H2SO4, Δ
Cl
NO2
NO2
NaOH, H2O, Δ
(Nucleophilic
aromatic subst.)

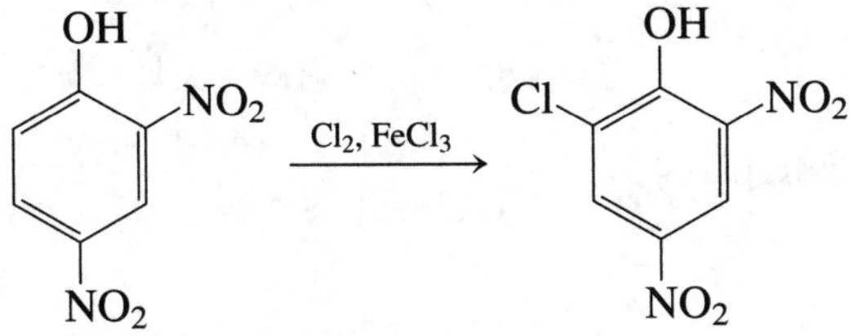
OH
NO2
NO2
Cl2, FeCl3
OH
Cl
NO2
NO2

56. **(a)**

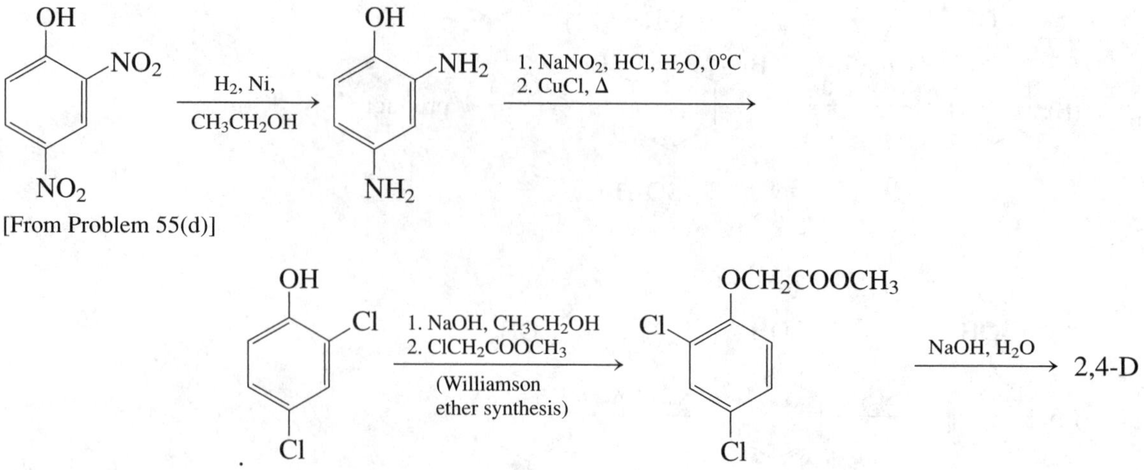

[From Problem 55(d)]

(b)

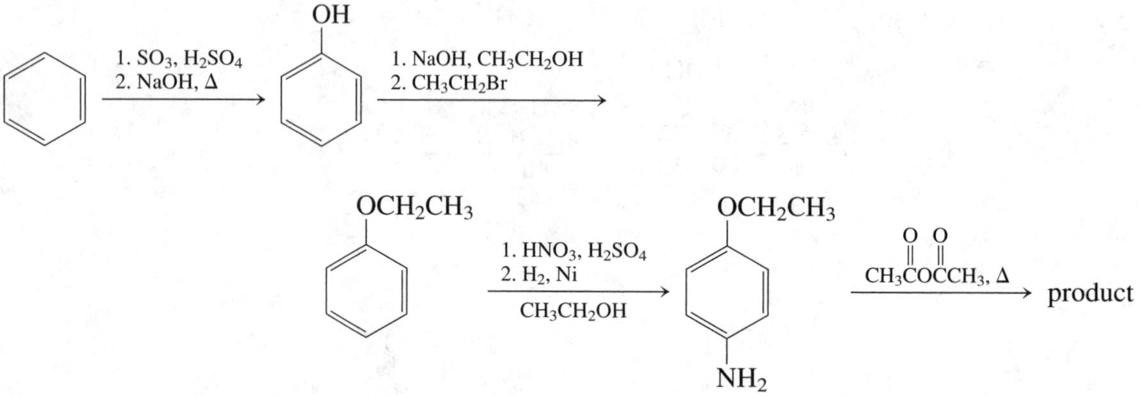

(c)

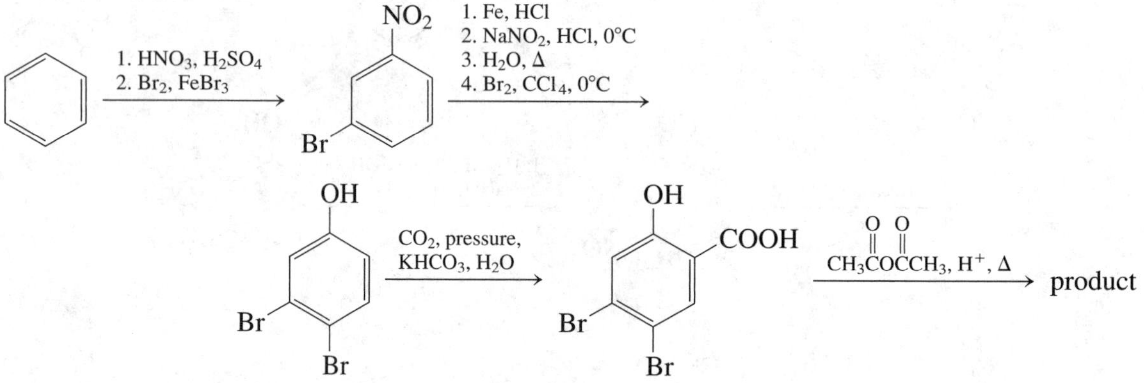

57. **(a)** 5-Bromo-2-chlorophenol; **(b)** 4-(hydroxymethyl)phenol;

(c) 2,4-dihydroxybenzenesulfonic acid; **(d)** 2-phenoxyphenol;

(e) 2-methylthio-2,5-cyclohexadiene-1,4-dione

58. **(a)**

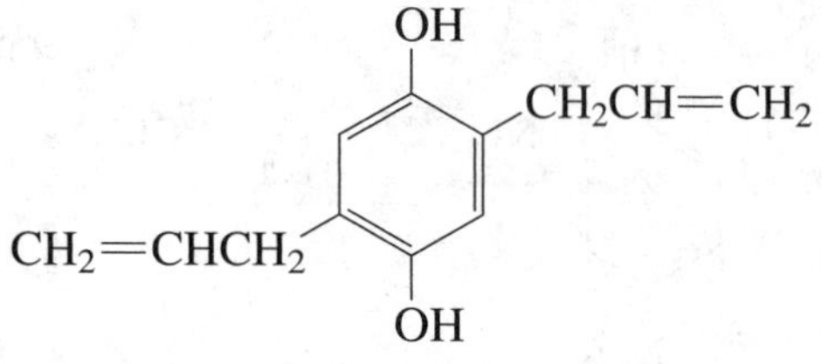

via **double** Claisen
rearrangement of

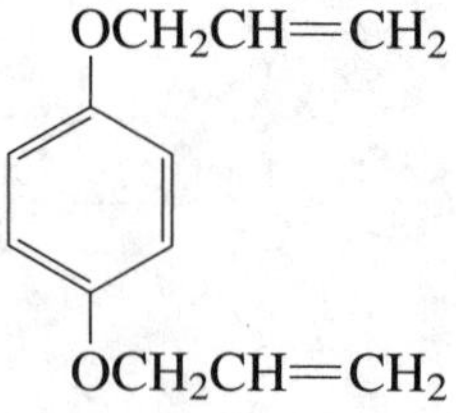

(b) Step-by-step:

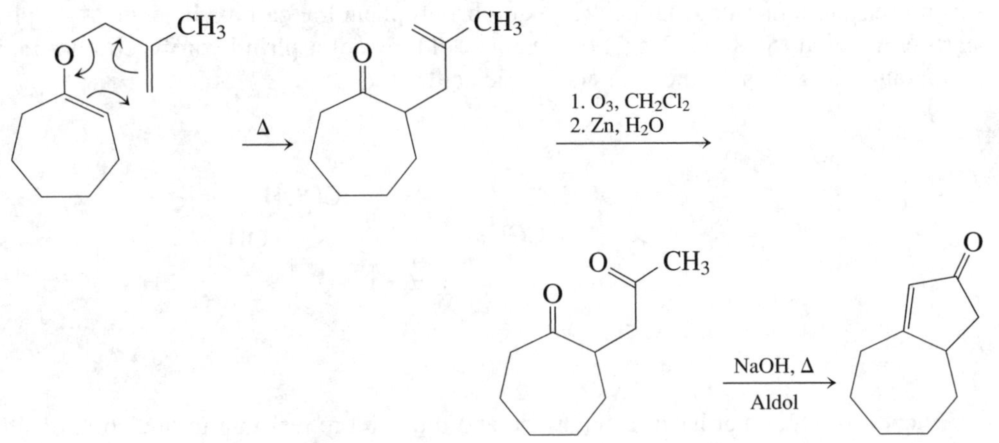
CH3
O
Δ
CH3
O
1. O3, CH2Cl2
2. Zn, H2O
O
CH3
O
NaOH, Δ
Aldol
O

(c)

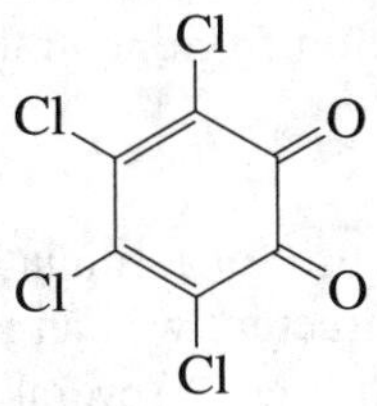
Cl
Cl
O
Cl
O
Cl

(d)

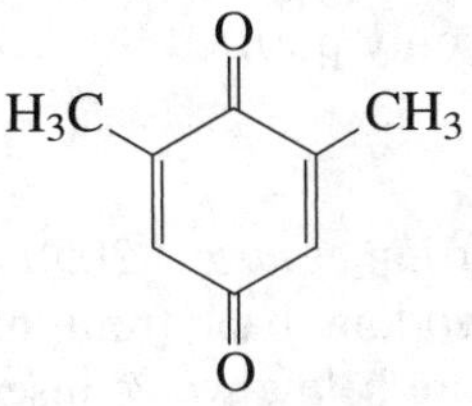
O
H3C
CH3
O

(e) The thiol is readily oxidized, so one likely reaction is

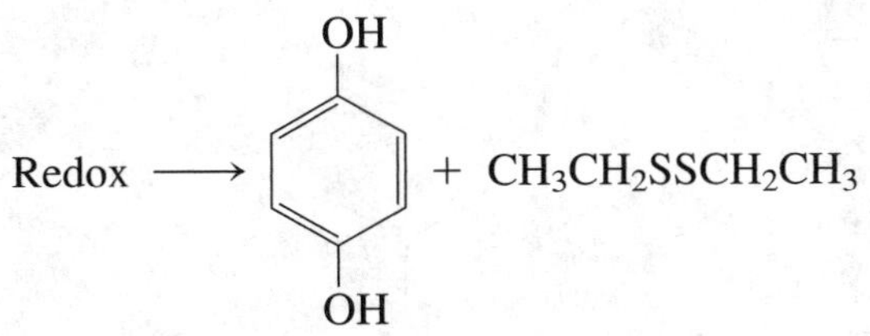
OH
Redox
+ CH3CH2SSCH2CH3
OH

Another possibility is

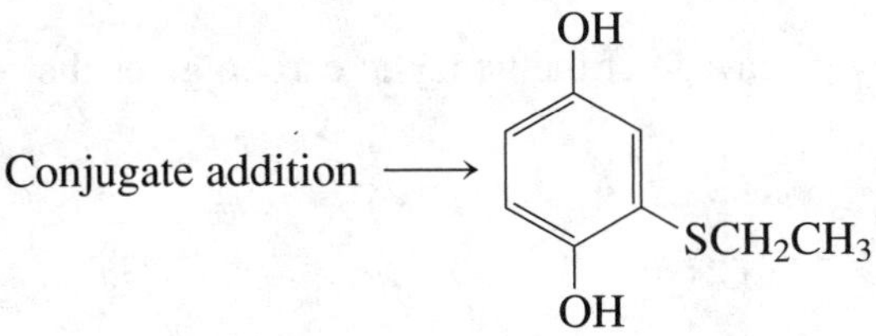
OH
Conjugate addition
SCH2CH3
OH

(f)

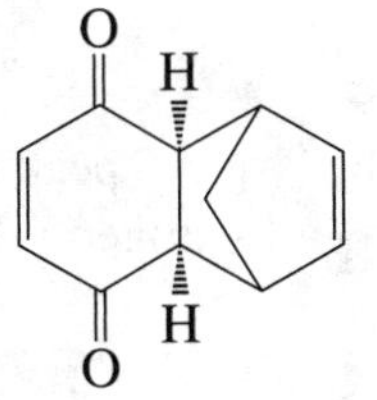
O
H
H
O

59. Aspirin is a *phenyl* ester. Phenyl esters are considerably more susceptible to hydrolysis than ordinary esters are, for two reasons: Delocalization of a lone electron pair from the phenol oxygen into the carbonyl group (ester resonance, see Section 20-1) is diminished, because this lone pair is also in resonance with the benzene ring. The result is a greater δ^+ on the carbonyl carbon, which facilitates nucleophilic attack. The second reason is thermodynamic: The equilibrium phenol + carboxylic acid ⇌ phenyl ester favors the starting material (Section 22-5). So, aqueous solutions of aspirin hydrolyze rather rapidly at room temperature to give salicylic acid and acetic acid.

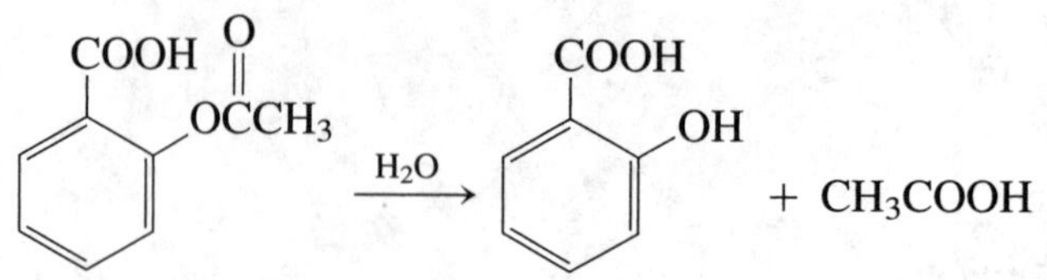

Tylenol, on the other hand, is an amide and much more resistant to nucleophilic attack. Hydrolysis would require extended heating, and strong acid or base is required. [Indeed, amides derived from benzenamine, such as *N*-acetylbenzeneamine (acetanilide), are commonly purified by recrystallization from neutral boiling water, further illustrating their stability.]

60. Without an exact mass measurement the molecular ion at $m/z = 305$ isn't much use for formula determination (there are far too many possibilities). However, working back from the actual structure we can see where the major fragmentation peaks come from. In the structure below we've inserted the masses of several portions of the molecule, obtained by just adding the integral masses of the atoms (H, 1; C, 12; N, 14; and O, 16).

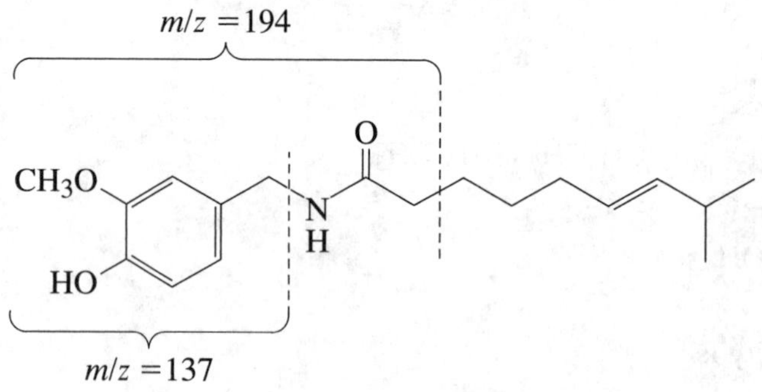

The base peak at $m/z = 137$ is recognizable right away: Cleavage at the benzylic carbon gives the resonance-stabilized cation

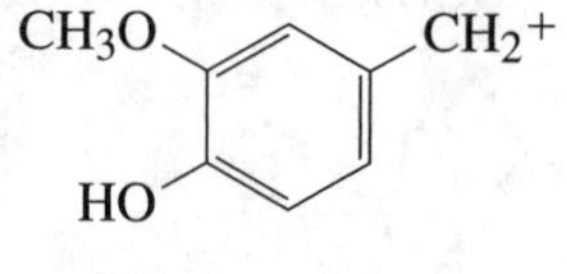

from which loss of a methyl group ($m/z = 15$) gives the $m/z = 122$ peak. Meanwhile the "tricky" peak at $m/z = 195$ is related to the $m/z = 194$ fragment identified above; in fact, it is that piece plus a hydrogen. That is,

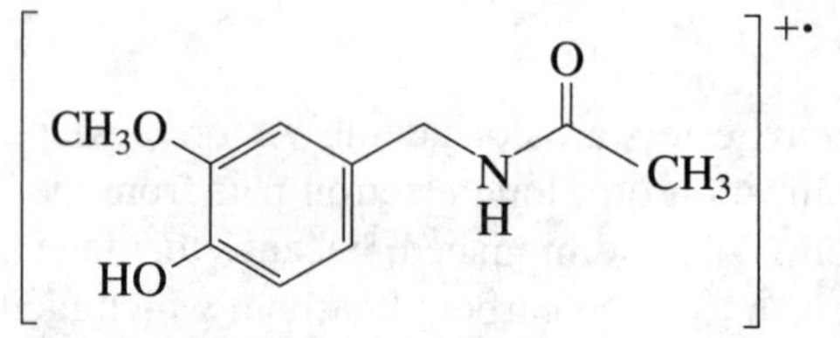

Where did that come from? Recall the McLafferty fragmentation of carbonyl compounds: It results in cleavage between the α and β carbons according to the following process:

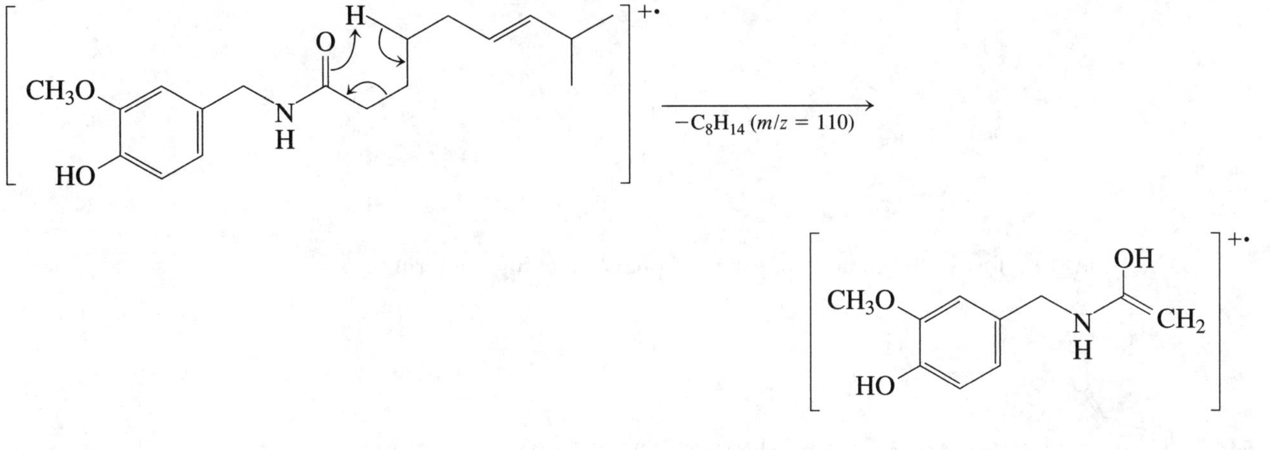

which is the enol isomer of the methyl ketone structure shown above.

The IR data is fairly straightforward to interpret. The band at 972 cm^{-1} derives from a bending motion of the trans alkene double bond, that at 1660 cm^{-1} from the amide C=O stretch, 3016 cm^{-1} from the alkene and arene C—H stretches, and 3445 and 3541 cm^{-1} from the amide N—H and phenol O—H stretches.

Finally, the NMR contains a wealth of information with virtually every individual group of hydrogens clearly resolved. The assignments are made easier if you pay attention to the integrations and the splittings, which conform to the $N + 1$ rule. Here they are (br = broad signal; qn = quintet):

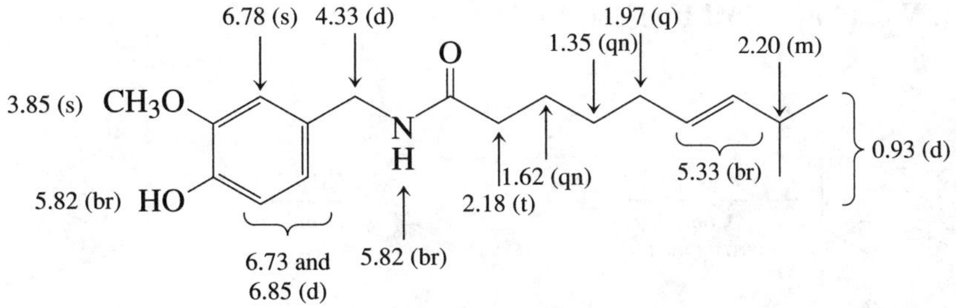

The OH and NH hydrogens are indistinguishable to the NMR in this molecule and combine to give a broad signal at δ 5.82 ppm. Unusually, the NH hydrogen splits the neighboring benzylic CH$_2$ group, whose signal at δ 4.33 ppm is a doublet (this assignment must be made by process of elimination—no other CH$_2$ group in the molecule can give a doublet in the NMR).

61. **(a)** Benzene rings containing only alkyl substituents are not very susceptible to attack by nucleophiles or radicals. Electrophilic attack is the most reasonable pathway. **(b)** Formation of the oxacyclopropane probably involves electrophilic attack of a reagent like H$_2$O$_2$ (derived from O$_2$) on the benzene ring, catalyzed by acid.

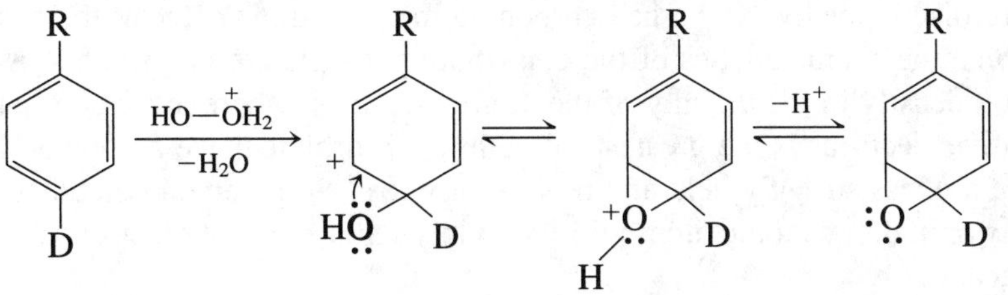

The final phenol probably arises from reversal of the last two steps. The oxacyclopropane ring can always close again, but eventually the carbocation reacts via an alternative pathway, involving D migration to give the rearranged aromatic product.

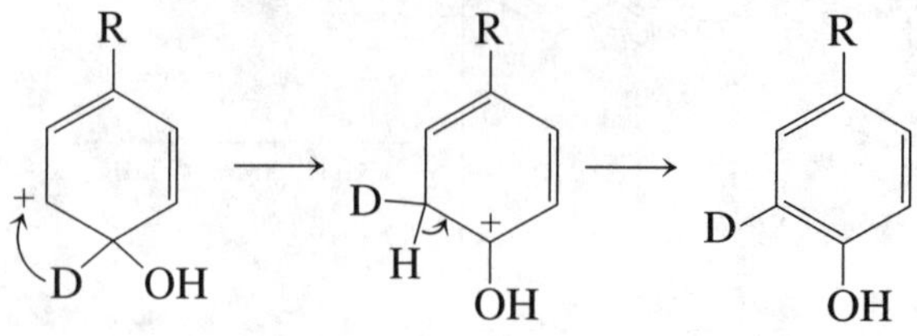

D^+ can also be lost in the final step, giving a phenol lacking deuterium.

62. (a) 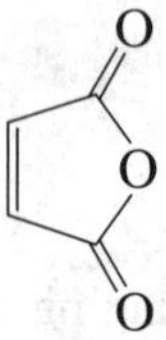 Δ (Diels-Alder)   **(b)** H_2, Pd/C, CH_3CH_2OH

(c) 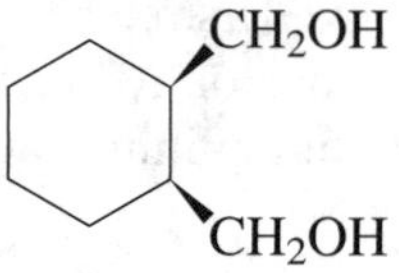   **(d)**

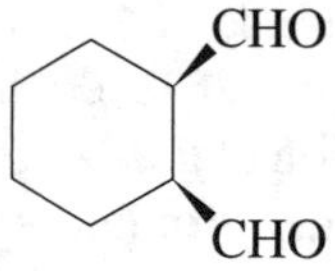

Now what? Work **backward** from the product.

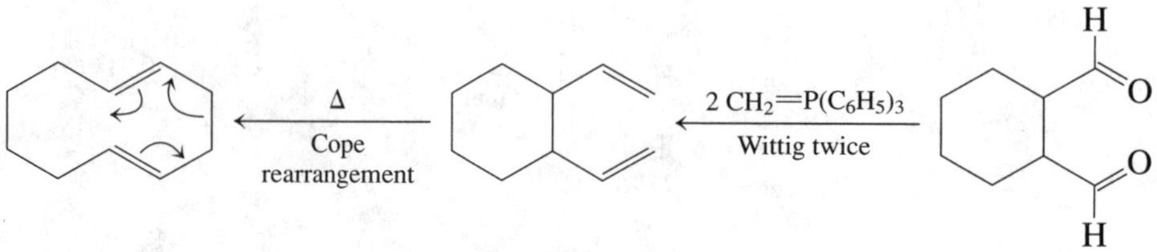

63. The normal "backside displacement" geometry associated with an S_N2 transition state is impossible to achieve at a carbon atom in a benzene ring. There is, in fact, no evidence to support the occurrence of the one-step S_N2 displacement mechanism at a benzene carbon.

The S_N1 pathway requires formation of a carbocation. We saw earlier (Section 13-9) that alkenyl halides are unreactive toward displacement reactions. The S_N2 pathway is poor for the same reason we stated above for benzene compounds. In addition, alkenyl cations are high-energy species because the placement of a positive charge on an sp^2-hybridized carbon is very unfavorable. The same issue arises in the case of benzene: The phenyl cation is a high-energy species and forms only with difficulty because it contains an sp^2-hybridized, positively charged carbon atom.

Why is an sp^2-hybridized, positively charged carbon atom a bad thing? Recall the basics of hybridization. Hybrid orbitals combine the characteristics of the contributing simple atomic orbitals. An occupied s orbital has maximum electron density in the vicinity of the atomic nucleus, where attraction between the positive nucleus and the negative electrons is the greatest. In contrast, p orbitals have a node at the nucleus; electrons in p orbitals therefore are less strongly held and more easily lost. So given a choice between having a vacant p orbital or a vacant s orbital, a cationic atom will invariably choose to put electrons in the s orbital and leave the p orbital vacant.

All carbocations we have seen since Chapter 7 have had unoccupied p orbitals, derived from cleavage of a bond with an sp^3 orbital, which is 1/4 s and 3/4 p in character:

sp³ hybridized   *sp²* hybridized
                          p orbital vacant

An sp^2 orbital is 1/3 s and 2/3 p in nature, more s and less p. Departure of a leaving group to leave it unoccupied and positively charged is disfavored by its greater partial s character, and also by the fact that in benzenes and alkenes no geometrical change to a more favorable orbital arrangement can occur.

64.

(One possibility)

It is an open question whether a phenyl cation is actually involved in reactions such as this. One alternative is addition–elimination (Section 22-4), but this option is not supported by kinetic evidence, which indicates that the displacement process is not bimolecular but unimolecular. Another possibility is a radical process, initiated by reduction of the phenyldiazonium cation to a radical by the iodide ion, which is a good reducing agent. This mechanism is more likely because it has a unimolecular rate-determining propagation step, loss of N_2 from the phenyldiazo radical to give a phenyl radical, a much easier species to form than the phenyl cation.

Although the exact mechanism for iodobenzene formation from the diazonium cation is unclear, the phenyl cation is not an *impossible* intermediate. It is still considered to be the best candidate for the intermediate in phenol formation from arenediazonium salts and hot water (Section 22-4). In addition, it is known that diazonium cations exchange their bonded N_2 with gaseous N_2 (detected in isotope studies), direct evidence that reversible dissociation of N_2 to leave behind a phenyl cation can take place. This system is an example of how a seemingly simple structure can exhibit a lot of complexity in its behavior.

65. Compounds (a) through (d) require nitrosation (also called *diazotization,* using $NaNO_2$, HCl, 0°C) to give the 3-methylbenzenediazonium salt, followed by treatment as appropriate for the final desired substituent: **(a)** H_3PO_2 (replaces N_2^+ with H); **(b)** CuBr, 100°C (for Br); **(c)** heat (for OH); **(d)** CuCN, KCN, 50°C (for CN). For **(e)** we are placing a substituent *on the nitrogen* rather than replacing the nitrogen with something else. Specifically, the problem asks for the conversion of a primary amine into a secondary amine, which is best done by reductive amination (Section 21-6):

66. Think about benzenediazonium salts as intermediates likely to be useful in these.

(a)

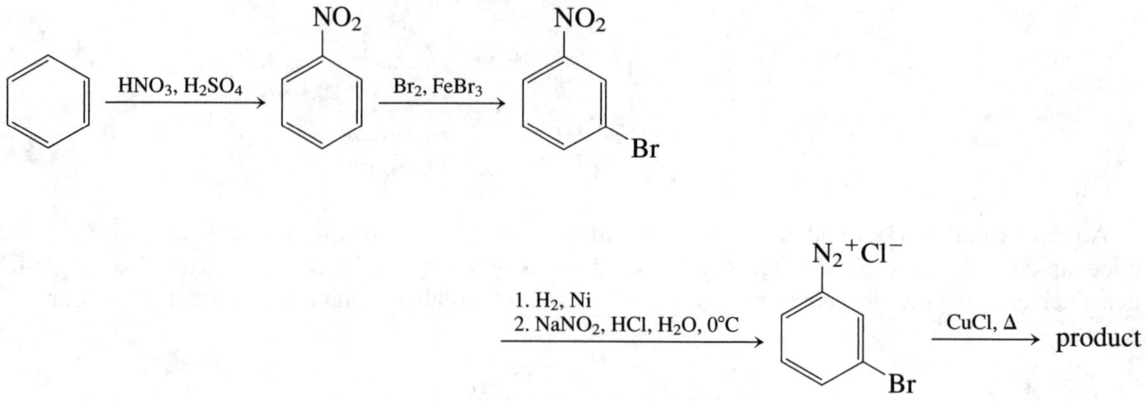

$$\text{benzene} \xrightarrow{\text{HNO}_3,\ \text{H}_2\text{SO}_4} \text{nitrobenzene} \xrightarrow{\text{Br}_2,\ \text{FeBr}_3} \text{3-bromonitrobenzene}$$

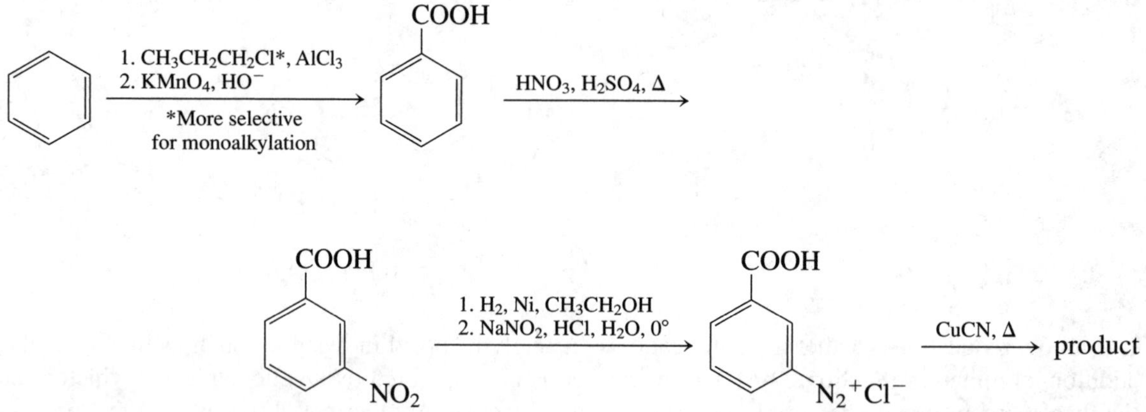

$$\text{3-bromo-1-(N}_2^+\text{Cl}^-)\text{benzene} \xrightarrow[\text{2. NaNO}_2,\ \text{HCl, H}_2\text{O, 0°C}]{\text{1. H}_2,\ \text{Ni}} \xrightarrow{\text{CuCl, }\Delta} \text{product}$$

(b)

$$\text{benzene} \xrightarrow[\substack{\text{2. KMnO}_4,\ \text{HO}^- \\ \text{*More selective} \\ \text{for monoalkylation}}]{\text{1. CH}_3\text{CH}_2\text{CH}_2\text{Cl*, AlCl}_3} \text{benzoic acid} \xrightarrow{\text{HNO}_3,\ \text{H}_2\text{SO}_4,\ \Delta}$$

$$\text{3-nitrobenzoic acid} \xrightarrow[\text{2. NaNO}_2,\ \text{HCl, H}_2\text{O, 0°}]{\text{1. H}_2,\ \text{Ni, CH}_3\text{CH}_2\text{OH}} \text{3-(N}_2^+\text{Cl}^-)\text{benzoic acid} \xrightarrow{\text{CuCN, }\Delta} \text{product}$$

(c)

$$\text{3-chloro-1-(N}_2^+\text{Cl}^-)\text{benzene} \xrightarrow{\text{H}_2\text{O, }\Delta} \text{3-chlorophenol} \xrightarrow{\text{HNO}_3,\ \text{H}_2\text{O, 0°C}} \text{product}$$

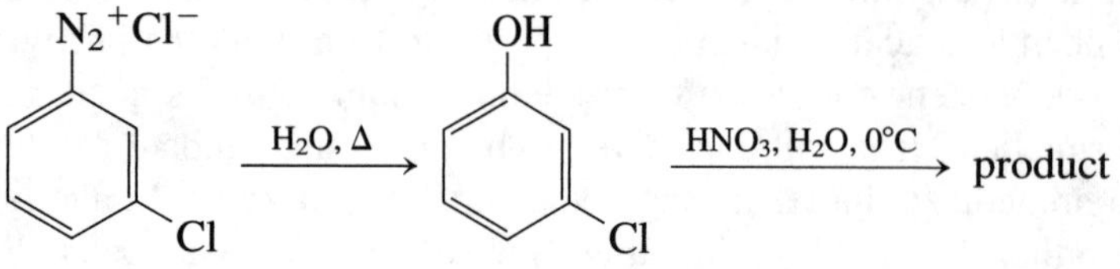

[See (a)]

(d)

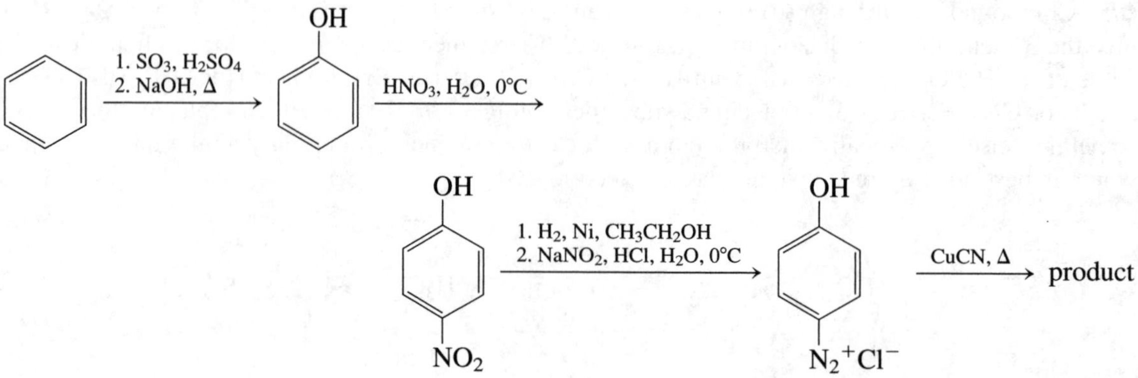

$$\text{benzene} \xrightarrow[\text{2. NaOH, }\Delta]{\text{1. SO}_3,\ \text{H}_2\text{SO}_4} \text{phenol} \xrightarrow{\text{HNO}_3,\ \text{H}_2\text{O, 0°C}}$$

$$\text{4-nitrophenol} \xrightarrow[\text{2. NaNO}_2,\ \text{HCl, H}_2\text{O, 0°C}]{\text{1. H}_2,\ \text{Ni, CH}_3\text{CH}_2\text{OH}} \text{4-(N}_2^+\text{Cl}^-)\text{phenol} \xrightarrow{\text{CuCN, }\Delta} \text{product}$$

(e)

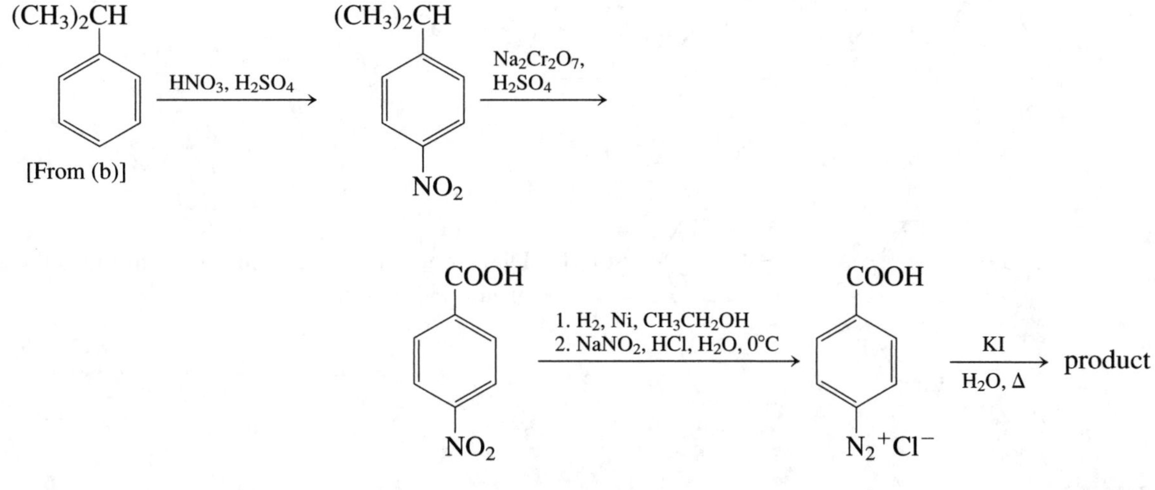

(CH₃)₂CH ⟶ HNO₃, H₂SO₄ ⟶ (CH₃)₂CH / NO₂ ⟶ Na₂Cr₂O₇, H₂SO₄ ⟶

[From (b)]

COOH / NO₂ ⟶ 1. H₂, Ni, CH₃CH₂OH 2. NaNO₂, HCl, H₂O, 0°C ⟶ COOH / N₂⁺Cl⁻ ⟶ KI / H₂O, Δ ⟶ product

Note concerning the first step: Nitration of methylbenzene gives mostly ortho product. The larger alkyl group favors para substitution.

(f)

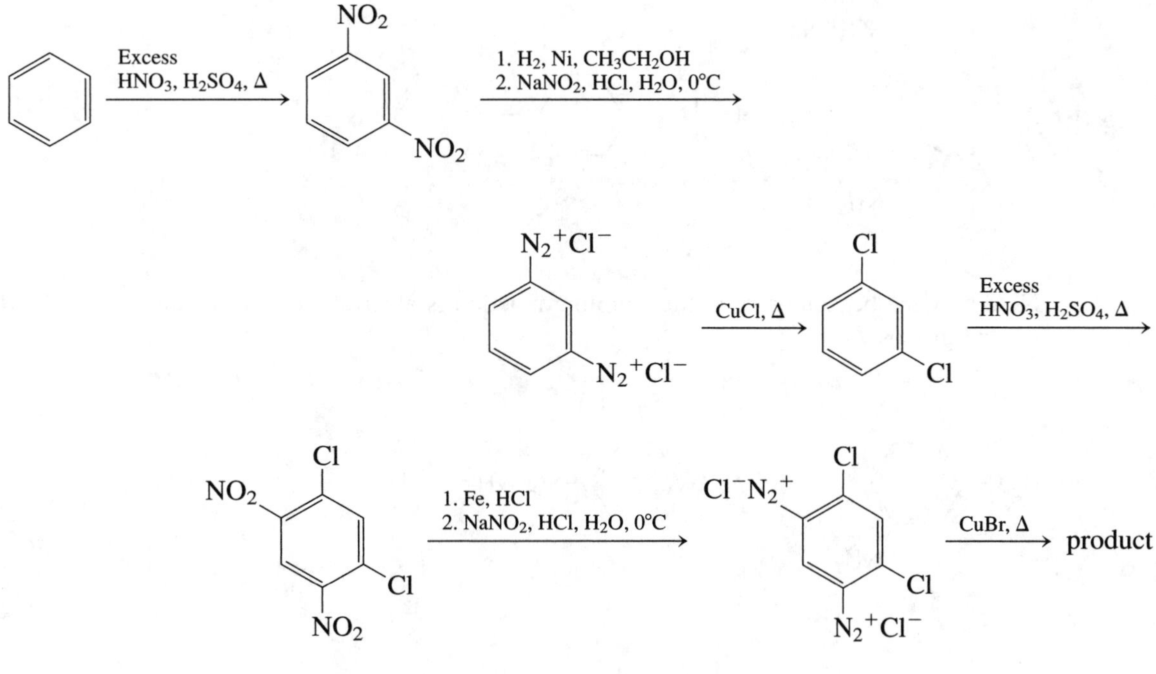

⟶ Excess HNO₃, H₂SO₄, Δ ⟶ NO₂ / NO₂ ⟶ 1. H₂, Ni, CH₃CH₂OH 2. NaNO₂, HCl, H₂O, 0°C ⟶

N₂⁺Cl⁻ / N₂⁺Cl⁻ ⟶ CuCl, Δ ⟶ Cl / Cl ⟶ Excess HNO₃, H₂SO₄, Δ ⟶

NO₂ / Cl / Cl / NO₂ ⟶ 1. Fe, HCl 2. NaNO₂, HCl, H₂O, 0°C ⟶ Cl⁻N₂⁺ / Cl / Cl / N₂⁺Cl⁻ ⟶ CuBr, Δ ⟶ product

(g)

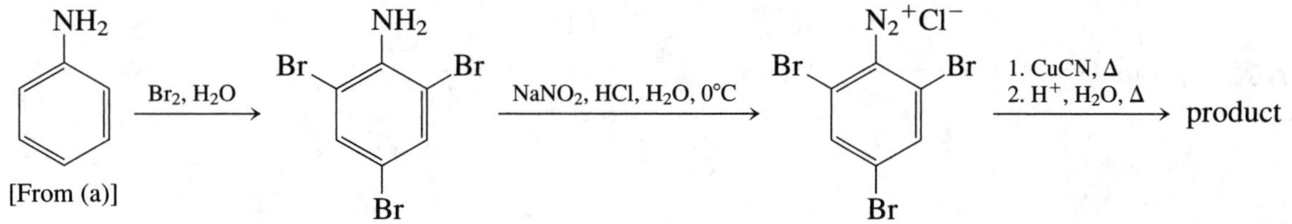

NH₂ ⟶ Br₂, H₂O ⟶ Br / NH₂ / Br / Br ⟶ NaNO₂, HCl, H₂O, 0°C ⟶ Br / N₂⁺Cl⁻ / Br / Br ⟶ 1. CuCN, Δ 2. H⁺, H₂O, Δ ⟶ product

[From (a)]

67. **(a)** HO— / —N=N— / —SO₃H / OH

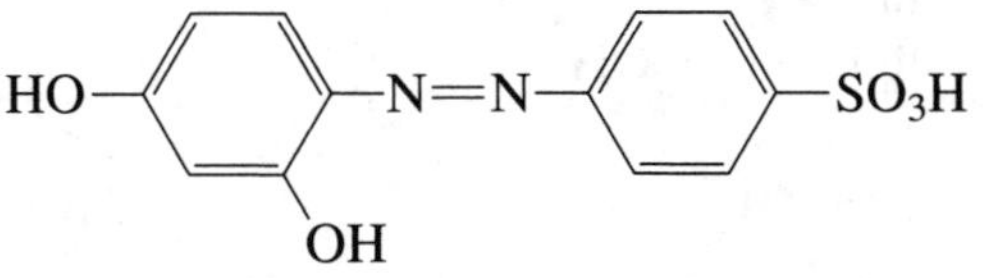

(b)

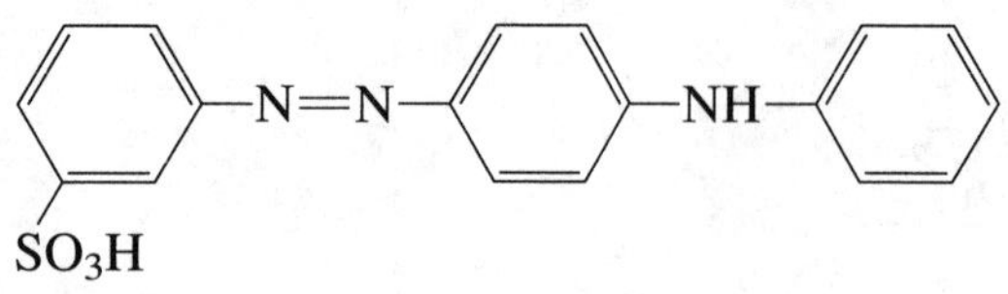

(c) 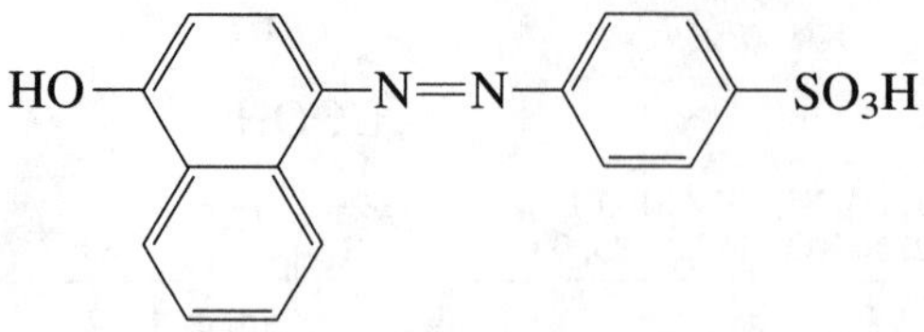 Diazo coupling is generally para to the activating group, if possible.

68. (a)

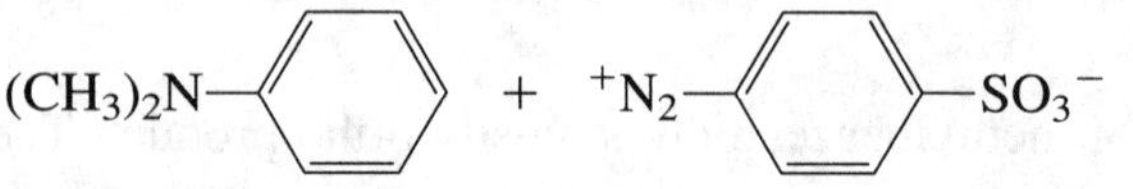

(b) 2

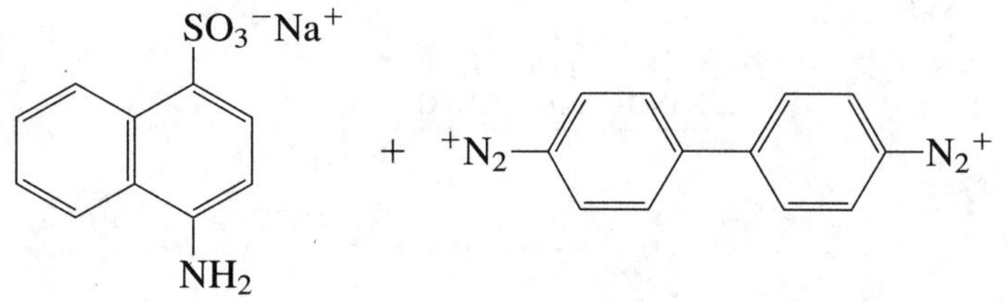

Note how the benzene ring in the coupling reaction is always strongly activated by OH, NH_2, or related groups.

(c)

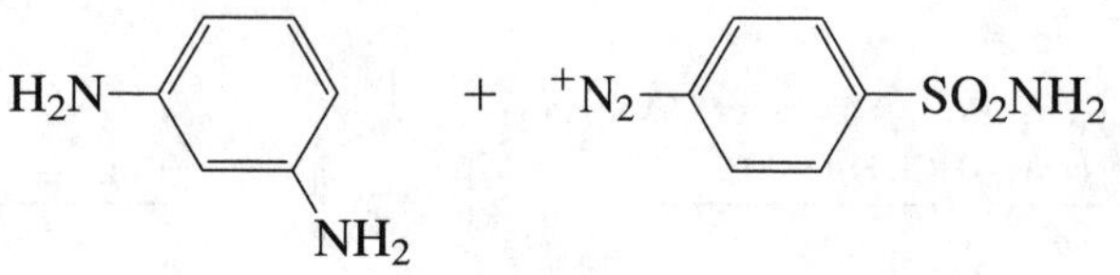

69. (a)

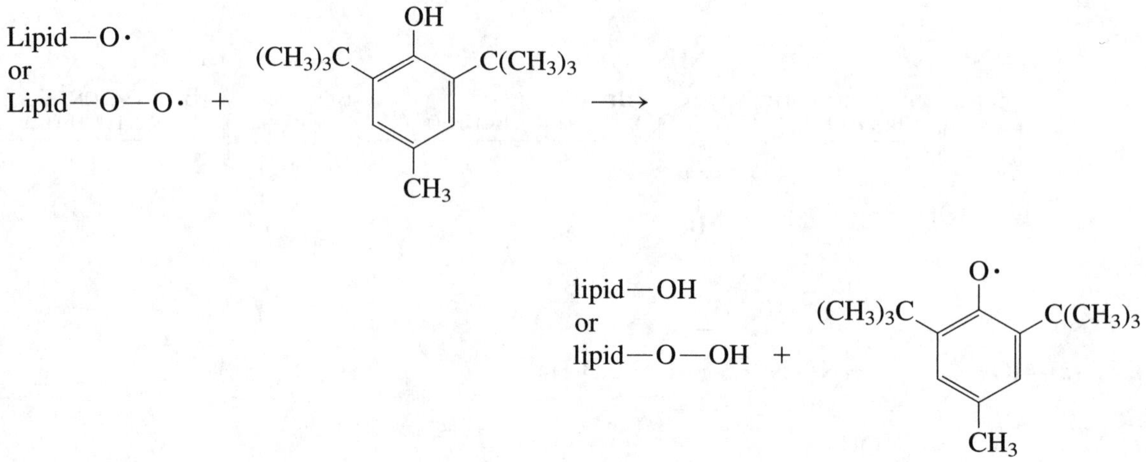

(b) Start with the lipid hydroperoxide of linoleic acid shown in Section 22-9 in the reaction labeled "propagation step 2." Form the alkoxyl radical and finish with β-scission.

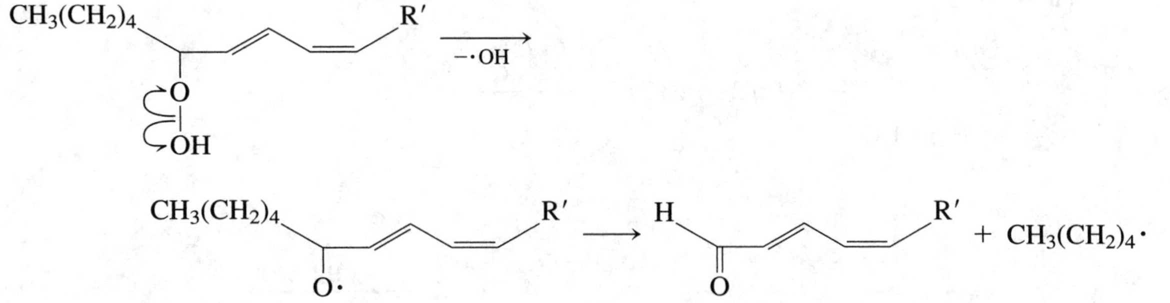

The $CH_3(CH_2)_4\cdot$ radical becomes pentane by abstracting a hydrogen atom from any reactive hydrogen donor, such as another lipid molecule.

70. There are several places to start: go step-by-step.

1. Degrees of unsaturation (Chapter 11).
Urushiol I, $H_{sat} = 42 + 2 = 44$; degrees of unsaturation $= \frac{(44 - 36)}{2} = 4$ π bonds or rings

Urushiol II, degrees of unsaturation $= \frac{(44 - 34)}{2} = 5$ π bonds or rings

2. Urushiol II contains only one double bond that is easily hydrogenated. The four degrees of unsaturation in urushiol I must be either rings or hard-to-hydrogenate π bonds (like those in a benzene ring).

3. Urushiol II contains the piece

$$CH_3CH_2CH_2CH_2CH_2CH_2CH{=}\underbrace{CHR}_{\text{Part of aldehyde A}}$$

4. Synthesis of aldehyde A is presented. Here are the structures of the intermediates.

(COOH attached via Kolbe reaction)

B **C** **D** **E**

F

(Double bond reduced and "benzyl" group removed)

Aldehyde A

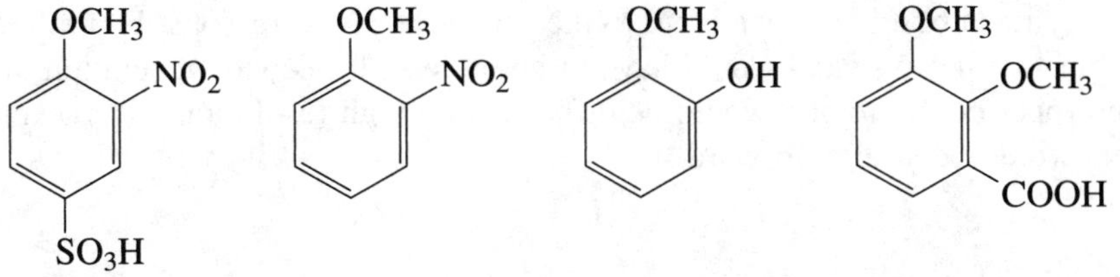

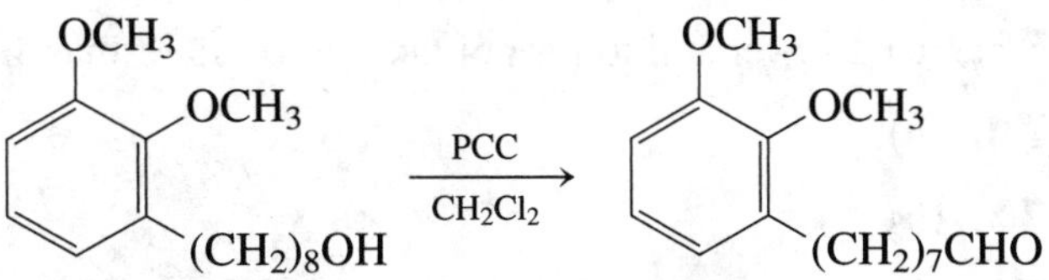

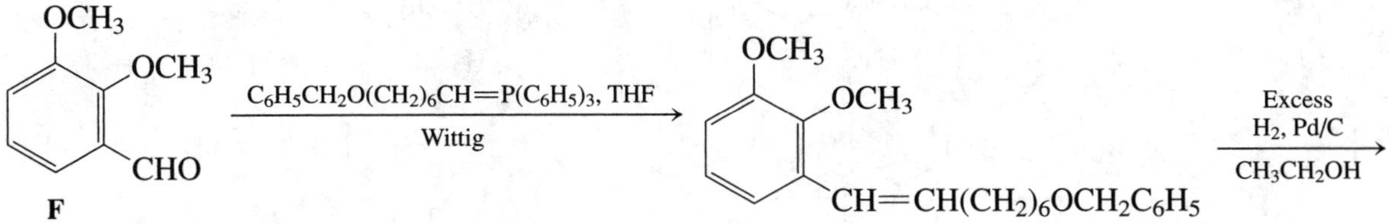

Going back to step 3, you can now work backward to the structure of urushiol II.

Dimethylurushiol II

Urushiol II

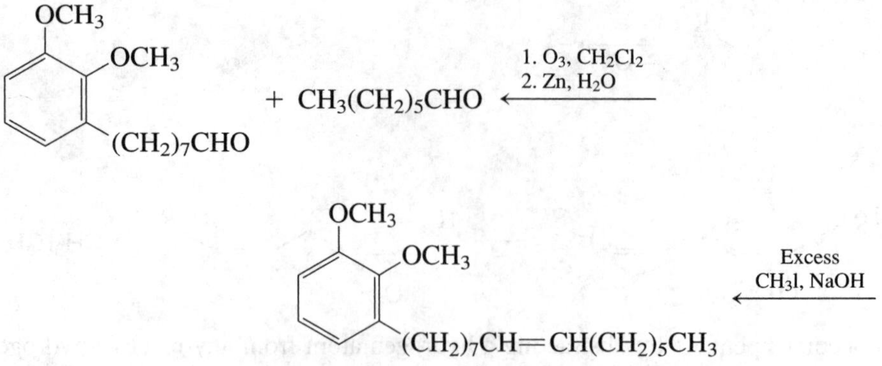

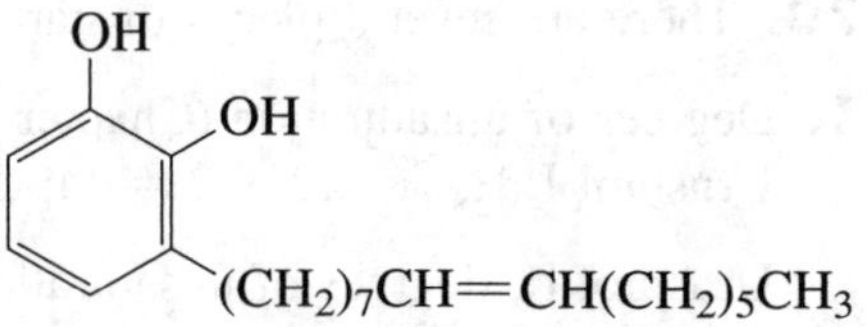

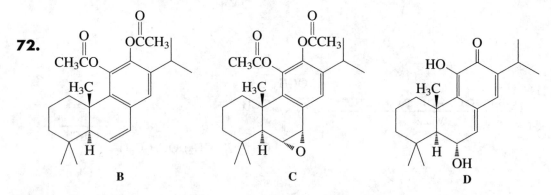

A reconsideration of step 2 shows that urushiol I must be

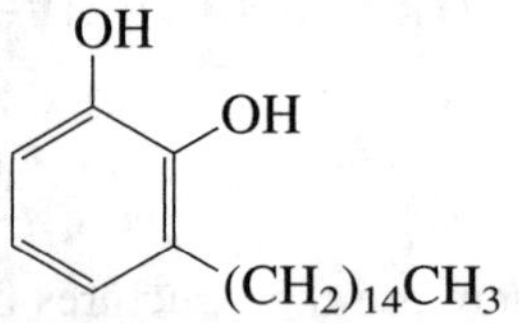

71. Yes to the first question: A benzylic carbon is oxidized. No to the second, for two reasons. First, the amine is a reactive and oxidizable functional group and would have to be protected. Second, generation of a new stereocenter in exclusively the proper configuration would be difficult, although resolution of a racemic product mixture into its enantiomers would be straightforward.

72.

B C D

73. (d) Conversion to phenol takes a *lot* more than boiling in water for two hours.

74. (b)

75. (b)

76. (a)

77. (a) Hydrogen bonding to the nitro group oxygen atom only occurs in the ortho isomer.

23

Ester Enolates and the Claisen Condensation: Synthesis of β-Dicarbonyl Compounds; Acyl Anion Equivalents

A large proportion of organic compounds of synthetic and biological importance contain more than one functional group. It is obviously necessary to be able to develop methods for the preparation of such substances. You have already learned quite a bit about the aldol condensation and the systems it forms, β-hydroxycarbonyl compounds and conjugated enones (Chapter 18). By presenting a related reaction of esters, the Claisen condensation, this chapter expands the concept of carbonyl condensation reactions. In addition to being a new method for carbon–carbon bond formation, the Claisen condensation gives rise to a class of compounds, β- or 1,3-dicarbonyl compounds, that have their own distinct reactivity and utility. The final chapter section presents a more advanced carbon–carbon bond forming method, showing how certain protected forms of carbonyl carbons may be made nucleophilic and used in displacement reactions.

Outline of the Chapter

23-1   β-Dicarbonyl Compounds: The Claisen Condensation
Construction of the most versatile combination of carbonyl groups.

23-2, 23-3   Synthetic Applications of β-Dicarbonyl Compounds

23-4   Acyl Anion Equivalents: Preparation of α-Hydroxyketones
Reversal of the normal polarity of a functionalized carbon in synthesis.

Keys to the Chapter

23-1. β-Dicarbonyl Compounds: The Claisen Condensation

The Claisen condensation is the main method for synthesizing 1,3-dicarbonyl compounds. Analyze this reaction on the basis of its similarities to the aldol condensation (Section 18-5): It is an enolate + carbonyl process, so bond formation occurs between the α-carbon of one carbonyl compound (which may be either an ester or a ketone) and the carbonyl carbon of another (an ester). Note the limitation: Under the conditions given, the reaction works only when the 1,3-dicarbonyl product still possesses a hydrogen on the carbon between the two carbonyl groups. Deprotonation of this acidic H by excess base allows the equilibrium to shift to the product.

Because there are several types of β-carbonyl compounds available from the Claisen condensation, Table 1 presents examples to help you keep them straight.

TABLE 1
Claisen condensations

Reaction Partners		Product	
Carbonyl	Enolate	New functional groups	Structure (New bond shown)
$\underset{\text{Carbonate}}{CH_3CH_2OCOCH_2CH_3}$ (O double bond)	$\underset{\text{General ester}}{CH_3COCH_2CH_3}$ (O double bond)	Ester + ester	$CH_3CH_2OC\text{—}CH_2COCH_2CH_3$ (two O double bonds) 1,3-Diester
$\underset{\text{General ester}}{CH_3COCH_2CH_3}$ (O double bond)	$\underset{\text{General ester}}{CH_3COCH_2CH_3}$ (O double bond)	Ketone + ester	$CH_3C\text{—}CH_2COCH_2CH_3$ (two O double bonds) 3-Ketoester
$\underset{\text{Formate}}{HCOCH_2CH_3}$ (O double bond)	$\underset{\text{General ester}}{CH_3COCH_2CH_3}$ (O double bond)	Aldehyde + ester	$HC\text{—}CH_2COCH_2CH_3$ (two O double bonds) Formyl ester
$\underset{\text{Carbonate}}{CH_3CH_2OCOCH_2CH_3}$ (O double bond)	$\underset{\text{Ketone}}{CH_3CCH_3}$ (O double bond)	Ester + ketone	$CH_3CH_2OC\text{—}CH_2CCH_3$ (two O double bonds) 3-Ketoester (The "hard" way)
$\underset{\text{General ester}}{CH_3COCH_2CH_3}$ (O double bond)	$\underset{\text{Ketone}}{CH_3CCH_3}$ (O double bond)	Ketone + ketone	$CH_3C\text{—}CH_2CCH_3$ (two O double bonds) 1,3-Diketone
$\underset{\text{Formate}}{HCOCH_2CH_3}$ (O double bond)	$\underset{\text{Ketone}}{CH_3CCH_3}$ (O double bond)	Aldehyde + ketone	$HC\text{—}CH_2CCH_3$ (two O double bonds) 3-Ketoaldehyde

23-2 and 23-3. Synthetic Applications of β-Dicarbonyl Compounds

In Section 23-2 two reactions of β-dicarbonyl compounds are presented. The first is alkylation of the readily deprotonated "carbon in the middle." The second pertains to β-dicarbonyl compounds where at least one of the carbonyl groups is an ester. Ester hydrolysis leads to a carboxylic acid that readily loses CO_2 (decarboxylation). When this sequence is carried out on a β-ketoester, the result is a ketone. When carried out on a diester of propanedioic (malonic) acid, the result is a carboxylic acid. In each case, groups attached in the preliminary alkylation step(s) wind up in the product.

Notice that the acetoacetic ester synthesis only makes **methyl** ketones, because the CH_3CO portion of the acetoacetic ester molecule is carried unchanged into the final product. To make other kinds of ketones, other 3-ketoesters must be prepared first. Problem 47 outlines the situation, and the general scheme below illustrates the process.

General 3-Ketoester Synthesis of Ketones

$$RCOCH_2CH_3 \; + \; CH_3COCH_2CH_3 \dashrightarrow RCCH_2COCH_2CH_3 \dashrightarrow RC\overset{R'}{\underset{R''}{—C—}}COCH_2CH_3 \dashrightarrow RCCH\overset{R'}{\underset{R''}{}}$$

Note that the necessary 3-ketoester comes from a crossed Claisen condensation. As a result, $RCO_2CH_2CH_3$ must be an ester that does not do Claisen condensations with itself (i.e., R must not have an α-CH_2 group). Otherwise the crossed condensation will be a disaster.

Section 23-3 reinforces the fact that anions of β-dicarbonyl compounds are still enolates. Thus, you will find that they do 1,4-additions to α, β-unsaturated carbonyl compounds (Michael additions). Furthermore, these additions can be followed by Robinson annulations, thereby resulting in six-membered rings.

23-4. Acyl Anion Equivalents

How are α-hydroxy ketones made? Because they contain alcohol groups, you might consider methods first introduced in Chapter 8 for the synthesis of alcohols: addition reactions of organometallic (carbanionic) reagents to aldehydes and ketones. However, if you tried to apply this approach, you would encounter a

problem: The necessary carbanion would have the structure $RC\overset{\displaystyle O}{\underset{\displaystyle \|}{}}:^-$, with a nucleophilic carbonyl carbon. Such a species is called an acyl anion and is not readily prepared. Indeed, ever since Chapter 8, you have had drilled into your brain the fact that carbonyl carbons are **electrophilic,** not nucleophilic. The whole idea of a nucleophilic acyl anion contradicts everything you've learned.

So, given all that, can anything be done to get around this limitation? Yes: **Carbonyl carbons can be chemically modified to become nucleophilic.** One way is to convert the carbonyl group of an aldehyde into a thioacetal and then deprotonate it with a strong base, forming a 1,3-dithiacyclohexane (1,3-dithiane) anion. Another way is to expose an aldehyde to thiazolium salts. Again, deprotonation of the original carbonyl carbon may follow, giving a nucleophilic anion. Either way, you have reversed the normal polarity of this carbon atom. Once such an anion (an acyl anion equivalent) has been formed, it can add to the normal, electrophilic carbonyl group of another molecule in the usual way. **The result is a new carbon–carbon bond between carbons that both started out with the same polarity (electrophilic).** This is **important.**

Application of polarity reversal in organic chemistry isn't magical. All you've done is turn carbonyl groups into new functions that are able to support negative charges. Then, after finishing with them, you've changed them back to carbonyl groups again. Still, this material can be troublesome to learn. You might try the following for study purposes: Write down several aldehydes and their (hard to make) acyl anions. Next, follow along one of the two general reaction sequences shown in the text section, drawing the corresponding acyl anion equivalent, adding it to another carbonyl group, and regenerating the original carbonyl carbon. The practice will be good for you.

Solutions to Problems

27. Begin by identifying any functional group that contains a hydrogen on a heteroatom (such as oxygen) and assigning to it (or just looking up) a reasonable pK_a value: **(b)** CH_3CO_2H, 4.8 (Table 19-3); **(c)** CH_3OH, 15.5 (Table 8-2). *Sooner or later (preferably sooner) you should memorize the facts that simple carboxylic acids typically have pK_a values around 4 to 5, and alcohols have pK_a values around 16 to 18.*

Next, look at all compounds for which the most acidic hydrogen is attached to an α-carbon of a carbonyl group. Especially, identify the ones flanked by multiple carbonyl groups: they will be more acidic. Estimate (or, again, look up—Table 23-1) pK_a values. Be sure to distinguish between the types of carbonyl functions (aldehyde, ketone, ester, etc.), because they differ in how much they acidify the α-hydrogens: aldehydes acidify the most, esters the least, and ketones are intermediate.

α-*Hydrogens between two carbonyl groups:*

(a) Between two ketone carbonyls; estimate pK_a of 9

(d) 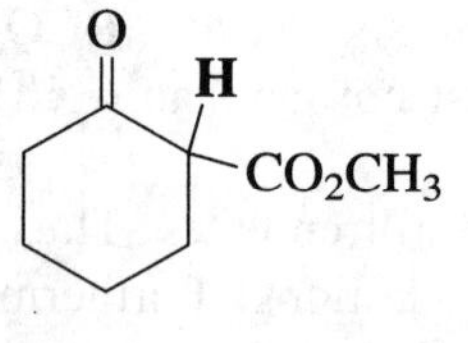 Between a ketone and an ester; estimate pK_a of 11

(g) $CH_3O_2CCH_2CO_2CH_3$ Between two ester carbonyls; estimate pK_a of 13

α-Hydrogens adjacent to only one carbonyl group:

(e) CH_3CHO Aldehyde; estimate pK_a of 17

(f) 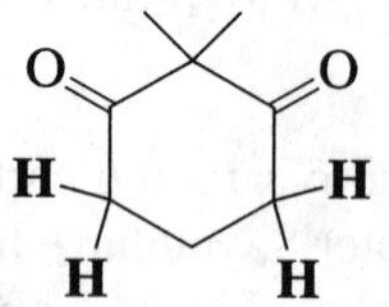 Ketone; estimate pK_a of 20

Compound **(h)** has no α-hydrogens. The hydrogens on its methyl groups are not appreciably acidified. It comes in last. So the final order of acidity is **(b)** > **(a)** > **(d)** > **(g)** > **(c)** > **(e)** > **(f)** > **(h)**.

28. Claisen condensations. (a), (b), (c) involve two identical molecules; (d), (e) are intramolecular examples; (f), (g), (h), (i) are mixed condensations. Make your new carbon–carbon bond (boldface) between the carbonyl carbon of one ester and the α-carbon of another.

(a) CH₃CH₂CH₂C—CHCOCH₂CH₃
            |
         CH₃CH₂

(b) C₆H₅CH(CH₃)CH₂C—CHCOCH₂CH₃
                      |
                   C₆H₅CHCH₃

(c) Unfavorable equilibrium. Claisen product is not stable; no reaction is observed.

(d)

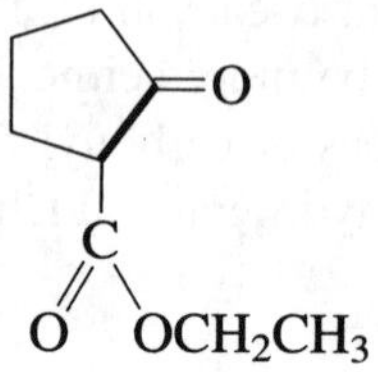

(e) 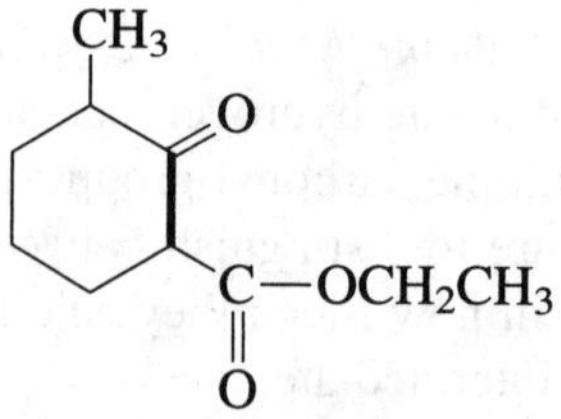 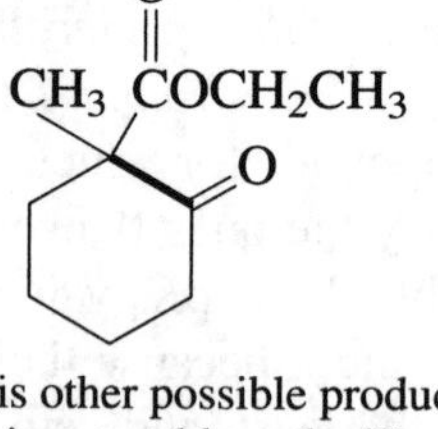
This other possible product is not stable and will not be isolated.

(f) HC—CHCOCH₂CH₃
          |
        C₆H₅

(g) C₆H₅C—CHCOCH₂CH₃
              |
           CH₃CH₂

(h) 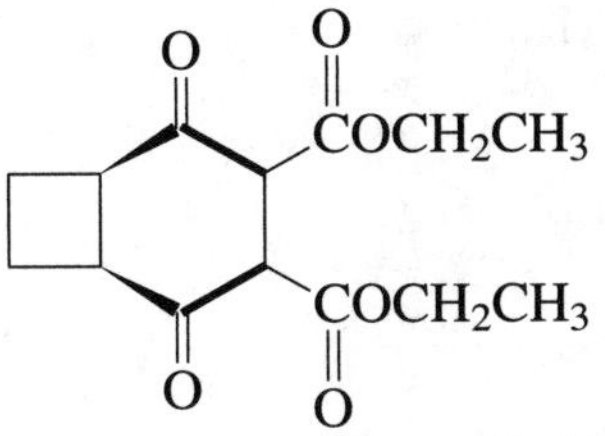 [structure: bicyclic compound with two ketone groups (O) and two $COCH_2CH_3$ ester groups]

(i) 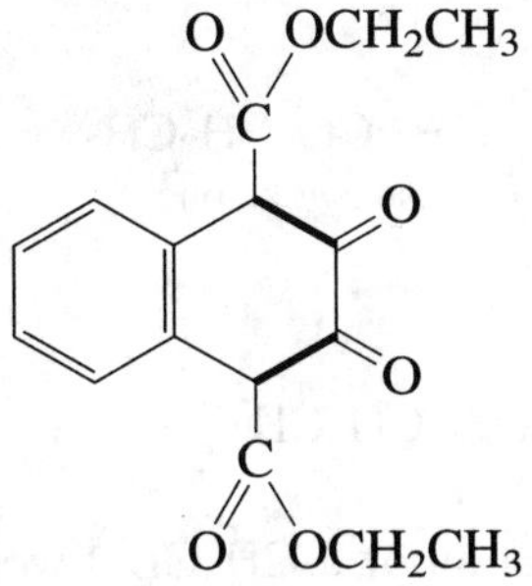 [structure: naphthalene-fused ring with two OCH_2CH_3 ester groups and two ketone O groups]

29. The second ester, $(CH_3)_2CHCOCH_3$, should be present in excess because (1) it does not form a stable product from Claisen condensation with itself and (2) it will be able to preferentially react with enolate ions from the first ester. Side reaction (condensation of first ester with itself):

$$2\ CH_3CH_2CO_2CH_3 \xrightarrow{NaOCH_3,\ CH_3OH} CH_3CH_2\overset{O}{\underset{}{C}}CHCO_2CH_3$$
$$\underset{CH_3}{|}$$

30. Analyze as you did for Problem 28. "Claisen" means 1. $NaOCH_2CH_3$, CH_3CH_2OH, 2. H^+, H_2O.

(a) 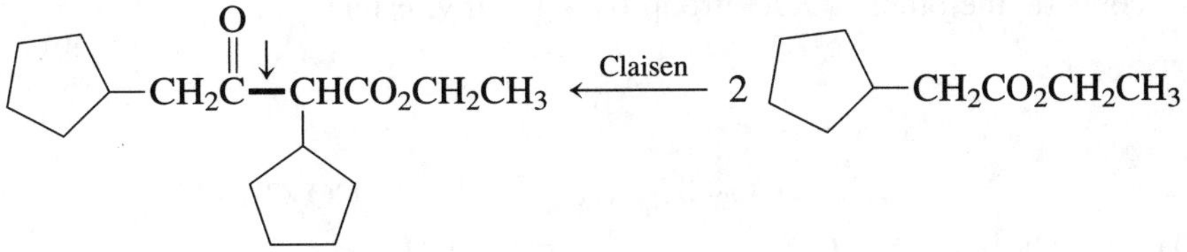 [cyclopentyl]$-CH_2\overset{O}{\underset{}{C}}\!\!\downarrow\!\!-CHCO_2CH_2CH_3$ (with cyclopentyl substituent) $\xleftarrow{\text{Claisen}}$ 2 [cyclopentyl]$-CH_2CO_2CH_2CH_3$

(b) $C_6H_5\overset{O}{\underset{}{C}}\!\!\downarrow\!\!-\underset{C_6H_5}{\underset{|}{CH}}CO_2CH_2CH_3 \xleftarrow{\text{Claisen}} C_6H_5CO_2CH_2CH_3 + C_6H_5CH_2CO_2CH_2CH_3$

(c) 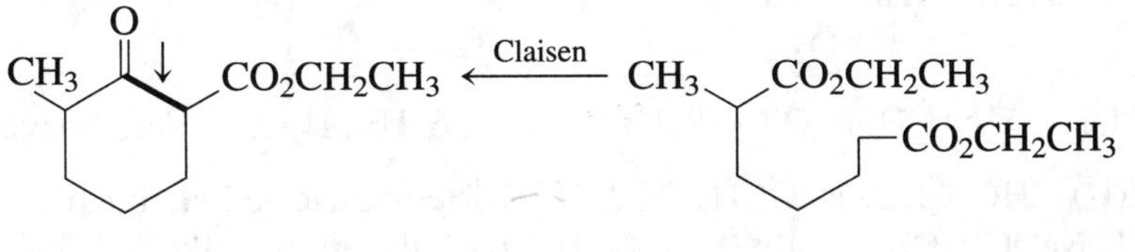 [cyclohexane ring with CH_3 and $\overset{O}{C}\!\!\downarrow\!\!CO_2CH_2CH_3$] $\xleftarrow{\text{Claisen}}$ CH_3 ... $CO_2CH_2CH_3$ chain with $-CO_2CH_2CH_3$ [Problem 28(e)!]

(d) $HC\overset{O\ O}{\underset{\uparrow}{C}}-CH_2CO_2CH_2CH_3 \xleftarrow{\text{Claisen}} HCCO_2CH_2CH_3 + CH_3CO_2CH_2CH_3$

(e) $C_6H_5\overset{O}{\underset{}{C}}-\underset{\uparrow}{CH_2}\overset{O}{\underset{}{C}}C_6H_5 \xleftarrow{\text{Claisen}} C_6H_5CO_2CH_2CH_3 + CH_3\overset{O}{\underset{}{C}}C_6H_5$
(Ketone + ester version)

(f) $CH_3CH_2O\overset{O}{\underset{}{C}}-\underset{\uparrow}{CH_2}\overset{O}{\underset{}{C}}OCH_2CH_3 \xleftarrow{\text{Claisen}} CH_3CH_2O\overset{O}{\underset{}{C}}OCH_2CH_3 + CH_3CO_2CH_2CH_3$
(Carbonate + ester version)

(g) $\triangleright\!\!-\!\!\overset{\overset{\displaystyle O}{\|}}{C}\!\!-\!\!CH_2\overset{\overset{\displaystyle O}{\|}}{C}CH_3$ $\xleftarrow{\text{Claisen}}$ $\triangleright\!\!-\!\!CO_2CH_2CH_3 + CH_3\overset{\overset{\displaystyle O}{\|}}{C}CH_3$
(Ester + ketone)

31. $H\overset{\overset{\displaystyle O}{\|}}{C}\!\!-\!\!CH_2\overset{\overset{\displaystyle O}{\|}}{C}H \Longrightarrow HCO_2CH_2CH_3 + CH_3\overset{\overset{\displaystyle O}{\|}}{C}H$?

Not likely to work because aldol condensation of 2 CH_3CHO would be a major competing process.

32. Analysis: $CH_3\overset{\overset{\displaystyle O}{\|}}{C}\underset{\underset{\displaystyle R'}{|}}{C}H\!\!-\!\!R \Longrightarrow CH_3\overset{\overset{\displaystyle O}{\|}}{C}\!\!-\!\!\underset{\underset{\displaystyle R'}{|}}{\overset{\overset{\displaystyle R}{|}}{C}}\!\!-\!\!CO_2CH_2CH_3 \Longrightarrow CH_3\overset{\overset{\displaystyle O}{\|}}{C}CH_2CO_2CH_2CH_3$
Starting material
for each synthesis.

The solvent for each reaction in this and the next problem can be ethanol.

(a) R = $-CH_2CH(CH_3)_2$, R′ = H. 1. $NaOCH_2CH_3$, 2. $(CH_3)_2CHCH_2Br$, 3. NaOH, H_2O, 4. H^+, H_2O, Δ; **(b)** R = R′ = $-CH_2CH_2CH_2-$. 1. 2 $NaOCH_2CH_3$, 2. $BrCH_2CH_2CH_2Br$, 3. NaOH, H_2O, 4. H^+, H_2O, Δ; **(c)** R = $-CH_2C_6H_5$, R′ = $-CH_2CH=CH_2$. 1. $NaOCH_2CH_3$, 2. $C_6H_5CH_2Br$, 3. $NaOCH_2CH_3$, 4. $CH_2=CHCH_2Br$, 5. NaOH, H_2O, 6. H^+, H_2O, Δ; **(d)** R = $-CH_2CH_3$, R′ = $-CH_2CO_2CH_2CH_3$. 1. $NaOCH_2CH_3$, 2. $BrCH_2CO_2CH_2CH_3$, 3. $NaOCH_2CH_3$, 4. CH_3CH_2Br, 5. NaOH, H_2O, 6. H^+, H_2O, Δ (decarboxylates only the COOH on the α-carbon to the ketone), 7. CH_3CH_2OH, H^+ (converts the other COOH group back to ethyl ester)

33. General pattern

$\underset{\underset{\displaystyle R'}{|}}{\overset{\overset{\displaystyle R}{|}}{CH}}\!\!-\!\!COOH \Longrightarrow \underset{\underset{\displaystyle R'}{|}}{\overset{\overset{\displaystyle R}{|}}{C}}\!\!\genfrac{}{}{0pt}{}{CO_2CH_2CH_3}{CO_2CH_2CH_3} \Longrightarrow CH_2\genfrac{}{}{0pt}{}{CO_2CH_2CH_3}{CO_2CH_2CH_3}$
Starting compound
for each synthesis.

(a) 1. $NaOCH_2CH_3$, 2. $CH_3CH_2CH_2CH_2I$, 3. $NaOCH_2CH_3$, 4. $\langle\!\!\!\!\bigcirc\!\!\!\!\rangle\!\!-\!\!CH_2Br$

(completes necessary alkylations), 5. NaOH, H_2O (hydrolyzes esters), 6. H^+, H_2O, Δ (decarboxylation);

(b) 1. $NaOCH_2CH_3$, 2. $(CH_3)_2CHCH_2I$, 3. $NaOCH_2CH_3$, 4. CH_3I (completes alkylations), 5. NaOH, H_2O, 6. H^+, H_2O, Δ; **(c)** 1. $NaOCH_2CH_3$, 2. $BrCH_2CO_2CH_2CH_3$ [alkylation, makes $CH_3CH_2O_2CCH_2CH(CO_2CH_2CH_3)_2$], 3. NaOH, H_2O, 4. H^+, H_2O, Δ;

(d) 1. 2 $NaOCH_2CH_3$, 2. $\begin{smallmatrix}CH_2Br\\ \langle\!\!\bigcirc\!\!\rangle\\ CH_2Br\end{smallmatrix}$, 3. NaOH, H_2O, 4. H^+, H_2O, Δ

34. (a) $\overset{O}{\underset{O}{\langle\text{cyclopentane-1,3-dione}\rangle}} + CH_2=CH\overset{\overset{\displaystyle O}{\|}}{C}CH_3 \xrightarrow[\substack{CH_3CH_2OH \\ \text{(Michael} \\ \text{addition)}}]{\text{Cat. } NaOCH_2CH_3}$ product

(b) [cycloheptenone structure] + $CH_2(CO_2CH_2CH_3)_2$ $\xrightarrow[\text{CH}_3\text{CH}_2\text{OH}]{\text{Cat. NaOCH}_2\text{CH}_3}$ product

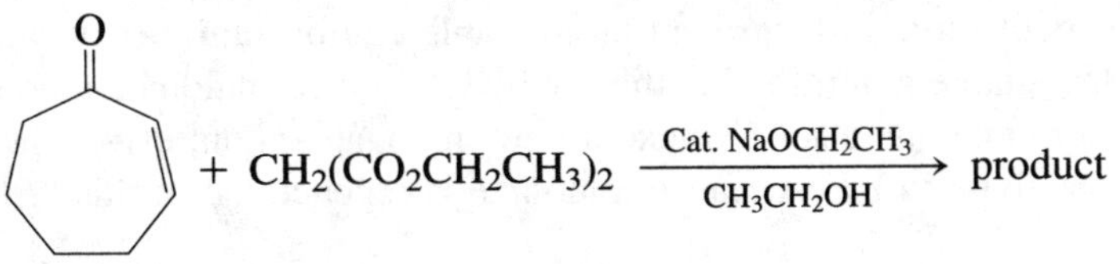

(c) [cyclopentenone structure] + $CH_3CCH_2CO_2CH_2CH_3$ $\xrightarrow[\substack{\text{CH}_3\text{CH}_2\text{OH} \\ \text{(Michael} \\ \text{addition)}}]{\substack{\text{Cat.} \\ \text{NaOCH}_2\text{CH}_3}}$ [cyclopentanone structure with CO$_2$CH$_2$CH$_3$ side group] $\xrightarrow[-\text{CO}_2]{\substack{\text{1. NaOH, H}_2\text{O} \\ \text{2. H}^+, \text{H}_2\text{O}, \Delta}}$ product

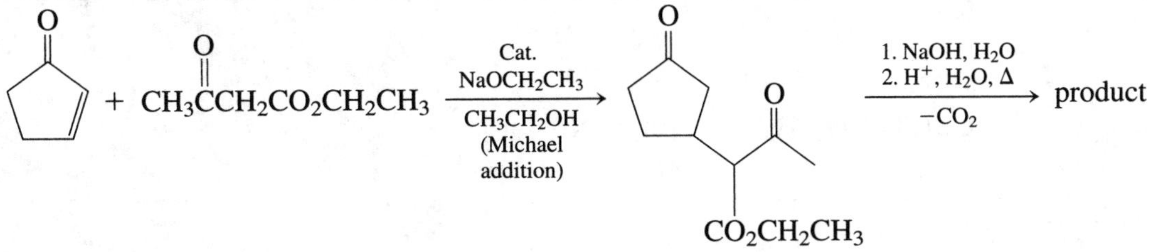

35. Here is one way that gets the job done:

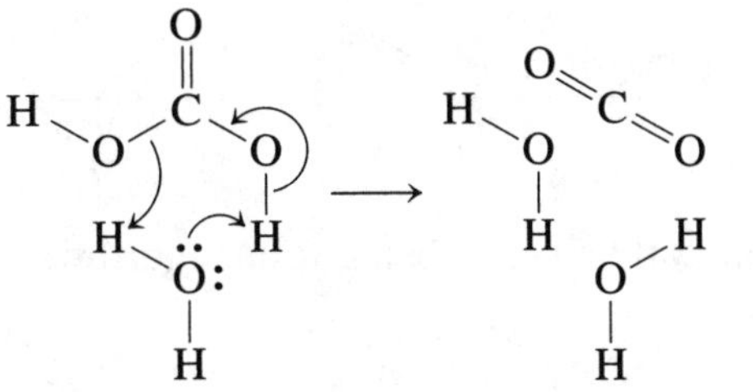

[mechanism structures]

36. **(a)** [mechanism structures with R]

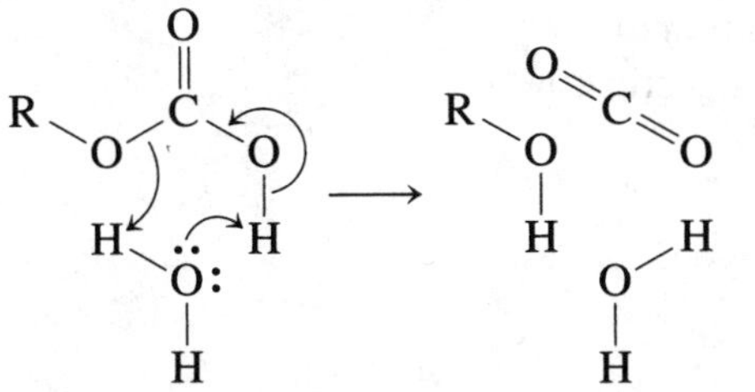

(b) H_2N [mechanism structures]

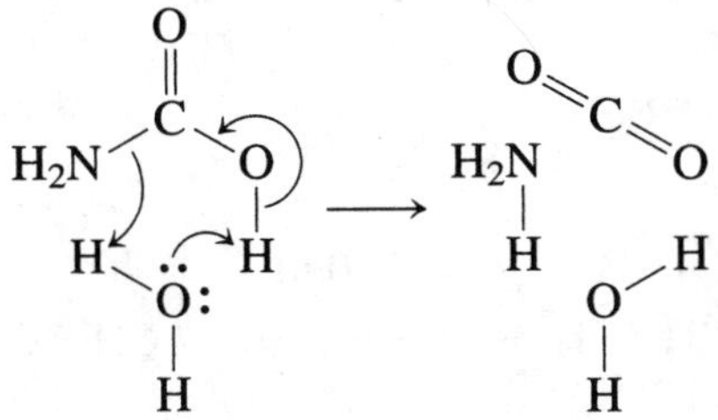

(c), (d) Does not work: An O—H bond is necessary (see the mechanisms above).

37. $(CH_3CH_2O_2C)_2CH$ $\xrightarrow[-\text{CH}_3\text{CH}_2\text{OH}]{^{-}:\ddot{O}CH_2CH_3}$ $(CH_3CH_2O_2C)_2\ddot{C}H$ + $CH_2=CHCCH_3$ $\underset{*}{\rightleftharpoons}$

$(CH_3CH_2O_2C)_2CHCH_2\ddot{C}HCOCH_3$ $\xrightarrow{\overset{H}{\underset{\curvearrowleft CH(CO_2CH_2CH_3)_2}{|}}}$

$(CH_3CH_2O_2C)_2CHCH_2CH_2COCH_3$ + $\ddot{C}H(CO_2CH_2CH_3)_2$
(Product) (Regenerated, to continue on)

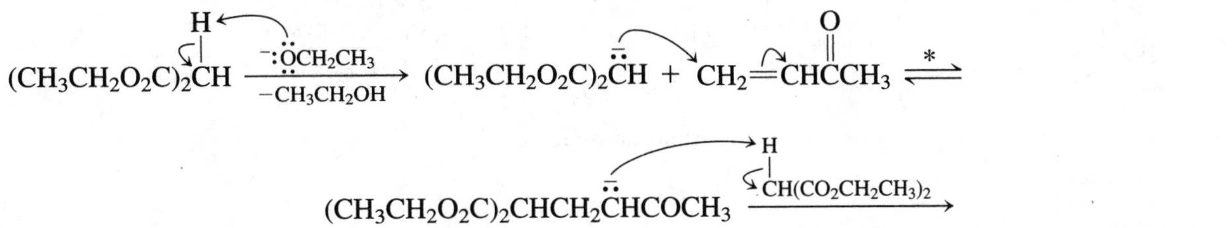

The step marked with an asterisk is reversible and, in fact, is an unfavorable equilibrium, because the product (a simple ketone enolate) is a less stable anion than is the doubly stabilized malonate anion. However, the next step, reaction with more malonic ester to make a new malonate anion, drives the equilibrium to product. The reaction is catalytic in base because malonate is regenerated in this last step.

38. These reactions form molecules that have the appearance of products of nucleophilic substitution.

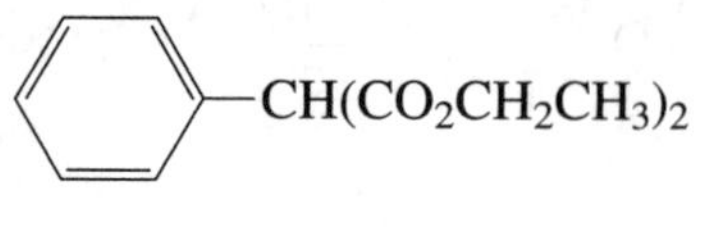

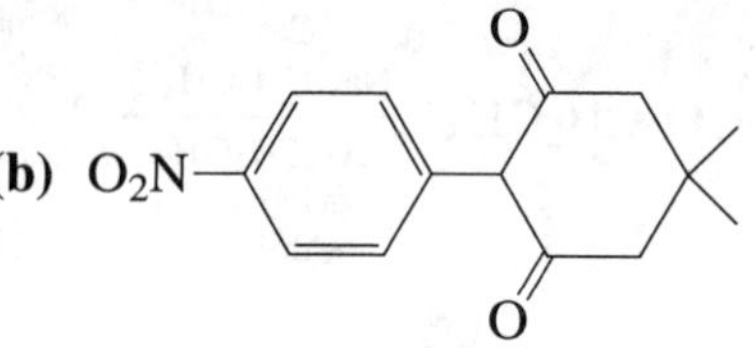

39.

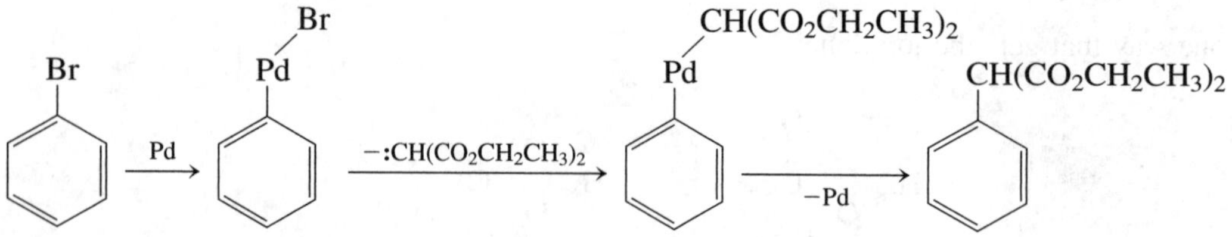

40. Work backward. Note carbon–carbon bonds being formed (arrows).

(a)

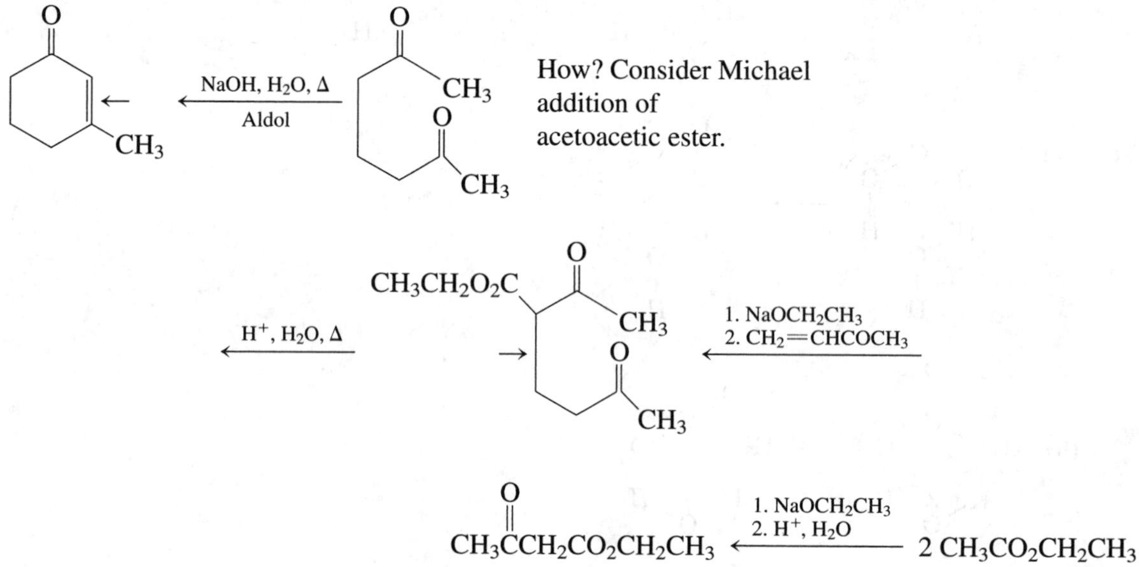

How? Consider Michael addition of acetoacetic ester.

(b)

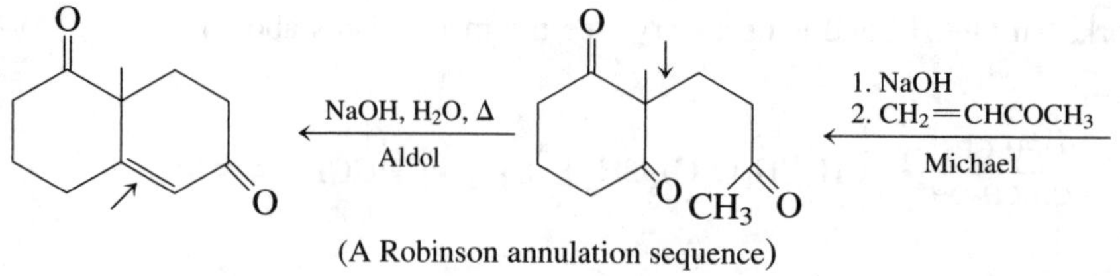

(A Robinson annulation sequence)

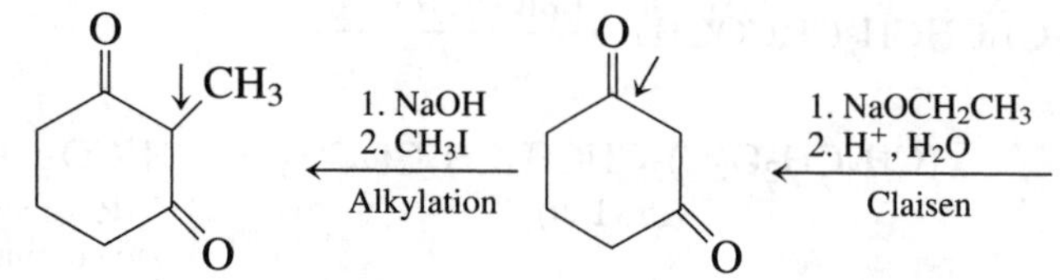

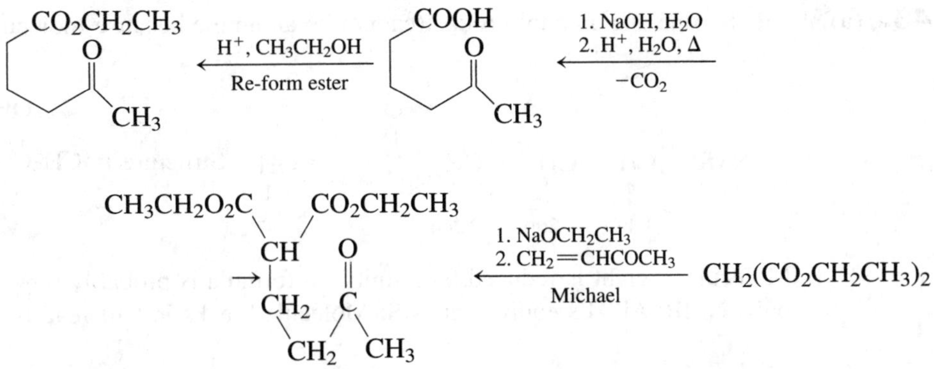

$$\underset{\overset{\displaystyle |}{O}}{\overset{\displaystyle CO_2CH_2CH_3}{\Big\|}} \quad \xleftarrow[\text{Re-form ester}]{H^+, CH_3CH_2OH} \quad \underset{\overset{\displaystyle |}{O}}{\overset{\displaystyle COOH}{\Big\|}} \quad \xleftarrow[-CO_2]{\substack{1.\ NaOH,\ H_2O \\ 2.\ H^+,\ H_2O,\ \Delta}}$$

(with CH_3 substituents)

$$CH_3CH_2O_2C \qquad CO_2CH_2CH_3$$
$$\underset{\underset{\underset{CH_2\quad CH_3}{|}}{\overset{|}{CH_2}}}{\overset{\displaystyle CH}{|}} \underset{C}{\overset{O}{\Big\|}}$$

$$\xleftarrow[\text{Michael}]{\substack{1.\ NaOCH_2CH_3 \\ 2.\ CH_2=CHCOCH_3}} CH_2(CO_2CH_2CH_3)_2$$

(c) A sequence identical to that of (b), but substitute two alkylations with $BrCH_2COCH_3$ for the two Michael additions to $CH_2=CHCOCH_3$.

41. (a) $(CH_3)_2CH-\underset{\overset{\displaystyle \|}{O}}{C}-\underset{\overset{\displaystyle |}{OH}}{CH}-CH(CH_3)_2$

(b) $C_6H_5-\underset{\overset{\displaystyle \|}{O}}{C}-\underset{\overset{\displaystyle |}{OH}}{CH}-C_6H_5$

(c) 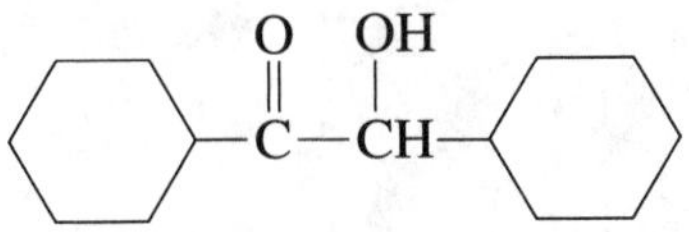 (cyclohexyl)$-\underset{\overset{\displaystyle \|}{O}}{C}-\underset{\overset{\displaystyle |}{OH}}{CH}-$(cyclohexyl)

(d) $C_6H_5CH_2-\underset{\overset{\displaystyle \|}{O}}{C}-\underset{\overset{\displaystyle |}{OH}}{CH}-CH_2C_6H_5$

42. (a) 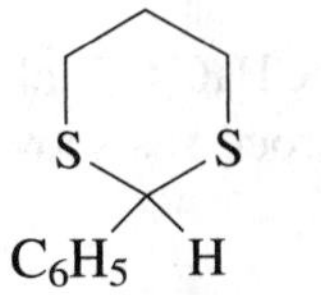 (1,3-dithiane ring with C_6H_5 and H on C-2) **(b)** 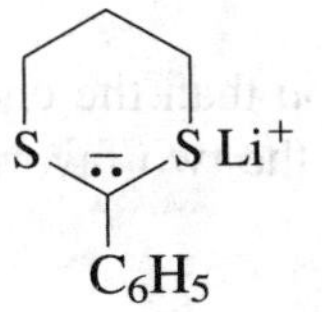 (1,3-dithiane ring with C_6H_5, S⁻ Li⁺)

In the same order as in Problem 41:

(a) $C_6H_5-\underset{\overset{\displaystyle \|}{O}}{C}-\underset{\overset{\displaystyle |}{OH}}{CH}-CH(CH_3)_2$

(b) $C_6H_5-\underset{\overset{\displaystyle \|}{O}}{C}-\underset{\overset{\displaystyle |}{OH}}{CH}-C_6H_5$ (same product!)

(c) 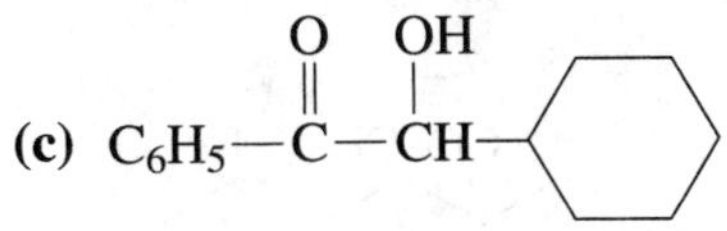 $C_6H_5-\underset{\overset{\displaystyle \|}{O}}{C}-\underset{\overset{\displaystyle |}{OH}}{CH}-$(cyclohexyl)

(d) $C_6H_5-\underset{\overset{\displaystyle \|}{O}}{C}-\underset{\overset{\displaystyle |}{OH}}{CH}-CH_2C_6H_5$

43. **(a)** A IR: Ketone and alcohol groups (cannot be an amine because molecular weight is an **even** number).

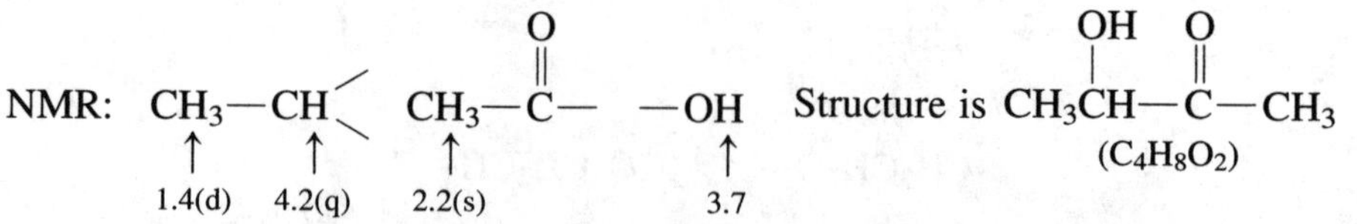

NMR: $CH_3-CH\Big\langle$ $CH_3-\overset{\overset{\displaystyle O}{\|}}{C}-$ $-OH$ Structure is $CH_3\overset{\overset{\displaystyle OH}{|}}{CH}-\overset{\overset{\displaystyle O}{\|}}{C}-CH_3$
                                                                                      ($C_4H_8O_2$)

          ↑   ↑   ↑   ↑
        1.4(d) 4.2(q) 2.2(s)   3.7

B Molecular weight is reduced by 2 units, so formula is probably now $C_4H_6O_2$. IR: Ketone signal only. NMR: All H's equivalent. MS: Molecule breaks in half readily, giving m/z 43,

C_2H_3O fragments. Simplest interpretation: $CH_3-\overset{\overset{\displaystyle O}{\|}}{C}-$, so molecule is $CH_3\overset{\overset{\displaystyle OO}{\| \|}}{CC}CH_3$

(b) Oxidation. Churning cream mixes it with air, thus allowing O_2 to react with ketoalcohol A to make diketone B. **(c)** You can synthesize A by reaction of acetaldehyde with catalytic *N*-dodecyl-thiazolium salt (Section 23-4). Oxidation gives B. **(d)** The diketone is conjugated.

44. Addition to carbonyl:

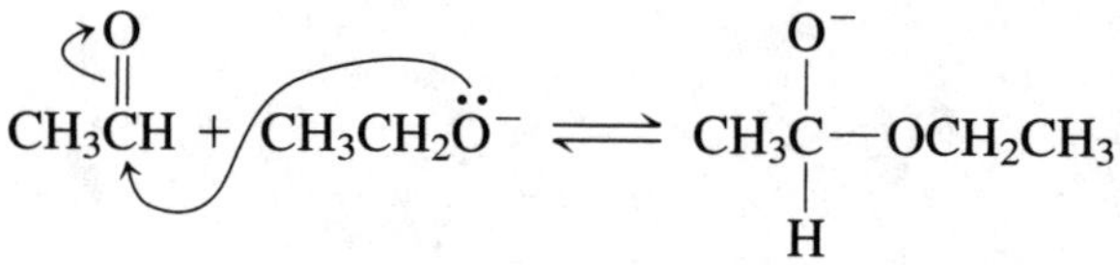

$$CH_3\overset{\overset{\displaystyle O}{\|}}{CH} + CH_3CH_2\ddot{O}^- \rightleftharpoons CH_3\overset{\overset{\displaystyle O^-}{|}}{\underset{\underset{\displaystyle H}{|}}{C}}-OCH_2CH_3$$

Deprotonation of α-carbon:

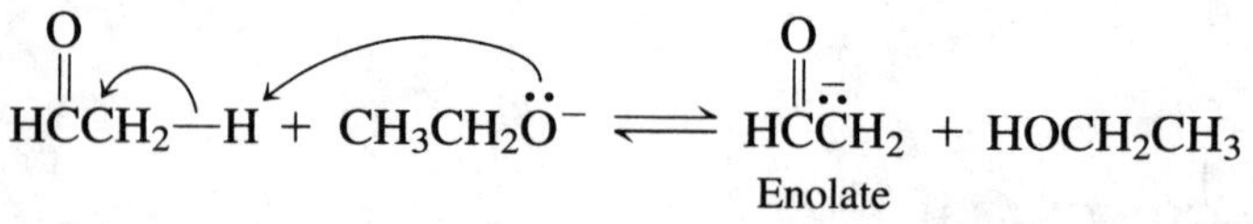

$$H\overset{\overset{\displaystyle O}{\|}}{C}CH_2-H + CH_3CH_2\ddot{O}^- \rightleftharpoons H\overset{\overset{\displaystyle O}{\|}}{C}\ddot{C}H_2 + HOCH_2CH_3$$
                                                      Enolate

Deprotonation of the aldehyde carbon leads to a much poorer anion than the enolate: $CH_3\overset{\overset{\displaystyle O}{\|}}{C}:^-$, electron pair in an sp^2 orbital, unable to be stabilized by resonance. Given the two favorable processes shown above, deprotonation of the $-\overset{\overset{\displaystyle O}{\|}}{C}H$ group is simply **not competitive.**

45.

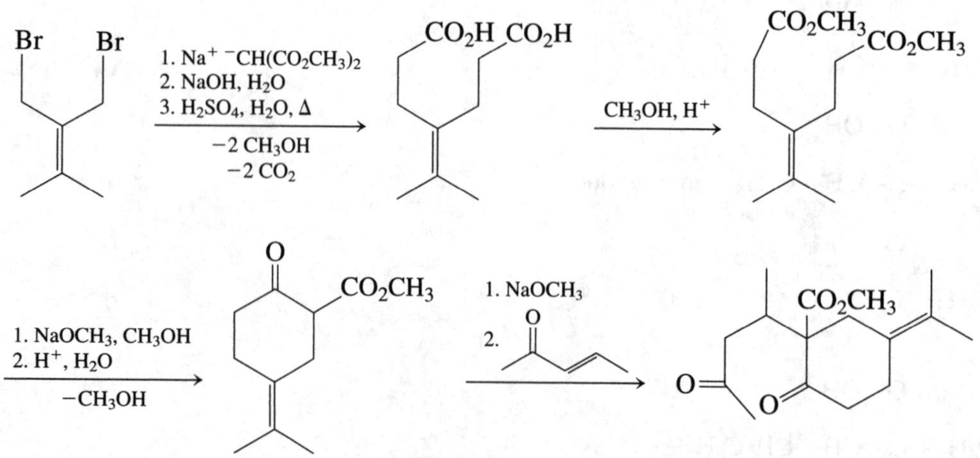

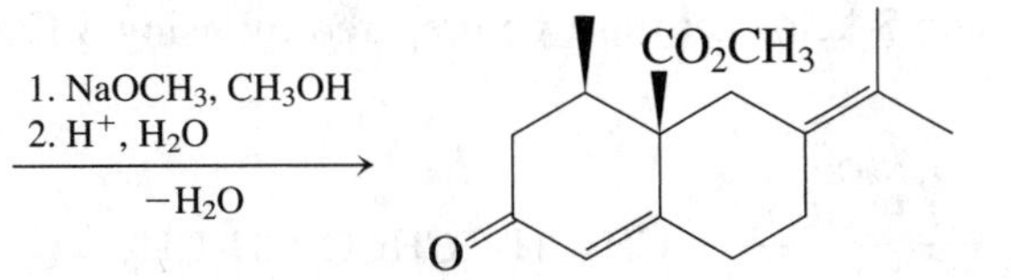

$$\xrightarrow[\text{$-H_2O$}]{\begin{array}{l}\text{1. NaOCH}_3,\text{ CH}_3\text{OH}\\\text{2. H}^+,\text{ H}_2\text{O}\end{array}}$$

46. The Knoevenagel condensation is a variation of aldol condensation using a β-dicarbonyl compound as the source of the enolate reaction partner. Its mechanism is the same as that of the aldol:

(a) $\overset{\overset{\displaystyle O}{\displaystyle\|}}{\text{CH}_3\text{C}}\text{CH}_2\text{CO}_2\text{CH}_2\text{CH}_3$

NaOCH$_2$CH$_3$

$\overset{\overset{\displaystyle O}{\displaystyle\|}}{\text{CH}_3\text{C}}\text{CHCO}_2\text{CH}_2\text{CH}_3$ $\longrightarrow$ $\longrightarrow$

$\longrightarrow$

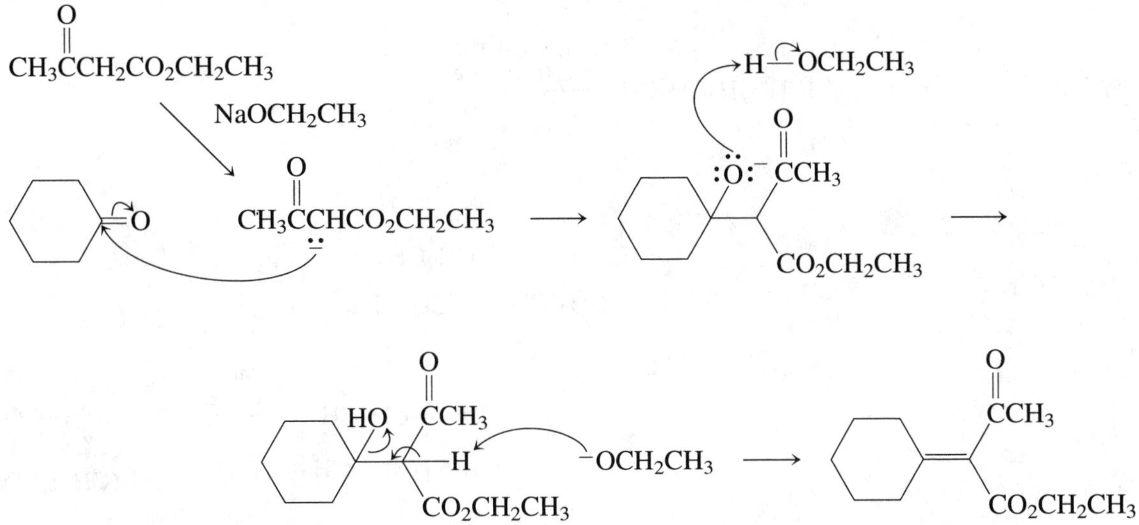

(b) As in the aldol, remove the elements of water (two hydrogen atoms from the carbon of the malonate ester and the oxygen atom from the aldehyde) and replace them with a carbon–carbon double bond:

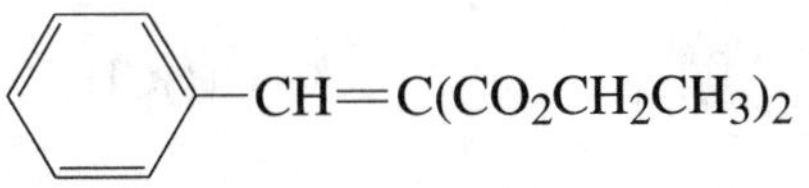

$$\text{—CH}{=}\text{C(CO}_2\text{CH}_2\text{CH}_3)_2$$

(c)

$$\begin{array}{l}\text{NaOCH}_2\text{CH}_3,\\\text{CH}_3\text{CH}_2\text{OH}\end{array}$$

$$=\!\text{O} \;+\; \text{CH}_2(\text{CO}_2\text{CH}_2\text{CH}_3)_2 \longrightarrow$$

$$\xrightarrow{\begin{array}{l}\text{1. LiAlH}_4\\\text{2. H}^+,\text{ H}_2\text{O}\end{array}}$$ CH$_2$OH $\xrightarrow{\text{PBr}_3}$ CH$_2$Br

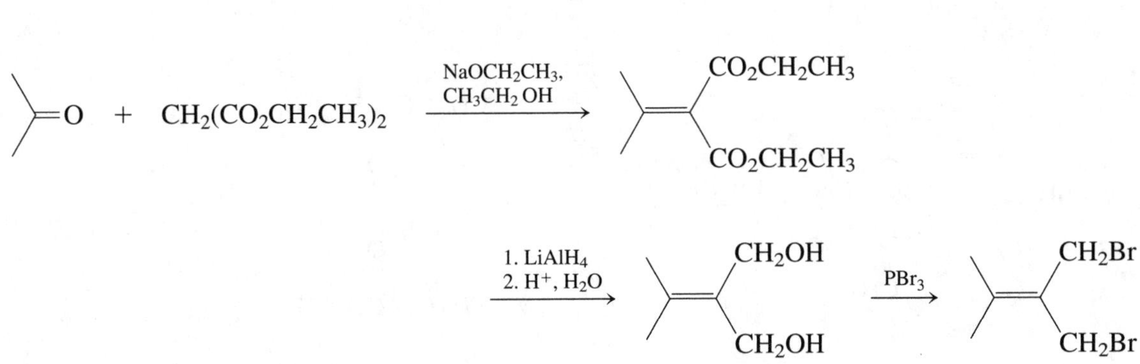

47. The acetoacetic ester ketone synthesis is only good for **methyl** ketones.

$$\overset{\overset{\displaystyle O}{\displaystyle\|}}{\text{CH}_3\text{C}}\text{—CHRR}' \text{ from } \overset{\overset{\displaystyle O}{\displaystyle\|}}{\text{CH}_3\text{C}}\text{—CH}_2\text{CO}_2\text{CH}_2\text{CH}_3$$

For other ketones, the appropriate 3-ketoester must be prepared by using a Claisen condensation.

(a)

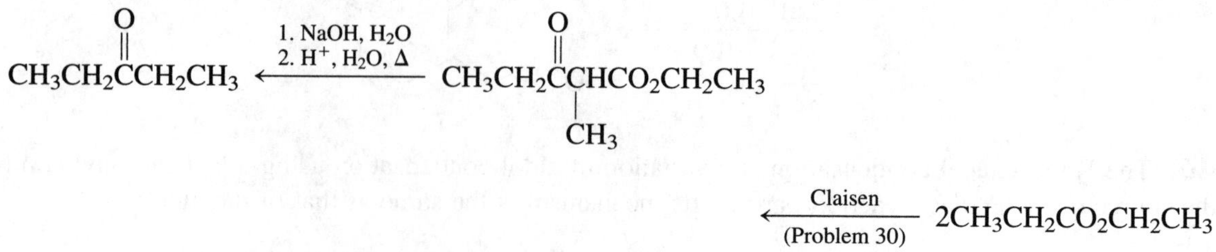

$$\underset{\text{CH}_3\text{CH}_2\overset{\overset{\displaystyle O}{\|}}{C}\text{CH}_2\text{CH}_3}{} \xleftarrow[\text{2. H}^+, \text{H}_2\text{O}, \Delta]{\text{1. NaOH, H}_2\text{O}} \underset{\underset{\displaystyle \text{CH}_3}{|}}{\text{CH}_3\text{CH}_2\overset{\overset{\displaystyle O}{\|}}{C}\text{CHCO}_2\text{CH}_2\text{CH}_3}$$

$$\xleftarrow[\text{(Problem 30)}]{\text{Claisen}} 2\text{CH}_3\text{CH}_2\text{CO}_2\text{CH}_2\text{CH}_3$$

(b)

$$\xleftarrow[\text{2. H}^+, \text{H}_2\text{O}, \Delta]{\text{1. NaOH, H}_2\text{O}}$$

$$\xrightarrow[\text{2. CH}_3\text{I}]{\text{1. NaOCH}_2\text{CH}_3}$$

$$\xleftarrow[]{\text{Mixed}\atop\text{Claisen}}$$

$$+$$
$$\text{CH}_3(\text{CH}_2)_4\text{CO}_2\text{CH}_2\text{CH}_3$$

(c)

$$\xleftarrow[\text{2. H}^+, \text{H}_2\text{O}, \Delta]{\text{1. NaOH, H}_2\text{O}}$$

$$\xleftarrow[\text{2. BrCH}_2\text{CH}=\text{CH}_2]{\text{1. NaOCH}_2\text{CH}_3}$$

$$\xleftarrow[]{\text{Dieckmann}}$$

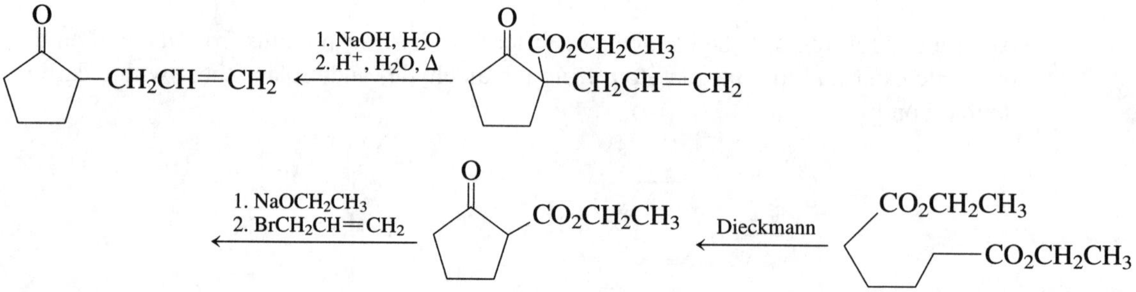

(d)

$$\xleftarrow[\text{2. H}^+, \text{H}_2\text{O}, \Delta]{\text{1. NaOH, H}_2\text{O}}$$

$$\xleftarrow[\text{2. 2}\text{CH}_2\text{Br}]{\text{1. 2 NaOCH}_2\text{CH}_3}$$

$$\xleftarrow[]{\text{Double}\atop\text{Claisen}}$$

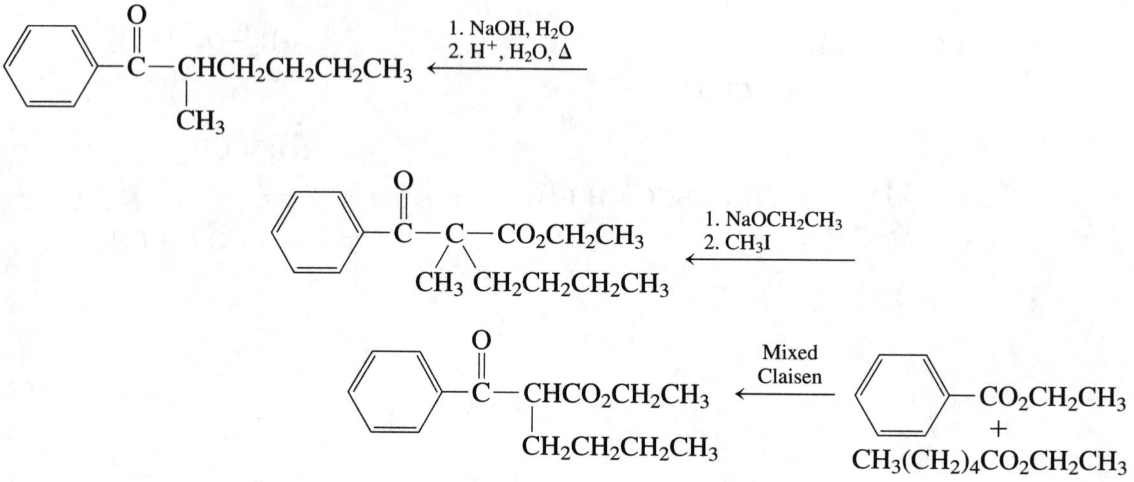

$$+$$

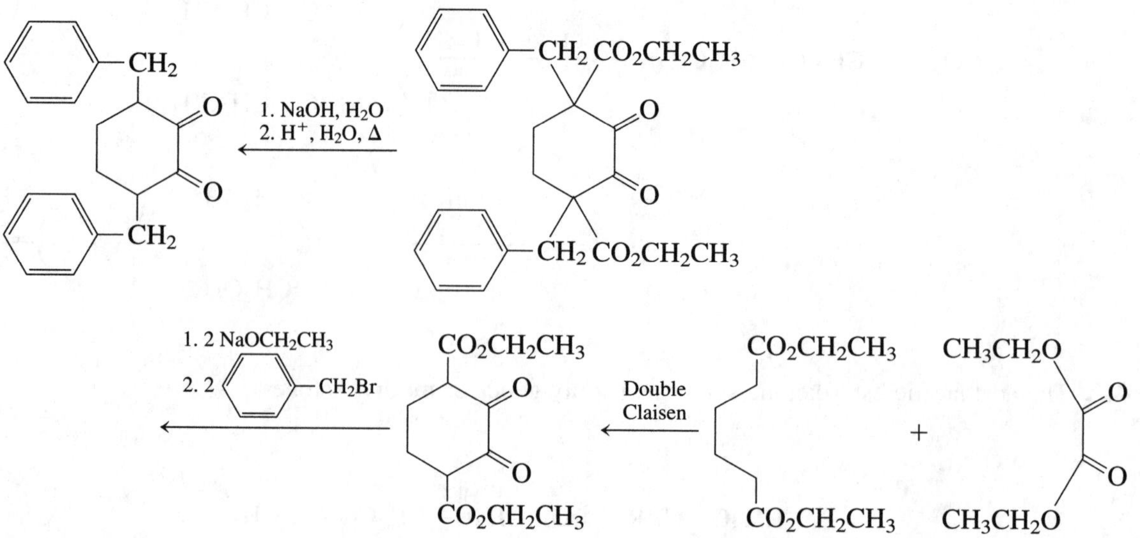

48. Ethanol is the reaction solvent unless specified otherwise.

Cyclopentanone

$$HCO_2CH_2CH_3 \; + \; CH_3CO_2CH_2CH_3 \xrightarrow{\text{Claisen}} \overset{\displaystyle O}{HCCH_2CO_2CH_2CH_3}$$

$$\overset{\displaystyle O}{CH_3CCH_3} \xrightarrow{\text{Br}_2,\ CH_3CO_2H} \overset{\displaystyle O}{BrCH_2CCH_3}$$

$$\overset{\displaystyle O}{HCCH_2CO_2CH_2CH_3} \xrightarrow[\text{2. BrCH}_2\text{COCH}_3]{\text{1. NaOCH}_2\text{CH}_3} \underset{\displaystyle CO_2CH_2CH_3}{\overset{\displaystyle O \qquad\quad O}{HCCHCH_2CCH_3}} \xrightarrow{H^+,\ H_2O,\ \Delta}$$

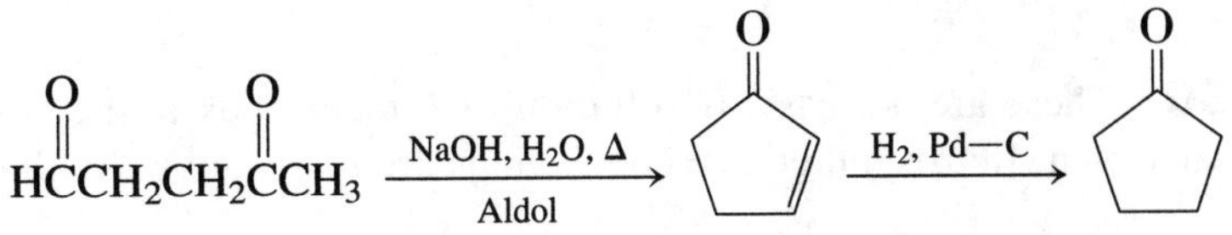

$$\overset{\displaystyle O \qquad\quad O}{HCCH_2CH_2CCH_3} \xrightarrow[\text{Aldol}]{\text{NaOH, H}_2\text{O},\ \Delta} \text{(cyclopentenone)} \xrightarrow{H_2,\ Pd\text{—}C} \text{(cyclopentanone)}$$

Cyclohexanone

$$\overset{\displaystyle O}{CH_3CCH_3} \; + \; CH_2{=}O \xrightarrow[\text{Aldol}]{\text{NaOH, H}_2\text{O}} \overset{\displaystyle O}{CH_3CCH_2CH_2OH} \xrightarrow{H^+,\ \Delta} \overset{\displaystyle O}{CH_3CCH{=}CH_2}$$

$$\underset{\text{From above}}{\overset{\displaystyle O}{HCCH_2CO_2CH_2CH_3}} \xrightarrow[\text{Michael}]{\substack{\text{1. NaOCH}_2\text{CH}_3 \\ \text{2. CH}_3\text{COCH}{=}\text{CH}_2}} \underset{\displaystyle CO_2CH_2CH_3}{\overset{\displaystyle O \qquad\qquad\quad O}{HCCHCH_2CH_2CCH_3}} \xrightarrow[\substack{\text{as above}}]{\text{Same steps}} \text{(cyclohexanone)}$$

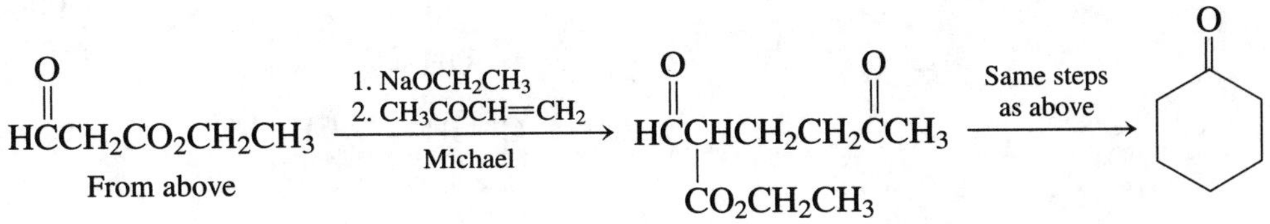

49.

Michael addition

$\xrightarrow[\substack{\text{Acid-catalyzed} \\ \text{aldol condensation}}]{H^+}$

(Enol as nucleophile)

$\xrightarrow{-H^+}$

$\xrightarrow[\substack{-H_2O \\ \text{Acid-catalyzed} \\ \text{dehydration}}]{H^+}$

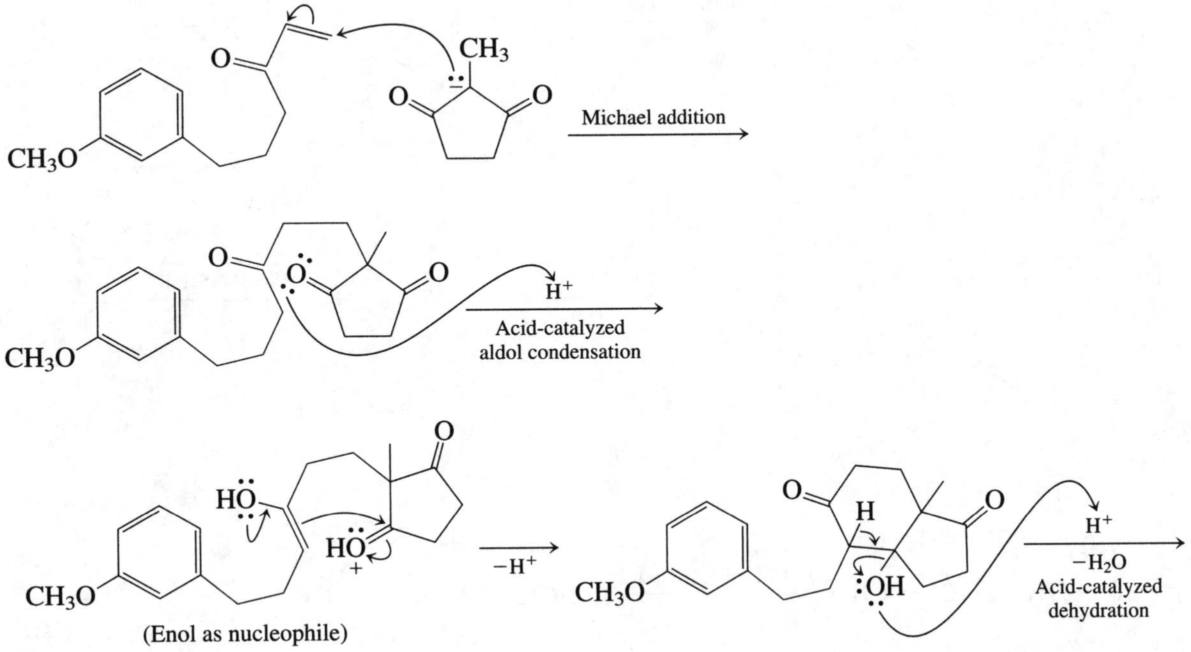

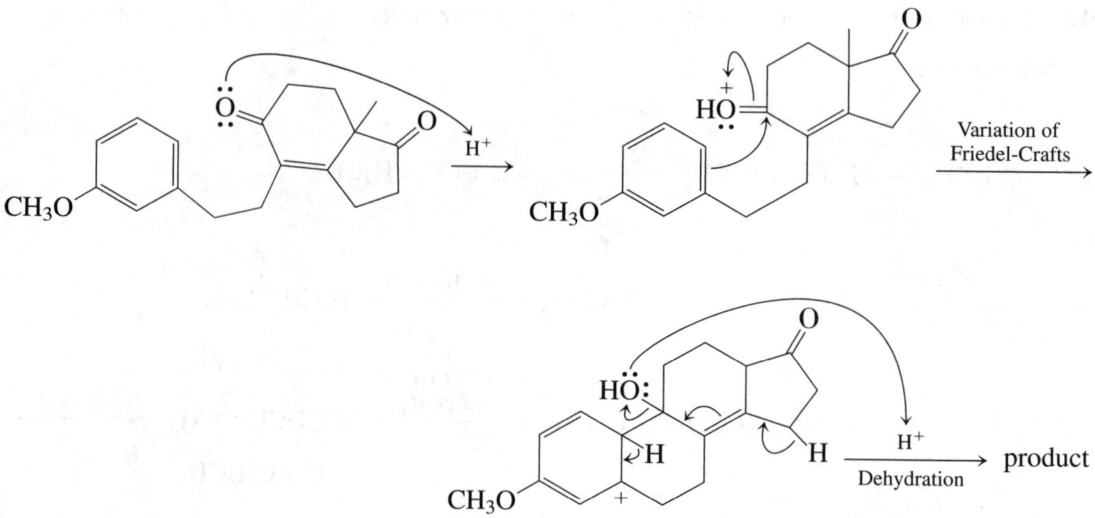

50. These are not easy. If you didn't get them, look at the answer for (a) and then try (b) and (c) again on your own. Retrosynthetic disconnections are indicated in boldface.

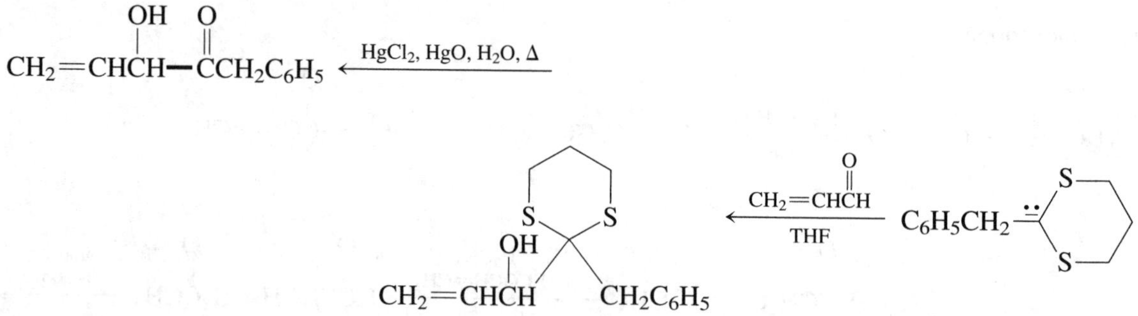

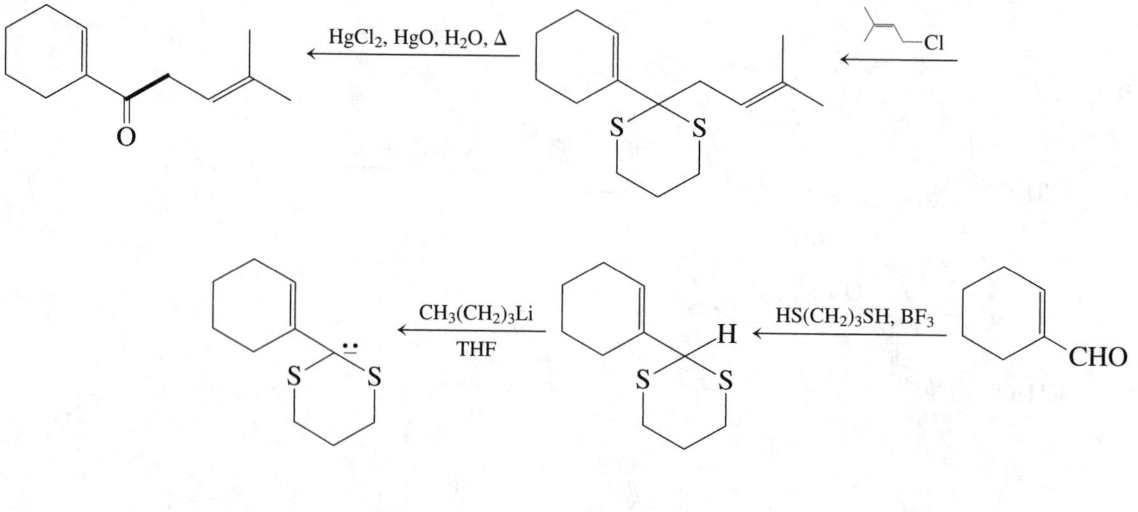

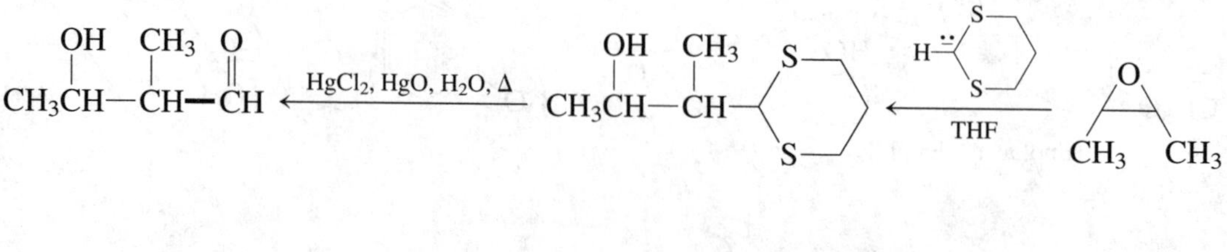

51. Identify the bond being made. It looks like the result of 1,4-addition to the α, β-unsaturated lactone by an alkanoyl anion equivalent.

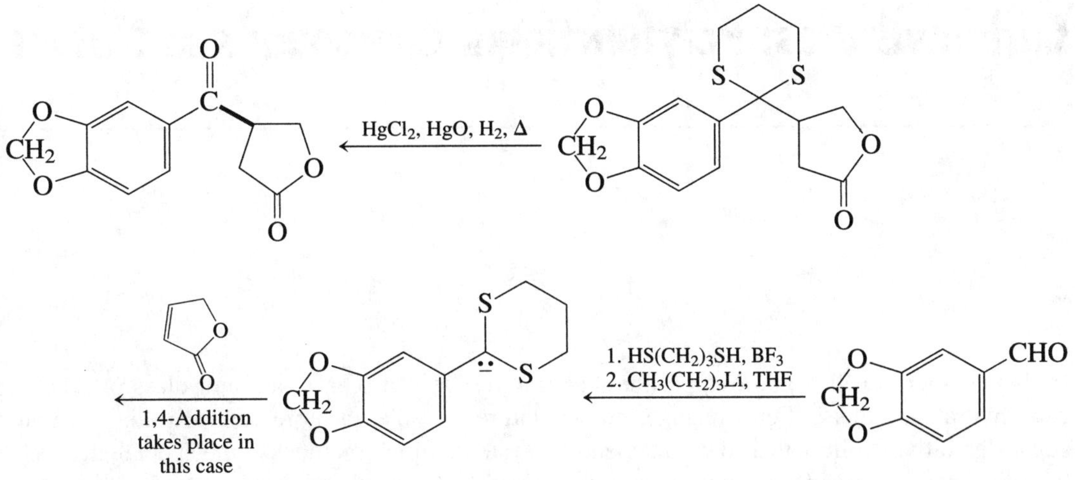

1,4-Addition
takes place in
this case

52. **(a)** Notice that the sequence begins by treating the starting material with *two* equivalents of strong base. The result is the formation of a dianion, which can be represented by the following Lewis structure.

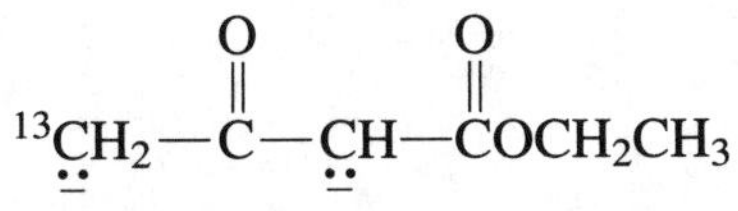

$$^{13}\ddot{C}H_2-\overset{O}{\overset{\|}{C}}-\underset{..}{C}H-\overset{O}{\overset{\|}{C}}OCH_2CH_3$$

Of the two anionic carbon atoms, the terminal one (^{13}C) is the more basic and therefore the stronger nucleophile because the charge is stabilized by only one adjacent carbonyl group. The remainder of the sequence is the completion of a β-ketoester ketone synthesis.

(b) Attempted alkylation of a β-dicarbonyl anion (an S_N2 process) with a tertiary haloalkane is doomed to give E2 elimination instead (textbook, Section 7-9).

Extra problem: The key is the need for *three* equivalents of strong base. The first two deprotonate the β-ketoamide CH_2 at C2 and the NH, respectively. What does the third do? Elimination of HCl from the benzene ring to give a benzyne, to which the ketoamide carbanion adds:

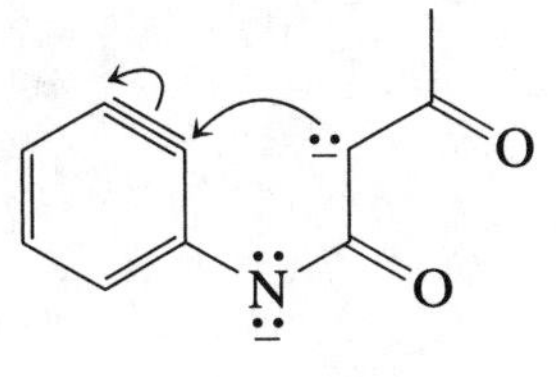

53. **(c)**

54. **(b)**

55. **(e)** From heat-induced dehydration of the open-chain diacid to give the cyclic anhydride (see text Section 19-8).

56. **(e)** The IR band at 2250 cm^{-1} is a dead giveaway for the nitrile triple bond. This molecule is the only one that matches the NMR (a 2H singlet and an ethyl group).

24

Carbohydrates: Polyfunctional Compounds in Nature

In this chapter you'll begin to apply the material you've just seen to a major class of "real world" molecules, carbohydrates (sugars). For a change, nomenclature will play a more central role: The names of sugars and sugar derivatives follow their own independent system, which includes some special stereochemical terms. On the other hand, most of the reactions are old ones and are needed for only a limited number of purposes such as structure determination, interconversion of derivatives, and synthesis of one sugar from another. You need to be good at deductive reasoning so that you can solve the puzzles posed by some of the problems. If you can, then this chapter should not be too hard for you.

Outline of the Chapter

24-1, 24-2, 24-3 Names and Structures of Carbohydrates
Be prepared for a lot of new terminology.

24-4 through 24-8 Polyfunctional Chemistry of Sugars
The basics: mostly (but not entirely) review material.

24-9, 24-10 Step-by-Step Buildup and Degradation of Sugars
Application in synthesis and structure determination.

24-11, 24-12 Complex Sugars in Nature

Keys to the Chapter

24-1. Names and Structures of Carbohydrates
The naming system presented in this section comes from an assortment of historically derived common names organized into a semiofficial framework that is universally used. So, sugars are all ketones or aldehydes containing alcohol groups, and their names all end in -*ose*. Almost all of them have stereocenters; and they are usually drawn in Fischer projections, with the carbon chain vertical and the carbonyl group nearest the top. If the stereocenter **nearest the bottom** of the Fischer projection has the OH on the right, it has an *R* configuration, and the sugar is said to belong to the D family. If this OH is on the left, you have an *S* configuration and an L-sugar. Look at Figures 24-1 and 24-2. Each horizontal row contains structures that are all **diastereomers** of one another. None of the mirror-image structures (**enantiomers**) of any of these is illustrated; the mirror image of any D-sugar is just an L-sugar with the same name.

24-2 and 24-3. Conformations and Cyclic Forms of Sugars; Mutarotation

Because sugars contain alcohols and carbonyl groups, they can (and usually do) form cyclic hemiacetals, typically with either five- or six-membered rings. Translating an open-chain structure into a picture of the cyclic hemiacetal is tricky: Here's a step-by-step way to do it.

1. Write in wedges and dotted lines and (for a D-series sugar) lay the structure on its side, with the top moved down to the right. Then, locate the OH group that you will use to form the cyclic hemiacetal with the carbonyl group.

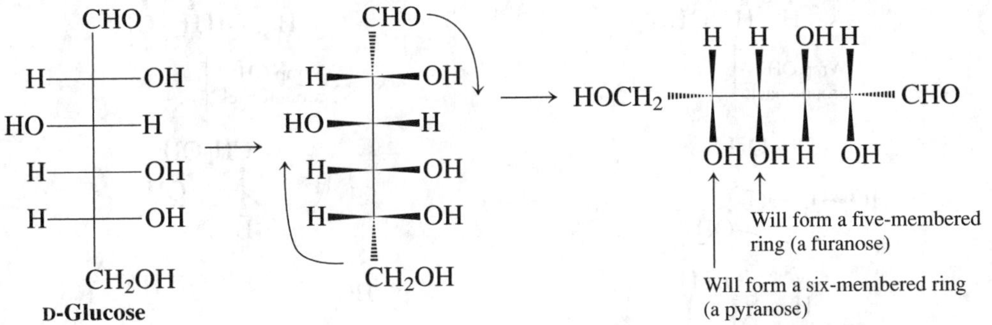

2. Rotate around the C—C bond to the right of the OH group you picked out until the OH is horizontal and pointing away from you. (Make a model!) Then wrap this left-hand end of the chain behind the plane of the paper to put the OH near to the carbonyl carbon. Finally, make the hemiacetal bond. (This sequence is shown step-by-step in the next paragraph.)

By convention, the procedure for L-series sugars is modified such that the top of the structure is rotated down **to the left** when it is laid on its side. That allows the L and D structures to look like mirror images when placed side-by-side.

β-D-Glucopyranose **β-L-Glucopyranose**

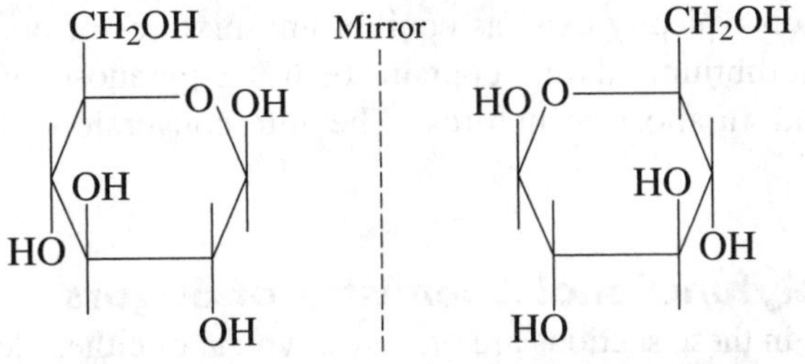

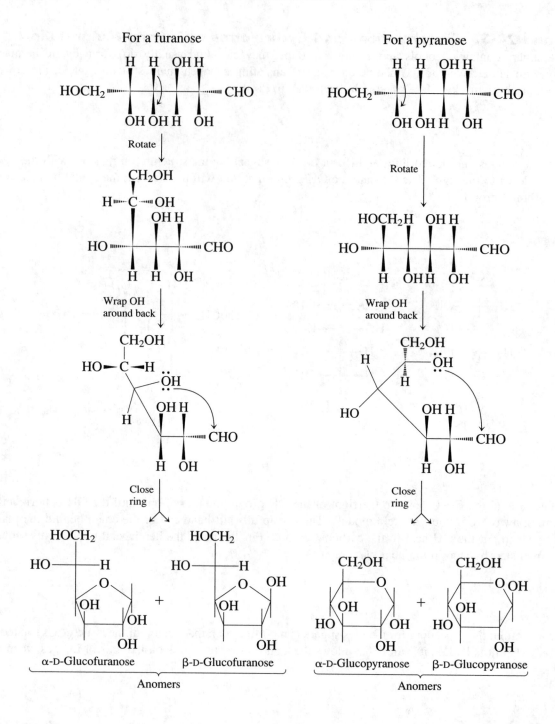

If you like, you can derive the cyclic structure of an L-sugar by doing the D-sugar first and then writing its mirror image.

Remember, sugars in solution typically exist as equilibrium mixtures of open chain plus cyclic hemiacetal structures. For glucose, this equilibrium mixture contains 63.6% β-pyranose, 36.4% α-pyranose, together with just traces of open chain and furanose structures. The interconversion of β and α anomers is called *mutarotation.*

24-4 though 24-8. Polyfunctional Chemistry of Sugars

Although most of the reactions in these sections are old ones, typical of either alcohol or aldehyde/ketone chemistry, a couple of new reagents are introduced. These mainly allow selective reactions to be carried out on the multifunctional sugar molecule. Examples are Br_2 in H_2O, which oxidizes only the aldehyde group of an aldose to CO_2H, and HNO_3, which oxidizes both the end carbons of an aldose, forming a dicarboxylic acid. The reactions in these sections have been chosen for their importance in practical aspects of sugar chemistry, the material in the latter half of this chapter.

24-9 and 24-10. Step-by-Step Buildup and Degradation of Sugars

Determining sugar structures was a major effort involving the development of reaction sequences to lengthen or shorten sugar chains and the use of some very clever logic dealing with the consequences of the stereochemistry of sugars and sugar derivatives. The "Fischer proof" illustrates the main techniques used. Note that the process repeatedly makes use of the synthesis of dicarboxylic acids, which are tested for the presence or absence of optical activity. An optically inactive diacid is assumed to be a meso compound, containing a plane of symmetry, and this kind of information is used to narrow down possible structures for unknown sugars.

24-11 and 24-12. Complex Sugars in Nature

These sections present simple extensions of the material just completed. Mother Nature has developed a convenient method for linking sugar molecules together: An alcohol group of one sugar forms an acetal by reactions with the hemiacetal group of another. This connection, called a *glycoside linkage,* is just a fancier version of a simple acetal formed by reaction of a sugar with an ordinary alcohol, like methanol (to form a "methyl glycoside," as in a previous section). Again, the determination of unknown structures represents a common type of problem. There are two important features that distinguish sugars containing free hemiacetal groups from those lacking them. In solution, hemiacetals are always in equilibrium with aldehydes. So, hemiacetal-containing sugars (1) undergo mutarotation and (2) are readily oxidized by mild oxidants like Ag^+. The latter feature is the basis for the Tollens's test for "reducing" (that is, oxidizable) sugars. Some of the sugars in this section contain acetal but not hemiacetal groups. Find them! These are examples of "nonreducing" sugars.

Solutions to Problems

33. You get

CHO
H————OH
H————OH
HO————H
CH$_2$OH

This compound is the mirror image (enantiomer) of **D-lyxose** (Figure 24-1). Therefore, this sugar is **L-lyxose,** a **diastereomer** of D-ribose.

34. **(a)** D-Aldopentose (Note: Only **one** stereocenter!); **(b)** L-aldohexose; **(c)** D-ketoheptose

35.

CHO
HO————H
HO————H
HO————H
CH$_2$OH

L-Ribose
Systematic name:
(2*S*,3*S*,4*S*)-2,3,4,5-
Tetrahydroxypentanal

CHO
HO————H
H————OH
HO————H
HO————H
CH$_2$OH

L-Glucose
Systematic name:
(2*S*,3*R*,4*S*,5*S*)-2,3,4,5,6-
Pentahydroxyhexanal

36. I know it sounds like an awful thing to make you do, but you might just have to review Sections 5-5 and 5-6 for this one. The *Study Guide* text for these sections may also help.

(a) L-Glyceraldehyde; (b) D-erythrulose; (c) just D-glucose (upside down!);
(d) L-xylose; (e) D-threose

37. Make models if you need to.

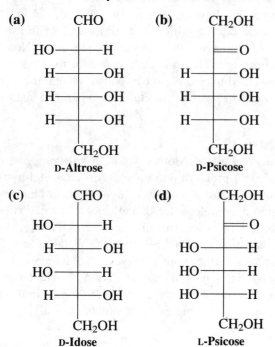

38. See procedure in the *Study Guide* text for this chapter. Careful—(b) and (c) are L-sugars.

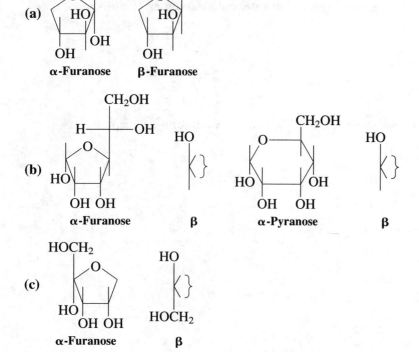

(d)

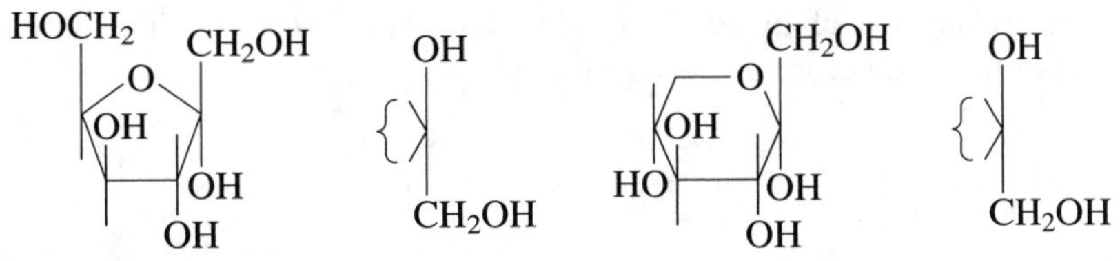

HOCH₂   CH₂OH
α-Furanose   β   α-Pyranose   β

(e)

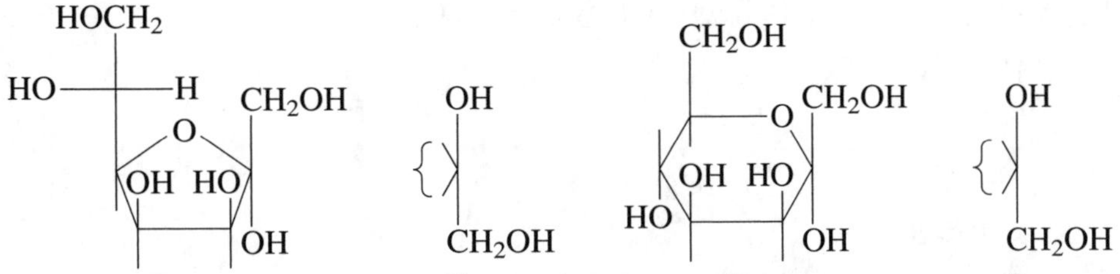

HOCH₂
α-Furanose   β   α-Pyranose   β

39. No. They are all hemiacetals and therefore are capable of readily interconverting their α and β anomers.

40. (a) 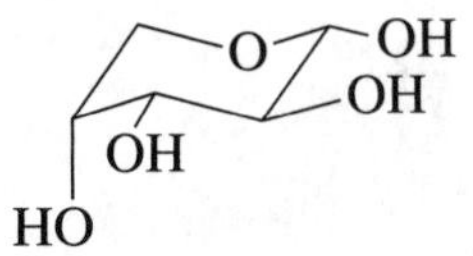   (b)

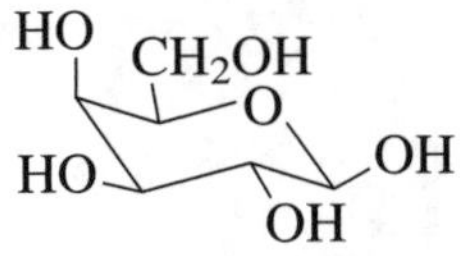

(c) 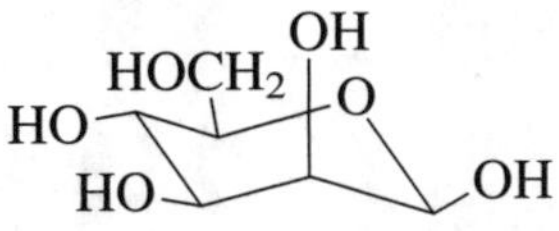   (d)

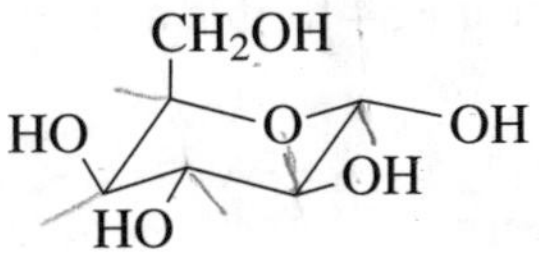

(d) is an unusual case where the CH₂OH is forced to be axial to allow all four OH's to be equatorial.

41. Base-catalyzed enolization allows the interconversion to take place. But notice that the product is not an ordinary enol. It has hydroxy groups on *both* carbons of the double bond: It is an *enediol*. Therefore, when it tautomerizes it has the option of losing a proton from either of two hydroxy groups, giving either the original ketone or the isomeric aldehyde.

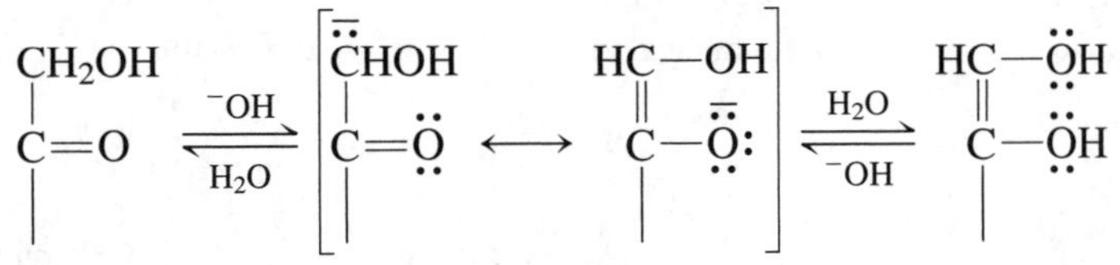

Ketose   **Enolate**   **Enediol**

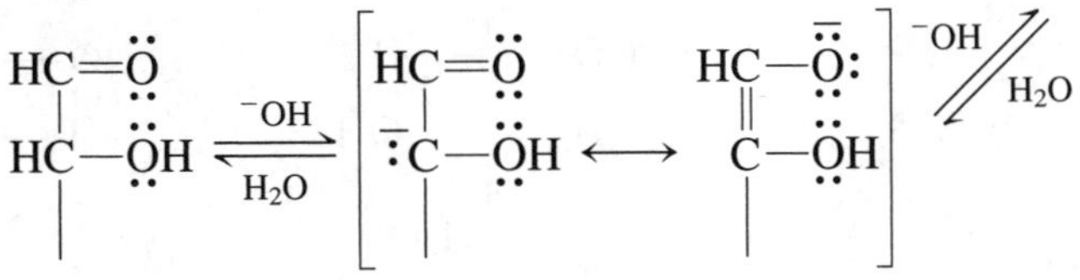

Aldose   **Other enolate**

42. Each carbon atom in the starting sugar is alongside its product after HIO_4 cleavage. The number of hydrogens on each carbon remains the same before and after cleavage.

(a)

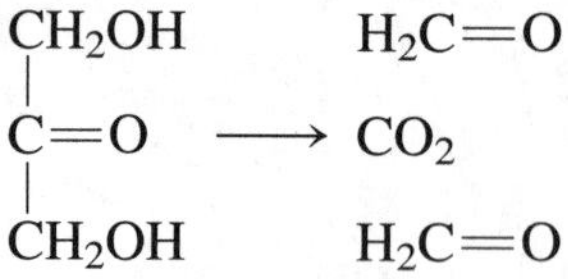

$$
\begin{array}{ll}
CH_2OH & H_2C=O \\
| & \\
C=O \longrightarrow & CO_2 \\
| & \\
CH_2OH & H_2C=O
\end{array}
\qquad \text{that is, 2 formaldehyde + 1 } CO_2
$$

(b)

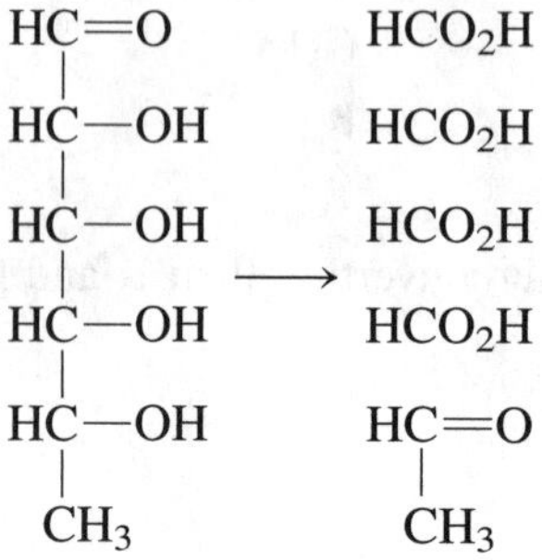

$$
\begin{array}{ll}
HC=O & HCO_2H \\
| & \\
HC-OH & HCO_2H \\
| & \\
HC-OH & HCO_2H \\
| \quad \longrightarrow & \\
HC-OH & HCO_2H \\
| & \\
HC-OH & HC=O \\
| & | \\
CH_3 & CH_3
\end{array}
\qquad \text{that is, 4 formic acid + 1 acetaldehyde}
$$

(c)

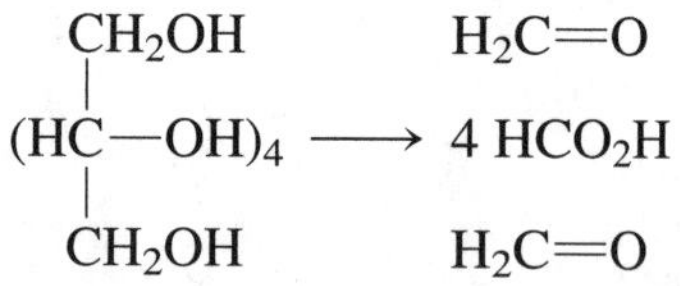

$$
\begin{array}{ll}
CH_2OH & H_2C=O \\
| & \\
(HC-OH)_4 \longrightarrow & 4\ HCO_2H \\
| & \\
CH_2OH & H_2C=O
\end{array}
\qquad \text{that is, 4 formic acid + 2 formaldehyde}
$$

43. (a) (i)

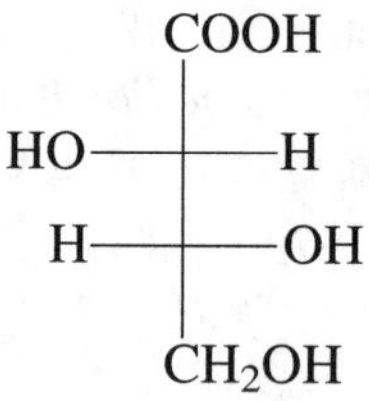

$$
\begin{array}{c}
COOH \\
HO-\!\!\!-H \\
H-\!\!\!-OH \\
CH_2OH
\end{array}
$$
D-Threonic acid

(ii)

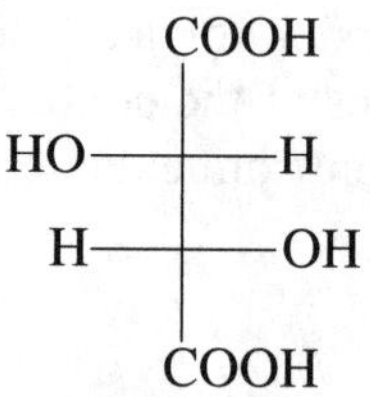

$$
\begin{array}{c}
COOH \\
HO-\!\!\!-H \\
H-\!\!\!-OH \\
COOH
\end{array}
$$
D-Tartaric acid

(iii)

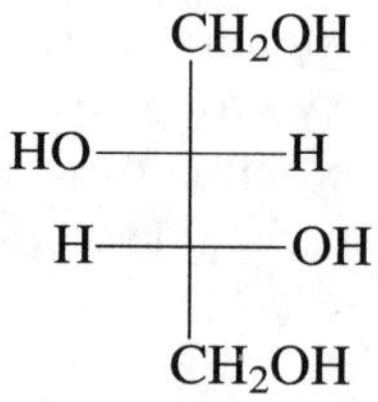

$$
\begin{array}{c}
CH_2OH \\
HO-\!\!\!-H \\
H-\!\!\!-OH \\
CH_2OH
\end{array}
$$
D-Threitol

(iv)

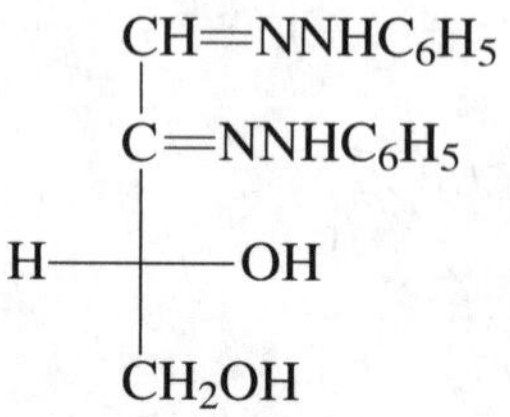

$$
\begin{array}{c}
CH=NNHC_6H_5 \\
| \\
C=NNHC_6H_5 \\
H-\!\!\!-OH \\
CH_2OH
\end{array}
$$
D-Threose
phenylosazone*

(b) (i)

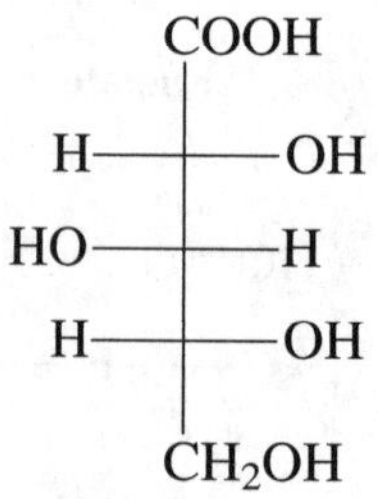

$$
\begin{array}{c}
COOH \\
H-\!\!\!-OH \\
HO-\!\!\!-H \\
H-\!\!\!-OH \\
CH_2OH
\end{array}
$$
D-Xylonic acid

(ii)

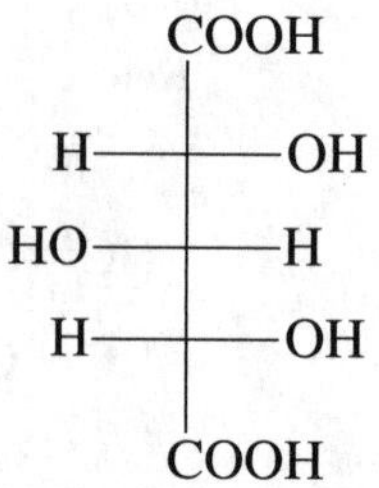

$$
\begin{array}{c}
COOH \\
H-\!\!\!-OH \\
HO-\!\!\!-H \\
H-\!\!\!-OH \\
COOH
\end{array}
$$
D-Xylaric acid

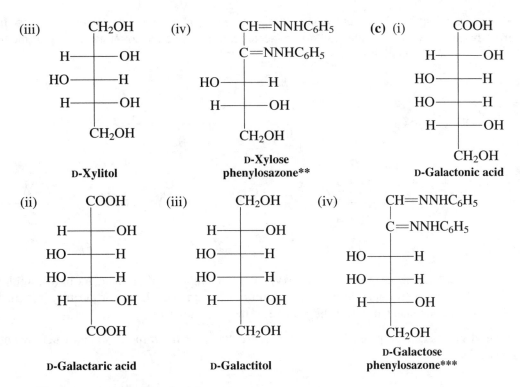

(iii) **D-Xylitol**

(iv) **D-Xylose phenylosazone****

(c) (i) **D-Galactonic acid**

(ii) **D-Galactaric acid**

(iii) **D-Galactitol**

(iv) **D-Galactose phenylosazone*****

*Same as D-erythrose phenylosazone.
**Same as D-lyxose phenylosazone.
***Same as D-talose phenylosazone.

44. **(a)** D-gulose (Figure 24-1);   **(b)** L-allose (all OH's on **left** side)

45. **(a)** Arabinose and lyxose. Ribitol and xylitol are meso compounds.

(b)

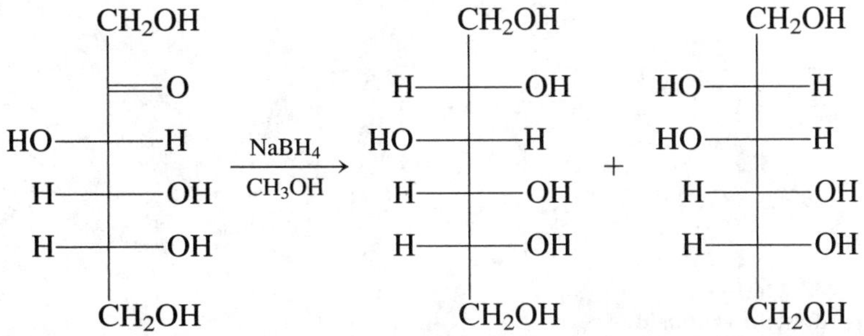

D-Fructose $\xrightarrow[\text{CH}_3\text{OH}]{\text{NaBH}_4}$ **D-Glucitol**   +   **D-Mannitol**

A new stereocenter is generated at C2, so two diastereomeric alditols are produced. In contrast, reduction of any aldose will be simpler because no new stereocenter will be generated; therefore, only a single product can form.

46. (a) and (d), because they still possess hemiacetal functionality. In (b) and (c), the OH at C1 in glucose has become OCH_3, and the molecule is now an acetal, incapable of mutarotation.

(e) is   Also with an acetal, not a hemiacetal, at C1.

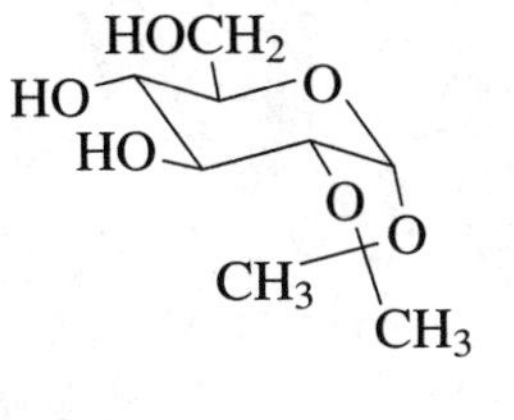

47. (a) The oxygen at C1 of an aldopyranose is a hemiacetal oxygen, not a simple alcohol oxygen. It can therefore be methylated the same way a hemiacetal can be converted to an acetal—with methanol and acid via a stabilized carbocation.

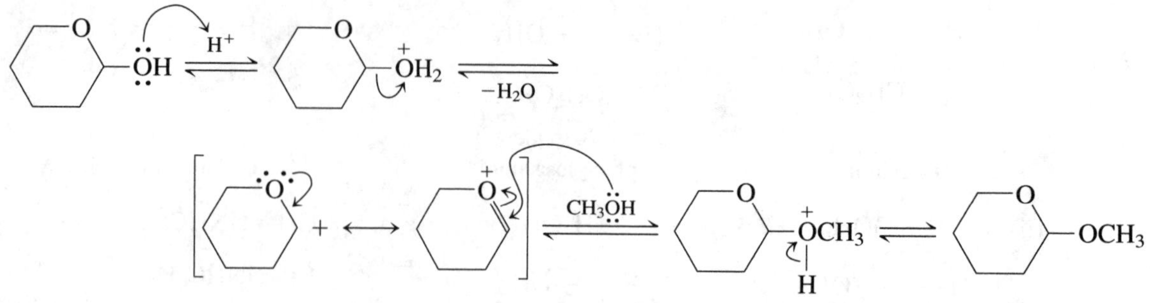

(b) The oxygen at C1 in this case is an acetal oxygen, not a simple methyl ether. As in (a), mild aqueous acid is sufficient for hydrolysis as a result of the same stabilized carbocation intermediate shown above (the mechanism is just the reverse of the one shown).

(c) Four methyl glycosides are possible (refer to the structures for fructofuranose and fructopyranose in Section 24-2).

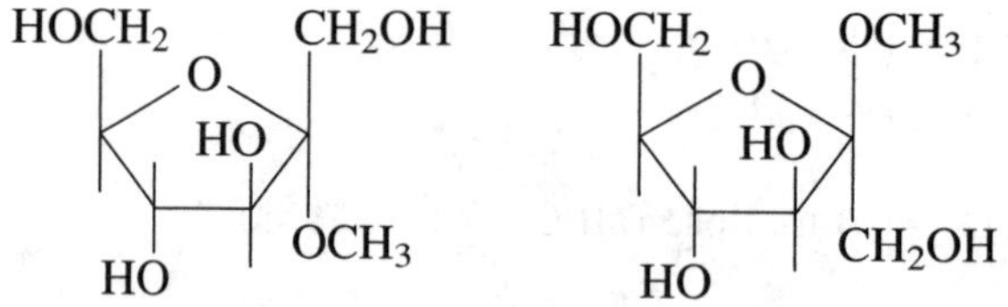

Methyl α-D-fructofuranoside

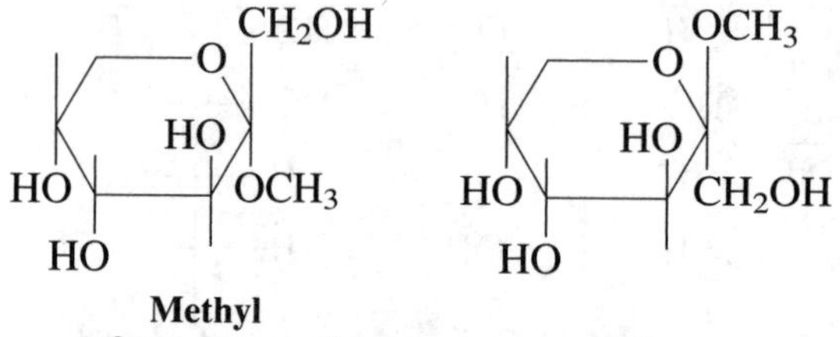

β

Methyl α-D-fructopyranoside

β

48. Arabinose (as a β-pyranose) forms a double acetal (Section 24-8). So does ribose, because in its α-pyranose form all four hydroxy groups are cis.

α-D-Ribopyranose

The best xylose and lyxose can do is have only one pair of adjacent cis hydroxy groups, so these readily form only monoacetals.

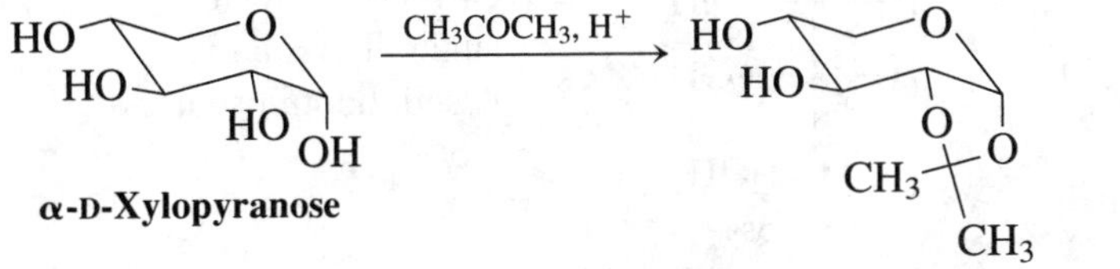

α-D-**Xylopyranose**

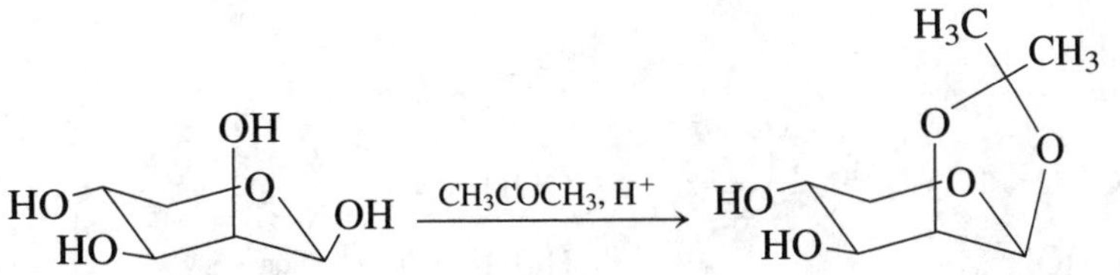

β-D-**Lyxopyranose**

49. (i) The sugar has seven carbons and is a *ketose,* because HIO_4 treatment produces a mole of CO_2 (see a similar reaction of D-fructose, Section 24-5). The sugar has two CH_2OH groups (leading to two formaldehydes) and four CHOH groups (leading to 4 mol of formic acid).

(ii) Because the sugar forms the same osazone as an **aldose,** its ketone must be at C2. So far, therefore, you have the following partial structure:

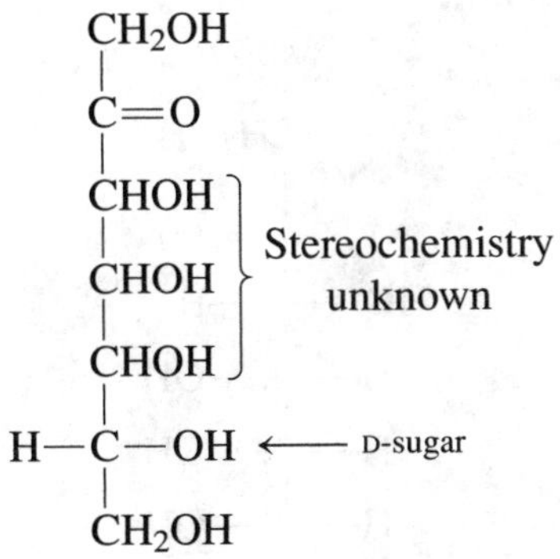

$$
\begin{array}{c}
CH_2OH \\
| \\
C{=}O \\
| \\
CHOH \\
| \\
CHOH \\
| \\
CHOH \\
| \\
H{-}C{-}OH \\
| \\
CH_2OH
\end{array}
$$

CHOH, CHOH, CHOH — Stereochemistry unknown

H—C—OH ← D-sugar

(iii) and (v) tell you that

$$
\begin{array}{c}
CHO \\
| \\
CHOH \\
| \\
CHOH \\
| \\
CHOH \\
| \\
CHOH \\
| \\
H{-}C{-}OH \\
| \\
CH_2OH
\end{array}
\xrightarrow{Ruff}
\begin{array}{c}
CHO \\
| \\
CHOH \\
| \\
CHOH \\
| \\
CHOH \\
| \\
H{-}C{-}OH \\
| \\
CH_2OH
\end{array}
\xrightarrow{Ruff}
$$

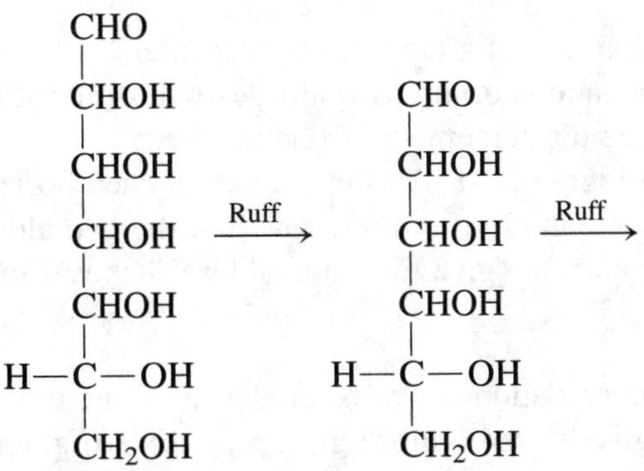

Aldoheptose A   **Aldohexose B**

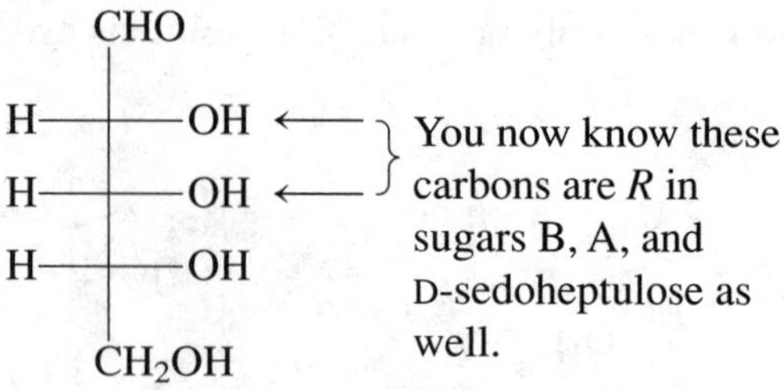

CHO

H——OH ←⎱ You now know these

H——OH ←⎰ carbons are *R* in

H——OH sugars B, A, and

CH$_2$OH D-sedoheptulose as
well.

D-Ribose

Next, (iv) tells you that

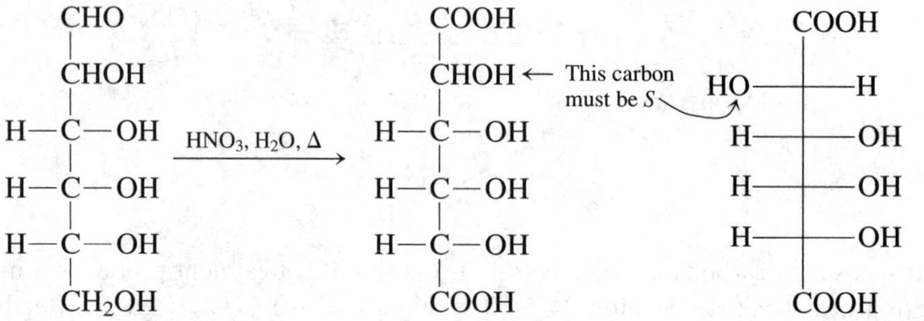

CHO

CHOH

H—C—OH $\xrightarrow{\text{HNO}_3,\ \text{H}_2\text{O},\ \Delta}$

H—C—OH

H—C—OH

CH$_2$OH

Aldohexose B

COOH

CHOH ← This carbon
must be *S*

H—C—OH

H—C—OH

H—C—OH

COOH

This is said to be
optically *active*

COOH

HO——H

H——OH

H——OH

H——OH

COOH

Otherwise the product
would be *meso*

From this information, you now can work backward toward the unknown: The stereocenters in D-sedoheptulose must be *3S, 4R, 5R,* and *6R*.

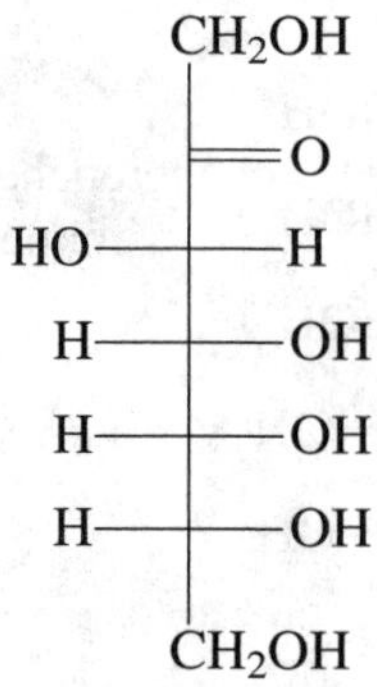

CH$_2$OH

==O

HO——H

H——OH

H——OH

H——OH

CH$_2$OH

D-Sedoheptulose

50. No. Kiliani-Fischer elongation does not alter the stereocenters in the starting material. The two pairs of products of Kiliani-Fischer elongations of any two aldoses will differ at the same stereocenters as did the starting materials. Thus they will be diastereomers of one another.

By way of illustration, refer to Figure 24-1. All the aldoses in each horizontal row are diasteromers of one another. The Kiliani-Fischer converts an aldose in one row into the two aldoses immediately below it. No two aldoses in one row (say, the pentoses) give the same aldose (hexose) in the next.

51. Refer to Figure 24-1. Ruff degradation converts an aldose in one row to the aldose immediately above it. **(a)** Galactose and talose both give lyxose (gulose gives xylose); **(b)** gulose and idose both give xylose (glucose gives arabinose); **(c)** allose and altrose both give ribose (mannose gives arabinose).

52. Two aldoheptoses are formed. After HNO_3 treatment, one gives an optically active diacid, the other an inactive meso compound.

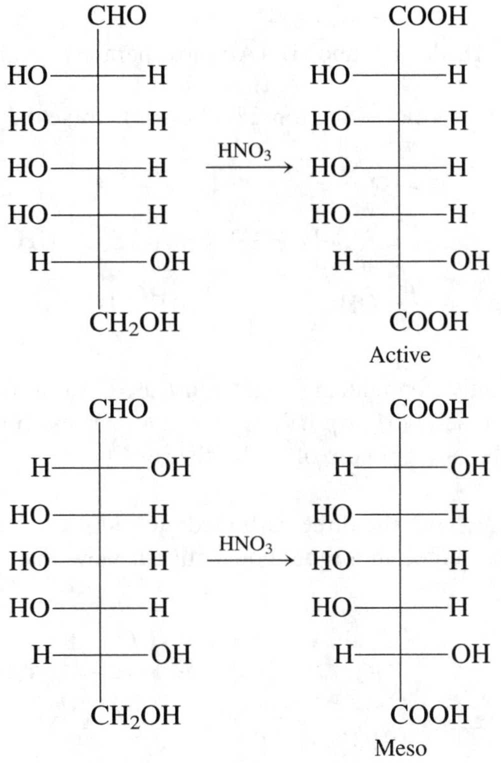

Active

Meso

53. **(a)**

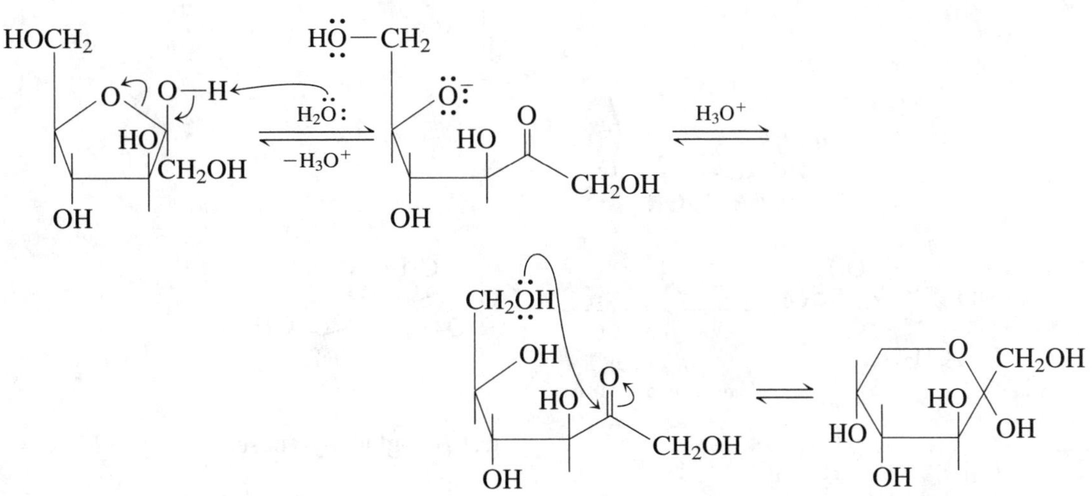

Product after protonation
(both anomers)

(b)

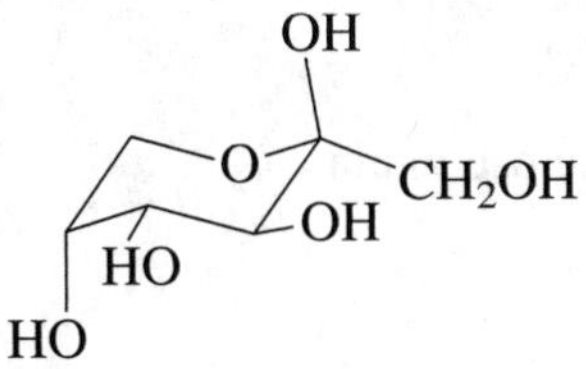

(Three equatorial and
two axial substituents)

(c) 0.65 kcal mol^{-1} (Table 2-1)

(d) An example of a "weighted average": $[\alpha]_{mixture} = X_A[\alpha]_A + X_B[\alpha]_B$ where X = mole fraction for each component. Let A = pyranose and B = furanose. So, $-92° = (0.68)(-132°) + (0.32)[\alpha]_B$, and $[\alpha]_B = -6°$.

54. Reducing: (a), (b), (c), (e), (f), (h), (i), and (j). (All have hemiacetal groups.)

55. Yes. At the lower right of the formula (Section 24-11) is a hemiacetal group.

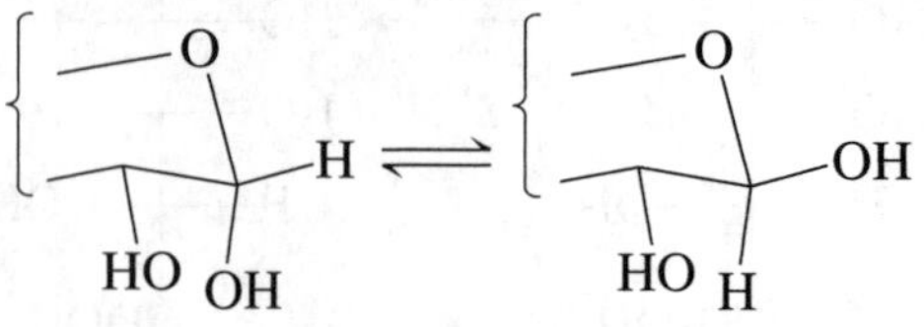

56. Trehalose must be (d), the only nonreducing sugar illustrated. Turanose is (c), the only one containing a ketose (the bottom half). Sophorose is (a) (top half is an α anomer and bottom half is a β anomer). Sugar (b) is composed of two aldoses that are epimers of each other at C4.

57. The structure is shown below, with the three carbohydrate units labeled. Redraw each one separately, and then rotate it such that it is presented in a more conventional view that can be compared with monosaccharides you know.

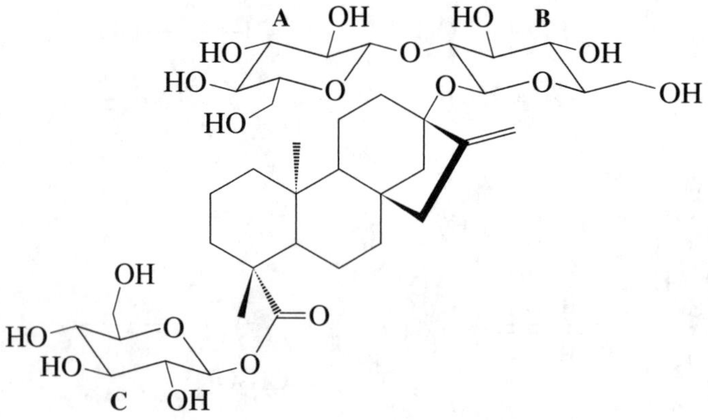

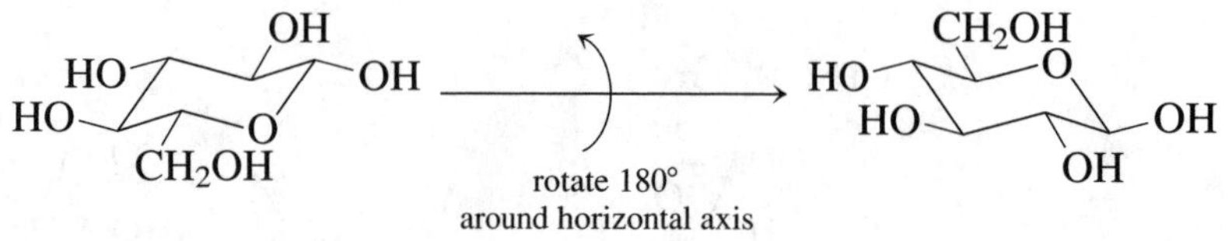

is just β-D-glucopyranose.

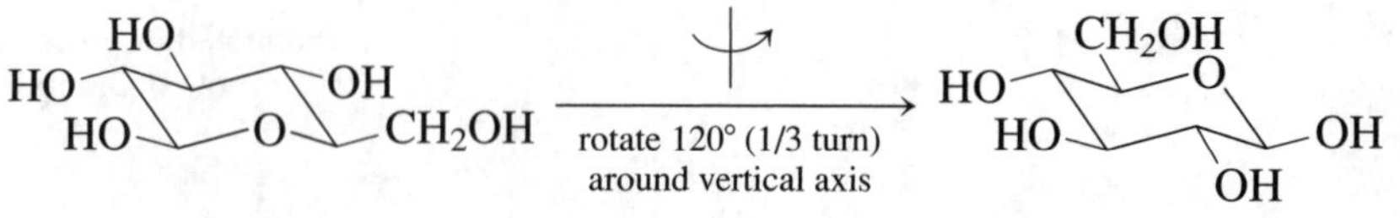

is just β-D-glucopyranose again,

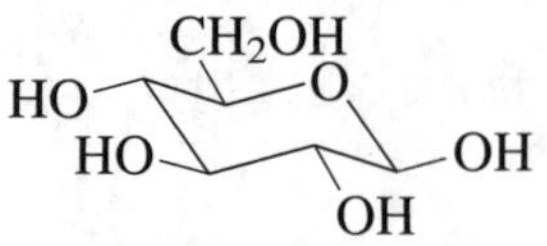

is also β-D-glucopyranose. Stevioside contains three molecules of glucose in its structure.

58. You can find a good conformational picture of β-D-galactose on page 1146 of the textbook, as the left-hand part of the structure of lactose. Switch stereochemistry at C1 for the α isomer. Notice in the answer below how the name, 3-(α-D-galactopyranos-1-yl)- β-D-galactopyranose, translates into the structure:

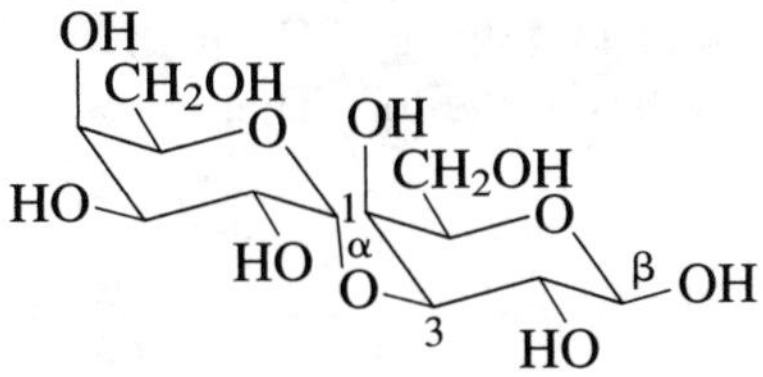

59. The hydroxyl group at C1 leaves most readily when protonated, because a resonance-stabilized carbocation is produced.

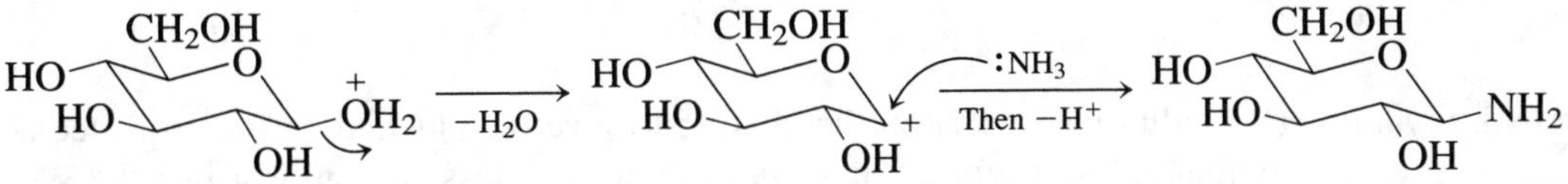

60. **(a)** Aldol condensations! Abbreviated mechanisms are shown below. Refer to Section 18-6 for more details, if necessary.

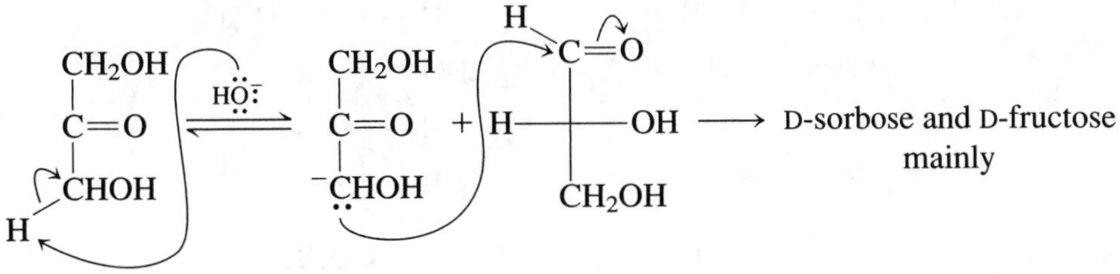

D-sorbose and D-fructose
mainly

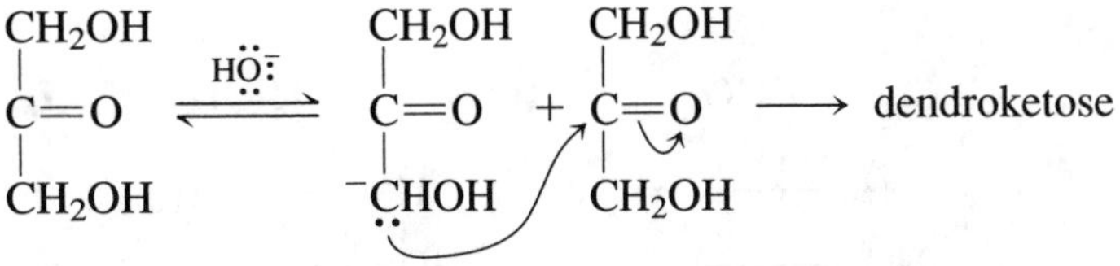

dendroketose

(b) The hint is supposed to make you think about enolate ions. Glyceraldehyde and 1,3-dihydroxy-propanone are readily **interconverted** in aqueous basic solution via enolates and enols.

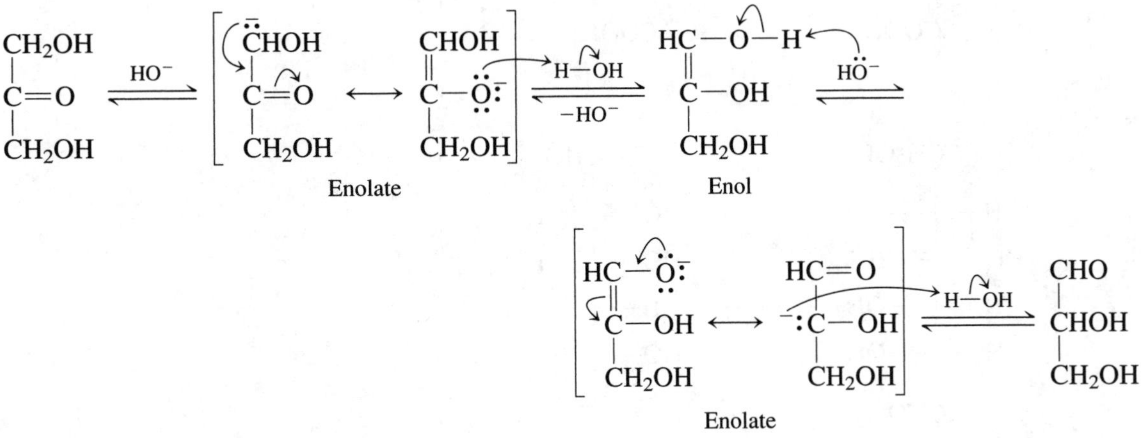

Enolate   Enol

Enolate

So a basic solution of either one rapidly becomes a mixture of the two, and the reactions in (a) can then occur. This aldose $\rightleftharpoons$ ketose interconversion is general. Glucose and fructose are interconverted by aqueous base, for example. We saw this same mechanism in Problem 41.

61. **(a)** Br_2, H_2O; **(b)** see answer to Problem 43b; **(c)** CH_3OH, H^+; **(d)** ester of acid in (b): —$COOCH_3$ at the top; **(e)** forms the amide: —$CONH_2$ at the top. Then

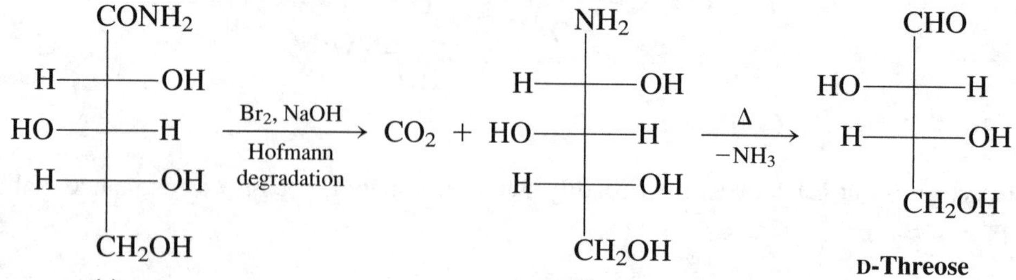

The hydroxyamine (f) readily loses NH_3 on the heating to give the aldehyde. This sequence achieves the removal of Cl from an aldose to form a new aldose with one less carbon, just like the Wohl and Ruff degradations.

62. **(a)** CH_3OH, H^+ **(b)** **(c)**

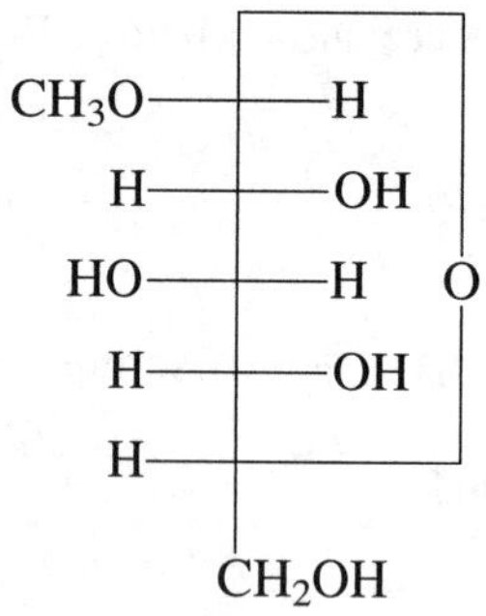

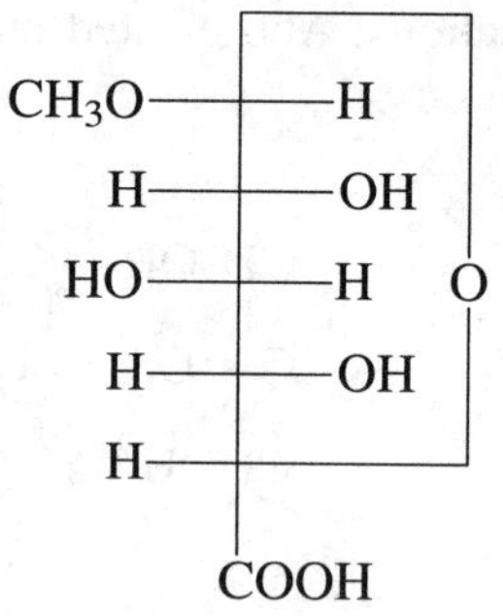

(d) 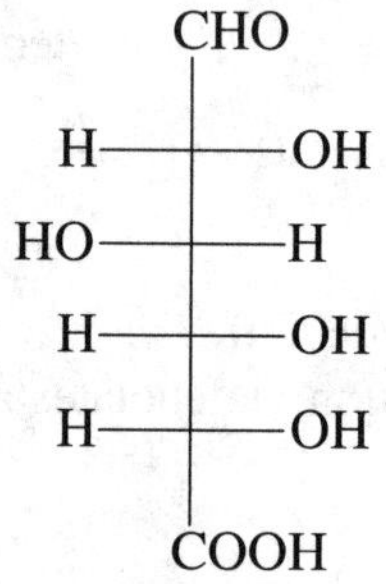 **(e)** 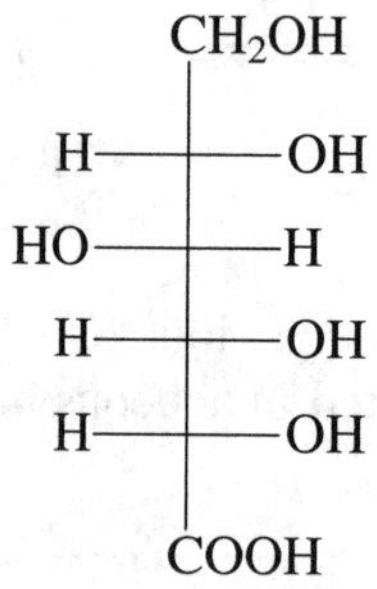 **(f)**

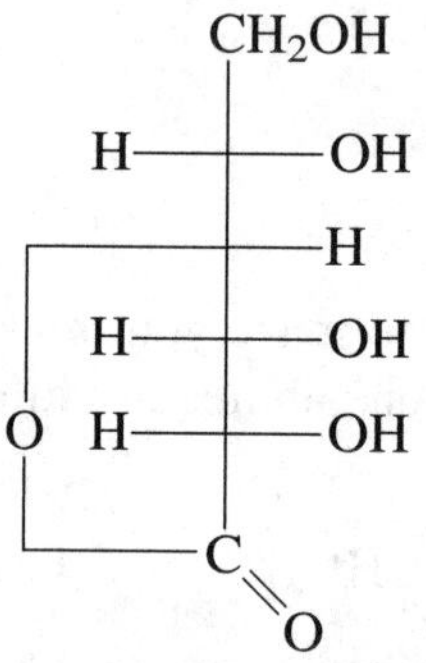

(g)

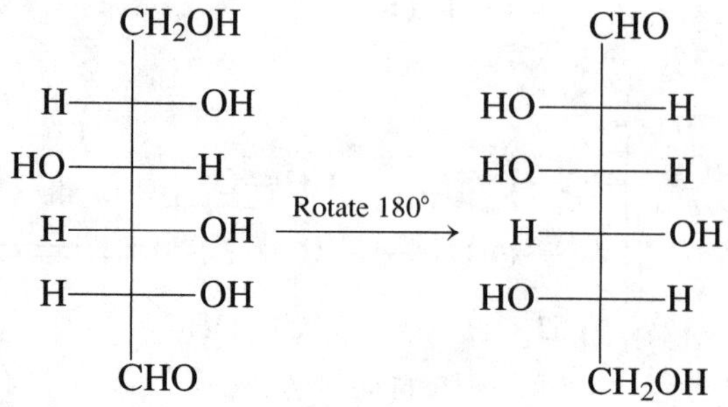

As you can see, the gulose Fischer synthesized was the L-enantiomer (hydroxy group on C5 is on the left).

63. First, D-glucuronic acid γ-lactone is

(Fischer projection with C=O at top, HO—H, HO—H, H—, HO—H, CHO at bottom; O bridging to form γ-lactone)

L-gulonic acid γ-lactone has already been illustrated [answer to (f) of Problem 62].

(a) NaBH$_4$ **(b)**

(Fischer projection: CH$_2$OH, H—OH, HO—H, H—OH, 5 =O, CH$_2$OH)

(c)

(cyclic furanose structure with O, OH, HO, HOH$_2$C, CH$_2$OH, OH)

(d) 1. 2 CH$_3$COCH$_3$, H$^+$, 2. KMnO$_4$ to oxidize unprotected primary alcohol to carboxylic acid;
(e) H$_2$O, H$^+$ (hydrolyze acetals); **(f)** Δ (−H$_2$O)

64. Follow along with the actual structure of D-lactose in Section 24-11.

1. Mild acid cleaves acetal linkages. You know what kind of functional linkage connects the two monosaccharides, but you know nothing about which carbon atoms bear the oxygen atoms of the acetal group.

2. Test to see if the "unknown" is a reducing sugar. It is, proving that one of the two component monosaccharides retains a hemiacetal function. (Otherwise, it would be like sucrose, in which both anomeric carbons are connected by an acetal oxygen.)

3. Complete methylation [with (CH$_3$)$_2$SO$_4$] of all free —OH groups followed by mild acid hydrolysis is revealing. The galactose portion adds four methyl groups, but the glucose adds only three. Therefore it is the galactose whose anomeric carbon is part of the acetal group linking the two.

4. and **5.** Comparison of the tri-*O*-methylglucose product of the methylation-hydrolysis sequence with known compounds would be one way to ascertain that the hydroxy groups at C4 and C5 were not methylated and, therefore, one must be responsible for the disaccharide (acetal) linkage, and the other is part of the cyclic hemiacetal, but we don't know which is which. We need to use other chemistry to open the hemiacetal ring and identify which hydroxy group is part of it. We have such a reaction: In Section 24-4 we learned that aldoses (which exist as cyclic hemiacetals) are oxidized at C1 by bromine in water to give (acyclic) aldonic acids. Starting with lactose, we therefore would have

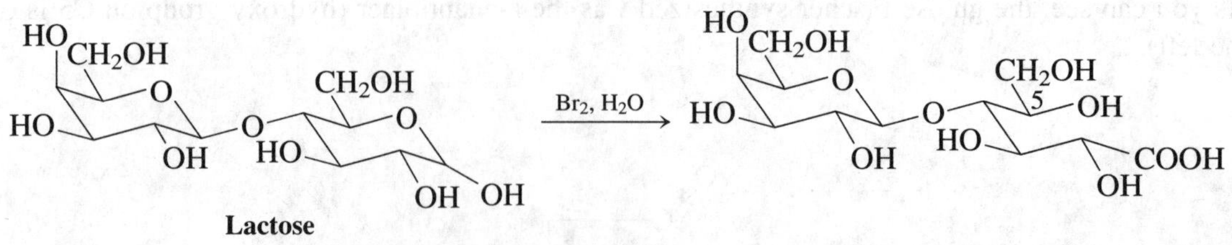

Lactose

This oxidation frees the hydroxyl group that was part of the cyclic hemiacetal but leaves the disaccharide acetal linkage unaffected. Now methylation will occur at the former hemiacetal hydroxy group—at C5 in the actual structure (see above). After mild acid hydrolysis, the presence of a methoxy group at C5 and a free —OH at C4 reveals conclusively that the disaccharide linkage utilized the C4 hydroxy group of the glucose unit, and that the latter was present as a pyranose ring, containing the oxygen attached to C5.

65. (a)

66. (a)

67. (d)

68. (d)

69. (a)

25

Heterocycles: Heteroatoms in Cyclic Organic Compounds

The material in this chapter falls into two broad categories: nonaromatic and aromatic heterocycles. You have already seen many examples of nonaromatic heterocycles (see the section references at the start of the text chapter). There will be new ones, especially compounds containing nitrogen atoms in three-, four-, or five-membered rings.

Heteroatoms may also be present in aromatic rings, and compounds with this feature are the subject of the last six sections of the chapter. In general, the properties of heterocyclic compounds will be predictable from principles you've already seen: They will be similar to acyclic compounds with similar heteroatoms unless (1) the ring is strained or (2) the ring is aromatic. When the ring is aromatic, the effect of the heteroatom on its chemistry will be important, and this will be a new topic to which you will need to apply your knowledge of inductive and resonance effects.

Outline of the Chapter

25-1   Naming the Heterocycles

25-2   Nonaromatic Heterocycles

25-3, 25-4   Aromatic Heterocyclopentadienes: Pyrrole, Furan, and Thiophene

25-5, 25-6   Pyridine, an Azabenzene
         The most common monocyclic heteroaromatic compounds.

25-7   Quinoline and Isoquinoline: The Benzpyridines

25-8   Alkaloids
         Even more rings!

Keys to the Chapter

25-1. Naming the Heterocycles
Note simply that the text sticks to strict systematic nomenclature for nonaromatic heterocycles but uses the universally accepted common names for the aromatic systems.

25-2. Nonaromatic Heterocycles
This section generalizes what you have previously learned concerning the preparation and reactions of cyclic ethers. Rings containing nitrogen and sulfur illustrate the principal similarity with their oxygen relatives: Ring opening to relieve strain governs the reactivity of the smaller rings (three or four atoms).

25-3 through 25-7. Aromatic Heterocycles

Five- and six-membered heterocyclic compounds make up the vast majority of aromatic heteroatom-containing systems. Counting electrons in the π systems of these rings is occasionally confusing. Here is a simple way to determine whether a lone pair of a heteroatom is part of the cyclic π system: If **only single bonds** link the heteroatom to its neighbors in the ring, then **one** lone pair from the heteroatom may be in a p orbital and become part of the cyclic π system. Examples are pyrrole, furan, and thiophene. Notice that in the latter two, only **one** of the two lone pairs on the heteroatom is part of the π system: The other is in an sp^2 orbital and has nothing to do with the molecule's aromaticity at all. If the heteroatom is **doubly bonded** to another ring atom, as is the case with pyridine, a lone pair on it will be in an sp^2 orbital and will **never** count toward the cyclic π system in the molecule. Try to apply these rules to as many structures illustrated in these sections as you can: They are all aromatic, so you know what the answers should be.

Typically, syntheses of the aromatic heterocycles utilize combinations of carbonyl condensation reactions and 1,2- or 1,4-additions of enolate and heteroatom nucleophiles to α, β-unsaturated carbonyl compounds—no fundamentally new chemistry here. Reactions of these systems show a blending of benzene–aromatic chemistry and the chemistry of nonaromatic analogs with the same heteroatom. Furan is a good example: It undergoes electrophilic substitution (aromatic chemistry), ring cleavage in acid (ether chemistry), and Diels-Alder cycloaddition (diene chemistry). Putting the large number of reactions here into their appropriate compartments should help you organize them for study and problem-solving.

Solutions to Problems

30. (a) (b)

(c) (d)

(e) 2-Formylfuran or furan-2-carbaldehyde; (f) *N*-methylpyrrole or 1-methylpyrrole;

(g) quinoline-4-carboxylic acid; (h) 2,3-dimethylthiophene

31. **1.** Pyrrole; **2.** pyridine and benzimidazole; **3.** pyridine and benzimidazole; **4.** none; **5.** quinoline; **6.** none; **7.** azacyclohexane (piperidine) fused to thiophene; **8.** oxacyclohexane (tetrahydropyran); **9.** aza-2,5-cyclohexadiene; **10.** benzene fused to 2,5-dihydrofuran (oxa-3-cyclopentene).

32. (a) (b)

(c) A bit tricky, so follow this mechanism.

33. (a)

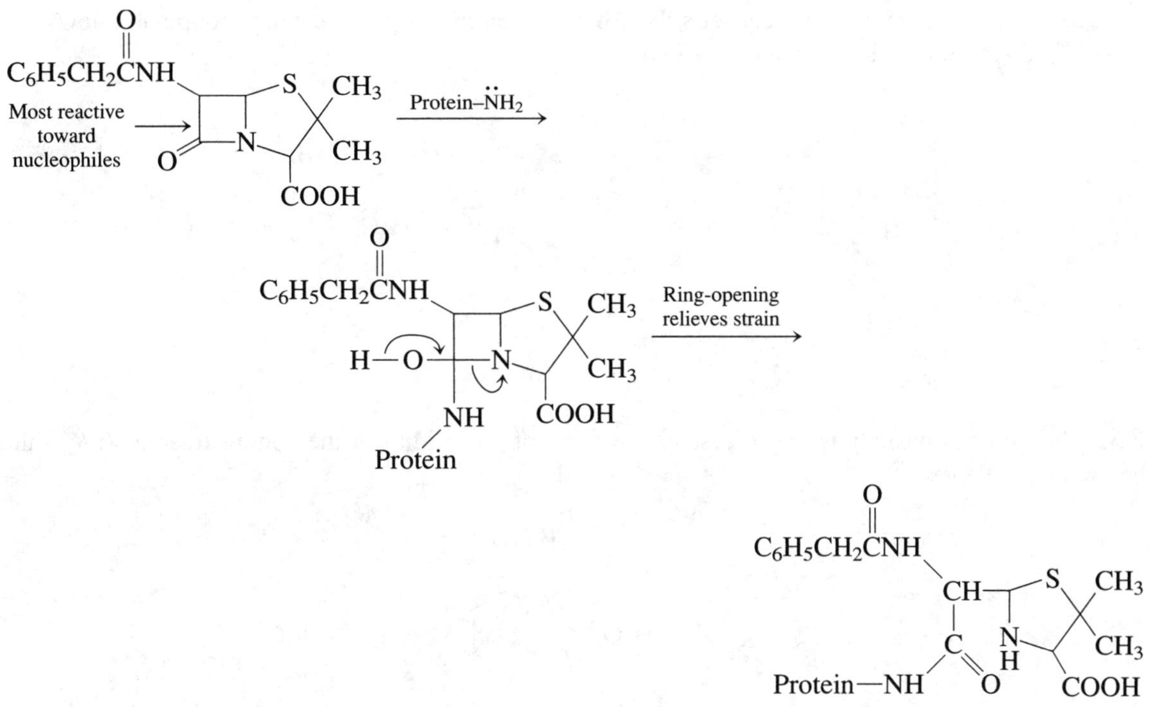

Most reactive toward nucleophiles

Penicilloyl protein

(b)

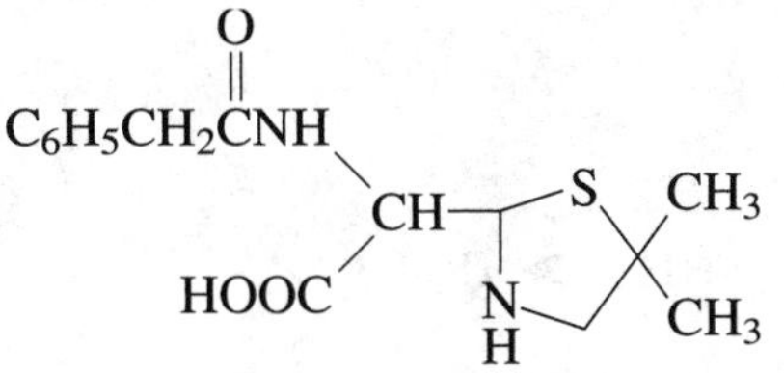

Formed via a parallel mechanism, with H_2O as the nucleophile. This product, penicilloic acid, no longer possesses the necessary strained azacyclobutanone ring for reaction with bacterial protein. It therefore lacks any antibiotic properties.

34. Use the Lewis acids to activate the ring oxygen in (a) and (b).

(a)

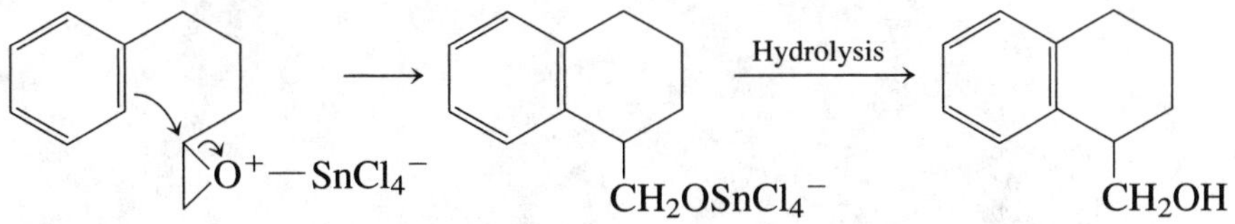

(A type of Friedel-Crafts reaction)

(b) $CH_3CH_2CH_2CH_2$—Li

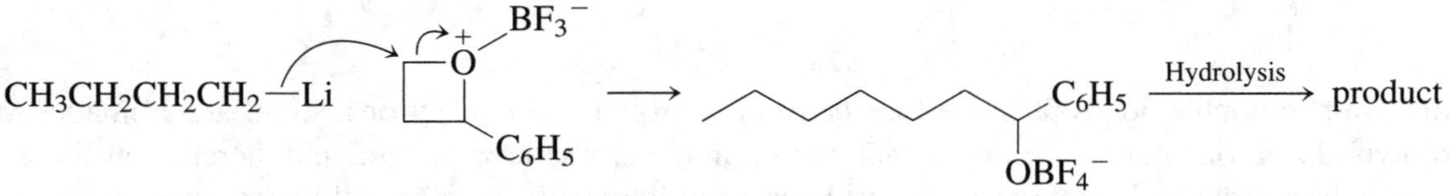

(c) Use the Lewis acid to activate the anhydride and form an acylium cation, similar to the first step in Friedel-Crafts alkanoylation.

$$CH_3\overset{O}{\overset{\|}{C}}-\overset{+}{\underset{\underset{MgBr_2^-}{|}}{O}}-\overset{O}{\overset{\|}{C}}CH_3 \rightleftharpoons CH_3CO^+ + CH_3\overset{O}{\overset{\|}{C}}-O-MgBr_2^- \quad \text{and}$$

$$CH_3\overset{O}{\overset{\|}{C}}-O-MgBr_2^- \rightleftharpoons CH_3\overset{O}{\overset{\|}{C}}-O-MgBr + Br^-$$

Then, the acylium ion converts the ether oxygen into a good leaving group and allows S_N2 displacement by bromide to occur.

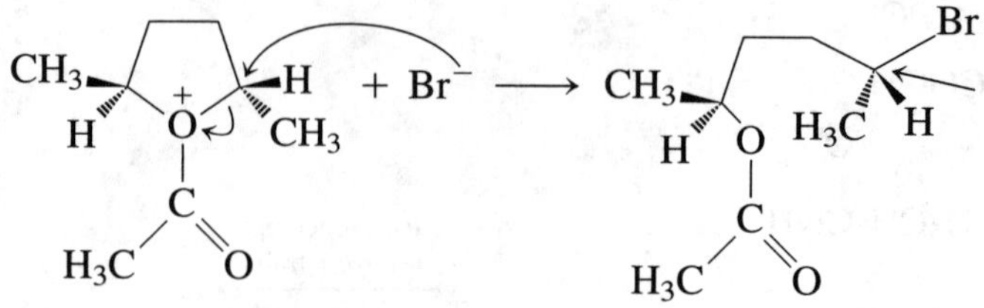

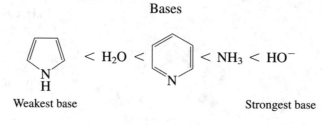

35. The order of basicity is the reverse of the order of the acidity of the conjugate acids (pK_a values shown below the structures).

Bases

$$\text{pyrrole} < H_2O < \text{pyridine} < NH_3 < HO^-$$

Weakest base   Strongest base

Conjugate acids

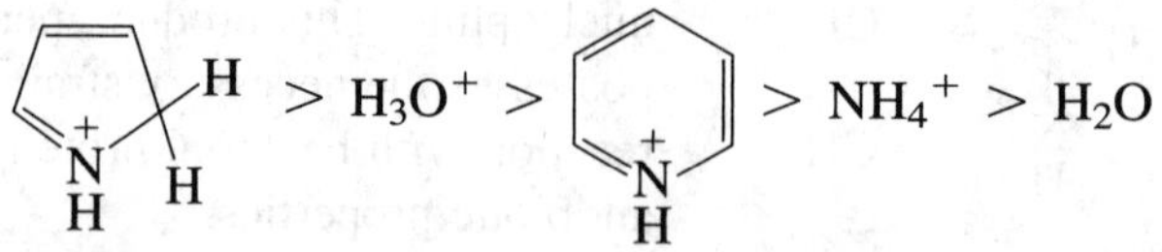

$$\text{protonated pyrrole} > H_3O^+ > \text{protonated pyridine} > NH_4^+ > H_2O$$

$pK_a = -4.4$ 0.0   5.3   9.2   15.7

Strongest acid   Weakest acid

36.

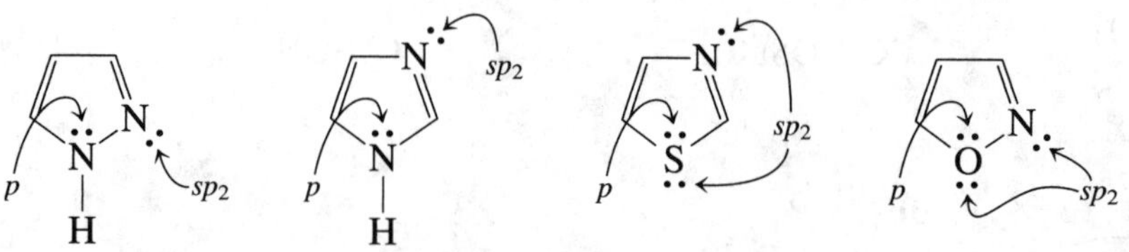

All have two double bonds plus one lone pair in a p orbital = 6 π electrons, so **all** are aromatic. All have sp^2-hybridized lone pairs on nitrogen, not tied up in the aromatic π system, and therefore available to act in a Lewis-base manner. Pyrrole lacks an sp^2-hybridized lone pair; therefore **all** the compounds above are stronger bases than pyrrole.

37. (a) 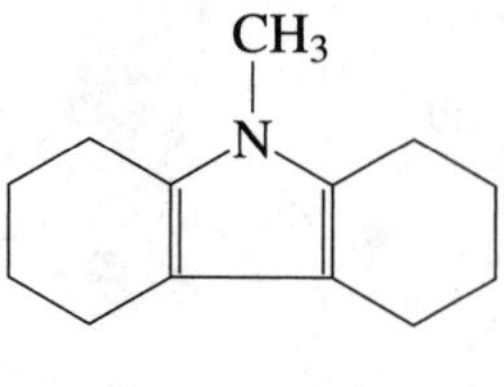   **(b)**

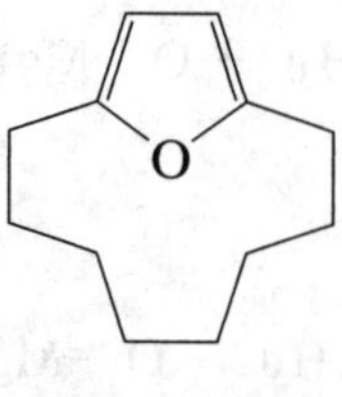

38. Abbreviated mechanism

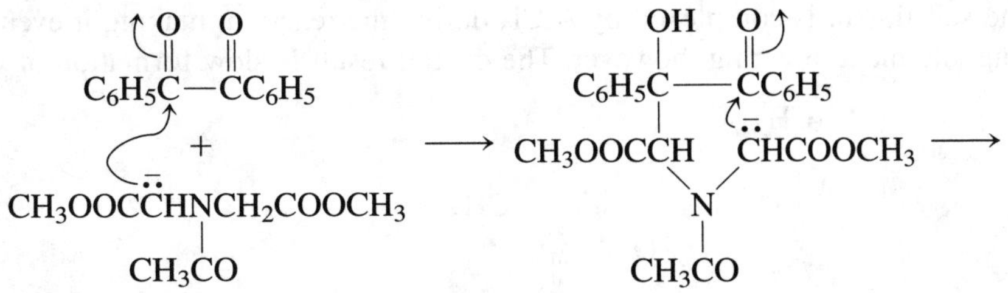

Under these reaction conditions ($NaOCH_3$, CH_3OH, heat) the amide is cleaved by methanolysis, a process similar to transesterification. See p. 939 and Chapter 20, problem 60.

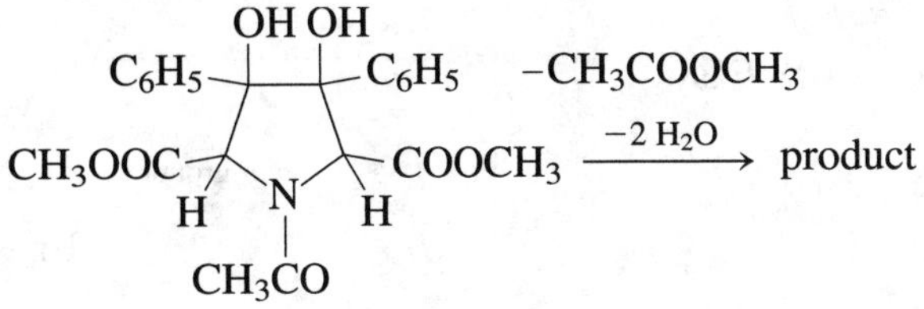

Synthesis

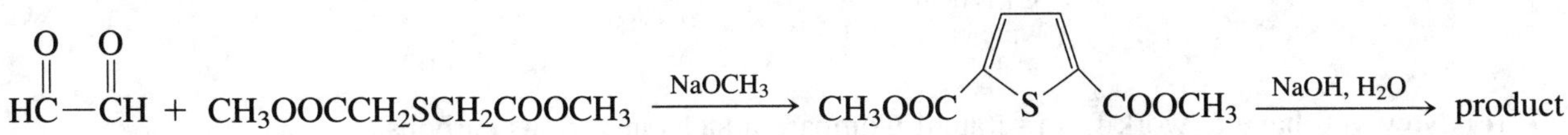

39. There are three factors to keep in mind: (1) the inherent preference of these compounds for substitution at C2 over C3, (2) the much greater reactivity of all of them compared with benzene, and (3) directing effects of substituent groups (which work the same way as in benzene). These are toughies!

(a) Two conflicting preferences

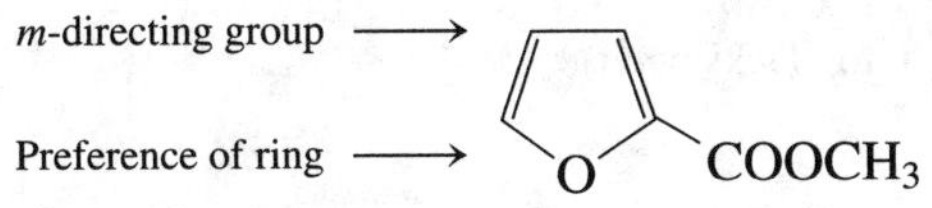

In this case, a mixture might be expected:

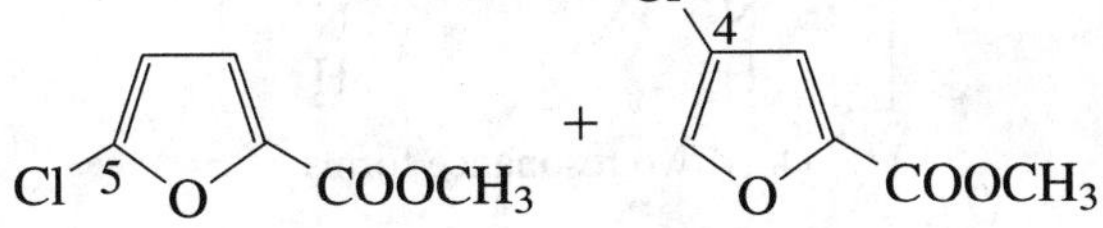

The first is actually the major product; the directing effect of the activating ring oxygen to C5 wins out over that of the moderately deactivating $COOCH_3$ group to C4.

(b) Easier

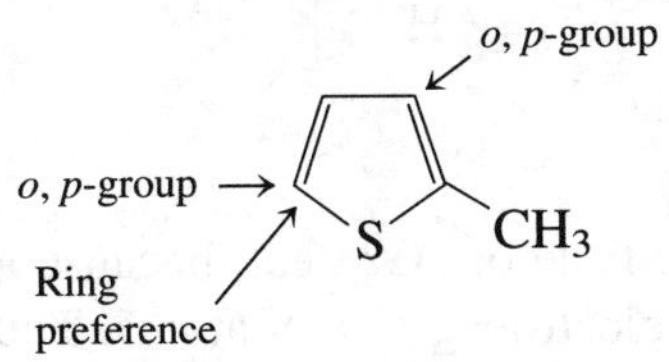

C5 is strongly activated

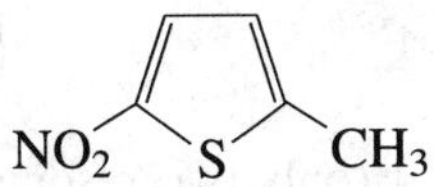

(c) Tricky. If this were benzene, the Friedel-Crafts reaction would not work at all because of the presence of the $COCH_3$ group. It does proceed here, because the heterocycle is much more reactive. The ketone substituent is complexed by $AlCl_3$ during the reaction, making it even more strongly deactivating and meta-directing, however. The overall result is slow formation of

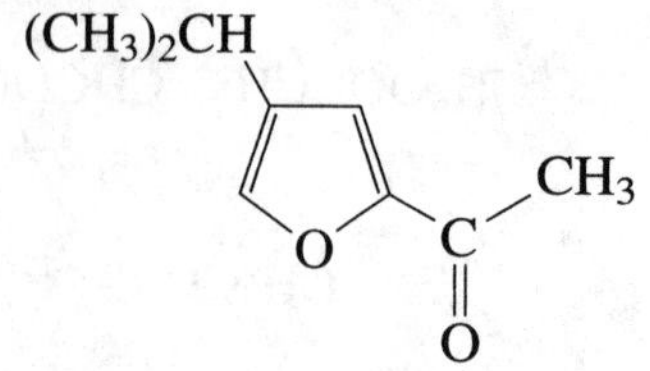

(d) Easy

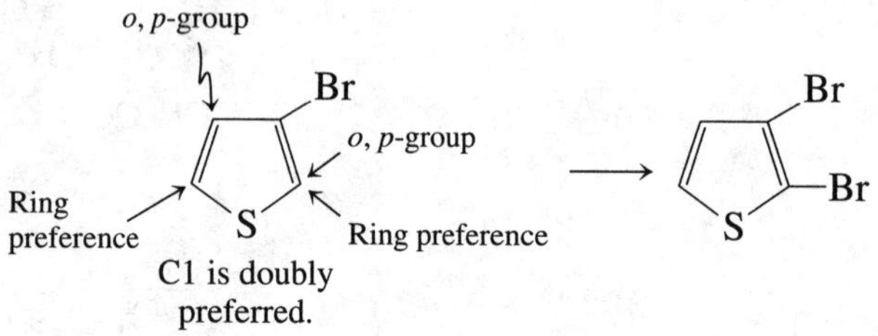

(e) Now you have to work from scratch! Compare attacks at various carbons.

C2

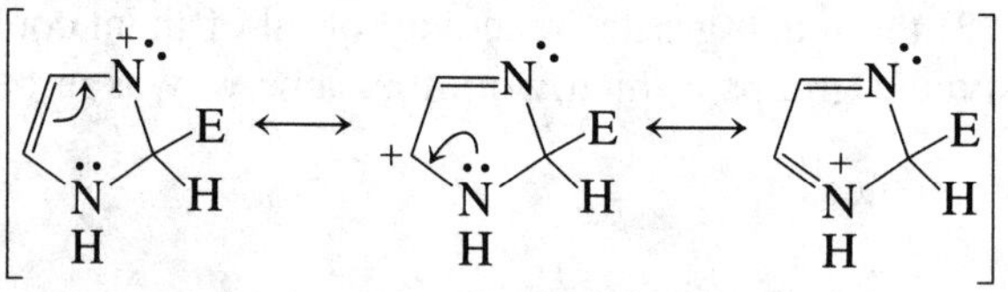

Poor! (N at top lacks octet)

C4

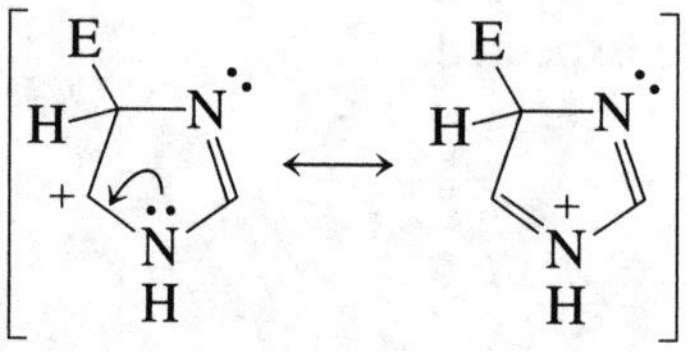

Only two resonance forms

C5

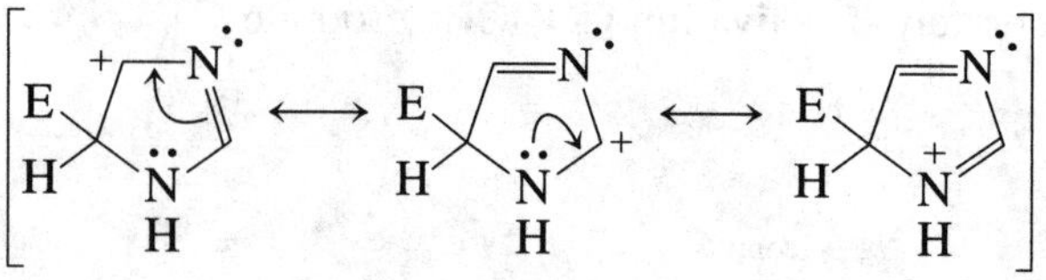

Three good resonance forms

Rule out C4 (only two resonance forms for cation). Rule out C2 next, because attack results in an electron sextet and a positive charge on one of the electronegative N atoms. Thus, C5 is the site of attack by typical electrophiles. In this particular example, however, the major product is due to diazo coupling at C2, because under the **basic** conditions the imidazole **anion** is attacked, and reaction at C2 gives a symmetrical intermediate with two equivalent resonance forms.

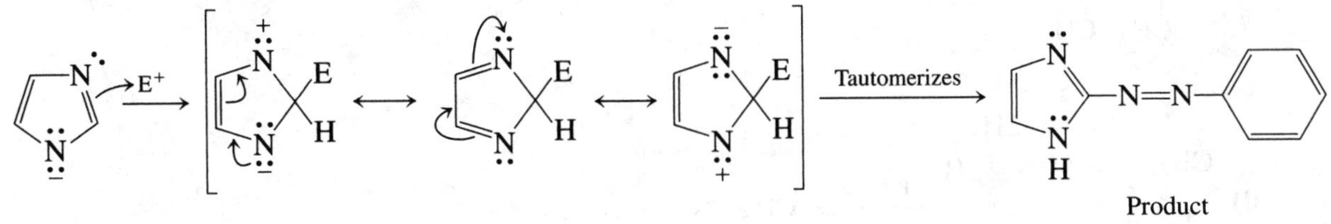

$E = C_6H_5N_2{}^+$

40. (a)

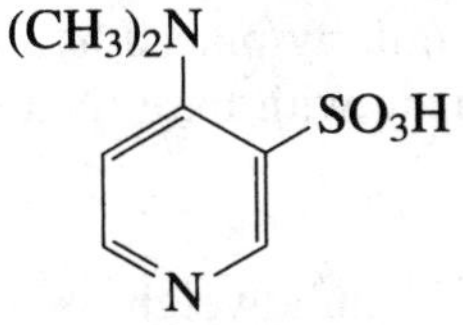

(b) Diels-Alder:

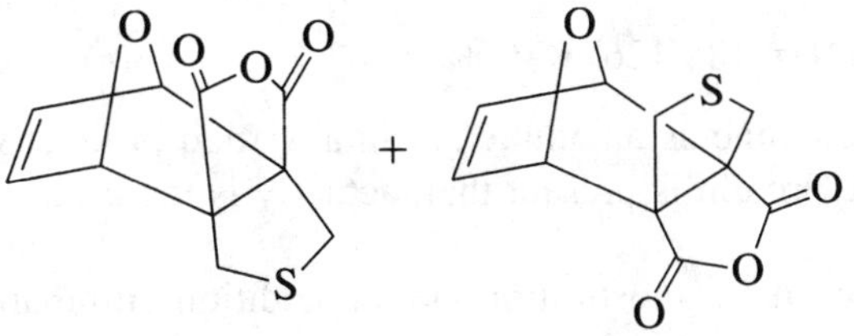

(c)

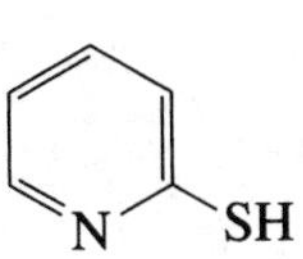

(d) $CH_3CH_2CH_2CH_2CC_6H_5$ (via)

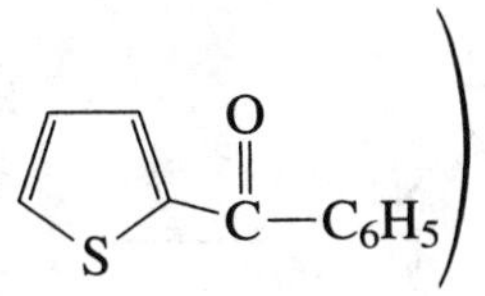

(e)

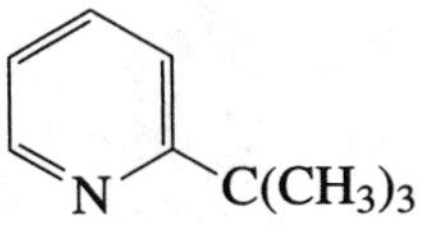

41. (a)

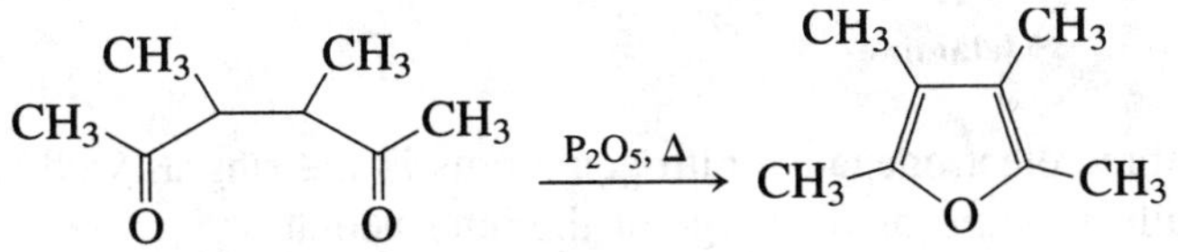

(b) Hantzsch $C_6H_5CCH_2COCH_2CH_3$ $\xrightarrow{H_2C=O, NH_3}$

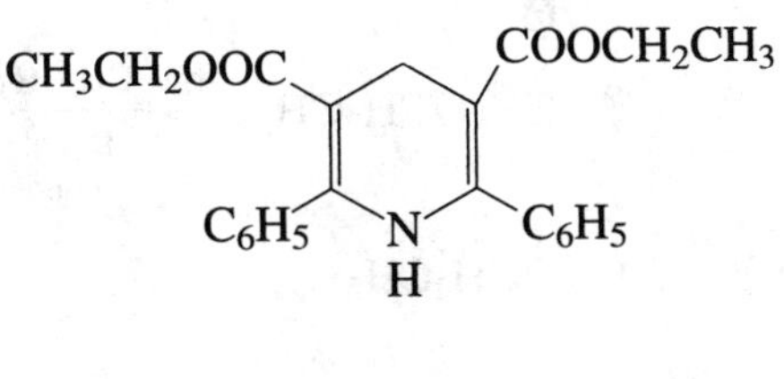

$\xrightarrow[\substack{\text{2. KOH, H}_2\text{O} \\ \text{3. CaO, } \Delta}]{\text{1. HNO}_3, \text{H}_2\text{SO}_4}$ product

(c) Paal-Knorr

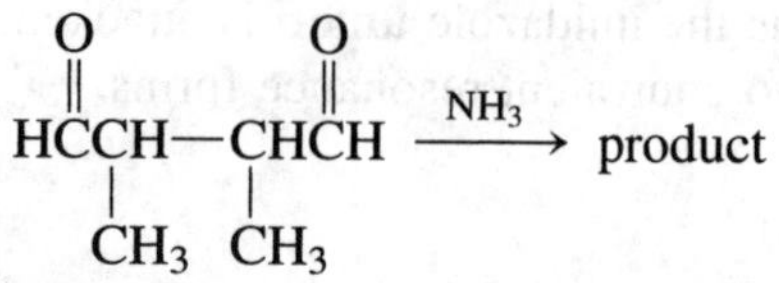

(d)

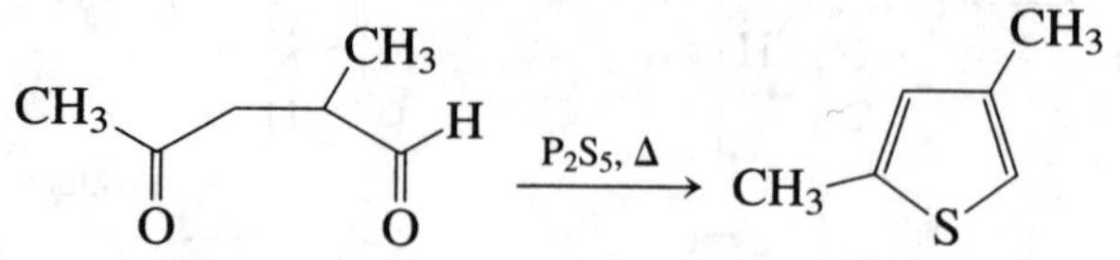

42. The only structural difference is the presence of a methyl group on a caffeine nitrogen, which in theobromine contains a hydrogen. Fortunately, this hydrogen should be usefully acidic because the nitrogen containing it lies between two carbonyl groups. Recall (Section 21-5) that the similarly placed hydrogen in 1,2-benzenedicarboximide (phthalimide) has a pK_a of about 8. Thus, 1. deprotonate with base (NaOH should be more than adequate) and 2. alkylate with CH_3I.

43. **(a)** I hope you haven't forgotten how to do this. Melamine's approximate formula weight is 36 (three C) + 6 (six H) + 84 (six N) = 126, of which 84 is nitrogen. Thus,

$$\% \ N = (84/126) \bullet 100\% = 67\%.$$

Clearly even a fairly small amount of melamine as an adulterant in any food product will increase the % N analysis, giving the impression that more protein is present than actually is the case.

(b) The reaction is a nucleophilic aromatic substitution via the addition-elimination mechanism (Recall Section 22-4):

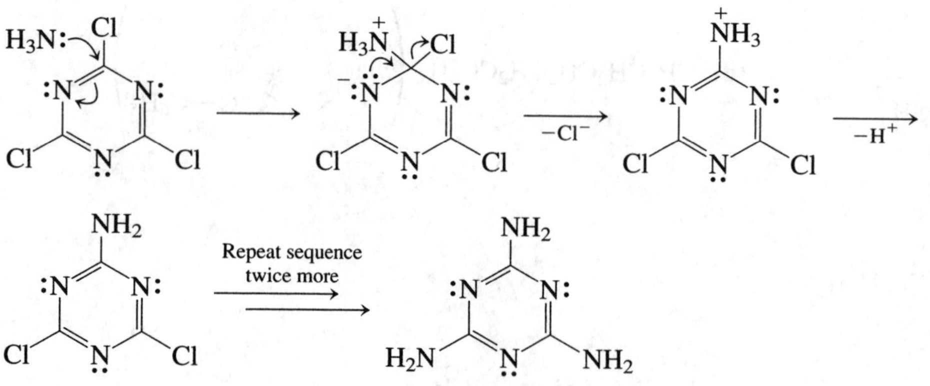

This process is favored by the presence of three electronegative nitrogen atoms in the ring as well as the chlorine substituents, all of which help stabilize the negative charge of the intermediates.

44.

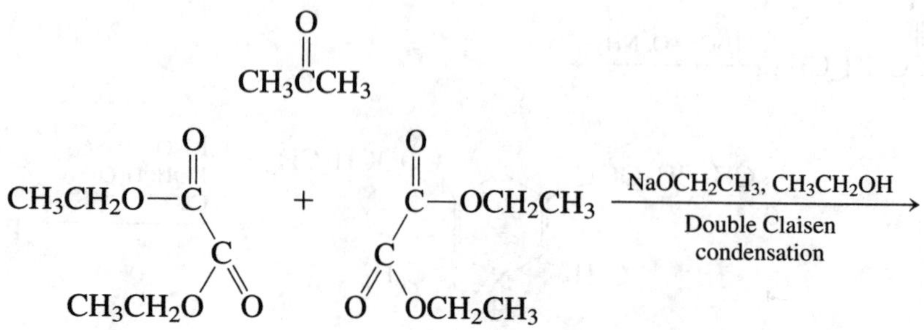

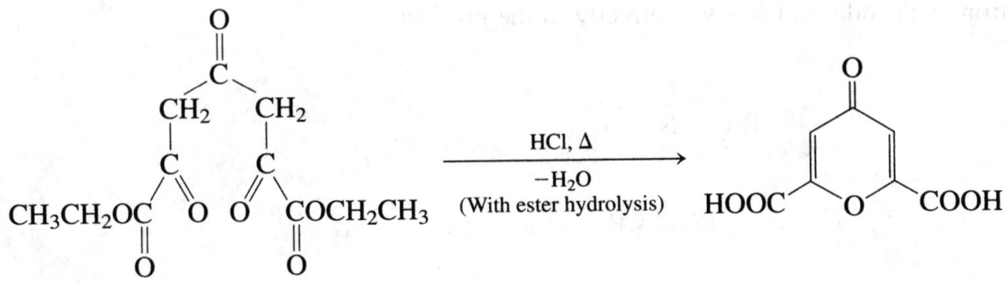

Compare synthesis of furans, etc., from 1,4-diketones.

45. Protonation of ketones and aldehydes can give rise to electrophiles capable of aromatic substitution:

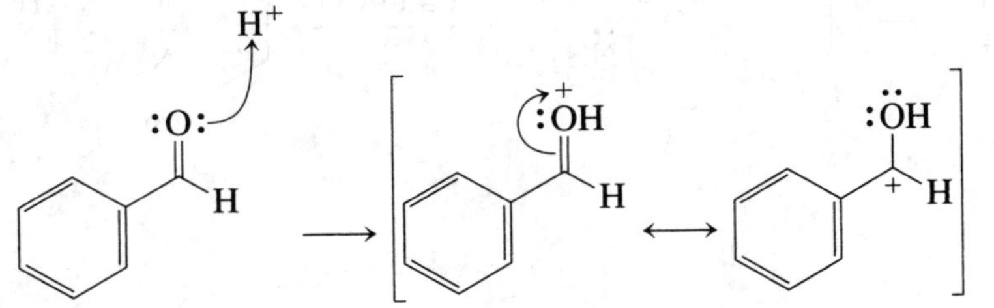

Then,

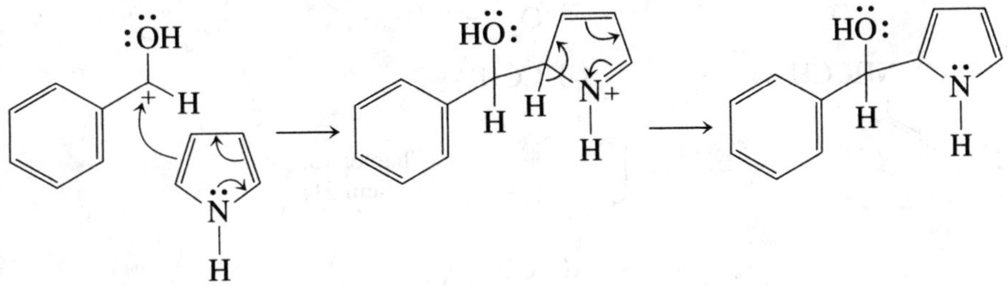

Next, protonation of the hydroxy group enables it to leave. The carbocation that results can undergo substitution with a second pyrrole ring:

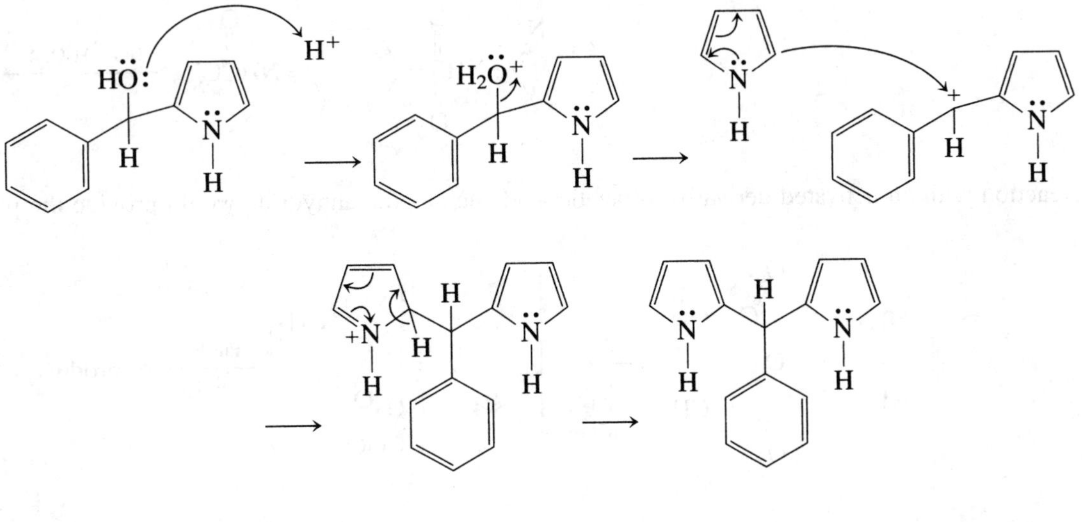

46. A six-electron cycloaddition takes you directly to the product:

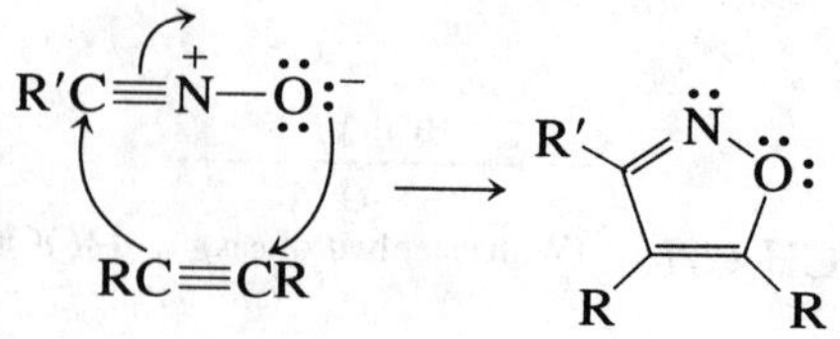

47. (a) H_2, Pt (b) Stepwise, ring opening of azacyclopropane first, followed by intramolecular amide (lactam) formation.

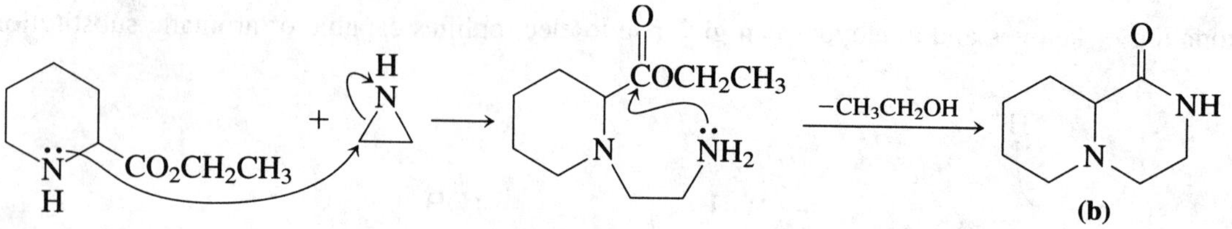

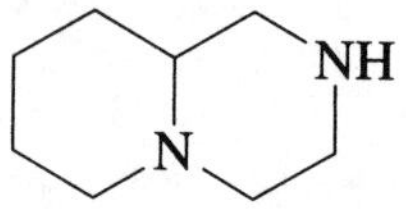

48.

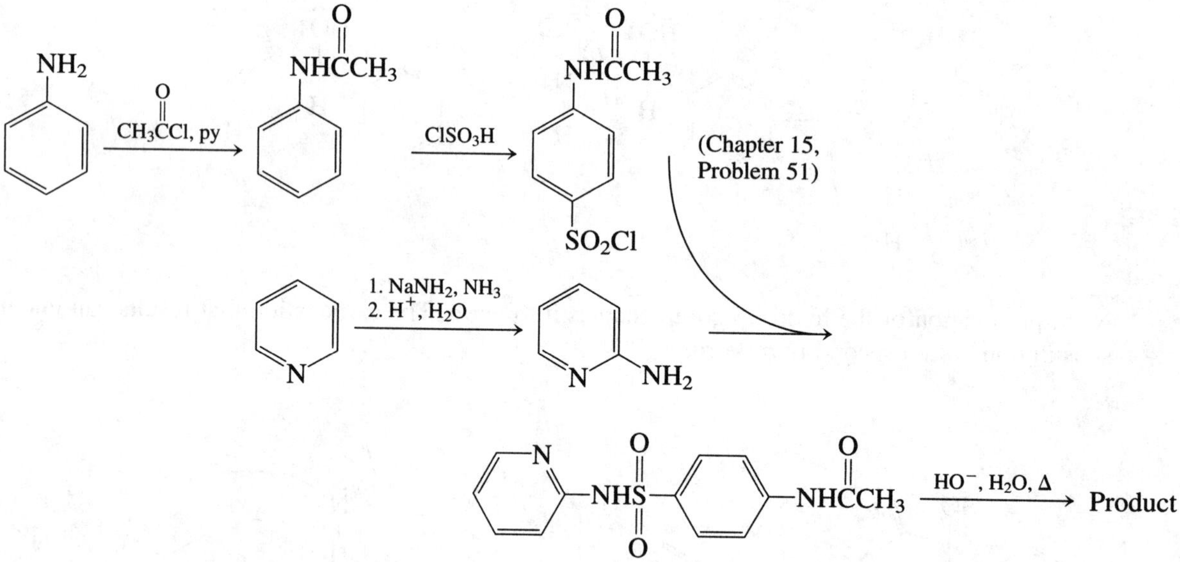

49. Reaction with an activated derivative of acetic acid such as the anhydride would provide the product.

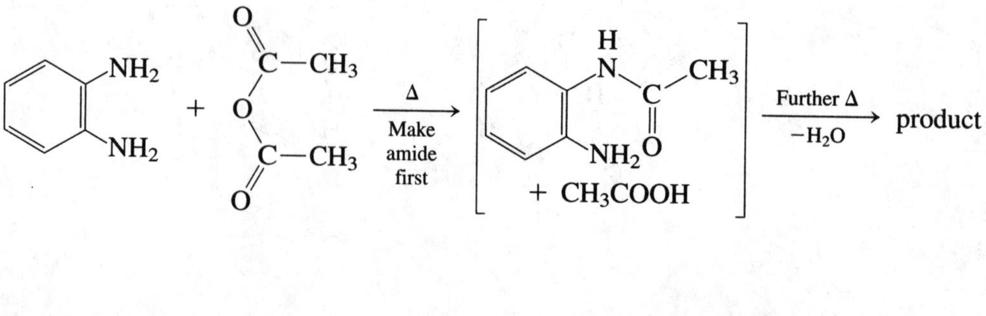

50. (a) 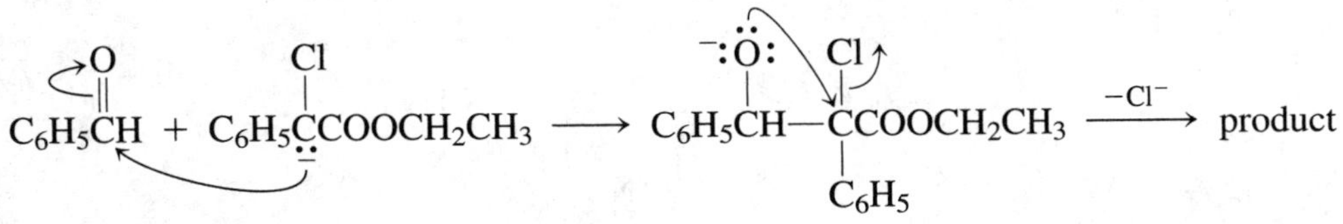 $C_6H_5CH + C_6H_5\overset{..}{\underset{..}{C}}COOCH_2CH_3 \longrightarrow C_6H_5CH-\overset{Cl}{\underset{C_6H_5}{C}}COOCH_2CH_3 \xrightarrow{-Cl^-}$ product

(b) 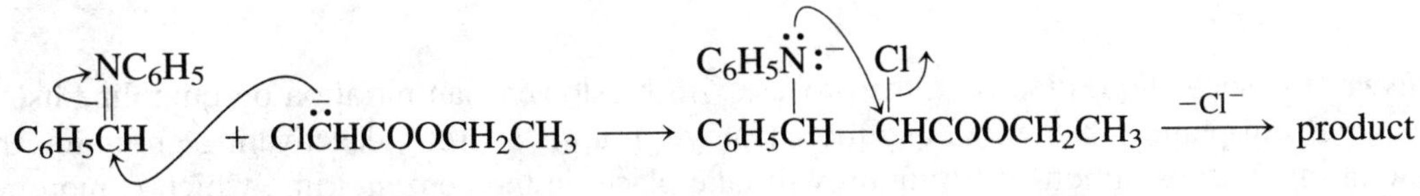 $C_6H_5\overset{NC_6H_5}{\underset{}{CH}} + Cl\overset{..}{\underset{..}{C}}HCOOCH_2CH_3 \longrightarrow C_6H_5CH-\overset{}{\underset{}{C}}HCOOCH_2CH_3 \xrightarrow{-Cl^-}$ product

51. (a) 1,3-Dibromo-5,5-dimethyl-1,3-diaza-2,4-cyclopentanedione.

(b) Thinking mechanistically, you have the following:

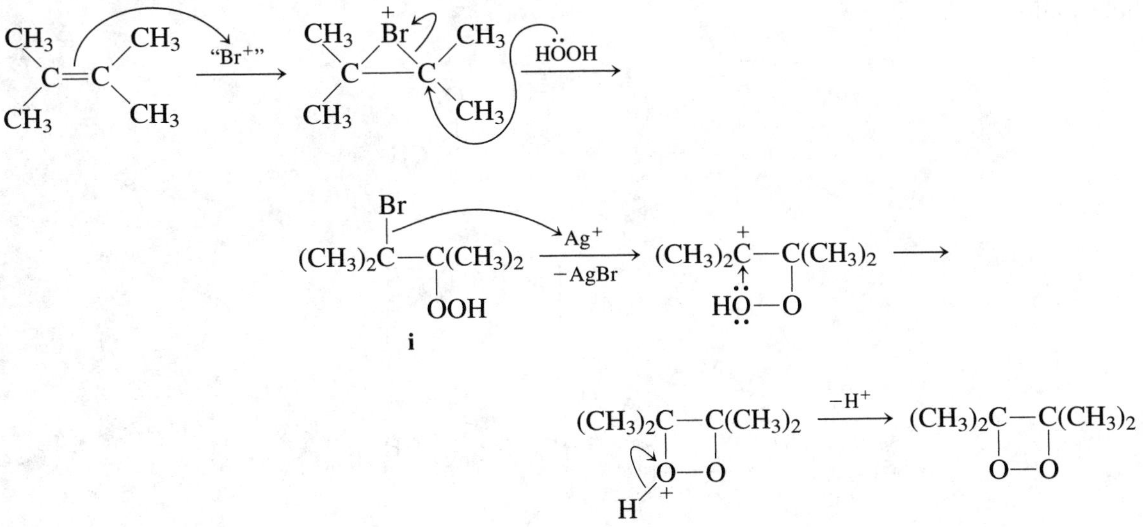

$$\underset{CH_3}{\overset{CH_3}{C}}=\underset{CH_3}{\overset{CH_3}{C}} \xrightarrow{\text{``}Br^+\text{''}} \underset{CH_3}{\overset{CH_3}{C}}-\underset{CH_3}{\overset{CH_3}{\overset{+}{Br}}}C \xrightarrow{H\overset{..}{O}OH}$$

$$\underset{\underset{\mathbf{i}}{OOH}}{(CH_3)_2C-C(CH_3)_2} \xrightarrow[-AgBr]{Ag^+} \underset{H\overset{..}{O}-O}{(CH_3)_2\overset{+}{C}-C(CH_3)_2} \longrightarrow$$

$$(CH_3)_2C-C(CH_3)_2 \xrightarrow{-H^+} (CH_3)_2C-C(CH_3)_2$$
$$\underset{\underset{H}{\overset{+}{O}-O}}{} \qquad \underset{O-O}{}$$
$$\qquad\qquad\qquad\qquad\qquad \mathbf{ii}$$
$$\qquad\qquad\qquad\qquad\qquad \textbf{3,3,4,4-Tetramethyl-}$$
$$\qquad\qquad\qquad\qquad\qquad \textbf{1,2-dioxacyclobutane}$$

52. Abbreviated mechanism (note the "double" Michael addition).

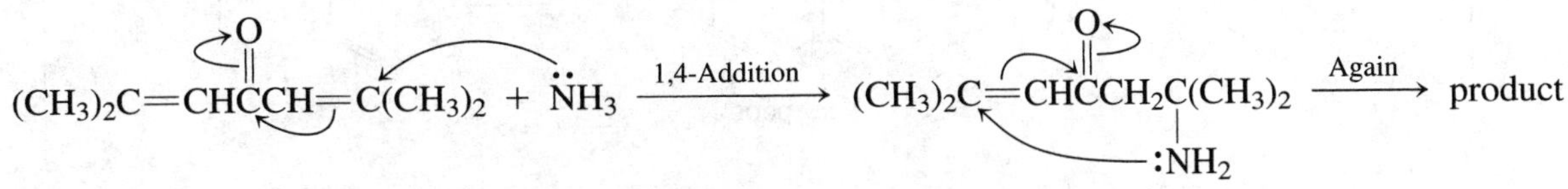

$$(CH_3)_2C=CHCCH=C(CH_3)_2 + \overset{..}{N}H_3 \xrightarrow{\text{1,4-Addition}} (CH_3)_2C=CHCCH_2C(CH_3)_2 \xrightarrow{\text{Again}} \text{product}$$
$$\qquad\qquad\qquad\qquad\qquad\qquad\qquad\qquad\qquad\qquad\qquad :NH_2$$

53. The hydrogen atoms on the nitrogen are relatively acidic because of resonance stabilization of the anion that results from deprotonation. Conjugate addition followed by intramolecular aldol condensation gives the product:

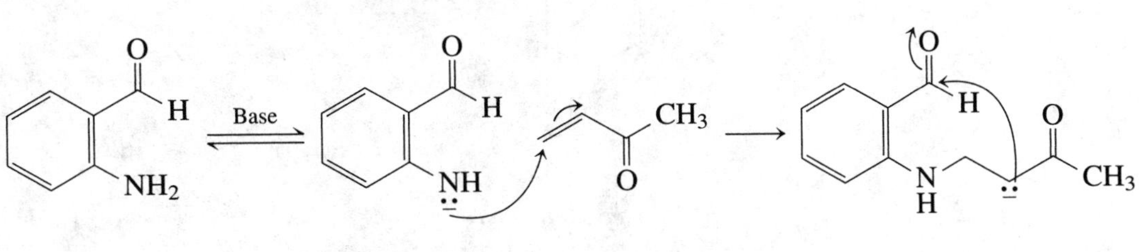

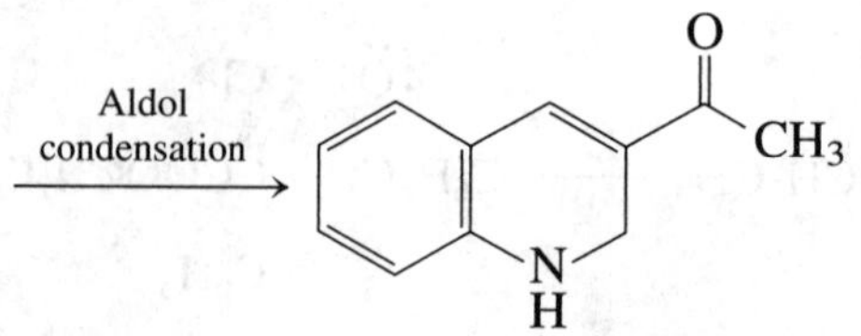

54. To answer the second question first, the process will be slower than nitration on quinoline itself. Nitration is an electrophilic substitution, and the alkanoyl group is a strong deactivating group toward electrophilic aromatic substitution. Substitution will take place on the benzene ring, which is more reactive toward electrophiles than is the pyridine ring. Without the extra ethanoyl group substitution would be expected to occur at positions 5 and 8, because the cationic intermediates from substitution at these positions do not disrupt the aromaticity of the pyridine ring. However, the extra substituent deactivates position 5 (see resonance forms), leaving substitution at position 8 as the most likely outcome.

Substitution at C5:

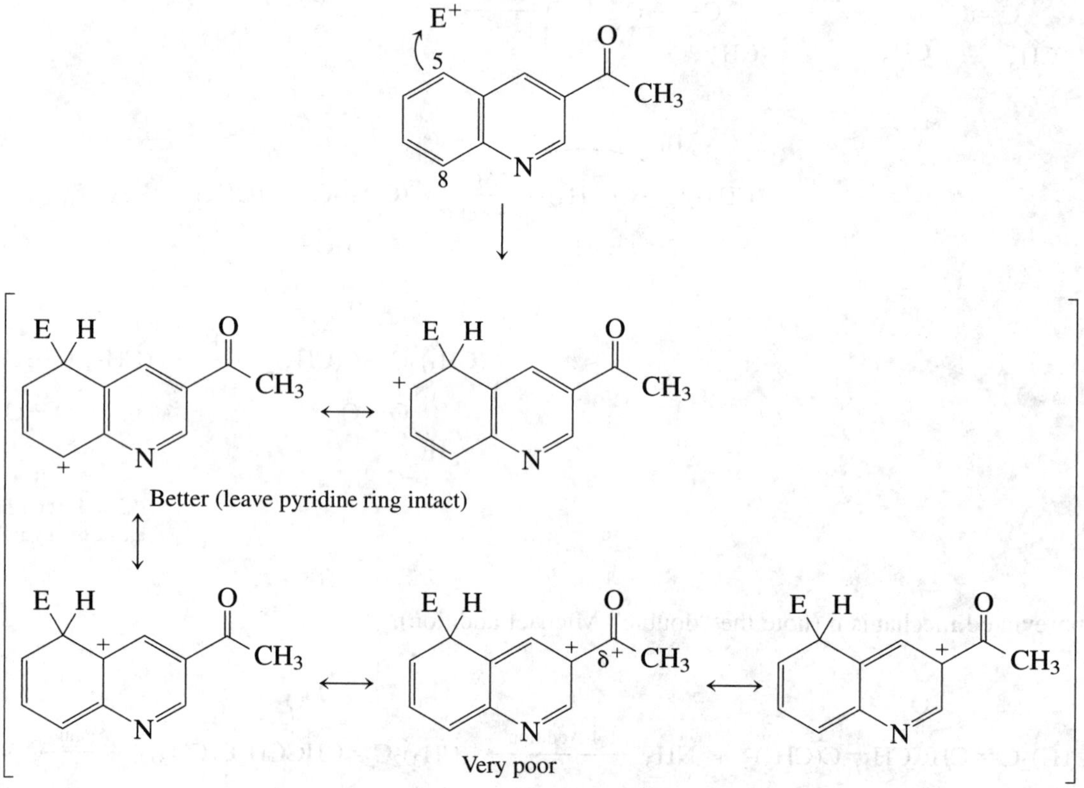

Substitution at C8:

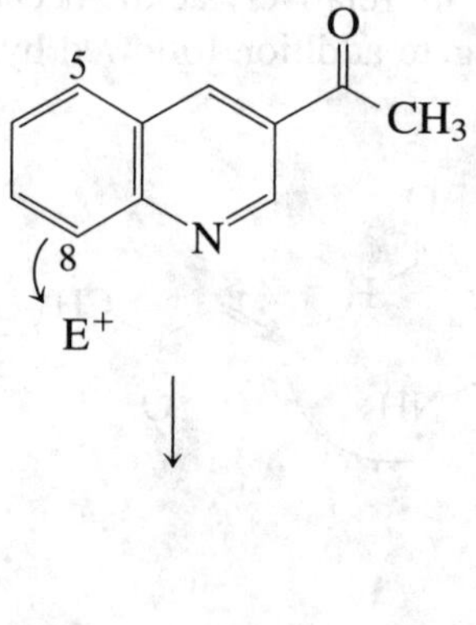

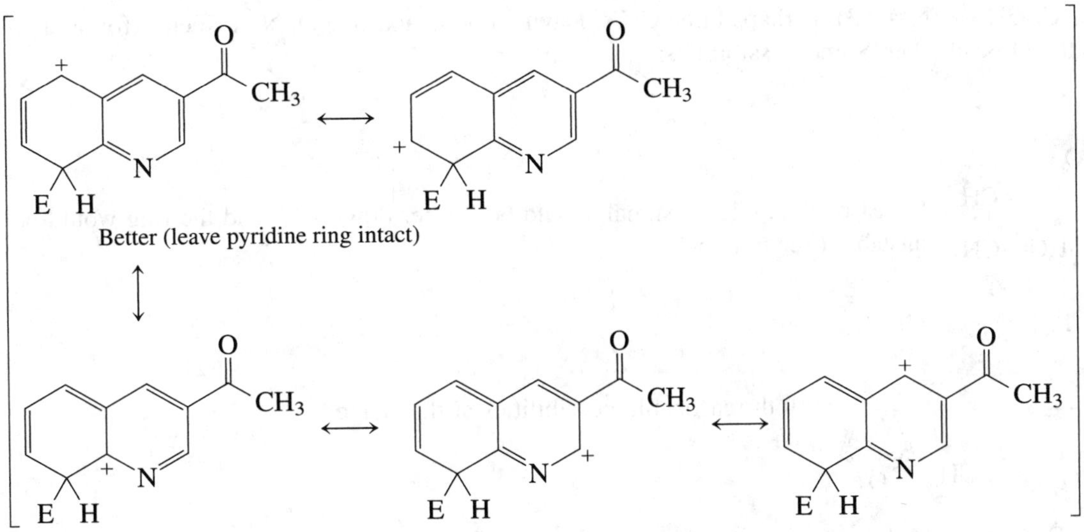

Better (leave pyridine ring intact)

55. $H_{sat} = 16 + 2 = 18$; degrees of unsaturation $= \frac{(18 - 8)}{2} = 5\ \pi$ bonds or rings. NMR: C_6H_5 group is present, accounting for 4 degrees of unsaturation. No alkene C—H signals, so the last degree of unsaturation is probably a ring. So far: C_6H_5—, 2 C, 3 H, O present, adding up to C_8H_8O. The three H's all couple to each other (all three signals are split), so an —OH is unlikely. Therefore, the O is an ether oxygen, giving as the only possible answer

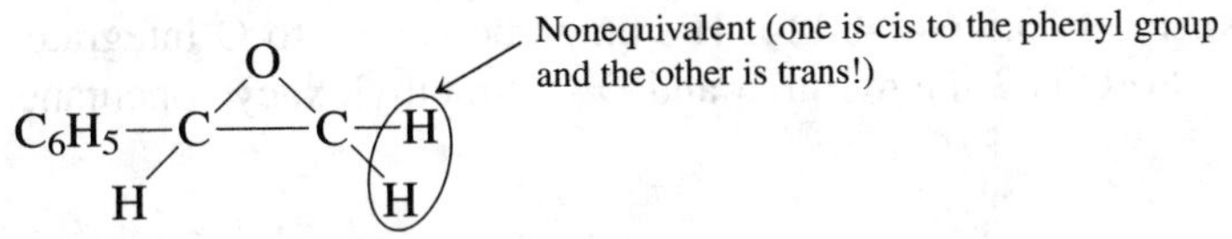

Nonequivalent (one is cis to the phenyl group and the other is trans!)

2-Phenyloxacyclopropane

Conc. aq. HCl causes ring opening to

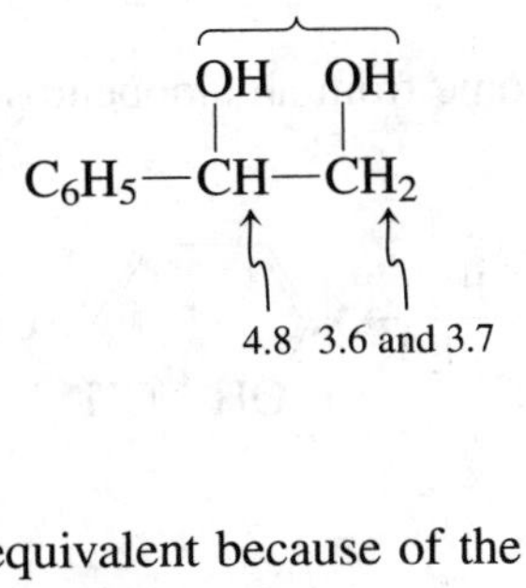

4.2

OH   OH

C_6H_5—CH—CH$_2$

4.8   3.6 and 3.7

The hydrogens of the CH$_2$ group are not equivalent because of the adjacent stereocenter.

56. $H_{sat} = 10 + 2 = 12$; degrees of unsaturation $= \frac{(12 - 6)}{2} = 3\ \pi$ bonds or rings in C. D has $1\ \pi$ bond or ring. Result of H_2 addition to C suggests that it has $2\ \pi$ bonds and 1 ring, and the π bonds are gone in D.

NMR of C: CH_3— ($\delta = 2.3$), perhaps three CH's, leaving one C and one O. No evidence for an alcohol, so assume the O is an ether. Some possibilities:

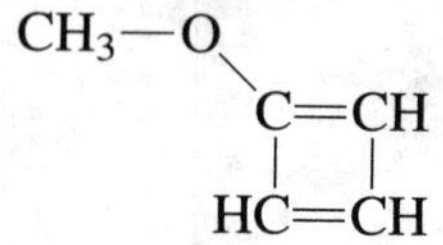

CH_3—O

Unreasonable. CH_3— signal should be further downfield, and the ring would be very unstable (Section 15-6).

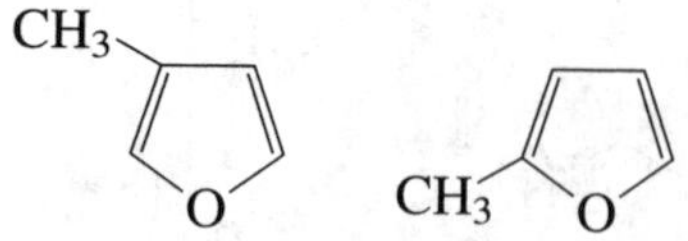

Both reasonable possibilities at this stage.

NMR of D: Complicated, but two pieces of information are extractable. First, the CH_3— is upfield ($\delta = 1.1$) and a doublet, consistent with either

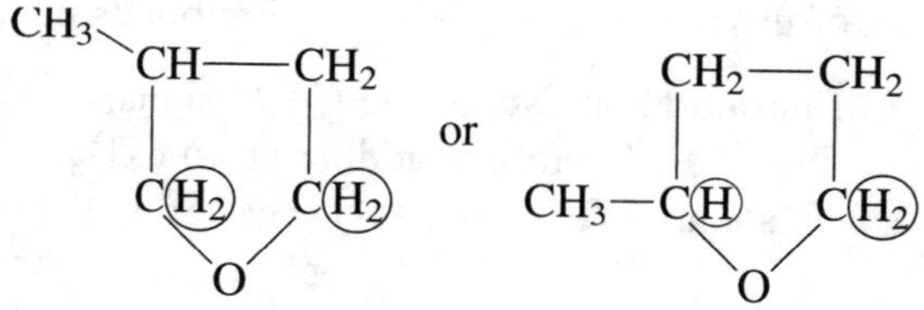

4 H near O   or   3 H near O

Second, the signals between $\delta = 3.5$ and 4.0 for H's on carbons next to O integrate to **3 H,** consistent only with the second structure. So C is 2-methylfuran and D is 2-methyloxacyclopentane.

57. $H_{sat} = 10 + 2 = 12$; degrees of unsaturation $= \frac{(12 - 4)}{2} = 4$ π bonds or rings. NMR: $\delta = 9.6$

suggests aldehyde (supported by IR), together with 3 CH's. Similar to Problem 49, try furans as possibilities:

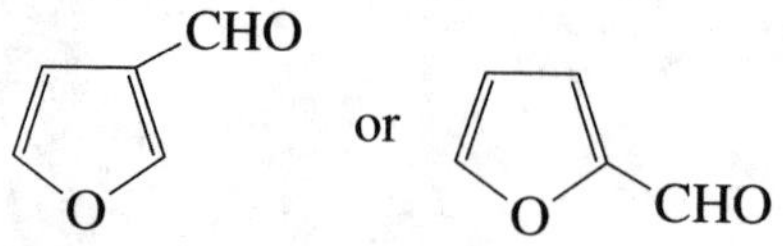

CHO   or   CHO

How to choose? Which is more likely to come from an aldopentose? A possible (abbreviated) mechanism:

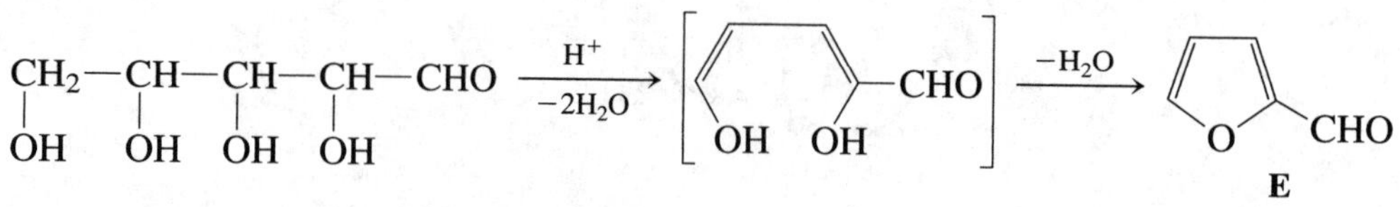

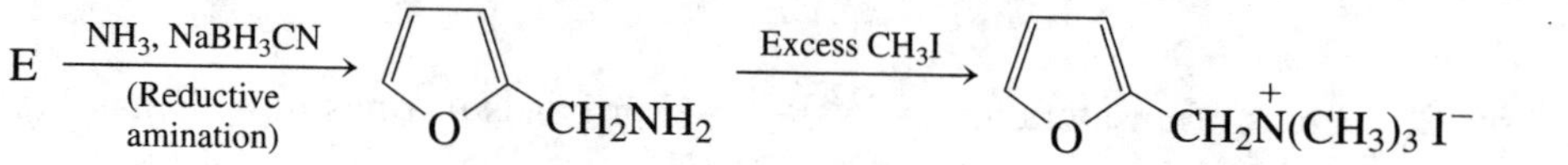

CH_2—CH—CH—CH—CHO $\xrightarrow[-2H_2O]{H^+}$ [CHO] $\xrightarrow{-H_2O}$ CHO
| | | | | |
OH   OH   OH   OH   OH   OH

E

E $\xrightarrow[\text{amination}]{\text{NH}_3, \text{NaBH}_3\text{CN}}$ CH_2NH_2 $\xrightarrow{\text{Excess CH}_3\text{I}}$ $CH_2\overset{+}{N}(CH_3)_3$ I^-
       (Reductive
        amination)

Furethonium

58.

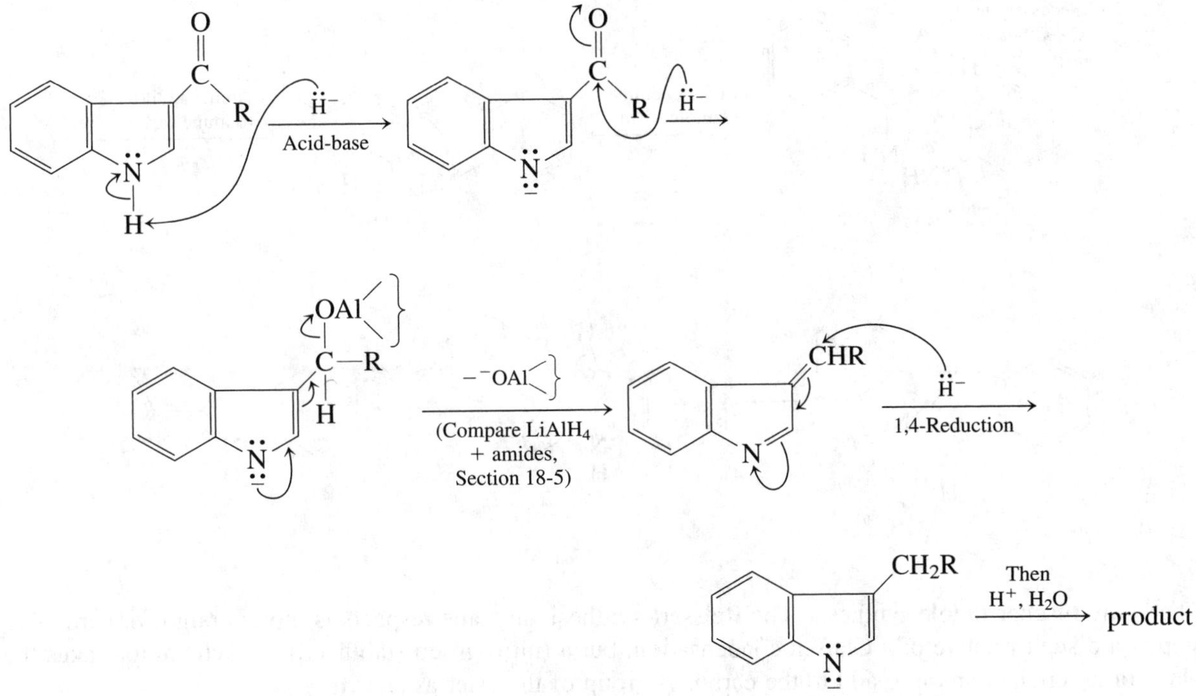

59. Stepwise:

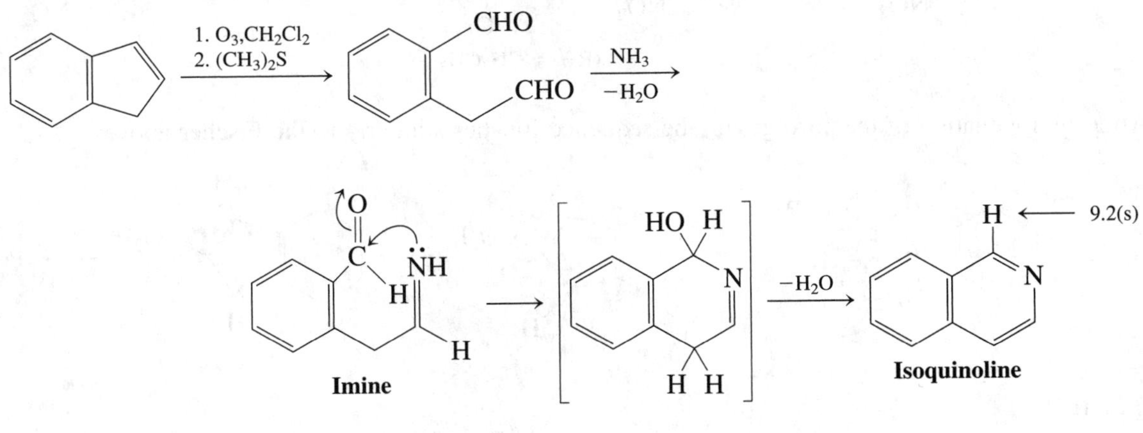

60. Begin with the hydrazone shown in the middle of the reaction scheme. Examine the structure of the product: A bond is needed between the methyl group and the benzene carbon *ortho* to the nitrogen. For brevity, catalytic protons are shown attaching in the same step as the reactions they catalyze. In fact, the protonations generally occur first, with tautomerism or addition taking place afterwards.

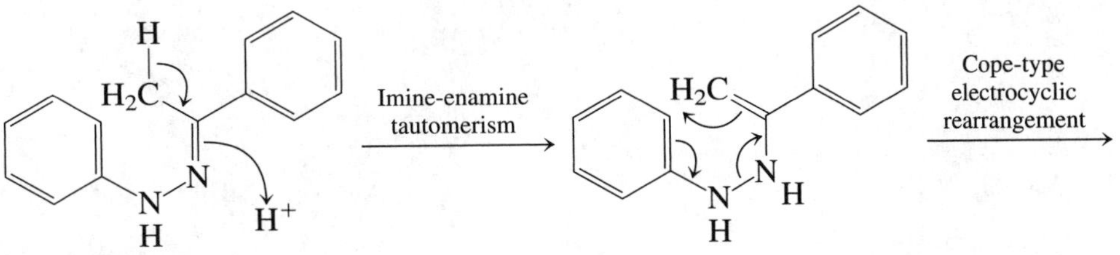

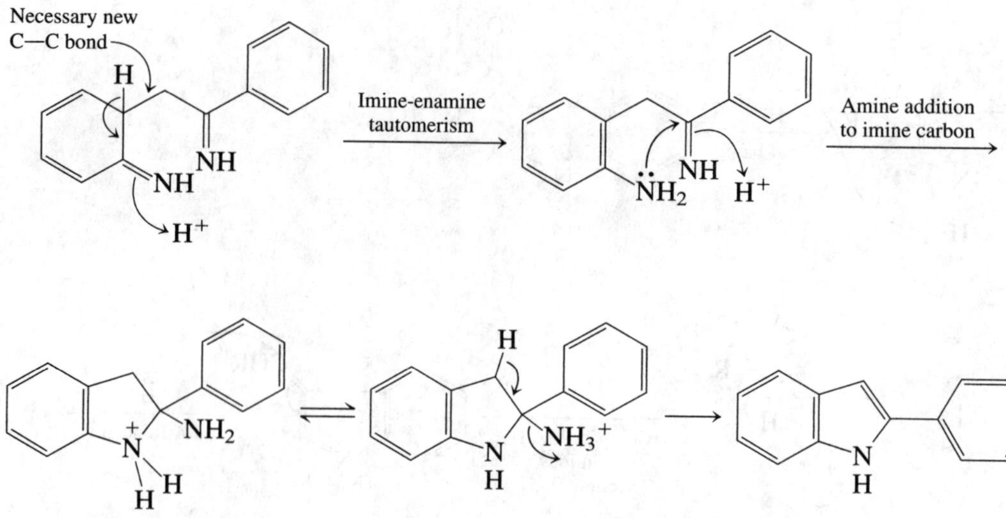
Necessary new
C—C bond
H
NH
NH
H+
Imine-enamine
tautomerism
NH
NH2
H+
Amine addition
to imine carbon
NH2
N
H
H
H
N
H
NH3+
N
H

That's the Fischer indole synthesis. The Reissert synthesis in some respects is more straightforward. The first step is a distant relative of a Claisen condensation, but a (nitro-group stabilized) benzylic anion takes the place of an enolate anion to add to the carbonyl group of the ester as shown:

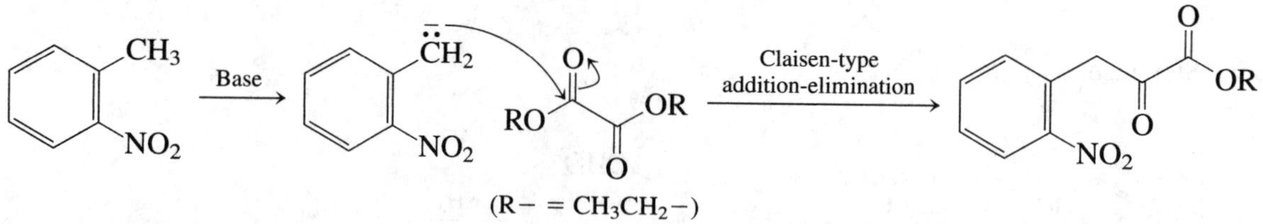
CH3
NO2
Base
CH2
NO2
O
RO
OR
O
(R– = CH3CH2–)
Claisen-type
addition-elimination
O
OR
O
NO2

After hydrogenation of the nitro group, the sequence finishes similarly to the Fischer above:

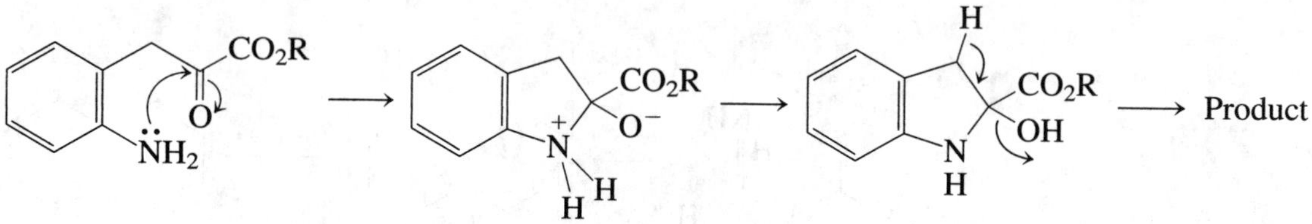
CO2R
O
NH2
CO2R
O−
N
H
H
H
CO2R
OH
N
H
Product

61. (c)

62. (e)

63. (d)

64. (c)

26

Amino Acids, Peptides, Proteins, and Nucleic Acids: Nitrogen-Containing Polymers in Nature

Here it is: the last chapter! It is, however, as much a beginning as an end. Chemistry in general, and organic chemistry in particular, are not isolated fields. Organic chemistry is the basic stuff of biology, and this chapter bridges the two. The basic principles that govern the behavior of organic molecules in general are shown here to be directly applicable to molecules of greater and greater complexity. You see here some of the most fundamental molecules of life viewed from the organic chemist's point of view. The next step in this direction is biochemistry.

Outline of the Chapter

26-1 Structure and Properties of Amino Acids
Nature's most versatile building blocks.

26-2, 26-3 Preparation of Amino Acids
Fancy footwork involving two very different functional groups.

26-4 Peptides and Proteins: Amino Acid Oligomers and Polymers
The whole is greater than the sum of its parts.

26-5 Determination of Polypeptide Primary Structure: Sequencing
Exercises in logic, plus a lot of hard work.

26-6, 26-7 Synthesis of Polypeptides: A Protecting Group Challenge
Even fancier footwork; how does Nature do it so easily??!!

26-8 Polypeptides in Nature: The Proteins Myoglobin and Hemoglobin

26-9, 26-10 Biosynthesis of Proteins: Nucleic Acids
How Nature *appears* to do it so easily.

26-11 DNA Sequencing and Synthesis: Gene Technology
The future is now!

Keys to the Chapter

26-1. Structure and Properties of Amino Acids

Read the introduction to the text chapter and this text section, and then come back here. All done? O.K., here's the big picture. The chemistry of life is complicated. Structures have to be built to hold things together, and a lot of chemical reactions have to be going on to perform the various functions that maintain life, such as energy storage and utilization. These all have to occur under a very constrained set of conditions: Water is the only available solvent, and in general only very narrow ranges of temperature and pH are acceptable. Otherwise everything falls apart. So how is it done? The answer begins with the amino acids.

Look first at the structures of the 20 most common examples (Table 26-1). They differ only in the group attached to the α-carbon. The variety in these groups establishes the versatility of amino acids. There are nonpolar groups, both small and large, capable of various degrees of steric interaction. There are uncharged but polar groups, capable of hydrogen bonding. There are nitrogen-containing groups of various base strengths, some of which will be protonated and positively charged at pH 7. There are oxygen and sulfur groups of various acid strengths, some of which will be deprotonated and negatively charged at pH 7. Because of this variety, Nature can choose from among these 20 compounds just the right one to fill any of a number of chemical needs. One feature that is emphasized in this section is the acid-base behavior of the amino acids. Table 26-1 lists pK_a values for all relevant groups. Notice, for one thing, that the way the amino acids have been drawn up to now, $H_2N—CHR—COOH$, is **wrong.** In solution, amino acids in fact **never** exist to any significant extent in this form, with neutral amino and carboxylic acid groups present at the same time. Depending on the pH, one or both of these groups is **always** charged, with the pH 7 structure being $^+H_3N—CHR—COO^-$. Because of this feature, amino acids are very good **buffers** at a variety of different pH's, depending on R. Obviously the effects of pH on amino acid structure are important, and Problem 32 will give you several examples to work on so you can get the feel for it.

26-2 and 26-3. Preparation of Amino Acids

Although none of the individual reactions in these sections are new, the sequences present them in powerful combinations aimed at solving the problem of introducing basic and acidic groups into the same molecule.

26-4 and 26-5. Peptides and Proteins: Amino Acid Oligomers and Polymers

Techniques for determining peptide and protein structure are extremely well worked out (and Problems 49–54 will give you plenty of chances to try them for yourself). The results, especially in the subtleties of folding of these polymeric chains, reveal the extent to which the characteristics of the different amino acids are combined and used in nature to generate large molecular assemblies perfectly suited for very specific biological roles. Problems 45–48 are intended to give you something to think about in this regard.

26-6 and 26-7. Synthesis of Polypeptides: A Protecting Group Challenge

Don't ever lose sight of the fact that the linkage between amino acids is nothing more than a simple amide bond: —HN—CO—. Nonetheless, the construction of peptide chains from simple amino acids is a major challenge for the same reasons that, say, mixed aldol or Claisen condensations (Sections 18-6 and 23-1) are tricky things to do: Each amino acid involved has **both** a potentially nucleophilic atom (the N) and a potentially electrophilic one (the carboxy C). Thus, an attempt at linking, say, the amine of amino acid 1 with the carboxy group of amino acid 2 is going to be complicated by the need to **prevent** the simultaneous linkage of either two molecules of a.a.1 to each other, two molecules of a.a.2 to each other, or a.a.1 to a.a.2 in the **wrong sense:** carboxy of 1 to amine of 2. The solution to the problem lies in, again, a very well worked out array of functional group protection–deprotection procedures, the simplest of which are presented here. Problems 55 and 56 will let you try them for yourself.

26-8 through 26-11. Proteins in Operation; Biosynthesis; DNA and Gene Technology

Obviously only a tiny taste of what's involved with these topics can be presented in the space of four chapter sections. Nonetheless, you should be able to sense the remarkable way in which the linkage of relatively small units gives rise to structures of such highly elaborate function. This stuff is really neat. You might like to read more about it some time.

Solutions to Problems

30.

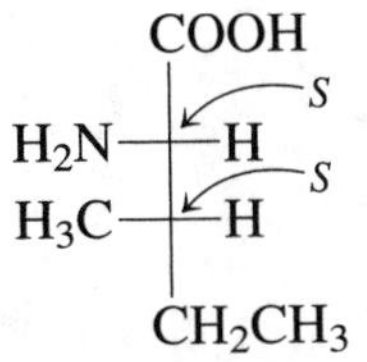

L-Isoleucine 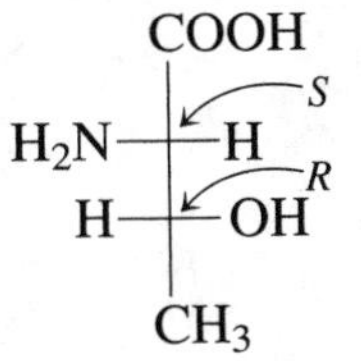   L-Threonine

Systematic name for L-threonine: (2S, 3R)-2-amino-3-hydroxybutanoic acid.

31.

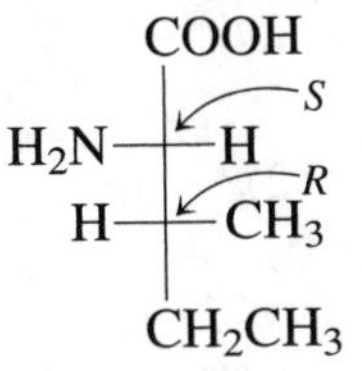

Allo-L-isoleucine

Systematic name for allo-L-isoleucine: (2S, 3R)-2-amino-3-methylpentanoic acid.

32. Structures are presented in order of increasing pH (in parentheses).

(a)
$$\begin{array}{c} COOH \\ H_3\overset{+}{N}-\!\!\!\!\!-H \ (1) \\ CH_3 \end{array} \quad \begin{array}{c} COO^- \\ H_3\overset{+}{N}-\!\!\!\!\!-H \ (7) \\ CH_3 \end{array} \quad \begin{array}{c} COO^- \\ H_2N-\!\!\!\!\!-H \ (12) \\ CH_3 \end{array}$$

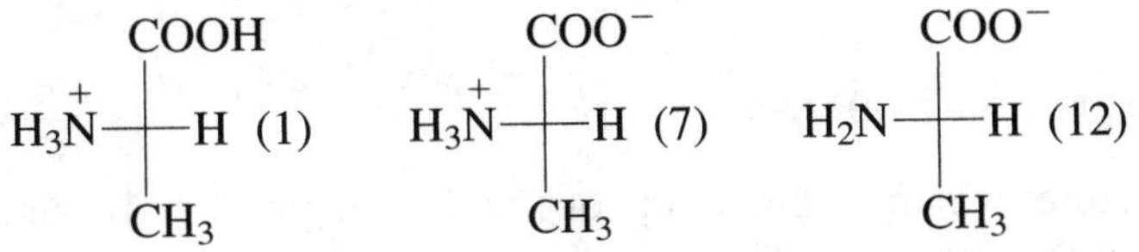

(b)
$$\begin{array}{c} COOH \\ H_3\overset{+}{N}-\!\!\!\!\!-H \ (1) \\ CH_2OH \end{array} \quad \begin{array}{c} COO^- \\ H_3\overset{+}{N}-\!\!\!\!\!-H \ (7) \\ CH_2OH \end{array} \quad \begin{array}{c} COO^- \\ H_2N-\!\!\!\!\!-H \ (12) \\ CH_2OH \end{array}$$

(c)
$$\begin{array}{c} COOH \\ H_3\overset{+}{N}-\!\!\!\!\!-H \ (1) \\ (CH_2)_4-\overset{+}{N}H_3 \end{array} \quad \begin{array}{c} COO^- \\ H_3\overset{+}{N}-\!\!\!\!\!-H \ (7) \\ (CH_2)_4-\overset{+}{N}H_3 \end{array} \quad \begin{array}{c} COO^- \\ H_2N-\!\!\!\!\!-H \ (9.5) \\ (CH_2)_4-\overset{+}{N}H_3 \end{array} \quad \begin{array}{c} COO^- \\ H_2N-\!\!\!\!\!-H \ (12) \\ (CH_2)_4-NH_2 \end{array}$$

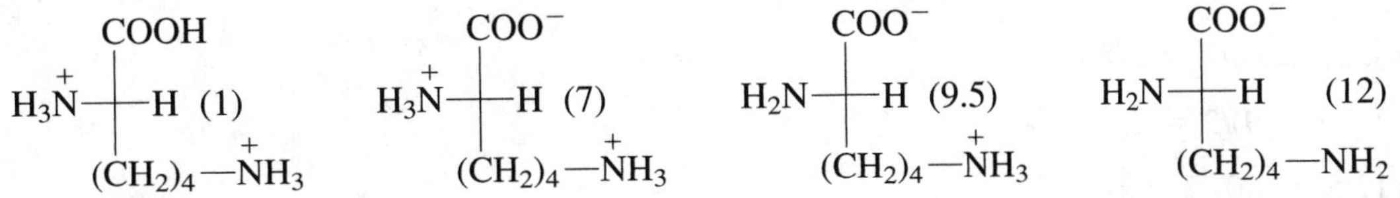

(d) imidazole side-chain structures (1), (5), (7), (12)

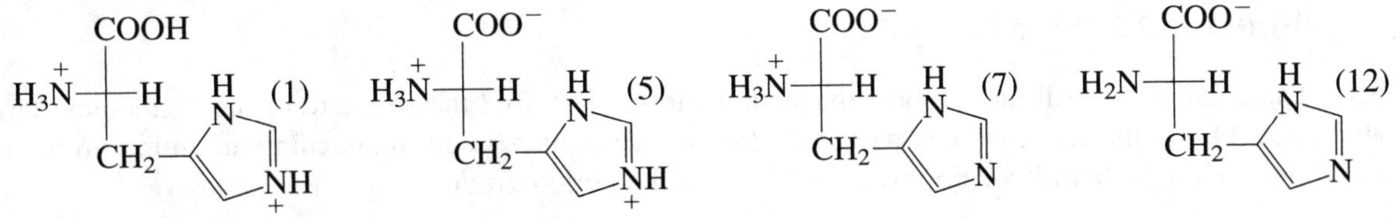

(e)
$$\begin{array}{c} COOH \\ H_3\overset{+}{N}-\!\!\!\!\!-H \ (1) \\ CH_2SH \end{array} \quad \begin{array}{c} COO^- \\ H_3\overset{+}{N}-\!\!\!\!\!-H \ (7) \\ CH_2SH \end{array} \quad \begin{array}{c} COO^- \\ H_3\overset{+}{N}-\!\!\!\!\!-H \ (9) \\ CH_2S^- \end{array} \quad \begin{array}{c} COO^- \\ H_2N-\!\!\!\!\!-H \ (12) \\ CH_2S^- \end{array}$$

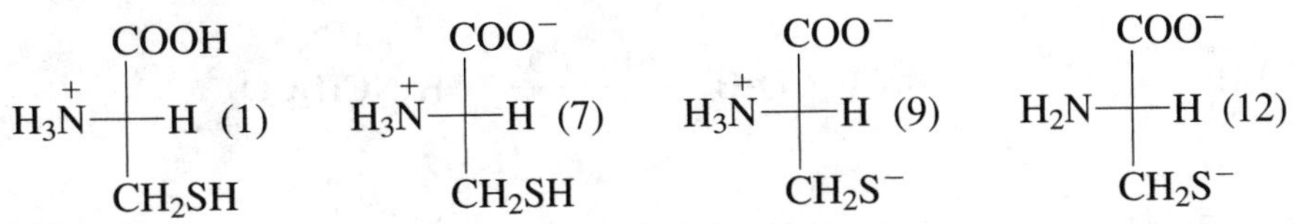

(f)
$$\begin{array}{c} COOH \\ H_3\overset{+}{N}-\!\!\!\!\!-H \ (1) \\ CH_2COOH \end{array} \quad \begin{array}{c} COO^- \\ H_3\overset{+}{N}-\!\!\!\!\!-H \ (3) \\ CH_2COOH \end{array} \quad \begin{array}{c} COO^- \\ H_3\overset{+}{N}-\!\!\!\!\!-H \ (7) \\ CH_2COO^- \end{array} \quad \begin{array}{c} COO^- \\ H_2N-\!\!\!\!\!-H \ (12) \\ CH_2COO^- \end{array}$$

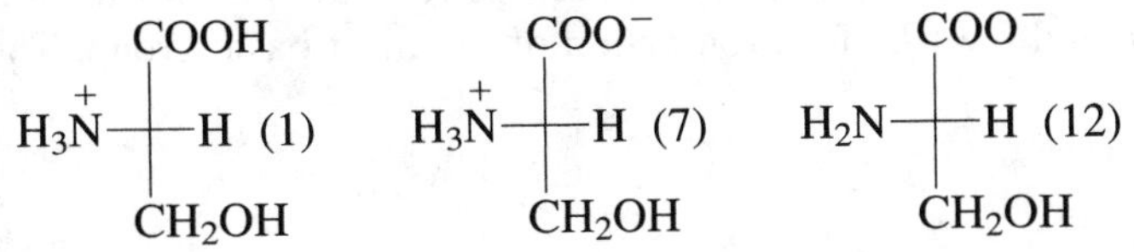

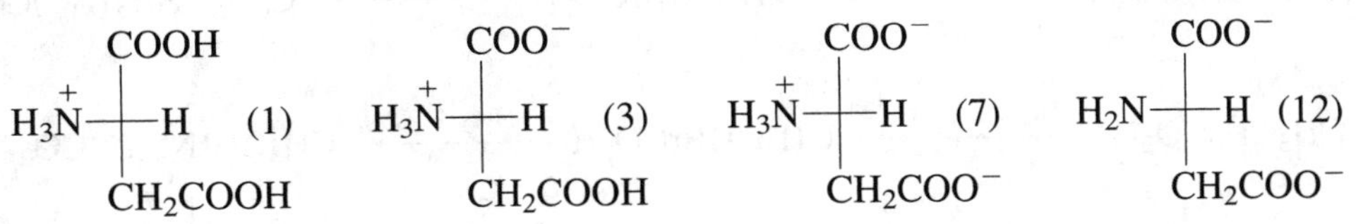

(g)

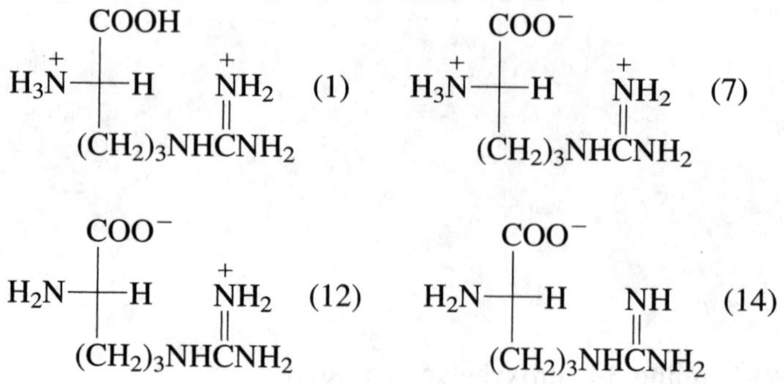

COOH
H$_3$N$^+$—H NH$_2^+$ (1)

(CH$_2$)$_3$NHCNH$_2$

COO$^-$
H$_3$N$^+$—H NH$_2^+$ (7)

(CH$_2$)$_3$NHCNH$_2$

COO$^-$
H$_2$N—H NH$_2^+$ (12)

(CH$_2$)$_3$NHCNH$_2$

COO$^-$
H$_2$N—H NH (14)

(CH$_2$)$_3$NHCNH$_2$

(h)

COOH
H$_3$N$^+$—H (1)
CH$_2$

OH

COO$^-$
H$_3$N$^+$—H (7)
CH$_2$

OH

COO$^-$
H$_2$N$^{..}$—H (9.5)
CH$_2$

OH

COO$^-$
H$_2$N$^{..}$—H (12)
CH$_2$

O$^-$

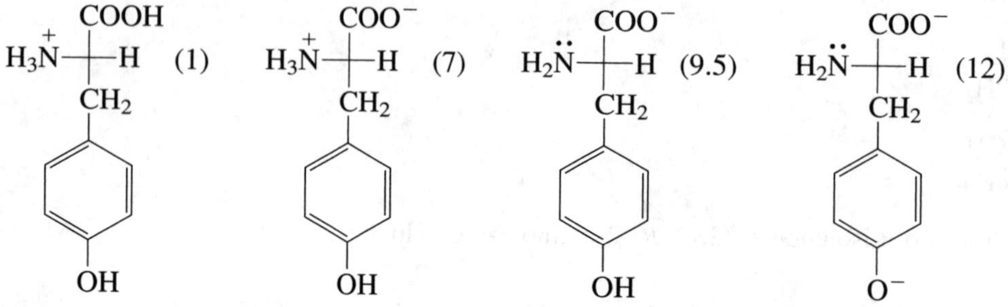

33. **(a)** Arg, Lys; **(b)** Ala, Ser, Tyr, His, Cys; **(c)** Asp

34. First identify the net charge-neutral structure, which is always the one with a single + and a single − charge. Choose the two pK_a values that bracket that structure. They are, repectively, the pK_a for deprotonation of the most acidic group in that structure and the pK_a for protonation of the most basic group. Their average is the pI.

(c) (9.0 + 10.5)/2 = 9.7

(d) (9.2 + 6.1)/2 = 7.6

(e) (8.2 + 2.0)/2 = 5.1

(f) (3.7 + 1.9)/2 = 2.8

(g) (12.5 + 9.0)/2 = 10.8

(h) (9.1 + 2.2)/2 = 5.7

35. The solutions for all three follow the same pattern. First, find the structure of the amino acid target. Then, based upon the reactions you have been told to use (which result in attaching an amino group to the α-carbon), begin each with a carboxylic acid lacking the amino group:

(a) For Gly, CH_3CO_2H $\xrightarrow{Br_2, \text{ cat. } PBr_3}$ $BrCH_2CO_2H$ $\xrightarrow{NH_3, H_2O}$ $^+H_3NCH_2CO_2^-$

(b) For Phe, $C_6H_5CH_2CO_2H$ $\xrightarrow{Br_2, \text{ cat. } PBr_3}$ $C_6H_5CHBrCO_2H$ $\xrightarrow{NH_3, H_2O}$ $C_6H_5CH(NH_3^+)CO_2^-$

(c) For Ala, $CH_3CH_2CO_2H$ $\xrightarrow{Br_2, \text{ cat. } PBr_3}$ $CH_3CHBrCO_2H$ $\xrightarrow{NH_3, H_2O}$ $CH_3CH(NH_3^+)CO_2^-$

36. This problem is related to the last, except that the starting material for a Strecker synthesis is an aldehyde, and the Strecker sequence adds a carbon atom (as a nitrile, CN, which becomes the carboxy group). So each starting material must be one carbon shorter than the intended target:

(a) For Gly, $H_2C{=}O$ $\xrightarrow[\text{2. HCN}]{\text{1. NH}_3}$ H_2NCH_2CN $\xrightarrow{H^+, H_2O}$ $^+H_3NCH_2CO_2{}^-$

(b) For Ile, $CH_3CH_2CH(CH_3)CHO$ $\xrightarrow[\text{2. HCN}]{\text{1. NH}_3}$ $CH_3CH_2CH(CH_3)CH(NH_2)CN$ $\xrightarrow{H^+, H_2O}$

$CH_3CH_2CH(CH_3)CH(NH_3{}^+)CO_2{}^-$

(c) For Ala, CH_3CHO $\xrightarrow[\text{2. HCN}]{\text{1. NH}_3}$ $CH_3CH(NH_2)CN$ $\xrightarrow{H^+, H_2O}$ $CH_3CH(NH_3{}^+)CO_2{}^-$

37. After steps 1 and 2 you have

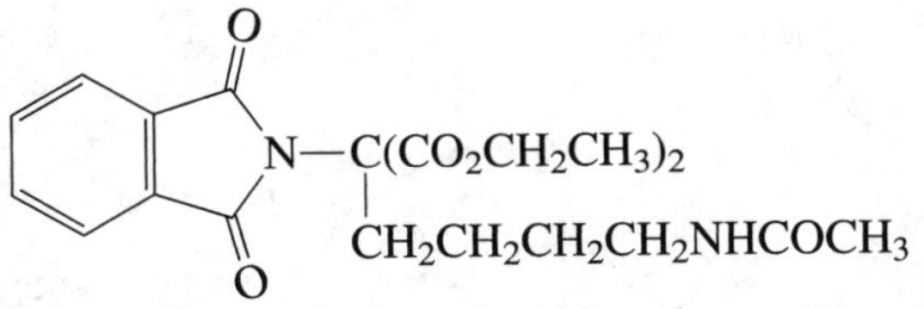

Acidic hydrolysis cleaves not only the phthalimido group from the ring nitrogen but also the acetyl group from the second nitrogen, giving $^-O_2CCH(NH_3{}^+)CH_2CH_2CH_2CH_2NH_3{}^+$, Lys.

38. (a) Because the R group is secondary, alkylation routes should be avoided. Use the Strecker synthesis.

$(CH_3)_2CHCHO$ $\xrightarrow[\text{2. HCN}]{\text{1. NH}_3}$ $(CH_3)_2CHCH\overset{\displaystyle NH_2}{\underset{\displaystyle |}{C}}N$ $\xrightarrow{H^+, H_2O}$ $(CH_3)_2CHCH\overset{\displaystyle ^+NH_3}{\underset{\displaystyle |}{C}}HCOO^-$

(b) The R group is primary; now you have a choice. Either the Strecker synthesis starting with $(CH_3)_2CHCH_2CHO$ or a Gabriel-based method will do.

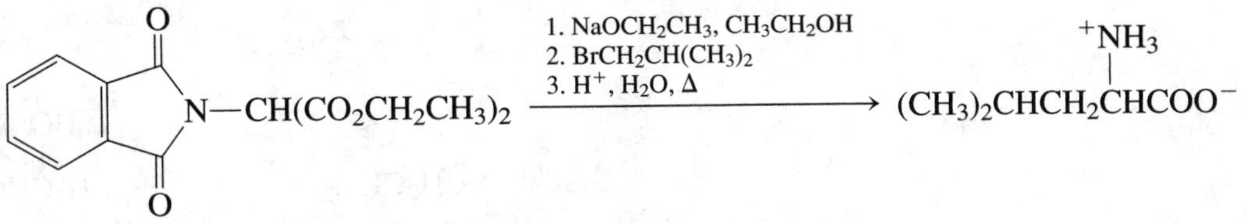

$\xrightarrow[\substack{\text{2. BrCH}_2\text{CH(CH}_3)_2 \\ \text{3. H}^+, \text{H}_2\text{O}, \Delta}]{\text{1. NaOCH}_2\text{CH}_3, \text{CH}_3\text{CH}_2\text{OH}}$ $(CH_3)_2CHCH_2\overset{\displaystyle ^+NH_3}{\underset{\displaystyle |}{C}}HCOO^-$

Even the Hell-Volhardt-Zelinsky–amination sequence works just fine here.

$(CH_3)_2CHCH_2CH_2COOH$ $\xrightarrow{Br_2, PBr_3}$ $(CH_3)_2CHCH_2\overset{\displaystyle Br}{\underset{\displaystyle |}{C}}HCOOH$

$\xrightarrow{NH_3, H_2O}$ $(CH_3)_2CHCH_2\overset{\displaystyle ^+NH_3}{\underset{\displaystyle |}{C}}HCOO^-$

(c) Several ways to go, but you have to first recognize the need for a three-carbon building block with leaving groups at each end to allow linkage to both the α-carbon and (later on) the amine nitrogen to form the ring. Start with a Gabriel-type sequence.

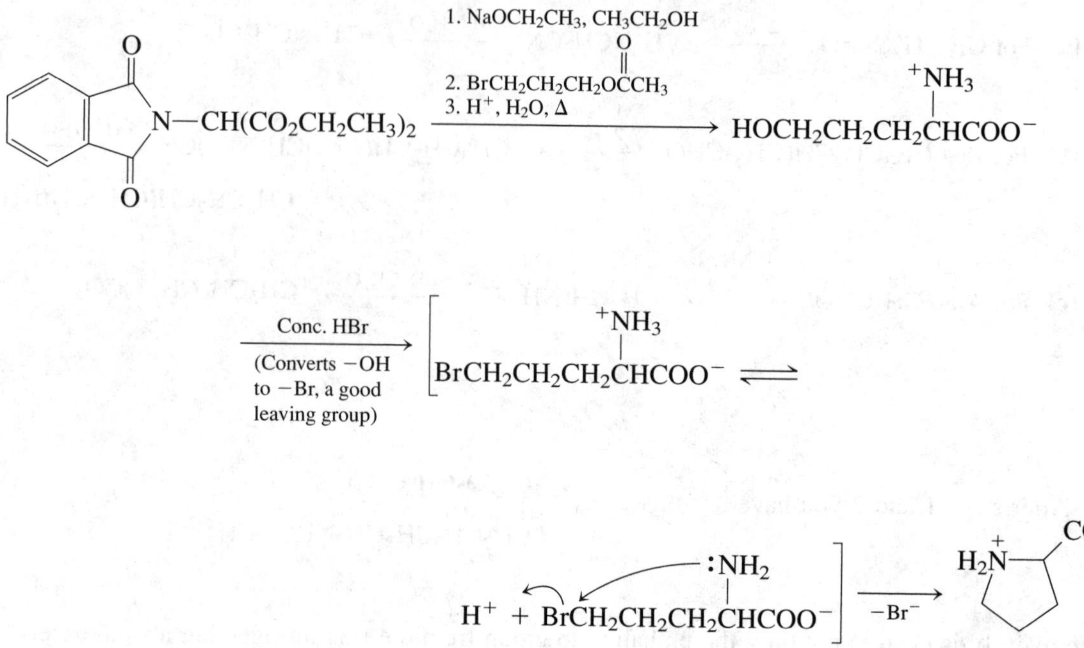

(d) Use a Gabriel-based method. Instead of forming the necessary carbon–carbon bond by S_N2 reaction with a haloalkane, use an aldol-type condensation with CH_3CHO.

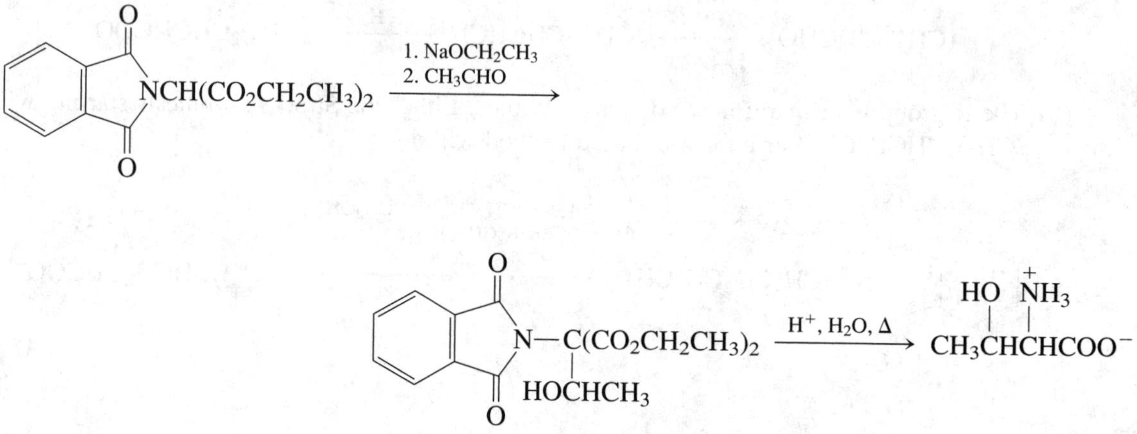

(e) The extra amine group must be present in protected form, irrespective of the method used. Here's a Gabriel-type sequence.

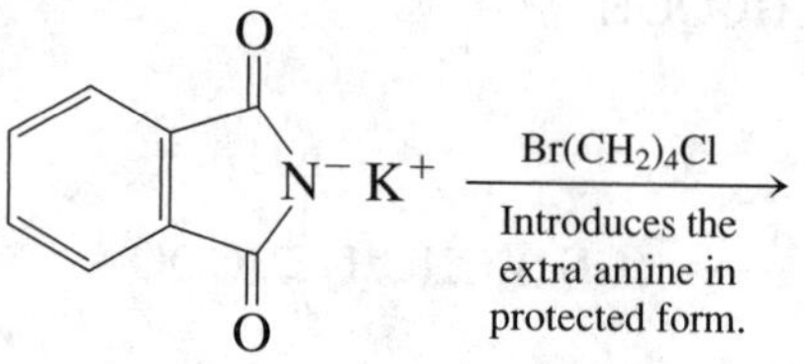

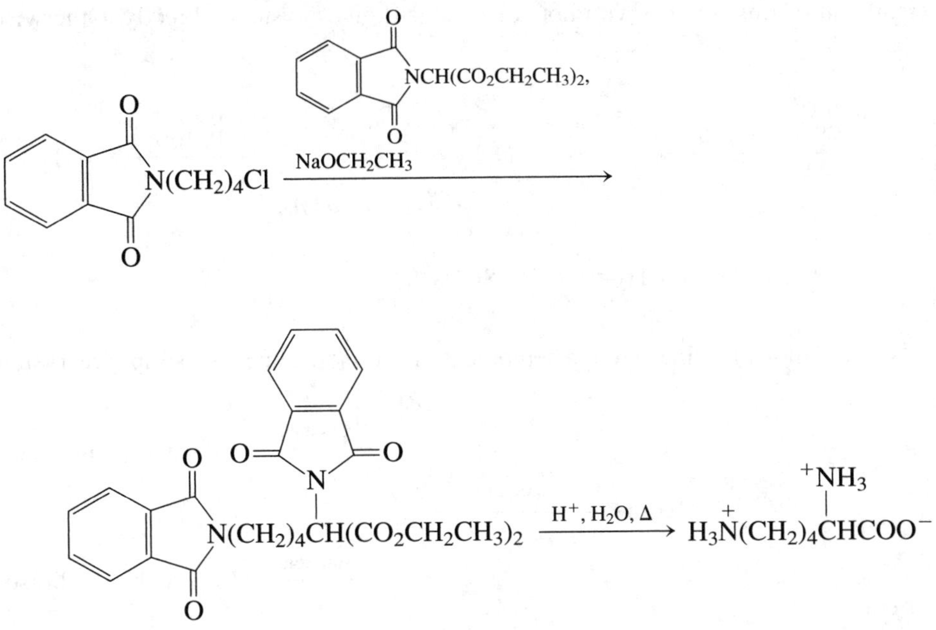

39. **(a)**

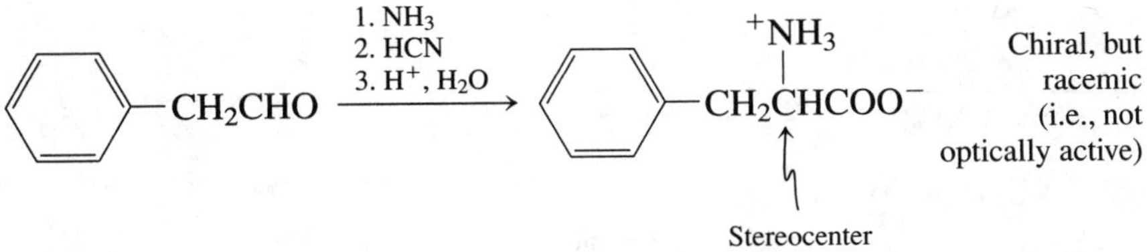

(b) The use of an optically active amine (an *S* enantiomer is shown) means that the addition product is actually a mixture, because a second stereocenter is generated, which can be either *R* or *S*. This is, therefore, a mixture of *R,S* and *S,S* products. Because these are **diastereomers** of each other, they don't necessarily form in identical yields. In fact, the *S,S* product (illustrated) greatly predominates, and, after hydrolysis and removal of the phenylmethyl group with H_2, mainly *S* amino acid is obtained.

40. Allicin is structurally related to cysteine, which should be readily available if you can first devise an approach to serine.

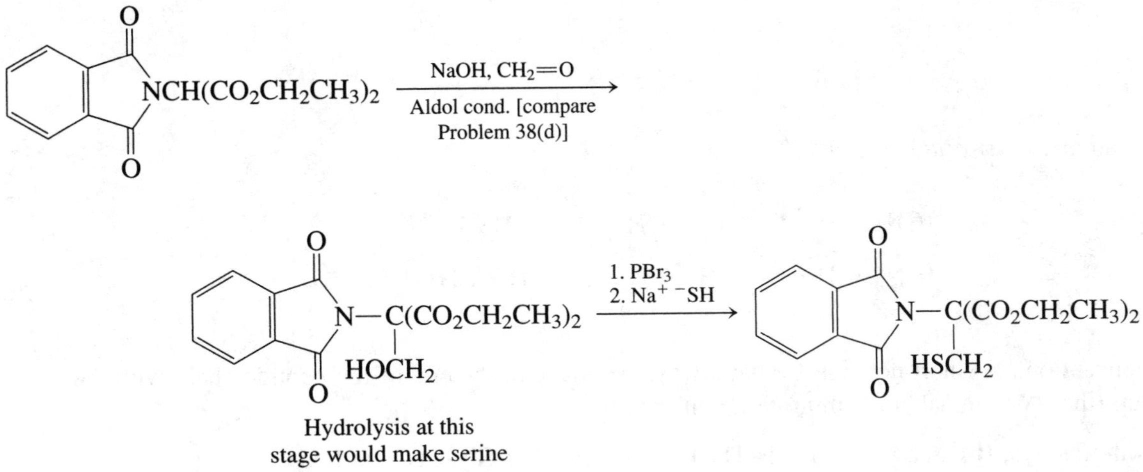

Treatment of this product with hot aqueous acid gives cysteine directly. Otherwise,

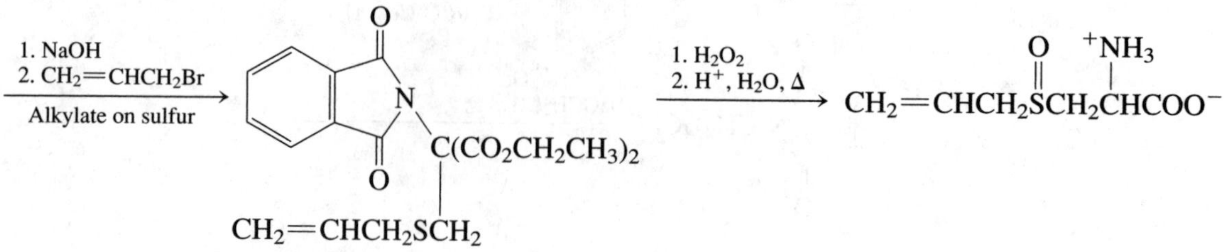

41. The alloisoleucines are diastereomers of the isoleucines, so simple recrystallization will separate them.

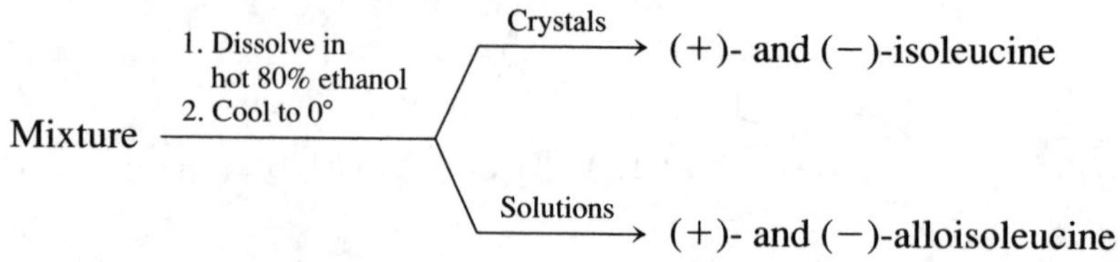

Continue with **each** mixture of enantiomers separately, making use of brucine as a resolving agent.

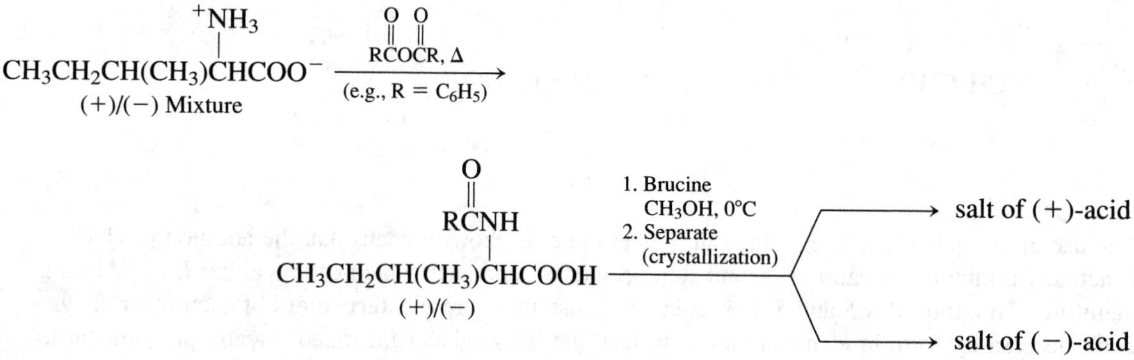

Finally, H^+, H_2O treatment releases each pure amino acid enantiomer in turn.

42. (a) tripeptide; (b) dipeptide; (c) tetrapeptide; (d) pentapeptide. The peptide bonds are simply the amide linkages,

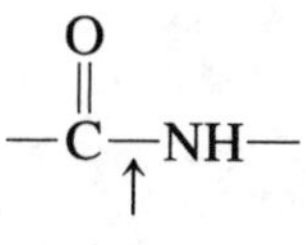

For example, in tripeptide (a):

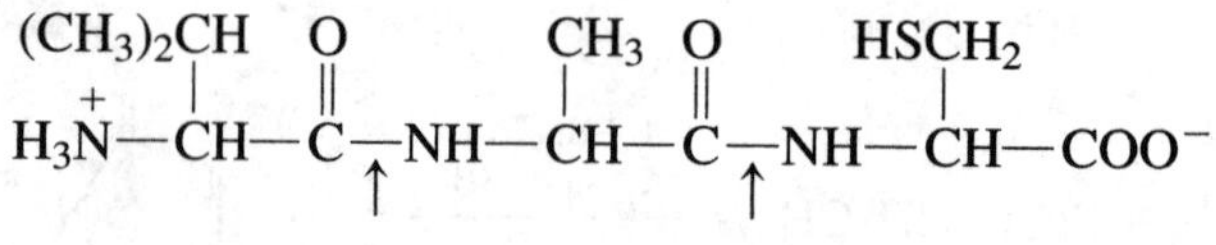

43. By convention, the short notation format always begins with the end of the peptide chain with the amino group (the "*N*-terminal" or "amino-terminal" end).

(a) Val-Ala-Cys; (b) Ser-Asp; (c) His-Thr-Pro-Lys; (d) Tyr-Gly-Gly-Phe-Leu

44. Determine the net charge on the amino acid or peptide at pH 7 and then recall that negative species migrate to the anode (A), positive to the cathode (C), and neutrals do not migrate at all (N).

Amino acids (Problem 32): (a), (b), (d), (e) N, (f) A, (c), (g) C

Peptides (Problem 42): (a) N, (b) A, (c) C, (d) N

45. The side chains are all small (—H, —CH₃, or —CH₂OH), and mostly nonpolar. In the illustrations, especially Figure 26-3, note that the sheet structure packs the R groups into small channels between layers of sheets, where only small groups will fit easily. The nonpolar nature of five of the six groups is also compatible with their location, a relatively nonpolar region with few hydrogen-bonding groups in the vicinity.

46. The α-helical stretches are fairly noticeable by their spiral shape (compare Figure 26-4). Myoglobin in fact contains eight significant α-helical stretches, which are labeled by the letters A–H:

α-Helix	Amino acid numbers	α-Helix	Amino acid numbers
A	3–18	E	58–77
B	20–35	F	86–94
C	36–42	G	100–118
D	51–57	H	125–148

In the figure, all but α-helix D (which is viewed on-end from this perspective) are fairly easy to pick out. The four prolines are located at or near the ends of α-helices and coincide with "kinks" in the overall tertiary structure of the molecule, a result of the conformational characteristics of the five-membered ring.

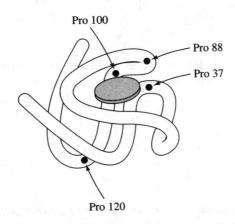

47. Except for the two histidines associated with the heme-bound iron atom, all the polar side chains are well positioned for hydrogen bonding with solvent molecules (water). In contrast, all the nonpolar side chains adopt interior positions, avoiding contact with polar solvent molecules.

48. **(a)** The sheet structure is favored by amino acids with small, nonpolar side chains and has very little ability to hydrogen bond to a polar solvent like water (Problem 45). **(b)** In globular proteins the polar side chains are exposed to the solvent, solubilizing the entire molecule (Problem 47). Similar effects are seen in micelles formed by soap molecules, in which polar groups are located on the surface, facilitating water solubility, whereas nonpolar groups are buried in the inside (Chemical Highlight 19-1). **(c)** If the tertiary structure of a globular protein is disrupted, its nonpolar amino acid side chains become exposed to the polar solvent, greatly reducing the overall solubility of the protein molecule.

49. (1) Purify and cleave the disulfide bridge (Section 9-10). (2) On a portion of the sample, degrade the entire chain by amide hydrolysis (6 N HCl, 110°C, 24 h) to determine amino acid composition by using an amino acid analyzer. (3) On another portion of this material, apply repetitive Edman degradation to determine the sequence of amino acids. Because only nine are present, the entire chain can be sequenced in this way.

50. (a)

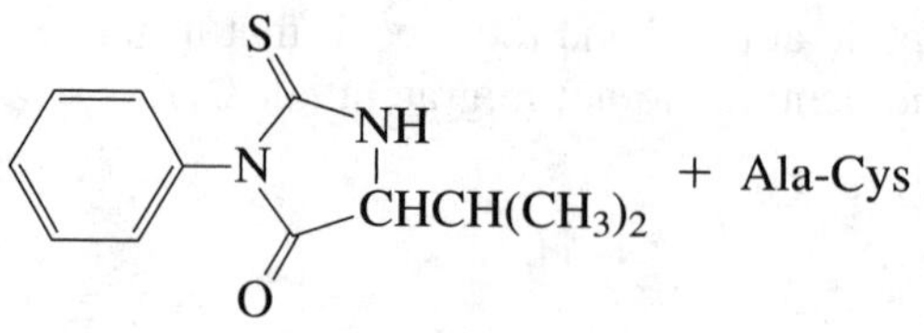

+ Ala-Cys

(b)

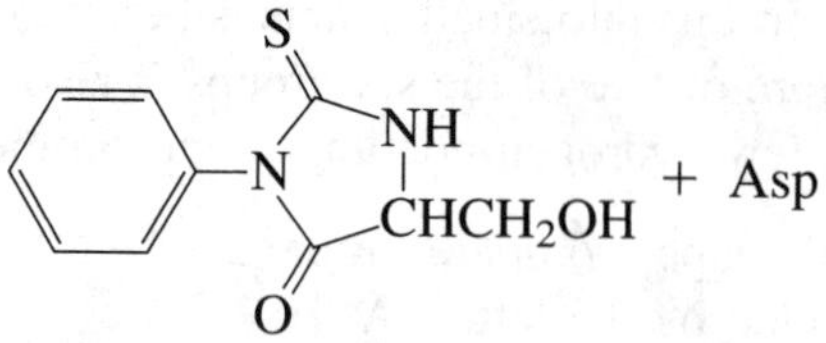

+ Asp

(c)

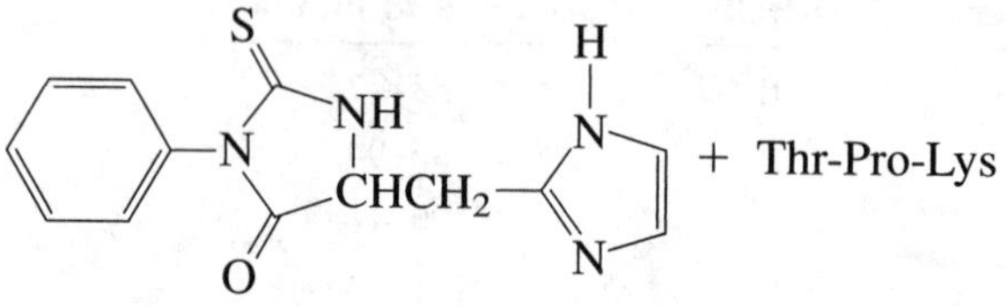

+ Thr-Pro-Lys

(d)

+ Gly-Gly-Phe-Leu

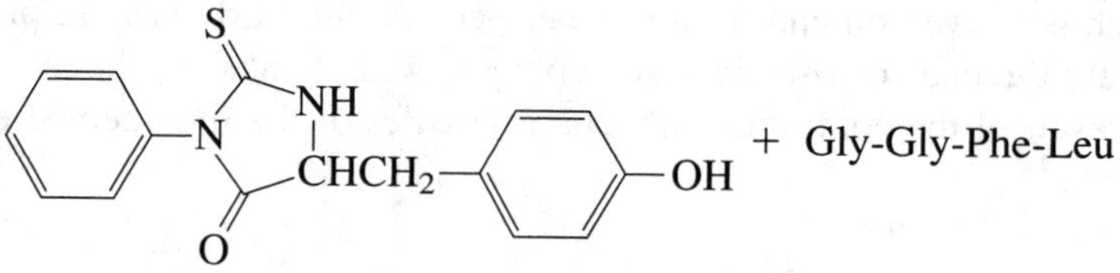

51. Because the peptide is cyclic, the Edman process will not give a normal result. It will simply form thiourea derivatives at the two "extra" Orn amino groups:

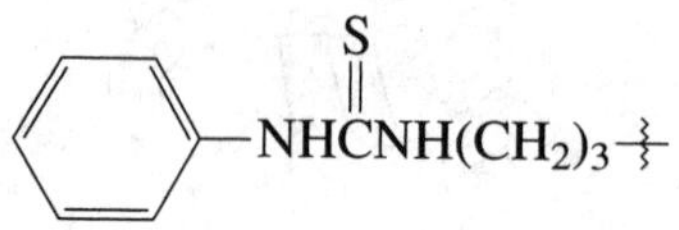

Because there is no α-amino group available to react, mild acid treatment will cleave no bonds in the product, and the cyclic polypeptide structure will remain intact.

52. Complete hydrolysis indicates that a total of nine amino acid units are present. Examine the four fragments of incomplete hydrolysis. You know that the peptide begins with Arg (given), so the fragment Arg-Pro-Pro-Gly must be first. There is only one Gly present, so the last Gly of this tetrapeptide must be the same one that is at the start of the tripeptide fragment Gly-Phe-Ser. You can use the same logic to start overlapping all the pieces to generate the whole solution. Thus, because only one Ser is present, the last Ser in the above tripeptide must be the same as the first one in Ser-Pro-Phe. So far, you have

```
1   2   3   4   5   6   7   8
Arg-Pro-Pro-Gly
            Gly-Phe-Ser
                    Ser-Pro-Phe
```

The final fragment, Phe-Arg, clearly is at the end, overlapping the Phe in position 8. So the answer is Arg-Pro-Pro-Gly-Phe-Ser-Pro-Phe-Arg.

53. Products of degradation, in order of appearance

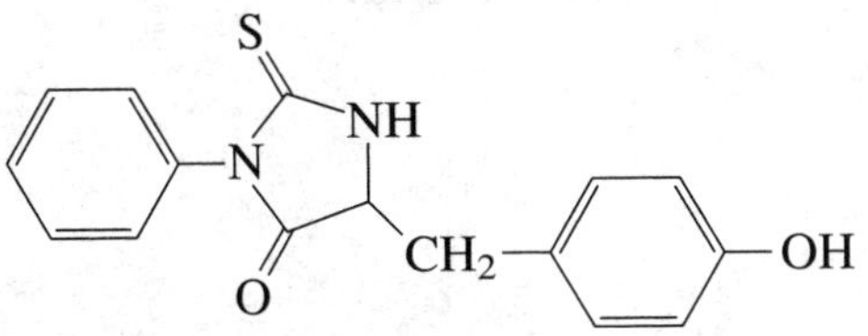

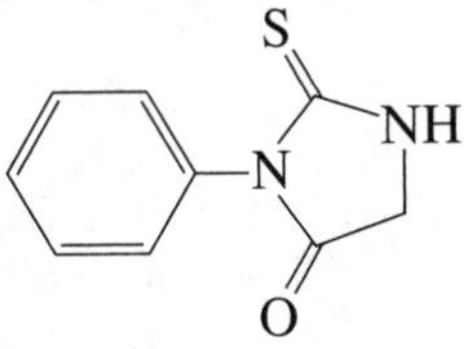

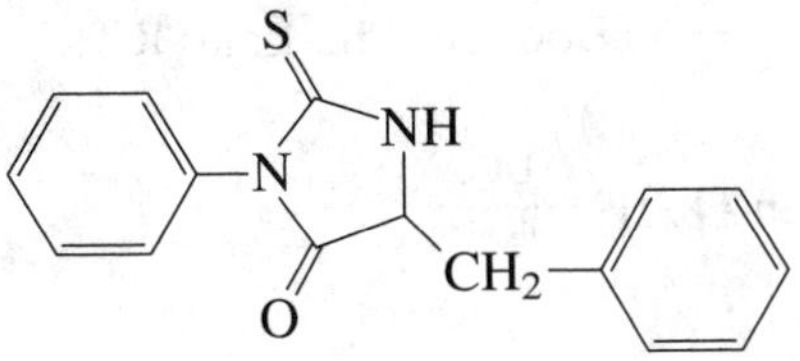

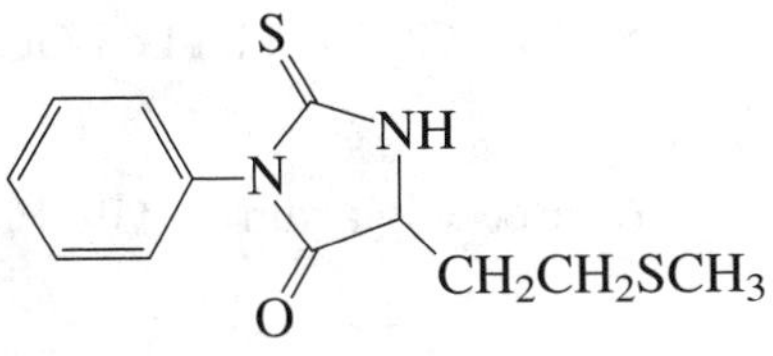

The last product to appear from leu-enkephalin is

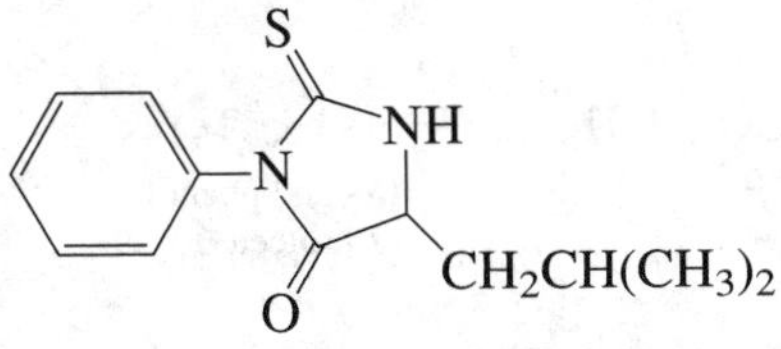

54. Look first for a piece that ends with an amino acid that should *not* be a site of cleavage by one of the enzymes. All the **chymotrypsin** fragments end in Phe, Trp, or Tyr, so that's no help. The **trypsin** results are more useful. It only cleaves after Arg or Lys, so the 18-amino acid fragment ending in Phe must be at the end of the intact hormone. Now it's a matter of matching up all the pieces. Start with this end piece from trypsin hydrolysis and overlap it with chymotrypsin fragments.

(Trypsin piece) Val-Tyr-Pro-Asp-Ala-Gly-Glu-Asp-Gln-Ser-Ala-Glu-Ala-Phe-Pro-Leu-Glu-Phe

(Chymotrypsin pieces) Pro-Asp-Ala-Gly-Glu-Asp-Gln-Ser-Ala-Glu-Ala-Phe-Pro Leu-Glu-Phe

Now identify a chymotrypsin piece to overlap with the Val-Tyr- front end of the trypsin piece and then continue the process, all the way to the *N*-terminal end (the "beginning") of the entire hormone.

Ser-Tyr-Ser-Met-Glu-His-Phe-Arg Trp-Gly-Lys Pro-Val-Gly-Lys   Pro-Val-Lys Val-Tyr-

Ser-Tyr Ser-Met-Glu-His-Phe Arg-Trp Gly-Lys-Pro-Val-Gly-Lys-Lys-Arg-Arg-Pro-Val-Lys-Val-Tyr

The complete answer is read directly, starting at Ser-Tyr- (just above), overlapping Val-Tyr with the large trypsin piece, and then on to the end: Ser-Tyr-Ser-Met-Gln-His-Phe-Arg-Trp-Gly-Lys-Pro-Val-Gly-Lys-Lys-Arg-Arg-Pro-Val-Lys-Val-Tyr-Pro-Asp-Ala-Gly-Glu-Asp-Gln-Ser-Ala-Glu-Ala-Phe-Pro-Leu-Glu-Phe

55. Start at the carboxy-terminal end.

1. Phe + $(CH_3)_3COCOCOC(CH_3)_3$ $\longrightarrow$ Boc-Phe
   (*N*-protected Phe)

2. Leu $\xrightarrow{CH_3OH,\ H^+}$ Leu-OCH_3
   (Methyl ester:
   carboxy-protected Leu)

3. Boc-Phe + Leu-OCH$_3$ $\xrightarrow{\text{DCC}}$ Boc-Phe-Leu-OCH$_3$ $\xrightarrow{\text{Dil. H}^+}$ Phe-Leu-OCH$_3$

4. Gly + (CH$_3$)$_3$COCOCOC(CH$_3$)$_3$ $\longrightarrow$ Boc-Gly
   $\qquad\qquad\qquad$ (*N*-protected Gly)

5. Boc-Gly + Phe-Leu-OCH$_3$ $\xrightarrow{\text{DCC}}$ Boc-Gly-Phe-Leu-OCH$_3$ $\xrightarrow{\text{Dil. H}^+}$ Gly-Phe-Leu-OCH$_3$

6. Boc-Gly again + Gly-Phe-Leu-OCH$_3$ $\xrightarrow{\text{DCC}}$

   Boc-Gly-Gly-Phe-Leu-OCH$_3$ $\xrightarrow{\text{Dil. H}^+}$ Gly-Gly-Phe-Leu-OCH$_3$

7. Tyr + excess (CH$_3$)$_3$COCOCOC(CH$_3$)$_3$ $\longrightarrow$ Boc-Tyr
   $\qquad\qquad\qquad\qquad\qquad\qquad$ (*N*- and phenolic
   $\qquad\qquad\qquad\qquad\qquad\qquad$ *O*-protected Tyr)

8. Boc-Tyr + Gly-Gly-Phe-Leu-OCH$_3$ $\xrightarrow{\text{DCC}}$

   Boc-Tyr-Gly-Gly-Phe-Leu-OCH$_3$ $\xrightarrow[\text{2. HO}^-\text{, H}_2\text{O}]{\text{1. H}^+\text{, H}_2\text{O}}$ Tyr-Gly-Gly-Phe-Leu
   $\qquad\qquad\qquad\qquad\qquad\qquad\qquad\qquad\qquad\qquad\qquad\qquad$ Leu-enkephalin

56. 1. His $\xrightarrow{\text{Cbz-Cl}}$ ring N-protected (Cbz) His $\xrightarrow{(\text{CH}_3)_3\text{COCOCOC(CH}_3)_3}$ amine *N*-protected Boc-(Cbz)His

   The purpose here is to block **both** reactive nitrogens of His differently. The Boc group will be later removed by acid to allow a peptide bond to be formed, while the ring N remains Cbz protected.

2. Pro $\xrightarrow{\text{CH}_3\text{OH, H}^+}$ Pro-OCH$_3$
   $\qquad\qquad$ (C-protected Pro)

3. Boc-(Cbz)His + Pro-OCH$_3$ $\xrightarrow{\text{DCC}}$ Boc-(Cbz)His-Pro-OCH$_3$ $\xrightarrow{\text{Dil. H}^+}$ (Cbz)His-Pro-OCH$_3$

4. Glu $\xrightarrow{135-140°C}$ pyroglutamic acid $\qquad$ Amine group is now an amide, so
   $\qquad\qquad\qquad\qquad\qquad\qquad\qquad\qquad\qquad$ no further protection is necessary.

5. Pyroglutamic acid + (Cbz)His-Pro-OCH$_3$ $\xrightarrow{\text{DCC}}$

   pyroglutamoyl-(Cbz)His-Pro-OCH$_3$ $\xrightarrow[\substack{\text{2. DCC} \\ \text{3. NH}_3}]{\text{1. HO}^-\text{, H}_2\text{O}}$

   pyroglutamoyl-(Cbz)His-Pro-NH$_2$ $\xrightarrow{\text{H}_2\text{, Pd}}$ TRH

57. (a) C:

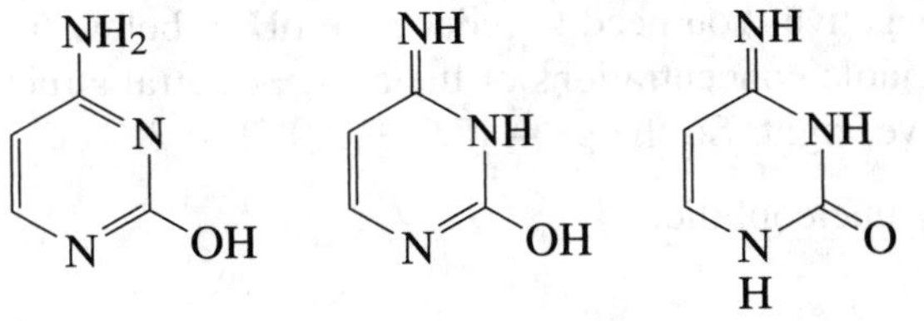

T:

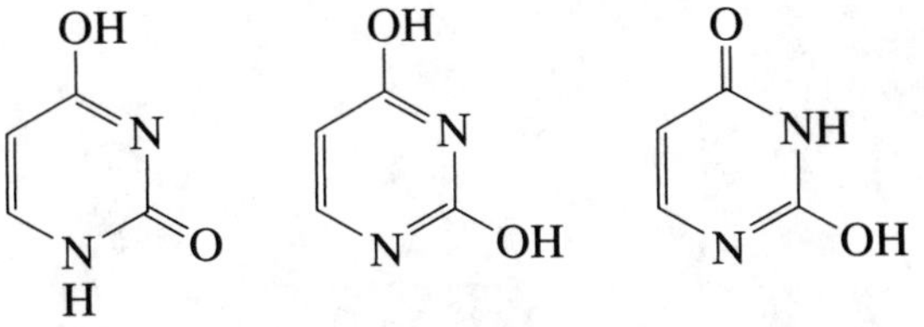

A:

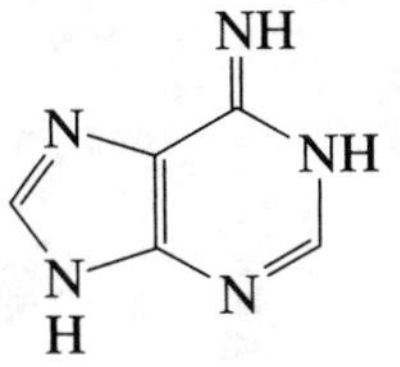

G:

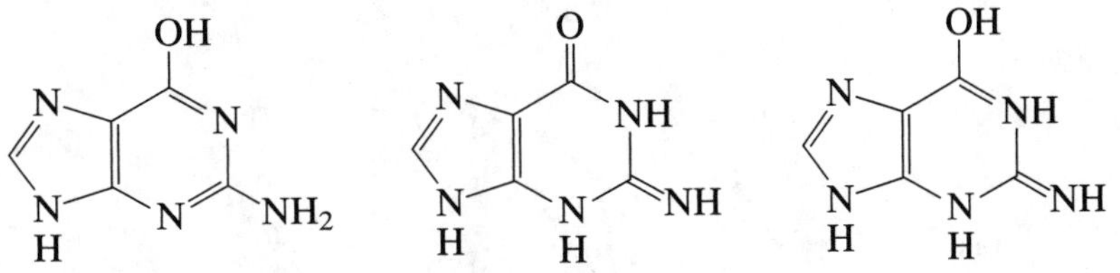

(b)

C:

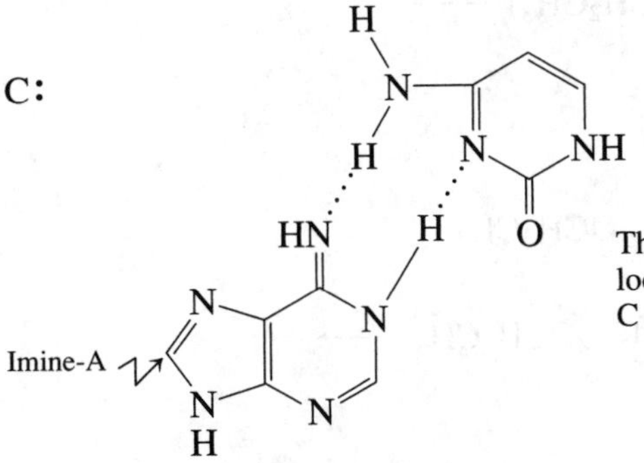

This mispairing makes the A look like a G (it pairs with C instead of a U)

Imine-A

(c) Amino acids: Tyr-Gly-Gly-Phe-Met. Possible codons: AUG-ⓊAC-GGA-GGA-UUU-AUG-UGA. If an A in the DNA strand were to mispair with C instead of U in the synthesis of this mRNA at the circled position, the result would be CAC instead of UAC, which codes for His instead of Tyr. So, the peptide ultimately synthesized would be His-Gly-Gly-Phe-Met.

58. $2332 \times 3 + 3$ (initiation codon) $+ 3$ (termination codon) $= 7002$.

59. (a) The more acidic SeH group will be deprotonated at pH 7:

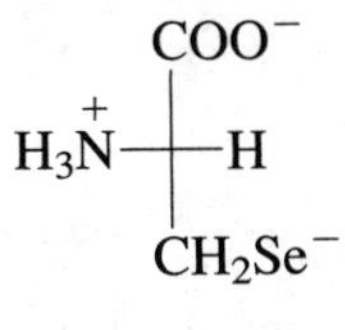

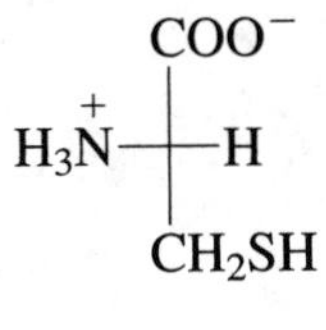

At pH 7 you are above the pK_a for the SeH group (5.2) but below the pK_a for the SH in Cys (8.2).

(b) The pH 7 structure for Sec above is negative. You need to reduce the pH to below 5.2 (the pK_a for the SeH group) to begin to see appreciable concentrations of the charge-neutral structure, which would resemble the Cys structure above, right. So the $pI = (5.2 + 2.0)/2 = 3.6$.

(c) More acidic, but also more powerfully nucleophilic.

60. **(a)**

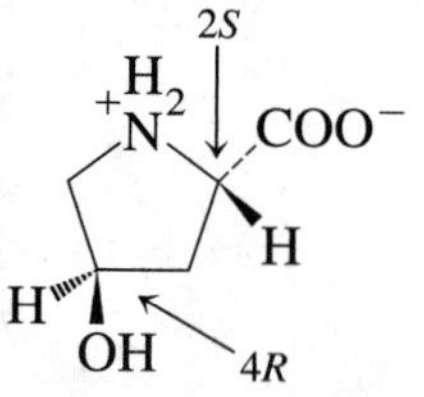

(b) **(i)**

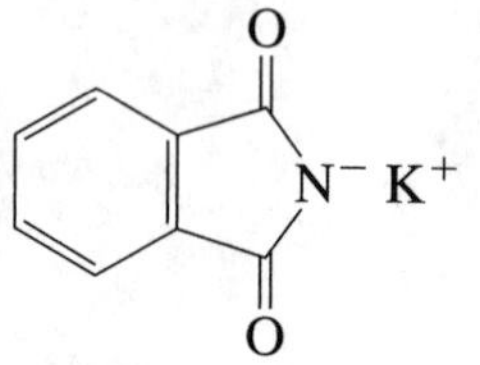

(ii) 1. H^+, H_2O, Δ; 2. HO^-, H_2O, Δ

Mechanisms:

(i)

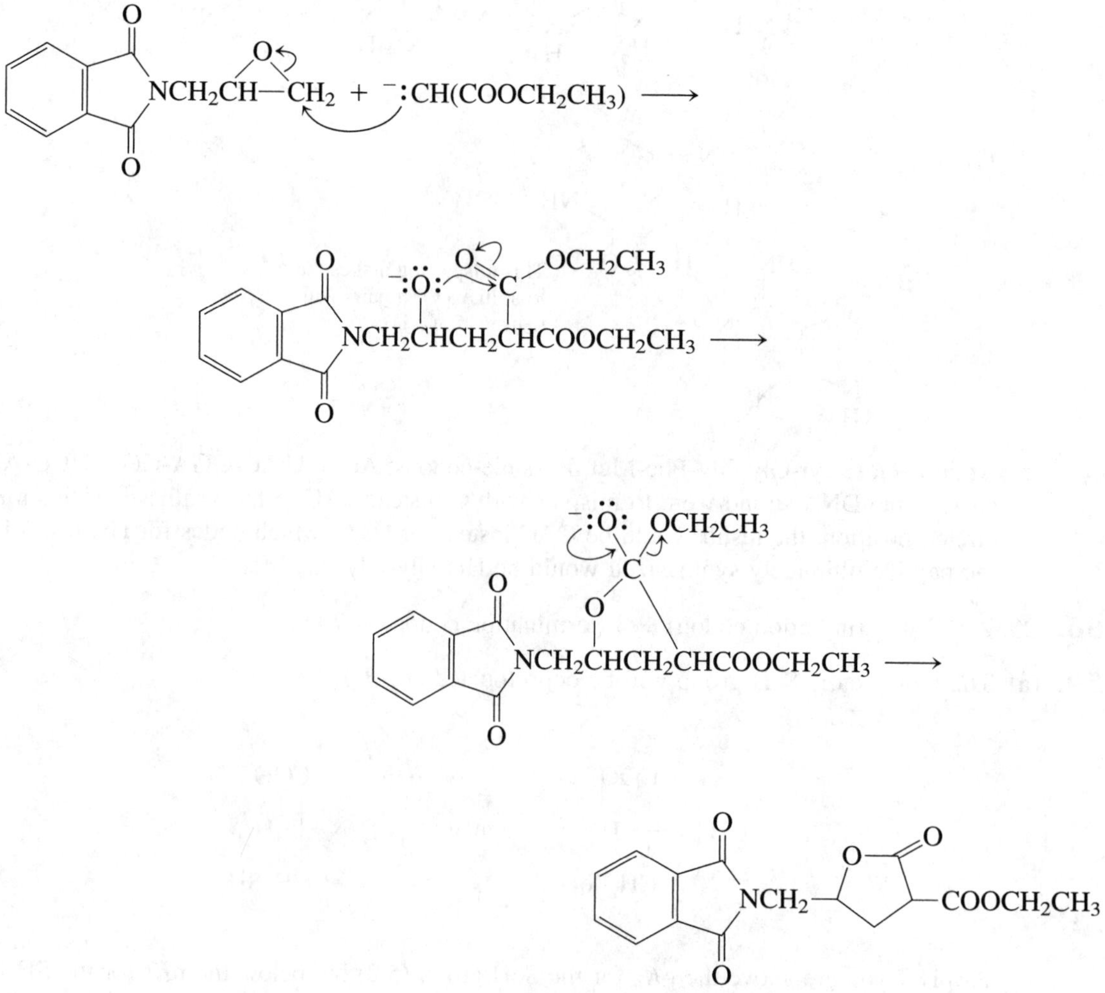

NCH_2CH-CH_2 + $^-$:$CH(COOCH_2CH_3)$ $\longrightarrow$

$NCH_2CHCH_2CHCOOCH_2CH_3$ $\longrightarrow$

$NCH_2CHCH_2CHCOOCH_2CH_3$ $\longrightarrow$

$NCH_2$$-$$COOCH_2CH_3$

(ii)

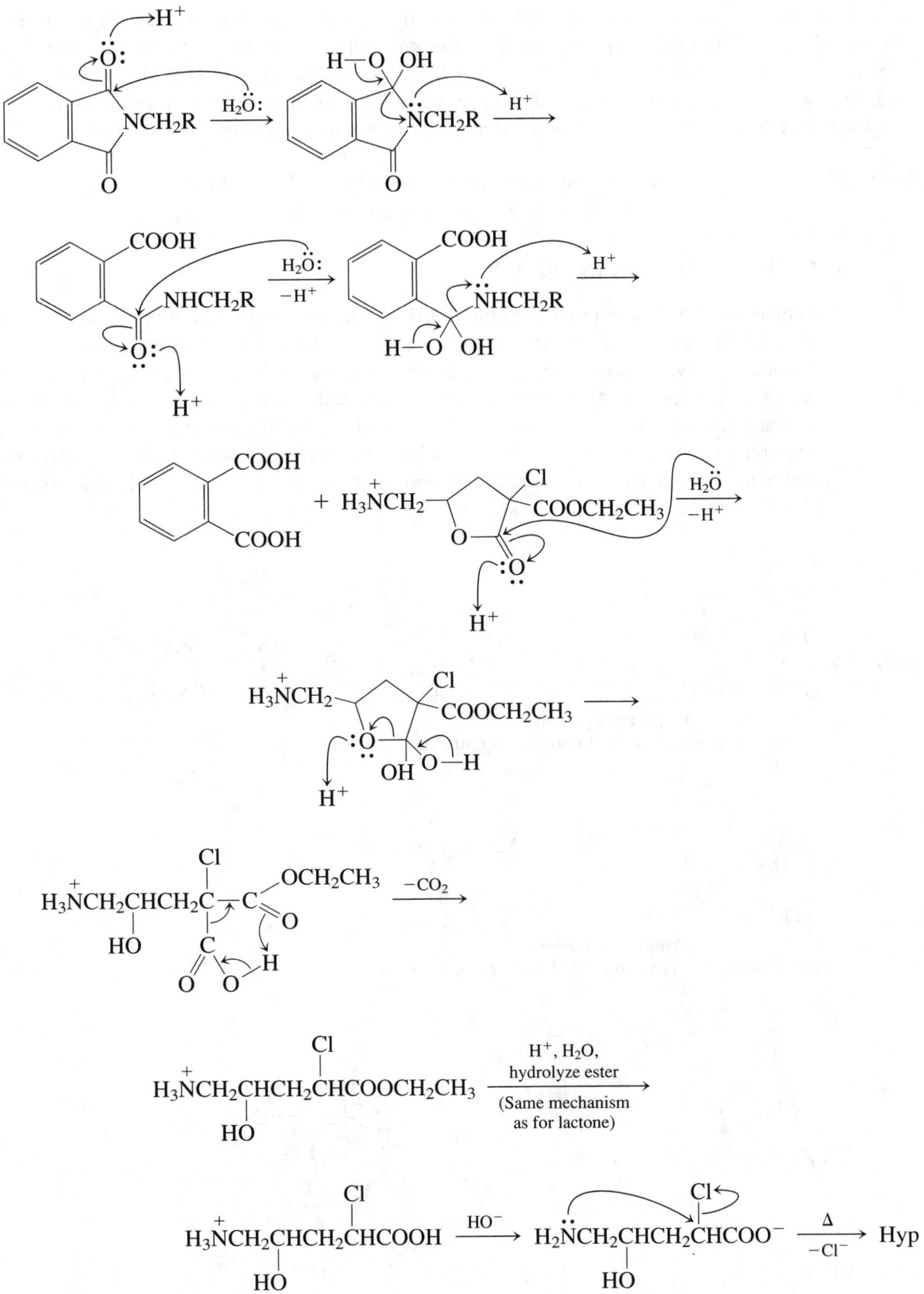

(c) There is no codon for Hyp in the table. Free Hyp is of no use to the body in the synthesis of collagen, because the body has no way of incorporating free Hyp into peptide chains. Gelatin does supply a lot of Pro, so it is useful in that regard, but it does not replace the indispensable need for vitamin C in collagen biosynthesis.

61. The process is a form of nucleophilic displacement. The reaction takes place at C1 of the galactose portion of the starting uridinediphosphate-galactose molecule. C1 is the anomeric carbon, the position where substitutions occur most easily, typically S_N1 under dilute acid conditions (recall glycoside formation and hydrolysis, Section 24-8). In this case the entire uridine diphosphate (UDP) molecule is the leaving group, and the C3 hydroxy group of the protein-bound *N*-acetylgalactosamine is the nucleophile.

62. (a)

AUG-GUG-CAC-CUG-ACU-CCU-GAG-GAG-AAG-etc.

Val- His- Leu- Thr- Pro- Glu- Glu- Lys- etc.

Initiation

(b) GAG → GUG, now codes for Val

(c) Replacement of Glu, a polar (anionic at pH 7) amino acid, by Val, a nonpolar (neutral at pH 7) one, reduces the overall polarity of the molecule, making it less water soluble. Because the nonpolar Val will prefer to reside in the interior of the molecule (and it replaces an amino acid that preferred to stick out into the water), the overall shape of the molecule (i.e., the tertiary structure) is changed. This change is especially disastrous because this substitution occurs very near the beginning of the first long stretch of α-helix in the molecule. The result is a defective hemoglobin that tends to aggregate into insoluble clumps that can block blood vessels and generally reduce the blood's capability to transport oxygen.

63. A:

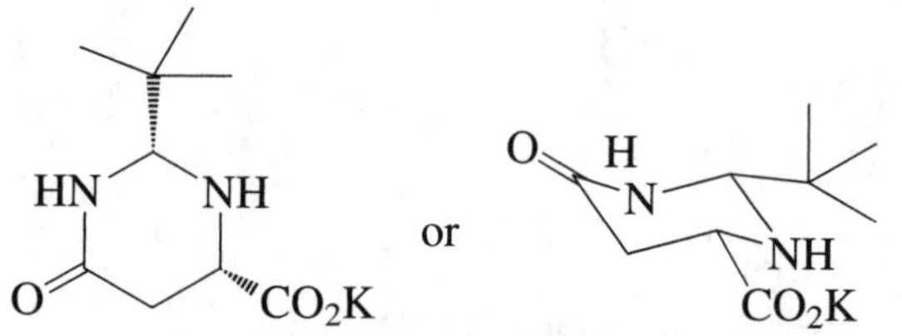

or

Major diastereomer:
Substituents are cis and both are equatorial

B:

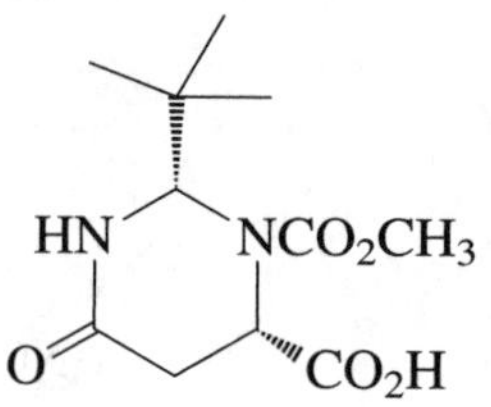

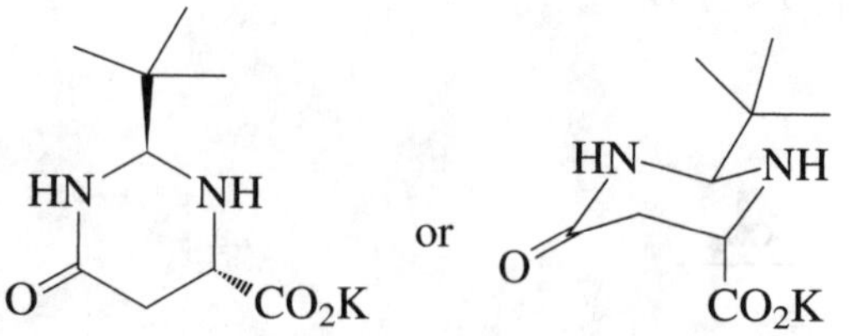

or

Major diastereomer:
Substituents are trans and only tert-butyl is equatorial

C:

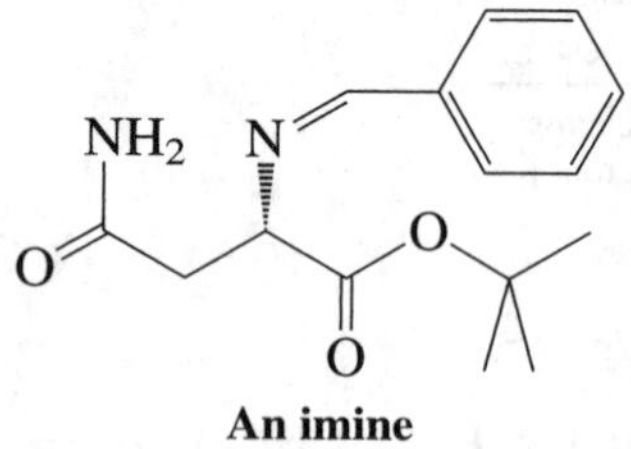

An imine

D:

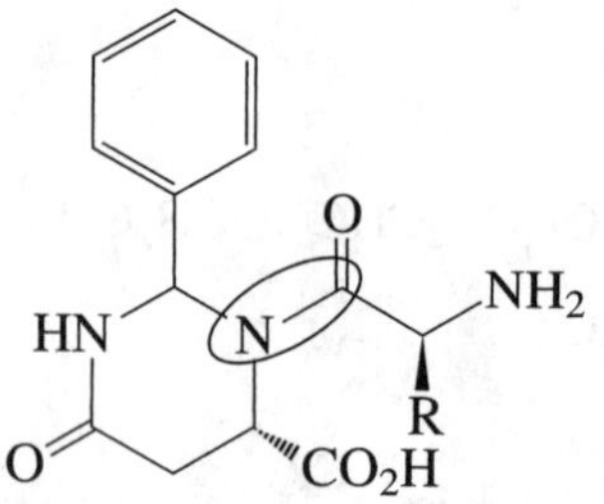

64. (b)

65. (c)

66. (c)

67. (d)

68. (c)

Glossary

Absolute configuration   The arrangement of substituents about a stereoisomer, labeled R or S according to sequence rules.

Abstraction   The removal of an atom.

Acene   A linear arrangement of fused benzenes, such as the three-ring system anthracene.

Acetal $R_2C(OR')_2$, the stable product of acid-catalyzed addition of excess alcohol to an aldehyde or ketone.

Acetaldehyde CH_3CHO, the aldehyde corresponding to acetic acid.

Acetal hydrolysis   Conversion of an acetal into a carbonyl compound and an alcohol in the presence of catalytic acid and excess water.

Acetic acid CH_3COOH, the carboxylic acid familiar as a dilute aqueous solution—vinegar.

Acetoacetic acid synthesis   Synthesis of a 3,3-disubstituted methyl ketone by alkylation of a β-ketoester.

Acetone   The common name of 2-propanone (dimethyl ketone), $(CH_3)_2C{=}O$.

Acetyl group   The group $CH_3CO{-}$.

Acetylene   The common name of ethyne, $HC{\equiv}CH$.

Achiral molecule   A molecule with reflection symmetry, and so superimposable on its mirror image.

Acid dissociation constant (K_a)   A mathematical measure of acid strength, determined from the equilibrium constant in reaction with water.

Activating group   Substituents speeding up a reaction, like electron-donating groups on benzene under electrophilic attack.

Acyl anion equivalent   A reagent bearing a negatively charged carbon atom that can be converted into a carbonyl group.

Acyl group $RCO{-}$, as in the acyl halides, carboxylic acid derivatives.

Acylium cation $RC{\equiv}O{:}^+$, an intermediate in Friedel-Crafts alkanoylation.

Adams's catalyst PtO_2, a catalyst that converts into colloidal platinum in the presence of hydrogen.

Addition   A reaction in which a π bond is broken and new single bonds form.

Addition–elimination   Nucleophilic substitution with a carboxylic acid or acid derivative by means of a tetrahedral intermediate.

Adenine (A)   One of the four bases in both DNA and RNA, a polyfunctional bicyclic nitrogen heterocycle.

Alcohol   An organic molecule whose functional group is the hydroxy group, $-OH$.

Aldaric acid   A dicarboxylic acid, also called saccharic acid, formed by vigorous oxidation of sugars with nitric acid.

Aldehyde   An organic molecule in which the carbonyl group, $C{=}O$, is bound to one hydrogen and one carbon atom.

Alditol   A polyhydroxy compound produced by sugar reduction.

Aldol condensation   Conversion of aldehydes to α, β-unsaturated aldehydes through an enolate by catalytic base and heating.

Aldonic acids   Carboxylic acid products of Fehling's and Tollen's tests or bromine oxidation on sugars.

Aldose   A sugar based on aldehydes—so that a three-carbon chain would be an aldotriose, for example.

Aliphatic compound   A nonaromatic compound.

Alkaloid   A physiologically active polycyclic compound containing nitrogen, often in a heterocycle—such as cocaine.

Alkanal   The systematic name for an aldehyde, such as methanal for formaldehyde.

Alkanamine   The *Chemical Abstracts* name for an aliphatic amine, such as methanamine (methylamine, aminomethane).

Alkane   A hydrocarbon containing only single bonds, and so an organic molecule lacking a functional group.

Alkanoate   An organic ester.

Alkanoic acid   The carboxylic acid derived from an alkane, such as butanoic acid (butyric acid).

Alkanol   The alcohol derived from an alkane.

Alkanone   The systematic name for a ketone, such as butanone.

Alkene   An organic molecule whose functional group is a carbon–carbon double bond.

Alkenol   An enol, or alcohol containing a carbon–carbon double bond.

Alkenyl   An alkene as a substituent—as in the alkenyl halides (haloalkenes), useful intermediates in synthesis.

Alkenyne   A hydrocarbon containing both double and triple bonds.

Alkoxide ion RO^-, the ion derived from deprotonation of an alcohol.

Alkoxy group   —OR, the functional group of the ethers.

Alkoxycarbonyl   An ester as a substituent, —COOR.

Alkyl   An alkane fragment as a substituent.

Alkyl azide $R—N_3$, the product of attack by an azide ion on a haloalkane.

Alkylene   The common name for an alkene, such as ethylene (ethene).

Alkylthio group   The RS group, called alkanethiolate as an anion.

Alkyne   An organic molecule whose functional group is a carbon–carbon triple bond.

Alkynol   An alkyne incorporating a hydroxy group.

Alkynyl   An alkyne as a substituent.

Allene   A diene in which the π bonds share a single carbon atom.

Allyl   The common name of 2-propenyl, a resonance-stabilized three-carbon intermediate.

Ambident species   A species subject to attack at two different sites to give two different products.

Amidate ion RCH_2CONH^-, the anion formed by removal of the NH hydrogen from an amide.

Amide   An organic molecule whose functional group is $-\overset{\overset{\text{O}}{\|}}{\text{C}}-\overset{|}{\text{N}}-$, singly bonded nitrogen in conjunction with a carbonyl group.

Amine   An organic molecule whose functional group is a singly bonded nitrogen atom.

Amino acid   A carboxylic acid with an amine group—at C2 in the $2S$-(α-) form used as building blocks in nature.

Amphoteric   Both acidic and basic.

Amylopectin   A branched polysaccharide component of starch.

Amylose   A helical polysaccharide component of starch.

Anhydride   An organic molecule in which two carbonyl groups are singly bonded to an oxygen atom.

Aniline   The common name of benzenamine, $C_6H_5—NH_2$.

[N]Annulene   A conjugated monocyclic hydrocarbon of N atoms, with aromatic character governed by Hückel's rule.

Anomers   Sugar hemiacetals, diastereomers with a new stereocenter called the anomeric carbon.

Anthracene   A linear arrangement of three fused benzene rings.

Anti conformation   The arrangement such that groups on adjacent carbon atoms are far apart—in contrast to *gauche*.

Anti dihydroxylation   An oxidation–hydrolysis sequence in which a diol forms from an alkene by attachment of two —OH groups to the opposite faces of the π bond.

Aprotic solvent   A solvent not containing positively polarized hydrogens.

Arene   A substituted benzene.

Arenediazonium salt   A derivative of the resonance-stabilized benzenediazonium cation, $C_6H_5N_2^+$.

Aromatic compound   (1) Benzene or a derivative—represented by a structure with three double bonds in a six-membered carbon ring; or (2) Any compound with a cyclic array of π-overlapping p orbitals containing $4n + 2$ π electrons.

Aromaticity   Resonance energy, the stabilizing energy difference between benzene and a nonaromatic cyclic triene.

Aryl (Ar)   An arene as a substituent, beginning with the phenyl group.

Asymmetric carbon   A stereocenter, or atom connected to four different substituents.

Asymmetric hydrogenation   A highly selective addition of hydrogen to alkenes used in the synthesis of amino acids.

Atomic orbital   In quantum mechanics, a region associated with the electrons surrounding an atom.

Aufbau principle   The assignment of electrons to atomic orbitals.

Axial bond   A bond parallel to the principal molecular axis of, usually, a cycloalkane ring.

Azacycloalkane   A saturated nitrogen heterocycle, such as azacyclohexane (piperidine), $C_5H_{11}N$.

Azide ion N_3^-, a nucleophile useful in introducing an amino group.

Azo group $-N=N-$, the functional group of several brilliant dyes, the azobenzenes.

Backside displacement   A concerted mechanism in which the nucleophile approaches from the side opposite the leaving group (S_N2 displacement, for example).

Baeyer-Villiger oxidation   Addition of peroxycarboxylic acid to a ketone followed by rearrangement to an ester.

Base   (1) The conjugate base of an acid; or (2) one of the four heterocyclic subunits of a nucleic acid.

Benzene   The aromatic compound C_6H_6, a remarkably stable conjugated cyclic system.

Benzoic acid   An aromatic carboxylic acid, C_6H_5—COOH.

Benzyl $C_6H_5CH_2-$, also called the phenylmethyl group—with a corresponding resonance-stabilized benzylic radical.

Benzylic resonance   Resonance stabilization of the phenylmethyl (benzyl) radical, $C_6H_5CH_2 \cdot$ (also cation and anion).

Benzyne   1,2-dehydrobenzene, a highly reactive intermediate, C_6H_4, with a bent triple bond.

Bimolecular reaction   A process, usually second-order, in which two molecules participate in the transition state.

Boat conformation   An unstable cyclohexane structure, with carbons 1 and 4 tilted out of the plane in the same direction.

Boc   The 1,1-dimethylethoxycarbonyl group, a protecting group in peptide synthesis.

Bond dissociation energy ($DH°$)   The energy required to break a chemical bond through homolytic cleavage, or the energy released in forming that bond.

Bond length   The separation between nuclei for maximum chemical bonding, in that a change would release no further energy.

Bond-angle strain   Relative instability of a molecule due to distortion of (for example) the tetrahedral carbon bond angle away from $109.5°$.

Bond-line notation   Molecular representation using straight lines for the main carbon chain and omitting all hydrogens.

Branched alkane   An alkane in which one or more methylene hydrogens have been replaced by an alkyl group.

Bridgehead carbon   A carbon atom shared by two rings.

Buckminsterfullerene   "Buckyball," or C_{60}, the most prevalent closed-shell allotrope of carbon.

Carbamic acid   RNH—COOH, an amide derivative of carbonic acid.

Carbamic ester   RNH—COOR′, an amide derivative of carbonic acid with common name urethane.

Carbanion   Negatively charged carbon.

Carbene $R_2C:$, a highly reactive species such as methylene ($H_2C:$).

Carbenoid   A reactive substance that, like a carbene, converts alkenes to cyclopropanes stereospecifically.

Carbocation   Positively charged carbon.

Carbocyclic compound   A ring of only carbon atoms.

Carbohydrate   A saccharide, which may be a sugar, starch, or cellulose.

α-Carbon   The carbon adjacent to a carbonyl group.

^{13}C NMR spectrometry   NMR spectrometry used to reveal molecular structure by observation of nuclei of the isotope carbon-13.

Carbonyl group   C=O, the functional group characteristic of aldehydes, ketones, and, together with an OH group, carboxylic acids.

Carboxamide   The amide derivative of carboxylic acids.

Carboxylate   An organic ester, or a salt of a carboxylic acid (see below).

Carboxylate salt   The salt of a carboxylic acid, such as sodium formate, $HCOO^-Na^+$.

Carboxylic acid   An organic molecule whose functional unit is —COOH, the carbonyl group in conjunction with a hydroxy group.

Catalytic hydrogenation   The addition of hydrogen to a double bond on the surface of a heterogenous catalyst.

Cbz   The phenylmethoxycarbonyl group, a protecting group in peptide synthesis.

Cellulose   A polysaccharide of glucose linked at C4.

Chair conformation   The most strain-free cyclohexane structure, with carbons 1 and 4 tilted out of the plane in opposite directions.

Chemical shift   The position of an NMR signal, related to molecular structure near the nucleus observed.

Chichibabin reaction   Nucleophilic substitution of an amine group in pyridine, using sodium amide in liquid ammonia.

Chiral molecule   A compound that lacks reflection symmetry, and so exists as nonsuperimposable enantiomers.

Chromic ester   The chromium analog to an organic ester and an intermediate in the Cr(VI) oxidation of alcohols.

Cis isomer   A stereoisomer in which substituents are on the same face, or side, of a ring or a double bond—in contrast to a trans isomer.

Claisen condensation   Addition–elimination of ester enolates with esters, giving β-dicarbonyl compounds.

Claisen rearrangement   An electrocyclic reaction of 2-propenyloxybenzene, giving 2-(2-propenyl) phenol.

Clemmensen reduction   Synthesis of an alkylbenzene by conversion of a carbonyl group to CH_2 in the presence of HCl and zinc.

Coal tar   The solid distillate of coal (leaving a residue called coke), a source of many aromatic compounds.

Codon   The three-nucleotide sequence in RNA that selects an amino acid in protein biosynthesis.

Common name   A traditional name, often derived from the molecule's source or discoverer—in contrast with systematic nomenclature.

Competition experiment   An experiment comparing the relative reactivity of two substrates.

Complex sugar   An oligomer or polymer of linked simple sugars.

Concerted reaction   A reaction in which bonds break and new bonds form "in concert," or simultaneously.

Condensation   The joining of two molecules with elimination of water.

Condensed formula   Molecular representation omitting all lone pairs and the bonds of the main carbon chain.

Conformational analysis   Study of the thermodynamic and kinetic behavior of conformations or conformers.

Conformations   Molecular structures differing only by rotation about carbon–carbon single bonds, also called conformers or rotamers.

Conjugate acid   The species HB^+ derived from a base B.

Conjugate addition   1,4-Addition to an α, β-unsaturated carbonyl compound.

Conjugate base   The species A^- derived from an acid HA.

Conjugated diene   A structure with two double bonds "in conjugation," or linked across a carbon–carbon single bond.

Conrotatory process   A reaction, such as a ring opening, in which carbon atoms rotate in the same direction.

Constitutional isomers   Molecules with the same molecular formula but with the atoms connected in a different order. Synonymous with structural isomers.

Coordination   The supplying of an electron pair to an atom by a Lewis base.

Cope rearrangement   A concerted reaction of compounds containing 1,5-diene units.

Coulomb's law   The mathematical law describing the force between electric charges, a main contribution to chemical bonding.

Coupling   The splitting of an NMR spectral peak by neighbors into more complex patterns.

Covalent bond   A bond formed by the sharing of electrons between atoms—in contrast to an ionic bond.

Cracking   The breaking of alkanes into smaller fragments, as in refining crude oil.

Cross-linkage   The connection of two or more polymer chains; gives rubber its familiar hardness and elasticity.

Crossed aldol condensation   An aldol condensation of two different aldehydes or ketones.

Cyano group   —CN, the functional group of the nitriles.

Cyanohydrin   NC—RR′COH, the product of hydrogen cyanide addition to the carbonyl group.

Cyclic bromonium ion   An intermediate in addition reactions, in which bromine bridges both carbons of a double bond.

Cycloaddition   An addition reaction, such as the Diels-Alder reaction, leading to a closed ring (the "cycloadduct").

Cycloalkane   An alkane whose carbons form a ring, also called a cyclic alkane or carbocycle.

Cycloalkanecarboxylic acid   A saturated cyclic carboxylic acid.

Cycloalkanone   The systematic name for a cyclic ketone.

Cytosine (C)   One of the four bases in both DNA and RNA, a polyfunctional nitrogen heterocycle.

D, L sugars R, S configurations based on the highest-numbered stereocenter, with all natural sugars D.

Dashed-wedge line notation   Molecular representation using dashed lines for bonds angled below the page, wedges for bonds angled above it.

DCC   Dicyclohexylcarbodiimide, a dehydrating agent in peptide synthesis.

Deactivating group   Substituents allowing only sluggish reaction, such as electron-withdrawing groups on benzene undergoing electrophilic attack.

Decoupling   Techniques that remove NMR coupling, such as proton decoupling and fast proton exchange.

Degenerate orbital   Orbitals of equal energy, such as the three p orbitals, p_x, p_y, and p_z.

Degree of unsaturation   The sum of the number of rings and π bonds in a molecule, determined from the molecular formula.

Dehydration   Elimination of a molecule of water.

Delocalization   The distribution of electrons over several nuclei, as in a resonance hybrid.

Deoxyribonucleic acid (DNA)   The dimeric helix of nucleotide chains that carries hereditary information—the genetic code.

Deshielding   An increase in the local magnetic field or removal of electrons from a nucleus, causing a downfield NMR chemical shift.

Dextrorotatory molecule   An enantiomer that rotates the plane of polarization clockwise as the viewer faces the light.

Diamine   A compound with two amine groups, such as 1,4-butanediamine (putrescine).

Diastereomers   Stereoisomers that are not mirror images of one another.

1,3-Diaxial interaction   Transannular strain due to steric crowding of axial substituents across a ring.

Diazo compound   A diazoalkane, $R_2C{=}N_2$, such as diazomethane (CH_2N_2), a useful synthetic intermediate.

Diazo coupling   Electrophilic substitution with arenediazonium salts, giving azo dyes.

Diazonium ion $R{-}N_2^+$, a highly reactive product of the reaction of a primary amine with nitrous acid.

Diazotization   Synthesis of arenediazonium salts from benzenamines (anilines) using cold nitrous acid.

β-Dicarbonyl compound   A useful intermediate bearing two carbonyl groups adjacent to the same carbon.

Dieckmann condensation   Intramolecular Claisen condensation, producing cyclic 3-keto esters.

Diels-Alder reaction   Concerted [4 + 2] cycloaddition of dienes, with four π electrons, and alkenes, with two, to form cyclohexenes.

Diene   A compound containing two double bonds, including conjugated and nonconjugated dienes as well as allenes.

Difunctional molecule   A compound with two functional groups, subject to the characteristic reactivity of both.

Dimer   Linkage of two identical subunits, often the first stage in polymerization.

Dioic acid   A dicarboxylic acid, such as propanedioic acid (malonic acid).

Disaccharide   A dimer of simple sugars, such as glucose.

Disrotatory process   A reaction, such as a ring closure, in which one carbon atom rotates clockwise, one counterclockwise.

Dissociation   The breaking of chemical bonds.

Disulfide   A molecule containing a sulfur–sulfur linkage.

DMSO   Dimethylsulfoxide, a useful polar aprotic solvent.

Double Claisen condensation   Intermolecular followed by intramolecular Claisen condensation, giving cyclic products.

E, Z system   Naming rules for diastereomers with double bonds, assigning priority separately to substituents on each carbon.

Eclipsed conformation   An arrangement such that, in a view along the carbon axis, each hydrogen or substituent group is aligned with one on the next carbon.

Edman degradation   Peptide sequencing by stepwise hydrolysis of *N*-terminal amino acids.

Electric dipole   A separation of charges giving each end of a neutral species a partial positive or negative charge.

Electrocyclic reaction   A ring closure that involves cyclic movement of electrons.

Electron affinity (EA)   The energy released when an electron attaches itself to an atom.

Electronegativity   The ability to accept electrons, highest in the rightmost elements of the periodic table.

Electrophile   A species (coded *blue* in this text) that tends to react with centers having unshared electron pairs.

Electrophilic addition   Addition to a double bond through attack by an electrophile followed by trapping of the resulting carbocation by a nucleophile.

Electrophilic aromatic substitution   Electrophilic attack on benzene in which hydrogen is replaced by the electrophilic group.

Elemental analysis   Determination of a chemical's empirical formula, and so the kinds and ratios of elements present.

Elimination (E)   A unimolecular (E1) or bimolecular (E2) reaction in which atoms, such as a halide and a hydrogen, are eliminated and π bonds form in their place.

Enamine   A product of amine–carbonyl condensation having both a carbon–carbon double bond and an amino group.

Enantiomers   Pairs of molecules that are nonsuperimposable mirror images.

Endo adduct   A bicyclic product with substituents placed trans with respect to the shorter bridge—in contrast to an exo adduct.

Endo rule   Selectivity favoring the endo adduct in the Diels-Alder reaction.

Endothermic reaction   A reaction absorbing heat, giving a positive enthalpy change.

Enol   An alcohol containing a double bond.

Enolate   A resonance-stabilized anion formed upon removal of the α-hydrogen from a carbonyl group.

Enone   An α, β-unsaturated carbonyl group.

Enthalpy change ($\Delta H°$)   The heat of a reaction at constant pressure—a negative value indicating that energy is released.

Entropy change ($\Delta S°$)   The change of a system toward greater dispersal of energy—a positive value favoring the reaction's progress.

Enzyme   A catalyst found in living systems.

Epimers   Diastereomers differing only at one stereocenter.

Epoxide   Common name for oxacyclopropane.

Equatorial bond   A bond approximately perpendicular to the principal molecular axis.

Equilibrium   The state of a reaction at which the concentrations of reactants and products no longer change.

Ester   In organic chemistry, a molecule whose functional group is $-\overset{\displaystyle O}{\overset{\|}{C}}-O-$, oxygen singly bonded to a carbonyl group.

Ester hydrolysis   The reverse of esterification, leading to a carboxylic acid and an alcohol in the presence of excess water.

Esterification   The production of esters and water from alcohols and carboxylic acids in the presence of a mineral acid.

Ether   An organic molecule whose functional group is the alkoxy group, $-OR$.

Exact mass   The precise mass of the peaks of mass spectra.

Excitation   Discrete changes in the state of a molecule when it absorbs radiation.

Exo adduct   A bicyclic product with substituents placed cis with respect to the shorter bridge—in contrast to an endo adduct.

Exothermic reaction   A reaction releasing heat, giving a negative enthalpy change.

Extended π system   A molecule with more than two conjugated double bonds, such as benzene.

Fat   An ester of long-chain carboxylic acids that is solid at room temperature—in contrast to liquid oils.

Fatty acid   The carboxylic acid parts of fats and oils.

Fehling's test   A chemical test for aldehydes by oxidation to a carboxylic acid and formation of cuprous oxide, a red precipitate.

Fingerprint region   The infrared spectrum between 600 and 1500 cm^{-1}.

Fischer projection   Representation of a configuration with horizontal lines for bonds pointed toward the viewer, vertical lines for bonds pointed away.

Fischer-Tropsch reaction   Catalytic synthesis of hydrocarbons from synthesis gas.

Formaldehyde $H_2C=O$, the parent aldehyde—the aldehyde corresponding to formic acid.

Formalin   An aqueous solution of formaldehyde, a common disinfectant.

Formic acid   HCOOH, the parent carboxylic acid.

Formyl group   The group HCO—.

Fragmentation pattern   A mass spectrum showing the abundance of the characteristic fragments of a molecular ion.

Friedel-Crafts reactions   Alkylation and alkanoylation (acylation) of benzene—electrophilic aromatic substitutions.

Frontside displacement   A (hypothetical) mechanism in which a nucleophile approaches from the same side as the leaving group.

Fructose   A ketohexose found in many fruits and honey.

Fullerene   A closed carbon shell, such as "buckyball," C_{60}.

Functional group   A group of atoms, also called a function, that controls an organic molecule's reactivity.

Functionalization   The introduction of a functional group, such as doubly or triply bonded carbon or other elements, into a molecule.

Furan C_4H_4O, an aromatic heterocyclopentadiene.

Furanose   A five-membered cyclic monosaccharide.

Fused ring   A structure in which two rings share two adjacent carbon atoms.

Gabriel synthesis   Synthesis of a primary amine using 1,2-benzenedicarboxylic imide (phthalimide).

Gauche conformation   A staggered arrangement such that groups on adjacent carbon atoms are close—in contrast to _anti_.

Geminal coupling   Coupling between nonequivalent hydrogens on the same carbon atom, such as terminal hydrogens at a double bond.

Geminal diol $RC(OH)_2R'$, a carbonyl hydrate.

Gibbs standard free energy change ($\Delta G°$)   The mathematical function of temperature that governs a reaction equilibrium.

Glucose   An aldohexose, also called dextrose or grape sugar.

Glutathione   A peptide that acts as an intracellular reducing agent.

Glycogen   A huge branched polysaccharide of glucose, a store of energy in humans and animals.

Glycoside   A sugar acetal, linking monosaccharides.

Glycosyl   Sugar as a substituent bound at its anomeric carbon.

Grignard reagent   RMgX, an organomagnesium compound useful in synthesis.

Guanadine $NH{=}C(NH_2)NH$—as a substituent, the guanadino group in amino acids like arginine.

Guanine (G)   One of the four bases in both DNA and RNA, a polyfunctional bicyclic nitrogen heterocycle.

Haloalkane   An organic molecule whose functional group is a carbon–halogen bond.

Halogenation–dehydrohalogenation   Synthesis of an alkyne through an alkenyl halide intermediate.

Hantzsch pyridine synthesis   Condensation of two β-dicarbonyl molecules, an aldehyde, and ammonia to an aromatic ring.

Haworth projection   Representation of D-sugars with the anomeric carbon at the right and substituents attached by vertical lines.

Heat of combustion (ΔH°_{comb})   The heat released in combustion, or the burning of a molecule, a useful measure of its stability.

Heck reaction   Coupling reaction between alkenyl halides and alkenes to produce dienes, catalyzed by metals such as Ni or Pd.

Hell-Vollhard-Zelinsky reaction   Bromination of a carboxylic acid at the α-carbon catalyzed by trace amounts of phosphorus.

Heme group   The oxygen-carrying porphyrin substituent in the natural polypeptides myoglobin and hemoglobin.

Hemiacetal   HRC(OH)OR, the product of addition of an alcohol to aldehydes or ketones.

Hemiaminal $HRC(NH_2)OH$, the nitrogen analog of a hemiacetal.

Heteroatom   An element other than carbon, most frequently nitrogen or oxygen.

Heterocycle   A cyclic molecule whose ring contains a heteroatom, or element other than carbon.

Heterogeneous catalyst   A catalyst insoluble in the reaction solvent.

Heterolytic cleavage   Dissociation in which the bonding electron pair is donated to one atom only—in contrast to homolytic cleavage.

Hofmann elimination   Alkene synthesis from quaternary ammonium salts, used to determine the structure of amines.

Hofmann rearrangement   Loss of the carbonyl group from a carboxamide by halogenation in the presence of base.

Hofmann rule   Regioselectivity whereby E2 elimination by a hindered base leads to the less stable terminal alkenes.

HOMO-LUMO   The gap between the highest occupied and lowest unoccupied molecular orbital, involved in electronic excitation.

Homologous series   The set of homologs, or molecules differing only by one or more methylene groups.

Homolytic cleavage   Dissociation in which the two bonding electrons are split between the resulting radicals—in contrast to heterolytic cleavage.

Hückel's rule   The rule that an aromatic annulene must contain $4n + 2$ π electrons.

Hund's rule   The sequence by which electrons fill degenerate orbitals, beginning with one electron of the same spin in each orbital.

Hybrid orbital   An orbital formed from the mixing of atomic orbitals on a single atom.

Hydration   Addition of the components of water to a double bond.

Hydrazine $H_2N{-}NH_2$.

Hydrazone $RR'C{=}N{-}NH_2$, a crystalline imine product used in identifying aldehydes and ketones.

Hydride ion $[H{:}]^-$, an anion of hydrogen possessing the helium electronic configuration.

Hydride shift   Rearrangement in which a hydrogen in a carbocation moves to a neighboring carbon.

Hydroboration   Addition of borane, BH_3, to a double bond.

Hydroboration–oxidation   A two-step addition of water to a double bond, with anti-Markovnikov regioselectivity.

α-Hydrogen   An acidic hydrogen on the carbon next to a carbonyl group, the α-carbon.

Hydrogen bond   A dipole-dipole attraction between a positively polarized hydrogen atom and an electronegative atom.

Hydrogen (^{1}H) NMR spectroscopy   NMR spectroscopy used to reveal molecular structure near a hydrogen nucleus.

Hydrogenation   Addition of gaseous hydrogen to a double bond.

Hydrogenolysis   Hydrogen addition with simultaneous bond breaking.

Hydrophilic   Enhancing solubility in water.

Hydrophobic   Insoluble in water.

Hydroxy group   —OH, the functional group of the alcohols.

Hyperconjugation   Delocalization of a bonding electron pair, into an empty or a partly empty atomic orbital, stabilizing the molecule.

Imidazole   Diazacyclopentadiene, an aromatic heterocycle found in amino acids like histidine.

Imide   The nitrogen analog of a cyclic anhydride, a ring in which two carbonyl carbons flank a nitrogen atom.

Imine   An organic compound, also called a Schiff base, whose functional group is a carbon–nitrogen double bond.

Iminium ion   [$R_2N—CH_2^+ \longleftrightarrow R_2N^+=CH_2$], the resonance-stabilized fragment left by loss of an alkyl group from an amine in mass spectrometry.

Indole   A benzpyrrole, or fused six- and five-membered aromatic rings with a nitrogen heteroatom.

Inductive effect   The stabilizing of a charge by its transmission through a chain of atoms.

Infrared (IR) radiation   Radiation just lower in energy than visible light, useful in spectroscopy for identification of functional groups in organic molecules.

Infrared spectroscopy   Determination of structure by assignment of vibrational excitations to specific functional groups.

Initiation   The initial step in a radical chain mechanism.

Intermediate   An unobserved species formed on the pathway between reactants and products, as described by the reaction mechanism.

Intramolecular aldol condensation   An aldol condensation between enolate ions and carbonyl groups in the same molecule.

Intramolecular esterification   The production of lactones from hydroxy acids on treatment with mineral acid catalyst.

Inversion   Acid- or enzyme-driven decrease in the specific rotation of sucrose, giving a mixture called invert sugar.

Inversion of configuration   A stereospecific reaction in which the reactant and product are of opposite configuration.

Ionic bond   A bond formed by the transfer of electrons between atoms—in contrast to a covalent bond.

Ionization potential (IP)   The energy needed to remove an electron from an atom.

Ipso substitution   The displacement of a group other than hydrogen from an aromatic ring.

Isoalkane   An alkane branched with a methyl substituent at C2.

Isoelectric point (pI)   The isoelectric pH, at which the charge-neutralized form of an amino acid predominates.

Isomers   Molecules with the same empirical formula, such as constitutional isomers and stereoisomers.

Isoquinoline   2-Azanaphthalene, an unsaturated heterocycle of fused six-membered rings.

IUPAC rules   Nomenclature adopted by the International Union of Pure and Applied Chemistry to reflect a molecule's structure.

Kekulé structure   Representation using a line notation for bonding electrons, with lone pairs shown by dots or omitted.

Keto   The carbonyl tautomer, which rapidly interconverts with its less stable enol form in the presence of acid or base.

Ketone   An organic molecule in which the carbonyl group, C=O, is bound to two carbon atoms.

Ketose   A sugar based on ketones—so that a three-carbon chain would be a ketotriose, for example.

Kiliani-Fischer synthesis   An older method for chain lengthening of sugars through a cyanohydrin intermediate.

Kinetic control   Conditions under which the major product of a reaction is the one that forms fastest.

Kolbe reaction   Attack of the phenoxide ion on carbon dioxide, giving a 2-hydroxybenzoic acid.

Lactam   A cyclic amide, or ring in which a carbonyl carbon is singly bonded to a nitrogen atom.

Lactone   A cyclic ester, the product of intramolecular esterification.

Lactose   The most abundant natural disaccharide, made of glucose and galactose units.

Leaving group   The group of atoms (coded *green* in this text) displaced in a substitution reaction.

Leaving-group ability   The ease with which a group can be displaced in a nucleophilic substitution, related inversely to basicity.

Levorotatory molecule   An enantiomer that rotates the plane of polarization counterclockwise as the viewer faces the light.

Lewis structure   Representation using dots to show valence electrons.

Lindlar's catalyst   A special deactivated form of palladium, used in partial hydrogenation of alkynes to form cis-alkenes.

Lipid   A water-insoluble biomolecule made of waxes and fats.

Lipid bilayer   A unimolecular sheet of phospholipids.

Lithium aluminum hydride $LiAlH_4$, a reducing agent in many organic syntheses.

London forces   Weak intermolecular forces between nonpolar molecules due to a correlation of electrons as they approach.

Magnetic resonance imaging (MRI)   A diagnostic tool that utilizes proton NMR to image the human body.

Major resonance contributor   The Lewis structures most nearly representative of the electron distribution in a delocalized molecule.

Malonic ester synthesis   Synthesis of 2,2-disubstituted carboxylic acids from diethyl propanedioate (malonic ester).

Maltose   A dimer of glucose with an acetal linkage.

Mannich base   The free amine whose salts are produced in the Mannich reaction.

Mannich reaction   Alkylation of enols by electrophilic iminium ions, giving β-aminocarbonyl compounds.

Markovnikov rule   Regioselectivity in electrophilic addition to alkenes whereby an electrophile attacks the less-substituted carbon.

Masked acyl anion   An acyl anion equivalent.

Mass spectrometry   Determination of structure from the deflection of molecular-ion fragments in a magnetic field.

McLafferty rearrangement   Decomposition in mass spectrometry of carbonyl compounds having a γ-hydrogen, giving an alkene and an enol.

Mechanism   The finer details of a reaction, including steps in which intermediates form and then rapidly change into products.

Mercapto group   The SH group.

Merrifield solid-state peptide synthesis   Automated synthesis using polystyrene to anchor a peptide chain.

Meso compound   A compound with at least two stereocenters yet superimposable on its mirror image.

Meta- (*m*-)   Common-name prefix for 1,3-disubstituted benzenes.

Meta directing   Allowing predominantly *meta* electrophilic substitution, like a trifluoromethyl group on benzene.

Metallation   The preparation of organometallic reagents.

Methylene   The highly reactive species $H_2C\colon$, the simplest carbene.

Methylene group $—CH_2—$, the basic unit of the alkanes.

Micelle   A spherical cluster in aqueous solution, such as the cluster of long-chain carboxylic acids in soap.

Michael addition   The conjugate addition of enolates to α,β-unsaturated aldehydes and ketones.

Mixed Claisen condensation   Claisen condensation involving two different esters and often giving product mixtures.

Molar extinction coefficient (ϵ)   The reported peak height in a UV-visible spectrum, also called molar absorptivity.

Molecular ion   A radical cation, $M^{+\cdot}$, formed by electron impact in a mass spectrometer.

Molecular orbital   An orbital formed of overlapping atomic orbitals from different atoms.

Monomer   The repeated building block in a dimer, oligomer, or polymer.

Monosaccharide   A simple sugar, or carbonyl compound bearing at least two additional hydroxy groups.

Mutarotation   Change in a sugar's optical rotation when it equilibrates with its anomer.

N + 1 rule   The rule that N equivalent neighbors produce $N + 1$ spectral peaks in NMR owing to spin–spin splitting.

Naphthalene   Two fused benzene rings, the simplest polycyclic benzenoid hydrocarbon.

Natural product   A compound produced by living organisms.

Neoalkane   A branched alkane in which C2 bears two methyl substituents.

Newman projection   Representation of an alkane along the carbon axis, so that the conformation is easily seen.

Nitrene   An intermediate in the Hofmann rearrangement having a neutral electron-deficient nitrogen atom.

Nitrile   An organic molecule whose functional group is a carbon–nitrogen triple bond.

N-Nitrosamine $R_2N—N{=}O$, the carcinogenic product of the reaction of a secondary amine with nitrous acid.

Nitrosyl cation NO^+, a resonance-stabilized electrophile derived from nitrous acid.

NMR spectrometry   Determination of molecular structure from how nuclei absorb radiation in a magnetic field (nuclear magnetic resonance).

NMR time scale   A limit to spectral resolution—peaks cannot be distinguished if changes are "fast on the NMR time scale."

Node   A surface separating portions of a wave function with opposite signs, indicating where an electron is never found.

Nuclear core   The nucleus of an atom together with all inner-shell electrons but not valence electrons.

Nuclear magnetic resonance   Characteristic excitations of a magnetic atomic nucleus in an external magnetic field, causing it to "flip," or alter, its spin state.

Nucleic acid   DNA or RNA, the nucleotide polymers that carry the genetic code and govern protein biosynthesis.

Nucleophile (Nu)   A species (coded *red* in this text) bearing lone electron pairs and attacking positively polarized centers.

Nucleophilic aromatic substitution   *Ipso* substitution, giving phenols, benzenamines (anilenes), or alkoxybenzenes.

Nucleophilic substitution   A bimolecular (S_N2) or unimolecular (S_N1) reaction in which a nucleophile displaces a leaving group.

Nucleophilicity   The reactivity of a nucleophile.

Nucleoside   The sugar–base subunit of a nucleotide.

Nucleotide   A phosphate-substituted nucleoside—a building block of DNA or RNA.

Octet   A group of eight electrons, as in the outermost shell of the noble gases after helium.

Oil   An ester of long-chain carboxylic acids that is liquid at room temperature—in contrast to solid fats.

Oligomer   A molecule with a subunit repeated only a few times, the result of attack on a dimer or trimer.

Optical activity   Rotation of the plane of polarization of incoming light by a chiral molecule, also called optical rotation.

Optical isomer   A molecule, such as an enantiomer, that is optically active, or rotates the plane of polarization of incoming light.

Optical purity   The percentage of optical activity observed relative to that of the pure enantiomer.

Organic molecules   Carbon compounds (traditionally excepting CO_2 and compounds where carbon enters solely in a CN subunit).

Organometallic reagent   A compound in which the carbon atom of an organic group is bound to a metal, and so a source of nucleophilic carbon.

Ortho- (o-)   Common-name prefix for 1,2- (adjacently) disubstituted benzenes.

Ortho and para directing   Allowing *ortho* and *para* electrophilic substitution, like a methyl group on benzene.

Oxacycloalkane   A cyclic ether, or ring in which a carbon atom is replaced by oxygen.

Oxalic acid   The common name of ethanedioic acid, the simplest dicarboxylic acid.

Oxidation   The addition of electronegative atoms such as oxygen or the removal of hydrogen from a molecule.

Oxime $RR'C=NOH$, a crystalline imine product used in identifying aldehydes and ketones.

Oxonium ion   The conjugate base of an alcohol.

Oxymercuration–demercuration   Hydration of a double bond by addition of a mercuric salt and water, then removal of mercury.

Ozonolysis   Cleaving of an alkene π bond leading to carbonyl compounds.

Paal-Knorr synthesis   Synthesis of pyrrole by cyclization of a γ-dicarbonyl compound.

Para- (p)   Common-name prefix for 1,4-disubstituted benzenes.

Pauli exclusion principle   The rule that no atomic orbital may be occupied by more than two electrons.

PCC   Pyridinium chlorochromate, an oxidizing agent.

Penicillin   An annulated β-lactam.

Peptide   An oligomer of amino acids joined by amide linkages—a dipeptide, tripeptide, and so on.

Peptide bond   The carboxylic acid–amine link in a peptide chain.

Peri fusion   Angular fusion of benzenes, as in the three-ring system phenanthrene.

Pericyclic reactions   Electrocyclic and other reactions with a transition state having a cyclic array of nuclei and electrons.

Peroxide   Any system with oxygens joined by a single bond.

Peroxycarboxylic acid RCO_3H, a carboxylic acid with an additional, electrophilic oxygen.

Phenol   The alcohol derived from benzene, formerly known as carbolic acid.

Phenoxide ion $C_6H_5O^-$, the resonance-stabilized anion accounting for the acidity of phenol.

Phenoxy C_6H_5O-, phenol as a substituent.

Phenyl   Benzene as a substituent, the simplest aryl, C_6H_5-.

Phenylmethyl group $C_6H_5CH_2-$, also called the benzyl group.

Phenylmethyl radical   The resonance-stabilized benzylic radical $C_6H_5CH_2\cdot$

Phenylosazone   A double phenylhydrazone produced by carbonyl condensation of sugars with phenylhydrazine.

Pheromone   A naturally occurring substance used for communication within a living species.

Phospholipid   A di- or triester of carboxylic acids and phosphoric acid, a component of cell membranes.

Photosynthesis   Plant synthesis of carbohydrates from carbon dioxide and water using energy from sunlight.

Pi (π) bond   A bond derived from overlap of parallel p orbitals, as in components of double and triple bonds.

$\pi \rightarrow \pi^*$ transitions   Excitation of electrons in π orbitals by visible or UV radiation.

Picric acid   2,4,6-Trinitrophenol, a highly explosive acid.

Plane-polarized light   Light whose electric field vectors lie in one plane, the plane of polarization.

Polarimeter   A device for measuring optical rotations.

Polarizability   The capacity of an atom's electrons to respond to a changing electric field.

Polarized bond   A covalent bond between atoms of different electronegativity, so that the bonding electrons are not shared equally.

Polycyclic benzenoid hydrocarbon   An extended π system of fused rings, also called a polycyclic aromatic hydrocarbon (PAH).

Polyethene   A branched polymer with common name polyethylene, familiar as plastic storage bags and containers.

Polyfunctional compound   A compound with many functional groups, such as a carbohydrate.

Polymer   A molecule with a repeated subunit, such as polyethene and the other synthetic compounds listed in Table 12-3.

Polymerization   The synthesis of polymers.

Polypeptide   An amino-acid polymer, such as a protein, with chains sometimes joined by disulfide bridges.

Polysaccharide   A polymeric complex sugar.

Polystyrene   A polymer derived from ethenylbenzene (styrene).

Porphyrin   A cyclic molecule, such as a heme group, of four linked pyrrole units.

Potential energy diagram   A graph of potential energy changes, as in a chemical reaction or rotation about a carbon–carbon bond.

Primary carbon   A carbon directly attached to only one other carbon atom.

Primary structure   The amino-acid sequence in a peptide chain—in contrast to secondary structure, the folding pattern.

Priority   A ranking of substituents at an asymmetric carbon following sequence rules.

Propagation   A stage subsequent to initiation in a radical chain mechanism, in which atoms or radicals attack other reactants.

Prostaglandin (PG)   A potent hormonelike substance containing alcohol groups, alkene bonds, and carbonyl groups.

Protecting group   A group that temporarily makes a functional group unreactive.

Proteins   Large natural polypeptides, such as globular proteins used as enzymes, fibrous proteins in muscle, and hemoglobin.

Protic solvent   A solvent containing positively polarized hydrogens, and so available to interact with anionic nucleophiles.

Pyranose   A six-membered cyclic monosaccharide.

Pyridine C_5H_5N, an azabenzene, or unsaturated nitrogen heterocycle with a six-membered ring.

Pyrolysis   The breaking of alkane bonds when a molecule is heated.

Pyrrole C_4H_5N, an aromatic heterocyclopentadiene.

Quanta   The discrete "packets" that make up electromagnetic radiation, each of energy dependent on wavelength.

Quaternary carbon   A carbon directly attached to four other carbon atoms.

Quinoline   1-Azanaphthalene, an unsaturated heterocycle of fused six-membered rings.

Quinomethane $CH_2C_6H_5O$, an intermediate produced following hydroxymethylation of phenol.

Quinone   The common name for derivatives of cyclohexadiendione (benzoquinone), a cyclic diketone.

Racemic mixture   A racemate, or 1:1 mixture of enantiomers.

Racemize   To undergo racemization, or equilibrate with the enantiomer's mirror image.

Radical   A fragment of a molecule possessing an unpaired electron as a result of homolytic cleavage.

Radical chain mechanism   A cycle in which propagation generates radical fragments, initiating further reaction until chain termination.

Raffinose   A nonreducing trisaccharide.

Raney nickel (Ra-Ni)   A catalyst prepared from finely dispersed nickel.

Rate-determining step   The step that determines the overall reaction rate because it is the slowest step in a sequence.

Reaction coordinate   The horizontal coordinate in a potential energy diagram, describing the progress of a reaction.

Reducing sugars   Sugars identifiable by Fehling's and Tollen's tests, oxidation to carboxylic acid.

Reduction   The removal of electronegative atoms such as oxygen or the addition of hydrogen to a molecule.

Reductive amination   Amine synthesis by condensation with a carbonyl compound to an imine intermediate, followed by reduction.

Reformate   A new hydrocarbon created, or reformed, from an old one.

Regioselectivity   The attack of a reagent preferentially at one of several possible centers.

Residue   An amino acid as a peptide subunit.

Resolution   The separation of enantiomers.

Resonance energy   The stabilizing energy difference between delocalized and nondelocalized structures.

Resonance hybrid   A molecule in which no one Lewis structure is correct, because electrons are delocalized, or shared, over several nuclei.

Retro-Claisen condensation   The reverse of Claisen condensation upon treatment of certain β-dicarbonyl compounds with base.

Retrosynthetic analysis   Planning a synthesis by reasoning backward from the desired product to identify important bonds.

Reverse polarization   The conversion of a carbon atom from an electrophilic center to a nucleophilic one, or vice versa.

Ribonucleic acid (RNA)   The chains of nucleotides, arranged in codons, that govern protein biosynthesis.

Ribose   An aldopentose, building block of the ribonucleic acids.

Ring strain   The relative instability of smaller rings due to torsional and bond-angle strain.

Robinson annulation   Michael addition followed by intramolecular aldol condensation, used in steroid synthesis.

Rotamers   Conformations or conformers.

Ruff degradation   Chain shortening of sugars by oxidative decarboxylation.

Saccharides   Sugars.

Sandmeyer reaction   Decomposition of arenediazonium salts to haloarenes in the presence of cuprous salts.

Saturated compound   A compound, such as an alkane, with only single bonds.

Saytzev rule   Regioselectivity whereby E2 elimination by a nonhindered base leads to the more stable internal alkenes.

Secondary carbon   A carbon directly attached to exactly two other carbon atoms.

Secondary structure   The folding pattern of a peptide chain—in contrast to primary structure, the sequence of residues.

Selectivity   The predominance of one of several possible products, so that the reagent is said to be selective.

Semicarbazone $RR'C{=}N{-}NH{-}CONH_2$, a crystalline imine product used in identifying aldehydes and ketones.

Sequence rules   Assignments of priority to substituents at a stereocenter, based on atomic mass and number close to the point of attachment.

Sequencing   Determination of a peptide's primary structure.

Shielding   A reduction in the local magnetic field, or donation of electrons to a nucleus, causing an upfield chemical shift.

Side chain   Substituents joined to the main chain, or sequence of residues in a polypeptide.

Sigma (σ) bond   A bond in which molecular orbitals are aligned in a colinear manner, as in a carbon–carbon single bond.

Simmons-Smith reagent ICH_2ZnI, a carbenoid used in cyclopropane synthesis from alkenes.

Simple sugar   A monosaccharide, or carbonyl compound bearing at least two additional hydroxy groups.

Single bond   A covalent bond in which the atoms gain octets by sharing just two electrons.

Singlet   A single sharp peak in an NMR spectrum.

Skew conformation   An arrangement of methyl groups between the extremes of a staggered or eclipsed conformation.

Specific rotation ([α])   A standard value for the optical rotation of a chiral compound.

Spectrometer   An instrument for recording spectra, by comparing radiation transmitted through a sample to a reference beam.

Spectrometry   A tool for determining molecular structure from how molecules absorb radiation.

Spectrum (*plural:* spectra)   A plot of intensity, showing peaks at wavelengths where a molecule absorbs radiation.

Spin   The intrinsic angular momentum of an electron, nucleus, or other quantum-mechanical object.

Spin–spin splitting   Coupling, or the splitting of an NMR spectral peak by neighbors into more complex patterns.

Staggered conformation   An arrangement such that, in a view along the carbon axis, each hydrogen or attached group is midway between two such groups on the next carbon.

Starch   A polysaccharide of glucose with α-acetal linkages.

Stereocenter   A carbon connected to four different substituents, so that molecules possessing just one such center are chiral.

Stereoisomers   Molecules with the same atoms connected in the same order but arranged differently in space.

Stereoselectivity   The predominance of one of several possible stereoisomeric products.

Stereospecific reaction   Transformation of a pure stereoisomer of starting material into a pure stereoisomer of product.

Steric hindrance   The repulsive interaction between atoms in close proximity, raising the molecule's energy.

Steroid   A compound, such as a sex hormone or cholesterol, with three fused six-membered rings fused to a five-membered ring.

Strecker synthesis   Synthesis of amino acids through an imine intermediate using hydrogen cyanide and ammonia.

Substitution   A reaction, such as nucleophilic substitution, in which a group is replaced by atoms from a reagent.

Substrate   The species attacked in a reaction, such as the target of a nucleophile in nucleophilic substitution.

Sucrose   A disaccharide of glucose and fructose.

Sugar   A saccharide, with empirical formula $C_n(H_2O)_n$.

Sulfate   A sulfur derivative based on the ion $ROSO_3{}^-$.

Sulfide   The sulfur analog of an ether, also called a thioether.

Sulfonamide   The product of reaction between a sulfonyl chloride and an amine, such as a sulfa drug.

Sulfonate   A sulfur derivative based on the ion $RSO_3{}^-$.

Sulfone   A molecule of the form RSO_2R.

Sulfonic acid   A molecule of the form RSO_3H.

Sulfonyl chloride   An acid chloride of a sulfonic acid, RSO_2Cl.

Sulfoxide   A molecule of the form $RSOR$, such as the solvent DMSO.

Synthesis   The making of molecules, with reagents, catalysts, and conditions chosen based on knowledge of the mechanism.

Synthesis gas   A pressurized mixture of carbon monoxide and hydrogen, useful starting material in hydrocarbon syntheses.

Systematic nomenclature   Nomenclature following IUPAC or Chemical Abstracts rules to reflect a molecule's structure.

Tautomers   Isomers differing by the shift of both a proton and a double bond, such as an alkenol and a carbonyl compound.

C-Terminal amino acid   The residue at the carboxy end of a peptide chain, placed to the right in drawing.

N-Terminal amino acid   The residue at the amino end of a peptide chain, placed to the left in drawing.

Terpene   One of the (often fragrant) natural compounds made of isoprene, or five-carbon, units.

Tertiary carbon   A carbon directly attached to exactly three other carbon atoms.

Tertiary structure   Further coiling or aggregation of the secondary structure of a peptide chain.

Tetrahedral intermediate   An intermediate in addition–elimination of a carboxylic derivative, named for its reactive carbon center.

Tetrahydrofuran (THF)   Oxacyclopentane, a common solvent.

Thermal isomerization   The interconversion of stereoisomers by heating, useful for measuring the strength of a π bond.

Thermodynamic control   Conditions under which the most stable of several possible reaction products forms to the greatest extent.

Thiacycloalkane   A saturated sulfur heterocycle.

Thiazolium ion   A five-membered ring with an acidic proton flanked by an NR group and a sulfur heteroatom.

Thioacetal   The sulfur analog of an acetal, which may be "desulfurized" by treatment with Raney nickel.

Thiol   A sulfur analog of an alcohol, RSH.

Thiophene C_4H_4S, an aromatic heterocyclopentadiene.

Thymine (T)   One of the four DNA bases, a polyfunctional nitrogen heterocycle.

Tollen's test   A chemical test for aldehydes by oxidation to a carboxylic acid and formation of a mirrorlike silver precipitate.

Toluene   Common name of methylbenzene, $C_6H_5CH_3$, a solvent.

Torsional strain   Relative instability of a molecule due to the eclipsing of neighboring hydrogens.

Total synthesis   The most efficient plan for obtaining a desired product, often by a series of reactions.

Trans isomer   A stereoisomer in which substituents are on opposite sides of a ring or a double bond—in contrast to a cis isomer.

Transannular strain   Relative instability of a molecule due to steric crowding of groups across a ring.

Transesterification   Interconversion of esters by acid- or base-catalyzed reaction with alcohols.

Transition state   The highest-energy point in a potential energy diagram between species in a process or reaction.

Trivial name   The common name of a molecule—in contrast with systematic nomenclature.

Twist-boat conformation   A structure obtained by partial removal of the transannular strain in boat cyclohexane.

Ubiquinone   One of a series of quinones, collectively called coenzyme Q (CoQ), used in biological reductions of oxygen.

Ultraviolet (UV) radiation   Radiation just higher in energy than visible light, used to produce electronic spectra.

Unimolecular reaction   A first-order process, one whose rate is proportional to the concentration of one species only.

Unsaturated compound   A compound, such as an alkene, containing double or triple bonds, and so able to undergo additions.

Uracil (U)   One of the four RNA bases, a polyfunctional nitrogen heterocycle.

Urea   RNH—CO—NHR', an amide derivative of carbonic acid.

UV-visible spectrometry   Determination of the structure of unsaturated systems from electronic excitations.

Valence electrons   Outer-shell electrons available to form a covalent bond.

van der Waals forces   The intermolecular forces that cause molecules to aggregate as solids and liquids.

Vibrational excitation   The characteristic excitation of atoms about the bonds between them as they absorb infrared radiation.

Vicinal coupling   Coupling of nonequivalent hydrogens on adjacent carbon atoms in NMR.

Vinyl   The common name of ethenyl, CH_2=CH—, as in vinyl chloride, the repeated subunit of the polymer PVC.

VSEPR method   The determining of molecular shapes by minimizing valence shell electron pair repulsions.

Wacker process   The catalytic conversion of ethene into acetaldehyde.

Wave function   In quantum mechanics, a mathematical expression used to predict the most likely location of an electron or other object.

Wax   An ester of long-chain carboxylic acids and long-chain alcohols.

Williamson ether synthesis S_N2 reaction of an alkoxide with a primary haloalkane or sulfonate ester, yielding an ether.

Wittig reaction   Nucleophilic attack of a phosphorus ylide on an aldehyde or ketone, leaving a doubly bonded carbon.

Wolff-Kishner reduction   The decomposition of hydrazones in the presence of a base, used to deoxygenate aldehydes and ketones.

Woodward-Hoffman rules   The stereochemistry of electrocyclic processes, based on the symmetry of molecular π orbitals.

Xylene   Common name of dimethylbenzene.

Ylide   Most commonly a phosphorus ylide $RHCP(C_6H_5)_3$, in which the heteroatom stabilizes a carbanion.

Zwitterion   A dipolar ion (or inner salt)—a structure that acts as both acid and base, such as an amino acid.